Intermediate ALGEBRA

Fourth Edition

Julie Miller

Daytona State College

Molly O'Neill

Daytona State College

Nancy Hyde

Broward College—Professor Emeritus

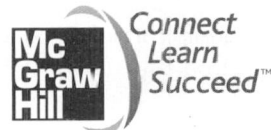

McGraw Hill

Connect
Learn
Succeed™

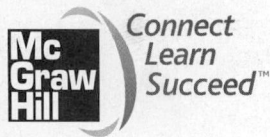

INTERMEDIATE ALGEBRA, FOURTH EDITION

Published by McGraw-Hill, a business unit of The McGraw-Hill Companies, Inc., 1221 Avenue of the Americas, New York, NY 10020. Copyright © 2014 by The McGraw-Hill Companies, Inc. All rights reserved. Printed in the United States of America. Previous editions © 2011, 2008, and 2004. No part of this publication may be reproduced or distributed in any form or by any means, or stored in a database or retrieval system, without the prior written consent of The McGraw-Hill Companies, Inc., including, but not limited to, in any network or other electronic storage or transmission, or broadcast for distance learning.

Some ancillaries, including electronic and print components, may not be available to customers outside the United States.

This book is printed on acid-free paper.

2 3 4 5 6 7 8 9 0 DOW/DOW 1 0 9 8 7 6 5 4

ISBN 978–0–07–338449–8
MHID 0–07–338449–6

ISBN 978–0–07–754871–1 (Annotated Instructor's Edition)
MHID 0–07–754871–X

Senior Vice President, Products & Markets: *Kurt L. Strand*
Vice President, General Manager, Products & Markets: *Marty Lange*
Vice President, Content Production & Technology Services: *Kimberly Meriwether David*
Director of Development: *Rose M. Koos*
Managing Director: *Ryan Blankenship*
Director, Developmental Mathematics: *Dawn R. Bercier*
Brand Manager: *Mary Ellen Rahn*
Director of Digital Content Development: *Emilie J. Berglund/Nicole Lloyd*
Developmental Editor: *Emily Williams*
Marketing Manager: *Peter A. Vanaria*

Project Manager: *Sherry Kane*
Buyer: *Nicole Baumgartner*
Senior Media Project Manager: *Sandra M. Schnee*
Senior Designer: *Laurie B. Janssen*
Cover Illustration: *Imagineering Media Services, Inc.*
Lead Content Licensing Specialist: *Carrie K. Burger*
Compositor: *Aptara®, Inc.*
Typeface: *10/12 Times Ten Roman*
Printer: *R. R. Donnelley*

All credits appearing on page or at the end of the book are considered to be an extension of the copyright page.

Photo Credits: Page 2: © Getty/Blend Images RF; p. 3: © Photodisc/Getty RF; p. 64: © Corbis RF; p. 66: © Rob Melnychuk/Getty RF; p. 68: © John Lund/Drew Kelly/Blend Images RF; p. 71: © Corbis RF; p. 85: © Creatas Images/MasterFile RF; p. 96: © Image Source RF; p. 97: © Don Hammond/Design Pics RF; p. 112: © BananaStock/Punchstock RF; p. 126: © Keith Brofsky/Getty RF; p. 135: © Corbis RF; p. 177: © PhotoDisc/Getty RF; p. 178: © Vol. 107/Corbis RF; p. 191: © Digital Vision/Getty RF; p. 227: © Andersen/Ross/Getty RF; p. 261: © imac/Alamy RF; p. 266(left): © The McGraw-Hill Companies, Inc./John Floumoy, photographer; p. 266(right): © Corbis RF; p. 269(left): © Comstock RF/JupiterImages; p. 269(right): © PhotoDisc/Getty RF; p. 291: © PhotoDisc/Getty RF; p. 311: © Dynamic Graphics/JupiterImages RF; p. 312: © Corbis RF; p. 325: © NASA/Corbis RF; p. 338: © Corbis RF; p. 402: © CMCD/Getty RF; p. 406: © LouAnn Wilson; p. 419: © PhotoDisc/Getty RF; p. 464: © Brand X Pictures RF; p. 466: © The McGraw-Hill Companies, Inc./Lars A. Niki, photographer; p. 467: © Corbis RF; p. 468: © Royalty-Free/Corbis RF; p. 471: © Corbis RF; p. 472: © Vol. 20/Corbis RF; p. 473(top): © Corbis RF; p. 473(bottom): © Corbis/Punchstock RF; p. 478: Courtesy of NOAA; p. 481: © Creatas/PunchStock RF; p. 492: © Vol. 59/Corbis RF; p. 545 © Getty RF; p. 551: © Vol. 145/Corbis RF; p. 555: © Image Source RF; p. 591: © Corbis RF; p. 630: © Royalty-Free/Corbis; p. 634: © The McGraw-Hill Companies, Inc./Jill Braaten, photographer; p. 655: © Medioimages/Superstock RF; p. 656: © Photodisc/Getty RF; p. 684: © Corbis; p. 701: © PhotoDisc/Getty RF; p. 774(top): © Getty/Photodisc RF; p. 774(bottom): © Royalty-Free/Corbis; p. 842: © Steve Mason/Getty RF.

Library of Congress Cataloging-in-Publication Data

Miller, Julie, 1962-
 Intermediate algebra / Julie Miller, Daytona State College, Daytona Beach, Molly O'Neill, Daytona State College, Daytona Beach, Nancy Hyde.— Fourth Edition.
  pages cm
 Includes index.
 ISBN 978–0–07–338449–8 — ISBN 0–07–338449–6 (hard copy : alk. paper) 1. Algebra—Textbooks. I. O'Neill, Molly, 1953- II. Hyde, Nancy. III. Title.

QA154.3.M554 2014
512.9—dc23
                    2012018138

The Internet addresses listed in the text were accurate at the time of publication. The inclusion of a website does not indicate an endorsement by the authors or McGraw-Hill, and McGraw-Hill does not guarantee the accuracy of the information presented at these sites.

Letter from the Authors

Dear Colleagues,

We originally embarked on this textbook project because we were seeing a lack of student success in our developmental math sequence. In short, we were not getting the results we wanted from our students with the materials and textbooks that we were using at the time. The primary goal of our project was to create teaching and learning materials that would get better results.

At Daytona State College, our students were instrumental in helping us develop the clarity of writing; the step-by-step examples; and the pedagogical elements, such as Avoiding Mistakes, Concept Connections, and Problem Recognition Exercises, found in our textbooks. They also helped us create the content for the McGraw-Hill lecture videos and dynamic Flash animations that accompany this text.

This project has been a true collaboration with our Board of Advisors and colleagues in developmental mathematics around the country. We are sincerely humbled by those of you who have used our books and by the over 400 colleagues around the country who have partnered with us providing valuable suggestions through reviews, symposia, focus groups, and being on our Board of Advisors. You partnered with us to create materials that will help students get better results. For that we are immeasurably grateful.

As an author team, we have an ongoing commitment to provide the best possible text materials for instructors and students. With your continued help and suggestions we will continue the quest to help all of our students get better results.

Sincerely,

Julie Miller
julie.miller.math@gmail.com

Molly O'Neill
molly.s.oneill@gmail.com

Nancy Hyde
nhyde@montanasky.com

About the Authors

Julie Miller is from Daytona State College, where she has taught developmental and upper-level mathematics courses for 20 years. Prior to her work at Daytona State College, she worked as a software engineer for General Electric in the area of flight and radar simulation. Julie earned a bachelor of science in applied mathematics from Union College in Schenectady, New York, and a master of science in mathematics from the University of Florida. In addition to this textbook, she has authored several course supplements for college algebra, trigonometry, and precalculus, as well as several short works of fiction and nonfiction for young readers.

"My father is a medical researcher, and I got hooked on math and science when I was young and would visit his laboratory. I can remember using graph paper to plot data points for his experiments and doing simple calculations. He would then tell me what the peaks and features in the graph meant in the context of his experiment. I think that applications and hands-on experience made math come alive for me and I'd like to see math come alive for my students."

—*Julie Miller*

Molly O'Neill is also from Daytona State College, where she has taught for 22 years in the School of Mathematics. She has taught a variety of courses from developmental mathematics to calculus. Before she came to Florida, Molly taught as an adjunct instructor at the University of Michigan–Dearborn, Eastern Michigan University, Wayne State University, and Oakland Community College. Molly earned a bachelor of science in mathematics and a master of arts and teaching from Western Michigan University in Kalamazoo, Michigan. Besides this textbook, she has authored several course supplements for college algebra, trigonometry, and precalculus and has reviewed texts for developmental mathematics.

"I differ from many of my colleagues in that math was not always easy for me. But in seventh grade I had a teacher who taught me that if I follow the rules of mathematics, even I could solve math problems. Once I understood this, I enjoyed math to the point of choosing it for my career. I now have the greatest job because I get to do math every day and I have the opportunity to influence my students just as I was influenced. Authoring these texts has given me another avenue to reach even more students."

—*Molly O'Neill*

Nancy Hyde served as a full-time faculty member of the Mathematics Department at Broward College for 24 years. During this time she taught the full spectrum of courses from developmental math through differential equations. She received a bachelor of science degree in math education from Florida State University and a master's degree in math education from Florida Atlantic University. She has conducted workshops and seminars for both students and teachers on the use of technology in the classroom. In addition to this textbook, she has authored a graphing calculator supplement for *College Algebra*.

"I grew up in Brevard County, Florida, where my father worked at Cape Canaveral. I was always excited by mathematics and physics in relation to the space program. As I studied higher levels of mathematics I became more intrigued by its abstract nature and infinite possibilities. It is enjoyable and rewarding to convey this perspective to students while helping them to understand mathematics."

—*Nancy Hyde*

Dedication
To Our Students
Julie Miller ❧ Molly O'Neill ❧ Nancy Hyde

Contents

Chapter R

Review of Basic Algebraic Concepts   1

R.1   Study Skills   2

Group Activity: Becoming a Successful Student   3

R.2   Sets of Numbers and Interval Notation   5

R.3   Operations on Real Numbers   16

R.4   Simplifying Algebraic Expressions   31

Chapter R Summary   39

Chapter R Review Exercises   41

Chapter R Test   42

Chapter 1

Linear Equations and Inequalities in One Variable   43

1.1   Linear Equations in One Variable   44

Problem Recognition Exercises: Equations versus Expressions   56

1.2   Applications of Linear Equations in One Variable   57

1.3   Applications to Geometry and Literal Equations   68

1.4   Linear Inequalities in One Variable   76

1.5   Compound Inequalities   85

1.6   Absolute Value Equations   97

1.7   Absolute Value Inequalities   103

Problem Recognition Exercises: Identifying Equations and Inequalities   113

Group Activity: Understanding the Symbolism of Mathematics   114

Chapter 1 Summary   115

Chapter 1 Review Exercises   122

Chapter 1 Test   125

Chapter 2

Linear Equations in Two Variables and Functions   127

2.1   Linear Equations in Two Variables   128

2.2   Slope of a Line and Rate of Change   145

2.3   Equations of a Line   157

Problem Recognition Exercises: Characteristics of Linear Equations   171

2.4   Applications of Linear Equations and Modeling   171

2.5   Introduction to Relations   183

2.6   Introduction to Functions   192

2.7   Graphs of Functions   204

Problem Recognition Exercises: Characteristics of Relations   217

Group Activity: Deciphering a Coded Message   218

Chapter 2 Summary   219

Chapter 2 Review Exercises   225

Chapter 2 Test   230

Chapters 1–2 Cumulative Review Exercises   233

Chapter 3   **Systems of Linear Equations and Inequalities   235**

3.1   Solving Systems of Linear Equations by the Graphing Method   236
3.2   Solving Systems of Linear Equations by the Substitution Method   246
3.3   Solving Systems of Linear Equations by the Addition Method   253
        Problem Recognition Exercises: Solving Systems of Linear Equations   260
3.4   Applications of Systems of Linear Equations in Two Variables   261
3.5   Linear Inequalities and Systems of Linear Inequalities in Two Variables   270
3.6   Systems of Linear Equations in Three Variables and Applications   283
3.7   Solving Systems of Linear Equations by Using Matrices   293
        Group Activity: Creating a Quadratic Model of the Form $y = at^2 + bt + c$ 302
        Chapter 3 Summary   303
        Chapter 3 Review Exercises   310
        Chapter 3 Test   314
        Chapters 1–3 Cumulative Review Exercises   316

Chapter 4   **Polynomials   319**

4.1   Properties of Integer Exponents and Scientific Notation   320
4.2   Addition and Subtraction of Polynomials and Polynomial Functions   329
4.3   Multiplication of Polynomials   340
4.4   Division of Polynomials   350
        Problem Recognition Exercises: Operations on Polynomials   359
4.5   Greatest Common Factor and Factoring by Grouping   360
4.6   Factoring Trinomials   368
4.7   Factoring Binomials   382
        Problem Recognition Exercises: Factoring Summary   392
4.8   Solving Equations by Using the Zero Product Rule   394
        Group Activity: Investigating Pascal's Triangle   408
        Chapter 4 Summary   409
        Chapter 4 Review Exercises   414
        Chapter 4 Test   418
        Chapters 1–4 Cumulative Review Exercises   419

Chapter 5   **Rational Expressions and Rational Equations   421**

5.1   Rational Expressions and Rational Functions   422
5.2   Multiplication and Division of Rational Expressions   432
5.3   Addition and Subtraction of Rational Expressions   437
5.4   Complex Fractions   447
        Problem Recognition Exercises: Operations on Rational Expressions   454
5.5   Solving Rational Equations   454
        Problem Recognition Exercises: Rational Equations vs. Expressions   462
5.6   Applications of Rational Equations and Proportions   463
5.7   Variation   474
        Group Activity: Computing the Future Value of an Investment   483
        Chapter 5 Summary   484
        Chapter 5 Review Exercises   489
        Chapter 5 Test   492
        Chapters 1–5 Cumulative Review Exercises   493

Chapter 6   Radicals and Complex Numbers   495

6.1   Definition of an *n*th Root   496
6.2   Rational Exponents   508
6.3   Simplifying Radical Expressions   515
6.4   Addition and Subtraction of Radicals   522
6.5   Multiplication of Radicals   528
     Problem Recognition Exercises: Simplifying Radical Expressions   536
6.6   Division of Radicals and Rationalization   536
6.7   Solving Radical Equations   546
6.8   Complex Numbers   556
     Group Activity: Margin of Error of Survey Results   566
     Chapter 6 Summary   568
     Chapter 6 Review Exercises   574
     Chapter 6 Test   577
     Chapters 1–6 Cumulative Review Exercises   578

Chapter 7   Quadratic Equations, Functions, and Inequalities   581

7.1   Square Root Property and Completing the Square   582
7.2   Quadratic Formula   592
7.3   Equations in Quadratic Form   606
     Problem Recognition Exercises: Quadratic and Quadratic Type Equations   611
7.4   Graphs of Quadratic Functions   612
7.5   Vertex of a Parabola: Applications and Modeling   626
7.6   Polynomial and Rational Inequalities   636
     Problem Recognition Exercises: Recognizing Equations and Inequalities   648
     Group Activity: Creating a Quadratic Model of the Form $y = a(x - h)^2 + k$ 649
     Chapter 7 Summary   650
     Chapter 7 Review Exercises   655
     Chapter 7 Test   658
     Chapters 1–7 Cumulative Review Exercises   660

Chapter 8   Exponential and Logarithmic Functions and Applications   663

8.1   Algebra of Functions and Composition   664
8.2   Inverse Functions   671
8.3   Exponential Functions   680
8.4   Logarithmic Functions   690
     Problem Recognition Exercises: Identifying Graphs of Functions   703
8.5   Properties of Logarithms   704
8.6   The Irrational Number *e* and Change of Base 712
     Problem Recognition Exercises: Logarithmic and Exponential Forms   725
8.7   Logarithmic and Exponential Equations and Applications   726
     Group Activity: Creating a Population Model   738
     Chapter 8 Summary   739
     Chapter 8 Review Exercises   744
     Chapter 8 Test   748
     Chapters 1–8 Cumulative Review Exercises   750

Chapter 9

Conic Sections 753

9.1   Distance Formula, Midpoint Formula, and Circles 754
9.2   More on the Parabola 765
9.3   The Ellipse and Hyperbola 774
      Problem Recognition Exercises: Formulas and Conic Sections 783
9.4   Nonlinear Systems of Equations in Two Variables 784
9.5   Nonlinear Inequalities and Systems of Inequalities 791
      Group Activity: Investigating the Graphs of Conic Sections on a Calculator 800
      Chapter 9 Summary 801
      Chapter 9 Review Exercises 806
      Chapter 9 Test 809
      Chapters 1–9 Cumulative Review Exercises 811

Chapter 10

Binomial Expansions, Sequences, and Series 813

10.1   Binomial Expansions 814
10.2   Sequences and Series 820
10.3   Arithmetic Sequences and Series 829
10.4   Geometric Sequences and Series 835
       Problem Recognition Exercises: Identifying Arithmetic and Geometric Series 843
       Group Activity: Investigating Mean and Standard Deviation 844
       Chapter 10 Summary 845
       Chapter 10 Review Exercises 848
       Chapter 10 Test 849
       Chapters 1–10 Cumulative Review Exercises 851

Additional Topics Appendix A-1
A.1   Determinants and Cramer's Rule A-1

Student Answer Appendix SA-1

Index I-1

Get Better Results

How Will Miller/O'Neill/Hyde Help Your Students *Get Better Results?*

Better Clarity, Quality, and Accuracy!

Julie Miller, Molly O'Neill, and Nancy Hyde know what students need to be successful in mathematics. Better results come from clarity in their exposition, quality of step-by-step worked examples, and accuracy of their exercises sets; but it takes more than just great authors to build a textbook series to help students achieve success in mathematics. Our authors worked with a strong mathematical team of instructors from around the country to ensure that the clarity, quality, and accuracy you expect from the Miller/O'Neill/Hyde series was included in this edition.

> "The most complete text at this level in its thoroughness, accuracy, and pedagogical soundness. The best developmental mathematics text I have seen."
> —Frederick Bakenhus, *Saint Phillips College*

Better Exercise Sets!

Comprehensive sets of exercises are available for every student level. Julie Miller, Molly O Neill, and Nancy Hyde worked with a board of advisors from across the country to offer the appropriate depth and breadth of exercises for your students. **Problem Recognition Exercises** were created to improve student performance while testing.

Practice exercise sets help students progress from skill development to conceptual understanding. Student tested and instructor approved, the Miller/O'Neill/Hyde exercise sets will help your students *get better results*.

- ▶ **Problem Recognition Exercises**
- ▶ **Skill Practice Exercises**
- ▶ **Study Skills Exercises**
- ▶ **Mixed Exercises**
- ▶ **Expanding Your Skills Exercises**
- ▶ **Vocabulary and Key Concepts Exercises**

> "This series was thoughtfully constructed with students' needs in mind. The Problem Recognition section was extremely well designed to focus on concepts that students often misinterpret."
> —Christine V. Wetzel-Ulrich, *Northampton Community College*

Better Step-By-Step Pedagogy!

Intermediate Algebra provides enhanced step-by-step learning tools to help students *get better results*.

- ▶ **Worked Examples** provide an "easy-to-understand" approach, clearly guiding each student through a step-by-step approach to master each practice exercise for better comprehension.
- ▶ **TIPs** offer students extra cautious direction to help improve understanding through hints and further insight.
- ▶ **Avoiding Mistakes** boxes alert students to common errors and provide practical ways to avoid them. Both of these learning aids will help students get better results by showing how to work through a problem using a clearly defined step-by-step methodology that has been class tested and student approved.

> "The book is designed with both instructors and students in mind. I appreciate that great care was used in the placement of 'Tips' and 'Avoiding Mistakes' as it creates a lot of teachable moments in the classroom."
> —Shannon Vinson, *Wake Tech Community College*

Get Better Results

Formula for Student Success

Step-by-Step Worked Examples

▶ Do you get the feeling that there is a disconnection between your students' class work and homework?

▶ Do your students have trouble finding worked examples that match the practice exercises?

▶ Do you prefer that your students see examples in the textbook that match the ones you use in class?

Miller/O'Neill/Hyde's *Worked Examples* offer a clear, concise methodology that replicates the mathematical processes used in the authors' classroom lectures!

Example 2 Solving a Linear Equation Involving Consecutive Integers

Three times the sum of two consecutive odd integers is 516. Find the integers.

Solution:

Step 1: Read the problem carefully.

Step 2: Label the unknown:

x represent the first odd integer.

$x + 2$ represents the next odd integer.

equation in words.

first odd integer) + (second odd integer)] = 516

$+ (x + 2)] = 516$ **Step 4:** Write a mathematical equation.

$3(2x + 2) = 516$ **Step 5:** Solve for x.

$6x + 6 = 516$

$6x = 510$

$x = 85$

rpret your results.

The first odd integer is x:

The second odd integer is $x + 2$:

$85 + 2$

Answer: The integers are 85 and 87.

Skill Practice

2. Four times the sum of three consecutive integers is 264. Find the integers.

> "As always, MOH's Worked Examples are so clear and useful for the students. All steps have wonderfully detailed explanations written with wording that the students can understand. MOH is also excellent with arrows and labels making the Worked Examples extremely clear and understandable."
> —Kelli Hammer, *Broward College–South*

> "Easy to read step-by-step solutions to sample textbook problems. The 'why' is provided for students, which is invaluable when working exercises without available teacher/tutor assistance."
> —Arcola Sullivan, *Copiah-Lincoln Community College*

Example 3 Solving a System of Linear Equations by Graphing

Solve the system by the graphing method.

$$x - 2y = -2$$
$$-3x + 2y = 6$$

30. $x + y = -1$
$2x - y = -5$

Classroom Examples

To ensure that the classroom experience also matches the examples in the text and the practice exercises, we have included references to even-numbered exercises to be used as Classroom Examples. These exercises are highlighted in the Practice Exercises at the end of each section.

Better Learning Tools

Chapter Openers

Tired of students not being prepared? The Miller/O'Neill/Hyde *Chapter Openers* help students get better results through engaging *Puzzles and Games* that introduce the chapter concepts and ask "Are You Prepared?"

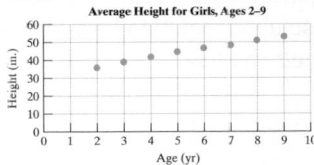

Chapter 2

In this chapter, we cover graphing and the applications of graphing. Graphs appear in magazines and newspapers and in other aspects of day-to-day life. In many fields of study such as the sciences and business, graphs are used to display data (information).

Are You Prepared?

Try answering questions 1–5 below. These problems will refresh your skills at reading and interpreting graphs. Match each answer with the correct letter and record the letter in the space below.

1. Use the graph to approximate the average height of a 5-yr-old girl.
2. Use the graph to estimate a girl's age if her height is 39 in. tall.
3. The equation $h = 2.5a + 31$ can be used to approximate a girl's height, h, in inches according to her age, a, in years. Use the formula $h = 2.5a + 31$ _____ girl's height at 6 yr old.
_____ ula $h = 2.5a + 31$ to estimate the age _____ height is 50 in.
_____ ula $h = 2.5a + 31$ to predict the height _____ d girl.

_____ ts the x-axis at an

$$\frac{}{3} \quad \frac{R}{5} \quad \frac{}{3} \quad \frac{P}{4}.$$

Average Height for Girls, Ages 2–9

T. 7.6 yr   E. 46 in.   I. 3 yr
X. 44 in.   C. 58.5 in.

> "I really like the chapter openers in MOH. The problems are a nice way to begin a class, and with the 'message' at the end, students can self check, and at some point can guess the answer similar to *Wheel of Fortune*. The activity is easy enough so that students feel confident, but at the same time, gives a nice review."
>
> —**Leonora Smook**, *Suffolk County Community College*

TIP and Avoiding Mistakes Boxes

TIP and **Avoiding Mistakes** boxes have been created based on the authors' classroom experiences—they have also been integrated into the **Worked Examples.** These pedagogical tools will help students get better results by learning how to work through a problem using a clearly defined step-by-step methodology.

Example 1   Multiplying Rational Expressions

Multiply. $\dfrac{5a - 5b}{10} \cdot \dfrac{2}{a^2 - b^2}$

Solution:

$$\frac{5a - 5b}{10} \cdot \frac{2}{a^2 - b^2}$$

$$= \frac{5(a - b)}{5 \cdot 2} \cdot \frac{2}{(a - b)(a + b)} \qquad \text{Factor numera... and denom}$$

$$= \frac{5(a - b) \cdot 2}{5 \cdot 2 \cdot (a - b)(a + b)} \qquad \text{Multiply.}$$

$$= \frac{5(a - b) \cdot 2}{5 \cdot 2 \cdot (a - b)(a + b)} = \frac{1}{c + b} \qquad \text{Reduce common factors and s}$$

Avoiding Mistakes
If all factors in the numerator simplify to 1, do not forget to write the factor of 1 in the numerator.

Avoiding Mistakes Boxes

Avoiding Mistakes boxes are integrated throughout the textbook to alert students to common errors and how to avoid them.

> "MOH presentation of reinforcement concepts builds students' confidence and provides easy to read guidance in developing basic skills and understanding concepts. I love the visual clue boxes 'Avoiding Mistakes.' Visual clue boxes provide tips and advice to assist students in avoiding common mistakes."
>
> —**Arcola Sullivan**, *Copiah-Lincoln Community College*

TIP Boxes

Teaching tips are usually revealed only in the classroom. Not anymore! TIP boxes offer students helpful hints and extra direction to help improve understanding and provide further insight.

TIP: The acronym, FOIL (first outer inner last) can be used as a memory device to multiply the two binomials.

Outer terms   First   Outer   Inner   Last

First terms

$(3y + 2)(7y - 6) = (3y)(7y) + (3y)(-6) + (2)(7y) + (2)(-6)$

Inner terms $= 21y^2 - 18y + 14y - 12$

Last terms $= 21y^2 - 4y - 12$

Note: It is important to realize that the acronym FOIL may only be used when finding the product of two binomials.

Get Better Results

Better Exercise Sets! Better Practice! Better Results!

► Do your students have trouble with problem solving?

► Do you want to help students overcome math anxiety?

► Do you want to help your students improve performance on math assessments?

Problem Recognition Exercises

Problem Recognition Exercises present a collection of problems that look similar to a student upon first glance, but are actually quite different in the manner of their individual solutions. Students sharpen critical thinking skills and better develop their "solution recall" to help them distinguish the method needed to solve an exercise—an essential skill in developmental mathematics.

Problem Recognition Exercises were tested in the authors' developmental mathematics classes and were created to improve student performance on tests.

> "The PREs are an excellent source of additional mixed problem sets. Frequently students have questions/comments like 'Where do I start?' or 'I know what to do once I get started, but I have trouble getting started.' Perhaps with these PREs, students will be able to overcome this obstacle."
> —Erika Blanken, *Daytona State College*

Problem Recognition Exercises

Recognizing Equations and Inequalities

At this point, you have learned how to solve a variety of equations and inequalities. Being able to distinguish the type of problem being posed is the first step in successfully solving it.

For Exercises 1–20,

a. Identify the problem type. Choose from

- linear equation
- quadratic equation
- rational equation
- absolute value equation
- radical equation
- equation quadratic in form

- polynomial equation
- linear inequality
- polynomial inequality
- rational inequality
- absolute value inequality
- compound inequality

b. Solve the equation or inequality. Write the solution to each inequality in interval notation if possible.

1. $(z^2 - 4)^2 - (z^2 - 4) - 12 = 0$ **2.** $3 + |4t - 1| < 6$ **3.** $2y(y - 4) \leq 5 + y$

4. $\sqrt[3]{11x - 3} + 4 = 6$ **5.** $-5 = -|w - 4|$ **6.** $\dfrac{5}{x - 2} + \dfrac{3}{x + 2} = 1$

7. $m^3 + 5m^2 - 4m - 20 \geq 0$ **8.** $-x - 4 > -5$ and $2x - 3 \leq 23$ **9.** $5 - 2[3 - (x - 4)] \leq 3x + 14$

10. $|2x - 6| = |x + 3|$ **11.** $\dfrac{3}{x - 2} \leq 1$ **12.** $9 < |x + 4|$

15. $-4 - x > 2$ or $8 < 2x$

18. $\dfrac{10}{x^2 + 1} < 0$

> "These are so important to test whether a student can recognize different types of problems and the method of solving each. They seem very unique—I have not noticed this feature in many other texts or at least your presentation of the problems is very organized and unique."
> —Linda Kuroski, *Erie Community College*

Student Centered Applications!

The Miller/O'Neill/Hyde Board of Advisors partnered with our authors to bring the *best applications* from every region in the country! These applications include real data and topics that are more relevant and interesting to today's student.

37. A rowing team trains on the Halifax River. It can row upstream 10 mi against the current in 2.5 hr and 16 mi downstream with the current in the same amount of time. Find the speed of the boat in still water and the speed of the current.

38. In her kayak, Bonnie can travel 31.5 mi downstream with the current in 7 hr. The return trip against the current takes 9 hr. Find the speed of the kayak in still water and the speed of the current.

Group Activities!

Each chapter concludes with a Group Activity to promote classroom discussion and collaboration—helping students not only to solve problems but to explain their solutions for better mathematical mastery. Group Activities are great for both full-time and adjunct instructors—bringing a more interactive approach to teaching mathematics! All required materials, activity time, and suggested group sizes are provided in the end-of-chapter material.

Group Activity

Computing the Future Value of an Investment

Materials: Calculator

Estimated time: 15 minutes

Group Size: 3

Suppose you are able to save $100 per month. If you invest the money in an account that pays 6% annual interest, how much money would you have at the end of 10 yr? This question can be answered by using the following formula.

$$S = R\left[\frac{(1 + i)^n - 1}{i}\right]$$

where

S is the future value of the investment.
R is the amount saved per period.
i is the interest rate per period.
n is the total number of periods.

In this example, $R = \$100$ (amount invested per month)

$i = \dfrac{0.06}{12} = 0.005$ (annual interest rate divided by 12 months)

$n = (12)(10) = 120$ (12 months per year times 10 years)

Therefore, $S = \$100\left[\dfrac{(1 + 0.005)^{120} - 1}{0.005}\right]$

$S = \$16,387.93$

1. Compute the future value of an account if you save $150 per month for 30 yr at an annual interest rate of 6%.

the monthly saving, interest rate, and ⋯⋯⋯⋯⋯⋯⋯⋯⋯⋯⋯⋯⋯ the value of the account varies with the interest rate and time the money is invested.

Monthly Savings	Annual Interest Rate	Years	Future Value

Get Better Results

Dynamic Math Animations

The Miller/O'Neill/Hyde author team has developed a series of Flash animations to illustrate difficult concepts where static images and text fall short. The animations leverage the use of on-screen movement and morphing shapes to enhance conceptual learning.

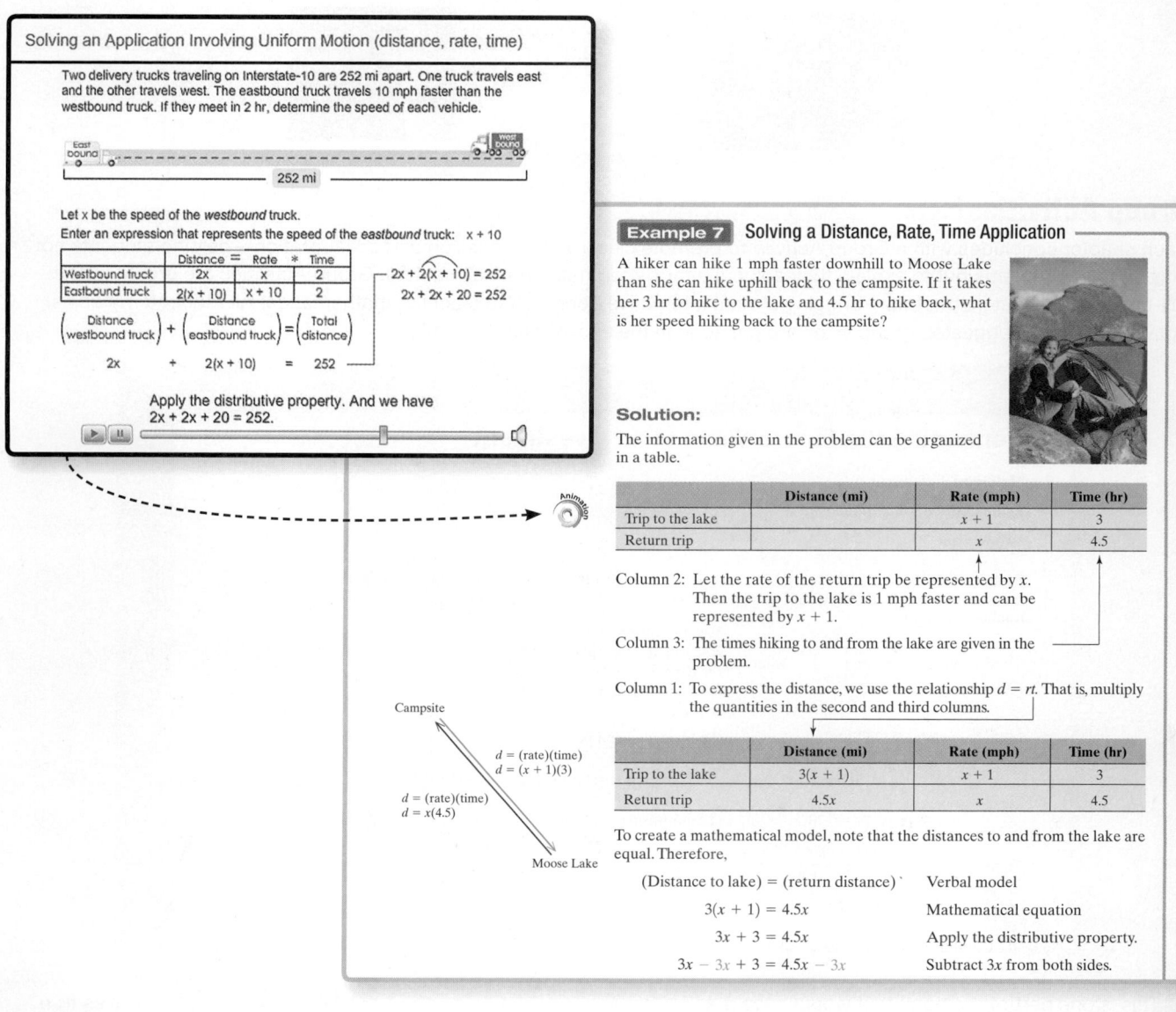

Solving an Application Involving Uniform Motion (distance, rate, time)

Two delivery trucks traveling on Interstate-10 are 252 mi apart. One truck travels east and the other travels west. The eastbound truck travels 10 mph faster than the westbound truck. If they meet in 2 hr, determine the speed of each vehicle.

252 mi

Let x be the speed of the *westbound* truck.
Enter an expression that represents the speed of the *eastbound* truck: x + 10

	Distance =	Rate *	Time
Westbound truck	2x	x	2
Eastbound truck	2(x + 10)	x + 10	2

$2x + 2(x + 10) = 252$
$2x + 2x + 20 = 252$

$$\left(\begin{array}{c}\text{Distance}\\\text{westbound truck}\end{array}\right) + \left(\begin{array}{c}\text{Distance}\\\text{eastbound truck}\end{array}\right) = \left(\begin{array}{c}\text{Total}\\\text{distance}\end{array}\right)$$

$$2x \quad + \quad 2(x + 10) \quad = \quad 252$$

Apply the distributive property. And we have
$2x + 2x + 20 = 252.$

Example 7 Solving a Distance, Rate, Time Application

A hiker can hike 1 mph faster downhill to Moose Lake than she can hike uphill back to the campsite. If it takes her 3 hr to hike to the lake and 4.5 hr to hike back, what is her speed hiking back to the campsite?

Solution:

The information given in the problem can be organized in a table.

	Distance (mi)	Rate (mph)	Time (hr)
Trip to the lake		$x + 1$	3
Return trip		x	4.5

Column 2: Let the rate of the return trip be represented by x. Then the trip to the lake is 1 mph faster and can be represented by $x + 1$.

Column 3: The times hiking to and from the lake are given in the problem.

Column 1: To express the distance, we use the relationship $d = rt$. That is, multiply the quantities in the second and third columns.

	Distance (mi)	Rate (mph)	Time (hr)
Trip to the lake	$3(x + 1)$	$x + 1$	3
Return trip	$4.5x$	x	4.5

Campsite

$d = (\text{rate})(\text{time})$
$d = (x + 1)(3)$

$d = (\text{rate})(\text{time})$
$d = x(4.5)$

Moose Lake

To create a mathematical model, note that the distances to and from the lake are equal. Therefore,

(Distance to lake) = (return distance)	Verbal model
$3(x + 1) = 4.5x$	Mathematical equation
$3x + 3 = 4.5x$	Apply the distributive property.
$3x - 3x + 3 = 4.5x - 3x$	Subtract $3x$ from both sides.

Through their classroom experience, the authors recognize that such media assets are great teaching tools for the classroom and excellent for online learning. The Miller/O'Neill/Hyde animations are interactive and quite diverse in their use. Some provide a virtual laboratory for which an application is simulated and where students can collect data points for analysis and modeling. Others provide interactive question-and-answer sessions to test conceptual learning. For word problem applications, the animations ask students to estimate answers and practice "number sense."

The animations were created by the authors based on over 75 years of combined teaching experience! To facilitate the use of the animations, the authors have placed icons in the text to indicate where animations are available. Students and instructors can access these assets online in either the ALEKS 360 Course product or Connect Math Hosted by ALEKS.

Modeling Using a Linear Equation in Two Variables

x Time (hr)	y Height (in.)
1	7.5
1.4	10.5
1.8	13.5
2.7	20.25
3.6	27
4.4	33
5.3	39.75
6.1	45.75

Water Level vs. Time

Animate

Get data point

Click here

Click on the "Get data point" button several times.

Copyright © 2008 Julie Miller

Example 1 Writing a Linear Model

Buffalo, New York, had 2 ft (24 in.) of snow on the
During the storm, snow fell at an average rate of $\frac{5}{8}$ i

a. Write a linear equation to compute the total sn
   the storm.

b. Graph the equation.

c. Use the equation to compute the depth of snow

d. If the snow depth was 31.5 in. at the end of the
   the storm lasted.

Solution:

a. The constant or base amount of snow before the storm began is 24 in. The
   rate of change is given by $\frac{5}{8}$ in. of snow per hour. If m is replaced by $\frac{5}{8}$ and b is
   replaced by 24, we have the linear equation

$$y = mx + b$$

$$y = \frac{5}{8}x + 24$$

b. The equation is in slope-
   intercept form, and the
   corresponding graph is shown
   in the figure.

Snow Depth Versus Time

$y = \frac{5}{8}x + 24$

5 in.

8 hr

c. $y = \frac{5}{8}x + 24$

$y = \frac{5}{8}(8) + 24$ Substitute $x = 8$.

$y = 5 + 24$ Solve for y.

$y = 29$ in.

The snow depth was 29 in. after 8 hr. The corresponding ordered pair is
(8, 29) and can be confirmed from the graph.

d. $y = \frac{5}{8}x + 24$

$31.5 = \frac{5}{8}x + 24$ Substitute $y = 31.5$.

$8(31.5) = 8\left(\frac{5}{8}x + 24\right)$ Multiply by 8 to clear fractions.

$252 = 5x + 192$ Clear parentheses.

$60 = 5x$ Solve for x.

$12 = x$

The storm lasted for 12 hr. The corresponding ordered pair is (12, 31.5) and
can be confirmed from the graph.

Avoiding Mistakes

A mathematical model *may be
reasonable only* for specific values
of the input variable. In Example 1,
the model $y = \frac{5}{8}x + 24$ is
reasonable for $0 \le x \le 12$. For
x values less than 0, the snow
storm hadn't started. For x values
greater than 12, the storm had
ended. We say that "model breakdown" occurs for values of
x outside the reasonable range.

Quality Content For Today's Online Learners

Online Exercises were carefully selected and developed to provide a seamless transition from textbook to technology.

For consistency, the guided solutions match the style and voice of the original text as though the author is guiding the students through the problems.

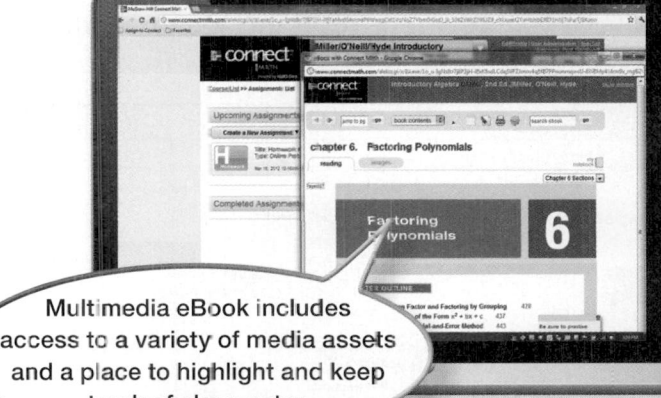

> Multimedia eBook includes access to a variety of media assets and a place to highlight and keep track of class notes

ALEKS Corporation's experience with algorithm development ensures a commitment to accuracy and a meaningful experience for students to demonstrate their understanding with a focus towards online learning.

The ALEKS® Initial Assessment is an artificially intelligent (AI), diagnostic assessment that identifies precisely what a student knows. Instructors can then use this information to make more informed decisions on what topics to cover in more detail with the class.

Hosted by **ALEKS Corp.**

ALEKS®

ALEKS is a unique, online program that significantly raises student proficiency and success rates in mathematics, while reducing faculty workload and office-hour lines. ALEKS uses artificial intelligence and adaptive questioning to assess precisely a student's knowledge, and deliver individualized learning tailored to the student's needs. With a comprehensive library of math courses, ALEKS delivers an unparalleled adaptive learning system that has helped millions of students achieve math success.

ALEKS Delivers a Unique Math Experience:

- **Research-Based, Artificial Intelligence** precisely measures each student's knowledge
- **Individualized Learning** presents the exact topics each student is most **ready to learn**
- **Adaptive, Open-Response Environment** includes comprehensive tutorials and resources
- **Detailed, Automated Reports** track student and class progress toward course mastery
- **Course Management Tools** include textbook integration, custom features, and more

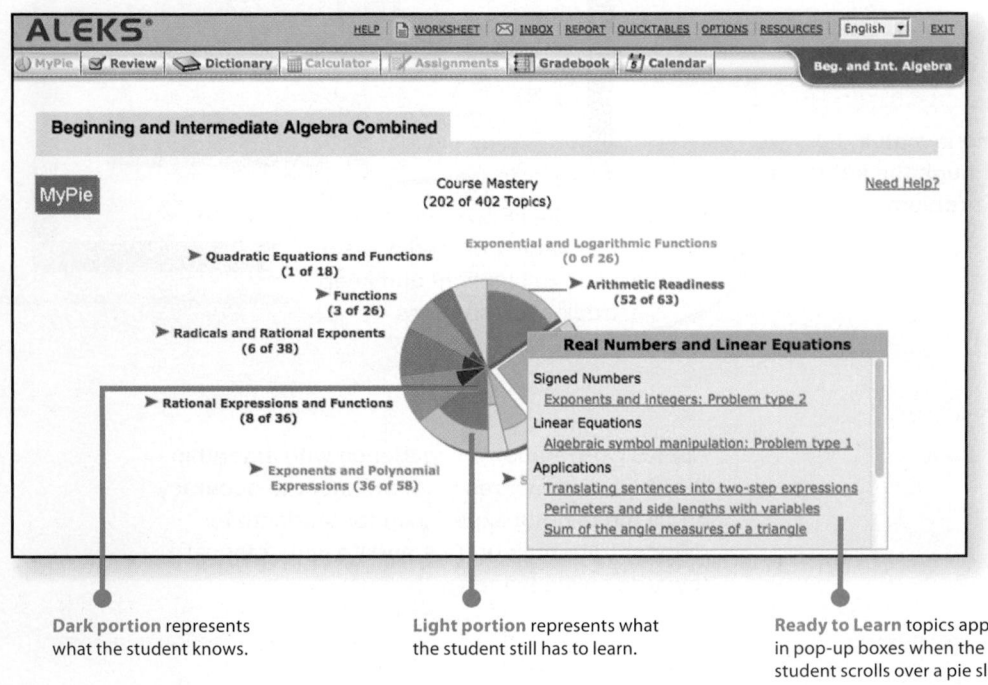

The ALEKS Pie summarizes a student's current knowledge, then delivers an individualized learning path with the exact topics the student is most ready to learn.

Dark portion represents what the student knows.

Light portion represents what the student still has to learn.

Ready to Learn topics appear in pop-up boxes when the student scrolls over a pie slice.

> "My experience with ALEKS has been effective, efficient, and eloquent. **Our students' pass rates improved from 49 percent to 82 percent with ALEKS.** We also saw student retention rates increase by 12% in the next course. Students feel empowered as they guide their own learning through ALEKS."
>
> —Professor Eden Donahou, *Seminole State College of Florida*

To learn more about ALEKS, please visit: **www.aleks.com/highered/math**

ALEKS® Prep Products

ALEKS Prep products focus on prerequisite and introductory material, and can be used during the first six weeks of the term to ensure student success in math courses ranging from Beginning Algebra through Calculus. ALEKS Prep quickly fills gaps in prerequisite knowledge by assessing precisely each student's preparedness and delivering individualized instruction on the exact topics students are most ready to learn. As a result, instructors can focus on core course concepts and see improved student performance with fewer drops.

> **"**ALEKS is wonderful. It is a professional product that takes very little time as an instructor to administer. Many of our students have taken Calculus in high school, but they have forgotten important algebra skills. ALEKS gives our students an opportunity to review these important skills.**"**
>
> —Professor Edward E. Allen, *Wake Forest University*

ALEKS®360 A Total Course Solution
With **eBook** Integration

A cost-effective total course solution: fully integrated, interactive eBook combined with the power of ALEKS adaptive learning and assessment.

Students can easily access the full eBook content, multimedia resources, and their notes from within their ALEKS Student Accounts.

To learn more about ALEKS, please visit: **www.aleks.com/highered/math**

Get Better Results

Additional Supplements

Student Resource Manual

The *Student Resource Manual (SRM),* created by the authors, is a printable, electronic supplement available to students through Connect Math Hosted by ALEKS Corp. Instructors can also choose to customize this manual and package with their course materials. With increasing demands on faculty schedules, this resource offers a convenient means for both full-time and adjunct faculty to promote active learning and success strategies in the classroom.

This manual supports the series in a variety of different ways:

- Discovery-based classroom activities written by the authors for each section
- Worksheets for extra practice written by the authors
- Lecture Notes designed to help students organize and take notes on key concepts
- Materials for a student portfolio

NEW Lecture Videos Created by the Authors

Julie Miller began creating these lecture videos for her own students to use when they were absent from class. The student response was overwhelmingly positive, prompting the author team to create the lecture videos for their entire developmental math book series. In these new videos, the authors walk students through the learning objectives using the same language and procedures outlined in the book. Students learn and review right alongside the author! Students can also access the written notes that accompany the videos.

- **Dynamic Math Animations**
  The authors have constructed a series of Flash animations to illustrate difficult concepts where static images and text fall short. The animations leverage the use of on-screen movement and morphing shapes to give students an interactive approach to conceptual learning. Some provide a virtual laboratory for which an application is simulated and where students can collect data points for analysis and modeling. Others provide interactive question-and-answer sessions to test conceptual learning.

- **Graphing Calculator Videos**
  The graphing calculator videos illustrate concepts using various operations on a graphing utility as well as help students master the most essential calculator skills used in the course. Topics include the explanation of various screens, avoiding common errors, and applying useful commands.

The videos are all closed-captioned for the hearing-impaired, and meet the Americans with Disabilities Act Standards for Accessible Design. These videos are available online through Connect Math Hosted by ALEKS Corp. as well as in ALEKS 360.

NEW Exercise Videos

The authors, along with a team of faculty who have used the Miller/O'Neill/Hyde textbooks for many years, have created new exercise videos for designated exercises in the textbook. These videos cover a representative sample of the main objectives in each section of the text. Each presenter works through selected problems, following the solution methodology employed in the text.

The video series is available online as part of Connect Math Hosted by ALEKS Corp. as well as in ALEKS 360. The videos are closed-captioned for the hearing impaired, and meet the Americans with Disabilities Act Standards for Accessible Design.

Annotated Instructor's Edition

In the *Annotated Instructor's Edition* (*AIE*), answers to all exercises appear adjacent to each exercise in a color used *only* for annotations. The *AIE* also contains Instructor Notes that appear in the margin. These notes offer instructors assistance with lecture preparation. In addition, there are Classroom Examples referenced in the text that are highlighted in the Practice Exercises. Also found in the *AIE* are icons within the Practice Exercises that serve to guide instructors in their preparation of homework assignments and lessons.

Powerpoints

The Powerpoints present key concepts and definitions with fully editable slides that follow the textbook. An instructor may project the slides in class or post to a website in an online course.

McGraw-Hill Connect Math Hosted by ALEKS Corp.

Connect Math Hosted by ALEKS Corp. is an exciting new assignment and assessment ehomework platform. Instructors can assign an AI-driven ALEKS Asssessment to identify the strengths and weaknesses of each student at the beginning of the term rather than after the first exam. Assignment creation and navigation is efficient and intuitive. The gradebook, based on instructor feedback, has a straightforward design and allows flexibility to import and export additional grades.

ALEKS Prep for Developmental Mathematics

ALEKS Prep for Beginning Algebra and Prep for Intermediate Algebra focus on prerequisite and introductory material for *Beginning Algebra* and *Intermediate Algebra*. These prep products can be used during the first 3 weeks of a course to prepare students for future success in the course and to increase retention and pass rates. Backed by two decades of National Science Foundation funded research, ALEKS interacts with students much like a human tutor, with the ability to precisely asses a student's preparedness and provide instruction on the topics the student is most likely to learn.

ALEKS Prep Course Products Feature:
- Artificial Intelligence Targeting Gaps in Individual Student's Knowledge
- Assessment and Learning Directed Toward Individual Student's Needs
- Open Response Environment with Realistic Input Tools
- Unlimited Online Access—PC and Mac Compatible

Free trial at www.aleks.com/free_trial/instructor

Instructor's Solutions Manual

The *Instructor's Solutions Manual* provides comprehensive, worked-out solutions to all exercises in the Chapter Openers, the Practice Exercises, the Problem Recognition Exercises, the end-of-chapter Review Exercises, the Chapter Tests, and the Cumulative Review Exercises.

Student's Solutions Manual

The *Student's Solutions Manual* provides comprehensive, worked-out solutions to the odd-numbered exercises in the Practice Exercise sets, the Problem Recognition Exercises, the end-of-chapter Review Exercises, the Chapter Tests, and the Cumulative Review Exercises. Answers to the Chapter Opener Puzzles are also provided.

Instructor's Test Bank

Among the supplements is a computerized test bank utilizing algorithm-based testing software to create customized exams quickly. Hundreds of text-specific, open-ended, and multiple-choice questions are included in the question bank. Sample chapter tests are also provided.

Get Better Results

Our Commitment to Market Development and Accuracy

McGraw-Hill's Development Process is an ongoing, never-ending, market-oriented approach to building accurate and innovative print and digital products. We begin developing a series by partnering with authors that desire to make an impact within their discipline to help students succeed. Next, we share these ideas and manuscript with instructors for review for feedback and to ensure that the authors' ideas represent the needs within that discipline. Throughout multiple drafts, we help our authors adapt to incorporate ideas and suggestions from reviewers to ensure that the series carries the same pulse as today's classrooms. With any new series, we commit to accuracy across the series and its supplements. In addition to involving instructors as we develop our content, we also utilize accuracy checks through our various stages of development and production. The following is a summary of our commitment to market development and accuracy:

1. 2 drafts of author manuscript
2. 2 rounds of manuscript review
3. 2 focus groups
4. 1 consultative, expert review
5. 2 accuracy checks
6. 2 rounds of proofreading and copyediting
7. Toward the final stages of production, we are able to incorporate additional rounds of quality assurance from instructors as they help contribute toward our digital content and print supplements

This process then will start again immediately upon publication in anticipation of the next edition. With our commitment to this process, we are confident that our series has the most developed content the industry has to offer, thus pushing our desire for quality and accurate content that meets the needs of today's students and instructors.

Acknowledgments and Reviewers

Paramount to the development of *Intermediate Algebra* was the invaluable feedback provided by the instructors from around the country that reviewed the manuscript or attended a market development event over the course of the several years the text was in development.

AMATYC 2011 Focus Group

Shanna Goff, *Grand Rapids Community College*
Betsy McKinney, *Grand Rapids Community College*
Michelle Hempton, *Lansing Community College*
Linda Schott, *Ozarks Technical Community College*

Jessica Lopez, *Saint Philips College*
Lorraine Lopez, *San Antonio College*
Peter Georgakis, *Santa Barbara City College*

NADE 2012 Focus Group

Beth Rountree, *Brevard College*
Kelly Jackson, *Camden Community College*
Sharonlyn Moses-Polk, *Central Carolina Technical College*
Sharhonda Elmore-Rivers, *Central Carolina Technical College*
Elizabeth Valentine, *Charleston Southern University*
Joseph Hernandez, *Delta College*
Randall Nicols, *Delta College*
Thomas Wells, *Delta College*
Teresa Hasenauer, *Indian River State College*

Tiffany Lewis, *Indian River State College*
Rita Lindsay, *Indian River State College*
Deanna Voehl, *Indian River State College*
Jennifer Manor, *Pierpont Community and Technical College*
Eden Donahou, *Seminole State College of Florida*
Dave DelRossi, *Tallahassee Community College*
Natalie Johnson, *Tarrant County College*
Claudia Martinez, *Valencia College*

Reviewers of Miller/O'Neill/Hyde Developmental Mathematics Series

Maryann Faller, *Adirondack Community College*
Albert Miller, *Ball State University*
Debra Pearson, *Ball State University*
Patricia Parkison, *Ball State University*
Robin Rufatto, *Ball State University*
Melanie Walker, *Bergen Community College*
Robert Fusco, *Bergen Community College*
Latonya Ellis, *Bishop State Community College*
Ana Leon, *Bluegrass Community College & Technical College*
Kaye Black, *Bluegrass Community College & Technical College*
Barbara Elzey, *Bluegrass Community College & Technical College*
Cheryl Grant, *Bowling Green State University*
Beth Rountree, *Brevard College*
Juliet Carl, *Broward College*
Lizette Foley, *Broward College*
Angie Matthews, *Broward College*
Mitchel Levy, *Broward College*
Jody Harris, *Broward College*
Michelle Carmel, *Broward College*
Antonnette Gibbs, *Broward College*
Elizabeth Valentine, *Charleston Southern University*
Adedoyin Adeyiga, *Cheyney University of Pennsylvania*
Dot French, *Community College of Philadelphia*
Brad Berger, *Copper Mountain College*
Donna Troy, *Cuyamaca College*
Brianna Kurtz, *Daytona State College–Daytona Beach*

Jennifer Walsh, *Daytona State College–Daytona Beach*
Marc Campbell, *Daytona State College–Daytona Beach*
Richard Rupp, *Del Mar College*
Joseph Hernandez, *Delta College*
Randall Nichols *Delta College*
Thomas Wells, *Delta College*
Paul Yun, *El Camino College*
Catherine Bliss, *Empire State College–Saratoga Springs*
Laurie Davis, *Erie Community College*
Linda Kuroski, *Erie Community College*
David Usinski, *Erie Community College*
Ron Bannon, *Essex County College*
David Platt, *Front Range Community College*
Alan Dinwiddie, *Front Range Community College*
Shanna Goff, *Grand Rapids Community College*
Betsy McKinney, *Grand Rapids Community College*
Cathy Gardner, *Grand Valley State University*
Jane Mays, *Grand Valley State University*
John Greene, *Henderson State University*
Fred Worth, *Henderson State University*
Ryan Baxter, *Illinois State University*
Angela Mccombs, *Illinois State University*
Elisha Van Meenen, *Illinois State University*
Teresa Hasenauer, *Indian River State College*
Tiffany Lewis, *Indian River State College*
Deanna Voehl, *Indian River State College*
Joe Jordan, *John Tyler Community College*

Get Better Results

Reviewers of the Miller/O'Neill/Hyde Developmental Mathematics Series

Sally Copeland, *Johnson County Community College*
Nancy Carpenter, *Johnson County Community College*
Susan Yellott, *Kilgore College*
Kim Miller, *Labette Community College*
Michelle Hempton, *Lansing Community College*
Michelle Whitmer, *Lansing Community College*
Kuen Lee, *Los Angeles Trade Tech*
Nic Lahue, *MCC-Longview Community College*
Jason Pallett, *MCC-Longview Community College*
Janet Wyatt, *MCC-Longview Community College*
Rene Barrientos, *Miami Dade College—Kendall*
Nelson De La Rosa, *Miami Dade College—Kendall*
Jody Balzer, *Milwaukee Area Technical College*
Shahla Razavi, *Mt. San Jacinto College*
Shawna Bynum, *Napa Valley College*
Tammy Ford, *North Carolina A & T University*
Ebrahim Ahmadizadeh, *Northampton Community College*
Christine Wetzel-Ulrich, *Northampton Community College*
Sharon Totten, *Northeast Alabama Community College*
Rodolfo Maglio, *Northeastern Illinios University*
Christine Copple, *Northwest State Community College*
Sumitana Chatterjee, *Nova Community College*
Charbel Fahed, *Nova Community College*
Ken Hirschel, *Orange County Community College*
Linda K. Schott, *Ozarks Technical Community College*
Matthew Harris, *Ozarks Technical Community College*
Daniel Kopsas, *Ozarks Technical Community College*
Andrew Aberle, *Ozarks Technical Community College*
Alan Papen, *Ozarks Technical Community College*
Angela Shreckhise, *Ozarks Technical Community College*
Jacob Lewellen, *Ozarks Technical Community College*
Marylynne Abbott, *Ozarks Technical Community College*
Jeffrey Gervasi, *Porterville College*
Stewart Hathaway, *Porterville College*
Lauran Johnson, *Richard Bland College*
Matthew Nickodemus, *Richard Bland College*
Cameron English, *Rio Hondo College*
Lydia Gonzalez, *Rio Hondo College*
Mark Littrell, *Rio Hondo College*
Matthew Pitassi, *Rio Hondo College*
Wayne Lee, *Saint Philips College*

Paula Looney, *Saint Philips College*
Fred Bakenhus, *Saint Philips College*
Lydia Casas, *Saint Philips College*
Gloria Guerra, *Saint Philips College*
Sounny Slitine, *Saint Philips College*
Jessica Lopez, *Saint Philips College*
Lorraine Lopez, *San Antonio College*
Peter Georgakis, *Santa Barbara City College*
Sandi Nieto-Navarro, *Santa Rosa Junior College*
Steve Drucker, *Santa Rosa Junior College*
Jean-Marie Magnier, *Springfield Tech Community College*
Dave Delrossi, *Tallahassee Community College*
Natalie Johnson, *Tarrant County College South*
Marilyn Peacock, *Tidewater Community College*
Yvonne Aucoin, *Tidewater Community College*
Cynthia Harris, *Triton College*
Jennifer Burkett, *Triton College*
Christyn Senese, *Triton College*
Jennifer Dale, *Triton College*
Patricia Hussey, *Triton College*
Glenn Jablonski, *Triton College*
Myrna La Rosa, *Triton College*
Michael Maltenfort, *Truman College*
Abdallah Shuaibi, *Truman College*
Marta Hidegkuti, *Truman College*
Sandra Wilder, *University of Akron*
Sandra Jovicic, *University of Akron*
Edward Migliore, *University of California–Santa Cruz*
Carolann Van Galder, *University of Wisconsin—Rock County*
Claudia Martinez, *Valencia College*
Stephen Toner, *Victor Valley Community College*
David Cooper, *Wake Tech Community College*
Karlata Elliott, *Wake Tech Community College*
Laura Kalbaugh, *Wake Tech Community College*
Kelly Vetter, *Wake Tech Community College*
Jacqui Fields, *Wake Tech Community College*
Jennifer Smeal, *Wake Tech Community College*
Shannon Vinson, *Wake Tech Community College*
Kim Walaski, *Wake Tech Community College*
Maziar Ouliaeinia, *Western Iowa Tech Community College*
Keith McCoy, *Wilbur Wright College*

Also, a special thanks to all instructors who have reviewed previous editions of this series.

Acknowledgments

The author team most humbly would like to thank all the people who have contributed to this project.

Special thanks to our team of digital authors for their thousands of hours of work: Jody Harris, Linda Schott, Lizette Hernandez Foley, Michael Larkin, Alina Coronel, and to the masters of ceremonies in the digital world: Donna Gerken, Nicole Lloyd, and Steve Toner. We also offer our sincerest appreciation to the video talent: Jody Harris, Alina Coronel, Didi Quesada, Tony Alfonso, and Brianna Kurtz. You folks are the best! To Mitchel Levy our exercise consultant, thank you for your watchful eye fine-tuning the exercise sets and for your ongoing valuable feedback. To Gene Rumann, thank you so much for ensuring accuracy in our manuscripts.

Finally, we greatly appreciate the many people behind the scenes at McGraw-Hill without whom we would still be on page 1. To Emily Williams, the best developmental editor in the business and steady rock that has kept the train on the track. To Mary Ellen Rahn, our executive editor and overall team leader. We're forever grateful for your support and innovative ideas. To our copy editor Pat Steele who has watched over our manuscript for many years and has been a long-time mentor for our writing. You're amazing. To the marketing team Alex Coleman, Peter Vaneria, Tim Cote, John Osgood, and Cheryl Sowa, thank you for your creative ideas in making our books come to life in the market. To the directors of digital content team, Emilie Berglund and Nicole Lloyd, we are most grateful for your long hours of work and innovation in a world that changes day-to-day. And many thanks to the team at ALEKS that oversees quality control in the digital content. To the production team Peggy Selle, Sherry Kane, Laurie Janssen, and Carrie Burger for making the manuscript beautiful and bringing it all together. And finally to Dawn Bercier, Ryan Blankenship, Marty Lange, and Kurt Strand, thank you for supporting our projects over the years and for the confidence you've always shown in us.

Most importantly, we give special thanks to the students and instructors who use our series in their classes.

New and Updated Content for Miller, O'Neill, and Hyde's *Intermediate Algebra*, Fourth Edition:

Updates throughout the Text:
- New Vocabulary and Key Concept Exercises
- New Tips and Avoiding Mistakes boxes
- Updated Applications and Data in all instances where appropriate

Chapter by Chapter Changes:

Chapter R
- 12 new exercises
- 23 new vocabulary exercises
- 4 new examples

Chapter 1
- 47 new exercises
- 40 new vocabulary exercises
- 7 new examples
- Reworked the definition of a Linear Equation in One Variable and the subsequent examples.
- Updated the Multiplication and Division Properties of Equality.
- Reworked the narrative on the introduction to the empty set and a linear inequality.
- New narrative on Inequalities and their solutions in the form of a graph, interval notation, and set builder notation. Examples and exercises in Section 1.4 now ask for set builder notation as well as interval notation.
- Split the Procedure Box on Solving a Compound Inequality into two parts for 'Or' and 'And'.

Chapter 2
- 38 new exercises
- 39 new vocabulary exercises
- Updated examples in Section 2.1 to show the corresponding coordinate pair next to the table of solutions.
- Reworked the narrative on introduction to slope.

Chapter 3
- 47 new exercises
- 27 new vocabulary exercises
- 1 new example
- Combined the objectives in Section 3.1 on Solutions to Systems of Linear Equations and Dependent and Inconsistent Systems of Linear Equations. Rewrote the narrative on the introduction to dependent and independent equations.
- In Section 3.1–3.3 changed the terminology from Dependent System to Dependent Equations.
- Changed directions to the exercise sets throughout Chapter 3 to more precisely explain the questions.

Chapter 4
- 34 new exercises
- 46 new vocabulary exercises
- 1 new example

Chapter 5
- 11 new exercises
- 17 new vocabulary exercises
- 2 new PRE exercises
- Rewrote narrative on the domain of a function and a complex fraction.
- Removed the objective and example in Section 5.4 on Using Complex Fractions in Applications.
- Swapped the Expanding Your Skills exercises from Section 5.5 and 5.6.

Chapter 6
- 43 new exercises
- 48 new vocabulary exercises
- 6 new examples
- Additional explanation provided in Examples of Section 6.1, in the Definition Box of an nth Root.
- Reworked Property Box on $\sqrt[n]{a}$ and the explanation following.
- Split Example 4 into two separate Examples in Section 6.3.
- Rewrote narrative and Procedure Box on a radical equation with extraneous solutions.
- Rewrote Definition of $\sqrt{-b}$ for $b > 0$.

Chapter 7
- 16 new exercises
- 39 new vocabulary exercises
- 4 new examples
- New title for the chapter to include 'Inequalities'. Also changed the title of Section 7.6 to 'Polynomial and Rational Inequalities'.
- Rewrote narrative introducing the square root property, vertical shift, and horizontal shift.

Chapter 8
- 10 new exercises
- 43 new vocabulary exercises
- 1 new example
- Updated definition of Composition of Functions.
- Included additional images and narrative to explain a One-to-One Function.
- New narrative on verifying if two functions are inverses.
- New definition on an Inverse Function.
- Reorganized Section 8.2 so that Introduction to the Inverse of a Function appears before finding the equation of the inverse of a function.

Chapter 9

- 7 new exercises
- 17 new vocabulary exercises
- 2 new PRE exercises
- 3 new examples
- Provides the standard form of a circle, an ellipse at the beginning of the explanation for each example where appropriate.
- New narrative on graphing an ellipse and a hyperbola at the origin and on the introduction to nonlinear inequalities in two variables.

Chapter 10

- 26 new exercises
- 21 new vocabulary exercises

- 4 new PRE exercises
- 4 new examples
- New Calculator Connections Box.
- Rewrote explanation on common difference.

Additional Topics Appendix

- 6 new exercises
- 3 new vocabulary exercises
- Changed terminology from dependent systems to dependent equations.

Review of Basic Algebraic Concepts

R

CHAPTER OUTLINE

R.1 Study Skills   2

   Group Activity: Becoming a Successful Student   3

R.2 Sets of Numbers and Interval Notation   5

R.3 Operations on Real Numbers   16

R.4 Simplifying Algebraic Expressions   31

Chapter R

In this chapter, we begin our study of algebra by reviewing the sets of numbers used in day-to-day life. We also review how to simplify numerical expressions and algebraic expressions. To prepare for this chapter, practice the following operations on whole numbers, decimals, fractions, and mixed numbers.

Are You Prepared?

For Exercises 1–12, simplify each expression. Then find the answer on the right and record the corresponding letter at the bottom of the page. When you are finished, you will have a key definition for this chapter.

Exercises		Answers	
1. $36{,}636 \div 43$	**2.** 0.25×6340	**I.** 389.7842	**B.** 1585
3. $24.0842 + 365.7$	**4.** $\dfrac{4}{7} \times \dfrac{21}{10}$	**R.** $\dfrac{1}{6}$	**D.** $\dfrac{6}{5}$
5. $\dfrac{7}{3} \div 14$	**6.** $\dfrac{5}{3} - \dfrac{1}{2}$	**T.** 852	**S.** 11.95
7. $\dfrac{2}{3} - \dfrac{7}{12}$	**8.** $4\dfrac{5}{9} - 1\dfrac{1}{3}$	**E.** $3\dfrac{2}{9}$	**P.** $\dfrac{1}{12}$
9. $3\dfrac{1}{5} \times 2\dfrac{1}{2}$	**10.** $3.75 + 8\dfrac{1}{5}$	**Y.** $\dfrac{7}{6}$	**O.** 8
11. $582 \div 0.01$	**12.** 582×0.01	**V.** $58{,}200$	**U.** 5.82

In this chapter we will show that $a(b + c) = ab + ac$. This important property is called the:

$\overline{}\ \overline{}\ \overline{}\ \overline{}\ \overline{}\ \overline{}\ \overline{}\ \overline{}\ \overline{}\ \overline{}\ \overline{}\ \overline{}\quad \overline{}\ \overline{}\ \overline{}\ \overline{}\ \overline{}\ \overline{}\ \overline{}\ .$
  4   3   10   1   5   3   2   12   1   3   11   8   7   5   9   7   8   5   1   6

Section R.1   Study Skills

Concepts

1. Before the Course
2. During the Course
3. Preparation for Exams
4. Where to Go for Help

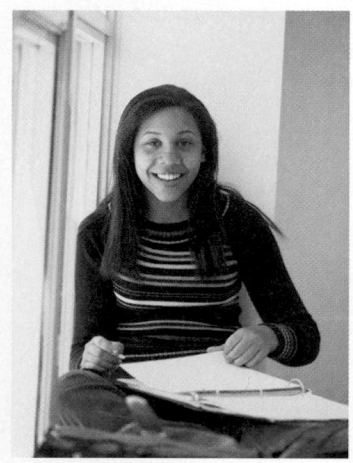

In taking a course in algebra, you are making a commitment to yourself, your instructor, and your classmates. Following some or all of the study tips below can help you be successful in this endeavor. The features of this text that will assist you are printed in blue.

1. Before the Course

- Purchase the necessary materials for the course before the course begins or on the first day.
- Obtain a three-ring binder to keep and organize your notes, homework, tests, and any other materials acquired in the class. We call this type of notebook a *portfolio*.
- Arrange your schedule so that you have enough time to attend class and to do homework. A common rule is to set aside at least 2 hours for homework for every hour spent in class. That is, if you are taking a 4-credit-hour course, plan on at least 8 hours a week for homework. If you experience difficulty in mathematics, plan for more time. A 4-credit-hour course will then take *at least* 12 hours each week—about the same as a part-time job.
- Communicate with your employer and family members the importance of your success in this course so that they can support you.
- Be sure to find out the type of calculator (if any) that your instructor requires. Also determine if there will be online homework or other computer requirements.

2. During the Course

- To prepare yourself for the next day's class, read the section in the text *before* coming to class. This will help you familiarize yourself with the material and terminology.
- Attend every class and be on time.
- Take notes in class. Write down all of the examples that the instructor presents. Read the notes after class, and add any comments to make your notes clearer to you. Use a tape recorder to record the lecture if the instructor permits the recording of lectures.
- Ask questions in class.
- Read the section in the text *after* the lecture, and pay special attention to the Tip boxes and Avoiding Mistakes boxes.
- After you read an example, try the accompanying Skill Practice problem. The skill practice problem mirrors the example and tests your understanding of what you have read.
- Do homework every night. Even if your class does not meet every day, you should still do some work every night to keep the material fresh in your mind.
- Check your homework with the answers that are supplied in the back of this text. Analyze what you did wrong and correct the exercises that do not match. Circle or star those that you cannot correct yourself. This way you can easily find them and ask your instructor the next day.
- Be sure to do the Vocabulary and Key Concepts exercises found at the beginning of the Practice Exercises.
- The Problem Recognition Exercises are found in Chapters 1–10. These provide additional practice distinguishing among a variety of problem types. Sometimes the most difficult part of learning mathematics is retaining all that you learn. These exercises are excellent tools for retention of material.

- Form a study group with fellow students in your class, and exchange phone numbers. You will be surprised by how much you can learn by talking about mathematics with other students.
- If you use a calculator in your class, read the Calculator Connections boxes to learn how and when to use your calculator.
- Ask your instructor where you might obtain extra help if necessary.

3. Preparation for Exams

- Look over your homework and rework exercises that gave you trouble. Pay special attention to the exercises you have circled or starred to be sure that you have learned that concept.
- Read through the Summary at the end of the chapter. Be sure that you understand each concept and example. If not, go to the section in the text and reread that section.
- Give yourself enough time to take the Chapter Test uninterrupted. Then check the answers. For each problem you answered incorrectly, go to the Review Exercises and do all of the problems that are similar.
- To prepare for the final exam, complete the Cumulative Review Exercises at the end of each chapter, starting with Chapter 2. If you complete the cumulative reviews after finishing each chapter, then you will be preparing for the final exam throughout the course. The Cumulative Review Exercises are another excellent tool for helping you retain material.

4. Where to Go for Help

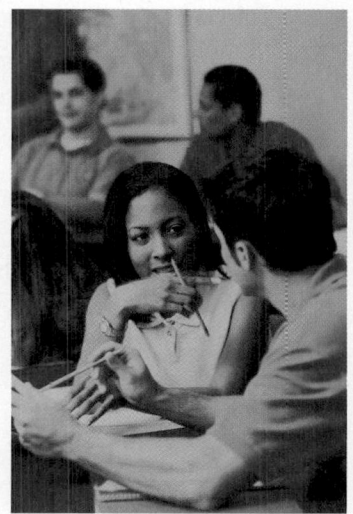

- At the first sign of trouble, see your instructor. Most instructors have specific office hours set aside to help students. Don't wait until after you have failed an exam to seek assistance.
- Get a tutor. Most colleges and universities have free tutoring available.
- When your instructor and tutor are unavailable, use the Student Solutions Manual for step-by-step solutions to the odd-numbered problems in the exercise sets.
- Work with another student from your class.
- Work on the computer. Many mathematics tutorial programs and websites are available on the Internet, including the website that accompanies this text.

Group Activity

Becoming a Successful Student

Materials: Computer with Internet access

Estimated time: 15 minutes

Group Size: 4

Good time management, good study skills, and good organization will help you be successful in this course. Answer the following questions and compare your answers with your group members.

1. To motivate yourself to complete a course, it is helpful to have clear reasons for taking the course. List your goals for taking this course and discuss them with your group.

2. Taking 12 credit-hours is the equivalent of a full-time job. Often students try to work too many hours while taking classes at school.

   a. Write down the number of hours you work per week and the number of credit-hours you are taking this term.

     Number of hours worked per week _____

     Number of credit-hours this term _____

Number of Credit-Hours	Maximum Number of Hours of Work per Week
3	40
6	30
9	20
12	10
15	0

   b. The table gives a recommended limit to the number of hours you should work for the number of credit-hours you are taking at school. (Keep in mind that other responsibilities in your life such as your family might also make it necessary to limit your hours at work even more.) How do your numbers from part (a) compare to those in the table? Are you working too many hours?

   c. It is often suggested that you devote two hours of study and homework time outside of class for each credit-hour you take at school. For example:

       12 credit-hours
     + 24 study hours
        36 total hours ⟵— full-time job!

     Based on the number of credit-hours you are taking, how many study hours should you plan for? _____

     What is the total number of hours (class time plus study time) that you should devote to school? _____

3. For the following week, write down the times each day that you plan to study math.

Monday	Tuesday	Wednesday	Thursday	Friday	Saturday	Sunday

4. Write down the date of your next math test. _____

5. Look through the book in Chapter 2 and find the page number corresponding to each feature in the book. Discuss with your group members how you might use each feature.

Problem Recognition Exercises: page _____

Chapter Summary: page _____

Chapter Review Exercises: page _____

Chapter Test: page _____

Cumulative Review Exercises: page _____

6. Look at the Skill Practice exercises after each example (for instance, find Skill Practice exercise 6 in Section 1.2). Where are the answers to these exercises located? Discuss with your group members how you might use the Skill Practice exercises.

7. Discuss with your group members places where you can go for extra help in math. Then write down three of the suggestions.

   _____

   _____

8. Do you keep an organized notebook for this class? Can you think of any suggestions that you can share with your group members to help them keep their materials organized?

9. Do you think that you have math anxiety? Read the following list for some possible solutions. Check the activities that you can realistically try to help you overcome this problem.

   _____ Read a book on math anxiety.

   _____ Search the Web for help tips on handling math anxiety.

   _____ See a counselor to discuss your anxiety.

   _____ Talk with your instructor to discuss strategies to manage math anxiety.

   _____ Evaluate your time management to see if you are trying to do too much. Then adjust your schedule accordingly.

10. Some students favor different methods of learning over others. For example, you might prefer:
    - Learning through listening and hearing.
    - Learning through seeing images, watching demonstrations, and visualizing diagrams and charts.
    - Learning by experience through a hands-on approach.
    - Learning through reading and writing.

    Most experts believe that the most effective learning comes when a student engages in *all* of these activities. However, each individual is different and may benefit from one activity more than another. You can visit a number of different websites to determine your "learning style." Try doing a search on the Internet with the key words *"learning styles assessment."* Once you have found a suitable website, answer the questionnaire and the site will give you feedback on what method of learning works best for you.

Sets of Numbers and Interval Notation   Section R.2

1. The Set of Real Numbers

Algebra is a powerful mathematical tool that is used to solve real-world problems in science, business, and many other fields. We begin our study of algebra with a review of basic definitions and notations used to express algebraic relationships.

Concepts

1. The Set of Real Numbers
2. Inequalities
3. Interval Notation
4. Translations Involving Inequalities

In mathematics, a collection of items (called elements) is called a **set**, and the set braces { } are used to enclose the elements of the set. For example, the set {a, e, i, o, u} represents the vowels in the English alphabet. The set {1, 3, 5, 7} represents the first four positive odd numbers. Another method to express a set is to *describe* the elements of the set by using **set-builder notation**. Consider the set {a, e, i, o, u} in set-builder notation.

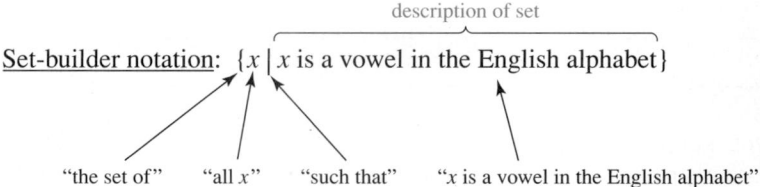

Consider the set {1, 3, 5, 7} in set-builder notation.

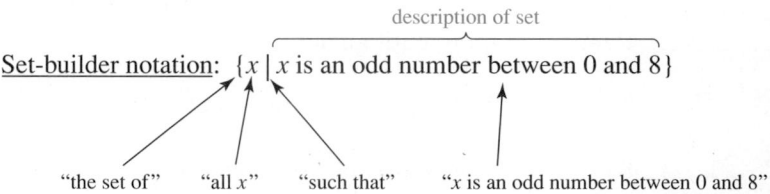

Several sets of numbers are used extensively in algebra. The numbers you are familiar with in day-to-day calculations are elements of the set of **real numbers**. These numbers can be represented graphically on a horizontal number line with a point labeled as 0. Positive real numbers are graphed to the right of 0, and negative real numbers are graphed to the left. Each point on the number line corresponds to exactly one real number, and for this reason, the line is called the **real number line** (Figure R-1).

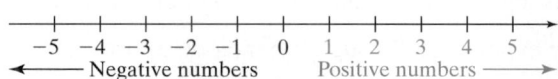

Figure R-1

Several sets of numbers are **subsets** (or part) of the set of real numbers. These are

  The set of natural numbers

  The set of whole numbers

  The set of integers

  The set of rational numbers

  The set of irrational numbers

Natural Numbers, Whole Numbers, and Integers

The set of **natural numbers** is {1, 2, 3, . . .}.
The set of **whole numbers** is {0, 1, 2, 3, . . .}.
The set of **integers** is { . . . , −3, −2, −1, 0, 1, 2, 3, . . .}.

The set of rational numbers consists of all the numbers that can be defined as a ratio of two integers.

Rational Numbers

The set of **rational numbers** is $\{\frac{p}{q} \mid p$ and q are integers and q does not equal zero$\}$.

Example 1 Identifying Rational Numbers

Show that each number is a rational number by finding two integers whose ratio equals the given number.

a. $\dfrac{-4}{7}$　　**b.** 8　　**c.** $0.\overline{6}$　　**d.** 0.87

Solution:

a. $\frac{-4}{7}$ is a rational number because it can be expressed as the ratio of the integers -4 and 7.

b. 8 is a rational number because it can be expressed as the ratio of the integers 8 and 1 $(8 = \frac{8}{1})$. In this example we see that *an integer is also a rational number.*

c. $0.\overline{6}$ represents the repeating decimal $0.6666666\ldots$ and can be expressed as the ratio of 2 and 3 $(0.\overline{6} = \frac{2}{3})$. In this example we see that *a repeating decimal is a rational number.*

d. 0.87 is the ratio of 87 and 100 $(0.87 = \frac{87}{100})$. In this example we see that *a terminating decimal is a rational number.*

Skill Practice Show that the numbers are rational by writing them as a ratio of integers.

1. $\dfrac{-9}{8}$　　**2.** 0　　**3.** $0.\overline{3}$　　**4.** 0.45

TIP: Any rational number can be represented by a terminating decimal or by a repeating decimal.

Some real numbers such as the number π (pi) cannot be represented by the ratio of two integers. In decimal form, an irrational number is a nonterminating, nonrepeating decimal. The value of π, for example, can be approximated as $\pi \approx 3.1415926535897932$. However, the decimal digits continue indefinitely with no pattern. Other examples of irrational numbers are the square roots of nonperfect squares, such as $\sqrt{3}$ and $\sqrt{10}$.

Irrational Numbers

The set of **irrational numbers** is a subset of the real numbers whose elements cannot be written as a ratio of two integers.

Note: An irrational number cannot be written as a terminating decimal or as a repeating decimal.

The set of real numbers consists of both the rational numbers and the irrational numbers. The relationships among the sets of numbers discussed thus far are illustrated in Figure R-2.

Answers

1. $\dfrac{-9}{8}$　**2.** $\dfrac{0}{1}$

3. $\dfrac{1}{3}$　**4.** $\dfrac{45}{100}$ or $\dfrac{9}{20}$

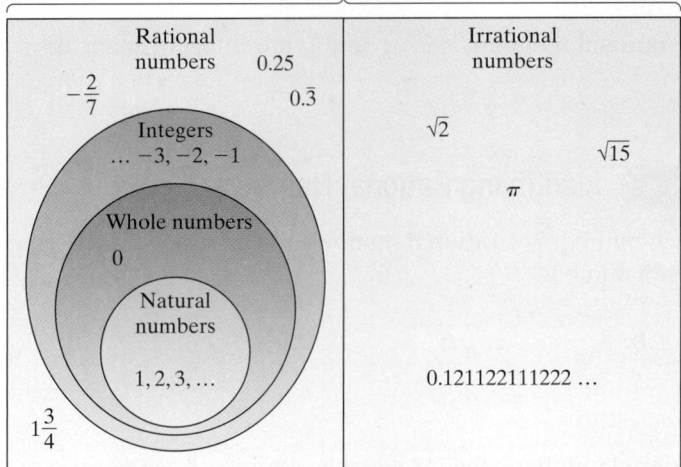

Real numbers

Rational numbers 0.25
$-\frac{2}{7}$ 0.$\overline{3}$

Irrational numbers

$\sqrt{2}$ $\sqrt{15}$
π

Integers
... −3, −2, −1

Whole numbers
0

Natural numbers
1, 2, 3, ...

0.121122111222 ...

$1\frac{3}{4}$

Figure R-2

Example 2 **Classifying Numbers by Set**

Check the set(s) to which each number belongs. The numbers may belong to more than one set.

	Natural Numbers	Whole Numbers	Integers	Rational Numbers	Irrational Numbers	Real Numbers
−6						
$\sqrt{23}$						
$-\frac{2}{7}$						
3						
2.$\overline{3}$						

Solution:

	Natural Numbers	Whole Numbers	Integers	Rational Numbers	Irrational Numbers	Real Numbers
−6			✓	✓		✓
$\sqrt{23}$					✓	✓
$-\frac{2}{7}$				✓		✓
3	✓	✓	✓	✓		✓
2.$\overline{3}$				✓		✓

Skill Practice

5. Check the set(s) to which each number belongs.

	1	0.47	$\sqrt{5}$	$-\frac{1}{2}$
Natural				
Whole				
Integer				
Rational				
Irrational				
Real				

Answer

5.

	1	0.47	$\sqrt{5}$	$-\frac{1}{2}$
Natural	✓			
Whole	✓			
Integer	✓			
Rational	✓	✓		✓
Irrational			✓	
Real	✓	✓	✓	✓

2. Inequalities

The relative value of two numbers can be compared by using the real number line. We say that a is less than b (written mathematically as $a < b$) if a lies to the left of b on the number line.

$$a < b$$

We say that a is greater than b (written mathematically as $a > b$) if a lies to the right of b on the number line.

$$a > b$$

From looking at the number line, note that $a > b$ is the same as $b < a$. Table R-1 summarizes the relational operators that compare two real numbers a and b.

Table R-1

Mathematical Expression	Translation	Other Meanings
$a < b$	a is less than b	b exceeds a
		b is greater than a
$a > b$	a is greater than b	a exceeds b
		b is less than a
$a \leq b$	a is less than or equal to b	a is at most b
		a is no more than b
$a \geq b$	a is greater than or equal to b	a is no less than b
		a is at least b
$a = b$	a is equal to b	
$a \neq b$	a is not equal to b	
$a \approx b$	a is approximately equal to b	

The symbols $<$, $>$, $\leq$, $\geq$, and $\neq$ are called inequality signs, and the expressions $a < b$, $a > b$, $a \leq b$, $a \geq b$, and $a \neq b$ are called **inequalities**.

Example 3 **Ordering Real Numbers**

Fill in the blank with the appropriate inequality sign: $<$ or $>$

a. -2 _____ -5 **b.** $\dfrac{4}{7}$ _____ $\dfrac{3}{5}$ **c.** -1.3 _____ $-1.\overline{3}$

Solution:

a. $-2 \underline{\ >\ } -5$

b. To compare $\frac{4}{7}$ and $\frac{3}{5}$, write the fractions as equivalent fractions with a common denominator.

$$\frac{4}{7} \cdot \frac{5}{5} = \frac{20}{35} \qquad \text{and} \qquad \frac{3}{5} \cdot \frac{7}{7} = \frac{21}{35}$$

Because $\dfrac{20}{35} < \dfrac{21}{35}$, then $\dfrac{4}{7} \underline{\ <\ } \dfrac{3}{5}$

Animation

c. -1.3 $\underline{\geq}$ $-1.33333\ldots$

Skill Practice Fill in the blanks with the appropriate sign, $<$ or $>$.

6. $2 \underline{\hspace{1cm}} -12$ **7.** $\dfrac{1}{4} \underline{\hspace{1cm}} \dfrac{2}{9}$ **8.** $-7.\overline{2} \underline{\hspace{1cm}} -7.2$

3. Interval Notation

The set $\{x \mid x \geq 3\}$ represents all real numbers greater than or equal to 3. This set can be illustrated graphically on the number line.

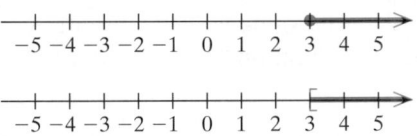

By convention, a closed circle ● or a square bracket [is used to indicate that an "endpoint" ($x = 3$) *is included* in the set.

The set $\{x \mid x > 3\}$ represents all real numbers strictly greater than 3. This set can be illustrated graphically on the number line.

By convention, an open circle ○ or a parenthesis (is used to indicate that an "endpoint" ($x = 3$) is *not* included in the set.

Notice that the sets $\{x \mid x \geq 3\}$ and $\{x \mid x > 3\}$ consist of an infinite number of elements that cannot all be listed. Another method to represent the elements of such sets is by using **interval notation**. To understand interval notation, first consider the real number line, which extends infinitely far to the left and right. The symbol ∞ is used to represent infinity. The symbol $-\infty$ is used to represent negative infinity.

To express a set of real numbers in interval notation, sketch the graph first, using the symbols ( ) or [ ]. Then use these symbols at the endpoints to define the interval.

| Example 4 | **Expressing Sets by Using Interval Notation** |

Graph each set on the number line, and express the set in interval notation.

  a. $\{x \mid x \geq -2\}$ **b.** $\{p \mid p > -2\}$

Solution:

  a. Set-builder notation: $\{x \mid x \geq -2\}$

    Graph:

    Interval notation: $[-2, \infty)$

The graph of the set $\{x \mid x \geq -2\}$ "begins" at -2 and extends infinitely far to the right. The corresponding interval notation "begins" at -2 and extends to ∞. Notice that a square bracket [is used at -2 for both the graph and the interval notation to include $x = -2$. *A parenthesis is always used at ∞ and at $-\infty$ because there is no endpoint.*

b. Set-builder notation: $\{p \mid p > -2\}$

Graph:

Number line from -5 to 5 with an open parenthesis at -2 extending right. Below, an arrow points to -2 and an arrow points to ∞.

Interval notation: $(-2, \infty)$

Skill Practice Graph each set, and express the set in interval notation.

9. $\{w \mid w \geq -7\}$ **10.** $\{x \mid x < 0\}$

In general, we use the following guidelines when applying interval notation.

Using Interval Notation

- The endpoints used in interval notation are always written from left to right. That is, the smaller number is written first, followed by a comma, followed by the larger number.
- Parentheses) or (indicate that an endpoint is *excluded* from the set.
- Square brackets] or [indicate that an endpoint is *included* in the set.
- Parentheses are always used with ∞ or $-\infty$.

Example 5   **Expressing Sets by Using Interval Notation**

Graph each set on the number line, and express the set in interval notation.

a. $\{z \mid z \leq -\frac{3}{2}\}$ **b.** $\{x \mid -4 < x \leq 2\}$

Solution:

a. Set-builder notation: $\{z \mid z \leq -\frac{3}{2}\}$

Graph:

Number line from -5 to 5 with a square bracket at $-\frac{3}{2}$ extending left. Labeled $-\infty$ on the left and ∞ on the right. Below, $(-\infty$ and $-\frac{3}{2}]$.

Interval notation: $(-\infty, -\frac{3}{2}]$

The graph of the set $\{z \mid z \leq -\frac{3}{2}\}$ extends infinitely far to the left. Interval notation is always written from left to right. Therefore, $-\infty$ is written first, followed by a comma, and then followed by the right-hand endpoint $-\frac{3}{2}$.

b. The inequality $-4 < x \leq 2$ means that x is greater than -4 and also less than or equal to 2. More concisely, we can say that x represents the real numbers *between* -4 and 2, including the endpoint, 2.

Set-builder notation: $\{x \mid -4 < x \leq 2\}$

Graph:

Number line from -5 to 5 with an open parenthesis at -4 and a square bracket at 2.

Interval notation: $(-4, 2]$

Answers

9.

Number line with a bracket at -7 extending right.

-7

$[-7, \infty)$

10.

Number line with a parenthesis at 0 extending left.

0

$(-\infty, 0)$

Skill Practice Graph the set on the number line, and express the set in interval notation.

11. $\{w \mid w \geq -\frac{5}{3}\}$ **12.** $\{y \mid -7 \leq y < 4\}$

Table R-2 summarizes interval notation.

Table R-2

Interval Notation	Graph	Interval Notation	Graph
(a, ∞)	$\xrightarrow{\quad\quad(\quad\longrightarrow}$ $\;\;\;\;a$	$[a, \infty)$	$\xrightarrow{\quad\quad[\quad\longrightarrow}$ $\;\;\;\;a$
$(-\infty, a)$	$\xleftarrow{\longleftarrow\quad) \quad\quad}$ $\;\;\;\;a$	$(-\infty, a]$	$\xleftarrow{\longleftarrow\quad] \quad\quad}$ $\;\;\;\;a$
(a, b)	$\xrightarrow{\quad(\quad\quad) \quad}$ $\;\;\;a \quad\quad b$	$[a, b]$	$\xrightarrow{\quad[\quad\quad] \quad}$ $\;\;\;a \quad\quad b$
$(a, b]$	$\xrightarrow{\quad(\quad\quad] \quad}$ $\;\;\;a \quad\quad b$	$[a, b)$	$\xrightarrow{\quad[\quad\quad) \quad}$ $\;\;\;a \quad\quad b$

4. Translations Involving Inequalities

In Table R-1, we learned that phrases such as *at least, at most, no more than, no less than,* and *between* can be translated into mathematical terms by using inequality signs.

Example 6 Translating Inequalities

The intensity of a hurricane is often defined according to its maximum sustained winds, for which wind speed is measured to the nearest mile per hour. Translate the italicized phrases into mathematical inequalities.

 a. A tropical storm is updated to hurricane status if the sustained wind speed, *w, is at least 74 mph.*

 b. Hurricanes are categorized according to intensity by the Saffir-Simpson scale. On a scale of 1 to 5, a category 5 hurricane is the most destructive. A category 5 hurricane has sustained winds, *w, exceeding 155 mph.*

 c. A category 4 hurricane has sustained winds, *w, of at least 131 mph but no more than 155 mph.*

Solution:

 a. $w \geq 74$ mph   **b.** $w > 155$ mph   **c.** 131 mph $\leq w \leq 155$ mph

Skill Practice Translate the italicized phrase to a mathematical inequality.

13. The gas mileage, m, for a Honda Civic is *at least 30 mpg.*

14. The gas mileage, m, for a Harley Davidson motorcycle is *more than 45 mpg.*

15. The gas mileage, m, for a Ford Explorer is *at least 10 mpg, but no more than 20 mpg.*

Answers

11.
$-\frac{5}{3}$
$[-\frac{5}{3}, \infty)$

12.
$-7 \quad\quad\quad 4$
$[-7, 4)$

13. $m \geq 30$
14. $m > 45$
15. $10 \leq m \leq 20$

Section R.2 Practice Exercises

Study Skills Exercises

In this text, we will provide skills for you to enhance your learning experience. Many of the Practice Exercises will begin with an activity that focuses on one of the following areas: learning about your course, using your text, taking notes, doing homework, and taking an exam. In subsequent chapters we will insert skills pertaining to the specific material in the chapter. Each activity requires only a few minutes and will help you pass this class and become a better math student.

To begin, write down the following information.

a. Instructor's name

b. Days of the week that the class meets

c. The room number in which the class meets

d. Is there a lab requirement for this course? If so, what is the requirement and what is the location of the lab?

e. Instructor's office number

f. Instructor's telephone number

g. Instructor's e-mail address

h. Instructor's office hours

Vocabulary and Key Concepts

1. a. In mathematics, a well-defined collection of elements is called a _____.

b. The statements $a < b$, $a > b$, and $a \neq b$ are examples of _____.

c. The statement $a < b$ is read as "_____."

d. The statement $c \geq d$ is read as "_____."

e. The statement $5 \neq 6$ is read as "_____."

f. The symbol ∞ represents _____ and $-\infty$ represents _____.

g. The set of real numbers greater than 5 can be written in set-builder notation as _____ and in _____ notation as $(5, \infty)$.

h. The interval $(-2, 5]$ (includes/excludes) the value -2 and (includes/excludes) the value 5.

i. When expressing interval notation, use a (parenthesis/bracket) with infinity.

Concept 1: The Set of Real Numbers

2. Determine the two consecutive integers between which the given number is located on the number line.

a. $\dfrac{19}{4}$

b. $-\dfrac{2}{3}$

c. -4.6

d. π

3. Plot the numbers on the number line.

$$\{1.7, \pi, -5, 4.\overline{2}\}$$

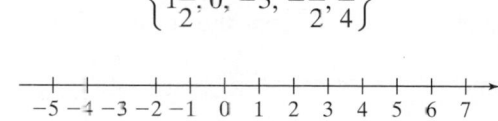

4. Plot the numbers on the number line.

$$\left\{1\dfrac{1}{2}, 0, -3, -\dfrac{1}{2}, \dfrac{3}{4}\right\}$$

For Exercises 5–10, show that each number is a rational number by finding a ratio of two integers equal to the given number. **(See Example 1.)**

5. -10

6. $7\dfrac{3}{4}$

7. $-\dfrac{3}{5}$

8. -0.1　　　　　　　　**9.** 0　　　　　　　　**10.** 0.35

11. Check the sets to which each number belongs. **(See Example 2.)**

	Real Numbers	Irrational Numbers	Rational Numbers	Integers	Whole Numbers	Natural Numbers
5						
$-\sqrt{9}$						
-1.7						
$\frac{1}{2}$						
$\sqrt{7}$						
$\frac{0}{4}$						
$0.\overline{2}$						

12. Check the sets to which each number belongs.

	Real Numbers	Irrational Numbers	Rational Numbers	Integers	Whole Numbers	Natural Numbers
$\frac{6}{8}$						
$1\frac{1}{2}$						
π						
0						
$-0.\overline{8}$						
$\frac{8}{2}$						
$4.\overline{2}$						

Concept 2: Inequalities

For Exercises 13–20, fill in the blanks with the appropriate symbol: $<$ or $>$. **(See Example 3.)**

13. -9 ___ -1　　　**14.** 0 ___ -6　　　**15.** $0.1\overline{5}$ ___ 0.15　　　**16.** $-2.\overline{5}$ ___ -2.5

17. $\frac{5}{3}$ ___ $\frac{10}{7}$　　　**18.** $-\frac{21}{5}$ ___ $-\frac{17}{4}$　　　**19.** $-\frac{5}{8}$ ___ $-\frac{1}{8}$　　　**20.** $-\frac{13}{15}$ ___ $-\frac{17}{12}$

Concept 3: Interval Notation

For Exercises 21–28, express the set in interval notation.

21.

22.

23.

24.

25.

26.

27.

28.

For Exercises 29–46, graph the sets and express each set in interval notation. **(See Examples 4–5.)**

29. $\{x \mid x > -1\}$

30. $\{x \mid x < 3\}$

31. $\{y \mid y \le -2\}$

32. $\{z \mid z \ge -4\}$

33. $\{w \mid w < \frac{9}{2}\}$

34. $\{p \mid p \ge -\frac{7}{3}\}$

35. $\{x \mid -2.5 < x \le 4.5\}$

36. $\{x \mid -6 \le x < 0\}$

37. All real numbers less than -3.

38. All real numbers greater than 2.34.

39. All real numbers greater than $\frac{5}{2}$.

40. All real numbers less than $\frac{4}{7}$.

41. All real numbers not less than 2.

42. All real numbers no more than 5.

43. All real numbers between -4 and 4.

44. All real numbers between -7 and -1.

45. All real numbers between -3 and 0, inclusive.

46. All real numbers between -1 and 6, inclusive.

For Exercises 47–54, write an expression in words that describes the set of numbers given by each interval. (Answers may vary.)

47. $(-\infty, -4)$

48. $[2, \infty)$

49. $(-2, 7]$

50. $(-3.9, 0)$

51. $[-180, 90]$

52. $(3.2, \infty)$

53. $(-\infty, \infty)$

54. $(-\infty, -1]$

Concept 4: Translations Involving Inequalities

For Exercises 55–64, write the expressions as an inequality. **(See Example 6.)**

55. The age, a, to get in to see a certain movie is at least 18 years old.

56. Winston is a cat that was picked up at the Humane Society. His age, a, at the time was no more than 2 years.

57. The cost, c, to have dinner at Jack's Café is at most $25.

58. The number of hours, h, that Katlyn spent studying was no less than 40.

59. The wind speed, s, for an F-5 tornado is no less than 261 mph.

60. The high temperature, t, for a certain December day in Albany is at most 26°F.

61. After a summer drought, the total rainfall, r, for June, July, and August was no more than 4.5 in.

62. Jessica works for a networking firm. Her salary, s, is at least $85,000 per year.

63. To play in a certain division of a tennis tournament, a player's age, a, must be at least 18 years but not more than 25 years.

64. The average age, a, of students at Central Community College is estimated to be between 25 years and 29 years.

The following chart defines the ranges for normal blood pressure, high normal blood pressure, and high blood pressure (*hypertension*). All values are measured in millimeters of mercury (mm Hg). (Source: American Heart Association.)

Normal	Systolic less than 130	Diastolic less than 85
High normal	Systolic 130–139, inclusive	Diastolic 85–89, inclusive
Hypertension	Systolic 140 or greater	Diastolic 90 or greater

For Exercises 65–68, write an inequality using the variable *p* that represents each condition.

65. Normal systolic blood pressure

66. Diastolic pressure in hypertension

67. High normal range for systolic pressure

68. Systolic pressure in hypertension

A pH scale determines whether a solution is acidic or alkaline. The pH scale runs from 0 to 14, with 0 being the most acidic and 14 being the most alkaline. A pH of 7 is neutral (distilled water has a pH of 7).

For Exercises 69–72, write the pH ranges as inequalities and label the substances as acidic or alkaline.

69. Lemon juice: 2.2 through 2.4, inclusive

70. Eggs: 7.6 through 8.0, inclusive

71. Carbonated soft drinks: 3.0 through 3.5, inclusive

72. Milk: 6.6 through 6.9, inclusive

Section R.3 Operations on Real Numbers

Concepts

1. Opposite and Absolute Value
2. Addition and Subtraction of Real Numbers
3. Multiplication and Division of Real Numbers
4. Exponential Expressions
5. Square Roots
6. Order of Operations
7. Evaluating Formulas

1. Opposite and Absolute Value

Several key definitions are associated with the set of real numbers and constitute the foundation of algebra. Two important definitions are the opposite of a real number and the absolute value of a real number.

> **Opposite of a Real Number**
>
> Two numbers that are the same distance from 0 but on opposite sides of 0 on the number line are called **opposites** of each other.
>
> Symbolically, we denote the opposite of a real number a as $-a$.

The numbers -4 and 4 are opposites of each other. Similarly, the numbers $\frac{3}{2}$ and $-\frac{3}{2}$ are opposites.

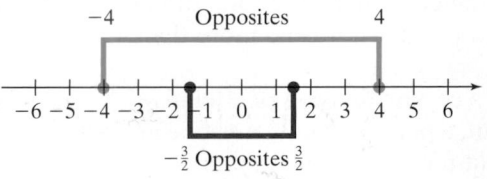

The **absolute value** of a real number a, denoted $|a|$, is the distance between a and 0 on the number line. *Note:* The absolute value of any real number is *nonnegative*.

For example: $|5| = 5$
and
$|-5| = 5$

Example 1 **Evaluating Absolute Value Expressions**

Simplify the expressions.

a. $|-2.5|$ **b.** $\left|\dfrac{5}{4}\right|$ **c.** $-|-4|$

Solution:

a. $|-2.5| = 2.5$

b. $\left|\dfrac{5}{4}\right| = \dfrac{5}{4}$

c. $-|-4| = -(4) = -4$

4 units　　$\frac{5}{2}$ units

2.5 units

Skill Practice Simplify.

1. $|-9.2|$ **2.** $\left|\dfrac{7}{6}\right|$ **3.** $-|-2|$

The absolute value of a number a is its distance from zero on the number line. The definition of $|a|$ may also be given algebraically depending on whether a is negative or nonnegative.

Absolute Value of a Real Number

Let a be a real number. Then

1. If a is nonnegative (that is, if $a \geq 0$), then $|a| = a$.

2. If a is negative (that is, if $a < 0$), then $|a| = -a$.

This definition states that if a is positive or zero, then $|a|$ equals a itself. If a is a negative number, then $|a|$ equals the opposite of a. For example,

$|9| = 9$ Because 9 is positive, $|9|$ equals the number 9 itself.

$|-7| = 7$ Because -7 is negative, $|-7|$ equals the opposite of -7, which is 7.

2. Addition and Subtraction of Real Numbers

Addition of Real Numbers

- To add two numbers with the *same sign*, add their absolute values and apply the common sign to the sum.
- To add two numbers with *different signs*, subtract the smaller absolute value from the larger absolute value. Then apply the sign of the number having the larger absolute value.

Answers

1. 9.2 **2.** $\dfrac{7}{6}$ **3.** -2

| Example 2 | Adding Real Numbers |

Perform the indicated operations.

a. $-2 + (-6)$ **b.** $-10.3 + 13.8$ **c.** $\dfrac{5}{6} + \left(-1\dfrac{1}{4}\right)$

Solution:

a. $-2 + (-6)$ First find the absolute value of the addends.
$|-2| = 2$ and $|-6| = 6$

$= -(2 + 6)$ Add their absolute values and apply the common sign. In this case, the common sign is negative.

Common sign is negative.

$= -8$ The sum is -8.

b. $-10.3 + 13.8$ First find the absolute value of the addends.
$|-10.3| = 10.3$ and $|13.8| = 13.8$

The absolute value of 13.8 is greater than the absolute value of -10.3. Therefore, the sum is positive.

$= +(13.8 - 10.3)$ Subtract the smaller absolute value from the larger absolute value.

Apply the sign of the number with the larger absolute value.

$= 3.5$

c. $\dfrac{5}{6} + \left(-1\dfrac{1}{4}\right)$

$= \dfrac{5}{6} + \left(-\dfrac{5}{4}\right)$ Write $-1\dfrac{1}{4}$ as a fraction.

$= \dfrac{5 \cdot 2}{6 \cdot 2} + \left(-\dfrac{5 \cdot 3}{4 \cdot 3}\right)$ The LCD is 12. Write each fraction with the LCD.

$= \dfrac{10}{12} + \left(-\dfrac{15}{12}\right)$ Find the absolute value of the addends.
$\left|\dfrac{10}{12}\right| = \dfrac{10}{12}$ and $\left|-\dfrac{15}{12}\right| = \dfrac{15}{12}$

The absolute value of $-\dfrac{15}{12}$ is greater than the absolute value of $\dfrac{10}{12}$. Therefore, the sum is negative.

$= -\left(\dfrac{15}{12} - \dfrac{10}{12}\right)$ Subtract the smaller absolute value from the larger absolute value.

Apply the sign of the number with the larger absolute value.

$= -\dfrac{5}{12}$

Skill Practice Perform the indicated operations.

4. $-4 + (-1)$ **5.** $-2.6 + 1.8$ **6.** $-1 + \left(-\dfrac{3}{7}\right)$

Answers

4. -5 **5.** -0.8 **6.** $-\dfrac{10}{7}$

Subtraction of real numbers is defined in terms of the addition process. To subtract two real numbers, add the opposite of the second number to the first number.

Subtraction of Real Numbers

If a and b are real numbers, then $a - b = a + (-b)$

| Example 3 | **Subtracting Real Numbers** |

Perform the indicated operations.

a. $-13 - 5$ **b.** $2.7 - (-3.8)$ **c.** $\dfrac{5}{2} - 4\dfrac{2}{3}$

Solution:

a. $-13 - 5$

$= -13 + (-5)$ Add the opposite of the second number to the first number.

$= -18$ Add.

b. $2.7 - (-3.8)$

$= 2.7 + (3.8)$ Add the opposite of the second number to the first number.

$= 6.5$ Add.

c. $\dfrac{5}{2} - 4\dfrac{2}{3}$

$= \dfrac{5}{2} + \left(-4\dfrac{2}{3}\right)$ Add the opposite of the second number to the first number.

$= \dfrac{5}{2} + \left(-\dfrac{14}{3}\right)$ Write the mixed number as a fraction.

$= \dfrac{5 \cdot 3}{2 \cdot 3} + \left(-\dfrac{14 \cdot 2}{3 \cdot 2}\right)$ The common denominator is 6.

$= \dfrac{15}{6} + \left(-\dfrac{28}{6}\right)$ Get a common denominator and add.

$= -\dfrac{13}{6}$ or $-2\dfrac{1}{6}$

Skill Practice Subtract.

7. $-9 - 8$ **8.** $1.1 - (-4.2)$ **9.** $\dfrac{1}{6} - 2\dfrac{1}{4}$

Answers

7. -17 **8.** 5.3 **9.** $-2\dfrac{1}{12}$ or $-\dfrac{25}{12}$

3. Multiplication and Division of Real Numbers

The sign of the product of two real numbers is determined by the signs of the factors.

> **Multiplication of Real Numbers**
> 1. The product of two real numbers with the *same* sign is *positive*.
> 2. The product of two real numbers with *different* signs is *negative*.
> 3. The product of any real number and zero is *zero*.

Example 4   **Multiplying Real Numbers**

Multiply the real numbers.

a. $(2)(-5.1)$ **b.** $-\dfrac{2}{3}\cdot\dfrac{9}{8}$ **c.** $\left(-3\dfrac{1}{3}\right)\left(-\dfrac{3}{10}\right)$

Solution:

a. $(2)(-5.1)$

$\quad = -10.2$ *Different* signs. The product is negative.

b. $-\dfrac{2}{3}\cdot\dfrac{9}{8}$

$\quad = -\dfrac{18}{24}$ *Different* signs. The product is negative.

$\quad = -\dfrac{3}{4}$ Simplify to lowest terms.

c. $\left(-3\dfrac{1}{3}\right)\left(-\dfrac{3}{10}\right)$

$\quad = \left(-\dfrac{10}{3}\right)\left(-\dfrac{3}{10}\right)$ Write the mixed number as a fraction.

$\quad = \dfrac{30}{30}$ *Same* signs. The product is positive.

$\quad = 1$ Simplify to lowest terms.

Skill Practice Multiply.

10. $(-5)(2.2)$ **11.** $\dfrac{5}{7}\cdot\left(-\dfrac{14}{15}\right)$ **12.** $\left(-5\dfrac{1}{4}\right)\left(-\dfrac{8}{3}\right)$

TIP: A number and its reciprocal have the same sign. For example:

$\left(-\dfrac{10}{3}\right)\left(-\dfrac{3}{10}\right) = 1$

and $3\cdot\dfrac{1}{3} = 1$

Notice from Example 4(c) that $\left(-\dfrac{10}{3}\right)\left(-\dfrac{3}{10}\right) = 1$. If the product of two numbers is 1, then the numbers are **reciprocals**. That is, the reciprocal of a real *number a* is $\dfrac{1}{a}$. Furthermore, $a\cdot\dfrac{1}{a} = 1$.

Answers

10. -11 **11.** $-\dfrac{2}{3}$ **12.** 14

Recall that subtraction of real numbers was defined in terms of addition. In a similar way, division of real numbers can be defined in terms of multiplication.

Procedure to Divide Real Numbers

To divide two real numbers, multiply the first number by the reciprocal of the second number. For example:

$$10 \div 5 = 2 \quad \text{or equivalently} \quad 10 \cdot \frac{1}{5} = 2$$

Because division of real numbers can be expressed in terms of multiplication, the sign rules that apply to multiplication also apply to division.

$$10 \div 2 = 10 \cdot \frac{1}{2} = 5$$

$$-10 \div (-2) = -10 \cdot \left(-\frac{1}{2}\right) = 5$$

Dividing two numbers of the same sign produces a *positive* quotient.

$$10 \div (-2) = 10 \cdot \left(-\frac{1}{2}\right) = -5$$

$$-10 \div 2 = -10 \cdot \frac{1}{2} = -5$$

Dividing two numbers of opposite signs produces a *negative* quotient.

Division of Real Numbers

Assume that a and b are real numbers such that $b \neq 0$.

1. If a and b have the *same* sign, then the quotient $\frac{a}{b}$ is *positive*.

2. If a and b have *different* signs, then the quotient $\frac{a}{b}$ is *negative*.

3. $\dfrac{0}{b} = 0$.

4. $\dfrac{b}{0}$ is undefined.

The relationship between multiplication and division can be used to investigate properties 3 and 4 from the preceding box. For example,

$$\frac{0}{6} = 0 \qquad \text{Because } 6 \cdot 0 = 0 \checkmark$$

$$\frac{6}{0} \text{ is undefined} \qquad \text{Because there is no number that when multiplied by 0 will equal 6}$$

Note: The quotient of 0 and 0 *cannot be determined.* Evaluating an expression of the form $\frac{0}{0} = ?$ is equivalent to asking, "What number times zero will equal 0?" That is, $(0)(?) = 0$. Any real number will satisfy this requirement; however, expressions involving $\frac{0}{0}$ are usually discussed in advanced mathematics courses.

Example 5   Dividing Real Numbers

Divide the real numbers. Write the answer as a fraction or whole number.

a. $\dfrac{-42}{7}$ **b.** $\dfrac{-96}{-144}$ **c.** $3\dfrac{1}{10} \div \left(-\dfrac{2}{5}\right)$ **d.** $\dfrac{-8}{-7}$

Solution:

a. $\dfrac{-42}{7} = -6$ *Different* signs. The quotient is negative.

b. $\dfrac{-96}{-144} = \dfrac{2}{3}$ *Same* signs. The quotient is positive. Simplify.

c. $3\dfrac{1}{10} \div \left(-\dfrac{2}{5}\right)$

$\quad = \dfrac{31}{10}\left(-\dfrac{5}{2}\right)$ Write the mixed number as an improper fraction, and multiply by the reciprocal of the second number.

$\quad = \dfrac{31}{\overset{}{\underset{2}{10}}}\left(-\dfrac{\overset{1}{5}}{2}\right)$

$\quad = -\dfrac{31}{4}$ *Different* signs. The quotient is negative.

d. $\dfrac{-8}{-7} = \dfrac{8}{7}$ *Same* signs. The quotient is positive.
Because 7 does not divide into 8 evenly, the answer can be left as a fraction.

TIP: Multiplication may be used to check a division problem.

$$\dfrac{-42}{7} = -6$$

Check: $(7)(-6) = -42$ ✓

Avoiding Mistakes

If the numerator and denominator of a fraction have opposite signs, then the quotient will be negative. Therefore, a fraction has the same value whether the negative sign is written in the numerator, in the denominator, or in front of the fraction.

$$-\dfrac{31}{4} = \dfrac{-31}{4} = \dfrac{31}{-4}$$

Skill Practice Divide.

13. $\dfrac{42}{-2}$ **14.** $\dfrac{-28}{-4}$ **15.** $-\dfrac{2}{3} \div 4$ **16.** $\dfrac{-1}{-2}$

4. Exponential Expressions

To simplify the process of repeated multiplication, exponential notation is often used. For example, the quantity $3 \cdot 3 \cdot 3 \cdot 3 \cdot 3$ can be written as 3^5 (3 to the fifth power).

Definition of b^n

Let b represent any real number and n represent a positive integer. Then

$$b^n = \underbrace{b \cdot b \cdot b \cdot b \cdot \ldots b}_{n \text{ factors of } b}$$

b^n is read as "b to the nth power."
b is called the **base** and n is called the **exponent**, or **power**.
b^2 is read as "b squared," and b^3 is read as "b cubed."

Answers

13. -21 **14.** 7
15. $-\dfrac{1}{6}$ **16.** $\dfrac{1}{2}$

Example 6 Evaluating Exponential Expressions

Simplify the expression.

a. 5^3 **b.** $(-2)^4$ **c.** -2^4 **d.** $\left(-\frac{1}{3}\right)^3$

Solution:

a. $5^3 = 5 \cdot 5 \cdot 5$ The base is 5, and the exponent is 3.

$= 125$

b. $(-2)^4 = (-2)(-2)(-2)(-2)$ The base is −2, and the exponent is 4.
The exponent 4 applies to the entire
$= 16$ contents of the parentheses.

c. $-2^4 = -[2 \cdot 2 \cdot 2 \cdot 2]$ The base is 2, and the exponent is 4.
Because no parentheses enclose the
$= -16$ negative sign, the exponent applies to only 2.

TIP: The quantity -2^4 can also be interpreted as $-1 \cdot 2^4$.

$$-2^4 = -1 \cdot 2^4 = -1 \cdot (2 \cdot 2 \cdot 2 \cdot 2) = -16$$

d. $\left(-\frac{1}{3}\right)^3 = \left(-\frac{1}{3}\right)\left(-\frac{1}{3}\right)\left(-\frac{1}{3}\right)$ The base is $-\frac{1}{3}$, and the exponent is 3.

$= -\frac{1}{27}$

Skill Practice Simplify.

17. 2^3 **18.** $(-10)^2$ **19.** -10^2 **20.** $\left(-\frac{3}{4}\right)^3$

Calculator Connections

Topic: Using the Exponent Keys

On many calculators, the x^2 key is used to square a number. The $\wedge$ key is used to raise a base to any power.

```
5^3
             125
(-2)^4
              16
-2^4
             -16
```

5. Square Roots

The inverse operation to squaring a number is to find its square roots. For example, finding a square root of 9 is equivalent to asking, "What number when squared equals 9?" One obvious answer is 3, because $(3)^2 = 9$. However, −3 is also a square root of 9 because $(-3)^2 = 9$. For now, we will focus on the **principal square root**, which is always taken to be nonnegative.

The symbol $\sqrt{\ }$, called a **radical sign**, is used to denote the principal square root of a number. Therefore, the principal square root of 9 can be written as $\sqrt{9}$. The expression $\sqrt{64}$ represents the principal square root of 64.

Answers
17. 8   **18.** 100
19. −100   **20.** $-\frac{27}{64}$

Example 7 Evaluating Square Roots

Evaluate the expressions, if possible.

a. $\sqrt{81}$ **b.** $\sqrt{\dfrac{25}{64}}$ **c.** $\sqrt{-16}$ **d.** $-\sqrt{16}$

Solution:

a. $\sqrt{81} = 9$ because $(9)^2 = 81$

b. $\sqrt{\dfrac{25}{64}} = \dfrac{5}{8}$ because $\left(\dfrac{5}{8}\right)^2 = \dfrac{25}{64}$

c. $\sqrt{-16}$ is *not a real number* because no real number when squared will be negative.

d. $-\sqrt{16} = -4$ because $-\sqrt{16} = -(\sqrt{16}) = -4$.

Skill Practice Evaluate the expressions, if possible.

21. $\sqrt{25}$ **22.** $\sqrt{\dfrac{49}{100}}$ **23.** $\sqrt{-4}$ **24.** $-\sqrt{9}$

Example 7(c) illustrates that the square root of a negative number is not a real number because no real number when squared will be negative.

Square Root of a Negative Number

Let a be a negative real number. Then $\sqrt{a}$ is not a real number.

---

6. Order of Operations

When algebraic expressions contain numerous operations, it is important to use the proper **order of operations**. Parentheses ( ), brackets [ ], and braces { } are used for grouping numbers and algebraic expressions. It is important to recognize that operations must be done first within parentheses and other grouping symbols.

Order of Operations

Step 1 First, simplify expressions within parentheses and other grouping symbols. These include absolute value bars, fraction bars, and radicals. If embedded parentheses are present, start with the innermost parentheses.

Step 2 Evaluate expressions involving exponents, radicals, and absolute values.

Step 3 Perform multiplication or division in the order in which they occur from left to right.

Step 4 Perform addition or subtraction in the order in which they occur from left to right.

Answers

21. 5   **22.** $\dfrac{7}{10}$

23. Not a real number

24. -3

Example 8 Applying the Order of Operations

Simplify the expression. $10 - [2 - 4(6 - 8)]^2 + \sqrt{16 - 7}$

Solution:

$10 - [2 - 4(6 - 8)]^2 + \sqrt{16 - 7}$

$= 10 - [2 - 4(-2)]^2 + \sqrt{9}$ Simplify inside the innermost parentheses and inside the radical.

$= 10 - [2 + 8]^2 + \sqrt{9}$ Simplify within square brackets. Perform multiplication before addition or subtraction.

$= 10 - [10]^2 + \sqrt{9}$

$= 10 - 100 + 3$ Simplify the exponential expression and the radical.

$= -90 + 3$ Perform addition or subtraction in the order in which they appear from left to right.

$= -87$

Avoiding Mistakes

Don't try to perform too many steps at once. Taking a shortcut may result in a careless error. For each step rewrite the entire expression, changing only the operation being evaluated.

Skill Practice Simplify the expression.

25. $36 \div 2^2 \cdot 3 - [(18 - 5) \cdot 2 + 6]$

Example 9 Applying the Order of Operations

Simplify the expression. $\dfrac{|(-3)^3 + (5^2 - 3)|}{-15 \div (-3)(2)}$

Solution:

$\dfrac{|(-3)^3 + (5^2 - 3)|}{-15 \div (-3)(2)}$ Simplify numerator and denominator separately.

$= \dfrac{|(-3)^3 + (25 - 3)|}{5(2)}$ *Numerator:* Simplify within the inner parentheses.
Denominator: Perform division and multiplication (left to right).

$= \dfrac{|(-3)^3 + (22)|}{10}$ *Numerator:* Simplify inner parentheses.
Denominator: Multiply.

$= \dfrac{|-27 + 22|}{10}$ Simplify exponent.

$= \dfrac{|-5|}{10}$ Add within the absolute value.

$= \dfrac{5}{10}$ or $\dfrac{1}{2}$ Evaluate the absolute value and simplify.

Calculator Connections

Topic: Applying the Order of Operations

To evaluate the expression
$$\dfrac{|(-3)^3 + (5^2 - 3)|}{-15 \div (-3)(2)}$$
on a graphing calculator, use parentheses to enclose the absolute value expression. Likewise, it is necessary to use parentheses to enclose the entire denominator.

Skill Practice Simplify the expression.

26. $\dfrac{-|5 - 7| + 11}{(-1 - 2)^2}$

Answers

25. -5 **26.** 1

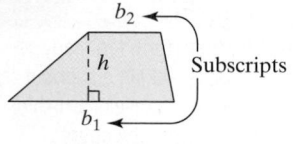

Figure R-3

7. Evaluating Formulas

An algebraic expression or formula involves operations on numbers and variables. A **variable** is a letter that may represent any numerical value. To evaluate an expression or formula, we substitute known values of the variables into the expression. Then we follow the order of operations.

A list of common geometry formulas is given in the inside back cover of the text. It is important to note that some geometric formulas use Greek letters (such as π) and some use variables with subscripts. A **subscript** is a number or letter written to the right of and below a variable. For example, the area of a trapezoid is given by $A = \frac{1}{2}(b_1 + b_2)h$. The values b_1 and b_2 (read as "b sub 1" and "b sub 2") represent two different bases of the trapezoid (Figure R-3).

Example 10 Evaluating a Formula

A homeowner in North Carolina wants to buy protective film for a trapezoid-shaped window. The film will adhere to shattered glass in the event that the glass breaks during a bad storm. Find the area of the window whose dimensions are given in Figure R-4.

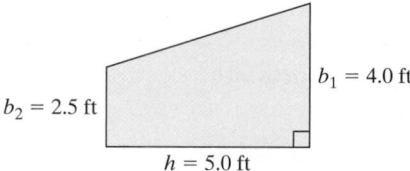

Figure R-4

Solution:

$$A = \frac{1}{2}(b_1 + b_2)h$$

$$= \frac{1}{2}(4.0 \text{ ft} + 2.5 \text{ ft})(5.0 \text{ ft}) \qquad \text{Substitute } b_1 = 4.0 \text{ ft, } b_2 = 2.5 \text{ ft, and } h = 5.0 \text{ ft.}$$

$$= \frac{1}{2}(6.5 \text{ ft})(5.0 \text{ ft}) \qquad \text{Simplify inside parentheses.}$$

$$= 16.25 \text{ ft}^2 \qquad \text{Multiply from left to right.}$$

The area of the window is 16.25 ft^2.

TIP: Subscripts should not be confused with *superscripts*, which are written above a variable. Superscripts are used to denote powers.

$$b_2 \neq b^2$$

Skill Practice

27. Use the formula given in Example 10 to find the area of the trapezoid.

$b_2 = 5$ in.

$h = 10$ in.

$b_1 = 12$ in.

Answer

27. The area is 85 in.^{2}

Section R.3   Practice Exercises

Study Skills Exercise

Sometimes you may run into a problem with homework, or you may find that you are having trouble keeping up with the pace of the class. A tutor can be a good resource. Answer the following questions.

a. Does your college offer tutoring?   **b.** Is it free?

c. Where should you go to sign up for a tutor?   **d.** Is there tutoring available online?

Vocabulary and Key Concepts

1. **a.** Two numbers that are the same distance from 0 but on opposite sides of 0 on the number line are called _____ .

   b. The absolute value of a real number, a, is denoted by _____ and is the distance between a and _____ on the number line.

   c. Given the expression b^n, the value b is called the _____ and _____ is called the exponent or power.

   d. The symbol $\sqrt{}$ is called a _____ sign and is used to find the principal _____ root of a nonnegative real number.

   e. If a and b are both negative, then $a + b$ will be (positive/negative) and ab will be (positive/negative).

   f. If $a < 0$ and $b > 0$, and if $|a| < |b|$, then the sign of $a + b$ will be (positive/negative) and the sign of $\frac{a}{b}$ will be (positive/negative).

   g. The expression $a - b = a +$ _____ . If $a > 0$ and $b < 0$, then the sign of $a - b$ is _____ .

   h. If a is a nonzero real number, then the reciprocal of a is _____ . The product of a number and its reciprocal is _____ .

   i. If either a or b is zero then $ab =$ _____ .

   j. If $a = 0$ and $b \neq 0$, then $\frac{a}{b} =$ _____ and $\frac{b}{a}$ is _____ .

Review Exercises

2. Write the set-builder notation representing the interval $[-5, 3)$.

For Exercises 3–6, express each set in interval notation.

3. $\{z \mid z < 4\}$ 4. $\{w \mid w \geq -2\}$ 5. $\{x \mid -3 \leq x < -1\}$ 6. $\{p \mid 0 < p \leq 8\}$

Concept 1: Opposite and Absolute Value

7. If the absolute value of a number can be thought of as its distance from zero, explain why an absolute value can never be negative.

8. If a number is negative, then its *opposite* will be

   a. Positive   **b.** Negative.

9. If a number is negative, then its *reciprocal* will be

   a. Positive   **b.** Negative.

10. If a number is negative, then its *absolute value* will be

    a. Positive   **b.** Negative.

11. Complete the table. **(See Example 1.)**

Number	Opposite	Reciprocal	Absolute Value
6			
	$-\frac{1}{11}$		
		$-\frac{1}{8}$	
	$\frac{13}{10}$		
0			
		$-0.\overline{3}$	

12. Complete the table.

Number	Opposite	Reciprocal	Absolute Value
-9			
	$\frac{2}{3}$		
		14	
-1			
0			
		$2\frac{1}{9}$	

For Exercises 13–20, fill in the blank with the appropriate symbol ($<$, $>$, $=$). **(See Example 1.)**

13. $-|6|$ _____ $|-6|$

14. $-(-5)$ _____ $-|-5|$

15. $|-4|$ _____ $|4|$

16. $-|2|$ _____ (-2)

17. $-|-1|$ _____ 1

18. -3 _____ $-|-7|$

19. $|2+(-5)|$ _____ $|2|+|-5|$

20. $|4+3|$ _____ $|4|+|3|$

Concept 2: Addition and Subtraction of Real Numbers

For Exercises 21–36, add or subtract as indicated. **(See Examples 2–3.)**

21. $-8+4$

22. $3+(-7)$

23. $-12+(-7)$

24. $-5+(-11)$

25. $-17-(-10)$

26. $-14-(-2)$

27. $5-(-9)$

28. $8-(-4)$

29. $-6.3-15.8$

30. $-21.9-4.7$

31. $1.5-9.6$

32. $4.8-10$

33. $\frac{2}{3}+\left(-2\frac{1}{3}\right)$

34. $-\frac{4}{7}+\left(1\frac{4}{7}\right)$

35. $-\frac{5}{9}-\frac{14}{15}$

36. $-6-\frac{2}{9}$

Concept 3: Multiplication and Division of Real Numbers

For Exercises 37–52, perform the indicated operation. **(See Examples 4–5.)**

37. $4(-8)$

38. $-21(3)$

39. $\frac{2}{9}\cdot\frac{12}{7}$

40. $\left(-\frac{5}{9}\right)\cdot\left(-1\frac{7}{11}\right)$

41. $\frac{-6}{-10}$

42. $\frac{-15}{-24}$

43. $-2\frac{1}{4}\div\frac{5}{8}$

44. $-\frac{2}{3}\div\left(-1\frac{5}{7}\right)$

45. $7\div0$

46. $\frac{1}{16}\div0$

47. $0\div(-3)$

48. $0\div11$

49. $(-1.2)(-3.1)$

50. $(4.6)(-2.25)$

51. $\frac{-5}{-11}$

52. $\frac{-3}{-13}$

Concept 4: Exponential Expressions

For Exercises 53–60, evaluate the expression. **(See Example 6.)**

53. 4^3

54. -2^3

55. -7^2

56. -3^4

57. $(-7)^2$

58. $(-5)^2$

59. $\left(\frac{5}{3}\right)^3$

60. $\left(\frac{10}{9}\right)^2$

Concept 5: Square Roots

For Exercises 61–68, evaluate the expression, if possible. **(See Example 7.)**

61. $\sqrt{9}$

62. $\sqrt{1}$

63. $\sqrt{-4}$

64. $\sqrt{-36}$

65. $\sqrt{\dfrac{1}{4}}$

66. $\sqrt{\dfrac{9}{4}}$

67. $-\sqrt{49}$

68. $-\sqrt{100}$

Concept 6: Order of Operations

For Exercises 69–96, simplify by using the order of operations. **(See Examples 8–9.)**

69. $5 + 3^3$

70. $10 - 2^4$

71. $5 \cdot 2^3$

72. $12 \div 2^2$

73. $(2 + 3)^2$

74. $(4 - 1)^3$

75. $2^2 + 3^2$

76. $4^3 - 1^3$

77. $6 + 10 \div 2 \cdot 3 - 4$

78. $12 \div 3 \cdot 4 - 18$

79. $4^2 - (5 - 2)^2 \cdot 3$

80. $5 - 3(8 \div 4)^2$

81. $2 - 5(9 - 4\sqrt{25})^2$

82. $5^2 - (\sqrt{9} + 4 \div 2)$

83. $\left(-\dfrac{3}{5}\right)^2 - \dfrac{3}{5} \cdot \dfrac{5}{9} + \dfrac{7}{10}$

84. $\dfrac{1}{2} - \left(\dfrac{2}{3} \div \dfrac{5}{9}\right) + \dfrac{5}{6}$

85. $1.75 \div 0.25 - (1.25)^2$

86. $5.4 - (0.3)^2 \div 0.09$

87. $\dfrac{\sqrt{10^2 - 8^2}}{3^2}$

88. $\dfrac{\sqrt{16 - 7} + 3^2}{\sqrt{16} - \sqrt{4}}$

89. $-|-11 + 5| + |7 - 2|$

90. $-|-8 - 3| - (-8 - 3)$

91. $25 - 2[(7 - 3)^2 \div 4] + \sqrt{18 - 2}$

92. $\sqrt{29 - 2^2} + [8 - 3(6 - 2)] \div 4 \cdot 5$

93. $\dfrac{|(10 - 7) - 2^3|}{6 - 16 \div 8 \cdot 3}$

94. $\dfrac{-12 - (7 - 3^2)^2|}{40 - 6^2 - 8 \div 2}$

95. $\left(\dfrac{1}{2}\right)^2 + \left(\dfrac{6 - 4}{5}\right)^2 + \left(\dfrac{5 + 2}{10}\right)^2$

96. $\left(\dfrac{2^3}{2^3 + 1}\right)^2 \div \left(\dfrac{8 - (-2)}{3^2}\right)^2$

For Exercises 97–98, find the average of the set of data values by adding the values and dividing by the number of values.

97. Find the average low temperature for a week in January in St. John's, Newfoundland. Round to the nearest tenth of a degree.

Day	Mon.	Tues.	Wed.	Thur.	Fri.	Sat.	Sun.
Low temperature	−18°C	−16°C	−20°C	−11°C	−4°C	−3°C	1°C

98. Find the average high temperature for a week in January in St. John's, Newfoundland. Round to the nearest tenth of a degree.

Day	Mon.	Tues.	Wed.	Thur.	Fri.	Sat.	Sun.
High temperature	−2°C	−6°C	−7°C	0°C	1°C	8°C	10°C

Concept 7: Evaluating Formulas

99. The formula $C = \dfrac{5}{9}(F - 32)$ converts temperatures in the Fahrenheit scale to the Celsius scale. Find the equivalent Celsius temperature for each Fahrenheit temperature.

  a. 77°F   **b.** 212°F   **c.** 32°F   **d.** −40°F

100. The formula $F = \dfrac{9}{5}C + 32$ converts Celsius temperatures to Fahrenheit temperatures. Find the equivalent Fahrenheit temperature for each Celsius temperature.

  a. −5°C   **b.** 0°C   **c.** 37°C   **d.** −40°C

The equation $G_H = \frac{1}{22}c + \frac{1}{30}h$ represents the amount of gasoline used (in gal) for a Honda Accord to drive c miles in the city and h miles on the highway. The equation $G_C = \frac{1}{12}c + \frac{1}{18}h$ represents the amount of gasoline used for a Cadillac Escalade to make the same trip. Use these formulas for Exercises 101–102.

101. Determine the amount of gas used by a Honda that travels 33 mi in the city and 80 mi on the highway.

102. Determine the amount of gas used by a Cadillac that travels 33 mi in the city and 80 mi on the highway.

Use the geometry formulas found in the inside back cover of the book to answer Exercises 103–112.

For Exercises 103–106, find the area. **(See Example 10.)**

103. Trapezoid

5 in.
2 in.
4 in.

104. Parallelogram

8.5 m
6 m

105. Triangle

3.1 cm
5.2 cm

106. Rectangle

$\frac{3}{4}$ yd
$7\frac{1}{6}$ yd

For Exercises 107–112, find the volume. (Use the π key on your calculator, and round the final answer to one decimal place.)

107. Sphere

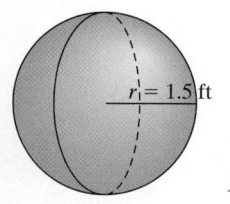
$r = 1.5$ ft

108. Sphere

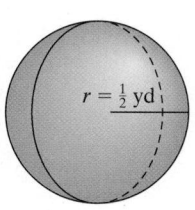
$r = \frac{1}{2}$ yd

109. Right circular cone

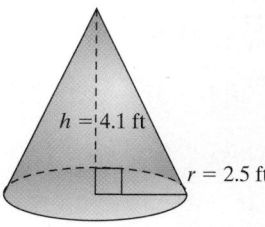
$h = 4.1$ ft
$r = 2.5$ ft

110. Right circular cone

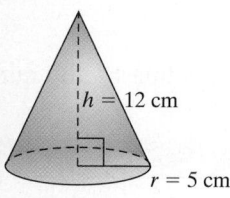
$h = 12$ cm
$r = 5$ cm

111. Right circular cylinder

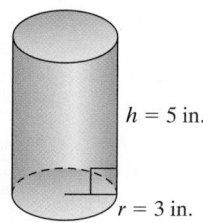
$h = 5$ in.
$r = 3$ in.

112. Right circular cylinder

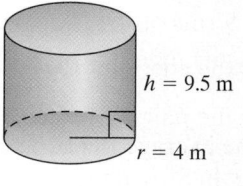
$h = 9.5$ m
$r = 4$ m

Graphing Calculator Exercises

113. Which expression when entered into a graphing calculator will yield the correct value of $\dfrac{12}{6-2}$?

$$12/6 - 2 \quad \text{or} \quad 12/(6 - 2)$$

114. Which expression when entered into a graphing calculator will yield the correct value of $\dfrac{24-6}{3}$?

$$(24 - 6)/3 \quad \text{or} \quad 24 - 6/3$$

115. Verify your solution to Exercise 87 by entering the expression into a graphing calculator:

$$(\sqrt{(10^2 - 8^2)})/3^2$$

116. Verify your solution to Exercise 88 by entering the expression into a graphing calculator:

$$(\sqrt{(16 - 7)} + 3^2)/(\sqrt{(16)} - \sqrt{(4)})$$

Simplifying Algebraic Expressions

1. Recognizing Terms, Factors, and Coefficients

A **term** is a constant or the product of a constant and one or more variables. An algebraic expression is a single term or a sum of two or more terms. For example, the expression

1. **Recognizing Terms, Factors, and Coefficients**
2. **Properties of Real Numbers**
3. **Simplifying Expressions**

$$-6x^2 + 5xyz - 11 \quad \text{or} \quad -6x^2 + 5xyz + (-11)$$

consists of the terms $-6x^2$, $5xyz$, and -11.

The terms $-6x^2$ and $5xyz$ are **variable terms**, and the term -11 is called a **constant term**. It is important to distinguish between a term and the **factors** within a term. For example, the quantity $5xyz$ is one term, but the values $5, x, y,$ and z are factors within the term. The constant factor in a term is called the numerical coefficient or simply **coefficient** of the term. In the terms $-6x^2$, $5xyz$, and -11, the coefficients are $-6, 5,$ and -11, respectively. A term containing only variables such as xy has a coefficient of 1.

Terms are called *like* **terms** if they each have the same variables and the corresponding variables are raised to the same powers. For example:

Like **Terms**			*Unlike* **Terms**			
$-6t$	and	$4t$	$-6t$	and	$4s$	(different variables)
$1.8ab$	and	$-3ab$	$1.8xy$	and	$-3x$	(different variables)
$\frac{1}{2}c^2d^3$	and	c^2d^3	$\frac{1}{2}c^2d^3$	and	c^2d	(different powers)
4	and	6	$4p$	and	4	(different variables)

Example 1 Identifying Terms, Factors, Coefficients, and *Like* Terms

a. List the terms of the expression. $-4x^2 - 7x + \frac{2}{3}$

b. Identify the coefficient of the term. yz^3

c. Identify the pair of *like* terms. $16b, 4b^2$ or $\frac{1}{2}c, -\frac{1}{6}c$

Solution:

a. The terms of the expression $-4x^2 - 7x + \frac{2}{3}$ are $-4x^2$, $-7x$, and $\frac{2}{3}$.

b. The term yz^3 can be written as $1yz^3$; therefore, the coefficient is 1.

c. $\frac{1}{2}c, -\frac{1}{6}c$ are *like* terms because they have the same variable raised to the same power.

Skill Practice Given: $-2x^2 + 5x + \dfrac{1}{2} - y^2$

1. List the terms of the expression.
2. Which term is the constant term?
3. Identify the coefficient of the term $-y^2$.

Answers

1. $-2x^2, \ 5x, \ \dfrac{1}{2}, \ -y^2$

2. $\dfrac{1}{2}$ 3. -1

2. Properties of Real Numbers

Simplifying algebraic expressions requires several important properties of real numbers that are stated in Table R-3. Assume that a, b, and c represent real numbers or real-valued algebraic expressions.

Table R-3

Property Name	Algebraic Representation	Example	Description/Notes
Commutative property of addition	$a + b = b + a$	$5 + 3 = 3 + 5$	The order in which two real numbers are added or multiplied does not affect the result.
Commutative property of multiplication	$a \cdot b = b \cdot a$	$(5)(3) = (3)(5)$	
Associative property of addition	$(a + b) + c$ $= a + (b + c)$	$(2 + 3) + 7$ $= 2 + (3 + 7)$	The manner in which real numbers are grouped under addition or multiplication does not affect the result.
Associative property of multiplication	$(a \cdot b)c = a(b \cdot c)$	$(2 \cdot 3)7 = 2(3 \cdot 7)$	
Distributive property of multiplication over addition	$a(b + c)$ $= ab + ac$	$3(5 + 2)$ $= 3 \cdot 5 + 3 \cdot 2$	A factor outside the parentheses is multiplied by each term inside the parentheses.
Identity property of addition	0 is the identity element for addition because $a + 0 = 0 + a = a$	$5 + 0 = 0 + 5 = 5$	Any number added to the identity element 0 will remain unchanged.
Identity property of multiplication	1 is the identity element for multiplication because $a \cdot 1 = 1 \cdot a = a$	$5 \cdot 1 = 1 \cdot 5 = 5$	Any number multiplied by the identity element 1 will remain unchanged.
Inverse property of addition	a and $(-a)$ are additive inverses because $a + (-a) = 0$ and $(-a) + a = 0$	$3 + (-3) = 0$	The sum of a number and its additive inverse (opposite) is the identity element 0.
Inverse property of multiplication	a and $\frac{1}{a}$ are multiplicative inverses because $a \cdot \frac{1}{a} = 1$ and $\frac{1}{a} \cdot a = 1$ (provided $a \neq 0$)	$5 \cdot \frac{1}{5} = 1$	The product of a number and its multiplicative inverse (reciprocal) is the identity element 1.

The properties of real numbers are used to multiply algebraic expressions. To multiply a term by an algebraic expression containing more than one term, we apply the distributive property of multiplication over addition.

Example 2   **Applying the Distributive Property**

Apply the distributive property.

a. $4(2x + 5)$

b. $-(-3.4q + 5.7r)$

c. $-3(a + 2b - 5c)$

d. $-\dfrac{2}{3}\left(-9x + \dfrac{3}{8}y - 5\right)$

Solution:

a. $4(2x + 5)$

$= 4(2x) + 4(5)$ Apply the distributive property.

$= 8x + 20$ Simplify, using the associative property of multiplication.

b. $-(-3.4q + 5.7r)$ The negative sign preceding the parentheses can be interpreted as a factor of -1.

$= -1(-3.4q + 5.7r)$

$= -1(-3.4q) + (-1)(5.7r)$ Apply the distributive property.

$= 3.4q - 5.7r$

c. $-3(a + 2b - 5c)$

$= -3(a) + (-3)(2b) + (-3)(-5c)$ Apply the distributive property.

$= -3a - 6b + 15c$ Simplify.

d. $-\dfrac{2}{3}\left(-9x + \dfrac{3}{8}y - 5\right)$

$= -\dfrac{2}{3}(-9x) + \left(-\dfrac{2}{3}\right)\left(\dfrac{3}{8}y\right) + \left(-\dfrac{2}{3}\right)(-5)$ Apply the distributive property.

$= \dfrac{18}{3}x - \dfrac{6}{24}y + \dfrac{10}{3}$ Simplify.

$= 6x - \dfrac{1}{4}y + \dfrac{10}{3}$ Simplify to lowest terms.

> **TIP:** When applying the distributive property, a negative factor preceding the parentheses will change the signs of the terms within the parentheses.
>
>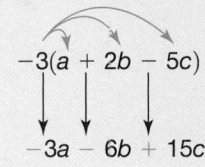

Skill Practice Apply the distributive property.

4. $10(30y - 40)$

5. $-(7t - 1.6s + 9.2)$

6. $-2(4x - 3y - 6)$

7. $-\dfrac{1}{2}(-4a + 7)$

Notice that the parentheses are removed after the distributive property is applied. Sometimes this is referred to as clearing parentheses.

Two terms can be added or subtracted only if they are *like* terms. To add or subtract *like* terms, we use the distributive property, as shown in Example 3.

Answers

4. $300y - 400$

5. $-7t + 1.6s - 9.2$

6. $-8x + 6y + 12$

7. $2a - \dfrac{7}{2}$

| Example 3 | Using the Distributive Property to Add and Subtract *Like* Terms |

Add and subtract as indicated.

a. $-8x + 3x$ **b.** $4.75y^2 - 9.25y^2 + y^2$

Solution:

a. $-8x + 3x$

$$= (-8 + 3)x \qquad \text{Apply the distributive property.}$$
$$= (-5)x \qquad \text{Simplify.}$$
$$= -5x$$

b. $4.75y^2 - 9.25y^2 + y^2$

$$= 4.75y^2 - 9.25y^2 + 1y^2 \qquad \text{Notice that } y^2 \text{ is interpreted as } 1y^2.$$
$$= (4.75 - 9.25 + 1)y^2 \qquad \text{Apply the distributive property.}$$
$$= (-3.5)y^2 \qquad \text{Simplify.}$$
$$= -3.5y^2$$

Skill Practice Combine *like* terms.

8. $-4y + 7y$ **9.** $a^2 - 6.2a^2 + 2.8a^2$

Although the distributive property is used to add and subtract *like* terms, it is tedious to write each step. Observe that adding or subtracting *like* terms is a matter of combining the coefficients and leaving the variable factors unchanged. This can be shown in one step. This shortcut will be used throughout the text. For example:

$$4w + 7w = 11w \qquad 8ab^2 + 10ab^2 - 5ab^2 = 13ab^2$$

3. Simplifying Expressions

Clearing parentheses and combining *like* terms are important tools to simplifying algebraic expressions. This is demonstrated in Example 4.

| Example 4 | Clearing Parentheses and Combining *Like* Terms |

Simplify by clearing parentheses and combining *like* terms.

a. $4 - 3(2x - 8) - 1$ **b.** $-(3s - 11t) - 5(2t + 8s) - 10s$

Solution:

a. $4 - 3(2x - 8) - 1$

$$= 4 - 6x + 24 - 1 \qquad \text{Apply the distributive property.}$$
$$= -6x + 4 + 24 - 1 \qquad \text{Group } like \text{ terms.}$$
$$= -6x + 27 \qquad \text{Combine } like \text{ terms.}$$

Answers

8. $3y$ **9.** $-2.4a^2$

b. $-(3s - 11t) - 5(2t + 8s) - 10s$

$\quad = -3s + 11t - 10t - 40s - 10s \qquad$ Apply the distributive property.

$\quad = -3s - 40s - 10s + 11t - 10t \qquad$ Group *like* terms.

$\quad = -53s + t \qquad$ Combine *like* terms.

Skill Practice Simplify by clearing parentheses and combining *like* terms.

10. $7 - 2(3x - 4) - 5$ **11.** $-(6z - 10y) - 4(3z + y) - 8y$

Example 5   Clearing Parentheses and Combining *Like* Terms

Simplify by clearing parentheses and combining *like* terms.

a. $2[1.5x + 4.7(x^2 - 5.2x) - 3x]$ **b.** $-\dfrac{1}{3}(3w - 6) - \left(\dfrac{1}{4}w + 4\right)$

Solution:

a. $2[1.5x + 4.7(x^2 - 5.2x) - 3x]$

$\quad = 2[1.5x + 4.7x^2 - 24.44x - 3x] \qquad$ Apply the distributive property to the inner parentheses.

$\quad = 2[1.5x - 24.44x - 3x + 4.7x^2] \qquad$ Group *like* terms.

$\quad = 2[-25.94x + 4.7x^2] \qquad$ Combine *like* terms.

$\quad = -51.88x + 9.4x^2 \qquad$ Apply the distributive property.

$\quad = 9.4x^2 - 51.88x$

> **TIP:** By using the commutative property of addition, the expression $-51.88x + 9.4x^2$ can also be written as $9.4x^2 + (-51.88x)$ or simply $9.4x^2 - 51.88x$. Although the expressions are all equal, it is customary to write the terms in descending order of the powers of the variable.

b. $-\dfrac{1}{3}(3w - 6) - \left(\dfrac{1}{4}w + 4\right)$

$\quad = -\dfrac{3}{3}w + \dfrac{6}{3} - \dfrac{1}{4}w - 4 \qquad$ Apply the distributive property.

$\quad = -w + 2 - \dfrac{1}{4}w - 4 \qquad$ Simplify fractions.

$\quad = -\dfrac{4}{4}w - \dfrac{1}{4}w + 2 - 4 \qquad$ Group *like* terms and find a common denominator.

$\quad = -\dfrac{5}{4}w - 2 \qquad$ Combine *like* terms.

Skill Practice Simplify by clearing parentheses and combining *like* terms.

12. $4[1.4a + 2.2(a^2 - 6a)] - 5.1a^2$ **13.** $-\dfrac{1}{2}(4p - 1) - \dfrac{5}{2}(p - 2)$

Answers

10. $-6x + 10$ **11.** $-2y - 18z$

12. $3.7a^2 - 47.2a$ **13.** $-\dfrac{9}{2}p + \dfrac{11}{2}$

Section R.4 Practice Exercises

Study Skills Exercise

It is very important to attend class every day. Math is cumulative in nature and you must master the material learned in the previous class to understand a new day's lesson. Because this is so important, many instructors tie attendance to the final grade. Write down the attendance policy for your class.

Vocabulary and Key Concepts

1. **a.** Given the expression $8x + cd - 3y + 90$, the terms $8x$, cd, and $-3y$ are variable terms, whereas 90 is a _____ term.

   b. The constant factor in a term is called the _____.

   c. Given the expression x, the value of the coefficient is _____, and the exponent is _____.

   d. Terms that have the same variables, with corresponding variables raised to the same powers are called _____ terms.

Review Exercises

For Exercises 2–4, simplify.

2. $6 - 2\sqrt{10^2 - 6^2} + 7$

3. $\dfrac{-3^2 - |5 + (-7)|}{26 - 2^2}$

4. $36 \div (-2) \cdot 3 - 4(6 - 8)$

For Exercises 5–8, write the set in interval notation.

5. $\{x \mid x > |-3|\}$

6. $\left\{x \mid x \le \left|-\dfrac{4}{3}\right|\right\}$

7. $\left\{w \mid -\dfrac{5}{2} < w \le \sqrt{9}\right\}$

8. $\left\{z \mid 2 \le z < \dfrac{11}{3}\right\}$

Concept 1: Recognizing Terms, Factors, and Coefficients

For Exercises 9–12:

    a. Determine the number of terms in the expression.

    b. Identify the constant term.

    c. List the coefficients of each term, separated by commas. **(See Example 1.)**

9. $2x^3 - 5xy + 6$

 10. $a^2 - 4ab - b^2 + 8$

11. $pq - 7 + q^2 - 4q + p$

12. $7x - 1 + 3xy$

Concept 2: Properties of Real Numbers

For Exercises 13–30, match each expression with the appropriate property.

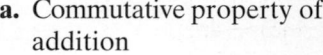

13. $3 + \dfrac{1}{2} = \dfrac{1}{2} + 3$

14. $7.2(4 + 1) = 7.2(4) + 7.2(1)$

    a. Commutative property of addition

15. $10 + 0 = 10$

16. $7 \cdot 1 = 7$

    b. Associative property of multiplication

17. $(6 + 8) + 2 = 6 + (8 + 2)$

18. $(4 + 19) + 7 = (19 + 4) + 7$

    c. Distributive property of multiplication over addition

19. $6 \cdot \dfrac{1}{6} = 1$

20. $2 + (-2) = 0$

d. Commutative property of multiplication

21. $9(4 \cdot 12) = (9 \cdot 4)12$

22. $\left(\dfrac{1}{4} + 2\right)20 = 5 + 40$

e. Associative property of addition

f. Identity property of addition

23. $42 \cdot 1 = 42$

24. $4 \cdot \dfrac{1}{4} = 1$

g. Identity property of multiplication

25. $(13 \cdot 41)6 = (41 \cdot 13)6$

26. $6(x + 3) = 6x + 18$

h. Inverse property of addition

27. $8 + (-8) = 0$

28. $21 + 0 = 21$

29. $3(y + 10) = 3(10 + y)$

30. $5(3 \cdot 7) = (5 \cdot 3)7$

i. Inverse property of multiplication

For Exercises 31–42, clear parentheses by applying the distributive property. **(See Example 2.)**

31. $2(x - 3y + 8)$

32. $5(-2a + 4b - 9c)$

33. $-10(4s - 9t - 3)$

34. $-4(-8x + 6y + 3z)$

35. $-(-7w + 5z)$

36. $-(-22a - 17b)$

37. $-\dfrac{1}{5}\left(-\dfrac{5}{2}a + 10b - 8\right)$

38. $-\dfrac{3}{4}\left(6x - 4y + \dfrac{4}{9}\right)$

39. $3(2.6x - 4.1)$

40. $5(-7.2y + 2.3)$

41. $2(7c - 8) - 5(6d - f)$

42. $-2(-3q + r) - 7(5s + 2t)$

Concept 3: Simplifying Expressions

For Exercises 43–80, clear parentheses and combine *like* terms. **(See Examples 3–5.)**

43. $8y - 2x + y + 5y$

44. $-9a + a - b + 5a$

45. $4p^2 - 2p + 3p - 6 + 2p^2$

46. $6q - 9 + 3q^2 - q^2 + 10$

47. $2p - 7p^2 - 5p + 6p^2$

48. $5a^2 - 2a - 7a^2 + 6a + 4$

49. $m - 4n^3 + 3 + 5n^3 - 9$

50. $x + 2y^3 - 2x - 8y^3$

51. $5ab + 2ab + 8a$

52. $-6m^2n - 3mn^2 - 2m^2n$

53. $14xy^2 - 5y^2 + 2xy^2$

54. $9uv + 3u^2 + 5uv + 4u^2$

55. $8(x - 3) + 1$

56. $-4(b + 2) - 3$

57. $-2(c + 3) - 2c$

58. $4(z - 4) - 3z$

59. $-(10w - 1) + 9 + w$

60. $-(2y + 7) - 4 + 3y$

61. $-9 - 4(2 - z) + 1$

62. $3 + 3(4 - w) - 11$

63. $4(2s - 7) - (s - 2)$

64. $2(t - 3) - (t - 7)$

65. $-3(-5 + 2w) - 8w + 2(w - 1)$

66. $5 - (-4t - 7) - t - 9$

67. $8x - 4(x - 2) - 2(2x + 1) - 6$

68. $6(y - 2) - 3(2y - 5) - 3$

69. $\dfrac{1}{2}(4 - 2c) + 5c$

70. $\dfrac{2}{3}(3d + 6) - 4d$

71. $3.1(2x + 2) - 4(1.2x - 1)$

72. $4.5(5 - y) + 3(1.9y + 1)$

73. $2\left[5\left(\dfrac{1}{2}a + 3\right) - (a^2 + a) + 4\right]$

74. $-3\left[3\left(b - \dfrac{2}{3}\right) - 2(b + 4) - 6b^2\right]$

75. $(2y - 5) - 2(y - y^2) - 3y$

76. $-(x + 6) + 3(x^2 + 1) + 2x$

77. $2.2\{4 - 8[6x - 1.5(x + 4) - 6] + 7.5x\}$

78. $-3.2 - \{6.1y - 4[9 - (2y + 2.5)] + 7y\}$

79. $\dfrac{1}{8}(24n - 16m) - \dfrac{2}{3}(3m - 18n - 2) + \dfrac{2}{3}$

80. $\dfrac{1}{5}(25a - 20b) - \dfrac{4}{7}(21a - 14b + 2) + \dfrac{1}{7}$

Expanding Your Skills

81. What is the identity element for addition? Use it in an example.

82. What is the identity element for multiplication? Use it in an example.

83. What is another name for a multiplicative inverse?

84. What is another name for an additive inverse?

85. Is the operation of subtraction commutative? If not, give an example.

86. Is the operation of division commutative? If not, give an example.

87. Given the rectangular regions:

 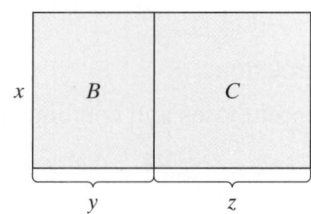

a. Write an expression for the area of region A. (Do not simplify.)

b. Write an expression for the area of region B.

c. Write an expression for the area of region C.

d. Add the expressions for the area of regions B and C.

e. Show that the area of region A is equal to the sum of the areas of regions B and C. What property of real numbers does this illustrate?

Chapter R   Summary

Section R.2   Sets of Numbers and Interval Notation

Key Concepts

Natural numbers: $\{1, 2, 3, \ldots\}$
Whole numbers: $\{0, 1, 2, 3, \ldots\}$
Integers: $\{\ldots, -3, -2, -1, 0, 1, 2, 3, \ldots\}$

Rational numbers: $\left\{ \dfrac{p}{q} \;\middle|\; p \text{ and } q \text{ are integers and } q \neq 0 \right\}$

Rational numbers include all terminating and repeating decimals.

Irrational numbers: A subset of the real numbers whose elements cannot be written as a ratio of two integers. Irrational numbers cannot be written as repeating or terminating decimals.

Real numbers: $\{x \mid x \text{ is rational or } x \text{ is irrational}\}$

$a < b$ "a is less than b"
$a > b$ "a is greater than b"
$a \leq b$ "a is less than or equal to b"
$a \geq b$ "a is greater than or equal to b"
$a < x < b$ "x is between a and b"

Examples

Example 1

Some rational numbers are:
$\frac{1}{7}, 0.5, 0.\overline{3}$

Some irrational numbers are:
$\sqrt{7}, \sqrt{2}, \pi$

Example 2

Set-Builder Notation	Interval Notation	Graph
$\{x \mid x > 3\}$	$(3, \infty)$	
$\{x \mid x \geq 3\}$	$[3, \infty)$	
$\{x \mid x < 3\}$	$(-\infty, 3)$	
$\{x \mid x \leq 3\}$	$(-\infty, 3]$	

Section R.3   Operations on Real Numbers

Key Concepts

The **reciprocal** of a number $a \neq 0$ is $\frac{1}{a}$.
The **opposite** of a number a is $-a$.
The **absolute value** of a, denoted $|a|$, is its distance from zero on the number line.

$|a| = a$ if $a \geq 0$
$|a| = -a$ if $a < 0$

Addition of Real Numbers

Same Signs: Add the absolute values of the numbers, and apply the common sign to the sum.
Unlike Signs: Subtract the smaller absolute value from the larger absolute value. Then apply the sign of the number having the larger absolute value.

Examples

Example 1

Given: -5
The reciprocal is $-\frac{1}{5}$. The opposite is 5.

The absolute value is 5.

Example 2

$-3 + (-4) = -7$
$-5 + 7 = 2$

Subtraction of Real Numbers

Add the opposite of the second number to the first number. $a - b = a + (-b)$

Example 3

$7 - (-5) = 7 + (5) = 12$

Multiplication and Division of Real Numbers

Same Signs: Product or quotient is positive.
Opposite Signs: Product or quotient is negative.

The product of any real number and 0 is 0.
The quotient of 0 and a nonzero number is 0.
The quotient of a nonzero number and 0 is undefined.

Example 4

$(-3)(-4) = 12$ $\qquad \dfrac{-15}{-3} = 5$

$(-2)(5) = -10$ $\qquad \dfrac{6}{-12} = -\dfrac{1}{2}$

$(-7)(0) = 0$ $\qquad 0 \div 9 = 0$

$-3 \div 0$ is undefined

Exponents and Radicals

$b^4 = b \cdot b \cdot b \cdot b$ (b is the **base**, 4 is the **exponent**)
$\sqrt{b}$ is the **principal square root** of b ($\sqrt{}$ is the radical sign).

Example 5

$6^3 = 6 \cdot 6 \cdot 6 = 216$

$\sqrt{100} = 10$

Order of Operations

1. Simplify expressions within parentheses and other grouping symbols first.
2. Evaluate expressions involving exponents, radicals, and absolute values.
3. Perform multiplication or division in order from left to right.
4. Perform addition or subtraction in order from left to right.

Example 6

$10 - 5(3 - 1)^2 + \sqrt{16}$

$= 10 - 5(2)^2 + \sqrt{16}$

$= 10 - 5(4) + 4$

$= 10 - 20 + 4$

$= -10 + 4$

$= -6$

Section R.4 Simplifying Algebraic Expressions

Key Concepts

A **term** is a constant or the product or quotient of a constant and one or more variables.

- A **variable term** contains at least one variable.
- A **constant term** has no variable.

The **coefficient** of a term is the numerical factor of the term.

Like **terms** have the same variables, and the corresponding variables are raised to the same powers.

Examples

Example 1

$-2x$ Variable term has coefficient -2.

x^2y Variable term has coefficient 1.

6 Constant term has coefficient 6.

Example 2

$4ab^3$ and $2ab^3$ are *like* terms.

Distributive Property of Multiplication over Addition

$a(b + c) = ab + ac$

Example 3

$2(x + 4y) = 2x + 8y$

$-(a + 6b - 5c) = -a - 6b + 5c$

Two terms can be added or subtracted if they are *like* terms. First clear parentheses before adding or subtracting *like* terms.

Example 4

$-4d - 12d + d$

$\quad = 9d$

Example 5

$-2[w - 4(w - 2)] + 3$

$\quad = -2[w - 4w + 8] + 3$

$\quad = -2[-3w + 8] + 3$

$\quad = 6w - 16 - 3$

$\quad = 6w - 13$

Chapter R   Review Exercises

Section R.2

1. Find a number that is a whole number but not a natural number.

For Exercises 2–3, answers may vary.

2. List three rational numbers that are not integers.

3. List five integers, two of which are not whole numbers.

For Exercises 4–9, write an expression in words that describes the set of numbers given by each interval. (Answers may vary.)

4. $(7, 16)$

5. $(0, 2.6]$

6. $[-6, -3]$

7. $(8, \infty)$

8. $(-\infty, 13]$

9. $(-\infty, \infty)$

For Exercises 10–12, graph each set and write the set in interval notation.

10. $\{x \mid x < 2\}$

11. $\{x \mid x \geq 0\}$

12. $\{x \mid -1 < x < 5\}$

13. True or false? $x < 3$ is equivalent to $3 > x$

Section R.3

For Exercises 14–15, find the opposite, reciprocal, and absolute value.

14. -8

15. $\dfrac{4}{9}$

For Exercises 16–17, simplify the exponents and the radicals.

16. $4^2, \sqrt{4}$

17. $25^2, \sqrt{25}$

For Exercises 18–33, perform the indicated operations.

18. $6 + (-8)$

19. $(-2) - (-5)$

20. $8(-2.7)$

21. $(-1.1)(7.41)$

22. $\dfrac{5}{8} \div \left(-\dfrac{13}{40}\right)$

23. $\left(-\dfrac{1}{4}\right) \div \left(-\dfrac{11}{16}\right)$

24. $2\dfrac{2}{5} - \left(1\dfrac{1}{10}\right)^2$

25. $4\dfrac{2}{3} - 3\left(1\dfrac{1}{6}\right)$

26. $\dfrac{2 - 4(3 - 7)}{-4 - 5(1 - 3)}$

27. $\dfrac{12(2) - 8}{4(-3) + 2(5)}$

28. $24 \div 8 \cdot 2$

29. $40 \div 5 \cdot 6$

30. $3^2 + 2(|-10 + 5| \div 5)$

31. $-91 + \sqrt{4}(\sqrt{25} - 13)^2$

32. $\dfrac{3(3 - 8)^2}{|8 - 3^2|}$

33. $\dfrac{4(5 - 2)^2}{|3 - 7 - 5|}$

34. Given $h = \frac{1}{2}gt^2 + v_0t + h_0$, find h if $g = -32$, $v_0 = 64$, $h_0 = 256$, and $t = 4$.

35. Find the area of a parallelogram with base 42 in. and height 18 in.

Section R.4

For Exercises 36–39, apply the distributive property and simplify.

36. $3(x + 5y)$

37. $\frac{1}{2}(x + 8y - 5)$

38. $-(-4x + 10y - z)$ **39.** $-(13a - b - 5c)$

For Exercises 40–43, clear parentheses if necessary, and combine *like* terms.

40. $5 - 6q + 13q - 19$ **41.** $18p + 3 - 17p + 8p$

42. $7 - 3(y + 4) - 3y$

43. $\frac{3}{4}(8x - 4) + \frac{1}{2}(6x + 4)$

For Exercises 44–45, answers may vary.

44. Write an example of the commutative property of addition.

45. Write an example of the associative property of multiplication.

Chapter R   Test

1. a. List the integers between -5 and 2, inclusive.

   b. List three rational numbers between 1 and 2. (Answers may vary.)

2. Write the opposite, reciprocal, and absolute value for each number.

   a. $-\frac{1}{2}$ **b.** 4 **c.** 0

3. Explain the difference between the interval $[4, \infty)$ and $(4, \infty)$.

4. Answer true or false: The set $\{x \mid x \geq 5\}$ is the same as $\{x \mid 5 \leq x\}$.

For Exercises 5–6, graph the inequality and express the set in interval notation.

5. $\{y \mid y < -\frac{4}{3}\}$ **6.** $\{p \mid 12 \leq p\}$

For Exercises 7–8, write each English phrase as an algebraic statement.

7. x is no more than 5.   **8.** p is at least 7.

For Exercises 9–12, simplify the expression.

9. $|-8| - 4(2 - 3)^2 \div \sqrt{4}$ **10.** $\dfrac{-6^2 - 10^2}{-1 + 3^2}$

11. $\left(-\dfrac{1}{6} + \sqrt{\dfrac{4}{9}}\right)^2$ **12.** $-8 \div 3 \cdot 2$

13. Given $z = \dfrac{x - \mu}{\sigma/\sqrt{n}}$, find z when $n = 16$, $x = 18$, $\sigma = 1.8$, and $\mu = 17.5$. (Round the answer to 1 decimal place.)

For Exercises 14–16, simplify the expressions.

14. $5b + 2 - 7b + 6 - 14$

15. $-3(4 - x) + 9(x - 1) - 5(2x - 4)$

16. $\frac{1}{2}(2x - 1) - \left(3x - \frac{3}{2}\right)$

For Exercises 17–20, answer true or false.

17. $(x + y) + 2 = 2 + (x + y)$ is an example of the associative property of addition.

18. $(2 \cdot 3) \cdot 5 = (3 \cdot 2) \cdot 5$ is an example of the commutative property of multiplication.

19. $\left(\frac{1}{2}x + \frac{1}{3}\right)6 = 3x + 2$ is an example of the distributive property of multiplication over addition.

20. $(10 + y) + z = 10 + (y + z)$ is an example of the associative property of addition.

Linear Equations and Inequalities in One Variable

1

CHAPTER OUTLINE

1.1 Linear Equations in One Variable   44

    Problem Recognition Exercises: Equations Versus Expressions   56

1.2 Applications of Linear Equations in One Variable   57

1.3 Applications to Geometry and Literal Equations   68

1.4 Linear Inequalities in One Variable   76

1.5 Compound Inequalities   85

1.6 Absolute Value Equations   97

1.7 Absolute Value Inequalities   103

    Problem Recognition Exercises: Identifying Equations and Inequalities   113

    Group Activity: Understanding the Symbolism of Mathematics   114

Chapter 1

In this chapter, we study linear equations and inequalities and their applications.

Are You Prepared?

To prepare yourself, try the crossword puzzle. The clues in the puzzle review formulas from geometry and other important mathematical facts that you will encounter in this chapter as you work through the application problems.

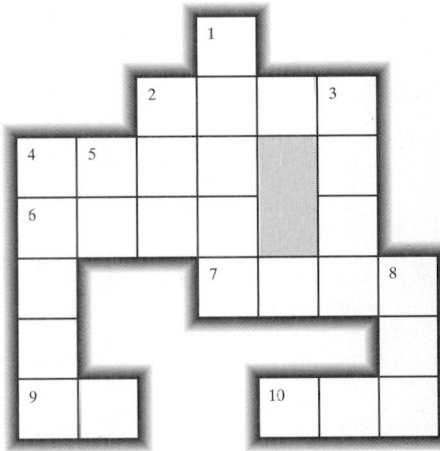

Across

2. What is the next consecutive integer after 1306?
4. Given that distance = rate × time, what is the distance between Atlanta and Los Angeles if it takes 33 hr, traveling 60 mph?
6. What is the next consecutive odd integer after 7803?
7. What is 10% of 64,780?
9. If an angle measures 41°, what is the complement?
10. If an angle measures 70°, what is the supplement?

Down

1. What is 40% of 32,640?
2. What is the sum of the measures of the angles in a triangle?
3. Evaluate $|-7729 + 262|$.
4. What number is 50 more than twice 8707?
5. If the area of a rectangle is 6370 ft^2, and the width is 65 ft, what is the length?
8. What is the amount of simple interest earned on $4000 at 5% interest for 4 yr?

Concepts

1. Linear Equations in One Variable
2. Solving Linear Equations
3. Clearing Fractions and Decimals
4. Conditional Equations, Contradictions, and Identities

1. Linear Equations in One Variable

An **equation** is a statement that indicates that two quantities are equal. The following are equations.

$$x = -4 \qquad p + 3 = 11 \qquad -2z = -20$$

All equations have an equal sign. Furthermore, notice that the equal sign separates the equation into two parts, the left-hand side and the right-hand side. A **solution to an equation** is a value of the variable that makes the equation a true statement. Substituting a solution to an equation for the variable makes the right-hand side equal to the left-hand side.

Equation	Solution	Check	
$p + 3 = 11$	8	$p + 3 = 11$ $\downarrow$ $8 + 3 = 11$ ✓	Substitute 8 for p. Right-hand side equals left-hand side.
$-2z = -20$	10	$-2z = -20$ $\downarrow$ $-2(10) = -20$ ✓	Substitute 10 for z. Right-hand side equals left-hand side.

The **solution set** to an equation is the set of *all* solutions to an equation. We write the solution set using set braces. For example:

Equation		Solution set
$p + 3 = 11$	This equation has one solution. $\longrightarrow$	$\{8\}$
$w^2 = 16$	This equation has two solutions. $\longrightarrow$	$\{4, -4\}$

Throughout this text we will learn to recognize and solve several different types of equations, but in this chapter, we will focus on the specific type of equation called a linear equation in one variable.

Definition of a Linear Equation in One Variable

Let a, b, and c be real numbers such that $a \neq 0$. A **linear equation in one variable** is an equation that can be written in the form

$$ax + b = c$$

Note: A linear equation in one variable is often called a first-degree equation because the variable x has an implied exponent of 1.

Examples	Notes
$3x + 5 = 20$	$a = 3, b = 5, c = 20$
$-2x - 4 = 6$ can be written as $-2x + (-4) = 6$	$a = -2, b = -4, c = 6$
$6x + 7 - 5x = 11$ can be written as $x + 7 = 11$	$a = 1, b = 7, c = 11$

2. Solving Linear Equations

To solve a linear equation, the goal is to simplify the equation to isolate the variable. Each step used in simplifying an equation results in an equivalent equation. *Equivalent equations* have the same solution set. For example, the equations $2x + 3 = 7$ and $2x = 4$ are equivalent because {2} is the solution set for both equations.

To solve an equation, we may use the addition, subtraction, multiplication, and division properties of equality. These properties state that adding, subtracting, multiplying, or dividing the same quantity on each side of an equation results in an equivalent equation.

Addition and Subtraction Properties of Equality

Let a, b, and c represent real numbers.

Addition property of equality: If $a = b$, then $a + c = b + c$.

*Subtraction property of equality: If $a = b$, then $a - c = b - c$.

*The subtraction property of equality follows directly from the addition property, because subtraction is defined in terms of addition.

$$\text{If} \quad a + (-c) = b + (-c)$$
$$\text{then,} \quad a - c = b - c$$

Multiplication and Division Properties of Equality

Let a, b, and c represent real numbers with $c \neq 0$.

Multiplication property of equality: If $a = b$, then $a \cdot c = b \cdot c$.

*Division property of equality: If $a = b$, then $\dfrac{a}{c} = \dfrac{b}{c}$.

*The division property of equality follows directly from the multiplication property, because division is defined as multiplication by the reciprocal.

$$\text{If} \quad a \cdot \frac{1}{c} = b \cdot \frac{1}{c}$$
$$\text{then,} \quad \frac{a}{c} = \frac{b}{c}$$

Example 1   Solving a Linear Equation

Solve the equation. $12 + x = 40$

Solution:

$$12 + x = 40$$
$$12 - 12 + x = 40 - 12 \qquad \text{To isolate } x, \text{ subtract 12 from both sides.}$$
$$x = 28 \qquad \text{Simplify.}$$

Check: $12 + x = 40$

$12 + (28) \stackrel{?}{=} 40$

$40 \stackrel{?}{=} 40 \checkmark$ True.

The solution set is $\{28\}$.

Skill Practice Solve the equation.

1. $x - 5 = -11$

Example 2 Solving Linear Equations

Solve each equation.

a. $-\dfrac{3}{5}p = \dfrac{4}{15}$ **b.** $4 = \dfrac{w}{2.2}$ **c.** $-x = 6$

Solution:

a.

$$-\dfrac{3}{5}p = \dfrac{4}{15}$$

$$\left(-\dfrac{5}{3}\right)\left(-\dfrac{3}{5}p\right) = \left(-\dfrac{5}{3}\right)\left(\dfrac{4}{15}\right)$$ To isolate p, multiply both sides by the reciprocal of $-\dfrac{3}{5}$.

$$p = \left(-\dfrac{\overset{1}{\cancel{5}}}{3}\right)\left(\dfrac{4}{\underset{3}{\cancel{15}}}\right)$$ Multiply fractions.

$$p = -\dfrac{4}{9}$$ The value $-\dfrac{4}{9}$ checks in the original equation.

The solution set is $\left\{-\dfrac{4}{9}\right\}$.

b.

$$4 = \dfrac{w}{2.2}$$

$$2.2(4) = 2.2 \cdot \left(\dfrac{w}{2.2}\right)$$ To isolate w, multiply both sides by 2.2.

$$8.8 = w$$ The value 8.8 checks in the original equation.

The solution set is $\{8.8\}$.

c.

$$-x = 6$$

$$-1(-x) = -1(6)$$ To isolate x, multiply both sides by -1.

$$x = -6$$ The value -6 checks in the original equation.

The solution set is $\{-6\}$.

Skill Practice Solve the equations.

2. $-\dfrac{6}{5}y = -\dfrac{3}{5}$ **3.** $5 = \dfrac{t}{16}$ **4.** $-a = -2$

Answers

1. $\{-6\}$ **2.** $\left\{\dfrac{1}{2}\right\}$

3. $\{80\}$ **4.** $\{2\}$

For more complicated linear equations, several steps are required to isolate the variable. These steps are listed below.

Solving a Linear Equation in One Variable

Step 1 Simplify both sides of the equation.
- Clear parentheses.
- Consider clearing fractions or decimals (if any are present) by multiplying both sides of the equation by a common denominator of all terms.
- Combine *like* terms.

Step 2 Use the addition or subtraction property of equality to collect the variable terms on one side of the equation.

Step 3 Use the addition or subtraction property of equality to collect the constant terms on the other side of the equation.

Step 4 Use the multiplication or division property of equality to make the coefficient of the variable term equal to 1.

Step 5 Check your answer and write the solution set.

Example 3 **Solving a Linear Equation**

Solve the linear equation and check the answer.

$$3x + 1 = -7$$

Solution:

$$3x + 1 = -7$$

$$3x + 1 - 1 = -7 - 1 \qquad \text{Subtract 1 from both sides.}$$

$$3x = -8 \qquad \text{Combine } like \text{ terms.}$$

$$\frac{3x}{3} = \frac{-8}{3} \qquad \text{To isolate } x, \text{ divide both sides of the equation by 3.}$$

$$x = -\frac{8}{3} \qquad \text{Simplify.}$$

$$\text{Check:} \qquad 3x + 1 \stackrel{?}{=} -7$$

$$3\left(-\frac{8}{3}\right) + 1 \stackrel{?}{=} -7$$

$$3\left(-\frac{8}{3}\right) + 1 \stackrel{?}{=} -7$$

$$-8 + 1 \stackrel{?}{=} -7$$

$$-7 \stackrel{?}{=} -7 \checkmark \quad \text{True.}$$

The solution set is $\left\{-\dfrac{8}{3}\right\}$.

Skill Practice Solve the linear equation and check the answer.

5. $5x - 19 = -23$

Answer

5. $\left\{-\dfrac{4}{5}\right\}$

Example 4   Solving a Linear Equation

Solve the linear equation and check the answer.

$$11z + 2 = 5(z - 2)$$

Solution:

$11z + 2 = 5(z - 2)$	
$11z + 2 = 5z - 10$	Apply the distributive property to clear parentheses.
$11z - 5z + 2 = 5z - 5z - 10$	Subtract $5z$ from both sides.
$6z + 2 = -10$	Combine *like* terms.
$6z + 2 - 2 = -10 - 2$	Subtract 2 from both sides.
$6z = -12$	Combine *like* terms.
$\dfrac{6z}{6} = \dfrac{-12}{6}$	To isolate z, divide both sides of the equation by 6.
$z = -2$	Simplify.

$$\begin{aligned}
\underline{\text{Check:}} \quad 11z + 2 &= 5(z - 2) \\
11(-2) + 2 &\overset{?}{=} 5(-2 - 2) \\
-22 + 2 &\overset{?}{=} 5(-4) \\
-20 &\overset{?}{=} -20 \ \checkmark \quad \text{True.}
\end{aligned}$$

The solution set is $\{-2\}$.

Skill Practice Solve the equations.

6. $7 + 2(y - 3) = 6y + 3$

Example 5   Solving a Linear Equation

Solve the equation. $-3(x - 4) + 2 = 7 - (x + 1)$

Solution:

$-3(x - 4) + 2 = 7 - (x + 1)$	
$-3x + 12 + 2 = 7 - x - 1$	Clear parentheses.
$-3x + 14 = -x + 6$	Combine *like* terms.
$-3x + x + 14 = -x + x + 6$	Add x to both sides of the equation.
$-2x + 14 = 6$	Combine *like* terms.
$-2x + 14 - 14 = 6 - 14$	Subtract 14 from both sides.

Answer

6. $\left\{ -\dfrac{1}{2} \right\}$

$$-2x = -8$$ Combine *like* terms.

$$\frac{-2x}{-2} = \frac{-8}{-2}$$ To isolate x, divide both sides by -2.

$$x = 4$$ Simplify. The solution checks in the original equation.

The solution set is $\{4\}$.

Skill Practice Solve the equation.

7. $4(2t + 2) - 6(t - 1) = 6 - t$

Example 6 **Solving a Linear Equation**

Solve the equation. $-4[y - 3(y - 5)] = 2(6 - 5y)$

Solution:

$$-4[y - 3(y - 5)] = 2(6 - 5y)$$

$$-4[y - 3y + 15] = 12 - 10y$$ Clear parentheses.

$$-4[-2y + 15] = 12 - 10y$$ Combine *like* terms.

$$8y - 60 = 12 - 10y$$ Clear parentheses.

$$8y + 10y - 60 = 12 - 10y + 10y$$ Add $10y$ to both sides of the equation.

$$18y - 60 = 12$$ Combine *like* terms.

$$18y - 60 + 60 = 12 + 60$$ Add 60 to both sides of the equation.

$$18y = 72$$

$$\frac{18y}{18} = \frac{72}{18}$$ To isolate y, divide both sides by 18.

$$y = 4$$ The solution checks.

The solution set is $\{4\}$.

Skill Practice Solve the equation.

8. $3[p + 2(p - 2)] = 4(p - 3)$

3. Clearing Fractions and Decimals

When an equation contains fractions or decimals, it is sometimes helpful to clear the fractions and decimals. This is accomplished by multiplying both sides of the equation by the least common denominator (LCD) of all terms within the equation. This is demonstrated in Example 7.

Answers

7. $\left\{-\dfrac{8}{3}\right\}$ **8.** $\{0\}$

Example 7   Solving a Linear Equation by Clearing Fractions

Solve the equation. $\dfrac{1}{4}w + \dfrac{1}{3}w - 1 = \dfrac{1}{2}(w - 4)$

Solution:

$$\dfrac{1}{4}w + \dfrac{1}{3}w - 1 = \dfrac{1}{2}(w - 4)$$

$$\dfrac{1}{4}w + \dfrac{1}{3}w - 1 = \dfrac{1}{2}w - 2 \qquad \text{Clear parentheses.}$$

$$12 \cdot \left(\dfrac{1}{4}w + \dfrac{1}{3}w - 1\right) = 12 \cdot \left(\dfrac{1}{2}w - 2\right) \qquad \begin{array}{l}\text{Multiply both sides of} \\ \text{the equation by the} \\ \text{LCD of all terms. In} \\ \text{this case, the LCD is 12.}\end{array}$$

$$12 \cdot \dfrac{1}{4}w + 12 \cdot \dfrac{1}{3}w + 12 \cdot (-1) = 12 \cdot \dfrac{1}{2}w + 12 \cdot (-2) \qquad \begin{array}{l}\text{Apply the distributive} \\ \text{property.}\end{array}$$

$$3w + 4w - 12 = 6w - 24$$

$$7w - 12 = 6w - 24$$

$$w - 12 = -24 \qquad \text{Subtract } 6w.$$

$$w = -12 \qquad \text{The solution checks.}$$

The solution set is $\{-12\}$.

Skill Practice Solve the equation by first clearing the fractions.

9. $\dfrac{3}{4}a + \dfrac{1}{2} = \dfrac{2}{3}a + \dfrac{1}{3}$

Example 8   Solving a Linear Equation by Clearing Fractions

Solve. $\dfrac{x - 2}{5} - \dfrac{x - 4}{2} = 2 + \dfrac{x + 4}{10}$

Solution:

$$\dfrac{x - 2}{5} - \dfrac{x - 4}{2} = \dfrac{2}{1} + \dfrac{x + 4}{10} \qquad \begin{array}{l}\text{The LCD of all} \\ \text{terms in the} \\ \text{equation is 10.}\end{array}$$

$$10\left(\dfrac{x - 2}{5} - \dfrac{x - 4}{2}\right) = 10\left(\dfrac{2}{1} + \dfrac{x + 4}{10}\right) \qquad \begin{array}{l}\text{Multiply both} \\ \text{sides by 10.}\end{array}$$

$$\dfrac{\overset{2}{10}}{1} \cdot \left(\dfrac{x - 2}{\underset{1}{5}}\right) - \dfrac{\overset{5}{10}}{1} \cdot \left(\dfrac{x - 4}{\underset{1}{2}}\right) = \dfrac{10}{1} \cdot \left(\dfrac{2}{1}\right) + \dfrac{\overset{1}{10}}{1} \cdot \left(\dfrac{x + 4}{\underset{1}{10}}\right) \qquad \begin{array}{l}\text{Apply the} \\ \text{distributive} \\ \text{property.}\end{array}$$

TIP: Clearing fractions is an application of the multiplication property of equality. We are multiplying both sides of the equation by the same number.

Answer

9. $\{-2\}$

$$2(x - 2) - 5(x - 4) = 20 + 1(x + 4) \quad \text{Clear fractions.}$$

$$2x - 4 - 5x + 20 = 20 + x + 4 \quad \begin{array}{l}\text{Apply the distributive} \\ \text{property.}\end{array}$$

$$-3x + 16 = x + 24 \quad \begin{array}{l}\text{Simplify both sides of the} \\ \text{equation.}\end{array}$$

$$-4x + 16 = 24 \quad \text{Subtract } x \text{ from both sides.}$$

$$-4x = 8 \quad \text{Subtract 16 from both sides.}$$

$$x = -2 \quad \begin{array}{l}\text{The value } -2 \text{ checks in the} \\ \text{original equation.}\end{array}$$

The solution set is $\{-2\}$.

Skill Practice Solve.

10. $\dfrac{1}{8} - \dfrac{x + 3}{4} = \dfrac{3x - 2}{2}$

The same procedure used to clear fractions in an equation can be used to clear decimals.

Example 9 **Solving a Linear Equation by Clearing Decimals**

Solve the equation. $0.55x - 0.6 = 2.05x$

Solution:

Recall that any terminating decimal can be written as a fraction. Therefore, the equation $0.55x - 0.6 = 2.05x$ is equivalent to

$$\frac{55}{100}x - \frac{6}{10} = \frac{205}{100}x$$

A convenient common denominator for all terms in this equation is 100. Multiplying both sides of the equation by 100 will have the effect of "moving" the decimal point 2 places to the right.

$$100(0.55x - 0.6) = 100(2.05x) \quad \text{Multiply both sides by 100 to clear decimals.}$$

$$55x - 60 = 205x$$

$$-60 = 150x \quad \text{Subtract } 55x \text{ from both sides.}$$

$$-\frac{60}{150} = x \quad \text{To isolate } x, \text{ divide both sides by 150.}$$

$$x = -\frac{2}{5} = -0.4 \quad \text{The solution checks.}$$

The solution set is $\{-0.4\}$.

Skill Practice Solve the equation by first clearing the decimals.

11. $2.2x + 0.5 = 1.6x + 0.2$

Answers

10. $\left\{\dfrac{3}{14}\right\}$ **11.** $\{-0.5\}$

4. Conditional Equations, Contradictions, and Identities

The solution to a linear equation is the value of x that makes the equation a true statement. While linear equations have one unique solution, some equations have no solution, and others have infinitely many solutions.

I. Conditional Equations

An equation that is true for some values of the variable but false for other values is called a **conditional equation**. The equation $x + 4 = 6$ is a conditional equation because it is true on the *condition* that $x = 2$. For other values of x, the statement $x + 4 = 6$ is false. The solution set is $\{2\}$.

II. Contradictions

Some equations have no solution, such as $x + 1 = x + 2$. There is no value of x that when increased by 1 will equal the same value increased by 2. If we tried to solve the equation by subtracting x from both sides, we get the contradiction $1 = 2$.

$$x + 1 = x + 2$$
$$x - x + 1 = x - x + 2$$
$$1 = 2 \quad \text{(contradiction)}$$

This indicates that the equation has no solution. An equation that has no solution is called a **contradiction**. The solution set for a contradiction is the empty set. The **empty set** is the set with no elements and is denoted by $\{\ \}$ or $\varnothing$.

III. Identities

An equation that is true for all real numbers is called an **identity**. For example, consider the equation $x + 4 = x + 4$. Because the left- and right-hand sides are *identical*, any real number substituted for x will result in equal quantities on both sides. If we solve the equation, we get the identity $4 = 4$. In such a case, the solution set is the set of real numbers.

$$x + 4 = x + 4$$
$$x - x + 4 = x - x + 4$$
$$4 = 4 \quad \text{(identity)} \qquad \text{The solution set is the set of real numbers.}$$

In set-builder notation, we have, $\{x \mid x \text{ is a real number}\}$.

Example 10   Identifying Conditional Equations, Contradictions, and Identities

Identify each equation as a conditional equation, a contradiction, or an identity. Then give the solution set.

a. $3[x - (x + 1)] = -2$

b. $5(3 + c) + 2 = 2c + 3c + 17$

c. $4x - 3 = 17$

Solution:

a. $3[x - (x + 1)] = -2$

$\quad\quad 3[x - x - 1] = -2$ Clear parentheses.

$\quad\quad\quad\quad 3[-1] = -2$ Combine *like* terms.

$\quad\quad\quad\quad\quad -3 = -2$ Contradiction

This equation is a contradiction. The solution set is { }.

b. $5(3 + c) + 2 = 2c + 3c + 17$

$\quad 15 + 5c + 2 = 5c + 17$ Clear parentheses and combine *like* terms.

$\quad\quad 5c + 17 = 5c + 17$ Identity

$\quad\quad\quad\quad\quad 0 = 0$

This equation is an identity. The solution set is $\{c \mid c$ is a real number$\}$.

c. $\quad 4x - 3 = 17$

$\quad\quad\quad 4x = 20$

$\quad\quad\quad\quad x = 5$

This equation is a conditional equation. The solution set is $\{5\}$.

> **TIP:** Interval notation can also be used to express the set of real numbers, $(-\infty, \infty)$.

Skill Practice Identify each equation as a conditional equation, an identity, or a contradiction. Then give the solution set.

12. $2(-5x - 1) = 2x - 12x + 6$ **13.** $2(3x - 1) = 6(x + 1) - 8$

14. $4x + 1 - x = 6x - 2$

Answers

12. Contradiction; { }
13. Identity; $\{x \mid x$ is a real number$\}$
14. Conditional equation; $\{1\}$

Section 1.1   Practice Exercises

Study Skills Exercise

> Some instructors allow the use of calculators. Does your instructor allow the use of a calculator? If so, what kind? Will you be allowed to use a calculator on tests or just for occasional calculator problems in the text?
>
> *Helpful Hint:* If you are not permitted to use a calculator on tests, you should do your homework in the same way, without the calculator.

Vocabulary and Key Concepts

1. **a.** An _____ is a statement that indicates that two quantities are equal.

     b. A _____ to an equation is a value of the variable that makes the equation a true statement.

     c. An equation that can be written in the form $ax + b = c$ $(a \neq 0)$ is called a _____ equation in one variable.

     d. A linear equation is also called a _____ -degree equation because the degree on the variable is 1.

     e. The set of all solutions to an equation is called the _____ _____.

     f. Two equations are equivalent if they have the same _____ set.

     g. A _____ equation is true for some values of the variable, but false for other values.

h. An equation that has no solution is called a _____.

i. The set containing no elements is called the _____ and is denoted by _____.

j. An equation that has all real numbers as its solution set is called an _____.

Review Exercises

2. Apply the distributive property. $-\dfrac{1}{2}(6x - 8y + 10)$

For Exercises 3–6, clear parentheses and combine *like* terms.

3. $8x - 3y + 2xy - 5x + 12xy$

4. $5ab + 5a - 13 - 2a + 17$

5. $2(3z - 4) - (z + 12)$

6. $-(6w - 5) + 3(4w - 5)$

Concept 1: Linear Equations in One Variable

For Exercises 7–12, label the equation as linear or nonlinear.

7. $2x + 1 = 5$

8. $10 = x + 6$

9. $x^2 + 7 = 9$

10. $3 + x^3 - x = 4$

11. $-3 = x$

12. $5.2 - 7x = 0$

13. Use substitution to determine which value is the solution to $2x - 1 = 5$.

   a. 2   **b.** 3   **c.** 0   **d.** -1

14. Use substitution to determine which value is the solution to $2y - 3 = -2$.

   a. 1   **b.** $\dfrac{1}{2}$ **c.** 0   **d.** $-\dfrac{1}{2}$

Concept 2: Solving Linear Equations

For Exercises 15–44, solve the equation and check the solution. **(See Examples 1–6.)**

15. $x + 7 = 19$

16. $-3 + y = -28$

17. $-x = 2$

18. $-t = \dfrac{3}{4}$

19. $-\dfrac{7}{8} = -\dfrac{5}{6}z$

20. $-\dfrac{12}{13} = \dfrac{4}{3}b$

21. $\dfrac{a}{5} = -8$

22. $\dfrac{x}{8} = \dfrac{1}{2}$

23. $2.53 = -2.3t$

24. $-4.8 = 6.1 + y$

25. $p - 2.9 = 3.8$

26. $-4.2a = 4.494$

27. $6q - 4 = 62$

28. $2w - 15 = 15$

29. $4y - 17 = 35$

30. $6z - 25 = 83$

31. $-b - 5 = 2$

32. $6 = -y + 1$

33. $3(x - 6) = 2x - 5$

34. $13y + 4 = 5(y - 4)$

35. $6 - (t + 2) = 5(3t - 4)$

36. $1 - 5(p + 2) = 2(p + 13)$

37. $6(a + 3) - 10 = -2(a - 4)$

38. $8(b - 2) + 3b = -9(b - 1)$

39. $-2[5 - (2z + 1)] - 4 = 2(3 - z)$

40. $3[w - (10 - w)] = 7(w + 1)$

41. $6(-y + 4) - 3(2y - 3) = -y + 5 + 5y$

42. $13 + 4w = -5(-w - 6) + 2(w + 1)$

43. $14 - 2x + 5x = -4(-2x - 5) - 6$

44. $8 - (p + 2) + 6p + 7 = p + 13$

Concept 3: Clearing Fractions and Decimals

For Exercises 45–56, solve the equations. **(See Examples 7–9.)**

45. $\dfrac{2}{3}x - \dfrac{1}{6} = -\dfrac{5}{12}x + \dfrac{3}{2} - \dfrac{1}{6}x$

46. $-\dfrac{1}{2}y + 4 = -\dfrac{9}{10}y + \dfrac{2}{5}$

47. $\dfrac{1}{5}(p - 5) = \dfrac{3}{5}p + \dfrac{1}{10}p + 1$

48. $\dfrac{5}{6}(q + 2) = -\dfrac{7}{9}q - \dfrac{1}{3} + 2$

49. $\dfrac{3x - 7}{2} + \dfrac{3 - 5x}{3} = \dfrac{3 - 6x}{5}$

50. $\dfrac{2y - 4}{5} = \dfrac{5y + 13}{4} + \dfrac{y}{2}$

51. $\dfrac{4}{3}(2q + 6) - \dfrac{5q - 6}{6} - \dfrac{q}{3} = 0$

52. $\dfrac{-3a + 9}{15} - \dfrac{2a - 5}{5} - \dfrac{a + 2}{10} = 0$

53. $6.3w - 1.5 = 4.8$

54. $0.2x + 53.6 = x$

55. $0.75(m - 2) + 0.25m = 0.5$

56. $0.4(n + 10) + 0.6n = 2$

Concept 4: Conditional Equations, Contradictions, and Identities

57. What is a conditional equation?

58. Explain the difference between a contradiction and an identity.

For Exercises 59–64, identify the equation as a conditional equation, a contradiction, or an identity. Then give the solution set. **(See Example 10.)**

59. $4x + 1 = 2(2x + 1) - 1$

60. $3x + 6 = 3x$

61. $-11x + 4(x - 3) = -2x - 12$

62. $5(x + 2) - 7 = 3$

63. $2x - 4 + 8x = 7x - 8 + 3x$

64. $-7x + 8 + 4x = -3(x - 3) - 1$

Mixed Exercises

For Exercises 65–96, solve the equations.

65. $-5b + 9 = -71$

66. $-3x + 18 = -66$

67. $16 = -10 + 13x$

68. $15 = -12 + 9x$

69. $10c + 3 = -3 + 12c$

70. $2w + 21 = 6w - 7$

71. $12b - 8b - 8 + 13 = 4b + 6 - 1$

72. $4z + 2 - 3z + 5 = 3 + z + 4$

73. $5(x - 2) - 2x = 3x + 7$

74. $2x + 3(x - 5) = 15$

75. $\dfrac{c}{2} - \dfrac{c}{4} + \dfrac{3c}{8} = 1$

76. $\dfrac{d}{5} - \dfrac{d}{10} + \dfrac{5d}{20} = \dfrac{7}{10}$

77. $0.75(8x - 4) = \dfrac{2}{3}(6x - 9)$

78. $-\dfrac{1}{2}(4z - 3) = -z$

79. $7(p + 2) - 4p = 3p + 14$

80. $6(z - 2) = 3z - 8 + 3z$

81. $4[3 + 5(3 - b) - 2b] = 6 - 2b$

82. $\dfrac{1}{3}(x + 3) - \dfrac{1}{6} = \dfrac{1}{6}(2x + 5)$

83. $3 - \dfrac{3}{4}x = 3\left(3 - \dfrac{1}{4}x\right)$

84. $\dfrac{9}{5} - 8w = 8\left(\dfrac{3}{5} - w\right)$

85. $\dfrac{5}{4} - \dfrac{y - 3}{8} = \dfrac{2y + 1}{2}$

86. $\dfrac{2}{3} - \dfrac{x + 2}{6} = \dfrac{5x - 2}{2}$

87. $\dfrac{2y - 9}{10} + \dfrac{3}{2} = y$

88. $\dfrac{2}{3}x - \dfrac{5}{6}x - 3 = \dfrac{1}{2}x - 5$

89. $0.48x - 0.08x = 0.12(260 - x)$

90. $0.07w + 0.06(140 - w) = 90$

91. $0.5x + 0.25 = \dfrac{1}{3}x + \dfrac{5}{4}$

92. $0.2b + \dfrac{1}{3} = \dfrac{7}{15}$

93. $0.3b - 1.5 = 0.25(b + 2) + 0.05b$ **94.** $0.7(a - 1) = 0.25 + 0.7a$

95. $-\dfrac{7}{8}y + \dfrac{1}{4} = \dfrac{1}{2}\left(5 - \dfrac{3}{4}y\right)$

96. $5x - (8 - x) = 2[-4 - (3 + 5x) - 13]$

Expanding Your Skills

97. A power company charges $0.12 per kilowatt-hour (kWh) and $14.89 in monthly taxes. The monthly charge C (in $) is given by $C = 0.12h + 14.89$ where h is the number of kilowatt-hours used. If a family's bill comes to $137.77, determine the number of kilowatt-hours used.

98. For a student's first semester at college, the college charges $105 per credit-hour plus a one-time $50 admissions fee. The cost C (in $) for the first semester is given by $C = 105h + 50$ where h is the number of credit-hours taken. If a student is billed $1415, how many credit-hours is she taking?

99. a. Simplify the expression. $-2(y - 1) + 3(y + 2)$

   b. Solve the equation. $-2(y - 1) + 3(y + 2) = 0$

   c. Explain the difference between simplifying an expression and solving an equation.

100. a. Simplify the expression. $4w - 8(2 + w)$

   b. Solve the equation. $4w - 8(2 + w) = 0$

   c. Explain the difference between simplifying an expression and solving an equation.

Problem Recognition Exercises

Equations Versus Expressions

For Exercises 1–20, identify each exercise as an expression or an equation. Then simplify the expressions and solve the equations.

1. $4x - 2 + 6 - 8x$

2. $-3y - 3 - 4y + 8$

3. $7b - 1 = 2b + 4$

4. $10t + 2 = 2 - 7t$

5. $4(a - 8) - 7(2a + 1)$

6. $10(2x + 3) - 8(5 - x)$

7. $7(2 - w) = 5w + 8$

8. $15(3 - 2y) = 21 + 2y$

9. $2(3x - 4) - 4(5x + 1) = -8x + 7$

10. $6(2 - 3a) - 2(8a + 3) = -12a - 19$

11. $\dfrac{1}{2}v + \dfrac{3}{5} - \dfrac{2}{3}v - \dfrac{7}{10}$

12. $-\dfrac{7}{8}t - \dfrac{4}{3}u - \dfrac{5}{4}t + \dfrac{11}{6}u$

13. $20x - 8 + 7x + 28 = 27x - 9$

14. $7 + 8w - 12 = 3w - 8 + 5w$

15. $\dfrac{5}{6}y - \dfrac{7}{8} = \dfrac{1}{2}y + \dfrac{3}{4}$

16. $\dfrac{4}{5} + 3z = \dfrac{1}{2}z + 1$

17. $0.29c + 4.495 - 0.12c$

18. $0.45k - 1.67 + 0.89 - 1.456k$

19. $0.125(2p - 8) = 0.25(p - 4)$

20. $0.5u + 1.2 - 0.74u = 0.8 - 0.24u + 0.4$

Applications of Linear Equations in One Variable	Section 1.2

1. Introduction to Problem Solving

One of the important uses of algebra is to develop mathematical models for understanding real-world phenomena. To solve an application problem, relevant information must be extracted from the wording of a problem and then translated into mathematical symbols. This is a skill that requires practice. The key is to stick with it and not to get discouraged.

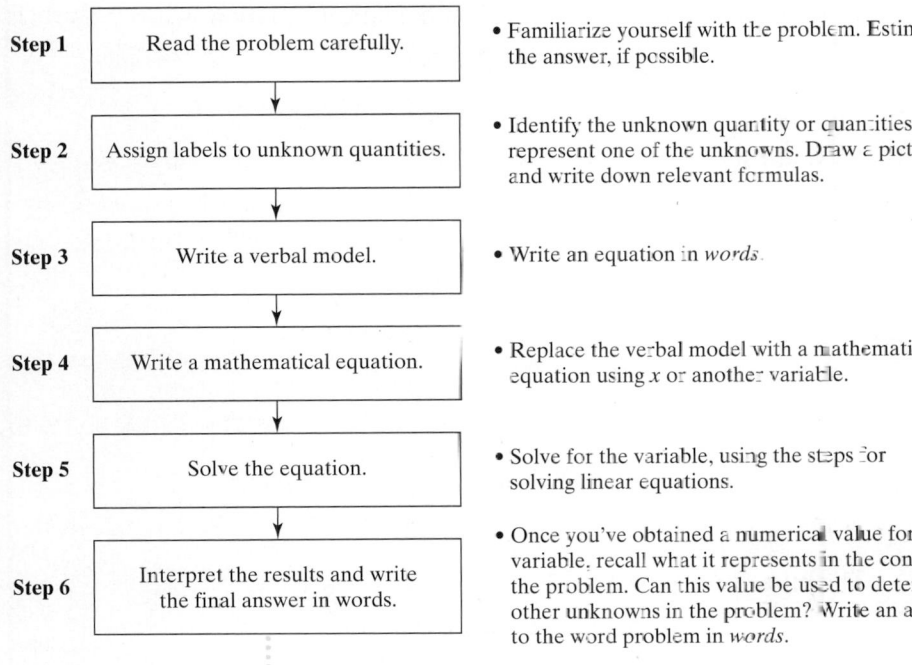

Problem-Solving Flowchart for Word Problems

Step 1 Read the problem carefully.
• Familiarize yourself with the problem. Estimate the answer, if possible.

Step 2 Assign labels to unknown quantities.
• Identify the unknown quantity or quantities. Let x represent one of the unknowns. Draw a picture and write down relevant formulas.

Step 3 Write a verbal model.
• Write an equation in *words*.

Step 4 Write a mathematical equation.
• Replace the verbal model with a mathematical equation using x or another variable.

Step 5 Solve the equation.
• Solve for the variable, using the steps for solving linear equations.

Step 6 Interpret the results and write the final answer in words.
• Once you've obtained a numerical value for the variable, recall what it represents in the context of the problem. Can this value be used to determine other unknowns in the problem? Write an answer to the word problem in *words*.

Concepts

1. Introduction to Problem Solving
2. Applications Involving Consecutive Integers
3. Applications Involving Percents and Rates
4. Applications Involving Principal and Interest
5. Applications Involving Mixtures
6. Applications Involving Distance, Rate, and Time

Avoiding Mistakes

Once you have reached a solution to a word problem, verify that it is reasonable in the context of the problem.

To write an English statement as an algebraic expression, review the list of key terms given in Table 1-1.

Table 1-1

Addition: $a + b$	Subtraction: $a - b$
• the **sum** of a and b	• the **difference** of a and b
• a plus b	• a minus b
• b added to a	• b subtracted from a
• b more than a	• a decreased by b
• a increased by b	• b less than a
• the total of a and b	
Multiplication: $a \cdot b$	**Division: $a \div b, \frac{a}{b}$**
• the **product** of a and b	• the **quotient** of a and b
• a times b	• a divided by b
• a multiplied by b	• b divided into a
	• the ratio of a and b
	• a over b
	• a per b

Example 1 **Translating and Solving a Linear Equation**

The sum of two numbers is 39. One number is 3 less than twice the other. What are the numbers?

Solution:

Step 1: Read the problem carefully.

Step 2: Let x represent one number.

Let $2x - 3$ represent the other number.

Step 3: (One number) + (other number) = 39

Step 4: Replace the verbal model with a mathematical equation.

(One number) + (other number) = 39

x + $(2x - 3)$ = 39

Step 5: Solve for x.

$$x + (2x - 3) = 39$$

$$3x - 3 = 39$$

$$3x = 42$$

$$\frac{3x}{3} = \frac{42}{3}$$

$$x = 14$$

Step 6: Interpret your results. Refer back to step 2.

One number is x: ⟶ 14

The other number is $2x - 3$:

$2(14) - 3$ ⟶ 25

Answer: The numbers are 14 and 25.

Skill Practice

1. One number is 5 more than 3 times another number. The sum of the numbers is 45. Find the numbers.

2. Applications Involving Consecutive Integers

The word *consecutive* means "following one after the other in order."

- The numbers −2, −1, 0, 1, 2, and so on are examples of consecutive integers. Notice that two consecutive integers differ by 1. Therefore, if x represents an integer, then $x + 1$ represents the next consecutive integer.
- The numbers 2, 4, 6, 8, and so on are consecutive *even* integers. Consecutive even integers differ by 2. Therefore, if x represents an even integer, then $x + 2$ represents the next consecutive even integer.
- The numbers 15, 17, 19, and so on are consecutive *odd* integers. Consecutive odd integers also differ by 2. Therefore, if x represents an odd integer, then $x + 2$ represents the next consecutive odd integer.

Answer

1. The numbers are 10 and 35.

| Example 2 | Solving a Linear Equation Involving Consecutive Integers |

Three times the sum of two consecutive odd integers is 516. Find the integers.

Solution:

Step 1: Read the problem carefully.

Step 2: Label the unknown:
Let x represent the first odd integer.
Then $x + 2$ represents the next odd integer.

Step 3: Write an equation in words.

$3[(\text{first odd integer}) + (\text{second odd integer})] = 516$

$3[x + (x + 2)] = 516$ **Step 4:** Write a mathematical equation.

$3(2x + 2) = 516$ **Step 5:** Solve for x.

$6x + 6 = 516$

$6x = 510$

$x = 85$

Step 6: Interpret your results.

The first odd integer is x: ⟶   85

The second odd integer is $x + 2$:

$85 + 2$ ⟶   87

Answer: The integers are 85 and 87.

Avoiding Mistakes

After completing a word problem, it is always a good idea to check that the answer is reasonable. Notice that 85 and 87 are consecutive odd integers, and three times their sum is $3(85 + 87)$, which equals 516.

Skill Practice

2. Four times the sum of three consecutive integers is 264. Find the integers.

3. Applications Involving Percents and Rates

In many real-world applications, percents are used to represent rates.

- The sales tax rate for a certain county is 6%.
- An ice cream machine is discounted 20%.
- A real estate sales broker receives a $4\frac{1}{2}$% commission on sales.
- A savings account earns 7% simple interest.

The following models are used to compute sales tax, commission, and simple interest. In each case the value is found by multiplying the base by the percentage.

Sales tax = (cost of merchandise)(tax rate)

Commission = (dollars in sales)(commission rate)

Simple interest = (principal)(annual interest rate)(time in years)

$\llcorner\!\!\longrightarrow I = Prt$

Answer

2. The integers are 21, 22, and 23.

| Example 3 | Solving a Percent Application |

A woman invests $5000 in an account that earns $5\frac{1}{4}$% simple interest. If the money is invested for 3 years (yr), how much money is in the account at the end of the 3-yr period?

Solution:

Let x represent the total money in the account.   Label the variables.

$P = \$5000$ (principal amount invested)

$r = 0.0525$ (interest rate)

$t = 3$ (time in years)

The total amount of money includes principal plus interest.

(Total money) = (principal) + (interest)	Verbal model
$x = P + Prt$	Mathematical equation
$x = \$5000 + (\$5000)(0.0525)(3)$	Substitute for P, r, and t.
$x = \$5000 + \$787.50$	
$x = \$5787.50$	Solve for x.

The total amount of money in the account is $5787.50.   Interpret the results.

TIP: Remember to use the decimal form of a percent when the number is used in a calculation.

Skill Practice

3. Markos earned $340 in 1 yr on an investment that paid a 4% dividend. Find the amount of money invested.

As consumers, we often encounter situations in which merchandise has been marked up or marked down from its original cost. It is important to note that percent increase and percent decrease are based on the original cost. For example, suppose a microwave oven originally priced at $305 is marked down 20%.

The discount is determined by 20% of the original price: $(0.20)(\$305) = \$61.00$. The new price is $\$305.00 - \$61.00 = \$244.00$.

| Example 4 | Solving a Percent Increase Application |

A college bookstore uses a standard markup of 40% on all books purchased wholesale from the publisher. If the bookstore sells a calculus book for $179.20, what was the original wholesale cost?

Answer

3. $8500

Solution:

Let x = original wholesale cost.   Label the variables.

The selling price of the book is based on the original cost of the book plus the bookstore's markup.

(Selling price) = (original price) + (markup)   Verbal model

(Selling price) = (original price) + (original price · markup rate)

$$179.20 = \qquad x \qquad + (x)(0.40)$$ Mathematical equation

$$179.20 = x + 0.40x$$

$$179.20 = 1.40x$$ Combine *like* terms.

$$\frac{179.20}{1.40} = x$$

$$x = 128$$ Simplify.

The original wholesale cost of the textbook was $128.00.   Interpret the results.

Skill Practice

4. An online bookstore gives a 20% discount on paperback books. Find the original price of a book that has a selling price of $5.28 after the discount.

4. Applications Involving Principal and Interest

Example 5   Solving an Investment Growth Application

Miguel had $10,000 to invest in two different mutual funds. One was a relatively safe bond fund that averaged 4% return on his investment at the end of 1 yr. The other fund was a riskier stock fund that averaged 7% return in 1 yr. If at the end of the year Miguel's portfolio grew to $10,625 ($625 above his $10,000 investment), how much money did Miguel invest in each fund?

Solution:

This type of word problem is sometimes categorized as a mixture problem. Miguel is "mixing" his money between two different investments. We have to determine how the money was divided to earn $625.

The information in this problem can be organized in a chart. (*Note:* There are two sources of money: the amount invested and the amount earned.)

	4% Bond Fund	7% Stock Fund	Total
Amount invested ($)	x	$(10{,}000 - x)$	10,000
Amount earned ($)	$0.04x$	$0.07(10{,}000 - x)$	625

Because the amount of principal is unknown for both accounts, we can let x represent the amount invested in the bond fund. If Miguel spends x dollars in the bond fund, then he has $(10{,}000 - x)$ left over to spend in the stock fund. The return for each fund is found by multiplying the principal and the percent growth rate.

To establish a mathematical model, we know that the total return ($625) must equal the earnings from the bond fund plus the earnings from the stock fund:

(Earnings from bond fund) + (earnings from stock fund) = (total earnings)

$$0.04x \quad + \quad 0.07(10{,}000 - x) \quad = \quad 625$$

$$0.04x + 0.07(10{,}000 - x) = 625 \qquad \text{Mathematical equation}$$

$$4x + 7(10{,}000 - x) = 62{,}500 \qquad \text{Multiply by 100 to clear decimals.}$$

$$4x + 70{,}000 - 7x = 62{,}500$$

$$-3x + 70{,}000 = 62{,}500 \qquad \text{Combine } like \text{ terms.}$$

$$-3x = -7500 \qquad \text{Subtract 70,000 from both sides.}$$

$$\frac{-3x}{-3} = \frac{-7500}{-3}$$

$$x = 2500 \qquad \text{Solve for } x \text{ and interpret the results.}$$

The amount invested in the bond fund is $2500.
The amount invested in the stock fund is $10,000 − x, or $7500.

Skill Practice

5. Jonathan borrowed $4000 in two loans. One loan charged 7% interest, and the other charged 1.5% interest. After 1 yr, Jonathan paid $225 in interest. Find the amount borrowed in each loan.

5. Applications Involving Mixtures

Example 6 Solving a Mixture Application

How many liters (L) of a 40% antifreeze solution must be added to 4 L of a 10% antifreeze solution to produce a 35% antifreeze solution?

Solution:

The given information is illustrated in Figure 1-1.

TIP: To understand the role of the concentration rate within a mixture problem, consider this example. Suppose you had 30 gal of a 10% antifreeze mixture. How much pure antifreeze is in the mixture?

pure antifreeze = 0.10(30 gal)

= 3 gal

Multiply the concentration rate by the amount of mixture.

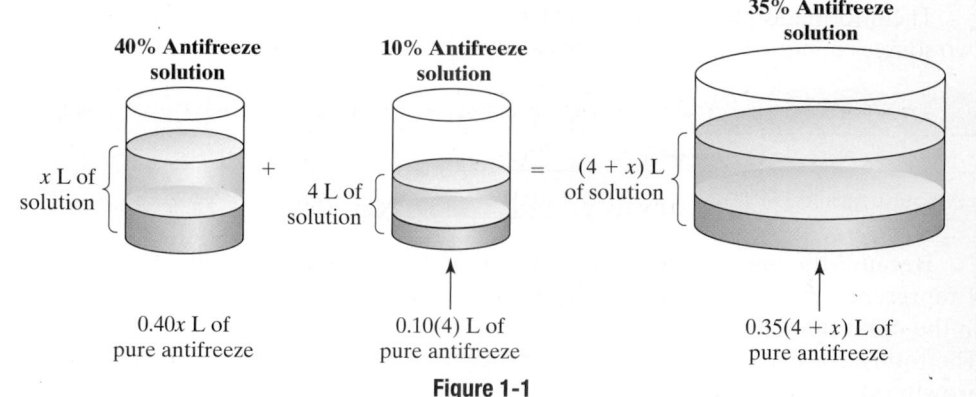

Figure 1-1

The information can also be organized in a table.

	40% Antifreeze	10% Antifreeze	Final Solution: 35% Antifreeze
Number of liters of solution	x	4	$(4 + x)$
Number of liters of pure antifreeze	$0.40x$	$0.10(4)$	$0.35(4 + x)$

Notice that an algebraic equation is obtained from the second row of the table relating the number of liters of pure antifreeze in each container.

$$\left(\begin{array}{c}\text{Pure antifreeze}\\\text{from solution 1}\end{array}\right) + \left(\begin{array}{c}\text{pure antifreeze}\\\text{from solution 2}\end{array}\right) = \left(\begin{array}{c}\text{pure antifreeze}\\\text{in the final solution}\end{array}\right)$$

$$0.40x \quad + \quad 0.10(4) \quad = \quad 0.35(4 + x)$$

$0.40x + 0.10(4) = 0.35(x + 4)$	Mathematical equation
$0.4x + 0.4 = 0.35x + 1.4$	Apply the distributive property.
$0.4x - 0.35x + 0.4 = 0.35x - 0.35x + 1.4$	Subtract $0.35x$ from both sides.
$0.05x + 0.4 = 1.4$	
$0.05x + 0.4 - 0.4 = 1.4 - 0.4$	Subtract 0.4 from both sides.
$0.05x = 1.0$	
$\dfrac{0.05x}{0.05} = \dfrac{1.0}{0.05}$	Divide both sides by 0.05.
$x = 20$	

Therefore, 20 L of a 40% antifreeze solution is needed.

Skill Practice

6. Find the number of ounces (oz) of 30% alcohol solution that must be mixed with 10 oz of a 70% solution to obtain a solution that is 40% alcohol.

6. Applications Involving Distance, Rate, and Time

The fundamental relationship among the variables distance, rate, and time is given by

$$\text{Distance} = (\text{rate})(\text{time}) \quad \text{or} \quad d = rt$$

For example, a motorist traveling 65 mph (miles per hour) for 3 hr (hours) will travel a distance of

$$d = \left(\frac{65 \text{ mi}}{\text{hr}}\right)(3 \text{ hr}) = 195 \text{ mi}$$

Answer

6. 30 oz of the 30% solution is needed.

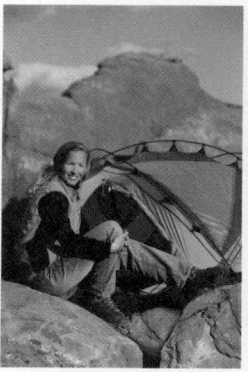

Example 7 Solving a Distance, Rate, Time Application

A hiker can hike 1 mph faster downhill to Moose Lake than she can hike uphill back to the campsite. If it takes her 3 hr to hike to the lake and 4.5 hr to hike back, what is her speed hiking back to the campsite?

Solution:

The information given in the problem can be organized in a table.

	Distance (mi)	Rate (mph)	Time (hr)
Trip to the lake		$x + 1$	3
Return trip		x	4.5

Column 2: Let the rate of the return trip be represented by x. Then the trip to the lake is 1 mph faster and can be represented by $x + 1$.

Column 3: The times hiking to and from the lake are given in the problem.

Column 1: To express the distance, we use the relationship $d = rt$. That is, multiply the quantities in the second and third columns.

	Distance (mi)	Rate (mph)	Time (hr)
Trip to the lake	$3(x + 1)$	$x + 1$	3
Return trip	$4.5x$	x	4.5

Campsite

$d = (\text{rate})(\text{time})$
$d = (x + 1)(3)$

$d = (\text{rate})(\text{time})$
$d = x(4.5)$

Moose Lake

To create a mathematical model, note that the distances to and from the lake are equal. Therefore,

(Distance to lake) = (return distance)	Verbal model
$3(x + 1) = 4.5x$	Mathematical equation
$3x + 3 = 4.5x$	Apply the distributive property.
$3x - 3x + 3 = 4.5x - 3x$	Subtract $3x$ from both sides.
$3 = 1.5x$	
$\dfrac{3}{1.5} = \dfrac{1.5x}{1.5}$	Divide by 1.5 to isolate the variable.
$2 = x$	Solve for x.

The hiker's speed on the return trip to the campsite is 2 mph.

Skill Practice

7. During a bad rainstorm, Jody drove 15 mph slower on a trip to her mother's house than she normally would when the weather is clear. If a trip to her mother's house takes 3.75 hr in clear weather and 5 hr in a bad storm, what is her normal driving speed during clear weather?

Answer

7. Jody normally drives 60 mph.

Section 1.2 Practice Exercises

Study Skills Exercise

After doing a section of homework, check the odd-numbered answers in the back of the text. Choose a method to identify the exercises that gave you trouble (i.e., circle the number or put a star by the number). List some reasons why it is important to label these problems.

Vocabulary and Key Concepts

1. **a.** Integers that follow one after the other without "gaps" are called _____ integers.

    b. The integers $-2, 0, 2$, and 4 are examples of consecutive _____ integers, and the integers $-3, -1$, 1, and 3 are examples of consecutive _____ integers.

    c. Two consecutive integers differ by _____, two consecutive odd integers differ by _____, and two consecutive even integers differ by _____.

    d. If x represents the smaller of two consecutive integers, then the expression _____ represents the next greater consecutive integer.

    e. If x represents the smaller of two consecutive odd integers, then the expression _____ represents the next greater consecutive odd integer.

    f. If x represents the smallest of three consecutive even integers, then the expressions _____ and _____ represent the next two greater consecutive even integers.

    g. Simple interest is interest computed on the principal invested or borrowed. Simple interest is computed as $I = $ _____ where P is the principal, r is the annual _____ rate, and t is the time in years.

    h. If \$5000 is borrowed at 6.5% simple interest for 4 yr, then the amount of simple interest is _____.

    i. Suppose that 12% of a solution is acid by volume and the remaining 88% is water. Then the amount of acid in a 4-L container of solution is _____. The amount of acid in a container containing $x + 8$ liters of solution is represented by the expression _____.

    j. If $d = rt$, then $r = \dfrac{\square}{\square}$ and $t = \dfrac{\square}{\square}$.

Review Exercises

2. Identify the smallest positive integer that could be used to clear fractions in the equation.
$$\frac{2}{3}x - \frac{4}{5} = 1 - \frac{1}{6}x$$

For Exercises 3–8, solve the equations.

3. $7a - 2 = 11$

4. $2z + 6 = -15$

5. $4(x - 3) + 7 = 19$

6. $-3(y - 5) + 4 = 1$

7. $\dfrac{3}{8}p + \dfrac{3}{4} = p - \dfrac{3}{2}$

8. $\dfrac{1}{4} - 2x = 5$

For the remaining exercises, follow the steps outlined in the Problem-Solving Flowchart found on page 57.

Concept 1: Introduction to Problem Solving

9. If x represents a number, write an expression for 5 more than the number.

10. If n represents a number, write an expression for 10 less than the number.

11. If *t* represents a number, write an expression for 7 less than twice the number.

12. If *y* represents a number, write an expression for 4 more than 3 times the number.

13. The larger of two numbers is 3 more than twice the smaller. The difference of the larger number and the smaller number is 8. Find the numbers. **(See Example 1.)**

14. One number is 3 less than another. Their sum is 15. Find the numbers.

15. The sum of 3 times a number and 2 is the same as the difference of the number and 4. Find the number.

16. Twice the sum of a number and 3 is the same as 1 subtracted from the number. Find the number.

Concept 2: Applications Involving Consecutive Integers

17. The sum of two consecutive page numbers in a book is 223. Find the page numbers. **(See Example 2.)**

18. The sum of the numbers on two consecutive raffle tickets is 808,455. Find the numbers on the tickets.

19. The sum of two consecutive odd integers is −148. Find the two integers.

20. The sum of three consecutive integers is −57. Find the integers.

21. Three times the smaller of two consecutive even integers is the same as −146 minus 4 times the larger integer. Find the integers.

22. Four times the smaller of two consecutive odd integers is the same as 73 less than 5 times the larger. Find the integers.

23. Two times the sum of three consecutive odd integers is the same as 23 more than 5 times the largest integer. Find the integers.

24. Five times the smallest of three consecutive even integers is 10 more than twice the largest. Find the integers.

Concept 3: Applications Involving Percents and Rates

25. Belle had the choice of taking out a 4-yr car loan at 8.5% simple interest or a 5-yr car loan at 7.75% simple interest. If she borrows $15,000, how much interest would she pay for each loan? Which option will require less interest? **(See Example 3.)**

26. Robert can take out a 3-yr loan at 8% simple interest or a 2-yr loan at $8\frac{1}{2}$% simple interest. If he borrows $7000, how much interest will he pay for each loan? Which option will require less interest?

27. An account executive earns $600 per month plus a 3% commission on sales. The executive's goal is to earn $2400 this month. How much must she sell to achieve this goal?

28. A salesperson earns $50 a day plus 12% commission on sales over $200. If her daily earnings are $76.88, how much money in merchandise did she sell?

29. J. W. is an artist and sells his pottery each year at a local Renaissance Festival. He keeps track of his sales and the 8.05% sales tax he collects by making notations in a ledger. Every evening he checks his records by counting the total money in his cash drawer. After a day of selling pottery, the cash totaled $1293.38. How much is from the sale of merchandise and how much is sales tax?

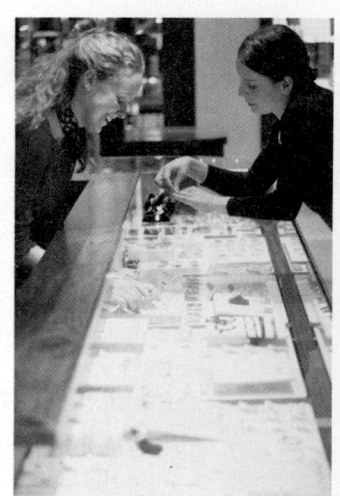

30. Wayne County has a sales tax rate of 7%. How much does Mike's used Honda Civic cost before tax if the total cost of the car *plus tax* is $13,888.60?

31. The price of a swimsuit after a 20% markup is $43.08. What was the price before the markup? **(See Example 4.)**

32. The price of a used textbook after a 35% markdown is $29.25. What was the original price?

Concept 4: Applications Involving Principal and Interest

33. Tony has a total of $12,500 in two accounts. One account pays 2% simple interest per year and the other pays 5% simple interest. If he earned $370 in interest in the first year, how much did he invest in each account? **(See Example 5.)**

34. Lillian had $15,000 invested in two accounts, one paying 9% simple interest and one paying 10% simple interest. How much was invested in each account if the interest after 1 yr is $1432?

35. Jason borrowed $18,000 in two loans. One loan charged 11% simple interest and the other charged 6% simple interest. After 1 yr, Jason paid a total of $1380. Find the amount borrowed in each loan.

36. Amanda borrowed $6000 from two sources: her parents and a credit union. Her parents charged 3% simple interest and the credit union charged 8% simple interest. If after 1 yr, Amanda paid $255 in interest, how much did she borrow from her parents, and how much did she borrow from the credit union?

37. Donna invested money in two accounts: one paying 4% simple interest and the other paying 3% simple interest. She invested $4000 more in the 4% account than in the 3% account. If she received $720 in interest at the end of 1 yr how much did she invest in each account?

38. Mr. Hall had some money in his bank earning 4.5% simple interest. He had $5000 more deposited in a credit union earning 6% simple interest. If his total interest for 1 yr was $1140, how much did he deposit in each account?

Concept 5: Applications Involving Mixtures

39. Ahmed mixes two plant fertilizers. How much fertilizer with 15% nitrogen should be mixed with 2 oz of fertilizer with 10% nitrogen to produce a fertilizer that is 14% nitrogen? **(See Example 6.)**

40. How much 8% saline solution should Kent mix with 80 cc (cubic centimeters) of an 18% saline solution to produce a 12% saline solution?

41. Jacque has 3 L of a 50% antifreeze mixture. How much 75% mixture should be added to get a mixture that is 60% antifreeze?

42. One fruit punch has 40% fruit juice and another is 70% fruit juice. How much of the 40% punch should be mixed with 10 gal of the 70% punch to create a fruit punch that is 45% fruit juice?

43. How many liters of an 18% alcohol solution must be added to a 10% alcohol solution to get 20 L of a 15% alcohol solution?

44. How many milliliters of a 2.5% bleach solution must be mixed with a 10% bleach solution to produce 600 mL of a 5% bleach solution?

45. Ronald has a 12% solution of the fertilizer Super Grow. How much pure Super Grow should he add to the mixture to get 32 oz of a 17.5% concentration?

46. How many ounces of water must be added to 20 oz of an 8% salt solution to make a 2% salt solution?

Concept 6: Applications Involving Distance, Rate, and Time

47. An airplane travels 60 mph faster from Atlanta to Fort Lauderdale than it travels on the return trip from Fort Lauderdale to Atlanta. If it takes 2 hr from Atlanta to Fort Lauderdale and 2.5 hr for the return trip, determine the speed of each trip. **(See Example 7.)**

48. A woman can hike 1 mph faster down a trail to Archuletta Lake than she can on the return trip uphill. It takes her 3 hr to get to the lake and 6 hr to return. What is her speed hiking down to the lake?

49. Two cars are 192 mi apart and travel toward each other on the same road. They meet in 2 hr. One car travels 4 mph faster than the other. What is the average speed of each car?

50. Two cars are 190 mi apart and travel toward each other along the same road. They meet in 2 hr. One car travels 5 mph slower than the other car. What is the average speed of each car?

Mixed Exercises

51. The sum of two integers is 30. Ten times one integer is 5 times the other integer. Find the integers. (*Hint:* If one number is x, then the other number is $30 - x$.)

52. The sum of two integers is 10. Three times one integer is 3 less than 8 times the other integer. Find the integers. (*Hint:* If one number is x, then the other number is $10 - x$.)

53. An older model of smart phone is marked down to $89.55 when newer technology comes on the market. If this is the new price after a 55% markdown, determine the original price.

54. A hardcover book is marked down to $15.60 once the book comes out in paperback. If this is the new price after a 35% markdown, determine the original price.

55. Two boats traveling the same direction leave a harbor at noon. After 3 hr they are 60 mi apart. If one boat travels twice as fast as the other, find the average rate of each boat.

56. Two canoes travel down a river, starting at 9:00. One canoe travels twice as fast as the other. After 3.5 hr, the canoes are 5.25 mi apart. Find the average rate of each canoe.

57. Ms. Riley deposited some money in an account paying 5% simple interest and twice that amount in an account paying 6% simple interest. If the total interest from the two accounts is $765 for 1 yr, how much was deposited into each account?

58. Sienna put some money in a certificate of deposit earning 4.2% simple interest. She deposited twice that amount in a money market account paying 4% simple interest. After 1 yr her total interest was $488. How much did Sienna deposit in her money market account?

59. Two different teas are mixed to make a blend that will be sold at a fair. Black tea sells for $2.20 per pound and green tea sells for $3.00 per pound. How much of each should be used to obtain 4 lb of a blend selling for $2.50?

60. A nut mixture consists of almonds and cashews. Almonds are $4.98 per pound, and cashews are $6.98 per pound. How many pounds of each type of nut should be mixed to produce 16 lb selling for $5.73 per pound?

61. After a recent crash of the housing market, the median price of a new home (including land) in the United States was $202,100. This represents a drop of 6% from the previous year. What was the median price the previous year?

62. According to the Bureau of Labor Statistics, the median weekly salary for an individual with a bachelor's degree is $1026 per week. This represents an increase of 35% over the salary of an individual with an associate's degree. What is the median weekly salary of an individual with an associate's degree?

Section 1.3   Applications to Geometry and Literal Equations

Concepts

1. **Applications Involving Geometry**
2. **Literal Equations**

1. Applications Involving Geometry

Some word problems involve the use of geometric formulas such as those listed in the inside back cover of this text.

Example 1   Solving an Application Involving Perimeter

The length of a rectangular corral is 2 ft more than 3 times the width. The corral is situated such that one of its shorter sides is adjacent to a barn and does not require fencing. If the total amount of fencing is 774 ft, find the dimensions of the corral.

Solution:

Read the problem and draw a sketch (Figure 1-2).

Figure 1-2

Let x represent the width.   Label variables.

Let $3x + 2$ represent the length.

TIP: In Example 1, the length of the field is given in terms of the width. Therefore, we let x represent the width.

To write a verbal model, we might consider using the formula for the perimeter of a rectangle. However, the formula $P = 2l + 2w$ incorporates all four sides of the rectangle. The formula must be modified to include only one factor of the width.

$$\begin{pmatrix} \text{Distance around} \\ \text{three sides} \end{pmatrix} = \begin{pmatrix} 2 \text{ times} \\ \text{the length} \end{pmatrix} + \begin{pmatrix} 1 \text{ times} \\ \text{the width} \end{pmatrix} \quad \text{Verbal model}$$

$$774 \quad = \quad 2(3x + 2) \quad + \quad x \quad \text{Mathematical equation}$$

$774 = 2(3x + 2) + x$	Solve for x.
$774 = 6x + 4 + x$	Apply the distributive property.
$774 = 7x + 4$	Combine *like* terms.
$770 = 7x$	Subtract 4 from both sides.
$110 = x$	Divide by 7 on both sides.
$x = 110$	

Because x represents the width, the width of the corral is 110 ft. The length is given by

$$3x + 2 \quad \text{or} \quad 3(110) + 2 = 332 \quad \text{Interpret the results.}$$

The width of the corral is 110 ft, and the length is 332 ft.

Avoiding Mistakes

To check the answer to Example 1, verify that the three sides add to 774 ft.

110 ft + 332 ft + 332 ft = 774 ft ✔

Skill Practice

1. The length of Karen's living room is 2 ft longer than the width. The perimeter is 80 ft. Find the length and width.

Recall some important facts involving angles.

- Two angles are complementary if the sum of their measures is 90°.
- Two angles are supplementary if the sum of their measures is 180°.
- The sum of the measures of the angles within a triangle is 180°.

Example 2   Solving an Application Involving Angles

Two angles are complementary. One angle measures 10° less than 4 times the other angle. Find the measure of each angle (Figure 1-3).

Figure 1-3

Answer

1. The length is 21 ft, and the width is 19 ft.

Solution:

Let x represent the measure of one angle.

Let $4x - 10$ represent the measure of the other angle.

Recall that two angles are complementary if the sum of their measures is 90°. Therefore, a verbal model is

(One angle) + (the complement of the angle) = 90°	Verbal model
$x + (4x - 10) = 90$	Mathematical equation
$5x - 10 = 90$	Solve for x.
$5x = 100$	
$x = 20$	

If $x = 20$, then $4x - 10 = 4(20) - 10 = 70$. The two angles are 20° and 70°.

Skill Practice

2. Two angles are supplementary, and the measure of one is 16° less than 3 times the other. Find their measures.

2. Literal Equations

Literal equations are equations that contain several variables. A formula is a literal equation with a specific application. For example, the perimeter of a rectangle can be found by the formula $P = 2l + 2w$. In this equation, P is expressed in terms of l and w. However, in science and other branches of applied mathematics, formulas may be more useful in alternative forms.

For example, the formula $P = 2l + 2w$ can be manipulated to solve for either l or w:

Solve for l		**Solve for w**	
$P = 2l + 2w$		$P = 2l + 2w$	
$P - 2w = 2l$	Subtract $2w$.	$P - 2l = 2w$	Subtract $2l$.
$\dfrac{P - 2w}{2} = l$	Divide by 2.	$\dfrac{P - 2l}{2} = w$	Divide by 2.
$l = \dfrac{P - 2w}{2}$		$w = \dfrac{P - 2l}{2}$	

To solve a literal equation for a specified variable, use the addition, subtraction, multiplication, and division properties of equality.

Example 3 **Applying a Literal Equation**

Buckingham Fountain is one of Chicago's most familiar landmarks. With 133 jets spraying a total of 14,000 gal of water per minute, Buckingham Fountain is one of the world's largest fountains. The circumference of the fountain is approximately 880 ft.

a. The circumference of a circle is given by $C = 2\pi r$. Solve the equation for r.

b. Use the equation from part (a) to find the radius and diameter of the fountain. Use 3.14 for π and round to the nearest foot.

Answer

2. 49° and 131°

Solution:

a. $C = 2\pi r$

$$\frac{C}{2\pi} = \frac{2\pi r}{2\pi}$$

$$\frac{C}{2\pi} = r$$

$$r = \frac{C}{2\pi}$$

Calculator Connections

Topic: Using the π Key
The π key on the calculator can also be used in the calculation for Example 3(b).

```
880/(2π)
        140.0563499
```

b. $r \approx \dfrac{880 \text{ ft}}{2(3.14)}$ Substitute 880 ft for C and 3.14 for π.

$\approx 140 \text{ ft}$

The radius is approximately 140 ft. The diameter is twice the radius $(d = 2r)$. Therefore, the diameter is 280 ft.

Skill Practice The formula to compute the surface area S of a sphere is given by $S = 4\pi r^2$.

3. Solve the equation for π.

4. A sphere has a surface area of 113 in.^{2} and a radius of 3 in. Use the formula found in part (a) to approximate π. Round to two decimal places.

Example 4   **Solving a Literal Equation**

The formula to find the area of a trapezoid is given by $A = \frac{1}{2}(b_1 + b_2)h$, where b_1 and b_2 are the lengths of the parallel sides and h is the height. (See Figure 1-4.)
    Solve this formula for b_1.

Figure 1-4

Solution:

$A = \frac{1}{2}(b_1 + b_2)h$ The goal is to isolate b_1.

$2A = 2 \cdot \frac{1}{2}(b_1 + b_2)h$ Multiply by 2 to clear fractions.

$2A = (b_1 + b_2)h$ Apply the distributive property.

$2A = b_1 h + b_2 h$

$2A - b_2 h = b_1 h$ Subtract $b_2 h$ from both sides.

$\dfrac{2A - b_2 h}{h} = \dfrac{b_1 h}{h}$ Divide by h.

$\dfrac{2A - b_2 h}{h} = b_1$

Skill Practice

5. The formula for the volume of a right circular cylinder is $V = \pi r^2 h$. Solve for h.

Answers

3. $\pi = \dfrac{S}{4r^2}$ **4.** 3.14   **5.** $h = \dfrac{V}{\pi r^2}$

TIP: When solving a literal equation for a specified variable, there is sometimes more than one way to express your final answer. This flexibility often presents difficulty for students. Students may leave their answer in one form, but the answer given in the text looks different. Yet both forms may be correct. To know if your answer is equivalent to the form given in the text you must try to manipulate it to look like the answer in the book, a process called *form fitting*.

The literal equation from Example 4 can be written in several different forms. The quantity $(2A - b_2h)/h$ can be split into two fractions.

$$b_1 = \frac{2A - b_2h}{h} = \frac{2A}{h} - \frac{b_2h}{h} = \frac{2A}{h} - b_2$$

Example 5   Solving a Literal Equation

Given $-2x + 3y = 5$, solve for y.

Solution:

$$-2x + 3y = 5$$

$$3y = 2x + 5 \qquad\qquad \text{Add } 2x \text{ to both sides.}$$

$$\frac{3y}{3} = \frac{2x + 5}{3} \qquad\qquad \text{Divide by 3 on both sides.}$$

$$y = \frac{2x + 5}{3} \quad \text{or} \quad y = \frac{2}{3}x + \frac{5}{3}$$

Skill Practice Solve for y.

6. $5x + 2y = 11$

Sometimes the variable we want to isolate may appear in more than one term in a literal equation. In such a case, isolate all terms with that variable on one side of the equation. Then apply the distributive property as demonstrated in Example 6.

Example 6   Solving a Literal Equation

Solve the equation for x. $ax - 3 = cx + 7$

Solution:

$$ax - 3 = cx + 7$$

$$ax - cx = 10 \qquad\qquad$$ Collect the terms containing x on one side of the equation. Collect the remaining terms on the other side.

The variable x appears twice in the equation. To isolate x, we want x to appear in only one term. To accomplish this, we apply the distributive property in reverse.

TIP: Applying the distributive property in reverse is called *factoring*. Factoring will be studied in detail in Chapter 4.

$$x(a - c) = 10 \qquad\qquad$$ Apply the distributive property. The variable x now appears one time in the equation.

$$\frac{x(a - c)}{(a - c)} = \frac{10}{(a - c)} \qquad\qquad$$ Divide both sides by $(a - c)$.

$$x = \frac{10}{a - c}$$

Answer

6. $y = \dfrac{11 - 5x}{2}$ or $y = -\dfrac{5}{2}x + \dfrac{11}{2}$

Skill Practice Solve for t.

7. $mt + 4 = nt + 9$

Answer

7. $t = \dfrac{5}{m - n}$

Section 1.3 Practice Exercises

Study Skills Exercise

In your next math class, take notes by drawing a vertical line about three-fourths of the way across the paper, as shown. On the left side, write down what your instructor puts on the board or overhead. On the right side, make your own comments about important words, procedures, or questions that you have.

Vocabulary and Key Concepts

1. **a.** For a rectangle having width w and length l, the perimeter is given by $P = $ _____.

   b. Two angles are complementary if the sum of their measures is _____.

   c. Two angles are _____ if the sum of their measures is 180°.

   d. The sum of the measures of the angles within a triangle is _____.

Review Exercises

For Exercises 2–6, solve the equations.

2. $7 + 5x - (2x - 6) = 6(x + 1) + 21$

3. $\dfrac{3}{5}y - 3 + 2y = 5$

4. $3[z - (2 - 3z) - 4] = z - 7$

5. $2a - 4 + 8a = 7a - 8 + 3a$

6. $3(t + 6) + t + 2 = 5(t + 4) - t$

Concept 1: Applications Involving Geometry

For Exercises 7–18, use the geometry formulas listed in the inside back cover of the text.

7. A volleyball court is twice as long as it is wide. If the perimeter is 177 ft, find the dimensions of the court. **(See Example 1.)**

8. The length of a rectangular picture frame is 4 in. less than twice the width. The perimeter is 112 in. Find the length and the width.

9. The lengths of the sides of a triangle are given by three consecutive even integers. The perimeter is 24 m. What is the length of each side?

10. A triangular garden has sides that can be represented by three consecutive integers. If the perimeter of the garden is 15 ft, what are the lengths of the sides?

11. Raoul would like to build a rectangular dog run in the rear of his backyard, away from the house. The width of the yard is $12\frac{1}{2}$ yd, and Raoul wants an area of 100 yd^2 for his dog.

    a. Find the dimensions of the dog run.

    b. How much fencing would Raoul need to enclose the dog run?

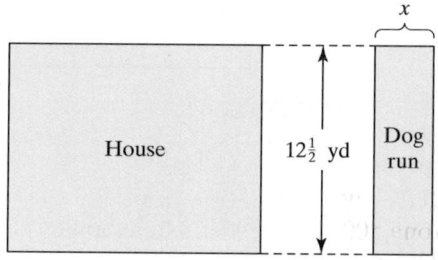

12. Joanne wants to plant a flower garden in her backyard in the shape of a trapezoid, adjacent to her house (see the figure). She also wants a front yard garden in the same shape, but with sides one-half as long. What should the dimensions be for each garden if Joanne has only a total of 60 ft of fencing?

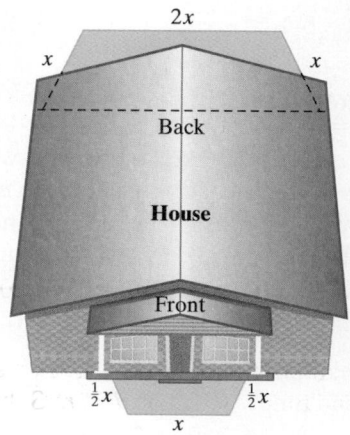

13. George built a rectangular pen for his rabbit such that the length is 7 ft less than twice the width. If the perimeter is 40 ft, what are the dimensions of the pen?

14. Antoine wants to put edging in the form of a square around a tree in his front yard. He has enough money to buy 18 ft of edging. Find the dimensions of the square that will use all the edging.

15. The measures of two angles in a triangle are equal. The third angle measures 2 times the sum of the equal angles. Find the measures of the three angles.

16. The smallest angle in a triangle is one-half the size of the largest. The middle angle measures 25° less than the largest. Find the measures of the three angles.

17. Two angles are complementary. One angle is 5 times as large as the other angle. Find the measure of each angle. **(See Example 2.)**

18. Two angles are supplementary. One angle measures 12° less than 3 times the other. Find the measure of each angle.

In Exercises 19–26, solve for x, and then find the measure of each angle.

19.

$(7x - 1)°$ $(2x + 1)°$

20.

$(10x + 36)°$

$[2(x + 15)]°$

21.

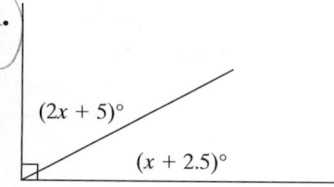

$(2x + 5)°$

$(x + 2.5)°$

22.

$[3(5x + 1)]°$

$(3x - 3)°$

23.

24.

25.

26.

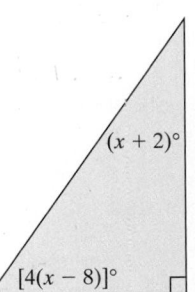

Concept 2: Literal Equations

 27. In 2010, Dario Franchitti won the Indianapolis 500 car race in 3 hr 5 min 37 sec ($\approx$ 3.094 hr). **(See Example 3.)**

   a. The relationship among the variables distance, rate, and time is given by $d = rt$. Solve for r.

   b. Determine Franchitti's average rate of speed if the total distance for the race is 500 mi. Round to one decimal place.

28. In 2011, Trevor Bayne won the 53rd running of the Daytona 500 car race with an average speed of 130.326 mph.

   a. Solve the formula $d = rt$ for t.

   b. Determine Bayne's total time of the race if the race is 500 mi long. Round to one decimal place.

29. The amount of simple interest earned or borrowed is the product of the principal, the annual interest rate, and the time invested (in years). This is given by $I = Prt$.

   a. Solve $I = Prt$ for t.

   b. Determine the amount of time necessary for the interest on $5000 invested at 4% to reach $1400.

30. The force of an object is equal to its mass times the acceleration, or $F = ma$.

   a. Solve $F = ma$ for m.

   b. The force on an object is 24.5 N (newtons), and the acceleration is 9.8 m/sec². Find the mass of the object (the answer will be in kilograms).

For Exercises 31–48, solve for the indicated variable. **(See Example 4.)**

31. $A = lw$ for l

32. $C_1 = \frac{5}{2}R$ for R

33. $I = Prt$ for P

34. $a + b + c = P$ for b

35. $W = K_2 - K_1$ for K_1

36. $y = mx + b$ for x

37. $F = \frac{9}{5}C + 32$ for C

38. $C = \frac{5}{9}(F - 32)$ for F

39. $K = \frac{1}{2}mv^2$ for v^2

40. $I = Prt$ for r

41. $v = v_0 + at$ for a

42. $a^2 + b^2 = c^2$ for b^2

43. $w = p(v_2 - v_1)$ for v_2

44. $A = lw$ for w

45. $ax + by = c$ for y

46. $P = 2L + 2W$ for L

47. $V = \frac{1}{3}Bh$ for B

48. $V = \frac{1}{3}\pi r^2 h$ for h

In Chapter 2 it will be necessary to change equations from the form $Ax + By = C$ to $y = mx + b$. For Exercises 49–60, express each equation in the form $y = mx + b$ by solving for y. **(See Example 5.)**

49. $3x + y = 6$

50. $x + y = -4$

51. $5x - 4y = 20$

52. $-4x - 5y = 25$

53. $-6x - 2y = 13$

54. $5x - 7y = 15$

55. $3x - 3y = 6$

56. $2x - 2y = 8$

 57. $9x + \dfrac{4}{3}y = 5$

58. $4x - \dfrac{1}{3}y = 5$

59. $-x + \dfrac{2}{3}y = 0$

60. $x - \dfrac{1}{4}y = 0$

In statistics, the z-score formula $z = \dfrac{x - \mu}{\sigma}$ is used in studying probability. Use this formula for Exercises 61–62.

61. a. Solve $z = \dfrac{x - \mu}{\sigma}$ for x.

    b. Find x when $z = 2.5$, $\mu = 100$, and $\sigma = 12$.

62. a. Solve $z = \dfrac{x - \mu}{\sigma}$ for μ.

    b. Find μ when $x = 150$, $z = 2.5$, and $\sigma = 16$.

63. Which expressions are equivalent to $\dfrac{-5}{x - 3}$?

    a. $-\dfrac{5}{x - 3}$ **b.** $\dfrac{5}{3 - x}$ **c.** $\dfrac{5}{-x + 3}$

64. Which expressions are equivalent to $\dfrac{z - 1}{-2}$?

    a. $\dfrac{1 - z}{2}$ **b.** $-\dfrac{z - 1}{2}$ **c.** $\dfrac{-z + 1}{2}$

65. Which expressions are equivalent to $\dfrac{-x - 7}{y}$?

    a. $-\dfrac{x + 7}{y}$ **b.** $\dfrac{x + 7}{-y}$ **c.** $\dfrac{-x - 7}{-y}$

66. Which expressions are equivalent to $\dfrac{-3w}{-x - y}$?

    a. $-\dfrac{3w}{-x - y}$ **b.** $\dfrac{3w}{x + y}$ **c.** $-\dfrac{3w}{x + y}$

For Exercises 67–75, solve for the indicated variable. **(See Example 6.)**

67. $6t - rt = 12$ for t

68. $5 = 4a + ca$ for a

69. $ax + 5 = 6x + 3$ for x

70. $cx - 4 = dx + 9$ for x

71. $A = P + Prt$ for P

72. $A = P + Prt$ for r

73. $T = mg - mf$ for m

74. $T = mg - mf$ for f

75. $ax + by = cx + z$ for x

Section 1.4 Linear Inequalities in One Variable

Concepts

1. Solving Linear Inequalities
2. Applications of Inequalities

1. Solving Linear Inequalities

In Sections 1.1–1.3, we learned how to solve linear equations and their applications. In this section, we will learn the process of solving linear *inequalities*. A **linear inequality** in one variable, x, is defined as any relationship of the form: $ax + b < c$, $ax + b \leq c$, $ax + b > c$, or $ax + b \geq c$, where $a \neq 0$.

The solution to the equation $x = 3$ can be graphed as a single point on the number line.

$$\begin{array}{ccccccccccc}
-5 & -4 & -3 & -2 & -1 & 0 & 1 & 2 & 3 & 4 & 5
\end{array}$$

Now consider the related *inequalities*.

Inequality	Solutions	Graph	Interval Notation	Set-Builder Notation
$x \le 3$	All real numbers less than or equal to 3.	⟵——————⟶ 3	$(-\infty, 3]$	$\{x \mid x \le 3\}$
$x < 3$	All real numbers less than 3.	⟵——————⟶ 3	$(-\infty, 3)$	$\{x \mid x < 3\}$
$x \ge 3$	All real numbers greater than or equal to 3.	——————⟶ 3	$[3, \infty)$	$\{x \mid x \ge 3\}$
$x > 3$	All real numbers greater than 3.	——————⟶ 3	$(3, \infty)$	$\{x \mid x > 3\}$

The solution sets to these inequalities have an infinite number of elements that cannot all be listed. Therefore, the solution set can be shown in a graph, or expressed in interval notation or in set-builder notation.

The addition and subtraction properties of equality indicate that a value added to or subtracted from both sides of an equation results in an equivalent equation. The same is true for inequalities.

Addition and Subtraction Properties of Inequality

Let a, b, and c represent real numbers.

*Addition property of inequality: If $a < b$
                                        then $a + c < b + c$

*Subtraction property of inequality: If $a < b$
                                        then $a - c < b - c$

*These properties may also be stated for $a \le b$, $a > b$, and $a \ge b$.

Example 1   **Solving a Linear Inequality**

Solve the inequality.

$$3x - 7 > 2(x - 4) - 1$$

Solution:

$3x - 7 > 2(x - 4) - 1$	
$3x - 7 > 2x - 8 - 1$	Apply the distributive property.
$3x - 7 > 2x - 9$	Combine *like* terms.
$3x - 2x - 7 > 2x - 2x - 9$	Subtract $2x$ from both sides.
$x - 7 > -9$	
$x - 7 + 7 > -9 + 7$	Add 7 to both sides.
$x > -2$	

Set-builder notation: $\{x \mid x > -2\}$ ⟵———————⟶
                                                      −2

Interval notation: $(-2, \infty)$

Skill Practice Solve the inequality.

1. $4(2x - 1) > 7x + 1$

Animation

Answer
1. $\{x \mid x > 5\}$
   $(5, \infty)$
   ——————⟶
                    5

Multiplying both sides of an equation by the same nonzero quantity results in an equivalent equation. However, the same is not always true for an inequality. If you multiply or divide an inequality by a *negative* quantity, the direction of the inequality symbol must be *reversed*.

For example, consider multiplying or dividing the inequality $4 < 5$ by -1.

Multiply/divide by -1: $4 < 5$

$-4 > -5$

$$-6 \;\; -5 \;\; -4 \;\; -3 \;\; -2 \;\; -1 \;\; 0 \;\; 1 \;\; 2 \;\; 3 \;\; 4 \;\; 5 \;\; 6$$

$-4 > -5$ $4 < 5$

The number 4 lies to the left of 5 on the number line. However, -4 lies to the right of -5. Changing the signs of two numbers changes their relative position on the number line. This is stated formally in the multiplication and division properties of inequality.

Multiplication and Division Properties of Inequality

Let a, b, and c represent real numbers.

*If c is *positive* and $a < b$, then $ac < bc$ and $\dfrac{a}{c} < \dfrac{b}{c}$

*If c is *negative* and $a < b$, then $ac > bc$ and $\dfrac{a}{c} > \dfrac{b}{c}$

The second statement indicates that if both sides of an inequality are multiplied or divided by a negative quantity, the inequality sign must be *reversed*.

*These properties may also be stated for $a \le b$, $a > b$, and $a \ge b$.

Example 2   Solving a Linear Inequality

Solve the inequality.

$$-2x - 5 < 2$$

Solution:

$$-2x - 5 < 2$$

$$-2x - 5 + 5 < 2 + 5 \qquad \text{Add 5 to both sides.}$$

$$-2x < 7$$

$$\frac{-2x}{-2} > \frac{7}{-2} \qquad \text{Divide by } -2 \;(\textit{reverse the inequality sign}).$$

$$x > -\frac{7}{2} \quad \text{or} \quad x > -3.5$$

Avoiding Mistakes

Do not forget to reverse the direction of the inequality sign when multiplying or dividing by a negative number.

Set-builder notation: $\{x \mid x > -\frac{7}{2}\}$

Interval notation: $(-\frac{7}{2}, \infty)$

$-\frac{7}{2}$

TIP: The inequality $-2x - 5 < 2$ could have been solved by isolating x on the right-hand side of the inequality. This creates a positive coefficient on the x term and eliminates the need to divide by a negative number.

$-2x - 5 < 2$

$\quad\quad -5 < 2x + 2$ Add $2x$ to both sides.

$\quad\quad -7 < 2x$ Subtract 2 from both sides.

$\quad\quad \dfrac{-7}{2} < \dfrac{2x}{2}$ Divide by 2 (because 2 is positive, do *not* reverse the inequality sign).

$\quad\quad -\dfrac{7}{2} < x$ (Note that the inequality $-\frac{7}{2} < x$ is equivalent to $x > -\frac{7}{2}$.)

Skill Practice Solve the inequality.

2. $-4x - 12 \geq 20$

Example 3   Solving a Linear Inequality

Solve the inequality.
$$-6(x - 3) \geq 2 - 2(x - 8)$$

Solution:

$\quad\quad -6(x - 3) \geq 2 - 2(x - 8)$

$\quad\quad\quad -6x + 18 \geq 2 - 2x + 16$ Apply the distributive property.

$\quad\quad\quad -6x + 18 \geq 18 - 2x$ Combine *like* terms.

$\quad -6x + 2x + 18 \geq 18 - 2x + 2x$ Add $2x$ to both sides.

$\quad\quad\quad -4x + 18 \geq 18$

$\quad\quad -4x + 18 - 18 \geq 18 - 18$ Subtract 18 from both sides.

$\quad\quad\quad\quad -4x \geq 0$

$\quad\quad\quad\quad \dfrac{-4x}{-4} \leq \dfrac{0}{-4}$ Divide by -4 (*reverse* the inequality sign).

$\quad\quad\quad\quad\quad x \leq 0$

Set-builder notation: $\{x \mid x \leq 0\}$

Interval notation: $(-\infty, 0]$

Skill Practice Solve the inequality.

3. $5(3x + 1) < 4(5x - 5)$

Answers

2. $\{x \mid x \leq -8\}$
   $(-\infty, -8]$

3. $\{x \mid x > 5\}$
   $(5, \infty)$

Example 4 Solving a Linear Inequality

Solve the inequality $\dfrac{-5x + 2}{-3} > x + 2$.

Solution:

$$\frac{-5x + 2}{-3} > x + 2$$

$$-3\left(\frac{-5x + 2}{-3}\right) < -3(x + 2) \qquad \text{Multiply by } -3 \text{ to clear fractions } (\textit{reverse the inequality sign}).$$

$$-5x + 2 < -3x - 6$$

$$-2x + 2 < -6 \qquad \text{Add } 3x \text{ to both sides.}$$

$$-2x < -8 \qquad \text{Subtract 2 from both sides.}$$

$$\frac{-2x}{-2} > \frac{-8}{-2} \qquad \text{Divide by } -2 \text{ (the inequality sign is reversed } \textit{again}\text{).}$$

$$x > 4 \qquad \text{Simplify.}$$

Set-builder notation: $\{x \mid x > 4\}$

Interval notation: $(4, \infty)$

Skill Practice Solve the inequality.

4. $\dfrac{x + 1}{-3} \geq -x + 1$

In Example 4, the inequality sign was reversed twice: once for multiplying the inequality by -3 and once for dividing by -2. If you are in doubt about whether you have the inequality sign in the correct direction, you can check your final answer by using the **test point method**. That is, pick a point in the proposed solution set, and verify that it makes the original inequality true. Furthermore, any test point picked outside the solution set should make the original inequality false.

Pick $x = 0$ as a test point

$$\frac{-5x + 2}{-3} > x + 2$$

$$\frac{-5(0) + 2}{-3} \overset{?}{>} (0) + 2$$

$$\frac{2}{-3} \overset{?}{>} 2 \qquad \text{False}$$

Pick $x = 5$ as a test point

$$\frac{-5x + 2}{-3} > x + 2$$

$$\frac{-5(5) + 2}{-3} \overset{?}{>} (5) + 2$$

$$\frac{-23}{-3} \overset{?}{>} 7$$

$$7\tfrac{2}{3} \overset{?}{>} 7 \qquad \text{True}$$

Answer

4. $\{x \mid x \geq 2\}$
  $[2, \infty)$

Because a test point to the right of $x = 4$ makes the inequality true, we have shaded the correct part of the number line.

2. Applications of Inequalities

Example 5   Solving a Linear Inequality Application

Beth received grades of 97, 82, 89, and 99 on her first four algebra tests. To earn an A in the course, she needs an average of 90 or more. What scores can she receive on the fifth test to earn an A?

Solution:

Let x represent the score on the fifth test.

The average of the five tests is given by $\dfrac{97 + 82 + 89 + 99 + x}{5}$

To earn an A, we have:

(Average of test scores) ≥ 90 Verbal model

$\dfrac{97 + 82 + 89 + 99 + x}{5} \geq 90$ Mathematical equation

$\dfrac{367 + x}{5} \geq 90$ Simplify the numerator.

$5\left(\dfrac{367 + x}{5}\right) \geq 5(90)$ Clear fractions.

$367 + x \geq 450$ Simplify.

$x \geq 83$

To earn an A, Beth would have to score at least 83 on her fifth test.

Skill Practice

5. Jamie is a salesman who works on commission, so his salary varies from month to month. To qualify for an automobile loan, his monthly salary must average at least $2100 for 6 months. His salaries for the past 5 months have been $1800, $2300, $1500, $2200, and $2800. What amount does he need to earn in the last month to qualify for the loan?

Example 6   Solving a Linear Inequality Application

The water level in a retention pond in northern California is 7.2 ft. During a time of drought, the water level decreases at a rate of 0.05 ft/day. The water level L (in ft) is given by $L = 7.2 - 0.05d$, where d is the number of days after the drought begins (Figure 1-5). For which days after the beginning of the drought will the water level be less than 6 ft?

Water Level vs. Days of Drought

$L = 7.2 - 0.05d$

Figure 1-5

Answer

5. Jamie's salary must be at least $2000.

Solution:

We require that $L < 6$.

$$\overbrace{7.2 - 0.05d}^{L} < 6 \qquad \text{Substitute } 7.2 - 0.05d \text{ for } L.$$

$$7.2 - 7.2 - 0.05d < 6 - 7.2 \qquad \text{Subtract 7.2 from both sides.}$$

$$-0.05d < -1.2$$

$$\frac{-0.05d}{-0.05} > \frac{-1.2}{-0.05} \qquad \begin{array}{l}\text{Divide both sides by } -0.05 \text{ and reverse}\\ \text{the direction of the inequality sign.}\end{array}$$

$$d > 24$$

The water level in the pond will be less than 6 ft after day 24 of the drought.

Skill Practice

6. The population of Alaska has steadily increased since 1950 according to the equation $P = 10t + 117$, where t represents the number of years after 1950 and P represents the population in thousands. For what years since 1950 was the population less than 417 thousand people?

Answer

6. The population was less than 417 thousand for $t < 30$. This corresponds to the years before 1980.

Section 1.4 Practice Exercises

Study Skills Exercise

Look over the notes that you took today. Do you understand what you wrote? If there were any rules, definitions, or formulas, highlight them so that they can be easily found when studying for the test. You may want to begin by highlighting the rule indicating when the direction of an inequality sign must be reversed.

Vocabulary and Key Concepts

1. **a.** A relationship of the form $ax + b > c$ or $ax + b < c \ (a \neq 0)$ is called a _____ _____ in one variable.

   b. When solving an inequality, the direction of the inequality sign must be reversed when multiplying or dividing both sides of the inequality by a _____ number.

   c. Which inequality statement represents the set of real numbers greater than 2? $x > 2$ or $2 < x$

Review Exercises

2. When using interval notation is it proper to use a parenthesis or a bracket with ∞ and $-\infty$?

For Exercises 3–8, complete the table.

	Set-Builder Notation	Interval Notation	Graph
3.	$\{x \mid x > 5\}$		───────────→
4.	$\{x \mid x \leq -2\}$		───────────→
5.		$(-3, 6]$	───────────→
6.		$[0, 4)$	───────────→
7.			──────[───→ 4
8.			←──────)────→ 10

Concept 1: Solving Linear Inequalities

For Exercises 9–10, solve the equation or inequality.

		Set Notation	Interval Notation	Graph
9. a.	$-2x + 4 = 10$			$\longrightarrow$
b.	$-2x + 4 < 10$			$\longrightarrow$
c.	$-2x + 4 > 10$			$\longrightarrow$
10. a.	$-4x + 2 = -6$			$\longrightarrow$
b.	$-4x + 2 < -6$			$\longrightarrow$
c.	$-4x + 2 > -6$			$\longrightarrow$

For Exercises 11–46, solve the inequality and graph the solution set. Write the solution set in (a) set-builder notation and (b) interval notation. **(See Examples 1–4.)**

11. $2y + 6 \le 4$

12. $3y + 11 > 5$

13. $-2x - 5 \le -25$

14. $-4z - 2 > -22$

15. $6z + 3 > 16$

16. $8w - 2 \le 13$

17. $-8 > \dfrac{2}{3}t$

18. $-4 \le \dfrac{1}{5}p$

19. $\dfrac{3}{4}(8y - 9) < 3$

20. $\dfrac{2}{5}(2x - 1) > 10$

21. $0.8a - 0.5 \le 0.3a - 11$

22. $0.2w - 0.7 < 0.4 - 0.9w$

23. $-5x + 7 < 22$

24. $-3w - 6 > 9$

25. $-\dfrac{5}{6}x \le -\dfrac{3}{4}$

26. $-\dfrac{3}{2}y > -\dfrac{21}{16}$

27. $\dfrac{3p - 1}{-2} > 5$

28. $\dfrac{3k - 2}{-5} \le 4$

29. $0.2t + 1 > 2.4t - 10$

30. $20 \le 8 - \dfrac{1}{3}x$

31. $3 - 4(y + 2) \le 6 + 4(2y + 1)$

32. $1 + 4(b - 2) < 2(b - 5) + 4$

33. $7.2k - 5.1 \ge 5.7$

34. $6h - 2.92 \le 16.58$

35. $\dfrac{3}{4}x - 8 \le 1$

36. $-\dfrac{2}{5}a - 3 > 5$

37. $-1.2b - 0.4 \ge -0.4b$

38. $-0.4t + 1.2 < -2$

39. $-\dfrac{3}{4}c - \dfrac{5}{4} \geq 2c$

40. $-\dfrac{2}{3}q - \dfrac{1}{3} > \dfrac{1}{2}q$

41. $4 - 4(y - 2) < -5y + 6$

42. $6 - 6(k - 3) \geq -4k + 12$

43. $-6(2x + 1) < 5 - (x - 4) - 6x$

44. $2(4p + 3) - p \leq 5 + 3(p - 3)$

45. $6a - (9a + 1) - 3(a - 1) \geq 2$

46. $8(q + 1) - (2q + 1) + 5 > 12$

Concept 2: Applications of Inequalities

47. Nadia received quiz grades of 80, 86, 73, and 91. **(See Example 5.)**

  a. What grade would she need to make on the fifth quiz to get a B average, that is, at least 80 but less than 90?

  b. Is it possible for Nadia to get an A average for her quizzes (at least 90)?

48. Ty received test grades of 78, 75, 71, 83, and 73.

  a. What grade would he need to make on the sixth test to get a C if a C is at least 75 but less than 80?

  b. Is it possible for Ty to get a B or better for his test average (at least 80)?

For Exercises 49–52, use the graph that shows the average height for boys based on age. Let a represent a boy's age (in years) and let h represent his height (in inches). **(See Example 6.)**

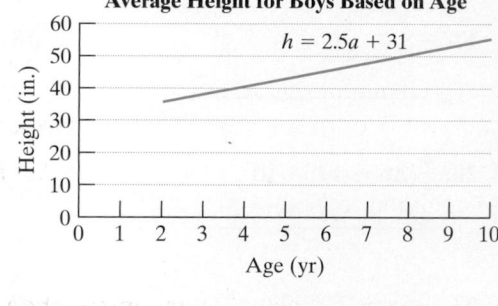

Average Height for Boys Based on Age

$h = 2.5a + 31$

Height (in.) / Age (yr)

49. Determine the age range for which the average height of boys is at least 51 in.

50. Determine the age range for which the average height of boys is greater than or equal to 41 in.

51. Determine the age range for which the average height of boys is no more than 46 in.

52. Determine the age range for which the average height of boys is at most 53.5 in.

53. Nolvia sells copy machines, and her salary is $25,000 plus a 4% commission on sales. The equation $S = 25{,}000 + 0.04x$ represents her salary S in dollars in terms of her total sales x in dollars.

  a. How much money in sales does Nolvia need to earn a salary that exceeds $40,000?

  b. How much money in sales does Nolvia need to earn a salary that exceeds $80,000?

  c. Why is the money in sales required to earn a salary of $80,000 more than twice the money in sales required to earn a salary of $40,000?

54. The amount of money A in a savings account depends on the principal P, the interest rate r, and the time in years t that the money is invested. The equation $A = P + Prt$ shows the relationship among the variables for an account earning simple interest. If an investor deposits $5000 at $6\frac{1}{2}\%$ simple interest, the account will grow according to the formula $A = 5000 + 5000(0.065)t$.

  a. How many years will it take for the investment to exceed $10,000? (Round to the nearest tenth of a year.)

  b. How many years will it take for the investment to exceed $15,000? (Round to the nearest tenth of a year.)

55. The revenue R for selling x fleece jackets is given by the equation $R = 49.95x$. The cost to produce x jackets is $C = 2300 + 18.50x$. Find the number of jackets that the company needs to sell to produce a profit. (*Hint:* A profit occurs when revenue exceeds cost.)

56. The revenue R for selling x mountain bikes is $R = 249.95x$. The cost to produce x bikes is $C = 56,000 + 140x$. Find the number of bikes that the company needs to sell to produce a profit.

Expanding Your Skills

For Exercises 57–60, assume $a > b$. Determine which inequality sign ($>$ or $<$) should be inserted to make a true statement. Assume $a \neq 0$ and $b \neq 0$.

57. $a + c$ _____ $b + c$,   for $c > 0$

58. $a + c$ _____ $b + c$,   for $c < 0$

59. ac _____ bc,   for $c < 0$

60. ac _____ bc,   for $c > 0$

Compound Inequalities

1. Union and Intersection of Sets

Two or more sets can be combined by the operations of union and intersection.

Concepts

1. Union and Intersection of Sets
2. Solving Compound Inequalities: And
3. Solving Inequalities of the Form $a < x < b$
4. Solving Compound Inequalities: Or
5. Applications of Compound Inequalities

Definition of *A* Union *B* and *A* Intersection *B*

The **union** of sets A and B, denoted $A \cup B$, is the set of elements that belong to set A or to set B or to both sets A and B.

The **intersection** of two sets A and B, denoted $A \cap B$, is the set of elements common to both A and B.

The concepts of the union and intersection of two sets are illustrated in Figures 1-6 and 1-7.

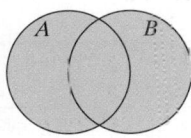

$A \cup B$
A union B
The elements in A *or* B *or* both
Figure 1-6

$A \cap B$
A intersection B
The elements in A *and* B
Figure 1-7

Example 1 **Finding the Union and Intersection of Sets**

Given the sets: $A = \{a, b, c, d, e, f\}$ $B = \{a, c, e, g, i, k\}$ $C = \{g, h, i, j, k\}$

Find: **a.** $A \cup B$ **b.** $A \cap B$ **c.** $A \cap C$

Solution:

a. $A \cup B = \{a, b, c, d, e, f, g, i, k\}$ The union of A and B includes all the elements of A along with all the elements of B. Notice that the elements a, c, and e are not listed twice.

b. $A \cap B = \{a, c, e\}$ The intersection of A and B includes only those elements that are common to both sets.

c. $A \cap C = \{\ \}$ (the empty set)   Because A and C share no common elements, the intersection of A and C is the empty set (also called the null set).

> **TIP:** The empty set may be denoted by the symbol $\{\ \}$ or by the symbol $\varnothing$.

Skill Practice Given: $A = \{r, s, t, u, v, w\}$ $B = \{s, v, w, y, z\}$ $C = \{x, y, z\}$

Find: **1.** $B \cup C$ **2.** $A \cap B$ **3.** $A \cap C$

Example 2   **Finding the Union and Intersection of Sets**

Given the sets: $A = \{x \mid x < 3\}$ $B = \{x \mid x \geq -2\}$ $C = \{x \mid x \geq 5\}$

Graph the following sets. Then express each set in interval notation.

a. $A \cap B$ **b.** $A \cup C$

Solution:

It is helpful to visualize the graphs of individual sets on the number line before taking the union or intersection.

a. Graph of $A = \{x \mid x < 3\}$

Graph of $B = \{x \mid x \geq -2\}$

Graph of $A \cap B$ (the "overlap")

Interval notation: $[-2, 3)$

Note that the set $A \cap B$ represents the real numbers greater than or equal to -2 and less than 3. This relationship can be written more concisely as a compound inequality: $-2 \leq x < 3$. We can interpret this inequality as "x is between -2 and 3, including $x = -2$."

b. Graph of $A = \{x \mid x < 3\}$

Graph of $C = \{x \mid x \geq 5\}$

Graph of $A \cup C$

Interval notation: $(-\infty, 3) \cup [5, \infty)$ $A \cup C$ includes all elements from set A along with the elements from set C.

Skill Practice Given the sets: $A = \{x \mid x \geq -1\}$, $B = \{x \mid x < 4\}$, and $C = \{x \mid x \geq 9\}$, determine the union or intersection and express the answer in interval notation.

4. $A \cap B$ **5.** $B \cup C$

Answers

1. $\{s, v, w, x, y, z\}$ **2.** $\{s, v, w\}$
3. $\{\ \}$ **4.** $[-1, 4)$
5. $(-\infty, 4) \cup [9, \infty)$

In Example 3, we find the union and intersection of sets expressed in interval notation.

Finding the Union and Intersection of Two Intervals

Find the union or intersection as indicated. Write the answer in interval notation.

a. $(-\infty, -2) \cup [-4, 3)$ **b.** $(-\infty, -2) \cap [-4, 3)$

Solution:

a. $(-\infty, -2) \cup [-4, 3)$ To find the union, graph each interval separately. The union is the collection of real numbers that lie in the first interval, the second interval, or both intervals.

$(-\infty, -2)$

$[-4, 3)$

The union consists of all real numbers in the red interval along with the real numbers in the blue interval: $(-\infty, 3)$

The union is $(-\infty, 3)$.

b. $(-\infty, -2) \cap [-4, 3)$

$(-\infty, -2)$

$[-4, 3)$

The intersection is the "overlap" of the two intervals: $[-4, -2)$.

The intersection is $[-4, -2)$.

Skill Practice Find the union or intersection. Write the answer in interval notation.

6. $(-\infty, -5] \cup (-7, 0)$ **7.** $(-\infty, -5] \cap (-7, 0)$

2. Solving Compound Inequalities: And

The solution to two inequalities joined by the word *and* is the intersection of their solution sets.

Solving a Compound Inequality: And

Step 1 Solve and graph each inequality separately.

Step 2 If the inequalities are joined by the word *and*, find the intersection of the two solution sets.

Step 3 Express the solution set in interval notation or in set-builder notation.

Answers
6. $(-\infty, 0)$ **7.** $(-7, -5]$

As you work through the examples in this section, remember that multiplying or dividing an inequality by a negative factor reverses the direction of the inequality sign.

Example 4 **Solving a Compound Inequality: And**

Solve the compound inequality.

$$-2x < 6 \quad \text{and} \quad x + 5 \le 7$$

Solution:

$-2x < 6$	and	$x + 5 \le 7$	Solve each inequality separately.

$$\frac{-2x}{-2} > \frac{6}{-2} \quad \text{and} \quad x \le 2 \qquad \text{Reverse the first inequality sign.}$$

$$x > -3 \quad \text{and} \quad x \le 2$$

     $\{x \mid x > -3\}$

     $\{x \mid x \le 2\}$

Take the intersection of the solution sets: $\{x \mid -3 < x \le 2\}$

The solution is $\{x \mid -3 < x \le 2\}$, or equivalently in interval notation, $(-3, 2]$.

Skill Practice Solve the compound inequality.

8. $5x + 2 \ge -8$ and $-4x > -24$

Example 5 **Solving a Compound Inequality: And**

Solve the compound inequality.

$$4.4a + 3.1 < -12.3 \quad \text{and} \quad -2.8a + 9.1 < -6.3$$

Solution:

$$4.4a + 3.1 < -12.3 \quad \text{and} \quad -2.8a + 9.1 < -6.3$$

$$4.4a < -15.4 \quad \text{and} \quad -2.8a < -15.4 \qquad \begin{array}{l}\text{Solve each} \\ \text{inequality} \\ \text{separately.}\end{array}$$

$$\frac{4.4a}{4.4} < \frac{-15.4}{4.4} \quad \text{and} \quad \frac{-2.8a}{-2.8} > \frac{-15.4}{-2.8} \qquad \begin{array}{l}\text{Reverse the} \\ \text{second} \\ \text{inequality sign.}\end{array}$$

$$a < -3.5 \quad \text{and} \quad a > 5.5$$

$\{a \mid a < -3.5\}$

$\{a \mid a > 5.5\}$

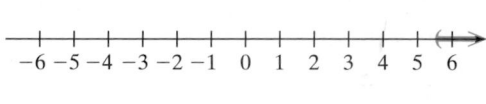   The intersection of the solution sets is the empty set: $\{\quad\}$

Answer

8. $\{x \mid -2 \le x < 6\}; [-2, 6)$

There are no real numbers that are simultaneously less than -3.5 and greater than 5.5. There is no solution.

The solution set is $\{\ \}$.

Skill Practice Solve the compound inequality.

9. $3.2y - 2.4 > 16.8$ and $-4.1y \geq 8.2$

Example 6 Solving a Compound Inequality: And

Solve the compound inequality.

$$-\frac{2}{3}x \leq 6 \quad \text{and} \quad -\frac{1}{2}x < 1$$

Solution:

$$-\frac{2}{3}x \leq 6 \qquad \text{and} \qquad -\frac{1}{2}x < 1$$

$$-\frac{3}{2}\left(-\frac{2}{3}x\right) \geq -\frac{3}{2}(6) \quad \text{and} \quad -2\left(-\frac{1}{2}x\right) > -2(1) \qquad \begin{array}{l}\text{Solve each} \\ \text{inequality} \\ \text{separately.}\end{array}$$

$$x \geq -9 \qquad \text{and} \qquad x > -2$$

$\{x \mid x \geq -9\}$

$\{x \mid x > -2\}$

Take the intersection of the solution sets:
$\{x \mid x > -2\}$

The solution set is $\{x \mid x > -2\}$, or in interval notation, $(-2, \infty)$.

Skill Practice Solve the compound inequality.

10. $-\frac{1}{4}z < \frac{5}{8}$ and $\frac{1}{2}z + 1 \geq 3$

---

3. Solving Inequalities of the Form $a < x < b$

An inequality of the form $a < x < b$ is a type of **compound inequality**, one that defines two simultaneous conditions on x.

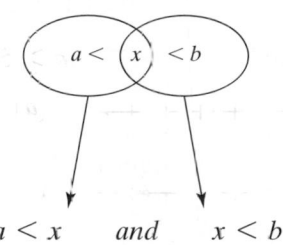

$$a < x \qquad \text{and} \qquad x < b$$

The solution set to the compound inequality $a < x < b$ is the *intersection* of the solution sets to the inequalities $a < x$ and $x < b$.

Answers
9. $\{\ \}$
10. $\{z \mid z \geq 4\}; [4, \infty)$

Example 7 **Solving an Inequality of the Form** $a < x < b$

Solve the inequality. $-4 < 3x + 5 \leq 10$

Solution:

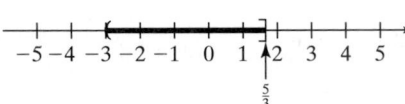

$-4 < 3x + 5$	and	$3x + 5 \leq 10$

Set up the intersection of two inequalities.

$-9 < 3x$	and	$3x \leq 5$

Solve each inequality.

$\dfrac{-9}{3} < \dfrac{3x}{3}$	and	$\dfrac{3x}{3} \leq \dfrac{5}{3}$

$-3 < x$	and	$x \leq \dfrac{5}{3}$

$$-3 < x \leq \frac{5}{3}$$

Take the intersection of the solution sets.

The solution is $\{x \mid -3 < x \leq \frac{5}{3}\}$, or equivalently in interval notation, $(-3, \frac{5}{3}]$.

Skill Practice Solve the inequality.

11. $-6 \leq 2x - 5 < 1$

To solve an inequality of the form $a < x < b$, we can also work with the inequality as a "three-part" inequality and isolate x. This is demonstrated in Example 8.

Example 8 **Solving an Inequality of the Form** $a < x < b$

Solve the inequality. $2 \geq \dfrac{p - 2}{-3} \geq -1$

Solution:

$$2 \geq \frac{p - 2}{-3} \geq -1$$

Isolate the variable in the middle part.

$$-3(2) \leq -3\left(\frac{p - 2}{-3}\right) \leq -3(-1)$$

Multiply all three parts by -3. Remember to reverse the inequality signs.

$$-6 \leq p - 2 \leq 3$$

Simplify.

$$-6 + 2 \leq p - 2 + 2 \leq 3 + 2$$

Add 2 to all three parts to isolate p.

$$-4 \leq p \leq 5$$

The solution is $\{p \mid -4 \leq p \leq 5\}$, or equivalently in interval notation $[-4, 5]$.

Skill Practice Solve the inequality.

12. $8 > \dfrac{t + 4}{-2} > -5$

Answers

11. $\{x \mid -\frac{1}{2} \leq x < 3\}$; $[-\frac{1}{2}, 3)$
12. $\{t \mid -20 < t < 6\}$; $(-20, 6)$

4. Solving Compound Inequalities: Or

In Examples 9 and 10, we solve compound inequalities that involve inequalities joined by the word *or*. In such a case, the solution to the compound inequality is the union of the solution sets of the individual inequalities.

Solving a Compound Inequality: Or

Step 1 Solve and graph each inequality separately.

Step 2 If the inequalities are joined by the word *or*, find the union of the two solution sets.

Step 3 Express the solution set in interval notation or in set-builder notation.

Example 9 Solving a Compound Inequality: Or

Solve the compound inequality.

$$-3y - 5 > 4 \quad \text{or} \quad 4 - y \le 6$$

Solution:

$$-3y - 5 > 4 \qquad \text{or} \qquad 4 - y \le 6$$

$$-3y > 9 \qquad \text{or} \qquad -y \le 2 \qquad \text{Solve each inequality separately.}$$

$$\frac{-3y}{-3} < \frac{9}{-3} \qquad \text{or} \qquad \frac{-y}{-1} \ge \frac{2}{-1} \qquad \text{Reverse the inequality signs.}$$

$$y < -3 \qquad \text{or} \qquad y \ge -2$$

 $\{y \,|\, y < -3\}$

$\{y \,|\, y \ge -2\}$

Take the union of the solution sets:
$\{y \,|\, y < -3 \text{ or } y \ge -2\}$

The solution is $\{y \,|\, y < -3 \text{ or } y \ge -2\}$ or, equivalently in interval notation, $(-\infty, -3) \cup [-2, \infty)$.

Skill Practice Solve the compound inequality.

13. $-10t - 8 \ge 12$ or $3t - 6 > 3$

Example 10 Solving a Compound Inequality: Or

Solve the compound inequality.

$$4x + 3 < 16 \quad \text{or} \quad -2x < 3$$

Solution:

$$4x + 3 < 16 \quad \text{or} \quad -2x < 3$$

$$4x < 13 \quad \text{or} \quad x > -\frac{3}{2} \qquad \text{Solve each inequality separately.}$$

$$x < \frac{13}{4} \quad \text{or} \quad x > -\frac{3}{2}$$

$$\{x \mid x < \tfrac{13}{4}\}$$

$$\{x \mid x > -\tfrac{3}{2}\}$$

Take the union of the solution sets.

The union of the solution sets is $\{x \mid x \text{ is a real number}\}$, or equivalently, $(-\infty, \infty)$.

Skill Practice Solve the compound inequality.

14. $x - 7 > -2$ or $-6x > -48$

5. Applications of Compound Inequalities

Compound inequalities are used in many applications, as shown in Examples 11 and 12.

Example 11 Translating Compound Inequalities

The normal level of thyroid-stimulating hormone (TSH) for adults ranges from 0.4 to 4.8 microunits per milliliter (μU/mL), inclusive. Let x represent the amount of TSH measured in microunits per milliliter.

  a. Write an inequality representing the normal range of TSH, inclusive.

  b. Write a compound inequality representing abnormal TSH levels.

Solution:

  a. $0.4 \le x \le 4.8$ **b.** $x < 0.4$ or $x > 4.8$

Skill Practice The length of a normal human pregnancy, w, is from 37 to 41 weeks, inclusive.

15. Write an inequality representing the normal length of a pregnancy.

16. Write a compound inequality representing an abnormal length for a pregnancy.

Answers

14. $\{x \mid x \text{ is a real number}\}; (-\infty, \infty)$
15. $37 \le w \le 41$
16. $w < 37$ or $w > 41$

TIP:

- In mathematics, the word "between" means strictly between two values. That is, the endpoints are *excluded*.

  Example: x is between 4 and 10 $\Rightarrow$ (4, 10).

- If the word "inclusive" is added to the statement, then we *include* the endpoints.

  Example: x is between 4 and 10, inclusive $\Rightarrow$ [4, 10].

Example 12   **Translating and Solving a Compound Inequality**

The sum of a number and 4 is between -5 and 12. Find all such numbers.

Solution:

Let x represent a number.

$$-5 < x + 4 < 12 \qquad \text{Translate the inequality.}$$

$$-5 - 4 < x + 4 - 4 < 12 - 4 \qquad \text{Subtract 4 from all three parts of the inequality.}$$

$$-9 < x < 8$$

The number may be any real number between -9 and 8: $\{x \mid -9 < x < 8\}$.

Skill Practice

17. The sum of twice a number and 11 is between 21 and 31. Find all such numbers.

Answer

17. Any real number between 5 and 10: $\{n \mid 5 < n < 10\}$

Section 1.5   Practice Exercises

Study Skills Exercise

Which activities might you try when working in a study group to help you learn and understand the material?

☐ Quiz one another by asking one another questions.

☐ Practice teaching one another.

☐ Share and compare class notes.

☐ Support and encourage one another.

☐ Work together on exercises and sample problems.

Vocabulary and Key Concepts

1. **a.** The _____ of two sets A and B, denoted by _____, is the set of elements that belong to A or B or both A and B.

   b. The _____ of two sets A and B, denoted by _____, is the set of elements common to both A and B.

   c. The solution set to the compound inequality $x < c$ and $x > d$ is the (union/intersection) of the solution sets of the individual inequalities.

   d. The compound inequality $a < x$ and $x < b$ can be written as the three-part inequality _____.

   e. The solution set to the compound inequality $x < a$ or $x > b$ is the (union/intersection) of the solution sets of the individual inequalities.

Review Exercises

2. Write the solution set to the inequality $4 < x + 1$ in set-builder notation and in interval notation.

For Exercises 3–6, solve the linear inequality. Write the solution in interval notation.

3. $-6u + 8 > 2$ **4.** $2 - 3z \geq -4$ **5.** $-12 \leq \frac{3}{4}p$ **6.** $5 > \frac{1}{3}w$

Concept 1: Union and Intersection of Sets

7. Given: $M = \{-3, -1, 1, 3, 5\}$ and
$N = \{-4, -3, -2, -1, 0\}$. **(See Example 1.)**

List the elements of the following sets:

a. $M \cap N$ **b.** $M \cup N$

8. Given: $P = \{a, b, c, d, e, f, g, h, i\}$ and
$Q = \{a, e, i, o, u\}$.

List the elements of the following sets.

a. $P \cap Q$ **b.** $P \cup Q$

For Exercises 9–20, refer to the sets $A, B, C,$ and D. Determine the union or intersection as indicated. Express the answer in interval notation, if possible. **(See Example 2.)**

$$A = \{x \mid x < -4\}, B = \{x \mid x > 2\}, C = \{x \mid x \geq -7\}, D = \{x \mid 0 \leq x < 5\}$$

9. $A \cap C$ **10.** $B \cap C$ **11.** $A \cup B$

12. $A \cup D$ **13.** $A \cap B$ **14.** $A \cap D$

15. $B \cup C$ **16.** $B \cup D$ **17.** $C \cap D$

18. $B \cap D$ **19.** $C \cup D$ **20.** $A \cup C$

For Exercises 21–26, find the intersection and union of sets as indicated. Write the answers in interval notation. **(See Example 3.)**

21. **a.** $(-2, 5) \cap [-1, \infty)$
b. $(-2, 5) \cup [-1, \infty)$

22. **a.** $(-\infty, 4) \cap [-1, 5)$
b. $(-\infty, 4) \cup [-1, 5)$

23. **a.** $\left(-\frac{5}{2}, 3\right) \cap \left(-1, \frac{9}{2}\right)$
b. $\left(-\frac{5}{2}, 3\right) \cup \left(-1, \frac{9}{2}\right)$

24. **a.** $(-3.4, 1.6) \cap (-2.2, 4.1)$
b. $(-3.4, 1.6) \cup (-2.2, 4.1)$

25. **a.** $(-4, 5] \cap (0, 2]$
b. $(-4, 5] \cup (0, 2]$

26. **a.** $[-1, 5) \cap (0, 3)$
b. $[-1, 5) \cup (0, 3)$

Concept 2: Solving Compound Inequalities: And

For Exercises 27–36, solve the compound inequality and graph the solution. Write the answer in interval notation. **(See Examples 4–6.)**

27. $y - 7 \geq -9$ and $y + 2 \leq 5$

28. $a + 6 > -2$ and $5a < 30$

29. $2t + 7 < 19$ and $5t + 13 > 28$

30. $5p + 2p \geq -21$ and $-9p + 3p \geq -24$

31. $2.1k - 1.1 \leq 0.6k + 1.9$ and
$0.3k - 1.1 < -0.1k + 0.9$

32. $0.6w + 0.1 > 0.3w - 1.1$ and
$2.3w + 1.5 \geq 0.3w + 6.5$

33. $\frac{2}{3}(2p - 1) \geq 10$ and $\frac{4}{5}(3p + 4) \geq 20$

34. $\frac{5}{2}(a + 2) < -6$ and $\frac{3}{4}(a - 2) < 1$

35. $-2 < -x - 12$ and $-14 < 5(x - 3) + 6x$

36. $-8 \geq -3y - 2$ and $3(y - 7) + 16 > 4y$

Concept 3: Solving Inequalities of the Form $a < x < b$

37. Write $-4 \leq t < \frac{3}{4}$ as two separate inequalities.

38. Write $-2.8 < y \leq 15$ as two separate inequalities.

39. Explain why $6 < x < 2$ has no solution.

40. Explain why $4 < t < 1$ has no solution.

41. Explain why $-5 > y > -2$ has no solution.

42. Explain why $-3 > w > -1$ has no solution.

For Exercises 43–54, solve the compound inequality and graph the solution set. Write the answer in interval notation.
(See Examples 7–8.)

43. $0 \leq 2b - 5 < 9$

44. $-6 < 3k - 9 \leq 0$

45. $-1 < \frac{a}{6} \leq 1$

46. $-3 \leq \frac{1}{2}x < 0$

47. $-\frac{2}{3} < \frac{y - 4}{-6} < \frac{1}{3}$

48. $\frac{1}{3} > \frac{t - 4}{-3} > -2$

can wait

49. $5 \leq -3x - 2 \leq 8$

50. $-1 < -2x + 4 \leq 5$

51. $12 > 6x + 3 \geq 0$

52. $-4 \geq 2x - 5 > -7$

53. $-0.2 < 2.6 + 7t < 4$

54. $-1.5 < 0.1x \leq 8.1$

Concept 4: Solving Compound Inequalities: Or

For Exercises 55–64, solve the compound inequality and graph the solution set. Write the answer in interval notation. **(See Examples 9–10.)**

55. $2y - 1 \geq 3$ or $y < -2$

56. $x < 0$ or $3x + 1 \geq 7$

57. $1 > 6z - 8$ or $8z - 6 \leq 10$

58. $22 > 4t - 10$ or $7 > 2t - 5$

59. $5(x - 1) \geq -5$ or $5 - x \leq 11$

60. $-p + 7 \geq 10$ or $3(p - 1) \leq 12$

61. $\frac{5}{3}v \leq 5$ or $-v - 6 < 1$

62. $\frac{3}{8}u + 1 > 0$ or $-2u \geq -4$

63. $0.5w + 5 < 2.5w - 4$ or $0.3w \leq -0.1w - 1.6$

64. $1.25a + 3 \leq 0.5a - 6$ or $2.5a - 1 \geq 9 - 1.5a$

Mixed Exercises

For Exercises 65–74, solve the compound inequality. Write the answer in interval notation.

65. a. $3x - 5 < 19$ and $-2x + 3 < 23$

    b. $3x - 5 < 19$ or $-2x + 3 < 23$

66. a. $0.5(6x + 8) > 0.8x - 7$ and $4(x + 1) < 7.2$

    b. $0.5(6x + 8) > 0.8x - 7$ or $4(x + 1) < 7.2$

67. a. $8x - 4 \geq 6.4$ or $0.3(x + 6) \leq -0.6$

    b. $8x - 4 \geq 6.4$ and $0.3(x + 6) \leq -0.6$

68. a. $-2r + 4 \leq -8$ or $3r + 5 \leq 8$

    b. $-2r + 4 \leq -8$ and $3r + 5 \leq 8$

69. $-4 \leq \dfrac{2 - 4x}{3} < 8$

70. $-1 < \dfrac{3 - x}{2} \leq 0$

71. $5 \geq -4(t - 3) + 3t$ or

    $6 < 12t + 8(4 - t)$

72. $3 > -(w - 3) + 4w$ or

    $-5 \geq -3(w - 5) + 6w$

73. $\dfrac{-x + 3}{2} > \dfrac{4 + x}{5}$ or $\dfrac{1 - x}{4} > \dfrac{2 - x}{3}$

74. $\dfrac{y - 7}{-3} < \dfrac{1}{4}$ or $\dfrac{y + 1}{-2} > -\dfrac{1}{3}$

Concept 5: Applications of Compound Inequalities

75. The normal number of white blood cells for human blood is between 4800 and 10,800 cells per cubic millimeter, inclusive. Let x represent the number of white blood cells per cubic millimeter. **(See Example 11.)**

    a. Write an inequality representing the normal range of white blood cells per cubic millimeter.

    b. Write a compound inequality representing abnormal levels of white blood cells per cubic millimeter.

76. Normal hemoglobin levels in human blood for adult males are between 13 and 16 grams per deciliter (g/dL), inclusive. Let x represent the level of hemoglobin measured in grams per deciliter.

    a. Write an inequality representing normal hemoglobin levels for adult males.

    b. Write a compound inequality representing abnormal levels of hemoglobin for adult males.

77. A polling company estimates that a certain candidate running for office will receive between 44% and 48% of the votes. Let x represent the percentage of votes for this candidate.

    a. Write a strict inequality representing the expected percentage of votes for this candidate.

    b. Write a compound inequality representing the percentage of votes that would fall outside the polling company's prediction.

78. A machine is calibrated to cut a piece of wood between 2.4 in. thick and 2.6 in. thick. Let x represent the thickness of the wood after it is cut.

    a. Write a strict inequality representing the expected range of thickness of the wood after it has been cut.

    b. Write a compound inequality representing the thickness of wood that would fall outside the normal range for this machine.

79. Twice a number is between -3 and 12. Find all such numbers. **(See Example 12.)**

80. The difference of a number and 6 is between 0 and 8. Find all such numbers.

81. One plus twice a number is either greater than 5 or less than -1. Find all such numbers.

82. One-third of a number is either less than -2 or greater than 5. Find all such numbers.

83. Amy knows from reading her syllabus in intermediate algebra that the average of her chapter tests accounts for 80% (0.8) of her overall course grade. She also knows that the final exam counts as 20% (0.2) of her grade.

   a. Suppose that the average of Amy's chapter tests is 92%. Determine the range of grades that she would need on her final exam to get an "A" in the class. (Assume that a grade of "A" is obtained if Amy's overall average is 90% or better.)

   b. Determine the range of grades that Amy would need on her final exam to get a "B" in the class. (Assume that a grade of "B" is obtained if Amy's overall average is at least 80% but less than 90%.)

84. Robert knows from reading his syllabus in intermediate algebra that the average of his chapter tests accounts for 60% (0.6) of his overall course grade. He also knows that the final exam counts as 40% (0.4) of his grade.

   a. Suppose that the average of Robert's chapter tests is 89%. Determine the range of grades that he would need on his final exam to get an "A" in the class. (Assume that a grade of "A" is obtained if Robert's overall average is 90% or better.)

   b. Determine the range of grades that Robert would need on his final exam to get a "B" in the class. (Assume that a grade of "B" is obtained if Robert's overall average is at least 80% but less than 90%.)

85. The average high and low temperatures for Vancouver, British Columbia, in January are 5.6°C and 0°C, respectively. The formula relating Celsius temperatures to Fahrenheit temperatures is given by $C = \frac{5}{9}(F - 32)$. Convert the inequality $0.0° \leq C \leq 5.6°$ to an equivalent inequality using Fahrenheit temperatures.

86. For a day in July, the temperatures in Austin, Texas, ranged from 20°C to 29°C. The formula relating Celsius temperatures to Fahrenheit temperatures is given by $C = \frac{5}{9}(F - 32)$. Convert the inequality $20° \leq C \leq 29°$ to an equivalent inequality using Fahrenheit temperatures.

Absolute Value Equations

1. Solving Absolute Value Equations

An equation of the form $|x| = a$ is called an **absolute value equation**. For example, consider the equation $|x| = 4$. From the definition of absolute value, the solutions are found by solving the equations $x = 4$ and $-x = 4$. This gives the equivalent equations $x = 4$ and $x = -4$.

   Also recall from Section R.3 that the absolute value of a number is its distance from zero on the number line. Therefore, geometrically, the solutions to the equation $|x| = 4$ are the values of x that are 4 units from zero on the number line (Figure 1-8).

Concepts

1. Solving Absolute Value Equations
2. Solving Equations Containing Two Absolute Values

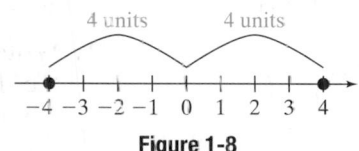

Figure 1-8

> ### Solving Absolute Value Equations of the Form $|x| = a$
> If a is a real number, then
> - If $a \geq 0$, the solutions to the equation $|x| = a$ are given by $x = a$ and $x = -a$.
> - If $a < 0$, there is no solution to the equation $|x| = a$.

To solve an absolute value equation of the form $|x| = a$ ($a \geq 0$), rewrite the equation as $x = a$ or $x = -a$.

Example 1 **Solving Absolute Value Equations**

Solve the absolute value equations.

   a. $|x| = 5$ **b.** $|w| - 2 = 12$ **c.** $|p| = 0$ **d.** $|x| = -6$

Solution:

a. $|x| = 5$ The equation is in the form $|x| = a$, where $a = 5$.

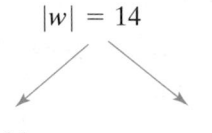

   $x = 5$ or $x = -5$ Rewrite the equation as $x = a$ or $x = -a$.
   The solution set is $\{5, -5\}$.

b. $|w| - 2 = 12$ Isolate the absolute value to write the equation in the form $|x| = a$.

          $|w| = 14$

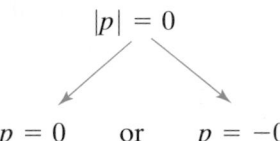

   $w = 14$ or $w = -14$ Rewrite the equation as $w = a$ or $w = -a$.
   The solution set is $\{14, -14\}$.

c. $|p| = 0$

   $p = 0$ or $p = -0$ Rewrite as two equations. Notice that the second equation $p = -0$ is the same as the first equation. Intuitively, $p = 0$ is the only number whose absolute value equals 0.
   The solution set is $\{0\}$.

d. $|x| = -6$ The equation is of the form $|x| = a$, but a is negative. There is no number whose absolute value is negative.
   No solution, $\{\ \}$

Skill Practice Solve the absolute value equations.

   1. $|y| = 7$ **2.** $|v| + 6 = 10$ **3.** $|w| = 0$ **4.** $|z| = -12$

We have solved absolute value equations of the form $|x| = a$. Notice that x can represent any algebraic quantity. For example, to solve the equation $|2w - 3| = 5$, we still rewrite the absolute value equation as two equations. In this case, we set the quantity $2w - 3$ equal to 5 and to -5, respectively.

$$|2w - 3| = 5$$

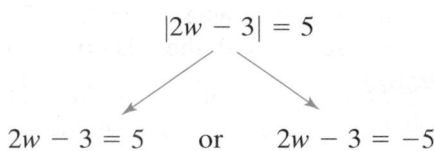

   $2w - 3 = 5$ or $2w - 3 = -5$

Answers
1. $\{7, -7\}$ **2.** $\{4, -4\}$
3. $\{0\}$ **4.** $\{\ \}$

Solving an Absolute Value Equation

Step 1   Isolate the absolute value. That is, write the equation in the form $|x| = a$, where a is a real number.

Step 2   If $a < 0$, there is no solution.

Step 3   Otherwise, if $a \geq 0$, rewrite the absolute value equation as $x = a$ or $x = -a$.

Step 4   Solve the individual equations from step 3.

Step 5   Check the answers in the original absolute value equation.

Example 2   **Solving an Absolute Value Equation**

Solve the equation. $|2w - 3| = 5$

Solution:

$|2w - 3| = 5$ The equation is already in the form $|x| = a$, where $x = 2w - 3$.

$2w - 3 = 5$ or $2w - 3 = -5$ Rewrite as two equations.

$2w = 8$ or $2w = -2$ Solve each equation.

$w = 4$ or $w = -1$

Check: $w = 4$ Check: $w = -1$ Check the values in the original equation.

$|2w - 3| = 5$ $|2w - 3| = 5$

$|2(4) - 3| \stackrel{?}{=} 5$ $|2(-1) - 3| \stackrel{?}{=} 5$

$|8 - 3| \stackrel{?}{=} 5$ $|-2 - 3| \stackrel{?}{=} 5$

$|5| \stackrel{?}{=} 5$ ✔ $|-5| \stackrel{?}{=} 5$ ✔

The solution set is $\{4, -1\}$.

Skill Practice Solve the equation.

5. $|4x + 1| = 9$

Example 3   **Solving an Absolute Value Equation**

Solve the equation. $|2c - 5| + 6 = 2$

Solution:

$|2c - 5| + 6 = 2$

$|2c - 5| = -4$ Isolate the absolute value. The equation is in the form $|x| = a$, where $x = 2c - 5$ and $a = -4$. Because $a < 0$, there is no solution.

No solution, $\{\ \}$ There are no numbers c that will make an absolute value equal to a negative number.

Avoiding Mistakes

Always isolate the absolute value first. Otherwise you will get answers that do not check.

Skill Practice Solve the equation.

6. $|3z + 10| + 3 = 1$

Answers

5. $\left\{2, -\dfrac{5}{2}\right\}$ **6.** No solution, $\{\ \}$

Example 4 Solving an Absolute Value Equation

Solve the equation. $-2\left|\dfrac{2}{5}p + 3\right| - 7 = -19$

Solution:

$$-2\left|\frac{2}{5}p + 3\right| - 7 = -19$$

$$-2\left|\frac{2}{5}p + 3\right| = -12 \qquad \text{Isolate the absolute value.}$$

$$\frac{-2\left|\frac{2}{5}p + 3\right|}{-2} = \frac{-12}{-2} \qquad \text{Divide both sides by } -2.$$

$$\left|\frac{2}{5}p + 3\right| = 6$$

$$\frac{2}{5}p + 3 = 6 \qquad \text{or} \qquad \frac{2}{5}p + 3 = -6 \qquad \text{Rewrite as two equations.}$$

$$2p + 15 = 30 \qquad \text{or} \qquad 2p + 15 = -30 \qquad \text{Multiply by 5 to clear fractions.}$$

$$2p = 15 \qquad \text{or} \qquad 2p = -45$$

$$p = \frac{15}{2} \qquad \text{or} \qquad p = -\frac{45}{2} \qquad \text{Both values check in the original equation.}$$

The solution set is $\left\{\dfrac{15}{2}, -\dfrac{45}{2}\right\}$.

Skill Practice Solve the equation.

7. $3\left|\dfrac{3}{2}a + 1\right| + 2 = 14$

Example 5 Solving an Absolute Value Equation

Solve the equation. $6.9 = |4.1 - p| + 6.9$

Solution:

$$6.9 = |4.1 - p| + 6.9$$

$$|4.1 - p| + 6.9 = 6.9 \qquad \text{First write the absolute value on the left. Then subtract 6.9 from both sides to write the equation in the form } |x| = a.$$

$$|4.1 - p| = 0 \qquad \text{Isolate the absolute value.}$$

$$4.1 - p = 0 \qquad \text{or} \qquad 4.1 - p = -0 \qquad \text{Rewrite as two equations. Notice that the equations are the same.}$$

Answer

7. $\left\{2, -\dfrac{10}{3}\right\}$

$$-p = -4.1 \qquad\qquad \text{Subtract 4.1 from both sides.}$$

$$p = 4.1 \qquad\qquad \underline{\text{Check: } p = 4.1}$$

$$|4.1 - p| + 6.9 = 6.9$$

$$|4.1 - 4.1| + 6.9 \stackrel{?}{=} 6.9$$

$$|0| + 6.9 \stackrel{?}{=} 6.9$$

The solution set is $\{4.1\}$. $\qquad\qquad 6.9 \stackrel{?}{=} 6.9 \ \checkmark$

Skill Practice Solve the equation.

8. $-3.5 = |1.2 + x| - 3.5$

2. Solving Equations Containing Two Absolute Values

Some equations have two absolute values such as $|x| = |y|$. If two quantities have the same absolute value, then the quantities are equal or the quantities are opposites.

Equality of Absolute Values

$$|x| = |y| \text{ implies that } x = y \text{ or } x = -y.$$

Example 6 Solving an Equation Having Two Absolute Values

Solve the equation. $\quad |2w - 3| = |5w + 1|$

Solution:

$$|2w - 3| = |5w + 1|$$

$2w - 3 = 5w + 1$ or	$2w - 3 = -(5w + 1)$	Rewrite as two equations, $x = y$ or $x = -y$.
$2w - 3 = 5w + 1$ or	$2w - 3 = -5w - 1$	Solve for w.
$-3w - 3 = 1$ or	$7w - 3 = -1$	
$-3w = 4$ or	$7w = 2$	
$w = -\dfrac{4}{3}$ or	$w = \dfrac{2}{7}$	Both values check in the original equation.

Avoiding Mistakes

To take the opposite of the quantity $5w + 1$, use parentheses and apply the distributive property.

The solution set is $\left\{ -\dfrac{4}{3}, \dfrac{2}{7} \right\}$.

Skill Practice Solve the equation.

9. $|3 - 2x| = |3x - 1|$

Answers

8. $\{-1.2\}$ **9.** $\left\{ \dfrac{4}{5}, -2 \right\}$

Example 7 Solving an Equation Having Two Absolute Values

Solve the equation. $|x - 4| = |x + 8|$

Solution:

$$|x - 4| = |x + 8|$$

$x - 4 = x + 8$ or $x - 4 = -(x + 8)$		Rewrite as two equations, $x = y$ or $x = -y$.
$-4 = 8$ or $x - 4 = -x - 8$		Solve for x.
$\uparrow$ $\quad\quad$ $2x - 4 = -8$		
contradiction		
$2x = -4$		
$x = -2$		$x = -2$ checks in the original equation.

The solution set is $\{-2\}$.

Answer

10. $\left\{\dfrac{1}{4}\right\}$

Skill Practice Solve the equation.

10. $|4t + 3| = |4t - 5|$

Section 1.6 Practice Exercises

Study Skills Exercise

One way to know that you really understand a concept is to explain it to someone else. In your own words, explain how to solve an absolute value equation.

Vocabulary and Key Concepts

1. a. An _____ value equation is an equation of the form $|x| = a$. If a is a positive real number then the solution set is _____.

b. What is the first step to solve the absolute value equation $|x| + 5 = 7$?

c. The absolute value equation $|x| = |y|$ implies that $x = $ _____ or $x = $ _____.

d. The solution set to the equation $|x + 4| = -2$ is _____. The solution set to the equation $|x + 4| = 0$ is _____.

Review Exercises

For Exercises 2–6, solve the compound inequalities. Write the answers in interval notation.

2. $3(a + 2) - 6 > 2$ and $-2(a - 3) + 14 > -3$

3. $3x - 5 \geq 7x + 3$ or $2x - 1 \leq 4x - 5$

4. $-5 < 2 - x \leq 6$

5. $5 \geq \dfrac{x - 4}{-2} > -3$

6. $4 \leq \dfrac{1}{3}x - 2 < 7$

Concept 1: Solving Absolute Value Equations

For Exercises 7–42, solve the absolute value equations. **(See Examples 1–5.)**

7. $|p| = 7$

8. $|q| = 10$

9. $|x| + 5 = 11$

10. $|x| - 3 = 20$

11. $|y| + 8 = 5$

12. $|x| + 12 = 6$

13. $|w| - 3 = -1$

14. $|z| - 14 = -10$

15. $|y| = \sqrt{2}$

16. $|y| = \sqrt{5}$

17. $|w| - 3 = -5$

18. $|w| + 4 = -8$

19. $|3q| = 0$ **20.** $|4p| = 0$ **21.** $|3x - 4| = 8$ **22.** $|4x + 1| = 6$

23. $5 = |2x - 4|$ **24.** $10 = |3x + 7|$ **25.** $\left|\dfrac{7z}{3} - \dfrac{1}{3}\right| + 3 = 6$ **26.** $\left|\dfrac{w}{2} + \dfrac{3}{2}\right| - 2 = 7$

27. $|0.2x - 3.5| = -5.6$ **28.** $|1.81 + 2x| = -2.2$ **29.** $1 = -4 + \left|2 - \dfrac{1}{4}w\right|$ **30.** $-12 = -6 - |6 - 2x|$

31. $10 = 4 + |2y + 1|$ **32.** $-1 = -|5x + 7|$ **33.** $-2|3b - 7| - 9 = -9$ **34.** $-3|5x + 1| + 4 = 4$

35. $-2|x + 3| = 5$ **36.** $-3|x - 5| = 7$ **37.** $0 = |6x - 9|$ **38.** $7 = |4k - 6| + 7$

39. $\left|-\dfrac{1}{5} - \dfrac{1}{2}k\right| = \dfrac{9}{5}$ **40.** $\left|-\dfrac{1}{6} - \dfrac{2}{9}h\right| = \dfrac{1}{2}$ **41.** $-3|2 - 6x| + 5 = -10$ **42.** $5|1 - 2x| - 7 = 3$

Concept 2: Solving Equations Containing Two Absolute Values

For Exercises 43–60, solve the absolute value equations. (See Examples 6–7.)

43. $|4x - 2| = |-8|$ **44.** $|3x + 5| = |-5|$ **45.** $|4w + 3| = |2w - 5|$

46. $|3y + 1| = |2y - 7|$ **47.** $|2y + 5| = |7 - 2y|$ **48.** $|9a + 5| = |9a - 1|$

49. $\left|\dfrac{4w - 1}{6}\right| = \left|\dfrac{2w}{3} + \dfrac{1}{4}\right|$ **50.** $\left|\dfrac{6p + 3}{8}\right| = \left|\dfrac{3}{4}p - 2\right|$ **51.** $|x + 2| = |-x - 2|$

52. $|2y - 3| = |-2y + 3|$ **53.** $|3.5m - 1.2| = |8.5m + 6|$ **54.** $|11.2n + 9| = |7.2n - 2.1|$

55. $|4x - 3| = -|2x - 1|$ **56.** $-|3 - 6y| = |8 - 2y|$ **57.** $|8 - 7w| = |7w - 8|$

58. $|4 - 3z| = |3z - 4|$ **59.** $|x + 2| + |x - 4| = 0$ **60.** $|t + 6| + |t - 1| = 0$

Expanding Your Skills

61. Write an absolute value equation whose solution is the set of real numbers 6 units from zero on the number line.

62. Write an absolute value equation whose solution is the set of real numbers $\frac{7}{2}$ units from zero on the number line.

63. Write an absolute value equation whose solution is the set of real numbers $\frac{4}{3}$ units from zero on the number line.

64. Write an absolute value equation whose solution is the set of real numbers 9 units from zero on the number line.

Absolute Value Inequalities

1. Solving Absolute Value Inequalities by Definition

In Section 1.6, we studied absolute value equations in the form $|x| = a$. In this section, we will solve absolute value *inequalities*. An inequality in any of the forms $|x| < a$, $|x| \le a$, $|x| > a$, or $|x| \ge a$ is called an **absolute value inequality**.

Recall that an absolute value represents distance from zero on the real number line. Consider the following absolute value equation and inequalities.

Concepts

1. Solving Absolute Value Inequalities by Definition
2. Solving Absolute Value Inequalities by the Test Point Method
3. Translating to an Absolute Value Expression

1. $|x| = 3$
   $x = 3$ or $x = -3$

Solution:

The set of all points 3 units from zero on the number line

2. $|x| > 3$
   $x < -3$ or $x > 3$

Solution:

The set of all points more than 3 units from zero

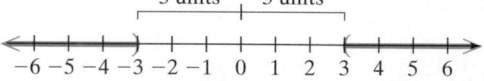

3. $|x| < 3$
   $-3 < x < 3$

Solution:

The set of all points less than 3 units from zero

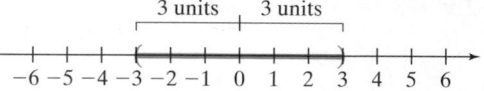

Solving Absolute Value Equations and Inequalities

Let a be a real number such that $a > 0$. Then

Equation/ Inequality	Solution (Equivalent Form)	Graph
$|x| = a$	$x = -a$ or $x = a$	
$|x| > a$	$x < -a$ or $x > a$	
$|x| < a$	$-a < x < a$	

To solve an absolute value inequality, first isolate the absolute value and then rewrite the absolute value inequality in its equivalent form.

Example 1   Solving an Absolute Value Inequality

Solve the inequality. $|3w + 1| - 4 < 7$

Solution:

$$|3w + 1| - 4 < 7$$

$|3w + 1| < 11$ ◄— Isolate the absolute value first.

The inequality is in the form $|x| < a$, where $x = 3w + 1$.

$-11 < 3w + 1 < 11$ Rewrite in the equivalent form $-a < x < a$.

$-12 < 3w < 10$ Solve for w.

$-4 < w < \dfrac{10}{3}$

TIP: Recall that a strict inequality (using the symbols > and <) will have parentheses at the endpoints of the interval form of the solution.

The solution is $\{w \mid -4 < w < \frac{10}{3}\}$, or equivalently in interval notation, $(-4, \frac{10}{3})$.

Answer
1. $[-7, 2]$

Skill Practice Solve the inequality. Write the solution in interval notation.

   1. $|2t + 5| + 2 \le 11$

Example 2 Solving an Absolute Value Inequality

Solve the inequality. $\quad 3 \leq 1 + \left| \dfrac{1}{2}t - 5 \right|$

Solution:

$3 \leq 1 + \left| \dfrac{1}{2}t - 5 \right|$

$1 + \left| \dfrac{1}{2}t - 5 \right| \geq 3 \qquad$ Write the inequality with the absolute value on the left.

$\left| \dfrac{1}{2}t - 5 \right| \geq 2 \qquad$ Isolate the absolute value.

The inequality is in the form $|x| \geq a$, where $x = \frac{1}{2}t - 5$.

$\dfrac{1}{2}t - 5 \leq -2 \qquad$ or $\qquad \dfrac{1}{2}t - 5 \geq 2 \qquad$ Rewrite in the equivalent form $x \leq -a$ or $x \geq a$.

$\dfrac{1}{2}t \leq 3 \qquad$ or $\qquad \dfrac{1}{2}t \geq 7 \qquad$ Solve the compound inequality.

$2\left(\dfrac{1}{2}t \right) \leq 2(3) \qquad$ or $\qquad 2\left(\dfrac{1}{2}t \right) \geq 2(7) \qquad$ Clear fractions.

$t \leq 6 \qquad$ or $\qquad t \geq 14 \qquad$

The solution is $\{t \mid t \leq 6 \text{ or } t \geq 14\}$ or, equivalently in interval notation, $(-\infty, 6] \cup [14, \infty)$.

> **TIP:** It is generally easier to solve an absolute value inequality if the absolute value appears on the left-hand side of the inequality.

Skill Practice Solve the inequality. Write the solution in interval notation.

2. $5 < 1 + \left| \dfrac{1}{3}c - 1 \right|$

By definition, the absolute value of a real number will always be nonnegative. Therefore, the absolute value of any expression will always be greater than a negative number. Similarly, an absolute value can never be less than a negative number. Let a represent a positive real number. Then

- The solution to the inequality $|x| > -a$ is all real numbers, $(-\infty, \infty)$.
- There is no solution to the inequality $|x| < -a$.

Example 3 Solving Absolute Value Inequalities

Solve the inequalities.

a. $|3d - 5| + 7 < 4 \qquad\qquad$ **b.** $|3d - 5| + 7 > 4$

Solution:

a. $|3d - 5| + 7 < 4 \qquad\qquad$ Isolate the absolute value. An absolute value expression cannot be less than a negative number. Therefore, there is no solution.

$|3d - 5| < -3$

No solution, $\{\ \}$

Answer

2. $(-\infty, -9) \cup (15, \infty)$

b. $|3d - 5| + 7 > 4$

$|3d - 5| > -3$

All real numbers, $(-\infty, \infty)$

Isolate the absolute value. The inequality is in the form $|x| > a$, where a is negative. An absolute value of any real number is greater than a negative number. Therefore, the solution is all real numbers.

Skill Practice Solve the inequalities.

3. $|4p + 2| + 6 < 2$ **4.** $|4p + 2| + 6 > 2$

Example 4   **Solving Absolute Value Inequalities**

Solve the inequalities.

a. $|4x + 2| \geq 0$ **b.** $|4x + 2| > 0$ **c.** $|4x + 2| \leq 0$

Solution:

a. $|4x + 2| \geq 0 \longleftarrow$ The absolute value is already isolated.

The absolute value of any real number is nonnegative. Therefore, the solution is all real numbers, $(-\infty, \infty)$.

b. $|4x + 2| > 0$

An absolute value will be greater than zero at all points *except where it is equal to zero*. That is, the point(s) for which $|4x + 2| = 0$ must be excluded from the solution set.

$|4x + 2| = 0$

$4x + 2 = 0$ or $4x + 2 = -0$ The second equation is the same as the first.

$4x = -2$

$x = -\dfrac{1}{2}$ Therefore, exclude $x = -\frac{1}{2}$ from the solution.

The solution is $\{x \mid x \neq -\frac{1}{2}\}$ or equivalently in interval notation, $(-\infty, -\frac{1}{2}) \cup (-\frac{1}{2}, \infty)$.

c. $|4x + 2| \leq 0$

An absolute value of a number cannot be less than zero. However, it can be *equal* to zero. Therefore, the only solutions to this inequality are the solutions to the related equation:

$|4x + 2| = 0$ From part (b), we see that the solution set is $\left\{-\dfrac{1}{2}\right\}$.

Answers

3. No solution, $\{\ \}$
4. All real numbers; $(-\infty, \infty)$
5. $(-\infty, \infty)$
6. $\left\{x \mid x \neq \dfrac{1}{3}\right\}$ or $\left(-\infty, \dfrac{1}{3}\right) \cup \left(\dfrac{1}{3}, \infty\right)$
7. $\left\{\dfrac{1}{3}\right\}$

Skill Practice Solve the inequalities.

5. $|3x - 1| \geq 0$ **6.** $|3x - 1| > 0$ **7.** $|3x - 1| \leq 0$

2. Solving Absolute Value Inequalities by the Test Point Method

For the problems in Examples 1 and 2, the absolute value inequality was converted to an equivalent compound inequality. However, sometimes students have difficulty setting up the appropriate compound inequality. To avoid this problem, you may want to use the test point method to solve absolute value inequalities.

Solving Inequalities by Using the Test Point Method

Step 1 Find the boundary points of the inequality. (Boundary points are the real solutions to the related equation and points where the inequality is undefined.)

Step 2 Plot the boundary points on the number line. This divides the number line into intervals.

Step 3 Select a test point from each interval and substitute it into the original inequality.
- If a test point makes the original inequality true, then that interval is part of the solution set.

Step 4 Test the boundary points in the original inequality.
- If the original inequality is strict ($<$ or $>$), do not include the boundary points in the solution set.
- If the original inequality is defined using $\leq$ or $\geq$, then include the boundary points that are well defined within the inequality.

Note: Any boundary point that makes an expression within the inequality undefined must *always* be excluded from the solution set.

To demonstrate the use of the test point method, we will repeat the absolute value inequalities from Examples 1 and 2. Notice that regardless of the method used, the absolute value is always isolated *first* before any further action is taken.

Example 5 — Solving an Absolute Value Inequality by the Test Point Method

Solve the inequality by using the test point method. $\quad |3w + 1| - 4 < 7$

Solution:

$|3w + 1| - 4 < 7$

$|3w + 1| < 11 \longleftarrow$ Isolate the absolute value.

$|3w + 1| = 11$ **Step 1:** Solve the related equation.

$3w + 1 = 11 \quad$ or $\quad 3w + 1 = -11$ Write as two equations.

$3w = 10 \quad$ or $\quad 3w = -12$

$w = \dfrac{10}{3} \quad$ or $\quad w = -4$ These are the only boundary points.

          Step 2: Plot the boundary points.

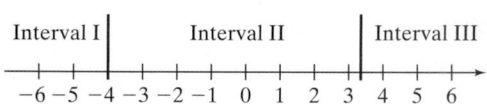

Step 3: Select a test point from each interval.

Test $w = -5$:

$|3(-5) + 1| - 4 \overset{?}{<} 7$

$|-14| - 4 \overset{?}{<} 7$

$14 - 4 \overset{?}{<} 7$

$10 \overset{?}{<} 7$ False

Test $w = 0$:

$|3(0) + 1| - 4 \overset{?}{<} 7$

$|1| - 4 \overset{?}{<} 7$

$1 - 4 \overset{?}{<} 7$

$-3 \overset{?}{<} 7$ True

Test $w = 4$:

$|3(4) + 1| - 4 \overset{?}{<} 7$

$|13| - 4 \overset{?}{<} 7$

$13 - 4 \overset{?}{<} 7$

$9 \overset{?}{<} 7$ False

False   True   False

-6 -5 -4 -3 -2 -1 0 1 2 3 4 5 6

$\frac{10}{3}$

Step 4: Because the original inequality is a strict inequality, the boundary points (where equality occurs) are not included.

The solution is $\{w \mid -4 < w < \frac{10}{3}\}$ or, equivalently in interval notation, $(-4, \frac{10}{3})$.

Skill Practice Solve the inequality.

8. $6 + |3t - 4| \le 10$

Example 6 **Solving an Absolute Value Inequality by the Test Point Method**

Solve the inequality by using the test point method. $3 \le 1 + \left|\frac{1}{2}t - 5\right|$

Solution:

$3 \le 1 + \left|\frac{1}{2}t - 5\right|$

$1 + \left|\frac{1}{2}t - 5\right| \ge 3$ Write the inequality with the absolute value on the left.

$\left|\frac{1}{2}t - 5\right| \ge 2$ ⟵————————— Isolate the absolute value.

$\left|\frac{1}{2}t - 5\right| = 2$ **Step 1:** Solve the related equation.

$\frac{1}{2}t - 5 = 2$ or $\frac{1}{2}t - 5 = -2$ Write as two equations.

$\frac{1}{2}t = 7$ or $\frac{1}{2}t = 3$

$t = 14$ or $t = 6$ These are the boundary points.

Interval I | Interval II | Interval III

6   14

Step 2: Plot the boundary points.

Step 3: Select a test point from each interval.

Answer

8. $\left[0, \frac{8}{3}\right]$

Test $t = 0$:

$3 \overset{?}{\le} 1 + \left|\dfrac{1}{2}(0) - 5\right|$

$3 \overset{?}{\le} 1 + |0 - 5|$

$3 \overset{?}{\le} 1 + |-5|$

$3 \overset{?}{\le} 6$ True

Test $t = 10$:

$3 \overset{?}{\le} 1 + \left|\dfrac{1}{2}(10) - 5\right|$

$3 \overset{?}{\le} 1 + |5 - 5|$

$3 \overset{?}{\le} 1 + |0|$

$3 \overset{?}{\le} 1$ False

Test $t = 16$:

$3 \overset{?}{\le} 1 + \left|\dfrac{1}{2}(16) - 5\right|$

$3 \overset{?}{\le} 1 + |8 - 5|$

$3 \overset{?}{\le} 1 + |3|$

$3 \overset{?}{\le} 4$ True

Step 4: The original inequality uses the sign $\ge$. Therefore, the boundary points (where equality occurs) must be part of the solution set.

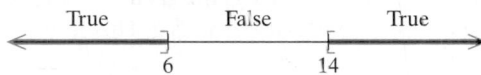

The solution is $\{t \mid t \le 6 \text{ or } t \ge 14\}$ or, equivalently in interval notation, $(-\infty, 6] \cup [14, \infty)$.

Skill Practice Solve the inequality.

9. $\left|\dfrac{1}{2}c + 4\right| + 1 > 6$

3. Translating to an Absolute Value Expression

Absolute value expressions can be used to describe distances. The distance between c and d is given by $|c - d|$. For example, the distance between -2 and 3 on the number line is $|(-2) - 3| = |-5| = 5$ as expected.

Example 7 Expressing Distances with Absolute Value

Write an absolute value inequality to represent the following phrases.

a. All real numbers x, whose distance from zero is greater than 5 units

b. All real numbers x, whose distance from -7 is less than 3 units

Solution:

a. All real numbers x, whose distance from zero is greater than 5 units

$|x - 0| > 5$ or simply $|x| > 5$

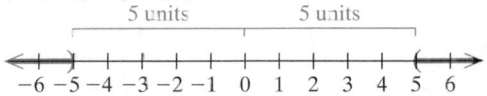

b. All real numbers x, whose distance from -7 is less than 3 units

$|x - (-7)| < 3$ or simply $|x + 7| < 3$

Skill Practice Write an absolute value inequality to represent the following phrases.

10. All real numbers whose distance from zero is greater than 10 units

11. All real numbers whose distance from 4 is less than 6 units

Answers

9. $(-\infty, -18) \cup (2, \infty)$

10. $|x| > 10$

11. $|x - 4| < 6$

Absolute value expressions can also be used to describe boundaries for measurement error.

Example 8   Expressing Measurement Error with Absolute Value

Latoya measured a certain compound on a scale in the chemistry lab at school. She measured 8 g of the compound, but the scale is only accurate to ± 0.1 g. Write an absolute value inequality to express an interval for the true mass, x, of the compound she measured.

TIP: The symbol $\pm$ means "plus or minus." The expression ± 0.1 is a short way of writing 0.1 or -0.1.

Solution:

Because the scale is only accurate to ± 0.1 g, the true mass, x, of the compound may deviate by as much as 0.1 g above or below 8 g. This may be expressed as an absolute value inequality:

$|x - 8.0| \le 0.1$ or equivalently $7.9 \le x \le 8.1$

Skill Practice

12. Vonzell molded a piece of metal in her machine shop. She measured the thickness at 12 mm. Her machine is accurate to ± 0.05 mm. Write an absolute value inequality to express an interval for the true measurement of the thickness, t, of the metal.

Answer

12. $|t - 12| \le 0.05$

Section 1.7   Practice Exercises

Study Skills Exercise

When you take a test, go through the test and do all the problems that you know first. Then go back and work on the problems that were more difficult. Give yourself a time limit for how much time you spend on each problem (maybe 3 to 5 min the first time through). Circle the importance of each statement.

	Not important	Somewhat important	Very important
a. Read through the entire test first.	1	2	3
b. If time allows, go back and check each problem.	1	2	3
c. Write out all steps instead of doing the work in your head.	1	2	3

Vocabulary and Key Concepts

1. a. If a is a positive real number, then the inequality $|x| < a$ is equivalent to _____ $< x <$ _____.

b. If a is a positive real number, then the inequality $|x| > a$ is equivalent to $x <$ _____ or x _____ a.

c. The solution set to the inequality $|x + 2| < -6$ is _____, whereas the solution set to the inequality $|x + 2| > -6$ is _____.

d. The solution set to the inequality $|x + 4| \le 0$ (includes/excludes) -4, whereas the solution set to the inequality $|x + 4| < 0$ (includes/excludes) -4.

Review Exercises

2. Solve the equation and inequalities. Write the solution sets to the inequalities in interval notation.

    a. $-4x - 3 = 5$ **b.** $-4x - 3 < 5$ **c.** $-4x - 3 > 5$

For Exercises 3–4, solve the equations.

3. $2 = |5 - 7x| + 1$ **4.** $|3x - 12| + 4 = 6 - 2$

For Exercises 5–8, solve the inequalities and graph the solution set. Write the solution in interval notation.

5. $-15 < 3w - 6 \le -9$ **6.** $5 - 2y \le 1$ and $3y + 2 \ge 14$

7. $m - 7 \le -5$ or $m - 7 \ge -10$ **8.** $3b - 2 < 7$ or $b - 2 > 4$

Concepts 1 and 2: Solving Absolute Value Inequalities

For Exercises 9–20, solve the equations and inequalities. For each inequality, graph the solution set and express the solution in interval notation. **(See Examples 1–6.)**

9. a. $|x| = 5$ **10. a.** $|a| = 4$ **11. a.** $|x - 3| = 7$

    b. $|x| > 5$ **b.** $|a| > 4$ **b.** $|x - 3| > 7$

    c. $|x| < 5$ **c.** $|a| < 4$ **c.** $|x - 3| < 7$

12. a. $|w + 2| = 6$ **13. a.** $|p| = -2$ **14. a.** $|x| = -14$

    b. $|w + 2| > 6$ **b.** $|p| > -2$ **b.** $|x| > -14$

    c. $|w + 2| < 6$ **c.** $|p| < -2$ **c.** $|x| < -14$

15. a. $|y + 1| = -6$ **16. a.** $|z - 4| = -3$ **17. a.** $|x| = 0$

    b. $|y + 1| > -6$ **b.** $|z - 4| > -3$ **b.** $|x| > 0$

    c. $|y + 1| < -6$ **c.** $|z - 4| < -3$ **c.** $|x| < 0$

18. a. $|p + 3| = 0$ **19. a.** $|k - 7| = 0$ **20. a.** $|2x + 4| + 3 = 2$

    b. $|p + 3| > 0$ **b.** $|k - 7| > 0$ **b.** $|2x + 4| + 3 > 2$

    c. $|p + 3| < 0$ **c.** $|k - 7| < 0$ **c.** $|2x + 4| + 3 < 2$

For Exercises 21–50, solve the absolute value inequalities. Graph the solution set and write the solution in interval notation. **(See Examples 1–6.)**

21. $|x| > 6$ **22.** $|x| \le 6$ **23.** $|t| \le 3$

24. $|p| > 3$ **25.** $|y + 2| \ge 0$ **26.** $0 \le |7n + 2|$

27. $5 \le |2x - 1|$

28. $7 \le |x - 2|$

29. $|k - 7| < -3$

30. $|h + 2| < -9$

31. $\left|\dfrac{w - 2}{3}\right| - 3 \le 1$

32. $\left|\dfrac{x + 3}{2}\right| - 2 \ge 4$

33. $12 \le |9 - 4y| - 2$

34. $5 > |2m - 7| + 4$

35. $4 > -1 + \left|\dfrac{2x + 1}{4}\right|$

36. $9 \ge 2 + \left|\dfrac{x - 4}{5}\right|$

37. $8 < |4 - 3x| + 12$

38. $-16 < |5x - 1| - 1$

39. $5 - |2m + 1| > 5$

40. $3 - |5x + 3| > 3$

41. $|p + 5| \le 0$

42. $|y + 1| - 4 \le -4$

43. $|z - 6| + 5 > 5$

44. $|2c - 1| - 4 > -4$

45. $5|2y - 6| + 3 \ge 13$

46. $7|y + 1| - 3 \ge 11$

47. $-3|6 - t| + 1 > -5$

48. $-4|8 - x| + 2 > -14$

49. $|0.02x + 0.06| - 0.1 < 0.05$

50. $|0.05x - 0.04| - 0.01 < 0.11$

Concept 3: Translating to an Absolute Value Expression

For Exercises 51–54, write an absolute value inequality equivalent to the expression given. **(See Example 7.)**

51. All real numbers whose distance from 0 is greater than 7

52. All real numbers whose distance from −3 is less than 4

53. All real numbers whose distance from 2 is at most 13

54. All real numbers whose distance from 0 is at least 6

55. A 32-oz jug of orange juice may not contain exactly 32 oz of juice. The possibility of measurement error exists when the jug is filled in the factory. If the maximum measurement error is ±0.05 oz, write an absolute value inequality representing the range of volumes, x, in which the orange juice jug may be filled. **(See Example 8.)**

56. The length of a board is measured to be 32.3 in. The maximum measurement error is ±0.2 in. Write an absolute value inequality that represents the range for the length of the board, x.

57. A bag of potato chips states that its weight is $6\frac{3}{4}$ oz. The maximum measurement error is ±$\frac{1}{8}$ oz. Write an absolute value inequality that represents the range for the weight, x, of the bag of chips.

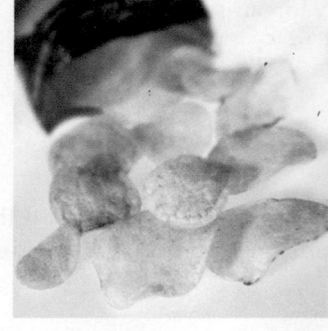

58. A $\frac{7}{8}$-in. bolt varies in length by at most ±$\frac{1}{16}$ in. Write an absolute value inequality that represents the range for the length, x, of the bolt.

59. The width, w, of a bolt is supposed to be 2 cm but may have a 0.01-cm margin of error. Solve $|w - 2| \le 0.01$, and interpret the solution to the inequality in the context of this problem.

60. In a political poll, the front-runner was projected to receive 53% of the votes with a margin of error of 3%. Solve $|p - 0.53| \leq 0.03$ and interpret the solution in the context of this problem.

Expanding Your Skills

For Exercises 61–64, match the graph with the inequality.

61.

62.

63.

64.

a. $|x - 2| < 4$ **b.** $|x - 1| > 4$ **c.** $|x - 3| < 2$ **d.** $|x - 5| > 1$

Problem Recognition Exercises

Identifying Equations and Inequalities

For Exercises 1–8, solve each equation or inequality. Express the solution in interval notation where appropriate.

1. a. $3x - 9 = 18$

   b. $|3x - 9| = 18$

   c. $|3x - 9| < 18$

   d. $|3x - 9| \geq 18$

2. a. $5y + 2 = -20$

   b. $|5y + 2| = -20$

   c. $|5y + 2| \leq -20$

   d. $|5y + 2| > -20$

3. a. $-2t - 14 = 0$

   b. $-2t - 14 > 0$

   c. $-2t - 14 \leq 0$

4. a. $\dfrac{x - 2}{3} = 9$

   b. $\dfrac{x - 2}{3} \geq 9$

   c. $\dfrac{x - 2}{3} < 9$

5. a. $|8t - 2| = |-2t + 3|$

  b. $8t - 2 = -2t + 3$

6. a. $-5 < x + 2$ and $x + 2 \leq 8$

   b. $-5 < x + 2 \leq 8$

7. a. $-4x - 9 < 11$ or $2 \leq x + 1$

   b. $-4x - 9 < 11$ and $2 \leq x + 1$

8. a. $4 < 2y$ or $-3(y + 2) > -2y + 1$

   b. $4 < 2y$ and $-3(y + 2) > -2y + 1$

For Exercises 9–28,

  a. Identify the type of equation or inequality. Choose from:
- linear equation
- absolute value equation
- linear inequality
- compound inequality
- absolute value inequality

  b. Solve the equation or inequality. Express the solution set in interval notation where appropriate.

9. $-0.5y + 0.7 = 3.7$

10. $3m - 9 = 18$

11. $|2t + 8| \leq 4$

12. $|1 - 3x| < -1$

13. $-11 < 2t + 1 < 19$

14. $2z - 3 \geq 11$ or $3z + 3 < 9$

15. $\left|\dfrac{1}{2}y + 3\right| = 5$

16. $|4x + 3| = |9 - 2x|$

17. $-\dfrac{3}{4}p \geq -9$

18. $8w + 4 \geq 5w + 1$

19. $\left|\dfrac{2x - 9}{3}\right| \geq 5$

20. $\left|\dfrac{10 - x}{5}\right| < 3$

21. $|2 - c| + 5 = 3$

22. $|10n + 2| + 7 = 7$

23. $\dfrac{w - 4}{5} - \dfrac{w + 1}{3} = 1$

24. $\dfrac{1}{3}y - \dfrac{5}{6} = \dfrac{1}{2}y + 1$

25. $2x - 7 > 9$ and $3x \leq 36$

26. $-3 + x > 2x$ and $2 \geq -\dfrac{1}{3}x$

27. $5(x - 2) + 7 = 2x + 3(x - 1)$

28. $7y - 4 = 3(y + 1) + 4y$

Group Activity

Understanding the Symbolism of Mathematics

Estimated time: 15 minutes

Group Size: 3

As you advance in your study of mathematics, you will notice that mathematics has its own language and syntax. For example, the statement $a > 0$ translates to "a is greater than zero." This also means that a is positive. Read and interpret the following conditions imposed on the variables a, b, c, d, and x. Then determine whether the statements in Exercises 1–12 are true or false. Work three exercises at a time individually. Then stop and check your answers with the other members of the group. Discuss any discrepancies and agree on a final answer.

$$a > 0 \qquad b < 0 \qquad -1 < c < 1 \qquad d > 1 \qquad x = 0$$

1. $ab > 0$ _____

2. $bd < 0$ _____

3. $a + d > 0$ _____

4. $b^2 < 0$ _____

5. $a^2 > 0$ _____

6. $c + d \geq 0$ _____

7. $c^2 < 1$ _____

8. $d^2 > 1$ _____

9. $|c| < 1$ _____

10. $bx < 0$ _____

11. $b + x > 0$ _____

12. $\dfrac{x}{d} = 0$ _____

For Exercises 13–24, suppose that m represents an *odd* integer and n represents an *even* integer. Determine whether the statements are true or false.

13. $m + 1$ is an odd integer. _____

14. $m + 2$ is an odd integer. _____

15. $m - 1$ is an odd integer. _____

16. $m - 2$ is an odd integer. _____

17. $m + n$ is an even integer. _____

18. $m - n$ is an even integer. _____

19. m^2 is an even integer. _____

20. n^2 is an even integer. _____

21. $(m + n)^2$ is an odd integer. _____

22. $(n + n)^2$ is an even integer. _____

23. $(m + n + 1)^3$ is an odd integer. _____

24. $(m + m - 2)^3$ is an odd integer. _____

Chapter 1   Summary

Section 1.1   Linear Equations in One Variable

Key Concepts

A **linear equation in one variable** can be written in the form $ax + b = c \ (a \neq 0)$.

Steps to Solve a Linear Equation in One Variable

1. Simplify both sides of the equation.
   - Clear parentheses.
   - Consider clearing fractions or decimals (if any are present) by multiplying both sides of the equation by a common denominator of all terms.
   - Combine *like* terms.
2. Use the addition or subtraction property of equality to collect the variable terms on one side of the equation.
3. Use the addition or subtraction property of equality to collect the constant terms on the other side.
4. Use the multiplication or division property of equality to make the coefficient on the variable term equal to 1.
5. Check your answer and write the solution set.

An equation that has no solution is called a **contradiction**.

An equation that has all real numbers as its solutions is called an **identity**.

Examples

Example 1

$$\frac{1}{2}(x - 4) - \frac{3}{4}(x + 2) = \frac{1}{4}$$

$$\frac{1}{2}x - 2 - \frac{3}{4}x - \frac{3}{2} = \frac{1}{4}$$

$$4\left(\frac{1}{2}x - 2 - \frac{3}{4}x - \frac{3}{2}\right) = 4\left(\frac{1}{4}\right)$$

$$2x - 8 - 3x - 6 = 1$$

$$-x - 14 = 1$$

$$-x = 15$$

$$x = -15$$

The solution -15 checks in the original equation.

The solution set is $\{-15\}$.

Example 2

$$3x + 6 = 3(x - 5)$$

$$3x + 6 = 3x - 15$$

$$6 = -15 \qquad \text{Contradiction}$$

There is no solution, $\{ \ \}$.

Example 3

$$-(5x + 12) - 3 = 5(-x - 3)$$

$$-5x - 12 - 3 = -5x - 15$$

$$-5x - 15 = -5x - 15$$

$$-15 = -15 \qquad \text{Identity}$$

All real numbers are solutions.

The solution set is $\{x \mid x \text{ is a real number}\}$.

Section 1.2 Applications of Linear Equations in One Variable

Key Concepts

Problem-Solving Steps for Word Problems

1. Read the problem carefully.
2. Assign labels to unknown quantities.
3. Write a verbal model.
4. Write a mathematical equation.
5. Solve the equation.
6. Interpret the results and write the final answer in words.

Sales tax: (cost of merchandise)(tax rate)

Commission: (dollars in sales)(commission rate)

Simple interest: $I = Prt$

Distance = (rate)(time) $d = rt$

Examples

Example 1

1. Estella needs to borrow $8500. She borrows part of the money from a friend and agrees to pay the friend 6% simple interest. She borrows the rest of the money from a bank that charges 10% simple interest. If she pays back the money at the end of 1 yr and also pays $750 in interest, find the amount that Estella borrowed from each source.

2. Let x represent the amount borrowed at 6%. Then $8500 - x$ is the amount borrowed at 10%.

	6% Account	10% Account	Total
Principal	x	$8500 - x$	8500
Interest	$0.06x$	$0.10(8500 - x)$	750

3. $\left(\begin{array}{c} \text{Interest} \\ \text{owed at } 6\% \end{array} \right) + \left(\begin{array}{c} \text{interest} \\ \text{owed at } 10\% \end{array} \right) = \left(\begin{array}{c} \text{total} \\ \text{interest} \end{array} \right)$

4. $0.06x + 0.10(8500 - x) = 750$

5. $6x + 10(8500 - x) = 75{,}000$

   $6x + 85{,}000 - 10x = 75{,}000$

   $-4x = -10{,}000$

   $x = 2500$

6. $x = 2500$

   $8500 - 2500 = 6000$

$2500 was borrowed at 6% and $6000 was borrowed at 10%.

Section 1.3 Applications to Geometry and Literal Equations

Key Concepts

Some useful formulas for word problems:

Perimeter

Rectangle: $P = 2l + 2w$

Area

Rectangle: $A = lw$

Square: $A = s^2$

Triangle: $A = \dfrac{1}{2}bh$

Trapezoid: $A = \dfrac{1}{2}(b_1 + b_2)h$

Angles

Two angles whose measures total 90° are complementary angles.

Two angles whose measures total 180° are supplementary angles.

The sum of the measures of the angles of a triangle is 180°.

 $x + y + z = 180°$

Literal equations (or formulas) are equations with several variables. To solve for a specific variable, follow the steps to solve a linear equation.

Examples

Example 1

A border of marigolds is to enclose a rectangular flower garden. If the length is twice the width and the perimeter is 25.5 ft, what are the dimensions of the garden?

$P = 2l + 2w$

$25.5 = 2(2x) + 2(x)$

$25.5 = 4x + 2x$

$25.5 = 6x$

$4.25 = x$

The width is 4.25 ft, and the length is 2(4.25) ft or 8.5 ft.

Example 2

Solve for y.

$4x - 5y = 20$

$-5y = -4x + 20$

$\dfrac{-5y}{-5} = \dfrac{-4x + 20}{-5}$

$y = \dfrac{-4x + 20}{-5}$ or $y = \dfrac{4}{5}x - 4$

Section 1.4   **Linear Inequalities in One Variable**

Key Concepts

A **linear inequality** is an inequality that can be written in one of the following forms, provided that $a \neq 0$.

$$ax + b < c, \qquad ax + b > c, \qquad ax + b \leq c, \quad \text{or}$$
$$ax + b \geq c$$

Properties of Inequalities

1. If $a < b$, then $a + c < b + c$.
2. If $a < b$, then $a - c < b - c$.
3. If c is positive and $a < b$, then $ac < bc$ and
   $\dfrac{a}{c} < \dfrac{b}{c}$.
4. If c is negative and $a < b$, then $ac > bc$ and
   $\dfrac{a}{c} > \dfrac{b}{c}$.

Properties 3 and 4 indicate that if we multiply or divide an inequality by a negative value, the direction of the inequality sign must be reversed.

Examples

Example 1

Solve.

$$\frac{14 - x}{-2} < -3x$$

$$-2\left(\frac{14 - x}{-2}\right) > -2(-3x) \qquad \text{(Reverse the inequality sign.)}$$

$$14 - x > 6x$$

$$-7x > -14$$

$$\frac{-7x}{-7} < \frac{-14}{-7} \qquad \text{(Reverse the inequality sign.)}$$

$$x < 2$$

Set-builder notation: $\{x \mid x < 2\}$
Interval notation: $(-\infty, 2)$

Section 1.5   **Compound Inequalities**

Key Concepts

$A \cup B$ is the **union** of A and B. This is the set of elements that belong to set A or set B or both sets A and B.

$A \cap B$ is the **intersection** of A and B. This is the set of elements common to both A and B.

Examples

Example 1

Union   **Intersection**

$A \cup B$

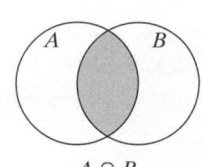

$A \cap B$

- Solve two or more inequalities joined by *and* by finding the intersection of their solution sets.

- Solve two or more inequalities joined by *or* by finding the union of their solution sets.

Example 2

$$-7x + 3 \geq -11 \quad \text{and} \quad 1 - x < 4.5$$

$$-7x \geq -14 \quad \text{and} \quad -x < 3.5$$

$$x \leq 2 \quad \text{and} \quad x > -3.5$$

$x \leq 2$

$x > -3.5$

The solution is $\{x \mid -3.5 < x \leq 2\}$ or equivalently $(-3.5, 2]$.

Inequalities of the form $a < x < b$:

The inequality $a < x < b$ is represented by

or, in interval notation, (a, b).

Example 3

$$5y + 1 \geq 6 \quad \text{or} \quad 2y - 5 \leq -11$$

$$5y \geq 5 \quad \text{or} \quad 2y \leq -6$$

$$y \geq 1 \quad \text{or} \quad y \leq -3$$

$y \geq 1$

$y \leq -3$

The solution is $\{y \mid y \geq 1 \text{ or } y \leq -3\}$ or equivalently $(-\infty, -3] \cup [1, \infty)$.

Example 4

Solve.

$$-13 \leq 3x - 1 < 5$$

$$-13 + 1 \leq 3x - 1 + 1 < 5 + 1$$

$$-12 \leq 3x < 6$$

$$\frac{-12}{3} \leq \frac{3x}{3} < \frac{6}{3}$$

$$-4 \leq x < 2$$

Interval notation: $[-4, 2)$

Section 1.6   Absolute Value Equations

Key Concepts

The equation $|x| = a$ is an **absolute value equation**. For $a \geq 0$, the solution to the equation $|x| = a$ is $x = a$ or $x = -a$.

Steps to Solve an Absolute Value Equation

1. Isolate the absolute value to write the equation in the form $|x| = a$.
2. If $a < 0$, there is no solution.
3. Otherwise, if $a \geq 0$, rewrite the equation $|x| = a$ as $x = a$ or $x = -a$.
4. Solve the equations from step 3.
5. Check answers in the original equation.

The equation $|x| = |y|$ implies $x = y$ or $x = -y$.

Examples

Example 1

$$|2x - 3| + 5 = 10$$

$$|2x - 3| = 5 \qquad \text{Isolate the absolute value.}$$

$$2x - 3 = 5 \quad \text{or} \quad 2x - 3 = -5$$

$$2x = 8 \quad \text{or} \qquad 2x = -2$$

$$x = 4 \quad \text{or} \qquad x = -1$$

The solution set is $\{4, -1\}$.

Example 2

$$|x + 2| + 5 = 1$$

$$|x + 2| = -4 \qquad \text{No solution, } \{\ \}$$

Example 3

$$|2x - 1| = |x + 4|$$

$$2x - 1 = x + 4 \quad \text{or} \quad 2x - 1 = -(x + 4)$$

$$x = 5 \qquad \text{or} \quad 2x - 1 = -x - 4$$

$$\text{or} \qquad 3x = -3$$

$$\text{or} \qquad x = -1$$

The solution set is $\{5, -1\}$.

Section 1.7   Absolute Value Inequalities

Key Concepts

Solutions to Absolute Value Inequalities

For $a > 0$, we have:

$|x| > a \Rightarrow x < -a \quad \text{or} \quad x > a$

$|x| < a \Rightarrow -a < x < a$

Examples

Example 1

$$|5x - 2| < 12$$

$$-12 < 5x - 2 < 12$$

$$-10 < 5x < 14$$

$$-2 < x < \frac{14}{5}$$

The solution is $\left(-2, \dfrac{14}{5}\right)$.

Test Point Method to Solve Inequalities

1. Find the boundary points of the inequality.
   (Boundary points are the real solutions to the
   related equation and points where the inequality
   is undefined.)
2. Plot the boundary points on the number line.
   This divides the number line into intervals.
3. Select a test point from each interval and
   substitute it into the original inequality.
   - If a test point makes the original inequality true,
     then that interval is part of the solution set.
4. Test the boundary points in the original
   inequality.
   - If the original inequality is strict ($<$ or $>$), do
     not include the boundary in the solution set.
   - If the original inequality is defined using $\leq$
     or $\geq$, then include the boundary points that are
     well defined within the inequality.

Note: Any boundary point that makes an expression
within the inequality undefined must *always* be excluded
from the solution set.

If *a* is *negative* ($a < 0$), then

1. $|x| < a$ has no solution.
2. $|x| > a$ is true for all real numbers.

Example 2

$|x - 3| \div 2 \geq 7$

$\quad |x - 3| \geq 5 \qquad$ Isolate the absolute value.

$\quad |x - 3| = 5 \qquad$ Solve the related equation.

$x - 3 = 5 \quad$ or $\quad x - 3 = -5$

$\quad x = 8 \quad$ or $\qquad x = -2 \qquad$ Boundary points

Interval I:

Test $x = -3$: $\qquad |(-3) - 3| + 2 \overset{?}{\geq} 7 \qquad$ True

Interval II:

Test $x = 0$: $\qquad |(0) - 3| + 2 \overset{?}{\geq} 7 \qquad$ False

Interval III:

Test $x = 9$: $\qquad |(9) - 3| + 2 \overset{?}{\geq} 7 \qquad$ True

The solution is $(-\infty, -2] \cup [8, \infty)$.

Example 3

$|x + 5| > -2$

The solution is all real numbers because an absolute
value will always be greater than a negative number.

$(-\infty, \infty)$

Example 4

$|x + 5| < -2$

There is no solution because an absolute value cannot
be less than a negative number.

The solution set is $\{ \ \}$.

Chapter 1 Review Exercises

Section 1.1

1. Describe the solution set for a contradiction.

2. Describe the solution set for an identity.

For Exercises 3–12, solve the equations and identify each as a conditional equation, a contradiction, or an identity.

3. $x - 27 = -32$ **4.** $y + \dfrac{7}{8} = 1$

5. $7.23 + 0.6x = 0.2x$ **6.** $0.1y + 1.122 = 5.2y$

7. $-(4 + 3m) = 9(3 - m)$

8. $-2(5n - 6) = 3(-n - 3)$

9. $\dfrac{x - 3}{5} - \dfrac{2x + 1}{2} = 1$

10. $3(x + 3) - 2 = 3x + 2$

11. $\dfrac{10}{8}m + 18 - \dfrac{7}{8}m = \dfrac{3}{8}m + 25$

12. $\dfrac{2}{3}m + \dfrac{1}{3}(m - 1) = -\dfrac{1}{3}m + \dfrac{1}{3}(4m - 1)$

Section 1.2

13. Explain how you would label three consecutive integers.

14. Explain how you would label two consecutive odd integers.

15. Explain what the formula $d = rt$ means.

16. Explain what the formula $I = Prt$ means.

17. a. Cory made $85,200 in taxable income. If he pays an average of 28% in taxes on his income, determine the amount of tax he must pay.

   b. What is his net income (after taxes)?

18. For a recent year, approximately 7.2 million men were in college in the United States. This represents an 8% increase over the number of men in college in 2000. Approximately how many men were in college in 2000? (Round to the nearest tenth of a million.)

19. For a recent year, there were 17,430 deaths due to alcohol-related accidents in the United States. This was a 5% increase over the number of alcohol-related deaths in 1999. How many such deaths were there in 1999?

20. Of three consecutive even integers, the sum of the smallest two integers is equal to 6 less than the largest. Find the integers.

21. To do a rope trick, a magician needs to cut a piece of rope so that one piece is one-third the length of the other piece. If she begins with a $2\frac{2}{3}$-ft rope, what will be the lengths of the two pieces of rope?

22. Sharyn invests $2000 more in an account that earns 9% simple interest than she invests in an account that earns 6% simple interest. How much did she invest in each account if her total interest is $405 after 1 yr?

23. How much 10% acid solution should be mixed with 1 L of 25% acid solution to produce a solution that is 15% acid?

24. Two friends plan to meet at a restaurant for lunch. They both leave their homes at 11:30 A.M. and between the two of them, they drive a total of 37.5 mi. Lynn drives in from a neighboring town and averages 15 mph faster than her friend Linda. If they meet at noon, find the average driving speed for each.

Section 1.3

25. The length of a rectangle is 2 ft more than the width. Find the dimensions if the perimeter is 40 ft.

For Exercises 26–27, solve for x, and then find the measure of each angle.

26.

$\left(\dfrac{x}{2} + 1\right)^{\circ}$

$(x - 25)^{\circ}$

27.

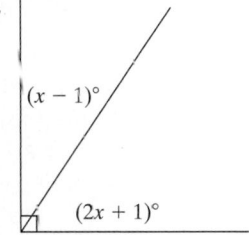

$(x-1)^\circ$

$(2x+1)^\circ$

For Exercises 28–31, solve for the indicated variable.

28. $3x - 2y = 4$ for y

29. $-6x + y = 12$ for y

30. $S = 2\pi r + \pi r^2 h$ for h

31. $A = \dfrac{1}{2}bh$ for b

32. a. The circumference of a circle is given by $C = 2\pi r$. Solve this equation for π.

b. Tom measures the radius of a circle to be 6 cm and the circumference to be 37.7 cm. Use these values to approximate π. (Round to 2 decimal places.)

Section 1.4

For Exercises 33–38, solve the inequality and graph the solution set. Write the solution set in (a) set-builder notation and (b) interval notation.

33. $-6x - 2 > 6$ ⟶

34. $-10x \le 15$ ⟶

35. $5 - 7(x + 3) > 19x$ ⟶

36. $4 - 3x \ge 10(-x + 5)$ ⟶

37. $\dfrac{5 - 4x}{8} \ge 9$ ⟶

38. $\dfrac{3 + 2x}{4} \le 8$ ⟶

39. Dave earned the following test scores in his biology class: 82, 88, 92, and 93. How high does he have to score on the fifth test to have an average of 90 or more?

Section 1.5

40. Explain the difference between the union and intersection of two sets. You may use the sets C and D in the following diagram to provide an example.

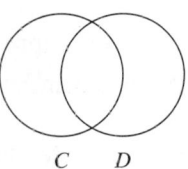

C D

Let $X = \{x \mid x \ge -10\}$, $Y = \{x \mid x < 1\}$, $Z = \{x \mid x > -1\}$, and $W = \{x \mid x \le -3\}$. For Exercises 41–46, find the intersection or union of the sets X, Y, Z, and W. Write the answers in interval notation.

41. $X \cap Y$ **42.** $X \cup Y$

43. $Y \cup Z$ **44.** $Y \cap Z$

45. $Z \cup W$ **46.** $Z \cap W$

For Exercises 47–56, solve the compound inequalities. Write the solutions in interval notation.

47. $4m > -11$ and $4m - 3 \le 13$

48. $4n - 7 < 1$ and $7 + 3n \ge -8$

49. $-3y + 1 \ge 10$ and $-2y - 5 \le -15$

50. $\dfrac{1}{2} - \dfrac{h}{12} \le -\dfrac{7}{12}$ and $\dfrac{1}{2} - \dfrac{h}{10} > -\dfrac{1}{5}$

51. $\dfrac{2}{3}t - 3 \le 1$ or $\dfrac{3}{4}t - 2 > 7$

52. $2(3x + 1) < -10$ or $3(2x - 4) \ge 0$

53. $-7 < -7(2w + 3)$ or $-2 < -4(3w - 1)$

54. $5(p + 3) + 4 > p - 1$ or $4(p - 1) + 2 > p + 8$

55. $2 \ge -(b - 2) - 5b \ge -6$

56. $-4 \le \dfrac{1}{2}(x - 1) < -\dfrac{3}{2}$

57. The product of $\frac{1}{3}$ and the sum of a number and 3 is between -1 and 5. Find all such numbers.

58. Normal levels of total cholesterol vary according to age. For adults between 25 and 40 yr old, the normal range is generally accepted to be between 140 and 225 mg/dL (milligrams per deciliter), inclusive.

a. Write an inequality representing the normal range for total cholesterol for adults between 25 and 40 yr old.

b. Write a compound inequality representing abnormal ranges for total cholesterol for adults between 25 and 40 yr old.

59. Normal levels of total cholesterol vary according to age. For adults younger than 25 yr old, the normal range is generally accepted to be between 125 and 200 mg/dL, inclusive.

a. Write an inequality representing the normal range for total cholesterol for adults younger than 25 yr old.

b. Write a compound inequality representing abnormal ranges for total cholesterol for adults younger than 25 yr old.

60. One method to approximate your maximum heart rate is to subtract your age from 220. To maintain an aerobic workout, it is recommended that you sustain a heart rate of between 60% and 75% of your maximum heart rate.

a. If the maximum heart rate h is given by the formula $h = 220 - A$, where A is a person's age, find your own maximum heart rate. (Answers will vary.)

b. Find the interval for your own heart rate that will sustain an aerobic workout. (Answers will vary.)

Section 1.6

For Exercises 61–74, solve the absolute value equations.

61. $|x| = 10$

62. $|x| = 17$

63. $|8.7 - 2x| = 6.1$

64. $|5.25 - 5x| = 7.45$

65. $16 = |x + 2| + 9$

66. $5 = |x - 2| + 4$

67. $|4x - 1| + 6 = 4$

68. $|3x - 1| + 7 = 3$

69. $\left|\dfrac{7x - 3}{5}\right| + 4 = 4$

70. $\left|\dfrac{4x + 5}{-2}\right| - 3 = -3$

71. $|3x - 5| = |2x + 1|$

72. $|8x + 9| = |8x - 1|$

73. $|2 + 7d| = |-7d - 2|$

74. $-|4y + 6| = |2y - 3|$

75. Which absolute value expression represents the distance between 3 and -2 on the number line?

$$|3 - (-2)| \qquad |-2 - 3|$$

Section 1.7

76. Write the compound inequality $x < -5$ or $x > 5$ as an absolute value inequality.

77. Write the compound inequality $-4 < x < 4$ as an absolute value inequality.

For Exercises 78–79, write an absolute value inequality that represents the solution sets graphed here.

78.

79.

For Exercises 80–93, solve the absolute value inequalities. Graph the solution set and write the solution in interval notation.

80. $|x + 6| \geq 8$

81. $|x + 8| \leq 3$

82. $2|7x - 1| + 4 > 4$

83. $4|5x + 1| - 3 > -3$

84. $|3x + 4| - 6 \leq -4$

85. $|5x - 3| + 3 \leq 6$

86. $\left|\dfrac{x}{2} - 6\right| < 5$

87. $\left|\dfrac{x}{3} + 2\right| < 2$

88. $|4 - 2x| + 8 \geq 8$

89. $|9 + 3x| + 1 \geq 1$ ⎯⎯⎯⎯⎯⎯⎯⎯→

90. $-2|5.2x - 7.8| < 13$ ⎯⎯⎯⎯⎯⎯→

91. $-|2.5x + 15| < 7$ ⎯⎯⎯⎯⎯⎯→

92. $|3x - 8| < -1$ ⎯⎯⎯⎯⎯→   **93.** $|x + 5| < -4$ ⎯⎯⎯⎯⎯→

94. State one possible situation in which an absolute value inequality will have no solution.

95. State one possible situation in which an absolute value inequality will have a solution of all real numbers.

96. The Neilsen ratings estimated that the percent, p, of the television viewing audience watching *American Idol* was 20% with a 3% margin of error. Solve the inequality $|p - 0.20| \leq 0.03$ and interpret the answer in the context of this problem.

97. The length, L, of a screw is supposed to be $3\frac{3}{8}$ in. Due to variation in the production equipment, there is a $\frac{1}{4}$-in. margin of error. Solve the inequality $|L - 3\frac{3}{8}| \leq \frac{1}{4}$ and interpret the answer in the context of this problem.

Chapter 1   Test

For Exercises 1–9, solve the equations.

1. $\dfrac{x}{7} + 1 = 20$

2. $8 - 5(4 - 3z) = 2(4 - z) - 8z$

3. $0.12(x) + 0.08(60,000 - x) = 10,500$

4. $\dfrac{5 - x}{6} - \dfrac{2x - 3}{2} = \dfrac{x}{3}$

5. $\left|\dfrac{1}{2}x + 3\right| - 4 = 4$

6. $|3x + 4| = |x - 12|$

7. $-5 = -8 + |2y - 3|$

8. $|3.7x - 5| + 7 = 6.2$

9. $|8x + 11| = |8x + 5|$

10. Label each equation as a conditional equation, an identity, or a contradiction.

   a. $(5x - 9) + 19 = 5(x + 2)$

   b. $2a - 2(1 + a) = 5$

   c. $(4w - 3) + 4 = 3(5 - w)$

11. The difference between two numbers is 72. If the larger is 5 times the smaller, find the two numbers.

12. Joëlle is determined to get some exercise and walks to the store at a brisk rate of 4.5 mph. She meets her friend Yun Ling at the store, and together they walk back at a slower rate of 3 mph. Joëlle's total walking time was 1 hr.

   a. How long did it take her to walk to the store?

   b. What is the distance to the store?

13. Shawnna banks at a credit union. Her money is distributed between two accounts: a certificate of deposit (CD) that earns 5% simple interest and a savings account that earns 3.5% simple interest. Shawnna has $100 less in her savings account than in the CD. If after 1 year her total interest is $81.50, how much did she invest in the CD?

14. A yield sign is in the shape of an equilateral triangle (all sides have equal length). Its perimeter is 81 in. Find the length of the sides.

15. The sum of three consecutive odd integers is 41 less than four times the largest. Find the numbers.

16. How many gallons of a 20% acid solution must be mixed with 6 gal of a 30% acid solution to make a 22% solution?

For Exercises 17–18, solve the equations for the indicated variable.

17. $4x + 2y = 6$ for y **18.** $x = \mu + z\sigma$ for z

For Exercises 19–21, solve the inequalities. Graph the solution and write the solution set in interval notation.

19. $x + 8 > 42$

20. $-\dfrac{3}{2}x + 6 \geq x - 3$

21. $-2 < 3x - 1 \leq 5$

For Exercises 22–32, solve the compound and absolute value inequalities. Write the answers in interval notation.

22. $-2 \leq 3x - 1 \leq 5$

23. $-4 \leq \dfrac{6 - 2x}{5} < 2$

24. $-\dfrac{3}{5}x - 1 \leq 8$ or $-\dfrac{2}{3}x \geq 16$

25. $-2x - 3 > -3$ and $x + 3 \geq 0$

26. $5x + 1 \leq 6$ or $2x + 4 > -6$

27. $2x - 3 > 1$ and $x + 4 < -1$

28. $|3 - 2x| + 6 < 2$

29. $|3x - 8| \geq 9$

30. $|0.4x + 0.3| - 0.2 < 7$

31. $|7 - 3x| + 1 > -3$

32. $6 \geq |2x - 5| - 5$

33. An elevator can accommodate a maximum weight of 2000 lb. If four passengers on the elevator have an average weight of 180 lb each, how many additional passengers of the same average weight can the elevator carry before the maximum weight capacity is exceeded?

34. The normal range in humans of the enzyme adenosine deaminase (ADA), is between 9 and 33 IU (international units), inclusive. Let x represent the ADA level in international units.

    a. Write an inequality representing the normal range for ADA.

    b. Write a compound inequality representing abnormal ranges for ADA.

35. The mass of a small piece of metal is measured to be 15.41 g. If the measurement error is at most ± 0.01 g, write an absolute value inequality that represents the possible mass, x, of the piece of metal.

Linear Equations in Two Variables and Functions

2

CHAPTER OUTLINE

2.1 Linear Equations in Two Variables   128

2.2 Slope of a Line and Rate of Change   145

2.3 Equations of a Line   157

   Problem Recognition Exercises: Characteristics of Linear Equations   171

2.4 Applications of Linear Equations and Modeling   171

2.5 Introduction to Relations   183

2.6 Introduction to Functions   192

2.7 Graphs of Functions   204

   Problem Recognition Exercises: Characteristics of Relations   217

   Group Activity: Deciphering a Coded Message   218

Chapter 2

In this chapter, we cover graphing and the applications of graphing. Graphs appear in magazines and newspapers and in other aspects of day-to-day life. In many fields of study such as the sciences and business, graphs are used to display data (information).

Are You Prepared?

Try answering questions 1–5 below. These problems will refresh your skills at reading and interpreting graphs. Match each answer with the correct letter and record the letter in the space below.

1. Use the graph to approximate the average height of a 5-yr-old girl.
2. Use the graph to estimate a girl's age if her height is 39 in. tall.
3. The equation $h = 2.5a + 31$ can be used to approximate a girl's height, h, in inches according to her age, a, in years. Use the formula $h = 2.5a + 31$ to estimate a girl's height at 6 yr old.
4. Use the formula $h = 2.5a + 31$ to estimate the age of a girl if her height is 50 in.
5. Use the formula $h = 2.5a + 31$ to predict the height of an 11-yr-old girl.

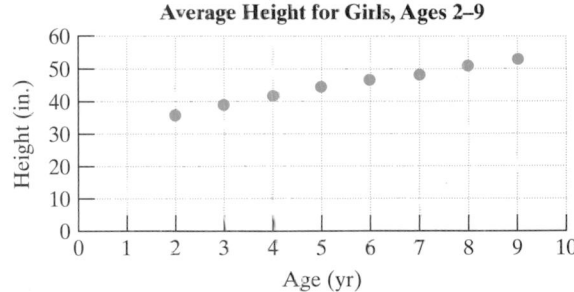

Average Height for Girls, Ages 2–9

A graph intersects the x-axis at an

___ - ___ N ___ ___ R ___ ___ P ___ .
  1   2   4   3   5   3   4

T. 7.6 yr   E. 46 in.   I. 3 yr
X. 44 in.   C. 58.5 in.

Concepts

1. **The Rectangular Coordinate System**
2. **Linear Equations in Two Variables**
3. **Graphing Linear Equations in Two Variables**
4. ***x*- and *y*-Intercepts**
5. **Horizontal and Vertical Lines**

1. The Rectangular Coordinate System

One application of algebra is the graphical representation of numerical information (or data). For example, Table 2-1 shows the percentage of individuals who participate in leisure sports activities according to the age of the individual.

Table 2-1

Age (years)	Percentage of Individuals Participating in Leisure Sports Activities
20	59%
30	52%
40	44%
50	34%
60	21%
70	18%

Source: U.S. National Endowment for the Arts

Information in table form is difficult to picture and interpret. However, when the data are presented in a graph, there appears to be a downward trend in the participation in leisure sports activities as age increases (Figure 2-1).

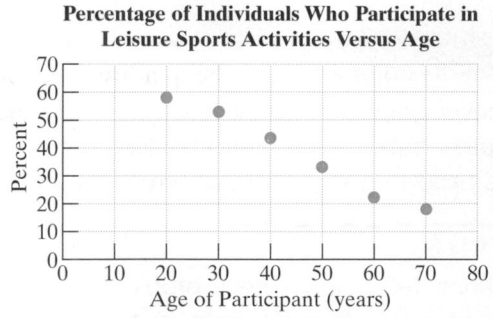

Percentage of Individuals Who Participate in Leisure Sports Activities Versus Age

Figure 2-1

In this example, two variables are related: age and the percentage of individuals who participate in leisure sports activities.

To picture two variables simultaneously, we use a graph with two number lines drawn at right angles to each other (Figure 2-2). This forms a **rectangular coordinate system**. The horizontal line is called the ***x*-axis**, and the vertical line is called the ***y*-axis**. The point where the lines intersect is called the **origin**. On the *x*-axis, the numbers to the right of the origin are positive, and the numbers to the left are negative. On the *y*-axis, the numbers above the origin are positive, and the numbers below are negative. The *x*- and *y*-axes divide the graphing area into four regions called **quadrants**.

Figure 2-2

Points graphed in a rectangular coordinate system are defined by two numbers as an **ordered pair** (x, y). The first number (called the **x-coordinate** or abscissa) is the horizontal position from the origin. The second number (called the **y-coordinate** or ordinate) is the vertical position from the origin. Example 1 shows how points are plotted in a rectangular coordinate system.

Example 1 **Plotting Points**

Plot each point and state the quadrant or axis where it is located.

a. $(4, 1)$ **b.** $(-3, 4)$ **c.** $(4, -3)$

d. $\left(-\frac{5}{2}, -2\right)$ **e.** $(0, 3)$ **f.** $(-4, 0)$

Solution:

a. The point $(4, 1)$ is in quadrant I.

b. The point $(-3, 4)$ is in quadrant II.

c. The point $(4, -3)$ is in quadrant IV.

d. The point $\left(-\frac{5}{2}, -2\right)$ can also be written as $(-2.5, -2)$. This point is in quadrant III.

e. The point $(0, 3)$ is on the y-axis.

f. The point $(-4, 0)$ is on the x-axis.

Figure 2-3

TIP: Notice that the points $(-3, 4)$ and $(4, -3)$ are in different quadrants. Changing the order of the coordinates changes the location of the point. That is why points are represented by *ordered* pairs (Figure 2-3).

Skill Practice Plot the point and state the quadrant or axis where it is located.

1. a. $(3, 5)$ **b.** $(-2, 0)$ **c.** $(2, -1)$

    d. $(0, 4)$ **e.** $(-2, -2)$ **f.** $(-5, 2)$

2. Linear Equations in Two Variables

Recall from Section 1.1 that an equation in the form $ax + b = c$ is called a linear equation in one variable. In this section, we will study linear equations in *two* variables.

Definition of a Linear Equation in Two Variables

Let A, B, and C be real numbers such that A and B are not both zero. A **linear equation in two variables** is an equation that can be written in the form

$$Ax + By = C \qquad \text{This form is called } standard\ form.$$

Answers

1. a. $(3, 5)$; quadrant I
   b. $(-2, 0)$; x-axis
   c. $(2, -1)$; quadrant IV
   d. $(0, 4)$; y-axis
   e. $(-2, -2)$; quadrant III
   f. $(-5, 2)$; quadrant II

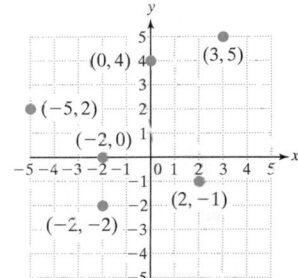

A *solution* to a linear equation in two variables is an ordered pair (x, y) that makes the equation a true statement.

> **Example 2** Determining Solutions to a Linear Equation
>
> For the linear equation $-2x + 3y = 8$, determine whether the ordered pair is a solution.
>
> **a.** $(-4, 0)$ **b.** $(2, -4)$ **c.** $\left(1, \dfrac{10}{3}\right)$
>
> **Solution:**
>
> **a.** $-2x + 3y = 8$ The ordered pair $(-4, 0)$ indicates that $x = -4$ and $y = 0$.
>
> $-2(-4) + 3(0) \overset{?}{=} 8$ Substitute $x = -4$ and $y = 0$ into the equation.
>
> $8 + 0 \overset{?}{=} 8 \checkmark$ (true)   The ordered pair $(-4, 0)$ makes the equation a true statement. The ordered pair is a solution to the equation.
>
> **b.** $-2x + 3y = 8$ Test the ordered pair $(2, -4)$.
>
> $-2(2) + 3(-4) \overset{?}{=} 8$ Substitute $x = 2$ and $y = -4$ into the equation.
>
> $-4 + (-12) \overset{?}{=} 8$
>
> $-16 \overset{?}{=} 8$ (false)   The ordered pair $(2, -4)$ does not make the equation a true statement. The ordered pair is *not* a solution to the equation.
>
> **c.** $-2x + 3y = 8$ Test the ordered pair $(1, \frac{10}{3})$.
>
> $-2(1) + 3\left(\dfrac{10}{3}\right) \overset{?}{=} 8$ Substitute $x = 1$ and $y = \frac{10}{3}$.
>
> $-2 + 10 \overset{?}{=} 8 \checkmark$ (true)   The ordered pair $(1, \frac{10}{3})$ is a solution to the equation.

Skill Practice Determine whether each ordered pair is a solution for the equation $x + 4y = -8$.

2. $(-2, -1)$ **3.** $(4, -3)$ **4.** $\left(-14, \dfrac{3}{2}\right)$

3. Graphing Linear Equations in Two Variables

Consider the linear equation $x - y = 3$. The solutions to the equation are ordered pairs such that the difference of x and y is 3. Several solutions are given in the following list:

Answers
2. Not a solution
3. Solution
4. Solution

Solution	Check
(x, y)	$x - y = 3$
$(3, 0)$	$(3) - (0) = 3$ ✔
$(4, 1)$	$(4) - (1) = 3$ ✔
$(0, -3)$	$(0) - (-3) = 3$ ✔
$(-1, -4)$	$(-1) - (-4) = 3$ ✔
$(2, -1)$	$(2) - (-1) = 3$ ✔

Figure 2-4

By graphing these ordered pairs, we see that the solution points form a line (see Figure 2-4). There are actually an infinite number of solutions to the equation $x - y = 3$. The graph of all solutions to a linear equation forms a line in the xy-plane. Conversely, each ordered pair on the line is a solution to the equation.

The Graph of an Equation in Two Variables

To graph a linear equation in two variables means that we will graph all ordered pair solutions to the equation.

To graph a linear equation, it is sufficient to find two solution points and draw the line through them. We will find three solution points and use the third point as a check point. This is demonstrated in Example 3.

Example 3 Graphing a Linear Equation in Two Variables

Graph the equation $3x + 5y = 15$.

Solution:

We will find three ordered pairs that are solutions to the equation. In the table, we have selected arbitrary values for x or y and must complete the ordered pairs.

x	y	
0		⟶ (0,)
	2	⟶ (, 2)
5		⟶ (5,)

From the first row, substitute $x = 0$.

$$3x + 5y = 15$$
$$3(0) + 5y = 15$$
$$5y = 15$$
$$y = 3$$

From the second row, substitute $y = 2$.

$$3x + 5y = 15$$
$$3x + 5(2) = 15$$
$$3x + 10 = 15$$
$$3x = 5$$
$$x = \frac{5}{3}$$

From the third row, substitute $x = 5$.

$$3x + 5y = 15$$
$$3(5) + 5y = 15$$
$$15 + 5y = 15$$
$$5y = 0$$
$$y = 0$$

The completed list of ordered pairs is shown. To graph the equation, plot the three solutions and draw the line through the points (Figure 2-5). Arrows on the ends of the line indicate that points on the line extend infinitely in both directions.

x	y
0	3
$\frac{5}{3}$	2
5	0

$\longrightarrow$ $(0, 3)$

$\longrightarrow$ $\left(\frac{5}{3}, 2\right)$

$\longrightarrow$ $(5, 0)$

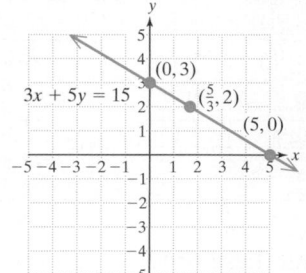

Figure 2-5

Skill Practice Given $2x - y = 1$, complete the table and graph the line through the points.

5.

x	y
0	
	5
1	

Example 4 Graphing a Linear Equation in Two Variables

Graph the equation $y = \dfrac{1}{2}x - 2$.

Solution:

Because the y-variable is isolated in the equation, it is easy to substitute a value for x and simplify the right-hand side to find y. Since any number can be used for x, choose numbers that are multiples of 2 that will simplify easily when multiplied by $\frac{1}{2}$.

x	y
0	
2	
4	

Substitute $x = 0$.

$y = \dfrac{1}{2}(0) - 2$

$y = 0 - 2$

$y = -2$

Substitute $x = 2$.

$y = \dfrac{1}{2}(2) - 2$

$y = 1 - 2$

$y = -1$

Substitute $x = 4$.

$y = \dfrac{1}{2}(4) - 2$

$y = 2 - 2$

$y = 0$

Answer

5.

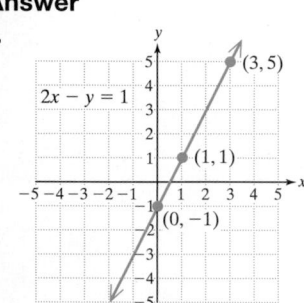

The completed list of ordered pairs is as follows. To graph the equation, plot the three solutions and draw the line through the points (Figure 2-6).

x	y
0	-2
2	-1
4	0

$\longrightarrow (0, -2)$
$\longrightarrow (2, -1)$
$\longrightarrow (4, 0)$

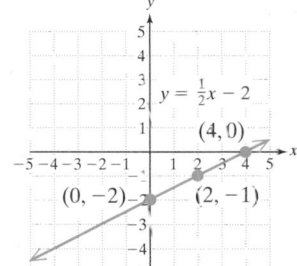

Figure 2-6

Skill Practice

6. Graph the equation $y = -\frac{1}{3}x + 1$.

   Hint: Select values of x that are multiples of 3.

4. *x*- and *y*-Intercepts

For many applications of graphing, it is advantageous to know the points where a graph intersects the *x*- or *y*-axis. These points are called the *x*- and *y*-intercepts.

In Figure 2-5, the *x*-intercept is $(5, 0)$. In Figure 2-6, the *x*-intercept is $(4, 0)$. In general, a point on the *x*-axis must have a *y*-coordinate of zero. In Figure 2-5, the *y*-intercept is $(0, 3)$. In Figure 2-6, the *y*-intercept is $(0, -2)$. In general, a point on the *y*-axis must have an *x*-coordinate of zero.

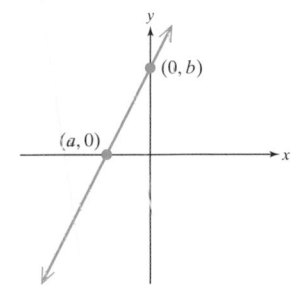

Figure 2-7

Definition of *x*- and *y*-Intercepts

An ***x*-intercept*** is a point $(a, 0)$ where a graph intersects the *x*-axis. (See Figure 2-7.)

A ***y*-intercept** is a point $(0, b)$ where a graph intersects the *y*-axis. (See Figure 2-7.)

*In some applications, an *x*-intercept is defined as the *x-coordinate* of a point of intersection that a graph makes with the *x*-axis. For example, if an *x*-intercept is at the point $(3, 0)$, it is sometimes stated simply as 3 (the *y*-coordinate is understood to be zero). Similarly, a *y*-intercept is sometimes defined as the *y-coordinate* of a point of intersection that a graph makes with the *y*-axis. For example, if a *y*-intercept is at the point $(0, 7)$, it may be stated simply as 7 (the *x*-coordinate is understood to be zero).

To find the *x*- and *y*-intercepts from an equation in *x* and *y*, follow these steps:

Answer

6.

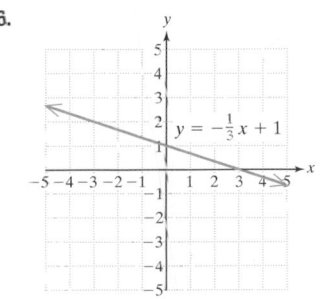

Determining the *x*- and *y*-Intercepts from an Equation

Given an equation in *x* and *y*,

- Find the *x*-intercept(s) by substituting $y = 0$ into the equation and solving for *x*.
- Find the *y*-intercept(s) by substituting $x = 0$ into the equation and solving for *y*.

Example 5 Finding the *x*- and *y*-Intercepts of a Line

Given $2x + 4y = 8$, find the *x*- and *y*-intercepts. Then graph the equation.

Solution:

To find the *x*-intercept, substitute $y = 0$.

$$2x + 4y = 8$$
$$2x + 4(0) = 8$$
$$2x = 8$$
$$x = 4$$

The *x*-intercept is $(4, 0)$.

To find the *y*-intercept, substitute $x = 0$.

$$2x + 4y = 8$$
$$2(0) + 4y = 8$$
$$4y = 8$$
$$y = 2$$

The *y*-intercept is $(0, 2)$.

In this case, the intercepts are two distinct points and may be used to graph the line. A third point can be found to verify that the points all fall on the same line (points that lie on the same line are said to be *collinear*). Choose a different value for either *x* or *y*, such as $y = 4$.

$$2x + 4y = 8$$
$$2x + 4(4) = 8 \qquad \text{Substitute } y = 4.$$
$$2x + 16 = 8 \qquad \text{Solve for } x.$$
$$2x = -8$$
$$x = -4 \qquad \text{The point } (-4, 4) \text{ lines up with the other two points}$$
$$\text{(Figure 2-8).}$$

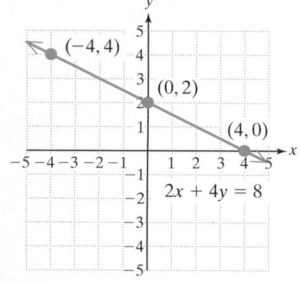

Figure 2-8

Skill Practice

7. Given $-2x + y = -4$, find the *x*- and *y*-intercepts. Then graph the equation.

Example 6 Finding the *x*- and *y*-Intercepts of a Line

Given $y = \frac{1}{4}x$, find the *x*- and *y*-intercepts. Then graph the equation.

Solution:

To find the *x*-intercept, substitute $y = 0$.

$$y = \frac{1}{4}x$$
$$(0) = \frac{1}{4}x$$
$$0 = x$$

The *x*-intercept is $(0, 0)$.

To find the *y*-intercept, substitute $x = 0$.

$$y = \frac{1}{4}x$$
$$y = \frac{1}{4}(0)$$
$$y = 0$$

The *y*-intercept is $(0, 0)$.

Answer

7.

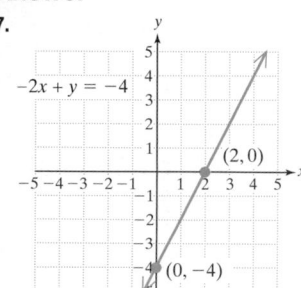

Notice that the x- and y-intercepts are both located at the origin $(0, 0)$. In this case, the intercepts do not yield two distinct points. Therefore, another point is necessary to draw the line. We may pick any value for either x or y. However, for this equation, it would be particularly convenient to pick a value for x that is a multiple of 4 such as $x = 4$.

$$y = \frac{1}{4}x$$

$$y = \frac{1}{4}(4) \qquad \text{Substitute } x = 4.$$

$$y = 1$$

The point $(4, 1)$ is a solution to the equation (Figure 2-9).

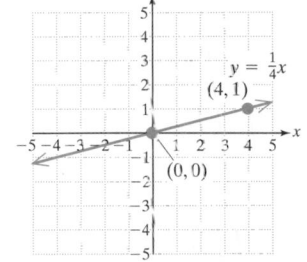

Figure 2-9

Skill Practice

8 Given $y = -5x$, find the x- and y-intercepts. Then graph the equation.

> **Avoiding Mistakes**
>
> You can always find a third point on a line to check the accuracy of your graph. For example, the point $(-4, -1)$ satisfies the equation $y = \frac{1}{4}x$ and lines up with the other points from Example 6.

Example 7   Interpreting the x- and y-Intercepts of a Line

Companies and corporations are permitted to depreciate assets that have a known useful life span. This accounting practice is called *straight-line depreciation*. In this procedure the useful life span of the asset is determined, and then the asset is depreciated by an equal amount each year until the taxable value of the asset is equal to zero.

The J. M. Gus trucking company purchases a new truck for $65,000. The truck will be depreciated at $13,000 per year. The equation that describes the depreciation line is

$$y = 65{,}000 - 13{,}000x$$

where y represents the value of the truck in dollars and x is the age of the truck in years.

a. Find the x- and y-intercepts. Plot the intercepts on a rectangular coordinate system, and draw the line that represents the straight-line depreciation.

b. What does the x-intercept represent in the context of this problem?

c. What does the y-intercept represent in the context of this problem?

Solution:

a. To find the x-intercept, substitute $y = 0$.

$$0 = 65{,}000 - 13{,}000x$$

$$13{,}000x = 65{,}000$$

$$x = 5$$

The x-intercept is $(5, 0)$.

To find the y-intercept, substitute $x = 0$.

$$y = 65{,}000 - 13{,}000(0)$$

$$y = 65{,}000$$

The y-intercept is $(0, 65{,}000)$.

Answer

8.

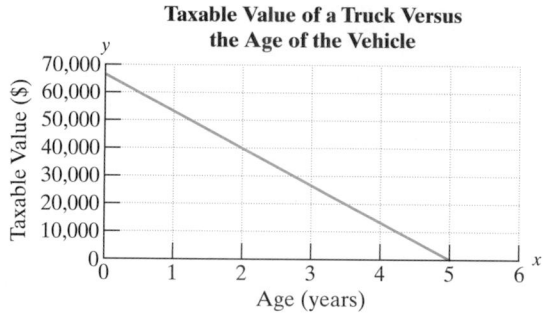

Taxable Value of a Truck Versus the Age of the Vehicle

In Example 7 we graphed the line only in the first quadrant where both the *x*- and *y*-coordinates are positive. (A negative *x*-coordinate would imply a negative age, and a negative *y*-coordinate would imply a negative value of a car, neither of which makes sense.)

b. The *x*-intercept $(5, 0)$ indicates that when the truck is 5 yr old, the taxable value of the truck will be \$0.

c. The *y*-intercept $(0, 65,000)$ indicates that when the truck was new (0 yr old), its taxable value was \$65,000.

Skill Practice

9. Acme Motor Company tests the engines of its trucks by running the engines in a laboratory. The engines burn 4 gal of fuel per hour. The engines begin the test with 30 gal of fuel. The equation $y = 30 - 4x$ represents the amount of fuel *y* left in the engine after *x* hours.
   a. Find the *x*- and *y*-intercepts.
   b. Interpret the *x*-intercept in the context of this problem.
   c. Interpret the *y*-intercept in the context of this problem.

5. Horizontal and Vertical Lines

Recall that a linear equation can be written in the form $Ax + By = C$, where *A* and *B* are not both zero. If either *A* or *B* is 0, then the resulting line is horizontal or vertical, respectively.

Vertical and Horizontal Lines

1. A **vertical line** is a line whose equation can be written in the form $x = k$, where *k* is a constant.
2. A **horizontal line** is a line whose equation can be written in the form $y = k$, where *k* is a constant.

Example 8   Graphing a Vertical Line

Graph the equation $x = 6$.

Answer

9. a. *x*-intercept: $(7.5, 0)$; *y*-intercept: $(0, 30)$
   b. The *x*-intercept $(7.5, 0)$ represents the amount of fuel in the truck after 7.5 hr. After 7.5 hr the tank contains 0 gal. It is empty.
   c. The *y*-intercept $(0, 30)$ represents the amount of fuel in the truck initially (after 0 hr). After 0 hr, the tank contains 30 gal of fuel.

Solution:

Because this equation is in the form $x = k$, the line is vertical and must cross the *x*-axis at $x = 6$. We can also construct a table of solutions to the equation $x = 6$. The choice for the *x*-coordinate must be 6, but *y* can be any real number (Figure 2-10).

x	y	
6	−4	⟶ $(6, -4)$
6	1	⟶ $(6, 1)$
6	5	⟶ $(6, 5)$

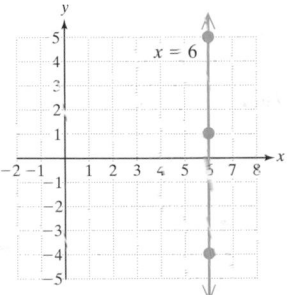

Figure 2-10

Skill Practice

10. Graph the equation $x = -4$.

Example 9   Graphing a Horizontal Line

Graph the equation $4y = -7$.

Solution:

The equation $4y = -7$ is equivalent to $y = -\frac{7}{4}$. Because the equation is in the form $y = k$, the line must be horizontal and must pass through the y-axis at $y = -\frac{7}{4}$ (Figure 2-11).

We can also construct a table of solutions to the equation $4y = -7$. The choice for the y-coordinate must be $-\frac{7}{4}$, but x can be any real number.

x	y	
0	$-\frac{7}{4}$	⟶ $(0, -\frac{7}{4})$
−3	$-\frac{7}{4}$	⟶ $(-3, -\frac{7}{4})$
2	$-\frac{7}{4}$	⟶ $(2, -\frac{7}{4})$

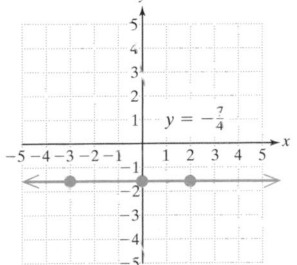

Figure 2-11

Skill Practice

11. Graph the equation $-2y = 9$.

TIP: Notice that horizontal and vertical lines that do not pass through the origin have only one intercept. For instance,

- The vertical line in Figure 2-10 has an x-intercept but no y-intercept.
- The horizontal line in Figure 2-11 has a y-intercept but no x-intercept.

Answers

10.

11.

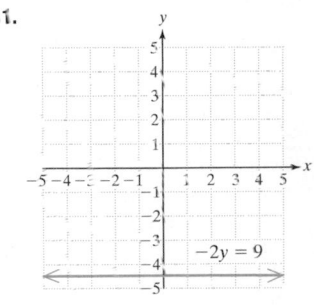

Calculator Connections

Topic: Using the *Table* and *Graph* Features

A viewing window of a graphing calculator shows a portion of a rectangular coordinate system. The *standard viewing window* for most calculators shows both the *x*- and *y*-axes between −10 and 10. Furthermore, the scale defined by the tick marks on both axes is usually set to 1.

The standard viewing window is shown here.

Linear equations can be analyzed with a graphing calculator.

- It is important to isolate the *y*-variable in the equation. Then enter the equation in the calculator. For example, to enter the equation from Example 5, we have:

$$2x + 4y = 8 \longrightarrow 4y = -2x + 8$$

$$\frac{4y}{4} = \frac{-2x}{4} + \frac{8}{4}$$

$$y = -\frac{1}{2}x + 2$$

- A *Table* feature can be used to find many solutions to an equation. Several solutions to $y = -\frac{1}{2}x + 2$ are shown here.

- A *Graph* feature can be used to graph a line.

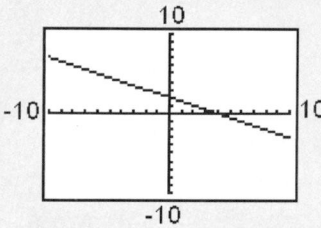

Sometimes the standard viewing window does not provide an adequate display for the graph of an equation. For example, in the standard viewing window, the graph of $y = -x + 15$ is visible only in a small portion of the upper right corner.

To see the *x*- and *y*-intercepts of this line, we can change the viewing window to accommodate larger values of *x* and *y*. Most calculators have a *Range* or *Window* feature that enables the user to change the minimum and maximum *x*- and *y*-values. In this case, we changed the values of *x* to range between −5 and 20, and the values of *y* to range between −10 and 20. We also changed the scaling for the *x*- and *y*-axes so that the tick marks appear in increments of 5 (Xscl = 5 and Yscl = 5).

Section 2.1   Practice Exercises

Study Skills Exercise

After getting a test back, it is a good idea to correct the test so that you do not make the same errors again. One recommended approach is to use a clean sheet of paper, and divide the paper down the middle vertically, as shown. For each problem that you missed on the test, rework the problem correctly on the left-hand side of the paper. Then give a written explanation on the right-hand side of the paper. To reinforce the correct procedure, return to the section of text from which the problem was taken and do several more problems.

Take the time this week to make corrections from your last test.

Perform the correct math here.	Explain the process here.
$2 + 4(5)$ $= 2 + 20$ $= 22$	Do multiplication before addition.

Vocabulary and Key Concepts

1. **a.** In a rectangular coordinate system, two number lines are drawn at right angles to each other. The horizontal line is called the _____ -axis, and the vertical line is called the _____ .

   b. A point in a rectangular coordinate system is defined by an _____ pair, (x, y).

   c. In a rectangular coordinate system, the point where the x- and y-axes intersect is called the _____ and is represented by the ordered pair _____ .

   d. The x- and y-axes divide the coordinate plane into four regions called _____ .

   e. A point with a positive x-coordinate and a _____ y-coordinate is located in quadrant IV.

   f. In quadrant _____ , both the x- and y-coordinates are negative.

   g. A linear equation in two variables is an equation that can be written in the form _____ where A and B are not both zero.

   h. A point where a graph intersects the x-axis is called a(n) _____ .

   i. A point where a graph intersects the y-axis is called a(n) _____ .

   j. A _____ line can be represented by an equation of the form $x = k$, where k is a constant.

   k. A _____ line can be represented by an equation of the form $y = k$, where k is a constant.

Concept 1: The Rectangular Coordinate System

2. Let a and b represent nonzero real numbers.

   a. An ordered pair of the form $(0, b)$ represents a point on which axis?

   b. An ordered pair of the form $(a, 0)$ represents a point on which axis?

3. Given the coordinates of a point, explain how to determine in which quadrant the point lies.

4. What is meant by the word *ordered* in the term *ordered pair*?

5. Plot the points on a rectangular coordinate system. **(See Example 1.)**

   a. $(-2, 1)$ **b.** $(0, 4)$

   c. $(0, 0)$ **d.** $(-3, 0)$

   e. $\left(\dfrac{3}{2}, -\dfrac{7}{3}\right)$ **f.** $(-4.1, -2.7)$

6. Plot the points on a rectangular coordinate system.

a. $(-2, 5)$ **b.** $\left(\dfrac{5}{2}, 0\right)$

c. $(4, -3)$ **d.** $(0, -2)$

e. $(2, 2)$ **f.** $(-3, -3)$

7. A point on the *x*-axis will have what *y*-coordinate?

8. A point on the *y*-axis will have what *x*-coordinate?

For Exercises 9–10, give the coordinates of the labeled points, and state the quadrant or axis where the point is located.

9.

10.

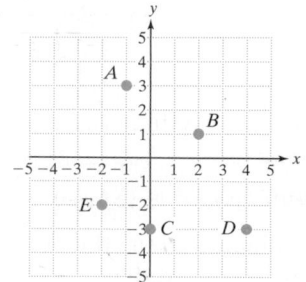

Concept 2: Linear Equations in Two Variables

For Exercises 11–14, determine if the ordered pair is a solution to the linear equation. **(See Example 2.)**

 11. $2x - 3y = 9$

a. $(0, -3)$

b. $(-6, 1)$

c. $\left(1, -\dfrac{7}{3}\right)$

12. $-5x - 2y = 6$

a. $(0, 3)$

b. $\left(-\dfrac{6}{5}, 0\right)$

c. $(-2, 2)$

13. $x = \dfrac{1}{3}y + 1$

a. $(-1, 0)$

b. $(2, 3)$

c. $(-6, 1)$

14. $y = -\dfrac{3}{2}x - 4$

a. $(0, -4)$

b. $(2, -7)$

c. $(-4, -2)$

Concept 3: Graphing Linear Equations in Two Variables

For Exercises 15–18, complete the table. Then graph the line defined by the points. **(See Examples 3–4.)**

15. $3x - 2y = 4$

x	y
0	
	4
−1	

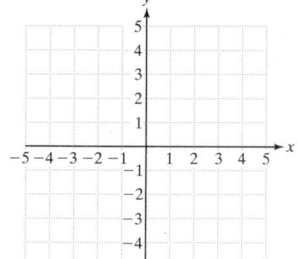

16. $4x + 3y = 6$

x	y
	2
3	
	−1

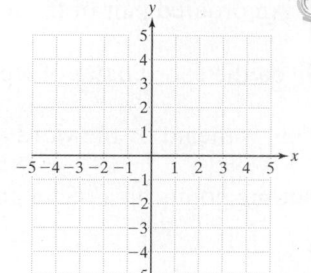

17. $y = -\dfrac{1}{5}x$

x	y
0	
5	
-5	

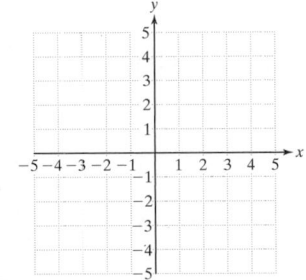

18. $y = \dfrac{1}{3}x$

x	y
0	
3	
6	

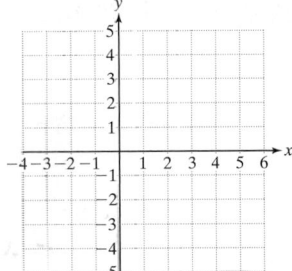

In Exercises 19–30, graph the linear equation by using a table of solutions. **(See Examples 3–4.)**

19. $x + y = 5$

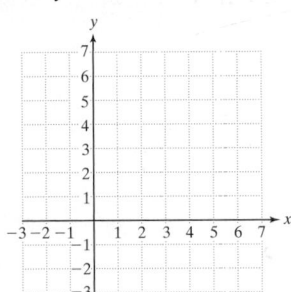

20. $x + y = -8$

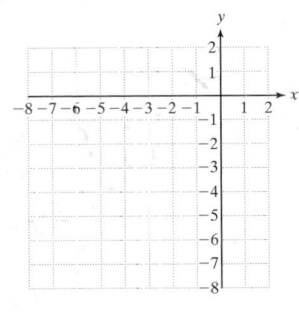

21. $3x - 4y = 12$

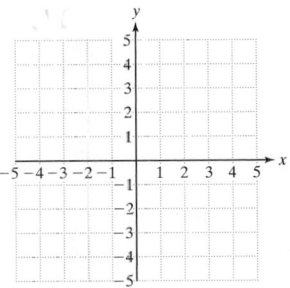

22. $5x + 3y = 15$

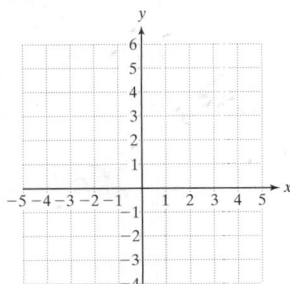

23. $y = -3x + 5$

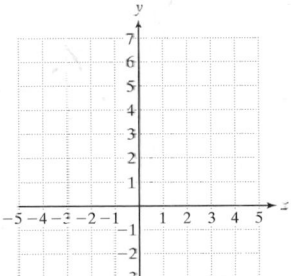

24. $y = -2x + 2$

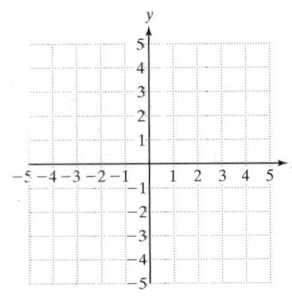

25. $y = \dfrac{2}{5}x - 1$

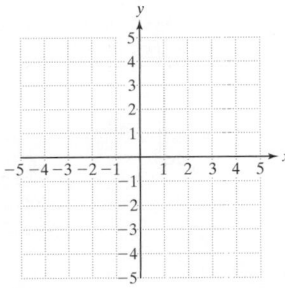

26. $y = \dfrac{5}{3}x + 1$

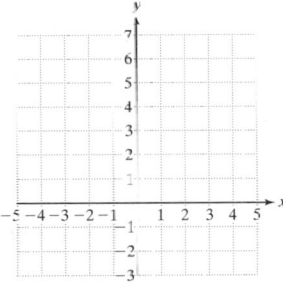

27. $x = -5y - 5$

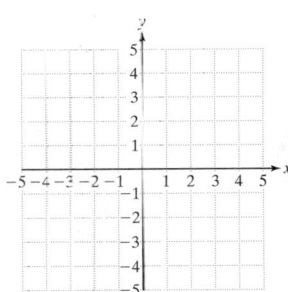

28. $x = 4y + 2$

29. $x = 2y$

30. $x = -3y$

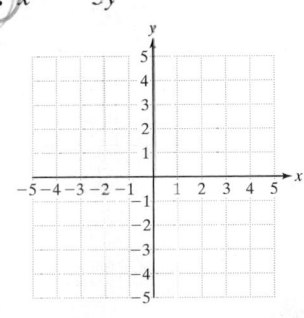

Concept 4: *x*- and *y*-Intercepts

31. Given a linear equation, how do you find an *x*-intercept? How do you find a *y*-intercept?

32. Can the point $(4, -1)$ be an *x*- or *y*-intercept? Why or why not?

For Exercises 33–44, **a.** find the *x*-intercept, **b.** find the *y*-intercept, and **c.** graph the equation. **(See Examples 5–6.)**

33. $2x + 3y = 18$

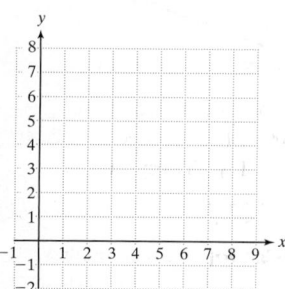

34. $2x - 5y = 10$

35. $x - 2y = 4$

36. $x + y = 8$

37. $5x = 3y$

38. $3y = -5x$

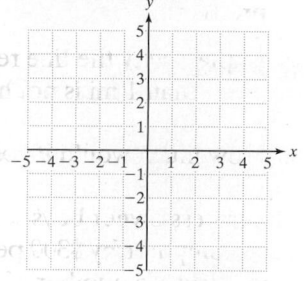

39. $y = 2x + 4$

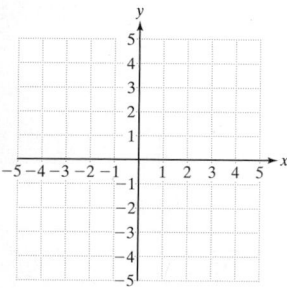

40. $y = -3x - 1$

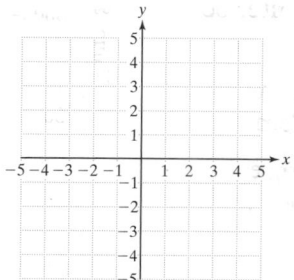

41. $y = -\dfrac{4}{3}x + 2$

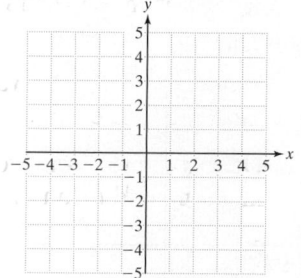

42. $y = -\dfrac{2}{5}x - 1$

43. $x = \dfrac{1}{4}y$

44. $x = \dfrac{2}{3}y$

45. A salesperson makes a base salary of $15,000 a year plus an 8% commission on total sales for the year. The yearly salary can be expressed as a linear equation. **(See Example 7.)**

$$y = 15,000 + 0.08x$$

a. What is the salesperson's salary for a year in which his sales total $500,000?

b. What is the salary for a year in which sales total $300,000?

c. What does the y-intercept mean in the context of this problem?

d. Why is it unreasonable to use negative values for x in this equation?

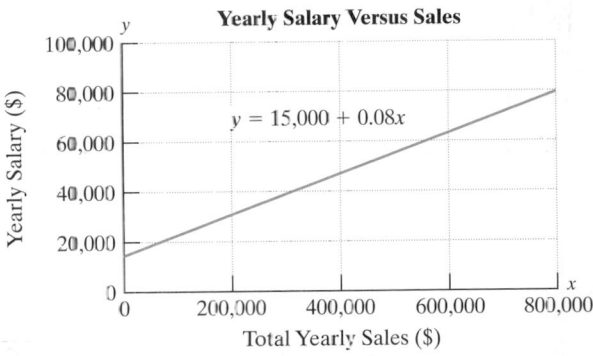

46. A taxi company in Portland charges $3.50 for any distance up to the first mile and $2.50 for every mile thereafter. The cost of a cab ride can be modeled graphically.

a. Explain why the first part of the model is represented by a horizontal line.

b. What does the y-intercept mean in the context of this problem?

c. Explain why the line representing the cost of traveling more than 1 mi is not horizontal.

d. How much would it cost to take a cab 3.5 mi?

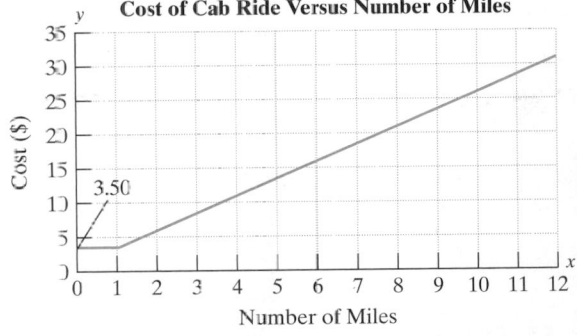

47. A business owner buys several new computers for the office for $1500 each. The accounting office depreciates each computer by $300 per year. The value y (in $) for each computer can be represented by $y = 1500 - 300x$, where x is the number of years after purchase.

a. How much will a computer be worth 1 yr after purchase?

b. After how many years will the computer be worth only $300?

c. Determine the y-intercept and interpret its meaning in the context of this problem.

d. Determine the x-intercept and interpret its meaning in the context of this problem.

48. The equation $y = -3.6x + 59$ can be used to approximate the air temperature y (in °F) at an altitude x (in 1000 ft).

a. Determine the air temperature at 10,000 ft.

b. At what altitude is the air temperature −5.8°F?

c. Determine the y-intercept and interpret its meaning in the context of this problem.

d. Determine the x-intercept and interpret its meaning in the context of this problem.

Concept 5: Horizontal and Vertical Lines

For Exercises 49–56, determine if the equation represents a horizontal line or a vertical line. Then graph the line and identify the x- and y-intercepts. **(See Examples 8–9.)**

49. $y = -1$

50. $y = 3$

51. $x = 2$

52. $x = -5$

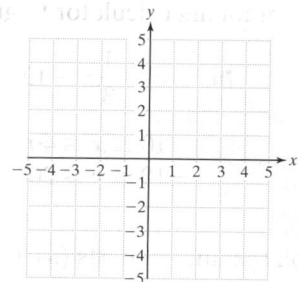

53. $2x + 6 = 5$

54. $-3x = 12$

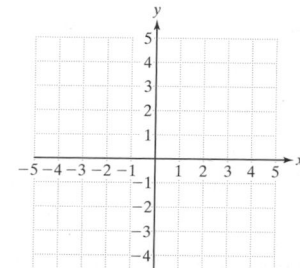

55. $-2y + 1 = 9$

56. $-5y = -10$

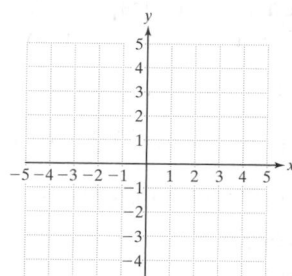

57. Explain why not every line has both an x-intercept and a y-intercept.

58. Is it possible for a line to have no x-intercept and no y-intercept?

59. Which of the lines defined here has only one unique intercept?

  a. $2x - 3y = 8$ **b.** $x = 7$ **c.** $3y = 9$ **d.** $-x + y = 0$

60. Which of the lines defined here has only one unique intercept?

    a. $y = -5$ **b.** $x + 2y = 0$ **c.** $3x - 4 = 2$ **d.** $x + 3y = 6$

Expanding Your Skills

For Exercises 61–64, find the x- and y-intercepts.

61. $\dfrac{x}{2} + \dfrac{y}{3} = 1$ **62.** $\dfrac{x}{7} + \dfrac{y}{4} = 1$ **63.** $\dfrac{x}{a} + \dfrac{y}{b} = 1$ **64.** $Ax + By = C$

Graphing Calculator Exercises

For Exercises 65–68, solve the equation for y. Use a graphing calculator to graph the equation on the standard viewing window.

65. $2x - 3y = 7$ **66.** $4x + 2y = -2$ **67.** $3y = 9$ **68.** $2y + 10 = 0$

For Exercises 69–72, use a graphing calculator to graph the lines on the suggested viewing window.

69. $y = -\dfrac{1}{2}x - 10$ **70.** $y = -\dfrac{1}{3}x + 12$ **71.** $-2x + 4y = 1$ **72.** $5y = 4x - 1$

    $-30 \le x \le 10$ $-10 \le x \le 40$ $-1 \le x \le 1$ $-0.5 \le x \le 0.5$

    $-15 \le y \le 5$ $-10 \le y \le 20$ $-1 \le y \le 1$ $-0.5 \le y \le 0.5$

For Exercises 73–74, graph the lines in parts (a)–(c) on the same viewing window. Compare the graphs. Are the lines exactly the same?

73. **a.** $y = x + 3$ **74.** **a.** $y = 2x + 1$

      b. $y = x + 3.1$ **b.** $y = 1.9x + 1$

      c. $y = x + 2.9$ **c.** $y = 2.1x - 1$

Slope of a Line and Rate of Change   Section 2.2

1. Introduction to the Slope of a Line

In Section 2.1, we learned how to graph a linear equation and to identify its x- and y-intercepts. In this section, we learn about another important feature of a line called the *slope* of a line. Geometrically, slope measures the "steepness" of a line.

    Figure 2-12 shows a set of stairs with a wheelchair ramp to the side. Notice that, even though the stairs and ramp both rise the same vertical distance, the stairs are steeper than the ramp. This is because the stairs rise 3 ft over a shorter horizontal distance than the ramp.

Concepts

1. Introduction to the Slope of a Line
2. The Slope Formula
3. Parallel and Perpendicular Lines
4. Applications and Interpretation of Slope

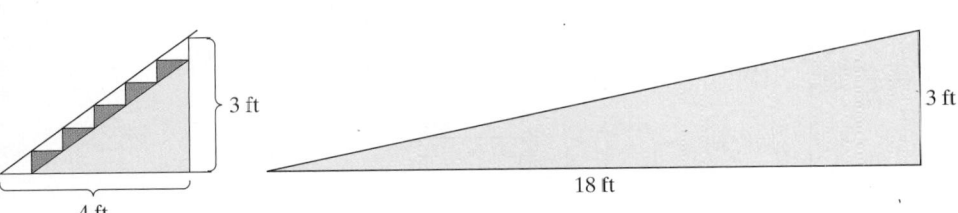

       3 ft   3 ft

                             18 ft

  4 ft

Figure 2-12

To measure the slope of a line quantitatively, consider two points on the line. The slope is the ratio of the vertical change between the two points to the horizontal change. That is, the **slope** is the ratio of the change in y to the change in x. As a memory device, we might think of the slope of a line as "rise over run."

$$\text{Slope} = \frac{\text{change in } y}{\text{change in } x} = \frac{\text{rise}}{\text{run}}$$

To move from point A to point B on the stairs, rise 3 ft and move to the right 4 ft (Figure 2-13).

Figure 2-13

$$\text{Slope} = \frac{\text{change in } y}{\text{change in } x} = \frac{3 \text{ ft}}{4 \text{ ft}} = \frac{3}{4}$$

To move from point A to point B on the wheelchair ramp, rise 3 ft and move to the right 18 ft (Figure 2-14).

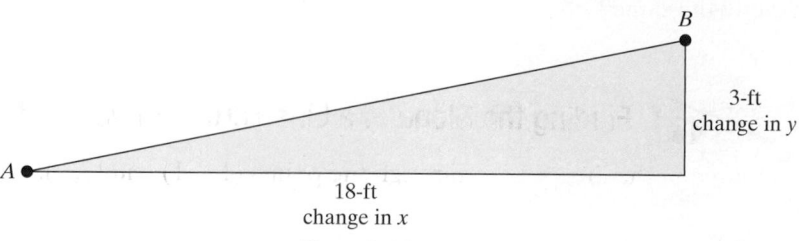

Figure 2-14

$$\text{Slope} = \frac{\text{change in } y}{\text{change in } x} = \frac{3 \text{ ft}}{18 \text{ ft}} = \frac{1}{6}$$

The slope of the stairs is $\frac{3}{4}$, which is greater than the slope of the ramp, which is $\frac{1}{6}$.

Example 1   Finding the Slope in an Application

Find the slope of the ladder against the wall.

Solution:

$$\text{Slope} = \frac{\text{change in } y}{\text{change in } x}$$

$$= \frac{15 \text{ ft}}{5 \text{ ft}}$$

$$= \frac{3}{1} \text{ or } 3$$

The slope is $\frac{3}{1}$, which indicates that a person climbs 3 ft vertically for every 1 ft traveled horizontally.

Skill Practice

1. Find the slope of the roof.

2. The Slope Formula

The slope of a line may be found by using *any* two points on the line—call these points (x_1, y_1) and (x_2, y_2). The change in y between the points can be found by taking the difference of the y-values: $y_2 - y_1$. The change in x can be found by taking the difference of the x-values in the same order: $x_2 - x_1$.

The slope of a line is often symbolized by the letter m and is given by the following formula.

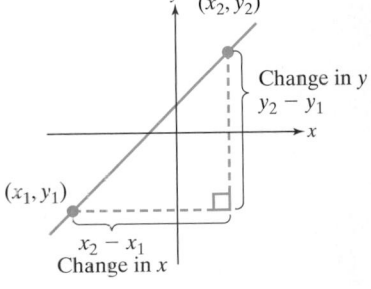

Formula for the Slope of a Line Given Two Distinct Points

The **slope** of a line passing through the distinct points (x_1, y_1) and (x_2, y_2) is

$$m = \frac{y_2 - y_1}{x_2 - x_1} \quad \text{provided} \quad x_2 - x_1 \neq 0$$

Example 2 **Finding the Slope of a Line Through Two Points**

Find the slope of the line passing through the points $(1, -1)$ and $(7, 2)$.

Solution:

To use the slope formula, first label the coordinates of each point, and then substitute their values into the slope formula.

$$
\begin{array}{ccc}
(1, -1) & \text{and} & (7, 2) \\
(x_1, y_1) & & (x_2, y_2)
\end{array}
\quad \text{Label the points.}
$$

$$m = \frac{y_2 - y_1}{x_2 - x_1} = \frac{2 - (-1)}{7 - 1} \quad \text{Apply the slope formula.}$$

$$= \frac{3}{6} \quad \text{Simplify.}$$

$$= \frac{1}{2}$$

The slope of the line can be verified from the graph (Figure 2-15).

> **Avoiding Mistakes**
>
> When applying the slope formula, y_2 and x_2 are taken from the same ordered pair. Likewise y_1 and x_1 are taken from the same ordered pair.

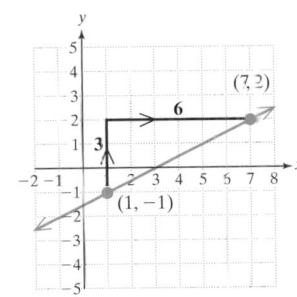

Figure 2-15

Answer

1. $\dfrac{2}{5}$

TIP: The slope formula does not depend on which point is labeled (x_1, y_1) and which point is labeled (x_2, y_2). For example, reversing the order in which the points are labeled in Example 2 results in the same slope:

$$\underset{(x_2,\, y_2)}{(1, -1)} \quad \text{and} \quad \underset{(x_1,\, y_1)}{(7, 2)} \quad \text{then} \quad m = \frac{-1 - 2}{1 - 7} = \frac{-3}{-6} = \frac{1}{2}$$

Skill Practice

2. Find the slope of the line that passes through the points $(-4, 5)$ and $(6, 8)$.

When you apply the slope formula, you will see that the slope of a line may be positive, negative, zero, or undefined.

- Lines that "increase," or "rise," from left to right have a *positive slope*.
- Lines that "decrease," or "fall," from left to right have a *negative slope*.
- Horizontal lines have a *zero slope*.
- Vertical lines have an *undefined slope*.

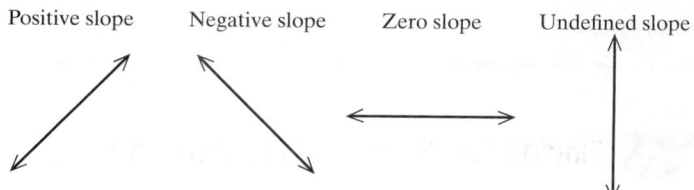

Positive slope   Negative slope   Zero slope   Undefined slope

Example 3   **Finding the Slope of a Line Through Two Points**

Find the slope of the line passing through the points $(3, -4)$ and $(-5, -1)$.

Solution:

$$\underset{(x_1,\, y_1)}{(3, -4)} \qquad \text{and} \qquad \underset{(x_2,\, y_2)}{(-5, -1)} \qquad \text{Label points.}$$

$$m = \frac{y_2 - y_1}{x_2 - x_1} = \frac{-1 - (-4)}{-5 - 3} \qquad \text{Apply the slope formula.}$$

$$= \frac{3}{-8} = -\frac{3}{8} \qquad \text{Simplify.}$$

The two points can be graphed to verify that $-\frac{3}{8}$ is the correct slope (Figure 2-16).

The line slopes downward from left to right.

$m = -\frac{3}{8}$

Figure 2-16

Skill Practice Find the slope of the line passing through the given points.

3. $(1, -8)$ and $(-5, -4)$

Answers

2. $\dfrac{3}{10}$ **3.** $-\dfrac{2}{3}$

Example 4 Finding the Slope of a Line Through Two Points

Find the slope of the line passing through the points $(-3, 4)$ and $(-3, -2)$.

Solution:

$$(-3, 4) \quad \text{and} \quad (-3, -2)$$
$$(x_1, y_1) \qquad\qquad (x_2, y_2) \qquad \text{Label points.}$$

$$m = \frac{y_2 - y_1}{x_2 - x_1} = \frac{-2 - 4}{-3 - (-3)} \qquad \text{Apply slope formula.}$$

$$= \frac{-6}{-3 + 3}$$

$$= \frac{-6}{0} \qquad\qquad \text{Undefined}$$

The slope is undefined. The points form a vertical line (Figure 2-17).

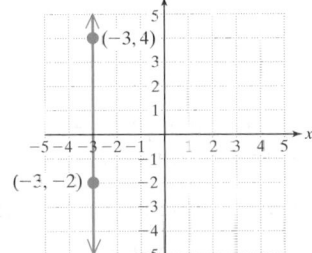

Figure 2-17

Skill Practice Find the slope of the line passing through the given points.

4. $(5, -2)$ and $(5, 5)$

Example 5 Finding the Slope of a Line Through Two Points

Find the slope of the line passing through the points $(0, 2)$ and $(-4, 2)$.

Solution:

$$(0, 2) \quad \text{and} \quad (-4, 2)$$
$$(x_1, y_1) \qquad\qquad (x_2, y_2) \qquad \text{Label the points.}$$

$$m = \frac{y_2 - y_1}{x_2 - x_1} = \frac{2 - 2}{-4 - 0} \qquad \text{Apply the slope formula.}$$

$$= \frac{0}{-4}$$

$$= 0 \qquad\qquad \text{Simplify.}$$

The slope is zero. The line through the two points is a horizontal line (Figure 2-18).

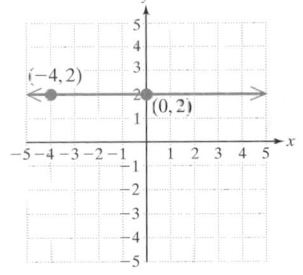

Figure 2-18

Skill Practice Find the slope of the line passing through the points.

5. $(1, 6)$ and $(-7, 6)$

3. Parallel and Perpendicular Lines

Lines in the same plane that do not intersect are *parallel*. Nonvertical parallel lines have the same slope and different *y*-intercepts (Figure 2-19).

Lines that intersect at a right angle are *perpendicular*. If two lines are perpendicular, then the slope of one line is the opposite of the reciprocal of the slope of the other (provided neither line is vertical) (Figure 2-20).

Answers

4. Undefined   **5.** 0

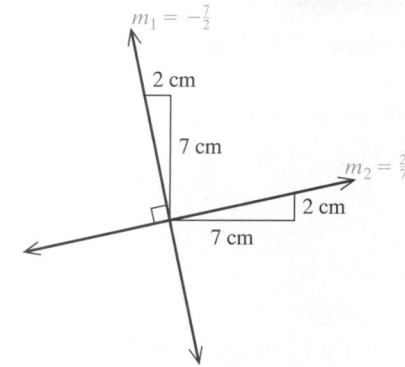

These lines are perpendicular.

$m_1 = -\frac{7}{2}$

2 cm

7 cm

$m_2 = \frac{2}{7}$

2 cm

7 cm

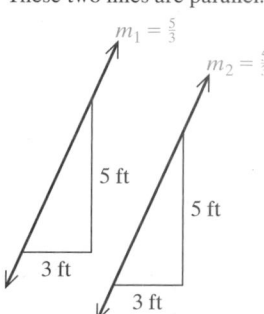

These two lines are parallel.

$m_1 = \frac{5}{3}$

$m_2 = \frac{5}{3}$

5 ft

5 ft

3 ft

3 ft

Figure 2-19   **Figure 2-20**

Slopes of Parallel Lines

If m_1 and m_2 represent the slopes of two parallel (nonvertical) lines, then

$$m_1 = m_2$$

See Figure 2-19.

Slopes of Perpendicular Lines

If $m_1 \neq 0$ and $m_2 \neq 0$ represent the slopes of two perpendicular lines, then

$$m_1 = -\frac{1}{m_2} \text{ or, equivalently, } m_1 \cdot m_2 = -1$$

See Figure 2-20.

Example 6   ## Determining the Slope of Parallel and Perpendicular Lines

Suppose a given line has a slope of -5.

a. Find the slope of a line parallel to the given line.

b. Find the slope of a line perpendicular to the given line.

Solution:

a. The slope of a line parallel to the given line is $m = -5$ (same slope).

b. The slope of a line perpendicular to the given line is $m = \frac{1}{5}$ (the opposite of the reciprocal of -5).

Skill Practice The slope of line L_1 is $-\frac{4}{3}$.

6. Find the slope of a line parallel to L_1.

7. Find the slope of a line perpendicular to L_1.

Answers

6. $-\frac{4}{3}$ **7.** $\frac{3}{4}$

Example 7   **Determining Whether Two Lines Are Parallel, Perpendicular, or Neither**

Two points are given from each of two lines: L_1 and L_2. Without graphing the points, determine if the lines are parallel, perpendicular, or neither.

L_1: $(2, -3)$ and $(4, 1)$

L_2: $(5, -6)$ and $(-3, -2)$

Solution:

First determine the slope of each line. Then compare the values of the slopes to determine if the lines are parallel or perpendicular.

For line 1:   For line 2:

L_1: $(2, -3)$ and $(4, 1)$ L_2: $(5, -6)$ and $(-3, -2)$ Label the points.
      (x_1, y_1) (x_2, y_2) (x_1, y_1) (x_2, y_2)

$$m = \frac{1 - (-3)}{4 - 2} \qquad\qquad m = \frac{-2 - (-6)}{-3 - 5} \qquad\quad \text{Apply the slope}$$
$$\text{formula.}$$

$$= \frac{4}{2} \qquad\qquad\qquad\qquad = \frac{4}{-8}$$

$$= 2 \qquad\qquad\qquad\qquad\quad = -\frac{1}{2}$$

The slope of L_1 is 2. The slope of L_2 is $-\frac{1}{2}$. The slope of L_1 is the opposite of the reciprocal of L_2. By comparing the slopes, the lines must be perpendicular.

TIP: You can also verify that the lines in Example 7 are perpendicular by noting that the product of their slopes is -1.

$$2 \cdot \left(-\frac{1}{2}\right) = -1$$

Skill Practice Two points are given for lines L_1 and L_2. Determine if the lines are parallel, perpendicular, or neither.

8. L_1: $(4, -1)$ and $(-3, 6)$
    L_2: $(-1, 3)$ and $(2, 0)$

4. Applications and Interpretation of Slope

In applications, the slope of a line represents a rate of change between the y variable and the x variable. For example, a hiker walking up a hill with a slope of $\frac{1}{6}$ means that 1 ft of elevation is gained for every 6 ft traveled horizontally. Using this rate of change, we can also say that a hiker gains 10 ft of elevation for 60 ft traveled horizontally. See Figure 2-21.

1 ft   10 ft

6 ft

60 ft

Figure 2-21

Example 8   **Interpreting the Slope of a Line in an Application**

The number of males 20 yr old or older who were employed full-time in the United States has grown linearly since 1970. Approximately 43.0 million males 20 yr old or older were employed full-time in 1970. By 2010, this number had grown to 69.0 million (Figure 2-22).

Answer

8. Parallel

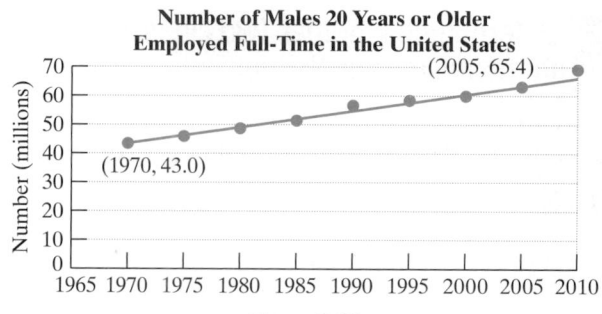

Figure 2-22

Source: U.S. Census Bureau

a. Find the slope of the line, using the points $(1970, 43.0)$ and $(2005, 65.4)$.

b. Interpret the meaning of the slope in the context of this problem.

Solution:

a. $(1970, 43.0)$ and $(2005, 65.4)$

(x_1, y_1) (x_2, y_2) Label the points.

$m = \dfrac{y_2 - y_1}{x_2 - x_1} = \dfrac{65.4 - 43.0}{2005 - 1970}$ Apply the slope formula.

$m = \dfrac{22.4}{35}$ or $m = 0.64$

> **TIP:** The slope, $m = \dfrac{22.4}{35}$, means that the workforce increased by 22.4 million men over 35 yr. This is the same rate as $\dfrac{0.64}{1}$, meaning that the workforce increased by 0.64 million men per year.

b. The slope is approximately 0.64, meaning that the full-time workforce has increased at a rate of 0.64 million men (or 640,000 men) per year between 1970 and 2005.

Answers

9. 0.018
10. The population density increased at a rate of 0.018 people per square mile per year.

Skill Practice The number of people per square mile (called population density) in Alaska has increased linearly since 1990. In 1990, there were 0.96 people per square mile. By 2010, this increased to 1.32 people per square mile.

9. Find the slope of the line passing through the points $(1990, 0.96)$ and $(2010, 1.32)$.
10. Interpret the meaning of the slope in the context of this problem.

Section 2.2   **Practice Exercises**

Study Skills Exercise

Instructors vary in what they emphasize on tests. For example, test material may come from the textbook, notes, handouts, and homework. What did your instructor emphasize on the last test?

Vocabulary and Key Concepts

1. a. The ratio of the vertical change and the horizontal change between two distinct points (x_1, y_1) and (x_2, y_2) on a line is called the _____ of the line. The slope can be computed from the formula $m = $ _____.

b. Lines in the same plane that do not intersect are called _____ lines. Such lines, if they are nonvertical, have the _____ slope and different y-intercepts.

c. Two lines are perpendicular if they intersect at a _____ angle.

d. If m_1 and m_2 represent the slopes of two nonvertical perpendicular lines, then $m_1 \cdot m_2 = $ _____.

Review Exercises

2. For each ordered pair, determine the quadrant in which the point lies.

   a. $(-5.73, 102.4)$ **b.** $(-1500, -2030)$ **c.** $\left(\dfrac{367}{1013}, \dfrac{41}{512}\right)$ **d.** $(\pi, -2.1718)$

3. Find the missing coordinate so that the ordered pairs are solutions to the equation $\frac{1}{2}x + y = 4$.

   a. $(0,\ \ )$ **b.** $(\ , 0)$ **c.** $(-4,\ \ )$

For Exercises 4–6, find the x- and y-intercepts (if possible) for each equation, and sketch the graph.

   4. $2x + 8 = 0$ **5.** $4 - 2y = 0$ **6.** $2x - 2y - 6 = 0$

      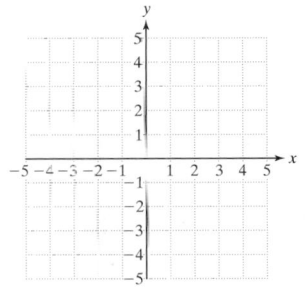

Concept 1: Introduction to the Slope of a Line

7. A 25-ft ladder is leaning against a house, as shown in the diagram. Find the slope of the ladder. **(See Example 1.)**

8. Find the slope shown in the figure.

9. Find the slope of the treadmill.

10. Find the average slope of the hill.

11. The road sign shown in the figure indicates the percent grade of a hill. This gives the slope of the road as the change in elevation per 100 horizontal ft. Given a 4% grade, write this as a slope in fractional form.

12. If a plane gains 1000 ft in altitude over a distance of 12,000 horizontal ft, what is the slope? Explain what this value means in the context of the problem.

Concept 2: The Slope Formula

For Exercises 13–30, use the slope formula to determine the slope of the line containing the two points. **(See Examples 2–5.)**

 13. $(6, 0)$ and $(0, -3)$

14. $(-5, 0)$ and $(0, 4)$

15. $(-2, 3)$ and $(4, -7)$

16. $(-5, -4)$ and $(1, -7)$

17. $(-2, 5)$ and $(2, -3)$

18. $(4, -2)$ and $(6, -8)$

19. $(0.3, -1.1)$ and $(-0.1, -0.8)$

20. $(0.4, -0.2)$ and $(0.3, -0.1)$

21. $(2, 3)$ and $(2, 7)$

22. $(-1, 5)$ and $(-1, 0)$

23. $(5, -1)$ and $(-3, -1)$

24. $(-8, 4)$ and $(1, 4)$

25. $(-4.6, 4.1)$ and $(0, 6.4)$

26. $(1.1, 4)$ and $(-3.2, -0.3)$

27. $\left(\dfrac{3}{2}, \dfrac{4}{3}\right)$ and $\left(\dfrac{7}{2}, 1\right)$

28. $\left(\dfrac{2}{3}, -\dfrac{1}{2}\right)$ and $\left(-\dfrac{1}{6}, -\dfrac{3}{2}\right)$

29. $\left(\dfrac{3}{4}, \dfrac{7}{3}\right)$ and $\left(\dfrac{1}{2}, 2\dfrac{1}{3}\right)$

30. $\left(\dfrac{9}{4}, \dfrac{2}{5}\right)$ and $\left(2\dfrac{1}{4}, \dfrac{1}{10}\right)$

31. Explain how to use the graph of a line to determine whether the slope of a line is positive, negative, zero, or undefined.

32. If the slope of a line is $\frac{4}{3}$, how many units of change in y will be produced by 6 units of change in x?

For Exercises 33–38, estimate the slope of the line from its graph.

33.

34.

35.

36.

37.

38.

Concept 3: Parallel and Perpendicular Lines

For Exercises 39–44, the slope of a line is given.

  a. Find the slope of a line parallel to the given line.

  b. Find the slope of a line perpendicular to the given line. **(See Example 6.)**

39. $m = 5$

40. $m = 3$

41. $m = -\dfrac{4}{7}$

42. $m = -\dfrac{2}{11}$

43. $m = 0$

44. m is undefined.

45. Can the slopes of two perpendicular lines both be positive? Explain your answer.

46. Suppose a line is defined by the equation $x = 2$. What is the slope of a line perpendicular to this line?

47. Suppose a line is defined by the equation $y = -5$. What is the slope of a line perpendicular to this line?

48. Suppose a line is defined by the equation $x = -3$. What is the slope of a line parallel to this line?

49. What is the slope of a line parallel to the x-axis?

50. What is the slope of a line perpendicular to the y-axis?

51. What is the slope of a line perpendicular to the x-axis?

52. What is the slope of a line parallel to the y-axis?

In Exercises 53–60, two points are given from each of two lines L_1 and L_2. Without graphing the points, determine if the lines are parallel, perpendicular, or neither. **(See Example 7.)**

53. L_1: $(2, 5)$ and $(4, 9)$
L_2: $(-1, 4)$ and $(3, 2)$

54. L_1: $(-3, -5)$ and $(-1, 2)$
L_2: $(0, 4)$ and $(7, 2)$

55. L_1: $(4, -2)$ and $(3, -1)$
L_2: $(-5, -1)$ and $(-10, -16)$

56. L_1: $(0, 0)$ and $(2, 3)$
L_2: $(-2, 5)$ and $(0, -2)$

57. L_1: $(5, 3)$ and $(5, 9)$
L_2: $(4, 2)$ and $(0, 2)$

58. L_1: $(3, 5)$ and $(2, 5)$
L_2: $(2, 4)$ and $(0, 4)$

59. L_1: $(-3, -2)$ and $(2, 3)$
L_2: $(-4, 1)$ and $(0, 5)$

60. L_1: $(7, 1)$ and $(0, 0)$
L_2: $(-10, -8)$ and $(4, -6)$

Concept 4: Applications and Interpretation of Slope

61. The graph shows the number of cellular phone subscriptions (in millions) purchased in the United States for selected years. **(See Example 8.)**

a. Use the coordinates of the given points to find the slope of the line, and express the answer in decimal form.

b. Interpret the meaning of the slope in the context of this problem.

62. The U. S. population (in millions) has grown approximately linearly since the year 2000.

a. Find the slope of the line defined by the two given points.

b. Interpret the meaning of the slope in the context of this problem.

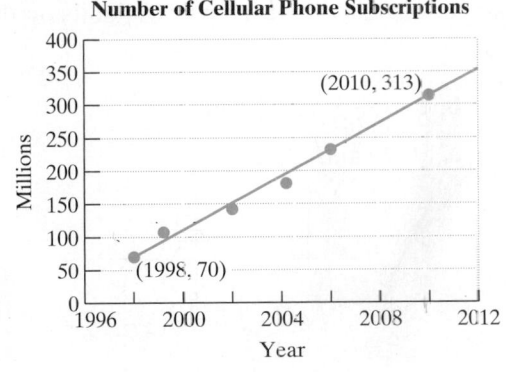

Number of Cellular Phone Subscriptions

(2010, 313)

(1998, 70)

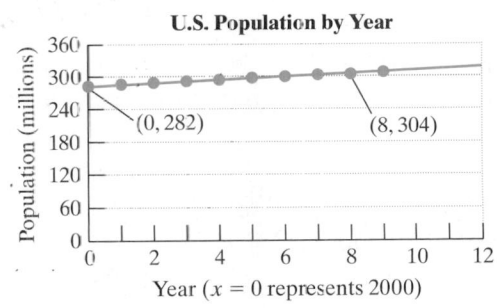

U.S. Population by Year

(0, 282)

(8, 304)

Year ($x = 0$ represents 2000)

63. The data in the graph show the average weight for boys based on age.

    a. Use the coordinates of the given points to find the slope of the line.

    b. Interpret the meaning of the slope in the context of this problem.

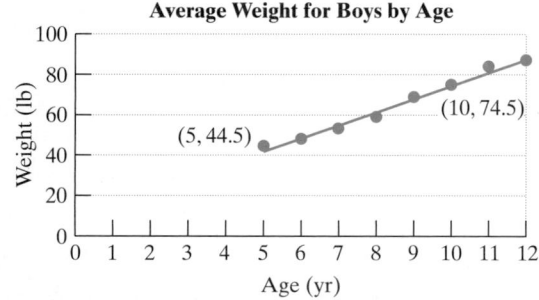

64. The data in the graph show the average weight for girls based on age.

    a. Use the coordinates of the given points to find the slope of the line, and write the answer in decimal form.

    b. Interpret the meaning of the slope in the context of this problem.

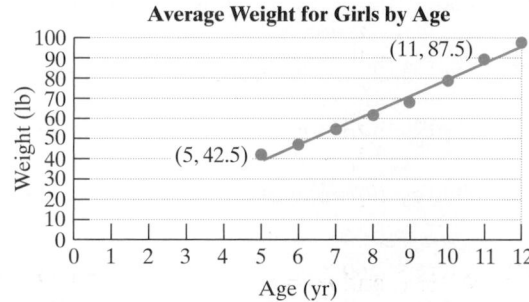

Mixed Exercises

The slope of a line is constant. This also means that the rate of change of y versus x is constant between any two points on the line. For Exercises 65–66, demonstrate this statement by computing the slope of the line using each pair of points on the line.

65. a. $(-1, -4)$ and $(0, -2)$

    b. $(0, -2)$ and $(3, 4)$

    c. $(-1, -4)$ and $(3, 4)$

66. a. $(-4, -1)$ and $(0, 1)$

    b. $(0, 1)$ and $(2, 2)$

    c. $(0, 1)$ and $(4, 3)$

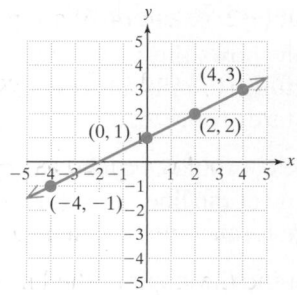

Given a nonlinear graph, the rate of change of y versus x may differ for different pairs of points on the graph. For Exercises 67–68, demonstrate this statement by computing the slope formula with each pair of points on the curve.

67. a. $(-2, 0)$ and $(0, -4)$

    b. $(0, -4)$ and $(2, 0)$

    c. $(2, 0)$ and $(3, 5)$

68. a. $(-3, -8)$ and $(-2, -3)$

    b. $(-2, -3)$ and $(0, 1)$

    c. $(0, 1)$ and $(2, -3)$

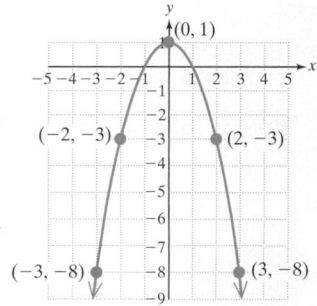

Expanding Your Skills

For Exercises 69–74, given a point P on a line and the slope m of the line, find a second point on the line (answers may vary). *Hint:* Graph the line to help you find the second point.

69. $P(0, 0)$ and $m = 2$

70. $P(-2, 1)$ and $m = -\dfrac{1}{3}$

71. $P(2, -3)$ and m is undefined

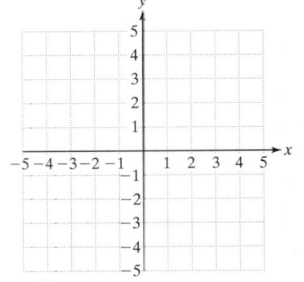

72. $P(-2, 4)$ and $m = 0$

73. $P(-1, 2)$ and $m = -\dfrac{2}{3}$

74. $P(-1, -4)$ and $m = \dfrac{4}{5}$

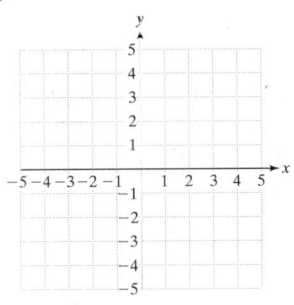

75. Given that $(-2, y)$ and $(4, 6)$ are points on a line whose slope is $-\frac{3}{2}$, find y.

76. Given that $(x, -4)$ and $(3, 2)$ are points on a line whose slope is $\frac{6}{7}$, find x.

77. The pitch of a roof is defined as $\dfrac{\text{rise}}{\text{rafter}}$.

   a. Determine the pitch of the roof shown.

   b. Determine the slope of the line segment from point D to point E.

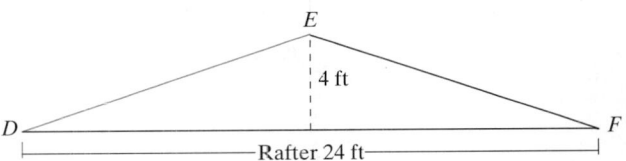

Equations of a Line

Section 2.3

1. Slope-Intercept Form

In Section 2.1, we learned that an equation of the form $Ax + By = C$ (where A and B are not both zero) represents a line in a rectangular coordinate system. An equation of a line written in this way is in **standard form**. In this section, we will learn a new form, called the **slope-intercept form**, which is useful in determining the slope and y-intercept of a line.

Concepts

1. **Slope-Intercept Form**
2. **The Point-Slope Formula**
3. **Equations of a Line: A Summary**

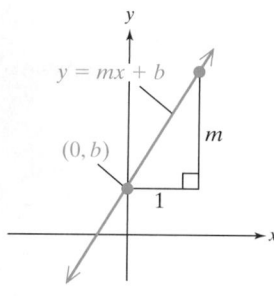

Let $(0, b)$ represent the y-intercept of a line. Let (x, y) represent any other point on the line where $x \neq 0$. Then the slope of the line through the two points is

$$m = \frac{y_2 - y_1}{x_2 - x_1} \quad \rightarrow \quad m = \frac{y - b}{x - 0} \qquad \text{Apply the slope formula.}$$

$$m = \frac{y - b}{x} \qquad\qquad\qquad \text{Simplify.}$$

$$m \cdot x = \left(\frac{y - b}{\cancel{x}}\right) \cdot \cancel{x} \qquad\qquad \text{Clear fractions.}$$

$$mx = y - b \qquad\qquad\qquad \text{Simplify.}$$

$$mx + b = y \quad \text{or} \quad y = mx + b \qquad \text{Solve for } y \text{: slope-intercept form}$$

Slope-Intercept Form of a Line

$y = mx + b$ is the slope-intercept form of a line.

m is the slope and the point $(0, b)$ is the y-intercept.

The equation $y = -4x + 7$ is written in slope-intercept form. By inspection, we can see that the slope of the line is -4 and the y-intercept is $(0, 7)$.

Example 1   **Finding the Slope and y-Intercept of a Line**

Given $3x + 4y = 4$, write the equation of the line in slope-intercept form. Then find the slope and y-intercept.

Solution:

Write the equation in slope-intercept form, $y = mx + b$, by solving for y.

$$3x + 4y = 4$$
$$4y = -3x + 4 \qquad \text{Subtract } 3x \text{ from both sides.}$$
$$\frac{4y}{4} = \frac{-3x}{4} + \frac{4}{4} \qquad \text{To isolate } y, \text{ divide both sides by 4.}$$
$$y = -\frac{3}{4}x + 1 \qquad \text{The slope is } -\frac{3}{4} \text{ and the } y\text{-intercept is } (0, 1).$$

Skill Practice Write the equation in slope-intercept form. Determine the slope and the y-intercept.

1. $2x - 4y = 3$

The slope-intercept form is a useful tool to graph a line. The y-intercept is a known point on the line, and the slope indicates the "direction" of the line and can be used to find a second point. Using slope-intercept form to graph a line is demonstrated in Example 2.

Answer

1. $y = \frac{1}{2}x - \frac{3}{4}$

Slope: $\frac{1}{2}$; y-intercept: $\left(0, -\frac{3}{4}\right)$

Example 2 Graphing a Line Using the Slope and *y*-Intercept

Graph the equation $y = -\frac{3}{4}x + 1$ using the slope and *y*-intercept.

Solution:

First plot the *y*-intercept $(0, 1)$. The slope $m = -\frac{3}{4}$ can be written as

$$m = \frac{-3}{4}$$

— The change in *y* is -3.

— The change in *x* is 4.

Figure 2-23

To find a second point on the line, start at the *y*-intercept and move *down* 3 *units* and to the *right* 4 *units*. Then draw the line through the two points (Figure 2-23).

Similarly, the slope can be written as

$$m = \frac{3}{-4}$$

— The change in *y* is 3.

— The change in *x* is -4.

To find a second point on the line, start at the *y*-intercept and move *up* 3 *units* and to the *left* 4 *units*. Then draw the line through the two points (see Figure 2-23).

Skill Practice

2. Graph the equation $y = \frac{1}{5}x - 2$ using the slope and *y*-intercept.

As we have seen earlier, two lines are parallel if they have the same slope and different *y*-intercepts. Two lines are perpendicular if the slope of one line is the opposite of the reciprocal of the slope of the other line. Otherwise, the lines are neither parallel nor perpendicular.

Example 3 Determining if Two Lines Are Parallel, Perpendicular, or Neither

Given the equations for two lines, L_1 and L_2, determine if the lines are parallel, perpendicular, or neither.

a. L_1: $y = -2x + 7$ **b.** L_1: $2y = -3x + 2$ **c.** L_1: $x + y = 6$

    L_2: $y = -2x - 1$ L_2: $-4x + 6y = -12$ L_2: $y = 6$

Solution:

a. The equations are written in slope-intercept form.

    L_1: $y = -2x + 7$ The slope is -2 and the *y*-intercept is $(0, 7)$.

    L_2: $y = -2x - 1$ The slope is -2 and the *y*-intercept is $(0, -1)$.

Because the slopes are the same and the *y*-intercepts are different, the lines are parallel.

Answer

2.

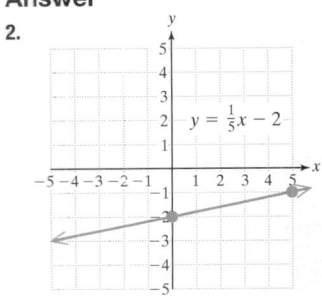

b. Write each equation in slope-intercept form by solving for y.

L_1: $2y = -3x + 2$ $\qquad\qquad\qquad$ L_2: $-4x + 6y = -12$

$$\frac{2y}{2} = \frac{-3x}{2} + \frac{2}{2} \quad \text{Divide by 2.}$$

$$6y = 4x - 12 \quad \text{Add } 4x \text{ to both sides.}$$

$$y = -\frac{3}{2}x + 1$$

$$\frac{6y}{6} = \frac{4}{6}x - \frac{12}{6} \quad \text{Divide by 6.}$$

$$y = \frac{2}{3}x - 2$$

The slope of L_1 is $-\dfrac{3}{2}$. $\qquad\qquad$ The slope of L_2 is $\dfrac{2}{3}$.

The value $-\frac{3}{2}$ is the opposite of the reciprocal of $\frac{2}{3}$. Therefore, the lines are perpendicular.

c. L_1: $x + y = 6$ is equivalent to $y = -x + 6$. The slope is -1.

L_2: $y = 6$ is a horizontal line, and the slope is 0.

The slopes are not the same. Therefore, the lines are not parallel. The slope of one line is not the opposite of the reciprocal of the other slope. Therefore, the lines are not perpendicular. The lines are neither parallel nor perpendicular.

Skill Practice Given the pair of equations, determine if the lines are parallel, perpendicular, or neither.

3. $y = -\dfrac{3}{4}x + 1$ $\qquad$ **4.** $3x + y = 4$ $\qquad$ **5.** $x - y = 7$

$\quad y = \dfrac{4}{3}x + 3$ $\qquad\qquad\quad 6x = 6 - 2y$ $\qquad\qquad x = 1$

Example 4 **Using Slope-Intercept Form to Find an Equation of a Line**

Use slope-intercept form to find an equation of the line with slope -3 and passing through the point $(1, -4)$.

Solution:

To find an equation of a line in slope-intercept form, $y = mx + b$, it is necessary to find the slope, m, and the y-intercept, b. The slope is given in the problem as $m = -3$. Therefore, the slope-intercept form becomes

$$y = mx + b$$

$$\downarrow$$

$$y = -3x + b$$

Answers
3. Perpendicular   **4.** Parallel
5. Neither

Furthermore, because the point $(1, -4)$ is on the line, it is a solution to the equation. Therefore, if we substitute $(1, -4)$ for x and y in the equation, we can solve for b.

$$-4 = -3(1) + b$$
$$-4 = -3 + b$$
$$-1 = b$$

Thus, the slope-intercept form is $y = -3x - 1$.

Skill Practice

6. Use slope-intercept form to find an equation of the line with slope 2 and passing through $(-3, -5)$.

Calculator Connections

Topic: Using the *Value* Feature

The *Value* feature of a graphing calculator prompts the user for a value of x, and then returns the corresponding y-value of an equation.

We can check the answer to Example 4 by graphing the equation, $y = -3x - 1$. Using the *Value* feature, we see that the line passes through the point $(1, -4)$ as expected.

2. The Point-Slope Formula

In Example 4, we used the slope-intercept form of a line to construct an equation of a line given its slope and a known point on the line. Here we provide another tool called the *point-slope formula* that (as its name suggests) can accomplish the same result.

Suppose a nonvertical line passes through a given point (x_1, y_1) and has slope m. If (x, y) is any other point on the line, then

$$m = \frac{y - y_1}{x - x_1} \qquad \text{Slope formula}$$

$$m(x - x_1) = \frac{y - y_1}{x - x_1}(x - x_1) \qquad \text{Clear fractions.}$$

$$m(x - x_1) = y - y_1$$

or

$$y - y_1 = m(x - x_1) \qquad \text{Point-slope formula}$$

Answer

6. $y = 2x + 1$

> ## The Point-Slope Formula
> The **point-slope formula** is given by
>
> $$y - y_1 = m(x - x_1)$$
>
> where m is the slope of the line and (x_1, y_1) is a known point on the line.

The point-slope formula is used specifically to find an equation of a line when a point on the line is known and the slope is known. To illustrate the point-slope formula, we will repeat the problem from Example 4.

Example 5   Using the Point-Slope Formula to Find an Equation of a Line

Use the point-slope formula to find an equation of the line having a slope of -3 and passing through the point $(1, -4)$. Write the answer in slope-intercept form.

Solution:

$m = -3$ and $(x_1, y_1) = (1, -4)$

$y - y_1 = m(x - x_1)$

$y - (-4) = -3(x - 1)$ Apply the point-slope formula.

$y + 4 = -3(x - 1)$ Simplify.

To write the answer in slope-intercept form, clear parentheses and solve for y.

$y + 4 = -3x + 3$ Clear parentheses.

$y = -3x - 1$ Solve for y. The answer is written in slope-intercept form. Notice that this is the same equation as in Example 4.

Skill Practice

7. Use the point-slope formula to write an equation for the line passing through the point $(-2, -6)$ and having a slope of -5. Write the answer in slope-intercept form.

> **TIP:** The solution to Example 5 can also be written in standard form, $Ax + By = C$.
>
> $y = -3x - 1$ Slope-intercept form
>
> $3x + y = -3x + 3x - 1$ Add $3x$ to both sides.
>
> $3x + y = -1$ Standard form
>
> In general, we will write the solution in slope-intercept form, because the slope and y-intercept can be easily identified.

Answer

7. $y = -5x - 16$

Find an equation of the line passing through the points $(5, -1)$ and $(3, 1)$. Write the answer in slope-intercept form.

Solution:

The slope formula can be used to compute the slope of the line between two points. Once the slope is known, the point-slope formula can be used to find an equation of the line.

First find the slope.

$$m = \frac{y_2 - y_1}{x_2 - x_1} = \frac{1 - (-1)}{3 - 5} = \frac{2}{-2} = -1$$

Next, apply the point-slope formula.

$y - y_1 = m(x - x_1)$

$y - 1 = -1(x - 3)$ Substitute $m = -1$ and use *either* point for (x_1, y_1). We will use $(3, 1)$ for (x_1, y_1).

$y - 1 = -x + 3$ Clear parentheses.

$y = -x + 4$ Solve for y. The final answer is in slope-intercept form.

TIP: In Example 6, once the slope is found from the two given points, we can alternatively use slope-intercept form to find an equation of the line. Substitute the coordinates from *either* known point for x and y. Using the point $(3, 1)$, we have:

$y = mx + b$
$1 = -1(3) + b$
$1 = -3 + b$
$4 = b$

Therefore, $y = -x + 4$.

Skill Practice

8. Use the point-slope formula to write an equation of the line that passes through the points $(-5, 2)$ and $(-1, -1)$. Write the answer in slope-intercept form.

TIP: In Example 6, the point $(3, 1)$ was used for (x_1, y_1) in the point-slope formula. However, either point could have been used. Using the point $(5, -1)$ for (x_1, y_1) produces the same final equation:

$y - (-1) = -1(x - 5)$
$y + 1 = -x + 5$
$y = -x + 4$

Find an equation of the line passing through the point $(-2, -3)$ and parallel to the line $4x + y = 8$. Write the answer in slope-intercept form.

Solution:

To find an equation of a line, we must know a point on the line and the slope. The known point is $(-2, -3)$. Because the line is parallel to $4x + y = 8$, the two lines must have the same slope. Writing the equation $4x + y = 8$ in slope-intercept form, we have $y = -4x + 8$. Therefore, the slope of both lines must be -4.

Answer

8. $y = -\frac{3}{4}x - \frac{7}{4}$

Now find an equation of the line passing through $(-2, -3)$ having a slope of -4.

$$y - y_1 = m(x - x_1) \qquad \text{Apply the point-slope formula.}$$

$$y - (-3) = -4[x - (-2)] \qquad \text{Substitute } m = -4 \text{ and } (-2, -3) \text{ for } (x_1, y_1).$$

$$y + 3 = -4(x + 2)$$

$$y + 3 = -4x - 8 \qquad \text{Clear parentheses.}$$

$$y = -4x - 11 \qquad \text{Write the answer in slope-intercept form.}$$

Skill Practice

9. Find an equation of a line containing $(4, -1)$ and parallel to $2x = y - 7$. Write the answer in slope-intercept form.

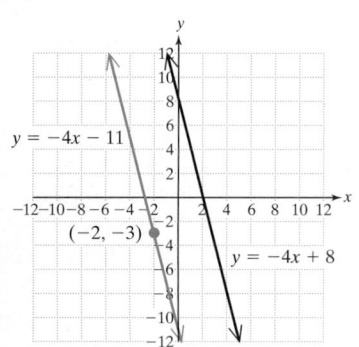

$y = -4x - 11$

$(-2, -3)$

$y = -4x + 8$

Figure 2-24

We can verify the answer to Example 7 by graphing both equations. We see that the line defined by $y = -4x - 11$ passes through the point $(-2, -3)$ and is parallel to the line $y = -4x + 8$. See Figure 2-24.

Example 8 Finding an Equation of a Line Perpendicular to Another Line

Find an equation of the line passing through the point $(4, 3)$ and perpendicular to the line $2x + 3y = 3$. Write the answer in slope-intercept form.

Solution:

The slope of the given line can be found from its slope-intercept form.

$$2x + 3y = 3$$

$$3y = -2x + 3 \qquad \text{Solve for } y.$$

$$\frac{3y}{3} = \frac{-2x}{3} + \frac{3}{3}$$

$$y = -\frac{2}{3}x + 1 \qquad \text{The slope is } -\frac{2}{3}.$$

The slope of a line *perpendicular* to this line must be the opposite of the reciprocal of $-\frac{2}{3}$; hence, $m = \frac{3}{2}$. Using $m = \frac{3}{2}$ and the known point $(4, 3)$, we can apply the point-slope formula to find an equation of the line.

$$y - y_1 = m(x - x_1) \qquad \text{Apply the point-slope formula.}$$

$$y - 3 = \frac{3}{2}(x - 4) \qquad \text{Substitute } m = \frac{3}{2} \text{ and } (4, 3) \text{ for } (x_1, y_1).$$

$$y - 3 = \frac{3}{2}x - 6 \qquad \text{Clear parentheses.}$$

$$y = \frac{3}{2}x - 3 \qquad \text{Solve for } y.$$

Skill Practice

10. Find an equation of the line passing through the point $(1, -6)$ and perpendicular to the line $x + 2y = 8$. Write the answer in slope-intercept form.

Answers

9. $y = 2x - 9$
10. $y = 2x - 8$

From Example 8, the line defined by $y = \frac{3}{2}x - 3$ should be perpendicular to the line $y = -\frac{2}{3}x + 1$ and should pass through the point $(4, 3)$.

   Note: In this example, we are using a *square window option*, which sets the scale to display distances on the x- and y-axes as equal units of measure.

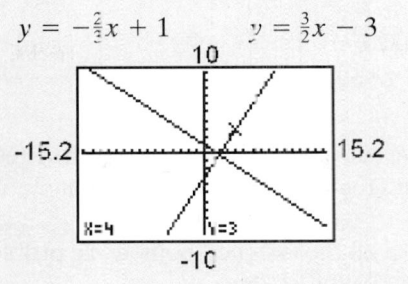

3. Equations of a Line: A Summary

A linear equation can be written in several different forms, as summarized in Table 2-2.

Table 2-2

Form	Example	Comments
Standard Form $Ax + By = C$	$2x + 3y = 6$	A and B must not *both* be zero.
Horizontal Line $y = k$ (k is constant)	$y = 3$	The slope is zero, and the y-intercept is $(0, k)$.
Vertical Line $x = k$ (k is constant)	$x = -2$	The slope is undefined and the x-intercept is $(k, 0)$.
Slope-Intercept Form $y = mx + b$ Slope is m. y-Intercept is $(0, b)$.	$y = -2x + 5$ Slope $= -2$. y-Intercept is $(0, 5)$.	Solving a linear equation for y results in slope-intercept form. The coefficient of the x-term is the slope, and the constant defines the location of the y-intercept.
Point-Slope Formula $y - y_1 = m(x - x_1)$ Slope is m and (x_1, y_1) is a point on the line.	$m = -2$ $(x_1, y_1) = (3, 1)$ $y - 1 = -2(x - 3)$	This formula is typically used to build an equation of a line when a point on the line is known and the slope is known.

   Although it is important to understand and apply slope-intercept form and the point-slope formula, they are not necessarily applicable to all problems. Example 9 illustrates how a little ingenuity may lead to a simple solution.

Example 9   Finding an Equation of a Line

Find an equation of the line passing through the point $(-4, 1)$ and perpendicular to the x-axis.

Solution:

Any line perpendicular to the x-axis must be *vertical*. Recall that all vertical lines can be written in the form $x = k$, where k is constant. A quick sketch can help find the value of the constant (Figure 2-25).

   Because the line must pass through a point whose x-coordinate is -4, the equation of the line is $x = -4$.

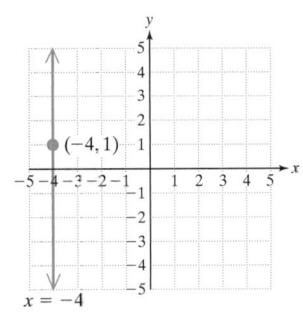

Figure 2-25

Skill Practice

11. Write an equation of the line through the point $(20, 50)$ and having a slope of 0.

Answer

11. $y = 50$

Section 2.3 Practice Exercises

Test yourself.

Yes _____ No _____ Did you have sufficient time to study for the test on the last chapter? If not, what could you have done to create more time for studying?

Yes _____ No _____ Did you work all the assigned homework problems in this chapter?

Yes _____ No _____ If you encountered difficulty in this chapter, did you see your instructor or tutor for help?

Yes _____ No _____ Have you taken advantage of the textbook supplements such as the *Student Solution Manual*?

Vocabulary and Key Concepts

1. **a.** Consider a line with slope m and y-intercept $(0, b)$. The slope-intercept form of an equation of the line is
   _____ .

   b. An equation of a line written in the form $Ax + By = C$, where A and B are not both zero, is said to be in _____ form.

   c. A line defined by an equation $y = k$, where k is a constant is a (horizontal/vertical) line.

   d. A line defined by an equation $x = k$, where k is a constant is a (horizontal/vertical) line.

   e. Given the slope-intercept form of an equation of a line, $y = mx + b$, the value of m is the _____ and b is the _____ .

   f. Given a point (x_1, y_1) on a line with slope m, the point-slope formula is given by _____ .

Review Exercises

2. Given $\dfrac{x}{2} + \dfrac{y}{3} = 1$

   a. Find the x-intercept.

   b. Find the y-intercept.

   c. Sketch the graph.

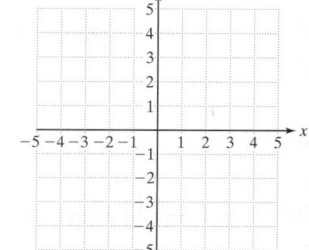

3. **a.** What is the slope of a horizontal line?

   b. The slope of a vertical line is _____ .

4. Using slopes, how do you determine whether two lines are parallel?

5. Using the slopes of two lines, how do you determine whether the lines are perpendicular?

6. Write the formula to find the slope of a line given two points (x_1, y_1) and (x_2, y_2).

Concept 1: Slope-Intercept Form

For Exercises 7–18, determine the slope and the y-intercept of the line. **(See Example 1.)**

7. $y = -\dfrac{2}{3}x - 4$ **8.** $y = \dfrac{3}{7}x - 1$ **9.** $y = 2 + 3x$ **10.** $y = 5 - 7x$

11. $17x + y = 0$ **12.** $x + y = 0$ **13.** $18 = 2y$ **14.** $-7 = \dfrac{1}{2}y$

15. $8x + 12y = 9$ **16.** $-9x + 10y = -4$ **17.** $y = 0.625x - 1.2$ **18.** $y = -2.5x + 1.8$

In Exercises 19–24, match the equation with the correct graph.

19. $y = \dfrac{3}{2}x - 2$ **20.** $y = -x + 3$ **21.** $y = \dfrac{13}{4}$ Animation

22. $y = x + \dfrac{1}{2}$ **23.** $x = -2$ **24.** $y = -\dfrac{1}{2}x + 2$

a.

b.

c.

d.

e.

f.
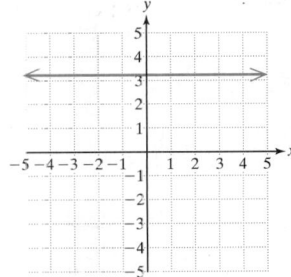

For Exercises 25–30, write the equations in slope-intercept form (if possible). Then graph each line, using the slope and y-intercept. **(See Example 2.)**

25. $y - 2 = 4x$ **26.** $3x = 5 - y$ **27.** $3x + 2y = 6$ Animation

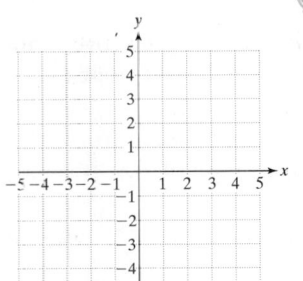

28. $x - 2y = 8$

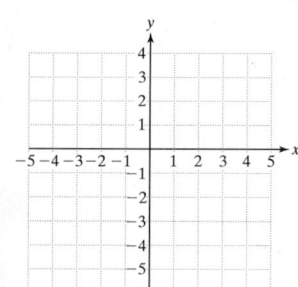

29. $2x - 5y = 0$

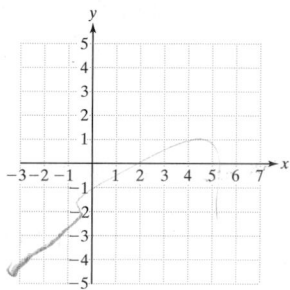

30. $3x - y = 0$

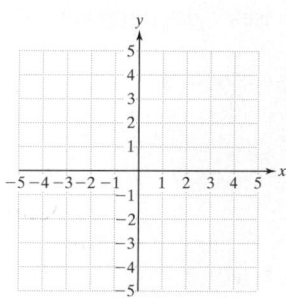

31. Given the standard form of a linear equation $Ax + By = C$, $B \neq 0$, solve for y and write the equation in slope-intercept form. What is the slope of the line? What is the y-intercept?

32. Use the result of Exercise 31 to determine the slope and y-intercept of the line $3x + 7y = 9$.

For Exercises 33–38, the equations of two lines are given. Determine if the lines are parallel, perpendicular, or neither. **(See Example 3.)**

33. $-3y = 5x - 1$
$6x = 10y - 12$

34. $x = 6y - 3$
$3x + \dfrac{1}{2}y = 0$

35. $3x - 4y = 12$
$\dfrac{1}{2}x - \dfrac{2}{3}y = 1$

36. $4.8x = 1.2y + 3.6$
$y - 1 = 4x$

37. $3y = 5x + 6$
$5x + 3y = 9$

38. $-y = 3x - 2$
$-6x + 2y = 6$

For Exercises 39–44, use the slope-intercept form of a line to find an equation of the line having the given slope and passing through the given point. **(See Example 4.)**

39. $m = 3, (0, 5)$

40. $m = -4, (0, 3)$

41. $m = 2, (4, -3)$

42. $m = 3, (-1, 5)$

43. $m = -\dfrac{4}{5}, (10, 0)$

44. $m = -\dfrac{2}{7}, (-3, 1)$

Concept 2: The Point-Slope Formula

For Exercises 45–74, write an equation of the line satisfying the given conditions. Write the answer in slope-intercept form or standard form.

45. The line passes through the point $(0, -2)$ and has a slope of 3.

46. The line passes through the point $(0, 5)$ and has a slope of $-\dfrac{1}{2}$.

47. The line passes through the point $(2, 7)$ and has a slope of 2. **(See Example 5.)**

48. The line passes through the point $(3, 10)$ and has a slope of -2.

49. The line passes through the point $(-2, -5)$ and has a slope of -3.

50. The line passes through the point $(-1, -6)$ and has a slope of 4.

51. The line passes through the point $(6, -3)$ and has a slope of $-\dfrac{4}{5}$.

52. The line passes through the point $(7, -2)$ and has a slope of $\dfrac{7}{2}$.

53. The line passes through $(0, 4)$ and $(3, 0)$. **(See Example 6.)**

54. The line passes through $(1, 1)$ and $(3, 7)$.

55. The line passes through $(6, 12)$ and $(4, 10)$.

56. The line passes through $(-2, -1)$ and $(3, -4)$.

57. The line passes through $(-5, 2)$ and $(-1, 2)$.

58. The line passes through $(-4, -1)$ and $(2, -1)$.

59. The line contains the point $(3, 2)$ and is parallel to a line with a slope of $-\frac{3}{4}$. **(See Example 7.)**

60. The line contains the point $(-1, 4)$ and is parallel to a line with a slope of $\frac{1}{2}$.

61. The line contains the point $(3, 2)$ and is perpendicular to a line with a slope of $-\frac{3}{4}$. **(See Example 8.)**

62. The line contains the point $(-2, 5)$ and is perpendicular to a line with a slope of $\frac{1}{2}$.

63. The line contains the point $(2, -5)$ and is parallel to $3x - 4y = -7$.

64. The line contains the point $(-6, -1)$ and is parallel to $2x + 3y = -12$.

65. The line contains the point $(-8, -1)$ and is perpendicular to $-15x + 3y = 9$.

66. The line contains the point $(4, -2)$ and is perpendicular to $4x + 3y = -6$.

67. The line contains the point $(4, 0)$ and is parallel to the line defined by $3x = 2y$.

68. The line contains the point $(-3, 0)$ and is parallel to the line defined by $-5x = 6y$.

69. The line is perpendicular to the line defined by $3y + 2x = 21$ and passes through the point $(2, 4)$.

70. The line is perpendicular to $7y - x = -21$ and passes through the point $(-14, 8)$.

71. The line is perpendicular to $\frac{1}{2}y = x$ and passes through $(-3, 5)$.

72. The line is perpendicular to $-\frac{1}{4}y = x$ and passes through $(-1, -5)$.

73. The line is parallel to the line $3x + y = 7$ and passes through the origin.

74. The line is parallel to the line $-2x + y = -5$ and passes through the origin.

Concept 3: Equations of a Line: A Summary

For Exercises 75–82, write an equation of the line satisfying the given conditions.

75. The line passes through $(2, -3)$ and has a zero slope.

76. The line contains the point $(\frac{5}{2}, 0)$ and has an undefined slope.

77. The line contains the point $(2, -3)$ and has an undefined slope. **(See Example 9.)**

78. The line contains the point $(\frac{5}{2}, 0)$ and has a zero slope.

79. The line is parallel to the x-axis and passes through $(4, 5)$.

80. The line is perpendicular to the x-axis and passes through $(4, 5)$.

81. The line is parallel to the line $x = 4$ and passes through $(5, 1)$.

82. The line is parallel to the line $y = -2$ and passes through $(-3, 4)$.

Expanding Your Skills

83. Is the equation $x = -2$ in slope-intercept form? Identify the slope and y-intercept.

84. Is the equation $x = 1$ in slope-intercept form? Identify the slope and y-intercept.

85. Is the equation $y = 3$ in slope-intercept form? Identify the slope and y-intercept.

86. Is the equation $y = -5$ in slope-intercept form? Identify the slope and y-intercept.

For Exercises 87–90, write an equation of the given line. Write the answer in slope-intercept form, if possible.

87.

88.

89.

90.

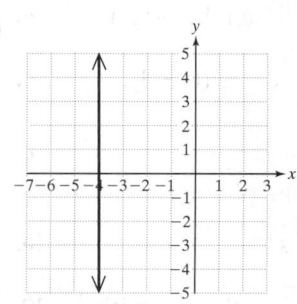

Graphing Calculator Exercises

For Exercises 91–94, use a graphing calculator to graph the lines on the same viewing window. Then explain how the lines are related.

91. $y_1 = \dfrac{1}{2}x + 4$

$y_2 = \dfrac{1}{2}x - 2$

92. $y_1 = -\dfrac{1}{3}x + 5$

$y_2 = -\dfrac{1}{3}x - 3$

93. $y_1 = x - 2$

$y_2 = 2x - 2$

$y_3 = 3x - 2$

94. $y_1 = -2x + 1$

$y_2 = -3x + 1$

$y_3 = -4x + 1$

For Exercises 95–96, use a graphing calculator to graph the lines on a square viewing window. Then explain how the lines are related.

95. $y_1 = 4x - 1$

$y_2 = -\dfrac{1}{4}x - 1$

96. $y_1 = \dfrac{1}{2}x - 3$

$y_2 = -2x - 3$

97. Use a graphing calculator to graph the equation from Exercise 53. Use the *Value* feature to verify that the line passes through the points $(0, 4)$ and $(3, 0)$.

98. Use a graphing calculator to graph the equation from Exercise 54. Use the *Value* feature to verify that the line passes through the points $(1, 1)$ and $(3, 7)$.

Problem Recognition Exercises

Characteristics of Linear Equations

For Exercises 1–20, choose the equation(s) from column B whose graph satisfies the condition described in column A. Give all possible answers.

Column A

1. Line whose slope is positive.

2. Line whose slope is negative.

3. Line that passes through the origin.

4. Line that contains the point $(2, 0)$.

5. Line whose y-intercept is $(0, -3)$.

6. Line whose y-intercept is $(0, 0)$.

7. Line whose slope is $-\dfrac{1}{3}$.

8. Line whose slope is $\dfrac{1}{2}$.

9. Line whose slope is 0.

10. Line whose slope is undefined.

11. Line that is parallel to the line with equation $x + 3y = 6$.

12. Line perpendicular to the line with equation $x - 4y = -4$.

13. Line that is vertical.

14. Line that is horizontal.

15. Line whose x-intercept is $(12, 0)$.

16. Line whose x-intercept is $\left(\dfrac{1}{5}, 0\right)$.

17. Line that has no x-intercept.

18. Line that is perpendicular to the x-axis.

19. Line with a negative slope and positive y-intercept.

20. Line with a positive slope and negative y-intercept.

Column B

a. $y = -4x$

b. $2x - 4y = 4$

c. $y = -\dfrac{1}{3}x - 3$

d. $3x + 5y = 10$

e. $3y = -9$

f. $y = 5x - 1$

g. $4x + 1 = 9$

h. $x + 3y = 12$

Applications of Linear Equations and Modeling

Section 2.4

1. Writing a Linear Model

A **mathematical model** is a formula or equation that represents a relationship between two or more variables in a real-world application. Algebra (or some other field of mathematics) can then be used to solve the problem. The use of mathematical models is found throughout the physical and biological sciences, sports, medicine, economics, business, and many other fields.

For an equation written in slope-intercept form, $y = mx + b$, the term mx is called the *variable term*. The value of this term changes with different values of x. The term b is called the *constant term* and it remains unchanged regardless of the value of x. The slope of the line, m, is called the *rate of change*. A linear equation can be created if the rate of change and the constant are known.

Concepts

1. Writing a Linear Model
2. Interpreting a Linear Model
3. Finding a Linear Model from Observed Data Points

Example 1 **Writing a Linear Model**

Buffalo, New York, had 2 ft (24 in.) of snow on the ground before a snowstorm. During the storm, snow fell at an average rate of $\frac{5}{8}$ in./hr.

a. Write a linear equation to compute the total snow depth y after x hr of the storm.

b. Graph the equation.

c. Use the equation to compute the depth of snow after 8 hr.

d. If the snow depth was 31.5 in. at the end of the storm, determine how long the storm lasted.

Solution:

a. The constant or base amount of snow before the storm began is 24 in. The rate of change is given by $\frac{5}{8}$ in. of snow per hour. If m is replaced by $\frac{5}{8}$ and b is replaced by 24, we have the linear equation

$$y = mx + b$$

$$y = \frac{5}{8}x + 24$$

b. The equation is in slope-intercept form, and the corresponding graph is shown in the figure.

c. $y = \frac{5}{8}x + 24$

$$y = \frac{5}{8}(8) + 24 \qquad \text{Substitute } x = 8.$$

$$y = 5 + 24 \qquad \text{Solve for } y.$$

$$y = 29 \text{ in.}$$

The snow depth was 29 in. after 8 hr. The corresponding ordered pair is $(8, 29)$ and can be confirmed from the graph.

Avoiding Mistakes

A mathematical model *may be reasonable only* for specific values of the input variable. In Example 1, the model $y = \frac{5}{8}x + 24$ is reasonable for $0 \leq x \leq 12$. For x values less than 0, the snow storm hadn't started. For x values greater than 12, the storm had ended. We say that "model breakdown" occurs for values of x outside the reasonable range.

d. $\qquad y = \frac{5}{8}x + 24$

$$31.5 = \frac{5}{8}x + 24 \qquad \text{Substitute } y = 31.5.$$

$$8(31.5) = 8\left(\frac{5}{8}x + 24\right) \qquad \text{Multiply by 8 to clear fractions.}$$

$$252 = 5x + 192 \qquad \text{Clear parentheses.}$$

$$60 = 5x \qquad \text{Solve for } x.$$

$$12 = x$$

The storm lasted for 12 hr. The corresponding ordered pair is $(12, 31.5)$ and can be confirmed from the graph.

Skill Practice When Joe graduated from college, he had $1000 in his savings account. When he began working, he decided he would add $120 per month to his savings account.

1. Write a linear equation to compute the amount of money y in Joe's account after x months of saving.
2. Use the equation to compute the amount of money in Joe's account after 6 months.
3. Joe needs $3160 for a down payment for a car. How long will it take for Joe's account to reach this amount?

2. Interpreting a Linear Model

Example 2 Interpreting a Linear Model

In 1938, President Franklin D. Roosevelt signed a bill enacting the Fair Labor Standards Act of 1938 (FLSA). In its final form, the act banned oppressive child labor and set the minimum hourly wage at 25 cents and the maximum workweek at 44 hr. Over the years, the minimum hourly wage has been increased by the government to meet the rising cost of living.

The minimum hourly wage y (in dollars per hour) in the United States since 1970 can be approximated by the equation

$$y = 0.14x + 1.60 \qquad x \geq 0$$

where x represents the number of years since 1970 ($x = 0$ corresponds to 1970, $x = 1$ corresponds to 1971, and so on) (Figure 2-26).

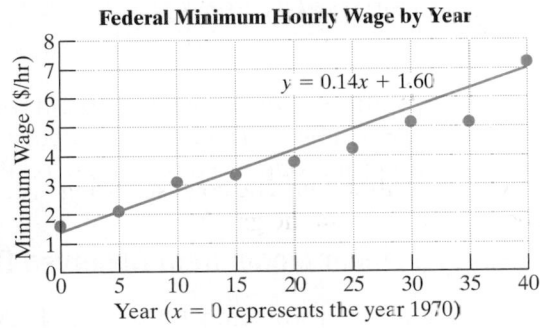

Federal Minimum Hourly Wage by Year

$y = 0.14x + 1.60$

Year ($x = 0$ represents the year 1970)

Figure 2-26

a. Determine the slope of the line and interpret the meaning of the slope as a rate of change.

b. Find the y-intercept of the line and interpret the meaning of the y-intercept in the context of this problem.

c. Use the linear equation to approximate the minimum wage in 1985.

Answers
1. $y = 120x + 1000$
2. $1720   3. 18 months

Solution:

a. The equation $y = 0.14x + 1.60$ is written in slope-intercept form. The slope is 0.14 and indicates that minimum hourly wage rose an average of $0.14 per year since 1970.

b. The y-intercept is $(0, 1.60)$. The y-intercept indicates that the minimum wage in the year 1970 ($x = 0$) was approximately $1.60 per hour.

c. The year 1985 is 15 years after the year 1970. Substitute $x = 15$ into the linear equation.

$$y = 0.14x + 1.60$$

$$y = 0.14(15) + 1.60 \qquad \text{Substitute } x = 15.$$

$$y = 2.1 + 1.60$$

$$y = 3.70$$

According to the linear model, the minimum wage in 1985 was approximately $3.70 per hour. (The actual minimum wage in 1985 was $3.35 per hour.)

Skill Practice One cell-phone plan charges a monthly fee plus a charge per minute for the number of minutes used beyond 400 min. If a customer exceeds the 400-min cap, then the monthly fee, y (in dollars), is given by $y = 0.40x + 49.99$, where x is the number of minutes used beyond 400 min.

4. What is the slope? Interpret its meaning in the context of the problem.

5. What is the y-intercept?

6. Use the equation to determine the cost of using 445 min in this plan.

3. Finding a Linear Model from Observed Data Points

Graphing a set of data points offers a visual method to determine whether the points follow a linear pattern. When two variables are related, it is often desirable to find a mathematical equation (or *model*) to describe the relationship.

Example 3   **Writing a Linear Model from Observed Data**

Figure 2-27 represents the number of women, y (in thousands), who graduated from law school in the United States by year. Let x represent the number of years since 1980.

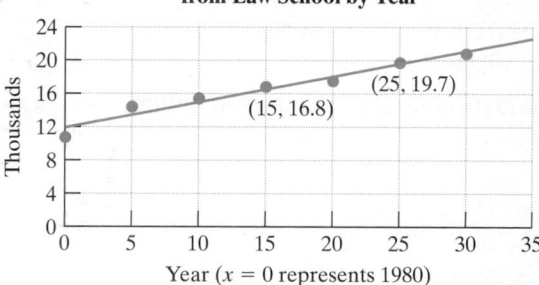

Source: U.S. National Center for Education Statistics

Figure 2-27

Answers

4. The slope is 0.40. The customer is charged $0.40 for each minute used beyond 400 min.

5. The y-intercept is $(0, 49.99)$. This means that if 0 min is used beyond 400 min, the customer is charged $49.99.

6. $67.99

a. Use the given ordered pairs to find a linear equation to model the number of women graduating from law school by year.

b. Determine the slope of the line and interpret the meaning of the slope as a rate of change.

c. Use the linear equation to predict the number of women who will graduate from law school in the year 2015.

d. Would it be practical to use the linear model to predict the number of women who would graduate from law school for the year 2050?

Solution:

a. The slope formula can be used to compute the slope of the line between the points.

$$(15, 16.8) \quad \text{and} \quad (25, 19.7)$$
$$(x_1, y_1) \qquad\qquad (x_2, y_2)$$
                                                   Label the points.

$$m = \frac{y_2 - y_1}{x_2 - x_1} = \frac{19.7 - 16.8}{25 - 15} = \frac{2.9}{10} = 0.29 \qquad \text{Apply the slope formula.}$$

$$y - y_1 = m(x - x_1) \qquad\qquad\qquad \text{Apply the point-slope formula.}$$

$$y - 16.8 = 0.29(x - 15) \qquad\qquad \text{Use the point } (15, 16.8)$$
                                                   and $m = 0.29$.

$$y - 16.8 = 0.29x - 4.35 \qquad\qquad \text{Clear parentheses.}$$

$$y = 0.29x + 12.45 \qquad\qquad\quad \text{Solve for } y. \text{ The equation is}$$
                                                 in slope-intercept form.

b. The slope is 0.29. This means that the number of women who graduate from law school increased at a rate of 0.29 thousand (290) per year during this time period.

c. The year 2015 is 35 years after the year 1980. Therefore, substitute $x = 35$.

$$y = 0.29(35) + 12.45$$

$$= 22.6$$

In the year 2015, we predict that approximately 22,600 women will graduate from law school.

d. It would not be practical to use this equation to predict the number of women who will graduate from law school in the year 2050. It is unreasonable to assume that the linear trend will continue this far beyond the observed data.

Skill Practice The figure shows data relating the cost of college textbooks (in dollars) to the number of pages in the book. Let y represent the cost of the book, and let x represent the number of pages.

7. Use the ordered pairs indicated in the figure to write a linear equation to model the cost of textbooks (in dollars) versus the number of pages.

8. What is the slope of this line and what does it mean in the context of this problem?

9. Use the equation to predict the cost of a textbook that has 360 pages.

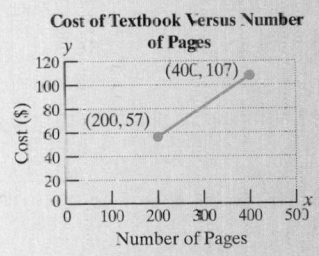

Cost of Textbook Versus Number of Pages

Answers

7. $y = 0.25x + 7$

8. $m = 0.25$; The cost of the textbook goes up by \$0.25 per page.

9. \$97

Section 2.4 Practice Exercises

Study Skills Exercise

On test day, take a look at any formulas or important points that you had to memorize before you enter the classroom. Then when you sit down to take your test, write these formulas on the test or on scrap paper. This is called a memory dump. Write down the formulas from Sections 2.1–2.4.

Vocabulary and Key Concepts

1. A mathematical _____ is a formula or equation that represents a relationship between two or more variables.

Review Exercises

2. Determine the x- and y-intercepts for the line defined by $30x - 40y = 1200$.

For Exercises 3–5,

    a. Find the slope (if possible) of the line passing through the two points.

    b. Find an equation of the line passing through the two points. Write the answer in slope-intercept form (if possible) and in standard form.

    c. Graph the line by using the slope and y-intercept. Verify that the line passes through the two given points.

3. $(-3, 0)$ and $(3, -2)$ **4.** $(1, -1)$ and $(3, -5)$ **5.** $(-4, 3)$ and $(-2, 3)$

          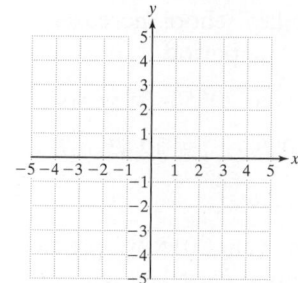

6. What is the slope of the line defined by $x = -2$?

Concept 1: Writing a Linear Model

7. A luxury car rental company charges $120 per day in addition to a flat fee of $65 for insurance. **(See Example 1.)**

    a. Write an equation that represents the cost y (in dollars) to rent the car for x days.

    b. Graph the equation.

    c. What is the y-intercept and what does it mean in the context of this problem?

    d. Use the model from part (a) to determine the cost of driving the rental car for 2 days, 5 days, and 7 days.

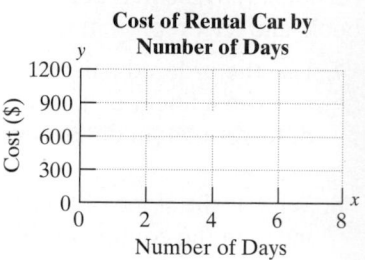

e. If the company offers a weekly rate of $799, is this a better deal than paying by the day for 7 days?

f. Find the total cost of driving the car for 4 days if the sales tax is 6%.

g. Is it reasonable to use negative values of x in the equation from part (a)? Why or why not?

8. Alex is a sales representative and earns a base salary of $1000 per month plus a 4% commission on his sales for the month.

a. Write a linear equation that expresses Alex's monthly salary y in terms of his sales x.

b. Graph the equation.

c. What is the y-intercept and what does it represent in the context of this problem?

d. What is the slope of the line and what does it represent in the context of this problem?

e. How much will Alex make if his sales for a given month are $30,000?

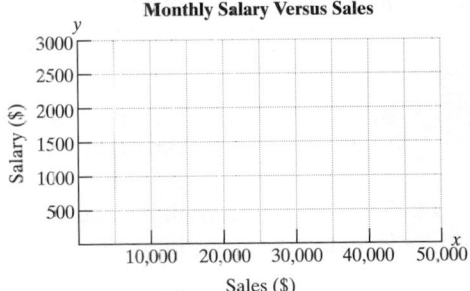

9. Ava recently purchased a home in Crescent Beach, Florida. Her property taxes for the first year are $2742. Ava estimates that her taxes will increase at a rate of $52 per year.

a. Write an equation to compute Ava's yearly property taxes. Let y be the amount she pays in taxes, and let x be the time in years.

b. Graph the line.

c. What is the slope of this line? What does the slope of the line represent in the context of this problem?

d. What is the y-intercept? What does the y-intercept represent in the context of this problem?

e. What will Ava's yearly property tax be in 10 years? In 15 years?

10. Millage rate is the amount per $1000 that is often used to calculate property tax. For example, a home with a $80,000 taxable value in a municipality with a 19 mil tax rate would require $(0.019)(\$80,000) = \$1520$ in property taxes.

a. In one county, homeowners pay a flat tax of $156 plus a rate of 19 mil on the taxable value of a home. Write a linear equation that represents the total property tax y (in $) for a home with a taxable value of x dollars.

b. Determine the amount of property tax on a home with a taxable value of $90,000.

c. If the property tax on a home is $2436, determine the taxable value of the home.

Concept 2: Interpreting a Linear Model

11. Sound travels at approximately one-fifth of a mile per second. Therefore, the difference in time, x (in seconds), between seeing lightning and hearing thunder can be used to estimate the distance y (in miles) between a storm and an observer. The distance of the storm can be approximated by the equation $y = 0.2x$, where $x \geq 0$. **(See Example 2.)**

    a. Use the linear model to determine the distance between a storm and an observer for the following times between seeing lightning and hearing thunder: 4 sec, 12 sec, and 16 sec.

    b. If a storm is 4.2 mi away, how many seconds will pass between seeing lightning and hearing thunder?

12. The force y (in pounds) required to stretch a particular spring x in. beyond its rest (or "equilibrium") position is given by the equation $y = 2.5x$, where $0 \leq x \leq 20$.

    a. Use the equation to determine the amount of force necessary to stretch the spring 6 in. from its rest position. How much force is necessary to stretch the spring twice as far?

    b. If 45 lb of force is exerted on the spring, how far will the spring be stretched?

13. The equation $y = 9.4x + 35.7$ represents the average amount y (in \$) spent on video games per person in the United States where x is the number of years since 2006.

    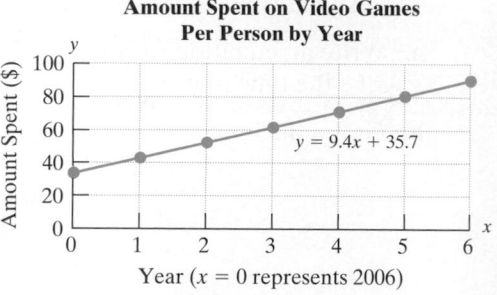

    a. Use the equation to approximate the average amount spent per person in the year 2010.

    b. Use the equation to approximate the average amount spent per person in the year 2008 and compare it with the actual amount spent of \$55.80.

    c. What is the slope of the line and what does it mean in the context of this problem?

    d. What is the y-intercept and what does it mean in the context of this problem?

14. Let y represent the average number of miles driven per year for passenger cars in the United States since 1980. Let x represent the year where $x = 0$ corresponds to 1980, $x = 1$ corresponds to 1981, and so on. The average yearly mileage for passenger cars can be approximated by the equation $y = 142x + 9060$, where $x \geq 0$.

    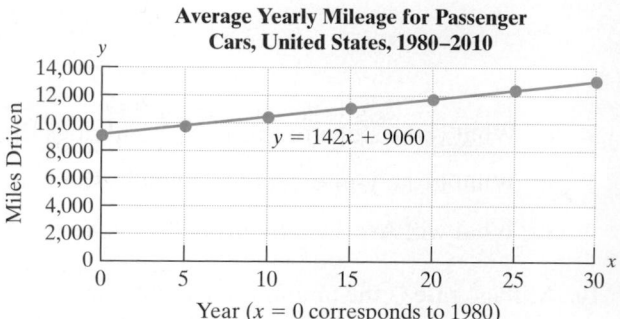

    a. Use the linear equation to approximate the average yearly mileage for passenger cars in the United States in the year 2005.

    b. Use the linear equation to approximate the average mileage for the year 1985, and compare it with the actual value of 9700 mi.

    c. What is the slope of the line and what does it mean in the context of this problem?

    d. What is the y-intercept and what does it mean in the context of this problem?

Concept 3: Finding a Linear Model from Observed Data Points

15. Windchill is the apparent temperature felt on exposed skin. This is a relationship between air temperature and wind speed. At a fixed air temperature of 5°F, windchill temperature is approximately linear for speeds between 10 mph and 110 mph. At 20 mph the windchill is −15°F and at 50 mph it is −24°F. **(See Example 3.)**

a. Make a graph with wind speed on the *x*-axis and windchill on the *y*-axis. Plot the points (20, −15) and (50, −24), and draw the line through the points.

b. Write an equation of the line through the given points. Write the equation in slope-intercept form.

c. Use the equation from part (b) to estimate the windchill for a wind speed of 40 mph.

d. Use the equation from part (b) to estimate the windchill for a wind speed of 46 mph.

e. What is the slope of the line and what does it mean in the context of this problem?

16. During a drought, the average water level in a retention pond decreased linearly with time.

a. Let *y* represent the water level (in feet) and *x* represent the number of days since the beginning of the drought. Use the ordered pairs given to write a linear equation to estimate the water level based on the number of days since the drought began.

b. Use the linear equation to approximate the water level after 15 days.

c. Use the linear equation to approximate the water level after 25 days.

d. What is the slope of the line and what does it mean in the context of this problem?

e. Interpret the meaning of the *x*-intercept. Do you think that this linear trend will continue indefinitely? Explain your answer.

17. The graph displays the number of associate degrees conferred in the United States at the end of selected academic years. The variable x represents the number of years since 1970, and the variable y represents the number of associate degrees in thousands.

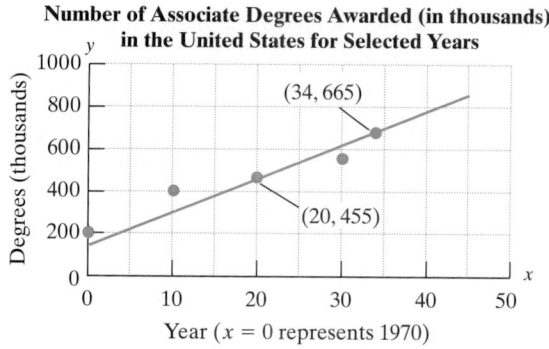

Number of Associate Degrees Awarded (in thousands) in the United States for Selected Years

a. Use the points $(20, 455)$ and $(34, 665)$ to create a linear model of the data.

b. What does the slope mean in the context of this problem?

c. If this linear trend continues, predict the number of associate degrees that will be conferred in the United States in the year 2015.

18. The number of prisoners in federal or state correctional facilities is shown in the figure by year. (*Source:* U.S. Department of Justice)

a. Use the given points to create a linear model of the data.

b. What does the slope mean in the context of this problem?

c. Use the equation in part (a) to predict the number of prisoners in federal or state correctional facilities for the year 2015.

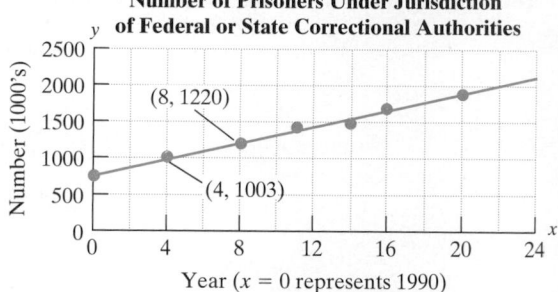

Number of Prisoners Under Jurisdiction of Federal or State Correctional Authorities

Year ($x = 0$ represents 1990)

19. At a concession stand at a high school football game, the owner notices that the relationship between the price of a hot dog and the number of hot dogs sold is linear. If the price is $2.50 per hot dog, then approximately 650 are sold each night. If the price is raised to $3.50, then the number sold drops to 475 per night.

a. Make a graph with the price of hot dogs on the x-axis and the number of hot dogs sold on the y-axis. Use the points $(2.50, 650)$ and $(3.50, 475)$. Then graph the line through the points with $x \geq 0$.

b. Find an equation of the line through the points. Write the equation in slope-intercept form.

c. Use the equation from part (b) to predict the number of hot dogs that would sell if the price were raised to $4.00. Round to the nearest whole unit.

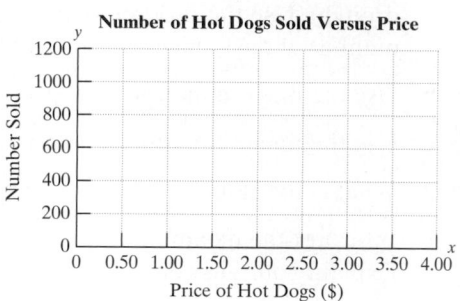

Number of Hot Dogs Sold Versus Price

Price of Hot Dogs ($)

20. Sales at a concession stand indicate that the relationship between the price of a drink and the number of drinks sold is linear. If drinks are sold at $1.00 each, then approximately 1020 are sold each night. If the price is raised to $1.50, then the number sold drops to 820 per night.

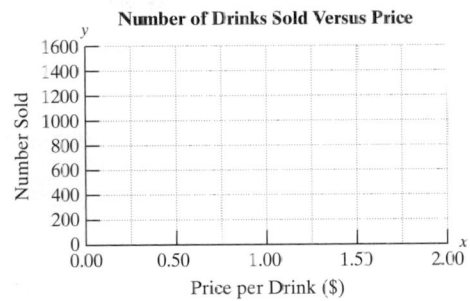

a. Make a graph with the price of drinks on the x-axis and the number of drinks sold on the y-axis. Graph the points $(1.00, 1020)$ and $(1.50, 820)$. Then graph the line through the points with $x \geq 0$.

b. Find an equation of the line through the points. Write the equation in slope-intercept form.

c. Use the equation from part (b) to predict the number of drinks that would sell if the price were $2.00 per drink.

21. In order to advise students properly, a college advisor is interested in the relationship between the number of hours a student studies in an average week and the student's GPA. The data are shown in the table.

a. Let x represent study time and let y represent GPA. Graph the points.

Grade Point Average Versus Weekly Study Time

Student	Study Time (in hours)	GPA
1	15	2.5
2	38	3.9
3	10	2.1
4	24	2.8
5	35	3.3
6	15	2.7
7	45	4.0
8	28	3.1
9	35	3.4
10	10	2.2
11	6	1.8

b. Does there appear to be a linear trend?

c. Use the data points $(28, 3.1)$ and $(10, 2.2)$ to find an equation of the line through these points.

d. Use the equation in part (c) to estimate the GPA of a student who studies for 30 hr a week.

e. Why would the linear model found in part (c) not be realistic for a student who studies more than 46 hr per week?

22. Loraine is enrolled in an algebra class that meets 5 days per week. Her instructor gives a test every Friday. Loraine has a study plan and keeps a portfolio with notes, homework, test corrections, and vocabulary. She also records the amount of time per day that she studies and does homework. The following data represent the amount of time she studied per day and her weekly test grades.

   a. Graph the points on a rectangular coordinate system. Do the data points appear to follow a linear trend?

Time Studied per Day (min) x	Weekly Test Grade (percent) y
60	69
70	74
80	79
90	84
100	89

   b. Find a linear equation that relates Loraine's weekly test score y to the amount of time she studied per day x. (*Hint:* Pick two ordered pairs from the observed data, and find an equation of the line through the points.)

   c. How many minutes should Loraine study per day in order to score at least 90% on her weekly examination? Would the equation used to determine the time Loraine needs to study to get 90% work for other students? Why or why not?

   d. If Loraine is only able to spend $\frac{1}{2}$ hr/day studying her math, predict her test score for that week.

Expanding Your Skills

Points are *collinear* if they lie on the same line. For Exercises 23–26, use the slope formula to determine if the points are collinear. (*Hint:* Three points are collinear if the slope calculated using one pair of points is equal to the slope calculated using a different pair of points.)

23. $(3, -4)\,(0, -5)\,(9, -2)$

24. $(4, 3)\,(-4, -1)\,(2, 2)$

25. $(0, 2)\,(-2, 12)\,(-1, 6)$

26. $(-2, -2)\,(0, -3)\,(-4, -1)$

Graphing Calculator Exercises

27. Use a *Table* feature to confirm your answers to Exercise 11(a).

28. Use a *Table* feature to confirm your answers to Exercise 12(a).

29. Graph the equation $y = -175x + 1087.5$ on the viewing window $0 \le x \le 5$ and $0 \le y \le 1200$. Use the *Value* feature to support your answer to Exercise 19 by showing that the line passes through the points $(2.5, 650)$ and $(3.5, 475)$.

30. Graph the equation $y = -400x + 1420$ on the viewing window $0 \le x \le 3$ and $0 \le y \le 1600$. Use the *Value* feature to support your answer to Exercise 20 by showing that the line passes through the points $(1, 1020)$ and $(1.5, 820)$.

Introduction to Relations

1. Definition of a Relation

In many naturally occurring phenomena, two variables may be linked by some other type of relationship. Table 2-3 shows a correspondence between the length of a woman's femur and her height. (The femur is the large bone in the thigh attached to the knee and hip.)

Concepts

1. **Definition of a Relation**
2. **Domain and Range of a Relation**
3. **Applications Involving Relations**

Table 2-3

Length of Femur (cm) x	Height (in.) y	Ordered Pair
45.5	65.5	→ (45.5, 65.5)
48.2	68.0	→ (48.2, 68.0)
41.8	62.2	→ (41.8, 62.2)
46.0	66.0	→ (46.0, 66.0)
50.4	70.0	→ (50.4, 70.0)

Each data point from Table 2-3 may be represented as an ordered pair. In this case, the first value represents the length of a woman's femur and the second, the woman's height. The set of ordered pairs {(45.5, 65.5), (48.2, 68.0), (41.8, 62.2), (46.0, 66.0), (50.4, 70.0)} defines a relation between femur length and height.

2. Domain and Range of a Relation

Definition of a Relation in x and y

A set of ordered pairs (x, y) is called a **relation in x and y**. Furthermore,

- The set of first components in the ordered pairs is called the **domain of the relation**.
- The set of second components in the ordered pairs is called the **range of the relation**.

Example 1   Finding the Domain and Range of a Relation

Find the domain and range of the relation linking the length of a woman's femur to her height {(45.5, 65.5), (48.2, 68.0), (41.8, 62.2), (46.0, 66.0), (50.4, 70.0)}.

Solution:

Domain: {45.5, 48.2, 41.8, 46.0, 50.4}   Set of first components

Range: {65.5, 68.0, 62.2, 66.0, 70.0}   Set of second components

Skill Practice Find the domain and range of the relation.

1. $\left\{ (0,0), (-8, 4), \left(\frac{1}{2}, 1\right), (-3, 4), (-8, 0) \right\}$

The x and y components that constitute the ordered pairs in a relation do not need to be numerical. This is demonstrated in Example 2.

Answer

1. Domain: $\left\{ 0, -8, \frac{1}{2}, -3 \right\}$;
   range: {0, 4, 1}

| Example 2 | Writing a Relation and Finding Its Domain and Range |

Table 2-4 gives five states in the United States and the corresponding number of representatives in the House of Representatives for a recent year.

Table 2-4

State *x*	Number of Representatives *y*
Alabama	7
California	53
Colorado	7
Florida	25
Kansas	4

a. The data in the table define a relation. Write a list of ordered pairs for this relation.

b. Write the domain and range.

Solution:

a. {(Alabama, 7), (California, 53), (Colorado, 7), (Florida, 25), (Kansas, 4)}

b. Domain: {Alabama, California, Colorado, Florida, Kansas}

Range: {7, 53, 25, 4} (*Note:* The element 7 is not listed twice.)

Skill Practice The table depicts six types of animals and their average longevity.

2. Write the ordered pairs indicated by the relation in the table.

3. Find the domain and range of the relation.

Animal *x*	Longevity (Years) *y*
Bear	22.5
Cat	11
Cow	20.5
Deer	12.5
Dog	11
Elephant	35

A relation may consist of a finite number of ordered pairs or an infinite number of ordered pairs. Furthermore, a relation may be defined by several different methods.

- A relation may be defined as a set of ordered pairs.

$$\{(1, 2), (-3, 4), (1, -4), (3, 4)\}$$

- A relation may be defined by a correspondence (Figure 2-28). The corresponding ordered pairs are $\{(1, 2), (1, -4), (-3, 4), (3, 4)\}$.

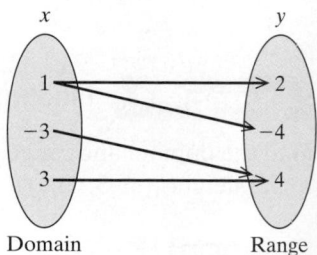

Figure 2-28

- A relation may be defined by a graph (Figure 2-29). The corresponding ordered pairs are $\{(1, 2), (-3, 4), (1, -4), (3, 4)\}$.

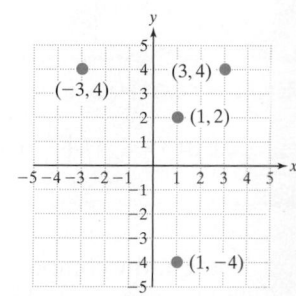

Figure 2-29

Answers

2. {(Bear, 22.5), (Cat, 11), (Cow, 20.5), (Deer, 12.5), (Dog, 11), (Elephant, 35)}

3. Domain: {Bear, Cat, Cow, Deer, Dog, Elephant}; range: {22.5, 11, 20.5, 12.5, 35}

- A relation may be expressed by an equation such as $x = y^2$. The solutions to this equation define an infinite set of ordered pairs of the form $\{(x, y) | x = y^2\}$. The solutions can also be represented by a graph in a rectangular coordinate system (Figure 2-30).

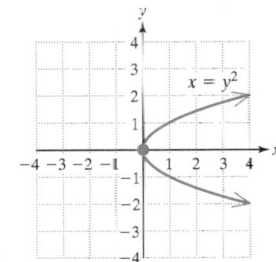

Figure 2-30

Example 3 **Finding the Domain and Range of a Relation**

Write the relation as a set of ordered pairs. Then find the domain and range.

a.

b.

Solution:

a. From the figure, the relation defines the set of ordered pairs:
   $\{(3, -9), (2, -9), (-7, -9)\}$

   Domain: $\{3, 2, -7\}$

   Range: $\{-9\}$

b. The points in the graph make up the set of ordered pairs:
   $\{(-2, -3), (-1, 0), (0, 1), (1, 0), (2, -3)\}$

   Domain: $\{-2, -1, 0, 1, 2\}$

   Range: $\{-3, 0, 1\}$

Skill Practice Find the domain and range of the relations.

4.

5.

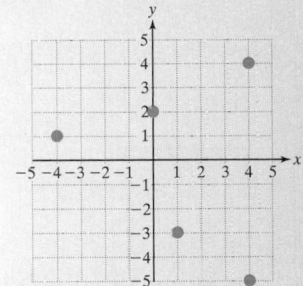

Example 4 Finding the Domain and Range of a Relation

Use interval notation to express the domain and range of the relation.

a.

b.

Solution:

Animation

a.

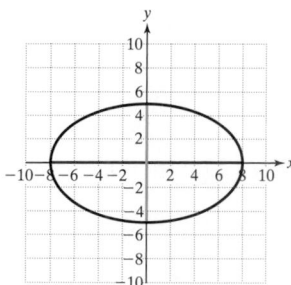

The domain consists of an infinite number of *x*-values extending from −8 to 8 (shown in red). The range consists of all *y*-values from −5 to 5 (shown in blue). Thus, the domain and range must be expressed in set-builder notation or in interval notation.

Domain: $[-8, 8]$

Range: $[-5, 5]$

b.

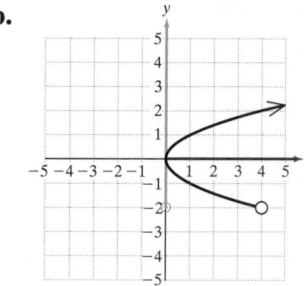

The arrow on the curve indicates that the graph extends infinitely far up and to the right. The open circle means that the graph will end at the point (4, −2), but not include that point.

Domain: $[0, \infty)$

Range: $(-2, \infty)$

Skill Practice Use interval notation to express the domain and range of the relations.

6.

7.

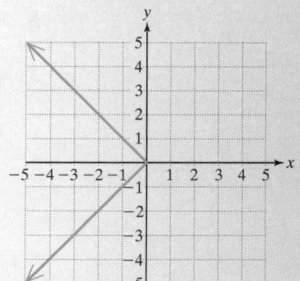

Answers

6. Domain: $[-4, 0]$
  Range: $[-2, 2]$

7. Domain: $(-\infty, 0]$
  Range: $(-\infty, \infty)$

3. Applications Involving Relations

> **Example 5** **Analyzing a Relation**

The data in Table 2-5 depict the length of a woman's femur and her corresponding height. Based on these data, a forensics specialist can find a linear relationship between height y (in inches) and femur length x (in centimeters):

$$y = 0.91x + 24 \qquad 40 \leq x \leq 55$$

From this type of relationship, the height of a woman can be inferred based on skeletal remains.

a. Find the height of a woman whose femur is 46.0 cm.

b. Find the height of a woman whose femur is 51.0 cm.

c. Why is the domain restricted to $40 \leq x \leq 55$?

Table 2-5

Length of Femur (cm) x	Height (in.) y
45.5	65.5
48.2	68.0
41.8	62.2
46.0	66.0
50.4	70.0

Solution:

a. $y = 0.91x + 24$

    $= 0.91(46.0) + 24$ Substitute $x = 46.0$ cm.

    $= 65.86$ The woman is approximately 65.9 in. tall.

b. $y = 0.91x + 24$

    $= 0.91(51.0) + 24$ Substitute $x = 51.0$ cm.

    $= 70.41$ The woman is approximately 70.4 in. tall.

c. The domain restricts femur length to values between 40 cm and 55 cm inclusive. These values are within the normal lengths for an adult female and are in the proximity of the observed data (Figure 2-31).

Figure 2-31

Skill Practice The linear equation, $y = -0.014x + 64.5$, for $1500 \leq x \leq 4000$, relates the weight of a car, x (in pounds), to its gas mileage, y (in mpg).

8. Find the gas mileage in miles per gallon for a car weighing 2550 lb.

9. Find the gas mileage for a car weighing 2850 lb.

10. Why is the domain restricted to $1500 \leq x \leq 4000$?

Answers

8. 28.8 mpg   **9.** 24.6 mpg

10. The relation is valid only for cars weighing between 1500 lb and 4000 lb, inclusive.

Section 2.5 Practice Exercises

Study Skills Exercise

A good way to determine what will be on a test is to look at both your notes and the exercises assigned by your instructor. List five kinds of problems that you think will be on the test for this chapter.

Vocabulary and Key Concepts

1. a. A set of ordered pairs (x, y) is called a _____ in x and y.

  b. The _____ of a relation is the set of first components in the ordered pairs.

  c. The _____ of a relation is the set of second components in the ordered pairs.

Review Exercises

2. Write an equation of a line passing through the points $(3, -4)$ and $(1, 6)$. Write the answer in slope-intercept form.

For Exercises 3–6,

  a. Determine if the equation represents a horizontal line, a vertical line, or a slanted line.

  b. Determine the slope of the line (if it exists).

  c. Determine the x-intercept (if it exists).

  d. Determine the y-intercept (if it exists).

3. $2x - 3 = 4$

4. $2x - 3y = 4$

5. $3x - 2y = 4$

6. $2 - 3y = 4$

Concept 2: Domain and Range of a Relation

For Exercises 7–14,

  a. Write the relation as a set of ordered pairs.

  b. Determine the domain and range. **(See Examples 1–3.)**

7.

Region	Number Living in Poverty (thousands)
Northeast	54.1
Midwest	65.6
South	110.7
West	70.7

8.

x	y
0	3
-2	$\frac{1}{2}$
-7	1
-2	8
5	1

9.

Country	Year of First Man or Woman in Space
USSR	1961
USA	1962
Poland	1978
Vietnam	1980
Cuba	1980

10.

State, x	Year of Statehood, y
Maine	1820
Nebraska	1823
Utah	1847
Hawaii	1959
Alaska	1959

11.

12.

13.

14.

For Exercises 15–30, find the domain and range of the relations. Use interval notation where appropriate.
(See Example 4.)

15.

16.

17.

18.

19.

20.

21.

22.

23.

24.

25.

26.

27.

28.

29.

30.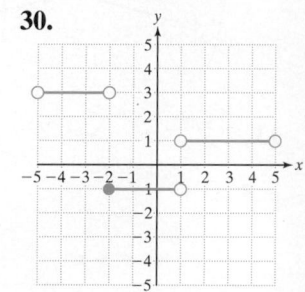

Concept 3: Applications Involving Relations

31. The table gives a relation between the month of the year and the average precipitation for that month for Miami, Florida. **(See Example 5.)**

    a. What is the range element corresponding to April?

    b. What is the range element corresponding to June?

    c. Which element in the domain corresponds to the least value in the range?

    d. Complete the ordered pair: (, 2.66)

    e. Complete the ordered pair: (Sept.,   )

    f. What is the domain of this relation?

Month x	Precipitation (in.) y	Month x	Precipitation (in.) y
Jan.	2.01	July	5.70
Feb.	2.08	Aug.	7.58
Mar.	2.39	Sept.	7.63
Apr.	2.85	Oct.	5.64
May	6.21	Nov.	2.66
June	9.33	Dec.	1.83

Source: U.S. National Oceanic and Atmospheric Administration

32. The table gives a relation between a person's age and the person's maximum recommended heart rate.

    a. What is the domain?

    b. What is the range?

    c. The range element 200 corresponds to what element in the domain?

    d. Complete the ordered pair: (50,   )

    e. Complete the ordered pair: (, 190)

Age (years) x	Maximum Recommended Heart Rate (Beats per Minute) y
20	200
30	190
40	180
50	170
60	160

33. The population of Canada, y (in millions), can be approximated by the relation $y = 0.146x + 31$, where x represents the number of years since 2000.

    a. Approximate the population of Canada in the year 2006.

    b. In what year did the population of Canada reach approximately 32,752,000?

34. World record times for women's track and field events are shown in the table. The women's world record time y (in seconds) required to run x meters can be approximated by the relation $y = 0.159x - 10.79$.

Distance (m)	Time (sec)	Winner's Name and Country
100	10.49	Florence Griffith Joyner (United States)
200	21.34	Florence Griffith Joyner (United States)
400	47.60	Marita Koch (East Germany)
800	113.28	Jarmila Kratochvilova (Czechoslovakia)
1000	148.98	Svetlana Masterkova (Russia)
1500	230.46	Qu Yunxia (China)

    a. Predict the time required for a 500-m race.

    b. Use this model to predict the time for a 1000-m race. Is this value exactly the same as the data value given in the table? Explain.

Expanding Your Skills

35. a. Define a relation with four ordered pairs such that the first element of the ordered pair is the name of a friend and the second element is your friend's place of birth.

    b. State the domain and range of this relation.

36. a. Define a relation with four ordered pairs such that the first element is a state and the second element is its capital.

    b. State the domain and range of this relation.

37. Use a mathematical equation to define a relation whose second component y is 1 less than 2 times the first component x.

38. Use a mathematical equation to define a relation whose second component y is 3 more than the first component x.

39. Use a mathematical equation to define a relation whose second component is the square of the first component.

40. Use a mathematical equation to define a relation whose second component is one-fourth the first component.

Section 2.6 Introduction to Functions

Concepts

1. Definition of a Function
2. Vertical Line Test
3. Function Notation
4. Finding Function Values from a Graph
5. Domain of a Function

1. Definition of a Function

In this section, we introduce a special type of relation called a function.

Definition of a Function

Given a relation in x and y, we say "y is a **function** of x" if, for each element x in the domain, there is exactly one value of y in the range.

Note: This means that no two ordered pairs may have the same first coordinate and different second coordinates.

To understand the difference between a relation that is a function and a relation that is not a function, consider Example 1.

Example 1 Determining Whether a Relation Is a Function

Determine which of the relations define y as a function of x.

a. **b.** **c.**

Solution:

a. This relation is defined by the set of ordered pairs

When $x = 1$, there are *two* possible range elements: $y = 3$ and $y = 4$. Therefore, this relation is *not* a function.

b. This relation is defined by the set of ordered pairs $\{(1, 4), (2, -1), (3, 2)\}$.

Notice that no two ordered pairs have the same value of x but different values of y. Therefore, this relation *is* a function.

c. This relation is defined by the set of ordered pairs $\{(1, 4), (2, 4), (3, 4)\}$.

Notice that no two ordered pairs have the same value of x but different values of y. Therefore, this relation *is* a function.

Skill Practice Determine if the relations define y as a function of x.

1.

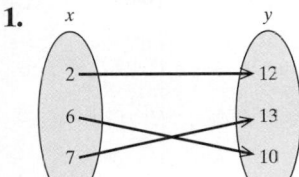

2. $\{(4, 2), (-5, 4), (0, 0), (8, 4)\}$

3. $\{(-1, 6), (8, 9), (-1, 4), (-3, 10)\}$

2. Vertical Line Test

A relation that is not a function has at least one domain element x paired with more than one range value y. For example, the set $\{(4, 2), (4, -2)\}$ does not define a function because two different y-values correspond to the same x. These two points are aligned vertically in the xy-plane, and a vertical line drawn through one point also intersects the other point (see Figure 2-32). If a vertical line drawn through a graph of a relation intersects the graph in more than one point, the relation cannot be a function. This idea is stated formally as the **vertical line test**.

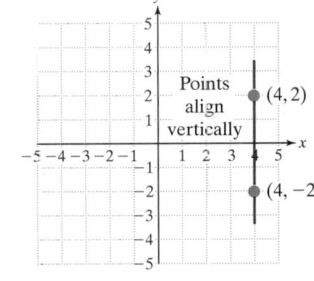

Figure 2-32

> ### The Vertical Line Test
>
> Consider a relation defined by a set of points (x, y) in a rectangular coordinate system. The graph defines y as a function of x if no vertical line intersects the graph in more than one point.

The vertical line test can be demonstrated by graphing the ordered pairs from the relations in Example 1.

a. $\{(1, 3), (1, 4), (2, -1), (3, -2)\}$ **b.** $\{(1, 4), (2, -1), (3, 2)\}$

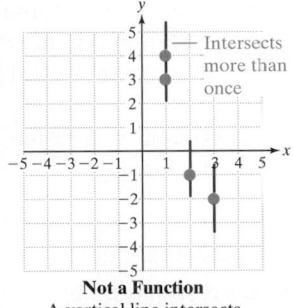

Not a Function
A vertical line intersects
in more than one point.

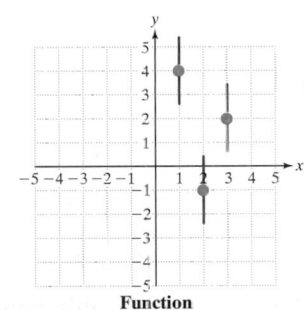

Function
No vertical line
intersects more than once.

Example 2 Using the Vertical Line Test

Use the vertical line test to determine whether the relations define y as a function of x.

a.

b.

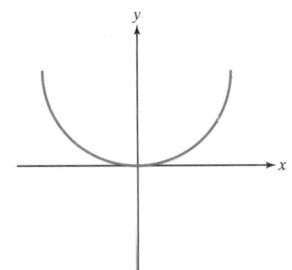

Answers

1. Yes   **2.** Yes   **3.** No

Solution:

a.

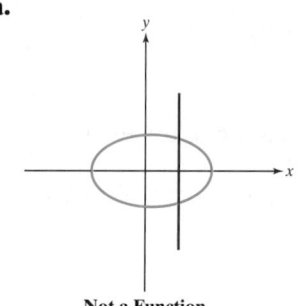

Not a Function
A vertical line intersects
in more than one point.

b.

Function
No vertical line intersects
in more than one point.

Skill Practice Use the vertical line test to determine whether the relations define y as a function of x.

4.

5.

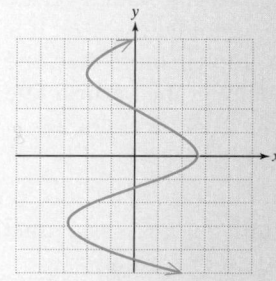

---

3. Function Notation

A function is defined as a relation with the added restriction that each value in the domain must have only one corresponding y value in the range. In mathematics, functions are often given by rules or equations to define the relationship between two or more variables. For example, the equation $y = 2x$ defines the set of ordered pairs such that the y value is twice the x value.

When a function is defined by an equation, we often use **function notation**. For example, the equation $y = 2x$ may be written in function notation as

$$f(x) = 2x$$

- f is the name of the function.
- x is an input value from the domain of the function.
- $f(x)$ is the function value (y value) corresponding to x.

The notation $f(x)$ is read as "f of x" or "the value of the function f at x."

A function may be evaluated at different values of x by substituting x-values from the domain into the function. For example, to evaluate the function defined by $f(x) = 2x$ at $x = 5$, substitute $x = 5$ into the function.

$$f(x) = 2x$$

$$f(5) = 2(5)$$

$$f(5) = 10$$

> **Avoiding Mistakes**
>
> Be sure to note that $f(x)$ is *not* $f \cdot x$.

> **TIP:** $f(5) = 10$ can be written as the ordered pair $(5, 10)$.

Answers

4. Yes   **5.** No

Thus, when $x = 5$, the corresponding function value is 10. We say:

- f of 5 is 10.
- f at 5 is 10.
- f evaluated at 5 is 10.

The names of functions are often given by either lowercase or uppercase letters, such as f, g, h, p, K, and M. The input variable may also be a letter other than x. For example, the function $y = P(t)$ might represent population as a function of time.

Example 3   **Evaluating a Function**

Given the function defined by $g(x) = \frac{1}{2}x - 1$, find the function values.

  a. $g(0)$ **b.** $g(2)$ **c.** $g(4)$ **d.** $g(-2)$

Solution:

a. $g(x) = \dfrac{1}{2}x - 1$ **b.** $g(x) = \dfrac{1}{2}x - 1$

    $g(0) = \dfrac{1}{2}(0) - 1$ $g(2) = \dfrac{1}{2}(2) - 1$

         $= 0 - 1$ $= 1 - 1$

         $= -1$ $= 0$

We say that "g of 0 is -1."   We say that "g of 2 is 0."
This is equivalent to the ordered   This is equivalent to the ordered
pair $(0, -1)$.   pair $(2, 0)$.

c. $g(x) = \dfrac{1}{2}x - 1$ **d.** $g(x) = \dfrac{1}{2}x - 1$

    $g(4) = \dfrac{1}{2}(4) - 1$ $g(-2) = \dfrac{1}{2}(-2) - 1$

         $= 2 - 1$ $= -1 - 1$

         $= 1$ $= -2$

We say that "g of 4 is 1."   We say that "g of -2 is -2."
This is equivalent to the ordered   This is equivalent to the ordered
pair $(4, 1)$.   pair $(-2, -2)$.

Skill Practice Given the function defined by $f(x) = -2x - 3$, find the function values.

  6. $f(1)$ **7.** $f(0)$ **8.** $f(-3)$ **9.** $f\left(\dfrac{1}{2}\right)$

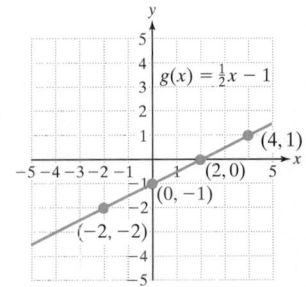

Figure 2-33

Notice that $g(0), g(2), g(4)$, and $g(-2)$ correspond to the ordered pairs $(0, -1)$, $(2, 0), (4, 1)$, and $(-2, -2)$. In the graph, these points "line up." The graph of *all* ordered pairs defined by this function is a line with a slope of $\frac{1}{2}$ and y-intercept of $(0, -1)$ (Figure 2-33). This should not be surprising because the function defined by $g(x) = \frac{1}{2}x - 1$ is equivalent to $y = \frac{1}{2}x - 1$.

Answers

6. -5 **7.** -3 **8.** 3   **9.** -4

Calculator Connections

Topic: Using the *Table* and *Value* Features

The values of $g(x)$ in Example 3 can be found using a *Table* feature.

$$Y_1 = \tfrac{1}{2}x - 1$$

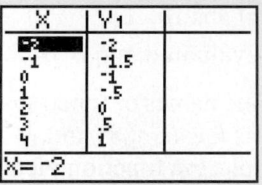

Function values can also be evaluated by using a *Value* feature. The value of $g(4)$ is shown here.

A function may be evaluated at numerical values or at algebraic expressions, as shown in Example 4.

Example 4 **Evaluating Functions**

Given the functions defined by $f(x) = x^2 - 2x$ and $g(x) = 3x + 5$, find the function values.

   a. $f(t)$ **b.** $g(w + 4)$ **c.** $f(-t)$

Solution:

a. $f(x) = x^2 - 2x$

      $f(t) = (t)^2 - 2(t)$ Substitute $x = t$ for all values of x in the function.

         $= t^2 - 2t$ Simplify.

b. $g(x) = 3x + 5$

   $g(w + 4) = 3(w + 4) + 5$ Substitute $x = w + 4$ for all values of x in the function.

          $= 3w + 12 + 5$

          $= 3w + 17$ Simplify.

c. $f(x) = x^2 - 2x$ Substitute $-t$ for x.

   $f(-t) = (-t)^2 - 2(-t)$

        $= t^2 + 2t$ Simplify.

Skill Practice Given the function defined by $g(x) = 4x - 3$, find the function values.

   10. $g(a)$ **11.** $g(x + h)$ **12.** $g(-x)$

Answers

10. $4a - 3$ **11.** $4x + 4h - 3$
12. $-4x - 3$

4. Finding Function Values from a Graph

We can find function values by looking at a graph of the function. The value of $f(a)$ refers to the y-coordinate of a point with x-coordinate a.

Example 5   Finding Function Values from a Graph

Consider the function pictured in Figure 2-34.

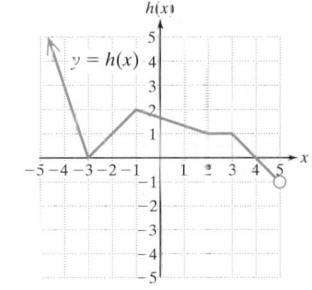

a. Find $h(-1)$.

b. Find $h(2)$.

c. Find $h(5)$.

d. For what value of x is $h(x) = 3$?

e. For what values of x is $h(x) = 0$?

Figure 2-34

Solution:

a. $h(-1) = 2$ This corresponds to the ordered pair $(-1, 2)$.

b. $h(2) = 1$ This corresponds to the ordered pair $(2, 1)$.

c. $h(5)$ is not defined.   The value 5 is not in the domain.

d. $h(x) = 3$ for $x = -4$ This corresponds to the ordered pair $(-4, 3)$.

e. $h(x) = 0$ for $x = -3$ and $x = 4$ These are the ordered pairs $(-3, 0)$ and $(4, 0)$.

Skill Practice Refer to the function graphed here.

13. Find $f(0)$.

14. Find $f(-2)$.

15. Find $f(5)$.

16. For what value(s) of x is $f(x) = 0$?

17. For what value(s) of x is $f(x) = -4$?

5. Domain of a Function

A function is a relation, and it is often necessary to determine its domain and range. Consider a function defined by the equation $y = f(x)$. The **domain** of f is the set of all x-values that when substituted into the function produce a real number. The **range** of f is the set of all y-values corresponding to the values of x in the domain.

To find the domain of a function defined by $y = f(x)$, keep these guidelines in mind.

- Exclude values of x that make the denominator of a fraction zero.
- Exclude values of x that make the expression within a square root negative.

Answers

13. 3   **14.** 1

15. Not defined

16. $x = -4$ and $x = 4$

17. $x = -5$

Example 6   **Finding the Domain of a Function**

Write the domain in interval notation.

a. $f(x) = \dfrac{x + 7}{2x - 1}$ b. $h(x) = \dfrac{x - 4}{x^2 + 9}$

c. $k(t) = \sqrt{t + 4}$ d. $g(t) = t^2 - 3t$

Solution:

a. $f(x) = \frac{x + 7}{2x - 1}$ will not be a real number when the denominator is zero, that is, when

$$2x - 1 = 0$$
$$2x = 1$$
$$x = \frac{1}{2} \qquad \text{The value } x = \tfrac{1}{2} \text{ must be } excluded \text{ from the domain.}$$

Interval notation: $\left(-\infty, \dfrac{1}{2}\right) \cup \left(\dfrac{1}{2}, \infty\right)$

b. For $h(x) = \frac{x - 4}{x^2 + 9}$ the quantity x^2 is greater than or equal to 0 for all real numbers x, and the number 9 is positive. The sum $x^2 + 9$ must be *positive* for all real numbers x. The denominator will never be zero; therefore, the domain is the set of all real numbers.

Interval notation: $(-\infty, \infty)$

c. The function defined by $k(t) = \sqrt{t + 4}$ will not be a real number when the radicand is negative. The domain is the set of all t values that make the radicand *greater than or equal to zero:*

$$t + 4 \geq 0$$
$$t \geq -4$$

Interval notation: $[-4, \infty)$

d. The function defined by $g(t) = t^2 - 3t$ has no restrictions on its domain because any real number substituted for t will produce a real number. The domain is the set of all real numbers.

Interval notation: $(-\infty, \infty)$

Skill Practice Write the domain in interval notation.

18. $f(x) = \dfrac{2x + 1}{x - 9}$ 19. $k(x) = \dfrac{-5}{4x^2 + 1}$

20. $g(x) = \sqrt{x - 2}$ 21. $h(x) = x + 6$

Answers
18. $(-\infty, 9) \cup (9, \infty)$ 19. $(-\infty, \infty)$
20. $[2, \infty)$ 21. $(-\infty, \infty)$

Section 2.6 Practice Exercises

Study Skills Exercise

> Look back over your notes for this chapter. Have you highlighted the important topics? Have you underlined the key terms? Have you indicated the places where you are having trouble? If you find that you have problems with a particular topic, write a question that you can ask your instructor either in class or in the instructor's office.

Vocabulary and Key Concepts

1. **a.** Given a relation in x and y, we say that y is a _____ of x if for each element x in the domain, there is exactly one value of y in the range.

   b. If a _____ line intersects the graph of a relation in more than one point, the relation is not a function.

   c. Function notation for the relation $y = 2x + 1$ is $f(x) =$ _____ .

   d. The set of x values in a function is called the _____ of the function, and the set of y values is called the _____ of the function.

   e. To find the domain of a function defined by $y = f(x)$, exclude all values of x that make the _____ of a fraction equal to zero. Also exclude all values of x that make the expression within a square root _____ .

   f. Given $f(x) = \dfrac{x - 4}{x + 2}$, the domain is restricted so that $x \neq$ _____ .

   g. Given $f(x) = \sqrt{x - 3}$, the domain is restricted so that $x \geq$ _____ .

   h. Given a function defined by $y = f(x)$, the statement $f(1) = 6$ is equivalent to what ordered pair?

Review Exercises

2. Simplify, if possible.

   a. $\dfrac{0}{6}$ **b.** $\dfrac{8}{0}$ **c.** $-\sqrt{16}$ **d.** $\sqrt{-16}$

For Exercises 3–4, **a.** write the relation as a set of ordered pairs, **b.** identify the domain, and **c.** identify the range.

3.

Parent, x	Child, y
Kevin	Kayla
Kevin	Kira
Kathleen	Katie
Kathleen	Kira

4.

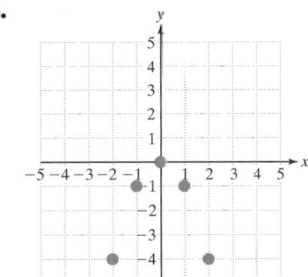

Concept 1: Definition of a Function

For Exercises 5–10, determine if the relation defines y as a function of x. **(See Example 1.)**

5.

6.

7.

8.
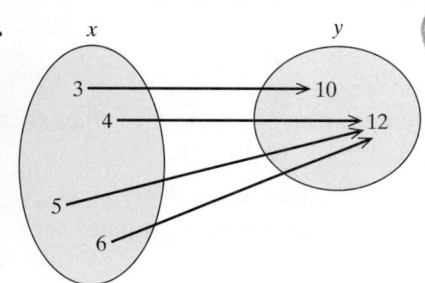

9. $\{(1, 2), (3, 4), (5, 4), (-9, 3)\}$

10. $\left\{(0, -1.1), \left(\frac{1}{2}, 8\right), (1.1, 8), \left(4, \frac{1}{2}\right)\right\}$

Concept 2: Vertical Line Test

For Exercises 11–16, use the vertical line test to determine whether the relation defines y as a function of x.
(See Example 2.)

11.

12.

13.

14.

15.

16.
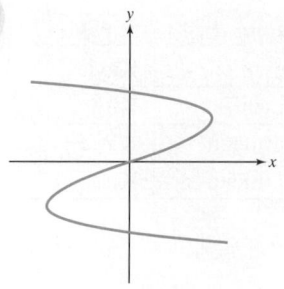

Concept 3: Function Notation

For Exercises 17–20, consider a function defined by $y = f(x)$.

17. Interpret the meaning of $f(2) = 5$.

18. Interpret the meaning of $f(-7) = 8$.

19. Write the ordered pair represented by $f(0) = -2$.

20. Write the ordered pair represented by $f(-10) = 11$.

Consider the functions defined by $f(x) = 6x - 2$, $g(x) = -x^2 - 4x + 1$, $h(x) = 7$, and $k(x) = |x - 2|$. For Exercises 21–52, find the following. **(See Examples 3–4.)**

21. $g(2)$ **22.** $k(2)$ **23.** $g(0)$ **24.** $h(0)$

25. $k(0)$ **26.** $f(0)$ **27.** $f(t)$ **28.** $g(a)$

29. $h(u)$ **30.** $k(v)$ **31.** $g(-3)$ **32.** $h(-5)$

33. $k(-2)$ **34.** $f(-6)$ **35.** $f(x + 1)$ **36.** $h(x + 1)$

37. $g(2x)$ **38.** $k(x - 3)$ **39.** $g(-\pi)$ **40.** $g(a^2)$

41. $h(a + b)$ **42.** $f(x + h)$ **43.** $f(-a)$ **44.** $g(-b)$

45. $k(-c)$ **46.** $h(-x)$ **47.** $f\left(\dfrac{1}{2}\right)$ **48.** $g\left(\dfrac{1}{4}\right)$

49. $h\left(\dfrac{1}{7}\right)$ **50.** $k\left(\dfrac{3}{2}\right)$ **51.** $f(-2.8)$ **52.** $k(-5.4)$

Consider the functions $p = \{(\frac{1}{2}, 6), (2, -7), (1, 0), (3, 2\pi)\}$ and $q = \{(6, 4), (2, -5), (\frac{2}{4}, \frac{1}{5}), (0, 9)\}$. For Exercises 53–60, find the function values.

53. $p(2)$ **54.** $p(1)$ **55.** $p(3)$ **56.** $p\left(\dfrac{1}{2}\right)$

57. $q(2)$ **58.** $q\left(\dfrac{3}{4}\right)$ **59.** $q(6)$ **60.** $q(0)$

Concept 4: Finding Function Values from a Graph

61. The graph of $y = f(x)$ is given. **(See Example 5.)**

    a. Find $f(0)$.

    b. Find $f(3)$.

    c. Find $f(-2)$.

    d. For what value(s) of x is $f(x) = -3$?

    e. For what value(s) of x is $f(x) = 3$?

    f. Write the domain of f.

    g. Write the range of f.

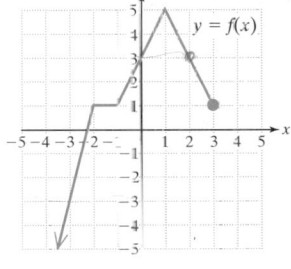

62. The graph of $y = g(x)$ is given.

    a. Find $g(-1)$.

    b. Find $g(1)$.

    c. Find $g(4)$.

    d. For what value(s) of x is $g(x) = 3$?

    e. For what value(s) of x is $g(x) = 0$?

    f. Write the domain of g.

    g. Write the range of g.

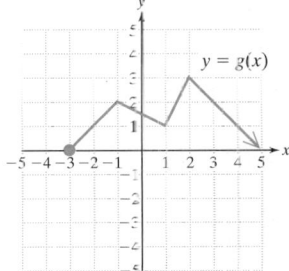

63. The graph of $y = H(x)$ is given.

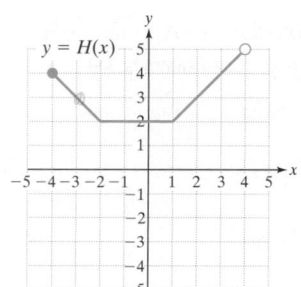

   a. Find $H(-3)$.

   b. Find $H(4)$.

   c. Find $H(3)$.

   d. For what value(s) of x is $H(x) = 3$?

   e. For what value(s) of x is $H(x) = 2$?

   f. Write the domain of H.

   g. Write the range of H.

64. The graph of $y = K(x)$ is given.

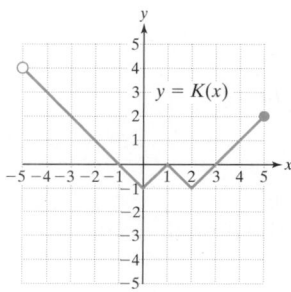

   a. Find $K(0)$.

   b. Find $K(-5)$.

   c. Find $K(1)$.

   d. For what value(s) of x is $K(x) = 0$?

   e. For what value(s) of x is $K(x) = 3$?

   f. Write the domain of K.

   g. Write the range of K.

65. The graph of $y = p(x)$ is given.

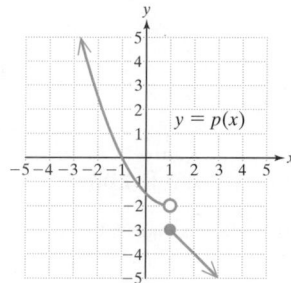

   a. Find $p(2)$.

   b. Find $p(-1)$.

   c. Find $p(1)$.

   d. For what value(s) of x is $p(x) = 0$?

   e. For what value(s) of x is $p(x) = -2$?

   f. Write the domain of p.

   g. Write the range of p.

66. The graph of $y = q(x)$ is given.

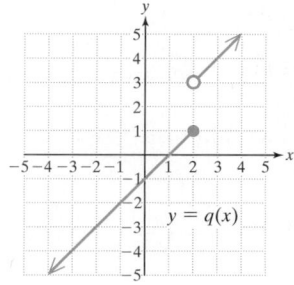

   a. Find $q(3)$.

   b. Find $q(-1)$.

   c. Find $q(2)$.

   d. For what value(s) of x is $q(x) = -4$?

   e. For what value(s) of x is $q(x) = 3$?

   f. Write the domain of q.

   g. Write the range of q.

For Exercises 67–76, refer to the functions $y = f(x)$ and $y = g(x)$, defined as follows:

$$f = \{(-3, 5), (-7, -3), (-\tfrac{3}{2}, 4), (1.2, 5)\}$$

$$g = \{(0, 6), (2, 6), (6, 0), (1, 0)\}$$

67. Identify the domain of f.

68. Identify the range of f.

69. Identify the range of g.

70. Identify the domain of g.

71. For what value(s) of x is $f(x) = 5$?

72. For what value(s) of x is $f(x) = -3$?

73. For what value(s) of x is $g(x) = 0$?

74. For what value(s) of x is $g(x) = 6$?

75. Find $f(-7)$.

76. Find $g(0)$.

Concept 5: Domain of a Function

77. Explain how to determine the domain of the function defined by $f(x) = \dfrac{x + 6}{x - 2}$.

78. Explain how to determine the domain of the function defined by $g(x) = \sqrt{x - 3}$.

For Exercises 79–94, find the domain. Write the answers in interval notation. **(See Example 6.)**

79. $k(x) = \dfrac{x - 3}{x + 6}$

80. $m(x) = \dfrac{x - 1}{x - 4}$

81. $f(t) = \dfrac{5}{t}$

82. $g(t) = \dfrac{t - 7}{t}$

83. $h(p) = \dfrac{p - 4}{p^2 + 1}$

84. $n(p) = \dfrac{p + 8}{p^2 + 2}$

85. $h(t) = \sqrt{t + 7}$

86. $k(t) = \sqrt{t - 5}$

87. $f(a) = \sqrt{a - 3}$

88. $g(a) = \sqrt{a + 2}$

89. $m(x) = \sqrt{1 - 2x}$

90. $n(x) = \sqrt{12 - 6x}$

91. $p(t) = 2t^2 + t - 1$

92. $q(t) = t^3 + t - 1$

93. $f(x) = x + 6$

94. $g(x) = 8x - \pi$

Mixed Exercises

95. The height (in feet) of a ball that is dropped from an 80-ft building is given by $h(t) = -16t^2 + 80$, where t is the time in seconds after the ball is dropped.

    a. Find $h(1)$ and $h(1.5)$

    b. Interpret the meaning of the function values found in part (a).

96. A ball is dropped from a 50-m building. The height (in meters) after t sec is given by $h(t) = -4.9t^2 + 50$.

    a. Find $h(1)$ and $h(1.5)$.

    b. Interpret the meaning of the function values found in part (a).

97. If Alicia rides a bike at an average speed of 11.5 mph, the distance that she rides can be represented by $d(t) = 11.5t$, where t is the time in hours.

    a. Find $d(1)$ and $d(1.5)$.

    b. Interpret the meaning of the function values found in part (a).

98. Brian's score on an exam is a function of the number of hours he spends studying. The function defined by

$$P(x) = \frac{100x^2}{50 + x^2} \ (x \geq 0)$$ indicates that he will achieve a score of $P\%$ if he studies for x hours.

Evaluate $P(0)$, $P(5)$, $P(10)$, $P(15)$, $P(20)$, and $P(25)$ and confirm the values on the graph. (Round to one decimal place.) Interpret $P(25)$ in the context of this problem.

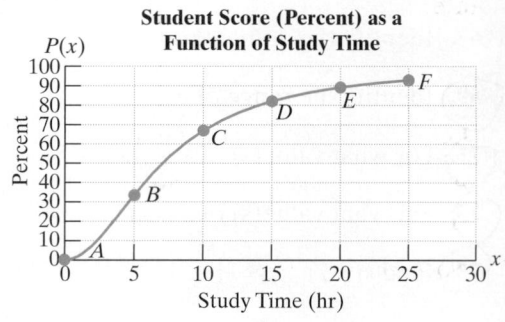

For Exercises 99–102, write a function defined by $y = f(x)$ subject to the conditions given.

99. The value of $f(x)$ is three more than two times x.

100. The value of $f(x)$ is four less than the square of x.

101. The value of $f(x)$ is ten less than the absolute value of x.

102. The value of $f(x)$ is sixteen times the square root of x.

Expanding Your Skills

For Exercises 103–104, write the domain in interval notation.

103. $q(x) = \dfrac{2}{\sqrt{x + 2}}$

104. $p(x) = \dfrac{8}{\sqrt{x - 4}}$

Graphing Calculator Exercises

105. Graph $k(t) = \sqrt{t - 5}$. Use the graph to support your answer to Exercise 86.

106. Graph $h(t) = \sqrt{t + 7}$. Use the graph to support your answer to Exercise 85.

Section 2.7   Graphs of Functions

Concepts

1. Linear and Constant Functions
2. Graphs of Basic Functions
3. Definition of a Quadratic Function
4. Finding the x- and y-Intercepts of a Graph Defined by $y = f(x)$

1. Linear and Constant Functions

A function may be expressed as a mathematical equation that relates two or more variables. In this section, we will look at several elementary functions.

We know from Section 2.1 that an equation in the form $y = k$, where k is a constant, is a horizontal line. In function notation, this can be written as $f(x) = k$. For example, the graph of the function defined by $f(x) = 3$ is a horizontal line, as shown in Figure 2-35.

We say that a function defined by $f(x) = k$ is a constant function because for any value of x, the function value is constant.

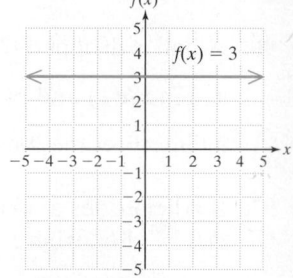

Figure 2-35

An equation of the form $y = mx + b$ is represented graphically by a line with slope m and y-intercept $(0, b)$. In function notation, this may be written as $f(x) = mx + b$. A function in this form is called a linear function. For example, the function defined by $f(x) = 2x - 3$ is a linear function with slope $m = 2$ and y-intercept $(0, -3)$ (Figure 2-36).

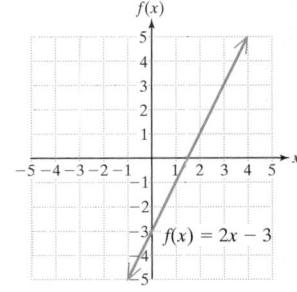

Figure 2-36

Linear Functions and Constant Functions

Let m and b represent real numbers such that $m \neq 0$. Then

A function that can be written in the form $f(x) = mx + b$ is a **linear function**.
A function that can be written in the form $f(x) = b$ is a **constant function**.

Note: The graphs of linear and constant functions are lines.

---

2. Graphs of Basic Functions

At this point, we are able to recognize the equations and graphs of linear and constant functions. In addition to linear and constant functions, the following equations define six basic functions that will be encountered in the study of algebra:

Equation	Function Notation	Type of Function
$y = x$	$f(x) = x$	Identity function
$y = x^2$	$f(x) = x^2$	Quadratic function
$y = x^3$	$f(x) = x^3$	Cubic function
$y = |x|$	$f(x) = |x|$	Absolute value function
$y = \sqrt{x}$	$f(x) = \sqrt{x}$	Square root function
$y = \dfrac{1}{x}$	$f(x) = \dfrac{1}{x}$	Reciprocal function

equivalent function notation

The graph of the function defined by $f(x) = x$ is linear, with slope $m = 1$ and y-intercept $(0, 0)$ (Figure 2-37).

To determine the shapes of the other basic functions, we can plot several points to establish the pattern of the graph. Analyzing the equation itself may also provide insight to the domain, range, and shape of the graph. To demonstrate this, we will graph $f(x) = x^2$ and $g(x) = \frac{1}{x}$.

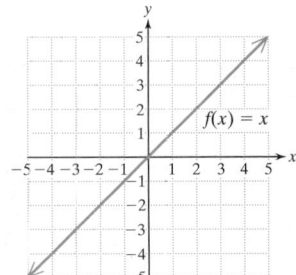

Figure 2-37

Example 1 Graphing Basic Functions

Graph the function defined by $f(x) = x^2$.

Solution:

The domain of the function given by $f(x) = x^2$ (or equivalently $y = x^2$) is all real numbers.

To graph the function, choose arbitrary values of x within the domain of the function. Be sure to choose values of x that are positive and values that are negative to determine the behavior of the function to the right and left of the origin (Table 2-6). The graph of $f(x) = x^2$ is shown in Figure 2-38.

The function values are equated to the square of x, so $f(x)$ will always be greater than or equal to zero. Hence, the y-coordinates on the graph will never be negative. The range of the function is $[0, \infty)$. The arrows on each branch of the graph imply that the pattern continues indefinitely.

Table 2-6

x	$f(x) = x^2$
0	0
1	1
2	4
3	9
−1	1
−2	4
−3	9

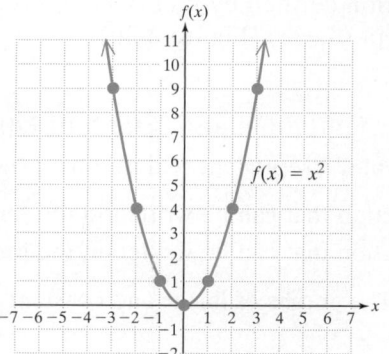

Figure 2-38

Skill Practice

1. Graph $f(x) = -x^2$ by first making a table of points.

Example 2 **Graphing Basic Functions**

Graph the function defined by $g(x) = \dfrac{1}{x}$.

Solution:

$g(x) = \dfrac{1}{x}$ Notice that $x = 0$ is not in the domain of the function. From the equation $y = \frac{1}{x}$, the y values will be the reciprocal of the x-values. The graph defined by $g(x) = \frac{1}{x}$ is shown in Figure 2-39.

x	$g(x) = \dfrac{1}{x}$
1	1
2	$\frac{1}{2}$
3	$\frac{1}{3}$
−1	−1
−2	$-\frac{1}{2}$
−3	$-\frac{1}{3}$

x	$g(x) = \dfrac{1}{x}$
$\frac{1}{2}$	2
$\frac{1}{3}$	3
$\frac{1}{4}$	4
$-\frac{1}{2}$	−2
$-\frac{1}{3}$	−3
$-\frac{1}{4}$	−4

Figure 2-39

Answers

1.

2.

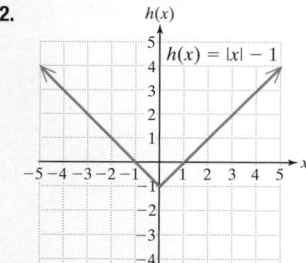

Notice that as x approaches ∞ and $-\infty$, the y values approach zero, and the graph approaches the x-axis. In this case, the x-axis is called a *horizontal asymptote*. Similarly, the graph of the function approaches the y-axis as x gets close to zero. In this case, the y-axis is called a *vertical asymptote*. In general terms, an asymptote is a line that a curve approaches more and more closely but never reaches.

Skill Practice

2. Graph $h(x) = |x| - 1$ by first making a table of points.

Calculator Connections

Topic: Graphing Nonlinear Functions

To enter a function, let $f(x) = y$. That is, to graph $f(x) = x^2$, enter $Y_1 = x^2$.

To graph $g(x) = \dfrac{1}{x}$, enter $Y_1 = 1/x$.

Summary of Six Basic Functions and Their Graphs

Function	Graph	Domain and Range
1. $f(x) = x$ Identity function		Domain $(-\infty, \infty)$ Range $(-\infty, \infty)$
2. $f(x) = x^2$ Quadratic function		Domain $(-\infty, \infty)$ Range $[0, \infty)$
3. $f(x) = x^3$ Cubic function		Domain $(-\infty, \infty)$ Range $(-\infty, \infty)$
4. $f(x) = |x|$ Absolute value function		Domain $(-\infty, \infty)$ Range $[0, \infty)$
5. $f(x) = \sqrt{x}$ Square root function		Domain $[0, \infty)$ Range $[0, \infty)$
6. $f(x) = \dfrac{1}{x}$ Reciprocal function		Domain $(-\infty, 0) \cup (0, \infty)$ Range $(-\infty, 0) \cup (0, \infty)$

The shapes of these six graphs will be developed in the homework exercises. These functions are used often in the study of algebra. Therefore, we recommend that you associate an equation with its graph and commit each to memory.

3. Definition of a Quadratic Function

In Example 1 we graphed the function defined by $f(x) = x^2$ by plotting points. This function belongs to a special category called quadratic functions.

> **Definition of a Quadratic Function**
>
> A **quadratic function** is a function defined by
>
> $$f(x) = ax^2 + bx + c \qquad \text{where } a, b, \text{ and } c \text{ are real numbers and } a \neq 0.$$

The graph of a quadratic function is in the shape of a **parabola**. The leading coefficient, a, determines the direction of the parabola.

- If $a > 0$, then the parabola opens upward, and the vertex is the minimum point on the parabola. For example, the graph of $f(x) = x^2$ is shown in Figure 2-40.
- If $a < 0$, then the parabola opens downward, and the vertex is the maximum point on the parabola. For example, the graph of $f(x) = -x^2$ is shown in Figure 2-41.

Figure 2-40

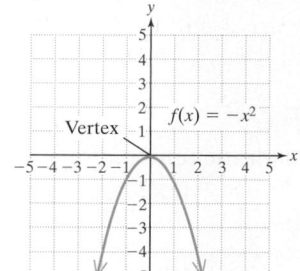

Figure 2-41

Example 3   Identifying Functions

Identify each function as linear, constant, quadratic, or none of these.

a. $f(x) = -4$ **b.** $f(x) = x^2 + 3x + 2$ **c.** $f(x) = 7 - 2x$

d. $f(x) = \dfrac{4x + 8}{8}$ **e.** $f(x) = \dfrac{6}{x} + 2$

Solution:

a. $f(x) = -4$ is a constant function. It is in the form $f(x) = b$, where $b = -4$.

b. $f(x) = x^2 + 3x + 2$ is a quadratic function. It is in the form $f(x) = ax^2 + bx + c$, where $a \neq 0$.

c. $f(x) = 7 - 2x$ is linear. Writing it in the form $f(x) = mx + b$, we get $f(x) = -2x + 7$, where $m = -2$ and $b = 7$.

d. $f(x) = \dfrac{4x + 8}{8}$ is linear. Writing it in the form $f(x) = mx + b$, we get

$$f(x) = \frac{4x}{8} + \frac{8}{8}$$

$$= \frac{1}{2}x + 1, \text{ where } m = \frac{1}{2} \text{ and } b = 1.$$

e. $f(x) = \dfrac{6}{x} + 2$ fits none of these categories.

Skill Practice Identify each function as constant, linear, quadratic, or none of these.

3. $m(x) = -2x^2 - 3x + 7$

4. $n(x) = -6$

5. $W(x) = \dfrac{4}{3}x - \dfrac{1}{2}$

6. $R(x) = \dfrac{4}{3x} - \dfrac{1}{2}$

4. Finding the *x*- and *y*-Intercepts of a Graph Defined by *y* = *f*(*x*)

In Section 2.1, we learned that to find the *x*-intercept, we substitute $y = 0$ and solve the equation for *x*. Using function notation, this is equivalent to finding the real solutions of the equation $f(x) = 0$. To find the *y*-intercept, substitute $x = 0$ and solve the equation for *y*. In function notation, this is equivalent to finding $f(0)$.

Finding Intercepts Using Function Notation

Given a function defined by $y = f(x)$,

Step 1 The *x*-intercepts are the real solutions to the equation $f(x) = 0$.

Step 2 The *y*-intercept is given by $f(0)$.

Example 4 Finding *x*- and *y*-Intercepts

Given the function defined by $f(x) = 2x - 4$:

a. Find the *x*-intercept(s).

b. Find the *y*-intercept.

c. Graph the function.

Solution:

a. To find the *x*-intercept(s), find the real solutions to the equation $f(x) = 0$.

$f(x) = 2x - 4$

$0 = 2x - 4$ Substitute $f(x) = 0$.

$4 = 2x$

$2 = x$ The *x*-intercept is $(2, 0)$.

Answers

3. Quadratic
4. Constant
5. Linear
6. None of these

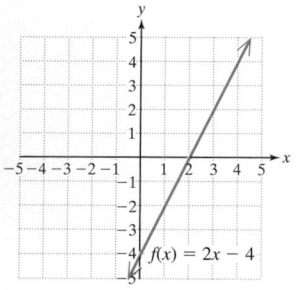

Figure 2-42

b. To find the y-intercept, evaluate $f(0)$.

$f(0) = 2(0) - 4$ Substitute $x = 0$.

$f(0) = -4$ The y-intercept is $(0, -4)$.

c. This function is linear, with a y-intercept of $(0, -4)$, an x-intercept of $(2, 0)$, and a slope of 2 (Figure 2-42).

Skill Practice Consider $f(x) = -5x + 1$.

7. Find the x-intercept.

8. Find the y-intercept.

9. Graph the function.

Calculator Connections

Topic: Finding *x*- and *y*-Intercepts

Refer to Example 4 with $f(x) = 2x - 4$. To find the y-intercept, let $x = 0$. We can do this using the *Value* key.

To find the x-intercept, use the *Zero* feature. To use this feature you must give bounds for the x-intercept. To find a left bound, place the curser to the left of the x-intercept and press Enter. For the right bound, place the curser to the right of the x-intercept and press Enter.

Then make a guess by placing the curser near the x-intercept. Press Enter and the x-intercept will be displayed.

Answers

7. $\left(\frac{1}{5}, 0\right)$ **8.** $(0, 1)$

9.

| Example 5 | Finding *x*- and *y*-Intercepts |

For the function pictured in Figure 2-43, estimate

a. The real values of *x* for which $f(x) = 0$.

b. The value of $f(0)$.

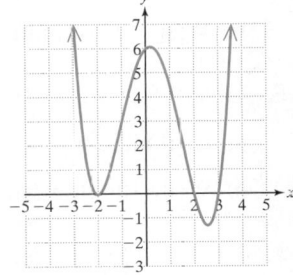

Solution:

a. The real values of *x* for which $f(x) = 0$ are the *x*-intercepts of the function. For this graph, the *x*-intercepts are located at $x = -2$, $x = 2$, and $x = 3$.

b. The value of $f(0)$ is the value of *y* at $x = 0$. That is, $f(0)$ is the *y*-intercept, $f(0) = 6$.

Figure 2-43

Skill Practice Use the function pictured.

10. a. Estimate the real value(s) of *x* for which $f(x) = 0$
b. Estimate the value of $f(0)$.

Answer
10. a. $x = -2$, $x = 1$, and $x = 4$
     b. $f(0) = 4$

Section 2.7 Practice Exercises

Study Skills Exercise

Make up a practice test for yourself for this chapter. Use examples or exercises from the text. Be sure to cover each concept that was presented.

Vocabulary and Key Concepts

1. a. A function that can be written in the form $f(x) = mx + b, m \neq 0$, is a _____ function.

b. A function that can be written in the form $f(x) = b$ is a _____ function.

c. A function that can be written in the form $f(x) = ax^2 + bx + c, a \neq 0$, is a _____ function.

d. The graph of a quadratic function is in the shape of a _____.

e. To find the *x*-intercept(s) of a graph defined by $y = f(x)$, find the real solutions to the equation $f(x) =$ _____.

f. To find the _____-intercept of a graph defined by $y = f(x)$, evaluate $f(0)$.

Review Exercises

2. Given: $g = \{(6, 1), (5, 2), (4, 3), (3, 4)\}$

   a. Is this relation a function?

   b. List the elements in the domain.

   c. List the elements in the range.

3. Given: $f = \{(7, 3), (2, 3), (-5, 3)\}$

   a. Is this relation a function?

   b. List the elements in the domain.

   c. List the elements in the range.

4. Given $h(x) = \dfrac{12}{x-1}$.

   a. Evaluate $h(0), h(1), h(2),$ and $h(3),$ if possible.

   b. Write the domain of h in interval notation.

5. Given: $f(x) = \sqrt{x+4}$

   a. Evaluate $f(0), f(-3), f(-4),$ and $f(-5),$ if possible.

   b. Write the domain of f in interval notation.

6. The force (measured in pounds) to stretch a certain spring x inches is given by $f(x) = 3x$. Evaluate $f(3)$ and $f(10),$ and interpret the results in the context of this problem.

Concept 1: Linear and Constant Functions

7. Fill in the blank with the word *vertical* or *horizontal*. The graph of a constant function is a _____ line.

8. For the linear function defined by $f(x) = mx + b$, identify the slope and y-intercept.

9. Graph the constant function $f(x) = 2$. Then use the graph to identify the domain and range of f.

10. Graph the linear function $g(x) = -2x + 1$. Then use the graph to identify the domain and range of g.

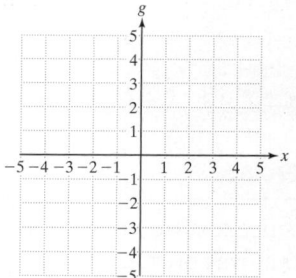

Concept 2: Graphs of Basic Functions

For Exercises 11–16, sketch a graph by completing the table and plotting the points. **(See Examples 1–2.)**

11. $f(x) = \dfrac{1}{x}$

x	$f(x)$
-2	
-1	
$-\frac{1}{2}$	
$-\frac{1}{4}$	

x	$f(x)$
$\frac{1}{4}$	
$\frac{1}{2}$	
1	
2	

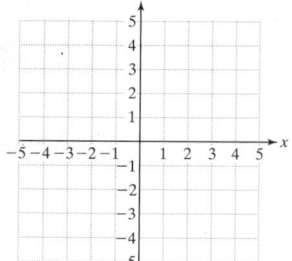

12. $g(x) = |x|$

x	$g(x)$
-2	
-1	
0	
1	
2	

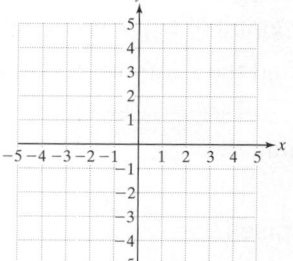

13. $h(x) = x^3$

x	$h(x)$
-2	
-1	
0	
1	
2	

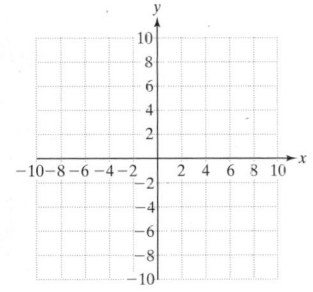

14. $k(x) =$ =

x	$k(x)$
-2	
-1	
0	
1	
2	

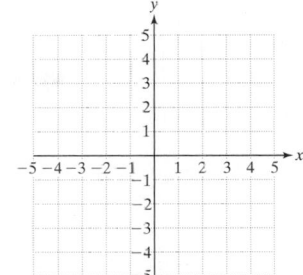

15. $q(x) = x^2$

x	$q(x)$
-2	
-1	
0	
1	
2	

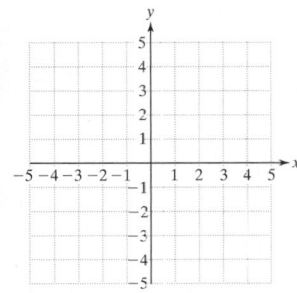

16. $p(x) = \sqrt{x}$

x	$p(x)$
0	
1	
4	
9	
16	

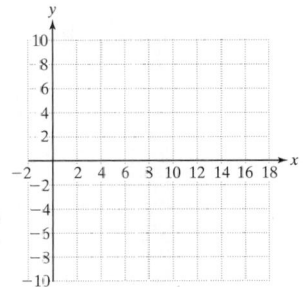

Concept 3: Definition of a Quadratic Function

For Exercises 17–28, determine if the function is constant, linear, quadratic, or none of these. **(See Example 3.)**

17. $f(x) = 2x^2 + 3x + 1$ **18.** $g(x) = -x^2 + 4x + 12$ **19.** $k(x) = -3x - 7$ **20.** $h(x) = -x - 3$

21. $m(x) = \dfrac{4}{3}$ **22.** $n(x) = 0.8$ **23.** $p(x) = \dfrac{2}{3x} + \dfrac{1}{4}$ **24.** $Q(x) = \dfrac{1}{5x} - 3$

25. $t(x) = \dfrac{2}{3}x + \dfrac{1}{4}$ **26.** $r(x) = \dfrac{1}{5}x - 3$ **27.** $w(x) = \sqrt{4 - x}$ **28.** $T(x) = -|x + 10|$

Concept 4: Finding the x- and y-Intercepts of a Graph Defined by $y = f(x)$

For Exercises 29–36, find the x- and y-intercepts, and graph the function. **(See Example 4.)**

29. $f(x) = 5x - 10$

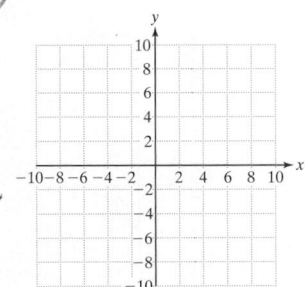

30. $f(x) = -3x + 12$

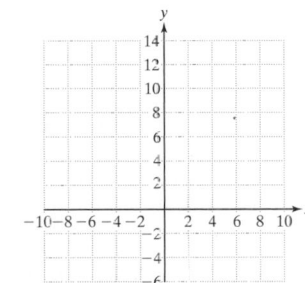

31. $g(x) = -6x + 5$

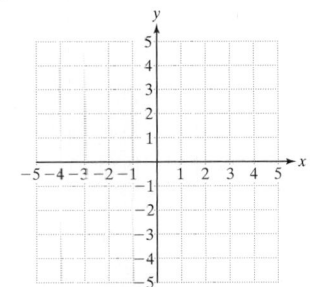

32. $h(x) = 2x + 9$

33. $f(x) = 18$

34. $g(x) = -7$

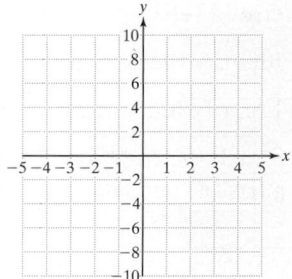

35. $g(x) = \dfrac{2}{3}x + 2$

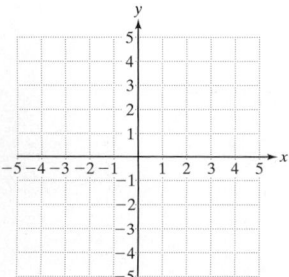

36. $h(x) = -\dfrac{3}{5}x - 3$

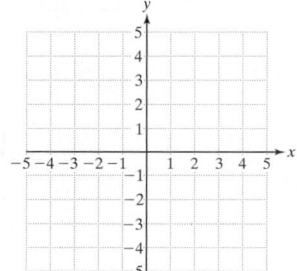

For Exercises 37–42, use the graph to estimate

a. The real values of x for which $f(x) = 0$.

b. The value of $f(0)$. **(See Example 5.)**

37.

38.

39.

40.

41.

42.

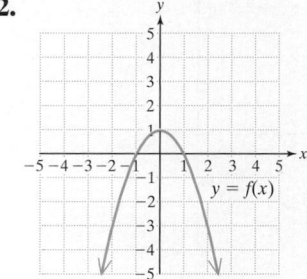

Mixed Exercises

For Exercises 43–52,

    a. Identify the domain of the function.

    b. Identify the *y*-intercept.

    c. Match the function with its graph by recognizing the basic shape of the graph and using the results from
    parts (a) and (b). Plot additional points if necessary.

43. $q(x) = 2x^2$
              44. $p(x) = -2x^2 + 1$
              45. $h(x) = x^3 + 1$

46. $k(x) = x^3 - 2$
          47. $r(x) = \sqrt{x} + 1$
              48. $s(x) = \sqrt{x + 4}$

49. $f(x) = \dfrac{1}{x - 3}$
           50. $g(x) = \dfrac{1}{x + 1}$
           51. $k(x) = |x + 2|$

52. $h(x) = |x - 1| + 2$

i.

ii.

iii.

iv.

v.

vi.

vii.

viii.

ix.

x.

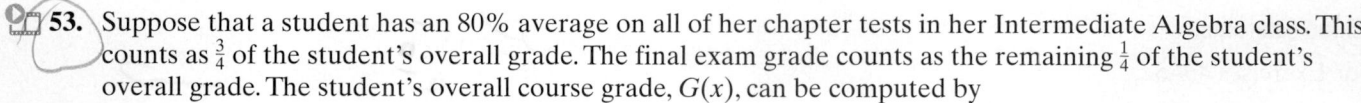

53. Suppose that a student has an 80% average on all of her chapter tests in her Intermediate Algebra class. This counts as $\frac{3}{4}$ of the student's overall grade. The final exam grade counts as the remaining $\frac{1}{4}$ of the student's overall grade. The student's overall course grade, $G(x)$, can be computed by

$$G(x) = \frac{3}{4}(80) + \frac{1}{4}x, \text{ where } x \text{ is the student's grade on the final exam.}$$

a. Is this function linear, quadratic, or neither?

b. Evaluate $G(90)$ and interpret its meaning in the context of the problem.

c. Evaluate $G(50)$ and interpret its meaning in the context of the problem.

54. The median weekly earnings, $E(x)$ in dollars, for women 16 yr and older working full time can be approximated by $E(x) = 0.14x^2 + 7.8x + 540$. For this function, x represents the number of years since 2000. (*Source:* U.S. Department of Labor)

a. Is this function linear, quadratic, or neither?

b. Evaluate $E(5)$ and interpret its meaning in the context of this problem.

c. Evaluate $E(10)$ and interpret its meaning in the context of this problem.

For Exercises 55–60, write a function defined by $y = f(x)$ under the given conditions.

55. The value of $f(x)$ is two more than the square of x.

56. The value of $f(x)$ is the square of the sum of two and x.

57. The function f is a constant function passing through the point $(4, 3)$.

58. The function f is a constant function with y-intercept $(0, 5)$.

59. The function f is a linear function with slope $\frac{1}{2}$ and y-intercept $(0, -2)$.

60. The function f is a linear function with slope -4 and y-intercept $(0, \frac{1}{3})$.

Graphing Calculator Exercises

For Exercises 61–66, use a graphing calculator to graph the basic functions. Verify your answers from the table on page 207.

61. $f(x) = x$

62. $f(x) = x^2$

63. $f(x) = x^3$

64. $f(x) = |x|$

65. $f(x) = \sqrt{x}$

66. $f(x) = \frac{1}{x}$

For Exercises 67–70, find the x- and y-intercepts using a graphing calculator and the *Value* and *Zero* features.

67. $y = -\frac{1}{8}x + 1$

68. $y = -\frac{1}{2}x - 3$

69. $y = \frac{4}{5}x + 4$

70. $y = \frac{7}{2}x - 7$

Problem Recognition Exercises

Characteristics of Relations

Exercises 1–15 refer to the sets of points (x, y) described in a–h.

a. $\{(0, 8), (1, 4), (\frac{1}{2}, 4), (-3, 5), (2, 1)\}$

b. $\{(-6, 4), (2, 3), (-9, 6), (2, 1), (0, 10)\}$

c. $c(x) = 3x^2 - 2x - 1$

d. $d(x) = 5x - 9$

e.

f.

g.

h.

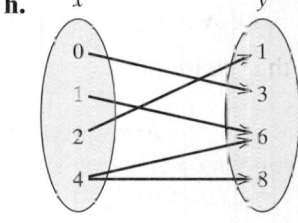

1. Which relations define y as a function of x?

2. Which relations contain the point $(2, 1)$?

3. Use relation (c) to find $c(-1)$.

4. Use relation (f) to find $f(-4)$.

5. Find the domain of relation (a).

6. Find the range of relation (b).

7. Find the domain of relation (g).

8. Find the range of relation (f).

9. Use relation (h) to complete the ordered pair $(\_\_, 3)$.

10. Find the x-intercept(s) of the graph of relation (d).

11. Find the y-intercept(s) of the graph shown in relation (e).

12. Use relation (g) to determine the value of x such that $g(x) = 2$.

13. Which relation describes a quadratic function?

14. Which relation describes a linear function?

15. Use relation (d) to find the value of x such that $d(x) = 6$.

Group Activity

Deciphering a Coded Message

Materials: A calculator

Estimated time: 20–25 minutes

Group Size: 4 (two pairs)

Cryptography is the study of coding and decoding messages. One type of coding process assigns a number to each letter of the alphabet and to the space character. For example:

A	B	C	D	E	F	G	H	I	J	K	L	M	N
1	2	3	4	5	6	7	8	9	10	11	12	13	14

O	P	Q	R	S	T	U	V	W	X	Y	Z	space
15	16	17	18	19	20	21	22	23	24	25	26	27

According to the numbers assigned to each letter, the message "*Decimals have a point*" would be coded as follows:

D E C I M A L S _ H A V E _ A _ P O I N T

4 5 3 9 13 1 12 19 27 8 1 22 5 27 1 27 16 15 9 14 20

Now suppose each letter is encoded by applying a function such as $f(x) = 2x + 5$, where x is the numerical value of each letter. For example:

The letter "a" would be coded as: $f(1) = 2(1) + 5 = 7$

The letter "b" would be coded as: $f(2) = 2(2) + 5 = 9$

Using this encoding function, we have

Message:	D	E	C	I	M	A	L	S	_	H	A	V	E	_	A	_	P	O	I	N	T
Original:	4	5	3	9	13	1	12	19	27	8	1	22	5	27	1	27	16	15	9	14	20
Coded Form:	13	15	11	23	31	7	29	43	59	21	7	49	15	59	7	59	37	35	23	33	45

To decode this message, the receiver would need to reverse the operations assigned by $f(x) = 2x + 5$. Since the function f multiplies x by 2 and then adds 5, we can reverse this process by subtracting 5 and dividing by 2. This is represented by $g(x) = \frac{x-5}{2}$.

1. a. One pair of students will encode the follow message according to $f(x) = 4x + 2$.

   MATH IS THE KEY TO THE SCIENCES

   b. The second pair of students will encode the follow message according to $f(x) = 3x - 1$.

   MATH IS NOT A SPECTATOR SPORT

2. With each message encoded, the pairs will exchange papers. Each pair will then decode the message.

Chapter 2 Summary

Section 2.1   Linear Equations in Two Variables

Key Concepts

A **linear equation in two variables** can be written in the form $Ax + By = C$, where A, B, and C are real numbers and A and B are not both zero.

    The graph of a linear equation in two variables is a line and can be represented in a rectangular coordinate system.

An **x-intercept** is a point $(a, 0)$ where a graph intersects the x-axis. Given an equation of a graph, to find an x-intercept, substitute 0 for y and solve for x.

A **y-intercept** is a point $(0, b)$ where a graph intersects the y-axis. Given an equation of a graph, to find a y-intercept, substitute 0 for x and solve for y.

An equation of a **vertical line** can be written in the form $x = k$.
An equation of a **horizontal line** can be written in the form $y = k$.

Examples

Example 1

To graph the equation $3x - 4y = 12$, we can construct a table of points.

x	y
0	-3
4	0
1	$-\dfrac{9}{4}$

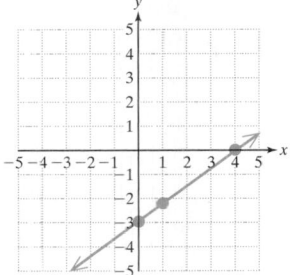

Example 2

Given $2x - 3y = 8$, find the x- and y-intercepts.

x-intercept: $2x + 3(0) = 8$

$$2x = 8$$

$$x = 4 \qquad (4, 0)$$

y-intercept: $2(0) + 3y = 8$

$$3y = 8$$

$$y = \frac{8}{3} \qquad \left(0, \frac{8}{3}\right)$$

Example 3

Graph $x = -2$.

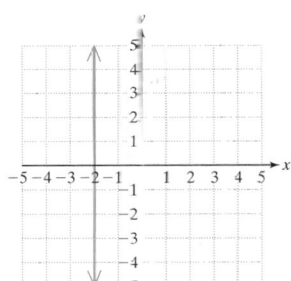

Example 4

Graph $y = 3$.

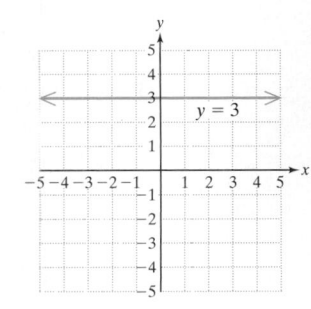

Section 2.2   Slope of a Line and Rate of Change

Key Concepts

The **slope** m of a line passing through two distinct points (x_1, y_1) and (x_2, y_2) is given by

$$m = \frac{y_2 - y_1}{x_2 - x_1}, \qquad x_2 - x_1 \neq 0$$

The slope of a line may be positive, negative, zero, or undefined.

Two parallel (nonvertical) lines have the same slope: $m_1 = m_2$.

Two lines are perpendicular if the slope of one line is the opposite of the reciprocal of the slope of the other line:

$$m_1 = -\frac{1}{m_2} \text{ or equivalently, } m_1 m_2 = -1$$

Examples

Example 1

The slope of the line passing through $(1, -3)$ and $(-3, 7)$ is

$$m = \frac{7 - (-3)}{-3 - 1} = \frac{10}{-4} = -\frac{5}{2}$$

Example 2

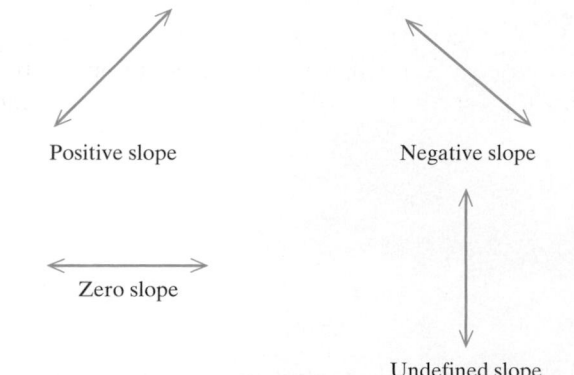

Positive slope   Negative slope

Zero slope

Undefined slope

Example 3

The slopes of two lines are given. Determine whether the lines are parallel, perpendicular, or neither.

a. $m_1 = -7$ and $m_2 = -7$ Parallel

b. $m_1 = -\dfrac{1}{5}$ and $m_2 = 5$ Perpendicular

c. $m_1 = -\dfrac{3}{2}$ and $m_2 = -\dfrac{2}{3}$ Neither

Section 2.3   Equations of a Line

Key Concepts

Standard Form: $Ax + By = C$ (A and B are not both zero)

Horizontal line: $y = k$

Vertical line: $x = k$

Slope-intercept form: $y = mx + b$

Point-slope formula: $y - y_1 = m(x - x_1)$

Slope-intercept form is used to identify the slope and y-intercept of a line when the equation is given. Slope-intercept form can also be used to graph a line.

The point-slope formula can be used to construct an equation of a line, given a point and a slope.

Examples

Example 1

Find the slope and y-intercept. Then graph the equation.

$$7x - 2y = 4 \qquad \text{Solve for } y.$$

$$-2y = -7x + 4$$

$$y = \frac{7}{2}x - 2 \qquad \text{Slope-intercept form}$$

The slope is $\frac{7}{2}$; the y-intercept is $(0, -2)$.

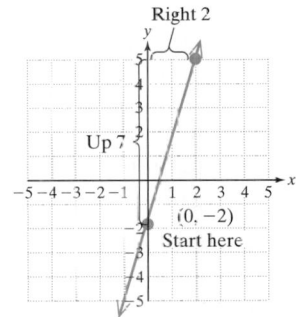

Example 2

Find an equation of the line passing through the point $(2, -3)$ and having slope $m = -4$.

Using the point-slope formula gives

$$y - y_1 = m(x - x_1)$$

$$y - (-3) = -4(x - 2)$$

$$y + 3 = -4x + 8$$

$$y = -4x + 5$$

Section 2.4   Applications of Linear Equations and Modeling

Key Concepts

A linear model can be constructed to describe data for a given situation.

- Given two points from the data, use the point-slope formula to find an equation of the line.

- Interpret the meaning of the slope and y-intercept in the context of the problem.

- Use the equation to predict values.

Examples

Example 1

The per capita income in the United States has been rising linearly since 1980. In the graph, x represents the number of years since 1980, and y represents average income in dollars.

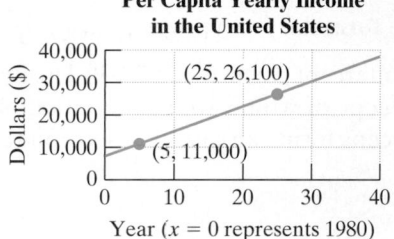

Per Capita Yearly Income in the United States

Year ($x = 0$ represents 1980)

Write an equation of the line, using the points $(5, 11{,}000)$ and $(25, 26{,}100)$.

Slope: $\dfrac{26{,}100 - 11{,}000}{25 - 5} = \dfrac{15{,}100}{20} = 755$

$y - 11{,}000 = 755(x - 5)$

$y - 11{,}000 = 755x - 3775$

$y = 755x + 7225$

- The slope 755 indicates that the average income has increased at a rate of \$755 per year.
- The y-intercept $(0, 7225)$ means that the average income in 1980 (year $x = 0$) was \$7225.

By substituting different values of x, the equation can be used to approximate the average income for that year. For the year 2010 ($x = 30$), we have:

$y = 755(30) + 7225$

$y = 29{,}875$

The average per capita income in 2010 would be approximately \$29,875.

Section 2.5   Introduction to Relations

Key Concepts

A set of ordered pairs (x, y) is called a **relation in x and y.**

The **domain** of a relation is the set of first components in the ordered pairs in the relation. The **range** of a relation is the set of second components in the ordered pairs.

Examples

Example 1

Let $A = \{(0, 0), (1, 1), (2, 4), (3, 9), (-1, 1), (-2, 4)\}$.

Domain of A: $\{0, 1, 2, 3, -1, -2\}$

Range of A: $\{0, 1, 4, 9\}$

Example 2

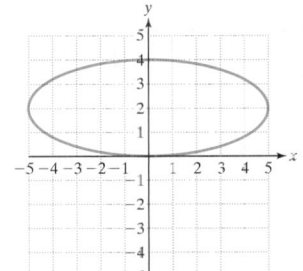

Domain: $[-5, 5]$
Range: $[0, 4]$

Section 2.6   Introduction to Functions

Key Concepts

Given a relation in x and y, we say "y is a **function** of x" if, for each element x in the domain, there is exactly one value of y in the range.

Note: This means that no two ordered pairs may have the same first coordinate and different second coordinates.

The Vertical Line Test for Functions

Consider a relation defined by a set of points (x, y) in a rectangular coordinate system. The graph defines y as a function of x if no vertical line intersects the graph in more than one point.

Examples

Example 1

Function $\{(1, 3), (2, 5), (6, 3)\}$
Not a function $\{(1, 3), (2, 5), (1, 4)\}$

Example 2

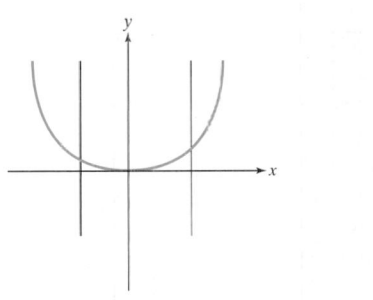

Function   Not a function

Function Notation

$f(x)$ is the value of the function f at x.

The domain of a function defined by $y = f(x)$ is the set of x-values that when substituted into the function produces a real number. In particular,

- Exclude values of x that make the denominator of a fraction zero.
- Exclude values of x that make the expression within a square root negative.

Example 3

Given $f(x) = -3x^2 + 5x$, find $f(-2)$.

$$f(-2) = -3(-2)^2 + 5(-2)$$
$$= -12 - 10$$
$$= -22$$

Example 4

Find the domain.

1. $f(x) = \dfrac{x + 4}{x - 5}$; Domain: $(-\infty, 5) \cup (5, \infty)$

2. $f(x) = \sqrt{x - 3}$; Domain: $[3, \infty)$

3. $f(x) = 3x^2 - 5$; Domain: $(-\infty, \infty)$

Section 2.7 Graphs of Functions

Key Concepts

A function of the form $f(x) = mx + b$ $(m \neq 0)$ is a **linear function**. Its graph is a line with slope m and y-intercept $(0, b)$.

A function of the form $f(x) = k$ is a **constant function**. Its graph is a horizontal line.

A function of the form $f(x) = ax^2 + bx + c$ $(a \neq 0)$ is a **quadratic function**. Its graph is a **parabola**.

Examples

Example 1

$f(x) = 2x - 3$ $f(x) = 3$

Graphs of basic functions:

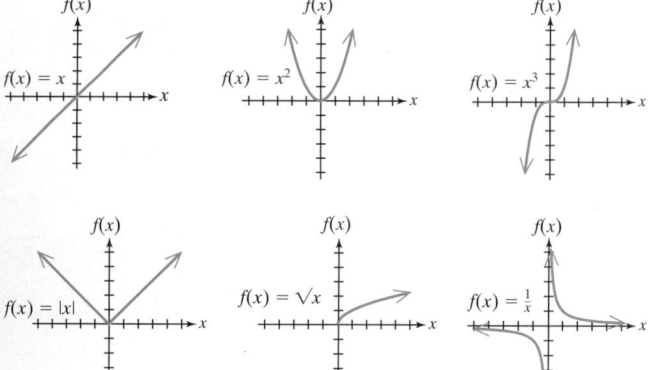

The x-intercepts of a function are determined by finding the real solutions to the equation $f(x) = 0$.

The y-intercept of a function is at $f(0)$.

Example 2

Find the x- and y-intercepts for the function pictured.

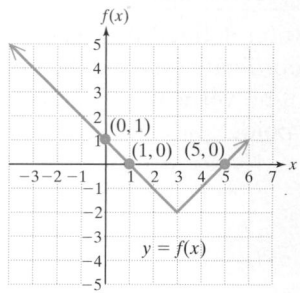

$f(x) = 0$, when $x = 1$ and $x = 5$.

The x-intercepts are $(1, 0)$ and $(5, 0)$.

$f(0) = 1$. The y-intercept is $(0, 1)$.

Chapter 2   Review Exercises

Section 2.1

1. Label the following on the diagram:

   a. Origin

   b. x-Axis

   c. y-Axis

   d. Quadrant I

   e. Quadrant II

   f. Quadrant III

   g. Quadrant IV

2. Determine if $(2, 5)$ is a solution to $-2x + 4y = -16$.

3. Determine if $(-3, 4)$ is a solution to $5x = -15$.

4. Determine the coordinates of the points labeled in the graph.

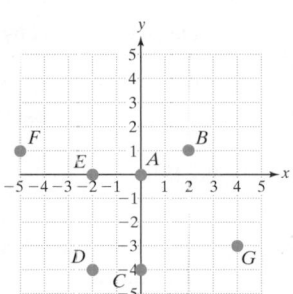

For Exercises 5–7, complete the table and graph the line defined by the points.

5. $3x - 2y = -6$

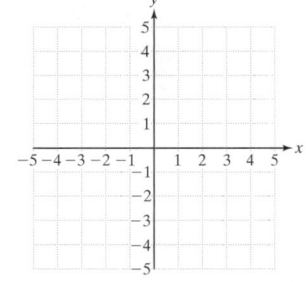

6. $2y - 3 = 10$

7. $6 - x = 2$

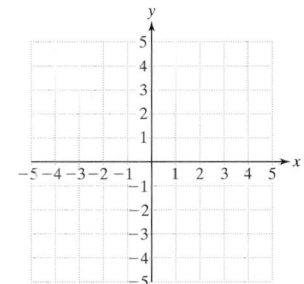

For Exercises 8–11, graph the lines. In each case find at least three points and identify the x- and y-intercepts (if possible).

8. $2x = 3y - 6$

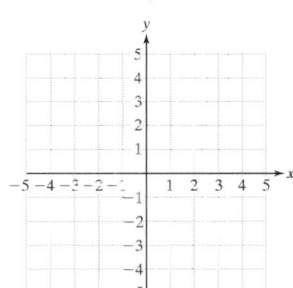

9. $5x - 2y = 0$

10. $2y = 6$

11. $-3x = 6$

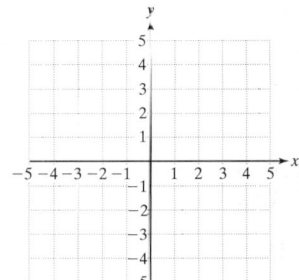

Section 2.2

12. Find the slope of the line.

a.

b.

c.

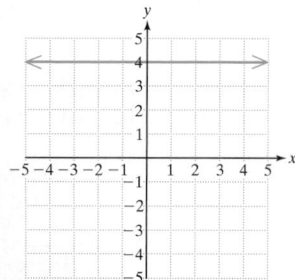

13. Draw a line with slope 2 (answers may vary).

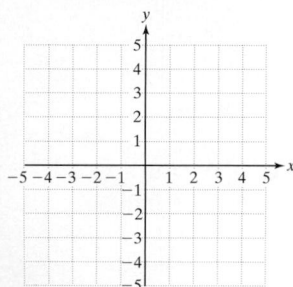

14. Draw a line with slope $-\frac{3}{4}$ (answers may vary).

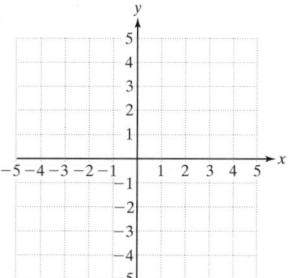

For Exercises 15–18, find the slope of the line that passes through each pair of points.

15. $(2, 6), (-1, 0)$

16. $(7, 2), (-3, -5)$

17. $(8, 2), (3, 2)$

18. $\left(-4, \frac{1}{2}\right), (-4, 1)$

19. Two points for each of two lines are given. Determine if the lines are parallel, perpendicular, or neither.

L_1: $(4, -6)$ and $(3, -2)$

L_2: $(3, -1)$ and $(7, 0)$

For Exercises 20–22, the slopes of two lines are given. Based on the slopes, are the lines parallel, perpendicular, or neither?

20. $m_1 = -\frac{1}{3}, m_2 = 3$

21. $m_1 = \frac{5}{4}, m_2 = \frac{4}{5}$

22. $m_1 = 7, m_2 = 7$

23. The graph indicates that the enrollment for a small college has been increasing linearly since 1990.

a. Use the two data points to find the slope of the line.

b. Interpret the meaning of the slope in the context of this problem.

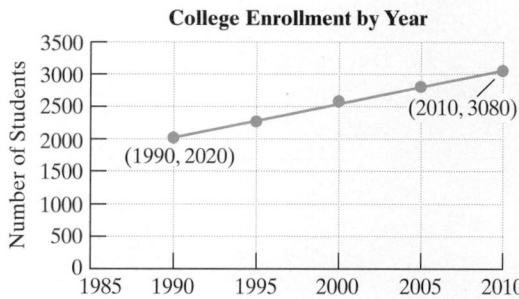

24. Approximate the slope of the stairway pictured here.

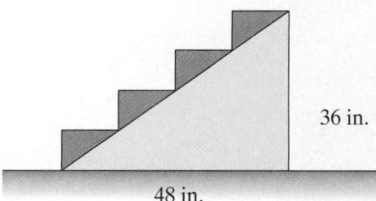

Section 2.3

25. Write a formula.

a. Horizontal line

b. Point-slope formula

c. Standard form

d. Vertical line

e. Slope-intercept form

For Exercises 26–30, write your answer in slope-intercept form or in standard form.

26. Write an equation of the line that has slope $\frac{1}{9}$ and y-intercept $(0, 6)$.

27. Write an equation of the line that has slope $-\frac{2}{3}$ and x-intercept $(3, 0)$.

28. Write an equation of the line that passes through the points $(-8, -1)$ and $(-5, 9)$.

29. Write an equation of the line that passes through the point $(6, -2)$ and is perpendicular to the line $y = -\frac{1}{3}x + 2$.

30. Write an equation of the line that passes through the point $(0, -3)$ and is parallel to the line $4x + 3y = -1$.

31. For each of the given conditions, find an equation of the line

    a. Passing through the point $(-3, -2)$ and parallel to the x-axis.

    b. Passing through the point $(-3, -2)$ and parallel to the y-axis.

    c. Passing through the point $(-3, -2)$ and having an undefined slope.

    d. Passing through the point $(-3, -2)$ and having a zero slope.

32. Are any of the lines in Exercise 31 the same?

Section 2.4

33. Ally loves the beach and decides to spend the summer selling various ice cream products on the beach. From her accounting course, she knows that her total cost is calculated as

    Total cost = fixed cost + variable cost

    She estimates that her fixed cost for the summer season is $50 per day. She also knows that each ice cream product costs her $0.75 from her distributor.

    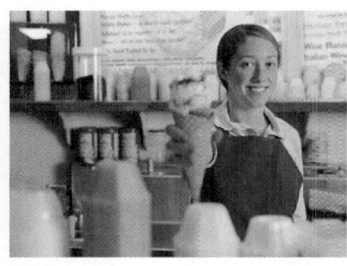

a. Write a relationship for the daily cost y in terms of the number of ice cream products sold per day x.

b. Graph the equation from part (a) by letting the horizontal axis represent the number of ice cream products sold per day and letting the vertical axis represent the daily cost.

c. What does the y-intercept represent in the context of this problem?

d. What is her cost if she sells 450 ice cream products?

e. What is the slope of the line?

f. What does the slope of the line represent in the context of this problem?

34. The margin of victory for a certain college football team seems to be linearly related to the number of rushing yards gained by the star running back. The table shows the statistics.

Yards Rushed	Margin of Victory
100	20
60	10
120	24
50	7

a. Graph the data to determine if a linear trend exists. Let x represent the number of yards rushed by the star running back and y represent the points in the margin of victory.

b. Find an equation for the line through the points $(50, 7)$ and $(100, 20)$.

c. Based on the equation, what would be the result of the football game if the star running back did not play?

Section 2.5

For Exercises 35–38, find the domain and range.

35. $\left\{ \left(\frac{1}{3}, 10 \right), \left(6, -\frac{1}{2} \right), \left(\frac{1}{4}, 4 \right), \left(7, \frac{2}{5} \right) \right\}$

36.

37.

38.

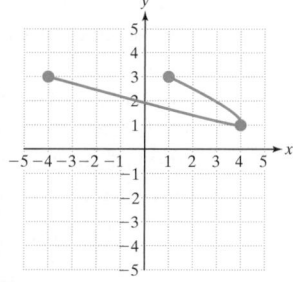

Section 2.6

39. Sketch a relation that is *not* a function. (Answers may vary.)

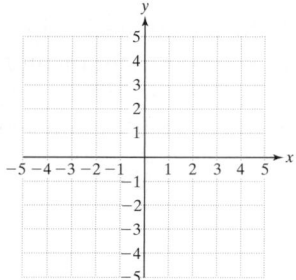

40. Sketch a relation that *is* a function. (Answers may vary.)

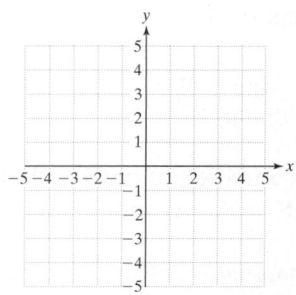

For Exercises 41–46:

a. Determine whether the relation defines y as a function of x.

b. Find the domain.

c. Find the range.

41.

42.

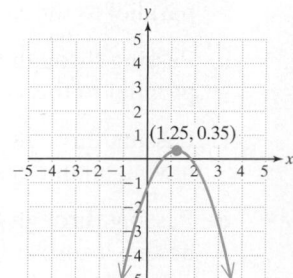

43. $\{(1, 3), (2, 3), (3, 3), (4, 3)\}$

44. $\{(0, 2), (0, 3), (4, 4), (0, 5)\}$

45.

46.

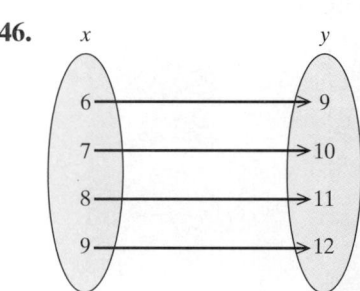

For Exercises 47–54, find the function values given $f(x) = 6x^2 - 4$.

47. $f(0)$ **48.** $f(1)$

49. $f(-1)$ **50.** $f(t)$

51. $f(b)$ **52.** $f(\pi)$

53. $f(a)$ **54.** $f(-2)$

For Exercises 55–58, write the domain of each function in interval notation.

55. $g(x) = 7x^3 + 1$ **56.** $h(x) = \dfrac{x + 10}{x - 11}$

57. $k(x) = \sqrt{x - 8}$ **58.** $w(x) = \sqrt{x + 2}$

59. Anita is a waitress and makes $6 per hour plus tips. Her tips average $5 per table. In one 8-hr shift, Anita's pay can be described by $p(x) = 48 + 5x$, where x represents the number of tables she waits on. Determine how much Anita will earn if she waits on

     a. 10 tables   **b.** 15 tables   **c.** 20 tables

Section 2.7

For Exercises 60–65, sketch the functions from memory.

60. $h(x) = x$ **61.** $f(x) = x^2$

 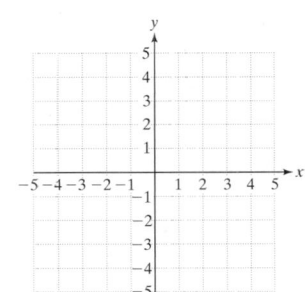

62. $g(x) = x^3$ **63.** $w(x) = |x|$

 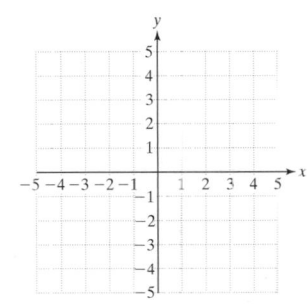

64. $s(x) = \sqrt{x}$ **65.** $r(x) = \dfrac{1}{x}$

 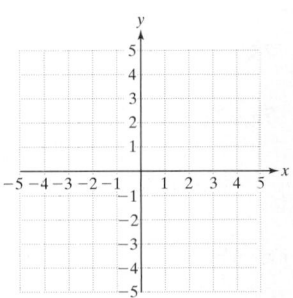

For Exercises 66–67, sketch the functions.

66. $q(x) = 3$ **67.** $k(x) = 2x + 1$

 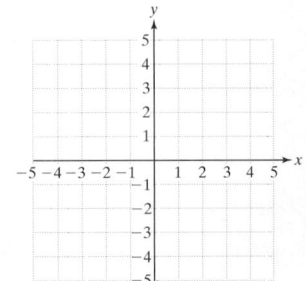

68. Given: $s(x) = (x - 2)^2$

     a. Find $s(4), s(-3), s(2), s(1)$, and $s(0)$.

     b. What is the domain of s?

69. Given: $r(x) = 2\sqrt{x - 4}$

     a. Find $r(2), r(4), r(5)$, and $r(8)$.

     b. What is the domain of r?

70. Given: $h(x) = \dfrac{3}{x - 3}$

     a. Find $h(-3), h(0), h(2)$, and $h(5)$.

     b. What is the domain of h?

71. Given: $k(x) = -|x + 3|$

     a. Find $k(-5), k(-4), k(-3)$, and $k(2)$.

     b. What is the domain of k?

For Exercises 72–73, find the x- and y-intercepts.

72. $p(x) = 4x - 7$

73. $q(x) = -2x + 9$

74. The function defined by $b(t) = 0.7t + 4.5$ represents the per capita consumption of bottled water in the United States since 1985. The values of $b(t)$ are measured in gallons, and $t = 0$ corresponds to the year 1985. (*Source:* U.S. Department of Agriculture)

  a. Evaluate $b(0)$ and $b(7)$ and interpret the results in the context of this problem.

  b. Determine the slope and interpret its meaning in the context of this problem.

For Exercises 75–80, refer to the graph.

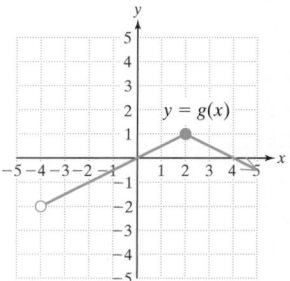

75. Find $g(-2)$.   **76.** Find $g(4)$.

77. For what value(s) of x is $g(x) = 0$?

78. For what value(s) of x is $g(x) = 4$?

79. Write the domain of g.

80. Write the range of g.

Chapter 2   Test

1. Given the equation $x - \frac{2}{3}y = 6$, complete the ordered pairs and graph the corresponding points. $(0, \quad)\ (\quad, 0)\ (\quad, -3)$

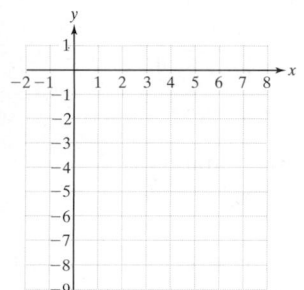

2. Determine whether the following statements are true or false and explain your answer.

  a. The product of the x- and y-coordinates is positive only for points in quadrant I.

  b. The quotient of the x- and y-coordinates is negative only for points in quadrant IV.

  c. The point $(-2, -3)$ is in quadrant III.

  d. The point $(0, 0)$ lies on the x-axis.

3. Determine if $(-4, -1)$ is a solution to
$$y = -\frac{1}{2}x - 3.$$

4. Explain the process for finding the x- and y-intercepts.

For Exercises 5–8, identify the x- and y-intercepts (if possible) and graph the line.

5. $6x - 8y = 24$ **6.** $x = -4$

7. $3x = 5y$

8. $2y = -6$

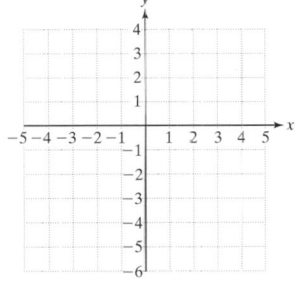

9. Find the slope of the line, given the following information:

    a. The line passes through the points $(7, -3)$ and $(-1, -8)$.

    b. The line is given by $6x - 5y = 1$.

10. Given the equation $-3x + 4y = 4$,

    a. Write the line in slope-intercept form.

    b. Determine the slope and y-intercept.

    c. Graph the line, using the slope and y-intercept.

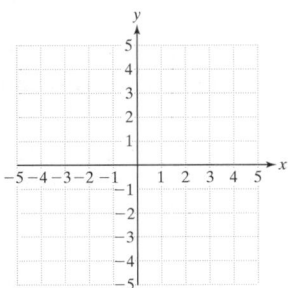

11. The slope of a line is -7.

    a. Find the slope of a line parallel to the given line.

    b. Find the slope of a line perpendicular to the given line.

12. Describe the relationship of the slopes of

    a. Two parallel lines

    b. Two perpendicular lines

13. Two points are given for each of two lines. Determine if the lines are parallel, perpendicular, or neither.

    L_1: $(4, -4)$ and $(1, -6)$

    L_2: $(-2, 0)$ and $(0, 3)$

14. Determine if the lines are parallel, perpendicular, or neither.

    a. $y = -x + 4$
        $y = x - 3$

    b. $9x - 3y = 1$
        $15x - 5y = 10$

    c. $3y = 6$
        $x = 0.5$

    d. $5x - 3y = 9$
        $3x - 5y = 10$

15. Write an equation that represents a line subject to the following conditions. (Answers may vary.)

    a. A line that does not pass through the origin and has a positive slope

    b. A line with an undefined slope

    c. A line perpendicular to the y-axis. What is the slope of such a line?

    d. A slanted line that passes through the origin and has a negative slope

16. Write an equation of the line containing the points $(2, -3)$ and $(4, 0)$.

17. Write an equation of a line containing $(4, -3)$ and parallel to $6x - 3y = 1$.

18. Write an equation of the line that passes through the point $(8, -\frac{1}{2})$ with slope -2. Write the answer in slope-intercept form.

19. Write an equation of the line that passes through the point $(-10, -3)$ and is perpendicular to $3x + y = 7$. Write the answer in slope-intercept form.

20. Jack sells used cars. He is paid $800 per month plus $300 commission for each automobile he sells.

    a. Write an equation that represents Jack's monthly earnings y in terms of the number of automobiles he sells x.

    b. Graph the linear equation you found in part (a).

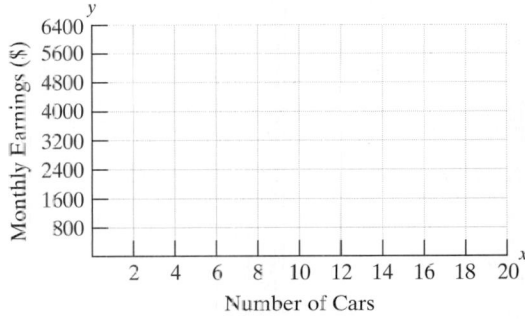

c. What does the *y*-intercept mean in the context of this problem?

d. How much will Jack earn in a month if he sells 17 automobiles?

21. The graph represents the life expectancy for females in the United States according to birth year. The value $x = 0$ represents a birth year of 1940.

Life Expectancy for Females in the United States According to Birth Year

Source: National Center for Health Statistics

a. Determine the *y*-intercept from the graph. What does the *y*-intercept represent in the context of this problem?

b. Using the two points $(0, 66)$ and $(30, 75)$, determine the slope of the line. What does the slope of the line represent in the context of this problem?

c. Use the *y*-intercept and the slope found in parts (a) and (b) to write an equation of the line by letting *x* represent the year of birth and *y* represent the corresponding life expectancy.

d. Using the linear equation from part (c), approximate the life expectancy for women born in the United States in 1994. How does your answer compare with the reported life expectancy of 79 yr?

For Exercises 22–25, graph the functions.

22. $f(x) = -3x - 1$

23. $k(x) = -2$

24. $p(x) = x^2$

25. $w(x) = |x|$

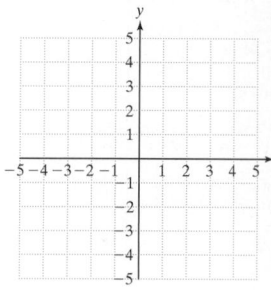

For Exercises 26–27, **a.** determine if the relation defines *y* as a function of *x*, **b.** identify the domain, and **c.** identify the range.

26.

27.

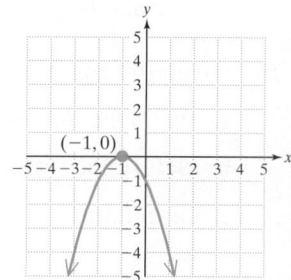

For Exercises 28–30, write the domain in interval notation.

28. $f(x) = \dfrac{x - 5}{x + 7}$

29. $f(x) = \sqrt{x + 7}$

30. $h(x) = (x + 7)(x - 5)$

31. Given: $r(x) = x^2 - 2x + 1$

a. Find $r(-2)$, $r(0)$, and $r(3)$.

b. What is the domain of *r*?

32. The function defined by $s(t) = 1.6t + 36$ approximates the per capita consumption of soft drinks in the United States since 1985. The values of $s(t)$ are measured in gallons, and $t = 0$ corresponds to the year 1985. (Source: U.S. Department of Agriculture)

a. Evaluate $s(0)$ and $s(7)$ and interpret the results in the context of this problem.

b. Determine the slope and interpret its meaning in the context of this problem.

For Exercises 33–36, determine if the function is constant, linear, quadratic, or none of these.

33. $f(x) = -3x^2$

34. $g(x) = -3x$

35. $h(x) = -3$

36. $k(x) = -\dfrac{3}{x}$

37. Explain how to find the x- and y-intercepts of the graph defined by $y = f(x)$.

38. Find the x- and y-intercepts for $f(x) = \dfrac{3}{4}x + 9$.

For Exercises 39–46, refer to the graph.

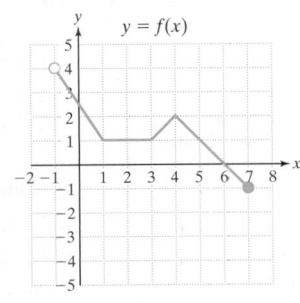

$y = f(x)$

39. Find $f(1)$.

40. Find $f(4)$.

41. Write the domain of f.

42. Write the range of f.

43. Answer true or false. The value $y = 5$ is in the range of f.

44. Find the x-intercept.

45. For what value(s) of x is $f(x) = 0$?

46. For what value(s) of x is $f(x) = 1$?

Chapters 1–2   Cumulative Review Exercises

For Exercises 1–3, simplify the expressions.

1. $\dfrac{5 - 2^3 \div 4 + 7}{-1 - 3(4 - 1)}$

2. $3 + \sqrt{25} - 8(\sqrt{9}) \div 6$

3. $4[-3x - 5(y - 2x) + 3] - 7(6y + x)$

For Exercises 4–6, solve the equation.

4. $\dfrac{2x - 3}{6} - \dfrac{x + 1}{4} = -2$

5. $z - (3 + 2z) + 5 = -z - 5$

6. $|x - 5| + 7 = 10$

For Exercises 7–11, solve the inequality. Write the answer in interval notation.

7. $-4 \le \dfrac{x - 1}{2} < 3$

8. $3x + 2 < 11$ and $-4 < 2x$

9. $3x + 2 < 11$ or $-4 < 2x$

10. $|2x - 1| + 3 < 12$

11. $\left|\dfrac{x - 3}{5}\right| \ge 2$

12. Given: $f(x) = \frac{1}{2}x - 1$ and $g(x) = 3x^2 - 2x$
   a. Find $f(4)$.   **b.** Find $g(-3)$.

13. Find the slope of the line that passes through the points $(4, -5)$ and $(-6, -3)$.

14. Determine if $(\frac{1}{3}, 7)$ is a solution to $y = 6x - 5$.

For Exercises 15–16, **a.** find the x- and y-intercepts, **b.** find the slope, and **c.** graph the line.

15. $3x - 5y = 10$

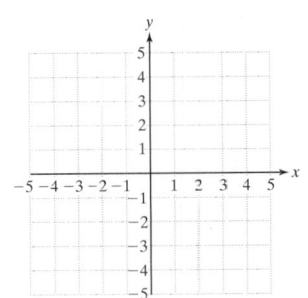

16. $2y + 4 = 10$

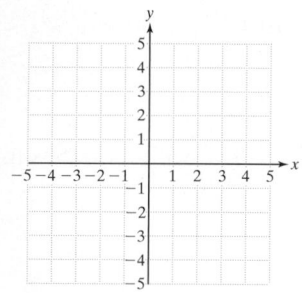

17. Find an equation for the vertical line that passes through the point (7, 11).

18. Find the slope of the line $5x - 2y = -10$.

19. State the domain and range of the relation. Is the relation a function? $\{(3, -1), (4, -5), (3, -8)\}$

20. Find an equation of the line passing through $(1, -4)$ and parallel to $2x + y = 6$. Write the answer in slope-intercept form.

21. Find an equation of the line passing through $(1, -4)$ and perpendicular to $y = \frac{1}{4}x - 2$. Write the answer in slope-intercept form.

22. At the movies, Laquita paid for drinks and popcorn for herself and her two children. She spent twice as much on popcorn as on drinks. If her total bill came to $17.94, how much did she spend on drinks and how much did she spend on popcorn?

23. Write the domain of the function in interval notation.

$$f(x) = \frac{1}{x - 15}$$

24. A chemist mixes a 20% salt solution with 15 L of a 50% salt solution to get a 38% salt solution. How much of the 20% solution does she use?

25. The yearly rainfall for Seattle, Washington, is 0.7 in. less than twice the yearly rainfall for Los Angeles, California. If the total yearly rainfall for the two cities is 50 in., how much rain does each city get per year?

Systems of Linear Equations and Inequalities

3

CHAPTER OUTLINE

3.1 Solving Systems of Linear Equations by the Graphing Method   236

3.2 Solving Systems of Linear Equations by the Substitution Method   246

3.3 Solving Systems of Linear Equations by the Addition Method   253

    Problem Recognition Exercises: Solving Systems of Linear Equations   260

3.4 Applications of Systems of Linear Equations in Two Variables   261

3.5 Linear Inequalities and Systems of Linear Inequalities in Two Variables   270

3.6 Systems of Linear Equations in Three Variables and Applications   283

3.7 Solving Systems of Linear Equations by Using Matrices   293

    Group Activity: Creating a Quadratic Model of the Form $y = at^2 + bt + c$ 302

Chapter 3

In this chapter, we solve systems of linear equations in two and three variables. One of the skills we need is to find the point of intersection between two lines.

Are You Prepared?

To practice, graph the five lines on the grid provided. Then answer the questions and place the letter corresponding to each answer in the space below. This will define an important vocabulary word for this chapter.

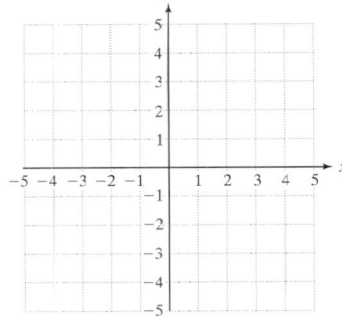

    Line 1. $y = x + 2$
    Line 2. $y = -x + 4$
    Line 3. $x - y = 1$
    Line 4. $y = -2$
    Line 5. $x + 4 = 1$

1. The point of intersection between lines 1 and 2.
2. The point of intersection between lines 3 and 4.
3. The point of intersection between lines 5 and 1.
4. The line that is parallel to line 3.
5. The line that is vertical.
6. The line that has (3, 1) as a solution.
7. The point of intersection between lines 1 and 3.

T $(-1, -2)$
I $(-3, -1)$
U Line 2
O $(1, 3)$
S Line 1
N no point of intersection
L Line 5

A __ __ __ __ __ __ __ __   to a system of linear equations is a point of
   4 1 5 6 2 3 1 7   intersection between two lines.

Section 3.1	**Solving Systems of Linear Equations by the Graphing Method**

Concepts

1. **Solutions to Systems of Linear Equations**
2. **Solving Systems of Linear Equations by Graphing**

1. Solutions to Systems of Linear Equations

A linear equation in two variables has an infinite number of solutions that form a line in a rectangular coordinate system. Two or more linear equations form a **system of linear equations**. For example:

$$x - 3y = -5$$
$$2x + 4y = 10$$

A **solution to a system of linear equations** is an ordered pair that is a solution to *both* individual linear equations.

Example 1 **Determining Solutions to a System of Linear Equations**

Determine whether the ordered pairs are solutions to the system.

$$x + y = -6$$
$$3x - y = -2$$

a. $(-2, -4)$ **b.** $(0, -6)$

Solution:

a. Substitute the ordered pair $(-2, -4)$ into both equations:

$$x + y = -6 \longrightarrow (-2) + (-4) \overset{?}{=} -6 \checkmark \text{ True}$$
$$3x - y = -2 \longrightarrow 3(-2) - (-4) \overset{?}{=} -2 \checkmark \text{ True}$$

Because the ordered pair $(-2, -4)$ is a solution to each equation, it is a solution to the *system* of equations.

b. Substitute the ordered pair $(0, -6)$ into both equations:

$$x + y = -6 \longrightarrow (0) + (-6) \overset{?}{=} -6 \checkmark \text{ True}$$
$$3x - y = -2 \longrightarrow 3(0) - (-6) \overset{?}{=} -2 \quad \text{ False}$$

Because the ordered pair $(0, -6)$ is not a solution to the second equation, it is *not* a solution to the system of equations.

Skill Practice Determine whether the ordered pairs are solutions to the system.

$$3x + 2y = -8$$
$$y = 2x - 18$$

1. $(-2, -1)$ **2.** $(4, -10)$

A solution to a system of two linear equations can be interpreted graphically as a point of intersection between the two lines.

Answers

1. No   **2.** Yes

Graphing the lines from Example 1 we see that the point of intersection is $(-2, -4)$. Therefore, we say that the solution set is $\{(-2, -4)\}$. See Figure 3-1.

When two lines are drawn in a rectangular coordinate system, three geometric relationships are possible:

1. Two lines may intersect at *exactly one point*.

2. Two lines may intersect at *no point*. This occurs if the lines are parallel.

3. Two lines may intersect at *infinitely many points* along the line. This occurs if the equations represent the same line (the lines are coinciding).

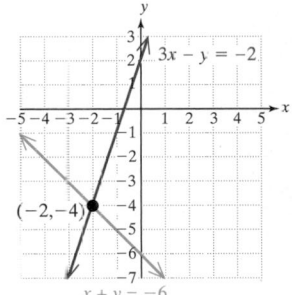

Figure 3-1

If a system of linear equations has one or more solutions, the system is said to be a **consistent system**. If a linear system has no solution, it is said to be an **inconsistent system**.

If two equations represent the same line, then all points along the line are solutions to the system of equations. In such a case, the equations are said to be **dependent**. If two linear equations represent different lines, then the equations are said to be **independent**. The different possibilities for solutions to systems of linear equations are given in Table 3-1.

Table 3-1   Solutions to Systems of Linear Equations in Two Variables

One Unique Solution	No Solution	Infinitely Many Solutions
One point of intersection	Parallel lines	Coinciding lines
System is consistent.	System is inconsistent.	System is consistent.
Equations are independent.	Equations are independent.	Equations are dependent.

2. Solving Systems of Linear Equations by Graphing

Example 2 **Solving a System of Linear Equations by Graphing**

Solve the system by graphing both linear equations and finding the point(s) of intersection.

$$y = \frac{1}{2}x - 2$$

$$4x + 2y = 6$$

Solution:

To graph each equation, write the equation in slope-intercept form $y = mx + b$.

First equation

$$y = \frac{1}{2}x - 2 \qquad \text{Slope: } \tfrac{1}{2}$$

Second equation

$$4x + 2y = 6$$
$$2y = -4x + 6$$
$$\frac{2y}{2} = \frac{-4x}{2} + \frac{6}{2}$$
$$y = -2x + 3 \qquad \text{Slope: } -2$$

From their slope-intercept forms, we see that the lines have different slopes, indicating that the lines must intersect at exactly one point. We can graph the lines using the slope and y-intercept to find the point of intersection (Figure 3-2).

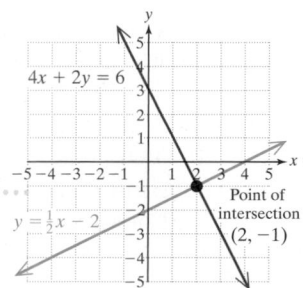

Figure 3-2

The point $(2, -1)$ appears to be the point of intersection. This can be confirmed by substituting $x = 2$ and $y = -1$ into both equations.

$$y = \frac{1}{2}x - 2$$
$$-1 \stackrel{?}{=} \frac{1}{2}(2) - 2$$
$$-1 \stackrel{?}{=} 1 - 2$$
$$-1 \stackrel{?}{=} -1 \ \checkmark \ \text{True}$$

$$4x + 2y = 6$$
$$4(2) + 2(-1) \stackrel{?}{=} 6$$
$$8 - 2 \stackrel{?}{=} 6$$
$$6 \stackrel{?}{=} 6 \ \checkmark \ \text{True}$$

The solution set is $\{(2, -1)\}$.

Skill Practice   Solve by using the graphing method.

3. $y = -3x - 5$
     $x - 2y = -4$

TIP: In Example 2, the lines could also have been graphed by using the x- and y-intercepts or by using a table of points. However, the advantage of writing the equations in slope-intercept form is that we can compare the slope *and* y-intercept of each line.

1. If the slopes differ, the lines are different and nonparallel and must cross in exactly one point.
2. If the slopes are the same and the y-intercepts are different, the lines are parallel and do not intersect.
3. If the slopes are the same and the y-intercepts are the same, the two equations represent the same line.

Answer
3. $\{(-2, 1)\}$

Example 3 Solving a System of Linear Functions by Graphing

Solve the system.

$$f(x) = 3$$

$$g(x) = 2x + 1$$

Solution:

This first function can be written as $y = 3$. This is an equation of a horizontal line. Writing the second equation as $y = 2x + 1$, we have a slope of 2 and a y-intercept of $(0, 1)$.

The graphs of the functions are shown in Figure 3-3. The point of intersection is $(1, 3)$. Therefore, the solution set is $\{(1, 3)\}$.

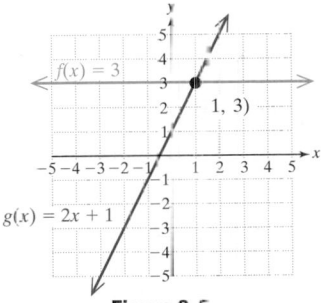

Figure 3-3

TIP: The equations in Example 3 are independent because they represent different lines. The system is consistent because it has a solution.

Skill Practice Solve the system by graphing.

4. $f(x) = 1$
$g(x) = -3x + 4$

Example 4 Solving a System of Linear Equations by Graphing

Solve the system by graphing.

$$-x + 3y = -6$$

$$6y = 2x + 6$$

Solution:

To graph the lines, write each equation in slope-intercept form.

$$-x + 3y = -6 \qquad\qquad 6y = 2x + 6$$

$$3y = x - 6$$

$$\frac{3y}{3} = \frac{x}{3} - \frac{6}{3} \qquad\qquad \frac{6y}{6} = \frac{2x}{6} + \frac{6}{6}$$

$$y = \frac{1}{3}x - 2 \qquad\qquad y = \frac{1}{3}x + 1$$

Because the lines have the same slope but different y-intercepts, they are parallel (Figure 3-4). Two parallel lines do not intersect, which implies that the system has no solution. The system is inconsistent.

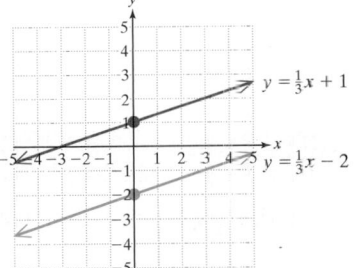

Figure 3-4

The solution set is the empty set, $\{\ \}$.

Skill Practice Solve the system by graphing.

5. $2y = 2x$
$-x + y = -3$

Answers

4. $\{(1, 1)\}$
5. The solution set is $\{\ \}$. The system is inconsistent.

Example 5   **Solving a System of Linear Equations by Graphing**

Solve the system by graphing.

$$x + 4y = 8$$

$$y = -\frac{1}{4}x + 2$$

Solution:

Write the first equation in slope-intercept form. The second equation is already in slope-intercept form.

First equation	**Second equation**
$x + 4y = 8$	$y = -\frac{1}{4}x + 2$
$4y = -x + 8$	
$\dfrac{4y}{4} = \dfrac{-x}{4} + \dfrac{8}{4}$	
$y = -\dfrac{1}{4}x + 2$	

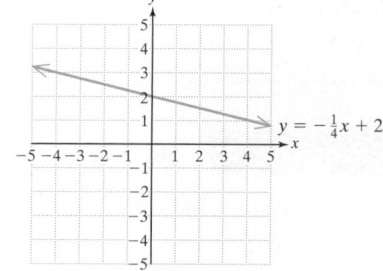

Notice that the slope-intercept forms of the two lines are identical. Therefore, the equations represent the same line (Figure 3-5). The equations are dependent, and the solution to the system of equations is the set of all points on the line.

Figure 3-5

Because the ordered pairs in the solution set cannot all be listed, we can write the solution in set-builder notation. Furthermore, the equations $x + 4y = 8$ and $y = -\frac{1}{4}x + 2$ represent the same line. Therefore, the solution set may be written as $\{(x, y) \mid y = -\frac{1}{4}x + 2\}$ or $\{(x, y) \mid x + 4y = 8\}$.

Skill Practice Solve the system by graphing.

6. $y = \dfrac{1}{2}x + 1$

 $x - 2y = -2$

Calculator Connections

Topic: Using the *Intersect* Feature

The solution to a system of equations can be found by using either a *Trace* feature or an *Intersect* feature on a graphing calculator to find the point of intersection between two curves.

For example, consider the system

$$-2x + y = 6$$

$$5x + y = -1$$

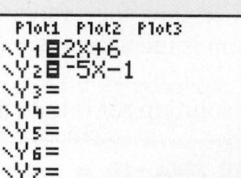

Answer

6. Infinitely many solutions;

$\{(x, y) \mid y = \dfrac{1}{2}x + 1\}$;

dependent equations

First graph the equations together on the same viewing window. Recall that to enter the equations into the calculator, the equations must be written with the y variable isolated.

Isolate y.

$$-2x + y = 6 \longrightarrow y = 2x + 6$$

$$5x + y = -1 \longrightarrow y = -5x - 1$$

By inspection of the graph, it appears that the solution is $(-1, 4)$. The *Trace* option on the calculator may come close to $(-1, 4)$ but may not show the exact solution (Figure 3-6). However, an *Intersect* feature on a graphing calculator may provide the exact solution (Figure 3-7). See your user's manual for further details.

Using *Trace*

Figure 3-6

Using *Intersect*

Figure 3-7

Section 3.1   Practice Exercises

Study Skills Exercise

Before you proceed further in Chapter 3, make your test corrections for the Chapter 2 test. See Exercise 1 of Section 2.1 for instructions.

Vocabulary and Key Concepts

1. a. A _____ of linear equations consists of two or more linear equations.

 b. A _____ to a system of linear equations is an ordered pair that is a solution to both individual equations in the system.

 c. Graphically, a solution to a system of linear equations in two variables is a point where the lines _____.

 d. A system of equations that has one or more solutions is said to be _____.

 e. The solution set to an inconsistent system of equations is _____.

 f. Two equations in a system of linear equations in two variables are said to be _____ if they represent the same line.

 g. Two equations in a system of linear equations in two variables are said to be _____ if they represent different lines.

Concept 1: Solutions to Systems of Linear Equations

2. From the graph shown, determine the solution to the system.

$$x + y = 4$$

$$y = 2x + 1$$

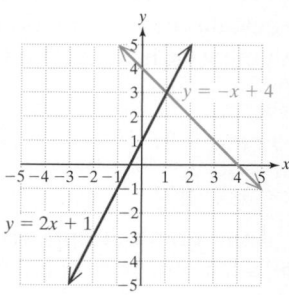

For Exercises 3–8, determine which points are solutions to the given system. **(See Example 1.)**

3. $y = 8x - 5$

$y = 4x + 3$

$(-1, 13), (-1, 1), (2, 11)$

4. $y = -\dfrac{1}{2}x - 5$

$y = \dfrac{3}{4}x - 10$

$(4, -7), (0, -10), \left(3, -\dfrac{9}{2}\right)$

5. $2x - 7y = -30$

$y = 3x + 7$

$(0, -30), \left(\dfrac{3}{2}, 5\right), (-1, 4)$

6. $x + 2y = 4$

$y = -\dfrac{1}{2}x + 2$

$(-2, 3), (4, 0), \left(3, \dfrac{1}{2}\right)$

7. $x - y = 6$

$4x + 3y = -4$

$(4, -2), (6, 0), (2, 4)$

8. $x - 3y = 3$

$2x - 9y = 1$

$(0, 1), (4, -1), (9, 2)$

For Exercises 9–14, the graph of a system of linear equations is given.

a. Identify whether the system is consistent or inconsistent.

b. Identify the equations as dependent or independent.

c. Identify the number of solutions to the system.

9. $y = x + 3$

$3x + y = -1$

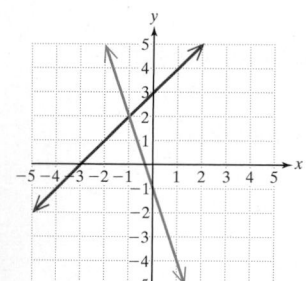

10. $5x - 3y = 6$

$3y = 2x + 3$

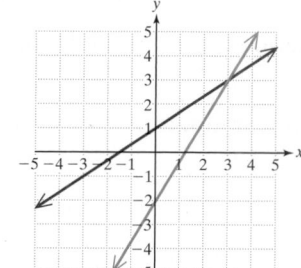

11. $2x = y + 4$

$-4x + 2y = 2$

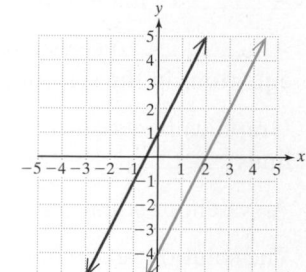

12. $y = -2x - 3$

$-4x - 2y = 0$

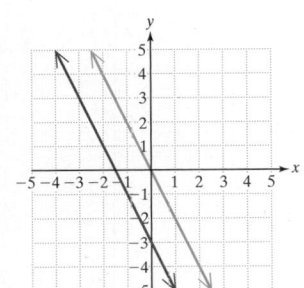

13. $y = \dfrac{1}{3}x + 2$

$-x + 3y = 6$

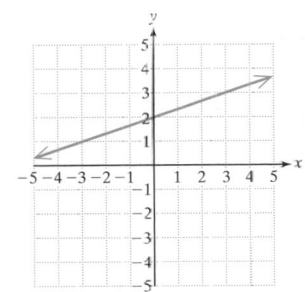

14. $y = -\dfrac{2}{3}x - 1$

$-4x - 6y = 6$

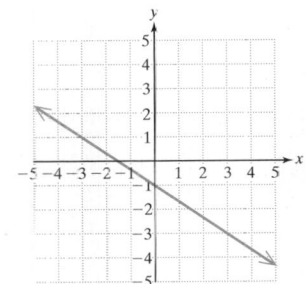

Concept 2: Solving Systems of Linear Equations by Graphing

For Exercises 15–32, solve the system by graphing. For systems that do not have one unique solution, also state the number of solutions and whether the system is inconsistent or the equations are dependent. **(See Examples 2–5.)**

15. $2x + y = -3$

$-x + y = 3$

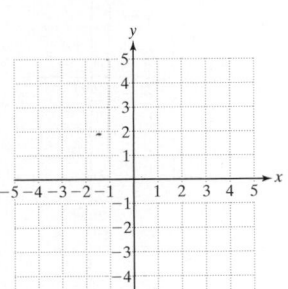

16. $4x - 3y = 12$

$3x + 4y = -16$

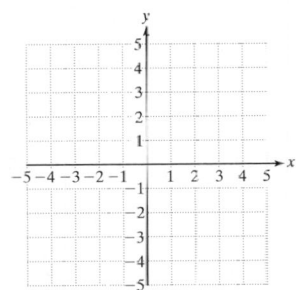

17. $f(x) = -2x + 3$

$g(x) = 5x - 4$

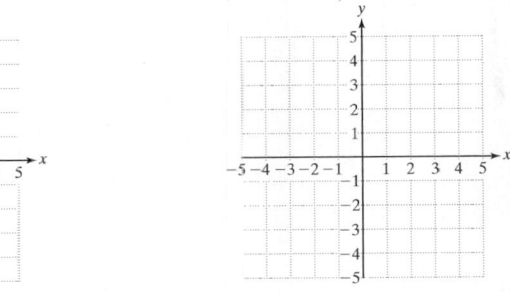

18. $h(x) = 2x + 5$

$g(x) = -x + 2$

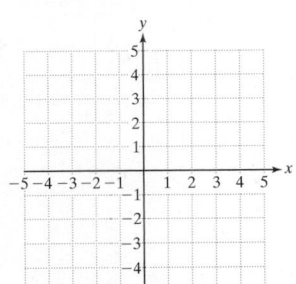

19. $k(x) = \dfrac{1}{3}x - 5$

$f(x) = -\dfrac{2}{3}x - 2$

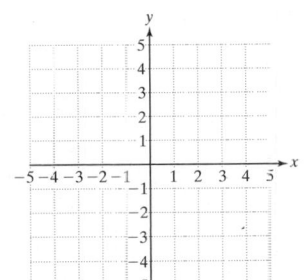

20. $f(x) = \dfrac{1}{2}x + 2$

$g(x) = \dfrac{5}{2}x - 2$

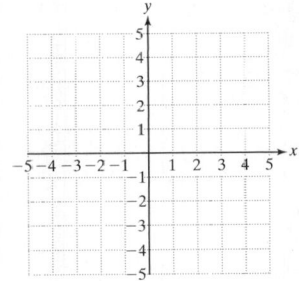

21. $x = 4$

$y = 2x - 3$

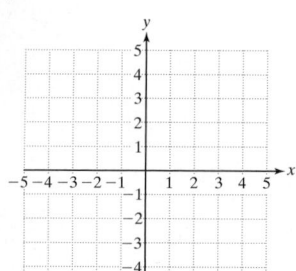

22. $3x + 2y = 6$

$y = -3$

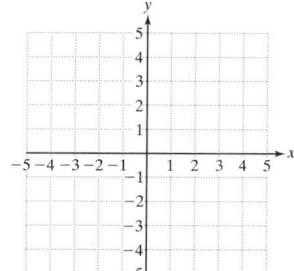

23. $y = -2x + 3$

$-2x = y + 1$

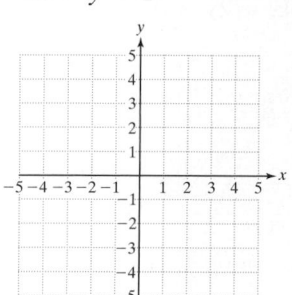

24. $y = \dfrac{1}{3}x - 2$

$x = 3y - 9$

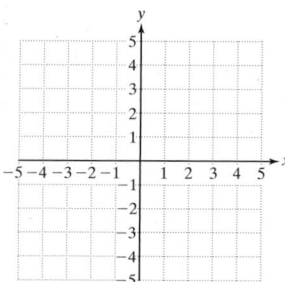

25. $y = \dfrac{2}{3}x - 1$

$2x = 3y + 3$

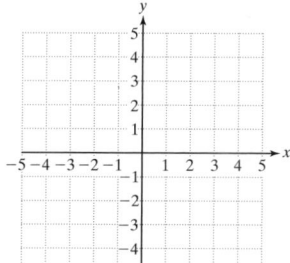

26. $4x = 16 - 8y$

$y = -\dfrac{1}{2}x + 2$

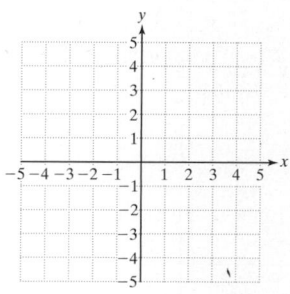

27. $2x = 4$

$\dfrac{1}{2}y = -1$

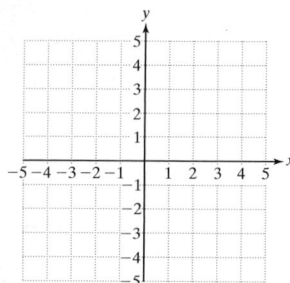

28. $y + 7 = 6$

$-5 = 2x$

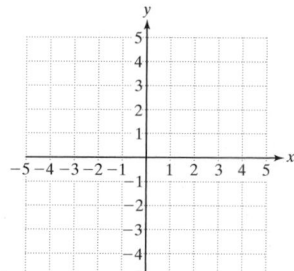

29. $-x + 3y = 6$

$6y = 2x + 12$

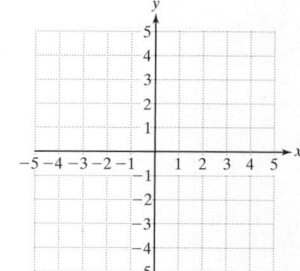

30. $3x = 2y - 4$

$-4y = -6x - 8$

31. $2x - y = 4$

$4x + 2 = 2y$

32. $x = 4y + 4$

$-2x + 8y = -16$

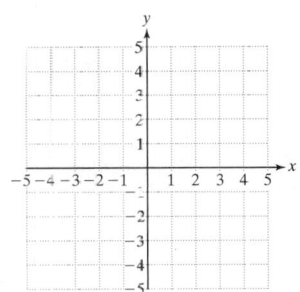

For Exercises 33–36, identify each statement as true or false.

33. A consistent system is a system that always has a unique solution.

34. Dependent equations form a system that has no solution.

35. If two lines coincide, the equations are dependent.

36. If two lines are parallel, the equations are independent.

Expanding Your Skills

37. Write a system of equations with solution set $\{(4, 5)\}$.

38. Write a system of equations with solution set $\{(-2, 6)\}$.

39. Find C and D such that the solution set to the system is $\{(1, 3)\}$.

$$Cx + 2y = 11$$

$$-3x + Dy = 9$$

40. Find M and N such that the solution set to the system is $\{(2, -4)\}$.

$$3x + My = -22$$

$$Nx + 4y = 6$$

Graphing Calculator Exercises

For Exercises 41–44, use a graphing calculator to graph each linear equation on the same viewing window. Use a *Trace* or *Intersect* feature to find the point(s) of intersection.

41. $y = 5.62x + 15.46$

$y = -1.96x - 11.07$

42. $y = -2.3x - 5.48$

$y = 4.62x + 26.352$

43. $2.4x - 4.8y = -9.36$

$-1.8x + 5.4y = 12.456$

44. $36x - 90y = -36$

$-15.5x - 5y = -80.75$

Section 3.2 Solving Systems of Linear Equations by the Substitution Method

Concepts

1. The Substitution Method
2. Solving Inconsistent Systems and Systems of Dependent Equations

1. The Substitution Method

Graphing a system of equations is one method to find the solution of the system. However, sometimes it is difficult to determine the solution using this method because of limitations in the accuracy of the graph. This is particularly true when the coordinates of a solution are not integer values or when the solution is a point not sufficiently close to the origin. Identifying the coordinates of the point $(\frac{3}{17}, -\frac{23}{9})$ or $(-251, 8349)$, for example, might be difficult from a graph.

In this section and Section 3.3, we will present two algebraic methods to solve a system of equations. The first is called the *substitution method*. This technique is particularly important because it can be used to solve more advanced problems including nonlinear systems of equations.

The first step in the substitution process is to isolate one of the variables from one of the equations. Consider the system

$$x + y = 16$$
$$x - y = 4$$

Solving the first equation for x yields $x = 16 - y$. Then, because x is equal to $16 - y$, the expression $16 - y$ may replace x in the second equation. This leaves the second equation in terms of y only.

First equation: $x + y = 16 \xrightarrow{\text{Solve for } x.} x = \underbrace{16 - y}$

Second equation: $(16 - y) - y = 4$ Substitute $x = 16 - y$.

$16 - 2y = 4$ Solve for y.

$-2y = -12$

$y = 6$

$x = 16 - y$ To find x, substitute $y = 6$ back into the expression $x = 16 - y$.

$x = 16 - (6)$

$x = 10$ Check the ordered pair $(10, 6)$ in both original equations.

$x + y = 16$ $x - y = 4$

$(10) + (6) \overset{?}{=} 16$ ✔ True $(10) - (6) \overset{?}{=} 4$ ✔ True

The solution set is $\{(10, 6)\}$.

Solving a System of Equations by the Substitution Method

Step 1   Isolate one of the variables from one equation.

Step 2   Substitute the quantity found in step 1 into the *other* equation.

Step 3   Solve the resulting equation.

Step 4   Substitute the value found in step 3 back into the equation in step 1 to find the value of the remaining variable.

Step 5   Check the solution in *both* equations, and write the answer as an ordered pair within set notation.

Example 1   Using the Substitution Method to Solve a System of Linear Equations

Solve the system by using the substitution method.

$$3x - 2y = -7$$
$$6x + y = 6$$

Solution:

The y variable in the second equation is the easiest variable to isolate because its coefficient is 1.

$3x - 2y = -7$

$6x + y = 6 \longrightarrow y = -6x + 6$ **Step 1:** Solve the second equation for y.

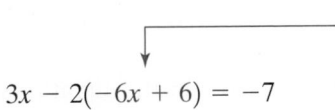

$3x - 2(-6x + 6) = -7$ **Step 2:** Substitute the quantity $-6x + 6$ for y in the *other* equation.

$3x + 12x - 12 = -7$

$15x - 12 = -7$

$15x = 5$ **Step 3:** Solve for x.

$x = \dfrac{1}{3}$

$y = -6x + 6$ **Step 4:** Substitute $x = \frac{1}{3}$ into the expression $y = -6x + 6$.

$y = -6\left(\dfrac{1}{3}\right) + 6$

$y = -2 + 6$

$y = 4$

$3x - 2y = -7 \qquad 6x + y = 6$ **Step 5:** Check the ordered pair $(\frac{1}{3}, 4)$ in each original equation.

$3\left(\dfrac{1}{3}\right) - 2(4) \stackrel{?}{=} -7 \qquad 6\left(\dfrac{1}{3}\right) + 4 \stackrel{?}{=} 6$

$1 - 8 \stackrel{?}{=} -7 \checkmark \qquad 2 + 4 \stackrel{?}{=} 6 \checkmark$

The solution set is $\{(\frac{1}{3}, 4)\}$.

> **Avoiding Mistakes**
>
> Do not substitute $y = -6x + 6$ into the same equation from which it came. This mistake will result in an identity:
>
> $$6x + y = 6$$
> $$6x + (-6x + 6) = 6$$
> $$6x - 6x + 6 = 6$$
> $$6 = 6$$

Skill Practice Solve by using the substitution method.

1. $3x + y = 8$
$\quad x - 2y = 12$

Answer

1. $\{(4, -4)\}$

> **Example 2** **Using the Substitution Method to Solve a System of Linear Equations**

Solve the system by using the substitution method.

$$-3x + 4y = 9$$

$$x = -\frac{1}{3}y + 2$$

Solution:

$$-3x + 4y = 9$$

$$x = -\frac{1}{3}y + 2$$ **Step 1:**   In the second equation, x is already isolated.

$$-3\left(-\frac{1}{3}y + 2\right) + 4y = 9$$ **Step 2:**   Substitute the quantity $-\frac{1}{3}y + 2$ for x in the *other* equation.

$$y - 6 + 4y = 9$$ **Step 3:**   Solve for y.

$$5y = 15$$

$$y = 3$$

Now use the known value of y to solve for the remaining variable x.

$$x = -\frac{1}{3}y + 2$$

$$x = -\frac{1}{3}(3) + 2$$ **Step 4:**   Substitute $y = 3$ into the equation $x = -\frac{1}{3}y + 2$.

$$x = -1 + 2$$

$$x = 1$$

Step 5:   Check the ordered pair $(1, 3)$ in each original equation.

$$-3x + 4y = 9 \qquad\qquad x = -\frac{1}{3}y + 2$$

$$-3(1) + 4(3) \overset{?}{=} 9 \qquad\qquad 1 \overset{?}{=} -\frac{1}{3}(3) + 2$$

$$-3 + 12 \overset{?}{=} 9 \ ✔ \text{ True} \qquad\qquad 1 \overset{?}{=} -1 + 2 \ ✔ \text{ True}$$

The solution set is $\{(1, 3)\}$.

Skill Practice   Solve by using the substitution method.

2. $x = 2y + 3$
     $4x - 2y = 0$

Answer

2. $\{(-1, -2)\}$

2. Solving Inconsistent Systems and Systems of Dependent Equations

> **Example 3** Solving an Inconsistent System

Solve the system by using the substitution method.

$$x = 2y - 4$$
$$-2x + 4y = 6$$

Solution:

$x = 2y - 4$	**Step 1:** The x variable is already isolated.
$-2x + 4y = 6$	
$-2(2y - 4) + 4y = 6$	**Step 2:** Substitute the quantity $x = 2y - 4$ into the *other* equation.
$-4y + 8 + 4y = 6$	**Step 3:** Solve for y.
$8 = 6$	The equation reduces to a contradiction, indicating that the system has no solution. The lines never intersect and must be parallel. The system is *inconsistent*.

There is no solution.
The system is inconsistent.
The solution set is { }.

TIP: The answer to Example 3 can be verified by writing each equation in slope-intercept form and graphing the equations.

Equation 1

$x = 2y - 4$

$2y = x + 4$

$\dfrac{2y}{2} = \dfrac{x}{2} + \dfrac{4}{2}$

$y = \dfrac{1}{2}x + 2$

Equation 2

$-2x + 4y = 6$

$4y = 2x + 6$

$\dfrac{4y}{4} = \dfrac{2x}{4} + \dfrac{6}{4}$

$y = \dfrac{1}{2}x + \dfrac{3}{2}$

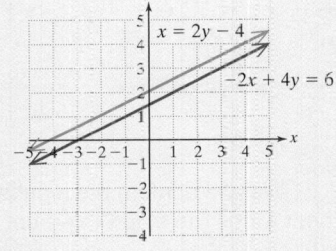

Notice that the equations have the same slope, but different y-intercepts; therefore, the lines must be parallel. There is no solution to this system of equations.

Skill Practice Solve by using the substitution method.

3. $8x - 16y = 3$

$y = \dfrac{1}{2}x + 1$

Answer

3. No solution; { }; inconsistent system

> ### Example 4 Solving a System of Dependent Equations
>
> Solve by using the substitution method.
>
> $$4x - 2y = -6$$
> $$y - 3 = 2x$$
>
> **Solution:**
>
> $$4x - 2y = -6$$
> $$y - 3 = 2x \longrightarrow y = \underline{2x + 3}$$ **Step 1:** Solve for one of the variables.
>
> $$4x - 2(2x + 3) = -6$$ **Step 2:** Substitute the quantity $2x + 3$ for y in the *other* equation.
>
> $$4x - 4x - 6 = -6$$ **Step 3:** Solve for x. Apply the distributive property to clear the parentheses.
>
> $$-6 = -6$$
>
> The system reduces to the identity $-6 = -6$. Therefore, the two equations are equivalent. The solution set consists of all points on the common line, giving us an infinite number of solutions. The equations are dependent and because the equations $4x - 2y = -6$ and $y - 3 = 2x$ represent the same line, the solution set can be written as
>
> $$\{(x, y) \mid 4x - 2y = -6\} \quad \text{or} \quad \{(x, y) \mid y - 3 = 2x\}$$
>
> **Skill Practice** Solve the system by using substitution.
>
> **4.** $3x + 6y = 12$
> $2y = -x + 4$

TIP: We can confirm the results of Example 4 by writing each equation in slope-intercept form. The slope-intercept forms are identical, indicating that the lines are the same.

slope-intercept form

$$4x - 2y = -6 \longrightarrow -2y = -4x - 6 \longrightarrow y = 2x + 3$$
$$y - 3 = 2x \longrightarrow y = 2x + 3$$

Answer

4. Infinitely many solutions;
$\{(x, y) \mid 3x + 6y = 12\}$;
dependent equations

Section 3.2   Practice Exercises

Review Exercises

For Exercises 1–4, use the slope-intercept form of the lines to determine the number of solutions to the system.

1. $y = 8x - 1$

   $2x - 16y = 3$

2. $4x + 6y = 1$

   $10x + 15y = \dfrac{5}{2}$

3. $2x - 4y = 0$

   $x - 2y = 9$

4. $6x - 3y = 8$

   $8x + 4y = -1$

5. Determine if the ordered pair $(-4, 3)$ is a solution to the system. $-x + 2y = 10$

   $2x - y = 11$

For Exercises 6–7, solve the system by graphing.

6. $x - y = 4$

   $3x + 4y = 12$

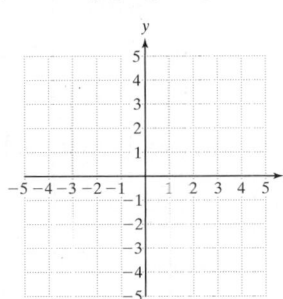

7. $y = 2x + 3$

   $6x + 3y = 9$

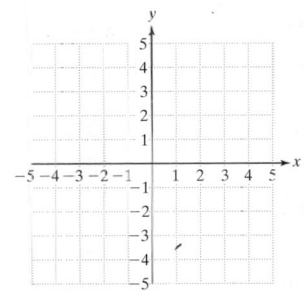

Concept 1: The Substitution Method

For Exercises 8–21, solve by using the substitution method. **(See Examples 1–2.)**

8. $4x + 12y = 4$

   $y = 5x + 11$

9. $y = -3x - 1$

   $2x - 3y = -8$

10. $10y + 34 = x$

   $-7x + y = -31$

11. $-3x + 8y = -1$

   $4x - 11 = y$

12. $12x - 2y = 0$

   $-7x + y = -1$

13. $3x + 12y = 36$

   $x - 5y = 12$

14. $x - 3y = -3$

   $2x + 3y = -6$

15. $x - y = 8$

   $3x + 2y = 9$

16. $5x - 2y = 10$

   $y = x - 1$

17. $2x - y = -1$

   $y = -2x$

18. $1 + 3y = 10$

   $5x + 2y = 6$

19. $2x + 3 = 7$

   $3x - 4y = 6$

20. $2x + 3y = 7$

   $-5x = 2y - 12$

21. $4x - 5y = 14$

   $3y = x - 7$

22. Describe the process of solving a system of linear equations by using substitution.

Concept 2: Solving Inconsistent Systems and Systems of Dependent Equations

For Exercises 23–30, solve the system using the substitution method. For systems that do not have one unique solution, also state the number of solutions and whether the system is inconsistent or the equations are dependent. **(See Examples 3–4.)**

23. $2x - 6y = -2$
$x = 3y - 1$

24. $-2x + 4y = 22$
$x = 2y - 11$

25. $y = \dfrac{1}{7}x + 3$
$x - 7y = -4$

26. $x = -\dfrac{3}{2}y + \dfrac{1}{2}$
$4x + 6y = 7$

27. $5x - y = 10$
$2y = 10x - 5$

28. $x + 4y = 8$
$3x = 3 - 12y$

29. $3x - y = 7$
$-14 + 6x = 2y$

30. $x = 4y + 1$
$-12y = -3x + 3$

31. When using the substitution method, explain how to determine whether two linear equations in a system are dependent.

32. When using the substitution method, explain how to determine whether a system of linear equations is inconsistent.

Mixed Exercises

For Exercises 33–56, solve the system using the substitution method. For systems that do not have one unique solution, also state the number of solutions and whether the system is inconsistent or the equations are dependent.

33. $x = 1.3y + 1.5$
$y = 1.2x - 4.6$

34. $y = 0.8x - 1.8$
$1.1x = -y + 9.6$

35. $y = \dfrac{2}{3}x - \dfrac{1}{3}$
$x = \dfrac{1}{4}y + \dfrac{17}{4}$

36. $x = \dfrac{1}{6}y - \dfrac{5}{3}$
$y = \dfrac{1}{5}x + \dfrac{21}{5}$

37. $-2x + y = 4$
$-\dfrac{1}{4}x + \dfrac{1}{8}y = \dfrac{1}{4}$

38. $8x - y = 8$
$\dfrac{1}{3}x - \dfrac{1}{24}y = \dfrac{1}{2}$

39. $3x + 2y = 6$
$y = x + 3$

40. $-x + 4y = -4$
$y = x - 1$

41. $-300x - 125y = 1350$
$y + 2 = 8$

42. $200y = 150x$
$y - 4 = 1$

43. $2x - y = 6$
$\dfrac{1}{6}x - \dfrac{1}{12}y = \dfrac{1}{2}$

44. $x - 4y = 8$
$\dfrac{1}{16}x - \dfrac{1}{4}y = \dfrac{1}{2}$

45. $y = -2.7x - 5.1$
$y = 3.1x - 63.1$

46. $y = 6.8x + 2.3$
$y = -4.1x + 56.8$

47. $4x + 4y = 5$
$x - 4y = -\dfrac{5}{2}$

48. $-2x + y = -6$
$6x - 13y = -12$

49. $2(x + 2y) = 12$
$-6x = 5y - 8$

50. $5x - 2y = -25$
$10x = 3(y - 10)$

51. $5(3y - 2) = x + 4$
$4y = 7x - 3$

52. $2x = -3(y + 3)$
$3x - 4y = -22$

53. $2x - 5 = 7$
$4 = 3y + 1$

54. $-2 = 4 - 2y$
$7x - 5 = -5$

55. $0.01y = 0.02x - 0.11$
$0.3x - 0.5y = 2$

56. $0.3x - 0.4y = 1.3$
$0.01x = 0.03y + 0.01$

57. The **centroid** of a region is the geometric center. For the region shown, the centroid is the point of intersection of the diagonals of the parallelogram.

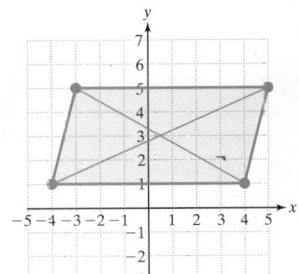

    a. Find the slope-intercept form of the line through the points $(-4, 1)$ and $(5, 5)$.

    b. Find the slope-intercept form of the line through the points $(-3, 5)$ and $(4, 1)$.

    c. Find the centroid of the region.

58. **a.** Find the slope-intercept form of the line through the points $(0, -3)$ and $(-2, 5)$.

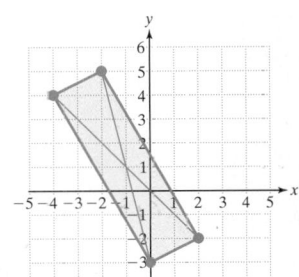

    b. Find the slope-intercept form of the line through the points $(-4, 4)$ and $(2, -2)$.

    c. Find the centroid of the region.

59. The cost to rent an apartment at Glendale Lakes is $800 per month plus a one-time fee of $250. An apartment at the Breakers requires a $500 pet fee plus $750 per month.

    a. Write the cost y to rent for x months for each apartment.

    b. Find the number of months for which the total amount spent for each apartment would be the same.

60. The Surfside Motel charges $159.50 per night plus a flat fee for parking of $24. The Tropical Winds Motel charges $165.50 per night with no parking fee.

    a. Write the cost y to stay for x nights for each motel. Assume that the patron needs parking.

    b. Find the number of nights for which the cost to stay at each motel would be the same.

Solving Systems of Linear Equations by the Addition Method

Section 3.3

1. The Addition Method

The next method we present to solve systems of linear equations is the *addition method* (sometimes called the elimination method). With the addition method, begin by writing both equations in standard form $Ax + By = C$. Then multiply one or both equations by appropriate constants to create opposite coefficients on either the x or the y variable. Next, add the equations to eliminate the variable having opposite coefficients. This process is demonstrated in Example 1.

Concepts

1. **The Addition Method**
2. **Solving Inconsistent Systems and Systems of Dependent Equations**

Example 1   Solving a System by the Addition Method

Solve the system by using the addition method.

$$3x - 4y = 2$$
$$4x + y = 9$$

Solution:

$$3x - 4y = 2 \qquad\qquad 3x - 4y = 2$$
$$4x + y = 9 \xrightarrow{\text{Multiply by 4.}} 16x + 4y = 36$$

Multiply the second equation by 4. This makes the coefficients of the y variables *opposite*.

$$3x - 4y = 2$$
$$\underline{16x + 4y = 36}$$
$$19x \qquad = 38$$
$$x = 2$$

Now if the equations are added, the y variable will be eliminated.

Solve for x.

$$3x - 4y = 2$$
$$3(2) - 4y = 2$$
$$6 - 4y = 2$$
$$-4y = -4$$
$$y = 1$$

Substitute $x = 2$ back into one of the original equations and solve for y.

Check the ordered pair $(2, 1)$ in each original equation:

$$3x - 4y = 2 \qquad\qquad 4x + y = 9$$
$$3(2) - 4(1) \overset{?}{=} 2 \checkmark \text{ True} \qquad 4(2) + (1) \overset{?}{=} 9 \checkmark \text{ True}$$

The solution set is $\{(2, 1)\}$.

Avoiding Mistakes

Be sure to multiply *both* sides of the equation by 4:

$$4(4x + 9) = 4(9)$$

TIP: Substituting $x = 2$ into the other equation, $4x + y = 9$, produces the same value for y.

$$4x + y = 9$$
$$4(2) + y = 9$$
$$8 + y = 9$$
$$y = 1$$

Skill Practice Solve by using the addition method.

1. $2x - 3y = 13$
   $x + 2y = 3$

The steps to solve a system of linear equations in two variables by the addition method is outlined in the following box.

Solving a System of Linear Equations by the Addition Method

Step 1 Write both equations in standard form: $Ax + By = C$.
Step 2 Clear fractions or decimals (optional).
Step 3 Multiply one or both equations by nonzero constants to create opposite coefficients for one of the variables.
Step 4 Add the equations from step 3 to eliminate one variable.
Step 5 Solve for the remaining variable.
Step 6 Substitute the known value found in step 5 into one of the original equations to solve for the other variable.
Step 7 Check the ordered pair in *both* equations and write the solution set.

Answer

1. $\{(5, -1)\}$

Example 2   **Solving a System by the Addition Method**

Solve the system by using the addition method.

$$4x + 5y = 2$$
$$3x = 1 - 4y$$

Solution:

$$4x + 5y = 2 \longrightarrow 4x + 5y = 2$$
$$3x = 1 - 4y \longrightarrow 3x + 4y = 1$$

Step 1: Write both equations in standard form. There are no fractions or decimals.

We may choose to eliminate either variable. To eliminate x, change the coefficients to 12 and -12.

$$4x + 5y = 2 \xrightarrow{\text{Multiply by 3.}} 12x + 15y = 6$$

Step 3: Multiply the first equation by 3.

$$3x + 4y = 1 \xrightarrow{\text{Multiply by } -4.} -12x - 16y = -4$$

Multiply the second equation by -4.

$$12x + 15y = 6$$
$$\underline{-12x - 16y = -4}$$
$$-y = 2$$
$$y = -2$$

Step 4: Add the equations.

Step 5: Solve for y.

$$4x + 5y = 2$$
$$4x + 5(-2) = 2$$
$$4x - 10 = 2$$
$$4x = 12$$
$$x = 3$$

Step 6: Substitute $y = -2$ back into one of the original equations and solve for x.

The solution set is $\{(3, -2)\}$.

Step 7: Check the ordered pair $(3, -2)$ in both original equations.

> **TIP:** The addition method works on the principle that adding the same quantity to both sides of an equation produces an equivalent equation. In step 4 of Example 2, the expressions $12x + 15y$ and 6 are equal. These expressions are added vertically with the equation below to produce an equivalent equation.

> **TIP:** To eliminate the x variable in Example 2, both equations were multiplied by appropriate constants to create $12x$ and $-12x$. We chose 12 because it is the *least common multiple* of 4 and 3.
>   We could have solved the system by eliminating the y variable. To eliminate y, we would multiply the top equation by 4 and the bottom equation by -5. This would make the coefficients of the y variable 20 and -20, respectively.
>
> $$4x + 5y = 2 \xrightarrow{\text{Multiply by 4.}} 16x + 20y = 8$$
> $$3x + 4y = 1 \xrightarrow{\text{Multiply by } -5.} -15x - 20y = -5$$

Skill Practice Solve by using the addition method.

2. $2y = 5x - 4$
   $3x - 4y = 1$

Answer

2. $\left\{ \left(1, \frac{1}{2}\right) \right\}$

Example 3 Solving a System by the Addition Method

Solve the system by using the addition method.

$$x - 2y = 6 + y$$
$$0.05y = 0.02x - 0.10$$

Solution:

$x - 2y = 6 + y \longrightarrow x - 3y = 6$

$0.05y = 0.02x - 0.10 \longrightarrow -0.02x + 0.05y = -0.10$

Step 1: Write both equations in standard form.

$x \quad - 3y = 6$

$-0.02x + 0.05y = -0.10 \xrightarrow{\text{Multiply by 100.}} -2x + 5y = -10$

Step 2: Clear decimals.

$x - 3y = 6 \xrightarrow{\text{Multiply by 2.}} 2x - 6y = 12$

$-2x + 5y = -10 \longrightarrow \underline{-2x + 5y = -10}$

Step 3: Create opposite coefficients.

$-y = 2$

Step 4: Add the equations.

$y = -2$

Step 5: Solve for y.

$x - 2y = 6 + y$

$x - 2(-2) = 6 + (-2)$

$x + 4 = 4$

Step 6: To solve for x, substitute $y = -2$ into one of the original equations.

$x = 0$

Step 7: The ordered pair $(0, -2)$ checks in each original equation.

The solution set is $\{(0, -2)\}$.

Skill Practice Solve by using the addition method.

3. $0.2x + 0.3y = 1.5$
$\quad 5x + \quad 3y = 20 - y$

2. Solving Inconsistent Systems and Systems of Dependent Equations

Example 4 Solving a System of Dependent Equations

Solve the system by using the addition method.

$$\frac{1}{5}x - \frac{1}{2}y = 1$$
$$-4x + 10y = -20$$

Solution:

$\frac{1}{5}x - \frac{1}{2}y = 1$ **Step 1:** Equations are in standard form.

$-4x + 10y = -20$

Answer

3. $\{(0, 5)\}$

$$10\left(\frac{1}{5}x - \frac{1}{2}y\right) = 10 \cdot 1 \xrightarrow{\text{Multiply by 10.}} 2x - 5y = 10$$

Step 2:
Clear fractions.

$$-4x + 10y = -20$$

$$2x - 5y = 10 \xrightarrow{\text{Multiply by 2.}} 4x - 10y = 20$$

$$-4x + 10y = -20 \xrightarrow{} \underline{-4x + 10y = -20}$$

$$0 = 0$$

Step 3: Multiply the first equation by 2.

Step 4:
Add the equations.

Notice that both variables were eliminated. The system of equations is reduced to the identity $0 = 0$. Therefore, the two original equations are dependent. The solution set consists of an infinite number of ordered pairs (x, y) that fall on the common line of intersection $-4x + 10y = -20$, or equivalently $\frac{1}{5}x - \frac{1}{2}y = 1$. The solution set is

$$\{(x, y) \mid -4x + 10y = -20\} \quad \text{or} \quad \left\{(x, y) \mid \frac{1}{5}x - \frac{1}{2}y = 1\right\}$$

Skill Practice Solve by the addition method.

4. $3x + y = 4$
$x + \dfrac{1}{3}y = \dfrac{4}{3}$

Example 5   Solving an Inconsistent System

Solve the system by using the addition method.

$$2y = -3x + 4$$
$$20(6x + 5y) = 40 + 20y$$

Solution:

Step 1:   Write the equations in standard form.

$$2y = -3x + 4 \xrightarrow{} 3x + 2y = 4$$

$$20(6x + 5y) = 40 + 20y \longrightarrow 120x + 100y = 40 + 20y \longrightarrow 120x + 80y = 40$$

With both equations now in standard form, we can proceed with the addition method.

Step 2:   The goal is to create opposite coefficients on either the x or y terms. The second equation can be divided by -40.

$$3x + 2y = 4 \qquad\qquad 3x + 2y = 4$$

$$120x + 80y = 40 \xrightarrow[\text{Divide by }-40.]{} \underline{-3x - 2y = -1}$$

$$0 = 3$$

Step 3:   Divide the second equation by -40.

Step 4:   Add the equations.

The equations reduce to a contradiction, indicating that the system has no solution. The system is inconsistent. The two equations represent parallel lines, as shown in Figure 3-8.

There is no solution, $\{\ \}$.

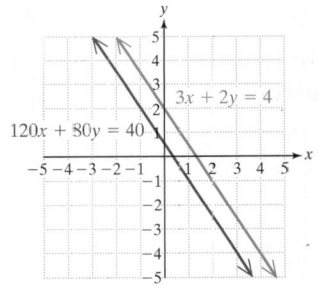

Figure 3-8

Skill Practice Solve by using the addition method.

5. $18 + 10x = 6y$
$5x - 3y = 9$

Answers

4. Infinitely many solutions;
$\{(x, y) \mid 3x + y = 4\}$;
dependent equations

5. No solution; $\{\ \}$; inconsistent system

Section 3.3   Practice Exercises

Vocabulary and Key Concepts

1. a. By what nonzero constant would the second equation be multiplied to eliminate the x variable from the system of equations?

$$3x - 4y = 8$$
$$x + 5y = -2$$

b. By what nonzero constant would the first equation be multiplied to eliminate the y variable from the system of equations?

$$3x - y = 8$$
$$-8x + 5y = 12$$

Review Exercises

For Exercises 2–4, use the slope-intercept form of the lines to determine the number of solutions for the system of equations.

2. $y = \dfrac{1}{2}x - 4$

$y = \dfrac{1}{2}x + 1$

3. $y = 2.32x - 8.1$

$y = 1.46x - 8.1$

4. $4x = y + 7$

$-2y = -8x + 14$

Concept 1: The Addition Method

For Exercises 5–16, solve the system by using the addition method. **(See Examples 1–3.)**

5. $3x - y = -1$

$-3x + 4y = -14$

6. $5x - 2y = 15$

$3x + 2y = -7$

7. $2x + 3y = 3$

$-10x + 2y = -32$

8. $2x - 5y = 7$

$3x - 10y = 13$

9. $3x + 7y = -20$

$-5x + 3y = -84$

10. $6x - 9y = -15$

$5x - 2y = -40$

11. $3x = 10y + 13$

$7y = 4x - 11$

12. $-5x = 6y - 4$

$5y = 1 - 3x$

13. $1.2x - 0.6y = 3$

$0.8x - 1.4y = 3$

14. $1.8x + 0.8y = 1.4$

$1.2x + 0.6y = 1.2$

15. $3x + 2 = 4y + 2$

$7x = 3y$

16. $-4y - 3 = 2x - 3$

$5y = 3x$

Concept 2: Solving Inconsistent Systems and Systems of Dependent Equations

For Exercises 17–24, solve the system. For systems that do not have one unique solution, also state the number of solutions and whether the system is inconsistent or the equations are dependent. **(See Examples 4–5.)**

17. $3x - 2y = 1$

$-6x + 4y = -2$

18. $3x - y = 4$

$6x - 2y = 8$

19. $6y = 14 - 4x$

$2x = -3y - 7$

20. $2x = 4 - y$

$-y = 2x - 2$

21. $12x - 4y = 2$

$6x = 1 + 2y$

22. $10x - 15y = 5$

$3y = 2x - 1$

23. $\dfrac{1}{2}x + y = \dfrac{7}{6}$

$x + 2y = 4.5$

24. $0.2x - 0.1y = -1.2$

$x - \dfrac{1}{2}y = 3$

Mixed Exercises

25. Describe a situation in which you would prefer to use the substitution method over the addition method.

26. If you used the addition method to solve the given system, would it be easier to eliminate the x or y variable? Explain.

$$3x - 5y = 4$$
$$7x + 10y = 31$$

For Exercises 27–32, answer True or False.

27. A system of dependent equations has exactly one solution.

28. An inconsistent system has infinitely many solutions.

29. If solving a system of equations results in the contradiction $0 = 4$, then the system is inconsistent.

30. If solving a system of equations results in the identity $-3 = -3$, then the equations are dependent.

31. If a system of linear equations represents two intersecting lines, then the system has exactly one solution.

32. If a system of linear equations represents two parallel lines, then the system has no solution.

For Exercises 33–56, solve the system. For systems that do not have one unique solution, also state the number of solutions and whether the system is inconsistent or the equations are dependent.

33. $2x - 4y = 8$
$y = 2x + 1$

34. $8x + 6y = -8$
$x = 6y - 10$

35. $2x + 5y = 9$
$4x - 7y = -16$

36. $0.1x + 0.5y = 0.7$
$0.2x + 0.7y = 0.8$

37. $0.2x - 0.1y = 0.8$
$0.1x - 0.1y = 0.4$

38. $y = \dfrac{1}{2}x - 3$
$4x + y = -3$

39. $4x - 6y = 5$
$2x - 3y = 7$

40. $3x + 6y = 7$
$2x + 4y = 5$

41. $\dfrac{1}{4}x - \dfrac{1}{6}y = -2$
$-\dfrac{1}{6}x + \dfrac{1}{5}y = 4$

42. $\dfrac{1}{3}x + \dfrac{1}{5}y = 7$
$\dfrac{1}{6}x - \dfrac{2}{5}y = -4$

43. $\dfrac{1}{3}x - \dfrac{1}{2}y = 0$
$x = \dfrac{3}{2}y$

44. $\dfrac{2}{5}x - \dfrac{2}{3}y = 0$
$y = \dfrac{3}{5}x$

45. $2(x + 2y) = 20 - y$
$-7(x - y) = 16 + 3y$

46. $-3(x + y) = 10 - 4y$
$4(x + 2y) = 50 + 3y$

47. $-4y = 10$
$4x + 3 = 1$

48. $-9x = 15$
$3y + 2 = 1$

49. $0.04x = -0.05y + 1.7$
$-0.03y = -2.4 + 0.07x$

50. $-0.01x = -0.06y + 3.2$
$0.08y = 0.03x + 4.6$

51. $3x - 2 = \dfrac{1}{3}(11 + 5y)$
$x + \dfrac{2}{3}(2y - 3) = -2$

52. $2(2y + 3) - 2x = 1 - x$
$x + y = \dfrac{1}{5}(7 + y)$

53. $\dfrac{1}{4}x + \dfrac{1}{2}y = \dfrac{11}{4}$
$\dfrac{2}{3}x + \dfrac{1}{3}y = \dfrac{7}{3}$

54. $\dfrac{1}{10}x - \dfrac{1}{2}y = -\dfrac{8}{5}$

$x + \dfrac{1}{4}y = -\dfrac{11}{2}$

55. $4x = 3y$

$y = \dfrac{4}{3}x + 2$

56. $4x - 2y = 6$

$x = \dfrac{1}{2}y + \dfrac{3}{2}$

57. A Lincoln Town Car gets 16 mpg in the city and 24 mpg on the highway. A Lincoln Navigator gets 14 mpg in the city and 20 mpg on the highway. Suppose that each vehicle travels c miles in the city and h miles on the highway. The Town Car uses 12 gal of gasoline and the Navigator uses 14 gal. Solve the system of equations to determine the number of city and number of highway miles driven. (*Source:* www.fueleconomy.gov)

$$\text{Town Car:} \quad \dfrac{1}{16}c + \dfrac{1}{24}h = 12$$

$$\text{Navigator:} \quad \dfrac{1}{14}c + \dfrac{1}{20}h = 14$$

58. A Subaru Forester gets 21 mpg in the city and 27 mpg on the highway. A Jeep Liberty gets 15 mpg in the city and 21 mpg on the highway. Suppose that each vehicle travels c miles in the city and h miles on the highway. The Subaru uses 12 gal of gasoline and the Jeep uses 16 gal. Solve the system of equations to determine the number of city and number of highway miles driven.

$$\text{Subaru:} \quad \dfrac{1}{21}c + \dfrac{1}{27}h = 12$$

$$\text{Jeep:} \quad \dfrac{1}{15}c + \dfrac{1}{21}h = 16$$

Expanding Your Skills

Sometimes the solution to a system of equations is an ordered pair containing fractions. In such a case, it is often easier to solve for each coordinate separately using the addition method. That is, use the addition method to solve for x. Then, rather than substituting x back into one of the original equations, repeat the addition method and solve for y. For Exercises 59–61, solve each system.

59. $9x + 11y = 47$

$-5x + 3y = 23$

60. $-6x + 7y = -4$

$4x - 9y = 31$

61. $4x - 10y = 19$

$5x + 12y = -41$

Problem Recognition Exercises

Solving Systems of Linear Equations

For Exercises 1–4, solve each system by using three different methods.

    a. Use the graphing method.   **b.** Use the substitution method.   **c.** Use the addition method.

1. $-3x + y = -2$

$4x - y = 4$

2. $3x - 2y = 4$

$x = \dfrac{2}{3}y + \dfrac{4}{3}$

3. $5x = 2y$

$y = \dfrac{5}{2}x + 1$

4. $2y = 3x + 1$

$4x = 4$

For Exercises 5–8, solve the system of equations by using the most efficient method.

5. $y = -4x - 9$

$8x + 3y = -29$

6. $5x - 2y = -17$

$x + 5y = 2$

7. $5x - 3y = 2$

$7x + 4y = -30$

8. $\dfrac{1}{10}x - \dfrac{2}{5}y = -\dfrac{3}{5}$

$\dfrac{3}{4}x + \dfrac{1}{3}y = \dfrac{13}{6}$

Applications of Systems of Linear Equations in Two Variables

1. Applications Involving Cost

In Chapter 1 we solved numerous application problems using equations that contained one variable. However, when an application has more than one unknown, sometimes it is more convenient to use multiple variables. In this section, we will solve applications containing two unknowns. When two variables are present, the goal is to set up a system of two independent equations.

Concepts

1. Applications Involving Cost
2. Applications Involving Mixtures
3. Applications Involving Principal and Interest
4. Applications Involving Uniform Motion
5. Applications Involving Geometry

Example 1 Solving a Cost Application

At an amusement park, five hot dogs and one drink cost $22. Two hot dogs and three drinks cost $14. Find the cost per hot dog and the cost per drink.

Solution:

Let h represent the cost per hot dog.   Label the variables.

Let d represent the cost per drink.

$\left(\begin{array}{c}\text{Cost of 5}\\\text{hot dogs}\end{array}\right) + \left(\begin{array}{c}\text{cost of 1}\\\text{drink}\end{array}\right) = \$22 \longrightarrow 5h + d = 22$ Write two equations.

$\left(\begin{array}{c}\text{Cost of 2}\\\text{hot dogs}\end{array}\right) + \left(\begin{array}{c}\text{cost of 3}\\\text{drinks}\end{array}\right) = \$14 \longrightarrow 2h + 3d = 14$

This system can be solved by either the substitution method or the addition method. We will solve by using the substitution method. The d variable in the first equation is the easiest variable to isolate.

$5h + d = 22 \longrightarrow d = -5h + 22$ Solve for d in the first equation.

$2h + 3d = 14$

$2h + 3(-5h + 22) = 14$ Substitute the quantity $-5h + 22$ for d in the *second* equation.

$2h - 15h + 66 = 14$ Clear parentheses.

$-13h + 66 = 14$ Solve for h.

$-13h = -52$

$h = 4$

$d = -5(4) + 22 \longrightarrow d = 2$ Substitute $h = 4$ in the equation $d = -5h + 22$.

Because $h = 4$, the cost per hot dog is $4.00.

Because $d = 2$, the cost per drink is $2.00.

Skill Practice

1. At the movie theater, Tom spent $15.50 on 3 soft drinks and 2 boxes of popcorn. Carly bought 5 soft drinks and 1 box of popcorn for a total of $16.50. Use a system of equations to find the cost of a soft drink and the cost of a box of popcorn.

Answer

1. Soft drinks cost $2.50 and popcorn costs $4.00.

> **TIP:** A word problem can be checked by verifying that the solution meets the conditions specified in the problem.
>
> 5 hot dogs + 1 drink = 5($4.00) + 1($2.00) = $22.00 ✔
> 2 hot dogs + 3 drinks = 2($4.00) + 3($2.00) = $14.00 ✔

2. Applications Involving Mixtures

Example 2 Solving an Application Involving Chemistry

One brand of cleaner used to etch concrete is 25% acid. A stronger industrial-strength cleaner is 50% acid. How many gallons of each cleaner should be mixed to produce 20 gal of a 40% acid solution?

Solution:

Let x represent the amount of 25% acid cleaner.

Let y represent the amount of 50% acid cleaner.

	25% Acid	50% Acid	40% Acid
Number of gallons of solution	x	y	20
Number of gallons of pure acid	$0.25x$	$0.50y$	0.40(20), or 8

From the first row of the table, we have

$$\left(\begin{array}{c}\text{Amount of}\\25\%\text{ solution}\end{array}\right) + \left(\begin{array}{c}\text{amount of}\\50\%\text{ solution}\end{array}\right) = \left(\begin{array}{c}\text{total amount}\\\text{of solution}\end{array}\right) \rightarrow x + y = 20$$

From the second row of the table we have

$$\left(\begin{array}{c}\text{Amount of}\\\text{pure acid in}\\25\%\text{ solution}\end{array}\right) + \left(\begin{array}{c}\text{amount of}\\\text{pure acid in}\\50\%\text{ solution}\end{array}\right) = \left(\begin{array}{c}\text{amount of}\\\text{pure acid in}\\\text{resulting solution}\end{array}\right) \rightarrow 0.25x + 0.50y = 8$$

$$x + y = 20 \longrightarrow x + y = 20$$
$$0.25x + 0.50y = 8 \longrightarrow 25x + 50y = 800 \quad \text{Multiply by 100 to clear decimals.}$$

Multiply by −25.
$$x + y = 20 \longrightarrow -25x - 25y = -500 \quad \text{Create opposite coefficients of } x.$$
$$\underline{25x + 50y = 800 \longrightarrow 25x + 50y = 800}$$
$$25y = 300 \quad \text{Add the equations to eliminate } x.$$
$$y = 12$$

$$x + y = 20$$
$$x + (12) = 20 \qquad \text{Substitute } y = 12 \text{ back into one of the original equations.}$$
$$x = 8$$

Therefore, 8 gal of 25% acid solution must be added to 12 gal of 50% acid solution to create 20 gal of a 40% acid solution.

Skill Practice

2. A pharmacist needs 8 ounces (oz) of a solution that is 50% saline. How many ounces of 60% saline solution and 20% saline solution must be mixed to obtain the mixture needed?

3. Applications Involving Principal and Interest

Example 3 **Solving an Application Involving Finance**

Serena invested money in two mutual funds. One had a return of 4.5% and the other had a return of 7%. Twice as much was invested at 7% as at 4.5%. If the amount earned on the original principal at the end of 1 yr was $1017.50, determine the amount invested in each fund.

Solution:

Let x represent the amount invested at 4.5%.

Let y represent the amount invested at 7%.

	4.5% Account	7% Account	Total
Principal	x	y	
Amount earned	$0.045x$	$0.07y$	1017.50

Because the amount invested at 7% was twice the amount invested at 4.5%, we have

$$\begin{pmatrix} \text{Amount} \\ \text{invested} \\ \text{at } 7\% \end{pmatrix} = 2\begin{pmatrix} \text{amount} \\ \text{invested} \\ \text{at } 4.5\% \end{pmatrix} \rightarrow y = 2x$$

From the second row of the table, we have

$$\begin{pmatrix} \text{Amount} \\ \text{earned from} \\ 4.5\% \text{ account} \end{pmatrix} + \begin{pmatrix} \text{amount} \\ \text{earned from} \\ 7\% \text{ account} \end{pmatrix} = \begin{pmatrix} \text{total} \\ \text{amount} \\ \text{earned} \end{pmatrix} \rightarrow 0.045x + 0.07y = 1017.50$$

$y = 2x$

$45x + 70y = 1{,}017{,}500$ Multiply by 1000 to clear decimals.

Because the y variable in the first equation is isolated, we will use the substitution method.

$45x + 70(2x) = 1{,}017{,}500$ Substitute the quantity $2x$ into the second equation.

$45x + 140x = 1{,}017{,}500$ Solve for x.

$185x = 1{,}017{,}500$

$x = \dfrac{1{,}017{,}500}{185}$

$x = 5500$

$y = 2x$

$y = 2(5500)$ Substitute $x = 5500$ into the equation $y = 2x$ to solve for y.

$y = 11{,}000$

Because $x = 5500$, the amount invested at 4.5% was $5500.

Because $y = 11{,}000$, the amount invested at 7% was $11,000.

Skill Practice

3. Seth invested money in two accounts, one paying 5% interest and the other paying 6% interest. The amount invested at 6% was $1000 more than the amount invested at 5%. He earned a total of $830 interest in 1 yr. Use a system of equations to find the amount invested in each account.

Answer

3. Seth invested $7000 at 5% and $8000 at 6%.

TIP: To check Example 3, note that $11,000 is twice $5500. Furthermore,

$$\begin{pmatrix} \text{Amount} \\ \text{earned from} \\ 4.5\% \text{ account} \end{pmatrix} + \begin{pmatrix} \text{amount} \\ \text{earned from} \\ 7\% \text{ account} \end{pmatrix} = \$5500(0.045) + \$11{,}000(0.07) = 1017.50 \checkmark$$

4. Applications Involving Uniform Motion

Example 4 **Solving a Distance, Rate, and Time Application**

A plane flies 660 mi from Atlanta to Miami in 1.2 hr when traveling with a tailwind. The return flight against the same wind takes 1.5 hr. Find the speed of the plane in still air and the speed of the wind.

Solution:

Let p represent the speed of the plane in still air.

Let w represent the speed of the wind.

The speed of the plane *with* the wind:

$$(\text{Plane's still airspeed}) + (\text{wind speed}) \rightarrow p + w$$

The speed of the plane *against* the wind:

$$(\text{Plane's still airspeed}) - (\text{wind speed}) \rightarrow p - w$$

Set up a chart to organize the given information:

	Distance	Rate	Time
With a tailwind	660	$p + w$	1.2
Against the wind	660	$p - w$	1.5

Two equations can be found by using the relationship $d = rt$.

$$\begin{pmatrix} \text{Distance} \\ \text{with} \\ \text{wind} \end{pmatrix} = \begin{pmatrix} \text{speed} \\ \text{with} \\ \text{wind} \end{pmatrix}\begin{pmatrix} \text{time} \\ \text{with} \\ \text{wind} \end{pmatrix} \longrightarrow 660 = (p + w)(1.2)$$

$$\begin{pmatrix} \text{Distance} \\ \text{against} \\ \text{wind} \end{pmatrix} = \begin{pmatrix} \text{speed} \\ \text{against} \\ \text{wind} \end{pmatrix}\begin{pmatrix} \text{time} \\ \text{against} \\ \text{wind} \end{pmatrix} \longrightarrow 660 = (p - w)(1.5)$$

TIP: In Section 3.3 we used the multiplication property of equality to create opposite coefficients. Example 4 demonstrates that we can also use the division property of equality to create opposite coefficients.

$660 = (p + w)(1.2)$ Notice that the first equation may be *divided* by 1.2
$660 = (p - w)(1.5)$ and still leave integer coefficients. Similarly, the second equation may be simplified by dividing by 1.5.

$$660 = (p + w)(1.2) \xrightarrow{\text{Divide by 1.2}} \frac{660}{1.2} = \frac{(p + w)\,\cancel{1.2}}{\cancel{1.2}} \longrightarrow 550 = p + w$$

$$660 = (p - w)(1.5) \xrightarrow{\text{Divide by 1.5}} \frac{660}{1.5} = \frac{(p - w)\,\cancel{1.5}}{\cancel{1.5}} \longrightarrow 440 = p - w$$

$$550 = p + w$$
$$\underline{440 = p - w}$$
$$990 = 2p \qquad \text{Add the equations.}$$
$$p = 495$$
$$550 = (495) + w \qquad \text{Substitute } p = 495 \text{ into the equation } 550 = p + w.$$
$$55 = w \qquad \text{Solve for } w.$$

The speed of the plane in still air is 495 mph, and the speed of the wind is 55 mph.

Skill Practice

4. A plane flies 1200 mi from Orlando to New York in 2 hr with a tailwind. The return flight against the same wind takes 2.5 hr. Find the speed of the plane in still air and the speed of the wind.

5. Applications Involving Geometry

Example 5 **Solving a Geometry Application**

The sum of the two acute angles in a right triangle is 90°. The measure of one angle is 6° less than 2 times the measure of the other angle. Find the measure of each angle.

Solution:

Let x represent one acute angle.

Let y represent the other acute angle.

The sum of the two acute angles is 90° ⟶ $x + y = 90$

One angle is 6° less than 2 times the other angle ⟶ $x = 2y - 6$

$$x + y = 90$$
$$x = \underbrace{2y - 6}$$

Because one variable is already isolated, we will use the substitution method.

$$(2y - 6) + y = 90 \qquad \text{Substitute } x = 2y - 6 \text{ into the first equation.}$$
$$3y - 6 = 90$$
$$3y = 96$$
$$y = 32$$
$$x = 2y - 6 \qquad \text{To find } x, \text{ substitute } y = 32 \text{ into the equation } x = 2y - 6.$$
$$x = 2(32) - 6$$
$$x = 64 - 6$$
$$x = 58$$

The two acute angles in the triangle measure 32° and 58°.

Skill Practice

5. Two angles are supplementary. The measure of one angle is 16° less than 3 times the measure of the other. Use a system of equations to find the measures of the angles.

Answers

4. The speed of the plane is 540 mph, and the speed of the wind is 60 mph.
5. The angles are 49° and 131°.

Section 3.4 Practice Exercises

Vocabulary and Key Concepts

1. **a.** If the cost of one T-shirt is $12, how much will five T-shirts cost? How much will x T-shirts cost?

   b. If a solution contains 10% bleach, how much pure bleach is in a 20-L container? How much pure bleach is in a container holding x liters?

   c. If $5000 is in a savings account that earns 4% simple interest, how much interest is earned at the end of 1 yr? If y dollars is invested at 4% simple interest for 1 yr, how much interest is earned?

   d. If a boat travels b km/hr in still water and the current is c km/hr, then what is the speed of the boat traveling against the current? What is the speed of the boat traveling with the current?

   e. The sum of the measures of the angles within a triangle is _____.

   f. If the measure of an angle is x and the measure of its supplement is y, then $x + y = $ _____.

   g. If the measure of an angle is x and the measure of its complement is y, then $x + y = $ _____.

   h. If the measure of one acute angle in a right triangle is a and the other acute angle has measure b, then $a + b = $ _____.

Review Exercises

2. State three methods that can be used to solve a system of linear equations in two variables.

For Exercises 3–4, state which method you would prefer to use to solve the system. Then solve the system.

3. $y = 9 - 2x$
   $3x - y = 16$

4. $7x - y = -25$
   $2x + 5y = 14$

Concept 1: Applications Involving Cost

5. A 1200-seat theater sells two types of tickets for a concert. Premium seats sell for $30 each and regular seats sell for $20 each. At one event $30,180 was collected in ticket sales with 10 seats left unsold. How many of each type of ticket was sold? **(See Example 1.)**

6. John and Ariana bought school supplies. John spent $10.65 on 4 notebooks and 5 pens. Ariana spent $7.50 on 3 notebooks and 3 pens. What is the cost of 1 notebook and what is the cost of 1 pen?

7. Mickey bought lunch for his fellow office workers on Monday. He spent $24.20 on 3 hamburgers and 2 fish sandwiches. Chloe bought lunch on Tuesday and spent $23.60 for 4 hamburgers and 1 fish sandwich. What is the price of 1 hamburger, and what is the price of 1 fish sandwich?

8. A group of four golfers paid $150 to play a round of golf. Of the golfers, one is a member of the club and three are nonmembers. Another group of golfers consists of two members and one nonmember. They paid a total of $75. What is the cost for a member to play a round of golf, and what is the cost for a nonmember?

9. Jen has 2 scoops of vanilla ice cream and 1 scoop of mud pie ice cream for a total of 40 g of fat. Jim has 1 scoop of vanilla and 2 scoops of mud pie for a total of 44 g of fat. How much fat is in 1 scoop of each ice cream?

10. Roselle has 2 cups of popcorn and 8 oz of soda for a total of 216 calories. Carmel has 1 cup of popcorn and 12 oz of soda for a total of 204 calories. Determine the number of calories per cup of popcorn and per ounce of soda.

Concept 2: Applications Involving Mixtures

11. A jar of one face cream contains 18% moisturizer, and another type contains 24% moisturizer. How many ounces of each should be combined to get 12 oz of a cream that is 22% moisturizer? **(See Example 2.)**

12. Logan wants to mix an 18% acid solution with a 45% acid solution to get 16 L of a 36% acid solution. How many liters of the 18% solution and how many liters of the 45% solution should be mixed?

13. How much fertilizer containing 8% nitrogen should be mixed with a fertilizer containing 12% nitrogen to get 8 L of a fertilizer containing 11% nitrogen?

14. How much 30% acid solution should be added to 10% acid solution to make 100 mL of a 12% acid solution?

15. How much pure bleach should Tim combine with a solution that is 4% bleach to make 12 oz of a 12% bleach solution? (*Hint:* Pure bleach is 100% bleach.)

16. A fruit punch that contains 25% fruit juice is combined with 100% fruit juice. How many ounces of each should be used to make 48 oz of a mixture that is 75% fruit juice? (*Hint:* Pure juice is 100% juice.)

Concept 3: Applications Involving Principal and Interest

17. Mr. Coté invested 3 times as much money in a stock fund that returned 8% interest after 1 yr as he did in a bond fund that earned 5% interest. If his total earnings came to $435 after 1 yr, how much did he invest in each fund? **(See Example 3.)**

18. Aliya deposited half as much money in a savings account earning 2.5% simple interest as she invested in a money market account that earns 3.5% simple interest. If the total interest after 1 yr is $247, how much did she invest in each account?

19. Mr. Levy borrowed money from two lenders. One lender charged 5.5% simple interest and the other charged 3.5% simple interest. Mr. Levy had to borrow $200 more at 5.5% than he did at 3.5%. If the total interest after the first year was $245, how much did he borrow at each rate?

20. Jody invested $5000 less in an account paying 3% simple interest than she did in an account paying 4% simple interest. At the end of the first year, the total interest from both accounts was $725. Find the amount invested in each account.

21. Alina borrowed a total of $15,000 from two banks to buy a new boat. Because of her excellent credit, one bank charged only 6% simple interest and the other charged 7% simple interest. At the end of 5 yr, the total amount of money she paid in interest was $4750. How much did she borrow from each bank?

22. Didi plans to take a trip to the Galapagos Islands in 4 yr and knows that she needs approximately $3500 for the trip. She invests a total of $15,500 in two funds. One fund has a 6% return, and the other has a 5% return. How much should she invest in each fund so that she earns $3500 after 4 yr?

Concept 4: Applications Involving Uniform Motion

23. It takes a boat 2 hr to travel 16 mi downstream with the current and 4 hr to return against the current. Find the speed of the boat in still water and the speed of the current. **(See Example 4.)**

24. A plane flew 720 mi in 3 hr with the wind. It would take 4 hr to travel the same distance against the wind. What is the speed of the plane in still air and the speed of the wind?

25. A plane flies from Atlanta to Los Angeles against the wind in 5 hr. The return trip back to Atlanta with the wind takes only 4 hr. If the distance between Atlanta and Los Angeles is 3200 km, find the speed of the plane in still air and the speed of the wind.

26. The Gulf Stream is a warm ocean current that extends from the eastern side of the Gulf of Mexico up through the Florida Straits and along the southeastern coast of the United States to Cape Hatteras, North Carolina. A boat travels with the current 100 mi from Miami, Florida, to Freeport, Bahamas, in 2.5 hr. The return trip against the same current takes $3\frac{1}{3}$ hr. Find the speed of the boat in still water and the speed of the current.

27. A moving sidewalk in the Atlanta airport moves people between gates. It takes Molly's 8-year-old son Stephen 20 sec to travel 100 ft walking with the sidewalk. It takes him 30 sec to travel 60 ft walking against the moving sidewalk (in the opposite direction). Find the speed of the moving sidewalk and Stephen's walking speed on nonmoving ground.

28. Kim rides a total of 48 km in the bicycle portion of a triathlon. The course is an "out and back" route. It takes her 3 hr on the way out against the wind. The ride back takes her 2 hr with the wind. Find the speed of the wind and Kim's speed riding her bike in still air.

Concept 5: Applications Involving Geometry

For Exercises 29–34, solve the applications involving geometry. If necessary, refer to the geometry formulas listed in the inside back cover of the text.

29. In a right triangle, one acute angle measures 6° more than 3 times the other. If the sum of the measures of the two acute angles must equal 90°, find the measures of the acute angles. **(See Example 5.)**

30. An isosceles triangle has two angles of the same measure. If the angle represented by y measures 3° less than the angle x, find the measures of all angles of the triangle.

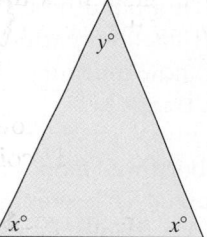

31. Two angles are supplementary. One angle measures 2° less than 3 times the other. What are the measures of the two angles?

32. The measure of one angle is 5 times the measure of another. If the two angles are supplementary, find the measures of the angles.

33. One angle measures 6° more than twice another. If the two angles are complementary, find the measures of the angles.

34. Two angles are complementary. One angle measures 15° more than 2 times the measure of the other. What are the measures of the two angles?

Mixed Exercises

35. How much pure gold (24K) must be mixed with 60% gold to get 20 grams (g) of 75% gold?

36. Connie is the head of maintenance at a large hospital. She received news of a new state mandate indicating that the minimum strength for disinfectant was to be 17%, up from the old requirement of 15%. Connie had plenty of barrels of 15% disinfectant left over, and also lots of the strong 55% disinfectant used in rooms for patients with highly contagious diseases. How many gallons of each disinfectant should be mixed to get 50 gal of 17% disinfectant?

37. A rowing team trains on the Halifax River. It can row upstream 10 mi against the current in 2.5 hr and 16 mi downstream with the current in the same amount of time. Find the speed of the boat in still water and the speed of the current.

38. In her kayak, Bonnie can travel 31.5 mi downstream with the current in 7 hr. The return trip against the current takes 9 hr. Find the speed of the kayak in still water and the speed of the current.

39. There are two types of tickets sold at the Canadian Formula One Grand Prix race. The price of 6 grandstand tickets and 2 general admissions tickets is $2330. The price of 4 grandstand tickets and 4 general admission tickets is $2020. What is the price of each type of ticket?

40. A basketball player scored 19 points by shooting two-point and three-point baskets. If she made a total of eight baskets, how many of each type did she make?

41. A bank offers two accounts, a money market account at 2% simple interest and a regular savings account at 1.3% interest. If Svetlana deposits $3000 between the two accounts and receives $51.25 in total interest in the first year, how much did she invest in each account?

42. Angelo invested $8000 in two accounts: one that pays 3% and one that pays 1.8%. At the end of the first year, his total interest earned was $222. How much did he deposit in the account that pays 3%?

43. The perimeter of a rectangle is 42 m. The length is 1 m longer than the width. Find the dimensions of the rectangle.

44. In a right triangle, the measure of one acute angle is one-fourth the measure of the other. Find the measures of the acute angles.

45. A coin collection consists of 50¢ pieces and $1 coins. If there are 21 coins worth $15.50, how many 50¢ pieces and $1 coins are there?

46. Jacob has a piggy bank consisting of nickels and dimes. If there are 30 coins worth $1.90, how many nickels and dimes are in the bank?

47. One phone plan charges $0.15 per text message. A second plan charges $0.10 per text message but adds a monthly fee of $4.95.

   a. Write a linear function representing the cost for the first plan for x text messages.

   b. Write a linear function representing the cost for the second plan for x text messages.

   c. Find the number of text messages for which the cost of either plan would be the same.

48. A rental car company rents a compact car for $20 a day, plus $0.25 per mile. A midsize car rents for $30 a day, plus $0.20 per mile.

   a. Write a linear function representing the cost to rent the compact car for x miles.

   b. Write a linear function representing the cost to rent a midsize car for x miles.

   c. Find the number of miles at which the cost to rent either car would be the same.

Section 3.5	Linear Inequalities and Systems of Linear Inequalities in Two Variables

Concepts

1. Graphing Linear Inequalities in Two Variables
2. Compound Linear Inequalities in Two Variables
3. Graphing a Feasible Region

1. Graphing Linear Inequalities in Two Variables

A **linear inequality in two variables** x and y is an inequality that can be written in one of the following forms: $Ax + By < C$, $Ax + By > C$, $Ax + By \leq C$, or $Ax + By \geq C$, provided A and B are not both zero.

A solution to a linear inequality in two variables is an ordered pair that makes the inequality true. For example, solutions to the inequality $x + y < 6$ are ordered pairs (x, y) such that the sum of the x- and y-coordinates is less than 6. This inequality has an infinite number of solutions, and therefore it is convenient to express the solution set as a graph.

To graph a linear inequality in two variables, we will follow these steps.

Graphing a Linear Inequality in Two Variables

Step 1 Solve for y, if possible.

Step 2 Graph the related equation. Draw a dashed line if the inequality is strict, $<$ or $>$. Otherwise, draw a solid line.

Step 3 Shade above or below the line as follows:
- Shade *above* the line if the inequality is of the form $y > mx + b$ or $y \geq mx + b$.
- Shade *below* the line if the inequality is of the form $y < mx + b$ or $y \leq mx + b$.

Note: A dashed line indicates that the line is *not* included in the solution set. A solid line indicates that the line *is* included in the solution set.

This process is demonstrated in Example 1.

Example 1   Graphing a Linear Inequality in Two Variables

Graph the solution set. $-3x + y \leq 1$

Solution:

$$-3x + y \leq 1$$

$$y \leq 3x + 1 \qquad \text{Solve for } y.$$

Next graph the line defined by the related equation $y = 3x + 1$.

Because the inequality is of the form $y \leq mx + b$, the solution to the inequality is the region *below* the line $y = 3x + 1$. See Figure 3-9.

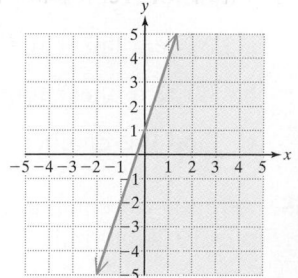

Figure 3-9

Skill Practice Graph the solution set.

1. $2x + y \geq -4$

Answer

1.

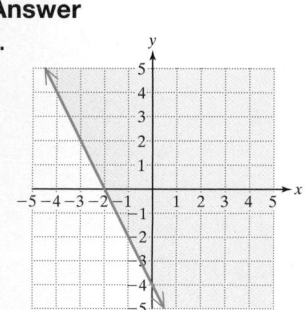

After graphing the solution to a linear inequality, we can verify that we have shaded the correct side of the line by using test points. In Example 1, we can pick an arbitrary ordered pair within the shaded region. Then substitute the x- and y-coordinates in the original inequality. If the result is a true statement, then that ordered pair is a solution to the inequality and suggests that other points from the same region are also solutions.

For example, the point $(0, 0)$ lies within the shaded region (Figure 3-10).

Figure 3-10

$$-3x + y \leq 1 \qquad \text{Substitute } (0, 0) \text{ in the original inequality.}$$

$$-3(0) + (0) \overset{?}{\leq} 1$$

$$0 + 0 \overset{?}{\leq} 1 \text{ ✔ True} \qquad \text{The point } (0, 0) \text{ from the shaded region is a solution.}$$

In Example 2, we will graph the solution set to a strict inequality. A strict inequality does not include equality and therefore is expressed with the symbols $<$ or $>$. In such a case, the boundary line will be drawn as a dashed line. This indicates that the boundary itself is *not* part of the solution set.

Example 2 Graphing a Linear Inequality in Two Variables

Graph the solution set. $-4y < 5x$

Solution:

$$-4y < 5x$$

$$\frac{-4y}{-4} > \frac{5x}{-4} \qquad \text{Solve for } y. \text{ Because we divide both sides by a negative number, reverse the inequality sign.}$$

$$y > -\frac{5}{4}x$$

Graph the line defined by the related equation, $y = -\frac{5}{4}x$. The boundary line is drawn as a dashed line because the inequality is strict. Also note that the line passes through the origin.

Because the inequality is of the form $y > mx + b$, the solution to the inequality is the region *above* the line. See Figure 3-11.

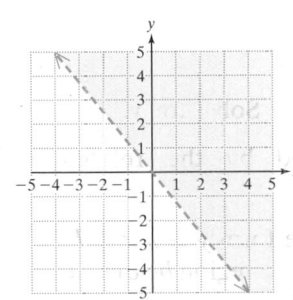

Figure 3-11

Skill Practice Graph the solution set.

2. $-3y < x$

Answer

2.

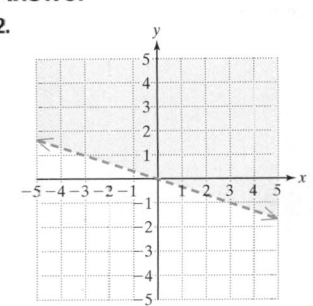

In Example 2, we cannot use the origin as a test point, because the point $(0, 0)$ is on the boundary line. Be sure to select a test point strictly within the shaded region. In this case, we choose $(2, 1)$. See Figure 3-12.

$$-4y < 5x$$
$$-4(1) \overset{?}{<} 5(2)$$
$$-4 \overset{?}{<} 10 \checkmark \text{ True}$$

Substitute $(2, 1)$ in the original inequality.

The point $(2, 1)$ from the shaded region is a solution to the original inequality.

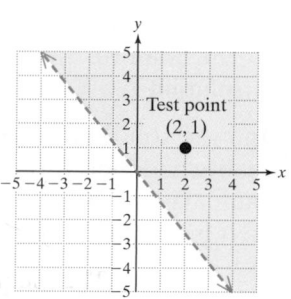

Figure 3-12

In Example 3, we encounter a situation in which we cannot solve for the y variable.

Example 3 Graphing a Linear Inequality

Graph the solution set. $4x \geq -12$

Solution:

$$4x \geq -12$$
$$x \geq -3$$

In this inequality, there is no y variable. However, we can simplify the inequality by solving for x.

Graph the related equation $x = -3$. This is a vertical line. The boundary is drawn as a solid line, because the inequality is not strict, $\geq$.

To shade the appropriate region, refer to the inequality, $x \geq -3$. The points for which x is greater than -3 are to the right of $x = -3$. Therefore, shade the region to the *right* of the line (Figure 3-13).

Selecting a test point such as $(0, 0)$ from the shaded region indicates that we have shaded the correct side of the line.

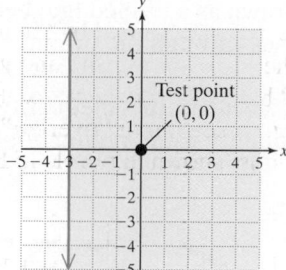

Figure 3-13

$$4x \geq -12$$
$$4(0) \overset{?}{\geq} -12 \checkmark$$

Substitute $x = 0$.

True

Answer

3.

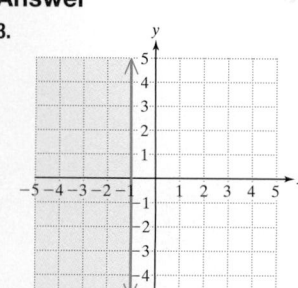

Skill Practice Graph the solution set.

3. $-2x \geq 2$

2. Compound Linear Inequalities in Two Variables

Some applications require us to find the union or intersection of the solution sets of two or more linear inequalities.

Example 4   Graphing a Compound Linear Inequality

Graph the solution set of the compound inequality.

$$y > \tfrac{1}{2}x + 1 \quad \text{and} \quad x + y < 1$$

Solution:

Solve each inequality for y.

<u>First inequality</u>

$y > \tfrac{1}{2}x + 1$

The inequality is of the form $y > mx + b$. Shade *above* the boundary line. See Figure 3-14.

<u>Second inequality</u>

$x + y < 1$

$y < -x + 1$

The inequality is of the form $y < mx + b$. Shade *below* the boundary line. See Figure 3-15.

Figure 3-14

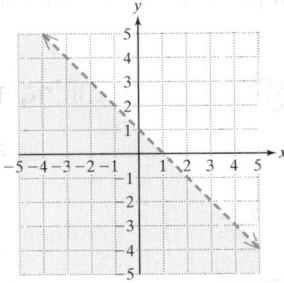

Figure 3-15

The region bounded by the inequalities is the region above the line $y = \tfrac{1}{2}x + 1$ and below the line $y = -x + 1$. This is the intersection or "overlap" of the two regions (shown in purple in Figure 3-16).

The intersection is the solution set to the system of inequalities. See Figure 3-17.

Figure 3-16

Figure 3-17

Answer

4.

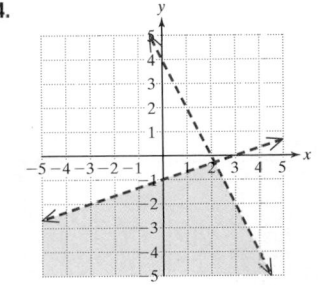

Skill Practice Graph the solution set.

4. $x - 3y > 3$ and $y < -2x + 4$

Example 5 demonstrates the union of the solution sets of two linear inequalities.

Example 5 Graphing a Compound Linear Inequality

Graph the solution set of the compound inequality.

$$3y \le 6 \quad \text{or} \quad y - x \le 0$$

Solution:

First inequality	Second inequality
$3y \le 6$	$y - x \le 0$
$y \le 2$	$y \le x$

The graph of $y \le 2$ is the region on and below the horizontal line $y = 2$. See Figure 3-18.

The inequality $y \le x$ is of the form $y \le mx + b$. Graph a solid line and the region below the line. See Figure 3-19.

Figure 3-18

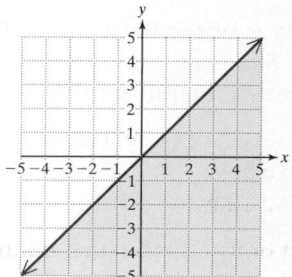

Figure 3-19

The solution to the compound inequality $3y \le 6$ or $y - x \le 0$ is the union of these regions (Figure 3-20).

Skill Practice Graph the solution set.

5. $2y \le 4 \quad \text{or} \quad y \le x + 1$

Figure 3-20

Answers

5.

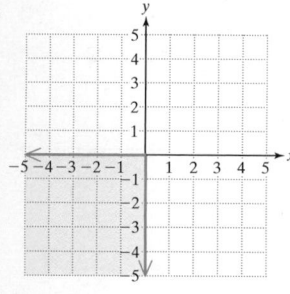

6.

Example 6 Graphing a Compound Linear Inequality

Describe the region of the plane defined by the system of inequalities.

$$x \le 0 \quad \text{and} \quad y \ge 0$$

Solution:

$x \le 0$ $x \le 0$ for points on the y-axis and in the second and third quadrants.

$y \ge 0$ $y \ge 0$ for points on the x-axis and in the first and second quadrants.

The intersection of these regions is the set of points in the second quadrant (with the boundaries included).

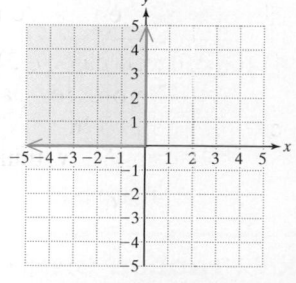

Skill Practice Graph the region defined by the system of inequalities.

6. $x \le 0 \quad \text{and} \quad y \le 0$

3. Graphing a Feasible Region

When two variables are related under certain constraints, a system of linear inequalities can be used to show a region of feasible values for the variables. The *feasible region* represents the ordered pairs that are true for each inequality in the system.

> **Example 7**   Graphing a Feasible Region
>
> Susan has two tests on Friday: one in chemistry and one in psychology. Because the two classes meet in consecutive hours, she has no study time between tests. Susan estimates that she has a maximum of 12 hr of study time before the tests, and she must divide her time between chemistry and psychology.
>
> Let x represent the number of hours Susan spends studying chemistry.
>
> Let y represent the number of hours Susan spends studying psychology.
>
> **a.** Find a set of inequalities to describe the constraints on Susan's study time.
>
> **b.** Graph the constraints to find the feasible region defining Susan's study time.

Solution:

a. Because Susan cannot study chemistry or psychology for a negative period of time, we have $x \geq 0$ and $y \geq 0$. Furthermore, her total time studying cannot exceed 12 hr: $x + y \leq 12$.

A system of inequalities that defines the constraints on Susan's study time is:

$$x \geq 0$$

$$y \geq 0$$

$$x + y \leq 12$$

b. The first two conditions $x \geq 0$ and $y \geq 0$ represent the set of points in the first quadrant. The third condition $x + y \leq 12$ represents the set of points below and including the line $x + y = 12$ (Figure 3-21).

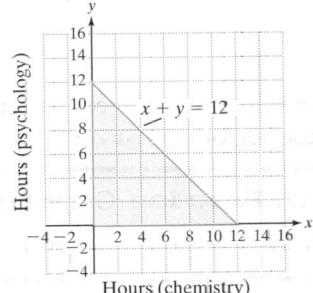

Figure 3-21

Discussion:

1. Refer to the feasible region. Is the ordered pair $(8, 5)$ part of the feasible region?

   No. The ordered pair $(8, 5)$ indicates that Susan spent 8 hr studying chemistry and 5 hr studying psychology. This is a total of 13 hr, which exceeds the constraint that Susan only had 12 hr to study. The point $(8, 5)$ lies outside the feasible region, above the line $x + y = 12$ (Figure 3-22).

2. Is the ordered pair $(7, 3)$ part of the feasible region?

   Yes. The ordered pair $(7, 3)$ indicates that Susan spent 7 hr studying chemistry and 3 hr studying psychology.

   This point lies within the feasible region and satisfies all three constraints.

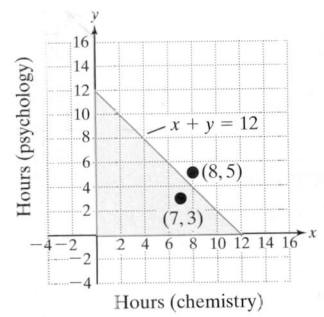

Figure 3-22

$$x \geq 0 \longrightarrow \quad 7 \geq 0 \quad \text{True}$$

$$y \geq 0 \longrightarrow \quad 3 \geq 0 \quad \text{True}$$

$$x + y \leq 12 \longrightarrow (7) + (3) \leq 12 \quad \text{True}$$

Notice that the ordered pair $(7, 3)$ corresponds to a point where Susan is not making full use of the 12 hr of study time.

3. Suppose there was one additional constraint imposed on Susan's study time. She knows she needs to spend at least twice as much time studying chemistry as she does studying psychology.

Because the time studying chemistry must be at least twice the time studying psychology, we have $x \geq 2y$.

This inequality may also be written as $y \leq \frac{1}{2}x$.

Figure 3-23 shows the first quadrant with the constraint $y \leq \frac{1}{2}x$.

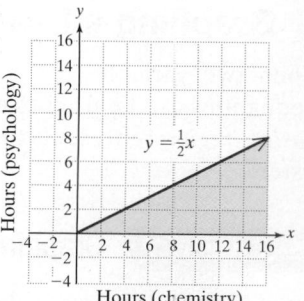

Figure 3-23

4. At what point in the feasible region is Susan making the most efficient use of her time for both classes?

First and foremost, Susan must make use of *all* 12 hr. This occurs for points along the line $x + y = 12$. Susan will also want to study for both classes with approximately twice as much time devoted to chemistry. Therefore, Susan will be deriving the maximum benefit at the point of intersection of the line

$x + y = 12$ and the line $y = \frac{1}{2}x$.

Using the substitution method, replace $y = \frac{1}{2}x$ into the equation $x + y = 12$.

$$x + \frac{1}{2}x = 12$$

$$2x + x = 24 \qquad \text{Clear fractions.}$$

$$3x = 24$$

$$x = 8 \qquad \text{Solve for } x.$$

$$y = \frac{(8)}{2} \qquad \text{To solve for } y, \text{substitute } x = 8$$

$$\text{into the equation } y = \frac{1}{2}x.$$

$$y = 4$$

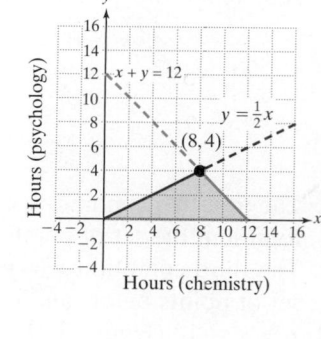

Therefore, Susan should spend 8 hr studying chemistry and 4 hr studying psychology.

Skill Practice

7. A local pet rescue group has a total of 30 cages that can be used to hold cats and dogs. Let x represent the number of cages used for cats, and let y represent the number used for dogs.

a. Write a set of inequalities to express the fact that the number of cat and dog cages cannot be negative.

b. Write an inequality to describe the constraint on the total number of cages for cats and dogs.

c. Graph the system of inequalities to find the feasible region describing the available cages.

Answer

7. a. $x \geq 0$ and $y \geq 0$
b. $x + y \leq 30$
c.

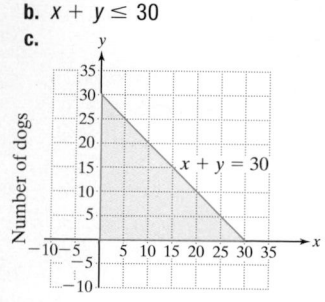

Section 3.5 Practice Exercises

Vocabulary and Key Concepts

1. a. An inequality that can be written in the form $Ax + By > C$ is called a _____ inequality in two variables.

b. Given the inequality $3x - 4y < 6$, the boundary line $3x - 4y = 6$ (is/is not) included in the solution set. However, given $3x - 4y \leq 6$, then the boundary line (is/is not) included in the solution set.

c. Given the inequality $5x + y > 5$, the boundary line $5x + y = 5$ should be drawn as a (solid/dashed) line to indicate that the line (is/is not) part of the solution set.

d. Given the inequality $y \geq -x + 4$, the boundary line $y = -x + 4$ should be drawn as a (solid/dashed) line to indicate that the line (is/is not) part of the solution set.

Review Exercises

For Exercises 2–5, solve the inequalities. Write the solution set in interval notation.

2. $5 < x + 1$ and $-2x + 6 \geq -6$

3. $5 - x \leq 4$ and $6 > 3x - 3$

4. $4 - y < 3y + 12$ or $-2(y + 3) \geq 12$

5. $-2x < 4$ or $3x - 1 \leq -13$

Concept 1: Graphing Linear Inequalities in Two Variables

For Exercises 6–9, decide if the given point is a solution to the inequality.

6. $2x - y > 8$

   a. $(3, -5)$ **b.** $(-1, -10)$

   c. $(4, -2)$ **d.** $(0, 0)$

7. $3y + x < 5$

   a. $(-1, 7)$ **b.** $(5, 0)$

   c. $(0, 0)$ **d.** $(2, -3)$

8. $y \leq -2$

   a. $(5, -3)$ **b.** $(-4, -2)$

   c. $(0, 0)$ **d.** $(3, 2)$

9. $x \geq 5$

   a. $(4, 5)$ **b.** $(5, -1)$

   c. $(8, 8)$ **d.** $(0, 0)$

10. When should you use a dashed line to graph the solution to a linear inequality?

For Exercises 11–16, decide which inequality symbol should be used ($<, >, \geq, \leq$) by looking at the graph.

11.

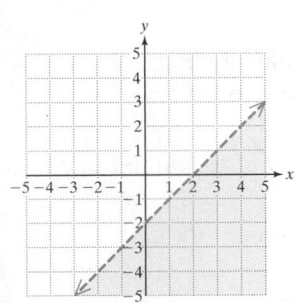

$x - y$ ___ 2

12.

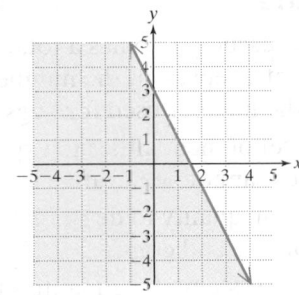

y ___ $-2x + 3$

13.

y ___ -4

14.

x ____ 3

15.

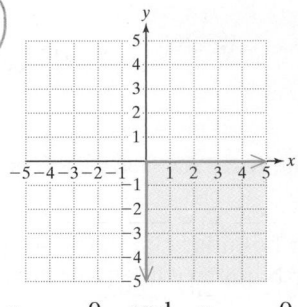

x ____ 0 and y ____ 0

16.

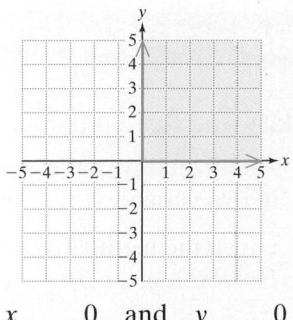

x ____ 0 and y ____ 0

For Exercises 17–40, graph the solution set. **(See Examples 1–3.)**

17. $x - 2y > 4$

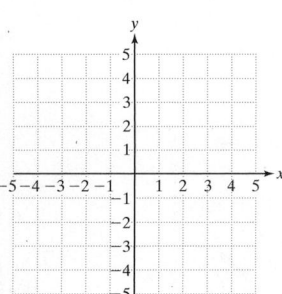

18. $x - 3y \geq 6$

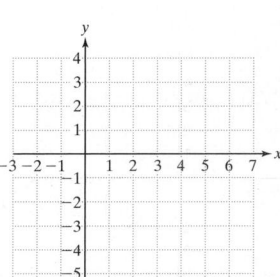

19. $5x - 2y < 10$

20. $x - 3y < 8$

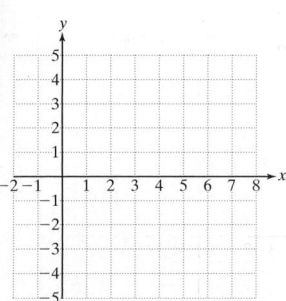

21. $2x \leq -6y + 12$

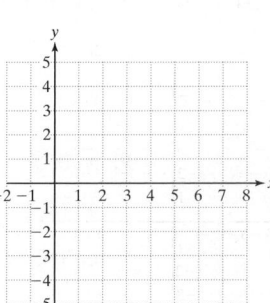

22. $4x < 3y + 12$

23. $2y \leq 4x$

24. $-6x < 2y$

25. $y \geq -2$

26. $y \geq 5$

27. $4x < 5$

28. $x + 6 < 7$

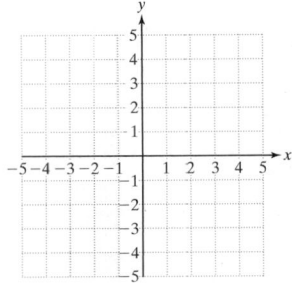

29. $y \geq \dfrac{2}{5}x - 4$

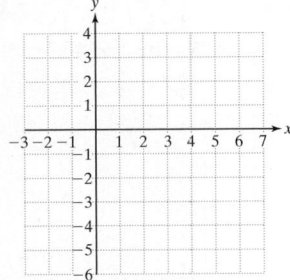

30. $y \geq -\dfrac{5}{2}x - 4$

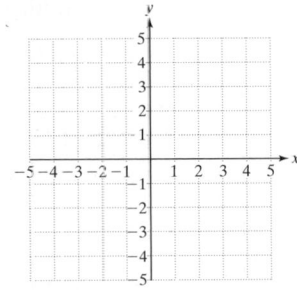

31. $y \leq \dfrac{1}{3}x + 6$

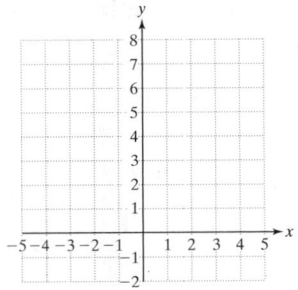

32. $y \leq -\dfrac{1}{4}x + 2$

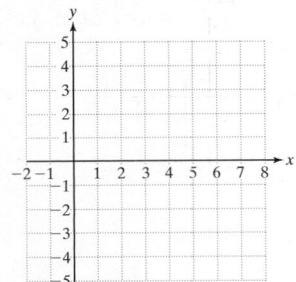

33. $y - 5x > 0$

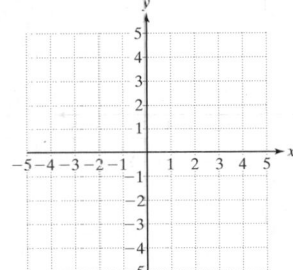

34. $y - \dfrac{1}{2}x > 0$

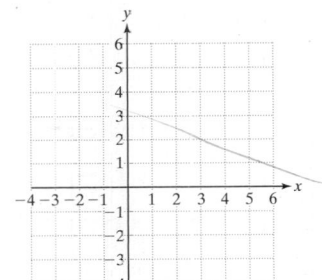

35. $\dfrac{x}{5} + \dfrac{y}{4} < 1$

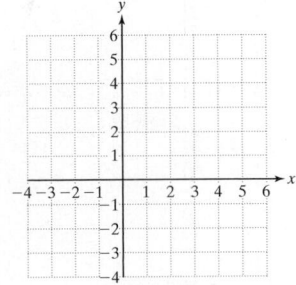

36. $x + \dfrac{y}{2} \geq 2$

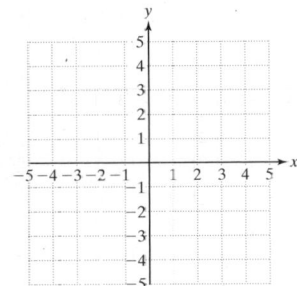

37. $0.1x + 0.2y \leq 0.6$

38. $0.3x - 0.2y < 0.6$

39. $x \leq -\dfrac{2}{3}y$

40. $x \geq -\dfrac{5}{4}y$

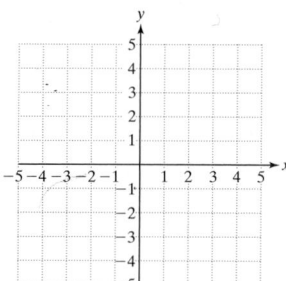

Concept 2: Compound Linear Inequalities in Two Variables

For Exercises 41–55, graph the solution set of each compound inequality. **(See Examples 4–6.)**

41. $y < 4$ and $y > -x + 2$

42. $y < 3$ and $x + 2y < 6$

43. $2x + y \leq 5$ or $x \geq 3$

44. $x + 3y \geq 3$ or $x \leq -2$

45. $x + y < 3$ and $4x + y < 6$

46. $x + y < 4$ and $3x + y < 9$

47. $2x - y \leq 2$ or $2x + 3y \geq 6$

48. $3x + 2y \geq 4$ or $x - y \leq 3$

49. $x > 4$ and $y < 2$

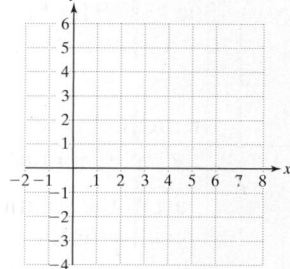

50. $x < 3$ and $y > 4$

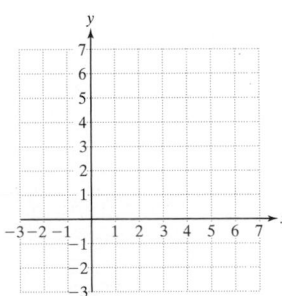

51. $x \leq -2$ or $y \leq 0$

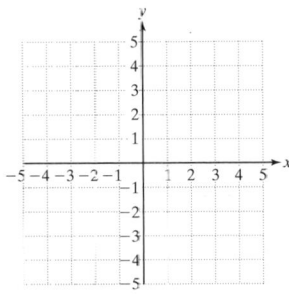

52. $x \geq 0$ or $y \geq -3$

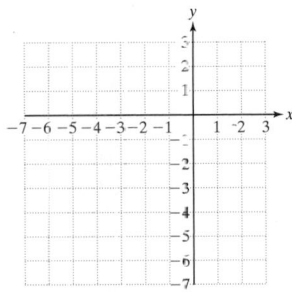

53. $x > 0$ and $x + y < 6$

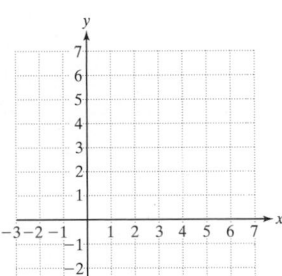

54. $x < 0$ and $x + y < 2$

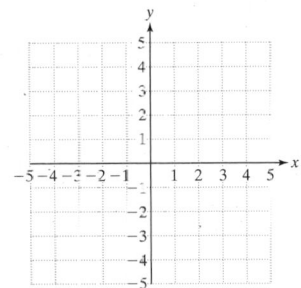

55. $y \leq 0$ or $x - y \leq -4$

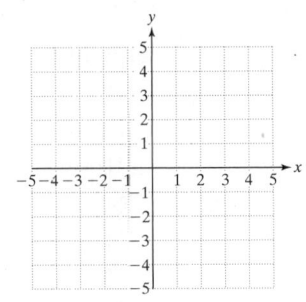

Concept 3: Graphing a Feasible Region

For Exercises 56–59, graph the feasible regions.

56. $x + y \leq 3$ and

   $x \geq 0, y \geq 0$

57. $x - y \leq 2$ and

   $x \geq 0, y \geq 0$

58. $x \geq 0, y \geq 0$

   $x + y \leq 8$ and

   $3x - 5y \leq 30$

59. $x \geq 0, y \geq 0$

   $x + y \leq 5$ and

   $x + 2y \leq 6$

   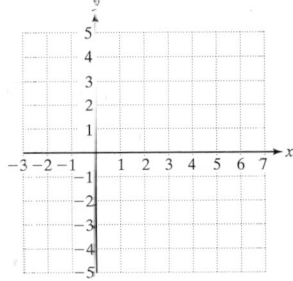

60. Suppose Sue has 50 ft of fencing with which she can build a rectangular dog run. Let x represent the length of the dog run and let y represent the width.

   a. Write an inequality representing the fact that the total perimeter of the dog run is at most 50 ft.

   b. Sketch part of the solution set for this inequality that represents all possible values for the length and width of the dog run. (*Hint:* Note that both the length and the width must be positive.)

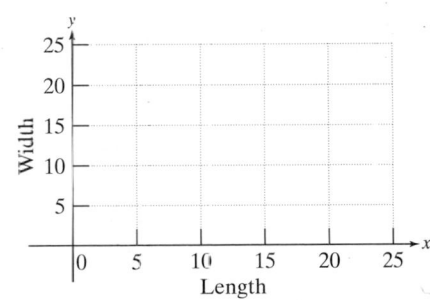

61. Suppose Rick has 40 ft of fencing with which he can build a rectangular garden. Let x represent the length of the garden and let y represent the width.

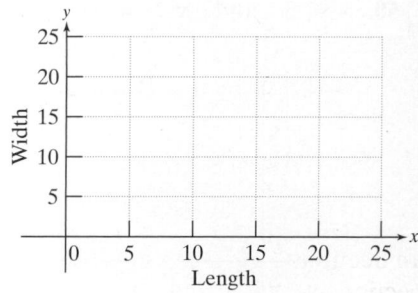

   a. Write an inequality representing the fact that the total perimeter of the garden is at most 40 ft.

   b. Sketch part of the solution set for this inequality that represents all possible values for the length and width of the garden. (*Hint:* Note that both the length and the width must be positive.)

62. A manufacturer produces two models of desks. Model A requires 4 hr to stain and finish and 3 hr to assemble. Model B requires 3 hr to stain and finish and 1 hr to assemble. The total amount of time available for staining and finishing is 24 hr and for assembling is 12 hr. Let x represent the number of Model A desks produced, and let y represent the number of Model B desks produced.

   a. Write two inequalities that express the fact that the number of desks to be produced cannot be negative.

   b. Write an inequality in terms of the number of Model A and Model B desks that can be produced if the total time for staining and finishing is at most 24 hr.

   c. Write an inequality in terms of the number of Model A and Model B desks that can be produced if the total time for assembly is no more than 12 hr.

   d. Graph the feasible region formed by graphing the preceding inequalities.

   e. Is the point $(3, 1)$ in the feasible region? What does the point $(3, 1)$ represent in the context of this problem?

   f. Is the point $(5, 4)$ in the feasible region? What does the point $(5, 4)$ represent in the context of this problem?

63. In scheduling two drivers for delivering pizza, James needs to have at least 65 hr scheduled this week. His two drivers, Karen and Todd, are not allowed to get overtime, so each one can work at most 40 hr. Let x represent the number of hours that Karen can be scheduled, and let y represent the number of hours Todd can be scheduled. **(See Example 7.)**

   a. Write two inequalities that express the fact that Karen and Todd cannot work a negative number of hours.

   b. Write two inequalities that express the fact that neither Karen nor Todd is allowed overtime (i.e., each driver can have at most 40 hr).

   c. Write an inequality that expresses the fact that the total number of hours from both Karen and Todd needs to be at least 65 hr.

   d. Graph the feasible region formed by graphing the inequalities.

   e. Is the point $(35, 40)$ in the feasible region? What does the point $(35, 40)$ represent in the context of this problem?

   f. Is the point $(20, 40)$ in the feasible region? What does the point $(20, 40)$ represent in the context of this problem?

Systems of Linear Equations in Three Variables and Applications

1. Solutions to Systems of Linear Equations in Three Variables

In Sections 3.1–3.3, we solved systems of linear equations in two variables. In this section, we will expand the discussion to solving systems involving three variables.

A **linear equation in three variables** can be written in the form $Ax + By + Cz = D$, where A, B, and C are not all zero. For example, the equation $2x + 3y + z = 6$ is a linear equation in three variables. Solutions to this equation are **ordered triples** of the form (x, y, z) that satisfy the equation. Some solutions to the equation $2x + 3y + z = 6$ are

Solution:		**Check:**
$(1, 1, 1)$	$\longrightarrow$	$2(1) + 3(1) + (1) \overset{?}{=} 6$ ✔ True
$(2, 0, 2)$	$\longrightarrow$	$2(2) + 3(0) + (2) \overset{?}{=} 6$ ✔ True
$(0, 1, 3)$	$\longrightarrow$	$2(0) + 3(1) + (3) \overset{?}{=} 6$ ✔ True

Infinitely many ordered triples serve as solutions to the equation $2x + 3y + z = 6$.

The set of all ordered triples that are solutions to a linear equation in three variables may be represented graphically by a plane in space. Figure 3-24 shows a portion of the plane $2x + 3y + z = 6$ in a 3-dimensional coordinate system.

An example of a system of three linear equations in three variables is shown here.

$$2x + y - 3z = -7$$
$$3x - 2y + z = 11$$
$$-2x - 3y - 2z = 3$$

A solution to a system of linear equations in three variables is an ordered triple that satisfies *each* equation. Geometrically, a solution is a point of intersection of the planes represented by the equations in the system.

A system of linear equations in three variables may have *one unique solution, infinitely many solutions,* or *no solution* (Table 3-2, Table 3-3, and Table 3-4).

Concepts

1. Solutions to Systems of Linear Equations in Three Variables
2. Solving Systems of Linear Equations in Three Variables
3. Applications of Linear Equations in Three Variables
4. Solving Inconsistent Systems and Systems of Dependent Equations

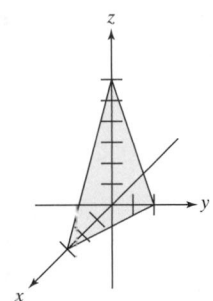

Figure 3-24

Table 3-2

One unique solution (planes intersect at one point)

- The system is consistent.
- The equations are independent.

Table 3-3

No solution (the three planes do not all intersect)

- The system is inconsistent.
- The equations are independent.

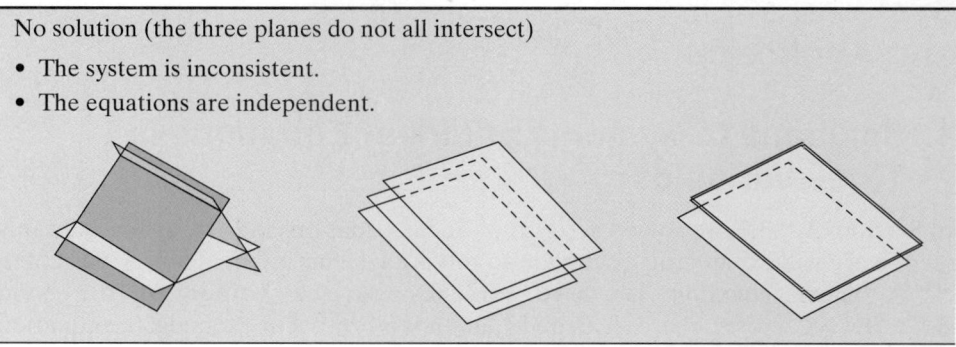

Table 3-4

Infinitely many solutions (planes intersect at infinitely many points)

- The system is consistent.
- The equations are dependent.

2. Solving Systems of Linear Equations in Three Variables

To solve a system involving three variables, the goal is to eliminate one variable. This reduces the system to two equations in two variables. One strategy for eliminating a variable is to pair up the original equations two at a time.

Solving a System of Three Linear Equations in Three Variables

Step 1 Write each equation in standard form $Ax + By + Cz = D$.

Step 2 Choose a pair of equations, and eliminate one of the variables by using the addition method.

Step 3 Choose a different pair of equations and eliminate the *same* variable.

Step 4 Once steps 2 and 3 are complete, you should have two equations in two variables. Solve this system by using the methods from Sections 3.2 and 3.3.

Step 5 Substitute the values of the variables found in step 4 into any of the three original equations that contain the third variable. Solve for the third variable.

Step 6 Check the ordered triple in each of the original equations. Then write the solution as an ordered triple within set notation.

Example 1   Solving a System of Linear Equations
in Three Variables

Solve the system.

$$2x + y - 3z = -7$$
$$3x - 2y + z = 11$$
$$-2x - 3y - 2z = 3$$

Solution:

$\boxed{A}$ $2x + y - 3z = -7$

$\boxed{B}$ $3x - 2y + z = 11$

$\boxed{C}$ $-2x - 3y - 2z = 3$

Step 1: The equations are already in standard form.

- It is often helpful to label the equations.
- The y variable can be easily eliminated from equations $\boxed{A}$ and $\boxed{B}$ and from equations $\boxed{A}$ and $\boxed{C}$. This is accomplished by creating opposite coefficients for the y terms and then adding the equations.

Step 2: Eliminate the y variable from equations $\boxed{A}$ and $\boxed{B}$.

$\boxed{A}$ $2x + y - 3z = -7$ $\xrightarrow{\text{Multiply by 2.}}$ $4x + 2y - 6z = -14$

$\boxed{B}$ $3x - 2y + z = 11$ $\longrightarrow$ $\underline{3x - 2y + z = 11}$

$$7x \quad - 5z = -3 \;\boxed{D}$$

Step 3: Eliminate the y variable again, this time from equations $\boxed{A}$ and $\boxed{C}$.

$\boxed{A}$ $2x + y - 3z = -7$ $\xrightarrow{\text{Multiply by 3.}}$ $6x + 3y - 9z = -21$

$\boxed{C}$ $-2x - 3y - 2z = 3$ $\longrightarrow$ $\underline{-2x - 3y - 2z = 3}$

$$4x \quad - 11z = -18 \;\boxed{E}$$

Step 4: Now equations $\boxed{D}$ and $\boxed{E}$ can be paired up to form a linear system in two variables. Solve this system.

$\boxed{D}$ $7x - 5z = -3$ $\xrightarrow{\text{Multiply by } -4.}$ $-28x + 20z = 12$

$\boxed{E}$ $4x - 11z = -18$ $\xrightarrow[\text{Multiply by 7.}]{}$ $\underline{28x - 77z = -126}$

$$-57z = -114$$
$$z = 2$$

Once one variable has been found, substitute this value into either equation in the two-variable system, that is, either equation $\boxed{D}$ or $\boxed{E}$.

$\boxed{D}$ $7x - 5z = -3$

$7x - 5(2) = -3$ Substitute $z = 2$ into equation $\boxed{D}$.

$7x - 10 = -3$

$7x = 7$

$x = 1$

> **TIP:** It is important to note that in steps 2 and 3, the *same* variable is eliminated.

$\boxed{A}$ $2x + y - 3z = -7$

$2(1) + y - 3(2) = -7$

$2 + y - 6 = -7$

$y - 4 = -7$

$y = -3$

Step 5: Now that two variables are known, substitute these values (x and z) into any of the original three equations to find the remaining variable y. Substitute $x = 1$ and $z = 2$ into equation $\boxed{A}$.

The solution set is $\{(1, -3, 2)\}$.

Step 6: Check the ordered triple in the three original equations.

Check: $2x + y - 3z = -7 \rightarrow 2(1) + (-3) - 3(2) \overset{?}{=} -7$ ✔ True

$3x - 2y + z = 11 \rightarrow 3(1) - 2(-3) + (2) \overset{?}{=} 11$ ✔ True

$-2x - 3y - 2z = 3 \rightarrow -2(1) - 3(-3) - 2(2) \overset{?}{=} 3$ ✔ True

Skill Practice Solve the system.

1. $x + 2y + z = 1$
    $3x - y + 2z = 13$
    $2x + 3y - z = -8$

3. Applications of Linear Equations in Three Variables

Example 2 Applying Systems of Linear Equations to Geometry

In a triangle, the smallest angle measures $10°$ more than one-half the measure of the largest angle. The middle angle measures $12°$ more than the measure of the smallest angle. Find the measure of each angle.

Solution:

Let x represent the measure of the smallest angle.
Let y represent the measure of the middle angle.
Let z represent the measure of the largest angle.

To solve for three variables, we need to establish three independent relationships among x, y, and z.

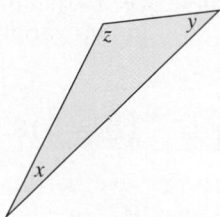

$\boxed{A}$ $x = \dfrac{1}{2}z + 10$

The smallest angle measures $10°$ more than one-half the measure of the largest angle.

$\boxed{B}$ $y = x + 12$

The middle angle measures $12°$ more than the measure of the smallest angle.

$\boxed{C}$ $x + y + z = 180$

The sum of the angles inscribed in a triangle is $180°$.

Answer

1. $\{(1, -2, 4)\}$

Clear fractions and write each equation in standard form.

Standard Form

A $\quad x \quad = \dfrac{1}{2}z + 10 \xrightarrow{\text{Multiply by 2.}} 2x = z + 20 \longrightarrow \quad 2x \quad - z = \quad 20$

B $\quad\quad y \quad = x + 12 \xrightarrow{\hspace{5cm}} \quad -x + y \quad = \quad 12$

C $\quad x + y + z = \quad 180 \xrightarrow{\hspace{5cm}} \quad x + y + z = 180$

Notice equation B is missing the z variable. Therefore, we can eliminate z again by pairing up equations A and C.

A $\quad 2x \quad\quad - z = \quad 20$
C $\quad \underline{x + y + z = 180}$
$\quad\quad 3x + y \quad\quad = 200 \quad$ D

B $\quad -x + y = 12 \xrightarrow{\text{Multiply by } -1.} \quad x - y = -12 \quad$ Pair up equations B and

D $\quad 3x + y = 200 \xrightarrow{\hspace{2cm}} \quad \underline{3x + y = \quad 200} \quad$ D to form a system of two variables.

$\quad\quad\quad\quad\quad\quad\quad\quad\quad\quad\quad\quad 4x \quad\quad = \quad 188$

$\quad\quad\quad\quad\quad\quad\quad\quad\quad\quad\quad\quad\quad x = 47 \quad$ Solve for x.

From equation B we have $-x + y = 12 \longrightarrow \quad -47 + y = \quad 12 \rightarrow y = 59$

From equation C we have $\quad x + y + z = 180 \rightarrow 47 + 59 + z = 180 \rightarrow z = 74$

The smallest angle measures 47°, the middle angle measures 59°, and the largest angle measures 74°.

Skill Practice

2. The perimeter of a triangle is 30 in. The shortest side is 4 in. shorter than the longest side. The longest side is 6 in. less than the sum of the other two sides. Find the length of each side.

Example 3 **Applying Systems of Linear Equations to Nutrition**

Doctors have become increasingly concerned about the sodium intake in the U.S. diet. Recommendations by the American Medical Association indicate that most individuals should not exceed 2400 mg of sodium per day.

Torie ate 1 slice of pizza, 1 serving of ice cream, and 1 glass of soda for a total of 1030 mg of sodium. David ate 3 slices of pizza, no ice cream, and 2 glasses of soda for a total of 2420 mg of sodium. Emilie ate 2 slices of pizza, 1 serving of ice cream, and 2 glasses of soda for a total of 1910 mg of sodium. How much sodium is in one serving of each item?

Solution:

Let x represent the sodium content of 1 slice of pizza.
Let y represent the sodium content of 1 serving of ice cream.
Let z represent the sodium content of 1 glass of soda.

Answer

2. The sides are 8 in., 10 in., and 12 in.

From Torie's meal we have: $\boxed{\text{A}}$ $x + y + z = 1030$

From David's meal we have: $\boxed{\text{B}}$ $3x \qquad + 2z = 2420$

From Emilie's meal we have: $\boxed{\text{C}}$ $2x + y + 2z = 1910$

Equation $\boxed{\text{B}}$ is missing the y variable. Eliminating y from equations $\boxed{\text{A}}$ and $\boxed{\text{C}}$, we have

$\boxed{\text{A}}$ $x + y + z = 1030$ $\xrightarrow{\text{Multiply by } -1.}$ $-x - y - z = -1030$

$\boxed{\text{C}}$ $2x + y + 2z = 1910$ $\xrightarrow{\hspace{2cm}}$ $\underline{2x + y + 2z = 1910}$

 $\boxed{\text{D}}$ $x \qquad + z = 880$

Solve the system formed by equations $\boxed{\text{B}}$ and $\boxed{\text{D}}$.

$\boxed{\text{B}}$ $3x + 2z = 2420$ $\xrightarrow{\hspace{2cm}}$ $3x + 2z = 2420$

$\boxed{\text{D}}$ $x + z = 880$ $\xrightarrow[\text{Multiply by } -2.]{}$ $\underline{-2x - 2z = -1760}$

 $x = 660$

From equation $\boxed{\text{D}}$ we have $x + z = 880 \longrightarrow 660 + z = 880 \longrightarrow z = 220$

From equation $\boxed{\text{A}}$ we have $x + y + z = 1030 \rightarrow 660 + y + 220 = 1030 \rightarrow y = 150$

Therefore, 1 slice of pizza has 660 mg of sodium, 1 serving of ice cream has 150 mg of sodium, and 1 glass of soda has 220 mg of sodium.

Skill Practice

3. Annette, Barb, and Carlita work in a clothing shop. One day the three had combined sales of $1480. Annette sold $120 more than Barb. Barb and Carlita combined sold $280 more than Annette. How much did each person sell?

4. Solving Inconsistent Systems and Systems of Dependent Equations

Example 4 Solving a System of Dependent Equations

Solve the system. If the system does not have one unique solution, state the number of solutions and whether the system is inconsistent or the equations are dependent.

$$\boxed{\text{A}} \quad 3x + y - z = 8$$
$$\boxed{\text{B}} \quad 2x - y + 2z = 3$$
$$\boxed{\text{C}} \quad x + 2y - 3z = 5$$

Solution:

The first step is to make a decision regarding the variable to eliminate. The y variable is particularly easy to eliminate because the coefficients of y in equations $\boxed{\text{A}}$ and $\boxed{\text{B}}$ are already opposites. The y variable can be eliminated from equations $\boxed{\text{B}}$ and $\boxed{\text{C}}$ by multiplying equation $\boxed{\text{B}}$ by 2.

$\boxed{\text{A}}$ $3x + y - z = 8$

$\boxed{\text{B}}$ $\underline{2x - y + 2z = 3}$

$5x + z = 11$ $\boxed{\text{D}}$

Pair up equations $\boxed{\text{A}}$ and $\boxed{\text{B}}$ to eliminate y.

Answer

3. Annette sold $600, Barb sold $480, and Carlita sold $400.

$\boxed{B}$ $2x - y + 2z = 3$ $\xrightarrow{\text{Multiply by 2.}}$ $4x - 2y + 4z = 6$ Pair up equations
$\boxed{C}$ $x + 2y - 3z = 5$ $\longrightarrow$ $\underline{x + 2y - 3z = 5}$ $\boxed{B}$ and $\boxed{C}$ to
$$5x \qquad + z = 11 \quad \boxed{E}$$ eliminate y.

Because equations $\boxed{D}$ and $\boxed{E}$ are equivalent equations, it appears that this is a dependent system. By eliminating variables we obtain the identity $0 = 0$.

$\boxed{D}$ $5x + z = 11$ $\xrightarrow{\text{Multiply by } -1.}$ $-5x - z = -11$
$\boxed{E}$ $5x + z = 11$ $\longrightarrow$ $\underline{5x + z = \quad 11}$
$$0 = \quad 0$$

The result $0 = 0$ indicates that there are infinitely many solutions and that the equations are dependent.

Skill Practice Solve the system. If the system does not have one unique solution, state the number of solutions and whether the system is inconsistent or the equations are dependent.

4. $x + y + z = 8$
$ 2x - y + z = 6$
$-5x - 2y - 4z = -30$

Example 5   Solving an Inconsistent System of Linear Equations

Solve the system. If the system does not have one unique solution, state the number of solutions and whether the system is inconsistent or the equations are dependent.

$$2x + 3y - 7z = 4$$
$$-4x - 6y + 14z = 1$$
$$5x + y - 3z = 6$$

Solution:

We will eliminate the x variable from equations $\boxed{A}$ and $\boxed{B}$.

$\boxed{A}$ $2x + 3y - 7z = 4$ $\xrightarrow{\text{Multiply by 2.}}$ $4x + 6y - 14z = 8$
$\boxed{B}$ $-4x - 6y + 14z = 1$ $\longrightarrow$ $\underline{-4x - 6y + 14z = 1}$
$$0 = 9 \qquad \text{(contradiction)}$$

The result $0 = 9$ is a contradiction, indicating that the system has no solution. The system is inconsistent.

Skill Practice

5. Solve the system. If the system does not have one unique solution, state the number of solutions and whether the system is inconsistent or the equations are dependent.

$$x - 2y + z = 5$$
$$x - 3y + 2z = -7$$
$$-2x + 4y - 2z = 6$$

Answers

4. Infinitely many solutions; dependent equations
5. No solution; inconsistent system

Section 3.6 Practice Exercises

Vocabulary and Key Concepts

1. **a.** An equation written in the form $Ax + By + Cz = D$, where A, B, and C are not all zero, is called a _____ equation in three variables.

   b. Solutions to a linear equation in three variables are of the form (x, y, z) and are called _____.

Review Exercises

2. Determine if the ordered pair $(-4, -7)$ is a solution to the system. $-5x + 3y = -1$
$$4x - 2y = -2$$

For Exercises 3–4, solve the systems by using two methods: **(a)** the substitution method and **(b)** the addition method.

3. $3x + y = 4$
   $4x + y = 5$

4. $2x - 5y = 3$
   $-4x + 10y = 3$

5. Marge can ride her bike 24 mi in $1\frac{1}{3}$ hr riding with the wind. Riding against the wind she can ride 24 mi in 2 hr. Find the speed at which Marge can ride in still air and the speed of the wind.

Concept 1: Solutions to Systems of Linear Equations in Three Variables

6. How many solutions are possible when solving a system of three equations with three variables?

7. Which of the following points are solutions to the system? $(2, 1, 7)$, $(3, -10, -6)$, $(4, 0, 2)$

$$2x - y + z = 10$$
$$4x + 2y - 3z = 10$$
$$x - 3y + 2z = 8$$

8. Which of the following points are solutions to the system? $(1, 1, 3)$, $(0, 0, 4)$, $(4, 2, 1)$

$$-3x - 3y - 6z = -24$$
$$-9x - 6y + 3z = -45$$
$$9x + 3y - 9z = 33$$

9. Which of the following points are solutions to the system? $(12, 2, -2)$, $(4, 2, 1)$, $(1, 1, 1)$

$$-x - y - 4z = -6$$
$$x - 3y + z = -1$$
$$4x + y - z = 4$$

10. Which of the following points are solutions to the system? $(0, 4, 3)$, $(3, 6, 10)$, $(3, 3, 1)$

$$x + 2y - z = 5$$
$$x - 3y + z = -5$$
$$-2x + y - z = -4$$

Concept 2: Solving Systems of Linear Equations in Three Variables

For Exercises 11–22, solve the system of equations. **(See Example 1.)**

11. $2x + y - 3z = -12$
    $3x - 2y - z = 3$
    $-x + 5y + 2z = -3$

12. $-3x - 2y + 4z = -15$
    $2x + 5y - 3z = 3$
    $4x - y + 7z = 15$

13. $x - 3y - 4z = -7$
    $5x + 2y + 2z = -1$
    $4x - y - 5z = -6$

14. $6x - 5y + z = 7$
    $5x + 3y + 2z = 0$
    $-2x + y - 3z = 11$

15. $4x + 2z = 12 + 3y$
    $2y = 3x + 3z - 5$
    $y = 2x + 7z + 8$

16. $y = 2x + z + 1$
    $-3x - 1 = -2y + 2z$
    $5x + 3z = 16 - 3y$

17.
$$x + y + z = 6$$
$$-x + y - z = -2$$
$$2x + 3y + z = 11$$

18.
$$x - y - z = -11$$
$$x + y - z = 15$$
$$2x - y + z = -9$$

19.
$$2x - 3y + 2z = -1$$
$$x + 2y = -4$$
$$x + z = 1$$

20.
$$x + y + z = 2$$
$$2x - z = 5$$
$$3y + z = 2$$

21.
$$4x + 9y = 8$$
$$8x + 6z = -1$$
$$6y + 6z = -1$$

22.
$$3x + 2z = 11$$
$$y - 7z = 4$$
$$x - 6y = 1$$

Concept 3: Applications of Linear Equations in Three Variables

23. A triangle has one angle that measures 5° more than twice the smallest angle, and the third angle measures 11° less than 3 times the measure of the smallest angle. Find the measures of the three angles. **(See Example 2.)**

24. The largest angle of a triangle measures 4° less than 5 times the measure of the smallest angle. The middle angle measures twice that of the smallest angle. Find the measures of the three angles.

25. The perimeter of a triangle is 55 cm. The measure of the shortest side is 8 cm less than the middle side. The measure of the longest side is 1 cm less than the sum of the other two sides. Find the lengths of the sides.

26. The perimeter of a triangle is 5 ft. The longest side of the triangle measures 20 in. more than the shortest side. The middle side is 3 times the measure of the shortest side. Find the lengths of the three sides in *inches*.

27. Sean kept track of his fiber intake from three sources for 3 weeks. The first week he had 3 servings of a fiber supplement, 1 serving of oatmeal, and 4 servings of cereal, which totaled 19 g of fiber. The second week he had 2 servings of the fiber supplement, 4 servings of oatmeal, and 2 servings of cereal totaling 25 g. The third week he had 5 servings of the fiber supplement, 3 servings of oatmeal, and 2 servings of cereal for a total of 30 g. Find the amount of fiber in one serving of each of the following: the fiber supplement, the oatmeal, and the cereal. **(See Example 3.)**

28. Natalie kept track of her calcium intake from three sources for 3 days. The first day she had 1 glass of milk, 1 serving of ice cream, and 1 calcium supplement in pill form which totaled 1180 mg of calcium. The second day she had 2 glasses of milk, 1 serving of ice cream, and 1 calcium supplement totaling 1680 mg. The third day she had 1 glass of milk, 2 servings of ice cream, and 1 calcium supplement for a total of 1260 mg. Find the amount of calcium in one glass of milk, in one serving of ice cream, and in one calcium supplement.

29. Goofie Golf has 18 holes that are par 3, par 4, or par 5. Most of the holes are par 4. In fact, there are 3 times as many par 4s as par 3s. There are 3 more par 5s than par 3s. How many of each type are there?

30. Combining peanuts, pecans, and cashews makes a party mixture of nuts. If the amount of peanuts equals the amount of pecans and cashews combined, and if there are twice as many cashews as pecans, how many ounces of each nut is used to make 48 oz of party mixture?

31. Souvenir hats, T-shirts, and jackets are sold at a rock concert. Three hats, two T-shirts, and one jacket cost $140. Two hats, two T-shirts, and two jackets cost $170. One hat, three T-shirts, and two jackets cost $180. Find the prices of the individual items.

32. Annie and Maria traveled overseas for 7 days and stayed in three different hotels in three different cities: Stockholm, Sweden; Oslo, Norway; and Paris, France.

The total bill for all seven nights (not including tax) was $1040. The total tax was $106. The nightly cost (excluding tax) to stay at the hotel in Paris was $80 more than the nightly cost (excluding tax) to stay in Oslo. Find the cost per night for each hotel excluding tax.

City	Number of Nights	Cost/Night ($)	Tax Rate
Paris, France	1	x	8%
Stockholm, Sweden	4	y	11%
Oslo, Norway	2	z	10%

33. Walter had $25,000 to invest. He split the money into three types of investment: small caps earning 6%, global market investments earning 10%, and a balanced fund earning 9%. He put twice as much money in the global account as he did in the balanced fund. If his earnings for the first year totaled $2160, how much did he invest in each account?

34. Raeann deposited $8000 into three accounts at her credit union: a checking account that pays 1.2% interest, a savings account that pays 2.5% interest, and a money market account that pays 3% interest. If she put 3 times more money in the 3% account than she did in the 1.2% account, and her total interest for 1 yr was $202, how much did she deposit into each account?

Concept 4: Solving Inconsistent Systems and Systems of Dependent Equations

For Exercises 35–46, solve the system. If the system does not have a unique solution, state whether the system is inconsistent or the equations are dependent. **(See Examples 1, 4, and 5.)**

35.
$$2x + y + 3z = 2$$
$$x - y + 2z = -4$$
$$-2x + 2y - 4z = 8$$

36.
$$x + y = z$$
$$2x + 4y - 2z = 6$$
$$3x + 6y - 3z = 9$$

37.
$$6x - 2y + 2z = 2$$
$$4x + 8y - 2z = 5$$
$$-2x - 4y + z = -2$$

38.
$$3x + 2y + z = 3$$
$$x - 3y + z = 4$$
$$-6x - 4y - 2z = 1$$

39.
$$\tfrac{1}{2}x + \tfrac{2}{3}y = \tfrac{5}{2}$$
$$\tfrac{1}{5}x - \tfrac{1}{2}z = -\tfrac{3}{10}$$
$$\tfrac{1}{3}y - \tfrac{1}{4}z = \tfrac{3}{4}$$

40.
$$\tfrac{1}{2}x + \tfrac{1}{4}y + z = 3$$
$$\tfrac{1}{8}x + \tfrac{1}{4}y + \tfrac{1}{4}z = \tfrac{9}{8}$$
$$x - y - \tfrac{2}{3}z = \tfrac{1}{3}$$

41.
$$-3x + y - z = 8$$
$$-4x + 2y + 3z = -3$$
$$2x + 3y - 2z = -1$$

42.
$$2x + 3y + 3z = 15$$
$$3x - 6y - 6z = -23$$
$$-9x - 3y + 6z = 8$$

43.
$$2x + y = 3(z - 1)$$
$$3x - 2(y - 2z) = 1$$
$$2(2x - 3z) = -6 - 2y$$

44.
$$2x + y = -3$$
$$2y + 16z = -10$$
$$-7x - 3y + 4z = 8$$

45.
$$-0.1y + 0.2z = 0.2$$
$$0.1x + 0.1y + 0.1z = 0.2$$
$$-0.1x - 0.3z = 0.2$$

46.
$$-0.4x - 0.3y = 0$$
$$0.3y + 0.1z = -0.1$$
$$0.4x - 0.1z = 1.2$$

Expanding Your Skills

The systems in Exercises 47–50 are called homogeneous systems because each system has $(0, 0, 0)$ as a solution. However, if the equations are dependent, the system will have infinitely many more solutions. For each system determine whether $(0, 0, 0)$ is the only solution or if the equations are dependent.

47.
$$2x - 4y + 8z = 0$$
$$-x - 3y + z = 0$$
$$x - 2y + 5z = 0$$

48.
$$2x - 4y + z = 0$$
$$x - 3y - z = 0$$
$$3x - y + 2z = 0$$

49.
$$4x - 2y - 3z = 0$$
$$-8x - y + z = 0$$
$$2x - y - \tfrac{3}{2}z = 0$$

50.
$$5x + y = 0$$
$$4y - z = 0$$
$$5x + 5y - z = 0$$

Solving Systems of Linear Equations by Using Matrices

1. Introduction to Matrices

In Sections 3.2, 3.3, and 3.6, we solved systems of linear equations by using the substitution method and the addition method. We now present a third method called the Gauss-Jordan method that uses matrices to solve a linear system.

A **matrix** is a rectangular array of numbers (the plural of *matrix* is *matrices*). The rows of a matrix are read horizontally, and the columns of a matrix are read vertically. Every number or entry within a matrix is called an element of the matrix.

The **order of a matrix** is determined by the number of rows and number of columns. A matrix with m rows and n columns is an $m \times n$ (read as "m by n") matrix. Notice that with the order of a matrix, the number of rows is given first, followed by the number of columns.

Concepts

1. **Introduction to Matrices**
2. **Solving Systems of Linear Equations by Using the Gauss-Jordan Method**

Example 1 Determining the Order of a Matrix

Determine the order of each matrix.

a. $\begin{bmatrix} 2 & -4 & 1 \\ 5 & \pi & \sqrt{7} \end{bmatrix}$ **b.** $\begin{bmatrix} 1.9 \\ 0 \\ 7.2 \\ -6.1 \end{bmatrix}$ **c.** $\begin{bmatrix} 1 & 0 & 0 \\ 0 & 1 & 0 \\ 0 & 0 & 1 \end{bmatrix}$ **d.** $[a \quad b \quad c]$

Solution:

a. This matrix has two rows and three columns. Therefore, it is a 2×3 matrix.

b. This matrix has four rows and one column. Therefore, it is a 4×1 matrix. A matrix with one column is called a **column matrix**.

c. This matrix has three rows and three columns. Therefore, it is a 3×3 matrix. A matrix with the same number of rows and columns is called a **square matrix**.

d. This matrix has one row and three columns. Therefore, it is a 1×3 matrix. A matrix with one row is called a **row matrix**.

Skill Practice Determine the order of the matrix.

1. $\begin{bmatrix} -5 & 2 \\ 1 & 3 \\ 8 & 9 \end{bmatrix}$ **2.** $[4\ -8]$ **3.** $\begin{bmatrix} 5 \\ 10 \\ 15 \end{bmatrix}$ **4.** $\begin{bmatrix} 2 & -0.5 \\ -1 & 6 \end{bmatrix}$

A matrix can be used to represent a system of linear equations written in standard form. To do so, we extract the coefficients of the variable terms and the constants within the equation. For example, consider the system

$$2x - y = 5$$
$$x + 2y = -5$$

The matrix **A** is called the **coefficient matrix**.

$$A = \begin{bmatrix} 2 & -1 \\ 1 & 2 \end{bmatrix}$$

Answers
1. 3×2 **2.** 1×2
3. 3×1 **4.** 2×2

If we extract both the coefficients and the constants from the equations, we can construct the **augmented matrix** of the system:

$$\left[\begin{array}{rr|r} 2 & -1 & 5 \\ 1 & 2 & -5 \end{array}\right]$$

A vertical bar is inserted into an augmented matrix to designate the position of the equal signs.

Example 2   Writing an Augmented Matrix for a System of Linear Equations

Write the augmented matrix for each linear system.

a. $-3x - 4y = 3$
$\quad\;\; 2x + 4y = 2$

b. $2x \qquad\;\; - 3z = 14$
$\qquad 2y + \;\; z = \;\; 2$
$\qquad x + y \qquad\;\;\; = \;\; 4$

Solution:

a. $\left[\begin{array}{rr|r} -3 & -4 & 3 \\ 2 & 4 & 2 \end{array}\right]$

b. $\left[\begin{array}{rrr|r} 2 & 0 & -3 & 14 \\ 0 & 2 & 1 & 2 \\ 1 & 1 & 0 & 4 \end{array}\right]$

> **TIP:** Notice that zeros are inserted to denote the coefficient of each missing term.

Skill Practice Write the augmented matrix for each system.

5. $-x + y = 4$
$\quad\; 2x - y = 1$

6. $\;\;\; 2x - \;\; y + \;\; z = 14$
$\quad -3x + 4y \qquad\;\;\; = \;\; 8$
$\qquad\; x - \;\; y + 5z = \;\; 0$

Example 3   Writing a Linear System from an Augmented Matrix

Write a system of linear equations represented by each augmented matrix.

a. $\left[\begin{array}{rr|r} 2 & -5 & -8 \\ 4 & 1 & 6 \end{array}\right]$

b. $\left[\begin{array}{rrr|r} 2 & -1 & 3 & 14 \\ 1 & 1 & -2 & -5 \\ 3 & 1 & -1 & 2 \end{array}\right]$

c. $\left[\begin{array}{rrr|r} 1 & 0 & 0 & 4 \\ 0 & 1 & 0 & -1 \\ 0 & 0 & 1 & 0 \end{array}\right]$

Solution:

a. $2x - 5y = -8$
$\quad\; 4x + \;\; y = \;\; 6$

b. $2x - y + 3z = \;\; 14$
$\quad\; x + y - 2z = -5$
$\quad 3x + y - \;\; z = \;\; 2$

c. $x + 0y + 0z = \;\; 4$
$\quad 0x + \;\; y + 0z = -1 \qquad$ or $\qquad$ $x = \;\; 4$
$\quad 0x + 0y + \;\; z = \;\; 0 \qquad\qquad\qquad$ $y = -1$
$\qquad\qquad\qquad\qquad\qquad\qquad\qquad\qquad$ $z = \;\; 0$

Skill Practice Write a system of linear equations represented by each augmented matrix.

7. $\left[\begin{array}{rr|r} 2 & 3 & 5 \\ -1 & 8 & 1 \end{array}\right]$

8. $\left[\begin{array}{rrr|r} -3 & 2 & 1 & 4 \\ 14 & 1 & 0 & 20 \\ -8 & 3 & 5 & 6 \end{array}\right]$

9. $\left[\begin{array}{rrr|r} 1 & 0 & 0 & -8 \\ 0 & 1 & 0 & 2 \\ 0 & 0 & 1 & 15 \end{array}\right]$

Answers

5. $\left[\begin{array}{rr|r} -1 & 1 & 4 \\ 2 & -1 & 1 \end{array}\right]$

6. $\left[\begin{array}{rrr|r} 2 & -1 & 1 & 14 \\ -3 & 4 & 0 & 8 \\ 1 & -1 & 5 & 0 \end{array}\right]$

7. $2x + 3y = 5$
$\quad -x + 8y = 1$

8. $-3x + 2y + \;\; z = \;\; 4$
$\quad\;\; 14x + \;\; y \qquad\;\;\; = 20$
$\quad -8x + 3y + 5z = \;\; 6$

9. $x = -8, y = 2, z = 15$

2. Solving Systems of Linear Equations by Using the Gauss-Jordan Method

We know that interchanging two equations results in an equivalent system of linear equations. Interchanging two rows in an augmented matrix results in an equivalent augmented matrix. Similarly, because each row in an augmented matrix represents a linear equation, we can perform the following elementary row operations that result in an equivalent augmented matrix.

Elementary Row Operations

The following *elementary row operations* performed on an augmented matrix produce an equivalent augmented matrix:

- Interchange two rows.
- Multiply every element in a row by a nonzero real number.
- Add a multiple of one row to another row.

When we are solving a system of linear equations by any method, the goal is to write a series of simpler but equivalent systems of equations until the solution is obvious. The *Gauss-Jordan method* uses a series of elementary row operations performed on the augmented matrix to produce a simpler augmented matrix. In particular, we want to produce an augmented matrix that has 1's along the diagonal of the matrix of coefficients and 0's for the remaining entries in the matrix of coefficients. A matrix written in this way is said to be written in **reduced row echelon form**. For example, the augmented matrix from Example 3(c) is written in reduced row echelon form.

$$\left[\begin{array}{ccc|c} 1 & 0 & 0 & 4 \\ 0 & 1 & 0 & -1 \\ 0 & 0 & 1 & 0 \end{array}\right]$$

The solution to the corresponding system of equations is easily recognized as $x = 4$, $y = -1$, and $z = 0$.

Similarly, matrix **B** represents a solution of $x = a$ and $y = b$.

$$\mathbf{B} = \left[\begin{array}{cc|c} 1 & 0 & a \\ 0 & 1 & b \end{array}\right]$$

Example 4	Solving a System by Using the Gauss-Jordan Method

Solve by using the Gauss-Jordan method.

$$2x - y = 5$$
$$x + 2y = -5$$

Solution:

$$\begin{bmatrix} 2 & -1 & | & 5 \\ 1 & 2 & | & -5 \end{bmatrix}$$ Set up the augmented matrix.

$R_1 \Leftrightarrow R_2$ → $\begin{bmatrix} 1 & 2 & | & -5 \\ 2 & -1 & | & 5 \end{bmatrix}$ Switch row 1 and row 2 to get a 1 in the upper left position.

$-2R_1 + R_2 \Rightarrow R_2$ → $\begin{bmatrix} 1 & 2 & | & -5 \\ 0 & -5 & | & 15 \end{bmatrix}$ Multiply row 1 by -2 and add the result to row 2. This produces an entry of 0 below the upper left position.

$-\frac{1}{5}R_2 \Rightarrow R_2$ → $\begin{bmatrix} 1 & 2 & | & -5 \\ 0 & 1 & | & -3 \end{bmatrix}$ Multiply row 2 by $-\frac{1}{5}$ to produce a 1 along the diagonal in the second row.

$-2R_2 + R_1 \Rightarrow R_1$ → $\begin{bmatrix} 1 & 0 & | & 1 \\ 0 & 1 & | & -3 \end{bmatrix}$ Multiply row 2 by -2 and add the result to row 1. This produces a 0 in the first row, second column.

$$C = \begin{bmatrix} 1 & 0 & | & 1 \\ 0 & 1 & | & -3 \end{bmatrix}$$

The matrix **C** is in reduced row echelon form. From the augmented matrix, we have $x = 1$ and $y = -3$. The solution set is $\{(1, -3)\}$.

Skill Practice

10. Solve by using the Gauss-Jordan method.
$$x - 2y = -21$$
$$2x + y = -2$$

The order in which we manipulate the elements of an augmented matrix to produce reduced row echelon form was demonstrated in Example 4. In general, the order is as follows.

• First produce a 1 in the first row, first column. Then use the first row to obtain 0's in the first column below this element.
• Next, if possible, produce a 1 in the second row, second column. Use the second row to obtain 0's above and below this element.
• Next, if possible, produce a 1 in the third row, third column. Use the third row to obtain 0's above and below this element.
• The process continues until reduced row echelon form is obtained.

Answer
10. $\{(-5, 8)\}$

Example 5	Solving a System by Using the Gauss-Jordan Method

Solve by using the Gauss-Jordan method.

$$x = -y + 5$$
$$-2x + 2z = y - 10$$
$$3x + 6y + 7z = 14$$

Solution:

First write each equation in the system in standard form.

$$x = -y + 5 \longrightarrow x + y = 5$$
$$-2x + 2z = y - 10 \longrightarrow -2x - y + 2z = -10$$
$$3x + 6y + 7z = 14 \longrightarrow 3x + 6y + 7z = 14$$

$$\begin{bmatrix} 1 & 1 & 0 & | & 5 \\ -2 & -1 & 2 & | & -10 \\ 3 & 6 & 7 & | & 14 \end{bmatrix}$$ Set up the augmented matrix.

$$2R_1 + R_2 \Rightarrow R_2 \longrightarrow$$
$$-3R_1 + R_3 \Rightarrow R_3 \longrightarrow$$
$$\begin{bmatrix} 1 & 1 & 0 & | & 5 \\ 0 & 1 & 2 & | & 0 \\ 0 & 3 & 7 & | & -1 \end{bmatrix}$$
Multiply row 1 by 2 and add the result to row 2. Multiply row 1 by -3 and add the result to row 3.

$$-1R_2 + R_1 \Rightarrow R_1 \longrightarrow$$
$$-3R_2 + R_3 \Rightarrow R_3 \longrightarrow$$
$$\begin{bmatrix} 1 & 0 & -2 & | & 5 \\ 0 & 1 & 2 & | & 0 \\ 0 & 0 & 1 & | & -1 \end{bmatrix}$$
Multiply row 2 by -1 and add the result to row 1. Multiply row 2 by -3 and add the result to row 3.

$$2R_3 + R_1 \Rightarrow R_1 \longrightarrow$$
$$-2R_3 + R_2 \Rightarrow R_2 \longrightarrow$$
$$\begin{bmatrix} 1 & 0 & 0 & | & 3 \\ 0 & 1 & 0 & | & 2 \\ 0 & 0 & 1 & | & -1 \end{bmatrix}$$
Multiply row 3 by 2 and add the result to row 1. Multiply row 3 by -2 and add the result to row 2.

From the reduced row echelon form of the matrix, we have $x = 3$, $y = 2$, and $z = -1$. The solution set is $\{(3, 2, -1)\}$.

Skill Practice Solve by using the Gauss-Jordan method.

11. $x + y + z = 2$
$x - y + z = 4$
$x + 4y + 2z = 1$

Answer

11. $\{(1, -1, 2)\}$

It is particularly easy to recognize a system of dependent equations or an inconsistent system of equations from the reduced row echelon form of an augmented matrix. This is demonstrated in Examples 6 and 7.

Example 6 **Solving a System of Dependent Equations by Using the Gauss-Jordan Method**

Solve by using the Gauss-Jordan method.

$$x - 3y = 4$$

$$\frac{1}{2}x - \frac{3}{2}y = 2$$

Solution:

$$\begin{bmatrix} 1 & -3 & | & 4 \\ \frac{1}{2} & -\frac{3}{2} & | & 2 \end{bmatrix}$$ Set up the augmented matrix.

$$\xrightarrow{-\frac{1}{2}R_1 + R_2 \Rightarrow R_2} \begin{bmatrix} 1 & -3 & | & 4 \\ 0 & 0 & | & 0 \end{bmatrix}$$ Multiply row 1 by $-\frac{1}{2}$ and add the result to row 2.

The second row of the augmented matrix represents the equation $0 = 0$. The equations are dependent. The solution set is $\{(x, y) \mid x - 3y = 4\}$.

Skill Practice Solve by using the Gauss-Jordan method.

12. $4x - 6y = 16$
$6x - 9y = 24$

Example 7 **Solving an Inconsistent System by Using the Gauss-Jordan Method**

Solve by using the Gauss-Jordan method.

$$x + 3y = 2$$

$$-3x - 9y = 1$$

Solution:

$$\begin{bmatrix} 1 & 3 & | & 2 \\ -3 & -9 & | & 1 \end{bmatrix}$$ Set up the augmented matrix.

$$\xrightarrow{3R_1 + R_2 \Rightarrow R_2} \begin{bmatrix} 1 & 3 & | & 2 \\ 0 & 0 & | & 7 \end{bmatrix}$$ Multiply row 1 by 3 and add the result to row 2.

The second row of the augmented matrix represents the contradiction $0 = 7$. The system is inconsistent. There is no solution, $\{\ \}$.

Skill Practice Solve by using the Gauss-Jordan method.

13. $6x + 10y = 1$
$15x + 25y = 3$

Answers

12. Infinitely many solutions;
$\{(x, y) \mid 4x - 6y = 16\}$;
dependent equations
13. No solution; $\{\ \}$; inconsistent system

Calculator Connections

Topic: Entering a Matrix into a Calculator

Many graphing calculators have a matrix editor in which the user defines the order of the matrix and then enters the elements of the matrix. For example, the 2×3 matrix

$$\mathbf{D} = \begin{bmatrix} 2 & -3 & -13 \\ 3 & 1 & 8 \end{bmatrix}$$

is entered as shown.

Once an augmented matrix has been entered into a graphing calculator, a *rref* function can be used to transform the matrix into reduced row echelon form.

```
rref([A])
      [[1 0 1]
       [0 1 5]]
```

Section 3.7 Practice Exercises

Vocabulary and Key Concepts

1. **a.** A _____ is a rectangular array of numbers. The order of a matrix is $m \times n$ where m is the number of _____ and n is the number of _____.

    b. A matrix that has exactly one column is called a _____ matrix. A matrix that has exactly _____ row is called a row matrix, and a matrix with the same number of rows and columns is called a _____ matrix.

    c. Given the system of equations shown, matrix **A** is called the _____ matrix. The matrix **B** is called the _____ matrix.

    $$\begin{array}{c} 3x - 2y = 6 \\ 4x + 5y = 9 \end{array} \quad \mathbf{A} = \begin{bmatrix} 3 & -2 \\ 4 & 5 \end{bmatrix} \quad \mathbf{B} = \begin{bmatrix} 3 & -2 & 6 \\ 4 & 5 & 9 \end{bmatrix}$$

    d. The matrix **A** is said to be written in reduced _____ form.

    $$\mathbf{A} = \begin{bmatrix} 1 & 0 & 8 \\ 0 & 1 & 2 \end{bmatrix}$$

Review Exercises

2. How much 50% acid solution should be mixed with pure acid to obtain 20 L of a mixture that is 70% acid?

For Exercises 3–5, solve the system by using any method.

3. $x - 6y = 9$
   $x + 2y = 13$

4. $x + y - z = 8$
   $x - 2y + z = 3$
   $x + 3y + 2z = 7$

5. $2x - y + z = -4$
   $-x + y + 3z = -7$
   $x + 3y - 4z = 22$

Concept 1: Introduction to Matrices

For Exercises 6–14, **(a)** determine the order of each matrix and **(b)** determine if the matrix is a row matrix, a column matrix, a square matrix, or none of these. **(See Example 1.)**

6. $\begin{bmatrix} 4 \\ 5 \\ -3 \\ 0 \end{bmatrix}$

7. $\begin{bmatrix} 5 \\ -1 \\ 2 \end{bmatrix}$

8. $\begin{bmatrix} -9 & 4 & 3 \\ -1 & -8 & 4 \\ 5 & 8 & 7 \end{bmatrix}$

9. $\begin{bmatrix} 3 & -9 \\ -1 & -3 \end{bmatrix}$

10. $\begin{bmatrix} 4 & -7 \end{bmatrix}$

11. $\begin{bmatrix} 0 & -8 & 11 & 5 \end{bmatrix}$

12. $\begin{bmatrix} 5 & -8.1 & 4.2 & 0 \\ 4.3 & -9 & 18 & 3 \end{bmatrix}$

13. $\begin{bmatrix} \frac{1}{3} & \frac{3}{4} & 6 \\ -2 & 1 & -\frac{7}{8} \end{bmatrix}$

14. $\begin{bmatrix} 5 & 1 \\ -1 & 2 \\ 0 & 7 \end{bmatrix}$

For Exercises 15–18, set up the augmented matrix. **(See Example 2.)**

15. $x - 2y = -1$
$2x + y = -7$

16. $x - 3y = 3$
$2x - 5y = 4$

17. $x - 2y = 5 - z$
$2x + 6y + 3z = -2$
$3x - y - 2z = 1$

18. $5x - 17 = -2z$
$8x + 6z = 26 + y$
$8x + 3y - 12z = 24$

For Exercises 19–22, write a system of linear equations represented by the augmented matrix. **(See Example 3.)**

19. $\left[\begin{array}{cc|c} 4 & 3 & 6 \\ 12 & 5 & -6 \end{array}\right]$

20. $\left[\begin{array}{cc|c} -2 & 5 & -15 \\ -7 & 15 & -45 \end{array}\right]$

21. $\left[\begin{array}{ccc|c} 1 & 0 & 0 & 4 \\ 0 & 1 & 0 & -1 \\ 0 & 0 & 1 & 7 \end{array}\right]$

22. $\left[\begin{array}{ccc|c} 1 & 0 & 0 & 0.5 \\ 0 & 1 & 0 & 6.1 \\ 0 & 0 & 1 & 3.9 \end{array}\right]$

Concept 2: Solving Systems of Linear Equations by Using the Gauss-Jordan Method

23. Given the matrix **E**

$$\mathbf{E} = \left[\begin{array}{cc|c} 3 & -2 & 8 \\ 9 & -1 & 7 \end{array}\right]$$

   a. What is the element in the second row and third column?

   b. What is the element in the first row and second column?

24. Given the matrix **F**

$$\mathbf{F} = \left[\begin{array}{cc|c} 1 & 8 & 0 \\ 12 & -13 & -2 \end{array}\right]$$

   a. What is the element in the second row and second column?

   b. What is the element in the first row and third column?

25. Given the matrix **Z**

$$\mathbf{Z} = \left[\begin{array}{cc|c} 2 & 1 & 11 \\ 2 & -1 & 1 \end{array}\right]$$

write the matrix obtained by multiplying the elements in the first row by $\frac{1}{2}$.

26. Given the matrix **J**

$$\mathbf{J} = \left[\begin{array}{cc|c} 1 & 1 & 7 \\ 0 & 3 & -6 \end{array}\right]$$

write the matrix obtained by multiplying the elements in the second row by $\frac{1}{3}$.

27. Given the matrix **K**

$$\mathbf{K} = \left[\begin{array}{cc|c} 5 & 2 & 1 \\ 1 & -4 & 3 \end{array}\right]$$

write the matrix obtained by interchanging rows 1 and 2.

28. Given the matrix **L**

$$\mathbf{L} = \left[\begin{array}{cc|c} 9 & 6 & 13 \\ -7 & 2 & 19 \end{array}\right]$$

write the matrix obtained by interchanging rows 1 and 2.

29. Given the matrix **M**

$$\mathbf{M} = \begin{bmatrix} 1 & 5 & | & 2 \\ -3 & -4 & | & -1 \end{bmatrix}$$

write the matrix obtained by multiplying the first row by 3 and adding the result to row 2.

30. Given the matrix **N**

$$\mathbf{N} = \begin{bmatrix} 1 & 3 & | & -5 \\ -2 & 2 & | & 12 \end{bmatrix}$$

write the matrix obtained by multiplying the first row by 2 and adding the result to row 2.

31. Given the matrix **R**

$$\mathbf{R} = \begin{bmatrix} 1 & 3 & 0 & | & -1 \\ 4 & 1 & -5 & | & 6 \\ -2 & 0 & -3 & | & 10 \end{bmatrix}$$

a. Write the matrix obtained by multiplying the first row by −4 and adding the result to row 2.

b. Using the matrix obtained from part (a), write the matrix obtained by multiplying the first row by 2 and adding the result to row 3.

32. Given the matrix **S**

$$\mathbf{S} = \begin{bmatrix} 1 & 2 & 0 & | & 10 \\ 5 & 1 & -4 & | & 3 \\ -3 & 4 & 5 & | & 2 \end{bmatrix}$$

a. Write the matrix obtained by multiplying the first row by −5 and adding the result to row 2.

b. Using the matrix obtained from part (a), write the matrix obtained by multiplying the first row by 3 and adding the result to row 3.

For Exercises 33–36, use the augmented matrices **A**, **B**, **C**, and **D** to answer true or false.

$$\mathbf{A} = \begin{bmatrix} 6 & -4 & | & 2 \\ 5 & -2 & | & 7 \end{bmatrix} \quad \mathbf{B} = \begin{bmatrix} 5 & -2 & | & 7 \\ 6 & -4 & | & 2 \end{bmatrix} \quad \mathbf{C} = \begin{bmatrix} 1 & -\frac{2}{3} & | & \frac{1}{3} \\ 5 & -2 & | & 7 \end{bmatrix} \quad \mathbf{D} = \begin{bmatrix} 5 & -2 & | & 7 \\ -12 & 8 & | & -4 \end{bmatrix}$$

33. The matrix **A** is a 2 × 3 matrix.

34. Matrix **B** is equivalent to matrix **A**.

35. Matrix **A** is equivalent to matrix **C**.

36. Matrix **B** is equivalent to matrix **D**.

37. What does the notation $R_2 \Leftrightarrow R_1$ mean when one is performing the Gauss-Jordan method?

38. What does the notation $2R_3 \Rightarrow R_3$ mean when one is performing the Gauss-Jordan method?

39. What does the notation $-3R_1 + R_2 \Rightarrow R_2$ mean when one is performing the Gauss-Jordan method?

40. What does the notation $4R_2 + R_3 \Rightarrow R_3$ mean when one is performing the Gauss-Jordan method?

For Exercises 41–56, solve the system by using the Gauss-Jordan method. For systems that do not have one unique solution, also state the number of solutions and whether the system is inconsistent or the equations are dependent. (See Example 4–7.)

41. $x - 2y = -1$
$2x + y = -7$

42. $x - 3y = 3$
$2x - 5y = 4$

43. $x + 3y = 6$
$-4x - 9y = 3$

44. $2x - 3y = -2$
$x + 2y = 13$

45. $x + 3y = 3$
$4x + 12y = 12$

46. $2x + 5y = 1$
$-4x - 10y = -2$

47. $x - y = 4$
$2x - y = 5$

48. $2x - y = 0$
$x + y = 3$

49. $x + 3y = -1$
$-3x - 6y = 12$

50. $x + y = 4$
$2x - 4y = -4$

51. $3x + y = -4$
$-6x - 2y = 3$

52. $2x + y = 4$
$6x + 3y = -1$

53. $x + y + z = 6$
$x - y + z = 2$
$x + y - z = 0$

54. $2x - 3y - 2z = 11$
$x + 3y + 8z = 1$
$3x - y + 14z = -2$

55. $x - 2y = 5 - z$
$2x - 6y + 3z = -10$
$3x - y - 2z = 5$

56. $5x = 10z + 15$
$x - y + 6z = 23$
$x + 3y - 12z = 13$

Graphing Calculator Exercises

For Exercises 57–62, use the matrix features on a graphing calculator to express each augmented matrix in reduced row echelon form. Compare your results to the solution you obtained in the indicated exercise.

57. $\begin{bmatrix} 1 & -2 & | & -1 \\ 2 & 1 & | & -7 \end{bmatrix}$

Compare with Exercise 41.

58. $\begin{bmatrix} 1 & -3 & | & 3 \\ 2 & -5 & | & 4 \end{bmatrix}$

Compare with Exercise 42.

59. $\begin{bmatrix} 1 & 3 & | & 6 \\ -4 & -9 & | & 3 \end{bmatrix}$

Compare with Exercise 43.

60. $\begin{bmatrix} 2 & -3 & | & -2 \\ 1 & 2 & | & 13 \end{bmatrix}$

Compare with Exercise 44.

61. $\begin{bmatrix} 1 & 1 & 1 & | & 6 \\ 1 & -1 & 1 & | & 2 \\ 1 & 1 & -1 & | & 0 \end{bmatrix}$

Compare with Exercise 53.

62. $\begin{bmatrix} 2 & -3 & -2 & | & 11 \\ 1 & 3 & 8 & | & 1 \\ 3 & -1 & 14 & | & -2 \end{bmatrix}$

Compare with Exercise 54.

Group Activity

Creating a Quadratic Model of the Form $y = at^2 + bt + c$

Estimated time: 20 minutes

Group Size: 3

Natalie Dalton was a player on the Daytona State College women's fast-pitch softball team. She threw a ball 120 ft from right field to make a play at third base. A photographer used strobe photography to follow the path of the ball. The height of the ball (in ft) at 0.2-sec intervals was recorded in the table.

Time (sec) t	0	0.2	0.4	0.6	0.8	1.0	1.2	1.4	1.6	1.8	2.0
Height (ft) y	5	11	16	19	21	22	21	19	16	12	6

1. Plot the points (t, y) from the data.

2. Select three data points and label
   them (t_1, y_1), (t_2, y_2), and (t_3, y_3).

   $(t_1, y_1) = $ _____

   $(t_2, y_2) = $ _____

   $(t_3, y_3) = $ _____

3. Substitute the values of t_1 and y_1 for t and y in the quadratic model $y = at^2 + bt + c$. The equation you are left with should only have the variables a, b, and c. Call this equation 1. Then repeat this step two more times, using the points (t_2, y_2) and (t_3, y_3). You should then be left with three equations involving a, b, and c only.

   Equation 1: _____

   Equation 2: _____

   Equation 3: _____

4. Solve the system of equations from step 3 for the variables a, b, and c.

5. Replace the values of a, b, and c into the quadratic function $y = at^2 + bt + c$. What does this equation represent?

6. Use the function found in step 5 to approximate the height of the ball after $\frac{1}{2}$ sec.

7. Use the function found in step 5 to approximate the height of the ball after 1.4 sec. How well does this value match the observed value of 19 ft?

Chapter 3 Summary

Section 3.1   Solving Systems of Linear Equations by the Graphing Method

Key Concepts

A **system of linear equations** in two variables can be solved by graphing.

   A **solution to a system of linear equations** is an ordered pair that satisfies each equation in the system. Graphically, this represents a point of intersection of the lines.

   There may be one solution, infinitely many solutions, or no solution.

One solution   Infinitely many   No solution
Consistent   solutions   Inconsistent
Independent   Consistent   Independent
                      Dependent

   A system of equations is **consistent** if there is at least one solution. A system is **inconsistent** if there is no solution.

   Two linear equations in x and y are **dependent** if the equations represent the same line. The solution set is the set of all points on the line.

   If two linear equations represent different lines, then the equations are **independent**.

Examples

Example 1

Solve by graphing. $x + y = 3$
                      $2x - y = 0$

Write each equation in slope-intercept form $(y = mx + b)$ to graph the lines.

$y = -x + 3$

$y = 2x$

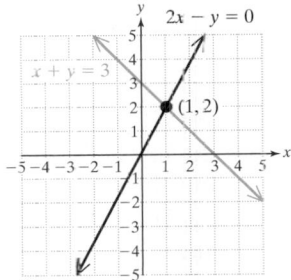

The solution is the point of intersection, $(1, 2)$.

The solution set is $\{(1, 2)\}$.

Section 3.2 Solving Systems of Linear Equations by the Substitution Method

Key Concepts

Substitution Method

1. Isolate one of the variables.
2. Substitute the quantity found in step 1 into the *other* equation.
3. Solve the resulting equation.
4. Substitute the value from step 3 back into the equation from step 1 to solve for the remaining variable.
5. Check the ordered pair in both equations, and write the answer as an ordered pair within set notation.

A system is consistent if there is at least one solution. A system is inconsistent if there is no solution. An inconsistent system is detected by a contradiction (such as $0 = 5$).

Two linear equations are independent if the equations represent different lines. The equations are dependent if they represent the same line. This produces infinitely many solutions. Two equations in a system of equations are dependent if the system reduces to an identity (such as $0 = 0$).

Examples

Example 1

$$2y = -6x + 14$$
$$2x + y = 5 \xrightarrow{\text{Isolate a variable.}} y = -2x + 5$$

Substitute

$$2(-2x + 5) = -6x + 14$$
$$-4x + 10 = -6x + 14$$
$$2x + 10 = 14$$
$$2x = 4$$
$$x = 2$$

$$y = -2x + 5 \qquad \text{Now solve for } y.$$
$$y = -2(2) + 5$$
$$y = 1$$

The ordered pair $(2, 1)$ checks in both equations.

The solution set is $\{(2, 1)\}$.

Example 2

$$y = -2x + 3$$
$$-4x - 2y = 1 \longrightarrow -4x - 2(-2x + 3) = 1$$
$$-4x + 4x - 6 = 1$$
$$-6 = 1$$

Contradiction. The system is inconsistent. There is no solution, $\{ \ \}$.

Example 3

$$x = -3y + 1$$
$$2x + 6y = 2 \longrightarrow 2(-3y + 1) + 6y = 2$$
$$-6y + 2 + 6y = 2$$
$$2 = 2$$

Identity. The equations are dependent. There are infinitely many solutions.

$$\{(x, y) | x = -3y + 1\}$$

Section 3.3 | Solving Systems of Linear Equations by the Addition Method

Key Concepts

Addition Method

1. Write both equations in standard form $Ax + By = C$.
2. Clear fractions or decimals (optional).
3. Multiply one or both equations by nonzero constants to create opposite coefficients for one of the variables.
4. Add the equations from step 3 to eliminate one variable.
5. Solve for the remaining variable.
6. Substitute the known value from step 5 back into one of the original equations to solve for the other variable.
7. Check the ordered pair in both equations and write the solution set.

Examples

Example 1

$$3x - 4y = 18 \xrightarrow{\text{Mult. by 3.}} 9x - 12y = 54$$
$$-5x - 3y = -1 \xrightarrow{\text{Mult. by } -4.} \underline{20x + 12y = 4}$$
$$29x = 58$$
$$x = 2$$

$$3(2) - 4y = 18$$
$$6 - 4y = 18$$
$$-4y = 12$$
$$y = -3$$

The ordered pair $(2, -3)$ checks in both equations.

The solution set is $\{(2, -3)\}$.

Section 3.4 Applications of Systems of Linear Equations in Two Variables

Key Concepts

Solve application problems by using systems of linear equations in two variables.

- Cost applications
- Mixture applications
- Principal and interest applications
- Uniform motion applications
- Geometry applications

Steps to Solve Applications:

1. Label two variables.
2. Construct two equations in words.
3. Write two equations.
4. Solve the system.
5. Write the answer.

Examples

Example 1

Mercedes borrowed $1500 more from a lender that charges 6.5% interest than she did from a lender that charges 4% interest. If the total interest owed at the end of 1 yr is $622.50, find the amount she borrowed at 6.5%.

Let x represent the amount borrowed at 6.5%.
Let y represent the amount borrowed at 4%.

$$\begin{pmatrix} \text{Amount borrowed} \\ \text{at } 6.5\% \end{pmatrix} = \begin{pmatrix} \text{amount borrowed} \\ \text{at } 4\% \end{pmatrix} + \$1500$$

$$\begin{pmatrix} \text{Interest owed} \\ \text{from } 6.5\% \\ \text{account} \end{pmatrix} + \begin{pmatrix} \text{interest owed} \\ \text{from } 4\% \\ \text{account} \end{pmatrix} = \$622.50$$

$$x = y + 1500$$
$$0.065x + 0.04y = 622.50$$

Using substitution gives

$$0.065(y + 1500) + 0.04y = 622.50$$
$$0.065y + 97.5 + 0.04y = 622.50$$
$$0.105y = 525$$
$$y = 5000$$

$$x = (5000) + 1500 = 6500$$

Mercedes borrowed $6500 at 6.5%.

Section 3.5   Linear Inequalities and Systems of Linear Inequalities in Two Variables

Key Concepts

A **linear inequality in two variables** is an inequality of the form $Ax + By < C$, $Ax + By > C$, $Ax + By \leq C$, or $Ax + By \geq C$.

Graphing a Linear Inequality in Two Variables

1. Solve for y, if possible.
2. Graph the related equation. Draw a dashed line if the inequality is strict, $<$ or $>$. Otherwise, draw a solid line.
3. Shade above or below the line according to the following convention.
   - Shade *above* the line if the inequality is of the form $y > mx + b$ or $y \geq mx + b$.
   - Shade *below* the line if the inequality is of the form $y < mx + b$ or $y \leq mx + b$.

You can use test points to check that you have shaded the correct region. Select an ordered pair from the proposed solution set and substitute the values of x and y in the original inequality. If the test point produces a true statement, then you have shaded the correct region.

The union or intersection of two or more linear inequalities is the union or intersection of the solution sets.

Examples

Example 1

Graph the solution to the inequality $2x - y < 4$.

Solve for y: $2x - y < 4$

$$-y < -2x + 4$$

$$y > 2x - 4$$

Graph the related equation, $y = 2x - 4$, with a dashed line.

Shade above the line.

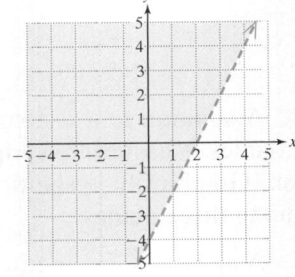

Example 2

Graph the solution set.

$x < 0$ and $y > 2$

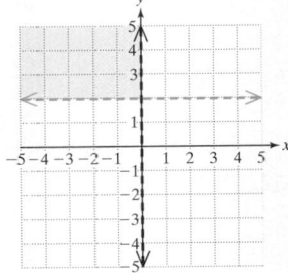

Example 3

Graph the solution set.

$x \leq 0$ or $y \geq 2$

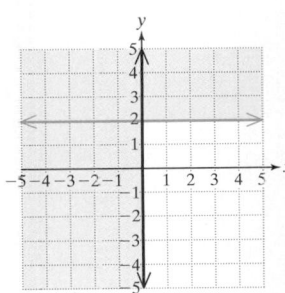

Section 3.6 Systems of Linear Equations in Three Variables and Applications

Key Concepts

A **linear equation in three variables** can be written in the form $Ax + By + Cz = D$, where A, B, and C are not all zero. The graph of a linear equation in three variables is a plane in space.

A solution to a system of linear equations in three variables is an **ordered triple** that satisfies each equation. Graphically, a solution is a point of intersection among three planes.

A system of linear equations in three variables may have one unique solution, infinitely many solutions (dependent equations), or no solution (inconsistent system).

Examples

Example 1

$\boxed{\text{A}}$ $x + 2y - z = 4$

$\boxed{\text{B}}$ $3x - y + z = 5$

$\boxed{\text{C}}$ $2x + 3y + 2z = 7$

$\boxed{\text{A}}$ $x + 2y - z = 4$

$\boxed{\text{B}}$ $\underline{3x - y + z = 5}$

$\quad\quad 4x + y \quad\quad = 9$ $\boxed{\text{D}}$

$2 \cdot \boxed{\text{A}}$ $2x + 4y - 2z = 8$

$\quad\boxed{\text{C}}$ $\underline{2x + 3y + 2z = 7}$

$\quad\quad 4x + 7y \quad\quad = 15$ $\boxed{\text{E}}$

$\boxed{\text{D}}$ $4x + y = 9 \xrightarrow{\text{Multiply by } -1.} -4x - y = -9$

$\boxed{\text{E}}$ $4x + 7y = 15 \quad\quad\quad\quad\quad \underline{4x + 7y = 15}$

$\quad\quad\quad\quad\quad\quad\quad\quad\quad\quad\quad\quad\quad 6y = 6$

$\quad\quad\quad\quad\quad\quad\quad\quad\quad\quad\quad\quad\quad\ y = 1$

Substitute $y = 1$ into either equation $\boxed{\text{D}}$ or $\boxed{\text{E}}$.

$\boxed{\text{D}}$ $4x + (1) = 9$

$\quad\quad\quad 4x = 8$

$\quad\quad\quad\ x = 2$

Substitute $x = 2$ and $y = 1$ into equation $\boxed{\text{A}}$, $\boxed{\text{B}}$, or $\boxed{\text{C}}$.

$\boxed{\text{A}}$ $(2) + 2(1) - z = 4$

$\quad\quad\quad\quad\quad\quad\quad z = 0$

The solution set is $\{(2, 1, 0)\}$.

Section 3.7 Solving Systems of Linear Equations by Using Matrices

Key Concepts

A **matrix** is a rectangular array of numbers displayed in rows and columns. Every number or entry within a matrix is called an element of the matrix.

The **order of a matrix** is determined by the number of rows and number of columns. A matrix with m rows and n columns is an $m \times n$ matrix.

A system of equations written in standard form can be represented by an **augmented matrix** consisting of the coefficients of the terms of each equation in the system.

The Gauss-Jordan method can be used to solve a system of equations by using the following elementary row operations on an augmented matrix.

1. Interchange two rows.
2. Multiply every element in a row by a nonzero real number.
3. Add a multiple of one row to another row.

These operations are used to write the matrix in **reduced row echelon form**.

$$\begin{bmatrix} 1 & 0 & | & a \\ 0 & 1 & | & b \end{bmatrix}$$

This represents the solution, $x = a$ and $y = b$.

Examples

Example 1

$\begin{bmatrix} 1 & 2 & 5 \end{bmatrix}$ is a 1×3 matrix (a row matrix).

$\begin{bmatrix} -1 & 8 \\ 1 & 5 \end{bmatrix}$ is a 2×2 matrix (a square matrix).

$\begin{bmatrix} 4 \\ 1 \end{bmatrix}$ is a 2×1 matrix (a column matrix).

Example 2

The augmented matrix for

$$\begin{aligned} 4x + y &= -12 \\ x - 2y &= 6 \end{aligned} \quad \text{is} \quad \begin{bmatrix} 4 & 1 & | & -12 \\ 1 & -2 & | & 6 \end{bmatrix}$$

Example 3

Solve the system from Example 2 by using the Gauss-Jordan method.

$R_1 \Leftrightarrow R_2$
$$\begin{bmatrix} 1 & -2 & | & 6 \\ 4 & 1 & | & -12 \end{bmatrix}$$

$\xrightarrow{-4R_1 + R_2 \Rightarrow R_2}$
$$\begin{bmatrix} 1 & -2 & | & 6 \\ 0 & 9 & | & -36 \end{bmatrix}$$

$\xrightarrow{\frac{1}{9}R_2 \Rightarrow R_2}$
$$\begin{bmatrix} 1 & -2 & | & 6 \\ 0 & 1 & | & -4 \end{bmatrix}$$

$\xrightarrow{2R_2 + R_1 \Rightarrow R_1}$
$$\begin{bmatrix} 1 & 0 & | & -2 \\ 0 & 1 & | & -4 \end{bmatrix}$$

From the reduced row echelon form of the matrix we have $x = -2$ and $y = -4$. The solution set is $\{(-2, -4)\}$.

Chapter 3   Review Exercises

Section 3.1

1. Determine if the ordered pair is a solution to the system.

$$-5x - 7y = 4$$

$$y = -\frac{1}{2}x - 1$$

a. $(2, 2)$ **b.** $(2, -2)$

For Exercises 2–4, answer true or false.

2. An inconsistent system has one solution.

3. Parallel lines form an inconsistent system.

4. Lines with different slopes intersect in one point.

For Exercises 5–8, solve the system by graphing. For systems that do not have one unique solution, also state the number of solutions and whether the system is inconsistent or the equations are dependent.

5. $f(x) = x - 1$

  $g(x) = 2x - 4$

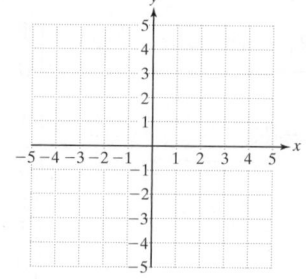

6. $y = 2x + 7$

  $y = -x - 5$

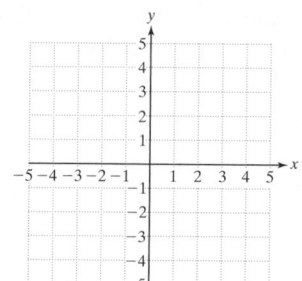

7. $6x + 2y = 4$

  $3x = -y + 2$

8. $y = \frac{1}{2}x - 2$

  $-4x + 8y = -8$

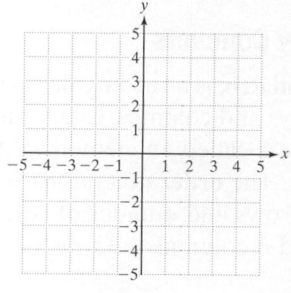

Section 3.2

For Exercises 9–14, solve the system by using the substitution method. For systems that do not have one unique solution, also state the number of solutions and whether the system is inconsistent or the equations are dependent.

9. $y = \frac{3}{4}x - 4$

  $-x + 2y = -6$

10. $6x + y = 5$

  $5x + y = 3$

11. $2(x + y) = 10 - 3y$

  $0.4x + y = 1.2$

12. $3x = 11y - 9$

  $y = \frac{3}{11}x + \frac{6}{11}$

13. $60(5x - y) = 90$

  $10x = 2y + 3$

14. $4x + y = 7$

  $x + \frac{1}{4}y = \frac{7}{4}$

15. The cost y (in \$) to rent a 5-ft by 5-ft storage space for x months is given for two companies.

  Company A: $y = 105 + 45x$
  Company B: $y = 48.50x$

  Determine the number of months for which the cost of renting a storage space would be the same for either company.

16. The cost y (in \$) to rent a car for x days is given for two rental car companies.

  Company A: $y = 44 + 81.50x$
  Company B: $y = 87x$

  Determine the number of days for which the cost of renting a car would be the same for each company.

Section 3.3

For Exercises 17–26, solve the system by using the addition method. For systems that do not have one unique solution, also state the number of solutions and whether the system is inconsistent or the equations are dependent.

17. $\frac{2}{5}x + \frac{3}{5}y = 1$

$x - \frac{2}{3}y = \frac{1}{3}$

18. $4x + 3y = 5$

$3x - 4y = 10$

19. $3x + 4y = 2$

$2x + 5y = -1$

20. $3x + y = 1$

$-x - \frac{1}{3}y = -\frac{1}{3}$

21. $2y = 3x - 8$

$-6x = -4y + 4$

22. $3x + y = 16$

$3(x + y) = y + 2x + 2$

23. $-(y + 4x) = 2x - 9$

$-2x + 2y = -10$

24. $-(4x - 35) = 3y$

$-(x - 15) = y$

25. $-0.4x + 0.3y = 1.8$

$0.6x - 0.2y = -1.2$

26. $0.02x - 0.01y = -0.11$

$0.01x + 0.04y = 0.26$

Section 3.4

27. Melinda invested twice as much money in an account paying 5% simple interest as she did in an account paying 3.5% simple interest. If her total interest at the end of 1 yr is $303.75, find the amount she invested in the 5% account.

28. A school carnival sold tickets to ride on a Ferris wheel. The charge was $1.50 for adults and $1.00 for students. If 54 tickets were sold for a total of $70.50, how many of each type of ticket were sold?

29. How many liters of 20% saline solution must be mixed with 50% saline solution to produce 16 L of a 31.25% saline solution?

30. It takes a pilot $1\frac{3}{4}$ hr to travel with the wind from Jacksonville, Florida, to Myrtle Beach, South Carolina. Her return trip takes 2 hr flying against the wind. What is the speed of the wind and the speed of the plane in still air if the distance between Jacksonville and Myrtle Beach is 280 mi?

31. Two phone companies offer discount rates to students.

Company 1: $9.95 per month, plus $0.10 per minute for long-distance calls

Company 2: $12.95 per month, plus $0.08 per minute for long-distance calls

a. Write a linear function describing the total cost for x min of long-distance calls from Company 1.

b. Write a linear function describing the total cost for x min of long-distance calls from Company 2.

c. How many minutes of long-distance calls would result in equal cost for both offers?

32. Two angles are complementary. One angle measures 6° more than 5 times the measure of the other. What are the measures of the two angles?

Section 3.5

For Exercises 33–38, graph the solution set.

33. $2x > -y + 5$

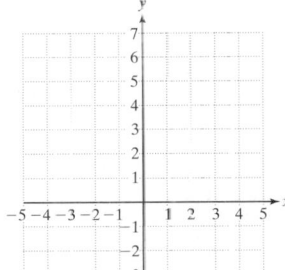

34. $2x \le -8 - 3y$

35. $x > -3$

36. $x \le 2$

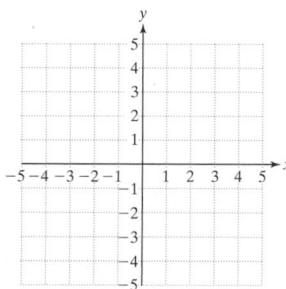

41. $y \ge x$ or $y \le -x$

37. $x \ge \frac{1}{2}y$

38. $x < \frac{2}{5}y$

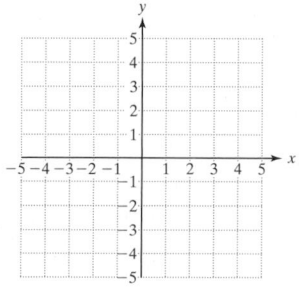

42. $x \ge 0$, $y \ge 0$,   and $y \le -\frac{2}{3}x + 4$

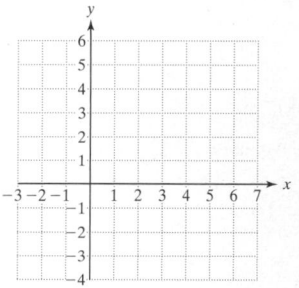

For Exercises 39–42, graph the system of inequalities.

39. $2x - y > -2$ and $2x - y \le 2$

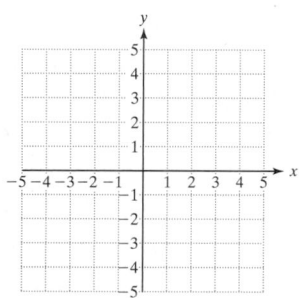

40. $3x + y \ge 6$ or $3x + y < -6$

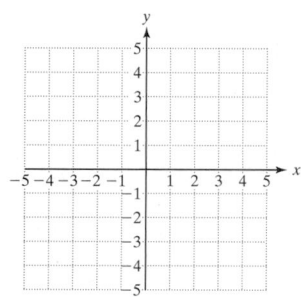

43. Suppose a farmer has 100 acres of land on which to grow oranges and grapefruit. Furthermore, because of demand from his customers, he wants to plant at least 4 times as many acres of orange trees as grapefruit trees.

Let x represent the number of acres of orange trees.

Let y represent the number of acres of grapefruit trees.

a. Write two inequalities that express the fact that the farmer cannot use a negative number of acres to plant orange and grapefruit trees.

b. Write an inequality that expresses the fact that the total number of acres used for growing orange and grapefruit trees is at most 100.

c. Write an inequality that expresses the fact that the farmer wants to plant at least 4 times as many orange trees as grapefruit trees.

d. Sketch the inequalities in parts (a)–(c) to find the feasible region for the farmer's distribution of orange and grapefruit trees.

Section 3.6

For Exercises 44–47, solve the system. If a system does not have a unique solution, state whether the system is inconsistent or the equations are dependent.

44. $5x + 5y + 5z = 30$

$-x + y + z = 2$

$10x + 6y - 2z = 4$

45. $5x + 3y - z = 5$

$x + 2y + z = 6$

$-x - 2y - z = 8$

46. $x + y + z = 4$

$-x - 2y - 3z = -6$

$2x + 4y + 6z = 12$

47. $3x \quad + 4z = 5$

$2y + 3z = 2$

$2x - 5y \quad = 8$

48. The perimeter of a right triangle is 30 ft. One leg is 2 ft longer than twice the shortest leg. The hypotenuse is 2 ft less than 3 times the shortest leg. Find the lengths of the sides of this triangle.

49. Three pumps are working to drain a construction site. Working together, the pumps can drain 950 gal/hr of water. The slowest pump drains 150 gal/hr less than the fastest pump. The fastest pump drains 150 gal/hr less than the sum of the other two pumps. How many gallons can each pump drain per hour?

50. The smallest angle in a triangle measures 9° less than the middle angle. The largest angle is 26° more than 3 times the measure of the smallest angle. Find the measure of the three angles.

Section 3.7

For Exercises 51–54, determine the order of each matrix.

51. $\begin{bmatrix} 2 & 4 & -1 \\ 5 & 0 & -3 \\ -1 & 6 & 10 \end{bmatrix}$

52. $\begin{bmatrix} -5 & 6 \\ 9 & 2 \\ 0 & -3 \end{bmatrix}$

53. $\begin{bmatrix} 0 & 13 & -4 & 16 \end{bmatrix}$

54. $\begin{bmatrix} 7 \\ 12 \\ -4 \end{bmatrix}$

For Exercises 55–56, set up the augmented matrix.

55. $x + y = 3$

$x - y = -1$

56. $x - y + z = 4$

$2x - y + 3z = 8$

$-2x + 2y - z = -9$

For Exercises 57–58, write a corresponding system of equations from the augmented matrix.

57. $\left[\begin{array}{cc|c} 1 & 0 & 9 \\ 0 & 1 & -3 \end{array}\right]$

58. $\left[\begin{array}{ccc|c} 1 & 0 & 0 & -5 \\ 0 & 1 & 0 & 2 \\ 0 & 0 & 1 & -8 \end{array}\right]$

59. Given the matrix **C**

$$\mathbf{C} = \left[\begin{array}{cc|c} 1 & 3 & 1 \\ 4 & -1 & 6 \end{array}\right]$$

a. What is the element in the second row and first column?

b. Write the matrix obtained by multiplying the first row by -4 and adding the result to row 2.

60. Given the matrix **D**

$$\mathbf{D} = \left[\begin{array}{ccc|c} 1 & 2 & 0 & -3 \\ 4 & -1 & 1 & 0 \\ -3 & 2 & 2 & 5 \end{array}\right]$$

a. Write the matrix obtained by multiplying the first row by -4 and adding the result to row 2.

b. Using the matrix obtained in part (a), write the matrix obtained by multiplying the first row by 3 and adding the result to row 3.

For Exercises 61–64, solve the system by using the Gauss-Jordan method.

61. $x + y = 3$
$x - y = -1$

62. $4x + 3y = 6$
$12x + 5y = -6$

63. $x - y + z = -4$
$2x + y - 2z = 9$
$x + 2y + z = 5$

64. $x - y + z = 4$
$2x - y + 3z = 8$
$-2x + 2y - z = -9$

Chapter 3   Test

1. Determine if the ordered pair $\left(\frac{1}{4}, 2\right)$ is a solution to the system.

$$4x - 3y = -5$$
$$12x + 2y = 7$$

For Exercises 2–4, match each figure with the appropriate description, a, b, or c.

a. The system is consistent.
The equations are dependent.
There are infinitely many solutions.

b. The system is consistent.
The equations are independent.
There is one solution.

c. The system is inconsistent.
The equations are independent.
There is no solution.

2.

3.

4.

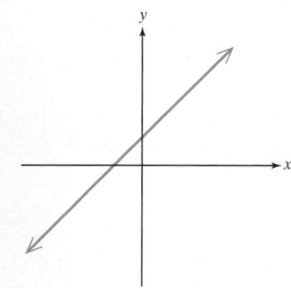

5. Solve the system by graphing.

$$4x - 2y = -4$$
$$3x + y = 7$$

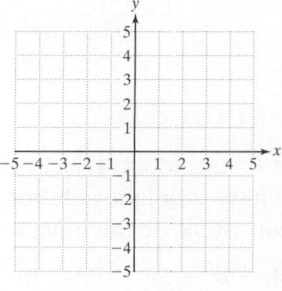

6. Solve the system by graphing.

$$f(x) = x + 3$$
$$g(x) = -\frac{3}{2}x - 2$$

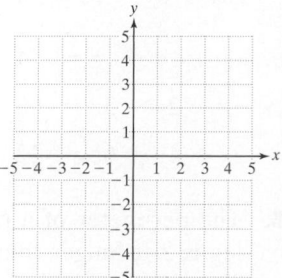

7. Solve the system by using the substitution method.

$$3x + 5y = 13$$
$$y = x + 9$$

8. Solve the system by using the addition method.

$$6x + 8y = 5$$
$$3x - 2y = 1$$

For Exercises 9–13, solve the system. For systems that do not have one unique solution, also state the number of solutions and whether the system is inconsistent or the equations are dependent.

9. $7y = 5x - 21$
$9y + 2x = -27$

10. $3x - 5y = -7$
$-18x + 30y = 42$

11. $\dfrac{1}{5}x = \dfrac{1}{2}y + \dfrac{17}{5}$

$\dfrac{1}{4}(x + 2) = -\dfrac{1}{6}y$

12. $4x = 5 - 2y$

$y = -2x + 4$

13. $-0.03y + 0.06x = 0.3$

$0.4x - 2 = -0.5y$

For Exercises 14–16, graph the solution set.

14. $2x - 5y \geq 10$

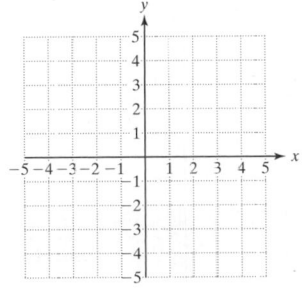

15. $x + y < 3$ and
$3x - 2y > -6$

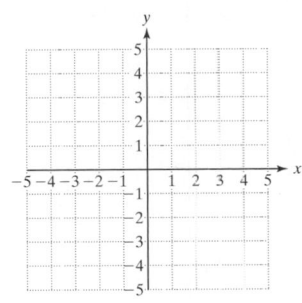

16. $5x \leq 5$ or $x + y \leq 0$

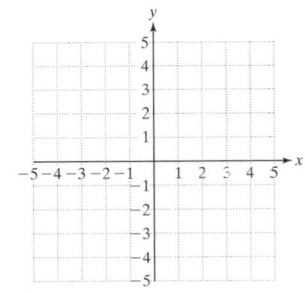

17. After menopause, women are at higher risk for hip fractures as a result of low calcium. As early as their teen years, women need at least 1200 mg of calcium per day (the USDA recommended daily allowance). One 8-oz glass of skim milk contains 300 mg of calcium, and one antacid tablet (regular strength) contains 400 mg of calcium. Let x represent the number of 8-oz glasses of milk that a woman drinks per day. Let y represent the number of antacid tablets (regular strength) that a woman takes per day.

   a. Write two inequalities that express the fact that the number of glasses of milk and the number of antacid tablets taken each day cannot be negative.

   b. Write a linear inequality in terms of x and y for which the daily calcium intake is at least 1200 mg.

   c. Graph the inequalities.

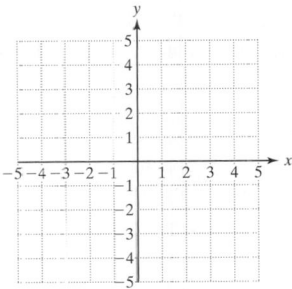

For Exercises 18–19, solve the system.

18. $2x + 2y + 4z = -6$

$3x + y + 2z = 29$

$x - y - z = 44$

19. $2(x + z) = 6 + x - 3y$

$2x = 11 + y - z$

$x + 2(y + z) = 8$

20. Aiko borrows a total of $5000 from two lenders. One charges 6.5% simple interest and the other charges 5% simple interest. If the total interest paid at the end of 1 yr is $268, how much did she borrow from each lender?

21. How many liters of a 20% acid solution should be mixed with a 60% acid solution to produce 200 L of a 44% acid solution?

22. Two angles are complementary. Two times the measure of one angle is 60° less than the measure of the other. Find the measure of each angle.

23. Working together, Joanne, Kent, and Geoff can process 504 orders per day for their business. Kent can process 20 more orders per day than Joanne can process. Geoff can process 104 fewer orders per day than Kent and Joanne combined. Find the number of orders that each person can process per day.

24. Write an example of a 3×2 matrix.

25. Given the matrix **A**

$$\mathbf{A} = \left[\begin{array}{rrr|r} 1 & 2 & 1 & -3 \\ 4 & 0 & 1 & -2 \\ -5 & -6 & 3 & 0 \end{array}\right]$$

a. Write the matrix obtained by multiplying the first row by -4 and adding the result to row 2.

b. Using the matrix obtained in part (a), write the matrix obtained by multiplying the first row by 5 and adding the result to row 3.

For Exercises 26–27, solve by using the Gauss-Jordan method.

26. $5x - 4y = 34$
$x - 2y = 8$

27. $x + y + z = 1$
$2x + y \phantom{{}+z} = 0$
$\phantom{2x + {}}-2y - z = 5$

Chapters 1–3 Cumulative Review Exercises

1. Simplify. $2[3^2 - 8(7 - 5)]$

2. Simplify. $7[4 - 2(w - 5) - 3(2w + 1)] + 20$

For Exercises 3–5, solve the equation.

3. $-5(2x - 1) - 2(3x + 1) = 7 - 2(8x + 1)$

4. $\dfrac{1}{2}(a - 2) - \dfrac{3}{4}(2a + 1) = -\dfrac{1}{6}$

5. $-4 = |2x - 3| - 9$

For Exercises 6–11, solve the inequality. Write the answer in interval notation if possible.

6. $-3y - 2(y + 1) < 8$

7. $4x > 16 \quad \text{or} \quad -6x - 3 \geq 9$

8. $4x > 16 \quad \text{and} \quad -6x - 3 \geq 9$

9. $0 \leq \dfrac{3x - 9}{6} \leq 5$

10. $|x - 4| + 1 < 11$

11. $4 < |2x + 4|$

12. Graph the solution set. $x - 5y \leq 5$

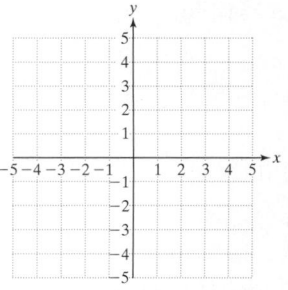

13. Identify the slope and the x- and y-intercepts of the line $5x - 2y = 15$.

For Exercises 14–15, graph the equations.

14. $y = -\dfrac{1}{3}x - 4$ **15.** $x = -2$

    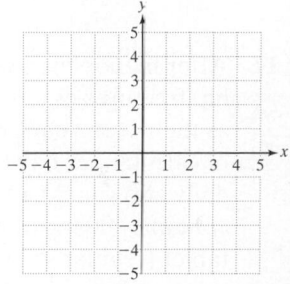

16. Find the slope of the line passing through the points $(4, -10)$ and $(6, -10)$.

17. Find an equation for the line that passes through the points $(3, -8)$ and $(2, -4)$. Write the answer in slope-intercept form.

18. Solve the system by using the addition method.

$$2x - 3y = 6$$

$$\frac{1}{2}x - \frac{3}{4}y = 1$$

19. Solve the system by using the substitution method.

$$2x + y = 4$$

$$y = 3x - 1$$

20. The perimeter of a triangle is 7 m. The length of the longest side is twice the length of the shortest side. The sum of the lengths of the shortest side and the middle side is 1 m more than the length of the longest side. Find the lengths of the sides.

21. Two video clubs rent DVDs according to the following fee schedules:

Club 1: $25 initiation fee plus $2.50 per DVD
Club 2: $10 initiation fee plus $3.00 per DVD

a. Write a linear function describing the total cost of renting x DVDs from Club 1.

b. Write a linear function describing the total cost of renting x DVDs from Club 2.

c. How many DVDs would have to be rented to make the cost for Club 1 the same as the cost for Club 2?

22. Solve the system.

$$3x = 3 - 2y - 3z$$

$$4x - 5y + 7z = 1$$

$$2x + 3y - 2z = 6$$

23. Determine the order of the matrix.

$$\begin{bmatrix} 4 & 5 & 1 \\ -2 & 6 & 0 \end{bmatrix}$$

24. Write an example of a 2×4 matrix.

25. Solve the system by using the Gauss-Jordan method.

$$2x - 4y = -2$$

$$4x + y = 5$$

Polynomials

4

CHAPTER OUTLINE

4.1 Properties of Integer Exponents and Scientific Notation   320

4.2 Addition and Subtraction of Polynomials and Polynomial Functions   329

4.3 Multiplication of Polynomials   340

4.4 Division of Polynomials   350

   Problem Recognition Exercises: Operations on Polynomials   359

4.5 Greatest Common Factor and Factoring by Grouping   360

4.6 Factoring Trinomials   368

4.7 Factoring Binomials   382

   Problem Recognition Exercises: Factoring Summary   392

4.8 Solving Equations by Using the Zero Product Rule   394

   Group Activity: Investigating Pascal's Triangle   408

Chapter 4

In this chapter, we study addition, subtraction, multiplication, and division of polynomials, along with an important operation called factoring.

Are You Prepared?

To prepare for this chapter, use the crossword puzzle to review the properties of real numbers and exponents from Chapter R.

Across

1. Simplify and give the word name for $(-2)^4$.

6. $a(b + c) = ab + ac$ illustrates the _____ property.

8. Simplify. $\left(\dfrac{1}{6}\right)^{-1} + \left(\dfrac{1}{3}\right)^{-2}$

9. Simplify and give the word name for y^0.

10. Simplify. $\left(\dfrac{3r^4}{r^5}\right)^2 \cdot (2r)^3$

Down

2. Simplify. $3(x + 4) - 2(x + 2)$

3. Simplify and give the word name for -2^2.

4. True or false:
   $(x + 2)(x + 3) = (x + 3)(x + 2)$

5. Simplify. $\dfrac{16d^4}{2d^3}$

7. Simplify. $-4(5 + t) + 5(t + 3) + 10$

319

Properties of Integer Exponents and Scientific Notation

Concepts

1. Simplifying Expressions with Exponents
2. Scientific Notation

1. Simplifying Expressions with Exponents

In Section R.3, we learned that exponents are used to represent repeated multiplication. The following properties of exponents (Table 4-1) are often used to simplify algebraic expressions.

Table 4-1   Properties of Exponents*

Description	Property	Example	Expanded Form
Multiplication of like bases	$b^m \cdot b^n = b^{m+n}$	$b^2 \cdot b^4 = b^{2+4}$ $= b^6$	$b^2 \cdot b^4 = (b \cdot b)(b \cdot b \cdot b \cdot b)$ $= b^6$
Division of like bases	$\dfrac{b^m}{b^n} = b^{m-n}$	$\dfrac{b^5}{b^2} = b^{5-2}$ $= b^3$	$\dfrac{b^5}{b^2} = \dfrac{b \cdot b \cdot b \cdot b \cdot b}{b \cdot b}$ $= b^3$
Power rule	$(b^m)^n = b^{m \cdot n}$	$(b^4)^2 = b^{4 \cdot 2}$ $= b^8$	$(b^4)^2 = (b \cdot b \cdot b \cdot b)(b \cdot b \cdot b \cdot b)$ $= b^8$
Power of a product	$(ab)^m = a^m b^m$	$(ab)^3 = a^3 b^3$	$(ab)^3 = (ab)(ab)(ab)$ $= (a \cdot a \cdot a)(b \cdot b \cdot b) = a^3 b^3$
Power of a quotient	$\left(\dfrac{a}{b}\right)^m = \dfrac{a^m}{b^m}$	$\left(\dfrac{a}{b}\right)^3 = \dfrac{a^3}{b^3}$	$\left(\dfrac{a}{b}\right)^3 = \left(\dfrac{a}{b}\right)\left(\dfrac{a}{b}\right)\left(\dfrac{a}{b}\right)$ $= \dfrac{a \cdot a \cdot a}{b \cdot b \cdot b} = \dfrac{a^3}{b^3}$

*Assume that a and b are real numbers ($b \neq 0$) and that m and n represent integers.

In addition to the properties of exponents, two definitions are used to simplify algebraic expressions.

Definition of b^0 and b^{-n}

Let n be an integer, and b be a real number such that $b \neq 0$.

*1. $b^0 = 1$ **Example:** $5^0 = 1$

2. $b^{-n} = \left(\dfrac{1}{b}\right)^n = \dfrac{1}{b^n}$ **Example:** $4^{-3} = \left(\dfrac{1}{4}\right)^3 = \dfrac{1}{4^3}$ or $\dfrac{1}{64}$

3. From the definition of b^{-n} we also have:

$$\left(\dfrac{a}{b}\right)^{-n} = \left(\dfrac{b}{a}\right)^n = \dfrac{b^n}{a^n} \quad \text{for } a \neq 0, b \neq 0.$$

     Example: $\left(\dfrac{3}{7}\right)^{-2} = \left(\dfrac{7}{3}\right)^2 = \dfrac{7^2}{3^2}$ or $\dfrac{49}{9}$

*Note: The value of 0^0 is not defined by definition 1 because the base, b, must not equal 0.

The definition of b^0 is consistent with the properties of exponents. For example, if b is a nonzero real number and n is an integer, then

$$\dfrac{b^n}{b^n} = 1$$

$$\dfrac{b^n}{b^n} = b^{n-n} = b^0$$

The expression $b^0 = 1$

The definition of b^{-n} is also consistent with the properties of exponents. If b is a nonzero real number, then

$$\frac{b^3}{b^5} = \frac{b \cdot b \cdot b}{b \cdot b \cdot b \cdot b \cdot b} = \frac{1}{b^2}$$

The expression $b^{-2} = \dfrac{1}{b^2}$

$$\frac{b^3}{b^5} = b^{3-5} = b^{-2}$$

Example 1 Simplifying Expressions with Exponents

Simplify the expressions.

a. $(-2)^4$ **b.** -2^4 **c.** -2^{-4} **d.** $(-7x)^0$ **e.** $-7x^0$

Solution:

a. $(-2)^4 = (-2)(-2)(-2)(-2)$

       $= 16$

b. $-2^4 = -1 \cdot 2^4$

       $= -1 \cdot (2 \cdot 2 \cdot 2 \cdot 2)$

       $= -16$

c. $-2^{-4} = -1 \cdot (2^{-4})$

       $= -1 \cdot \left(\dfrac{1}{2^4}\right)$

       $= -1 \cdot \dfrac{1}{2 \cdot 2 \cdot 2 \cdot 2}$

       $= -\dfrac{1}{16}$

d. $(-7x)^0 = 1$ because $b^0 = 1$

e. $-7x^0 = -7 \cdot x^0$

       $= -7 \cdot 1$

       $= -7$

Skill Practice Simplify the expressions.

1. $(-3)^2$ **2.** -3^2 **3.** -3^{-2} **4.** $(-8y)^0$ **5.** $-8y^0$

Example 2 Using Properties of Exponents

Simplify the expressions. Write the final answer with positive exponents only.

a. $x^3 x^5 x^{-2}$ **b.** $\dfrac{y^7}{y^4}$ **c.** $(b^2)^{-5}$

Answers

1. 9   **2.** -9 **3.** $-\dfrac{1}{9}$

4. 1   **5.** -8

Solution:

a. $x^3 x^5 x^{-2}$

$\quad = x^{3+5+(-2)}$ Multiply like bases by adding exponents.

$\quad = x^6$ Simplify.

b. $\dfrac{y^7}{y^4}$

$\quad = y^{7-4}$ Divide like bases by subtracting exponents.

$\quad = y^3$ Simplify.

c. $(b^2)^{-5}$

$\quad = b^{2(-5)}$ Apply the power rule.

$\quad = b^{-10}$ Multiply the exponents.

$\quad = \dfrac{1}{b^{10}}$ Write the answer with positive exponents.

Skill Practice Simplify the expressions. Write the final answer with positive exponents only.

6. $w^7 w^{-3} w$ **7.** $\dfrac{t^{11}}{t^6}$ **8.** $(p^{-4})^2$

Example 3   **Simplifying an Expression with Exponents**

Simplify the expression. $\left(\dfrac{1}{5}\right)^{-3} - (2)^{-2} + 3^0$

Solution:

$\left(\dfrac{1}{5}\right)^{-3} - (2)^{-2} + 3^0$

$\quad = 5^3 - \left(\dfrac{1}{2}\right)^2 + 3^0$ Simplify negative exponents.

$\quad = 125 - \dfrac{1}{4} + 1$ Evaluate expressions with exponents.

$\quad = \dfrac{500}{4} - \dfrac{1}{4} + \dfrac{4}{4}$ Write the expressions with a common denominator.

$\quad = \dfrac{503}{4}$ Simplify.

Skill Practice Simplify the expression.

9. $\left(\dfrac{2}{3}\right)^{-1} + 4^{-1} - \left(\dfrac{1}{4}\right)^0$

Answers

6. w^5 **7.** t^5 **8.** $\dfrac{1}{p^8}$ **9.** $\dfrac{3}{4}$

Example 4 **Simplifying an Expression with Exponents**

Simplify the expression. Write the answer with positive exponents only.

$$\frac{(2a^7b^{-4})^3}{(4a^3b^{-2})^2}$$

Solution:

$$\frac{(2a^7b^{-4})^3}{(4a^3b^{-2})^2}$$

$$= \frac{2^3a^{21}b^{-12}}{4^2a^6b^{-4}} \qquad \text{Apply the power rule.}$$

$$= \frac{8a^{21}b^{-12}}{16a^6b^{-4}} \qquad \text{Simplify the coefficients.}$$

$$= \frac{8a^{21-6}b^{-12-(-4)}}{16} \qquad \text{Divide like bases by subtracting exponents.}$$

$$= \frac{\overset{1}{8}a^{15}b^{-8}}{\underset{2}{16}} \qquad \text{Simplify.}$$

$$= \frac{a^{15}}{2b^8} \qquad \text{Simplify negative exponents.}$$

Skill Practice Simplify the expression. Write the final answer with positive exponents only.

10. $\dfrac{(3x^3y^{-4})^2}{(x^{-2}y)^{-4}}$

Example 5 **Simplifying an Expression with Exponents**

Simplify the expression. Write the answer with positive exponents only.

$$4xy^{-3}\left(\frac{8x^2}{3x^5y^2}\right)^{-2}$$

Solution:

$$4xy^{-3}\left(\frac{8x^2}{3x^5y^2}\right)^{-2} = 4xy^{-3}\cdot\left(\frac{8}{3x^3y^2}\right)^{-2} \qquad \text{Simplify within parentheses.}$$

$$= \frac{4xy^{-3}}{1}\cdot\frac{8^{-2}}{3^{-2}x^{-6}y^{-4}} \qquad \begin{array}{l}\text{Raise the expression in}\\\text{parentheses to the }-2\text{ power.}\end{array}$$

$$= \frac{4x}{y^3}\cdot\frac{3^2x^6y^4}{8^2} \qquad \text{Simplify negative exponents.}$$

$$= \frac{4\cdot9\cdot x\cdot x^6\cdot y^4}{64y^3} \qquad \begin{array}{l}\text{Multiply the fractions and simplify}\\\text{the expressions }3^2\text{ and }8^2.\end{array}$$

$$= \frac{\overset{1}{4}\cdot9\cdot x^{1+6}y^{4-3}}{\underset{16}{64}} \qquad \begin{array}{l}\text{Add the exponents on }x.\\\text{Subtract the exponents on }y.\end{array}$$

$$= \frac{9x^7y}{16} \qquad \text{Simplify.}$$

Answer

10. $\dfrac{9}{x^2y^4}$

Skill Practice Simplify the expression. Write the answer with positive exponents only.

11. $-9m^5t\left(\dfrac{3m^{-2}t^3}{4t^6}\right)^{-3}$

2. Scientific Notation

Scientists in a variety of fields often work with very large or very small numbers. For instance, the distance between the Earth and the Sun is approximately 93,000,000 mi. The mass of an electron is 0.000 000 000 000 000 000 000 000 000 000 911 kg.

   Scientific notation was devised as a shortcut method of expressing very large and very small numbers. The principle behind scientific notation is to use a power of 10 to express the magnitude of the number. For example, a number such as 50,000 can be written as $5 \times 10{,}000$ or equivalently as 5×10^4. Similarly, the number 0.0035 is equal to $3.5 \times \frac{1}{1000}$ or, equivalently, 3.5×10^{-3}.

> **Definition of Scientific Notation**
>
> A number expressed in the form $a \times 10^n$, where $1 \le |a| < 10$ and n is an integer, is said to be written in **scientific notation**.

Consider the following numbers in scientific notation:

The distance between the Sun and the Earth: $93{,}000{,}000 \text{ mi} = 9.3 \times 10^7 \text{ mi}$

7 places

The mass of an electron: $0.000\ 000\ 000\ 000\ 000\ 000\ 000\ 000\ 000\ 000\ 911 \text{ kg}$

$= 9.11 \times 10^{-31} \text{ kg}$ 31 places

In each case, the power of 10 corresponds to the number of place positions that the decimal point is moved. The power of 10 is sometimes called the order of magnitude (or simply the magnitude) of the number. The order of magnitude of the distance between the Earth and Sun is 10^7 mi (tens of millions). The mass of an electron has an order of magnitude of 10^{-31} kg.

Example 6 Writing Numbers in Scientific Notation

Fill in the table by writing the numbers in scientific notation or standard notation as indicated.

Quantity	Standard Notation	Scientific Notation
Number of NASCAR fans	75,000,000 people	
Width of an influenza virus	0.000000001 m	
Cost of hurricane Katrina		$\$1.25 \times 10^{11}$
Probability of winning the Florida state lottery		$4.35587878 \times 10^{-8}$

Solution:

Quantity	Standard Notation	Scientific Notation
Number of NASCAR fans	75,000,000 people	7.5×10^7 people
Width of an influenza virus	0.000000001 m	1×10^{-9} m
Cost of hurricane Katrina	$125,000,000,000	1.25×10^{11}
Probability of winning the Florida state lottery	0.0000000435587878	$4.35587878 \times 10^{-8}$

Skill Practice Rewrite each number in either scientific notation or standard notation.

12. 2,600,000   **13.** 0.00088   **14.** -5.7×10^{-8} **15.** 1.9×10^5

Calculator Connections

Topic: Using Scientific Notation on a Calculator

Calculators use scientific notation to display very large or very small numbers. To enter scientific notation in a calculator, try using the **EE** key or the **EXP** key to express the power of 10.

   Notice that E in the display of the calculator introduces the power of 10.

```
9.3E7
            93000000
7.25E-2
               .0725
```

Example 7   Applying Scientific Notation

a. During a recent economic crisis, the U.S. government lent money to troubled financial institutions that had a large number of mortgage-related assets. The government committed an estimated $750,000,000,000. How much money does this represent per person if the U.S. population was 300,000,000 at that time?

b. The mean distance between the Earth and the Andromeda Galaxy is approximately 1.8×10^6 light-years. Assuming that 1 light-year is 6×10^{12} mi, what is the distance in miles to the Andromeda Galaxy?

Solution:

a. Express each value in scientific notation. Then divide the total amount to be paid off by the number of people.

$$\frac{7.5 \times 10^{11}}{3 \times 10^8}$$

$$= \left(\frac{7.5}{3}\right) \times \left(\frac{10^{11}}{10^8}\right) \qquad \text{Divide 7.5 by 3 and subtract the powers of 10.}$$

$$= 2.5 \times 10^3$$

In standard notation, this amounts to $2500 per person.

Answers

12. 2.6×10^6 **13.** 8.8×10^{-4}
14. -0.000000057 **15.** 190,000

b. Multiply the number of light-years by the number of miles per light-year.

$$(1.8 \times 10^6)(6 \times 10^{12})$$

$$= (1.8)(6) \times (10^6)(10^{12})$$

$$= 10.8 \times 10^{18}$$ Multiply 1.8 and 6 and add the powers of 10.

The number 10.8×10^{18} is not in "proper" scientific notation because 10.8 is not between 1 and 10.

$$= (1.08 \times 10^1) \times 10^{18}$$ Rewrite 10.8 as 1.08×10^1.

$$= 1.08 \times (10^1 \times 10^{18})$$ Apply the associative property of multiplication.

$$= 1.08 \times 10^{19}$$

The distance between the Earth and the Andromeda Galaxy is 1.08×10^{19} mi.

Skill Practice

16. The thickness of a penny is 6×10^{-2} in. How many pennies would have to be stacked to equal the height of the Empire State Building (1.5×10^4 in.)?

17. The distance from the Earth to the "nearby" star, Barnard's Star, is 7.6 light-years (where 1 light-year $= 6 \times 10^{12}$ mi). How many miles away is Barnard's Star?

Answers

16. Approximately 2.5×10^5 pennies

17. 4.56×10^{13} mi

Section 4.1   Practice Exercises

Vocabulary and Key Concepts

1. a. A(n) _____ is used to show repeated multiplication of the base.

   b. For $b \neq 0$, the expression b^0 is defined to be _____.

   c. For $b \neq 0$, the expression b^{-n} is defined as _____.

   d. A number expressed in the form $a \times 10^n$, where $1 \leq |a| < 10$ and n is an integer is said to be written in _____ _____.

Concept 1: Simplifying Expressions with Exponents

2. Write the expressions in expanded form and simplify.

$$b^4 \cdot b^3 \text{ and } (b^4)^3$$

3. Write the expressions in expanded form and simplify.

$$ab^3 \text{ and } (ab)^3$$

For Exercises 4–9, write an example of each property. (Answers may vary.)

4. $b^n \cdot b^m = b^{n+m}$

5. $(ab)^n = a^n b^n$

6. $(b^n)^m = b^{nm}$

7. $\dfrac{b^n}{b^m} = b^{n-m}$ $(b \neq 0)$

8. $\left(\dfrac{a}{b}\right)^n = \dfrac{a^n}{b^n}$ $(b \neq 0)$

9. $b^0 = 1$ $(b \neq 0)$

For Exercises 10–28, simplify. **(See Example 1.)**

10. $\left(\dfrac{2}{3}\right)^{-1}$

11. $\left(\dfrac{1}{3}\right)^{-1}$

12. 3^{-1}

13. 5^{-2}

14. 8^{-2}

15. -5^{-2}

16. -8^{-2}

17. $(-5)^{-2}$

18. $(-8)^{-2}$

19. $\left(-\dfrac{1}{4}\right)^{-3}$

20. $\left(-\dfrac{3}{8}\right)^{-1}$

21. $\left(-\dfrac{3}{2}\right)^{-4}$

22. $\left(-\dfrac{1}{9}\right)^{-2}$

23. $-\left(\dfrac{2}{5}\right)^{-3}$

24. $-\left(\dfrac{1}{2}\right)^{-5}$

25. $(10ab)^0$

26. $(13x)^0$

27. $10ab^0$

28. $13x^0$

For Exercises 29–80, simplify and write the answer with positive exponents only. **(See Examples 2–5.)**

29. $y^3 \cdot y^5$

30. $x^4 \cdot x^8$

31. $\dfrac{13^8}{13^6}$

32. $\dfrac{5^7}{5^3}$

33. $(y^2)^4$

34. $(z^3)^4$

35. $(3x^2)^4$

36. $(2y^5)^3$

37. p^{-3}

38. q^{-5}

39. $7^{10} \cdot 7^{-13}$

40. $11^{-9} \cdot 11^7$

41. $\dfrac{w^3}{w^5}$

42. $\dfrac{t^4}{t^8}$

43. $a^{-2}c^{-5}$

44. $b^{-1}b^{-8}$

45. $\dfrac{r}{r^{-1}}$

46. $\dfrac{s^{-1}}{s}$

47. $\dfrac{z^{-6}}{z^{-2}}$

48. $\dfrac{w^{-8}}{w^{-3}}$

49. $\dfrac{a^3}{b^{-2}}$

50. $\dfrac{c^4}{d^{-1}}$

51. $(6xyz^2)^0$

52. $(-7ab^3)^0$

53. $2^4 + 2^{-2}$

54. $3^2 + 3^{-1}$

55. $1^{-2} + 5^{-2}$

56. $4^{-2} + 2^{-2}$

57. $\left(\dfrac{2}{3}\right)^{-2} - \left(\dfrac{1}{2}\right)^2 + \left(\dfrac{1}{3}\right)^0$

58. $\left(\dfrac{1}{6}\right)^{-1} + \left(\dfrac{2}{3}\right)^0 - \left(\dfrac{1}{4}\right)^{-2}$

59. $\left(\dfrac{4}{5}\right)^{-1} + \left(\dfrac{3}{2}\right)^2 - \left(\dfrac{2}{7}\right)^0$

60. $\left(\dfrac{4}{5}\right)^0 - \left(\dfrac{2}{3}\right)^2 + \left(\dfrac{9}{5}\right)^{-1}$

61. $\dfrac{p^2q}{p^5q^{-1}}$

62. $\dfrac{m^{-1}n^3}{m^4n^{-2}}$

63. $\dfrac{-48ab^{10}}{32a^4b^3}$

64. $\dfrac{25x^2y^{12}}{10x^5y^7}$

65. $(-3x^{-4}y^5z^2)^{-4}$

66. $(-6a^{-2}b^3c)^{-2}$

67. $(4m^{-2}n)(-m^6n^{-3})$

68. $(-6pq^{-3})(2p^4q)$

69. $(p^{-2}q)^3(2pq^4)^2$

70. $(mn^3)^2(5m^{-2}n^2)$

71. $\left(\dfrac{x^2}{y}\right)^3(5x^2y)$

72. $\left(\dfrac{a}{b^2}\right)^2(3a^2b^3)$

73. $\dfrac{(-8a^2b^2)^4}{(16a^3b^7)^2}$

74. $\dfrac{(-3x^2y^3)^2}{(-2xy^4)^3}$

75. $\left(\dfrac{-2x^6y^{-5}}{3x^{-2}y^6}\right)^{-3}$

76. $\left(\dfrac{-6a^2b^{-3}}{5a^{-1}b}\right)^{-2}$

77. $\left(\dfrac{2x^{-3}y^0}{4x^6y^{-5}}\right)^{-2}$

78. $\left(\dfrac{a^3b^2c^0}{a^{-1}b^{-2}c^{-3}}\right)^{-2}$

79. $3xy^5\left(\dfrac{2x^4y}{6x^5y^3}\right)^{-2}$

80. $7x^{-3}y^{-4}\left(\dfrac{3x^{-1}y^5}{9x^3y^{-2}}\right)^{-3}$

Concept 2: Scientific Notation

81. The European Organization for Nuclear Research (known by the acronym CERN) has built the world's largest high-energy particle accelerator, called the Large Hadron Collider (LHC). Scientists hope the LHC will answer many open questions in physics. Write the following numbers in scientific notation. **(See Example 6.)**

   a. The LHC cost $8,000,000,000 to build.

   b. 3,000,000 DVDs worth of data will be produced each year.

   c. 14,000,000,000,000 electron volts (eV) of energy will be produced to smash the protons together.

   d. 1 eV is equivalent to 0.000 000 000 000 000 000 1602 joules (J).

82. Write the numbers in scientific notation.

   a. The estimated population of the United States in 2011 was 311,000,000.

   b. The size of the smallest visible object in an optical microscope is 0.0000002 m.

   c. A trillion is defined as 1,000,000,000,000.

83. Write the numbers in standard notation.

   a. The Andromeda Galaxy contains at least 2×10^{11} stars.

   b. The diameter of a capillary is 4×10^{-6} m.

   c. The mean distance of Venus from the Sun is 1.082×10^{11} m.

84. Write the numbers in standard notation.

   a. At the end of a recent year, the Department of Energy's inventory of high-level radioactive waste was approximately 3.784×10^{5} m^3.

   b. The diameter of a water molecule is 3×10^{-10} m.

   c. The distance a bullet will travel in 1 sec when fired from a 0.22 caliber gun is 4.1×10^{2} m.

For Exercises 85–90, determine which numbers are in "proper" scientific notation. If the number is not in "proper" scientific notation, correct it.

85. 35×10^{4}

86. 0.469×10^{-7}

87. 7×10^{0}

88. 8.12×10^{1}

89. 9×10^{1}

90. 6.9×10^{0}

For Exercises 91–98, perform the indicated operations and write the answer in scientific notation.

91. $(6.5 \times 10^{3})(5.2 \times 10^{-8})$

92. $(3.26 \times 10^{-6})(8.2 \times 10^{9})$

93. $(0.0000024)(6,700,000,000)$

94. $(3,400,000,000)(70,000,000,000,000)$

95. $(8.5 \times 10^{-2}) \div (2.5 \times 10^{-15})$

96. $(3 \times 10^{9}) \div (1.5 \times 10^{13})$

97. $(900,000,000) \div (360,000)$

98. $(0.0000000002) \div (8,000,000)$

99. If one H_2O molecule contains 2 hydrogen atoms and 1 oxygen atom, and 10 H_2O molecules contain 20 hydrogen atoms and 10 oxygen atoms, how many hydrogen atoms and oxygen atoms are contained in 6.02×10^{23} H_2O molecules? **(See Example 7.)**

100. The star named Alpha Centauri is 4.3 light-years from the Earth. If 1 light-year is approximately 6×10^{9} mi, how far is Alpha Centauri?

101. If the county of Queens, New York, has a population of approximately 2,200,000 and the area is 110 mi^2, how many people are there per square mile? **(See Example 7.)**

102. If the county of Catawba, North Carolina, has a population of approximately 150,000 and the area is 400 mi^2, how many people are there per square mile?

103. According to the Federal Emergency Management Agency (FEMA), the annual loss due to earthquakes in California is approximately 3.5×10^9. If this is representative as a yearly average, find the loss over 15 yr.

104. Avogadro's number $N_a = 6.02 \times 10^{23}$ is the number of atoms in 1 *mole* of an element.

    a. How many atoms are in 5 moles of carbon-12?

    b. If 75 g of carbon-12 has 4.515×10^{25} atoms, how many moles is this?

Expanding Your Skills

105. A 20-yr-old starts a savings plan for her retirement. She will put $20 per month into a mutual fund that she hopes will average 6% growth annually.

    a. If she plans to retire at age 65, for how many months will she be depositing money?

    b. By age 65, how much money will she have deposited?

    c. The value of an account built in this fashion is given by

$$A = P \cdot \left[\left(1 + \frac{r}{12} \right)^N - 1 \right] \cdot \left(1 + \frac{12}{r} \right)$$

    where A is the final amount of money in the account, P is the amount of the monthly deposit, and N is the number of months. Use a calculator to find the total amount in the woman's retirement account at age 65.

For Exercises 106–111, simplify each expression. Assume that a and b represent positive integers and x and y are nonzero real numbers.

106. $x^{a+1}x^{a+5}$

107. $y^{a-5}y^{a+7}$

108. $\dfrac{y^{2a+1}}{y^{a-1}}$

109. $\dfrac{x^{3a-3}}{x^{a+1}}$

110. $\dfrac{x^{3b-2}y^{b+1}}{x^{2b+1}y^{2b+2}}$

111. $\dfrac{x^{2a-2}y^{a+3}}{x^{a+4}y^{a-3}}$

Addition and Subtraction of Polynomials and Polynomial Functions

Section 4.2

1. Polynomials: Basic Definitions

One commonly used algebraic expression is called a polynomial. A **polynomial** in x is defined as a finite sum of terms of the form ax^n, where a is a real number and the exponent n is a whole number. For each term, a is called the **coefficient**, and n is called the **degree of the term**. For example:

Concepts

1. **Polynomials: Basic Definitions**
2. **Addition of Polynomials**
3. **Subtraction of Polynomials**
4. **Polynomial Functions**

Term (Expressed in the Form ax^n)	Coefficient	Degree
$3x^5$	3	5
x^{14} → rewrite as $1x^{14}$	1	14
7 → rewrite as $7x^0$	7	0
$\frac{1}{2}p$ → rewrite as $\frac{1}{2}p^1$	$\frac{1}{2}$	1

If a polynomial has exactly one term, it is called a **monomial**. A two-term polynomial is called a **binomial**, and a three-term polynomial is called a **trinomial**. Usually the terms of a polynomial are written in descending order according to degree. In descending order, the highest-degree term is written first and is called the **leading term**. Its coefficient is called the **leading coefficient**. The **degree of a polynomial** is the greatest degree of all its terms. Thus, the leading term determines the degree of the polynomial.

	Expression	Descending Order	Leading Coefficient	Degree of Polynomial
Monomials	$2x^9$	$2x^9$	2	9
	-49	-49	-49	0
Binomials	$10y - 7y^2$	$-7y^2 + 10y$	-7	2
	$6 - \dfrac{2}{3}b$	$-\dfrac{2}{3}b + 6$	$-\dfrac{2}{3}$	1
Trinomials	$w + 2w^3 + 9w^6$	$9w^6 + 2w^3 + w$	9	6
	$2.5a^4 - a^8 + 1.3a^3$	$-a^8 + 2.5a^4 + 1.3a^3$	-1	8

Polynomials may have more than one variable. In such a case, the degree of a term is the sum of the exponents of the variables contained in the term. For example, the term $2x^3y^4z$ has degree 8 because the exponents applied to x, y, and z are 3, 4, and 1, respectively.

The following polynomial has a degree of 12 because the highest degree of its terms is 12.

$$11x^4y^3z \quad - \quad 5x^3y^2z^7 \quad + \quad 2x^2y \quad + \quad 7$$

degree	degree	degree	degree
8	12	3	0

2. Addition of Polynomials

To add or subtract two polynomials, we combine *like* terms. Recall that two terms are *like* **terms** if they each have the same variables and the corresponding variables are raised to the same powers.

Example 1 **Adding Polynomials**

Add the polynomials.

a. $(3t^3 + 2t^2 - 5t) + (t^3 - 6t)$ **b.** $\left(\dfrac{2}{3}w^2 - w + \dfrac{1}{8}\right) + \left(\dfrac{4}{3}w^2 + 8w - \dfrac{1}{4}\right)$

Solution:

a. $(3t^3 + 2t^2 - 5t) + (t^3 - 6t)$

$\quad = 3t^3 + t^3 + 2t^2 + (-5t) + (-6t)$ Group *like* terms.

$\quad = 4t^3 + 2t^2 - 11t$ Add *like* terms.

b. $\left(\frac{2}{3}w^2 - w + \frac{1}{8}\right) + \left(\frac{4}{3}w^2 + 8w - \frac{1}{4}\right)$

$= \frac{2}{3}w^2 + \frac{4}{3}w^2 + (-w) + 8w + \frac{1}{8} + \left(-\frac{1}{4}\right)$ Group *like* terms.

$= \frac{6}{3}w^2 + 7w + \left(\frac{1}{8} - \frac{2}{8}\right)$ Add fractions with common denominators.

$= 2w^2 + 7w - \frac{1}{8}$ Simplify.

Skill Practice Add the polynomials.

1. $(2x^2 + 5x - 2) + (6x^2 - 8x - 8)$

2. $\left(-\frac{1}{4}m^2 - 2m + \frac{1}{3}\right) + \left(\frac{3}{4}m^2 + 7m - \frac{1}{12}\right)$

Example 2   **Adding Polynomials**

Add the polynomials. $(a^2b + 7ab + 6) + (5a^2b - 2ab - 7)$

Solution:

Polynomials can be added vertically. Be sure to line up the *like* terms.

$$
\begin{aligned}
a^2b &+ 7ab + 6 \\
+ \; 5a^2b &- 2ab - 7 \\
\hline
6a^2b &+ 5ab - 1
\end{aligned}
$$
    Add *like* terms.

Skill Practice Add the polynomials.

3. $(-5a^2b - 6ab^2 + 2) + (2a^2b + ab^2 - 3)$

---

3. Subtraction of Polynomials

Subtraction of two polynomials is similar to subtracting real numbers. Add the opposite of the second polynomial to the first polynomial.

The opposite (or additive inverse) of a real number a is $-a$. Similarly, if A is a polynomial, then $-A$ is its opposite.

Example 3   **Finding the Opposite of a Polynomial**

Find the opposite of the polynomials.

a. $5a - 2b - c$ **b.** $-5.5y^4 - 2.4y^3 + 1.1y$

TIP: Notice that the sign of each term is changed when finding the opposite of a polynomial.

Solution:

Expression	Opposite	Simplified Form
a. $5a - 2b - c$	$-(5a - 2b - c)$	$-5a + 2b + c$
b. $-5.5y^4 - 2.4y^3 + 1.1y$	$-(-5.5y^4 - 2.4y^3 + 1.1y)$	$5.5y^4 + 2.4y^3 - 1.1y$

Skill Practice Find the opposite of the polynomials.

4. $-7z + 6w$ **5.** $2p - 3q + r + 1$

Answers

1. $8x^2 - 3x - 10$

2. $\frac{1}{2}m^2 + 5m + \frac{1}{4}$

3. $-3a^2b - 5ab^2 - 1$

4. $7z - 6w$

5. $-2p + 3q - r - 1$

> **Subtraction of Polynomials**
> If A and B are polynomials, then $A - B = A + (-B)$.

Example 4 Subtracting Polynomials

Subtract the polynomials. $(3x^2 + 2x - 5) - (4x^2 - 7x + 2)$

Solution:

$(3x^2 + 2x - 5) - (4x^2 - 7x + 2)$

$= (3x^2 + 2x - 5) + (-4x^2 + 7x - 2)$ Add the opposite of the second polynomial.

$= 3x^2 + (-4x^2) + 2x + 7x + (-5) + (-2)$ Group *like* terms.

$= -x^2 + 9x - 7$ Combine *like* terms.

Skill Practice Subtract the polynomials.

6. $(6a^2 - 2a) - (-3a^2 + 2a + 3)$

Example 5 Subtracting Polynomials

Subtract the polynomials. $(6x^2y - 2xy + 5) - (x^2y - 3)$

Solution:

$(6x^2y - 2xy + 5) - (x^2y - 3)$

Subtraction of polynomials can be performed vertically by vertically aligning *like* terms. Then add the opposite of the second polynomial. "Placeholders" (shown in red) may be used to help line up *like* terms.

$$
\begin{array}{l}
6x^2y - 2xy + 5 \\
\underline{-(x^2y + 0xy - 3)}
\end{array}
\quad\xrightarrow[\text{opposite.}]{\text{Add the}}\quad
\begin{array}{l}
6x^2y - 2xy + 5 \\
\underline{+\ -x^2y - 0xy + 3} \\
5x^2y - 2xy + 8
\end{array}
$$

Skill Practice Subtract the polynomials.

7. $(7p^2q - 6) - (2p^2q + 4pq + 4)$

Example 6 Subtracting Polynomials

Subtract $\dfrac{1}{2}x^4 - \dfrac{3}{4}x^2 + \dfrac{1}{5}$ from $\dfrac{3}{2}x^4 + \dfrac{1}{2}x^2 - 4x$

Solution:

In general, to subtract a from b, we write $b - a$. Therefore, to subtract

$\dfrac{1}{2}x^4 - \dfrac{3}{4}x^2 + \dfrac{1}{5}$ from $\dfrac{3}{2}x^4 + \dfrac{1}{2}x^2 - 4x$

Answers

6. $9a^2 - 4a - 3$
7. $5p^2q - 4pq - 10$

we have

$$\left(\frac{3}{2}x^4 + \frac{1}{2}x^2 - 4x\right) - \left(\frac{1}{2}x^4 - \frac{3}{4}x^2 + \frac{1}{5}\right)$$

$$= \frac{3}{2}x^4 + \frac{1}{2}x^2 - 4x - \frac{1}{2}x^4 + \frac{3}{4}x^2 - \frac{1}{5}$$
Subtract the polynomials by adding the opposite of the second polynomial.

$$= \frac{3}{2}x^4 - \frac{1}{2}x^4 + \frac{1}{2}x^2 + \frac{3}{4}x^2 - 4x - \frac{1}{5}$$
Group *like* terms.

$$= \frac{3}{2}x^4 - \frac{1}{2}x^4 + \frac{2}{4}x^2 + \frac{3}{4}x^2 - 4x - \frac{1}{5}$$
Write *like* terms with a common denominator.

$$= \frac{2}{2}x^4 + \frac{5}{4}x^2 - 4x - \frac{1}{5}$$
Combine *like* terms.

$$= x^4 + \frac{5}{4}x^2 - 4x - \frac{1}{5}$$
Simplify.

Skill Practice

8. Subtract $\frac{1}{2}p^3 + \frac{1}{3}p^2 + \frac{1}{2}p$ from $\frac{1}{3}p^3 + \frac{3}{4}p^2 - p$

4. Polynomial Functions

A **polynomial function** is a function defined by a finite sum of terms of the form ax^n, where a is a real number and n is a whole number. For example, the functions defined here are polynomial functions:

$$f(x) = 3x - 8$$
$$g(x) = 4x^5 - 2x^3 + 5x - 3$$
$$h(x) = -\frac{1}{2}x^4 + \frac{3}{5}x^3 - 4x^2 + \frac{5}{9}x - 1$$
$$k(x) = 7 \quad (7 = 7x^0, \text{ which is of the form } ax^n, \text{ where } n = 0 \text{ is a whole number})$$

The following functions are *not* polynomial functions:

$$m(x) = \frac{1}{x} - 8 \quad \left(\frac{1}{x} = x^{-1}, \text{ the exponent } -1 \text{ is not a whole number}\right)$$
$$q(x) = |x| \quad (|x| \text{ is not of the form } ax^n)$$

Example 7 Evaluating a Polynomial Function

Given $P(x) = x^3 + 2x^2 - x - 2$, find the function values.

a. $P(-3)$ **b.** $P(-1)$ **c.** $P(0)$ **d.** $P(2)$

Answer

8. $-\frac{1}{6}p^3 + \frac{5}{12}p^2 - \frac{3}{2}p$

Solution:

a. $P(x) = x^3 + 2x^2 - x - 2$

$P(-3) = (-3)^3 + 2(-3)^2 - (-3) - 2$

$\qquad = -27 + 2(9) + 3 - 2$

$\qquad = -27 + 18 + 3 - 2$

$\qquad = -8$

b. $P(-1) = (-1)^3 + 2(-1)^2 - (-1) - 2$

$\qquad = -1 + 2(1) + 1 - 2$

$\qquad = -1 + 2 + 1 - 2$

$\qquad = 0$

c. $P(0) = (0)^3 + 2(0)^2 - (0) - 2$

$\qquad = -2$

d. $P(2) = (2)^3 + 2(2)^2 - (2) - 2$

$\qquad = 8 + 2(4) - 2 - 2$

$\qquad = 8 + 8 - 2 - 2$

$\qquad = 12$

The function values can be confirmed from the graph of $y = P(x)$ (Figure 4-1).

Figure 4-1

Skill Practice Given: $P(x) = -2x^3 - 4x + 6$

9. Find $P(0)$.   **10.** Find $P(-2)$.

11. Find $P(-1)$.   **12.** Find $P(2)$.

Example 8   **Using Polynomial Functions in an Application**

The length of a rectangle is 4 m less than 3 times the width. Let x represent the width. Write a polynomial function P that represents the perimeter of the rectangle and simplify the result.

Solution:

Let x represent the width. Then $3x - 4$ is the length. The perimeter of a rectangle is given by $P = 2L + 2W$. Thus,

$$P(x) = 2(3x - 4) + 2(x)$$

$$= 6x - 8 + 2x$$

$$= 8x - 8$$

Skill Practice

13. The longest side of a triangle is 2 ft less than 4 times the shortest side. The middle side is 3 ft more than twice the shortest side. Let x represent the shortest side. Find a polynomial function P that represents the perimeter of the triangle, and simplify the result.

Answers

9. $P(0) = 6$ **10.** $P(-2) = 30$

11. $P(-1) = 12$ **12.** $P(2) = -18$

13. $P(x) = x + (4x - 2) + (2x + 3)$

$\qquad = 7x + 1$

Using a Polynomial Function in an Application

The percent of male smokers in the United States has varied by year since 1990. The function defined by $M(x) = -0.003x^2 - 0.257x + 28.0$ approximates the percent of male smokers, $M(x)$, where x is the number of years since 1990. See Figure 4-2.

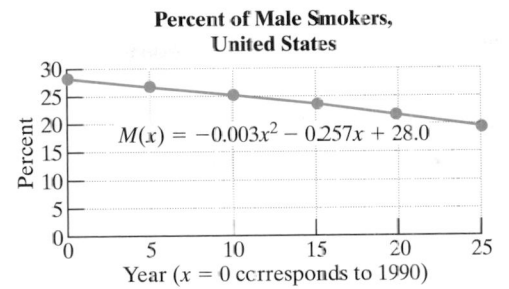

Percent of Male Smokers, United States

$M(x) = -0.003x^2 - 0.257x + 28.0$

Year ($x = 0$ corresponds to 1990)

Source: U.S. National Center for Health Statistics.

Figure 4-2

a. Evaluate $M(5)$ and interpret the meaning in the context of this problem.

b. Determine the percent of male smokers in the year 2008.

Solution:

a. $M(5) = -0.003(5)^2 - 0.257(5) + 28.0$ Substitute $x = 5$ into the function.

≈ 26.6

$M(5) \approx 26.6$ means that in the year 1995, 26.6% of males in the United States smoked.

b. The year 2008 is 18 yr since 1990. Therefore, evaluate $M(18)$.

$M(18) = -0.003(18)^2 - 0.257(18) + 28.0$ Substitute $x = 18$ into the function.

≈ 22.4

$M(18) \approx 22.4$ means that in the year 2008, 22.4% of males in the United States smoked.

Skill Practice The population of Kenya can be modeled by $P(t) = 0.017t^2 + 1.04t + 38$. The value t is the number of years since 2007 and $P(t)$ is the population of Kenya (in millions). (*Source:* Central Intelligence Agency)

14. Evaluate $P(3)$ and interpret its meaning.
15. If this trend continues, predict the population of Kenya for the year 2015.

Answers

14. $M(3) = 41.273$ means that in the year 2010 ($t = 3$ years since 2007), the population of Kenya was 41.273 million.
15. $M(8) = 47.408$; In the year 2015, the population of Kenya will be 47.408 million.

Section 4.2 Practice Exercises

Vocabulary and Key Concepts

1. a. A _____ in the variable, x, is a single term or a sum of terms of the form ax^n, where a is a real number and n is a nonnegative integer.

b. Given the term ax^n, a is called the _____ , and _____ is called the degree of the term.

c. Given the term x, the coefficient of the term is _____ and the degree is _____ .

d. A monomial is a polynomial with exactly _____ term(s).

e. A _____ is a polynomial with exactly two terms.

f. A _____ is a polynomial with exactly three terms.

g. The term with the highest degree is called the _____ term and its coefficient is called the _____ _____ .

h. The degree of a polynomial is the _____ degree of all of its terms.

i. The degree of a nonzero constant such as 7 is _____.

j. If a term of a polynomial has more than one variable, then the degree of the term is the sum of the _____ of the variables contained in the term.

k. A _____ function is a function defined by a finite sum of terms of the form ax^n, where a is a real number and n is a whole number.

Review Exercises

For Exercises 2–6, simplify the expression.

2. $\left(\dfrac{6tv^2}{15t^3}\right)^{-1}$

3. $(2ac^{-2})(5a^{-1}c^4)$

4. $\left(\dfrac{1}{4}\right)^{-2} - \left(\dfrac{2}{3}\right)^0$

5. $(3.4 \times 10^5)(5.0 \times 10^{-2})$

6. $\dfrac{6.2 \times 10^3}{3.1 \times 10^5}$

Concept 1: Polynomials: Basic Definitions

For Exercises 7–12, write the polynomial in descending order. Then identify the leading coefficient and the degree.

7. $a^2 - 6a^3 - a$

8. $2b - b^4 + 5b^2$

9. $6x^2 - x + 3x^4 - 1$

10. $8 - 4y + y^5 - y^2$

11. $100 - t^2$

12. $-51 + s^2$

For Exercises 13–18, write a polynomial in one variable that is described by the following. (Answers may vary.)

13. A monomial of degree 5

14. A monomial of degree 4

15. A trinomial of degree 2

16. A trinomial of degree 3

17. A binomial of degree 4

18. A binomial of degree 2

Concept 2: Addition of Polynomials

For Exercises 19–30, add the polynomials and simplify. **(See Examples 1–2.)**

19. $(-4m^2 + 4m) + (5m^2 + 6m)$

20. $(3n^3 + 5n) + (2n^3 - 2n)$

21. $(3x^4 - x^3 - x^2) + (3x^3 - 7x^2 + 2x)$

22. $(6x^3 - 2x^2 - 12) + (x^2 + 3x + 9)$

23. $\left(\dfrac{1}{2}w^3 + \dfrac{2}{9}w^2 - 1.8w\right) + \left(\dfrac{3}{2}w^3 - \dfrac{1}{9}w^2 + 2.7w\right)$

24. $\left(2.9t^4 - \dfrac{7}{8}t + \dfrac{5}{3}\right) + \left(-8.1t^4 - \dfrac{1}{8}t - \dfrac{1}{3}\right)$

25. Add $(9x^2y - 5xy + 1)$ to $(8x^2y + xy - 15)$.

26. Add $(-x^3y^2 + 5xy)$ to $(10x^3y^2 + x^2y - 10)$.

27. Add $(-7a + 6a^2 + 1)$ to $(-8 - 4a - 2a^2)$.

28. Add $(1 - 12p + 8p^3)$ to $(6p^2 + p^3 - 14)$.

29.
$$\begin{array}{r} 12x^3 \qquad\ + 6x - 8 \\ +\ (-3x^3 - 5x^2 - 4x\quad) \\ \hline \end{array}$$

30.
$$\begin{array}{r} -8y^4 - 8y^3 - 6y^2 \qquad\ - 9 \\ +\ (4y^4 + 5y^3 \qquad\ - 10y - 3) \\ \hline \end{array}$$

Concept 3: Subtraction of Polynomials

For Exercises 31–36, write the opposite of the given polynomial. **(See Example 3.)**

31. $-30y^3$

32. $-2x^2$

33. $4p^3 + 2p - 12$

34. $8t^2 - 4t - 3$

35. $-11ab^2 + a^2b$

36. $-23rs - 4r + 9s$

For Exercises 37–46, subtract the polynomials and simplify. **(See Examples 4–5.)**

37. $(13z^5 - z^2) - (7z^5 + 5z^2)$

38. $(8w^4 + 3w^2) - (12w^4 - w^2)$

39. $(-3x^3 + 3x^2 - x + 6) - (1 - x - x^2 - x^3)$

40. $(-8x^3 + 6x + 7) - (-4 - 2x - 5x^3)$

41. $(-3xy^3 + 3x^2y - x + 6) - (-xy^3 - xy - x + 1)$

42. $(-8x^2y^2 + 6xy^2 + 7xy) - (5xy^2 - 2xy - 4)$

43. $\quad 4t^3 - 6t^2 \qquad - 18$
$\quad - (3t^3 + 7t^2 + 9t - \;\; 5)$

44. $\quad 5w^3 - 9w^2 + \;\; 6w + 13$
$\quad - (7w^3 \qquad\quad - 10w - \;\; 8)$

45. $\left(\dfrac{1}{5}a^2 - \dfrac{1}{2}ab + \dfrac{1}{10}b^2 + 3\right) - \left(-\dfrac{3}{10}a^2 + \dfrac{2}{5}ab - \dfrac{1}{2}b^2 - 5\right)$

46. $\left(\dfrac{4}{7}a^2 - \dfrac{1}{7}ab + \dfrac{1}{14}b^2 - 7\right) - \left(\dfrac{1}{2}a^2 - \dfrac{2}{7}ab - \dfrac{9}{14}b^2 + 1\right)$

47. Subtract $(9x^2 - 5x + 1)$ from $(8x^2 + x - 15)$. **(See Example 6.)**

48. Subtract $(-x^3 + 5x)$ from $(10x^3 + x^2 - 10)$.

49. Find the difference of $(3x^5 - 2x^3 + 4)$ and $(x^4 + 2x^5 - 7)$.

50. Find the difference of $(7x^{10} - 2x^4 - 3x)$ and $(-4x^2 - 5x^4 + x + 5)$.

Mixed Exercises

For Exercises 51–74, add or subtract as indicated. Write the answers in descending order, if possible.

51. $(8y^2 - 4y^3) - (3y^2 - 8y^3)$

52. $(-9y^2 - 8) - (4y^2 + 3)$

53. $(-2r - 6r^4) + (-r^4 - 9r)$

54. $(-8s^9 + 7s^2) + (7s^9 - s^2)$

55. $(5xy + 13x^2 + 3y) - (4x^2 - 8y)$

56. $(6p^2q - 2q) - (-2p^2q + 13)$

57. $(11ab - 23b^2) + (7ab - 19b^2)$

58. $(-4x^2y + 9) + (8x^2y - 12)$

59. $[2p - (3p + 5)] + (4p - 6) + 2$

60. $-(q - 2) - [4 - (2q - 3) + 5]$

61. $5 - [2m^2 - (4m^2 + 1)]$

62. $[4n^3 - (n^3 + 4)] + 3n^3$

63. $(6x^3 - 5) - (-3x^3 + 2x) - (2x^3 - 6x)$

64. $(9p^4 - 2) + (7p^4 + 1) - (8p^4 - 10)$

65. $(-ab + 5a^2b) - [7ab^2 - 2ab - (7a^2b + 2ab^2)]$

66. $(m^3n^2 + 4m^2n) - [-5m^3n^2 - 4mn - (7m^2n - 6mn)]$

67. $(8x^3 - x^2 + 3) - [5x^2 + x - (4x^3 + x - 2)]$

68. $(y^2 + 6y - 6) - [(2y^3 - 4y) - (3y^2 + y + 1)]$

69. $\quad 12a^2b - 4ab^2 - \;\; ab$
$\quad -(4a^2b + \;\; ab^2 - 5ab)$

70. $\quad 2x^2 - \;\; 7xy + 3y^2$
$\quad - (9x^2 - 10xy - \;\; y^2)$

71. $\quad -5x^4 \qquad - 11x^2 \qquad + 6$
$\quad - (-5x^4 + 3x^3 + \;\; 5x^2 - 10x + 5)$

72. $\quad 9z^4 \qquad\quad + 2z^2 \qquad + 11$
$\quad - (9z^4 - 4z^3 + 8z^2 - 9z - \;\; 4)$

73. $\quad -2.2p^5 - 9.1p^4 \qquad\quad + 5.3p^2 - 7.9p$
$\quad + \qquad\qquad - 6.4p^4 - 8.5p^3 - 10.3p^2$

74. $\quad 5.5w^4 \qquad\quad + 4.6w^2 - 9.3w - 8.3$
$\quad + 0.4w^4 - 7.3w^3 \qquad\qquad - 5.8w + 4.6$

For Exercises 75–76, find the perimeter.

75.

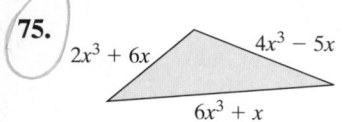

$2x^3 + 6x$ $4x^3 - 5x$

$6x^3 + x$

76.

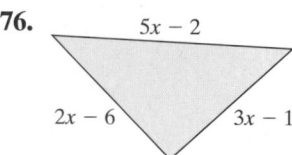

$5x - 2$

$2x - 6$ $3x - 1$

Concept 4: Polynomial Functions

For Exercises 77–84, determine whether the given function is a polynomial function. If it is a polynomial function, state the degree. If not, state the reason why.

77. $h(x) = \dfrac{2}{3}x^2 - 5$

78. $k(x) = -7x^4 - 0.3x + x^3$

79. $p(x) = 8x^3 + 2x^2 - \dfrac{3}{x}$

80. $q(x) = x^2 - 4x^{-3}$

81. $g(x) = -7$

82. $g(x) = 4x$

83. $M(x) = |x| + 5x$

84. $N(x) = x^2 + |x|$

85. Given $P(x) = -x^4 + 2x - 5$, find the function values. **(See Example 7.)**

    a. $P(2)$ **b.** $P(-1)$ **c.** $P(0)$ **d.** $P(1)$

86. Given $N(x) = -x^2 + 5x$, find the function values.

    a. $N(1)$ **b.** $N(-1)$ **c.** $N(2)$ **d.** $N(0)$

87. Given $H(x) = \frac{1}{2}x^3 - x + \frac{1}{4}$, find the function values.

    a. $H(0)$ **b.** $H(2)$ **c.** $H(-2)$ **d.** $H(-1)$

88. Given $K(x) = \frac{2}{3}x^2 + \frac{1}{9}$, find the function values.

    a. $K(0)$ **b.** $K(3)$ **c.** $K(-3)$ **d.** $K(-1)$

89. A rectangular garden is designed to be 3 ft longer than it is wide. Let x represent the width of the garden. Find a function P that represents the perimeter in terms of x. **(See Example 8.)**

90. Pauline measures a rectangular conference room and finds that the length is 4 yd greater than twice the width. Let x represent the width. Find a function P that represents the perimeter in terms of x.

91. The cost in dollars of producing x calendars is $C(x) = 5.40x + 99$. The revenue for selling x calendars is $R(x) = 12x$. To calculate profit, subtract the cost from the revenue.

    a. Write and simplify a function P that represents profit in terms of x.

    b. Find the profit of producing and selling 50 calendars.

92. The cost in dollars of producing x lawn chairs is $C(x) = 4.5x + 10.1$. The revenue for selling x chairs is $R(x) = 12.99x$. To calculate profit, subtract the cost from the revenue.

    a. Write and simplify a function P that represents profit in terms of x.

    b. Find the profit of producing and selling 100 lawn chairs.

93. The function defined by $D(x) = 5.2x^2 + 40.4x + 1636$ approximates the average yearly dormitory charge for 4-yr universities x years since 1990. $D(x)$ is the cost in dollars, and x represents the number of years since 1990. **(See Example 9.)**

a. Evaluate $D(0)$ and $D(18)$ and interpret their meaning in the context of this problem.

b. If this trend continues, what will the annual dormitory charge be in the year 2015?

Yearly Dormitory Cost, Four-Year Universities

$D(x) = 5.2x^2 + 40.4x + 1636$

Source: U.S. National Center for Education Statistics

94. The population of bacteria in a culture can be modeled by $P(t) = -0.01t^3 + 12.96t + 10$, where t is the time in hours after the culture was started and $P(t)$ is the population in thousands.

a. Evaluate $P(0)$ and $P(14)$ and interpret their meaning in the context of this problem.

b. Predict the population of bacteria 24 hr after the culture was started.

Population of Bacteria Versus Time

$P(t) = -0.01t^3 + 12.96t + 10$

Time (hr)

95. The number of women, W, to be paid child support in the United States can be approximated by

$$W(t) = 143t + 6580$$

where t is the number of years since 2000, and $W(t)$ is the yearly total measured in thousands. (*Source:* U.S. Bureau of the Census)

a. Evaluate $W(0)$, $W(5)$, and $W(10)$.

b. Interpret the meaning of the function value $W(10)$.

96. The total yearly amount of child support due (in billions of dollars) in the United States can be approximated by

$$D(t) = 0.925t + 4.625$$

where t is the number of years since 2000, and $D(t)$ is the amount due (in billions of dollars).

a. Evaluate $D(0)$, $D(4)$, and $D(8)$.

b. Interpret the meaning of the function value of $D(8)$.

Expanding Your Skills

97. A toy rocket is shot from ground level at an angle of 60° from the horizontal. See the figure. The x- and y-positions of the rocket (measured in feet) vary with time t according to

$$x(t) = 25t$$
$$y(t) = -16t^2 + 43.3t$$

a. Evaluate $x(0)$ and $y(0)$, and write the values as an ordered pair. Interpret the meaning of these function values in the context of this problem. Match the ordered pair with a point on the graph.

b. Evaluate $x(1)$ and $y(1)$ and write the values as an ordered pair. Interpret the meaning of these function values in the context of this problem. Match the ordered pair with a point on the graph.

c. Evaluate $x(2)$ and $y(2)$, and write the values as an ordered pair. Match the ordered pair with a point on the graph.

Path of Rocket

Horizontal Distance (ft)

Section 4.3 Multiplication of Polynomials

Concepts

1. Multiplying Polynomials
2. Special Case Products: Difference of Squares and Perfect Square Trinomials
3. Translations Involving Polynomials
4. Applications Involving a Product of Polynomials

1. Multiplying Polynomials

The properties of exponents covered in Section 4.1 can be used to simplify many algebraic expressions including the multiplication of monomials. To multiply monomials, first use the associative and commutative properties of multiplication to group coefficients and like bases. Then simplify the result by using the properties of exponents.

Example 1 Multiplying Monomials

Multiply the monomials. $(3x^2y^7)(5x^3y)$

Solution:

$(3x^2y^7)(5x^3y)$

$= (3 \cdot 5)(x^2 \cdot x^3)(y^7 \cdot y)$ Group coefficients and like bases.

$= 15x^5y^8$ Add exponents and simplify.

Skill Practice Multiply the monomials.

1. $(-8r^3s)(-4r^4s^4)$

The distributive property is used to multiply polynomials: $a(b + c) = ab + ac$.

Example 2 Multiplying a Polynomial by a Monomial

Multiply the polynomials.

a. $5y^3(2y^2 - 7y + 6)$ **b.** $-4a^3b^7c\left(2ab^2c^4 - \frac{1}{2}a^5b\right)$

Solution:

a. $5y^3(2y^2 - 7y + 6)$

$= (5y^3)(2y^2) + (5y^3)(-7y) + (5y^3)(6)$ Apply the distributive property.

$= 10y^5 - 35y^4 + 30y^3$ Simplify each term.

b. $-4a^3b^7c\left(2ab^2c^4 - \frac{1}{2}a^5b\right)$

$= (-4a^3b^7c)(2ab^2c^4) + (-4a^3b^7c)\left(-\frac{1}{2}a^5b\right)$ Apply the distributive property.

$= -8a^4b^9c^5 + 2a^8b^8c$ Simplify each term.

Skill Practice Multiply the polynomials.

2. $-6b^2(2b^2 + 3b - 8)$ **3.** $8st^3\left(\frac{1}{2}s^2 - \frac{1}{4}st\right)$

Answers

1. $32r^7s^5$
2. $-12b^4 - 18b^3 + 48b^2$
3. $4s^3t^3 - 2s^2t^4$

Thus far, we have illustrated polynomial multiplication involving monomials. Next, the distributive property will be used to multiply polynomials with more than one term. For example:

$$(x + 3)(x + 5) = (x + 3)x + (x + 3)5 \qquad \text{Apply the distributive property.}$$

$$= (x + 3)x + (x + 3)5 \qquad \text{Apply the distributive property again.}$$

$$= x^2 + 3x + 5x + 15$$

$$= x^2 + 8x + 15 \qquad \text{Combine } like \text{ terms.}$$

Note: Using the distributive property results in multiplying each term of the first polynomial by each term of the second polynomial:

$$(x + 3)(x + 5) = x^2 + 5x + 3x + 15$$

$$= x^2 + 8x + 15$$

Example 3 Multiplying Polynomials

Multiply the polynomials. $(2x^2 + 4)(3x^2 - x + 5)$

Solution:

$$(2x^2 + 4)(3x^2 - x + 5) \qquad \text{Multiply each term in the first polynomial by each term in the second.}$$

$$= (2x^2)(3x^2) + (2x^2)(-x) + (2x^2)(5) \qquad \text{Apply the distributive}$$
$$+ (4)(3x^2) + (4)(-x) + (4)(5) \qquad \text{property.}$$

$$= 6x^4 - 2x^3 + 10x^2 + 12x^2 - 4x + 20 \qquad \text{Simplify each term.}$$

$$= 6x^4 - 2x^3 + 22x^2 - 4x + 20 \qquad \text{Combine } like \text{ terms.}$$

TIP: Multiplication of polynomials can be performed vertically by a process similar to column multiplication of real numbers.

$$(2x^2 + 4)(3x^2 - x + 5) \longrightarrow \quad 3x^2 - x + 5$$
$$\underline{\times \quad 2x^2 \qquad + \quad 4}$$
$$12x^2 - 4x + 20$$
$$\underline{6x^4 - 2x^3 + 10x^2}$$
$$6x^4 - 2x^3 + 22x^2 - 4x + 20$$

Note: When multiplying by the column method, it is important to align *like* terms vertically before adding terms.

Skill Practice Multiply the polynomials.

4. $(2y - 1)(3y^2 - 2y - 1)$

Answer

4. $6y^3 - 7y^2 + 1$

Example 4   **Multiplying Polynomials**

Multiply the polynomials. $(3y + 2)(7y - 6)$

Solution:

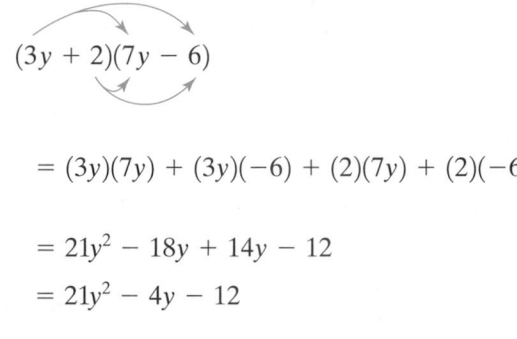

$(3y + 2)(7y - 6)$ Multiply each term in the
                                                    first polynomial by each
                                                    term in the second.

$= (3y)(7y) + (3y)(-6) + (2)(7y) + (2)(-6)$ Apply the distributive
                                                    property.

$= 21y^2 - 18y + 14y - 12$ Simplify each term.

$= 21y^2 - 4y - 12$ Combine *like* terms.

TIP: The acronym, FOIL (first outer inner last) can be used as a memory device to multiply the two binomials.

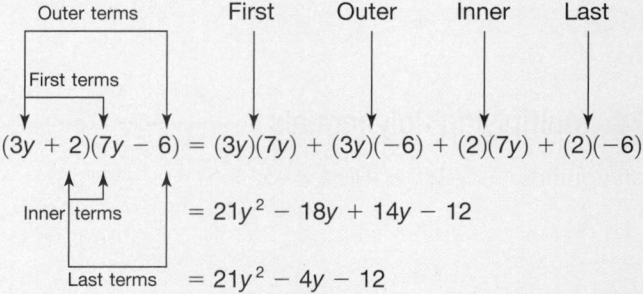

Note: It is important to realize that the acronym FOIL may only be used when finding the product of two *binomials*.

Skill Practice Multiply the polynomials.

5. $(4t + 5)(2t + 3)$

---

2. Special Case Products: Difference of Squares and Perfect Square Trinomials

In some cases, the product of two binomials takes on a special pattern.

I.   The first special case occurs when multiplying the sum and difference of the same two terms. For example:

$(2x + 3)(2x - 3)$

$= 4x^2 - 6x + 6x - 9$

$= 4x^2 - 9$

Notice that the "middle terms" are opposites. This leaves only the difference between the square of the first term and the square of the second term. For this reason, the product is called a *difference of squares*.

Note: The binomials $2x + 3$ and $2x - 3$ are called **conjugates**. In one expression, $2x$ and 3 are added, and in the other, $2x$ and 3 are subtracted. In general, $a + b$ and $a - b$ are conjugates of each other.

Answer

5. $8t^2 + 22t + 15$

II. The second special case involves the square of a binomial. For example:

$(3x + 7)^2$

$= (3x + 7)(3x + 7)$

$= 9x^2 + 21x + 21x + 49$

$= 9x^2 + 42x + 49$

$= (3x)^2 + 2(3x)(7) + (7)^2$

When squaring a binomial, the product will be a trinomial called a *perfect square trinomial*. The first and third terms are formed by squaring the terms of the binomial. The middle term is twice the product of the terms in the binomial.

Note: The expression $(3x - 7)^2$ also results in a perfect square trinomial, but the middle term is negative.

$$(3x - 7)(3x - 7) = 9x^2 - 21x - 21x + 49 = 9x^2 - 42x + 49$$

The following summarizes these special case products.

Special Case Product Formulas

1. $(a + b)(a - b) = a^2 - b^2$ The product is called a **difference of squares**.

2. $(a + b)^2 = a^2 + 2ab + b^2$

$(a - b)^2 = a^2 - 2ab + b^2$ The product is called a **perfect square trinomial**.

It is advantageous for you to become familiar with these special case products because they will be presented again when we factor polynomials.

Example 5   Using Special Products

Use the special product formulas to multiply the polynomials.

a. $(6c - 7d)(6c + 7d)$ **b.** $(5x - 2)^2$ **c.** $(4x^3 + 3y^2)^2$

Solution:

a. $(6c - 7d)(6c + 7d)$ $a = 6c, b = 7d$

$= (6c)^2 - (7d)^2$ Apply the formula $a^2 - b^2$.

$= 36c^2 - 49d^2$ Simplify each term.

b. $(5x - 2)^2$ $a = 5x, b = 2$

$= (5x)^2 - 2(5x)(2) + (2)^2$ Apply the formula $a^2 - 2ab + b^2$.

$= 25x^2 - 20x + 4$ Simplify each term.

c. $(4x^3 + 3y^2)^2$ $a = 4x^3, b = 3y^2$

$= (4x^3)^2 + 2(4x^3)(3y^2) + (3y^2)^2$ Apply the formula $a^2 + 2ab + b^2$.

$= 16x^6 + 24x^3y^2 + 9y^4$ Simplify each term.

Skill Practice Multiply the polynomials.

6. $(5x - 4y)(5x + 4y)$ **7.** $(c - 3)^2$ **8.** $(7s^2 + 2t)^2$

Answers

6. $25x^2 - 16y^2$

7. $c^2 - 6c + 9$

8. $49s^4 + 28s^2t + 4t^2$

The special case products can be used to simplify more complicated algebraic expressions.

Example 6 Using Special Products

Multiply. $[x + (y + z)][x - (y + z)]$

Solution:

$[x + (y + z)][x - (y + z)]$	This product is in the form $(a + b)(a - b)$, where $a = x$ and $b = (y + z)$.
$= (x)^2 - (y + z)^2$	Apply the formula $a^2 - b^2$.
$= (x)^2 - (y^2 + 2yz + z^2)$	Expand $(y + z)^2$ by using the special case product formula.
$= x^2 - y^2 - 2yz - z^2$	Apply the distributive property.

Skill Practice Multiply.

9. $[a + (c + 5)][a - (c + 5)]$

Example 7 Using Special Products

Multiply. $(x + y)^3$

Solution:

$(x + y)^3$

$= (x + y)^2(x + y)$	Rewrite as the square of a binomial and another factor.
$= (x^2 + 2xy + y^2)(x + y)$	Expand $(x + y)^2$ by using the special case product formula.
$= (x^2)(x) + (x^2)(y) + (2xy)(x)$ $+ (2xy)(y) + (y^2)(x) + (y^2)(y)$	Apply the distributive property.
$= x^3 + x^2y + 2x^2y + 2xy^2 + xy^2 + y^3$	Simplify each term.
$= x^3 + 3x^2y + 3xy^2 + y^3$	Combine *like* terms.

Skill Practice Multiply.

10. $(t + 2)^3$

3. Translations Involving Polynomials

Example 8 Translating Between English Form and Algebraic Form

Complete the table.

English Form	Algebraic Form
The square of the sum of x and y	
	$x^2 + y^2$
The square of the product of 3 and x	

Answers
9. $a^2 - c^2 - 10c - 25$
10. $t^3 + 6t^2 + 12t + 8$

Solution:

English Form	Algebraic Form	Notes
The square of the sum of x and y	$(x + y)^2$	The *sum* is squared, not the individual terms.
The sum of the squares of x and y	$x^2 + y^2$	The individual terms x and y are squared first. Then the sum is taken.
The square of the product of 3 and x	$(3x)^2$	The product of 3 and x is taken. Then the result is squared.

Skill Practice Translate to algebraic form:

11. The square of the difference of a and b

12. The difference of the square of a and the square of b

13. Translate to English form: $a - b^2$

4. Applications Involving a Product of Polynomials

Example 9   Applying a Product of Polynomials

A box is created from a sheet of cardboard 20 in. on a side by cutting a square from each corner and folding up the sides (Figures 4-3 and 4-4). Let x represent the length of the sides of the squares removed from each corner.

a. Write a function V that represents the volume of the box in terms of x.

b. Find the volume if a 4-in. square is removed.

Figure 4-3

Figure 4-4

Solution:

a. The volume of a rectangular box is given by the formula $V = lwh$. The length and width can both be expressed as $20 - 2x$. The height of the box is x. The volume is given by

$$V = l \cdot w \cdot h$$
$$V(x) = (20 - 2x)(20 - 2x)x$$
$$= (20 - 2x)^2 x$$
$$= (400 - 80x + 4x^2)x$$
$$= 400x - 80x^2 + 4x^3$$
$$= 4x^3 - 80x^2 + 400x$$

Answers

11. $(a - b)^2$ **12.** $a^2 - b^2$

13. The difference of a and the square of b

b. If a 4-in. square is removed from the corners of the box, we have $x = 4$. Thus, the volume is

$$V(x) = 4x^3 - 80x^2 + 400x$$
$$V(4) = 4(4)^3 - 80(4)^2 + 400(4)$$
$$= 4(64) - 80(16) + 400(4)$$
$$= 256 - 1280 + 1600$$
$$= 576$$

The volume is 576 in.^{3}

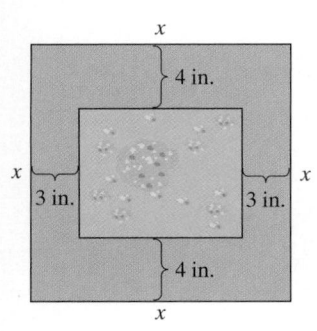

Answers

14. $A(x) = (x - 8)(x - 6)$;
     $A(x) = x^2 - 14x + 48$
15. 24 in.^{2}

Skill Practice A rectangular photograph is mounted on a square piece of cardboard whose sides have length x. The border that surrounds the photo is 3 in. on each side and 4 in. on both top and bottom.

14. Write a function A for the area of the photograph in terms of x.

15. Determine the area of the photograph if x is 12 in.

Section 4.3 Practice Exercises

Vocabulary and Key Concepts

1. a. To multiply $2(4x - 5)$, apply the _____ property.

b. The conjugate of $4x + 7$ is _____.

c. When two conjugates are multiplied the resulting binomial is a difference of _____. This is given by the formula $(a + b)(a - b) = $ _____.

d. When a binomial is squared, the resulting trinomial is a _____ square trinomial. This is given by the formula $(a + b)^2 = $ _____.

Review Exercises

2. Simplify. $(-4x^2y - 2xy + 3xy^2) - (2x^2y - 4xy^2) + (6x^2y + 5xy)$

3. Simplify. $(-2 - 3x) - [5 - (6x^2 + 4x + 1)]$

4. Given $f(x) = 4x^3 - 5$, find the function values.

   a. $f(3)$ **b.** $f(0)$ **c.** $f(-2)$

5. Given $g(x) = x^4 - x^2 - 3$, find the function values.

   a. $g(-1)$ **b.** $g(2)$ **c.** $g(0)$

6. Write the distributive property of multiplication over addition. Give an example of the distributive property. (Answers may vary.)

Concept 1: Multiplying Polynomials

For Exercises 7–40, multiply the polynomials. **(See Examples 1–4.)**

7. $(7x^4y)(-6xy^5)$

8. $(-4a^3b^7)(-2ab^3)$

9. $(2.2a^6b^4c^7)(5ab^4c^3)$

10. $(8.5c^4d^5e)(6cd^2e)$

11. $\frac{1}{5}(2a - 3)$

12. $\frac{1}{3}(6b + 4)$

13. $2m^3n^2(m^2n^3 - 3mn^2 + 4n)$

14. $3p^2q(p^3q^3 - pq^2 - 4p)$

15. $6xy^2\left(\frac{1}{2}x - \frac{2}{3}xy\right)$

16. $12ab\left(\dfrac{5}{6}a + \dfrac{1}{4}ab^2\right)$ **17.** $(x + y)(x - 2y)$ **18.** $(3a + 5)(a - 2)$

19. $(6x - 1)(5 + 2x)$ **20.** $(7 + 3x)(x - 8)$ **21.** $(y^2 - 12)(2y^2 + 3)$

22. $(4p^2 - 1)(2p^2 + 5)$ **23.** $(5s + 3t)(5s - 2t)$ **24.** $(4a + 3b)(4a - b)$

25. $(n^2 + 10)(5n + 3)$ **26.** $(m^2 + 8)(3m + 7)$ **27.** $(1.3a - 4b)(2.5a + 7b)$

28. $(2.1x - 3.5y)(4.7x + 2y)$ **29.** $(2x + y)(3x^2 + 2xy + y^2)$ **30.** $(h - 5k)(h^2 - 2hk + 3k^2)$

31. $(x - 7)(x^2 + 7x + 49)$ **32.** $(x + 3)(x^2 - 3x + 9)$ **33.** $(4a - b)(a^3 - 4a^2b + ab^2 - b^3)$

34. $(3m + 2n)(m^3 + 2m^2n - mn^2 + 2n^3)$ **35.** $\left(\dfrac{1}{2}a - 2b + c\right)(a + 6b - c)$

36. $\left(\dfrac{1}{2}a^2 - 2ab + b^2\right)(2a + b)$ **37.** $(-x^2 + 2x + 1)(3x - 5)$ **38.** $(x + y - 2z)(5x - y + z)$

39. $\left(\dfrac{1}{5}y - 10\right)\left(\dfrac{1}{2}y - 15\right)$ **40.** $\left(\dfrac{2}{3}x + 6\right)\left(\dfrac{1}{2}x - 9\right)$

Concept 2: Special Case Products: Difference of Squares and Perfect Square Trinomials

For Exercises 41–60, multiply by using the special case products. **(See Example 5.)**

41. $(a - 8)(a + 8)$ **42.** $(b + 2)(b - 2)$ **43.** $(3p + 1)(3p - 1)$ **44.** $(5q - 3)(5q + 3)$

45. $\left(x - \dfrac{1}{3}\right)\left(x + \dfrac{1}{3}\right)$ **46.** $\left(\dfrac{1}{2}x + \dfrac{1}{3}\right)\left(\dfrac{1}{2}x - \dfrac{1}{3}\right)$ **47.** $(3h - k)(3h + k)$ **48.** $(x - 7y)(x + 7y)$

49. $(3h - k)^2$ **50.** $(x - 7y)^2$ **51.** $(t - 7)^2$ **52.** $(w + 9)^2$

53. $(u + 3v)^2$ **54.** $(a - 4b)^2$ **55.** $\left(h + \dfrac{1}{6}k\right)^2$ **56.** $\left(\dfrac{2}{5}x + 1\right)^2$

57. $(2z^2 - w^3)(2z^2 + w^3)$ **58.** $(a^4 - 2b^3)(a^4 + 2b^3)$ **59.** $(5x^2 - 3y)^2$ **60.** $(4p^3 - 2m)^2$

For Exercises 61–62, the product of two binomials is shown. Determine if the binomials are conjugates.

61. a. $(-5x + 4)(5x + 4)$ **62. a.** $(-3 - 7x)(3 + 7x)$

    b. $(-5x + 4)(5x - 4)$ **b.** $(-3 + 7x)(3 + 7x)$

63. Multiply the expressions. Explain their similarities.   **64.** Multiply the expressions. Explain their similarities.

    a. $(A - B)(A + B)$ **a.** $(A + B)(A - B)$

    b. $[(x + y) - B][(x + y) + B]$ **b.** $[A - (3h + k)][A - (3h + k)]$

For Exercises 65–70, multiply the expressions. **(See Example 6.)**

65. $[(w + v) - 2][(w + v) + 2]$ **66.** $[(x + y) - 6][(x + y) + 6]$ **67.** $[2 - (x + y)][2 + (x + y)]$

68. $[a - (b + 1)][a + (b + 1)]$ **69.** $[(3a - 4) + b][(3a - 4) - b]$ **70.** $[(5p - 7) - q][(5p - 7) + q]$

71. Explain how to multiply $(x + y)^3$.

72. Explain how to multiply $(a - b)^3$.

For Exercises 73–76, multiply the expressions. **(See Example 7.)**

73. $(2x + y)^3$ **74.** $(x - 5y)^3$ **75.** $(4a - b)^3$ **76.** $(3a + 4b)^3$

77. Explain how you would multiply the binomials

$$(x - 2)(x + 6)(2x + 1)$$

78. Explain how you would multiply the binomials

$$(a + b)(a - b)(2a + b)(2a - b)$$

For Exercises 79–86, simplify the expressions.

79. $2a^2(a + 5)(3a + 1)$ **80.** $-5y(2y - 3)(y + 3)$ **81.** $(x + 3)(x - 3)(x + 5)$ **82.** $(t + 2)(t - 3)(t + 1)$

83. $-3(2x + 7) - (4x - 1)^2$ **84.** $(p + 10)^2 - 4(p + 6)^2$ **85.** $(y + 1)^2 - (2y + 3)^2$ **86.** $(b - 3)^2 - (3b - 1)^2$

Concept 3: Translations Involving Polynomials

For Exercises 87–90, translate from English form to algebraic form. **(See Example 8.)**

87. The square of the sum of r and t

88. The square of a plus the cube of b

89. The difference of x squared and y cubed

90. The square of the product of 3 and a

For Exercises 91–94, translate from algebraic form to English form. **(See Example 8.)**

91. $p^3 + q^2$ **92.** $a^3 - b^3$ **93.** xy^2 **94.** $(c + d)^3$

Concept 4: Applications Involving a Product of Polynomials

95. A rectangular garden has a walk around it of width x. The garden is 20 ft by 15 ft. Write a function representing the combined area $A(x)$ of the garden and walk. Simplify the result.

96. An 8-in. by 10-in. photograph is in a frame of width x. Write a function that represents the area $A(x)$ of the frame alone. Simplify the result.

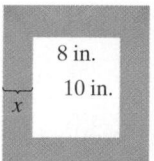

97. A box is created from a square piece of cardboard 8 in. on a side by cutting a square from each corner and folding up the sides. Let x represent the length of the sides of the squares removed from each corner. **(See Example 9.)**

a. Write a function representing the volume of the box.

b. Find the volume if 1-in. squares are removed from the corners.

98. A box is created from a rectangular piece of metal with dimensions 12 in. by 9 in. by removing a square from each corner of the metal sheet and folding up the sides. Let x represent the length of the sides of the squares removed from each corner.

    a. Write a function representing the volume of the box.

    b. Find the volume if 2-in. squares are removed from the corners.

For Exercises 99–104, write an expression for the area and simplify your answer.

99. Square

$x - 2$

100. Square

$x + 3$

101. Rectangle

$x - 2$

$x + 2$

102. Rectangle

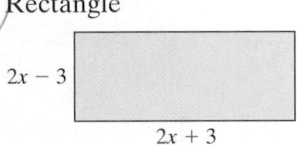

$2x - 3$

$2x + 3$

103. Triangle

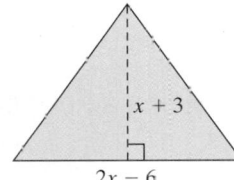

$x + 3$

$2x - 6$

104. Triangle

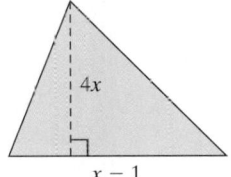

$4x$

$x - 1$

For Exercises 105–106, write an expression for the volume and simplify your answer.

105.

$3x + 10$

$3x$

x

106.

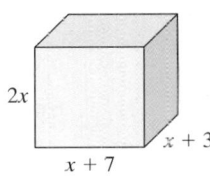

$2x$

$x + 7$

$x + 3$

Expanding Your Skills

For Exercises 107–108, simplify completely.

107. $\dfrac{[(x + h)^2 - 3(x + h) - 5] - (x^2 - 3x - 5)}{h}$

108. $\dfrac{[(x + h)^2 - 4(x + h) + 2] - (x^2 - 4x + 2)}{h}$

109. Explain how to multiply $(x + 2)^4$.

110. Explain how to multiply $(y - 3)^4$.

111. $(2x - 3)$ multiplied by what binomial will result in the trinomial $10x^2 - 27x + 18$? Check your answer by multiplying the binomials.

112. $(4x + 1)$ multiplied by what binomial will result in the trinomial $12x^2 - 5x - 2$? Check your answer by multiplying the binomials.

113. $(4y + 3)$ multiplied by what binomial will result in the trinomial $8y^2 + 2y - 3$? Check your answer by multiplying the binomials.

114. $(3y - 2)$ multiplied by what binomial will result in the trinomial $3y^2 - 17y + 10$? Check your answer by multiplying the binomials.

Concepts

1. Division by a Monomial
2. Long Division
3. Synthetic Division

Section 4.4 Division of Polynomials

1. Division by a Monomial

Division of polynomials is presented in this section as two separate cases. The first case illustrates division by a monomial divisor. The second case illustrates division by a polynomial with two or more terms.

To divide a polynomial by a monomial, divide each individual term in the polynomial by the divisor and simplify the result.

> **Dividing a Polynomial by a Monomial**
> If a, b, and c are polynomials such that $c \neq 0$, then
> $$\frac{a+b}{c} = \frac{a}{c} + \frac{b}{c} \qquad \text{Similarly,} \qquad \frac{a-b}{c} = \frac{a}{c} - \frac{b}{c}$$

Example 1 Dividing a Polynomial by a Monomial

Divide the polynomials.

a. $\dfrac{3x^4 - 6x^3 + 9x}{3x}$

b. $(10c^3d - 15c^2d^2 + 2cd^3) \div (-5c^2d^2)$

Solution:

a. $\dfrac{3x^4 - 6x^3 + 9x}{3x}$

$= \dfrac{3x^4}{3x} - \dfrac{6x^3}{3x} + \dfrac{9x}{3x}$ Divide each term in the numerator by $3x$.

$= x^3 - 2x^2 + 3$ Simplify each term, using the properties of exponents.

b. $(10c^3d - 15c^2d^2 + 2cd^3) \div (-5c^2d^2)$

$= \dfrac{10c^3d - 15c^2d^2 + 2cd^3}{-5c^2d^2}$

$= \dfrac{10c^3d}{-5c^2d^2} - \dfrac{15c^2d^2}{-5c^2d^2} + \dfrac{2cd^3}{-5c^2d^2}$ Divide each term in the numerator by $-5c^2d^2$.

$= -\dfrac{2c}{d} + 3 - \dfrac{2d}{5c}$ Simplify each term.

Skill Practice Divide.

1. $\dfrac{18y^3 - 6y^2 - 12y}{6y}$

2. $(-24a^3b^2 - 16a^2b^3 + 8ab) \div (-8ab)$

2. Long Division

If the divisor has two or more terms, a long division process similar to the division of whole numbers is used.

Example 2 **Using Long Division to Divide Polynomials**

Divide the polynomials by using long division.

$$(3x^2 - 14x - 10) \div (x - 2)$$

Solution:

$$x - 2 \overline{)3x^2 - 14x - 10}$$

Divide the leading term in the dividend by the leading term in the divisor.

$\dfrac{3x^2}{x} = 3x$. This is the first term in the quotient.

The terms in the quotient are aligned with the *like* terms in the dividend.

$$\begin{array}{r} 3x \\ x - 2 \overline{)3x^2 - 14x - 10} \\ -(3x^2 - 6x) \end{array}$$

Multiply $3x$ by the divisor and record the result: $3x(x - 2) = 3x^2 - 6x$.

$$\begin{array}{r} 3x \\ x - 2 \overline{)3x^2 - 14x - 10} \\ -3x^2 + 6x \\ \hline -8x \end{array}$$

Next, subtract the quantity $3x^2 - 6x$. To do this, add its opposite.

$$\begin{array}{r} 3x - 8 \\ x - 2 \overline{)3x^2 - 14x - 10} \\ -3x^2 + 6x \\ \hline -8x - 10 \\ -(-8x + 16) \end{array}$$

Bring down the next column and repeat the process.

Divide the leading term by x: $\dfrac{-8x}{x} = -8$

Multiply the divisor by -8 and record the result: $-8(x - 2) = -8x + 16$.

$$\begin{array}{r} 3x - 8 \\ x - 2 \overline{)3x^2 - 14x - 10} \\ -3x^2 + 6x \\ \hline -8x - 10 \\ +8x - 16 \\ \hline -26 \end{array}$$

Subtract the quantity $-8x + 16$ by adding its opposite.

The remainder is -26. We do not continue because the degree of the remainder is less than the degree of the divisor.

TIP: Take a minute to compare the long division process with numbers to the process of dividing polynomials.

$$\begin{array}{r} 148 \\ 31 \overline{)4602} \\ -31 \\ \hline 150 \\ -124 \\ \hline 262 \\ -248 \\ \hline 14 \end{array}$$

Summary:

The quotient is $3x - 8$

The remainder is -26

The divisor is $x - 2$

The dividend is $3x^2 - 14x - 10$

The solution to a long division problem is often written in the form:

Quotient $+ \dfrac{\text{remainder}}{\text{divisor}}$.

This answer is $3x - 8 + \dfrac{-26}{x - 2}$.

Skill Practice Divide.

3. $(4x^2 + 6x - 8) \div (x + 3)$

The division of polynomials can be checked in the same fashion as the division of real numbers. To check we use the division algorithm.

Division Algorithm

$$\text{Dividend} = (\text{divisor})(\text{quotient}) + \text{remainder}$$

$$3x^2 - 14x - 10 \overset{?}{=} (x - 2)(3x - 8) + (-26)$$
$$\overset{?}{=} 3x^2 - 8x - 6x + 16 + (-26)$$
$$\overset{?}{=} 3x^2 - 14x - 10 \checkmark \quad \text{True}$$

Example 3 **Using Long Division to Divide Polynomials**

Divide the polynomials by using long division: $(-2x^3 - 10x^2 + 56) \div (2x - 4)$

Solution:

First note that the dividend has a missing power of x and can be written as $-2x^3 - 10x^2 + 0x + 56$. The term $0x$ is a placeholder for the missing term. It is helpful to use the placeholder to keep the powers of x lined up.

> **TIP:** To get the first term in the quotient, you can also ask, "What can be multiplied by $2x$ to get $-2x^3$?"

$$\begin{array}{r} -x^2 \\ 2x - 4 \overline{\smash{)}-2x^3 - 10x^2 + 0x + 56} \\ -(-2x^3 + 4x^2) \end{array}$$

Leave space for the missing power of x.

Divide $\dfrac{-2x^3}{2x} = -x^2$ to get the first term of the quotient.

$$\begin{array}{r} -x^2 - 7x \\ 2x - 4 \overline{\smash{)}-2x^3 - 10x^2 + 0x + 56} \\ \underline{2x^3 - 4x^2} \\ -14x^2 + 0x \\ -(-14x^2 + 28x) \end{array}$$

Subtract by adding the opposite.
Bring down the next column.

Divide $\dfrac{-14x^2}{2x} = -7x$ to get the next term in the quotient.

$$\begin{array}{r} -x^2 - 7x - 14 \\ 2x - 4 \overline{\smash{)}-2x^3 - 10x^2 + 0x + 56} \\ \underline{2x^3 - 4x^2} \\ -14x^2 + 0x \\ \underline{14x^2 - 28x} \\ -28x + 56 \\ -(-28x + 56) \end{array}$$

Subtract by adding the opposite.
Bring down the next column.

Divide $\dfrac{-28x}{2x} = -14$ to get the next term in the quotient.

Answer

3. $4x - 6 + \dfrac{10}{x + 3}$

$$\begin{array}{r} -x^2 - 7x - 14 \\ 2x-4\overline{)-2x^3 - 10x^2 + 0x + 56} \\ \underline{2x^3 - 4x^2} \\ -14x^2 + 0x \\ \underline{14x^2 - 28x} \\ -28x + 56 \\ \underline{28x - 56} \\ 0 \end{array}$$

Subtract by adding the opposite.

← The remainder is 0.

The answer is $-x^2 - 7x - 14$.

Skill Practice Divide.

4. $\dfrac{4y^3 - 2y + 2}{2y + 2}$

In Example 3, the quotient is $-x^2 - 7x - 14$ and the remainder is 0. Because the remainder is zero, $2x - 4$ divides *evenly* into $-2x^3 - 10x^2 + 56$. For this reason, the divisor and quotient are *factors* of $-2x^3 - 10x^2 + 56$. To check, we have

$$\text{Dividend} = (\text{divisor})\,(\text{quotient}) + \text{remainder}$$

$$-2x^3 - 10x^2 + 56 \overset{?}{=} (2x-4)(-x^2 - 7x - 14) + 0$$

$$\overset{?}{=} -2x^3 - 14x^2 - 28x + 4x^2 + 28x + 56$$

$$\overset{?}{=} -2x^3 - 10x^2 + 56 \ \checkmark \ \text{True}$$

Example 4 Using Long Division to Divide Polynomials

Divide. $(15x^3 - 4x + 9x^4 - 5) \div (3x^2 - 4)$

Solution:

Write the dividend in descending powers of x. $\quad 9x^4 + 15x^3 - 4x - 5$
The dividend has a missing power of x^2. $\quad 9x^4 + 15x^3 + 0x^2 - 4x - 5$

The divisor has a missing power of x and can be written as $3x^2 + 0x - 4$.

$$\begin{array}{r} 3x^2 + 5x + 4 \\ 3x^2 + 0x - 4\overline{)9x^4 + 15x^3 + 0x^2 - 4x - 5} \\ \underline{-(9x^4 + 0x^3 - 12x^2)} \\ 15x^3 + 12x^2 - 4x \\ \underline{-(15x^3 + 0x^2 - 20x)} \\ 12x^2 + 16x - 5 \\ \underline{-(12x^2 + 0x - 16)} \\ 16x + 11 \end{array}$$

The remainder is $16x + 11$. The degree of the remainder is less than the degree of the divisor.

TIP: Both the divisor and dividend must be written in descending order before performing polynomial division.

The answer is $3x^2 + 5x + 4 + \dfrac{16x + 11}{3x^2 - 4}$.

Skill Practice Divide.

5. $(x^3 + 1 + 2x^2) \div (x^2 + 1)$

Answers

4. $2y^2 - 2y + 1$

5. $x + 2 + \dfrac{-x - 1}{x^2 + 1}$

3. Synthetic Division

In this section, we introduced the process of long division to divide two polynomials. Next, we will learn another technique, called **synthetic division**, to divide two polynomials. Synthetic division can be used when dividing a polynomial by a first-degree divisor of the form $x - r$, where r is a constant. Synthetic division is considered a "shortcut" because it uses the coefficients of the divisor and dividend without writing the variables.

Consider dividing the polynomials $(3x^2 - 14x - 10) \div (x - 2)$.

$$
\begin{array}{r}
3x - 8 \\
x - 2 \overline{)\,3x^2 - 14x - 10} \\
\underline{-(3x^2 - 6x)} \\
-8x - 10 \\
\underline{-(-8x + 16)} \\
-26
\end{array}
$$

Avoiding Mistakes

It is important to check that the divisor is in the form $(x - r)$ before applying synthetic division. The variable x in the divisor must be of first degree, and its coefficient must be 1.

First note that the divisor $x - 2$ is in the form $x - r$, where $r = 2$. Therefore, synthetic division can be used to find the quotient and remainder.

Step 1: Write the value of r in a box. $\longrightarrow$

$$
\underline{2}\,\vert\; 3 \quad -14 \quad -10 \;\longleftarrow
$$

Step 2: Write the coefficients of the dividend to the right of the box.

$$3$$

Step 3: Skip a line and draw a horizontal line below the list of coefficients.

Step 4: Bring down the leading coefficient from the dividend and write it below the line.

$$
\underline{2}\,\vert\; 3 \quad -14 \quad -10
$$
$$
6
$$
$$
3 \quad -8 \;\longleftarrow
$$

Step 5: Multiply the value of r by the number below the line $(2 \times 3 = 6)$. Write the result in the next column above the line.

Step 6: Add the numbers in the column above the line $(-14 + 6)$, and write the result below the line.

Repeat steps 5 and 6 until all columns have been completed.

Step 7: To get the final result, we use the numbers below the line. The number in the last column is the remainder. The other numbers are the coefficients of the quotient.

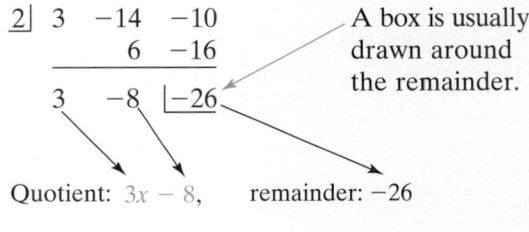

$$
\underline{2}\,\vert\; 3 \quad -14 \quad -10
$$
$$
6 \quad -16
$$
$$
3 \quad -8 \;\boxed{-26}
$$

A box is usually drawn around the remainder.

Quotient: $3x - 8$,   remainder: -26

The degree of the quotient will always be 1 less than that of the dividend. Because the dividend is a second-degree polynomial, the quotient will be a first-degree polynomial. In this case, the quotient is $3x - 8$ and the remainder is -26.

Example 5 **Using Synthetic Division to Divide Polynomials**

Divide the polynomials $(5x + 4x^3 - 6 + x^4) \div (x + 3)$ by using synthetic division.

Solution:

As with long division, the terms of the dividend and divisor should be written in descending order. Furthermore, missing powers must be accounted for by using placeholders (shown here in red).

$$5x + 4x^3 - 6 + x^4$$
$$= x^4 + 4x^3 + 0x^2 + 5x - 6$$

To use synthetic division, the divisor must be in the form $(x - r)$. The divisor $x + 3$ can be written as $x - (-3)$. Hence, $r = -3$.

Step 1: Write the value of r in a box.

Step 2: Write the coefficients of the dividend to the right of the box.

$$-3 \underline{|\ 1\ \ 4\ \ 0\ \ 5\ \ -6}$$
$$1$$

Step 3: Skip a line and draw a horizontal line below the list of coefficients.

Step 4: Bring down the leading coefficient from the dividend and write it below the line.

Step 5: Multiply the value of r by the number below the line $(-3 \times 1 = -3)$. Write the result in the next column above the line.

$$-3 \underline{|\ 1\ \ 4\ \ 0\ \ 5\ \ -6}$$
$$\underline{\ \ \ \ -3}$$
$$1\ \ 1$$

Step 6: Add the numbers in the column above the line: $4 + (-3) = 1$.

Repeat steps 5 and 6:

$$-3 \underline{|\ 1\ \ \ 4\ \ \ 0\ \ \ 5\ \ \ \ -6}$$
$$\underline{\ \ \ \ -3\ \ -3\ \ \ 9\ \ \ -42}$$
$$1\ \ \ 1\ \ -3\ \ 14\ \ \underline{|-48} \longleftarrow \text{remainder}$$

constant

x-term coefficient

x^2-term coefficient

x^3-term coefficient

The quotient is
$x^3 + x^2 - 3x + 14$.

The remainder is -48.

The answer is $x^3 + x^2 - 3x + 14 + \dfrac{-48}{x + 3}$.

Skill Practice Divide the polynomials by using synthetic division.

6. $(5y^2 - 4y + 2y^3 - 5) \div (y + 3)$

Answer

6. $2y^2 - y - 1 + \dfrac{-2}{y + 3}$

TIP: It is interesting to compare the long division process to the synthetic division process. For Example 5, long division is shown on the left, and synthetic division is shown on the right. Notice that the same pattern of coefficients used in long division appears in the synthetic division process.

$$
\begin{array}{r}
x^3 + x^2 - 3x + 14 \\
x + 3 \overline{)x^4 + 4x^3 + 0x^2 + 5x - 6} \\
-(x^4 + 3x^3) \\
\hline
x^3 + 0x^2 \\
-(x^3 + 3x^2) \\
\hline
-3x^2 + 5x \\
-(-3x^2 - 9x) \\
\hline
14x - 6 \\
-(14x + 42) \\
\hline
-48
\end{array}
$$

$$
\begin{array}{r|rrrrr}
-3 & 1 & 4 & 0 & 5 & -6 \\
   &   & -3 & -3 & 9 & -42 \\
\hline
   & 1 & 1 & -3 & 14 & \underline{|-48} \\
   & \downarrow & \downarrow & \downarrow & \downarrow & \downarrow \\
   & x^3 & x^2 & x & \text{constant} & \text{remainder}
\end{array}
$$

Quotient: $x^3 + x^2 - 3x + 14$

Remainder: -48

Example 6   Using Synthetic Division to Divide Polynomials

Divide the polynomials by using synthetic division. $(p^4 - 81) \div (p - 3)$

Solution:

$(p^4 - 81) \div (p - 3)$

$(p^4 + 0p^3 + 0p^2 + 0p - 81) \div (p - 3)$ Insert placeholders (red) for missing powers of p.

$$
\begin{array}{r|rrrrr}
3 & 1 & 0 & 0 & 0 & -81 \\
  &   & 3 & 9 & 27 & 81 \\
\hline
  & 1 & 3 & 9 & 27 & \underline{|0}
\end{array}
$$

Quotient: $p^3 + 3p^2 + 9p + 27$

Remainder: 0

The answer is $p^3 + 3p^2 + 9p + 27$.

Skill Practice   Divide the polynomials by using synthetic division.

7. $(x^3 + 1) \div (x + 1)$

Answer

7. $x^2 - x + 1$

Section 4.4   Practice Exercises

Vocabulary and Key Concepts

1. **a.** The _____ algorithm states that: Dividend = (divisor)(_____) + (_____).

   b. _____ division or long division can be used when dividing a polynomial by a divisor of the form $x - r$, where r is a constant.

Review Exercises

2. **a.** Add $(3x + 1) + (2x - 5)$.

   b. Multiply $(3x + 1)(2x - 5)$.

3. **a.** Subtract $(a - 10b) - (5a + b)$.

   b. Multiply $(a - 10b)(5a + b)$.

4. a. Subtract $(2y^2 + 1) - (y^2 - 5y + 1)$.

  b. Multiply $(2y^2 + 1)(y^2 - 5y + 1)$.

5. a. Add $(x^2 - x) + (6x^2 + x + 2)$.

  b. Multiply $(x^2 - x)(6x^2 + x + 2)$.

For Exercises 6–8, answers may vary.

6. Write an example of a product of two binomials and simplify.

7. Write an example of the square of a binomial and simplify.

8. Write an example of the product of conjugates and simplify

Concept 1: Division by a Monomial

For Exercises 9–24, divide the polynomials. Check your answer by multiplication. **(See Example 1.)**

9. $\dfrac{16t^4 - 4t^2 + 20t}{-4t}$

10. $\dfrac{2x^3 - 8x^2 - 2x}{-2x}$

11. $(36y + 24y^2 + 6y^3) \div (3y)$

12. $(6p^2 - 18p^4 + 30p^5) \div (6p)$

13. $(4x^3y + 12x^2y^2 - 4xy^3) \div (4xy)$

14. $(25m^2n - 10m^4n + m^3n) \div (5m^2n)$

15. $(-8y^4 - 12y^3 + 32y^2) \div (-4y^2)$

16. $(-2y^5 - 8y^6 + 16y^4 - 10y^3) \div (2y^3)$

17. $(3p^4 - 6p^3 + 2p^2 - p) \div (-6p)$

18. $(-4q^3 + 8q^2 - q) \div (-12q)$

19. $(a^3 + 5a^2 + a - 5) \div (a)$

20. $(2m^5 - 3m^4 + m^3 - m^2 + 9m) \div (m^2)$

21. $\dfrac{6s^3t^5 - 8s^2t^4 + 10st^2}{-2st^4}$

22. $\dfrac{-8r^4w^2 - 4r^3w + 2w^3}{-4r^3w}$

23. $(8p^4q^7 - 9p^5q^6 - 11p^3q - 4) \div (p^2q)$

24. $(20a^5b^5 - 20a^3b^2 + 5a^2b + 6) \div (a^2b)$

Concept 2: Long Division

25. a. Divide $(2x^3 - 7x^2 + 5x - 1) \div (x - 2)$, and identify the divisor, quotient, and remainder.

  b. Explain how to check by using multiplication.

26. a. Divide $(x^3 + 4x^2 + 7x - 3) \div (x + 3)$, and identify the divisor, quotient, and remainder.

  b. Explain how to check by using multiplication.

For Exercises 27–46, divide the polynomials by using long division. Check your answer by multiplication.
(See Examples 2–4.)

27. $(x^2 + 11x + 19) \div (x + 4)$

28. $(x^3 - 7x^2 - 13x + 3) \div (x + 2)$

29. $(3y^3 - 7y^2 - 4y + 3) \div (y - 3)$

30. $(z^3 - 2z^2 + 2z - 5) \div (z - 4)$

31. $(-12a^2 + 77a - 121) \div (3a - 11)$

32. $(28x^2 - 29x + 6) \div (4x - 3)$

33. $(9y + 18y^2 - 20) \div (3y + 4)$

34. $(-2y + 3y^2 - 1) \div (y - 1)$

35. $(18x^3 + 7x + 12) \div (3x - 2)$

36. $(8x^3 - 6x + 22) \div (2x - 1)$

37. $(8a^3 + 1) \div (2a + 1)$

38. $(81x^4 - 1) \div (3x + 1)$

39. $(x^4 - x^3 - x^2 + 4x - 2) \div (x^2 + x - 1)$

40. $(2a^5 - 7a^4 + 11a^3 - 22a^2 + 29a - 10) \div (2a^2 - 5a + 2)$

41. $(2x^3 - 10x + x^4 - 25) \div (x^2 - 5)$

42. $(-5x^3 + x^4 - 4 - 10x) \div (x^2 + 2)$

43. $(x^4 - 3x^2 + 10) \div (x^2 - 2)$

44. $(3y^4 - 25y^2 - 18) \div (y^2 - 3)$

45. $(n^4 - 16) \div (n - 2)$

46. $(m^3 + 27) \div (m + 3)$

Concept 3: Synthetic Division

47. Explain the conditions under which you may use synthetic division to divide polynomials.

48. Can synthetic division be used directly to divide $(4x^4 + 3x^3 - 7x + 9)$ by $(2x + 5)$? Explain why or why not.

49. Can synthetic division be used to divide $(6x^5 - 3x^2 + 2x - 14)$ by $(x^2 - 3)$? Explain why or why not.

50. Can synthetic division be used to divide $(3x^4 - x + 1)$ by $(x - 5)$? Explain why or why not.

51. The following table represents the result of a synthetic division.

$$
\begin{array}{r|rrrr}
5 & 1 & -2 & -4 & 3 \\
  &   & 5 & 15 & 55 \\
\hline
  & 1 & 3 & 11 & \underline{58}
\end{array}
$$

Use x as the variable.

a. Identify the divisor.

b. Identify the quotient.

c. Identify the remainder.

52. The following table represents the result of a synthetic division.

$$
\begin{array}{r|rrrrr}
-2 & 2 & 3 & 0 & -1 & 6 \\
   &   & -4 & 2 & -4 & 10 \\
\hline
   & 2 & -1 & 2 & -5 & \underline{16}
\end{array}
$$

Use x as the variable.

a. Identify the divisor.

b. Identify the quotient.

c. Identify the remainder.

For Exercises 53–70, divide by using synthetic division. Check your answer by multiplication. **(See Examples 5–6.)**

53. $(x^2 - 2x - 48) \div (x - 8)$

54. $(x^2 - 4x - 12) \div (x - 6)$

55. $(t^2 - 3t - 4) \div (t + 1)$

56. $(h^2 + 7h + 12) \div (h + 3)$

57. $(5y^2 + 5y + 1) \div (y - 1)$

58. $(3w^2 + w - 5) \div (w + 2)$

59. $(3 + 7y^2 - 4y + 3y^3) \div (y + 3)$

60. $(2z - 2z^2 + z^3 - 5) \div (z + 3)$

61. $(x^3 - 3x^2 + 4) \div (x - 2)$

62. $(3y^4 - 25y^2 - 18) \div (y - 3)$

63. $(a^5 - 32) \div (a - 2)$

64. $(b^3 + 27) \div (b + 3)$

65. $(x^3 - 216) \div (x - 6)$

66. $(y^4 - 16) \div (y + 2)$

67. $(6t^3 + 7t^2 - t + 3) \div \left(t + \dfrac{2}{3}\right)$

68. $(5z^3 - 6z^2 - 4z + 4) \div \left(z - \dfrac{1}{5}\right)$

69. $(4w^4 - w^2 + 6w - 3) \div \left(w - \dfrac{1}{2}\right)$

70. $(-12y^4 - 5y^3 - y^2 + y + 3) \div \left(y + \dfrac{3}{4}\right)$

Mixed Exercises

For Exercises 71–82, divide the polynomials by using an appropriate method.

71. $(-x^3 - 8x^2 - 3x - 2) \div (x + 4)$

72. $(8xy^2 - 9x^2y + 6x^2y^2) \div (x^2y^2)$

73. $(22x^2 - 11x + 33) \div (11x)$

74. $(2m^3 - 4m^2 + 5m - 33) \div (m - 3)$

75. $(12y^3 - 17y^2 - 30y - 10) \div (3y^2 - 2y + 5)$

76. $(90h^{12} - 63h^9 + 45h^8 - 36h^7) \div (9h^9)$

77. $(4x^4 + 6x^3 + 3x - 1) \div (2x^2 + 1)$

78. $(y^4 - 3y^3 - 5y^2 - 2y + 5) \div (y + 2)$

79. $(16k^{11} - 32k^{10} + 8k^8 - 40k^4) \div (8k^8)$

80. $(4m^3 - 18m^2 + 22m - 10) \div (2m^2 - 4m + 3)$

81. $(5x^3 + 9x^2 + 10x) \div (5x^2)$

82. $(15k^4 + 3k^3 + 4k^2 + 4) \div (3k^2 - 1)$

Expanding Your Skills

83. Given $P(x) = 4x^3 + 10x^2 - 8x - 20$,

   a. Evaluate $P(-4)$.

   b. Divide. $(4x^3 + 10x^2 - 8x - 20) \div (x + 4)$

   c. Compare the value found in part (a) to the remainder found in part (b).

84. Given $P(x) = -3x^3 - 12x^2 + 5x - 8$,

   a. Evaluate $P(-6)$.

   b. Divide. $(-3x^3 - 12x^2 + 5x - 8) \div (x + 6)$

   c. Compare the value found in part (a) to the remainder found in part (b).

85. Based on your solutions to Exercises 81–82, make a conjecture about the relationship between the value of a polynomial function $P(x)$ at $x = r$, and the value of the remainder of $P(x) \div (x - r)$.

86. **a.** Use synthetic division to divide. $(7x^2 - 16x + 9) \div (x - 1)$

   b. Based on your solution to part (a), is $x - 1$ a *factor* of $7x^2 - 16x + 9$?

87. **a.** Use synthetic division to divide. $(8x^2 + 13x + 5) \div (x + 1)$

   b. Based on your solution to part (a), is $x + 1$ a *factor* of $8x^2 + 13x + 5$?

Problem Recognition Exercises

Operations on Polynomials

Perform the indicated operations.

1. **a.** $(3x + 1)^2$

   b. $(3x + 1)(3x - 1)$

   c. $(3x + 1) - (3x - 1)$

3. **a.** $\dfrac{4x^2 + 8x - 10}{2x}$

   b. $\dfrac{4x^2 + 8x - 10}{2x - 1}$

   c. $(4x^2 + 8x - 10) \div (x - 1)$

2. **a.** $(9m - 5) - (9m + 5)$

   b. $(9m - 5)(9m + 5)$

   c. $(9m - 5)^2$

4. **a.** $\dfrac{3y^2 - 15y + 4}{3y}$

   b. $\dfrac{3y^2 - 15y + 4}{3y + 6}$

   c. $(3y^2 - 15y + 4) \div (y + 6)$

5. **a.** $(p - 5)(p + 5) - (p^2 + 5)$

 b. $(p - 5)(p + 5) - (p + 5)^2$

 c. $(p - 5)(p + 5) - (p^2 - 25)$

6. **a.** $(x + 4)(x - 4) - (x + 4)^2$

 b. $(x + 4)(x - 4) - (x^2 + 4)$

 c. $(x + 4)(x - 4) - (x^2 - 16)$

7. $(5t^2 - 6t + 2) - (3t^2 - 7t + 3)$

8. $-5x^2(3x^2 + x - 2)$

9. $(6z + 5)(6z - 5)$

10. $(6y^3 + 2y^2 + y - 2) + (3y^3 - 4y + 3)$

11. $(3b - 4)(2b - 1)$

12. $(5a + 2)(2a^2 + 3a + 1)$

13. $(t^3 - 4t^2 + t - 9) + (t + 12) - (2t^2 - 6t)$

14. $(2b^3 - 3b - 10) \div (b - 2)$

15. $(k + 4)^2 + (-4k + 9)$

16. $(3x^4 - 11x^3 - 4x^2 - 5x + 20) \div (x - 4)$

17. $-2t(t^2 + 6t - 3) + t(3t + 2)(3t - 2)$

18. $\dfrac{7x^2y^3 - 14xy^2 - x^2}{-7xy}$

19. $\left(\dfrac{1}{4}p^3 - \dfrac{1}{6}p^2 + 5\right) - \left(-\dfrac{2}{3}p^3 + \dfrac{1}{3}p^2 - \dfrac{1}{5}p\right)$

20. $-6w^3(1.2w - 2.6w^2 + 5.1w^3)$

21. $(6a^2 - 4b)^2$

22. $\left(\dfrac{1}{2}z^2 - \dfrac{1}{3}\right)\left(\dfrac{1}{2}z^2 + \dfrac{1}{3}\right)$

23. $(m - 3)^2 - 2(m + 8)$

24. $(2x - 5)(x + 1) - (x - 3)^2$

25. $(m^2 - 6m + 7)(2m^2 + 4m - 3)$

26. $(x^3 - 64) \div (x - 4)$

27. $[5 - (a + b)]^2$

28. $[a - (x - y)][a + (x - y)]$

29. $(x + y)^2 - (x - y)^2$

30. $(a - 4)^3$

31. $\left(-\dfrac{1}{2}x + \dfrac{1}{3}\right)\left(\dfrac{1}{4}x - \dfrac{1}{2}\right)$

32. $-3x^2y^3z^4\left(\dfrac{1}{6}x^4yzw^3\right)$

Section 4.5   Greatest Common Factor and Factoring by Grouping

Concepts

1. Factoring Out the Greatest Common Factor
2. Factoring Out a Negative Factor
3. Factoring Out a Binomial Factor
4. Factoring by Grouping

1. Factoring Out the Greatest Common Factor

Sections 4.5 through 4.7 are devoted to a mathematical operation called factoring. To factor an integer means to write the integer as a product of two or more integers. To factor a polynomial means to express the polynomial as a product of two or more polynomials. For example, in the product $5 \cdot 7 = 35$, the numbers 5 and 7 are factors of 35. In the product $(2x + 1)(x - 6) = 2x^2 - 11x - 6$, the quantities $(2x + 1)$ and $(x - 6)$ are factors of $2x^2 - 11x - 6$.

The **greatest common factor (GCF)** of a polynomial is the greatest factor that divides each term of the polynomial evenly. For example, the greatest common factor of $9x^4 + 18x^3 - 6x^2$ is $3x^2$. To factor out the greatest common factor from a polynomial, follow these steps:

Factoring Out the Greatest Common Factor

Step 1   Identify the greatest common factor of all terms of the polynomial.

Step 2   Write each term as the product of the GCF and another factor.

Step 3   Use the distributive property to factor out the greatest common factor.

Note: To check the factorization, multiply the polynomials.

Example 1   Factoring Out the Greatest Common Factor

Factor out the greatest common factor.

a. $12x^3 + 30x^2$ **b.** $12c^2d^3 - 30c^3d^2 - 3cd$

Solution:

a. $12x^3 + 30x^2$ The GCF is $6x^2$.

$= 6x^2(2x) + 6x^2(5)$ Write each term as the product of the GCF and another factor.

$= 6x^2(2x + 5)$ Factor out $6x^2$ by using the distributive property.

TIP: Any factoring problem can be checked by multiplying the factors.

Check: $6x^2(2x + 5) = 12x^3 + 30x^2$ ✔

> ### Avoiding Mistakes
> In Example 1(b), the GCF of $3cd$ is equal to one of the terms of the polynomial. In such a case, you must leave a 1 in place of that term after the GCF is factored out.
> $$3cd(4cd^2 - 10c^2d - 1)$$

b. $12c^2d^3 - 30c^3d^2 - 3cd$ The GCF is $3cd$.

$= 3cd(4cd^2) - 3cd(10c^2d) - 3cd(1)$ Write each term as the product of the GCF and another factor.

$= 3cd(4cd^2 - 10c^2d - 1)$ Factor out $3cd$ by using the distributive property.

Check: $3cd(4cd^2 - 10c^2d - 1) = 12c^2d^3 - 30c^3d^2 - 3cd$ ✔

Skill Practice Factor out the greatest common factor.

1. $45y^5 - 15y^2 + 30y$ **2.** $16a^2b^5 + 12a^3b^3 + 4a^3b^2$

---

2. Factoring Out a Negative Factor

Sometimes it is advantageous to factor out the *opposite* of the GCF, particularly when the leading coefficient of the polynomial is negative. This is demonstrated in Example 2. Notice that this *changes the signs* of the remaining terms inside the parentheses.

Answers
1. $15y(3y^4 - y + 2)$
2. $4a^2b^2(4b^3 + 3ab + a)$

Example 2   Factoring Out a Negative Factor

Factor out the quantity $-5a^2b$ from the polynomial $-5a^4b - 10a^3b^2 + 15a^2b^3$.

Solution:

$$-5a^4b - 10a^3b^2 + 15a^2b^3 \quad \text{The GCF is } 5a^2b. \text{ However, in this case we will factor out the opposite of the GCF, } -5a^2b.$$

$$= -5a^2b(a^2) + -5a^2b(2ab) + -5a^2b(-3b^2) \quad \text{Write each term as the product of } -5a^2b \text{ and another factor.}$$

$$= -5a^2b(a^2 + 2ab - 3b^2) \quad \text{Factor out } -5a^2b \text{ by using the distributive property.}$$

Skill Practice Factor out the quantity $-6xy$ from the polynomial.

3. $24x^4y^3 - 12x^2y + 18xy^2$

3. Factoring Out a Binomial Factor

The distributive property may also be used to factor out a common factor that consists of more than one term. This is shown in Example 3.

Example 3   Factoring Out a Binomial Factor

Factor out the greatest common factor.

$$x^3(x + 2) - x(x + 2) - 9(x + 2)$$

Solution:

$$x^3(x + 2) - x(x + 2) - 9(x + 2) \quad \text{The GCF is the quantity } (x + 2).$$

$$= (x + 2)(x^3) - (x + 2)(x) - (x + 2)(9) \quad \text{Write each term as the product of } (x + 2) \text{ and another factor.}$$

$$= (x + 2)(x^3 - x - 9) \quad \text{Factor out } (x + 2) \text{ by using the distributive property.}$$

Skill Practice Factor out the greatest common factor.

4. $a^2(b + 2) + 5(b + 2)$

Answers

3. $-6xy(-4x^3y^2 + 2x - 3y)$
4. $(b + 2)(a^2 + 5)$

4. Factoring by Grouping

When two binomials are multiplied, the product before simplifying contains four terms. For example:

$$(3a + 2)(2b - 7) = (3a + 2)(2b) + (3a + 2)(-7)$$

$$= (3a + 2)(2b) + (3a + 2)(-7)$$

$$= 6ab + 4b - 21a - 14$$

In Example 4, we learn how to reverse this process. That is, given a four-term polynomial, we will factor it as a product of two binomials. The process is called **factoring by grouping**.

> ### Factoring by Grouping
>
> To factor a four-term polynomial by grouping:
>
> **Step 1**   Identify and factor out the GCF from all four terms.
> **Step 2**   Factor out the GCF from the first pair of terms. Factor out the GCF from the second pair of terms. (Sometimes it is necessary to factor out the *opposite* of the GCF.)
> **Step 3**   If the two terms share a common binomial factor, factor out the binomial factor.

Example 4   Factoring by Grouping

Factor by grouping.

$$6ab - 21a + 4b - 14$$

Solution:

$6ab - 21a + 4b - 14$

$= 6ab - 21a \; | \; + 4b - 14$

$= 3a(2b - 7) + 2(2b - 7)$

$= (2b - 7)(3a + 2)$

Step 1:   Identify and factor out the GCF from all four terms. In this case the GCF is 1.

Group the first pair of terms and the second pair of terms.

Step 2:   Factor out the GCF from each pair of terms.

Note: The two terms now share a common binomial factor of $(2b - 7)$.

Step 3:   Factor out the common binomial factor.

Check: $(2b - 7)(3a + 2) = 2b(3a) + 2b(2) - 7(3a) - 7(2)$

$$= 6ab + 4b - 21a - 14 \; ✔$$

Avoiding Mistakes

In step 2, the expression $3a(2b - 7) + 2(2b - 7)$ is not yet factored because it is a *sum*, not a product. To factor the expression, you must carry it one step further.

$$3a(2b - 7) + 2(2b - 7)$$

$$= (2b - 7)(3a + 2)$$

The factored form must be represented as a product.

Skill Practice   Factor by grouping.

5. $7c^2 + cd + 14c + 2d$

Answer

5. $(7c + d)(c + 2)$

Example 5 Factoring by Grouping

Factor by grouping.

$$x^3 + 3x^2 - 3x - 9$$

Solution:

$x^3 + 3x^2 - 3x - 9$ **Step 1:** Identify and factor out the GCF from all four terms. In this case the GCF is 1.

$= x^3 + 3x^2 \;\vdots\; - 3x - 9$ Group the first pair of terms and the second pair of terms.

$= x^2(x + 3) - 3(x + 3)$ **Step 2:** Factor out x^2 from the first pair of terms.

Factor out -3 from the second pair of terms (this causes the signs to change in the second parentheses). The terms now contain a common binomial factor.

$= (x + 3)(x^2 - 3)$ **Step 3:** Factor out the common binomial $(x + 3)$.

TIP: One frequent question is, can the order be switched between factors? The answer is yes. Because multiplication is commutative, the order in which two or more factors are written does not matter. Thus, the following factorizations are equivalent:

$$(x + 3)(x^2 - 3) = (x^2 - 3)(x + 3)$$

Skill Practice Factor by grouping.

6. $a^3 - 4a^2 - 3a + 12$

Example 6 Factoring by Grouping

Factor by grouping.

$$24p^2q^2 - 18p^2q + 60pq^2 - 45pq$$

Solution:

$24p^2q^2 - 18p^2q + 60pq^2 - 45pq$

$= 3pq(8pq - 6p + 20q - 15)$ **Step 1:** Remove the GCF $3pq$ from all four terms.

$= 3pq(8pq - 6p \;\vdots\; + 20q - 15)$ Group the first pair of terms and the second pair of terms.

$$= 3pq[2p(4q - 3) + 5(4q - 3)]$$

Step 2:   Factor out the GCF from each pair of terms. The terms share the binomial factor $(4q - 3)$.

$$= 3pq(4q - 3)(2p + 5)$$

Step 3:   Factor out the common binomial $(4q - 3)$.

Skill Practice   Factor the polynomial.

7. $24x^2y - 12x^2 + 20xy - 10x$

Notice that in step 3 of factoring by grouping, a common binomial is factored from the two terms. These binomials must be *exactly* the same in each term. If the two binomial factors differ, try rearranging the original four terms.

Example 7   Factoring by Grouping Where Rearranging Terms Is Necessary

Factor the polynomial.

$$4x + 6pa - 8a - 3px$$

Solution:

$$4x + 6pa - 8a - 3px$$

Step 1:   Identify and factor out the GCF from all four terms. In this case the GCF is 1.

$$= 4x + 6pa \mid - 8a - 3px$$

$$= 2(2x + 3pa) - 1(8a + 3px)$$

Step 2:   The binomial factors in each term are different.

$$= 4x - 8a \mid - 3px + 6pa$$

Try rearranging the original four terms in such a way that the first pair of coefficients is in the same ratio as the second pair of coefficients. Notice that the ratio 4 to -8 is the same as the ratio -3 to 6.

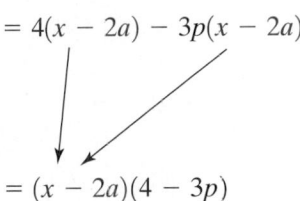

$$= 4(x - 2a) - 3p(x - 2a)$$

Step 2:   Factor out 4 from the first pair of terms.

Factor out $-3p$ from the second pair of terms.

$$= (x - 2a)(4 - 3p)$$

Step 3:   Factor out the common binomial factor.

> **Avoiding Mistakes**
>
> Remember that when factoring by grouping, the binomial factors must be *exactly* the same.

Skill Practice   Factor the polynomial.

8. $3ry + 2s + sy + 6r$

Section 4.5   Practice Exercises

Vocabulary and Key Concepts

1. a. Factoring a polynomial means to write it as a _____ of two or more polynomials.

   b. The _____ _____ _____ (GCF) of a polynomial is the greatest factor that divides each term of the polynomial evenly.

   c. The first step toward factoring a polynomial is to factor out the _____ _____
   _____ .

   d. To factor a four-term polynomial, we try the process of factoring by _____ .

Review Exercises

For Exercises 2–8, perform the indicated operation.

2. $(-4a^3b^5c)(-2a^7c^2)$

3. $(7t^4 + 5t^3 - 9t) - (-2t^4 + 6t^2 - 3t)$

4. $(5x^3 - 9x + 5) + (4x^3 + 3x^2 - 2x + 1) - (6x^3 - 3x^2 + x + 1)$

5. $(5y^2 - 3)(y^2 + y + 2)$

6. $(a + 6b)^2$

7. $\dfrac{6v^3 - 12v^2 + 2v}{-2v}$

8. $\dfrac{3x^3 + 2x^2 - 4}{x + 2}$

Concept 1: Factoring Out the Greatest Common Factor

For Exercises 9–24, factor out the greatest common factor. **(See Example 1.)**

9. $3x + 12$

10. $15x - 10$

11. $6z^2 + 4z$

12. $49y^3 - 35y^2$

13. $4p^6 - 4p$

14. $5q^2 - 5q$

15. $12x^4 - 36x^2$

16. $51w^4 - 34w^3$

17. $9st^2 + 27t$

18. $8a^2b^3 + 12a^2b$

19. $9a^4b^3 + 27a^3b^4 - 18a^2b^5$

20. $3x^5y^4 - 15x^4y^5 + 9x^2y^7$

21. $10x^2y + 15xy^2 - 5xy$

22. $12c^3d - 15c^2d + 3cd$

23. $13b^2 - 11a^2b - 12ab$

24. $6a^3 - 2a^2b + 5a^2$

Concept 2: Factoring Out a Negative Factor

For Exercises 25–30, factor out the indicated quantity. **(See Example 2.)**

25. $-x^2 - 10x + 7$: Factor out -1.

26. $-5y^2 + 10y + 3$: Factor out -1.

27. $-12x^3y - 6x^2y - 3xy$: Factor out $-3xy$.

28. $-32a^4b^2 + 24a^3b + 16a^2b$: Factor out $-8a^2b$.

29. $-2t^3 + 11t^2 - 3t$: Factor out $-t$.

30. $-7y^2z - 5yz - z$: Factor out $-z$.

Concept 3: Factoring Out a Binomial Factor

For Exercises 31–38, factor out the GCF. **(See Example 3.)**

31. $2a(3z - 2b) - 5(3z - 2b)$

32. $5x(3x + 4) + 2(3x + 4)$

33. $2x^2(2x - 3) + (2x - 3)$

34. $z(w - 9) + (w - 9)$　　**35.** $y(2x + 1)^2 - 3(2x + 1)^2$　　**36.** $a(b - 7)^2 + 5(b - 7)^2$

37. $3y(x - 2)^2 + 6(x - 2)^2$　　**38.** $10z(z + 3)^2 - 2(z + 3)^2$

39. Construct a polynomial that has a greatest common factor of $3x^2$. (Answers may vary.)

40. Construct two different trinomials that have a greatest common factor of $5x^2y^3$. (Answers may vary.)

41. Construct a binomial that has a greatest common factor of $(c + d)$. (Answers may vary.)

Concept 4: Factoring by Grouping

42. If a polynomial has four terms, what technique would you use to factor it?

43. Factor the polynomials by grouping.

　a. $2ax - ay + 6bx - 3by$

　b. $10w^2 - 5w - 6bw + 3b$

　c. Explain why you factored out $3b$ from the second pair of terms in part (a) but factored out the quantity $-3b$ from the second pair of terms in part (b).

44. Factor the polynomials by grouping.

　a. $3xy + 2bx + 6by + 4b^2$

　b. $15ac + 10ab - 6bc - 4b^2$

　c. Explain why you factored out $2b$ from the second pair of terms in part (a) but factored out the quantity $-2b$ from the second pair of terms in part (b).

For Exercises 45–64, factor each polynomial by grouping (if possible). **(See Examples 4–7.)**

45. $y^3 + 4y^2 + 3y + 12$　　　　　**46.** $ab + b + 2a + 2$

47. $6p - 42 + pq - 7q$　　　　　**48.** $2t - 8 + st - 4s$

49. $2mx + 2nx + 3my + 3ny$　　　**50.** $4x^2 + 6xy - 2xy - 3y^2$

51. $10ax - 15ay - 8bx + 12by$　　**52.** $35a^2 - 15a + 14a - 6$

53. $x^3 - x^2 - 3x + 3$　　　　　**54.** $2rs + 4s - r - 2$

55. $6p^2q + 18pq - 30p^2 - 90p$　　**56.** $5s^2t + 20st - 15s^2 - 60s$

57. $100x^3 - 300x^2 + 200x - 600$　**58.** $2x^5 - 10x^4 + 6x^3 - 30x^2$

59. $6ax - by + 2bx - 3ay$　　　　**60.** $5pq - 12 - 4q + 15p$

61. $4a - 3b - ab + 12$　　　　　**62.** $x^2y + 6x - 2x^3 - 2y$

63. $7y^3 - 21y^2 + 5y - 10$　　　　**64.** $5ax + 10bx - 2ac + 4bc$

65. Explain why the grouping method failed for Exercise 63.

66. Explain why the grouping method failed for Exercise 64.

Mixed Exercises

67. Solve the equation $U = Av + Acw$ for A by first factoring out A.

68. Solve the equation $S = rt + wt$ for t by first factoring out t.

69. Solve the equation $ay + bx = cy$ for y.

70. Solve the equation $cd + 2x = ac$ for c.

71. The area of a rectangle of width w is given by $A = 2w^2 + w$. Factor the right-hand side of the equation to find an expression for the length of the rectangle.

72. The amount in a savings account bearing simple interest at an annual interest rate r for t years is given by

$A = P + Prt$ where P is the principal amount invested.

a. Solve the equation for P.

b. Compute the amount of principal originally invested if the account is worth \$12,705 after 3 yr at a 7% interest rate.

Expanding Your Skills

For Exercises 73–80, factor out the greatest common factor and simplify.

73. $(a + 3)^4 + 6(a + 3)^5$

74. $(4 - b)^4 - 2(4 - b)^3$

75. $24(3x + 5)^3 - 30(3x + 5)^2$

76. $10(2y + 3)^2 + 15(2y + 3)^3$

77. $(t + 4)^2 - (t + 4)$

78. $(p + 6)^2 - (p + 6)$

79. $15w^2(2w - 1)^3 + 5w^3(2w - 1)^2$

80. $8z^4(3z - 2)^2 + 12z^3(3z - 2)^3$

Section 4.6   Factoring Trinomials

Concepts

1. Factoring Trinomials: AC-Method
2. Factoring Trinomials: Trial-and-Error Method
3. Factoring Perfect Square Trinomials
4. Factoring by Using Substitution

1. Factoring Trinomials: AC-Method

In Section 4.5, we learned how to factor out the greatest common factor from a polynomial and how to factor a four-term polynomial by grouping. In this section, we present two methods to factor trinomials. The first method is called the ac-method. The second method is called the trial-and-error method.

The product of two binomials results in a four-term expression that can sometimes be simplified to a trinomial. To factor the trinomial, we want to reverse the process.

Multiply: $(2x + 3)(x + 2) = $ — Multiply the binomials. → $2x^2 + 4x + 3x + 6$

$= $ — Add the middle terms. → $2x^2 + 7x + 6$

Factor: $2x^2 + 7x + 6 = $ — Rewrite the middle term as a sum or difference of terms. → $2x^2 + 4x + 3x + 6$

$= $ — Factor by grouping. → $(2x + 3)(x + 2)$

To factor a trinomial $ax^2 + bx + c$ by the ac-method, we rewrite the middle term bx as a sum or difference of terms. The goal is to produce a four-term polynomial that can be factored by grouping. The process is outlined as follows.

The AC-Method to Factor $ax^2 + bx + c$ $(a \neq 0)$

Step 1　After factoring out the GCF, multiply the coefficients of the first and last terms, ac.

Step 2　Find two integers whose product is ac and whose sum is b. (If no pair of integers can be found, then the trinomial cannot be factored further and is called a **prime polynomial**.)

Step 3　Rewrite the middle term bx as the sum of two terms whose coefficients are the integers found in step 2.

Step 4　Factor by grouping.

The ac-method for factoring trinomials is illustrated in Example 1. Before we begin, however, keep these two important guidelines in mind.

- For any factoring problem you encounter, always factor out the GCF from all terms first.
- To factor a trinomial, write the trinomial in the form $ax^2 + bx + c$.

Example 1　Factoring a Trinomial by the AC-Method

Factor by using the ac-method.　$12x^2 - 5x - 2$

Solution:

$12x^2 - 5x - 2$ 　　　　　　The GCF is 1.

$a = 12$　$b = -5$　$c = -2$　　**Step 1:** The expression is written in the form $ax^2 + bx + c$. Find the product $ac = 12(-2) = -24$.

Factors of –24	**Factors of –24**
$(1)(-24)$	$(-1)(24)$
$(2)(-12)$	$(-2)(12)$
$(3)(-8)$	$(-3)(8)$
$(4)(-6)$	$(-4)(6)$

Step 2: List all the factors of -24, and find the pair whose sum equals -5.

The numbers 3 and -8 produce a product of -24 and a sum of -5.

$12x^2 - 5x - 2$

Step 3: Write the middle term of the trinomial as two terms whose coefficients are the selected numbers 3 and -8.

$= 12x^2 + 3x - 8x - 2$

$= 12x^2 + 3x \mid -8x - 2$ 　　**Step 4:** Factor by grouping.

$= 3x(4x + 1) - 2(4x + 1)$

$= (4x + 1)(3x - 2)$ 　　　　The check is left for the reader.

Skill Practice

1. Factor by using the ac-method.　$10x^2 + x - 3$.

TIP: One frequently asked question is whether the order matters when we rewrite the middle term of the trinomial as two terms (step 3). The answer is no. From Example 1, the two middle terms in step 3 could have been reversed.

$$12x^2 - 5x - 2 = 12x^2 - 8x + 3x - 2$$

$$= 4x(3x - 2) + 1(3x - 2)$$

$$= (3x - 2)(4x + 1)$$

This example also shows that the order in which two factors are written does not matter. The expression $(3x - 2)(4x + 1)$ is equivalent to $(4x + 1)(3x - 2)$ because multiplication is a commutative operation.

Example 2 Factoring a Trinomial by the AC-Method

Factor the trinomial by using the ac-method. $-20c^3 + 34c^2d - 6cd^2$

Solution:

$$-20c^3 + 34c^2d - 6cd^2$$

$$= -2c(10c^2 - 17cd + 3d^2) \qquad \text{Factor out } -2c.$$

Step 1: Find the product
$a \cdot c = (10)(3) = 30$

Factors of 30	Factors of 30
$1 \cdot 30$	$(-1)(-30)$
$2 \cdot 15$	$(-2)(-15)$
$3 \cdot 10$	$(-3)(-10)$
$5 \cdot 6$	$(-5)(-6)$

Step 2: The numbers -2 and -15 form a product of 30 and a sum of -17.

$$= -2c(10c^2 - 17cd + 3d^2)$$

$$= -2c(10c^2 - 2cd \; \vdots \; - 15cd + 3d^2) \qquad \textbf{Step 3:} \text{ Write the middle term of the trinomial as two terms whose coefficients are } -2 \text{ and } -15.$$

$$= -2c[2c(5c - d) - 3d(5c - d)] \qquad \textbf{Step 4:} \text{ Factor by grouping.}$$

$$= -2c(5c - d)(2c - 3d)$$

Skill Practice

2. Factor by using the ac-method. $-4wz^3 - 2w^2z^2 + 20w^3z$

TIP: In Example 2, removing the GCF from the original trinomial produced a new trinomial with smaller coefficients. This makes the factoring process simpler because the product ac is smaller.

Original trinomial	With the GCF factored out
$-20c^3 + 34c^2d - 6cd^2$	$-2c(10c^2 - 17cd + 3d^2)$
$ac = (-20)(-6) = 120$	$ac = (10)(3) = 30$

Answer

2. $-2wz(2z + 5w)(z - 2w)$

2. Factoring Trinomials: Trial-and-Error Method

Another method that is widely used to factor trinomials of the form $ax^2 + bx + c$ is the trial-and-error method. To understand how the trial-and-error method works, first consider the multiplication of two binomials:

Product of $2 \cdot 1$ Product of $3 \cdot 2$

$$(2x + 3)(1x + 2) = 2x^2 + 4x + 3x + 6 = 2x^2 + 7x - 6$$

sum of products
of inner terms
and outer terms

In Example 3, we will factor this trinomial by reversing this process.

Example 3 **Factoring a Trinomial by the Trial-and-Error Method**

Factor by the trial-and-error method. $2x^2 + 7x + 6$

Solution:

To factor by the trial-and-error method, we must fill in the blanks to create the correct product.

Factors of 2

$$2x^2 + 7x + 6 = (\Box x \quad \Box)(\Box x \quad \Box)$$

Factors of 6

- The first terms in the binomials must be $2x$ and x. This creates a product of $2x^2$, which is the first term in the trinomial.
- The second terms in the binomials must form a product of 6. This means that the factors must both be positive or both be negative. Because the middle term of the trinomial is positive, we will consider only *positive* factors of 6. The options are $1 \cdot 6$, $2 \cdot 3$, $6 \cdot 1$, and $3 \cdot 2$.
- Test each combination of factors until the correct product of binomials is found.

$(2x + 1)(x + 6) = 2x^2 + 12x + 1x + 6$
$\qquad\qquad\qquad = 2x^2 + 13x + 6$ *Incorrect.* Wrong middle term.

$(2x + 2)(x + 3) = 2x^2 + 6x + 2x + 6$
$\qquad\qquad\qquad = 2x^2 + 8x + 6$ *Incorrect.* Wrong middle term.

$(2x + 6)(x + 1) = 2x^2 + 2x + 6x + 6$
$\qquad\qquad\qquad = 2x^2 + 8x + 6$ *Incorrect.* Wrong middle term.

$(2x + 3)(x + 2) = 2x^2 + 4x + 3x + 6$
$\qquad\qquad\qquad = 2x^2 + 7x + 6$ *Correct.*

The factored form of $2x^2 + 7x + 6$ is $(2x + 3)(x + 2)$.

Skill Practice Factor by the trial-and-error method.

3. $5y^2 - 9y + 4$

When applying the trial-and-error method, sometimes it is not necessary to test all possible combinations of factors. For the trinomial, $2x^2 + 7x + 6$, the GCF is 1. Therefore, any binomial factor that shares a common factor greater than 1 will not work and does not need to be tested. For example, the following binomials cannot work:

$$(2x + 2)(x + 3) \qquad\qquad (2x + 6)(x + 1)$$

Common factor of 2 $\qquad\qquad\qquad$ Common factor of 2

Although the trial-and-error method is tedious, its principle is generally easy to remember. We reverse the process of multiplying binomials.

The Trial-and-Error Method to Factor $ax^2 + bx + c$

Step 1 Factor out the greatest common factor.

Step 2 List all pairs of positive factors of a and pairs of positive factors of c. Consider the reverse order for either list of factors.

Step 3 Construct two binomials of the form

Factors of a

$$(\Box x \quad \Box)(\Box x \quad \Box)$$

Factors of c

Step 4 Test each combination of factors and signs until the correct product is found.

Step 5 If no combination of factors produces the correct product, the trinomial cannot be factored further and is a **prime polynomial**.

Example 4 — Factoring a Trinomial by the Trial-and-Error Method

Factor by the trial-and-error method. $\quad 13y - 6 + 8y^2$

Solution:

$8y^2 + 13y - 6$ $\qquad\qquad$ Write in the form $ax^2 + bx + c$.

$\qquad\qquad$ **Step 1:** The GCF is 1.

Factors of 8	Factors of 6
$1 \cdot 8$	$1 \cdot 6$
$2 \cdot 4$	$2 \cdot 3$
	$3 \cdot 2$
	$6 \cdot 1$

$\Big\}$ (reverse order)

Step 2: List the positive factors of 8 and positive factors of 6. Consider the reverse order in one list of factors.

$(2y \quad 1)(4y \quad 6)$
$(2y \quad 2)(4y \quad 3)$
$(2y \quad 3)(4y \quad 2)$
$(2y \quad 6)(4y \quad 1)$
$(1y \quad 1)(8y \quad 6)$
$(1y \quad 3)(8y \quad 2)$

Step 3: Construct all possible binomial factors by using different combinations of the factors of 8 and 6.

Without regard to signs, these factorizations cannot work because the terms in the binomial share a common factor greater than 1.

Test the remaining factorizations. Keep in mind that to produce a product of −6, the signs within the parentheses must be opposite (one positive and one negative). Also, the sum of the products of the inner terms and outer terms must be combined to form 13y.

$(1y \quad 6)(8y \quad 1)$ *Incorrect.*   Wrong middle term.

Regardless of signs, the product of inner terms 48y and the product of outer terms 1y cannot be combined to form the middle term 13y.

$(1y \quad 2)(8y \quad 3)$ *Correct.*   The terms 16y and 3y can be combined to form the middle term 13y, provided the signs are applied correctly. We require $+16y$ and $-3y$.

The correct factorization of $8y^2 + 13y - 6$ is $(y + 2)(8y - 3)$.

Skill Practice Factor by the trial-and-error method.

4. $5t - 6 + 4t^2$

In Example 4, the factors of −6 must have opposite signs to produce a negative product. Therefore, one binomial factor is a sum and one is a difference. Determining the correct signs is an important aspect of factoring trinomials. We suggest the following guidelines:

TIP: Given the trinomial $ax^2 + bx + c$ ($a > 0$), the signs can be determined as follows:

1. If c is *positive*, then the signs in the binomials must be the same (either both positive or both negative). The correct choice is determined by the middle term. If the middle term is positive, then both signs must be positive. If the middle term is negative, then both signs must be negative.

c is positive.	c is positive.
Example: $20x^2 + 43x + 21$	Example: $20x^2 - 43x + 21$
$(4x + 3)(5x + 7)$	$(4x - 3)(5x - 7)$
same signs	same signs

2. If c is *negative*, then the signs in the binomials must be different. The middle term in the trinomial determines which factor gets the positive sign and which factor gets the negative sign.

c is negative.	c is negative.
Example: $x^2 + 3x - 28$	Example: $x^2 - 3x - 28$
$(x + 7)(x - 4)$	$(x - 7)(x + 4)$
different signs	different signs

Answer

4. $(4t - 3)(t + 2)$

Example 5   Factoring a Trinomial by the
                    Trial-and-Error Method

Factor by the trial-and-error method.

$$-80x^3y + 208x^2y^2 - 20xy^3$$

Solution:

$-80x^3y + 208x^2y^2 - 20xy^3$

$= -4xy(20x^2 - 52xy + 5y^2)$ **Step 1:** Factor out $-4xy$.

$= -4xy(\square x\ \square y)(\square x\ \square y)$

Factors of 20	Factors of 5
$1 \cdot 20$	$1 \cdot 5$
$2 \cdot 10$	$5 \cdot 1$
$4 \cdot 5$	

Step 2: List the positive factors of 20 and positive factors of 5. Consider the reverse order in one list of factors.

Step 3: Construct all possible binomial factors by using different combinations of the factors of 20 and factors of 5. The signs in the parentheses must both be negative.

$\left.\begin{array}{l} -4xy(1x - 1y)(20x - 5y) \\ -4xy(2x - 1y)(10x - 5y) \\ -4xy(4x - 1y)(5x - 5y) \end{array}\right\}$ *Incorrect.*   These binomials contain a common factor.

$-4xy(1x - 5y)(20x - 1y)$ *Incorrect.*   Wrong middle term.
$\qquad\qquad\qquad\qquad\qquad -4xy(x - 5y)(20x - 1y)$
$\qquad\qquad\qquad\qquad\qquad\quad = -4xy(20x^2 - 101xy + 5y^2)$

$-4xy(4x - 5y)(5x - 1y)$ *Incorrect.*   Wrong middle term.
$\qquad\qquad\qquad\qquad\qquad -4xy(4x - 5y)(5x - 1y)$
$\qquad\qquad\qquad\qquad\qquad\quad = -4xy(20x^2 - 29x + 5y^2)$

$-4xy(2x - 5y)(10x - 1y)$ **Correct.** $-4xy(2x - 5y)(10x - 1y)$
$\qquad\qquad\qquad\qquad\qquad\quad = \mathbf{-4xy(20x^2 - 52xy + 5y^2)}$
$\qquad\qquad\qquad\qquad\qquad\quad = -80x^3y + 208x^2y^2 - 20xy^3$

The correct factorization of $-80x^3y + 208x^2y^2 - 20xy^3$ is $-4xy(2x - 5y)(10x - y)$.

Skill Practice   Factor by the trial-and-error method.

5. $-4z^3 - 22z^2 - 30z$

Answer

5. $-2z(2z + 5)(z + 3)$

Example 6 | **Factoring a Trinomial by the Trial-and-Error Method**

Factor completely.

$$2x^2 + 9x + 14$$

Solution:

$2x^2 + 9x + 14$	The GCF is 1 and the trinomial is written in the form $ax^2 + bx + c$.
$(2x + 14)(x + 1)$	*Incorrect.* $(2x + 14)$ contains a common factor of 2.
$(2x + 2)(x + 7)$	*Incorrect.* $(2x + 2)$ contains a common factor of 2.
$(2x + 1)(x + 14) = 2x^2 + 15x + 14$	*Incorrect.* Wrong middle term.
$(2x + 7)(x + 2) = 2x^2 + 11x + 14$	*Incorrect.* Wrong middle term.

No combination of factors results in the correct product. Therefore, the trinomial is prime (cannot be factored).

Skill Practice Factor completely.

6. $6r^2 - 13r + 10$

If a trinomial has a leading coefficient of 1, the factoring process simplifies significantly. Consider the trinomial $x^2 + bx + c$. To produce a leading term of x^2, we can construct binomials of the form $(x + \square)(x + \square)$. The remaining terms may be satisfied by two numbers p and q whose product is c and whose sum is b:

$$\overbrace{(x + p)(x + q)}^{\text{Factors of } c} = x^2 + qx + px + pq = x^2 + \underbrace{(p + q)}_{\text{Sum} = b}x + \underbrace{pq}_{\text{Product} = c}$$

This process is demonstrated in Example 7.

Example 7 | **Factoring a Trinomial with a Leading Coefficient of 1**

Factor completely.

$$x^2 - 10x + 16$$

Solution:

$x^2 - 10x + 16$	Factor out the GCF from all terms. In this case, the GCF is 1.
$= (x \quad \square)(x \quad \square)$	The trinomial is written in the form $x^2 + bx + c$. To form the product x^2, use the factors x and x.

Answer

6. Prime

Next, look for two numbers whose product is 16 and whose sum is -10. Because the middle term is negative, we will consider only the negative factors of 16.

Factors of 16	Sum
$-1(-16)$	$-1 + (-16) = -17$
$-2(-8)$	$-2 + (-8) = -10$
$-4(-4)$	$-4 + (-4) = -8$

The numbers are -2 and -8.

Therefore, $x^2 - 10x + 16 = (x - 2)(x - 8)$.

Skill Practice Factor completely.

7. $c^2 + 6c - 27$

3. Factoring Perfect Square Trinomials

Recall from Section 4.3 that the square of a binomial always results in a **perfect square trinomial**.

$$(a + b)^2 = (a + b)(a + b) = a^2 + ab + ab + b^2 = a^2 + 2ab + b^2$$
$$(a - b)^2 = (a - b)(a - b) = a^2 - ab - ab + b^2 = a^2 - 2ab + b^2$$

For example, $(2x + 7)^2 = (2x)^2 + 2(2x)(7) + (7)^2 = 4x^2 + 28x + 49$

$a = 2x \quad b = 7 \qquad\qquad a^2 + 2ab + b^2$

To factor the trinomial $4x^2 + 28x + 49$, the ac-method or the trial-and-error method can be used. However, recognizing that the trinomial is a perfect square trinomial, we can use one of the following patterns to reach a quick solution.

> **Factored Form of a Perfect Square Trinomial**
> $$a^2 + 2ab + b^2 = (a + b)^2$$
> $$a^2 - 2ab + b^2 = (a - b)^2$$

TIP: The following are perfect squares.

$1^2 = 1$	$(x^1)^2 = x^2$
$2^2 = 4$	$(x^2)^2 = x^4$
$3^2 = 9$	$(x^3)^2 = x^6$
$4^2 = 16$	$(x^4)^2 = x^8$
$\vdots$	$\vdots$

Any expression raised to an even power (multiple of 2) is a perfect square.

TIP: To determine if a trinomial is a perfect square trinomial, follow these steps:

1. Check if the first and third terms are both perfect squares with positive coefficients.
2. If this is the case, identify a and b, and determine if the middle term equals $2ab$ or $-2ab$.

Answer

7. $(c + 9)(c - 3)$

Example 8 Factoring a Perfect Square Trinomial

Factor completely. $x^2 + 12x + 36$

Solution:

$x^2 + 12x + 36$ The GCF is 1.

- The first and third terms are positive.
- The first term is a perfect square:
$$x^2 = (x)^2$$

Perfect squares

$= x^2 + 12x + 36$

- The third term is a perfect square:
$$36 = (6)^2$$
- The middle term is twice the product of x and 6:

$$12x = 2(x)(6)$$

$= (x)^2 + 2(x)(6) + (6)^2$ The trinomial is in the form
$a^2 + 2ab + b^2$, where $a = x$ and $b = 6$.

$= (x + 6)^2$ Factor as $(a + b)^2$.

Skill Practice Factor completely.

8. $x^2 + 2x + 1$

Example 9 Factoring a Perfect Square Trinomial

Factor completely. $4x^2 - 36xy + 81y^2$

Solution:

$4x^2 - 36xy + 81y^2$ The GCF is 1.

- The first and third terms are positive.
- The first term is a perfect square:
$$4x^2 = (2x)^2.$$

Perfect squares

$= 4x^2 - 36xy + 81y^2$

- The third term is a perfect square:
$$81y^2 = (9y)^2.$$
- The middle term:

$$-36xy = -2(2x)(9y)$$

$= (2x)^2 - 2(2x)(9y) + (9y)^2$ The trinomial is in the form
$a^2 - 2ab + b^2$, where $a = 2x$ and $b = 9y$.

$= (2x - 9y)^2$ Factor as $(a - b)^2$.

Skill Practice Factor completely.

9. $9y^2 - 12yz + 4z^2$

Answers

8. $(x + 1)^2$
9. $(3y - 2z)^2$

4. Factoring by Using Substitution

Sometimes it is convenient to use substitution to convert a polynomial into a simpler form before factoring.

Example 10 Using Substitution to Factor a Polynomial

Factor by using substitution. $(2x - 7)^2 - 3(2x - 7) - 40$

Solution:

$(2x - 7)^2 - 3(2x - 7) - 40$

$= u^2 - 3u - 40$	Substitute $u = 2x - 7$. The trinomial is simpler in form.
$= (u - 8)(u + 5)$	Factor the trinomial.
$= [(2x - 7) - 8][(2x - 7) + 5]$	Reverse substitute. Replace u by $2x - 7$.
$= (2x - 7 - 8)(2x - 7 + 5)$	Simplify.
$= (2x - 15)(2x - 2)$	The second binomial has a GCF of 2.
$= (2x - 15)(2)(x - 1)$	Factor out the GCF from the second binomial.
$= 2(2x - 15)(x - 1)$	

Skill Practice Factor by using substitution.

10. $(3x + 1)^2 + 2(3x + 1) - 15$

Example 11 Using Substitution to Factor a Polynomial

Factor by using substitution. $6y^6 - 5y^3 - 4$

Solution:

$6y^6 - 5y^3 - 4$

$= 6(y^3)^2 - 5(y^3) - 4$	Let $u = y^3$.
$= 6u^2 - 5u - 4$	Substitute u for y^3 in the trinomial.
$= (2u + 1)(3u - 4)$	Factor the trinomial.
$= (2y^3 + 1)(3y^3 - 4)$	Reverse substitute. Replace u with y^3.

TIP: The ac-method or trial-and-error method can also be used for Example 11 without using substitution.

Skill Practice Factor by using substitution.

11. $2x^4 + 7x^2 + 3$

Answers

10. $3(3x - 2)(x + 2)$
11. $(2x^2 + 1)(x^2 + 3)$

As you work through the exercises in this section, keep these guidelines in mind to factor trinomials.

Factoring Trinomials of the Form $ax^2 + bx + c$ $(a \neq 0)$

When factoring trinomials, the following guidelines should be considered:

Step 1 Factor out the greatest common factor.

Step 2 Check to see if the trinomial is a perfect square trinomial. If so, factor it as either $(a + b)^2$ or $(a - b)^2$. (With a perfect square trinomial, you do not need to use the ac-method or trial-and-error method.)

Step 3 If the trinomial is not a perfect square, use either the ac-method or the trial-and-error method to factor.

Step 4 Check the factorization by multiplication.

Note: Consider using substitution if a trinomial is in the form $au^2 + bu + c$, where u is an algebraic expression.

Section 4.6   Practice Exercises

Vocabulary and Key Concepts

1. **a.** Given a trinomial $x^2 + bx + c$, if c is positive, then the signs in the binomial factors are either both _____ or both negative.

   b. Given a trinomial $x^2 + bx + c$, if c is negative, then the signs in the binomial factors are (choose one: both positive, both negative, opposite).

   c. Which is the correct factored form of $2x^2 - 5x - 12$, the product $(2x + 3)(x - 4)$ or $(x - 4)(2x + 3)$?

   d. Which is the complete factorization of $6x^2 - 4x - 10$, the product $(3x - 5)(2x + 2)$ or $2(3x - 5)(x + 1)$?

   e. A perfect square trinomial $a^2 + 2ab + b^2$ factors as _____. Likewise $a^2 - 2ab + b^2$ factors as _____.

Review Exercises

2. Explain how to check a factoring problem.

For Exercises 3–8, factor the polynomial completely.

3. $36c^2d^7e^{11} + 12c^3d^5e^{15} - 6c^2d^4e^7$

4. $5x^3y^3 + 15x^4y^2 - 35x^2y^4$

5. $2x(3a - b) - (3a - b)$

6. $6(v - 8) - 3u(v - 8)$

7. $wz^2 + 2wz - 33az - 66a$

8. $3a^2x + 9ab - abx - 3b^2$

Concepts 1–2: Factoring Trinomials

In Exercises 9–46, factor the trinomial completely by using any method. Remember to look for a common factor first. **(See Examples 1–7.)**

9. $b^2 - 12b + 32$

10. $a^2 - 12a + 27$

11. $y^2 + 10y - 24$

12. $w^2 + 3w - 54$

13. $x^2 + 13x + 30$

14. $t^2 + 9t + 8$

15. $c^2 - 6c - 16$ **16.** $z^2 - 3z - 28$ **17.** $2x^2 - 7x - 15$

18. $2y^2 - 13y + 15$ **19.** $a + 6a^2 - 5$ **20.** $10b^2 - 3 - 29b$

21. $s^2 + st - 6t^2$ **22.** $p^2 - pq - 20q^2$ **23.** $3x^2 - 60x + 108$

24. $4c^2 + 12c - 72$ **25.** $2c^2 - 2c - 24$ **26.** $3x^2 + 12x - 15$

27. $2x^2 + 8xy - 10y^2$ **28.** $20z^2 + 26zw - 28w^2$ **29.** $33t^2 - 18t + 2$

30. $5p^2 - 10p + 7$ **31.** $3x^2 + 14xy + 15y^2$ **32.** $2a^2 + 15ab - 27b^2$

33. $5u^3v - 30u^2v^2 + 45uv^3$ **34.** $3a^3 + 30a^2b + 75ab^2$ **35.** $x^3 - 5x^2 - 14x$

36. $p^3 + 2p^2 - 24p$ **37.** $-23z - 5 + 10z^2$ **38.** $3 + 16y^2 + 14y$

39. $b^2 + 2b + 15$ **40.** $x^2 - x - 1$ **41.** $-2t^2 + 12t + 80$

42. $-3c^2 + 33c - 72$ **43.** $14a^2 + 13a - 12$ **44.** $12x^2 - 16x + 5$

45. $6a^2b + 22ab + 12b$ **46.** $6cd^2 + 9cd - 42c$

Concept 3: Factoring Perfect Square Trinomials

47. a. Multiply the binomials $(x + 5)(x + 5)$.

    b. Factor $x^2 + 10x + 25$.

48. a. Multiply the binomials $(2w - 5)(2w - 5)$.

    b. Factor $4w^2 - 20w + 25$.

49. a. Multiply the binomials $(3x - 2y)^2$.

    b. Factor $9x^2 - 12xy + 4y^2$.

50. a. Multiply the binomials $(x + 7y)^2$.

    b. Factor $x^2 + 14xy + 49y^2$.

For Exercises 51–54, fill in the blank to make the trinomial a perfect square trinomial.

51. $9x^2 + (\underline{\hspace{1cm}}) + 25$ **52.** $16x^4 - (\underline{\hspace{1cm}}) + 1$

53. $64z^4 + (\underline{\hspace{1cm}}) + t^2$ **54.** $9m^4 - (\underline{\hspace{1cm}}) + 49n^2$

For Exercises 55–66, factor out the greatest common factor, if necessary. Then determine if the polynomial is a perfect square trinomial. If it is, factor it. **(See Examples 8–9.)**

55. $y^2 - 8y + 16$ **56.** $x^2 + 10x + 25$ **57.** $64m^2 + 80m + 25$

58. $100c^2 - 140c + 49$ **59.** $w^2 - 5w + 9$ **60.** $2a^2 + 14a + 98$

61. $9a^2 - 30ab + 25b^2$ **62.** $16x^4 - 48x^2y + 9y^2$ **63.** $16t^2 - 80tv + 20v^2$

64. $12x^2 - 12xy + 3y^2$ **65.** $5b^4 - 20b^2 + 20$ **66.** $a^4 + 12a^2 + 36$

Concept 4: Factoring by Using Substitution

For Exercises 67–70, factor the polynomial in part (a). Then use substitution to help factor the polynomials in parts (b) and (c).

67. a. $u^2 - 10u + 25$

    b. $x^4 - 10x^2 + 25$

    c. $(a + 1)^2 - 10(a + 1) + 25$

68. a. $u^2 + 12u + 36$

    b. $y^4 + 12y^2 + 36$

    c. $(b - 2)^2 + 12(b - 2) + 36$

69. a. $u^2 + 11u - 26$

    b. $w^6 + 11w^3 - 26$

    c. $(y - 4)^2 + 11(y - 4) - 26$

70. a. $u^2 + 17u + 30$

    b. $z^6 + 17z^3 + 30$

    c. $(x + 3)^2 + 17(x + 3) + 30$

For Exercises 71–82, factor by using substitution. **(See Examples 10–11.)**

71. $(3x - 1)^2 - (3x - 1) - 6$

72. $(2x + 5)^2 - (2x + 5) - 12$

73. $2(x - 5)^2 + 9(x - 5) + 4$

74. $4(x - 3)^2 + 7(x - 3) + 3$

75. $3(y + 4)^2 + 5(y + 4) - 2$

76. $(3t - 2)^2 - (3t - 2) - 20$

77. $3y^6 + 11y^3 + 6$

78. $3x^4 - 5x^2 - 12$

79. $4p^4 + 5p^2 + 1$

80. $t^4 + 3t^2 + 2$

81. $x^4 + 15x^2 + 36$

82. $t^6 - 16t^3 + 63$

Mixed Exercises

83. A student factored $4y^2 - 10y + 4$ as $(2y - 1)(2y - 4)$ on her factoring test. Why did her professor deduct several points, even though $(2y - 1)(2y - 4)$ does multiply out to $4y^2 - 10y + 4$?

84. A student factored $9w^2 + 36w + 36$ as $(3w + 6)^2$ on his factoring test. Why did his instructor deduct several points, even though $(3w + 6)^2$ does multiply out to $9w^2 + 36w + 36$?

For Exercises 85–105, factor completely by using an appropriate method. (Be sure to note the number of terms in the polynomial.)

85. $w^4 + 12w^2 + 36$

86. $9 - 6t^2 + t^4$

87. $81w^2 + 90w + 25$

88. $49a^2 - 28ab + 4b^2$

89. $3x(a + b) - 6(a + b)$

90. $4p(t - 8) + 2(t - 8)$

91. $12a^2bc^2 + 4ab^2c^2 - 6abc^3$

92. $18x^2z - 6xyz + 30xz^2$

93. $-20x^3 + 74x^2 - 60x$

94. $-24y^3 + 90y^2 - 75y$

95. $2y^2 - 9y - 4$

96. $3w^2 - 12w + 4$

97. $2(w^2 - 5)^2 + (w^2 - 5) - 15$

98. $5(t^2 + 3)^2 + 21(t^2 + 3) + 4$

99. $1 - 4d + 3d^2$

100. $2 - 5a + 2a^2$

101. $ax - 5a^2 + 2bx - 10ab$

102. $my + y^2 - 3xm - 3xy$

103. $8z^2 + 24zw - 224w^2$

104. $9x^2 - 18xy - 135y^2$

105. $ay + ax - 5cy - 5cx$

For Exercises 106–114, factor the expression that defines each function.

106. $f(x) = 2x^2 + 13x - 7$

107. $g(x) = 3x^2 + 14x + 8$

108. $m(t) = t^2 - 22t + 121$

109. $n(t) = t^2 + 20t + 100$

110. $P(x) = x^3 + 4x^2 + 3x$

111. $Q(x) = x^4 + 6x^3 + 8x^2$

112. $h(a) = a^3 + 5a^2 - 6a - 30$

113. $k(a) = a^3 - 4a^2 + 2a - 8$

114. $f(x) = 3x^3 - 9x^2 + 5x - 15$

Section 4.7 Factoring Binomials

Concepts

1. Difference of Squares
2. Using a Difference of Squares in Grouping
3. Sum and Difference of Cubes
4. Summary of Factoring Binomials
5. Factoring Binomials of the Form $x^6 - y^6$

1. Difference of Squares

Up to this point we have learned how to

- Factor out the greatest common factor from a polynomial.
- Factor a four-term polynomial by grouping.
- Recognize and factor perfect square trinomials.
- Factor trinomials by the ac-method and by the trial-and-error method.

Next, we will learn how to factor binomials that fit the pattern of a difference of squares. Recall from Section 4.3 that the product of two conjugates results in a **difference of squares**

$$(a + b)(a - b) = a^2 - b^2$$

Therefore, to factor a difference of squares, the process is reversed. Identify a and b and construct the conjugate factors.

> **Factored Form of a Difference of Squares**
> $$a^2 - b^2 = (a + b)(a - b)$$

Example 1 Factoring a Difference of Squares

Factor the binomial completely. $16x^2 - 9$

Solution:

$16x^2 - 9$ — The GCF is 1. The binomial is a difference of squares.

$= (4x)^2 - (3)^2$ — Write in the form $a^2 - b^2$, where $a = 4x$ and $b = 3$.

$= (4x + 3)(4x - 3)$ — Factor as $(a + b)(a - b)$.

Skill Practice Factor completely.

1. $4z^2 - 1$

Example 2 Factoring a Difference of Squares

Factor the binomial completely. $98c^2d - 50d^3$

Solution:

$98c^2d - 50d^3$

$= 2d(49c^2 - 25d^2)$ — The GCF is $2d$. The resulting binomial is a difference of squares.

$= 2d[(7c)^2 - (5d)^2]$ — Write in the form $a^2 - b^2$, where $a = 7c$ and $b = 5d$.

$= 2d(7c + 5d)(7c - 5d)$ — Factor as $(a + b)(a - b)$.

Skill Practice Factor completely.

2. $7y^3z - 63yz^3$

Answers
1. $(2z - 1)(2z + 1)$
2. $7yz(y + 3z)(y - 3z)$

Example 3 Factoring a Difference of Squares

Factor the binomial completely. $z^4 - 81$

Solution:

$z^4 - 81$	The GCF is 1. The binomial is a difference of squares.
$= (z^2)^2 - (9)^2$	Write in the form $a^2 - b^2$, where $a = z^2$ and $b = 9$.
$= (z^2 + 9)(z^2 - 9)$	Factor as $(a + b)(a - b)$.
	$z^2 - 9$ is also a difference of squares.
$= (z^2 + 9)(z + 3)(z - 3)$	

Skill Practice Factor completely.

3. $b^4 - 16$

The difference of squares $a^2 - b^2$ factors as $(a - b)(a + b)$. However, the *sum* of squares is not factorable.

Sum of Squares

Suppose a and b have no common factors. Then the **sum of squares** $a^2 + b^2$ is *not* factorable over the real numbers.

That is, $a^2 + b^2$ is prime over the real numbers.

To see why $a^2 + b^2$ is not factorable, consider the product of binomials:

$$(a \quad b)(a \quad b) \stackrel{?}{=} a^2 + b^2$$

If all possible combinations of signs are considered, none produces the correct product.

$(a + b)(a - b) = a^2 - b^2$	Wrong sign
$(a + b)(a + b) = a^2 + 2ab + b^2$	Wrong middle term
$(a - b)(a - b) = a^2 - 2ab + b^2$	Wrong middle term

After exhausting all possibilities, we see that if a and b share no common factors, then the sum of squares $a^2 + b^2$ is a prime polynomial.

2. Using a Difference of Squares in Grouping

Sometimes a difference of squares can be used along with other factoring techniques.

Answer

3. $(b^2 + 4)(b - 2)(b + 2)$

| Example 4 | **Using a Difference of Squares in Grouping** |

Factor completely. $y^3 - 6y^2 - 4y + 24$

Solution:

$$y^3 - 6y^2 - 4y + 24 \qquad \text{The GCF is 1.}$$

$$= y^3 - 6y^2 \mid - 4y + 24 \qquad \begin{array}{l}\text{The polynomial has four terms.}\\\text{Factor by grouping.}\end{array}$$

$$= y^2(y - 6) - 4(y - 6)$$

$$= (y - 6)(y^2 - 4) \qquad y^2 - 4 \text{ is a difference of squares.}$$

$$= (y - 6)(y + 2)(y - 2)$$

Skill Practice Factor completely.

4. $a^3 + 5a^2 - 9a - 45$

| Example 5 | **Factoring a Four-Term Polynomial by Grouping Three Terms** |

Factor completely. $x^2 - y^2 - 6y - 9$

Solution:

Grouping "2 by 2" will not work to factor this polynomial. However, if we factor out -1 from the last three terms, the resulting trinomial will be a perfect square trinomial.

$$x^2 \mid - y^2 - 6y - 9 \qquad \text{Group the last three terms.}$$

$$= x^2 - 1(y^2 + 6y + 9) \qquad \text{Factor out } -1 \text{ from the last three terms.}$$

$$= x^2 - (y + 3)^2 \qquad \begin{array}{l}\text{Factor the perfect square trinomial}\\y^2 + 6y + 9 \text{ as } (y + 3)^2.\end{array}$$

$$\qquad\qquad \begin{array}{l}\text{The quantity } x^2 - (y + 3)^2 \text{ is a difference}\\\text{of squares, } a^2 - b^2, \text{ where } a = x \text{ and}\\b = (y + 3).\end{array}$$

$$= [x - (y + 3)][x + (y + 3)] \qquad \text{Factor as } a^2 - b^2 = (a + b)(a - b).$$

$$= (x - y - 3)(x + y + 3) \qquad \begin{array}{l}\text{Apply the distributive property to clear}\\\text{the inner parentheses.}\end{array}$$

Avoiding Mistakes

When factoring the expression $x^2 - (y + 3)^2$ as a difference of squares, be sure to use parentheses around the quantity $(y + 3)$. This will help you remember to "distribute the negative" in the expression $[x - (y + 3)]$.

$[x - (y + 3)] = (x - y - 3)$

Skill Practice Factor completely.

5. $x^2 + 10x + 25 - y^2$

Answers

4. $(a + 5)(a - 3)(a + 3)$
5. $(x + 5 - y)(x + 5 + y)$

TIP: From Example 5, the expression $x^2 - (y + 3)^2$ can also be factored by using substitution. Let $u = y + 3$.

$x^2 - (y + 3)^2$

$= x^2 - u^2$ Substitution $u = y + 3$.

$= (x - u)(x + u)$ Factor as a difference of squares.

$= [x - (y + 3)][x + (y + 3)]$ Substitute back.

$= (x - y - 3)(x + y + 3)$ Apply the distributive property.

3. Sum and Difference of Cubes

For binomials that represent the sum or difference of cubes, factor by using the following formulas.

Factored Form of a Sum and Difference of Cubes

Sum of cubes: $a^3 + b^3 = (a + b)(a^2 - ab + b^2)$

Difference of cubes: $a^3 - b^3 = (a - b)(a^2 + ab + b^2)$

TIP: The following are perfect cubes.

$1^3 = 1$	$(x^1)^3 = x^3$
$2^3 = 8$	$(x^2)^3 = x^6$
$3^3 = 27$	$(x^3)^3 = x^9$
$4^3 = 64$	$(x^4)^3 = x^{12}$
$\vdots$	$\vdots$

Any expression raised to a multiple of 3 is a perfect cube.

Multiplication can be used to confirm the formulas for factoring a sum or difference of cubes.

$(a + b)(a^2 - ab + b^2) = a^3 - a^2b + ab^2 + a^2b - ab^2 + b^3 = a^3 + b^3$ ✔

$(a - b)(a^2 + ab + b^2) = a^3 + a^2b + ab^2 - a^2b - ab^2 - b^3 = a^3 - b^3$ ✔

To help you remember the formulas for factoring a sum or difference of cubes, keep the following guidelines in mind.

- The factored form is the product of a binomial and a trinomial.
- The first and third terms in the trinomial are the squares of the terms within the binomial factor. Therefore, these terms are always positive.
- Without regard to sign, the middle term in the trinomial is the product of terms in the binomial factor.

Square the first term of the binomial.   Product of terms in the binomial

$x^3 + 8 = (x)^3 + (2)^3 = (x + 2)[(x)^2 - (x)(2) + (2)^2]$

Square the last term of the binomial.

- The sign within the binomial factor is the same as the sign of the original binomial.
- The first and third terms in the trinomial are always positive.
- The sign of the middle term in the trinomial is opposite the sign within the binomial.

TIP: To help remember the placement of the signs in factoring the sum or difference of cubes, remember SOAP: Same sign, Opposite signs, Always Positive.

Same sign   Positive

$x^3 + 8 = (x)^3 + (2)^3 = (x + 2)[(x)^2 - (x)(2) + (2)^2]$

Opposite signs

Example 6 Factoring a Difference of Cubes

Factor. $8x^3 - 27$

Solution:

$$8x^3 - 27 \qquad \text{$8x^3$ and 27 are perfect cubes.}$$

$$= (2x)^3 - (3)^3 \qquad \text{Write as } a^3 - b^3, \text{ where } a = 2x \text{ and } b = 3.$$

$$a^3 - b^3 = (a - b)(a^2 + ab + b^2)$$
$$(2x)^3 - (3)^3 = (2x - 3)[(2x)^2 + (2x)(3) + (3)^2] \qquad \text{Apply the difference of cubes formula.}$$

$$= (2x - 3)(4x^2 + 6x + 9) \qquad \text{Simplify.}$$

Skill Practice Factor completely.

6. $125p^3 - 8$

Example 7 Factoring a Sum of Cubes

Factor. $125t^3 + 64z^6$

Solution:

$$125t^3 + 64z^6 \qquad \text{$125t^3$ and $64z^6$ are perfect cubes.}$$

$$= (5t)^3 + (4z^2)^3 \qquad \text{Write as } a^3 + b^3, \text{ where } a = 5t \text{ and } b = 4z^2.$$

$$a^3 + b^3 = (a + b)(a^2 - ab + b^2) \qquad \text{Apply the sum of cubes formula.}$$

$$(5t)^3 + (4z^2)^3 = [(5t) + (4z^2)][(5t)^2 - (5t)(4z^2) + (4z^2)^2]$$
$$= (5t + 4z^2)(25t^2 - 20tz^2 + 16z^4) \qquad \text{Simplify.}$$

Skill Practice Factor completely.

7. $x^3 + 1000y^6$

4. Summary of Factoring Binomials

After factoring out the greatest common factor, the next step in any factoring problem is to recognize what type of pattern it follows. Exponents that are divisible by 2 are perfect squares, and those divisible by 3 are perfect cubes. The formulas for factoring binomials are summarized here.

Answers

6. $(5p - 2)(25p^2 + 10p + 4)$
7. $(x + 10y^2)(x^2 - 10xy^2 + 100y^4)$

Summary of Factoring Binomials
- Difference of squares: $\quad a^2 - b^2 = (a + b)(a - b)$
- Difference of cubes: $\quad a^3 - b^3 = (a - b)(a^2 + ab + b^2)$
- Sum of cubes: $\quad\quad\ \ a^3 + b^3 = (a + b)(a^2 - ab + b^2)$

Example 8   **Review of Factoring Binomials**

Factor the binomials.

a. $m^3 - \dfrac{1}{8}$ **b.** $9k^2 + 24m^2$ **c.** $128y^6 + 54x^3$ **d.** $50y^6 - 8x^2$

Solution:

a. $m^3 - \dfrac{1}{8}$ | m^3 is a perfect cube: $m^3 = (m)^3$. $\frac{1}{8}$ is a perfect cube: $\frac{1}{8} = (\frac{1}{2})^3$.

$= (m)^3 - \left(\dfrac{1}{2}\right)^3$ | This is a difference of cubes, where $a = m$ and $b = \frac{1}{2}$:

$= \left(m - \dfrac{1}{2}\right)\left(m^2 + \dfrac{1}{2}m + \dfrac{1}{4}\right)$ | $a^3 - b^3 = (a - b)(a^2 + ab + b^2)$.

b. $9k^2 + 24m^2$ | Factor out the GCF.

$= 3(3k^2 + 8m^2)$ | The resulting binomial is not a difference of squares or a sum or difference of cubes. It cannot be factored further over the real numbers.

c. $128y^6 + 54x^3$ | Factor out the GCF.

$= 2(64y^6 + 27x^3)$ | Both 64 and 27 are perfect cubes, and the exponents of both x and y are multiples of 3. This is a sum of cubes, where $a = 4y^2$ and $b = 3x$.

$= 2[(4y^2)^3 + (3x)^3]$

$= 2(4y^2 + 3x)(16y^4 - 12xy^2 + 9x^2)$ | $a^3 + b^3 = (a + b)(a^2 - ab + b^2)$.

d. $50y^6 - 8x^2$ | Factor out the GCF.

$= 2(25y^6 - 4x^2)$ | Both 25 and 4 are perfect squares. The exponents of both x and y are multiples of 2. This is a difference of squares, where $a = 5y^3$ and $b = 2x$.

$= 2[(5y^3)^2 - (2x)^2]$

$= 2(5y^3 + 2x)(5y^3 - 2x)$ | $a^2 - b^2 = (a + b)(a - b)$.

Skill Practice Factor the binomials.

8. $x^2 - \dfrac{1}{25}$ **9.** $16y^3 + 4y$ **10.** $24a^7 - 3a$ **11.** $18p^4 - 50t^2$

Answers
8. $\left(x + \dfrac{1}{5}\right)\left(x - \dfrac{1}{5}\right)$
9. $4y(4y^2 + 1)$
10. $3a(2a^2 - 1)(4a^4 + 2a^2 + 1)$
11. $2(3p^2 + 5t)(3p^2 - 5t)$

5. Factoring Binomials of the Form $x^6 - y^6$

Example 9   Factoring Binomials

Factor the binomial $x^6 - y^6$ as

a. A difference of cubes

b. A difference of squares

Solution:

Notice that the expressions x^6 and y^6 are both perfect squares and perfect cubes because the exponents are both multiples of 2 and of 3. Consequently, $x^6 - y^6$ can be interpreted initially as either a difference of cubes or a difference of squares.

a. $x^6 - y^6$

$$= (x^2)^3 - (y^2)^3$$

Write as $a^3 - b^3$, where $a = x^2$ and $b = y^2$.

$$= (x^2 - y^2)[(x^2)^2 + (x^2)(y^2) + (y^2)^2]$$

Apply the formula $a^3 - b^3 = (a - b)(a^2 + ab + b^2)$.

$$= (x^2 - y^2)(x^4 + x^2y^2 + y^4)$$

Factor $x^2 - y^2$ as a difference of squares.

$$= (x + y)(x - y)(x^4 + x^2y^2 + y^4)$$

The expression $x^4 + x^2y^2 + y^4$ cannot be factored by using the skills learned thus far.

b. $x^6 - y^6$

$$= (x^3)^2 - (y^3)^2$$

Write as $a^2 - b^2$, where $a = x^3$ and $b = y^3$.

$$= (x^3 + y^3)(x^3 - y^3)$$

Apply the formula $a^2 - b^2 = (a + b)(a - b)$.

Factor $x^3 + y^3$ as a sum of cubes. Factor $x^3 - y^3$ as a difference of cubes.

$$= (x + y)(x^2 - xy + y^2)(x - y)(x^2 + xy + y^2)$$

TIP: Notice that the expressions x^6 and y^6 are both perfect squares and perfect cubes because both exponents are multiples of 2 and of 3. Consequently, $x^6 - y^6$ can be factored initially as either the difference of squares or as the difference of cubes. In such a case, it is recommended that you factor the expression as a difference of squares first because it factors more completely into polynomials of lower degree.

$$x^6 - y^6 = (x + y)(x^2 - xy + y^2)(x - y)(x^2 + xy + y^2)$$

Answer

12. $(a - 2)(a + 2)(a^2 + 2a + 4)$
$(a^2 - 2a + 4)$

Skill Practice Factor completely.

12. $a^6 - 64$

Section 4.7 Practice Exercises

Study Skills Exercise

Multiplying polynomials and factoring polynomials are inverse operations. That is, to check a factoring problem you can multiply, and to check a multiplication problem you can factor. To practice both operations, write a factored polynomial on one side of a 3×5 card with the directions, *Multiply*. On the other side of the card, write the expanded form of the polynomial with the directions, *Factor*. Now you can mix up the cards and get a good sense of what is meant by the directions: *Factor* and *Multiply*.

Vocabulary and Key Concepts

1. **a.** The binomial $x^2 - 36$ is an example of a _____ of squares. A difference of squares $a^2 - b^2$ factors as _____.

   b. The binomial $y^2 + 9$ is an example of a _____ of squares.

   c. A sum of squares with greatest common factor 1 (is/is not) factorable over the real numbers.

   d. The square of a binomial always results in a perfect _____ trinomial.

   e. The binomial $x^3 + 64$ is an example of a _____ of _____.

   f. The binomial $c^3 - 27$ is an example of a _____ of _____.

   g. A difference of cubes $a^3 - b^3$ factors as (   )(   ).

   h. A sum of cubes $a^3 + b^3$ factors as (   )(   ).

Review Exercises

2. **a.** Identify which expressions represent perfect squares.
   $2,\ 4,\ 8,\ 16,\ 25,\ 64,\ x^2,\ x^3,\ x^4,\ x^5,\ x^9$

   b. Identify which expressions represent perfect cubes.
   $2,\ 4,\ 8,\ 16,\ 25,\ 64,\ x^2,\ x^3,\ x^4,\ x^5,\ x^9$

For Exercises 3–8, factor completely.

3. $4x^2 - 20x + 25$

4. $9t^2 - 42t + 49$

5. $10x + 6xy + 5 + 3y$

6. $21a + 7ab - 3b - b^2$

7. $32p^2 - 28p - 4$

8. $6q^2 + 37q - 35$

Concept 1: Difference of Squares

9. Explain how to identify and factor a difference of squares.

10. Can you factor $25x^2 + 4$?

For Exercises 11–22, factor the binomials. Identify the binomials that are prime. **(See Examples 1–3.)**

11. $x^2 - 9$

12. $y^2 - 25$

13. $16 - 49w^2$

14. $81 - 64b^2$

15. $8a^2 - 162b^2$

16. $50c^2 - 72d^2$

17. $25u^2 + 1$

18. $w^2 + 4$

19. $2a^4 - 32$

20. $5y^4 - 5$

21. $49 - k^6$

22. $4 - h^5$

Concept 2: Using a Difference of Squares in Grouping

For Exercises 23–36, use the difference of squares along with factoring by grouping. **(See Examples 4–5.)**

23. $x^3 - x^2 - 16x + 16$ **24.** $x^3 + 5x^2 - x - 5$ **25.** $4x^3 + 12x^2 - x - 3$

26. $5x^3 - x^2 - 45x + 9$ **27.** $9y^3 + 7y^2 - 36y - 28$ **28.** $9z^3 - 5z^2 - 36z + 20$

29. $49x^2 + 28x + 4 - y^2$ **30.** $100y^2 + 140y + 49 - z^2$ **31.** $w^2 - 9n^2 + 6n - 1$

32. $m^2 - 25c^2 + 20c - 4$ **33.** $p^4 - 10p^2 + 25 - t^4$ **34.** $m^4 - 14m^2 + 49 - z^2$

35. $9u^4 - 4v^4 + 20v^2 - 25$ **36.** $x^4 - 9y^4 - 42y^2 - 49$

Concept 3: Sum and Difference of Cubes

37. Explain how to identify and factor a sum of cubes.

38. Explain how to identify and factor a difference of cubes.

For Exercises 39–52, factor the sum or difference of cubes. **(See Examples 6–7.)**

39. $8x^3 - 1$ (Check by multiplying.)   **40.** $y^3 + 64$ (Check by multiplying.)

41. $125c^3 + 27$ **42.** $216u^3 - v^3$ **43.** $x^3 - 1000$ **44.** $y^3 - 27$

45. $64t^6 + 1$ **46.** $125r^6 + 1$ **47.** $2000y^6 + 2x^3$ **48.** $3a^6 + 24b^3$

49. $16z^4 - 54z$ **50.** $x^5 - 64x^2$ **51.** $p^{12} - 125$ **52.** $t^9 - 8$

Concept 4: Summary of Factoring Binomials

For Exercises 53–80, factor completely. **(See Example 8.)**

53. $36y^2 - \dfrac{1}{25}$ **54.** $16p^2 - \dfrac{1}{9}$ **55.** $18d^{12} - 32$ **56.** $3z^8 - 12$

57. $242v^2 + 32$ **58.** $8p^2 + 200$ **59.** $4x^2 - 16$ **60.** $9m^2 - 81n^2$

61. $25 - 49q^2$ **62.** $1 - 25p^2$ **63.** $(t + 2s)^2 - 36$ **64.** $(5x + 4)^2 - y^2$

65. $27 - t^3$ **66.** $8 + y^3$ **67.** $27a^3 + \dfrac{1}{8}$ **68.** $b^3 + \dfrac{27}{125}$

69. $2m^3 + 16$ **70.** $3x^3 - 375$ **71.** $x^4 - y^4$ **72.** $81u^4 - 16v^4$

73. $a^9 + b^9$ **74.** $27m^9 - 8n^9$ **75.** $\dfrac{1}{8}p^3 - \dfrac{1}{125}$ **76.** $1 - \dfrac{1}{27}d^3$

77. $4w^2 + 25$ **78.** $64 + a^2$ **79.** $\dfrac{1}{25}x^2 - \dfrac{1}{4}y^2$ **80.** $\dfrac{1}{100}a^2 - \dfrac{4}{49}b^2$

Concept 5: Factoring Binomials of the Form $x^6 - y^6$

For Exercises 81–88, factor completely. **(See Example 9.)**

81. $a^6 - b^6$ (*Hint:* First factor as a difference of squares.)

82. $64x^6 - y^6$ **83.** $64 - y^6$

84. $1 - p^6$

85. $h^6 + k^6$ (*Hint:* Factor as a sum of cubes.)

86. $27q^6 + 125p^6$

87. $8x^6 + 125$

88. $t^6 + 1$

Mixed Exercises

89. Find a difference of squares that has $(2x + 3)$ as one of its factors.

90. Find a difference of squares that has $(4 - p)$ as one of its factors.

91. Find a difference of cubes that has $(4a^2 + 6a + 9)$ as its trinomial factor.

92. Find a sum of cubes that has $(25c^2 - 10cd + 4d^2)$ as its trinomial factor.

93. Find a sum of cubes that has $(4x^2 + y)$ as its binomial factor.

94. Find a difference of cubes that has $(3t - r^2)$ as its binomial factor.

95. Consider the shaded region:

a. Find an expression that represents the area of the shaded region.

b. Factor the expression found in part (a).

c. Find the area of the shaded region if $x = 6$ in. and $y = 4$ in.

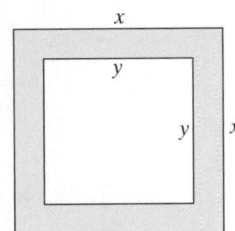

96. A manufacturer needs to know the area of a metal washer. The outer radius of the washer is R and the inner radius is r.

a. Find an expression that represents the area of the washer.

b. Factor the expression found in part (a).

c. Find the area of the washer if $R = \frac{1}{2}$ in. and $r = \frac{1}{4}$ in. (Round to the nearest 0.01 in.^{2})

Expanding Your Skills

For Exercises 97–102, factor the polynomials by using the difference of squares, sum of cubes, or difference of cubes with grouping.

97. $x^2 - y^2 + x + y$

98. $64m^2 - 25n^2 + 8m + 5n$

99. $x^3 + y^3 + x + y$

100. $4pu^3 - 4pv^3 - 7yu^3 + 7yv^3$

101. $576a^5 - 9a^2 - 64a^3c^2 + c^2$

102. $32t^5 - 108t^2 - 72t^3v^2 + 243v^2$

Problem Recognition Exercises

Factoring Summary

We now review the techniques of factoring presented thus far along with a general strategy for factoring polynomials.

Factoring Strategy

Step 1 Factor out the greatest common factor (Section 4.5).

Step 2 Identify whether the polynomial has two terms, three terms, or more than three terms.

Step 3 If the polynomial has more than three terms, try factoring by grouping (Section 4.5 and Section 4.7).

Step 4 If the polynomial has three terms, check first for a perfect square trinomial. Otherwise, factor the trinomial with the ac-method or the trial-and-error method (Section 4.6).

Step 5 If the polynomial has two terms, determine if it fits the pattern for a difference of squares, difference of cubes, or sum of cubes. Remember, a sum of squares is not factorable over the real numbers (Section 4.7).

Step 6 Be sure to factor the polynomial completely.

Step 7 Check by multiplying.

---

1. What is meant by a prime polynomial?

2. What is the first step in factoring any polynomial?

3. When factoring a binomial, what patterns do you look for?

4. When factoring a trinomial, what pattern do you look for first?

5. What do you look for when factoring a four-term polynomial?

6. How would you use substitution to factor $3(4x^2 + 1)^2 + 20(4x^2 + 1) + 12$.

For Exercises 7–66,

   a. Factor out the GCF from each polynomial and identify the category in which the remaining polynomial best fits. Choose from

   - difference of squares
   - difference of cubes
   - perfect square trinomial
   - four terms—grouping
   - sum of squares
   - sum of cubes
   - trinomial (ac-method or trial-and-error)
   - none of these

   b. Factor the polynomial completely.

7. $6x^2 - 21x - 45$

8. $8m^3 - 10m^2 - 3m$

9. $8a^2 - 50$

10. $ab + ay - b^2 - by$

11. $14u^2 - 11uv + 2v^2$

12. $9p^2 - 12pq + 4q^2$

13. $16x^3 - 2$

14. $9m^2 + 16n^2$

15. $27y^3 + 125$

16. $3x^2 - 16$

17. $128p^6 + 54q^3$

18. $5b^2 - 30b + 45$

19. $16a^4 - 1$

20. $81u^2 - 90uv + 25v^2$

21. $p^2 - 12p + 36 - c^2$

22. $4x^2 + 16$

23. $12ax - 6ay + 4bx - 2by$

24. $125y^3 - 8$

25. $5y^2 + 14y - 3$

26. $2m^4 - 128$

27. $t^2 - 100$

28. $4m^2 - 49n^2$

29. $y^3 + 27$

30. $x^3 + 1$

31. $d^2 + 3d - 28$

32. $c^2 + 5c - 24$

33. $x^2 - 12x + 36$

34. $p^2 + 16p + 64$

35. $2ax^2 - 5ax + 2bx - 5b$

36. $8x^2 - 4bx + 2ax - ab$

37. $10y^2 + 3y - 4$

38. $12z^2 + 11z + 2$

39. $10p^2 - 640$

40. $50a^2 - 72$

41. $z^4 - 64z$

42. $t^4 - 8t$

43. $b^3 - 4b^2 - 45b$

44. $y^3 - 14y^2 + 40y$

45. $9w^2 + 24wx + 16x^2$

46. $4k^2 - 20kp + 25p^2$

47. $60x^2 - 20x + 30ax - 10a$

48. $50x^2 - 200x + 10cx - 40c$

49. $w^4 - 16$

50. $k^4 - 81$

51. $t^6 - 8$

52. $p^6 + 27$

53. $8p^2 - 22p + 5$

54. $9m^2 - 3m - 20$

55. $36y^2 - 12y + 1$

56. $9a^2 + 42a + 49$

57. $2x^2 + 50$

58. $4y^2 + 64$

59. $12r^2s^2 + 7rs^2 - 10s^2$

60. $7z^2w^2 - 10zw^2 - 8w^2$

61. $x^2 + 8xy - 33y^2$

62. $s^2 - 9st - 36t^2$

63. $m^6 + n^3$

64. $a^3 - b^6$

65. $x^2 - 4x$

66. $y^2 - 9y$

For Exercises 67–101, factor completely using the strategy found on page 392.

67. $x^2(x + y) - y^2(x + y)$

68. $u^2(u - v) - v^2(u - v)$

69. $(a + 3)^4 + 6(a + 3)^5$

70. $(4 - b)^4 - 2(4 - b)^3$

71. $24(3x + 5)^3 - 30(3x + 5)^2$

72. $10(2y + 3)^2 + 15(2y + 3)^3$

73. $\frac{1}{100}x^2 + \frac{1}{35}x + \frac{1}{49}$

74. $\frac{1}{25}a^2 + \frac{1}{15}a + \frac{1}{36}$

75. $(5x^2 - 1)^2 - 4(5x^2 - 1) - 5$

76. $(x^3 + 4)^2 - 10(x^3 + 4) + 24$

77. $16p^4 - q^4$

78. $s^4t^4 - 81$

79. $y^3 + \frac{1}{64}$

80. $z^3 + \frac{1}{125}$

81. $6a^3 + a^2b - 6ab^2 - b^3$

82. $4p^3 + 12p^2q - pq^2 - 3q^3$

83. $\frac{1}{9}t^2 + \frac{1}{6}t + \frac{1}{16}$

84. $\frac{1}{25}y^2 + \frac{1}{5}y + \frac{1}{4}$

85. $x^2 + 12x + 36 - a^2$

86. $a^2 + 10a + 25 - b^2$

87. $p^2 + 2pq + q^2 - 81$

88. $m^2 - 2mn + n^2 - 9$

89. $b^2 - (x^2 + 4x + 4)$

90. $p^2 - (y^2 - 6y + 9)$

91. $4 - u^2 + 2uv - v^2$

92. $25 - a^2 - 2ab - b^2$

93. $6ax - by + 2bx - 3ay$

94. $5pq - 12 - 4q + 15p$

95. $u^6 - 64$ [*Hint:* Factor first as a difference of squares, $(u^3)^2 - (8)^2$.]

96. $1 - v^6$

97. $x^8 - 1$

98. $y^8 - 256$

99. $a^2 - b^2 + a + b$

100. $25c^2 - 9d^2 + 5c - 3d$

101. $5wx^3 + 5wy^3 - 2zx^3 - 2zy^3$

Section 4.8   Solving Equations by Using the Zero Product Rule

Concepts

1. Solving Equations by Using the Zero Product Rule
2. Applications of Quadratic Equations
3. Definition of a Quadratic Function
4. Applications of Quadratic Functions

1. Solving Equations by Using the Zero Product Rule

In Section 1.1, we defined a linear equation in one variable as an equation of the form $ax + b = c \, (a \neq 0)$. A linear equation in one variable is sometimes called a first-degree polynomial equation because the highest degree of all its terms is 1. A second-degree polynomial equation is called a quadratic equation.

> **Definition of a Quadratic Equation in One Variable**
>
> If a, b, and c are real numbers such that $a \neq 0$, then a **quadratic equation** is an equation that can be written in the form
>
> $$ax^2 + bx + c = 0$$

The following equations are quadratic because they can each be written in the form $ax^2 + bx + c = 0 \, (a \neq 0)$.

$$-4x^2 + 4x = 1 \qquad x(x - 2) = 3 \qquad (x - 4)(x + 4) = 9$$
$$-4x^2 + 4x - 1 = 0 \qquad x^2 - 2x = 3 \qquad x^2 - 16 = 9$$
$$x^2 - 2x - 3 = 0 \qquad x^2 - 25 = 0$$
$$x^2 + 0x - 25 = 0$$

One method to solve a quadratic equation is to factor and apply the zero product rule. The **zero product rule** states that if the product of two factors is zero, then one or both of its factors is equal to zero.

> **The Zero Product Rule**
>
> If $ab = 0$, then $a = 0$ or $b = 0$.

For example, the quadratic equation $x^2 - x - 12 = 0$ can be written in factored form as $(x - 4)(x + 3) = 0$. By the zero product rule, one or both factors must be zero: $x - 4 = 0$ or $x + 3 = 0$. Therefore, to solve the quadratic equation, set each factor to zero and solve for x.

$$(x - 4)(x + 3) = 0 \qquad\qquad \text{Apply the zero product rule.}$$

$$x - 4 = 0 \quad \text{or} \quad x + 3 = 0 \qquad \text{Set each factor to zero.}$$

$$x = 4 \quad \text{or} \qquad\quad x = -3 \qquad \text{Solve each equation for } x.$$

Quadratic equations, like linear equations, arise in many applications of mathematics, science, and business. The following steps summarize the factoring method to solve a quadratic equation.

Solving a Quadratic Equation by Factoring

Step 1 Write the equation in the form $ax^2 + bx + c = 0$.

Step 2 Factor completely.

Step 3 Apply the zero product rule. That is, set each factor equal to zero and solve the resulting equations.*

*The solution(s) found in step 3 may be checked by substitution in the original equation.

Example 1 Solving a Quadratic Equation

Solve. $2x^2 - 5x = 12$

Solution:

$$2x^2 - 5x = 12$$

$$2x^2 - 5x - 12 = 0 \qquad\qquad \text{Write the equation in the form } ax^2 + bx + c = 0.$$

$$(2x + 3)(x - 4) = 0 \qquad\qquad \text{Factor completely.}$$

$$2x + 3 = 0 \quad \text{or} \quad x - 4 = 0 \qquad \text{Set each factor equal to zero.}$$

$$2x = -3 \quad \text{or} \qquad\quad x = 4 \qquad \text{Solve each equation.}$$

$$x = -\frac{3}{2} \quad \text{or} \qquad\quad x = 4$$

Check: $x = -\dfrac{3}{2}$ \qquad\qquad Check: $x = 4$

$$2x^2 - 5x = 12 \qquad\qquad 2x^2 - 5x = 12$$

$$2\left(-\frac{3}{2}\right)^2 - 5\left(-\frac{3}{2}\right) \overset{?}{=} 12 \qquad 2(4)^2 - 5(4) \overset{?}{=} 12$$

$$2\left(\frac{9}{4}\right) + \frac{15}{2} \overset{?}{=} 12 \qquad\qquad 2(16) - 20 \overset{?}{=} 12$$

$$\frac{9}{2} + \frac{15}{2} \overset{?}{=} 12 \qquad\qquad 32 - 20 \overset{?}{=} 12 \checkmark$$

$$\frac{24}{2} \overset{?}{=} 12 \checkmark$$

Avoiding Mistakes

The zero product rule tells us that if $ab = 0$, then $a = 0$ or $b = 0$. This property does not hold for other numbers. For example if $ab = 12$, then it is not necessary that a or b must equal 12.

The solution set is $\left\{-\dfrac{3}{2}, 4\right\}$.

Example 2 Solving a Quadratic Equation

Solve. $6x^2 + 8x = 0$

Solution:

$$6x^2 + 8x = 0$$
$$2x(3x + 4) = 0 \qquad \text{Factor completely.}$$

$$2x = 0 \quad \text{or} \quad 3x + 4 = 0 \qquad \text{Set each factor equal to zero.}$$
$$x = 0 \qquad\qquad 3x = -4 \qquad \text{Solve each equation for } x.$$
$$x = -\dfrac{4}{3}$$

The solution set is $\left\{0, -\dfrac{4}{3}\right\}$.   The solutions check.

Example 3 Solving a Quadratic Equation

Solve. $9x(4x + 2) - 10x = 8x + 25$

Solution:

$$9x(4x + 2) - 10x = 8x + 25$$
$$36x^2 + 18x - 10x = 8x + 25 \qquad \text{Clear parentheses.}$$
$$36x^2 + 8x = 8x + 25 \qquad \text{Combine } like \text{ terms.}$$
$$36x^2 - 25 = 0 \qquad \text{Make one side of the equation equal to zero. The equation is in the form } ax^2 + bx + c = 0. \text{ (\textit{Note: } b = 0.)}$$
$$(6x - 5)(6x + 5) = 0 \qquad \text{Factor completely.}$$

$$6x - 5 = 0 \quad \text{or} \quad 6x + 5 = 0 \qquad \text{Set each factor equal to zero.}$$
$$6x = 5 \quad \text{or} \quad 6x = -5 \qquad \text{Solve each equation.}$$
$$x = \dfrac{5}{6} \quad \text{or} \quad x = -\dfrac{5}{6} \qquad \text{The check is left to the reader.}$$

The solution set is $\left\{\dfrac{5}{6}, -\dfrac{5}{6}\right\}$.

Skill Practice Solve.

3. $5a(2a - 3) + 4(a + 1) = 3a(3a - 2)$

Example 4 Solving an Equation

Solve. $2x(x + 5) + 3 = 2x^2 - 5x + 1$

Solution:

$2x(x + 5) + 3 = 2x^2 - 5x + 1$

$2x^2 + 10x + 3 = 2x^2 - 5x + 1$ Clear parentheses.

$15x + 2 = 0$ Make one side of the equation equal to

$15x = -2$ zero. The equation is not quadratic. It is

     in the form $ax + b = 0$, which is linear.

$x = -\dfrac{2}{15}$ Solve by using the method for linear

     equations.

The solution set is $\left\{-\dfrac{2}{15}\right\}$.   The check is left to the reader.

Skill Practice Solve.

4. $t^2 - 3t + 1 = t^2 + 2t + 11$

The zero product rule can be used to solve higher-degree polynomial equations provided one side of the equation is zero and the other is written in factored form.

Example 5 Solving a Higher-Degree Polynomial Equation

Solve. $-2(y + 7)(y - 1)(10y + 3) = 0$

Solution:

$-2(y + 7)(y - 1)(10y + 3) = 0$

One side of the equation is zero, and the other side is already factored.

$-2 = 0$ or $y + 7 = 0$ or $y - 1 = 0$ or $10y + 3 = 0$ Set each factor equal to zero.

No solution $y = -7$ or $y = 1$ or $y = -\dfrac{3}{10}$ Solve each equation for y.

Notice that when the constant factor is set to zero, the result is the contradiction $-2 = 0$. The constant factor does not produce a solution to the equation. Therefore, the only solutions are -7, 1, and $-\frac{3}{10}$. Each solution can be checked in the original equation.

The solution set is $\left\{-7, 1, -\dfrac{3}{10}\right\}$.

Skill Practice Solve.

5. $3(w + 2)(2w + 1)(w - 8) = 0$

Answers

3. $\{4, 1\}$ **4.** $\{-2\}$

5. $\left\{-2, -\dfrac{1}{2}, 8\right\}$

Example 6 Solving a Higher-Degree Polynomial Equation

Solve. $z^3 + 3z^2 - 4z - 12 = 0$

Solution:

$$z^3 + 3z^2 - 4z - 12 = 0$$ This is a higher-degree polynomial equation.

$$z^3 + 3z^2 \mid - 4z - 12 = 0$$ One side of the equation is zero. Now factor. Because there are four terms, try factoring

$$z^2(z + 3) - 4(z + 3) = 0$$ by grouping.

$$(z + 3)(z^2 - 4) = 0$$ $z^2 - 4$ can be factored further as a difference of squares.

$$(z + 3)(z - 2)(z + 2) = 0$$

$z + 3 = 0$ or $z - 2 = 0$ or $z + 2 = 0$ Set each factor equal to zero.

$z = -3$ or $z = 2$ or $z = -2$ Solve each equation.

The solution set is $\{-3, 2, -2\}$.

Skill Practice Solve.

6. $x^3 + x^2 - 9x - 9 = 0$

2. Applications of Quadratic Equations

Example 7 Solving an Application of a Quadratic Equation

The product of two consecutive odd integers is 35. Find the integers.

Solution:

Let x represent the smaller odd integer and $x + 2$ represent the next consecutive odd integer.

$$\left(\begin{array}{c}\text{First odd}\\ \text{integer}\end{array}\right) \cdot \left(\begin{array}{c}\text{next odd}\\ \text{integer}\end{array}\right) = 35 \qquad \text{Verbal model}$$

$x \cdot (x + 2) = 35$ Mathematical equation

$x^2 + 2x = 35$ Clear parentheses.

$x^2 + 2x - 35 = 0$ Set the equation equal to zero.

$(x + 7)(x - 5) = 0$ Factor.

$x + 7 = 0$ or $x - 5 = 0$ Set each factor equal to zero.

$x = -7$ or $x = 5$ Solve each equation.

If $x = -7$ then the next odd integer is $x + 2 = -5$.
If $x = 5$ then the next odd integer is $x + 2 = 7$.

There are two pairs of odd integers that are solutions, $-7, -5$ and $5, 7$.

Answers

6. $\{-1, 3, -3\}$
7. The integers are -8 and -6 or 6 and 8.

Skill Practice

7. The product of two consecutive even integers is 48. Find the integers.

Example 8 Solving an Application of a Quadratic Equation

The length of a basketball court is 6 ft less than 2 times the width. If the total area is 4700 ft², find the dimensions of the court.

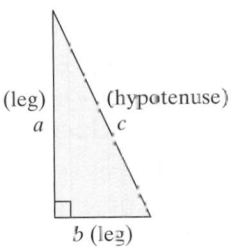

Figure 4-5

Solution:

If the width of the court is represented by w, then the length can be represented by $2w - 6$ (Figure 4-5).

$$A = (\text{length})(\text{width}) \qquad \text{Area of a rectangle}$$

$$4700 = (2w - 6)w \qquad \text{Mathematical equation}$$

$$4700 = 2w^2 - 6w$$

$$2w^2 - 6w - 4700 = 0 \qquad \begin{array}{l}\text{Set the equation equal to} \\ \text{zero and factor.}\end{array}$$

$$2(w^2 - 3w - 2350) = 0 \qquad \text{Factor out the GCF.}$$

$$2(w - 50)(w + 47) = 0 \qquad \text{Factor the trinomial.}$$

$$2 \ne 0 \quad \text{or} \quad w - 50 = 0 \quad \text{or} \quad w + 47 = 0 \qquad \begin{array}{l}\text{Set each factor} \\ \text{equal to zero.}\end{array}$$

contradiction

$$w = 50 \quad \text{or} \quad w \ne -47 \qquad \begin{array}{l}\text{A negative} \\ \text{width is not} \\ \text{possible.}\end{array}$$

The width is 50 ft.

The length is $2w - 6 = 2(50) - 6 = 94$ ft.

Skill Practice

8. The width of a rectangle is 5 in. less than 3 times the length. The area is 2 in.²
   Find the length and width.

A right triangle is a triangle that contains a 90° angle. Furthermore, the sum of the squares of the two legs (the shorter sides) of a right triangle equals the square of the hypotenuse (the longest side). This important fact is known as the Pythagorean theorem. For the right triangle shown in Figure 4-6, the Pythagorean theorem is stated as

$$a^2 + b^2 = c^2$$

In this formula, a and b are the legs and c is the hypotenuse. Notice that the hypotenuse is the longest side and is opposite the right angle.

(leg) a · (hypotenuse) c · b (leg)

Figure 4-6

TIP: When applying the Pythagorean theorem, it does not matter which leg you label a and which you label b. Since the lengths of the legs are interchangeable you can also write the Pythagorean theorem as $\text{leg}^2 + \text{leg}^2 = \text{hyp}^2$.

The triangle given in Figure 4-7 is a right triangle. We have

$$a^2 \;+\; b^2 \;=\; c^2$$

$$(5 \text{ ft})^2 + (12 \text{ ft})^2 = (13 \text{ ft})^2$$

$$25 \text{ ft}^2 + 144 \text{ ft}^2 = 169 \text{ ft}^2$$

$$169 \text{ ft}^2 = 169 \text{ ft}^2 \ ✔$$

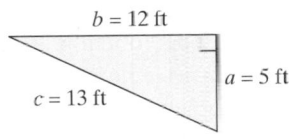

$b = 12$ ft · $a = 5$ ft · $c = 13$ ft

Figure 4-7

Answer

8. The width is 1 in., and the length is 2 in.

Example 9   **Application of a Quadratic Equation**

A region of coastline off Biscayne Bay is approximately in the shape of a right angle. The corresponding triangular area has sandbars and is marked off on navigational charts as being shallow water. If one leg of the triangle is 0.5 mi shorter than the other leg, and the hypotenuse is 2.5 mi, find the lengths of the legs of the triangle (Figure 4-8).

Figure 4-8

Solution:

Let x represent the longer leg.

Then $x - 0.5$ represents the shorter leg.

$$a^2 + b^2 = c^2 \qquad \text{Pythagorean theorem}$$

$$x^2 + (x - 0.5)^2 = (2.5)^2$$

> **TIP:** Recall that the square of a binomial results in a perfect square trinomial.
>
> $$(a - b)^2 = a^2 - 2ab + b^2$$
>
> $$(x - 0.5)^2 = (x)^2 - 2(x)(0.5) + (0.5)^2$$
>
> $$\qquad\qquad = x^2 - x + 0.25$$

$$x^2 + (x)^2 - 2(x)(0.5) + (0.5)^2 = 6.25$$

$$x^2 + x^2 - x + 0.25 = 6.25$$

$$2x^2 - x - 6 = 0 \qquad \text{Write the equation in the form } ax^2 + bx + c = 0.$$

$$(2x + 3)(x - 2) = 0 \qquad \text{Factor.}$$

$$2x + 3 = 0 \quad \text{or} \quad x - 2 = 0 \qquad \text{Set both factors to zero.}$$

$$x \cancel{=} -\frac{3}{2} \quad \text{or} \quad x = 2 \qquad \text{Solve both equations for } x.$$

The side of a triangle cannot be negative, so we reject the solution $x = -\frac{3}{2}$.

Therefore, one leg of the triangle is 2 mi.

The other leg is $x - 0.5 = 2 - 0.5 = 1.5$ mi.

Skill Practice

9. The longer leg of a right triangle measures 7 ft more than the shorter leg. The hypotenuse is 8 ft longer than the shorter leg. Find the lengths of the sides of the triangle.

3. Definition of a Quadratic Function

In Section 2.7, we graphed several basic functions by plotting points, including $f(x) = x^2$. This function is called a quadratic function, and its graph is in the shape of a **parabola**. In general, any second-degree polynomial function is a quadratic function.

Definition of a Quadratic Function

Let a, b, and c represent real numbers such that $a \neq 0$. Then a function defined by $f(x) = ax^2 + bx + c$ is called a **quadratic function**.

Answer

9. The sides are 5 ft, 12 ft, and 13 ft.

The graph of a quadratic function is a parabola that opens upward or downward. The leading coefficient a determines the direction of the parabola. For the quadratic function defined by $f(x) = ax^2 + bx + c$:

If $a > 0$, the parabola opens upward. For example, $f(x) = x^2$

If $a < 0$, the parabola opens downward. For example, $g(x) = -x^2$

Recall from Section 2.7 that the x-intercepts of a function $y = f(x)$ are the real solutions to the equation $f(x) = 0$. The y-intercept is found by evaluating $f(0)$.

Example 10 Finding the x- and y-Intercepts of a Quadratic Function

Find the x- and y-intercepts.

$$f(x) = x^2 - x - 12$$

Solution:

To find the x-intercept, substitute $f(x) = 0$.

$f(x) = x^2 - x - 12$

$0 = x^2 - x - 12$ Substitute 0 for $f(x)$. The result is a quadratic equation.

$0 = (x - 4)(x + 3)$ Factor.

$x - 4 = 0$ or $x + 3 = 0$ Set each factor equal to zero.

$x = 4$ or $x = -3$ Solve each equation.

The x-intercepts are $(4, 0)$ and $(-3, 0)$.

To find the y-intercept, find $f(0)$.

$f(x) = x^2 - x - 12$

$f(0) = (0)^2 - (0) - 12$ Substitute $x = 0$.

$= -12$

The y-intercept is $(0, -12)$.

Skill Practice

10. Find the x- and y-intercepts of the function defined by
$f(x) = x^2 + 8x + 12$.

Answer

10. x-intercepts: $(-6, 0)$ and $(-2, 0)$;
y-intercept: $(0, 12)$

Topic: Using the Zero Feature

The graph of $f(x) = x^2 - x - 12$ supports the solution to Example 10. Use the *zero* feature to confirm that the graph crosses the x-axis at -3 and 4. The y-intercept is given as $(0, -12)$.

4. Applications of Quadratic Functions

Example 11   Application of a Quadratic Function

A model rocket is shot vertically upward with an initial velocity of 288 ft/sec. The function given by $h(t) = -16t^2 + 288t$ relates the rocket's height $h(t)$ (in feet) to the time t after launch (in seconds).

   a. Find $h(0)$, $h(5)$, $h(10)$, and $h(15)$, and interpret the meaning of these function values in the context of the rocket's height and time after launch.

   b. Find the t-intercepts of the function, and interpret their meaning in the context of the rocket's height and time after launch.

   c. Find the time(s) at which the rocket is at a height of 1152 ft.

Solution:

   a. $h(t) = -16t^2 + 288t$

       $h(0) = -16(0)^2 + 288(0) = 0$

       $h(5) = -16(5)^2 + 288(5) = 1040$

      $h(10) = -16(10)^2 + 288(10) = 1280$

      $h(15) = -16(15)^2 + 288(15) = 720$

      $h(0) = 0$ means that at $t = 0$ sec, the height of the rocket is 0 ft.

      $h(5) = 1040$ means that 5 sec after launch, the height is 1040 ft.

      $h(10) = 1280$ means that 10 sec after launch, the height is 1280 ft.

      $h(15) = 720$ means that 15 sec after launch, the height is 720 ft.

   b. The t-intercepts of the function are represented by the real solutions of the equation $h(t) = 0$.

$-16t^2 + 288t = 0$	Set $h(t) = 0$.
$-16t(t - 18) = 0$	Factor.
$-16t = 0$ or $t - 18 = 0$	Apply the zero product rule.
$t = 0$ or $t = 18$	

The rocket is at ground level initially (at $t = 0$ sec) and then again after 18 sec when it hits the ground.

c. Set $h(t) = 1152$ and solve for t.

$$h(t) = -16t^2 + 288t$$

$$1152 = -16t^2 + 288t \qquad \text{Substitute 1152 for } h(t).$$

$$16t^2 - 288t + 1152 = 0 \qquad \text{Set the equation equal to zero.}$$

$$16(t^2 - 18t + 72) = 0 \qquad \text{Factor out the GCF.}$$

$$16(t - 6)(t - 12) = 0 \qquad \text{Factor.}$$

$$t = 6 \quad \text{or} \quad t = 12$$

The rocket will reach a height of 1152 ft after 6 sec (on the way up) and after 12 sec (on the way down). (See Figure 4-9.)

Height of Rocket Versus Time After Launch

(6, 1152) $\quad h(t) = -16t^2 + 288t$

(12, 1152)

Time (sec)

Figure 4-9

Skill Practice An object is dropped from the top of a building that is 144 ft high. The function given by $h(t) = -16t^2 + 144$ relates the height $h(t)$ of the object (in feet) to the time t in seconds after it is dropped.

11. a. Find $h(0)$ and interpret the meaning of the function value in the context of this problem.

   b. Find the t-intercept(s) and interpret the meaning in the context of this problem.

   c. When will the object be 128 ft above ground level?

Answer

11a. $h(0) = 144$, which is the initial height of the object (after 0 sec).

   b. The t-intercept is (3, 0), which means the object is at ground level (0 ft high) after 3 sec. The intercept $(-3, 0)$ does not make sense for this problem since time cannot be negative.

   c. One second after release, the object will be 128 ft above ground level.

Section 4.8 Practice Exercises

Vocabulary and Key Concepts

1. a. An equation that can be written in the form $ax^2 + bx + c = 0$, $a \neq 0$, is called a _____ equation.

   b. The zero product rule states that if $ab = 0$, then $a =$ _____ or $b =$ _____.

   c. The _____ theorem states that given a right triangle with legs a and b and hypotenuse c, then $a^2 + b^2 =$ _____.

   d. A function defined by $f(x) = ax^2 + bx + c$, $a \neq 0$, is called a _____ function.

   e. Given a quadratic function $f(x) = ax^2 + bx + c$, $a \neq 0$, find the x-intercept(s) by solving the equation _____. Find the _____-intercept by evaluating $f(0)$.

   f. If x is an integer, then _____ represents the next greater integer. If x is an odd integer, then _____ represents the next greater odd integer. Likewise if x is an even integer, then _____ represents the next greater even integer.

   g. The area of a rectangle of length l and width w is given by $A =$ _____.

   h. The area of a triangle with base b and height h is given by the formula $A =$ _____.

Review Exercises

2. Write the factored form for each binomial, if possible.

    a. $x^2 - y^2$ **b.** $x^2 + y^2$ **c.** $x^3 - y^3$ **d.** $x^3 + y^3$

For Exercises 3–8, factor completely.

3. $10x^2 + 3x$ **4.** $7x^2 - 28$ **5.** $2p^2 - 9p - 5$

6. $3q^2 - 4q - 4$ **7.** $t^3 - 1$ **8.** $z^2 - 11z + 30$

Concept 1: Solving Equations by Using the Zero Product Rule

9. What conditions are necessary to solve an equation by using the zero product rule?   **10.** State the zero product rule.

For Exercises 11–16, determine which of the equations are written in the correct form to apply the zero product rule directly. If an equation is not in the correct form, explain what is wrong.

11. $2x(x - 3) = 0$ **12.** $(u + 1)(u - 3) = 10$ **13.** $3p^2 - 7p + 4 = 0$

14. $t^2 - t - 12 = 0$ **15.** $a(a + 3)^2 = 5$ **16.** $\left(\dfrac{2}{3}x - 5\right)\left(x + \dfrac{1}{2}\right) = 0$

For Exercises 17–20, factor the polynomial or solve the equation as indicated.

17. a. Factor. $w^2 - 81$
      b. Solve. $w^2 - 81 = 0$

18. a. Factor. $p^2 - 25$
      b. Solve. $p^2 - 25 = 0$

19. a. Factor. $3x^2 + 14x - 5$
      b. Solve. $3x^2 + 14x - 5 = 0$

20. a. Factor. $2y^2 - y - 3$
      b. Solve. $2y^2 - y - 3 = 0$

For Exercises 21–56, solve the equation. **(See Examples 1–6.)**

21. $(x + 3)(x + 5) = 0$ **22.** $(x + 7)(x - 4) = 0$ **23.** $(2w + 9)(5w - 1) = 0$

24. $(3a + 1)(4a - 5) = 0$ **25.** $x(x + 4)(10x - 3) = 0$ **26.** $t(t - 6)(3t - 11) = 0$

27. $0 = 5(y - 0.4)(y + 2.1)$ **28.** $0 = -4(z - 7.5)(z - 9.3)$ **29.** $x^2 + 6x - 27 = 0$

30. $2x^2 + x - 15 = 0$ **31.** $2x^2 + 5x = 3$ **32.** $-11x = 3x^2 - 4$

33. $10x^2 = 15x$ **34.** $5x^2 = 7x$ **35.** $6(y - 2) - 3(y + 1) = 8$

36. $4x + 3(x - 9) = 6x + 1$ **37.** $-9 = y(y + 6)$ **38.** $-62 = t(t - 16) + 2$

39. $9p^2 - 15p - 6 = 0$ **40.** $6y^2 + 2y = 48$ **41.** $(x + 1)(2x - 1)(x - 3) = 0$

42. $2x(x - 4)^2(4x + 3) = 0$ **43.** $(y - 3)(y + 4) = 8$ **44.** $(t + 10)(t + 5) = 6$

45. $(2a - 1)(a - 1) = 6$ **46.** $w(6w + 1) = 2$ **47.** $p^2 + (p + 7)^2 = 169$

48. $x^2 + (x + 2)^2 = 100$ **49.** $3t(t + 5) - t^2 = 2t^2 + 4t - 1$ **50.** $a^2 - 4a - 2 = (a + 3)(a - 5)$

51. $2x^3 - 8x^2 - 24x = 0$ **52.** $2p^3 + 20p^2 + 42p = 0$ **53.** $w^3 = 16w$

54. $12x^3 = 27x$ **55.** $0 = 2x^3 + 5x^2 - 18x - 45$ **56.** $0 = 3y^3 + y^2 - 48y - 16$

Concept 2: Applications of Quadratic Equations

57. If 5 is added to the square of a number, the result is 30. Find all such numbers.

58. Four less than the square of a number is 77. Find all such numbers.

59. The square of a number is equal to 12 more than the number. Find all such numbers.

60. The square of a number is equal to 20 more than the number. Find all such numbers.

61. The product of two consecutive integers is 42. Find the integers.

62. The product of two consecutive integers is 110. Find the integers.

63. The product of two consecutive odd integers is 63. Find the integers. **(See Example 7.)**

64. The product of two consecutive even integers is 120. Find the integers.

65. A rectangular pen is to contain 35 ft^2 of area. If the width is 2 ft less than the length, find the dimensions of the pen. **(See Example 8.)**

 66. The length of a rectangular photograph is 7 in. more than the width. If the area is 78 in.^{2}, what are the dimensions of the photograph?

67. The length of a rectangular room is 5 yd more than the width. If the area is 300 yd^2, find the length and the width of the room.

68. The top of a rectangular dining room table is twice as long as it is wide. Find the dimensions of the table if the area is 18 ft^2.

69. The height of a triangle is 1 in. more than the base. If the height is increased by 2 in. while the base remains the same, the new area becomes 20 in.^{2}

    a. Find the base and height of the original triangle.

    b. Find the area of the original triangle.

70. The base of a triangle is 2 cm more than the height. If the base is increased by 4 cm while the height remains the same, the new area is 56 cm^2.

    a. Find the base and height of the original triangle.

    b. Find the area of the original triangle.

71. The area of a triangular garden is 25 ft^2. The base is twice the height. Find the base and the height of the triangle.

72. The height of a triangle is 1 in. more than twice the base. If the area is 18 in.^{2}, find the base and height of the triangle.

73. The sum of the squares of two consecutive positive integers is 41. Find the integers.

74. The sum of the squares of two consecutive, positive even integers is 164. Find the integers.

75. Justin must travel from Summersville to Clayton. He can drive 10 mi through the mountains at 40 mph. Or he can drive east and then north on superhighways at 60 mph. The alternative route forms a right angle as shown in the diagram. The eastern leg is 2 mi less than the northern leg. **(See Example 9.)**

    a. Find the total distance Justin would travel in going the alternative route.

    b. If Justin wants to minimize the time of the trip, which route should he take?

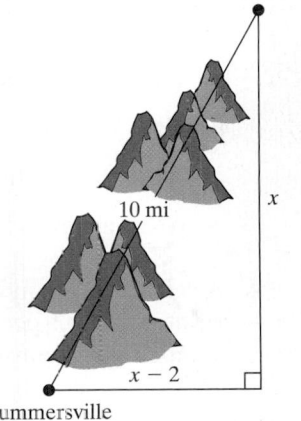

76. A 17-ft ladder is standing up against a wall. The distance between the base of the ladder and the wall is 7 ft less than the distance between the top of the ladder and the base of the wall. Find the distance between the base of the ladder and the wall.

77. A right triangle has side lengths represented by three consecutive even integers. Find the lengths of the three sides, measured in meters.

78. The hypotenuse of a right triangle is 3 m more than twice the short leg. The longer leg is 2 m more than twice the shorter leg. Find the lengths of the sides.

79. Determine the length of the radius of a circle whose area is numerically equal to its circumference.

80. Determine the length of the radius of a circle whose area is numerically twice its circumference.

Concept 3: Definition of a Quadratic Function

For Exercises 81–84, **a.** Find the values of x for which $f(x) = 0$. **b.** Find $f(0)$.

81. $f(x) = x^2 - 3x$

82. $f(x) = 4x^2 + 2x$

 83. $f(x) = x^2 - 6x - 7$

84. $f(x) = 2x^2 + 11x + 5$

For Exercises 85–88, find the x- and y-intercepts for the functions defined by $y = f(x)$. **(See Example 10.)**

85. $f(x) = \dfrac{1}{2}(x - 2)(x + 1)(2x)$

86. $f(x) = (x + 1)(x - 2)(x + 3)^2$

87. $f(x) = x^2 - 2x + 1$

88. $f(x) = x^2 + 4x + 4$

For Exercises 89–92, find the x-intercepts of each function and use that information to match the function with its graph.

89. $g(x) = x^2 - 9$

90. $h(x) = x(x - 2)(x + 4)$

91. $f(x) = 4(x + 1)$

92. $k(x) = (x + 1)(x + 3)(x - 2)(x - 1)$

a.   **b.**   **c.**   **d.**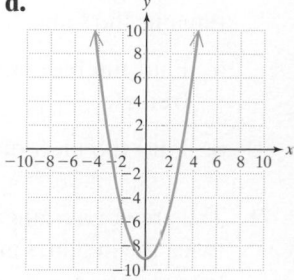

Concept 4: Applications of Quadratic Functions

93. A rocket is fired upward from ground level with an initial velocity of 490 m/sec. The height of the rocket $s(t)$ in meters is a function of the time t in seconds after launch. **(See Example 11.)**

$$s(t) = -4.9t^2 + 490t$$

a. What characteristics of s indicate that it is a quadratic function?

b. Find the t-intercepts of the function.

c. What do the t-intercepts mean in the context of this problem?

d. At what times is the rocket at a height of 485.1 m?

94. A certain company makes water purification systems. The factory can produce x water systems per year. The profit $P(x)$ the company makes is a function of the number of systems x it produces.

$$P(x) = -2x^2 + 1000x$$

a. Is this function linear or quadratic?

b. Find the number of water systems x that would produce a zero profit.

c. What points on the graph do the answers in part (b) represent?

d. Find the number of systems for which the profit is $80,000.

For Exercises 95–100, factor the expressions represented by $f(x)$. Explain how the factored form relates to the graph of the function. Can the graph of the function help you determine the factors?

95. $f(x) = x^2 - 7x + 10$

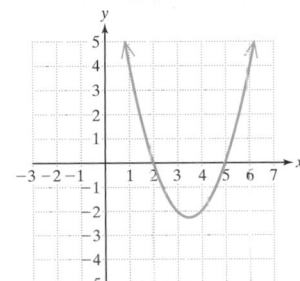

96. $f(x) = x^2 - 2x - 3$

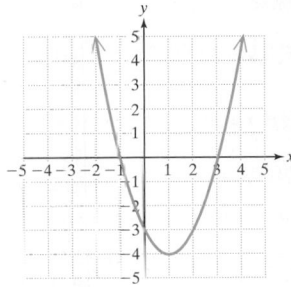

97. $f(x) = x^2 + 2x + 1$

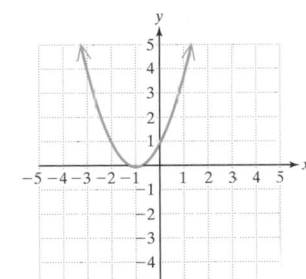

98. $f(x) = x^2 - 8x + 16$

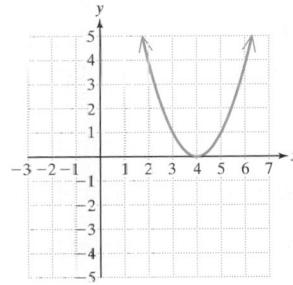

99. $f(x) = -x^2 - 6x - 5$

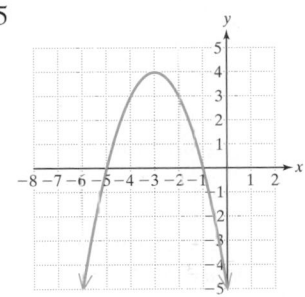

100. $f(x) = -x^2 + 6x - 8$

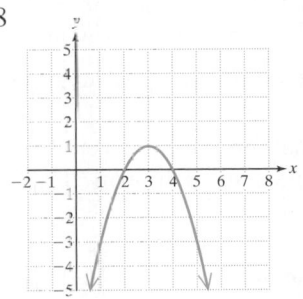

Expanding Your Skills

101. The surface area of a right circular cylinder is represented by $SA = 2\pi r^2 + 2\pi rh$. If the surface area is 156π ft^2 and the height is 7 ft, determine the radius of the cylinder.

102. Determine the length and width of a rectangle with a perimeter of 20 yd and an area of 16 yd^2.

103. Determine the length and width of a rectangle with a perimeter of 28 ft and an area of 48 ft^2.

For Exercises 104–107, find an equation that has the given solutions. For example, 2 and -1 are solutions to $(x - 2)(x + 1) = 0$ or $x^2 - x - 2 = 0$. In general, x_1 and x_2 are solutions to the equation $a(x - x_1)(x - x_2) = 0$, where a can be any nonzero real number. For each problem, there is more than one correct answer depending on your choice of a.

104. $x = -3$ and $x = 1$

105. $x = 2$ and $x = -2$

106. $x = 0$ and $x = -5$

107. $x = 0$ and $x = -3$

Graphing Calculator Exercises

For Exercises 108–111, graph Y_1. Use the *Zero* feature to approximate the x-intercepts. Then solve $Y_1 = 0$ and compare the solutions to the x-intercepts.

108. $Y_1 = -x^2 + x + 2$

109. $Y_1 = -x^2 - x + 20$

110. $Y_1 = x^2 - 6x + 9$

111. $Y_1 = x^2 + 4x + 4$

Group Activity

Investigating Pascal's Triangle

Estimated time: 15 minutes

Group Size: 2 or 3

1. Determine the value of $(a + b)^0$. Write your answer here: _____

2. Write the binomials in expanded form. [For example, for $(a + b)^2$, multiply the binomials $(a + b)(a + b)$ and write the result in the space provided.] Check your answers with another group.

    a. $(a + b)^1$ _____

    b. $(a + b)^2$ _____

    c. $(a + b)^3$ _____

3. a. How many terms were in the expansion of $(a + b)^2$? _____

    b. How many terms were in the expansion of $(a + b)^3$? _____

    c. How many terms would you expect in the expansion of $(a + b)^4$? _____

    d. How many terms would you expect in the expansion of $(a + b)^5$? _____

    e. How many terms would you expect in the expansion of $(a + b)^n$? _____

4. Based on your answers from question 2, do you see a pattern regarding the exponents on the factors of a and b in each term? Explain.

5. The coefficients of each expansion also follow a pattern. The triangle of numbers shown here is called **Pascal's triangle**. Pascal's triangle gives the coefficients of a binomial expansion of the form $(a + b)^n$. Work with your group members to see if you can figure out the pattern for the coefficients of $(a + b)^4$, $(a + b)^5$, $(a + b)^6$, and $(a + b)^7$.

$(a + b)^n$	Coefficients of the Expansion
$(a + b)^0$	
$(a + b)^1$	
$(a + b)^2$	
$(a + b)^3$	
$(a + b)^4$	
$(a + b)^5$	
$(a + b)^6$	

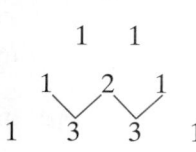

```
           1
         1   1
       1   2   1
     1   3   3   1
```

6. Using the expected number of terms found in question 3, the pattern of the exponents found in question 4, and the pattern for the coefficients in question 5, write the expansions for the following.

    a. $(a + b)^4 =$ _____

    b. $(a + b)^5 =$ _____

    c. $(a + b)^6 =$ _____

Chapter 4   Summary

Section 4.1   Properties of Integer Exponents and Scientific Notation

Key Concepts

Let a and b ($b \neq 0$) represent real numbers and m and n represent positive integers.

$$b^m \cdot b^n = b^{m+n} \qquad \frac{b^m}{b^n} = b^{m-n}$$

$$(b^m)^n = b^{mn} \qquad (ab)^m = a^m b^m$$

$$\left(\frac{a}{b}\right)^m = \frac{a^m}{b^m} \qquad b^0 = 1$$

$$b^{-n} = \left(\frac{1}{b}\right)^n = \frac{1}{b^n}$$

A number expressed in the form $a \times 10^n$, where $1 \leq |a| < 10$ and n is an integer, is written in **scientific notation**.

Examples

Example 1

$$\left(\frac{2x^2 y}{z^{-1}}\right)^{-3} (x^{-4} y^0)$$

$$= \left(\frac{2^{-3} x^{-6} y^{-3}}{z^3}\right)(x^{-4} \cdot 1)$$

$$= \frac{2^{-3} x^{-10} y^{-3}}{z^3}$$

$$= \frac{1}{2^3 x^{10} y^3 z^3} \quad \text{or} \quad \frac{1}{8 x^{10} y^3 z^3}$$

Example 2

$$0.0000002 \times 35{,}000$$

$$= (2 \times 10^{-7})(3.5 \times 10^4)$$

$$= 7 \times 10^{-3} \text{ or } 0.007$$

Section 4.2   Addition and Subtraction of Polynomials and Polynomial Functions

Key Concepts

A **polynomial** in x is defined by a sum of terms of the form ax^n, where a is a real number and n is a whole number.

- a is the **coefficient** of the term.
- n is the **degree of the term**.

The **degree of a polynomial** is the greatest degree of its terms.

The term of a polynomial with the greatest degree is the **leading term**. Its coefficient is the **leading coefficient**.

A one-term polynomial is a **monomial**.
A two-term polynomial is a **binomial**.
A three-term polynomial is a **trinomial**.

To add or subtract polynomials, add or subtract *like* terms.

Examples

Example 1

$$7y^4 - 2y^2 + 3y + 8$$

is a polynomial with leading coefficient 7 and degree 4.

Example 2

$$f(x) = 4x^3 - 6x - 11$$

f is a polynomial function with leading term $4x^3$ and leading coefficient 4. The degree of f is 3.

Example 3

For $f(x) = 4x^3 - 6x - 11$, find $f(-1)$.
$$f(-1) = 4(-1)^3 - 6(-1) - 11$$
$$= -9$$

Example 4

$$(-4x^3 y + 3x^2 y^2) - (7x^3 y - 5x^2 y^2)$$
$$= -4x^3 y + 3x^2 y^2 - 7x^3 y + 5x^2 y^2$$
$$= -11x^3 y + 8x^2 y^2$$

Section 4.3 Multiplication of Polynomials

Key Concepts

To multiply polynomials, multiply each term in the first polynomial by each term in the second polynomial.

Special Products

1. Multiplication of **conjugates**

$$(x + y)(x - y) = x^2 - y^2$$

The product is called a **difference of squares**.

2. Square of a binomial

$$(x + y)^2 = x^2 + 2xy + y^2$$
$$(x - y)^2 = x^2 - 2xy + y^2$$

The product is called a **perfect square trinomial**.

Examples

Example 1

$$(x - 2)(3x^2 - 4x + 11)$$
$$= 3x^3 - 4x^2 + 11x - 6x^2 + 8x - 22$$
$$= 3x^3 - 10x^2 + 19x - 22$$

Example 2

$$(3x + 5)(3x - 5)$$
$$= (3x)^2 - (5)^2$$
$$= 9x^2 - 25$$

Example 3

$$(4y + 3)^2$$
$$= (4y)^2 + (2)(4y)(3) + (3)^2$$
$$= 16y^2 + 24y + 9$$

Section 4.4 Division of Polynomials

Key Concepts

Division of polynomials:

1. For division by a monomial, use the properties

$$\frac{a + b}{c} = \frac{a}{c} + \frac{b}{c} \quad \text{and} \quad \frac{a - b}{c} = \frac{a}{c} - \frac{b}{c}$$

for $c \neq 0$.

2. If the divisor has more than one term, use long division.
   - First write each polynomial in descending order.
   - Insert placeholders for missing powers.

Examples

Example 1

$$\frac{-12a^2 - 6a + 9}{-3a}$$

$$= \frac{-12a^2}{-3a} - \frac{6a}{-3a} + \frac{9}{-3a}$$

$$= 4a + 2 - \frac{3}{a}$$

Example 2

$$(3x^2 - 5x + 1) \div (x + 2)$$

$$
\begin{array}{r}
3x - 11 \\
x + 2 \overline{)3x^2 - 5x + 1} \\
\underline{-(3x^2 + 6x)} \\
-11x + 1 \\
\underline{-(-11x - 22)} \\
23
\end{array}
$$

Answer: $3x - 11 + \dfrac{23}{x + 2}$

3. **Synthetic division** may be used to divide a polynomial by a binomial in the form $x - r$, where r is a constant.

Example 3

$(3x^2 - 5x + 1) \div (x + 2)$

$$\begin{array}{r} -2\,\underline{|\,3 \quad -5 \quad 1} \\ \quad\;\; -6 \quad 22 \\ \hline 3 \quad -11 \quad \underline{|\,23} \end{array}$$

Answer: $3x - 11 + \dfrac{23}{x + 2}$

Section 4.5   Greatest Common Factor and Factoring by Grouping

Key Concepts

The **greatest common factor (GCF)** is the largest factor common to all terms of a polynomial. To factor out the GCF from a polynomial, use the distributive property.

A four-term polynomial may be **factored by grouping**.

Steps to Factor by Grouping

1. Identify and factor out the GCF from all four terms.
2. Factor out the GCF from the first pair of terms. Factor out the GCF from the second pair of terms. (Sometimes it is necessary to factor out the *opposite* of the GCF.)
3. If the two pairs of terms share a common binomial factor, factor out the binomial factor.

Examples

Example 1

$3x^2(a - b) - 6x(a - b)$

$= 3x(a + b)x - 3x(a + b)(2)$

$= 3x(a + b)(x - 2)$

Example 2

$60xa - 30xb - 80ya + 40yb$

$= 10[6xa - 3xb - 8ya + 4yb]$

$= 10[3x(2a - b) - 4y(2a - b)]$

$= 10(2a - b)(3x - 4y)$

Section 4.6   Factoring Trinomials

Key Concepts

AC-Method

To factor trinomials of the form $ax^2 + bx + c$:

1. Factor out the GCF. Find the product ac.
2. Find two integers whose product is ac and whose sum is b. (If no pair of numbers can be found, then the trinomial is prime.)
3. Rewrite the middle term bx as the sum of two terms whose coefficients are the numbers found in step 2.
4. Factor the polynomial by grouping.

Examples

Example 1

$10y^2 + 35y - 20 = 5(2y^2 + 7y - 4)$

$$ac = (2)(-4) = -8$$

Find two integers whose product is -8 and whose sum is 7. The numbers are 8 and -1.

$5[2y^2 + 8y - 1y - 4]$

$= 5[2y(y + 4) - 1(y + 4)]$

$= 5(y + 4)(2y - 1)$

Trial-and-Error Method

To factor trinomials in the form $ax^2 + bx + c$:

1. Factor out the GCF.
2. List the pairs of factors of a and the pairs of factors of c. Consider the reverse order in either list.
3. Construct two binomials of the form

Factors of a

$(\square x \ \ \square)(\square x \ \ \square)$

Factors of c

4. Test each combination of factors until the product of the outer terms and the product of inner terms add to the middle term.
5. If no combination of factors works, the polynomial is prime.

The factored form of a **perfect square trinomial** is the square of a binomial:

$a^2 + 2ab + b^2 = (a + b)^2$

$a^2 - 2ab + b^2 = (a - b)^2$

Sometimes it is easier to factor a polynomial after making a substitution.

Example 2

$10y^2 + 35y - 20 = 5(2y^2 + 7y - 4)$

The pairs of factors of 2 are $2 \cdot 1$.
The pairs of factors of -4 are

$$
\begin{array}{cc}
-1 \cdot 4 & 1 \cdot (-4) \\
-2 \cdot 2 & 2 \cdot (-2) \\
-4 \cdot 1 & 4 \cdot (-1)
\end{array}
$$

$(2y - 2)(y + 2) = 2y^2 + 2y - 4$	No
$(2y - 4)(y + 1) = 2y^2 - 2y - 4$	No
$(2y + 1)(y - 4) = 2y^2 - 7y - 4$	No
$(2y + 2)(y - 2) = 2y^2 - 2y - 4$	No
$(2y + 4)(y - 1) = 2y^2 + 2y - 4$	No
$(2y - 1)(y + 4) = 2y^2 + 7y - 4$	Yes

Therefore, $10y^2 + 35y - 20$ factors as $5(2y - 1)(y + 4)$.

Example 3

$9w^2 - 30wz + 25z^2$

$= (3w)^2 - 2(3w)(5z) + (5z)^2$

$= (3w - 5z)^2$

Example 4

$(7v^2 - 1)^2 - (7v^2 - 1) - 12$	Let $u = (7v^2 - 1)$.
$= u^2 - u - 12$	Substitute.
$= (u + 3)(u - 4)$	Factor.
$= (7v^2 - 1 + 3)(7v^2 - 1 - 4)$	Back substitute.
$= (7v^2 + 2)(7v^2 - 5)$	Simplify.

Section 4.7   Factoring Binomials

Key Concepts

Factoring Binomials: Summary

Difference of squares:

$a^2 - b^2 = (a + b)(a - b)$

Sum of squares:

If a and b share no common factors, then $a^2 + b^2$ is prime.

Examples

Example 1

$25u^2 - 9v^4 = (5u)^2 - (3v^2)^2$

$= (5u + 3v^2)(5u - 3v^2)$

Example 2

$32 + 2w^2 = 2(16 + w^2)$ cannot be factored further because $16 + w^2$ is a sum of squares.

Difference of cubes:

$$a^3 - b^3 = (a - b)(a^2 + ab + b^2)$$

Sum of cubes:

$$a^3 + b^3 = (a + b)(a^2 - ab + b^2)$$

Sometimes it is necessary to group three terms with one term.

Example 3

$$8c^3 - d^6 = (2c - d^2)(4c^2 + 2cd^2 + d^4)$$

Example 4

$$27w^9 + 64x^3$$
$$= (3w^3 + 4x)(9w^6 - 12w^3x + 16x^2)$$

Example 5

$$4a^2 - 12ab + 9b^2 - c^2$$

$$= 4a^2 - 12ab + 9b^2 \mid - c^2 \qquad \text{Group 3 by 1.}$$

$$= (2a - 3b)^2 - c^2 \qquad \text{Perfect square trinomial.}$$

$$= (2a - 3b - c)(2a - 3b + c) \qquad \text{Difference of squares.}$$

Section 4.8 — Solving Equations by Using the Zero Product Rule

Key Concepts

An equation of the form $ax^2 + bx + c = 0$, where $a \neq 0$, is a **quadratic equation**.

The **zero product rule** states that if $ab = 0$, then $a = 0$ or $b = 0$. The zero product rule can be used to solve a quadratic equation or higher-degree polynomial equation that is factored and equal to zero.

$f(x) = ax^2 + bx + c \, (a \neq 0)$ defines a **quadratic function**. The x-intercepts of a function defined by $y = f(x)$ are determined by finding the real solutions to the equation $f(x) = 0$. The y-intercept of $y = f(x)$ is at $f(0)$.

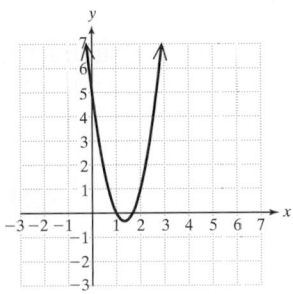

Examples

Example 1

$$0 = x(2x - 3)(x + 4)$$

$$x = 0 \quad \text{or} \quad 2x - 3 = 0 \quad \text{or} \quad x + 4 = 0$$

$$x = \frac{3}{2} \quad \text{or} \quad x = -4$$

The solution set is $\left\{0, \dfrac{3}{2}, -4\right\}$.

Example 2

Find the x-intercepts.

$$f(x) = 3x^2 - 8x + 5$$
$$0 = 3x^2 - 8x + 5$$
$$0 = (3x - 5)(x - 1)$$

$$3x - 5 = 0 \quad \text{or} \quad x - 1 = 0$$

$$x = \frac{5}{3} \quad \text{or} \quad x = 1$$

The x-intercepts are $(\frac{5}{3}, 0)$ and $(1, 0)$.

Find the y-intercept.

$$f(x) = 3x^2 - 8x + 5$$
$$f(0) = 3(0)^2 - 8(0) + 5$$
$$f(0) = 5$$

The y-intercept is $(0, 5)$.

Chapter 4   Review Exercises

Section 4.1

For Exercises 1–8, simplify the expression and write the answer with positive exponents.

1. $(3x)^3(3x)^2$

2. $(-6x^{-4})(3x^{-8})$

3. $\dfrac{24x^5y^3}{-8x^4y}$

4. $\dfrac{-18x^{-2}y^3}{-12x^{-5}y^5}$

5. $(-2a^2b^{-5})^{-3}$

6. $(-4a^{-2}b^3)^{-2}$

7. $\left(\dfrac{-4x^4y^{-2}}{5x^{-1}y^4}\right)^{-4}$

8. $\left(\dfrac{25x^2y^{-3}}{5x^4y^{-2}}\right)^{-5}$

9. Write the numbers in scientific notation.

  a. The population of Asia was 3,686,600,000 in 2000.

  b. A nanometer is 0.000001 of a millimeter.

10. Write the numbers in scientific notation.

  a. A millimeter is 0.001 of a meter.

  b. The population of Asia is predicted to be 5,155,700,000 by 2040.

11. Write the numbers in standard form.

  a. A micrometer is 1×10^{-3} of a millimeter.

  b. A nanometer is 1×10^{-9} of a meter.

12. Write the numbers in standard form.

  a. The total square footage of shopping centers in the United States is approximately 5.23×10^9 ft^2.

  b. The total yearly sales of those shopping centers is $\$1.091 \times 10^{12}$. (*Source:* International Council of Shopping Centers)

For Exercises 13–16, perform the indicated operations. Write the answer in scientific notation.

13. $\dfrac{2,500,000}{0.0004}$

14. $\dfrac{0.0005}{25,000}$

15. $(3.6 \times 10^8)(9 \times 10^{-2})$

16. $(7 \times 10^{-12})(5.2 \times 10^3)$

Section 4.2

For Exercises 17–18, identify the polynomial as a monomial, binomial, or trinomial; then give the degree of the polynomial.

17. $6x^4 + 10x - 1$ **18.** 18

19. Given the polynomial function defined by $g(x) = 4x - 7$, find the function values.

  a. $g(0)$ **b.** $g(-4)$ **c.** $g(3)$

20. Given the polynomial function defined by $p(x) = -x^4 - x + 12$, find the function values.

  a. $p(0)$ **b.** $p(1)$ **c.** $p(-2)$

21. The amount, $A(x)$, of bottled water consumed per capita in the United States can be approximated by

$$A(x) = 0.047x^2 + 1.46x + 16.8$$

where x is the number of years since the year 2000 and $A(x)$ is measured in gallons.

  a. Evaluate $A(5)$ and interpret the meaning in the context of the problem.

  b. Interpret the meaning of $A(15)$.

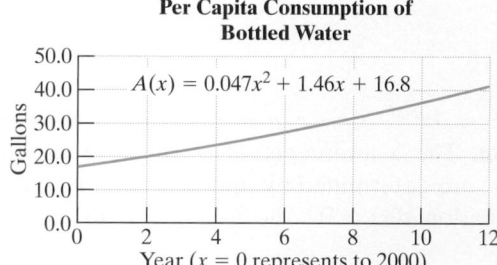

Per Capita Consumption of Bottled Water

$A(x) = 0.047x^2 + 1.46x + 16.8$

Gallons / Year ($x = 0$ represents to 2000)

For Exercises 22–33, add or subtract the polynomials as indicated.

22. $(x^2 - 2x - 3xy - 7) + (-3x^2 - x + 2xy + 6)$

23. $\begin{aligned} &7xy - 3xz + 5yz \\ +&13xy - 15xz - 8yz \end{aligned}$

24. $\begin{aligned} &-4a^3 + 8a^2 - 3a \\ -&(-7a^3 + 3a^2 - 9a) \end{aligned}$

25. $(3a^2 - 2a - a^3) - (5a^2 - a^3 - 8a)$

26. $\left(\dfrac{5}{8}x^4 - \dfrac{1}{4}x^2 - \dfrac{1}{2}\right) - \left(-\dfrac{3}{8}x^4 + \dfrac{3}{4}x^2 + \dfrac{1}{2}\right)$

27. $\left(\dfrac{5}{6}x^4 + \dfrac{1}{2}x^2 - \dfrac{1}{3}\right) - \left(-\dfrac{1}{6}x^4 - \dfrac{1}{4}x^2 - \dfrac{1}{3}\right)$

28. $(7x - y) - [-(2x + y) - (-3x - 6y)]$

29. $-(4x - 4y) - [(4x + 2y) - (3x + 7y)]$

30. Add $-4x + 6$ to $-7x - 5$.

31. Add $2x^2 - 4x$ to $2x^2 - 7x$.

32. Subtract $-4x + 6$ from $-7x - 5$.

33. Subtract $2x^2 - 4x$ from $2x^2 - 7x$.

Section 4.3

For Exercises 34–51, multiply the polynomials.

34. $2x(x^2 - 7x - 4)$ **35.** $-3x(6x^2 - 5x + 4)$

36. $(x + 6)(x - 7)$ **37.** $(x - 2)(x - 9)$

38. $\left(\dfrac{1}{2}x + 1\right)\left(\dfrac{1}{2}x - 5\right)$ **39.** $\left(-\dfrac{1}{5} + 2y\right)\left(\dfrac{1}{5} + y\right)$

40. $(3x + 5)(9x^2 - 15x + 25)$

41. $(x - y)(x^2 + xy + y^2)$

42. $(2x - 5)^2$ **43.** $\left(\dfrac{1}{2}x + 4\right)^2$

44. $(3y - 11)(3y + 11)$ **45.** $(6w - 1)(6w - 1)$

46. $\left(\dfrac{2}{3}t + 4\right)\left(\dfrac{2}{3}t - 4\right)$ **47.** $\left(z + \dfrac{1}{4}\right)\left(z - \dfrac{1}{4}\right)$

48. $[(x + 2) - b][(x + 2) + b]$

49. $[c - (w + 3)][c + (w + 3)]$

50. $(2x + 1)^3$ **51.** $(y^2 - 3)^3$

52. A square garden is surrounded by a walkway of uniform width x. If the sides of the garden are given by the expression $2x + 3$, find and simplify a polynomial for the following.

   a. The area of the garden.

   b. The area of the walkway and garden.

   c. The area of the walkway only.

53. The length of a rectangle is 2 ft more than 3 times the width. Let x represent the width of the rectangle.

   a. Write a function P that represents the perimeter of the rectangle.

   b. Write a function A that represents the area of the rectangle.

54. In parts (a) and (b), one of the statements is true and the other is false. Identify the true statement and explain why the false statement is incorrect.

   a. $2x^2 + 5x = 7x^3$ $(2x^2)(5x) = 10x^3$

   b. $4x - 7x = -3x$ $4x - 7x = -3$

Section 4.4

For Exercises 55–56, divide the polynomials.

55. $(6x^3y + 12x^2y^2 - 9xy^3) \div (3xy)$

56. $(10x^4 + 15x^3 - 20x^2) \div (-5x^2)$

57. a. Divide $(9y^4 + 14y^2 - 8) \div (3y + 2)$.

   b. Identify the quotient and the remainder.

   c. Explain how you can check your answer.

For Exercises 58–61, divide the polynomials by using long division.

58. $(x^2 + 7x + 10) \div (x + 5)$

59. $(x^2 + 8x - 16) \div (x + 4)$

60. $(2x^5 - 4x^4 + 2x^3 - 4) \div (x^2 - 3x)$

61. $(2x^5 + 3x^3 + x^2 - 4) \div (x^2 + x)$

62. Explain the conditions under which you may use synthetic division.

63. The following table is the result of a synthetic division.

$$\begin{array}{r|rrrrr} 3| & 2 & 5 & -2 & 6 & 1 \\ & & 6 & 33 & 93 & 297 \\ \hline & 2 & 11 & 31 & 99 & |298 \end{array}$$

Use x as the variable.

a. Identify the divisor.

b. Identify the quotient.

c. Identify the remainder.

For Exercises 64–68, divide the polynomials by using synthetic division.

64. $(t^3 - 3t^2 + 8t - 12) \div (t - 2)$

65. $(x^2 + 7x + 14) \div (x + 5)$

66. $(x^2 + 8x + 20) \div (x + 4)$

67. $(w^3 - 6w^2 + 8) \div (w - 3)$

68. $(p^4 - 16) \div (p - 2)$

Section 4.5

For Exercises 69–72, factor by removing the greatest common factor.

69. $-x^3 - 4x^2 + 11x$

70. $21w^3 - 7w + 14$

71. $5x(x - 7) - 2(x - 7)$

72. $3t(t + 4) + 5(t + 4)$

For Exercises 73–76, factor by grouping (remember to take out the GCF first).

73. $m^3 - 8m^2 + m - 8$

74. $24x^3 - 36x^2 + 72x - 108$

75. $4ax^2 + 2bx^2 - 6ax - 3xb$

76. $y^3 - 6y^2 + y - 6$

Section 4.6

77. What characteristics determine a perfect square trinomial?

For Exercises 78–87, factor the polynomials by using any method.

78. $18x^2 + 27xy + 10y^2$ **79.** $3m^2 + mt - 10t^2$

80. $60a^2 + 65a^3 - 20a^4$ **81.** $2k^2 + 7k^3 + 6k^4$

82. $49x^2 + 36 - 84x$ **83.** $80z + 32 + 50z^2$

84. $(9w + 2)^2 + 4(9w + 2) - 5$

85. $(4x + 3)^2 - 12(4x + 3) + 36$

86. $18a^4 + 39a^2 - 15$ **87.** $3w^4 - 2w^2 - 5$

Section 4.7

For Exercises 88–94, factor the binomials.

88. $25 - y^2$ **89.** $x^3 - \dfrac{1}{27}$

90. $b^2 + 64$

91. $h^3 + 9h$ **92.** $a^3 + 64$

93. $k^4 - 16$ **94.** $9y^3 - 4y$

For Exercises 95–98, factor by grouping and by using the difference of squares.

95. $x^2 - 8xy + 16y^2 - 9$ (*Hint:* Group three terms that make up a perfect square trinomial, then factor as a difference of squares.)

96. $a^2 + 12a + 36 - b^2$

97. $t^2 + 16t + 64 - 25c^2$ **98.** $y^2 - 6y + 9 - 16x^2$

Section 4.8

99. How do you determine if an equation is quadratic?

100. What shape is the graph of a quadratic function?

For Exercises 101–104, label the equation as quadratic or linear.

101. $x^2 + 6x = 7$

102. $(x - 3)(x + 4) = 9$

103. $2x - 5 = 3$

104. $x + 3 = 5x^2$

105. a. Factor. $5x^2 + 6x - 8$

   b. Solve. $5x^2 + 6x - 8 = 0$

106. a. Factor. $3x^2 - 19x + 28$

   b. Solve. $3x^2 - 19x + 28 = 0$

For Exercises 107–110, use the zero product rule to solve the equations.

107. $x^2 - 2x - 15 = 0$

108. $8x^2 = 59x - 21$

109. $2t(t + 5) + 1 = 3t - 3 - t^2$

110. $3(x - 1)(x + 5)(2x - 9) = 0$

For Exercises 111–114, find the x- and y-intercepts of the function. Then match the function with its graph.

111. $f(x) = -4x^2 + 4$

112. $g(x) = 2x^2 - 2$

113. $h(x) = 5x^3 - 10x^2 - 20x + 40$

114. $k(x) = -\dfrac{1}{8}x^2 + \dfrac{1}{2}$

a.

b.

c.

d.

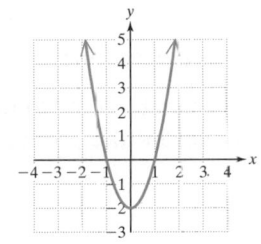

115. A moving van has the capacity to hold 1200 ft³ in volume. If the van is 10 ft high and the length is 1 ft less than twice the width, find the dimensions of the van.

116. A missile is shot upward from a submarine 1280 ft below sea level. The initial velocity of the missile is 672 ft/sec. A function that approximates the height of the missile (relative to sea level) is given by

$$h(t) = -16t^2 + 672t - 1280$$

where $h(t)$ is the height in feet and t is the time in seconds.

a. Complete the table to determine the height of the missile for the given values of t.

Time t (sec)	Height $h(t)$ (ft)
0	
1	
3	
10	
20	
30	
42	

b. Interpret the meaning of a negative value of $h(t)$.

c. Factor the function to find the time required for the missile to emerge from the water and the time required for the missile to reenter the water. (*Hint:* The height of the missile will be zero at sea level.)

Chapter 4   Test

For Exercises 1–4, simplify the expression, and write the answer with positive exponents only.

1. $\dfrac{20a^7}{4a^{-6}}$

2. $\dfrac{x^6x^3}{x^{-2}}$

3. $\left(\dfrac{-3x^6}{5y^7}\right)^2$

4. $\dfrac{(2^{-1}xy^{-2})^{-3}(x^{-4}y)}{(x^0y^5)^{-1}}$

5. Multiply. $(8 \times 10^{-6})(7.1 \times 10^5)$

6. Divide. (Write the answer in scientific notation.) $(9{,}200{,}000) \div (0.004)$

7. For the function defined by $F(x) = 5x^3 - 2x^2 + 8$, find the function values $F(-1)$, $F(2)$, and $F(0)$.

8. Perform the indicated operations. Write the answer in descending order.
$$(5x^2 - 7x + 3) - (x^2 + 5x - 25) + (4x^2 + 4x - 20)$$

For Exercises 9–11, multiply the polynomials.

9. $(2a - 5)(a^2 - 4a - 9)$

10. $\left(\dfrac{1}{3}x - \dfrac{3}{2}\right)(6x + 4)$

11. $(5x - 4y^2)(5x + 4y^2)$

12. Explain why $(5x + 7)^2 \neq 25x^2 + 49$.

13. Write and simplify an expression that describes the area of the square.

$7x - 4$

14. Divide the polynomials.
$$(2x^3y^4 + 5x^2y^2 - 6xy^3 - xy) \div (2xy)$$

15. Divide the polynomials.
$$(10p^3 + 13p^2 - p + 3) \div (2p + 3)$$

16. Divide the polynomials by using synthetic division. $(y^4 - 2y + 5) \div (y - 2)$

17. Explain the strategy for factoring a polynomial expression.

18. Explain the process to solve a polynomial equation by the zero product rule.

For Exercises 19–32, factor completely.

19. $3a^2 + 27ab + 54b^2$ **20.** $c^4 - 1$

21. $xy - 7x + 3y - 21$ **22.** $49 + p^2$

23. $-10u^2 + 30u - 20$ **24.** $12t^2 - 75$

25. $5y^2 - 50y + 125$ **26.** $21q^2 + 14q$

27. $2x^3 + x^2 - 8x - 4$ **28.** $y^3 - 125$

29. $x^2 + 8x + 16 - y^2$ **30.** $r^6 - 256r^2$

31. $(x^2 + 1)^2 + 3(x^2 + 1) + 2$

32. $12a - 6ac + 2b - bc$

For Exercises 33–38, solve the equation.

33. $(2x - 3)(x + 5) = 0$

34. $w^2 - 7w = 0$

35. $y^2 - 6y = 16$

36. $x(5x + 4) = 1$

37. $4p - 64p^3 = 0$

38. $t^2 + \dfrac{1}{2}t + \dfrac{1}{16} = 0$

39. A child launches a toy rocket from the ground. The height, $h(x)$, of the rocket can be determined by its horizontal distance, x, from the launch pad by

$$h(x) = -\dfrac{x^2}{256} + x$$

where x and h are in feet and $x \geq 0$ and $h \geq 0$.

How many feet from the launch pad will the rocket hit the ground?

40. The population of Japan, $P(t)$ (in millions), can be approximated by:

$$P(t) = -0.01t^2 - 0.062t + 127.7,$$

where $t = 0$ represents the year 2000.

a. Evaluate $p(4)$ and interpret in the context of the problem.

b. Approximate the number of people in Japan in the year 2006.

c. If the trend continues, predict the population of Japan in the year 2015.

For Exercises 41–44, find the x- and y-intercepts of the function. Then match the function with its graph.

41. $f(x) = x^2 - 6x + 8$

42. $k(x) = x^3 + 4x^2 - 9x - 36$

43. $p(x) = -2x^2 - 8x - 6$

44. $q(x) = x^3 - x^2 - 12x$

a.

b.

c.

d.

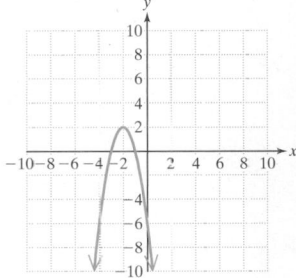

Chapters 1–4   Cumulative Review Exercises

1. Graph the inequality and express the set in interval notation: All real numbers at least 5, but not more than 12

2. Simplify the expression $3x^2 - 5x + 2 - 4(x^2 + 3)$.

3. Graph from memory.

a. $y = x^2$ **b.** $y = |x|$

    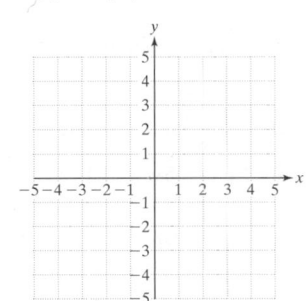

4. Simplify the expression $\left(\dfrac{1}{3}\right)^{-2} - \left(\dfrac{1}{2}\right)^3$.

5. The recent population of Mexico was approximately 9.85×10^7. At the current growth rate of 1.7%, this number is expected to double after 42 yr. How many people does this represent? Express the answer in scientific notation.

6. In the 2006 Orange Bowl football championship, Penn State scored 3 points more than Florida State in a three-overtime thriller. The total number of points scored was 49. Find the number of points scored by each team.

7. Find the value of each angle in the triangle.

8. Divide $(x^3 + 64) \div (x + 4)$.

9. Determine the slope and y-intercept of the line $4x - 3y = -9$, and graph the line.

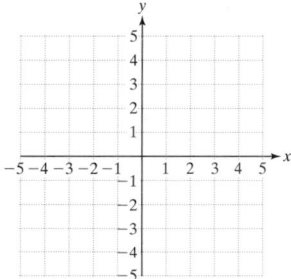

10. Write an equation of the line that is perpendicular to the y-axis and passes through the point $(-2, -5)$.

11. Simplify the expression.

$$\left(\frac{36a^{-2}b^4}{18b^{-6}}\right)^{-3}$$

12. Solve the system.

$$2x - y + 2z = 1$$
$$-3x + 5y - 2z = 11$$
$$x + y - 2z = -1$$

13. Determine whether the relation is a function.

a. $\{(2, 1), (3, 1), (-8, 1), (5, 1)\}$

b.

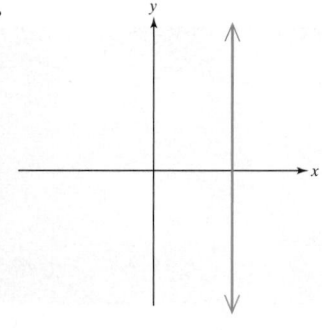

14. A telephone pole is leaning after a storm (see figure). What is the slope of the pole?

15. Given $P(x) = \frac{1}{6}x^2 + x - 5$, find the function value $P(6)$.

16. Solve. $\frac{1}{3}x - \frac{1}{6} = \frac{1}{2}(x - 3)$

17. Given $3x - 2y = 5$, solve for y.

18. A student scores $76, 85$, and 92 on her first three algebra tests.

a. Is it possible for her to score high enough on the fourth test to bring her test average up to 90? Assume that each test is weighted equally and that the maximum score on a test is 100 points.

b. What is the range of values required for the fourth test so that the student's test average will be between 80 and 89, inclusive?

19. How many liters of a 40% acid solution and how many liters of a 15% acid solution must be mixed to obtain 25 L of a 30% acid solution?

20. Multiply the polynomials $(4b - 3)(2b^2 + 1)$.

21. Add the polynomials.

$$(5a^2 + 3a - 1) + (3a^3 - 5a + 6)$$

22. Divide the polynomials $(6w^3 - 5w^2 - 2w) \div (2w^2)$.

For Exercises 23–24, solve the equations.

23. $y^2 - 5y = 14$

24. $a^3 + 9a^2 + 20a = 0$

25. Solve the inequality and write the solution set in interval notation. $-5 < -8 + |2x - 5|$

Rational Expressions and Rational Equations

CHAPTER OUTLINE

5.1 Rational Expressions and Rational Functions   422

5.2 Multiplication and Division of Rational Expressions   432

5.3 Addition and Subtraction of Rational Expressions   437

5.4 Complex Fractions   447

   Problem Recognition Exercises: Operations on Rational Expressions   454

5.5 Solving Rational Equations   454

   Problem Recognition Exercises: Rational Equations vs. Expressions   462

5.6 Applications of Rational Equations and Proportions   463

5.7 Variation   474

   Group Activity: Computing the Future Value of an Investment   483

Chapter 5

In this chapter, we define a rational expression as a ratio of two polynomials. The operations on rational expressions mirror the operations on fractions.

Are You Prepared?

To prepare for this chapter, work Exercises 1–8 to practice simplifying expressions containing fractions and solving equations containing fractions. Fill in the words corresponding to the solutions to Exercises 1–8.

1. $\dfrac{4}{9} \cdot \dfrac{15}{40}$ **2.** $\dfrac{6}{35} \div \dfrac{2}{7}$ **3.** $\dfrac{8}{3} - \dfrac{7}{6}$

4. $-\dfrac{6}{5} + \dfrac{2}{3}$ **5.** $-\dfrac{16}{3} \div 6$ **6.** $-4 \div \left(-\dfrac{2}{3}\right)$

Solve the equations.

7. $\dfrac{x-5}{2} = \dfrac{x}{3}$ **8.** $\dfrac{1}{2}x - \dfrac{2}{3} = 1 + \dfrac{1}{6}x$

To $\underline{}_{\;2}$ or $\underline{}_{\;4}$ $\underline{}_{\;8}$ $\underline{}_{\;3}$ $\underline{}_{\;1}$ $\underline{}_{\;6}$ a $\underline{}_{\;5}$ $\underline{}_{\;7}$.

Solution	Word
$\dfrac{3}{2}$	expressions
$\{15\}$	denominator
$\{5\}$	rational
$-\dfrac{8}{9}$	common
$-\dfrac{8}{15}$	subtract
6	need
$\dfrac{3}{5}$	add
$\dfrac{1}{6}$	we

Concepts

1. **Rational Functions**
2. **Simplifying Rational Expressions**
3. **Simplifying Ratios of −1**

Section 5.1   Rational Expressions and Rational Functions

1. Rational Functions

In Chapter 4, we introduced polynomials and polynomial functions. The ratio of two polynomials defines a **rational expression**. This leads to the following definition of a rational function.

> #### Definition of a Rational Function
>
> A function is a **rational function** if it can be written in the form $f(x) = \dfrac{p(x)}{q(x)}$, where p and q are polynomial functions and $q(x) \neq 0$.

For example, the functions $f, g, h,$ and k are rational functions.

$$f(x) = \frac{1}{x}, \qquad g(x) = \frac{2}{x-3}, \qquad h(a) = \frac{a+7}{a^2-5}, \qquad k(t) = \frac{t+4}{2t^2-11t+5}$$

In Section 2.7, we introduced the rational function defined by $f(x) = \frac{1}{x}$. Recall that $f(x) = \frac{1}{x}$ has a restriction on its domain that $x \neq 0$ and the graph of $f(x) = \frac{1}{x}$ has a vertical asymptote at $x = 0$ (Figure 5-1).

In this course, we restrict our discussion of functions defined by $y = f(x)$ to those whose domain and range are subsets of the set of real numbers. Therefore, the domain of a function is the set of real numbers that when substituted into the function produce a real number. For a rational function, we must exclude values that make the denominator zero. To find the domain of a rational function, we offer these guidelines.

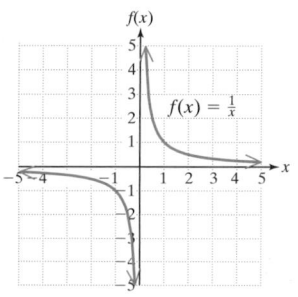

Figure 5-1

> #### Finding the Domain of a Rational Function
>
> **Step 1**   Set the denominator equal to zero and solve the resulting equation.
>
> **Step 2**   The domain is the set of all real numbers *excluding* the values found in step 1.

Example 1   **Evaluating a Rational Function**

Given $g(x) = \dfrac{2}{x-3}$

a. Determine the function values (if they exist).
$g(0), g(1), g(2), g(2.5), g(2.9), g(3), g(3.1), g(3.5), g(4),$ and $g(5)$.

b. Write the domain of the function.

Solution:

a. $g(0) = \dfrac{2}{(0)-3} = -\dfrac{2}{3}$ $g(3) = \dfrac{2}{(3)-3} = \dfrac{2}{0}$ (undefined)

$g(1) = \dfrac{2}{(1)-3} = -1$ $g(3.1) = \dfrac{2}{(3.1)-3} = \dfrac{2}{0.1} = 20$

$g(2) = \dfrac{2}{(2)-3} = -2$ $g(3.5) = \dfrac{2}{(3.5)-3} = \dfrac{2}{0.5} = 4$

$g(2.5) = \dfrac{2}{(2.5)-3} = \dfrac{2}{-0.5} = -4$ $g(4) = \dfrac{2}{(4)-3} = 2$

$g(2.9) = \dfrac{2}{(2.9)-3} = \dfrac{2}{-0.1} = -20$ $g(5) = \dfrac{2}{(5)-3} = 1$

b. $g(x) = \dfrac{2}{x-3}$

The value of the function is undefined when the denominator equals zero.

$x - 3 = 0$

Set the denominator equal to zero and solve for x.

$x = 3$

The value $x = 3$ must be excluded from the domain.

The domain can be written in set-builder notation or in interval notation.

Set-builder notation: $\{x \mid x \text{ is a real number}, x \neq 3\}$

Interval notation: $(-\infty, 3) \cup (3, \infty)$

Skill Practice Determine the function values.

$$h(x) = \frac{x+2}{x+6}$$

1. $h(0)$ **2.** $h(2)$ **3.** $h(-2)$ **4.** $h(-6)$

5. Write the domain in set-builder notation.

6. Write the domain in interval notation.

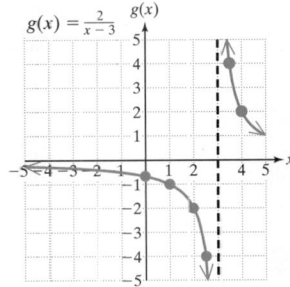

Figure 5-2

The graph of $g(x) = \dfrac{2}{x-3}$ is shown in Figure 5-2. Notice that the function has a vertical asymptote at $x = 3$ where the function is undefined (Figure 5-2).

Calculator Connections

Topic: Using *Dot* Mode on a Graphing Calculator

A *Table* feature can be used to check the function values in Example 1, $g(x) = \dfrac{2}{x-3}$.

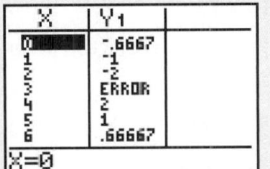

The graph of a rational function may be misleading on a graphing calculator. For example, $g(x) = 2/(x-3)$ has a vertical asymptote at $x = 3$, but the vertical asymptote is *not* part of the graph (Figure 5-3). A graphing calculator may try to "connect the dots" between two consecutive points, one to the left of $x = 3$ and one to the right of $x = 3$. This creates a line that is nearly vertical and appears to be part of the graph.

Figure 5-3

To show that this line is not part of the graph, we can graph the function in a *Dot* mode (see the owner's manual for your calculator). The graph of $g(x) = 2/(x-3)$ in dot mode indicates that the line $x = 3$ is not part of the function (Figure 5-4).

Figure 5-4

Answers

1. $\dfrac{1}{3}$ **2.** $\dfrac{1}{2}$

3. 0   **4.** Undefined

5. $\{x \mid x \text{ is a real number}, x \neq -6\}$

6. $(-\infty, -6) \cup (-6, \infty)$

Example 2 Finding the Domain of a Rational Function

Write the domain in set-builder notation and in interval notation.

a. $f(x) = \dfrac{x + 4}{2x^2 - 11x + 5}$

b. $g(x) = \dfrac{x}{x^2 + 4}$

Solution:

a. $f(x) = \dfrac{x + 4}{2x^2 - 11x + 5}$

$2x^2 - 11x + 5 = 0$ Set the denominator equal to zero and solve for x.

$(2x - 1)(x - 5) = 0$ This is a factorable quadratic equation.

$2x - 1 = 0$ or $x - 5 = 0$

$x = \dfrac{1}{2}$ or $x = 5$

The domain written in set-builder notation is
$\{x \mid x \text{ is a real number and } x \neq \frac{1}{2}, x \neq 5\}$.

The domain written in interval notation is $(-\infty, \frac{1}{2}) \cup (\frac{1}{2}, 5) \cup (5, \infty)$.

b. $g(x) = \dfrac{x}{x^2 + 4}$

Because the quantity x^2 is nonnegative for any real number x, the denominator $x^2 + 4$ cannot equal zero; therefore, no real numbers are excluded from the domain.

The domain written in set-builder notation is $\{x \mid x \text{ is a real number}\}$.

The domain written in interval notation is $(-\infty, \infty)$.

TIP: The domain of a rational function excludes values for which the denominator is zero.

Skill Practice Write the domain in set-builder notation and in interval notation.

7. $h(x) = \dfrac{x + 10}{2x^2 - x - 1}$

8. $g(t) = \dfrac{t}{t^2 + 1}$

2. Simplifying Rational Expressions

A rational expression is an expression in the form $\frac{p}{q}$, where p and q are polynomials and $q \neq 0$. As with fractions, it is often advantageous to simplify rational expressions to lowest terms.

The method for simplifying rational expressions mirrors the process to simplify fractions. In each case, factor the numerator and denominator. Common factors in the numerator and denominator form a ratio of 1 and can be simplified.

Answers
7. $\{x \mid x$ is a real number and $x \neq -\frac{1}{2}, x \neq 1\}$; $(-\infty, -\frac{1}{2}) \cup (-\frac{1}{2}, 1) \cup (1, \infty)$
8. $\{t \mid t$ is a real number$\}$; $(-\infty, \infty)$

Simplifying a fraction: $\dfrac{15}{35}$ $\xrightarrow{\text{factor}}$ $\dfrac{3 \cdot 5}{7 \cdot 5} = \dfrac{3}{7} \cdot \dfrac{5}{5} = \dfrac{3}{7} \cdot 1 = \dfrac{3}{7}$

Simplifying a rational expression: $\dfrac{x^2 - x - 12}{x^2 - 16}$ $\xrightarrow{\text{factor}}$ $\dfrac{(x + 3)(x - 4)}{(x + 4)(x - 4)} = \dfrac{(x + 3)}{(x + 4)} \cdot \dfrac{(x - 4)}{(x - 4)}$

$$= \dfrac{(x + 3)}{(x + 4)} \cdot 1$$

$$= \dfrac{(x + 3)}{(x + 4)}$$

This process is stated formally as the fundamental principle of rational expressions.

Fundamental Principle of Rational Expressions

Let p, q, and r represent polynomials. Then

$$\frac{pr}{qr} = \frac{p}{q} \cdot \frac{r}{r} = \frac{p}{q} \cdot 1 = \frac{p}{q} \qquad \text{for } q \neq 0 \text{ and } r \neq 0$$

A rational expression is not defined for the values of the variable that make the denominator equal to zero. We refer to these values as *restrictions* on the variable.

Example 3 **Simplifying a Rational Expression**

Given the expression $\dfrac{2x^3 + 12x^2 + 16x}{6x + 24}$

a. Factor the numerator and denominator.

b. Determine the restrictions on x.

c. Simplify the expression.

Solution:

a. $\dfrac{2x^3 + 12x^2 + 16x}{6x + 24}$

$$= \dfrac{2x(x^2 + 6x + 8)}{6(x + 4)} \qquad \text{Factor the numerator and denominator.}$$

$$= \dfrac{2x(x + 4)(x + 2)}{6(x + 4)}$$

> **Avoiding Mistakes**
>
> Always determine the restrictions on the variable *before* simplifying an expression.

b. The expression is not defined for all values of x for which the denominator is equal to zero.

$6(x + 4) = 0$ Solve the equation.

$x = -4$

The restriction on x is that $x \neq -4$.

c. $\dfrac{2x(x+4)(x+2)}{2\cdot 3(x+4)} = \dfrac{x(x+2)}{3}\cdot\dfrac{\overset{1}{2\cancel{(x+4)}}}{2\cancel{(x+4)}}$ Simplify the ratio of common factors.

$\qquad\qquad\qquad = \dfrac{x(x+2)}{3}$ (provided $x \neq -4$)

Skill Practice Given $\dfrac{x^2+3x-28}{2x+14}$

9. **a.** Factor the numerator and denominator.
  b. Determine the restrictions on x.
  c. Simplify the expression.

It is important to note that the expressions

$$\dfrac{2x^3+12x^2+16x}{6x+24} \qquad \text{and} \qquad \dfrac{x(x+2)}{3}$$

are equal for all values of x that make each expression a real number. Therefore,

$$\dfrac{2x^3+12x^2+16x}{6x+24} = \dfrac{x(x+2)}{3}$$

for all values of x *except* $x = -4$. (At $x = -4$, the original expression is undefined.) From this point forward, when we simplify rational expressions we will not explicitly write the restrictions by the simplified form. These will be implied from the original expression.

Example 4 involves simplifying a quotient of two monomials. This was also covered in Section 4.1.

Example 4 **Simplifying a Rational Expression**

Simplify. $\dfrac{2x^2y^8}{8x^4y^3}$

Solution:

$\dfrac{2x^2y^8}{8x^4y^3}$ This expression has the restriction that $x \neq 0$ and $y \neq 0$.

$= \dfrac{2x^2y^8}{2^3x^4y^3}$ Factor the denominator.

$= \dfrac{y^5}{2^2x^2}\cdot\dfrac{\overset{1}{2x^2y^3}}{2x^2y^3}$ Simplify common factors whose ratio is 1.

$= \dfrac{y^5}{4x^2}$

Skill Practice Simplify.

10. $\dfrac{9a^5b^3}{18a^8b}$

Answers

9. **a.** $\dfrac{(x+7)(x-4)}{2(x+7)}$ **b.** $x \neq -7$

  c. $\dfrac{x-4}{2}$ provided $x \neq -7$

10. $\dfrac{b^2}{2a^3}$

Example 5 Simplifying a Rational Expression

Simplify. $\dfrac{t^3 + 8}{t^2 + 4t + 4}$

Solution:

$$\dfrac{t^3 + 8}{t^2 + 4t + 4} = \dfrac{(t + 2)(t^2 - 2t + 4)}{(t + 2)^2}$$

Factor the numerator and denominator. The numerator is a sum of cubes. The denominator is a perfect square trinomial. The restriction on t is $t \neq -2$.

$$= \dfrac{(t^2 - 2t + 4)}{(t + 2)} \cdot \dfrac{\overset{1}{\cancel{(t + 2)}}}{\cancel{(t + 2)}}$$

Simplify common factors whose ratio is 1.

$$= \dfrac{t^2 - 2t + 4}{t + 2}$$

> **TIP:** Recall the formula to factor a sum of cubes.
>
> $a^3 + b^3$
> $= (a + b)(a^2 - ab + b^2)$
>
> Recall the formula to factor a perfect square trinomial.
>
> $a^2 + 2ab + b^2 = (a + b)^2$

Skill Practice Simplify.

11. $\dfrac{p^3 - 27}{p^2 - 6p + 9}$

Avoiding Mistakes

Because the fundamental property of rational expressions is based on the identity property of multiplication, reducing applies only to factors (remember that factors are multiplied). Therefore, terms that are added or subtracted cannot be reduced. For example:

$$\dfrac{3x}{3y} = \dfrac{3}{3} \cdot \dfrac{x}{y} = (1) \cdot \dfrac{x}{y} = \dfrac{x}{y} \qquad \text{However, } \dfrac{x + 3}{y + 3} \text{ cannot be simplified.}$$

↑ Reduce common factor. ↑ Cannot reduce common terms.

3. Simplifying Ratios of −1

When two factors are identical in the numerator and denominator, they form a ratio of 1 and can be simplified. Sometimes we encounter two factors that are *opposites* and form a ratio equal to −1. For example:

Simplified Form	**Details/Notes**
$\dfrac{-5}{5} = -1$	The ratio of a number and its opposite is **−1**.
$\dfrac{100}{-100} = -1$	The ratio of a number and its opposite is **−1**.

$$\dfrac{x + 7}{-x - 7} = -1 \qquad \dfrac{x + 7}{-x - 7} = \dfrac{x + 7}{-1(x + 7)} = \dfrac{\overset{1}{\cancel{x + 7}}}{-1\cancel{(x + 7)}} = \dfrac{1}{-1} = -1$$

Factor out −1.

$$\dfrac{2 - x}{x - 2} = -1 \qquad \dfrac{2 - x}{x - 2} = \dfrac{-1(-2 + x)}{x - 2} = \dfrac{-1\overset{1}{\cancel{(x - 2)}}}{\cancel{x - 2}} = \dfrac{-1}{1} = -1$$

Answer

11. $\dfrac{p^2 + 3p + 9}{p - 3}$

Recognizing factors that are opposites is useful when simplifying rational expressions. For example, $a - b$ and $b - a$ are opposites because the opposite of $a - b$ can be written $-(a - b) = -a + b = b - a$. Therefore, in general, $\frac{a - b}{b - a} = -1$.

Example 6 **Simplifying a Rational Expression**

Simplify the rational expression to lowest terms. $\dfrac{x - 5}{25 - x^2}$

Solution:

$\dfrac{x - 5}{25 - x^2}$

$= \dfrac{x - 5}{(5 - x)(5 + x)}$ Factor.

Notice that $x - 5$ and $5 - x$ are opposites and form a ratio of -1.

$= \dfrac{1}{(5 + x)} \cdot \dfrac{\overset{-1}{\cancel{(x - 5)}}}{\cancel{(5 - x)}}$ In general, $\dfrac{a - b}{b - a} = -1$.

$= \dfrac{1}{5 + x}(-1)$

$= -\dfrac{1}{5 + x}$ or $-\dfrac{1}{x + 5}$

TIP: The factor of -1 may be applied in front of the rational expression, or it may be applied to the numerator or to the denominator. Therefore, the final answer may be written in different forms.

$$-\dfrac{1}{x + 5} \quad \text{or} \quad \dfrac{-1}{x + 5} \quad \text{or} \quad \dfrac{1}{-(x + 5)}$$

Answer

12. $-\dfrac{5}{x + 3}$

Skill Practice Simplify the expression.

12. $\dfrac{20 - 5x}{x^2 - x - 12}$

Section 5.1 Practice Exercises

Vocabulary and Key Concepts

1. **a.** A _____ function is a function that can be written as $f(x) = \dfrac{p(x)}{q(x)}$, where $p(x)$ and $q(x)$ are polynomial functions and $q(x) \neq 0$.

   b. The domain of a rational function is all real numbers excluding the values of the variable that make the _____ equal to zero.

   c. For polynomials p, q, and r, where ($q \neq 0$ and $r \neq 0$), $\dfrac{pr}{qr}$ simplifies to _____.

   d. The ratio $\dfrac{a - b}{a - b} =$ _____, whereas the ratio $\dfrac{a - b}{b - a} =$ _____ provided that $a \neq b$.

Concept 1: Rational Functions

For Exercises 2–6, determine the function values, if possible. **(See Example 1.)**

2. $f(x) = \dfrac{5}{x+1}$; $f(0), f(2), f(-1), f(-6)$

3. $k(x) = \dfrac{-3}{x+4}$; $k(0), k(-1), k(2), k(-4)$

4. $m(x) = \dfrac{x-4}{x+6}$; $m(-6), m(-4), m(0), m(4)$

5. $n(a) = \dfrac{3a+1}{a^2+1}$; $n(1), n(0), n\left(-\dfrac{1}{3}\right), n(-1)$

6. $f(t) = \dfrac{2t-8}{t^2+9}$; $f(4), f(-4), f(3), f(-3)$

For Exercises 7–20,

    a. Write the domain in set-builder notation.

    b. Write the domain in interval notation. **(See Example 2.)**

7. $f(x) = \dfrac{9}{x}$

8. $g(a) = -\dfrac{10}{a}$

9. $h(v) = \dfrac{v+1}{v-7}$

10. $p(t) = \dfrac{t+9}{t+3}$

11. $k(x) = \dfrac{3x-1}{2x-5}$

12. $n(t) = \dfrac{6t+5}{3t+8}$

13. $f(q) = \dfrac{q+1}{q^2+6q-27}$

14. $k(a) = \dfrac{a^2}{2a^2+3a-5}$

15. $h(c) = \dfrac{c}{c^2+25}$

16. $m(x) = \dfrac{x}{x^2+16}$

17. $f(x) = \dfrac{x+5}{x^2-25}$

18. $g(t) = \dfrac{t+4}{t^2-16}$

19. $p(x) = \dfrac{x-5}{3}$

20. $r(x) = \dfrac{x+2}{8}$

For Exercises 21–24, write the domain of each function in interval notation and use that information to match the function with its graph.

21. $m(x) = \dfrac{1}{x + 4}$ **22.** $n(x) = \dfrac{1}{x + 1}$ **23.** $q(x) = \dfrac{1}{x - 4}$ **24.** $p(x) = \dfrac{1}{x - 1}$

a.   **b.**   **c.**   **d.**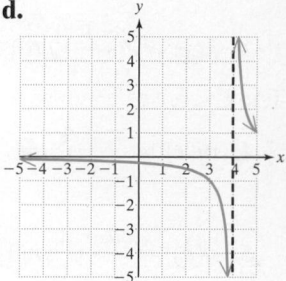

Concept 2: Simplifying Rational Expressions

For Exercises 25–26, simplify the expression if possible.

25. a. $\dfrac{8x}{4y}$ **b.** $\dfrac{8 + x}{4 + y}$ **26. a.** $\dfrac{a - 21}{14 + b}$ **b.** $\dfrac{-21a}{14b}$

For Exercises 27–30,

    a. Factor the numerator and denominator.

    b. Determine the restrictions on x.

    c. Simplify the expression. **(See Example 3.)**

27. $\dfrac{x^2 + 6x + 8}{x^2 + 3x - 4}$ **28.** $\dfrac{x^2 - 6x}{2x^2 - 11x - 6}$

29. $\dfrac{x^2 - 18x + 81}{x^2 - 81}$ **30.** $\dfrac{x^2 + 14x + 49}{x^2 - 49}$

For Exercises 31–52, simplify the rational expressions. **(See Examples 4–5.)**

31. $\dfrac{100x^3y^5}{36xy^8}$ **32.** $\dfrac{48ab^3c^2}{6a^7bc^0}$ **33.** $\dfrac{7w^{11}z^6}{14w^3z^3}$ **34.** $\dfrac{12r^9s^3}{24r^8s^4}$

35. $\dfrac{-3m^4n}{12m^6n^4}$ **36.** $\dfrac{-5x^3y^2}{20x^4y^2}$ **37.** $\dfrac{6a + 18}{9a + 27}$ **38.** $\dfrac{5y - 15}{3y - 9}$

39. $\dfrac{x - 5}{x^2 - 25}$ **40.** $\dfrac{3z - 6}{3z^2 - 12}$ **41.** $\dfrac{-7c}{21c^2 - 35c}$ **42.** $\dfrac{2p + 3}{2p^2 + 7p + 6}$

43. $\dfrac{2t^2 + 7t - 4}{-2t^2 - 5t + 3}$ **44.** $\dfrac{y^2 + 8y - 9}{y^2 - 5y + 4}$ **45.** $\dfrac{(p + 1)(2p - 1)^4}{(p + 1)^2(2p - 1)^2}$ **46.** $\dfrac{r(r - 3)^5}{r^3(r - 3)^2}$

47. $\dfrac{9 - z^2}{2z^2 + z - 15}$ **48.** $\dfrac{2c^2 + 2c - 12}{-8 + 2c + c^2}$ **49.** $\dfrac{2z^3 + 128}{16 + 8z + z^2}$ **50.** $\dfrac{p^3 - 1}{5 - 10p + 5p^2}$

51. $\dfrac{10x^3 - 25x^2 + 4x - 10}{-4 - 10x^2}$ **52.** $\dfrac{8x^3 - 12x^2 + 6x - 9}{16x^4 - 9}$

Concept 3: Simplifying Ratios of -1

For Exercises 53–64, simplify the rational expressions. **(See Example 6.)**

53. $\dfrac{r + 6}{6 + r}$ **54.** $\dfrac{a + 2}{2 + a}$ **55.** $\dfrac{b + 8}{-b - 8}$ **56.** $\dfrac{7 + w}{-7 - w}$

57. $\dfrac{10 - x}{x - 10}$ **58.** $\dfrac{y - 14}{14 - y}$ **59.** $\dfrac{2t - 2}{1 - t}$ **60.** $\dfrac{5p - 10}{2 - p}$

61. $\dfrac{c + 4}{c - 4}$ **62.** $\dfrac{b + 2}{b - 2}$ **63.** $\dfrac{y - x}{12x^2 - 12y^2}$ **64.** $\dfrac{4w^2 - 49z^2}{14z - 4w}$

Mixed Exercises

For Exercises 65–84, simplify the rational expression.

65. $\dfrac{t^2 - 1}{t^2 + 7t + 6}$ **66.** $\dfrac{x^2 + 4x + 4}{x^2 - 4}$ **67.** $\dfrac{8p + 8}{2p^2 - 4p - 6}$ **68.** $\dfrac{15y - 15}{3y^2 + 9y - 12}$

69. $\dfrac{-16a^2bc^4}{8ab^2c^4}$ **70.** $\dfrac{-9x^3yz^2}{27x^4yz}$ **71.** $\dfrac{x^2 - y^2}{8y - 8x}$ **72.** $\dfrac{p^2 - 49}{14 - 2p}$

73. $\dfrac{b + 4}{2b^2 + 5b - 12}$ **74.** $\dfrac{c - 6}{3c^2 - 17c - 6}$ **75.** $\dfrac{-2x + 34}{-4x + 6}$ **76.** $\dfrac{-9w - 3}{3w + 12}$

77. $\dfrac{(a - 2)^2(a - 5)^3}{(a - 2)^3(a - 5)}$ **78.** $\dfrac{t^2(t - 11)^4}{t^5(t - 11)^2}$ **79.** $\dfrac{4x - 2x^2}{5x - 10}$ **80.** $\dfrac{2y - 6}{3y^2 - y^3}$

81. $\dfrac{x^3 - 2x^2 - 25x + 50}{x^3 + 5x^2 - 4x - 20}$ **82.** $\dfrac{4y^3 + 12y^2 - y - 3}{2y^3 + y^2 - 18y - 9}$ **83.** $\dfrac{t^3 + 8}{3t^2 + t - 10}$ **84.** $\dfrac{w^3 - 27}{4w^2 - 5w - 21}$

Expanding Your Skills

85. Write a rational expression whose domain is $(-\infty, 2) \cup (2, \infty)$. (Answers may vary.)

86. Write a rational expression whose domain is $(-\infty, 3) \cup (3, \infty)$. (Answers may vary.)

87. Write a rational function whose domain is $(-\infty, -5) \cup (-5, \infty)$. (Answers may vary.)

88. Write a rational function whose domain is $(-\infty, -6) \cup (-6, \infty)$. (Answers may vary.)

Section 5.2   Multiplication and Division of Rational Expressions

Concepts

1. Multiplication of Rational Expressions
2. Division of Rational Expressions

1. Multiplication of Rational Expressions

Recall that to multiply fractions, we multiply the numerators and multiply the denominators. The same is true for multiplying rational expressions.

> **Multiplication Property of Rational Expressions**
>
> Let p, q, r, and s represent polynomials, such that $q \neq 0$ and $s \neq 0$. Then
>
> $$\frac{p}{q} \cdot \frac{r}{s} = \frac{pr}{qs}$$

For example:

Multiply the Fractions	Multiply the Rational Expressions
$\dfrac{2}{3} \cdot \dfrac{5}{7} = \dfrac{10}{21}$	$\dfrac{2x}{3y} \cdot \dfrac{5z}{7} = \dfrac{10xz}{21y}$

Sometimes it is possible to simplify a ratio of common factors to 1 *before* multiplying. To do so, we must first factor the numerators and denominators of each fraction.

$$\frac{7}{10} \cdot \frac{15}{21} \xrightarrow{\text{Factor.}} \frac{7}{2 \cdot 5} \cdot \frac{3 \cdot 5}{3 \cdot 7} = \frac{\overset{1}{\cancel{7}} \cdot \overset{1}{\cancel{3}} \cdot \overset{1}{\cancel{5}}}{2 \cdot \cancel{5} \cdot \cancel{3} \cdot \cancel{7}} = \frac{1}{2}$$

The same process is also used to multiply rational expressions.

> **Multiplying Rational Expressions**
>
> **Step 1** Factor the numerator and denominator of each expression.
> **Step 2** Multiply the numerators, and multiply the denominators.
> **Step 3** Reduce the ratios of common factors to 1 or -1 and simplify.

Example 1   Multiplying Rational Expressions

Multiply. $\dfrac{5a - 5b}{10} \cdot \dfrac{2}{a^2 - b^2}$

Solution:

$$\frac{5a - 5b}{10} \cdot \frac{2}{a^2 - b^2}$$

$$= \frac{5(a - b)}{5 \cdot 2} \cdot \frac{2}{(a - b)(a + b)} \qquad \text{Factor numerator and denominator.}$$

$$= \frac{5(a - b) \cdot 2}{5 \cdot 2 \cdot (a - b)(a + b)} \qquad \text{Multiply.}$$

$$= \frac{\overset{1}{\cancel{5}}(\overset{1}{\cancel{a - b}}) \cdot \overset{1}{\cancel{2}}}{\cancel{5} \cdot 2 \cdot (\cancel{a - b})(a + b)} = \frac{1}{a + b} \qquad \text{Reduce common factors and simplify.}$$

Avoiding Mistakes

If all factors in the numerator simplify to 1, do not forget to write the factor of 1 in the numerator.

Skill Practice Multiply.

1. $\dfrac{3y - 6}{6y} \cdot \dfrac{y^2 + 3y + 2}{y^2 - 4}$

Example 2 **Multiplying Rational Expressions**

Multiply. $\dfrac{4w - 20p}{2w^2 - 50p^2} \cdot \dfrac{2w^2 + 7wp - 15p^2}{3w + 9p}$

Solution:

$\dfrac{4w - 20p}{2w^2 - 50p^2} \cdot \dfrac{2w^2 + 7wp - 15p^2}{3w + 9p}$

$= \dfrac{4(w - 5p)}{2(w^2 - 25p^2)} \cdot \dfrac{(2w - 3p)(w + 5p)}{3(w + 3p)}$ Factor numerator and denominator.

$= \dfrac{2 \cdot 2(w - 5p)}{2(w - 5p)(w + 5p)} \cdot \dfrac{(2w - 3p)(w + 5p)}{3(w + 3p)}$ Factor further.

$= \dfrac{2 \cdot 2(w - 5p)(2w - 3p)(w + 5p)}{2(w - 5p)(w + 5p) \cdot 3(w + 3p)}$ Multiply.

$= \dfrac{2 \cdot 2\overset{1}{(w - 5p)}(2w - 3p)\overset{1}{(w + 5p)}}{2\underset{}{(w - 5p)}(w + 5p) \cdot 3(w + 3p)}$ Simplify common factors.

$= \dfrac{2(2w - 3p)}{3(w + 3p)}$

Notice that the expression is left in factored form to show that it has been simplified to lowest terms.

Skill Practice Multiply.

2. $\dfrac{p^2 + 8p + 16}{10p + 10} \cdot \dfrac{2p + 6}{p^2 + 7p + 12}$

---

2. Division of Rational Expressions

Recall that to divide fractions, multiply the first fraction by the reciprocal of the second fraction.

Divide: $\dfrac{15}{14} \div \dfrac{10}{49}$ $\xrightarrow[\text{of the second fraction.}]{\text{Multiply by the reciprocal}}$ $\dfrac{15}{14} \cdot \dfrac{49}{10} = \dfrac{3 \cdot \overset{1}{5} \cdot \overset{1}{7} \cdot 7}{2 \cdot 7 \cdot 2 \cdot 5} = \dfrac{21}{4}$

The same process is used for dividing rational expressions.

Division Property of Rational Expressions

Let p, q, r, and s represent polynomials, such that $q \neq 0, r \neq 0, s \neq 0$. Then

$$\dfrac{p}{q} \div \dfrac{r}{s} = \dfrac{p}{q} \cdot \dfrac{s}{r} = \dfrac{ps}{qr}$$

Answers

1. $\dfrac{y + 1}{2y}$ 2. $\dfrac{p + 4}{5(p + 1)}$

Example 3 **Dividing Rational Expressions**

Divide. $\dfrac{8t^3 + 27}{9 - 4t^2} \div \dfrac{4t^2 - 6t + 9}{2t^2 - t - 3}$

Solution:

$\dfrac{8t^3 + 27}{9 - 4t^2} \div \dfrac{4t^2 - 6t + 9}{2t^2 - t - 3}$

$= \dfrac{8t^3 + 27}{9 - 4t^2} \cdot \dfrac{2t^2 - t - 3}{4t^2 - 6t + 9}$

Multiply the first fraction by the reciprocal of the second.

$= \dfrac{(2t + 3)(4t^2 - 6t + 9)}{(3 - 2t)(3 + 2t)} \cdot \dfrac{(2t - 3)(t + 1)}{4t^2 - 6t + 9}$

Factor numerator and denominator. Notice $8t^3 + 27$ is a sum of cubes. Furthermore, $4t^2 - 6t + 9$ does not factor over the real numbers.

$= \dfrac{\overset{1}{\cancel{(2t + 3)}}\,\overset{1}{\cancel{(4t^2 - 6t + 9)}}\,\overset{-1}{\cancel{(2t - 3)}}(t + 1)}{\cancel{(3 - 2t)}\,\cancel{(3 + 2t)}\,\cancel{(4t^2 - 6t + 9)}}$

Simplify to lowest terms.

$= (-1)\dfrac{(t + 1)}{1}$

$= -(t + 1) \qquad \text{or} \qquad -t - 1$

> **Avoiding Mistakes**
>
> When dividing rational expressions, your first step should be to take the reciprocal of the second fraction (divisor). Do this first so that you do not forget.

Skill Practice Divide.

3. $\dfrac{x^2 + x}{5x^3 - x^2} \div \dfrac{10x^2 + 12x + 2}{25x^2 - 1}$

> **TIP:** In Example 3, the factors $(2t - 3)$ and $(3 - 2t)$ are opposites and form a ratio of -1. The factors $(2t + 3)$ and $(3 + 2t)$ are equal and form a ratio of 1.
>
> $$\dfrac{2t - 3}{3 - 2t} = -1 \quad \text{whereas} \quad \dfrac{2t + 3}{3 + 2t} = 1$$

Answer

3. $\dfrac{1}{2x}$

Section 5.2 Practice Exercises

Study Skills Exercise

Write an example of how to multiply two fractions, divide two fractions, add two fractions, and subtract two fractions. For example:

a. $\dfrac{5}{12} \cdot \dfrac{9}{10}$ **b.** $\dfrac{3}{4} \div \dfrac{15}{8}$ **c.** $\dfrac{5}{6} + \dfrac{7}{9}$ **d.** $\dfrac{5}{4} - \dfrac{2}{7}$

As you learn about rational expressions, compare the operations on rational expressions to those on fractions. This is a great place to use 3×5 cards again. Write an example of an operation with fractions on one side and the same operation with rational expressions on the other side.

Vocabulary and Key Concepts

1. a. Given polynomials $p, q, r,$ and s such that $q \neq 0$ and $s \neq 0, \dfrac{p}{q} \cdot \dfrac{r}{s} = \dfrac{\square}{\square}$.

   b. Given polynomials $p, q, r,$ and s such that $q \neq 0, r \neq 0,$ and $s \neq 0$ $\dfrac{p}{q} \div \dfrac{r}{s} = \dfrac{\square}{\square}$.

Review Exercises

For Exercises 2–9, simplify the rational expression to lowest terms.

2. $\dfrac{3w^2 - 12}{w^2 + 5w + 6}$
    3. $\dfrac{t^2 - 5t - 6}{t^2 - 7t + 6}$
    4. $\dfrac{5y + 15}{5y^2 + 16y + 3}$
    5. $\dfrac{2 - p}{p^2 - p - 2}$

6. $\dfrac{5x^2yz^3}{20xyz}$
    7. $\dfrac{7x + 14}{7x^2 - 7x - 42}$
    8. $\dfrac{25 - x^2}{x^2 - 10x + 25}$
    9. $\dfrac{a^3b^2c^5}{2a^3bc^2}$

Concept 1: Multiplication of Rational Expressions

For Exercises 10–21, multiply the rational expressions. **(See Examples 1–2.)**

10. $\dfrac{8w^2}{9} \cdot \dfrac{3}{2w^4}$
    11. $\dfrac{16}{z^7} \cdot \dfrac{z^4}{8}$
    12. $\dfrac{5p^2q^4}{12pq^3} \cdot \dfrac{6p^2}{20q^2}$

13. $\dfrac{27r^5}{7s} \cdot \dfrac{28rs^3}{9r^3s^2}$
    14. $\dfrac{3z + 12}{8z^3} \cdot \dfrac{16z^3}{9z + 36}$
    15. $\dfrac{x^2y}{x^2 - 4x - 5} \cdot \dfrac{2x^2 - 13x + 15}{xy^3}$

16. $\dfrac{3y^2 + 18y + 15}{6y + 6} \cdot \dfrac{y - 5}{y^2 - 25}$
    17. $\dfrac{10w - 8}{w + 2} \cdot \dfrac{3w^2 - w - 14}{25w^2 - 16}$
    18. $\dfrac{x - 5y}{x^2 + xy} \cdot \dfrac{y^2 - x^2}{10y - 2x}$

19. $\dfrac{3x - 15}{4x^2 - 2x} \cdot \dfrac{10x - 20x^2}{5 - x}$
    20. $x(x + 5)^2 \cdot \dfrac{2}{x^2 - 25}$
    21. $y(y^2 - 4) \cdot \dfrac{y}{y + 2}$

Concept 2: Division of Rational Expressions

For Exercises 22–35, divide the rational expressions. **(See Example 3.)**

22. $\dfrac{5x}{7} \div \dfrac{10x^2}{21}$
    23. $\dfrac{2a}{7b^3} \div \dfrac{10a^5}{77}$

24. $\dfrac{6x^2y^2}{(x - 2)} \div \dfrac{3xy^2}{(x - 2)^2}$
    25. $\dfrac{(r + 3)^2}{4r^3s} \div \dfrac{r + 3}{rs}$

26. $\dfrac{t^2 + 5t}{t + 1} \div (t + 5)$
    27. $\dfrac{6p + 7}{p + 2} \div (36p^2 - 49)$

28. $\dfrac{a}{a - 10} \div \dfrac{a^3 + 6a^2 - 40a}{a^2 - 100}$
    29. $\dfrac{b^2 - 6b + 9}{b^2 - b - 6} \div \dfrac{b^2 - 9}{4}$

30. $\dfrac{2x^2 + 5xy + 2y^2}{4x^2 - y^2} \div \dfrac{x^2 + xy - 2y^2}{2x^2 + xy - y^2}$
    31. $\dfrac{6s^2 + st - 2t^2}{6s^2 - 5st + t^2} \div \dfrac{3s^2 + 17st + 10t^2}{6s^2 + 13st - 5t^2}$

32. $\dfrac{x^4 - x^3 + x^2 - x}{2x^3 + 2x^2 + x + 1} \div \dfrac{x^3 - 4x^2 + x - 4}{2x^3 - 8x^2 + x - 4}$
    33. $\dfrac{a^3 + a - a^2 + 1}{a^3 + a^2 + ab^2 + b^2} \div \dfrac{a^3 + a + a^2b + b}{2a^2 + 2ab + ab^2 + b^3}$

34. $\dfrac{3y - y^2}{y^3 - 27} \div \dfrac{y}{y^2 + 3y + 9}$
    35. $\dfrac{8x - 4x^2}{xy - 2y + 3x - 6} \div \dfrac{3x + 6}{y + 3}$

Mixed Exercises

For Exercises 36–55, perform the indicated operations.

36. $\dfrac{8a^4b^3}{3c} \div \dfrac{a^7b^2}{9c}$

37. $\dfrac{3x^5}{2x^2y^7} \div \dfrac{4x^3y}{6y^6}$

38. $\dfrac{2}{25x^2} \cdot \dfrac{5x}{12} \div \dfrac{2}{15x}$

39. $\dfrac{4y}{7} \div \dfrac{y^2}{14} \cdot \dfrac{3}{y}$

40. $\dfrac{10x^2 - 13xy - 3y^2}{8x^2 - 10xy - 3y^2} \cdot \dfrac{2y + 8x}{2x^2 + 2y^2}$

41. $\dfrac{6a^2 + ab - b^2}{10a^2 + 5ab} \cdot \dfrac{2a^3 + 4a^2b}{3a^2 + 5ab - 2b^2}$

42. $(3m^2 - 12m) \div \dfrac{m^2 - 4m}{m^2 - 6m + 8}$

43. $(2x^2 + 8) \div \dfrac{x^4 - 16}{x^2 + x - 6}$

44. $\dfrac{(a + b)^2}{a - b} \cdot \dfrac{a^3 - b^3}{a^2 - b^2} \div \dfrac{a^2 + ab + b^2}{(a - b)^2}$

45. $\dfrac{m^2 - n^2}{(m - n)^2} \div \dfrac{m^2 - 2mn + n^2}{m^2 - mn + n^2} \cdot \dfrac{(m - n)^4}{m^3 + n^3}$

46. $\dfrac{x^2 - 4y^2}{x + 2y} \div (x + 2y) \cdot \dfrac{2y}{x - 2y}$

47. $\dfrac{x^2 - 6xy + 9y^2}{x^2 - 4y^2} \cdot \dfrac{x^2 - 5xy + 6y^2}{3y - x} \div \dfrac{x^2 - 9y^2}{x + 2y}$

48. $\dfrac{8x^3 - 27y^3}{4x^2 - 9y^2} \div \dfrac{8x^2 + 12xy + 18y^2}{2x + 3y}$

49. $\dfrac{25m^2 - 1}{125m^3 - 1} \div \dfrac{5m + 1}{25m^2 + 5m + 1}$

50. $\dfrac{m^3 + 2m^2 - mn^2 - 2n^2}{m^3 - m^2 - 20m} \cdot \dfrac{m^3 - 25m}{m^3 + m^2n - 4m - 4n}$

51. $\dfrac{2a^2 + ab - 8a - 4b}{2a^2 - 6a + ab - 3b} \cdot \dfrac{a^2 - 6a + 9}{a^2 - 16}$

52. $\dfrac{7}{3x + 15} \cdot (x + 5) \div \dfrac{14}{9x - 27}$

53. $\dfrac{45}{2x + 1} \cdot (8x + 4) \div \dfrac{27}{4x + 2}$

54. $\dfrac{12y + 3}{6y^2 - y - 12} \div \dfrac{4y^2 - 19y - 5}{2y^2 - y - 3}$

55. $\dfrac{2x^2 - 11x - 6}{3x - 2} \div \dfrac{2x^2 - 5x - 3}{3x^2 - 7x - 6}$

Expanding Your Skills

For Exercises 56–59, write an expression for the area of the figure and simplify.

56.

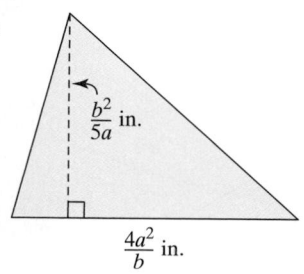

$\dfrac{b^2}{5a}$ in.

$\dfrac{4a^2}{b}$ in.

57.

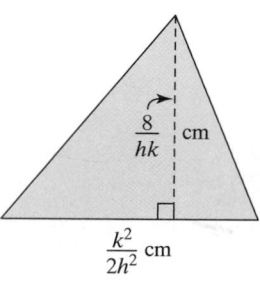

$\dfrac{8}{hk}$ cm

$\dfrac{k^2}{2h^2}$ cm

58.

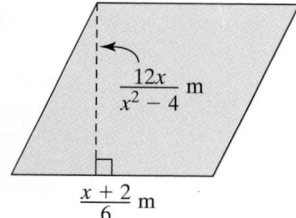

$\dfrac{12x}{x^2 - 4}$ m

$\dfrac{x + 2}{6}$ m

59.

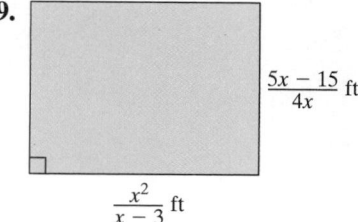

$\dfrac{5x - 15}{4x}$ ft

$\dfrac{x^2}{x - 3}$ ft

Addition and Subtraction of Rational Expressions | **Section 5.3**

1. Addition and Subtraction of Rational Expressions with Like Denominators

Concepts

1. Addition and Subtraction of Rational Expressions with Like Denominators
2. Least Common Denominator
3. Equivalent Rational Expressions
4. Addition and Subtraction of Rational Expressions with Unlike Denominators

To add or subtract rational expressions, the expressions must have the same denominator. As with fractions, we add or subtract rational expressions with the same denominator by combining the terms in the numerator and then writing the result over the common denominator. Then, if possible, we simplify the expression to lowest terms.

> **Addition and Subtraction Properties of Rational Expressions**
>
> Let p, q, and r represent polynomials where $q \neq 0$. Then
>
> **1.** $\dfrac{p}{q} + \dfrac{r}{q} = \dfrac{p+r}{q}$ **2.** $\dfrac{p}{q} - \dfrac{r}{q} = \dfrac{p-r}{q}$

Example 1 **Adding and Subtracting Rational Expressions with Like Denominators**

Add or subtract as indicated.

a. $\dfrac{1}{8} + \dfrac{3}{8}$ **b.** $\dfrac{5x}{2x-1} + \dfrac{3}{2x-1}$ **c.** $\dfrac{x^2}{x-4} - \dfrac{x+12}{x-4}$

Solution:

a. $\dfrac{1}{8} + \dfrac{3}{8} = \dfrac{1-3}{8}$ Add the terms in the numerator.

$= \dfrac{4}{8}$

$= \dfrac{1}{2}$ Simplify the fraction.

b. $\dfrac{5x}{2x-1} + \dfrac{3}{2x-1} = \dfrac{5x+3}{2x-1}$ Add the terms in the numerator. The answer is already in lowest terms.

c. $\dfrac{x^2}{x-4} - \dfrac{x+12}{x-4}$ Combine the terms in the numerator. Use parentheses to group the terms in the numerator that follow the subtraction sign. This will help you remember to apply the distributive property.

$= \dfrac{x^2 - (x+12)}{x-4}$

$= \dfrac{x^2 - x - 12}{x-4}$ Apply the distributive property

$= \dfrac{(x-4)(x+3)}{(x-4)}$ Factor the numerator and denominator.

$= \dfrac{\overset{1}{\cancel{(x-4)}}(x+3)}{\cancel{(x-4)}}$ Simplify the rational expression.

$= x + 3$

1. $\dfrac{5}{12} - \dfrac{1}{12}$ **2.** $\dfrac{4c - 3}{c - 2} + \dfrac{8}{c - 2}$ **3.** $\dfrac{t^2}{t - 7} - \dfrac{5t + 14}{t - 7}$

2. Least Common Denominator

If two rational expressions have different denominators, each expression must be rewritten with a common denominator before adding or subtracting the expressions. The **least common denominator (LCD)** of two or more rational expressions is defined as the least common multiple of the denominators.

For example, consider the fractions $\frac{1}{20}$ and $\frac{1}{8}$. By inspection, the least common denominator is 40. To understand why, find the prime factorization of both denominators.

$$20 = 2^2 \cdot 5 \quad \text{and} \quad 8 = 2^3$$

A common multiple of 20 and 8 must be a multiple of 5, a multiple of 2^2, and a multiple of 2^3. However, any number that is a multiple of $2^3 = 8$ is automatically a multiple of $2^2 = 4$. Therefore, it is sufficient to construct the least common denominator as the product of unique prime factors, where each factor is raised to its highest power.

The LCD of $\dfrac{1}{2^2 \cdot 5}$ and $\dfrac{1}{2^3}$ is $2^3 \cdot 5 = 40$.

> **Finding the LCD of Two or More Rational Expressions**
>
> **Step 1** Factor all denominators completely.
> **Step 2** The LCD is the product of unique prime factors from the denominators, where each factor is raised to the highest power to which it appears in any denominator.

Example 2 **Finding the LCD of Rational Expressions**

Find the LCD of the rational expressions.

a. $\dfrac{1}{12}, \dfrac{5}{18}, \dfrac{7}{30}$ **b.** $\dfrac{1}{2x^3y}, \dfrac{5}{16xy^2z}$

Solution:

a. $\dfrac{1}{12}, \dfrac{5}{18}, \dfrac{7}{30}$

$\dfrac{1}{2^2 \cdot 3}, \dfrac{5}{2 \cdot 3^2}, \dfrac{7}{2 \cdot 3 \cdot 5}$ Factor the denominators completely.

$\text{LCD} = 2^2 \cdot 3^2 \cdot 5 = 180$ The LCD is the product of the factors 2, 3, and 5. Each factor is raised to its highest power.

Answers

1. $\dfrac{1}{3}$ **2.** $\dfrac{4c + 5}{c - 2}$ **3.** $t + 2$

b. $\dfrac{1}{2x^3y}, \dfrac{5}{16xy^2z}$

$\dfrac{1}{2x^3y}, \dfrac{5}{2^4xy^2z}$ Factor the denominators completely.

$LCD = 2^4x^3y^2z$ The LCD is the product of the factors $2, x, y,$

$\quad\;\; = 16x^3y^2z$ and z. Each factor is raised to its highest power.

Skill Practice Find the LCD of the rational expressions.

4. $\dfrac{7}{40}, \dfrac{1}{15}, \dfrac{5}{6}$ **5.** $\dfrac{1}{9a^3b^2}, \dfrac{5}{18a^4b}$

| **Example 3** | **Finding the LCD of Rational Expressions** |

Find the LCD of the rational expressions.

a. $\dfrac{x^2 + 3}{x^2 + 9x + 20}, \dfrac{6}{x^2 + 8x + 16}$ **b.** $\dfrac{x + 4}{x - 3}, \dfrac{1}{3 - x}$

Solution:

a. $\dfrac{x^2 + 3}{x^2 + 9x + 20}, \dfrac{6}{x^2 + 8x + 16}$

$\dfrac{x^2 + 3}{(x + 4)(x + 5)}, \dfrac{6}{(x + 4)^2}$ Factor the denominators completely.

$LCD = (x + 5)(x + 4)^2$ The LCD is the product of the factors $(x + 5)$ and $(x + 4)$. Each factor is raised to its highest power.

b. $\dfrac{x + 4}{x - 3}, \dfrac{1}{3 - x}$ The denominators are already factored.

Notice that $x - 3$ and $3 - x$ are opposite factors. If -1 is factored from either expression, the binomial factors will be the same.

$\dfrac{x + 4}{x - 3}, \dfrac{1}{-1(-3 + x)}$ $\qquad$ $\dfrac{x + 4}{-1(-x + 3)}, \dfrac{1}{3 - x}$

$\qquad$ Factor $\qquad\qquad$ Factor
$\qquad$ out -1. $\qquad\qquad$ out -1.

$\qquad$ same binomial factors $\qquad\qquad$ same binomial factors

$LCD = (x - 3)(-1)$ $\qquad\qquad$ $LCD = (-1)(3 - x)$

$\quad\;\; = -x + 3$ $\qquad\qquad\qquad\;\; = -3 + x$

$\quad\;\; = 3 - x$ $\qquad\qquad\qquad\;\; = x - 3$

The LCD can be expressed as either $(3 - x)$ or $(x - 3)$.

Skill Practice Find the LCD of the rational expressions.

6. $\dfrac{5}{x^2 + 4x + 4}, \dfrac{x + 1}{x^2 - x - 6}$ **7.** $\dfrac{6}{z - 7}, \dfrac{1}{7 - z}$

Answers

4. 120 **5.** $18a^4b^2$
6. $(x + 2)^2(x - 3)$
7. $z - 7$ or $7 - z$

3. Equivalent Rational Expressions

Rational expressions can be added if they have common denominators. Once the LCD has been determined, each rational expression must be converted to an **equivalent rational expression** with the indicated denominator.

Using the identity property of multiplication, we know that for $q \neq 0$ and $r \neq 0$,

$$\frac{p}{q} = \frac{p}{q} \cdot 1 = \frac{p}{q} \cdot \frac{r}{r} = \frac{pr}{qr}$$

This principle is used to convert a rational expression to an equivalent expression with a different denominator. For example, $\frac{1}{2}$ can be converted to an equivalent expression with a denominator of 12 as follows:

$$\frac{1}{2} = \frac{1}{2} \cdot 1 = \frac{1}{2} \cdot \frac{6}{6} = \frac{1 \cdot 6}{2 \cdot 6} = \frac{6}{12}$$

In this example, we multiplied $\frac{1}{2}$ by a convenient form of 1. The ratio $\frac{6}{6}$ was chosen so that the product produced a new denominator of 12. Notice that multiplying $\frac{1}{2}$ by $\frac{6}{6}$ is equivalent to multiplying the numerator and denominator of the original expression by 6.

Example 4   **Creating Equivalent Rational Expressions**

Convert each expression to an equivalent rational expression with the indicated denominator.

a. $\dfrac{7}{5p^2} = \dfrac{}{20p^6}$ **b.** $\dfrac{w}{w + 5} = \dfrac{}{w^2 + 3w - 10}$

Solution:

a.
$$\frac{7}{5p^2} = \frac{}{20p^6}$$

$$\frac{7}{5p^2} \cdot \frac{4p^4}{4p^4} = \frac{28p^4}{20p^6}$$ $5p^2$ must be multiplied by $4p^4$ to create $20p^6$. Multiply numerator and denominator by $4p^4$.

b. $\dfrac{w}{w + 5} = \dfrac{}{w^2 + 3w - 10}$

$$= \frac{}{(w + 5)(w - 2)}$$ Factor the denominator.

$$\frac{w}{w + 5} = \frac{w}{w + 5} \cdot \frac{w - 2}{w - 2}$$ Multiply the numerator and denominator by the missing factor $(w - 2)$.

$$= \frac{w^2 - 2w}{(w + 5)(w - 2)}$$

TIP: Notice that in Example 4 we multiplied the polynomials in the numerator but left the denominator in factored form. This convention is followed because when we add and subtract rational expressions, the terms in the numerators must be combined.

Skill Practice   Convert each expression to an equivalent rational expression with the indicated denominator.

8. $\dfrac{1}{8xy} = \dfrac{}{16x^3y^2}$ **9.** $\dfrac{5b}{b - 3} = \dfrac{}{b^2 - 9}$

Answers

8. $\dfrac{2x^2y}{16x^3y^2}$ **9.** $\dfrac{5b^2 + 15b}{b^2 - 9}$

4. Addition and Subtraction of Rational Expressions with Unlike Denominators

To add or subtract rational expressions with unlike denominators, we must convert each expression to an equivalent expression with the same denominator. For example, consider adding the expressions $\frac{3}{x-2} + \frac{5}{x+1}$. The LCD is $(x-2)(x+1)$. For each expression, identify the factors from the LCD that are missing in the denominator. Then multiply the numerator and denominator of the expression by the missing factor(s):

$$\frac{(3)}{(x-2)} \cdot \frac{(x+1)}{(x+1)} + \frac{(5)}{(x+1)} \cdot \frac{(x-2)}{(x-2)}$$
The rational expressions now have the same denominator and can be added.

$$= \frac{3(x+1) + 5(x-2)}{(x-2)(x+1)}$$
Combine terms in the numerator.

$$= \frac{3x + 3 + 5x - 10}{(x-2)(x+1)}$$
Clear parentheses and simplify.

$$= \frac{8x - 7}{(x-2)(x+1)}$$

Adding and Subtracting Rational Expressions

Step 1 Factor the denominator of each rational expression.
Step 2 Identify the LCD.
Step 3 Rewrite each rational expression as an equivalent expression with the LCD as its denominator.
Step 4 Add or subtract the numerators, and write the result over the common denominator.
Step 5 Simplify, if possible.

Example 5 Adding Rational Expressions with Unlike Denominators

Add. $\frac{3}{7b} + \frac{4}{b^2}$

Solution:

$\frac{3}{7b} + \frac{4}{b^2}$

Step 1: The denominators are already factored.
Step 2: The LCD is $7b^2$.

$= \frac{3}{7b} \cdot \frac{b}{b} + \frac{4}{b^2} \cdot \frac{7}{7}$ **Step 3:** Write each expression with the LCD.

$= \frac{3b}{7b^2} + \frac{28}{7b^2}$ **Step 4:** Add the numerators, and write the result over the LCD.

$= \frac{3b + 28}{7b^2}$ **Step 5:** Simplify.

Skill Practice Add.

10. $\frac{4}{5y} + \frac{1}{3y^3}$

Answer

10. $\frac{12y^2 + 5}{15y^3}$

Example 6   Subtracting Rational Expressions
with Unlike Denominators

Subtract. $\dfrac{3t - 2}{t^2 + 4t - 12} - \dfrac{5}{2t + 12}$

Solution:

$$\dfrac{3t - 2}{t^2 + 4t - 12} - \dfrac{5}{2t + 12}$$

$$= \dfrac{3t - 2}{(t + 6)(t - 2)} - \dfrac{5}{2(t + 6)}$$ **Step 1:** Factor the denominators.

Step 2: The LCD is $2(t + 6)(t - 2)$.

$$= \dfrac{(2)}{(2)} \cdot \dfrac{(3t - 2)}{(t + 6)(t - 2)} - \dfrac{5}{2(t + 6)} \cdot \dfrac{(t - 2)}{(t - 2)}$$ **Step 3:** Write each expression with the LCD.

$$= \dfrac{2(3t - 2) - 5(t - 2)}{2(t + 6)(t - 2)}$$ **Step 4:** Add the numerators and write the result over the LCD.

$$= \dfrac{6t - 4 - 5t + 10}{2(t + 6)(t - 2)}$$ **Step 5:** Simplify.

$$= \dfrac{t + 6}{2(t + 6)(t - 2)}$$ Combine *like* terms.

$$= \dfrac{\overset{1}{\cancel{t + 6}}}{2(\cancel{t + 6})(t - 2)}$$ Simplify.

$$= \dfrac{1}{2(t - 2)}$$

Skill Practice Subtract.

11. $\dfrac{2x + 3}{x^2 + x - 2} - \dfrac{5}{3x - 3}$

Example 7   Adding and Subtracting Rational Expressions
with Unlike Denominators

Add and subtract as indicated. $\dfrac{2}{x} + \dfrac{x}{x + 3} - \dfrac{3x + 18}{x^2 + 3x}$

Solution:

$$\dfrac{2}{x} + \dfrac{x}{x + 3} - \dfrac{3x + 18}{x^2 + 3x}$$

$$= \dfrac{2}{x} + \dfrac{x}{x + 3} - \dfrac{3x + 18}{x(x + 3)}$$ **Step 1:** Factor the denominators.

Step 2: The LCD is $x(x + 3)$.

$$= \dfrac{2}{x} \cdot \dfrac{(x + 3)}{(x + 3)} + \dfrac{x}{(x + 3)} \cdot \dfrac{x}{x} - \dfrac{3x + 18}{x(x + 3)}$$ **Step 3:** Write each expression with the LCD.

Answer

11. $\dfrac{1}{3(x + 2)}$

$$= \frac{2(x+3) + x^2 - (3x+18)}{x(x+3)}$$

Step 4: Add the numerators, and write the result over the LCD.

$$= \frac{2x + 6 + x^2 - 3x - 18}{x(x+3)}$$

Step 5: Simplify.

$$= \frac{x^2 - x - 12}{x(x+3)}$$

Combine *like* terms.

$$= \frac{(x-4)(x+3)}{x(x+3)}$$

Factor the numerator.

$$= \frac{(x-4)\overset{1}{\cancel{(x+3)}}}{x\cancel{(x+3)}}$$

Simplify.

$$= \frac{x-4}{x}$$

Avoiding Mistakes

It is important to insert parentheses around the quantity being subtracted.

Skill Practice Add.

12. $\dfrac{a^2 + a + 24}{a^2 - 9} + \dfrac{5}{a+3}$

Example 8 Subtracting Rational Expressions with Unlike Denominators

Subtract. $\quad \dfrac{6}{w} - \dfrac{4}{-w}$

Solution:

$$\frac{6}{w} - \frac{4}{-w}$$

Step 1: The denominators are already factored.

Step 2: The denominators are opposites and differ by a factor of -1. The LCD can either be taken as w or $-w$. We will use an LCD of w.

$$= \frac{6}{w} - \frac{4}{-w} \cdot \frac{(-1)}{(-1)}$$

Step 3: Write each expression with the LCD. Note that $(-w)(-1) = w$.

$$= \frac{6}{w} - \frac{-4}{w}$$

$$= \frac{6 - (-4)}{w}$$

Step 4: Subtract the numerators, and write the result over the LCD.

$$= \frac{10}{w}$$

Step 5: Simplify.

Skill Practice Subtract.

13. $\dfrac{3}{-y} - \dfrac{5}{y}$

Answers

12. $\dfrac{a+3}{a-3}$ **13.** $-\dfrac{8}{y}$

> **Example 9**   **Adding Rational Expressions with Unlike Denominators**

Add. $\dfrac{x^2}{x-y} + \dfrac{y^2}{y-x}$

Solution:

$\dfrac{x^2}{x-y} + \dfrac{y^2}{y-x}$

Step 1: The denominators are already factored.

Step 2: The denominators are opposites and differ by a factor of -1. The LCD can be taken as either $(x-y)$ or $(y-x)$. We will use an LCD of $(x-y)$.

$= \dfrac{x^2}{(x-y)} + \dfrac{y^2}{(y-x)} \cdot \dfrac{(-1)}{(-1)}$ **Step 3:** Write each expression with the LCD. Note that $(y-x)(-1) = -y + x = x - y$.

$= \dfrac{x^2}{x-y} + \dfrac{-y^2}{x-y}$

$= \dfrac{x^2 - y^2}{x-y}$ **Step 4:** Combine the numerators, and write the result over the LCD.

$= \dfrac{(x+y)\overset{1}{\cancel{(x-y)}}}{\cancel{x-y}}$ **Step 5:** Factor and simplify to lowest terms.

$= x + y$

Skill Practice   Add.

14. $\dfrac{3a}{a-5} + \dfrac{15}{5-a}$

Answer

14. 3

Section 5.3   Practice Exercises

Vocabulary and Key Concepts

1. a. Given polynomials p, q, and r such that $q \neq 0$, $\dfrac{p}{q} + \dfrac{r}{q} = \dfrac{\Box}{\Box}$ and $\dfrac{p}{q} - \dfrac{r}{q} = \dfrac{\Box}{\Box}$

   b. The _____ _____ _____ (LCD) of two rational expressions is defined as the least common multiple of their denominators.

Review Exercises

For Exercises 2–6, perform the indicated operation.

2. $\dfrac{x}{x-y} \div \dfrac{x^2}{y-x}$

3. $\dfrac{9b+9}{4b+8} \cdot \dfrac{2b+4}{3b-3}$

4. $\dfrac{8a^2}{7b^3} \div \dfrac{4a}{21b}$

5. $\dfrac{(5-a)^2}{10a-2} \cdot \dfrac{25a^2-1}{a^2-10a+25}$

6. $\dfrac{x^2-z^2}{14x^2z^4} \div \dfrac{x^2+2xz+z^2}{3xz^3}$

Concept 1: Addition and Subtraction of Rational Expressions with Like Denominators

For Exercises 7–14, add or subtract as indicated and simplify if possible. **(See Example 1.)**

7. $\dfrac{3}{5x} + \dfrac{7}{5x}$

8. $\dfrac{1}{2x^2} - \dfrac{5}{2x^2}$

9. $\dfrac{x}{x^2 - 2x - 3} - \dfrac{3}{x^2 - 2x - 3}$

10. $\dfrac{x}{x^2 + 4x - 12} + \dfrac{6}{x^2 + 4x - 12}$

11. $\dfrac{5x - 1}{(2x + 9)(x - 6)} - \dfrac{3x - 6}{(2x + 9)(x - 6)}$

12. $\dfrac{4 - x}{8x + 1} - \dfrac{5x - 6}{8x + 1}$

13. $\dfrac{x + 2}{x - 5} + \dfrac{x - 12}{x - 5}$

14. $\dfrac{2x - 1}{x - 2} + \dfrac{x - 5}{x - 2}$

Concept 2: Least Common Denominator

For Exercises 15–26, find the least common denominator (LCD). **(See Examples 2–3.)**

15. $\dfrac{5}{8}; \ \dfrac{3}{20x}$

16. $\dfrac{y}{15a}; \ \dfrac{y^2}{35}$

17. $\dfrac{-5}{6m^4}; \ \dfrac{1}{15mn^7}$

18. $\dfrac{13}{12cd^5}; \ \dfrac{9}{8c^3}$

19. $\dfrac{6}{(x - 4)(x + 2)}; \ \dfrac{-8}{(x - 4)(x - 6)}$

20. $\dfrac{x}{(2x - 1)(z - 7)}; \ \dfrac{2}{(2x - 1)(x + 1)}$

21. $\dfrac{3}{x(x - 1)(x + 7)^2}; \ \dfrac{-1}{x^2(x + 7)}$

22. $\dfrac{14}{(x - 2)^2(x + 9)}; \ \dfrac{41}{x(x - 2)(x + 9)}$

23. $\dfrac{5}{x - 6}; \ \dfrac{x - 5}{x^2 - 8x + 12}$

24. $\dfrac{7a}{a + 4}; \ \dfrac{a + 12}{a^2 - 16}$

25. $\dfrac{3a}{a - 4}; \ \dfrac{5}{4 - a}$

26. $\dfrac{10}{x - 6}; \ \dfrac{z + 1}{6 - x}$

Concept 3: Equivalent Rational Expressions

For Exercises 27–32, fill in the blank to make an equivalent fraction with the given denominator. **(See Example 4.)**

27. $\dfrac{5}{3x} = \dfrac{}{9x^2y}$

28. $\dfrac{-5}{xy} = \dfrac{}{4x^2y^3}$

29. $\dfrac{2x}{x - 1} = \dfrac{}{x(x - 1)(x + 2)}$

30. $\dfrac{5x}{2x - 5} = \dfrac{}{(2x - 5)(x + 8)}$

31. $\dfrac{y}{y + 6} = \dfrac{}{y^2 + 5y - 6}$

32. $\dfrac{t^2}{t - 8} = \dfrac{}{t^2 - 6t - 16}$

Concept 4: Addition and Subtraction of Rational Expressions with Unlike Denominators

For Exercises 33–58, add or subtract as indicated. **(See Examples 5–9.)**

33. $\dfrac{4}{3p} - \dfrac{5}{2p^2}$

34. $\dfrac{6}{5a^2b} - \dfrac{1}{10ab}$

35. $\dfrac{s - 1}{s} - \dfrac{t + 1}{t}$

36. $\dfrac{x + 2}{x} - \dfrac{y - 2}{y}$

37. $\dfrac{4a - 2}{3a + 12} - \dfrac{a - 2}{a + 4}$

38. $\dfrac{6y + 5}{5y - 25} - \dfrac{y + 2}{y - 5}$

39. $\dfrac{10}{b(b + 5)} + \dfrac{2}{b}$

40. $\dfrac{6}{w(w - 2)} + \dfrac{3}{w}$

41. $\dfrac{x - 2}{x - 6} - \dfrac{x + 2}{6 - x}$

42. $\dfrac{x - 10}{x - 8} - \dfrac{x + 10}{8 - x}$

43. $\dfrac{6b}{b - 4} - \dfrac{1}{b + 1}$

44. $\dfrac{a}{a - 3} - \dfrac{5}{a + 6}$

45. $\dfrac{2}{2x + 1} + \dfrac{4}{x - 2}$

46. $\dfrac{3}{y + 6} + \dfrac{1}{3y + 1}$

47. $\dfrac{y - 2}{y - 4} + \dfrac{2y^2 - 15y + 12}{y^2 - 16}$

48. $\dfrac{x^2 + 13x + 18}{x^2 - 9} + \dfrac{x + 1}{x + 3}$

49. $\dfrac{x + 2}{x^2 - 36} - \dfrac{x}{x^2 + 9x + 18}$

50. $\dfrac{7}{x^2 - x - 2} + \dfrac{x}{x^2 + 4x + 3}$

51. $\dfrac{5}{w} + \dfrac{8}{-w}$

52. $\dfrac{4}{y} + \dfrac{5}{-y}$

53. $\dfrac{n}{5 - n} + \dfrac{2n - 5}{n - 5}$

54. $\dfrac{c}{7 - c} + \dfrac{2c - 7}{c - 7}$

55. $\dfrac{2}{3x - 15} + \dfrac{x}{25 - x^2}$

56. $\dfrac{5}{9 - x^2} - \dfrac{4}{x^2 + 4x + 3}$

57. $\dfrac{m}{20 + 9m + m^2} - \dfrac{4}{12 + 7m + m^2}$ **58.** $\dfrac{t}{6 + 5t + t^2} - \dfrac{2}{2 + 3t + t^2}$

Mixed Exercises

For Exercises 59–80, simplify.

59. $\dfrac{x + 3}{x^2} + \dfrac{x + 5}{2x}$

60. $\dfrac{x + 2}{5x^2} + \dfrac{x + 4}{15x}$

61. $w + 2 + \dfrac{1}{w - 2}$

62. $h - 3 + \dfrac{1}{h + 3}$

63. $\dfrac{9}{x^2 - 2x + 1} - \dfrac{x - 3}{x^2 - x}$

64. $\dfrac{2}{4z^2 - 12z + 9} - \dfrac{z + 1}{2z^2 - 3z}$

65. $\dfrac{t + 1}{t + 3} - \dfrac{t - 2}{t - 3} + \dfrac{6}{t^2 - 9}$

66. $\dfrac{y - 3}{y - 2} - \dfrac{y + 1}{2y - 5} + \dfrac{-4y + 7}{2y^2 - 9y + 10}$

67. $(x - 1) \cdot \left[\dfrac{3}{x^2 - 1} + \dfrac{x}{2x - 2} \right]$

68. $(3x - 2) \cdot \left[\dfrac{x}{3x^2 + x - 2} + \dfrac{2}{x + 1} \right]$

69. $\dfrac{3z}{z - 3} - \dfrac{z}{z + 4}$

70. $\dfrac{2p}{p - 5} - \dfrac{p}{p + 6}$

71. $\dfrac{2x}{x^2 - y^2} - \dfrac{1}{x - y} + \dfrac{1}{y - x}$

72. $\dfrac{3w - 1}{2w^2 + w - 3} - \dfrac{2 - w}{w - 1} - \dfrac{w}{1 - w}$

73. $(2p + 1) \cdot \left[\dfrac{2p}{6p + 3} - \dfrac{1}{p + 4} \right]$

74. $(y + 8) \cdot \left[\dfrac{4}{2y + 1} - \dfrac{y}{2y^2 + 17y + 8} \right]$

75. $\dfrac{1}{x + 5} + \dfrac{3}{(x + 5)^2} - \dfrac{2}{(x + 5)^3}$

76. $\dfrac{1}{x - 2} + \dfrac{4}{(x - 2)^2} - \dfrac{3}{(x - 2)^3}$

77. $\dfrac{-10}{z^2 - 6z + 5} + \dfrac{15}{z^2 - 4z - 5}$

78. $\dfrac{-4}{n^2 + 6n + 5} + \dfrac{3}{n^2 + 7n + 10}$

79. $\dfrac{5}{x^2 - 4} + \dfrac{2}{x^3 - 8}$

80. $\dfrac{-2}{x^2 - 9} + \dfrac{3}{x^3 - 27}$

For Exercises 81–84, write an expression that represents the perimeter of the figure and simplify.

81.

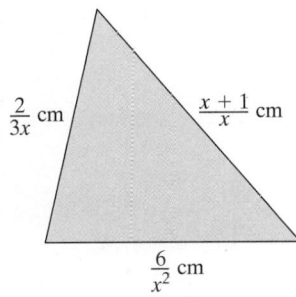

$\frac{2}{3x}$ cm $\frac{x+1}{x}$ cm

$\frac{6}{x^2}$ cm

82.

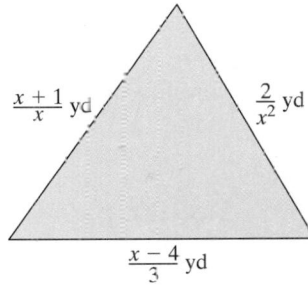

$\frac{x+1}{x}$ yd $\frac{2}{x^2}$ yd

$\frac{x-4}{3}$ yd

83.

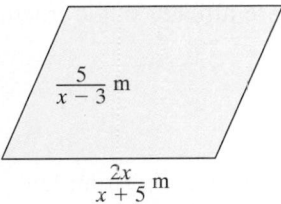

$\frac{5}{x-3}$ m

$\frac{2x}{x+5}$ m

84.

$\frac{3}{x+2}$ ft

$\frac{x}{x+1}$ ft

Complex Fractions

1. Simplifying Complex Fractions by Method I

Concepts

1. Simplifying Complex Fractions by Method I
2. Simplifying Complex Fractions by Method II

A **complex fraction** is an expression containing one or more fractional expressions in the numerator, denominator, or both. For example:

$$\frac{\dfrac{5x^2}{y}}{\dfrac{10x}{y^2}} \quad \text{and} \quad \frac{2 + \dfrac{1}{2} - \dfrac{1}{3}}{\dfrac{3}{4} + \dfrac{1}{6}}$$

are complex fractions.

Two methods will be presented to simplify complex fractions. The first method (Method I) follows the order of operations to simplify the numerator and denominator separately before dividing. The process is summarized as follows.

Simplifying a Complex Fraction—Method I

Step 1 Add or subtract expressions in the numerator to form a single fraction. Add or subtract expressions in the denominator to form a single fraction.

Step 2 Divide the rational expressions from step 1.

Step 3 Simplify to lowest terms, if possible.

Example 1 Simplifying a Complex Fraction by Method I

Simplify the expression. $\dfrac{\dfrac{5x^2}{y}}{\dfrac{10x}{y^2}}$

Solution:

$\dfrac{\dfrac{5x^2}{y}}{\dfrac{10x}{y^2}}$

Step 1: The numerator and denominator of the complex fraction are already single fractions.

$= \dfrac{5x^2}{y} \div \dfrac{10x}{y^2}$

$= \dfrac{5x^2}{y} \cdot \dfrac{y^2}{10x}$

Step 2: Multiply the numerator of the complex fraction by the reciprocal of the denominator.

$= \dfrac{5x^2 y^2}{10xy}$

$= \dfrac{1}{2} x^{2-1} y^{2-1}$ **Step 3:** Simplify.

$= \dfrac{1}{2}xy$ or $\dfrac{xy}{2}$

Skill Practice Simplify the expression.

1. $\dfrac{\dfrac{18a^3}{b^2}}{\dfrac{6a^2}{b}}$

Sometimes it is necessary to simplify the numerator and denominator of a complex fraction before the division is performed. This is illustrated in Example 2.

Example 2 Simplifying a Complex Fraction by Method I

Simplify the expression. $\dfrac{\dfrac{1}{4x} - \dfrac{3}{2}}{3 - \dfrac{1}{2x}}$

Solution:

$\dfrac{\dfrac{1}{4x} - \dfrac{3}{2}}{3 - \dfrac{1}{2x}}$

Step 1: Combine fractions in the numerator and denominator separately.

$= \dfrac{\dfrac{1}{4x} - \dfrac{2x}{2x}\cdot\dfrac{3}{2}}{\dfrac{2x}{2x}\cdot\dfrac{3}{1} - \dfrac{1}{2x}}$

The LCD in the numerator is $4x$.
The LCD in the denominator is $2x$.

Answer

1. $\dfrac{3a}{b}$

$$= \frac{\dfrac{1-6x}{4x}}{\dfrac{6x-1}{2x}}$$

Step 2: Divide the expression in the numerator of the complex fraction by the expression in the denominator.

$$= \frac{\overset{-1}{1-6x}}{\underset{2}{4x}} \cdot \frac{\overset{1}{2x}}{6x-1}$$

Multiply by the reciprocal of the divisor.

$$= -\frac{1}{2}$$

Step 3: Simplify to lowest terms.

Skill Practice Simplify the expression.

2. $\dfrac{\dfrac{1}{9m}-\dfrac{4}{3}}{4-\dfrac{1}{3m}}$

2. Simplifying Complex Fractions by Method II

We will now use a second method to simplify complex fractions—Method II. Recall that multiplying the numerator and denominator of a rational expression by the same quantity does not change the value of the expression. This is the basis for Method II.

Simplifying a Complex Fraction—Method II

Step 1 Multiply the numerator and denominator of the complex fraction by the LCD of *all* individual fractions within the expression.

Step 2 Apply the distributive property, and simplify the numerator and denominator.

Step 3 Simplify to lowest terms, if possible.

Example 3 Simplifying a Complex Fraction by Method II

Simplify by using Method II. $\dfrac{4-\dfrac{6}{x}}{\dfrac{2}{x}-\dfrac{3}{x^2}}$

Solution:

$$\frac{4-\dfrac{6}{x}}{\dfrac{2}{x}-\dfrac{3}{x^2}}$$

The LCD of all individual terms is x^2.

$$= \frac{x^2 \cdot \left(4-\dfrac{6}{x}\right)}{x^2 \cdot \left(\dfrac{2}{x}-\dfrac{3}{x^2}\right)}$$

Step 1: Multiply the numerator and denominator of the complex fraction by the LCD of x^2.

Answer

2. $-\dfrac{1}{3}$

$$= \frac{x^2 \cdot (4) - x^2 \cdot \left(\frac{6}{x}\right)}{x^2 \cdot \left(\frac{2}{x}\right) - x^2 \cdot \left(\frac{3}{x^2}\right)}$$

Step 2: Apply the distributive property.

$$= \frac{4x^2 - 6x}{2x - 3}$$

$$= \frac{2x(2x - 3)}{2x - 3}$$

Step 3: Factor and simplify.

$$= 2x$$

Skill Practice Simplify by using Method II.

3. $\dfrac{y - \dfrac{1}{y}}{1 - \dfrac{1}{y^2}}$

Example 4   Simplifying a Complex Fraction by Method II

Simplify by using Method II. $\dfrac{x^{-1} - x^{-2}}{1 + 2x^{-1} - 3x^{-2}}$

Solution:

$$\frac{x^{-1} - x^{-2}}{1 + 2x^{-1} - 3x^{-2}}$$

TIP: When writing $2x^{-1}$ with positive exponents recall that

$$2x^{-1} = 2 \cdot \frac{1}{x}$$

$$= \frac{2}{1} \cdot \frac{1}{x} = \frac{2}{x}$$

$$= \frac{\dfrac{1}{x} - \dfrac{1}{x^2}}{1 + \dfrac{2}{x} - \dfrac{3}{x^2}}$$

Rewrite the expression with positive exponents. The LCD of all individual terms is x^2.

$$= \frac{x^2 \cdot \left(\dfrac{1}{x} - \dfrac{1}{x^2}\right)}{x^2 \cdot \left(1 + \dfrac{2}{x} - \dfrac{3}{x^2}\right)}$$

Step 1: Multiply the numerator and denominator of the complex fraction by the LCD x^2.

$$= \frac{x^2\left(\dfrac{1}{x}\right) - x^2\left(\dfrac{1}{x^2}\right)}{x^2(1) + x^2\left(\dfrac{2}{x}\right) - x^2\left(\dfrac{3}{x^2}\right)}$$

Step 2: Apply the distributive property.

$$= \frac{x - 1}{x^2 + 2x - 3}$$

Answer

3. y

$$= \frac{x - 1}{(x + 3)(x - 1)}$$ **Step 3:** Factor and simplify to lowest terms.

$$= \frac{\overset{1}{\cancel{x - 1}}}{(x + 3)(\cancel{x - 1})}$$

$$= \frac{1}{x + 3}$$

Skill Practice Simplify by using Method II.

4. $\dfrac{c^{-1}b - b^{-1}c}{b^{-1} + c^{-1}}$

Example 5 **Simplifying a Complex Fraction by Method II**

Simplify the expression by Method II.
$$\frac{\dfrac{1}{w + 3} - \dfrac{1}{w - 3}}{1 + \dfrac{9}{w^2 - 9}}$$

Solution:

$$\frac{\dfrac{1}{w + 3} - \dfrac{1}{w - 3}}{1 + \dfrac{9}{w^2 - 9}}$$

$$= \frac{\dfrac{1}{w + 3} - \dfrac{1}{w - 3}}{1 + \dfrac{9}{(w + 3)(w - 3)}}$$ Factor all denominators to find the LCD.

The LCD of $\dfrac{1}{1}, \dfrac{1}{w + 3}, \dfrac{1}{w - 3},$ and $\dfrac{9}{(w + 3)(w - 3)}$ is $(w + 3)(w - 3)$.

$$= \frac{(w + 3)(w - 3)\left(\dfrac{1}{w + 3} - \dfrac{1}{w - 3}\right)}{(w + 3)(w - 3)\left[1 + \dfrac{9}{(w + 3)(w - 3)}\right]}$$ **Step 1:** Multiply the numerator and denominator of the complex fraction by $(w + 3)(w - 3)$.

$$= \frac{\cancel{(w + 3)}(w - 3)\left(\dfrac{1}{\cancel{w + 3}}\right) - (w + 3)\cancel{(w - 3)}\left(\dfrac{1}{\cancel{w - 3}}\right)}{(w + 3)(w - 3)1 + \cancel{(w + 3)}\cancel{(w - 3)}\left[\dfrac{9}{\cancel{(w + 3)}\cancel{(w - 3)}}\right]}$$ **Step 2:** Distributive property.

$$= \frac{(w - 3) - (w + 3)}{(w + 3)(w - 3) + 9}$$ **Step 3:** Simplify.

$$= \frac{w - 3 - w - 3}{w^2 - 9 + 9}$$ Apply the distributive property.

$$= \frac{-6}{w^2}$$

$$= -\frac{6}{w^2}$$

Answer

4. $b - c$

Skill Practice Simplify by using Method II.

5. $\dfrac{\dfrac{2}{x+1} - \dfrac{1}{x-1}}{\dfrac{x}{x-1} - \dfrac{1}{x+1}}$

Answer

5. $\dfrac{x-3}{x^2+1}$

Section 5.4 Practice Exercises

Vocabulary and Key Concepts

1. A _____ fraction is an expression containing one or more fractional expressions in the numerator, denominator, or both.

Review Exercises

For Exercises 2–4, simplify to lowest terms.

2. $\dfrac{x^3+y^3}{5x+5y}$

3. $\dfrac{25a^3b^3c}{15a^4bc}$

4. $\dfrac{6t^2-27t+30}{12t-30}$

For Exercises 5–8, perform the indicated operations.

5. $\dfrac{5}{x^2} + \dfrac{3}{2x}$

6. $\dfrac{2y-4}{y+1} \cdot \dfrac{y^2+3y+2}{y^2-4}$

7. $\dfrac{3}{a-5} - \dfrac{1}{a+1}$

8. $\dfrac{7}{12-6b} \div \dfrac{14b}{b^2+b-6}$

Concept 1: Simplifying Complex Fractions by Method I

For Exercises 9–16, simplify the complex fractions by using Method I. **(See Examples 1–2.)**

9. $\dfrac{\dfrac{5x^2}{9y^2}}{\dfrac{3x}{y^2x}}$

10. $\dfrac{\dfrac{3w^2}{4rs}}{\dfrac{15wr}{s^2}}$

11. $\dfrac{\dfrac{x-6}{3x}}{\dfrac{3x-18}{9}}$

12. $\dfrac{\dfrac{a+4}{6}}{\dfrac{16-a^2}{3}}$

13. $\dfrac{\dfrac{2}{3} + \dfrac{1}{6}}{\dfrac{1}{2} - \dfrac{1}{4}}$

14. $\dfrac{\dfrac{7}{8} + \dfrac{3}{4}}{\dfrac{1}{3} - \dfrac{5}{6}}$

15. $\dfrac{8 - \dfrac{5}{2x}}{\dfrac{5}{8x} - 2}$

16. $\dfrac{10 - \dfrac{3}{5x}}{\dfrac{3}{10x} - 5}$

Concept 2: Simplifying Complex Fractions by Method II

For Exercises 17–48, simplify the complex fractions by using Method II. **(See Examples 3–5.)**

17. $\dfrac{\dfrac{7y}{y+3}}{\dfrac{1}{4y+12}}$

18. $\dfrac{\dfrac{6x}{x-5}}{\dfrac{1}{4x-20}}$

19. $\dfrac{1 + \dfrac{1}{3}}{\dfrac{5}{6} - 1}$

20. $\dfrac{2 + \dfrac{4}{5}}{-1 + \dfrac{3}{10}}$

21. $\dfrac{\dfrac{3q}{p} - q}{q - \dfrac{q}{p}}$

22. $\dfrac{\dfrac{b}{a} + 3b}{b + \dfrac{2b}{a}}$

23. $\dfrac{\dfrac{2}{a} + \dfrac{3}{a^2}}{\dfrac{4}{a^2} - \dfrac{9}{a}}$

24. $\dfrac{\dfrac{2}{y^2} + \dfrac{1}{y}}{\dfrac{4}{y^2} - \dfrac{1}{y}}$

25. $\dfrac{t^{-1} - 1}{1 - t^{-2}}$

26. $\dfrac{d^{-2} - c^{-2}}{c^{-1} - d^{-1}}$

27. $\dfrac{-8}{\dfrac{6w}{w - 1} - 4}$

28. $\dfrac{6}{2z - \dfrac{10}{z - 4}}$

29. $\dfrac{\dfrac{y}{y + 3}}{\dfrac{y}{y + 3} + y}$

30. $\dfrac{\dfrac{4}{w - 4}}{\dfrac{4}{w - 4} - 1}$

31. $\dfrac{1 - \dfrac{1}{x} - \dfrac{6}{x^2}}{1 - \dfrac{4}{x} + \dfrac{3}{x^2}}$

32. $\dfrac{1 + \dfrac{1}{x} - \dfrac{12}{x^2}}{\dfrac{9}{x^2} + \dfrac{3}{x} - 2}$

33. $\dfrac{2 - \dfrac{2}{t + 1}}{2 + \dfrac{2}{t}}$

34. $\dfrac{3 + \dfrac{3}{p - 1}}{3 - \dfrac{3}{p}}$

35. $\dfrac{\dfrac{2}{a} - \dfrac{3}{a + 1}}{\dfrac{2}{a + 1} - \dfrac{3}{a}}$

36. $\dfrac{\dfrac{5}{b} + \dfrac{4}{b + 1}}{\dfrac{4}{b} - \dfrac{5}{b + 1}}$

37. $\dfrac{\dfrac{1}{y + 2} + \dfrac{4}{y - 3}}{\dfrac{2}{y - 3} - \dfrac{7}{y + 2}}$

38. $\dfrac{\dfrac{1}{t - 4} + \dfrac{1}{t + 5}}{\dfrac{6}{t + 5} + \dfrac{2}{t - 4}}$

39. $\dfrac{\dfrac{2}{x + h} - \dfrac{2}{x}}{h}$

40. $\dfrac{\dfrac{1}{2x + 2h} - \dfrac{1}{2x}}{h}$

41. $\dfrac{x^{-2}}{x + 3x^{-1}}$

42. $\dfrac{x^{-1} + x^{-2}}{5x^{-2}}$

43. $\dfrac{2a^{-1} + 3b^{-2}}{a^{-1} - b^{-1}}$

44. $\dfrac{2m^{-1} + n^{-1}}{m^{-2} - 4n^{-1}}$

45. $\dfrac{\dfrac{1}{4 + h} - \dfrac{1}{4}}{h}$

46. $\dfrac{\dfrac{1}{3 + 3h} - \dfrac{1}{3}}{h}$

47. $\dfrac{\dfrac{6}{x + h} - \dfrac{6}{x}}{h}$

48. $\dfrac{\dfrac{-3}{x + h} + \dfrac{3}{x}}{h}$

Expanding Your Skills

49. The slope formula is used to find the slope of the line passing through the points (x_1, y_1) and (x_2, y_2). Write the slope formula from memory.

For Exercises 50–53, find the slope of the line that passes through the given points.

50. $\left(1\dfrac{1}{2}, \dfrac{2}{5}\right), \left(\dfrac{1}{4}, -2\right)$ **51.** $\left(-\dfrac{3}{7}, \dfrac{3}{5}\right), (-1, -3)$ **52.** $\left(\dfrac{5}{8}, \dfrac{9}{10}\right), \left(-\dfrac{1}{16}, -\dfrac{1}{5}\right)$ **53.** $\left(\dfrac{1}{4}, \dfrac{1}{3}\right), \left(\dfrac{1}{8}, \dfrac{1}{6}\right)$

54. Show that $(x + x^{-1})^{-1} = \dfrac{x}{x^2 + 1}$ by writing the expression on the left without negative exponents and simplifying.

55. Show that $(x^{-1} + y^{-1})^{-1} = \dfrac{xy}{x + y}$ by writing the expression on the left without negative exponents and simplifying.

56. Simplify. $\dfrac{x}{1 - \left(1 + \dfrac{1}{x}\right)^{-1}}$

57. Simplify. $\dfrac{x}{1 - \left(1 - \dfrac{1}{x}\right)^{-1}}$

Problem Recognition Exercises

Operations on Rational Expressions

For Exercises 1–24, identify the operation (addition, subtraction, multiplication, or division), then simplify the expression. In each case, be sure to ask yourself if you need a common denominator.

1. $\dfrac{2}{2y-3} - \dfrac{3}{2y} + 1$

2. $(x+5) + \left(\dfrac{7}{x-4}\right)$

3. $\dfrac{5x^2-6x+1}{x^2-1} \div \dfrac{16x^2-9}{4x^2+7x+3} - \dfrac{x}{4x-3}$

4. $\dfrac{a^2-25}{3a^2+3ab} \cdot \dfrac{a^2+4a+ab+4b}{a^2+9a+20}$

5. $\dfrac{4}{y+1} + \dfrac{y+2}{y^2-1} - \dfrac{3}{y-1}$

6. $\dfrac{8w^2}{w^3-16w} - \dfrac{4w}{w^2-4w}$

7. $\dfrac{a^2-16}{2x+6} \cdot \dfrac{x+3}{a-4}$

8. $\dfrac{t^2-9}{t} \div \dfrac{t+3}{t+2}$

9. $\dfrac{2+\dfrac{1}{a}}{4-\dfrac{1}{a^2}}$

10. $\dfrac{\dfrac{6x^2y}{5}}{\dfrac{3x}{y}}$

11. $\dfrac{6xy}{x^2-y^2} + \dfrac{x+y}{y-x}$

12. $(x^2-6x+8) \cdot \left(\dfrac{3}{x-2}\right)$

13. $\dfrac{3}{x-2} - \dfrac{x-2}{6}$

14. $\dfrac{5}{x+7} + \dfrac{x+7}{10}$

15. $\dfrac{1}{w-1} - \dfrac{w+2}{3w-3}$

16. $\dfrac{3y+6}{y^2-3y-10} \div \dfrac{27}{y-5}$

17. $\dfrac{y+\dfrac{2}{y}-3}{1-\dfrac{2}{y}}$

18. $\dfrac{2}{t-3} - \dfrac{3}{t+2} + 5$

19. $\dfrac{4x^2+22x+24}{4x+4} \cdot \dfrac{6x+6}{4x^2-9}$

20. $\dfrac{12x^3y^5z}{5x^4} \div \dfrac{16xy^7}{10z^2}$

21. $\dfrac{3x-1}{4} + \dfrac{7}{6x-2}$

22. $\dfrac{2x^{-1}+3x^{-2}}{x^{-2}-5x^{-1}}$

23. $(y+2) \cdot \dfrac{2y+1}{y^2-4} - \dfrac{y-2}{y+3}$

24. $\dfrac{a^2}{a-10} - \dfrac{100-20a}{10-a}$

Section 5.5   Solving Rational Equations

Concepts

1. Solving Rational Equations
2. Formulas Involving Rational Equations

1. Solving Rational Equations

Thus far, we have studied two types of equations in one variable: linear equations and quadratic equations. In this section, we will study another type of equation called a rational equation.

> **Definition of a Rational Equation**
> An equation with one or more rational expressions is called a **rational equation**.

The following equations are rational equations:

$$\frac{1}{2}x + \frac{1}{3} = \frac{1}{4}x \qquad \frac{3}{5} + \frac{1}{x} = \frac{2}{3} \qquad 3 - \frac{6w}{w+1} = \frac{6}{w+1}$$

To understand the process of solving a rational equation, first review the procedure of clearing fractions from Section 1.1.

Example 1   **Solving an Equation with Fractions**

Solve the equation. $\dfrac{1}{2}x + \dfrac{1}{3} = \dfrac{1}{4}x$

Solution:

$$\frac{1}{2}x + \frac{1}{3} = \frac{1}{4}x \qquad \text{The LCD of all terms in the equation is 12.}$$

$$12\left(\frac{1}{2}x + \frac{1}{3}\right) = 12\left(\frac{1}{4}x\right) \qquad \text{Multiply both sides by 12 to clear fractions.}$$

$$12 \cdot \frac{1}{2}x + 12 \cdot \frac{1}{3} = 12 \cdot \frac{1}{4}x \qquad \text{Apply the distributive property.}$$

$$6x + 4 = 3x \qquad \text{Solve the resulting equation.}$$

$$3x = -4$$

$$x = -\frac{4}{3}$$

Check: $\qquad \dfrac{1}{2}x + \dfrac{1}{3} = \dfrac{1}{4}x$

$$\frac{1}{2}\left(-\frac{4}{3}\right) + \frac{1}{3} \stackrel{?}{=} \frac{1}{4}\left(-\frac{4}{3}\right)$$

$$-\frac{2}{3} + \frac{1}{3} \stackrel{?}{=} -\frac{1}{3}$$

$$-\frac{1}{3} \stackrel{?}{=} -\frac{1}{3} \checkmark$$

The solution set is $\left\{-\dfrac{4}{3}\right\}$.

Skill Practice Solve the equation.

1. $\dfrac{1}{2}x + \dfrac{21}{20} = \dfrac{1}{5}x$

The same process of clearing fractions is used to solve rational equations when variables are present in the denominator.

Answer

1. $\left\{-\dfrac{7}{2}\right\}$

Example 2 Solving a Rational Equation

Solve the equation. $\dfrac{3}{5} + \dfrac{1}{x} = \dfrac{2}{3}$

Solution:

$$\frac{3}{5} + \frac{1}{x} = \frac{2}{3}$$

The LCD of all terms in the equation is $15x$. Note that in this equation there is a restriction that $x \neq 0$.

$$15x\left(\frac{3}{5} + \frac{1}{x}\right) = 15x\left(\frac{2}{3}\right)$$

Multiply by $15x$ to clear fractions.

$$15x \cdot \frac{3}{5} + 15x \cdot \frac{1}{x} = 15x \cdot \frac{2}{3}$$

Apply the distributive property.

$$9x + 15 = 10x$$

Solve the resulting equation.

$$15 = x$$

Check: $x = 15$ $\dfrac{3}{5} + \dfrac{1}{x} = \dfrac{2}{3}$

$$\frac{3}{5} + \frac{1}{(15)} \stackrel{?}{=} \frac{2}{3}$$

$$\frac{9}{15} + \frac{1}{15} \stackrel{?}{=} \frac{2}{3}$$

$$\frac{10}{15} \stackrel{?}{=} \frac{2}{3} \quad ✔$$

The solution set is $\{15\}$.

Skill Practice Solve the equation.

2. $\dfrac{3}{y} + \dfrac{4}{3} = -1$

Example 3 Solving a Rational Equation

Solve the equation. $3 - \dfrac{6w}{w + 1} = \dfrac{6}{w + 1}$

Solution:

$$3 - \frac{6w}{w + 1} = \frac{6}{w + 1}$$

The LCD of all terms in the equation is $w + 1$. Note that in this equation there is a restriction that $w \neq -1$.

$$(w + 1)(3) - (w + 1)\left(\frac{6w}{w + 1}\right) = (w + 1)\left(\frac{6}{w + 1}\right)$$

Multiply by $(w + 1)$ on both sides to clear fractions.

$$(w + 1)(3) - (w + 1)\left(\frac{6w}{w + 1}\right) = (w + 1)\left(\frac{6}{w + 1}\right)$$

Apply the distributive property.

$$3w + 3 - 6w = 6$$

Solve the resulting equation.

$$-3w = 3$$

$$w = -1$$

Answer

2. $\left\{-\dfrac{9}{7}\right\}$

Check:
$$3 - \frac{6w}{w+1} = \frac{6}{w+1}$$

$$3 - \frac{6(-1)}{(-1)+1} \overset{?}{=} \frac{6}{(-1)+1}$$

The denominator is 0
for the value of $w = -1$.

The value -1 is one of the restrictions on w found in the first step. As expected, the value $w = -1$ does not check. Since no other potential solution exists, the equation has no solution.

The solution set is the empty set, { }.

Skill Practice Solve the equation.

3. $5 - \dfrac{8}{x+2} = \dfrac{4x}{x+2}$

Examples 1–3 show that the steps to solve a rational equation mirror the process of clearing fractions from Section 1.1. However, we must check whether the potential solutions are defined in each expression in the original equation. A potential solution that does not check is called an **extraneous solution**.

The steps for solving a rational equation are summarized as follows.

Solving a Rational Equation

Step 1 Factor the denominators of all rational expressions. Identify any values of the variable for which any expression is undefined.

Step 2 Identify the LCD of all terms in the equation.

Step 3 Multiply both sides of the equation by the LCD.

Step 4 Solve the resulting equation.

Step 5 Check the potential solutions in the original equation. Note that any value from step 1 for which the equation is undefined cannot be a solution to the equation.

Example 4 Solving a Rational Equation

Solve the equation. $\quad 1 + \dfrac{3}{x} = \dfrac{28}{x^2}$

Solution:

$1 + \dfrac{3}{x} = \dfrac{28}{x^2}$ The LCD of all terms in the equation is x^2.
Expressions will be undefined for $x = 0$.

$x^2\left(1 + \dfrac{3}{x}\right) = x^2\left(\dfrac{28}{x^2}\right)$ Multiply both sides by x^2 to clear fractions.

$x^2 \cdot 1 + x^2 \cdot \dfrac{3}{x} = x^2 \cdot \dfrac{28}{x^2}$ Apply the distributive property.

$x^2 + 3x = 28$ The resulting equation is quadratic.

$x^2 + 3x - 28 = 0$ Set the equation equal to zero and factor.

$(x+7)(x-4) = 0$

$x = -7 \quad$ or $\quad x = 4$

Answer

3. { } (The value -2 does not check.)

Check: $x = -7$ Check: $x = 4$

$$1 + \frac{3}{x} = \frac{28}{x^2} \qquad\qquad 1 + \frac{3}{x} = \frac{28}{x^2}$$

$$1 + \frac{3}{-7} \overset{?}{=} \frac{28}{(-7)^2} \qquad\qquad 1 + \frac{3}{4} \overset{?}{=} \frac{28}{(4)^2}$$

$$\frac{49}{49} - \frac{21}{49} \overset{?}{=} \frac{28}{49} \qquad\qquad \frac{16}{16} + \frac{12}{16} \overset{?}{=} \frac{28}{16}$$

$$\frac{28}{49} \overset{?}{=} \frac{28}{49} \checkmark \qquad\qquad \frac{28}{16} \overset{?}{=} \frac{28}{16} \checkmark$$

The solution set is $\{-7, 4\}$.

Skill Practice Solve the equation.

4. $1 + \dfrac{6}{x} = \dfrac{16}{x^2}$

Example 5   Solving a Rational Equation

Solve. $\dfrac{36}{p^2 - 9} = \dfrac{2p}{p + 3} - 1$

Solution:

$$\frac{36}{p^2 - 9} = \frac{2p}{p + 3} - 1$$

$$\frac{36}{(p + 3)(p - 3)} = \frac{2p}{p + 3} - 1 \qquad \begin{array}{l} \text{The LCD is } (p + 3)(p - 3). \\ \text{Expressions will be undefined for } p = 3 \\ \text{and } p = -3. \end{array}$$

Multiply both sides by the LCD to clear fractions.

$$(p + 3)(p - 3)\left[\frac{36}{(p + 3)(p - 3)}\right] = (p + 3)(p - 3)\left(\frac{2p}{p + 3}\right) - (p + 3)(p - 3)1$$

$$\cancel{(p + 3)}\cancel{(p - 3)}\left[\frac{36}{\cancel{(p + 3)}\cancel{(p - 3)}}\right] = \cancel{(p + 3)}(p - 3)\left(\frac{2p}{\cancel{p + 3}}\right) - (p + 3)(p - 3)1$$

$$36 = 2p(p - 3) - (p + 3)(p - 3) \qquad \text{Solve the resulting equation.}$$

$$36 = 2p^2 - 6p - (p^2 - 9) \qquad\qquad \text{The equation is quadratic.}$$

$$36 = 2p^2 - 6p - p^2 + 9$$

$$36 = p^2 - 6p + 9$$

$$0 = p^2 - 6p - 27 \qquad\qquad \text{Set the equation equal to zero and factor.}$$

$$0 = (p - 9)(p + 3)$$

$$p = 9 \quad \text{or} \quad p = -3$$

Answer

4. $\{-8, 2\}$

Check: $p = 9$

$$\frac{36}{p^2 - 9} = \frac{2p}{p + 3} - 1$$

$$\frac{36}{(9)^2 - 9} \overset{?}{=} \frac{2(9)}{(9) + 3} - 1$$

$$\frac{36}{72} \overset{?}{=} \frac{18}{12} - 1$$

$$\frac{1}{2} \overset{?}{=} \frac{3}{2} - 1$$

$$\frac{1}{2} \overset{?}{=} \frac{1}{2} \checkmark$$

Check: $p = -3$

$$\frac{36}{p^2 - 9} = \frac{2p}{p + 3} - 1$$

$$\frac{36}{(-3)^2 - 9} \overset{?}{=} \frac{2(-3)}{(-3) + 3} - 1$$

Denominator is zero.

Here the value -3 is *not* a solution to the original equation because it is restricted in the original equation. However, 9 checks in the original equation.

The solution set is $\{9\}$.

Skill Practice Solve.

5. $\dfrac{6}{x + 2} - \dfrac{20x}{x^2 - x - 6} = \dfrac{x}{x + 2}$

---

2. Formulas Involving Rational Equations

Example 6 Solving a Literal Equation Involving Rational Expressions

Solve for the indicated variable. $V = \dfrac{mv}{m + M}$ for m

Solution:

$$V = \frac{mv}{m + M} \quad \text{for } m$$

$$V(m + M) = \left(\frac{mv}{m + M}\right)(m + M) \qquad \text{Multiply by the LCD and clear fractions.}$$

$$V(m + M) = mv$$

$$Vm + VM = mv \qquad \text{Use the distributive property to clear parentheses.}$$

$$Vm - mv = -VM \qquad \text{Collect all } m \text{ terms on one side.}$$

$$m(V - v) = -VM \qquad \text{Factor out } m.$$

$$\frac{m(V - v)}{(V - v)} = \frac{-VM}{(V - v)} \qquad \text{Divide by } (V - v).$$

$$m = \frac{-VM}{V - v}$$

Avoiding Mistakes

Variables in algebra are case-sensitive. For example, M and m are different variables.

Answer

5. $\{-9\}$ (The value -2 does not check.)

TIP: The factor of -1 that appears in the numerator may be written in the denominator or out in front of the expression. The following expressions are equivalent:

$$m = \frac{-VM}{V - v} \quad \text{or}$$

$$\frac{VM}{-(V - v)} = \frac{VM}{v - V} \quad \text{or} \quad -\frac{VM}{V - v}$$

Skill Practice Solve the equation for x.

6. $y = \dfrac{ax + b}{x + d}$

Answer

6. $x = \dfrac{b - yd}{y - a}$ or $x = \dfrac{yd - b}{a - y}$

Section 5.5 Practice Exercises

Vocabulary and Key Concepts

1. **a.** The equation $\dfrac{5}{x + 2} + \dfrac{1}{2} = \dfrac{4}{5}$ is an example of a _____ equation.

  b. After solving a rational equation, check each potential solution to determine if it makes the _____ equal to zero in one or more of the rational expressions. If so, that potential solution is not part of the solution set.

  c. Given $\dfrac{3}{2x + 1} + \dfrac{36}{2x^2 - 7x - 4} = \dfrac{4}{x - 4}$, is it possible for 4 to be a solution to the equation?

Review Exercises

For Exercises 2–7, perform the indicated operations.

2. $\dfrac{1}{x^2 - 16} + \dfrac{1}{x^2 + 8x + 16}$

3. $\dfrac{3}{y^2 - 1} - \dfrac{2}{y^2 - 2y + 1}$

4. $\dfrac{m^2 - 9}{m^2 - 3m} \div (m^2 - m - 12)$

5. $\dfrac{2t^2 + 7t + 3}{4t^2 - 1} \div (t + 3)$

6. $\dfrac{1 + x^{-1}}{1 - x^{-2}}$

7. $\dfrac{x + y}{x^{-1} + y^{-1}}$

Concept 1: Solving Rational Equations

8. Why is it important to check your answer when solving a rational equation?

For Exercises 9–42, solve the rational equations. **(See Examples 1–5.)**

9. $\dfrac{x + 2}{3} - \dfrac{x - 4}{4} = \dfrac{1}{2}$

10. $\dfrac{x + 6}{3} - \dfrac{x + 8}{5} = 0$

11. $\dfrac{3y}{4} - 2 = \dfrac{5y}{6}$

12. $\dfrac{2w}{5} - 8 = \dfrac{4w}{2}$

13. $\dfrac{5}{4p} - \dfrac{7}{6} + 3 = 0$

14. $\dfrac{7}{15w} - \dfrac{3}{10} - 2 = 0$

15. $\dfrac{1}{2} - \dfrac{3}{2x} = \dfrac{4}{x} - \dfrac{5}{12}$

16. $\dfrac{2}{3x} + \dfrac{1}{4} = \dfrac{11}{6x} - \dfrac{1}{3}$

17. $\dfrac{3}{x - 4} + 2 = \dfrac{5}{x - 4}$

18. $\dfrac{5}{x + 3} - 2 = \dfrac{7}{x + 3}$

19. $\dfrac{1}{3} + \dfrac{2}{w - 3} = 1$

20. $\dfrac{3}{5} + \dfrac{7}{p + 2} = 2$

21. $\dfrac{12}{x} - \dfrac{12}{x-5} = \dfrac{2}{x}$ **22.** $\dfrac{25}{y} - \dfrac{25}{y-2} = \dfrac{2}{y}$ **23.** $\dfrac{3}{a^2} - \dfrac{4}{a} = -1$ **24.** $\dfrac{3}{w^2} = 2 + \dfrac{1}{w}$

25. $\dfrac{1}{4}a - 4a^{-1} = 0$ **26.** $\dfrac{1}{3}t - 12t^{-1} = 0$ **27.** $\dfrac{y}{y+3} + \dfrac{2}{y^2+3y} = \dfrac{6}{y}$ **28.** $\dfrac{-8}{t^2-6t} + \dfrac{t}{t-6} = \dfrac{1}{t}$

29. $\dfrac{4}{t-2} - \dfrac{8}{t^2-2t} = -2$ **30.** $\dfrac{x}{x+6} = \dfrac{72}{x^2-36} + 4$

31. $\dfrac{6}{5y+10} - \dfrac{1}{y-5} = \dfrac{4}{y^2-3y-10}$ **32.** $\dfrac{-3}{x^2-7x+12} - \dfrac{2}{x^2+x-12} = \dfrac{10}{x^2-16}$

33. $\dfrac{x}{x-5} + \dfrac{1}{5} = \dfrac{5}{x-5}$ **34.** $\dfrac{x}{x-2} + \dfrac{2}{3} = \dfrac{2}{x-2}$

35. $\dfrac{6}{x^2-4x+3} - \dfrac{1}{x-3} = \dfrac{1}{4x-4}$ **36.** $\dfrac{1}{4x^2-36} - \dfrac{5}{x+3} + \dfrac{2}{x-3} = 0$

37. $\dfrac{1}{k+2} - \dfrac{4}{k-2} - \dfrac{k^2}{4-k^2} = 0$ **38.** $\dfrac{h}{2} - \dfrac{h}{h-4} = \dfrac{4}{4-h}$

39. $\dfrac{5}{x^2-7x+12} = \dfrac{2}{x-3} + \dfrac{5}{x-4}$ **40.** $\dfrac{9}{x^2+7x+10} = \dfrac{5}{x+2} - \dfrac{3}{x+5}$

41. $\dfrac{4}{x^2+7x+12} - \dfrac{7}{x^2+8x+15} = \dfrac{1}{x^2+9x+20}$ **42.** $\dfrac{5}{x^2-6x+8} - \dfrac{2}{x^2+3x-10} = \dfrac{8}{x^2+x-20}$

Concept 2: Formulas Involving Rational Equations

For Exercises 43–60, solve the formula for the indicated variable. **(See Example 6.)**

43. $K = \dfrac{ma}{F}$ for m **44.** $K = \dfrac{ma}{F}$ for a **45.** $K = \dfrac{IR}{E}$ for E

46. $K = \dfrac{IR}{E}$ for R **47.** $I = \dfrac{E}{R+r}$ for R **48.** $I = \dfrac{E}{R+r}$ for r

49. $h = \dfrac{2A}{B+b}$ for B **50.** $\dfrac{V}{\pi h} = r^2$ for h **51.** $x = \dfrac{at+b}{t}$ for t

52. $\dfrac{T+mf}{m} = g$ for m **53.** $\dfrac{x-y}{xy} = z$ for x **54.** $\dfrac{w-n}{wn} = P$ for w

55. $a + b = \dfrac{2A}{h}$ for h **56.** $1 + rt = \dfrac{A}{P}$ for P **57.** $\dfrac{1}{R} = \dfrac{1}{R_1} + \dfrac{1}{R_2}$ for R

58. $\dfrac{b+a}{ab} = \dfrac{1}{f}$ for b **59.** $v = \dfrac{s_2 - s_1}{t_2 - t_1}$ for t_2 **60.** $a = \dfrac{v_2 - v_1}{t_2 - t_1}$ for v_1

Mixed Exercises

For Exercises 61–74, solve the equations.

61. $\dfrac{3}{x+2} + \dfrac{2}{x} = \dfrac{-4}{x^2 + 2x}$

62. $\dfrac{1}{y^2 - 3y} + \dfrac{8}{y} = \dfrac{2}{y-3}$

63. $4c(c+1) = 3(c^2 + 4)$

64. $3t(2t - 2) = 5(t^2 - 1)$

65. $\dfrac{2}{v-1} - \dfrac{4}{v+5} = \dfrac{3}{v^2 + 4v - 5}$

66. $\dfrac{-2}{a+4} - \dfrac{3}{a-5} = \dfrac{6}{a^2 - a - 20}$

67. $5(x - 9) = 3(x + 4) - 2(4x + 1)$

68. $4z - 3(5z - 3) = z - 12$

69. $\dfrac{3y}{10} - \dfrac{5}{2y} = \dfrac{y}{5}$

70. $\dfrac{2h}{3} - \dfrac{8}{3h} = \dfrac{h}{2}$

71. $\dfrac{1}{2}(4d - 1) + \dfrac{2}{3}(2d + 2) = \dfrac{5}{6}(4d + 1)$

72. $\dfrac{2}{5}(2b + 5) + \dfrac{1}{10}(7b - 10) = \dfrac{1}{2}(3b + 2)$

73. $8t^{-1} + 2 = 3t^{-1}$

74. $6z^{-2} - 5z^{-1} = 0$

Expanding Your Skills

75. Find the value of y so that the slope of the line between the points $(3, 1)$ and $(11, y)$ is $\frac{1}{2}$.

76. Find the value of x so that the slope of the line between the points $(-2, -5)$ and $(x, 10)$ is 3.

77. Find the value of x so that the slope of the line between the points $(4, -2)$ and $(x, 2)$ is 4.

78. Find the value of y so that the slope of the line between the points $(3, 2)$ and $(-1, y)$ is $-\frac{3}{4}$.

Problem Recognition Exercises

Rational Equations vs. Expressions

1. **a.** Simplify. $\dfrac{3}{w-5} + \dfrac{10}{w^2 - 25} - \dfrac{1}{w+5}$

b. Solve. $\dfrac{3}{w-5} + \dfrac{10}{w^2 - 25} - \dfrac{1}{w+5} = 0$

c. Identify each problem in parts (a) and (b) as either an equation or an expression.

2. **a.** Simplify. $\dfrac{x}{2x+4} + \dfrac{2}{3x+6} - 1$

b. Solve. $\dfrac{x}{2x+4} + \dfrac{2}{3x+6} = 1$

c. Identify each problem in parts (a) and (b) as either an equation or an expression.

For Exercises 3–20, first ask yourself whether the problem is an expression to simplify or an equation to solve. Then simplify or solve as indicated.

3. $\dfrac{2}{a^2 + 4a + 3} + \dfrac{1}{a+3}$

4. $\dfrac{1}{c+6} + \dfrac{4}{c^2 + 8c + 12}$

5. $\dfrac{7}{y^2 - y - 2} + \dfrac{1}{y+1} - \dfrac{3}{y-2} = 0$

6. $\dfrac{3}{b+2} - \dfrac{1}{b-1} - \dfrac{5}{b^2 + b - 2} = 0$

7. $\dfrac{x}{x-1} - \dfrac{12}{x^2 - x}$

8. $\dfrac{3}{5t - 20} + \dfrac{4}{t-4}$

9. $\dfrac{3}{w} - 5 = \dfrac{7}{w} - 1$

10. $\dfrac{-3}{y^2} - \dfrac{1}{y} = -2$

11. $\dfrac{4p + 1}{8p - 12} + \dfrac{p - 3}{2p - 3}$

12. $\dfrac{x + 1}{2x + 4} - \dfrac{x^2}{x + 2}$

13. $\dfrac{1}{2x^2} + \dfrac{1}{6x}$

14. $\dfrac{5}{4a} + \dfrac{1}{6a^2}$

15. $\dfrac{3}{2t} + \dfrac{2}{3t^2} = \dfrac{-1}{t}$

16. $\dfrac{-3}{b^2} + \dfrac{1}{5b} = \dfrac{1}{2b}$

17. $\dfrac{3}{c^2 + 4c + 3} - \dfrac{2}{c^2 + 6c + 9}$

18. $\dfrac{1}{y^2 - 10y + 25} - \dfrac{3}{y^2 - 7y + 10}$

19. $\dfrac{4}{w - 4} - \dfrac{36}{2w^2 - 7w - 4} = \dfrac{3}{2w + 1}$

20. $\dfrac{2}{x - 3} - \dfrac{5}{x + 2} = \dfrac{25}{x^2 - x - 6}$

Applications of Rational Equations and Proportions

1. Solving Proportions

A proportion is a rational equation that equates two ratios.

Definition of Ratio and Proportion

1. The **ratio** of a to b is $\dfrac{a}{b}$ $(b \neq 0)$ and can also be expressed as $a:b$ or $a \div b$.
2. An equation that equates two ratios or rates is called a **proportion**. Therefore, if $b \neq 0$ and $d \neq 0$, then $\dfrac{a}{b} = \dfrac{c}{d}$ is a proportion.

Concepts

1. Solving Proportions
2. Applications of Proportions
3. Similar Triangles
4. Applications of Rational Equations

The process for solving rational equations can be used to solve proportions.

Example 1 — Solving a Proportion

Solve the proportion. $\quad \dfrac{5}{19} = \dfrac{95}{y}$

Solution:

$$\dfrac{5}{19} = \dfrac{95}{y} \qquad \text{The LCD is } 19y. \text{ Note that } y \neq 0.$$

$$19y\left(\dfrac{5}{19}\right) = 19y\left(\dfrac{95}{y}\right) \qquad \text{Multiply both sides by the LCD.}$$

$$\cancel{19}y\left(\dfrac{5}{\cancel{19}}\right) = 19\cancel{y}\left(\dfrac{95}{\cancel{y}}\right) \qquad \text{Clear fractions.}$$

$$5y = 1805 \qquad \text{Solve the resulting equation.}$$

$$\dfrac{5y}{5} = \dfrac{1805}{5}$$

$$y = 361 \qquad \text{The solution 361 checks in the original equation.}$$

The solution set is $\{361\}$.

Skill Practice Solve the proportion.

1. $\dfrac{8}{5} = \dfrac{12}{x}$

2. Applications of Proportions

Example 2 Solving a Proportion

The recommended ratio of total cholesterol to HDL cholesterol is 7 to 2. If Rich's blood test revealed that he has a total cholesterol level of 210 mg/dL (milligrams per deciliter), what should his HDL level be to fit within the recommendations?

Solution:

One method of solving this problem is to set up a proportion. Write two equivalent ratios depicting the amount of total cholesterol to HDL cholesterol. Let x represent the unknown amount of HDL cholesterol.

$$\boxed{\text{Given ratio}} \longrightarrow \frac{7}{2} = \frac{210}{x} \longleftarrow \boxed{\text{Amount of total cholesterol}} \atop \longleftarrow \boxed{\text{Amount of HDL cholesterol}}$$

$$2x\left(\frac{7}{2}\right) = 2x\left(\frac{210}{x}\right) \qquad \text{Multiply both sides by the LCD } 2x.$$

$$7x = 420 \qquad \text{Clear fractions.}$$

$$x = 60$$

Rich's HDL cholesterol level should be 60 mg/dL to fit within the recommended level.

Skill Practice

2. The ratio of cats to dogs at an animal rescue facility is 8 to 5. How many dogs are in the facility if there are 400 cats?

Example 3 Solving a Proportion

The ratio of male to female police officers in a certain town is 11 : 3. If the total number of officers is 112, how many are men and how many are women?

Solution:

Let x represent the number of male police officers.

Then $112 - x$ represents the number of female police officers.

Answers

1. $\left\{\dfrac{15}{2}\right\}$ 2. There are 250 dogs.

$$\frac{\text{Male}}{\text{Female}} \longrightarrow \frac{11}{3} = \frac{x}{112 - x} \longleftarrow \frac{\text{Number of males}}{\text{Number of females}}$$

$$3(112 - x)\left(\frac{11}{3}\right) = 3(112 - x)\left(\frac{x}{112 - x}\right) \quad \begin{array}{l}\text{Multiply both sides by} \\ 3(112 - x).\end{array}$$

$$11(112 - x) = 3x \qquad\qquad \begin{array}{l}\text{The resulting equation is} \\ \text{linear.}\end{array}$$

$$1232 - 11x = 3x$$

$$1232 = 14x$$

$$\frac{1232}{14} = \frac{14x}{14}$$

$$x = 88$$

Then $112 - x = 112 - 88 = 24$

There are 88 male police officers and 24 female officers.

Skill Practice

3. Professor Wolfe has a ratio of passing students to failing students of 5 to 4. One semester he had a total of 207 students. How many students passed and how many failed?

3. Similar Triangles

Proportions are used in geometry with **similar triangles**. Two triangles are similar if their corresponding angles are equal. In such a case, the lengths of the corresponding sides are proportional. In Figure 5-5, triangle ABC is similar to triangle XYZ. Therefore, the following ratios are equivalent.

$$\frac{a}{x} = \frac{b}{y} = \frac{c}{z}$$

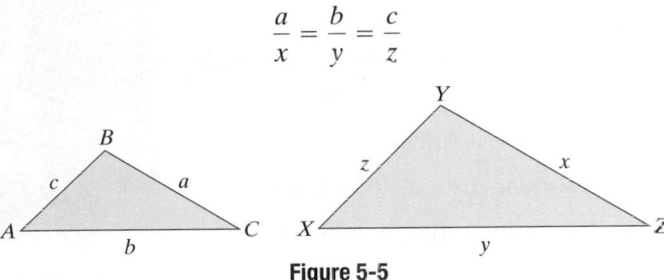

Figure 5-5

Example 4   Using Similar Triangles in an Application

The shadow cast by a yardstick is 2 ft long. The shadow cast by a tree is 11 ft long. Find the height of the tree.

Solution:

Let x represent the height of the tree.

We will assume that the measurements were taken at the same time of day. Therefore, the angle of the sun is the same on both objects, and we can set up similar triangles (Figure 5-6).

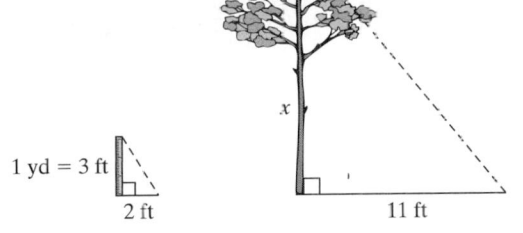

1 yd = 3 ft

2 ft

11 ft

Figure 5-6

Answer

3. 115 passed and 92 failed.

Height of yardstick	$\rightarrow$	$\dfrac{3 \text{ ft}}{x \text{ ft}}$	$=$	$\dfrac{2}{11 \text{ ft}}$	$\leftarrow$	Length of yardstick's shadow
Height of tree	$\rightarrow$				$\leftarrow$	Length of tree's shadow

$$\frac{3}{x} = \frac{2}{11} \qquad \text{Write an equation.}$$

$$11x \cdot \left(\frac{3}{x}\right) = 11x \cdot \left(\frac{2}{11}\right) \qquad \text{Multiply by the LCD.}$$

$$33 = 2x \qquad \text{Solve the equation.}$$

$$16.5 = x \qquad \text{Interpret the results.}$$

The tree is 16.5 ft high.

Skill Practice

4. Triangle XYZ is similar to triangle ABC. Solve for x.

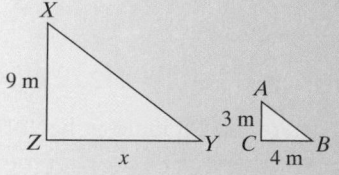

4. Applications of Rational Equations

Example 5 **Solving an Application Involving Distance, Rate, and Time**

An athlete's average speed on her bike is 14 mph faster than her average speed running. She can bike 31.5 mi in the same time that it takes her to run 10.5 mi. Find her speed running and her speed biking.

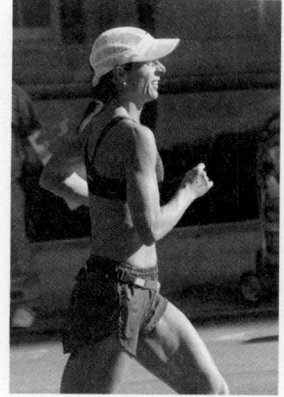

Solution:

Because the speed biking is given in terms of the speed running, let x represent the running speed.

    Let x represent the speed running.

    Then $x + 14$ represents the speed biking.

    Organize the given information in a chart.

	Distance	**Rate**	**Time**
Running	10.5	x	$\dfrac{10.5}{x}$
Biking	31.5	$x + 14$	$\dfrac{31.5}{x + 14}$

$$\text{Because } d = rt, \text{ then } t = \frac{d}{r}$$

The time required to run 10.5 mi is the same as the time required to bike 31.5 mi, so we can equate the two expressions for time:

$$\frac{10.5}{x} = \frac{31.5}{x + 14}$$ The LCD is $x(x + 14)$.

$$x(x + 14)\left(\frac{10.5}{x}\right) = x(x + 14)\left(\frac{31.5}{x + 14}\right)$$ Multiply by $x(x + 14)$ to clear fractions.

$$10.5(x + 14) = 31.5x$$ The resulting equation is linear.

$$10.5x + 147 = 31.5x$$ Solve for x.

$$-21x = -147$$

$$x = 7$$

Then $x + 14 = 7 + 14 = 21$.

The athlete runs 7 mph and bikes 21 mph.

Skill Practice

5. Devon can cross-country ski 5 km/hr faster than his sister, Shanelle. Devon skis 45 km in the same amount of time that Shanelle skis 30 km. Find their speeds.

Example 6   ## Solving an Application Involving Distance, Rate, and Time

Valentina travels 70 km to Rome by train, and then takes a bus 30 km to the Coliseum. The bus travels 24 km/hr slower than the train. If the total time traveling on the bus and train is 2 hr, find the average speed of the train and the average speed of the bus.

Solution:

Because the speed of the bus is given in terms of the speed of the train, let x represent the speed of the train.

Let x represent the speed of the train.

Let $x - 24$ represent the speed of the bus.

Organize the given information in a chart.

	Distance	Rate	Time
Train	70	x	$\dfrac{70}{x}$
Bus	30	$x - 24$	$\dfrac{30}{x - 24}$

Fill in the last column with $t = \dfrac{d}{r}$.

In this problem, we are given that the total time is 2 hr. So we add the two times to equal 2.

$$\frac{70}{x} + \frac{30}{x-24} = 2 \qquad \text{The LCD is } x(x-24).$$

$$x(x-24)\left(\frac{70}{x}\right) = x(x-24)\frac{30}{x-24} = x(x-24)(2) \qquad \begin{array}{l}\text{Multiply by the} \\ \text{LCD to clear the} \\ \text{fractions.}\end{array}$$

$$70(x-24) + 30x = 2x(x-24)$$

$$70x - 1680 + 30x = 2x^2 - 48x \qquad \begin{array}{l}\text{The resulting} \\ \text{equation is} \\ \text{quadratic.}\end{array}$$

$$0 = 2x^2 - 148x + 1680 \qquad \begin{array}{l}\text{Set the equation} \\ \text{equal to 0.}\end{array}$$

$$0 = 2(x^2 - 74x + 840) \qquad \text{Factor.}$$

$$0 = 2(x-60)(x-14) \qquad \text{Solve for } x.$$

$$x - 60 = 0 \quad \text{or} \quad x - 14 = 0 \qquad \begin{array}{l}\text{Set each factor} \\ \text{equal to 0.}\end{array}$$

$$x = 60 \quad \text{or} \quad x = 14$$

If $x = 14$, then the rate of the bus would be $14 - 24 = -10$. Because this is not reasonable, the solution is $x = 60$. That is, the average rate of the train is 60 km/hr and the average rate of the bus is $60 - 24$ or 36 km/hr.

Skill Practice

6. Jason drives 50 mi to a train station and then continues his trip with a 210-mi train ride. The car travels 20 mph slower than the train. If the total travel time is 4 hr, find the average speed of the car and the average speed of the train.

Example 7 **Solving an Application Involving "Work"**

JoAn can wallpaper a bathroom in 3 hr. Bonnie can wallpaper the same bathroom in 5 hr. How long would it take them if they worked together?

Solution:

Let x represent the amount of time required for both people working together to complete the job.

Answer

6. The car's average speed is 50 mph and the train's average speed is 70 mph.

One method to approach this problem is to add the rates of speed at which each person works.

$$\left(\begin{array}{c}\text{JoAn's}\\\text{speed}\end{array}\right) + \left(\begin{array}{c}\text{Bonnie's}\\\text{speed}\end{array}\right) = \left(\begin{array}{c}\text{speed working}\\\text{together}\end{array}\right)$$

$$\frac{1 \text{ job}}{3 \text{ hr}} + \frac{1 \text{ job}}{5 \text{ hr}} = \frac{1 \text{ job}}{x \text{ hr}}$$

$$\frac{1}{3} + \frac{1}{5} = \frac{1}{x}$$

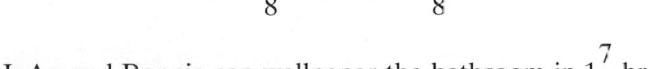

$$15x \cdot \left(\frac{1}{3} + \frac{1}{5}\right) = 15x \cdot \left(\frac{1}{x}\right)$$

$$5x + 3x = 15$$

$$8x = 15$$

$$x = \frac{15}{8} \quad \text{or} \quad x = 1\frac{7}{8}$$

JoAn and Bonnie can wallpaper the bathroom in $1\frac{7}{8}$ hr.

Skill Practice

7. Antonio can install a new roof in 4 days. Carlos can install the same size roof in 6 days. How long will it take them to install a roof if they work together?

TIP: An alternative approach to solving a "work" problem is to determine the portion of the job that each person can complete in 1 hr. Let x represent the amount of time required to complete the job working together. Then

- JoAn completes $\frac{1}{3}$ of the job in 1 hr, and $\frac{1}{3}x$ jobs in x hours.
- Bonnie completes $\frac{1}{5}$ of the job in 1 hr and $\frac{1}{5}x$ jobs in x hours.

$$\left(\begin{array}{c}\text{Portion of the job}\\\text{completed by JoAn}\end{array}\right) + \left(\begin{array}{c}\text{portion of the job}\\\text{completed by Bonnie}\end{array}\right) = \left(\begin{array}{c}\text{1 whole}\\\text{job}\end{array}\right)$$

$$\frac{1}{3}x \quad + \quad \frac{1}{5}x \quad = \quad 1$$

$$15 \cdot \left(\frac{1}{3}x + \frac{1}{5}x\right) = 15 \cdot (1)$$

$$5x + 3x = 15$$

$$8x = 15$$

$$x = \frac{15}{8} \quad \text{or} \quad x = 1\frac{7}{8}$$

The time working together is $1\frac{7}{8}$ hr.

Answer

7. It will take them $\frac{12}{5}$ days or $2\frac{2}{5}$ days.

Section 5.6 Practice Exercises

Vocabulary and Key Concepts

1. a. An equation that equates two rates or ratios is called a _____.

   b. Given similar triangles, the lengths of corresponding sides are _____.

Review Exercises

For Exercises 2–10, perform the indicated operation and simplify, or solve the equation.

2. $3 - \dfrac{6}{x} = x + 8$

3. $2 + \dfrac{6}{x} = x + 7$

4. $\dfrac{5}{3x - 6} - \dfrac{3}{4x - 8}$

5. $\dfrac{4}{5t - 1} + \dfrac{1}{10t - 2}$

6. $\dfrac{2}{y - 1} - \dfrac{5}{4} = \dfrac{-1}{y + 1}$

7. $\dfrac{5}{w - 2} = 7 - \dfrac{10}{w + 2}$

8. $\dfrac{ab}{6} \div \dfrac{a^2}{12} \cdot \dfrac{a + 1}{ab + b}$

9. $\dfrac{8p^2 - 32}{p^2 - 4p + 4} \cdot \dfrac{3p^2 - 3p - 6}{2p^2 + 20p + 32}$

10. $\dfrac{3t}{t + 6} = t + \dfrac{2t}{t + 6}$

Concept 1: Solving Proportions

For Exercises 11–26, solve the proportions. **(See Example 1.)**

11. $\dfrac{y}{6} = \dfrac{20}{15}$

12. $\dfrac{12}{18} = \dfrac{14}{x}$

13. $\dfrac{9}{75} = \dfrac{m}{50}$

14. $\dfrac{n}{15} = \dfrac{12}{45}$

15. $\dfrac{p - 1}{4} = \dfrac{p + 3}{3}$

16. $\dfrac{q - 5}{2} = \dfrac{q + 2}{3}$

17. $\dfrac{x + 1}{5} = \dfrac{4}{15}$

18. $\dfrac{t - 1}{7} = \dfrac{2}{21}$

19. $\dfrac{5 - 2x}{x} = \dfrac{1}{4}$

20. $\dfrac{2y + 3}{y} = \dfrac{3}{2}$

21. $\dfrac{2}{y - 1} = \dfrac{y - 3}{4}$

22. $\dfrac{1}{x - 5} = \dfrac{x - 3}{3}$

23. $\dfrac{1}{49w} = \dfrac{w}{9}$

24. $\dfrac{1}{4z} = \dfrac{z}{25}$

25. $\dfrac{x + 3}{5x + 26} = \dfrac{2}{x + 4}$

26. $\dfrac{-2}{x - 2} = \dfrac{x - 3}{8x + 11}$

Concept 2: Applications of Proportions

27. A preschool advertises that it has a 3-to-1 ratio of children to adults. If 18 children are enrolled, how many adults must be on the staff? **(See Example 2.)**

28. An after-school care facility tries to maintain a 4-to-1 ratio of children to adults. If the facility hired five adults, what is the maximum number of children that can enroll?

29. A 3.5-oz box of candy has a total of 21.0 g of fat. How many grams of fat would a 14-oz box of candy contain?

30. A 6-oz box of candy has 350 calories. How many calories would a 10-oz box contain?

31. A fisherman in the North Atlantic catches eight swordfish for a total of 1840 lb. How many swordfish were caught if a commercial fishing boat arrives in port with 230,000 lb of swordfish?

32. If a 64-oz bottle of laundry detergent costs $12.00, how much would an 80-oz bottle cost?

33. Pam drives her Toyota Prius 243 mi in city driving on 4.5 gal of gas. At this rate how many gallons of gas are required to drive 621 mi?

34. On a map, the distance from Sacramento, California, to San Francisco, California, is 8 cm. The legend gives the actual distance as 96 mi. On the same map, Fatima measured 7 cm from Sacramento to Modesto, California. What is the actual distance?

35. Yellowstone National Park in Wyoming has the largest population of free-roaming bison. To approximate the number of bison, 200 are captured and tagged and then left free to roam. Later, a sample of 120 bison is observed and 6 have tags. Approximate the population of bison in the park.

36. Laws have been instituted in Florida to help save the manatee. To establish the number of manatees in Florida, 150 manatees were tagged. A new sample was taken later, and among the 40 manatees in the sample, 3 were tagged. Approximate the number of manatees in Florida.

37. The ratio of men to women enrolled in Math for Elementary Teachers at Indiana University-Purdue University Indianapolis is 1 to 5. If the total enrollment in these classes is approximately 186 students per semester, how many men are enrolled? **(See Example 3.)**

38. The ratio of Hank's income spent on rent to his income spent on car payments is 3 to 1. If he spends a total of $1640 per month on the rent and car payment, how much does he spend on each item?

39. The ratio of single men in their 20s to single women in their 20s is 119 to 100 (*Source:* U.S. Census). In a random group of 1095 single college students in their 20s, how many are men and how many are women?

40. A chemist mixes water and alcohol in a 7 to 8 ratio. If she makes a 450-L solution, how much is water and how much is alcohol?

Concept 3: Similar Triangles

For Exercises 41–44, triangle *ABC* is similar to triangle *XYZ*. Find the lengths of the missing sides. **(See Example 4.)**

41.

42.

43.

44.

Concept 4: Applications of Rational Equations

45. If 5 is added to the reciprocal of a number, the result is $\frac{16}{3}$. Find the number.

46. If $\frac{2}{3}$ is added to the reciprocal of a number, the result is $\frac{17}{3}$. Find the number.

47. If 7 is decreased by the reciprocal of a number, the result is $\frac{9}{2}$. Find the number.

48. If a number is added to its reciprocal, the result is $\frac{13}{6}$. Find the number.

For Exercises 49–50, use the fact that distance = (rate)(time).

49. A truck travels 7 mph faster than a car. Let x represent the speed of the car.

   a. Write an expression for the speed of the truck.

   b. Write an expression for the time it takes the car to travel 48 mi.

   c. Write an expression for the time it takes the truck to travel 83 mi.

50. A car travels 4 mph slower than a motorcycle. Let x represent the speed of the motorcycle.

   a. Write an expression for the speed of the car.

   b. Write an expression for the time it takes the motorcycle to travel 50 mi.

   c. Write an expression for the time it takes the car to travel 145 mi.

51. A motorist travels 80 mi while driving in a bad rainstorm. In sunny weather, the motorist drives 20 mph faster and covers 120 mi in the same amount of time. Find the speed of the motorist in the rainstorm and the speed in sunny weather. **(See Example 5.)**

52. Brooke walks 2 km/hr slower than her older sister Adrianna. If Brooke can walk 12 km in the same amount of time that Adrianna can walk 18 km, find their speeds.

53. Two out-of-town firefighting crews have been called to a wildfire in the mountains. The Wescott Fire Station is 96 mi from the fire, and the Broadmoor Fire Station is 88 mi from the fire. The fire truck from the Wescott Fire Station travels 6.4 mph faster than the Broadmoor fire truck. If it takes the trucks the same amount of time to reach the fire, what is the average speed of each truck?

54. Kathy can run 3 mi to the beach in the same time Dennis can ride his bike 7 mi to work. Kathy runs 8 mph slower than Dennis rides his bike. Find their speeds.

55. A bicyclist rides 30 mi against a wind and returns 30 mi with the wind. His average speed for the return trip is 5 mph faster. How fast did the cyclist ride against the wind if the total time of the trip was 5 hr? **(See Example 6.)**

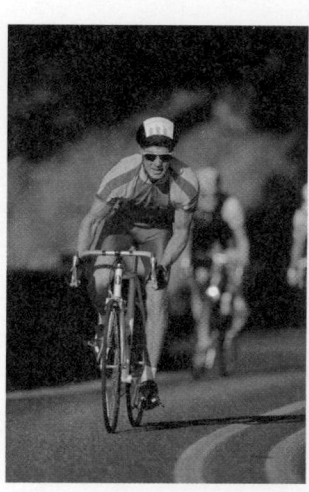

56. A boat travels 60 mi to an island and 60 mi back again. Changes in the wind and tide made the average speed on the return trip 3 mph slower than the speed on the way out. If the total time of the trip took 9 hr, find the speed going to the island and the speed of the return trip.

57. Celeste walked 140 ft on a moving walkway at the airport. Then she walked on the ground for 100 ft. She travels 2 ft/sec faster on the walkway than she does on the ground. If the time it takes her to travel the total distance of 240 ft is 40 sec, how fast does Celeste travel on and off the moving walkway?

58. Julio rides his bike for 6 mi and gets a flat tire. Then he has to walk with the bike for another mile. His speed walking is 6 mph less than his speed riding the bike. If the total time is 1 hr, find his speed riding the bike and his speed walking.

59. Beatrice does professional triathlons. She runs 2 mph faster than her friend Joe, a weekend athlete. If they each run 12 mi, Beatrice finishes 30 min ($\frac{1}{2}$ hr) ahead of Joe. Determine how fast each person runs.

60. A bus leaves a terminal at 9:00. A car leaves 1 hr later and averages 10 mph faster than the bus. If the car overtakes the bus after 200 mi, find the average speed of the bus and the average speed of the car.

61. One painter can paint a room in 6 hr. Another painter can paint the same room in 8 hr. How long would it take them working together? **(See Example 7.)**

62. Karen can wax her SUV in 2 hr. Clarann can wax the same SUV in 3 hr. If they work together, how long will it take them to wax the SUV?

63. A new housing development offers fenced-in yards that all have the same dimensions. Joel can fence a yard in 12 hr, and Michael can fence a yard in 15 hr. How long will it take if they work together?

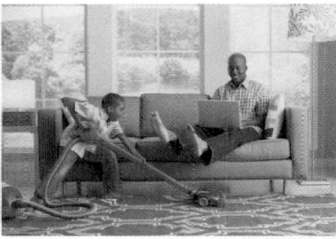

64. Ted can change an advertisement on a billboard in 4 hr. Marie can do the same job in 5 hr. How long would it take them if they worked together?

65. A swimming pool takes 30 hr to fill using an old pump. When a new pump was installed, it took only 12 hr to fill the pool with both pumps. However, the old pump had to be repaired.

   a. Determine how long it would take for the new pump to fill the pool alone.

   b. If the new pump begins filling the empty pool at 4 P.M. on Thursday, when should the technician return to stop the pump?

66. One carpenter can complete a kitchen in 8 days. With the help of another carpenter, they can do the job together in 4 days. How long would it take the second carpenter if he worked alone?

67. Gus works twice as fast as Sid. Together they can dig a garden in 4 hr. How long would it take each person working alone?

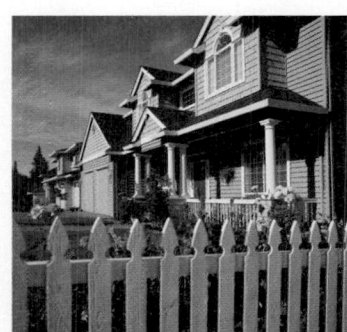

68. It takes a child 3 times longer to vacuum a house than an adult. If it takes 1 hr for one adult and one child working together to vacuum a house, how long would it take each person working alone?

Concepts

1. Definition of Direct and Inverse Variation
2. Translations Involving Variation
3. Applications of Variation

Section 5.7 Variation

1. Definition of Direct and Inverse Variation

In this section, we introduce the concept of variation. Direct and inverse variation models can show how one quantity varies in proportion to another.

> **Definition of Direct and Inverse Variation**
>
> Let k be a nonzero constant real number. Then the following statements are equivalent:
>
> **1.** y varies **directly** as x. $\left.\begin{array}{l}\ \\ \ \end{array}\right\}$ $y = kx$
> y is directly proportional to x.
>
> **2.** y varies **inversely** as x. $\left.\begin{array}{l}\ \\ \ \end{array}\right\}$ $y = \dfrac{k}{x}$
> y is inversely proportional to x.
>
> *Note:* The value of k is called the constant of variation.

For a car traveling at 30 mph, the equation $d = 30t$ indicates that the distance traveled is *directly proportional* to the time of travel. For positive values of k, when two variables are directly related, as one variable increases, the other variable will also increase. Likewise, if one variable decreases, the other will decrease. In the equation $d = 30t$, the longer the time of the trip, the greater the distance traveled. The shorter the time of the trip, the shorter the distance traveled.

For positive values of k, when two positive variables are *inversely related*, as one variable increases, the other will decrease, and vice versa. Consider a car traveling between Toronto and Montreal, a distance of 500 km. The time required to make the trip is inversely proportional to the speed of travel: $t = 500/r$. As the rate of speed r increases, the quotient $500/r$ will decrease. Hence, the time will decrease. Similarly, as the rate of speed decreases, the trip will take longer.

2. Translations Involving Variation

The first step in using a variation model is to translate an English phrase to an equivalent mathematical equation.

Example 1   Translating to a Variation Model

Translate each expression to an equivalent mathematical model.

a. The circumference of a circle varies directly as the radius.

b. At a constant temperature, the volume of a gas varies inversely as the pressure.

c. The length of time of a meeting is directly proportional to the *square* of the number of people present.

Solution:

a. Let C represent circumference and r represent radius. The variables are directly related, so use the model $C = kr$.

b. Let V represent volume and P represent pressure. Because the variables are inversely related, use the model $V = \dfrac{k}{P}$.

c. Let t represent time and let N be the number of people present at a meeting. Because t is directly related to N^2, use the model $t = kN^2$.

Skill Practice Translate to a variation model.

1. The time t it takes to drive a particular distance is inversely proportional to the speed s.
2. The amount of your paycheck P varies directly with the number of hours h that you work.
3. q varies inversely as the square of t.

Sometimes a variable varies directly as the product of two or more other variables. In this case, we have joint variation.

Definition of Joint Variation

Let k be a nonzero constant real number. Then the following statements are equivalent:

$\left.\begin{array}{l} y \text{ varies } \textbf{jointly} \text{ as } w \text{ and } z. \\ y \text{ is jointly proportional to } w \text{ and } z. \end{array}\right\} \quad y = kwz$

Example 2   Translating to a Variation Model

Translate each expression into an equivalent mathematical model.

a. y varies jointly as u and the square root of v.

b. The gravitational force of attraction between two planets varies jointly as the product of their masses and inversely as the square of the distance between them.

Solution:

a. $y = ku\sqrt{v}$

b. Let m_1 and m_2 represent the masses of the two planets. Let F represent the gravitational force of attraction and d represent the distance between the planets. The variation model is $F = \dfrac{km_1m_2}{d^2}$.

Skill Practice Translate to a variation model.

4. a varies jointly as b and c.
5. x varies directly as the square root of y and inversely as z.

Answers

1. $t = \dfrac{k}{s}$ 2. $P = kh$

3. $q = \dfrac{k}{t^2}$ 4. $a = kbc$

5. $x = \dfrac{k\sqrt{y}}{z}$

3. Applications of Variation

Consider the variation models $y = kx$ and $y = k/x$. In either case, if values for x and y are known, we can solve for k. Once k is known, we can use the variation equation to find y if x is known, or to find x if y is known. This concept is the basis for solving many problems involving variation.

Finding a Variation Model

Step 1 Write a general variation model that relates the variables given in the problem. Let k represent the constant of variation.

Step 2 Solve for k by substituting known values of the variables into the model from step 1.

Step 3 Substitute the value of k into the original variation model from step 1.

Example 3 Solving an Application Involving Direct Variation

The variable z varies directly as w. When w is 16, z is 56.

a. Write a variation model for this situation. Use k as the constant of variation.

b. Solve for the constant of variation.

c. Find the value of z when w is 84.

Solution:

a. $z = kw$

b. $z = kw$

$56 = k(16)$ Substitute known values for z and w. Then solve for the unknown value of k.

$\dfrac{56}{16} = \dfrac{k(16)}{16}$ To isolate k, divide both sides by 16.

$\dfrac{7}{2} = k$ Simplify $\dfrac{56}{16}$ to $\dfrac{7}{2}$.

c. With the value of k known, the variation model can now be written as $z = \dfrac{7}{2}w$.

$z = \dfrac{7}{2}(84)$ To find z when $w = 84$, substitute $w = 84$ into the equation.

$z = 294$

Skill Practice The variable q varies directly as the square of v. When v is 2, q is 40.

6. Write a variation model for this relationship.

7. Solve for the constant of variation.

8. Find q when $v = 7$.

Answers

6. $q = kv^2$ **7.** $k = 10$
8. $q = 490$

Example 4   Solving an Application Involving Direct Variation

The speed of a racing canoe in still water varies directly as the square root of the length of the canoe.

a. If a 16-ft canoe can travel 6.2 mph in still water, find a variation model that relates the speed of a canoe to its length.

b. Find the speed of a 25-ft canoe.

Solution:

a. Let s represent the speed of the canoe and L represent the length. The general variation model is $s = k\sqrt{L}$. To solve for k, substitute the known values for s and L.

$$s = k\sqrt{L}$$

$6.2 = k\sqrt{16}$ Substitute $s = 6.2$ mph and $L = 16$ ft.

$6.2 = k \cdot 4$

$\dfrac{6.2}{4} = \dfrac{4k}{4}$ Solve for k.

$k = 1.55$

$s = 1.55\sqrt{L}$ Substitute $k = 1.55$ into the model $s = k\sqrt{L}$.

b. $s = 1.55\sqrt{L}$

$= 1.55\sqrt{25}$ Find the speed when $L = 25$ ft.

$= 7.75$ mph   The speed of a 25-ft canoe is 7.75 mph.

Skill Practice The amount of water needed by a mountain hiker varies directly as the time spent hiking. The hiker needs 2.4 L for a 3-hr hike.

9. Write a model that relates the amount of water needed to the time of the hike.

10. How much water will be needed for a 5-hr hike?

Example 5   Solving an Application Involving Inverse Variation

The loudness of sound measured in decibels (dB) varies inversely as the square of the distance between the listener and the source of the sound. If the loudness of sound is 17.92 dB at a distance of 10 ft from a stereo speaker, what is the decibel level 20 ft from the speaker?

Solution:

Let L represent the loudness of sound in decibels and d represent the distance in feet. The inverse relationship between decibel level and the square of the distance is modeled by

$$L = \frac{k}{d^2}$$

$17.92 = \dfrac{k}{(10)^2}$ Substitute $L = 17.92$ dB and $d = 10$ ft.

$17.92 = \dfrac{k}{100}$

Answers

9. $w = 0.8t$ **10.** 4 L

$$(17.92)100 = \frac{k}{100} \cdot \cancel{100} \qquad \text{Solve for } k \text{ (clear fractions).}$$

$$k = 1792$$

$$L = \frac{1792}{d^2} \qquad \text{Substitute } k = 1792 \text{ into the original model } L = \frac{k}{d^2}.$$

With the value of k known, we can find L for any value of d.

$$L = \frac{1792}{(20)^2} \qquad \text{Find the loudness when } d = 20 \text{ ft.}$$

$$= 4.48 \text{ dB}$$

Notice that the loudness of sound is 17.92 dB at a distance 10 ft from the speaker. When the distance from the speaker is increased to 20 ft, the decibel level decreases to 4.48 dB. This is consistent with an inverse relationship. For $k > 0$, as one variable is increased, the other is decreased. It also seems reasonable that the farther one moves away from the source of a sound, the softer the sound becomes.

Skill Practice

11. The yield on a bond varies inversely as the price. The yield on a particular bond is 4% when the price is $100. Find the yield when the price is $80.

Example 6 **Solving an Application Involving Joint Variation**

In the early morning hours of August 29, 2005, Hurricane Katrina plowed into the Gulf Coast of the United States, bringing unprecedented destruction to southern Louisiana, Mississippi, and Alabama. The winds of a hurricane are strong enough to send a piece of plywood through a tree.

The kinetic energy of an object varies jointly as the weight of the object at sea level and as the square of its velocity. During a hurricane, a 0.5-lb stone traveling at 60 mph has 81 joules (J) of kinetic energy. Suppose the wind speed doubles to 120 mph. Find the kinetic energy.

Solution:

Let E represent the kinetic energy, let w represent the weight, and let v represent the velocity of the stone. The variation model is

$$E = kwv^2$$

$$81 = k(0.5)(60)^2 \qquad \text{Substitute } E = 81 \text{ J}, w = 0.5 \text{ lb, and } v = 60 \text{ mph.}$$

$$81 = k(0.5)(3600) \qquad \text{Simplify exponents.}$$

$$81 = k(1800)$$

$$\frac{81}{1800} = \frac{k(\cancel{1800})}{\cancel{1800}} \qquad \text{Divide by 1800.}$$

$$0.045 = k \qquad \text{Solve for } k.$$

Answer

11. 5%

With the value of k known, the model $E = kwv^2$ can be written as $E = 0.045wv^2$. We now find the kinetic energy of a 0.5-lb stone traveling at 120 mph.

$$E = 0.045(0.5)(120)^2$$

$$= 324$$

The kinetic energy of a 0.5-lb stone traveling at 120 mph is 324 J.

Skill Practice

12. The amount of simple interest earned in an account varies jointly as the interest rate and time of the investment. An account earns \$40 in 2 yr at 4% interest. How much interest would be earned in 3 yr at a rate of 5%?

In Example 6, when the velocity increased by 2 times, the kinetic energy increased by 4 times (note that 324 J = 4·81 J). This factor of 4 occurs because the kinetic energy is proportional to the *square* of the velocity. When the velocity increased by 2 times, the kinetic energy increased by 2^2 times.

Answer

12. \$75

Section 5.7   Practice Exercises

Study Skills Exercise

It is not too early to think about your final exam. Write the page number of the cumulative review for Chapters 1–5. Make this exercise set part of your homework this week.

Vocabulary and Key Concepts

1. a. Let k be a nonzero constant. If y varies directly as x, then $y = $ _____ , where k is the constant of variation.

   b. Let k be a nonzero constant. If y varies inversely as x, then $y = $ _____ , where k is the constant of variation.

   c. Let k be a nonzero constant. If y varies jointly as x and w, then $y = $ _____ , where k is the constant of variation.

Review Exercises

For Exercises 2–7, solve the proportion.

2. $\dfrac{x}{4} = \dfrac{13}{10}$

3. $\dfrac{8}{y} = \dfrac{6}{11}$

4. $\dfrac{3}{8} = \dfrac{w + 2}{6}$

5. $\dfrac{2}{3} = \dfrac{x - 4}{2}$

6. $\dfrac{p - 5}{p} = \dfrac{2}{7}$

7. $\dfrac{k}{k + 1} = \dfrac{1}{9}$

Concept 1: Definition of Direct and Inverse Variation

8. In the equation $r = kt$, does r vary directly or inversely with t?

9. In the equation $w = \frac{k}{v}$, does w vary directly or inversely with v?

10. a. Given $y = 12x$, if x increases, then y will (increase/decrease).

   b. Given $y = \dfrac{12}{x}$, if $x > 0$ increases, then y will (increase/decrease).

Concept 2: Translations Involving Variation

For Exercises 11–22, write a variation model. Use k as the constant of variation. **(See Examples 1–2.)**

11. T varies directly as q.

12. W varies directly as z.

13. b varies inversely as c.

14. m varies inversely as t.

15. Q is directly proportional to x and inversely proportional to y.

16. d is directly proportional to p and inversely proportional to n.

17. c varies jointly as s and t.

18. w varies jointly as p and f.

19. L varies jointly as w and the square root of v.

20. q varies jointly as v and the square root of w.

21. x varies directly as the square of y and inversely as z.

22. a varies directly as n and inversely as the square of d.

Concept 3: Applications of Variation

For Exercises 23–28, find the constant of variation k. **(See Example 3.)**

23. y varies directly as x, and when x is 4, y is 18.

24. m varies directly as x and when x is 8, m is 22.

25. p is inversely proportional to q and when q is 16, p is 32.

26. T is inversely proportional to x and when x is 40, T is 200.

27. y varies jointly as w and v. When w is 50 and v is 0.1, y is 8.75.

28. N varies jointly as t and p. When t is 1 and p is 7.5, N is 330.

Solve Exercises 29–40 by using the steps found on page 476. **(See Examples 3–4.)**

29. x varies directly as p. If $x = 50$ when $p = 10$, find x when p is 14.

30. y is directly proportional to z. If $y = 12$ when $z = 36$, find y when z is 21.

31. b is inversely proportional to c. If b is 4 when c is 3, find b when $c = 2$.

32. q varies inversely as w. If q is 8 when w is 50, find q when w is 125.

33. Z varies directly as the square of w, and $Z = 14$ when $w = 4$. Find Z when $w = 8$.

34. m varies directly as the square of x. If $m = 200$ when $x = 20$, find m when x is 32.

35. Q varies inversely as the square of p, and $Q = 4$ when $p = 3$. Find Q when $p = 2$.

36. z is inversely proportional to the square of t. If $z = 15$ when $t = 4$, find z when $t = 10$.

37. L varies jointly as a and the square root of b, and $L = 72$ when $a = 8$ and $b = 9$. Find L when $a = \frac{1}{2}$ and $b = 36$.

38. Y varies jointly as the cube of x and the square root of w, and $Y = 128$ when $x = 2$ and $w = 16$. Find Y when $x = \frac{1}{2}$ and $w = 64$.

39. B varies directly as m and inversely as n, and $B = 20$ when $m = 10$ and $n = 3$. Find B when $m = 15$ and $n = 12$.

40. R varies directly as s and inversely as t, and $R = 14$ when $s = 2$ and $t = 9$. Find R when $s = 4$ and $t = 3$.

For Exercises 41–58, use a variation model to solve for the unknown value.

41. The amount of medicine that a physician prescribes for a patient varies directly as the weight of the patient. A physician prescribes 3 grams (g) of a medicine for a 150-lb person.

   a. How many grams should be prescribed for a 180-lb person?

   b. How many grams should be prescribed for a 225-lb person?

   c. How many grams should be prescribed for a 120-lb person?

42. The number of people that a turkey can serve varies directly as the weight of the turkey. A 15-lb turkey can serve 10 people.

   a. How many people will a 12-lb turkey serve?

   b. How many people will an 18-lb turkey serve?

   c. How many people will a 21-lb turkey serve?

43. The unit cost of producing CDs is inversely proportional to the number of CDs produced. If 5000 CDs are produced, the cost per CD is $0.48.

   a. What would be the unit cost if 6000 CDs were produced?

   b. What would be the unit cost if 8000 CDs were produced?

   c. What would be the unit cost if 2400 CDs were produced?

44. An author self-publishes a book and finds that the number of books she can sell per month varies inversely as the price of the book. The author can sell 1500 books per month when the price is set at $8 per book.

   a. How many books would she expect to sell if the price were $12?

   b. How many books would she expect to sell if the price were $15?

   c. How many books would she expect to sell if the price were $6?

45. The amount of pollution entering the atmosphere varies directly as the number of people living in an area. If 80,000 people create 56,800 tons of pollutants, how many tons enter the atmosphere in a city with a population of 500,000?

46. The area of a picture projected on a wall varies directly as the square of the distance from the projector to the wall. If a 10-ft distance produces a 16-ft^2 picture, what is the area of a picture produced when the projection unit is moved to a distance 20 ft from the wall?

47. The intensity of light from a light source varies inversely as the square of the distance from the source. If the intensity of a light bulb is 400 lumen/m^2 (lux) at a distance of 5 m, determine the intensity at 8 m. (See Example 5.)

48. The frequency of a vibrating string is inversely proportional to its length. A 24-in. piano string vibrates at 252 cycles/sec. What is the frequency of an 18-in. piano string?

49. The current in a wire varies directly as the voltage and inversely as the resistance. If the current is 9 amperes (A) when the voltage is 90 volts (V) and the resistance is 10 ohms (Ω), find the current when the voltage is 185 V and the resistance is 10 Ω.

50. The resistance of a wire varies directly as its length and inversely as the square of its diameter. A 40-ft wire with 0.1-in. diameter has a resistance of 4 Ω. What is the resistance of a 50-ft wire with a diameter of 0.20 in.?

51. The amount of simple interest earned in an account varies jointly as the amount of principal invested and the amount of time the money is invested. If $2500 in principal earns $500 in interest after 4 yr, then how much interest will be earned on $7000 invested for 10 yr? (See Example 6.)

52. The amount of simple interest earned in an account varies jointly as the amount of principal invested and the amount of time the money is invested. If $6000 in principal earns $840 in interest after 2 yr, then how much interest will be earned on $4500 invested for 8 yr?

53. The stopping distance of a car is directly proportional to the square of the speed of the car. If a car traveling at 40 mph has a stopping distance of 109 ft, find the stopping distance of a car that is traveling at 25 mph. (Round your answer to one decimal place.)

54. The weight of a medicine ball varies directly as the cube of its radius. A ball with a radius of 3 in. weighs 4.32 lb. How much would a medicine ball weigh if its radius were 5 in.?

55. The surface area of a cube varies directly as the square of the length of an edge. The surface area is 24 ft^2 when the length of an edge is 2 ft. Find the surface area of a cube with an edge that is 5 ft.

56. The period of a pendulum is the length of time required to complete one swing back and forth. The period varies directly as the square root of the length of the pendulum. If it takes 1.8 sec for a 0.81-m pendulum to complete one period, what is the period of a 1-m pendulum?

57. The power in an electric circuit varies jointly as the current and the square of the resistance. If the power is 144 watts (W) when the current is 4 A and the resistance is 6 Ω, find the power when the current is 3 A and the resistance is 10 Ω.

58. The strength of a wooden beam varies jointly as the width of the beam and the square of the thickness of the beam and inversely as the length of the beam. A beam that is 48 in. long, 6 in. wide, and 2 in. thick can support a load of 417 lb. Find the maximum load that can be safely supported by a board that is 12 in. wide, 72 in. long, and 4 in. thick.

Expanding Your Skills

59. The area A of a square varies directly as the square of the length l of its sides.

   a. Write a general variation model with k as the constant of variation.

   b. If the length of the sides is doubled, what effect will that have on the area?

   c. If the length of the sides is tripled, what effect will that have on the area?

60. In a physics laboratory, a spring is fixed to the ceiling. With no weight attached to the end of the spring, the spring is said to be in its equilibrium position. As weights are applied to the end of the spring, the force stretches the spring a distance d from its equilibrium position. A student in the laboratory collects the following data:

Force F (lb)	2	4	6	8	10
Distance d (cm)	2.5	5.0	7.5	10.0	12.5

   a. Based on the data, do you suspect a direct relationship between force and distance or an inverse relationship?

   b. Find a variation model that describes the relationship between force and distance.

Group Activity

Computing the Future Value of an Investment

Materials: Calculator

Estimated time: 15 minutes

Group Size: 3

Suppose you are able to save $100 per month. If you invest the money in an account that pays 6% annual interest, how much money would you have at the end of 10 yr? This question can be answered by using the following formula.

$$S = R\left[\frac{(1 + i)^n - 1}{i}\right]$$

where

S is the future value of the investment.
R is the amount saved per period.
i is the interest rate per period.
n is the total number of periods.

In this example, $R = \$100$ (amount invested per month)

$$i = \frac{0.06}{12} = 0.005$$ (annual interest rate divided by 12 months)

$$n = (12)(10) = 120$$ (12 months per year times 10 years)

Therefore, $S = \$100\left[\dfrac{(1 + 0.005)^{120} - 1}{0.005}\right]$

$$S = \$16,387.93$$

1. Compute the future value of an account if you save $150 per month for 30 yr at an annual interest rate of 6%.

   $R = $ _____

   $i = $ _____ $S = $ _____

   $n = $ _____

2. Compute the future value of an account given the monthly saving, interest rate, and time period given in the table. With your group members, discuss how the value of the account varies with the interest rate and time the money is invested.

Monthly Savings	Annual Interest Rate	Years	Future Value
$100	6%	20	
$150	6%	20	
$200	6%	20	

Monthly Savings	Annual Interest Rate	Years	Future Value
$200	2.4%	20	
$200	3.6%	20	
$200	4.8%	20	

Monthly Savings	Annual Interest Rate	Years	Future Value
$200	6%	25	
$200	6%	30	
$200	6%	35	

3. Suppose you need to accumulate $50,000 in 15 yr to pay for your child's college tuition. You decide to invest the money in a mutual fund and the average rate of return is 9% per year. How much should you save each month to reach your goal? Round up to the nearest dollar.

4. Suppose you invest $5000 in an account on July 1 each year for 20 yr. If the *annual* growth on your investment is 8%, how much will the account be worth in 20 yr? (*Hint:* In this example, the savings is yearly, not monthly.)

$R = $ _____

$i = $ _____ $S = $ _____

$n = $ _____

Chapter 5   Summary

Section 5.1   Rational Expressions and Rational Functions

Key Concepts

A **rational expression** is in the form $\dfrac{p}{q}$, where p and q are polynomials and $q \neq 0$.

A **rational function** is a function of the form $f(x) = \dfrac{p(x)}{q(x)}$, where $p(x)$ and $q(x)$ are polynomial functions and $q(x) \neq 0$.

The *domain* of a rational function excludes the values for which the denominator is zero.

To simplify a rational expression to lowest terms, factor the numerator and denominator completely. Then simplify factors whose ratio is 1 or −1. A rational expression written in lowest terms will still have the same restrictions on the domain as the original expression.

Examples

Example 1

Find the domain of the function.

$$f(x) = \frac{x - 3}{(x + 4)(2x - 1)}$$

Domain: $\left\{ x \mid x \text{ is a real number and } x \neq -4, x \neq \dfrac{1}{2} \right\}$

or $(-\infty, -4) \cup \left(-4, \tfrac{1}{2}\right) \cup \left(\tfrac{1}{2}, \infty\right)$

Example 2

Simplify to lowest terms.

$$\frac{t^2 - 6t - 16}{5t + 10}$$

$$= \frac{(t - 8)(t + 2)}{5(t + 2)}$$

$$= \frac{(t - 8)}{5} \cdot \frac{\overset{1}{\cancel{(t + 2)}}}{\cancel{(t + 2)}}$$

$$= \frac{(t - 8)}{5} \cdot (1) = \frac{t - 8}{5}$$

Section 5.2   Multiplication and Division of Rational Expressions

Key Concepts

To multiply rational expressions, factor the numerators and denominators completely. Then simplify factors whose ratio is 1 or -1.

Examples

Example 1

$$\frac{b^2 - a^2}{a^2 - 2ab + b^2} \cdot \frac{a^2 - 3ab + 2b^2}{2a + 2b}$$

$$= \frac{\overset{-1}{(b - a)}\overset{1}{(b + a)} \cdot (a - 2b)\overset{1}{(a - b)}}{(a - b)(a - b) \cdot 2(a - b)} \qquad \text{Factor.}$$

$$= -\frac{a - 2b}{2} \quad \text{or} \quad \frac{2b - a}{2} \qquad \text{Simplify.}$$

To divide rational expressions, multiply by the reciprocal of the divisor.

Example 2

$$\frac{9x + 3}{x^2 - 4} \div \frac{3x + 1}{4x + 8}$$

$$= \frac{3\overset{1}{(3x + 1)}}{(x - 2)(x + 2)} \cdot \frac{4\overset{1}{(x + 2)}}{3x + 1} \qquad \begin{array}{l}\text{Factor. Multiply by} \\ \text{the reciprocal.}\end{array}$$

$$= \frac{12}{x - 2} \qquad \text{Simplify.}$$

| Section 5.3 | Addition and Subtraction of Rational Expressions |

Key Concepts

To add or subtract rational expressions, the expressions must have the same denominator.

The **least common denominator (LCD)** is the product of unique factors from the denominators, in which each factor is raised to its highest power.

Steps to Add or Subtract Rational Expressions

1. Factor the denominator of each rational expression.
2. Identify the LCD.
3. Rewrite each rational expression as an equivalent expression with the LCD as its denominator. [This is accomplished by multiplying the numerator and denominator of each rational expression by the missing factor(s) from the LCD.]
4. Add or subtract the numerators, and write the result over the common denominator.
5. Simplify, if possible.

Examples

Example 1

For $\dfrac{1}{3(x-1)^3(x+2)}$ and $\dfrac{-5}{6(x-1)(x+7)^2}$

$\text{LCD} = 6(x-1)^3(x+2)(x+7)^2$

Example 2

$\dfrac{c}{c^2 - c - 12} - \dfrac{1}{2c - 8}$

$= \dfrac{c}{(c-4)(c+3)} - \dfrac{1}{2(c-4)}$ Factor the denominators.

The LCD is $2(c-4)(c+3)$

$\dfrac{2}{2} \cdot \dfrac{c}{(c-4)(c+3)} - \dfrac{1}{2(c-4)} \cdot \dfrac{(c+3)}{(c+3)}$ Write equivalent fractions with LCD.

$= \dfrac{2c - (c+3)}{2(c-4)(c+3)}$ Subtract.

$= \dfrac{2c - c - 3}{2(c-4)(c+3)}$ Simplify.

$= \dfrac{c - 3}{2(c-4)(c+3)}$

Section 5.4 Complex Fractions

Key Concepts

Complex fractions can be simplified by using Method I or Method II.

Method I uses the order of operations to simplify the numerator and denominator separately before multiplying by the reciprocal of the denominator of the complex fraction.

To use Method II, multiply the numerator and denominator of the complex fraction by the LCD of all the individual fractions. Then simplify the result.

Examples

Example 1

Simplify by using Method I.

$$\frac{\dfrac{5x+15}{7}}{\dfrac{x+3}{2}}$$

$$=\frac{5x+15}{7}-\frac{x+3}{2}$$

$$=\frac{5x+15}{7}\cdot\frac{2}{x+3}$$

$$=\frac{5(\overset{1}{\cancel{x+3}})}{7}\cdot\frac{2}{\cancel{x+3}}$$

$$=\frac{10}{7}$$

Example 2

Simplify by using Method II.

$$\frac{1-\dfrac{4}{w^2}}{1-\dfrac{1}{w}-\dfrac{6}{w^2}}\qquad\text{The LCD is }w^2.$$

$$=\frac{w^2\left(1-\dfrac{4}{w^2}\right)}{w^2\left(1-\dfrac{1}{w}-\dfrac{6}{w^2}\right)}$$

$$=\frac{w^2-4}{w^2-w-6}$$

$$=\frac{(w-2)(\overset{1}{\cancel{w+2}})}{(w-3)(\cancel{w+2})}$$

$$=\frac{w-2}{w-3}$$

Section 5.5 Solving Rational Equations

Key Concepts

Steps to Solve a Rational Equation

1. Factor the denominators of all rational expressions. Identify any restrictions on the variable.
2. Identify the LCD of all expressions in the equation.
3. Multiply both sides of the equation by the LCD.
4. Solve the resulting equation.
5. Check each potential solution.

Examples

Example 1

$$\frac{1}{w} - \frac{1}{2w-1} = \frac{-2w}{2w-1}$$

The LCD is $w(2w-1)$.

$$w(2w-1)\frac{1}{w} - w(2w-1) \cdot \frac{1}{2w-1}$$

$$= w(2w-1) \cdot \frac{-2w}{2w-1}$$

$$(2w-1)1 - w(1) = w(-2w)$$

$$2w - 1 - w = -2w^2 \quad \text{(quadratic equation)}$$

$$2w^2 + w - 1 = 0$$

$$(2w-1)(w+1) = 0$$

$$w \neq \frac{1}{2} \quad \text{or} \quad w = -1$$

The solution set is $\{-1\}$. (The value $\frac{1}{2}$ does not check.)

Section 5.6 Applications of Rational Equations and Proportions

Key Concepts

An equation that equates two **ratios** is called a **proportion**.

$$\frac{a}{b} = \frac{c}{d} \quad \text{provided } b \neq 0, d \neq 0$$

Example 1

A sample of 85 g of a particular ice cream contains 17 g of fat. How much fat does 324 g of the same ice cream contain?

$$\text{fat (g)} \longrightarrow \frac{17}{85} = \frac{x}{324} \longleftarrow \text{fat (g)}$$
$$\text{ice cream (g)} \longrightarrow \qquad \qquad \longleftarrow \text{ice cream (g)}$$

$$(85 \cdot 324) \cdot \frac{17}{85} = (85 \cdot 324) \cdot \frac{x}{324} \quad \text{Multiply by the LCD.}$$

$$5508 = 85x$$

$$x = 64.8 \text{ g}$$

There would be 64.8 g of fat in 324 g of ice cream.

Examples

Example 2

An old water pump can fill a tank in 6 hr, and a new pump can fill the tank in 4 hr. How long will it take to fill the tank if both pumps are working?

Let x represent the time working together.

$$\begin{pmatrix} \text{Speed of} \\ \text{old pump} \end{pmatrix} + \begin{pmatrix} \text{speed of} \\ \text{new pump} \end{pmatrix} = \begin{pmatrix} \text{speed working} \\ \text{together} \end{pmatrix}$$

$$\frac{1}{6} + \frac{1}{4} = \frac{1}{x}$$

$$12x \cdot \left(\frac{1}{6} + \frac{1}{4}\right) = 12x \cdot \left(\frac{1}{x}\right)$$

$$2x + 3x = 12$$

$$5x = 12$$

$$x = \frac{12}{5} \quad \text{It will take } \frac{12}{5} \text{ hr.}$$

Section 5.7   Variation

Key Concepts

Direct Variation

y varies **directly** as x.
y is directly proportional to x. $\Big\}\ y = kx$

Inverse Variation

y varies **inversely** as x.
y is inversely proportional to x. $\Big\}\ y = \dfrac{k}{x}$

Joint Variation

y varies **jointly** as w and z.
y is jointly proportional to w and z. $\Big\}\ y = kwz$

Steps to Find a Variation Model

1. Write a general variation model that relates the variables given in the problem. Let k represent the constant of variation.

2. Solve for k by substituting known values of the variables into the model from step 1.

3. Substitute the value of k into the original variation model from step 1.

Examples

Example 1

t varies directly as the square root of x.

$t = k\sqrt{x}$

Example 2

W is inversely proportional to the cube of x.

$W = \dfrac{k}{x^3}$

Example 3

y is jointly proportional to x and to the square of z.

$y = kxz^2$

Example 4

C varies directly as the square root of d and inversely as t. If $C = 12$ when d is 9 and t is 6, find C if d is 16 and t is 12.

Step 1: $C = \dfrac{k\sqrt{d}}{t}$

Step 2: $12 = \dfrac{k\sqrt{9}}{6} \Rightarrow 12 = \dfrac{k \cdot 3}{6} \Rightarrow k = 24$

Step 3: $C = \dfrac{24\sqrt{d}}{t} \Rightarrow C = \dfrac{24\sqrt{16}}{12} \Rightarrow C = 8$

Chapter 5   Review Exercises

Section 5.1

1. Let $k(y) = \dfrac{y}{y^2 - 1}$.

   a. Find the function values (if they exist): $k(2)$, $k(0)$, $k(1)$, $k(-1)$, $k\left(\dfrac{1}{2}\right)$.

   b. Identify the domain for k. Write the answer in interval notation.

2. Let $h(x) = \dfrac{x}{x^2 + 1}$.

   a. Find the function values (if they exist): $h(1)$, $h(0)$, $h(-1)$, $h(-3)$, $h\left(\dfrac{1}{2}\right)$.

   b. Identify the domain for h. Write the answer in interval notation.

For Exercises 3–10, simplify the rational expressions.

3. $\dfrac{28a^3b^3}{14a^2b^3}$

4. $\dfrac{25x^2yz^3}{125xyz}$

5. $\dfrac{x^2 - 4x + 3}{x - 3}$

6. $\dfrac{k^2 + 3k - 10}{k^2 - 5k + 6}$

7. $\dfrac{x^3 - 27}{9 - x^2}$

8. $\dfrac{a^4 - 81}{3 - a}$

9. $\dfrac{2t^2 + 3t - 5}{7 - 6t - t^2}$

10. $\dfrac{y^3 - 4y}{y^2 - 5y + 6}$

For Exercises 11–14, write the domain of each function in interval notation. Use that information to match the function with its graph.

11. $f(x) = \dfrac{1}{x - 3}$

12. $m(x) = \dfrac{1}{x + 2}$

13. $k(x) = \dfrac{6}{x^2 - 3x}$

14. $p(x) = \dfrac{-2}{x^2 + 4}$

a.

b.

c.

d.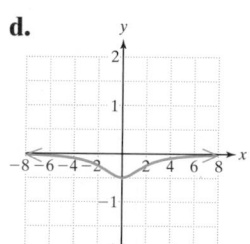

Section 5.2

For Exercises 15–26, multiply or divide as indicated.

15. $\dfrac{3a + 9}{a^2} \cdot \dfrac{a^3}{6a + 18}$

16. $\dfrac{4 - y}{5} \div \dfrac{2y - 8}{15}$

17. $\dfrac{x - 4y}{x^2 + xy} \div \dfrac{20y - 5x}{x^2 - y^2}$

18. $(x^2 + 5x - 24)\left(\dfrac{x + 8}{x - 3}\right)$

19. $\dfrac{7k + 28}{2k + 4} \cdot \dfrac{k^2 - 2k - 8}{k^2 + 2k - 8}$

20. $\dfrac{ab + 2a + b + 2}{ab - 3b + 2a - 6} \cdot \dfrac{ab - 3b + 4a - 12}{ab - b + 4a - 4}$

21. $\dfrac{x^2 + 8x - 20}{x^2 + 6x - 16} \div \dfrac{x^2 - 6x - 40}{x^2 - 3x - 40}$

22. $\dfrac{2b - b^2}{b^3 - 8} \cdot \dfrac{b^2 + 2b + 4}{b^2}$

23. $\dfrac{2w}{21} \div \dfrac{3w^2}{7} \cdot \dfrac{4}{w}$

24. $\dfrac{5y^2 - 20}{y^3 + 8} \div \dfrac{7y^2 - 14y}{y^3 + y}$

25. $\dfrac{x^2 + x - 20}{x^2 - 4x + 4} \cdot \dfrac{x^2 + x - 6}{12 + x - x^2} \div \dfrac{2x + 10}{10 - 5x}$

26. $(9k^2 - 25) \cdot \left(\dfrac{k + 5}{3k - 5}\right)$

Section 5.3

For Exercises 27–38, add or subtract as indicated.

27. $\dfrac{1}{x} + \dfrac{1}{x^2} - \dfrac{1}{x^3}$

28. $\dfrac{1}{x + 2} + \dfrac{5}{x - 2}$

29. $\dfrac{y}{2y - 1} + \dfrac{3}{1 - 2y}$

30. $\dfrac{a + 2}{2a + 6} - \dfrac{3}{a + 3}$

31. $\dfrac{4k}{k^2 + 2k + 1} + \dfrac{3}{k^2 - 1}$

32. $4x + 3 - \dfrac{2x + 1}{x + 4}$

33. $\dfrac{2}{a + 3} + \dfrac{2a^2 - 2a}{a^2 - 2a - 15}$

34. $\dfrac{6}{x^2 + 4x + 3} + \dfrac{7}{x^2 + 5x + 6}$

35. $\dfrac{2}{3x - 5} - 8$

36. $\dfrac{7}{4k^2 - k - 3} + \dfrac{1}{4k^2 - 7k + 3}$

37. $\dfrac{6a}{3a^2 - 7a + 2} + \dfrac{2}{1 - 3a} + \dfrac{3a}{a - 2}$

38. $4 + \dfrac{2y - 5}{y + 2} + \dfrac{y}{3 - y}$

Section 5.4

For Exercises 39–46, simplify the complex fraction.

39. $\dfrac{\dfrac{2x}{3x^2 - 3}}{\dfrac{4x}{6x - 6}}$

40. $\dfrac{\dfrac{k + 2}{3}}{\dfrac{5}{k - 2}}$

41. $\dfrac{\dfrac{2}{x} + \dfrac{1}{xy}}{\dfrac{4}{x^2}}$

42. $\dfrac{\dfrac{4}{y} - 1}{\dfrac{1}{y} - \dfrac{4}{y^2}}$

43. $\dfrac{\dfrac{1}{a - 1} + 1}{\dfrac{1}{a + 1} - 1}$

44. $\dfrac{\dfrac{3}{x - 1} - \dfrac{1}{1 - x}}{\dfrac{2}{x - 1} - \dfrac{2}{x}}$

45. $\dfrac{1 + xy^{-1}}{x^2 y^{-2} - 1}$

46. $\dfrac{5a^{-1} + (ab)^{-1}}{3a^{-2}}$

For Exercises 47–48, find the slope of the line containing the two points.

47. $\left(\dfrac{2}{3}, -\dfrac{7}{4}\right)$ and $\left(\dfrac{13}{6}, -\dfrac{5}{3}\right)$

48. $\left(\dfrac{8}{15}, -\dfrac{1}{3}\right)$ and $\left(\dfrac{13}{10}, \dfrac{9}{5}\right)$

Section 5.5

For Exercises 49–54, solve the equation.

49. $\dfrac{x + 3}{x^2 - x} - \dfrac{8}{x^2 - 1} = 0$

50. $\dfrac{y}{y + 3} - \dfrac{3}{3 - y} = \dfrac{18}{y^2 - 9}$

51. $x - 9 = \dfrac{72}{x - 8}$

52. $\dfrac{3x + 1}{x + 5} = \dfrac{x - 1}{x + 1} + 2$

53. $5y^{-2} + 1 = 6y^{-1}$

54. $1 + \dfrac{7}{6}m^{-1} = \dfrac{13}{6}m^{-1}$

55. Solve for x. $c = \dfrac{ax + b}{x}$

56. Solve for P. $\dfrac{A}{rt} = P + \dfrac{P}{rt}$

Section 5.6

For Exercises 57–60, solve the proportions.

57. $\dfrac{5}{4} = \dfrac{x}{6}$

58. $\dfrac{x}{36} = \dfrac{6}{7}$

59. $\dfrac{x + 2}{3} = \dfrac{5(x + 1)}{4}$

60. $\dfrac{x}{x + 2} = \dfrac{-3}{5}$

61. In one game Peyton Manning completed 34 passes for 357 yd. At this rate how many yards would be gained for 22 passes?

62. Erik bought $108 Canadian with $100 American. At this rate, how many Canadian dollars can he buy with $235 American?

63. Tony rode 175 mi on a 2-day bicycle ride to benefit the Multiple Sclerosis Foundation. The second day he rode 5 mph slower than the first day because of a strong headwind. If Tony rode 100 mi on the first day and 75 mi on the second day in a total time of 10 hr, how fast did he ride each day?

64. Stephen drove his car 45 mi. He ran out of gas and had to walk 3 mi to a gas station. His speed driving is 15 times his speed walking. If the total time for the drive and walk was $1\frac{1}{2}$ hr, what was his speed driving?

65. Doug and Jean work as phone solicitors. They work in batches of 400 calls. Doug can finish a batch in an average of 8 hr, and Jean can finish a batch in 10 hr. How long would it take them to finish a batch if they worked together?

66. Two pipes can fill a tank in 6 hr. The larger pipe works twice as fast as the smaller pipe. How long would it take each pipe to fill the tank if they worked separately?

Section 5.7

67. The force applied to a spring varies directly with the distance that the spring is stretched.

  a. Write a variation model using k as the constant of variation.

  b. When 6 lb of force is applied, the spring stretches 2 ft. Find k.

  c. How much force is required to stretch the spring 4.2 ft?

68. Suppose y varies inversely with x and $y = 32$ when $x = 2$. Find y when $x = 4$.

69. Suppose y varies jointly with x and the square root of z, and $y = 3$ when $x = 3$ and $z = 4$. Find y when $x = 8$ and $z = 9$.

 70. The distance d that one can see to the horizon varies directly as the square root of the height above sea level. If a person 25 m above sea level can see 30 km, how far can a person see if she is 64 m above sea level?

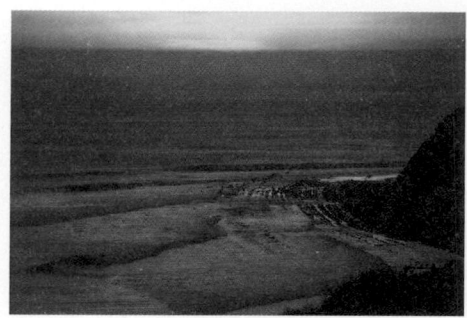

Chapter 5   Test

1. For the function $h(x) = \dfrac{2x - 14}{x^2 - 49}$

  a. Evaluate $h(0)$, $h(5)$, $h(7)$, and $h(-7)$, if possible.

  b. Write the domain of h in interval notation.

2. Write the domain of $k(x) = \dfrac{5x - 3}{7}$ in interval notation.

3. For the function $f(x) = \dfrac{2x + 6}{x^2 - x - 12}$

  a. Write the domain in set-builder notation.

  b. Simplify to lowest terms.

For Exercises 4–5, simplify to lowest terms.

4. $\dfrac{12m^3n^7}{18mn^8}$

5. $\dfrac{9x^2 - 9}{3x^2 + 2x - 5}$

6. Find the slope of the line containing the points $\left(\dfrac{1}{12}, -\dfrac{3}{4}\right)$ and $\left(\dfrac{5}{6}, -\dfrac{8}{3}\right)$.

For Exercises 7–13, simplify.

7. $\dfrac{2x - 5}{25 - 4x^2} \cdot (2x^2 - x - 15)$

8. $\dfrac{x^2}{x - 4} - \dfrac{8x - 16}{x - 4}$

9. $\dfrac{4x}{x + 1} + x + \dfrac{2}{x + 1}$

10. $\dfrac{3 + \dfrac{3}{k}}{4 + \dfrac{4}{k}}$

11. $\dfrac{2u^{-1} + 2v^{-1}}{4u^{-3} + 4v^{-3}}$

12. $\dfrac{ax + bx + 2a + 2b}{ax - 3a + bx - 3b} \cdot \dfrac{x - 3}{5 - x} \div \dfrac{x + 2}{ax - 5a}$

13. $\dfrac{3}{x^2 + 8x + 15} - \dfrac{1}{x^2 + 7x + 12} - \dfrac{1}{x^2 + 9x + 20}$

For Exercises 14–16, solve the equation.

14. $\dfrac{7}{z + 1} - \dfrac{z - 5}{z^2 - 1} = \dfrac{6}{z}$

15. $\dfrac{3}{y^2 - 9} + \dfrac{4}{y + 3} = 1$

16. $\dfrac{4x}{x - 4} = 3 + \dfrac{16}{x - 4}$

17. Solve for T. $\dfrac{1 + Tv}{T} = p$

18. Solve for m_1. $F = \dfrac{Gm_1m_2}{r^2}$

19. If the reciprocal of a number is added to 3 times the number, the result is $\frac{13}{2}$. Find the number.

20. Triangle ABC is similar to triangle XYZ. Find the lengths of the missing sides.

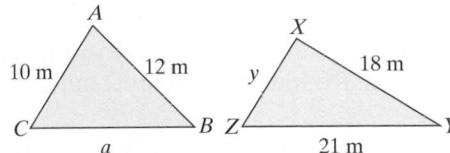

21. On a certain map, the distance between New York and Los Angeles is 8.2 in., and the actual distance is 2820 mi. What is the distance between two cities that are 5.7 in. apart on the same map? Round to the nearest mile.

22. Lance can ride 48 mi on his bike against the wind. With the wind at his back, he rides 4 mph faster and can ride 60 mi in the same amount of time. Find his speed riding against the wind and his speed riding with the wind.

23. Barbara can type a chapter in a book in 4 hr. Jack can type a chapter in a book in 10 hr. How long would it take them to type a chapter if they worked together?

24. Write a variation model using k as the constant of variation. The variable x varies directly as y and inversely as the square of t.

25. The period of a pendulum varies directly as the square root of the length of the pendulum. If the period of the pendulum is 2.2 sec when the length is 4 ft, find the period when the length is 9 ft.

Chapters 1–5   Cumulative Review Exercises

1. Check the sets to which each number belongs.

Set \ Number	-22	π	6	$-\sqrt{2}$
Real numbers				
Irrational numbers				
Rational numbers				
Integers				
Whole numbers				
Natural numbers				

2. Solve the system of equations.
$$3x + 4y = -4$$
$$2x - 8y = -8$$

3. Perform the indicated operations.
$(2x - 3)(x - 4) - (x - 5)^2$

4. The area of a trapezoid is given by $A = \frac{1}{2}h(b_1 + b_2)$.

   a. Solve for b_1.

   b. Find b_1 when $h = 4$ cm, $b_2 = 6$ cm, and $A = 32$ cm^2.

5. The dimensions of a rectangular swimming pool are such that the length is 10 m less than twice the width. If the perimeter is 160 m, find the length and width.

6. Solve the system of equations.
$$3x - 3y + z = -13$$
$$2x - y - 2z = 4$$
$$x + 2y - 3z = 15$$

7. Find an equation of the line through $(-3, 5)$ that is perpendicular to the line $y = 3x$. Write the answer in slope-intercept form.

8. Graph the line $2x - y = 3$.

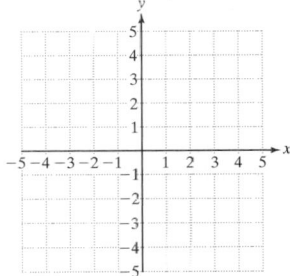

9. The speed of a car varies inversely as the time to travel a fixed distance. A car traveling the speed limit of 60 mph travels between two points in 10 sec. How fast is a car moving if it takes only 8 sec to cover the same distance?

10. Find the x-intercepts. $f(x) = -12x^3 + 17x^2 - 6x$

11. Factor $64y^3 - 8z^6$ completely over the real numbers.

12. Factor. $10x^2 - x - 2$

13. Perform the indicated operations.

$$\frac{2x^2 + 11x - 21}{4x^2 - 10x + 6} \div \frac{2x^2 - 98}{x^2 - x + xa - a}$$

14. Perform the indicated operations.

$$\frac{x}{x^2 + 5x - 50} - \frac{1}{x^2 - 7x + 10} + \frac{1}{x^2 + 8x - 20}$$

15. Simplify the complex fraction.

$$\frac{1 - \dfrac{49}{c^2}}{\dfrac{7}{c} + 1}$$

16. Solve. $y^3 - 5y^2 - y = -5$

17. Solve the equation.

$$\frac{4y}{y + 2} - \frac{y}{y - 1} = \frac{9}{y^2 + y - 2}$$

 18. Max knows that the distance between Roanoke, Virginia, and Washington, D.C., is 195 mi. On a certain map, the distance between the two cities is 6.5 in. On the same map, the distance between Roanoke and Cincinnati, Ohio, is 9.25 in. Find the distance in miles between Roanoke and Cincinnati. Round to the nearest mile.

19. Determine whether the equation represents a horizontal or vertical line. Then identify the slope of the line.

    a. $x = -5$ **b.** $2y = 8$

20. Simple interest varies jointly as the interest rate and as the time the money is invested. If an investment yields $1120 interest at 8% for 2 yr, how much interest will the investment yield at 10% for 5 yr?

Radicals and Complex Numbers

6

CHAPTER OUTLINE

6.1 Definition of an *n*th Root   496

6.2 Rational Exponents   508

6.3 Simplifying Radical Expressions   515

6.4 Addition and Subtraction of Radicals   522

6.5 Multiplication of Radicals   528

    Problem Recognition Exercises: Simplifying Radical Expressions   536

6.6 Division of Radicals and Rationalization   536

6.7 Solving Radical Equations   546

6.8 Complex Numbers   556

    Group Activity: Margin of Error of Survey Results   566

Chapter 6

In this chapter, we study radical expressions. This includes operations on square roots, cube roots, fourth roots, and so on. We also revisit the Pythagorean theorem and its applications as well as introduce the set of complex numbers.

Are You Prepared?

Take a minute to review your knowledge of square roots and operations on polynomials. These skills will help you be successful in this chapter.

    For Exercises 1–10, perform the indicated operations. Then write the letter corresponding to each answer in the space at the bottom of the page.

1. $(4x - 3)(x + 5)$ **2.** $4x^2(3x^2 + 5x + 1)$ **3.** $(4x - 3)(4x + 3)$ **4.** $(4x - 3)^2$

5. $\sqrt{4}$ **6.** 4^2 **7.** $\sqrt{-4}$ **8.** $\sqrt{10^2 - 6^2}$

9. $(-3x^4 + 9x^2 - 15x) - (-3x^4 - 7x^2 + 9x + 9)$ **10.** $(2x)(6x^3)$

R	16	N	$16x^2 - 9$	I	$16x^2 - 24x + 9$	M 8
Y	$16x^2 - 24x - 9$	A	not a real number	E	$12x^4 + 20x^3 + 4x^2$	U 2
G	$4x^2 + 17x - 15$	B	$12x^4$			

In this chapter, we also introduce a new type of number called an

$\underline{\quad}\,\underline{\quad}\,\underline{\quad}\,\underline{\quad}\,\underline{\quad}\,\underline{\quad}\,\underline{\quad}\,\underline{\quad}\,\underline{\quad}$ $\underline{\quad}\,\underline{\quad}\,\underline{\quad}\,\underline{\quad}\,\underline{\quad}\,\underline{\quad}$.

  4 8 7 1 4 3 7 6 9   3 5 8 10 2 6

Concepts

1. **Definition of a Square Root**
2. **Definition of an *n*th Root**
3. **Roots of Variable Expressions**
4. **Pythagorean Theorem**
5. **Radical Functions**

1. Definition of a Square Root

The reverse operation to squaring a number is to find its square roots. For example, finding a square root of 36 is equivalent to asking, "what number when squared equals 36?"

One obvious answer to this question is 6 because $(6)^2 = 36$, but -6 will also work, because $(-6)^2 = 36$.

Definition of a Square Root

b is a **square root** of a if $b^2 = a$.

Example 1   Identifying Square Roots

Identify the square roots of the real numbers.

  a. 25   **b.** 49   **c.** 0   **d.** -9

Solution:

  a. 5 is a square root of 25 because $(5)^2 = 25$.

    -5 is a square root of 25 because $(-5)^2 = 25$.

  b. 7 is a square root of 49 because $(7)^2 = 49$.

    -7 is a square root of 49 because $(-7)^2 = 49$.

  c. 0 is a square root of 0 because $(0)^2 = 0$.

  d. There are no real numbers that when squared will equal a negative number; therefore, there are no real-valued square roots of -9.

Skill Practice   Identify the square roots of the real numbers.

  1. 64   **2.** 16   **3.** 1   **4.** -100

TIP:

- All positive real numbers have two real-valued square roots (one positive and one negative).
- Zero has only one square root, which is zero itself.
- A negative number has no real-valued square roots.

Recall from Section R.3 that the positive square root of a real number can be denoted with a **radical sign** $\sqrt{}$.

Positive and Negative Square Roots

Let a represent a positive real number. Then

  1. $\sqrt{a}$ is the *positive* square root of a. The positive square root is also called the **principal square root**.

  2. $-\sqrt{a}$ is the *negative* square root of a.

  3. $\sqrt{0} = 0$

Answers

1. -8 and 8
2. -4 and 4
3. -1 and 1
4. No real-valued square roots

Example 2 Simplifying Square Roots

Simplify the square roots.

a. $\sqrt{36}$ **b.** $\sqrt{\dfrac{4}{9}}$ **c.** $\sqrt{0.04}$

Solution:

a. $\sqrt{36}$ denotes the positive square root of 36.

$\sqrt{36} = 6$ because $(6)^2 = 36$

b. $\sqrt{\dfrac{4}{9}}$ denotes the positive square root of $\dfrac{4}{9}$.

$\sqrt{\dfrac{4}{9}} = \dfrac{2}{3}$ because $\left(\dfrac{2}{3}\right)^2 = \dfrac{4}{9}$

c. $\sqrt{0.04}$ denotes the positive square root of 0.04.

$\sqrt{0.04} = 0.2$ because $(0.2)^2 = 0.04$

Skill Practice Simplify the square roots.

5. $\sqrt{81}$ **6.** $\sqrt{\dfrac{36}{49}}$ **7.** $\sqrt{0.09}$

The numbers 36, $\frac{4}{9}$, and 0.04 are **perfect squares** because their square roots are rational numbers.

Radicals that cannot be simplified to rational numbers are irrational numbers. Recall that an irrational number cannot be written as a terminating or repeating decimal. For example, the symbol $\sqrt{13}$ is used to represent the *exact* value of the square root of 13. The symbol $\sqrt{42}$ is used to represent the *exact* value of the square root of 42. These values can be approximated by a rational number by using a calculator.

$$\sqrt{13} \approx 3.605551275 \qquad \sqrt{42} \approx 6.480740698$$

TIP: Before using a calculator to evaluate a square root, try estimating the value first.

$\sqrt{13}$ must be a number between 3 and 4 because $\sqrt{9} < \sqrt{13} < \sqrt{16}$.
$\sqrt{42}$ must be a number between 6 and 7 because $\sqrt{36} < \sqrt{42} < \sqrt{49}$.

Calculator Connections

Topic: Approximating Square Roots

Use a calculator to approximate the values of $\sqrt{13}$ and $\sqrt{42}$.

```
√(13)
         3.605551275
√(42)
         6.480740698
```

A negative number cannot have a real number as a square root because no real number when squared is negative. For example, $\sqrt{-25}$ is *not* a real number because there is no real number b for which $(b)^2 = -25$.

Answers

5. 9   **6.** $\dfrac{6}{7}$ **7.** 0.3

Example 3 Simplifying Square Roots

Simplify the square roots, if possible.

a. $\sqrt{-144}$ b. $-\sqrt{144}$ c. $\sqrt{-0.01}$ d. $-\sqrt{\dfrac{1}{9}}$

Solution:

a. $\sqrt{-144}$ is *not* a real number. No real number when squared equals -144.

b. $-\sqrt{144}$

$$= -1 \cdot \sqrt{144}$$
$$= -1 \cdot \quad 12$$
$$= -12$$

> **TIP:** For the expression $-\sqrt{144}$, the factor of -1 is *outside* the radical.

c. $\sqrt{-0.01}$ is *not* a real number. No real number when squared equals -0.01.

d. $-\sqrt{\dfrac{1}{9}}$

$$= -1 \cdot \sqrt{\dfrac{1}{9}}$$
$$= -1 \cdot \dfrac{1}{3}$$
$$= -\dfrac{1}{3}$$

Skill Practice Simplify the square roots, if possible.

8. $\sqrt{-81}$ 9. $-\sqrt{64}$ 10. $-\sqrt{0.25}$ 11. $\sqrt{-\dfrac{1}{4}}$

2. Definition of an *n*th Root

Finding a square root of a number is the reverse process of squaring a number. This concept can be extended to finding a third root (called a cube root), a fourth root, and in general an **nth root**.

> ### Definition of an *n*th Root
> b is an *n*th root of a if $b^n = a$.
>
> Example: 2 is a square root of 4 because 2^2 is 4.
>
> Example: 2 is a third root of 8 because 2^3 is 8.
>
> Example: 2 is a fourth root of 16 because 2^4 is 16.

The radical sign $\sqrt{\ }$ is used to denote the principal square root of a number. The symbol $\sqrt[n]{\ }$ is used to denote the principal *n*th root of a number. In the expression $\sqrt[n]{a}$, n is called the **index** of the radical, and a is called the **radicand**. For a square root, the index is 2, but it is usually not written ($\sqrt[2]{a}$ is denoted simply as $\sqrt{a}$). A radical with an index of 3 is called a **cube root**, denoted by $\sqrt[3]{a}$.

Answers
8. Not a real number   9. -8
10. -0.5 11. Not a real number

Evaluating $\sqrt[n]{a}$

1. If $n > 1$ is an *even* integer and $a > 0$, then $\sqrt[n]{a}$ is the principal (positive) *n*th root of a. Example: $\sqrt{9} = 3$
2. If $n > 1$ is an *odd* integer, then $\sqrt[n]{a}$ is the *n*th root of a. Example: $\sqrt[3]{8} = 2$, $\sqrt[3]{-8} = -2$
3. If $n > 1$ is an integer, then $\sqrt[n]{0} = 0$.

For the purpose of simplifying *n*th roots, it is helpful to know numbers that are perfect squares as well as the following common perfect cubes, fourth powers, and fifth powers.

Perfect Cubes	Perfect Fourth Powers	Perfect Fifth Powers
$1^3 = 1$	$1^4 = 1$	$1^5 = 1$
$2^3 = 8$	$2^4 = 16$	$2^5 = 32$
$3^3 = 27$	$3^4 = 81$	$3^5 = 243$
$4^3 = 64$	$4^4 = 256$	$4^5 = 1024$
$5^3 = 125$	$5^4 = 625$	$5^5 = 3125$

Example 4 **Identifying the *n*th Root of a Real Number**

Simplify the expressions, if possible.

a. $\sqrt{4}$ b. $\sqrt[3]{64}$ c. $\sqrt[5]{-32}$ d. $\sqrt[4]{81}$

e. $\sqrt[6]{1,000,000}$ f. $\sqrt{-100}$ g. $\sqrt[4]{-16}$

Solution:

a. $\sqrt{4} = 2$ because $(2)^2 = 4$

b. $\sqrt[3]{64} = 4$ because $(4)^3 = 64$

c. $\sqrt[5]{-32} = -2$ because $(-2)^5 = -32$

d. $\sqrt[4]{81} = 3$ because $(3)^4 = 81$

e. $\sqrt[6]{1,000,000} = 10$ because $(10)^6 = 1,000,000$

f. $\sqrt{-100}$ is not a real number.   No real number when squared equals -100.

g. $\sqrt[4]{-16}$ is not a real number.   No real number when raised to the fourth power equals -16.

Skill Practice Simplify if possible.

12. $\sqrt[4]{16}$ 13. $\sqrt[3]{1000}$ 14. $\sqrt[5]{-1}$ 15. $\sqrt[5]{32}$

16. $\sqrt[5]{100,000}$ 17. $\sqrt{-36}$ 18. $\sqrt[3]{-27}$

Calculator Connections

Topic: Approximating *n*th Roots

A calculator can be used to approximate *n*th roots by using the $\sqrt[\ ]{\ }$ function. On most calculators, the index is entered first.

```
3 ×√(64)
              4
5 ×√(-32)
             -2
6 ×√(1000000)
             10
```

Examples 4(f) and 4(g) illustrate that an *n*th root of a negative quantity is not a real number if the index is even. This is because no real number raised to an even power is negative.

3. Roots of Variable Expressions

Finding an *n*th root of a variable expression is similar to finding an *n*th root of a numerical expression. For roots with an even index, however, particular care must be taken to obtain a nonnegative result.

Answers

12. 2   13. 10   14. -1
15. 2   16. 10
17. Not a real number   18. -3

Evaluating $\sqrt[n]{a^n}$

1. If n is a positive *odd* integer, then $\sqrt[n]{a^n} = a$.
2. If n is a positive *even* integer, then $\sqrt[n]{a^n} = |a|$.

The absolute value bars are necessary for roots with an even index because the variable a may represent a positive quantity or a negative quantity. By using absolute value bars, $\sqrt[n]{a^n} = |a|$ is nonnegative and represents the principal nth root of a^n.

Example 5 Simplifying Expressions of the Form $\sqrt[n]{a^n}$

Simplify the expressions.

a. $\sqrt[4]{(-3)^4}$ **b.** $\sqrt[5]{(-3)^5}$ **c.** $\sqrt{(x+2)^2}$ **d.** $\sqrt[3]{(a+b)^3}$ **e.** $\sqrt{y^4}$

Solution:

a. $\sqrt[4]{(-3)^4} = |-3| = 3$ Because this is an *even*-indexed root, absolute value bars are necessary to make the answer positive.

b. $\sqrt[5]{(-3)^5} = -3$ This is an *odd*-indexed root, so absolute value bars are not necessary.

c. $\sqrt{(x+2)^2} = |x+2|$ Because this is an *even*-indexed root, absolute value bars are necessary. The sign of the quantity $x + 2$ is unknown; however, $|x + 2| \geq 0$ regardless of the value of x.

d. $\sqrt[3]{(a+b)^3} = a + b$ This is an *odd*-indexed root, so absolute value bars are not necessary.

e. $\sqrt{y^4} = \sqrt{(y^2)^2}$

$= |y^2|$ Because this is an even-indexed root, use absolute value bars.

$= y^2$ However, because y^2 is nonnegative, the absolute value bars are not necessary.

Skill Practice Simplify the expressions.

19. $\sqrt{(-4)^2}$ **20.** $\sqrt[3]{(-4)^3}$ **21.** $\sqrt{(y+9)^2}$ **22.** $\sqrt[3]{(t+1)^3}$ **23.** $\sqrt[4]{v^8}$

If n is an even integer, then $\sqrt[n]{a^n} = |a|$; however, if the variable a is assumed to be *nonnegative*, then the absolute value bars may be dropped. That is, $\sqrt[n]{a^n} = a$ provided $a \geq 0$. In many examples and exercises, we will make the assumption that the variables within a radical expression are positive real numbers. In such a case, the absolute value bars are not needed to evaluate $\sqrt[n]{a^n}$.

In Chapter 4 you became familiar with the patterns associated with perfect squares and perfect cubes. In general, any expression raised to an even power is a perfect square. An expression raised to a power that is a multiple of three is a perfect cube.

Perfect Squares	Perfect Cubes
$(x^1)^2 = x^2$	$(x^1)^3 = x^3$
$(x^2)^2 = x^4$	$(x^2)^3 = x^6$
$(x^3)^2 = x^6$	$(x^3)^3 = x^9$
$(x^4)^2 = x^8$	$(x^4)^3 = x^{12}$

These patterns may be extended to higher powers.

Example 6 Simplifying *n*th Roots

Simplify the expressions. Assume that all variables are positive real numbers.

a. $\sqrt{y^8}$ **b.** $\sqrt[3]{27a^3}$ **c.** $\sqrt[5]{\dfrac{a^5}{b^5}}$ **d.** $-\sqrt[4]{\dfrac{81x^4y^8}{16}}$

Solution:

a. $\sqrt{y^8} = \sqrt{(y^4)^2}$ **b.** $\sqrt[3]{27a^3} = \sqrt[3]{(3a)^3}$

    $= y^4$ $= 3a$

c. $\sqrt[5]{\dfrac{a^5}{b^5}} = \sqrt[5]{\left(\dfrac{a}{b}\right)^5}$ **d.** $-\sqrt[4]{\dfrac{81x^4y^8}{16}} = -\sqrt[4]{\left(\dfrac{3xy^2}{2}\right)^4}$

    $= \dfrac{a}{b}$ $= -\dfrac{3xy^2}{2}$

> **TIP:** In Example 6, the variables are assumed to represent positive numbers. Therefore, absolute value bars are not necessary in the simplified form.

Skill Practice Simplify the expressions. Assume all variables represent positive real numbers.

24. $\sqrt{t^6}$ **25.** $\sqrt[3]{64y^{12}}$ **26.** $\sqrt[4]{\dfrac{x^4}{y^4}}$ **27.** $-\sqrt[5]{\dfrac{32a^5}{b^{10}}}$

---

4. Pythagorean Theorem

In Section 4.8, we used the Pythagorean theorem in several applications. For the triangle shown in Figure 6-1, the **Pythagorean theorem** may be stated as $a^2 + b^2 = c^2$. In this formula, *a* and *b* are the legs of the right triangle, and *c* is the hypotenuse.

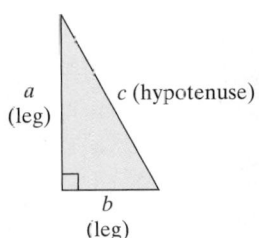

Figure 6-1

Example 7 Applying the Pythagorean Theorem

Use the Pythagorean theorem and the definition of the principal square root to find the length of the unknown side.

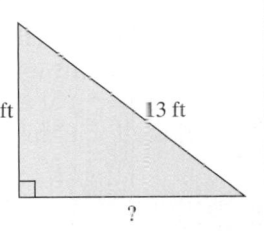

Solution:

Label the sides of the triangle.

$a^2 + b^2 = c^2$ Apply the Pythagorean theorem.

$(5)^2 + b^2 = (13)^2$

$25 + b^2 = 169$ Simplify.

$b^2 = 169 - 25$ Isolate b^2.

$b^2 = 144$ By definition, *b* must be one of the square roots of 144. Because *b* represents the length of a side of a triangle, choose the positive square root of 144.

$b = 12$

The third side is 12 ft long.

Skill Practice Use the Pythagorean theorem and the definition of the principal square root to find the length of the unknown side of the right triangle.

28.

Answers

24. t^3 **25.** $4y^4$ **26.** $\dfrac{x}{y}$

27. $-\dfrac{2a}{b^2}$ **28.** 9 cm

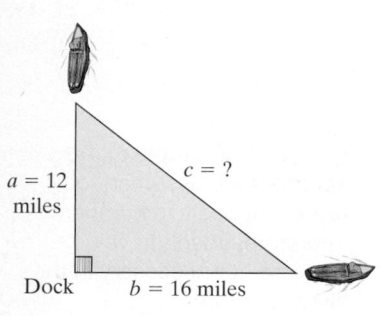

$a = 12$ miles

$c = ?$

Dock $b = 16$ miles

Figure 6-2

Example 8 **Applying the Pythagorean Theorem**

Two boats leave a dock at 12:00 noon. One travels due north at 6 mph, and the other travels due east at 8 mph (Figure 6-2). How far apart are the two boats after 2 hr?

Solution:

The boat traveling north travels a distance of (6 mph)(2 hr) = 12 mi. The boat traveling east travels a distance of (8 mph)(2 hr) = 16 mi. The course of the boats forms a right triangle where the hypotenuse represents the distance between them.

$$a^2 + b^2 = c^2$$

$$(12)^2 + (16)^2 = c^2 \qquad \text{Apply the Pythagorean theorem.}$$

$$144 + 256 = c^2 \qquad \text{Simplify.}$$

$$400 = c^2$$

$$\sqrt{400} = c \qquad \text{By definition, } c \text{ must be one of the square roots of } 400. \text{ Choose the positive square root of } 400 \text{ to}$$

$$20 = c \qquad \text{represent the distance between the two boats.}$$

The boats are 20 mi apart.

Skill Practice

29. Two cars leave from the same place at the same time. One travels west at 40 mph, and the other travels north at 30 mph. How far apart are they after 2 hr?

5. Radical Functions

If n is an integer greater than 1, then a function written in the form $f(x) = \sqrt[n]{x}$ is called a **radical function**. Note that if n is an even integer, then the function will be a real number only if the radicand is nonnegative. Therefore, the domain is restricted to nonnegative real numbers, or equivalently, $[0, \infty)$. If n is an odd integer, then the domain is all real numbers.

Example 9 **Determining the Domain of Radical Functions**

For each function, write the domain in interval notation.

   a. $g(t) = \sqrt[4]{t - 2}$ **b.** $h(a) = \sqrt[3]{a - 3}$ **c.** $k(x) = \sqrt{3 - 5x} + 2$

Solution:

   a. $g(t) = \sqrt[4]{t - 2}$ The index is even. The radicand must be nonnegative.

        $t - 2 \geq 0$ Set the radicand greater than or equal to zero.

           $t \geq 2$ Solve for t.

    The domain is $[2, \infty)$.

   b. $h(a) = \sqrt[3]{a - 3}$ The index is odd; therefore, the domain is all real numbers.

    The domain is $(-\infty, \infty)$.

Answer

29. 100 mi

c. $k(x) = \sqrt{3 - 5x} + 2$ The index is even; therefore, the radicand must be nonnegative.

$3 - 5x \geq 0$ Set the radicand greater than or equal to zero.

$-5x \geq -3$ Solve for x.

$\dfrac{-5x}{-5} \leq \dfrac{-3}{-5}$ Reverse the inequality sign.

$x \leq \dfrac{3}{5}$

The domain is $\left(-\infty, \frac{3}{5}\right]$.

Skill Practice For each function, write the domain in interval notation.

30. $f(x) = \sqrt{x + 5}$ **31.** $g(t) = \sqrt[3]{t - 9}$ **32.** $h(a) = \sqrt{1 - 2a}$

Calculator Connections

Topic: Graphing Radical Functions

We can graph the functions defined in Example 9. The graphs support the answers we obtained for the domain of each function.

$g(t) = \sqrt[4]{t - 2}$ $h(a) = \sqrt[3]{a - 3}$ $k(x) = \sqrt{3 - 5x} + 2$

Example 10   Graphing a Radical Function

Given $f(x) = \sqrt{3 - x}$

a. Write the domain of f in interval notation.

b. Graph f by making a table of ordered pairs.

Solution:

a. $f(x) = \sqrt{3 - x}$

$3 - x \geq 0$ The index is even. The radicand must be greater than or equal to zero.

$-x \geq -3$

$x \leq 3$ Reverse the inequality sign.

The domain is $(-\infty, 3]$.

Answers

30. $[-5, \infty)$ **31.** $(-\infty, \infty)$

32. $\left(-\infty, \frac{1}{2}\right]$

b. Create a table of ordered pairs where x values are taken to be less than or equal to 3.

x	$f(x)$
3	0
2	1
-1	2
-6	3

$f(3) = \sqrt{3 - 3} = \sqrt{0} = 0$
$f(2) = \sqrt{3 - 2} = \sqrt{1} = 1$
$f(-1) = \sqrt{3 - (-1)} = \sqrt{4} = 2$
$f(-6) = \sqrt{3 - (-6)} = \sqrt{9} = 3$

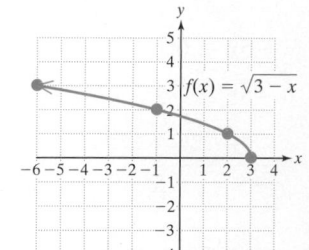

Answers

33. a. $[-4, \infty)$
  b.

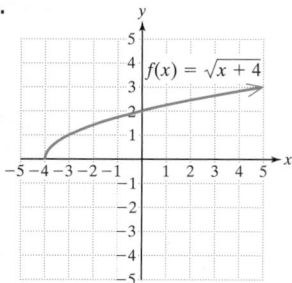

Skill Practice

33. Given $f(x) = \sqrt{x + 4}$
  a. Write the domain of f in interval notation.
  b. Graph f by making a table of ordered pairs.

Section 6.1   Practice Exercises

Vocabulary and Key Concepts

1. a. If $b^2 = a$, then _____ is a square root of _____.

  b. The symbol $\sqrt{a}$ denotes the positive or _____ square root of a.

  c. b is an nth root of a if _____ = _____.

  d. Given the symbol $\sqrt[n]{a}$, the value n is called the _____ and a is called the _____.

  e. The symbol $\sqrt[3]{a}$ denotes the _____ root of a.

  f. The expression $\sqrt{-4}$ (is/is not) a real number. The expression $-\sqrt{4}$ (is/is not) a real number.

  g. The expression $\sqrt[n]{a^n} = |a|$ if n is (even/odd). The expression $\sqrt[n]{a^n} = a$ if n is (even/odd).

  h. Given a right triangle with legs a and b and hypotenuse c, the _____ theorem is stated as $a^2 + b^2 = $ _____.

  i. In interval notation, the domain of $f(x) = \sqrt{x}$ is _____, whereas the domain of $g(x) = \sqrt[3]{x}$ is _____.

  j. Which of the following values of x are *not* in the domain of $h(x) = \sqrt{x + 3}$: $x = -5, x = -4, x = -3,$ $x = -2, x = -1, x = 0$?

Concept 1: Definition of a Square Root

2. Simplify the expression $\sqrt[3]{8}$. Explain how you can check your answer.

3. a. Find the square roots of 64. **(See Example 1.)**

  b. Find $\sqrt{64}$.

  c. Explain the difference between the answers in part (a) and part (b).

4. a. Find the square roots of 121.

  b. Find $\sqrt{121}$.

  c. Explain the difference between the answers in part (a) and part (b).

5. a. What is the principal square root of 81?

  b. What is the negative square root of 81?

6. a. What is the principal square root of 100?

  b. What is the negative square root of 100?

7. Using the definition of a square root, explain why $\sqrt{-36}$ is not a real number.

For Exercises 8–19, evaluate the roots without using a calculator. Identify those that are not real numbers.
(See Examples 2–3.)

8. $\sqrt{25}$ **9.** $\sqrt{49}$ **10.** $-\sqrt{25}$ **11.** $-\sqrt{49}$

12. $\sqrt{-25}$ **13.** $\sqrt{-49}$ **14.** $\sqrt{\dfrac{100}{121}}$ **15.** $\sqrt{\dfrac{64}{9}}$

16. $\sqrt{0.64}$ **17.** $\sqrt{0.81}$ **18.** $-\sqrt{0.0144}$ **19.** $-\sqrt{0.16}$

Concept 2: Definition of an *n*th Root

20. Using the definition of an *n*th root, explain why $\sqrt[4]{-16}$ is not a real number.

For Exercises 21–38, evaluate the roots without using a calculator. Identify those that are not real numbers.
(See Example 4.)

21. a. $\sqrt{64}$ **b.** $\sqrt[3]{64}$ **c.** $-\sqrt{64}$
       d. $-\sqrt[3]{64}$ **e.** $\sqrt{-64}$ **f.** $\sqrt[3]{-64}$

22. a. $\sqrt{16}$ **b.** $\sqrt[4]{16}$ **c.** $-\sqrt{16}$
       d. $-\sqrt[4]{16}$ **e.** $\sqrt{-16}$ **f.** $\sqrt[4]{-16}$

23. $\sqrt[3]{-27}$ **24.** $\sqrt[3]{-125}$ **25.** $\sqrt[3]{\dfrac{1}{8}}$ **26.** $\sqrt[5]{\dfrac{1}{32}}$

27. $\sqrt[5]{32}$ **28.** $\sqrt[4]{1}$ **29.** $\sqrt[3]{-\dfrac{125}{64}}$ **30.** $\sqrt[3]{-\dfrac{8}{27}}$

31. $\sqrt[4]{-1}$ **32.** $\sqrt[6]{-1}$ **33.** $\sqrt[6]{1,000,000}$ **34.** $\sqrt[4]{10,000}$

35. $-\sqrt[3]{0.008}$ **36.** $-\sqrt[4]{0.0016}$ **37.** $\sqrt[4]{0.0625}$ **38.** $\sqrt[3]{0.064}$

Concept 3: Roots of Variable Expressions

For Exercises 39–58, simplify the radical expressions. **(See Example 5.)**

39. $\sqrt{a^2}$ **40.** $\sqrt[4]{a^4}$ **41.** $\sqrt[3]{a^3}$ **42.** $\sqrt[5]{a^5}$

43. $\sqrt[6]{a^6}$ **44.** $\sqrt[7]{a^7}$ **45.** $\sqrt{(x+1)^2}$ **46.** $\sqrt[3]{(y+3)^3}$

47. $\sqrt{x^2 y^4}$ **48.** $\sqrt[3]{(u+v)^3}$ **49.** $-\sqrt[3]{\dfrac{x^3}{y^3}}, \ y \neq 0$ **50.** $\sqrt[4]{\dfrac{a^4}{b^8}}, \ b \neq 0$

51. $\dfrac{2}{\sqrt[4]{x^4}}, \ x \neq 0$ **52.** $\sqrt{(-5)^2}$ **53.** $\sqrt[3]{(-92)^3}$ **54.** $\sqrt[6]{(50)^6}$

55. $\sqrt[10]{(-2)^{10}}$ **56.** $\sqrt[5]{(-2)^5}$ **57.** $\sqrt[7]{(-923)^7}$ **58.** $\sqrt[6]{(-417)^6}$

For Exercises 59–74, simplify the expressions. Assume all variables are positive real numbers. **(See Example 6.)**

59. $\sqrt{y^8}$ **60.** $\sqrt{x^4}$ **61.** $\sqrt{\dfrac{a^6}{b^2}}$ **62.** $\sqrt{\dfrac{w^2}{z^4}}$

63. $-\sqrt{\dfrac{25}{q^2}}$ **64.** $-\sqrt{\dfrac{p^6}{81}}$ **65.** $\sqrt{9x^2 y^4 z^2}$ **66.** $\sqrt{4a^4 b^2 c^6}$

67. $\sqrt{\dfrac{h^2 k^4}{16}}$ **68.** $\sqrt{\dfrac{4x^2}{y^8}}$ **69.** $-\sqrt[3]{\dfrac{t^3}{27}}$ **70.** $\sqrt[4]{\dfrac{16}{w^4}}$

71. $\sqrt[5]{32y^{10}}$ **72.** $\sqrt[3]{64x^6 y^3}$ **73.** $\sqrt[6]{64p^{12}q^{18}}$ **74.** $\sqrt[4]{16r^{12}s^8}$

Concept 4: Pythagorean Theorem

For Exercises 75–78, find the length of the third side of each triangle by using the Pythagorean theorem. **(See Example 7.)**

75.

76.

77.

78.

For Exercises 79–82, use the Pythagorean theorem.

79. Roberto and Sherona began running from the same place at the same time. They ran along two different paths that formed right angles with each other. Roberto ran 4 mi and stopped, while Sherona ran 3 mi and stopped. How far apart were they when they stopped? **(See Example 8.)**

80. Leine and Laura began hiking from their campground. Laura headed south while Leine headed east. Laura walked 12 mi and Leine walked 5 mi. How far apart were they when they stopped walking?

81. Two mountain bikers take off from the same place at the same time. One travels north at 4 mph, and the other travels east at 3 mph. How far apart are they after 5 hr?

82. Professor Ortiz leaves campus on her bike, heading west at 12 ft/sec. Professor Wilson leaves campus at the same time and walks south at 5 ft/sec. How far apart are they after 40 sec?

Concept 5: Radical Functions

For Exercises 83–86, evaluate the function for the given values of x. Then write the domain of the function in interval notation. **(See Example 9.)**

83. $h(x) = \sqrt{x-2}$
   a. $h(0)$
   b. $h(1)$
   c. $h(2)$
   d. $h(3)$
   e. $h(6)$

84. $k(x) = \sqrt{x+1}$
   a. $k(-3)$
   b. $k(-2)$
   c. $k(-1)$
   d. $k(0)$
   e. $k(3)$

85. $g(x) = \sqrt[3]{x-2}$
   a. $g(-6)$
   b. $g(1)$
   c. $g(2)$
   d. $g(3)$

86. $f(x) = \sqrt[3]{x+1}$
   a. $f(-9)$
   b. $f(-2)$
   c. $f(0)$
   d. $f(7)$

For each function defined in Exercises 87–94, write the domain in interval notation. **(See Example 9.)**

87. $f(x) = \sqrt{5-2x}$ **88.** $g(x) = \sqrt{3-4x}$ **89.** $k(x) = \sqrt[3]{4x-7}$ **90.** $R(x) = \sqrt[3]{x+1}$

91. $M(x) = \sqrt{x-5}+3$ **92.** $N(x) = \sqrt{x+3}-1$ **93.** $F(x) = \sqrt[3]{x+7}-2$ **94.** $G(x) = \sqrt[3]{x-10}+4$

For Exercises 95–102,

a. Write the domain of *f* in interval notation.

b. Graph *f* by making a table of ordered pairs. **(See Example 10.)**

95. $f(x) = \sqrt{1 - x}$ **96.** $f(x) = \sqrt{2 - x}$ **97.** $f(x) = \sqrt{x + 3}$ **98.** $f(x) = \sqrt{x + 1}$

      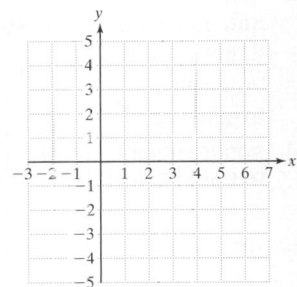

99. $f(x) = \sqrt{x} + 2$ **100.** $f(x) = \sqrt{x} - 1$ **101.** $f(x) = \sqrt[3]{x - 1}$ **102.** $f(x) = \sqrt[3]{x + 2}$

      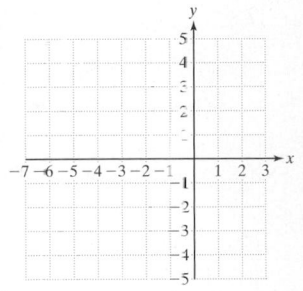

Mixed Exercises

For Exercises 103–106, write the English phrase as an algebraic expression.

103. The sum of *q* and the square of *p*

104. The product of 11 and the cube root of *x*

105. The quotient of 6 and the cube root of *x*

106. The difference of *y* and the principal square root of *x*

107. If a square has an area of 64 in.², then what are the lengths of the sides?

108. If a square has an area of 121 m², then what are the lengths of the sides?

Graphing Calculator Exercises

For Exercises 109–116, use a calculator to evaluate the expressions to four decimal places.

109. $\sqrt{69}$ **110.** $\sqrt{5798}$ **111.** $2 + \sqrt[3]{5}$ **112.** $3 - 2\sqrt[4]{10}$

113. $7\sqrt[4]{25}$ **114.** $-3\sqrt[3]{9}$ **115.** $\dfrac{3 - \sqrt{19}}{11}$ **116.** $\dfrac{5 + 2\sqrt{15}}{12}$

117. Graph $h(x) = \sqrt{x - 2}$. Use the graph to confirm the domain found in Exercise 83.

118. Graph $k(x) = \sqrt{x + 1}$. Use the graph to confirm the domain found in Exercise 84.

119. Graph $g(x) = \sqrt[3]{x - 2}$. Use the graph to confirm the domain found in Exercise 85.

120. Graph $f(x) = \sqrt[3]{x + 1}$. Use the graph to confirm the domain found in Exercise 86.

Section 6.2 Rational Exponents

Concepts

1. **Definition of $a^{1/n}$ and $a^{m/n}$**
2. **Converting Between Rational Exponents and Radical Notation**
3. **Properties of Rational Exponents**
4. **Applications Involving Rational Exponents**

1. Definition of $a^{1/n}$ and $a^{m/n}$

In Section 4.1, the properties for simplifying expressions with integer exponents were presented. In this section, the properties are expanded to include expressions with rational exponents. We begin by defining expressions of the form $a^{1/n}$.

> **Definition of $a^{1/n}$**
>
> Let a be a real number, and let n be an integer such that $n > 1$. If $\sqrt[n]{a}$ is a real number, then
>
> $$a^{1/n} = \sqrt[n]{a}$$

Example 1 Evaluating Expressions of the Form $a^{1/n}$

Convert the expression to radical form and simplify, if possible.

 a. $(-8)^{1/3}$ **b.** $81^{1/4}$ **c.** $-100^{1/2}$ **d.** $(-100)^{1/2}$ **e.** $16^{-1/2}$

Solution:

 a. $(-8)^{1/3} = \sqrt[3]{-8} = -2$

 b. $81^{1/4} = \sqrt[4]{81} = 3$

 c. $-100^{1/2} = -1 \cdot 100^{1/2}$ The exponent applies only to the base of 100.

$\qquad\qquad = -1\sqrt{100}$

$\qquad\qquad = -10$

 d. $(-100)^{1/2}$ is not a real number because $\sqrt{-100}$ is not a real number.

 e. $16^{-1/2} = \dfrac{1}{16^{1/2}}$ Write the expression with a positive exponent.

$\qquad\qquad\qquad$ Recall that $b^{-n} = \dfrac{1}{b^n}$.

$\qquad = \dfrac{1}{\sqrt{16}}$

$\qquad = \dfrac{1}{4}$

Skill Practice Convert the expression to radical form and simplify, if possible.

 1. $(-64)^{1/3}$ **2.** $16^{1/4}$ **3.** $-36^{1/2}$ **4.** $(-36)^{1/2}$ **5.** $64^{-1/3}$

If $\sqrt[n]{a}$ is a real number, then we can define an expression of the form $a^{m/n}$ in such a way that the multiplication property of exponents still holds true. For example:

$$(16^{1/4})^3 = (\sqrt[4]{16})^3 = (2)^3 = 8$$

$16^{3/4}$

$$(16^3)^{1/4} = \sqrt[4]{16^3} = \sqrt[4]{4096} = 8$$

Answers

1. -4 **2.** 2 **3.** -6

4. Not a real number **5.** $\dfrac{1}{4}$

Definition of $a^{m/n}$

Let a be a real number, and let m and n be positive integers such that m and n share no common factors and $n > 1$. If $\sqrt[n]{a}$ is a real number, then

$$a^{m/n} = (a^{1/n})^m = (\sqrt[n]{a})^m \quad \text{and} \quad a^{m/n} = (a^m)^{1/n} = \sqrt[n]{a^m}$$

The rational exponent in the expression $a^{m/n}$ is essentially performing two operations. The numerator of the exponent raises the base to the mth power. The denominator takes the nth root.

Example 2   Evaluating Expressions of the Form $a^{m/n}$

Convert each expression to radical form and simplify.

a. $8^{2/3}$ **b.** $100^{5/2}$ **c.** $\left(\dfrac{1}{25}\right)^{3/2}$ **d.** $4^{-3/2}$ **e.** $(-81)^{3/4}$

Solution:

a. $8^{2/3} = (\sqrt[3]{8})^2$ Take the cube root of 8 and square the result.

$\phantom{8^{2/3}} = (2)^2$ Simplify.

$\phantom{8^{2/3}} = 4$

b. $100^{5/2} = (\sqrt{100})^5$ Take the square root of 100 and raise the result to the fifth power.

$\phantom{100^{5/2}} = (10)^5$ Simplify.

$\phantom{100^{5/2}} = 100{,}000$

c. $\left(\dfrac{1}{25}\right)^{3/2} = \left(\sqrt{\dfrac{1}{25}}\right)^3$ Take the square root of $\dfrac{1}{25}$ and cube the result.

$\phantom{\left(\dfrac{1}{25}\right)^{3/2}} = \left(\dfrac{1}{5}\right)^3$ Simplify.

$\phantom{\left(\dfrac{1}{25}\right)^{3/2}} = \dfrac{1}{125}$

d. $4^{-3/2} = \left(\dfrac{1}{4}\right)^{3/2} = \dfrac{1}{4^{3/2}}$ Write the expression with positive exponents.

$\phantom{4^{-3/2}} = \dfrac{1}{(\sqrt{4})^3}$ Take the square root of 4 and cube the result.

$\phantom{4^{-3/2}} = \dfrac{1}{2^3}$ Simplify.

$\phantom{4^{-3/2}} = \dfrac{1}{8}$

e. $(-81)^{3/4}$ is not a real number because $\sqrt[4]{-81}$ is not a real number.

Skill Practice Convert each expression to radical form and simplify.

6. $9^{3/2}$ **7.** $8^{5/3}$ **8.** $\left(\dfrac{1}{27}\right)^{4/3}$ **9.** $32^{-4/5}$ **10.** $(-4)^{3/2}$

2. Converting Between Rational Exponents and Radical Notation

Example 3   **Using Radical Notation and Rational Exponents**

Convert each expression to radical notation. Assume all variables represent positive real numbers.

a. $a^{3/5}$ **b.** $5^{1/3}x^{2/3}$ **c.** $3y^{1/4}$ **d.** $9z^{-3/4}$

Solution:

a. $a^{3/5} = \sqrt[5]{a^3}$ or $\left(\sqrt[5]{a}\right)^3$

b. $5^{1/3}x^{2/3} = (5x^2)^{1/3} = \sqrt[3]{5x^2}$

c. $3y^{1/4} = 3\sqrt[4]{y}$ Note that the coefficient 3 is not raised to the $\frac{1}{4}$ power.

d. $9z^{-3/4} = 9 \cdot \dfrac{1}{z^{3/4}} = \dfrac{9}{\sqrt[4]{z^3}}$ Note that the coefficient 9 has an implied exponent of 1, not $-\frac{3}{4}$.

Skill Practice Convert each expression to radical notation. Assume all variables represent positive real numbers.

11. $t^{4/5}$ **12.** $2^{1/4}y^{3/4}$ **13.** $10p^{1/2}$ **14.** $11q^{-2/3}$

Example 4   **Using Radical Notation and Rational Exponents**

Convert each expression to an equivalent expression by using rational exponents. Assume that all variables represent positive real numbers.

a. $\sqrt[4]{b^3}$ **b.** $\sqrt{7a}$ **c.** $7\sqrt{a}$

Solution:

a. $\sqrt[4]{b^3} = b^{3/4}$ **b.** $\sqrt{7a} = (7a)^{1/2}$ **c.** $7\sqrt{a} = 7a^{1/2}$

Skill Practice Convert to an equivalent expression using rational exponents. Assume all variables represent positive real numbers.

15. $\sqrt[3]{x^2}$ **16.** $\sqrt{5y}$ **17.** $5\sqrt{y}$

Answers

6. 27   **7.** 32   **8.** $\dfrac{1}{81}$

9. $\dfrac{1}{16}$ **10.** Not a real number

11. $\sqrt[5]{t^4}$ **12.** $\sqrt[4]{2y^3}$ **13.** $10\sqrt{p}$

14. $\dfrac{11}{\sqrt[3]{q^2}}$ **15.** $x^{2/3}$ **16.** $(5y)^{1/2}$

17. $5y^{1/2}$

3. Properties of Rational Exponents

In Section 4.1, several properties and definitions were introduced to simplify expressions with integer exponents. These properties also apply to rational exponents.

Definitions and Properties of Exponents

Let a and b be nonzero real numbers. Let m and n be rational numbers such that a^m, a^n, and b^m are real numbers.

Description	Property	Example
1. Multiplying like bases	$a^m a^n = a^{m+n}$	$x^{1/3} x^{4/3} = x^{5/3}$
2. Dividing like bases	$\dfrac{a^m}{a^n} = a^{m-n}$	$\dfrac{x^{5/5}}{x^{1/5}} = x^{2/5}$
3. The power rule	$(a^m)^n = a^{mn}$	$(2^{1/3})^{1/2} = 2^{1/6}$
4. Power of a product	$(ab)^m = a^m b^m$	$(9y)^{1/2} = 9^{1/2} y^{1/2} = 3y^{1/2}$
5. Power of a quotient	$\left(\dfrac{a}{b}\right)^m = \dfrac{a^m}{b^m}$	$\left(\dfrac{4}{25}\right)^{1/2} = \dfrac{4^{1/2}}{25^{1/2}} = \dfrac{2}{5}$

Description	Definition	Example
1. Negative exponents	$a^{-m} = \left(\dfrac{1}{a}\right)^m = \dfrac{1}{a^m}$	$(8)^{-1/3} = \left(\dfrac{1}{8}\right)^{1/3} = \dfrac{1}{2}$
2. Zero exponent	$a^0 = 1$	$5^0 = 1$

Example 5   Simplifying Expressions with Rational Exponents

Use the properties of exponents to simplify the expressions. Assume all variables represent positive real numbers.

a. $y^{2/5} y^{3/5}$ **b.** $\dfrac{6a^{-1/2}}{a^{3/2}}$ **c.** $\left(\dfrac{s^{1/2} t^{1/3}}{w^{3/4}}\right)^4$

Solution:

a. $y^{2/5} y^{3/5} = y^{(2/5)+(3/5)}$ Multiply like bases by adding exponents.

$\qquad\qquad = y^{5/5}$ Simplify.

$\qquad\qquad = y$

b. $\dfrac{6a^{-1/2}}{a^{3/2}}$ Divide like bases by subtracting exponents.

$\quad = 6a^{(-1/2)-(3/2)}$ Simplify: $-\dfrac{4}{2} = -2$

$\quad = 6a^{-2}$

$\quad = \dfrac{6}{a^2}$ Simplify the negative exponent.

c. $\left(\dfrac{s^{1/2} t^{1/3}}{w^{3/4}}\right)^4 = \dfrac{s^{(1/2)\cdot 4} t^{(1/3)\cdot 4}}{w^{(3/4)\cdot 4}}$ Apply the power rule. Multiply exponents.

$\qquad\qquad = \dfrac{s^2 t^{4/3}}{w^3}$ Simplify.

Skill Practice Use the properties of exponents to simplify the expressions. Assume all variables represent positive real numbers.

18. $x^{1/2} \cdot x^{3/4}$ **19.** $\dfrac{4k^{-2/3}}{k^{1/3}}$ **20.** $\left(\dfrac{a^{1/3} b^{1/2}}{c^{5/8}}\right)^6$

Answers

18. $x^{5/4}$ **19.** $\dfrac{4}{k}$ **20.** $\dfrac{a^2 b^3}{c^{15/4}}$

4. Applications Involving Rational Exponents

Example 6 Applying Rational Exponents

Suppose P dollars in principal is invested in an account that earns interest annually. If after t years the investment grows to A dollars, then the annual rate of return r on the investment is given by

$$r = \left(\frac{A}{P}\right)^{1/t} - 1$$

Find the annual rate of return on $5000 which grew to $6894.21 after 6 yr.

Solution:

$$r = \left(\frac{A}{P}\right)^{1/t} - 1$$

$$= \left(\frac{6894.21}{5000}\right)^{1/6} - 1 \quad \text{where } A = \$6894.21, P = \$5000, \text{ and } t = 6$$

$$\approx 0.055 \text{ or } 5.5\%$$

The annual rate of return is 5.5%.

Skill Practice

21. The formula for the radius of a sphere is

$$r = \left(\frac{3V}{4\pi}\right)^{1/3}$$

where V is the volume. Find the radius of a sphere whose volume is 113.04 in.^{3} (Use 3.14 for π.)

Answer

21. 3 in.

Section 6.2 Practice Exercises

Study Skills Exercise

Before you do your homework for this section, go back to Section 4.1 and review the properties of exponents. Do several problems from the Section 4.1 exercises. This will help you with the concepts in Section 6.2.

For the exercises in this set, assume that all variables represent positive real numbers unless otherwise stated.

Vocabulary and Key Concepts

1. a. If n is an integer greater than 1, then radical notation for $a^{1/n}$ is _____.

b. If m and n are positive integers that share no common factors and $n > 1$, then in radical notation $a^{m/n} =$ _____.

c. The radical notation for $x^{-1/2}$ is _____.

d. $8^{1/3} =$ _____ and $8^{-1/3} =$ _____.

Review Exercises

2. Given $\sqrt[3]{27}$, identify the index and radicand.

For Exercises 3–6, evaluate the radicals.

3. $\sqrt{25}$ **4.** $\sqrt[3]{8}$ **5.** $\sqrt[4]{81}$ **6.** $(\sqrt[4]{16})^3$

Concept 1: Definition of $a^{1/n}$ and $a^{m/n}$

For Exercises 7–18, convert the expressions to radical form and simplify. **(See Example 1.)**

7. $144^{1/2}$ **8.** $16^{1/4}$ **9.** $-144^{1/2}$ **10.** $-16^{1/4}$

11. $(-144)^{1/2}$ **12.** $(-16)^{1/4}$ **13.** $(-64)^{1/3}$ **14.** $(-32)^{1/5}$

15. $25^{-1/2}$ **16.** $27^{-1/3}$ **17.** $-49^{-1/2}$ **18.** $-64^{-1/2}$

19. Explain how to interpret the expression $a^{m/n}$ as a radical.

20. Explain why $(\sqrt[3]{8})^4$ is easier to evaluate than $\sqrt[3]{8^4}$.

For Exercises 21–24, simplify the expression, if possible. **(See Example 2.)**

21. a. $16^{3/4}$ **22. a.** $81^{3/4}$ **23. a.** $25^{3/2}$ **24. a.** $4^{3/2}$

  b. $-16^{3/4}$ **b.** $-81^{3/4}$ **b.** $-25^{3/2}$ **b.** $-4^{3/2}$

  c. $(-16)^{3/4}$ **c.** $(-81)^{3/4}$ **c.** $(-25)^{3/2}$ **c.** $(-4)^{3/2}$

  d. $16^{-3/4}$ **d.** $81^{-3/4}$ **d.** $25^{-3/2}$ **d.** $4^{-3/2}$

  e. $-16^{-3/4}$ **e.** $-81^{-3/4}$ **e.** $-25^{-3/2}$ **e.** $-4^{-3/2}$

  f. $(-16)^{-3/4}$ **f.** $(-81)^{-3/4}$ **f.** $(-25)^{-3/2}$ **f.** $(-4)^{-3/2}$

For Exercises 25–48, simplify the expression. **(See Example 2.)**

25. $64^{-3/2}$ **26.** $81^{-3/2}$ **27.** $243^{3/5}$ **28.** $1^{5/3}$

29. $-27^{-4/3}$ **30.** $-16^{-5/4}$ **31.** $\left(\dfrac{100}{9}\right)^{-3/2}$ **32.** $\left(\dfrac{49}{100}\right)^{-1/2}$

33. $(-4)^{-3/2}$ **34.** $(-49)^{-3/2}$ **35.** $(-8)^{1/3}$ **36.** $(-9)^{1/2}$

37. $-8^{1/3}$ **38.** $-9^{1/2}$ **39.** $\dfrac{1}{36^{-1/2}}$ **40.** $\dfrac{1}{16^{-1/2}}$

41. $\dfrac{1}{1000^{-1/3}}$ **42.** $\dfrac{1}{81^{-3/4}}$ **43.** $\left(\dfrac{1}{8}\right)^{2/3} - \left(\dfrac{1}{4}\right)^{-1/2}$ **44.** $\left(\dfrac{1}{8}\right)^{-2/3} + \left(\dfrac{1}{4}\right)^{-1/2}$

45. $\left(\dfrac{1}{16}\right)^{-3/4} - \left(\dfrac{1}{49}\right)^{-1/2}$ **46.** $\left(\dfrac{1}{16}\right)^{1/4} - \left(\dfrac{1}{49}\right)^{1/2}$ **47.** $\left(\dfrac{1}{4}\right)^{1/2} + \left(\dfrac{1}{64}\right)^{-1/3}$ **48.** $\left(\dfrac{1}{36}\right)^{1/2} + \left(\dfrac{1}{64}\right)^{-5/6}$

Concept 2: Converting Between Rational Exponents and Radical Notation

For Exercises 49–56, convert each expression to radical notation. **(See Example 3.)**

49. $q^{2/3}$ **50.** $t^{3/5}$ **51.** $6y^{3/4}$ **52.** $8b^{4/9}$

53. $x^{2/3}y^{1/3}$ **54.** $c^{2/5}d^{3/5}$ **55.** $6r^{-2/5}$ **56.** $7x^{-3/4}$

For Exercises 57–64, write each expression by using rational exponents rather than radical notation. **(See Example 4.)**

57. $\sqrt[3]{x}$

58. $\sqrt[4]{a}$

59. $10\sqrt{b}$

60. $-2\sqrt[3]{t}$

61. $\sqrt[3]{y^2}$

62. $\sqrt[6]{z^5}$

63. $\sqrt[4]{a^2b^3}$

64. $\sqrt{abc}$

Concept 3: Properties of Rational Exponents

For Exercises 65–88, simplify the expressions by using the properties of rational exponents. Write the final answer using positive exponents only. **(See Example 5.)**

65. $x^{1/4}x^{-5/4}$

66. $2^{2/3}2^{-5/3}$

67. $\dfrac{p^{5/3}}{p^{2/3}}$

68. $\dfrac{q^{5/4}}{q^{1/4}}$

69. $(y^{1/5})^{10}$

70. $(x^{1/2})^8$

71. $6^{-1/5}6^{3/5}$

72. $a^{-1/3}a^{2/3}$

73. $\dfrac{4t^{-1/3}}{t^{4/3}}$

74. $\dfrac{5s^{-1/3}}{s^{5/3}}$

75. $(a^{1/3}a^{1/4})^{12}$

76. $(x^{2/3}x^{1/2})^6$

77. $(5a^2c^{-1/2}d^{1/2})^2$

78. $(2x^{-1/3}y^2z^{5/3})^3$

79. $\left(\dfrac{x^{-2/3}}{y^{-3/4}}\right)^{12}$

80. $\left(\dfrac{m^{-1/4}}{n^{-1/2}}\right)^{-4}$

81. $\left(\dfrac{16w^{-2}z}{2wz^{-8}}\right)^{1/3}$

82. $\left(\dfrac{50p^{-1}q}{2pq^{-3}}\right)^{1/2}$

83. $(25x^2y^4z^6)^{1/2}$

84. $(8a^6b^3c^9)^{2/3}$

85. $(x^2y^{-1/3})^6(x^{1/2}yz^{2/3})^2$ **86.** $(a^{-1/3}b^{1/2})^4(a^{-1/2}b^{3/5})^{10}$ **87.** $\left(\dfrac{x^{3m}y^{2m}}{z^{5m}}\right)^{1/m}$

88. $\left(\dfrac{a^{4n}b^{3n}}{c^n}\right)^{1/n}$

Concept 4: Applications Involving Rational Exponents

89. If P dollars in principal grows to A dollars after t years with annual interest, then the interest rate is given by $r = \left(\dfrac{A}{P}\right)^{1/t} - 1$. **(See Example 6.)**

  a. In one account, \$10,000 grows to \$16,802 after 5 yr. Compute the interest rate. Round your answer to a tenth of a percent.

  b. In another account \$10,000 grows to \$18,000 after 7 yr. Compute the interest rate. Round your answer to a tenth of a percent.

  c. Which account produced a higher average yearly return?

90. If the area A of a square is known, then the length of its sides, s, can be computed by the formula $s = A^{1/2}$.

  a. Compute the length of the sides of a square having an area of 100 in.^{2}

  b. Compute the length of the sides of a square having an area of 72 in.^{2} Round your answer to the nearest 0.1 in.

91. The radius r of a sphere of volume V is given by $r = \left(\dfrac{3V}{4\pi}\right)^{1/3}$. Find the radius of a sphere having a volume of 85 in.^{3} Round your answer to the nearest 0.1 in.

92. Is $(a + b)^{1/2}$ the same as $a^{1/2} + b^{1/2}$? If not, give a counterexample.

Expanding Your Skills

For Exercises 93–104, write the expression using rational exponents. Then simplify and convert back to radical notation.

Example: $\sqrt[15]{x^{10}}$ $\xrightarrow[\text{exponents}]{\text{Rational}}$ $x^{10/15}$ $\xrightarrow{\text{Simplify}}$ $x^{2/3}$ $\xrightarrow[\text{notation}]{\text{Radical}}$ $\sqrt[3]{x^2}$

93. $\sqrt[6]{y^3}$ 94. $\sqrt[4]{w^2}$ 95. $\sqrt[12]{z^3}$ 96. $\sqrt[18]{t^3}$

97. $\sqrt[9]{x^6}$ 98. $\sqrt[12]{p^9}$ 99. $\sqrt[6]{x^3y^6}$ 100. $\sqrt[8]{m^2p^8}$

101. $\sqrt{16x^8y^6}$ 102. $\sqrt{81a^{12}b^{20}}$ 103. $\sqrt[3]{8x^3y^2z}$ 104. $\sqrt[3]{64m^2n^3p}$

For Exercises 105–108, write the expression as a single radical.

105. $\sqrt{\sqrt[3]{x}}$ 106. $\sqrt[3]{\sqrt{x}}$ 107. $\sqrt[5]{\sqrt[3]{w}}$ 108. $\sqrt[3]{\sqrt[4]{w}}$

For Exercises 109–116, use a calculator to approximate the expressions. Round to four decimal places, if necessary.

109. $9^{1/2}$ 110. $125^{-1/3}$ 111. $50^{-1/4}$ 112. $(172)^{3/5}$

113. $\sqrt[3]{5^2}$ 114. $\sqrt[4]{6^3}$ 115. $\sqrt{10^3}$ 116. $\sqrt[3]{16}$

Simplifying Radical Expressions

1. Multiplication Property of Radicals

You may have already noticed certain properties of radicals involving a product or quotient.

Concepts

1. Multiplication Property of Radicals
2. Simplifying Radicals by Using the Multiplication Property of Radicals
3. Simplifying Radicals by Using the Order of Operations

Multiplication Property of Radicals

Let a and b represent real numbers such that $\sqrt[n]{a}$ and $\sqrt[n]{b}$ are both real. Then

$$\sqrt[n]{ab} = \sqrt[n]{a} \cdot \sqrt[n]{b}$$

The multiplication property of radicals follows from the property of rational exponents.

$$\sqrt[n]{ab} = (ab)^{1/n} = a^{1/n}b^{1/n} = \sqrt[n]{a} \cdot \sqrt[n]{b}$$

The multiplication property of radicals indicates that a product within a radicand can be written as a product of radicals, provided the roots are real numbers. For example:

$$\sqrt{144} = \sqrt{16} \cdot \sqrt{9}$$

The reverse process is also true. A product of radicals can be written as a single radical provided the roots are real numbers and they have the same indices.

$$\sqrt{3} \cdot \sqrt{12} = \sqrt{36}$$

2. Simplifying Radicals by Using the Multiplication Property of Radicals

In algebra, it is customary to simplify radical expressions.

Simplified Form of a Radical

Consider any radical expression where the radicand is written as a product of prime factors. The expression is in *simplified form* if all the following conditions are met:

1. The radicand has no factor raised to a power greater than or equal to the index.
2. The radicand does not contain a fraction.
3. There are no radicals in the denominator of a fraction.

For example, the following radicals are not simplified.

1. The expression $\sqrt[3]{x^5}$ fails condition 1.

2. The expression $\sqrt{\dfrac{1}{4}}$ fails condition 2.

3. The expression $\dfrac{1}{\sqrt[3]{8}}$ fails condition 3.

The expressions $\sqrt{x^2}$, $\sqrt{x^4}$, $\sqrt{x^6}$, and $\sqrt{x^8}$ are not simplified because they fail condition 1 (the exponents are not less than the index). However, each radicand is a perfect square and is easily simplified for $x \geq 0$.

$$\sqrt{x^2} = x$$
$$\sqrt{x^4} = x^2$$
$$\sqrt{x^6} = x^3$$
$$\sqrt{x^8} = x^4$$

However, how is an expression such as $\sqrt{x^9}$ simplified? This and many other radical expressions are simplified by using the multiplication property of radicals. We demonstrate the process in Examples 1–4.

Example 1   **Using the Multiplication Property to Simplify a Radical Expression**

Simplify the expression assuming that $x \geq 0$. $\sqrt{x^9}$

Solution:

The expression $\sqrt{x^9}$ is equivalent to $\sqrt{x^8 \cdot x}$.

$$\sqrt{x^9} = \sqrt{x^8 \cdot x}$$
$$= \sqrt{x^8} \cdot \sqrt{x} \qquad \text{Apply the multiplication property of radicals.}$$
$$\qquad\qquad\qquad \text{Note that } x^8 \text{ is a perfect square because } x^8 = (x^4)^2.$$
$$= x^4\sqrt{x} \qquad \text{Simplify.}$$

Skill Practice Simplify the expression. Assume $a \geq 0$.

1. $\sqrt{a^{11}}$

Answer

1. $a^5\sqrt{a}$

In Example 1, the expression x^9 is not a perfect square. Therefore, to simplify $\sqrt{x^9}$, it was necessary to write the expression as the product of the largest perfect square and a remaining or "left-over" factor: $\sqrt{x^9} = \sqrt{x^8 \cdot x}$. This process also applies to simplifying nth roots, as shown in Example 2.

Example 2   **Using the Multiplication Property to Simplify Radical Expressions**

Simplify each expression. Assume all variables represent positive real numbers.

a. $\sqrt[4]{b^7}$ **b.** $\sqrt[3]{w^7 z^9}$

Solution:

The goal is to rewrite each radicand as the product of the greatest perfect square (perfect cube, perfect fourth power, and so on) and a leftover factor.

a. $\sqrt[4]{b^7} = \sqrt[4]{b^4 \cdot b^3}$ b^4 is the greatest perfect fourth power in the radicand.

$\phantom{\sqrt[4]{b^7}} = \sqrt[4]{b^4} \cdot \sqrt[4]{b^3}$ Apply the multiplication property of radicals.

$\phantom{\sqrt[4]{b^7}} = b\sqrt[4]{b^3}$ Simplify.

b. $\sqrt[3]{w^7 z^9} = \sqrt[3]{(w^6 z^9) \cdot (w)}$ $w^6 z^9$ is the greatest perfect cube in the radicand.

$\phantom{\sqrt[3]{w^7 z^9}} = \sqrt[3]{w^6 z^9} \cdot \sqrt[3]{w}$ Apply the multiplication property of radicals.

$\phantom{\sqrt[3]{w^7 z^9}} = w^2 z^3 \sqrt[3]{w}$ Simplify.

Skill Practice Simplify the expressions. Assume all variables represent positive real numbers.

2. $\sqrt[4]{v^{25}}$ **3.** $\sqrt[3]{p^{17} q^{10}}$

Each expression in Example 2 involves a radicand that is a product of variable factors. If a numerical factor is present, sometimes it is necessary to factor the coefficient before simplifying the radical.

Example 3   **Using the Multiplication Property to Simplify a Radical**

Simplify the expression. $\sqrt{56}$

Solution:

$\sqrt{56} = \sqrt{2^3 \cdot 7}$ Factor the radicand.

$\phantom{\sqrt{56}} = \sqrt{(2^2) \cdot (2 \cdot 7)}$ 2^2 is the greatest perfect square in the radicand.

$\phantom{\sqrt{56}} = \sqrt{2^2} \cdot \sqrt{2 \cdot 7}$ Apply the multiplication property of radicals.

$\phantom{\sqrt{56}} = 2\sqrt{14}$ Simplify.

$$
\begin{array}{r|l}
2 & 56 \\
2 & 28 \\
2 & 14 \\
  & 7
\end{array}
$$

> **TIP:** It may be easier to visualize the greatest perfect square factor within the radicand as follows:
>
> $\sqrt{56} = \sqrt{4 \cdot 14}$
> $\phantom{\sqrt{56}} = \sqrt{4} \cdot \sqrt{14}$
> $\phantom{\sqrt{56}} = 2\sqrt{14}$

Skill Practice Simplify.

4. $\sqrt{24}$

Answers

2. $v^6 \sqrt[4]{v}$ **3.** $p^5 q^3 \sqrt[3]{p^2 q}$ **4.** $2\sqrt{6}$

Topic: Verifying a Simplified Radical

A calculator can be used to support the solution to Example 3. The decimal approximation for $\sqrt{56}$ and $2\sqrt{14}$ agree for the first 10 digits. This in itself does not make $\sqrt{56} = 2\sqrt{14}$. It is the multiplication property of radicals that guarantees that the expressions are equal.

```
√(56)
         7.483314774
2*√(14)
         7.483314774
```

Example 4 Using the Multiplication Property to Simplify Radicals

Simplify. $6\sqrt{50}$

TIP: The radical can also be simplified as:

$6\sqrt{50} = 6\sqrt{25 \cdot 2}$
$= 6\sqrt{25} \cdot \sqrt{2}$
$= 6 \cdot 5\sqrt{2}$
$= 30\sqrt{2}$

Solution:

$6\sqrt{50} = 6\sqrt{5^2 \cdot 2}$ — Factor the radicand.
$= 6 \cdot \sqrt{5^2} \cdot \sqrt{2}$ — Apply the multiplication property of radicals.
$= 6 \cdot 5 \cdot \sqrt{2}$ — Simplify.
$= 30\sqrt{2}$ — Simplify.

Skill Practice Simplify.

5. $5\sqrt{18}$

Example 5 Using the Multiplication Property to Simplify Radicals

Simplify the expression. Assume that x, y, and z represent positive real numbers.

$\sqrt[3]{40x^3y^5z^7}$

TIP: In Example 5, the numerical coefficient within the radicand can be written $8 \cdot 5$ because 8 is the greatest perfect cube factor of 40:

$\sqrt[3]{40x^3y^5z^7} = \sqrt[3]{8 \cdot 5x^3y^5z^7}$
$= \sqrt[3]{(8x^3y^3z^6)(5y^2z)}$
$= \sqrt[3]{8x^3y^3z^6} \cdot \sqrt[3]{5y^2z}$
$= 2xyz^2 \cdot \sqrt[3]{5y^2z}$

Solution:

$\sqrt[3]{40x^3y^5z^7}$
$= \sqrt[3]{2^3 5x^3y^5z^7}$ — Factor the radicand.
$= \sqrt[3]{(2^3x^3y^3z^6) \cdot (5y^2z)}$ — $2^3x^3y^3z^6$ is the greatest perfect cube.
$= \sqrt[3]{2^3x^3y^3z^6} \cdot \sqrt[3]{5y^2z}$ — Apply the multiplication property of radicals.
$= 2xyz^2\sqrt[3]{5y^2z}$ — Simplify.

$\begin{array}{r} 2|40 \\ 2|20 \\ 2|10 \\ 5 \end{array}$

Skill Practice Simplify. Assume that $a > 0$ and $b > 0$.

6. $\sqrt[4]{32a^{10}b^{19}}$

3. Simplifying Radicals by Using the Order of Operations

Often a radical can be simplified by applying the order of operations. In Example 6, the first step will be to simplify the expression within the radicand.

Answers
5. $15\sqrt{2}$ **6.** $2a^2b^4\sqrt[4]{2a^2b^3}$

Example 6 **Using the Order of Operations to Simplify Radicals**

Use the order of operations to simplify the expressions. Assume $a > 0$.

a. $\sqrt{\dfrac{a^7}{a^3}}$ **b.** $\sqrt[3]{\dfrac{3}{81}}$

Solution:

a. $\sqrt{\dfrac{a^7}{a^3}}$ The radicand contains a fraction. However, the fraction can be reduced to lowest terms.

$= \sqrt{a^4}$

$= a^2$ Simplify the radical.

b. $\sqrt[3]{\dfrac{3}{81}}$ The radical contains a fraction that can be simplified.

$= \sqrt[3]{\dfrac{1}{27}}$ Reduce to lowest terms.

$= \dfrac{1}{3}$ Simplify.

Skill Practice Use the order of operations to simplify the expressions. Assume $v > 0$.

7. $\sqrt{\dfrac{v^{21}}{v^5}}$ **8.** $\sqrt[5]{\dfrac{64}{2}}$

Example 7 **Simplifying a Radical Expression**

Simplify. $\dfrac{7\sqrt{50}}{10}$

Solution:

$\dfrac{7\sqrt{50}}{10} = \dfrac{7\sqrt{25 \cdot 2}}{10}$ 25 is the greatest perfect square in the radicand.

$= \dfrac{7 \cdot 5\sqrt{2}}{10}$ Simplify the radical.

$= \dfrac{7 \cdot \overset{1}{5}\sqrt{2}}{\underset{2}{10}}$ Simplify the fraction to lowest terms.

$= \dfrac{7\sqrt{2}}{2}$

Avoiding Mistakes

The expression $\frac{7\sqrt{2}}{2}$ cannot be simplified further because one factor of 2 is in the radicand and the other is outside the radical.

Skill Practice Simplify.

9. $\dfrac{2\sqrt{300}}{30}$

Answers

7. v^8 **8.** 2   **9.** $\dfrac{2\sqrt{3}}{3}$

Section 6.3 Practice Exercises

Study Skills Exercise

The final exam is just around the corner. Your old tests and quizzes provide good material to study for the final exam. Use your old tests to make a list of the concepts on which you need to concentrate. Ask your professor for help if there are still concepts that you do not understand.

For the exercises in this set, assume that all variables represent positive real numbers unless otherwise stated.

Vocabulary and Key Concepts

1. **a.** The multiplication property of radicals indicates that if $\sqrt[n]{a}$ and $\sqrt[n]{b}$ are real numbers, then
   $\sqrt[n]{ab} = $ _____ · _____ .

   b. Explain why the following radical is not in simplified form. $\sqrt{x^3}$

   c. The radical expression $\sqrt[3]{x^{10}}$ (is/is not) in simplified form.

   d. The radical expression $\sqrt{18}$ simplifies to __$\sqrt{2}$.

   e. To simplify the radical expression $\sqrt[3]{t^{14}}$ the radicand is rewritten as $\sqrt[3]{\underline{} \cdot t^2}$.

   f. On a calculator, $\sqrt{2}$ is given as 1.414213562. Is this decimal number the exact value of $\sqrt{2}$?

Review Exercises

For Exercises 2–4, simplify the expression. Write the answer with positive exponents only.

2. $(a^2 b^{-4})^{1/2} \left(\dfrac{a}{b^{-3}} \right)$

3. $\left(\dfrac{p^4}{q^{-6}} \right)^{-1/2} (p^3 q^{-2})$

4. $(x^{1/3} y^{5/6})^{-6}$

5. Write $x^{4/7}$ in radical notation.

6. Write $y^{2/5}$ in radical notation.

7. Write $\sqrt{y^9}$ by using rational exponents.

8. Write $\sqrt[3]{x^2}$ by using rational exponents.

Concept 2: Simplifying Radicals by Using the Multiplication Property of Radicals

For Exercises 9–32, simplify the radicals. **(See Examples 1–5.)**

9. $\sqrt{x^5}$

10. $\sqrt{p^{15}}$

11. $\sqrt[3]{q^7}$

12. $\sqrt[3]{r^{17}}$

13. $\sqrt{a^5 b^4}$

14. $\sqrt{c^9 d^6}$

15. $-\sqrt[4]{x^8 y^{13}}$

16. $-\sqrt[4]{p^{16} q^{17}}$

17. $\sqrt{28}$

18. $\sqrt{63}$

19. $\sqrt{20}$

20. $\sqrt{50}$

21. $5\sqrt{18}$

22. $2\sqrt{24}$

23. $\sqrt[3]{54}$

24. $\sqrt[3]{250}$

25. $\sqrt{25ab^3}$

26. $\sqrt{64m^5 n^{20}}$

27. $\sqrt[3]{40x^7}$

28. $\sqrt[3]{81y^{17}}$

29. $\sqrt[3]{-16x^6 yz^3}$

30. $\sqrt[3]{-192a^6 bc^2}$

31. $\sqrt[4]{80w^4 z^7}$

32. $\sqrt[4]{32p^8 qr^5}$

Concept 3: Simplifying Radicals by Using the Order of Operations

For Exercises 33–44, simplify the radical expressions. **(See Examples 6–7.)**

33. $\sqrt{\dfrac{x^3}{x}}$

34. $\sqrt{\dfrac{y^5}{y}}$

35. $\sqrt{\dfrac{p^7}{p^3}}$

36. $\sqrt{\dfrac{q^{11}}{q^5}}$

37. $\sqrt{\dfrac{50}{2}}$

38. $\sqrt{\dfrac{98}{2}}$

39. $\sqrt[3]{\dfrac{3}{24}}$

40. $\sqrt[3]{\dfrac{2}{250}}$

41. $\dfrac{5\sqrt[3]{16}}{6}$

42. $\dfrac{7\sqrt{18}}{9}$

43. $\dfrac{5\sqrt[3]{72}}{12}$

44. $\dfrac{3\sqrt[3]{250}}{10}$

Mixed Exercises

For Exercises 45–72, simplify the radical expressions.

45. $\sqrt{80}$

46. $\sqrt{108}$

47. $-6\sqrt{75}$

48. $-8\sqrt{8}$

49. $\sqrt{25x^4y^3}$

50. $\sqrt{125p^3q^2}$

51. $\sqrt[3]{27x^2y^3z^4}$

52. $\sqrt[3]{108a^3bc^2}$

53. $\sqrt{\dfrac{12w^5}{3w}}$

54. $\sqrt{\dfrac{64x^9}{4x^3}}$

55. $\sqrt{\dfrac{3y^3}{300y^{15}}}$

56. $\sqrt{\dfrac{4h}{100h^5}}$

57. $\sqrt[3]{\dfrac{16a^2b}{2a^2b^4}}$

58. $\sqrt[3]{\dfrac{-27a^4}{8a}}$

59. $\sqrt{2^3a^{14}b^8c^{31}d^{22}}$

60. $\sqrt{7^5u^{12}v^{20}w^{65}x^{80}}$

61. $\sqrt[3]{54a^6b^4}$

62. $\sqrt[3]{72m^5n^3}$

63. $-5a\sqrt{12a^2b^4c}$

64. $-7y\sqrt{75xy^5z^6}$

65. $\sqrt[4]{7x^5y}$

66. $\sqrt[4]{10cd^7}$

67. $\sqrt{54a^4b^2}$

68. $\sqrt{48r^6s^2}$

69. $\dfrac{2\sqrt{27}}{3}$

70. $\dfrac{7\sqrt{24}}{2}$

71. $\dfrac{3\sqrt{125}}{20}$

72. $\dfrac{10\sqrt{63}}{12}$

For Exercises 73–76, write a mathematical expression for the English phrase and simplify.

73. The quotient of 1 and the cube root of w^6

74. The principal square root of the quotient of h^2 and 49

75. The principal square root of the quantity k raised to the third power

76. The cube root of $2x^4$

For Exercises 77–80, determine the length of the third side of the right triangle. Write the answer as a simplified radical.

77.
10 ft, ?, 8 ft

78.
?, 4 in., 12 in.

79.
?, 18 m, 12 m

80.
3 cm, 7 cm, ?

81. On a baseball diamond, the bases are 90 ft apart. Find the exact distance from home plate to second base. Then round to the nearest tenth of a foot.

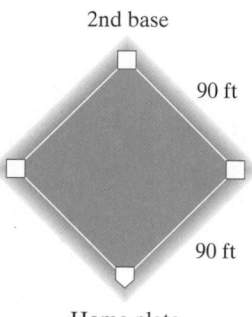

2nd base

90 ft

90 ft

Home plate

82. Linda is at the beach flying a kite. The kite is directly over a sand castle 60 ft away from Linda. If 100 ft of kite string is out (ignoring any sag in the string), how high is the kite? (Assume that Linda is 5 ft tall.) See figure.

100 ft

60 ft

5 ft

Expanding Your Skills

83. Tom has to travel from town *A* to town *C* across a small mountain range. He can travel one of two routes. He can travel on a four-lane highway from *A* to *B* and then from *B* to *C* at an average speed of 55 mph. Or he can travel on a two-lane road directly from town *A* to town *C*, but his average speed will be only 35 mph. If Tom is in a hurry, which route will take him to town *C* faster?

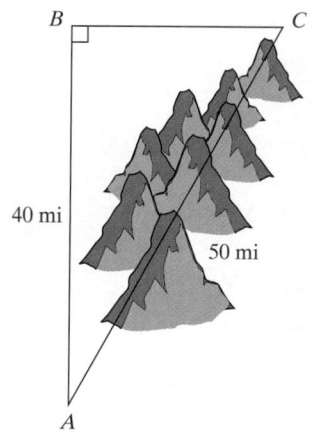

B

C

40 mi

50 mi

A

84. One side of a rectangular pasture is 80 ft in length. The diagonal distance is 110 yd. If fencing costs $3.29 per foot, how much will it cost to fence the pasture?

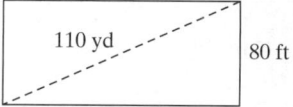

110 yd

80 ft

Section 6.4 Addition and Subtraction of Radicals

Concept

1. Addition and Subtraction of Radicals

1. Addition and Subtraction of Radicals

Definition of *Like* Radicals

Two radical terms are called *like* radicals if they have the same index and same radicand.

Like radicals can be added or subtracted by using the distributive property.

Same index

Distributive property

$$3\sqrt{x} + 7\sqrt{x} = (3 + 7)\sqrt{x} = 10\sqrt{x}$$

Same radicand

Example 1 Adding and Subtracting Radicals

Add or subtract as indicated.

a. $6\sqrt{11} - 2\sqrt{11}$ **b.** $\sqrt{3} + \sqrt{3}$

Solution:

a. $6\sqrt{11} - 2\sqrt{11}$

$\quad = (6 - 2)\sqrt{11}$ Apply the distributive property.

$\quad = 4\sqrt{11}$ Simplify.

b. $\sqrt{3} + \sqrt{3}$

$\quad = 1\sqrt{3} + 1\sqrt{3}$ Note that $\sqrt{3} = 1\sqrt{3}$.

$\quad = (1 + 1)\sqrt{3}$ Apply the distributive property.

$\quad = 2\sqrt{3}$ Simplify.

Avoiding Mistakes

The process of adding *like* radicals with the distributive property is similar to adding *like* terms. The end result is that the numerical coefficients are added and the radical factor is unchanged.

$$\sqrt{3} + \sqrt{3} = 1\sqrt{3} + 1\sqrt{3} = 2\sqrt{3}$$

Be careful: $\sqrt{3} + \sqrt{3} \neq \sqrt{6}$

In general: $\sqrt{x} + \sqrt{y} \neq \sqrt{x + y}$

Skill Practice Add or subtract as indicated.

1. $5\sqrt{6} - 8\sqrt{6}$ **2.** $\sqrt{10} + \sqrt{10}$

Example 2 Adding and Subtracting Radicals

Add or subtract as indicated.

a. $-2\sqrt[3]{ab} + 7\sqrt[3]{ab} - \sqrt[3]{ab}$ **b.** $\frac{1}{4}x\sqrt{3y} - \frac{3}{2}x\sqrt{3y}$

Solution:

a. $-2\sqrt[3]{ab} + 7\sqrt[3]{ab} - \sqrt[3]{ab}$

$\quad = (-2 + 7 - 1)\sqrt[3]{ab}$ Apply the distributive property.

$\quad = 4\sqrt[3]{ab}$ Simplify.

b. $\frac{1}{4}x\sqrt{3y} - \frac{3}{2}x\sqrt{3y}$

$\quad = \left(\frac{1}{4} - \frac{3}{2}\right)x\sqrt{3y}$ Apply the distributive property.

$\quad = \left(\frac{1}{4} - \frac{6}{4}\right)x\sqrt{3y}$ Get a common denominator.

$\quad = -\frac{5}{4}x\sqrt{3y}$ Simplify.

Answers

1. $-3\sqrt{6}$ **2.** $2\sqrt{10}$

3. $5\sqrt[3]{xy} - 3\sqrt[3]{xy} + 7\sqrt[3]{xy}$

4. $\dfrac{5}{6}y\sqrt{2} + \dfrac{1}{4}y\sqrt{2}$

Example 3 shows that it is often necessary to simplify radicals before adding or subtracting.

Example 3 Adding and Subtracting Radicals

Simplify the radicals and add or subtract as indicated. Assume all variables represent positive real numbers.

a. $3\sqrt{8} + \sqrt{2}$ **b.** $8\sqrt{x^3y^2} - 3y\sqrt{x^3}$

c. $\sqrt{50x^2y^5} - 13y\sqrt{2x^2y^3} + xy\sqrt{98y^3}$

Solution:

a. $3\sqrt{8} + \sqrt{2}$ The radicands are different. Try simplifying the radicals first.

$= 3 \cdot 2\sqrt{2} + \sqrt{2}$ Simplify: $\sqrt{8} = \sqrt{2^3} = 2\sqrt{2}$

$= 6\sqrt{2} + \sqrt{2}$

$= (6 + 1)\sqrt{2}$ Apply the distributive property.

$= 7\sqrt{2}$ Simplify.

b. $8\sqrt{x^3y^2} - 3y\sqrt{x^3}$ The radicands are different. Simplify the radicals first.

$= 8xy\sqrt{x} - 3xy\sqrt{x}$ Simplify $\sqrt{x^3y^2} = xy\sqrt{x}$ and $\sqrt{x^3} = x\sqrt{x}$.

$= (8 - 3)xy\sqrt{x}$ Apply the distributive property.

$= 5xy\sqrt{x}$ Simplify.

c. $\sqrt{50x^2y^5} - 13y\sqrt{2x^2y^3} + xy\sqrt{98y^3}$ Simplify each radical.

$$\begin{cases} \sqrt{50x^2y^5} = \sqrt{25 \cdot 2x^2y^5} \\ \qquad = 5xy^2\sqrt{2y} \\ -13y\sqrt{2x^2y^3} = -13xy^2\sqrt{2y} \\ xy\sqrt{98y^3} = xy\sqrt{49 \cdot 2y^3} \\ \qquad = 7xy^2\sqrt{2y} \end{cases}$$

$= 5xy^2\sqrt{2y} - 13xy^2\sqrt{2y} + 7xy^2\sqrt{2y}$

$= (5 - 13 + 7)xy^2\sqrt{2y}$ Apply the distributive property.

$= -xy^2\sqrt{2y}$

5. $\sqrt{75} + 2\sqrt{3}$

6. $4\sqrt{a^2b} - 6a\sqrt{b}$

7. $-3\sqrt{2y^3} + 5y\sqrt{18y} - 2\sqrt{50y^3}$

Answers

3. $9\sqrt[3]{xy}$ **4.** $\dfrac{13}{12}y\sqrt{2}$

5. $7\sqrt{3}$ **6.** $-2a\sqrt{b}$

7. $2y\sqrt{2y}$

In some cases, when two radicals are added, the resulting sum is written in factored form. This is demonstrated in Example 4.

Example 4 — Adding Radical Expressions

Add the radicals. Assume that $x \geq 0$. $3\sqrt{2x^2} + \sqrt{8}$

Solution:

$3\sqrt{2x^2} + \sqrt{8}$

$\quad = 3x\sqrt{2} + 2\sqrt{2}$ Simplify each radical. Notice that the radicands are the same, but the terms are not *like* terms. The first term has a factor of x and the second does not.

$\quad = (3x + 2)\sqrt{2}$ Apply the distributive property. The expression cannot be simplified further because $3x$ and 2 are not *like* terms.

Skill Practice Add the radicals. Assume that $y \geq 0$.

8. $4\sqrt{45} - \sqrt{5y^4}$

Answer

8. $(12 - y^2)\sqrt{5}$

Section 6.4 Practice Exercises

For the exercises in this set, assume that all variables represent positive real numbers, unless otherwise stated.

Vocabulary and Key Concepts

1. **a.** Two radical terms are called *like* radicals if they have the same _____ and the same _____.

   b. The expression $\sqrt{3x} + \sqrt{3x}$ simplifies to _____.

   c. The expression $\sqrt{2} + \sqrt{3}$ (can/cannot) be simplified further whereas the expression $\sqrt{2} \cdot \sqrt{3}$ (can/cannot) be simplified further.

   d. The expression $\sqrt{2} + \sqrt{18}$ simplifies to _____.

Review Exercises

For Exercises 2–5, simplify the radicals.

2. $\sqrt[3]{-16s^4t^9}$

3. $-\sqrt[4]{x^7y^4}$

4. $\sqrt{36a^2b^3}$

5. $\sqrt[3]{\dfrac{7b^8}{56b^2}}$

6. Write $(4x^2)^{1/3}$ in radical notation.

7. Write $\sqrt[4]{x^3y}$ using rational exponents.

8. Simplify. $32^{-1/5}$

For Exercises 9–10, simplify the expressions. Write the answer with positive exponents only.

9. $y^{2/3}y^{1/4}$

10. $(x^{1/2}y^{-3/4})^{-4}$

Concept 1: Addition and Subtraction of Radicals

For Exercises 11–12, determine if the radical terms are *like*.

11. a. $\sqrt{2}$ and $\sqrt[3]{2}$

b. $\sqrt{2}$ and $3\sqrt{2}$

c. $\sqrt{2}$ and $\sqrt{5}$

12. a. $7\sqrt[3]{x}$ and $\sqrt[3]{x}$

b. $\sqrt[3]{x}$ and $\sqrt[4]{x}$

c. $2\sqrt[4]{x}$ and $x\sqrt[4]{2}$

13. Explain the similarities between the pairs of expressions.

a. $7\sqrt{5} + 4\sqrt{5}$ and $7x + 4x$

b. $-2\sqrt{6} - 9\sqrt{3}$ and $-2x - 9y$

14. Explain the similarities between the pairs of expressions.

a. $-4\sqrt{3} + 5\sqrt{3}$ and $-4z + 5z$

b. $13\sqrt{7} - 18$ and $13a - 18$

For Exercises 15–32, add or subtract the radical expressions, if possible. **(See Examples 1–2.)**

15. $3\sqrt{5} + 6\sqrt{5}$

16. $5\sqrt{a} + 3\sqrt{a}$

17. $3\sqrt[3]{tw} - 2\sqrt[3]{tw} + \sqrt[3]{tw}$

18. $6\sqrt[3]{7} - 2\sqrt[3]{7} + \sqrt[3]{7}$

19. $6\sqrt{10} - \sqrt{10}$

20. $13\sqrt{11} - \sqrt{11}$

21. $\sqrt[4]{3} + 7\sqrt[4]{3} - \sqrt[4]{14}$

22. $2\sqrt{11} + 3\sqrt{13} + 5\sqrt{11}$

23. $8\sqrt{x} + 2\sqrt{y} - 6\sqrt{x}$

24. $10\sqrt{10} - 8\sqrt{10} + \sqrt{2}$

25. $\sqrt[3]{ab} + a\sqrt[3]{b}$

26. $x\sqrt[4]{y} - y\sqrt[4]{x}$

27. $\sqrt{2t} + \sqrt[3]{2t}$

28. $\sqrt[4]{5c} + \sqrt[3]{5c}$

29. $\dfrac{5}{6}z\sqrt[3]{6} + \dfrac{7}{9}z\sqrt[3]{6}$

30. $\dfrac{3}{4}a\sqrt[4]{b} + \dfrac{1}{6}a\sqrt[4]{b}$

31. $0.81x\sqrt{y} - 0.11x\sqrt{y}$

32. $7.5\sqrt{pq} - 6.3\sqrt{pq}$

33. Explain the process for adding the two radicals. Then find the sum. $3\sqrt{2} + 7\sqrt{50}$

34. Explain the process for subtracting two radicals. Then find the difference. $\sqrt{12x} - \sqrt{75x}$

For Exercises 35–64, add or subtract the radical expressions as indicated. **(See Examples 3–4.)**

35. $\sqrt{36} + \sqrt{81}$

36. $3\sqrt{80} - 5\sqrt{45}$

37. $2\sqrt{12} + \sqrt{48}$

38. $5\sqrt{32} + 2\sqrt{50}$

39. $4\sqrt{7} + \sqrt{63} - 2\sqrt{28}$

40. $8\sqrt{3} - 2\sqrt{27} + \sqrt{75}$

41. $5\sqrt{18} + \sqrt{27} - 4\sqrt{50}$

42. $7\sqrt{40} - \sqrt{8} + 4\sqrt{50}$

43. $\sqrt[3]{81} - \sqrt[3]{24}$

44. $17\sqrt[3]{81} - 2\sqrt[3]{24}$

45. $3\sqrt{2a} - \sqrt{8a} - \sqrt{72a}$

46. $\sqrt{12t} - \sqrt{27t} + 5\sqrt{3t}$

47. $2s^2\sqrt[3]{s^2t^6} + 3t^2\sqrt[3]{8s^8}$

48. $4\sqrt[3]{x^4} - 2x\sqrt[3]{x}$

49. $7\sqrt[3]{x^4} - x\sqrt[3]{x}$

50. $6\sqrt[3]{y^{10}} - 3y^2\sqrt[3]{y^4}$

51. $5p\sqrt{20p^2} + p^2\sqrt{80}$

52. $2q\sqrt{48q^2} - \sqrt{27q^4}$

53. $\sqrt[3]{a^2b} - \sqrt[3]{8a^2b}$

54. $w\sqrt{80} - 3\sqrt{125w^2}$

55. $5x\sqrt{x} + 6\sqrt{x}$

56. $9y^2\sqrt{2} + 4\sqrt{2}$

57. $\sqrt{50x^2} - 3\sqrt{8}$

58. $\sqrt{9x^3} - \sqrt{25x}$

59. $11\sqrt[3]{54cd^3} - 2\sqrt[3]{2cd^3} + d\sqrt[3]{16c}$

60. $x\sqrt[3]{64x^5y^2} - x^2\sqrt[3]{x^2y^2} + 5\sqrt[3]{x^8y^2}$

61. $\frac{3}{2}ab\sqrt{24a^3} + \frac{4}{3}\sqrt{54a^5b^2} - a^2b\sqrt{150a}$

62. $mn\sqrt{72n} + \frac{2}{3}n\sqrt{8m^2n} - \frac{5}{6}\sqrt{50m^2n^3}$

63. $x\sqrt[3]{16} - 2\sqrt[3]{27x} + \sqrt[3]{54x^3}$

64. $5\sqrt[4]{y^5} - 2y\sqrt[4]{y} + \sqrt[4]{16y^7}$

Mixed Exercises

For Exercises 65–72, answer true or false. If an answer is false, explain why or give a counterexample.

65. $\sqrt{x} + \sqrt{y} = \sqrt{x + y}$

66. $\sqrt{x} + \sqrt{x} = 2\sqrt{x}$

67. $5\sqrt[3]{x} + 2\sqrt[3]{x} = 7\sqrt[3]{x}$

68. $6\sqrt{x} + 5\sqrt[3]{x} = 11\sqrt{x}$

69. $\sqrt{y} + \sqrt{y} = \sqrt{2y}$

70. $\sqrt{c^2 + d^2} = c + d$

71. $2w\sqrt{5} + 4w\sqrt{5} = 6w^2\sqrt{5}$

72. $7x\sqrt{3} - 2\sqrt{3} = (7x - 2)\sqrt{3}$

For Exercises 73–76, write the English phrase as an algebraic expression. Simplify each expression.

73. The sum of the principal square root of 48 and the principal square root of 12

74. The sum of the cube root of 16 and the cube root of 2

75. The difference of 5 times the cube root of x^6 and the square of x

76. The sum of the cube of y and the principal fourth root of y^{12}

For Exercises 77–80, write an English phrase that translates the mathematical expression. (Answers may vary.)

77. $\sqrt{18} - 5^2$

78. $4^3 - \sqrt[3]{4}$

79. $\sqrt[4]{x} + y^3$

80. $a^4 - \sqrt{a}$

For Exercises 81–82, find the exact value of the perimeter, and then approximate the value to one decimal place.

81.

82.

83. The figure has perimeter $14\sqrt{2}$ ft. Find the value of x.

84. The figure has perimeter $12\sqrt{7}$ cm. Find the value of x.

Expanding Your Skills

85. a. An irregularly shaped garden is shown in the figure. All distances are expressed in yards. Find the perimeter. (*Hint*: Use the Pythagorean theorem to find the length of each side.) Write the final answer in radical form.

b. Approximate your answer to two decimal places.

c. If edging costs $1.49 per foot and sales tax is 6%, find the total cost of edging the garden.

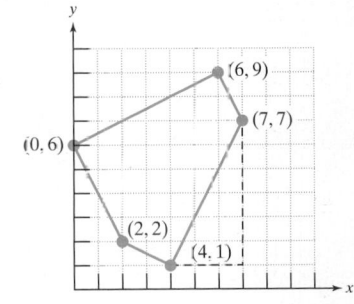

Concepts

1. **Multiplication Property of Radicals**
2. **Expressions of the Form $(\sqrt[n]{a})^n$**
3. **Special Case Products**
4. **Multiplying Radicals with Different Indices**

Section 6.5 Multiplication of Radicals

1. Multiplication Property of Radicals

In this section, we will learn how to multiply radicals by using the multiplication property of radicals first introduced in Section 6.3.

> **The Multiplication Property of Radicals**
>
> Let a and b represent real numbers such that $\sqrt[n]{a}$ and $\sqrt[n]{b}$ are both real. Then
> $$\sqrt[n]{ab} = \sqrt[n]{a} \cdot \sqrt[n]{b}$$

To multiply two radical expressions, we use the multiplication property of radicals along with the commutative and associative properties of multiplication.

Example 1 Multiplying Radical Expressions

Multiply each expression and simplify the result. Assume all variables represent positive real numbers.

a. $(3\sqrt{2})(5\sqrt{6})$ **b.** $(2x\sqrt{y})(-7\sqrt{xy})$ **c.** $(15c\sqrt[3]{cd})\left(\frac{1}{3}\sqrt[3]{cd^2}\right)$

Solution:

a. $(3\sqrt{2})(5\sqrt{6})$

$\quad = (3 \cdot 5)(\sqrt{2} \cdot \sqrt{6})$ Commutative and associative properties of multiplication

$\quad = 15\sqrt{12}$ Multiplication property of radicals

$\quad = 15\sqrt{2^2 3}$

$\quad = 15 \cdot 2\sqrt{3}$ Simplify the radical.

$\quad = 30\sqrt{3}$

b. $(2x\sqrt{y})(-7\sqrt{xy})$

$\quad = (2x)(-7)(\sqrt{y} \cdot \sqrt{xy})$ Commutative and associative properties of multiplication

$\quad = -14x\sqrt{xy^2}$ Multiplication property of radicals

$\quad = -14xy\sqrt{x}$ Simplify the radical.

c. $(15c\sqrt[3]{cd})\left(\frac{1}{3}\sqrt[3]{cd^2}\right)$

$\quad = \left(15c \cdot \frac{1}{3}\right)(\sqrt[3]{cd} \cdot \sqrt[3]{cd^2})$ Commutative and associative properties of multiplication

$\quad = 5c\sqrt[3]{c^2 d^3}$ Multiplication property of radicals

$\quad = 5cd\sqrt[3]{c^2}$ Simplify the radical.

Skill Practice Multiply the expressions and simplify the results. Assume all variables represent positive real numbers.

1. $(4\sqrt{6})(-2\sqrt{3})$
2. $(3ab\sqrt{b})(-2\sqrt{ab})$
3. $(2\sqrt[3]{4ab})(5\sqrt[3]{2a^2b})$

When multiplying radical expressions with more than one term, we use the distributive property.

Example 2 Multiplying Radical Expressions

Multiply. $3\sqrt{11}(2 + \sqrt{11})$

Solution:

$3\sqrt{11}(2 + \sqrt{11})$

$= 3\sqrt{11} \cdot (2) + 3\sqrt{11} \cdot \sqrt{11}$ Apply the distributive property.

$= 6\sqrt{11} + 3\sqrt{11^2}$ Multiplication property of radicals

$= 6\sqrt{11} + 3 \cdot 11$ Simplify the radical.

$= 6\sqrt{11} + 33$

Skill Practice Multiply.

4. $5\sqrt{5}(2\sqrt{5} - 2)$

Example 3 Multiplying Radical Expressions

Multiply.

a. $(\sqrt{5} + 3\sqrt{2})(2\sqrt{5} - \sqrt{2})$ **b.** $(-10a\sqrt{b} + 7b)(a\sqrt{b} + 2b)$

Solution:

a. $(\sqrt{5} + 3\sqrt{2})(2\sqrt{5} - \sqrt{2})$

$= 2\sqrt{5^2} - \sqrt{10} + 6\sqrt{10} - 3\sqrt{2^2}$ Apply the distributive property.

$= 2 \cdot 5 + 5\sqrt{10} - 3 \cdot 2$ Simplify radicals and combine *like* radicals.

$= 10 + 5\sqrt{10} - 6$

$= 4 + 5\sqrt{10}$ Combine *like* terms.

b. $(-10a\sqrt{b} + 7b)(a\sqrt{b} + 2b)$

$= -10a^2\sqrt{b^2} - 20ab\sqrt{b} + 7ab\sqrt{b} + 14b^2$ Apply the distributive property.

$= -10a^2b - 13ab\sqrt{b} + 14b^2$ Simplify and combine *like* terms.

Answers

1. $-24\sqrt{2}$ 2. $-6ab^2\sqrt{a}$
3. $20a\sqrt[3]{b^2}$ 4. $50 - 10\sqrt{5}$

Skill Practice Multiply.

5. $(2\sqrt{3} - 3\sqrt{10})(\sqrt{3} + 2\sqrt{10})$

6. $(x\sqrt{y} + y)(3x\sqrt{y} - 2y)$

Example 4 **Multiplying Radical Expressions**

Multiply. $(2\sqrt{x} + \sqrt{y})(6 - \sqrt{x} + 8\sqrt{y})$

Solution:

$(2\sqrt{x} + \sqrt{y})(6 - \sqrt{x} + 8\sqrt{y})$

$= 12\sqrt{x} - 2\sqrt{x^2} + 16\sqrt{xy} + 6\sqrt{y} - \sqrt{xy} + 8\sqrt{y^2}$ Apply the distributive property.

$= 12\sqrt{x} - 2x + 16\sqrt{xy} + 6\sqrt{y} - \sqrt{xy} + 8y$ Simplify the radicals.

$= 12\sqrt{x} - 2x + 15\sqrt{xy} + 6\sqrt{y} + 8y$ Combine *like* terms.

Skill Practice Multiply.

7. $(\sqrt{t} + 3\sqrt{w})(4 - \sqrt{t} - \sqrt{w})$

---

2. Expressions of the Form $(\sqrt[n]{a})^n$

The multiplication property of radicals can be used to simplify an expression of the form $(\sqrt{a})^2$, where $a \ge 0$.

$$(\sqrt{a})^2 = \sqrt{a} \cdot \sqrt{a} = \sqrt{a^2} = a \qquad \text{where } a \ge 0$$

This logic can be applied to *n*th roots.

- If $\sqrt[n]{a}$ is a real number, then $(\sqrt[n]{a})^n = a$.

Example 5 **Simplifying Radical Expressions**

Simplify the expressions. Assume all variables represent positive real numbers.

a. $(\sqrt{11})^2$ **b.** $(\sqrt[5]{z})^5$ **c.** $(\sqrt[3]{pq})^3$

Solution:

a. $(\sqrt{11})^2 = 11$ **b.** $(\sqrt[5]{z})^5 = z$ **c.** $(\sqrt[3]{pq})^3 = pq$

Skill Practice Simplify.

8. $(\sqrt{14})^2$ **9.** $(\sqrt[7]{q})^7$ **10.** $(\sqrt[5]{3z})^5$

Answers

5. $-54 + \sqrt{30}$
6. $3x^2y + xy\sqrt{y} - 2y^2$
7. $4\sqrt{t} - t - 4\sqrt{tw} + 12\sqrt{w} - 3w$
8. 14   **9.** q **10.** $3z$

3. Special Case Products

From Examples 2–4, you may have noticed a similarity between multiplying radical expressions and multiplying polynomials.

In Section 4.3 we learned that the square of a binomial results in a perfect square trinomial:

$$(a + b)^2 = a^2 + 2ab + b^2$$
$$(a - b)^2 = a^2 - 2ab + b^2$$

The same patterns occur when squaring a radical expression with two terms.

Example 6 Squaring a Two-Term Radical Expression

Square the radical expressions. Assume all variables represent positive real numbers.

a. $(\sqrt{d} + 3)^2$ **b.** $(5\sqrt{y} - \sqrt{2})^2$

Solution:

a. $(\sqrt{d} + 3)^2$ This expression is in the form $(a + b)^2$, where $a = \sqrt{d}$ and $b = 3$.

$\qquad a^2 + 2ab + b^2$

$= (\sqrt{d})^2 + 2(\sqrt{d})(3) + (3)^2$ Apply the formula $(a + b)^2 = a^2 + 2ab + b^2$.

$= d + 6\sqrt{d} + 9$ Simplify.

b. $(5\sqrt{y} - \sqrt{2})^2$ This expression is in the form $(a - b)^2$, where $a = 5\sqrt{y}$ and $b = \sqrt{2}$.

$\qquad a^2 - 2ab + b^2$

$= (5\sqrt{y})^2 - 2(5\sqrt{y})(\sqrt{2}) + (\sqrt{2})^2$ Apply the formula $(a - b)^2 = a^2 - 2ab + b^2$.

$= 25y - 10\sqrt{2y} + 2$ Simplify.

TIP: The product $(\sqrt{d} + 3)^2$ can also be found by using the distributive property.

$(\sqrt{d} + 3)^2 = (\sqrt{d} + 3)(\sqrt{d} + 3)$

Skill Practice Square the radical expressions. Assume all variables represent positive real numbers.

11. $(\sqrt{a} - 5)^2$ **12.** $(4\sqrt{x} + \sqrt{3})^2$

Recall from Section 4.3 that the product of two conjugate binomials results in a difference of squares.

$$(a + b)(a - b) = a^2 - b^2$$

The same pattern occurs when multiplying two conjugate radical expressions.

Example 7 Multiplying Conjugate Radical Expressions

Multiply the radical expressions. Assume all variables represent positive real numbers.

a. $(\sqrt{3} + 2)(\sqrt{3} - 2)$ **b.** $\left(\dfrac{1}{3}\sqrt{s} - \dfrac{3}{4}\sqrt{t}\right)\left(\dfrac{1}{3}\sqrt{s} + \dfrac{3}{4}\sqrt{t}\right)$

Answers
11. $a - 10\sqrt{a} + 25$
12. $16x + 8\sqrt{3x} + 3$

TIP: The product $(\sqrt{3} + 2)(\sqrt{3} - 2)$ can also be found by using the distributive property.

$$(\sqrt{3} + 2)(\sqrt{3} - 2)$$

Solution:

a. $(\sqrt{3} + 2)(\sqrt{3} - 2)$ The expression is in the form $(a + b)(a - b)$, where $a = \sqrt{3}$ and $b = 2$.

$$\overset{a^2 - b^2}{\downarrow \quad \downarrow}$$

$$= (\sqrt{3})^2 - (2)^2 \quad\quad \text{Apply the formula } (a + b)(a - b) = a^2 - b^2.$$

$$= 3 - 4 \quad\quad\quad\quad\quad \text{Simplify.}$$

$$= -1$$

b. $\left(\dfrac{1}{3}\sqrt{s} - \dfrac{3}{4}\sqrt{t}\right)\left(\dfrac{1}{3}\sqrt{s} + \dfrac{3}{4}\sqrt{t}\right)$ This expression is in the form $(a - b)(a + b)$, where $a = \frac{1}{3}\sqrt{s}$ and $b = \frac{3}{4}\sqrt{t}$.

$$\overset{a^2 - b^2}{\downarrow \quad\quad \downarrow}$$

$$= \left(\dfrac{1}{3}\sqrt{s}\right)^2 - \left(\dfrac{3}{4}\sqrt{t}\right)^2 \quad\quad \begin{array}{l}\text{Apply the formula}\\ (a + b)(a - b) = a^2 - b^2.\end{array}$$

$$= \dfrac{1}{9}s - \dfrac{9}{16}t \quad\quad\quad\quad\quad\quad \text{Simplify.}$$

Skill Practice Multiply the conjugates. Assume all variables represent positive real numbers.

13. $(\sqrt{5} + 3)(\sqrt{5} - 3)$ **14.** $\left(\dfrac{1}{2}\sqrt{a} + \dfrac{2}{5}\sqrt{b}\right) \cdot \left(\dfrac{1}{2}\sqrt{a} - \dfrac{2}{5}\sqrt{b}\right)$

4. Multiplying Radicals with Different Indices

The product of two radicals can be simplified provided the radicals have the same index. If the radicals have different indices, then we can use the properties of rational exponents to obtain a common index.

Example 8   **Multiplying Radicals with Different Indices**

Multiply the expressions. Write the answers in radical form.

a. $\sqrt[3]{5} \cdot \sqrt[4]{5}$ **b.** $\sqrt[3]{7} \cdot \sqrt{2}$

Solution:

a. $\sqrt[3]{5} \cdot \sqrt[4]{5}$

$$= 5^{1/3}5^{1/4} \quad\quad\quad\quad\quad \text{Rewrite each expression with rational exponents.}$$

$$= 5^{(1/3)+(1/4)} \quad\quad\quad\quad \text{Because the bases are equal, we can add the exponents.}$$

$$= 5^{(4/12)+(3/12)} \quad\quad\quad \text{Write the fractions with a common denominator.}$$

$$= 5^{7/12} \quad\quad\quad\quad\quad\quad \text{Simplify the exponent.}$$

$$= \sqrt[12]{5^7} \quad\quad\quad\quad\quad\quad \text{Rewrite the expression as a radical.}$$

Answers

13. -4 **14.** $\dfrac{1}{4}a - \dfrac{4}{25}b$

b. $\sqrt[3]{7} \cdot \sqrt{2}$

$= 7^{1/3}2^{1/2}$ Rewrite each expression with rational exponents.

$= 7^{2/6}2^{3/6}$ Write the rational exponents with a common denominator.

$= (7^2 2^3)^{1/6}$ Apply the power rule of exponents.

$= \sqrt[6]{7^2 2^3}$ Rewrite the expression as a single radical.

$= \sqrt[6]{392}$ Simplify.

Skill Practice Multiply the expressions. Write the answers in radical form. Assume all variables represent positive real numbers.

15. $\sqrt{x} \cdot \sqrt[3]{x}$ **16.** $\sqrt[4]{a^3} \cdot \sqrt[3]{b}$

Answers

15. $\sqrt[6]{x^5}$ **16.** $\sqrt[12]{a^9 b^4}$

Section 6.5   Practice Exercises

For the exercises in this set, assume that all variables represent positive real numbers, unless otherwise stated.

Vocabulary and Key Concepts

1. a. If $\sqrt[n]{a}$ and $\sqrt[n]{b}$ are real numbers, then $\sqrt[n]{a} \cdot \sqrt[n]{b} =$ _____.

   b. If $x \geq 0$, then $\sqrt{x} \cdot \sqrt{x} =$ _____.

   c. If $\sqrt[n]{a}$ is a real number, then $(\sqrt[n]{a})^n =$ _____.

   d. Two binomials $(x + \sqrt{2})$ and $(x - \sqrt{2})$ are called _____ of each other, and their product is $(x)^2 - (\sqrt{2})^2$.

   e. If $m \geq 0$ and $n \geq 0$, then $(\sqrt{m} + \sqrt{n})(\sqrt{m} - \sqrt{n}) =$ _____.

   f. Which is the correct simplification of $(\sqrt{c} + 4)^2$?

       $c + 16$ or $c + 8\sqrt{c} + 16$

Review Exercises

For Exercises 2–4, simplify the radicals.

2. $\sqrt[3]{-16x^5y^6z^7}$ **3.** $-\sqrt{20a^2b^3c}$ **4.** $\sqrt{\dfrac{8y^3z^5}{y}}$

For Exercises 5–6, simplify the expressions. Write the answer with positive exponents only.

5. $x^{1/3}y^{1/4}x^{-1/6}y^{1/3}$ **6.** $\dfrac{b^{1/4}}{b^{3/2}}$

For Exercises 7–8, add or subtract as indicated.

7. $-2\sqrt[3]{7} + 4\sqrt[3]{7}$ **8.** $4\sqrt{8x^3} - x\sqrt{50x}$

Concept 1: Multiplication Property of Radicals

For Exercises 9–44, multiply the radical expressions. **(See Examples 1–4.)**

9. $\sqrt[3]{7} \cdot \sqrt[3]{3}$

10. $\sqrt[4]{6} \cdot \sqrt[4]{2}$

11. $\sqrt{2} \cdot \sqrt{10}$

12. $\sqrt[3]{4} \cdot \sqrt[3]{12}$

13. $\sqrt[4]{16} \cdot \sqrt[4]{64}$

14. $\sqrt{5x^3} \cdot \sqrt{10x^4}$

15. $(4\sqrt[3]{4})(2\sqrt[3]{5})$

16. $(2\sqrt{5})(3\sqrt{7})$

17. $(8a\sqrt{b})(-3\sqrt{ab})$

18. $(p\sqrt[4]{q^3})(\sqrt[4]{pq})$

19. $\sqrt{30} \cdot \sqrt{12}$

20. $\sqrt{20} \cdot \sqrt{54}$

21. $\sqrt{6x} \cdot \sqrt{12x}$

22. $(\sqrt{3ab^2})(\sqrt{21a^2b})$

23. $\sqrt{5a^3b^2} \cdot \sqrt{20a^3b^3}$

24. $\sqrt[3]{m^2n^2} \cdot \sqrt[3]{48m^4n^2}$

25. $(4\sqrt{3xy^3})(-2\sqrt{6x^3y^2})$

26. $(2\sqrt[4]{3x})(4\sqrt[4]{27x^6})$

27. $(\sqrt[3]{4a^2b})(\sqrt[3]{2ab^3})(\sqrt[3]{54a^2b})$

28. $(\sqrt[3]{9x^3y})(\sqrt[3]{6xy})(\sqrt[3]{8x^2y^5})$

29. $\sqrt{3}(4\sqrt{3} - 6)$

30. $3\sqrt{5}(2\sqrt{5} + 4)$

31. $\sqrt{2}(\sqrt{6} - \sqrt{3})$

32. $\sqrt{5}(\sqrt{3} + \sqrt{7})$

33. $-\frac{1}{3}\sqrt{x}(6\sqrt{x} + 7)$

34. $-\frac{1}{2}\sqrt{y}(8 - 3\sqrt{y})$

35. $(\sqrt{3} + 2\sqrt{10})(4\sqrt{3} - \sqrt{10})$

36. $(8\sqrt{7} - \sqrt{5})(\sqrt{7} + 3\sqrt{5})$

37. $(\sqrt{x} + 4)(\sqrt{x} - 9)$

38. $(\sqrt{w} - 2)(\sqrt{w} - 9)$

39. $(\sqrt[3]{y} + 2)(\sqrt[3]{y} - 3)$

40. $(4 + \sqrt[5]{p})(5 + \sqrt[5]{p})$

41. $(\sqrt{a} - 3\sqrt{b})(9\sqrt{a} - \sqrt{b})$

42. $(11\sqrt{m} + 4\sqrt{n})(\sqrt{m} + \sqrt{n})$

43. $(\sqrt{p} + 2\sqrt{q})(8 + 3\sqrt{p} - \sqrt{q})$

44. $(5\sqrt{s} - \sqrt{t})(\sqrt{s} + 5 + 6\sqrt{t})$

Concept 2: Expressions of the Form $(\sqrt[n]{a})^n$

For Exercises 45–52, simplify the expressions. Assume all variables represent positive real numbers. **(See Example 5.)**

45. $(\sqrt{15})^2$

46. $(\sqrt{58})^2$

47. $(\sqrt{3y})^2$

48. $(\sqrt{19yz})^2$

49. $(\sqrt[3]{6})^3$

50. $(\sqrt[5]{24})^5$

51. $\sqrt{709} \cdot \sqrt{709}$

52. $\sqrt{401} \cdot \sqrt{401}$

Concept 3: Special Case Products

53. a. Write the formula for the product of two conjugates. $(x + y)(x - y) =$

b. Multiply $(x + 5)(x - 5)$.

54. a. Write the formula for squaring a binomial. $(x + y)^2 =$

b. Multiply $(x + 5)^2$.

For Exercises 55–66, multiply the radical expressions. **(See Examples 6–7.)**

55. $(\sqrt{13} + 4)^2$

56. $(6 - \sqrt{11})^2$

57. $(\sqrt{p} - \sqrt{7})^2$

58. $(\sqrt{q} + \sqrt{2})^2$

59. $(\sqrt{2a} - 3\sqrt{b})^2$

60. $(\sqrt{3w} + 4\sqrt{z})^2$

61. $(\sqrt{3} + x)(\sqrt{3} - x)$

62. $(y + \sqrt{6})(y - \sqrt{6})$

63. $(\sqrt{6} + \sqrt{2})(\sqrt{6} - \sqrt{2})$

64. $(\sqrt{15} + \sqrt{5})(\sqrt{15} - \sqrt{5})$

65. $\left(\frac{2}{3}\sqrt{x} + \frac{1}{2}\sqrt{y}\right)\left(\frac{2}{3}\sqrt{x} - \frac{1}{2}\sqrt{y}\right)$

66. $\left(\frac{1}{4}\sqrt{s} + \frac{1}{5}\sqrt{t}\right)\left(\frac{1}{4}\sqrt{s} - \frac{1}{5}\sqrt{t}\right)$

For Exercises 67–68, multiply the expressions.

67. a. $(\sqrt{3} + \sqrt{x})(\sqrt{3} - \sqrt{x})$

  b. $(\sqrt{3} + \sqrt{x})(\sqrt{3} + \sqrt{x})$

  c. $(\sqrt{3} - \sqrt{x})(\sqrt{3} - \sqrt{x})$

68. a. $(\sqrt{5} + \sqrt{y})(\sqrt{5} - \sqrt{y})$

  b. $(\sqrt{5} + \sqrt{y})(\sqrt{5} + \sqrt{y})$

  c. $(\sqrt{5} - \sqrt{y})(\sqrt{5} - \sqrt{y})$

Mixed Exercises

For Exercises 69–76, identify each statement as true or false. If an answer is false, explain why.

69. $\sqrt{3} \cdot \sqrt{2} = \sqrt{6}$

70. $\sqrt{5} \cdot \sqrt[3]{2} = \sqrt{10}$

71. $(x - \sqrt{5})^2 = x - 5$

72. $3(2\sqrt{5x}) = 6\sqrt{5x}$

73. $5(3\sqrt{4x}) = 15\sqrt{20x}$

74. $\dfrac{\sqrt{5x}}{5} = \sqrt{x}$

75. $\dfrac{3\sqrt{x}}{3} = \sqrt{x}$

76. $(\sqrt{t} - 1)(\sqrt{t} + 1) = t - 1$

For Exercises 77–84, perform the indicated operations.

77. $(-\sqrt{6x})^2$

78. $(-\sqrt{8a})^2$

79. $(\sqrt{3x + 1})^2$

80. $(\sqrt{x} - 1)^2$

81. $(\sqrt{x + 3} - 4)^2$

82. $(\sqrt{x + 1} + 3)^2$

83. $(\sqrt{2t - 3} + 5)^2$

84. $(\sqrt{3w - 2} - 4)^2$

For Exercises 85–88, find the exact area.

85.

$\sqrt{40}$ ft

$3\sqrt{2}$ ft

86.

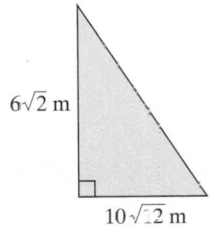

$6\sqrt{2}$ m

$10\sqrt{2}$ m

87.

$3\sqrt{5}$ in.

$6\sqrt{12}$ in.

88.

$2\sqrt{18}$ yd

$7\sqrt{6}$ yd

Concept 4: Multiplying Radicals with Different Indices

For Exercises 89–100, multiply or divide the radicals with different indices. Write the answers in radical form and simplify. **(See Example 8.)**

89. $\sqrt{x} \cdot \sqrt[4]{x}$

90. $\sqrt[3]{y} \cdot \sqrt{y}$

91. $\sqrt[5]{2z} \cdot \sqrt[3]{2z}$

92. $\sqrt[3]{5w} \cdot \sqrt[4]{5w}$

93. $\sqrt[3]{p^2} \cdot \sqrt{p^3}$

94. $\sqrt[4]{q^3} \cdot \sqrt[3]{q^2}$

95. $\dfrac{\sqrt{u^3}}{\sqrt[3]{u}}$

96. $\dfrac{\sqrt{v^5}}{\sqrt[4]{v}}$

97. $\sqrt[3]{x} \cdot \sqrt[6]{y}$

98. $\sqrt{a} \cdot \sqrt[6]{b}$

99. $\sqrt[4]{8} \cdot \sqrt{3}$

100. $\sqrt{11} \cdot \sqrt[6]{2}$

Expanding Your Skills

For Exercises 101–106, multiply.

101. $\sqrt[3]{2xy} \cdot \sqrt[4]{5xy}$

102. $\sqrt{6ab} \cdot \sqrt[3]{7ab}$

103. $\sqrt[3]{4m^2n} \cdot \sqrt{6mn}$

104. $\sqrt[4]{5xy^3} \cdot \sqrt[3]{10x^2y}$

105. $(\sqrt[3]{a} + \sqrt[3]{b})(\sqrt[3]{a^2} - \sqrt[3]{ab} + \sqrt[3]{b^2})$

106. $(\sqrt[3]{a} - \sqrt[3]{b})(\sqrt[3]{a^2} + \sqrt[3]{ab} + \sqrt[3]{b^2})$

Problem Recognition Exercises

Simplifying Radical Expressions

For Exercises 1–20, simplify the expressions.

1. a. $\sqrt{24}$
  b. $\sqrt[3]{24}$

2. a. $\sqrt{54}$
  b. $\sqrt[3]{54}$

3. a. $\sqrt{200y^6}$
  b. $\sqrt[3]{200y^6}$

4. a. $\sqrt{32z^{15}}$
  b. $\sqrt[3]{32z^{15}}$

5. a. $\sqrt{80}$
  b. $\sqrt[3]{80}$
  c. $\sqrt[4]{80}$

6. a. $\sqrt{48}$
  b. $\sqrt[3]{48}$
  c. $\sqrt[4]{48}$

7. a. $\sqrt{x^5y^6}$
  b. $\sqrt[3]{x^5y^6}$
  c. $\sqrt[4]{x^5y^6}$

8. a. $\sqrt{a^{10}b^9}$
  b. $\sqrt[3]{a^{10}b^9}$
  c. $\sqrt[4]{a^{10}b^9}$

9. a. $\sqrt[3]{32s^5t^6}$
  b. $\sqrt[4]{32s^5t^6}$
  c. $\sqrt[5]{32s^5t^6}$

10. a. $\sqrt[3]{96v^7w^{20}}$
  b. $\sqrt[4]{96v^7w^{20}}$
  c. $\sqrt[5]{96v^7w^{20}}$

11. a. $\sqrt{5} + \sqrt{5}$
  b. $\sqrt{5} \cdot \sqrt{5}$

12. a. $\sqrt{10} + \sqrt{10}$
  b. $\sqrt{10} \cdot \sqrt{10}$

13. a. $2\sqrt{6} - 5\sqrt{6}$
  b. $2\sqrt{6} \cdot 5\sqrt{6}$

14. a. $3\sqrt{7} - 10\sqrt{7}$
  b. $3\sqrt{7} \cdot 10\sqrt{7}$

15. a. $\sqrt{8} + \sqrt{2}$
  b. $\sqrt{8} \cdot \sqrt{2}$

16. a. $\sqrt{12} + \sqrt{3}$
  b. $\sqrt{12} \cdot \sqrt{3}$

17. a. $5\sqrt{18} - 4\sqrt{8}$
  b. $5\sqrt{18} \cdot 4\sqrt{8}$

18. a. $\sqrt{50} - \sqrt{72}$
  b. $\sqrt{50} \cdot \sqrt{72}$

19. a. $4\sqrt[3]{24} + 6\sqrt[3]{3}$
  b. $4\sqrt[3]{24} \cdot 6\sqrt[3]{3}$

20. a. $2\sqrt[3]{54} - 5\sqrt[3]{2}$
  b. $2\sqrt[3]{54} \cdot 5\sqrt[3]{2}$

Section 6.6   Division of Radicals and Rationalization

Concepts

1. Simplified Form of a Radical
2. Division Property of Radicals
3. Rationalizing the Denominator—One Term
4. Rationalizing the Denominator—Two Terms

1. Simplified Form of a Radical

Recall that for a radical expression to be in simplified form the following three conditions must be met.

> **Simplified Form of a Radical**
>
> Consider any radical expression in which the radicand is written as a product of prime factors. The expression is in simplified form if all the following conditions are met:
>
> 1. The radicand has no factor raised to a power greater than or equal to the index.
> 2. The radicand does not contain a fraction.
> 3. No radicals are in the denominator of a fraction.

In the previous sections, we have concentrated on the first condition in the simplification process. Next, we will demonstrate how to satisfy the second and third conditions involving radicals and fractions.

2. Division Property of Radicals

The multiplication property of radicals makes it possible to write a product within a radical to be separated as a product of radicals. We now state a similar property for radicals involving quotients.

Division Property of Radicals

Let a and b represent real numbers such that $\sqrt[n]{a}$ and $\sqrt[n]{b}$ are both real. Then,

$$\sqrt[n]{\frac{a}{b}} = \frac{\sqrt[n]{a}}{\sqrt[n]{b}} \quad b \neq 0$$

The division property of radicals indicates that a quotient within a radicand can be written as a quotient of radicals provided the roots are real numbers. For example:

$$\sqrt{\frac{25}{36}} = \frac{\sqrt{25}}{\sqrt{36}}$$

The reverse process is also true. A quotient of radicals can be written as a single radical provided that the roots are real numbers and that they have the same index.

$$\text{Same index} \longrightarrow \frac{\sqrt[3]{125}}{\sqrt[3]{8}} = \sqrt[3]{\frac{125}{8}}$$

In Examples 1 and 2, we will apply the division property of radicals to simplify radical expressions.

Example 1 **Using the Division Property to Simplify Radicals**

Use the division property of radicals to simplify the expressions. Assume the variables represent positive real numbers.

a. $\sqrt{\dfrac{a^6}{b^4}}$ **b.** $\sqrt[3]{\dfrac{81y^5}{x^3}}$

Solution:

a. $\sqrt{\dfrac{a^6}{b^4}}$ The radicand contains an irreducible fraction.

$= \dfrac{\sqrt{a^6}}{\sqrt{b^4}}$ Apply the division property to rewrite as a quotient of radicals.

$= \dfrac{a^3}{b^2}$ Simplify the radicals.

b. $\sqrt[3]{\dfrac{81y^5}{x^3}}$ The radicand contains an irreducible fraction.

$= \dfrac{\sqrt[3]{81y^5}}{\sqrt[3]{x^3}}$ Apply the division property to rewrite as a quotient of radicals.

$= \dfrac{\sqrt[3]{3^4 \cdot y^5}}{\sqrt[3]{x^3}}$ Factor the radicand in the numerator to simplify the radical.

$= \dfrac{3y\sqrt[3]{3y^2}}{x}$ Simplify the radicals in the numerator and the denominator. The expression is simplified because it now satisfies all conditions.

Skill Practice Simplify the expressions.

1. $\sqrt{\dfrac{x^4}{y^{10}}}$ **2.** $\sqrt[3]{\dfrac{w^7}{64}}$

Example 2   Using the Division Property to Simplify a Radical

Use the division property of radicals to simplify the expressions. Assume the variables represent positive real numbers.

$$\frac{\sqrt[4]{8p^7}}{\sqrt[4]{p^3}}$$

Solution:

$\dfrac{\sqrt[4]{8p^7}}{\sqrt[4]{p^3}}$ There is a radical in the denominator of the fraction.

$= \sqrt[4]{\dfrac{8p^7}{p^3}}$ Apply the division property to write the quotient under a single radical.

$= \sqrt[4]{8p^4}$ Simplify the fraction.

$= p\sqrt[4]{8}$ Simplify the radical.

Skill Practice Simplify the expression.

3. $\dfrac{\sqrt[3]{16y^4}}{\sqrt[3]{y}}$

---

3. Rationalizing the Denominator—One Term

The third condition restricts radicals from the denominator of an expression. The process to remove a radical from the denominator is called **rationalizing the denominator**. In many cases, rationalizing the denominator creates an expression that is computationally simpler. For example,

$$\frac{6}{\sqrt{3}} = 2\sqrt{3} \qquad \text{and} \qquad \frac{-2}{2 + \sqrt{6}} = 2 - \sqrt{6}$$

Answers

1. $\dfrac{x^2}{y^5}$ **2.** $\dfrac{w^2\sqrt[3]{w}}{4}$ **3.** $2y\sqrt[3]{2}$

We will demonstrate the process to rationalize the denominator as two separate cases:

- Rationalizing the denominator (one term)
- Rationalizing the denominator (two terms involving square roots)

To begin the first case, recall that the nth root of a perfect nth power simplifies completely.

$$\sqrt{x^2} = x \qquad x \geq 0$$
$$\sqrt[3]{x^3} = x$$
$$\sqrt[4]{x^4} = x \qquad x \geq 0$$
$$\sqrt[5]{x^5} = x$$
$$\cdots$$

Therefore, to rationalize a radical expression, use the multiplication property of radicals to create an nth root of an nth power.

Example 3 **Rationalizing Radical Expressions**

Fill in the missing radicand to rationalize the radical expressions. Assume all variables represent positive real numbers.

a. $\sqrt{a} \cdot \sqrt{?} = \sqrt{a^2} = a$ **b.** $\sqrt[3]{y} \cdot \sqrt[3]{?} = \sqrt[3]{y^3} = y$

c. $\sqrt[4]{2z^3} \cdot \sqrt[4]{?} = \sqrt[4]{2^4 z^4} = 2z$

Solution:

a. $\sqrt{a} \cdot \sqrt{?} = \sqrt{a^2} = a$ What multiplied by $\sqrt{a}$ will equal $\sqrt{a^2}$?

    $\sqrt{a} \cdot \sqrt{a} = \sqrt{a^2} = a$

b. $\sqrt[3]{y} \cdot \sqrt[3]{?} = \sqrt[3]{y^3} = y$ What multiplied by $\sqrt[3]{y}$ will equal $\sqrt[3]{y^3}$?

    $\sqrt[3]{y} \cdot \sqrt[3]{y^2} = \sqrt[3]{y^3} = y$

c. $\sqrt[4]{2z^3} \cdot \sqrt[4]{?} = \sqrt[4]{2^4 z^4} = 2z$ What multiplied by $\sqrt[4]{2z^3}$ will equal $\sqrt[4]{2^4 z^4}$?

    $\sqrt[4]{2z^3} \cdot \sqrt[4]{2^3 z} = \sqrt[4]{2^4 z^4} = 2z$

Skill Practice Fill in the missing radicand to rationalize the radical expression.

4. $\sqrt{7} \cdot \sqrt{?}$ **5.** $\sqrt[5]{t^2} \cdot \sqrt[5]{?}$ **6.** $\sqrt[3]{5x^2} \cdot \sqrt[3]{?}$

    To rationalize the denominator of a radical expression, multiply the numerator and denominator by an appropriate expression to create an nth root of an nth power in the denominator.

Answers

4. 7   **5.** t^3 **6.** $5^2 x$

Example 4   **Rationalizing the Denominator—One Term**

Simplify the expression. $\dfrac{5}{\sqrt[3]{a}}$ $a \neq 0$

Solution:

To remove the radical from the denominator, a cube root of a perfect cube is needed in the denominator. Multiply numerator and denominator by $\sqrt[3]{a^2}$ because $\sqrt[3]{a} \cdot \sqrt[3]{a^2} = \sqrt[3]{a^3} = a$.

$$\frac{5}{\sqrt[3]{a}} = \frac{5}{\sqrt[3]{a}} \cdot \frac{\sqrt[3]{a^2}}{\sqrt[3]{a^2}}$$

$$= \frac{5\sqrt[3]{a^2}}{\sqrt[3]{a^3}} \qquad \text{Multiply the radicals.}$$

$$= \frac{5\sqrt[3]{a^2}}{a} \qquad \text{Simplify.}$$

Skill Practice Simplify the expression. Assume $y > 0$.

7. $\dfrac{2}{\sqrt[4]{y}}$

Note that for $a \neq 0$, the expression $\dfrac{\sqrt[3]{a^2}}{\sqrt[3]{a^2}} = 1$. In Example 4, multiplying the

expression $\dfrac{5}{\sqrt[3]{a}}$ by this ratio does not change its value.

Example 5   **Rationalizing the Denominator—One Term**

Simplify the expression. $\sqrt{\dfrac{y^5}{7}}$

Solution:

$\sqrt{\dfrac{y^5}{7}}$ The radical contains an irreducible fraction.

$= \dfrac{\sqrt{y^5}}{\sqrt{7}}$ Apply the division property of radicals.

$= \dfrac{y^2\sqrt{y}}{\sqrt{7}}$ Simplify the radical in the numerator.

$= \dfrac{y^2\sqrt{y}}{\sqrt{7}} \cdot \dfrac{\sqrt{7}}{\sqrt{7}}$ Rationalize the denominator.

               Note: $\sqrt{7} \cdot \sqrt{7} = \sqrt{7^2} = 7.$

$= \dfrac{y^2\sqrt{7y}}{\sqrt{7^2}}$

$= \dfrac{y^2\sqrt{7y}}{7}$ Simplify.

Avoiding Mistakes

A factor within a radicand cannot be simplified with a factor outside the radicand. For example, $\dfrac{\sqrt{7y}}{7}$ cannot be simplified.

Answer

7. $\dfrac{2\sqrt[4]{y^3}}{y}$

Skill Practice Simplify the expression.

8. $\sqrt{\dfrac{8}{3}}$

Example 6 Rationalizing the Denominator—One Term

Simplify the expression. $\dfrac{15}{\sqrt[3]{25s}}$

Solution:

$\dfrac{15}{\sqrt[3]{25s}}$

$= \dfrac{15}{\sqrt[3]{5^2 s}} \cdot \dfrac{\sqrt[3]{5s^2}}{\sqrt[3]{5s^2}}$ Because $25 = 5^2$, one additional factor of 5 is needed to form a perfect cube. Two additional factors of s are needed to make a perfect cube. Multiply numerator and denominator by $\sqrt[3]{5s^2}$.

$= \dfrac{15\sqrt[3]{5s^2}}{\sqrt[3]{5^3 s^3}}$

$= \dfrac{15\sqrt[3]{5s^2}}{5s}$ Simplify the perfect cube.

$= \dfrac{\overset{3}{\cancel{15}}\sqrt[3]{5s^2}}{\underset{1}{\cancel{5}s}}$ Reduce to lowest terms.

$= \dfrac{3\sqrt[3]{5s^2}}{s}$

TIP: In the expression $\dfrac{15\sqrt[3]{5s^2}}{5s}$, the factor of 15 and the factor of 5 may be reduced because both are outside the radical.

$\dfrac{15\sqrt[3]{5s^2}}{5s} = \dfrac{15}{5} \cdot \dfrac{\sqrt[3]{5s^2}}{s}$

$= \dfrac{3\sqrt[3]{5s^2}}{s}$

Skill Practice Simplify the expression.

9. $\dfrac{18}{\sqrt[3]{3y^2}}$

4. Rationalizing the Denominator—Two Terms

Example 7 demonstrates how to rationalize a two-term denominator involving square roots.

First recall from the multiplication of polynomials that the product of two conjugates results in a difference of squares.

$$(a + b)(a - b) = a^2 - b^2$$

If either a or b has a square root factor, the expression will simplify without a radical. That is, the expression is *rationalized*. For example:

$$(2 + \sqrt{6})(2 - \sqrt{6}) = (2)^2 - (\sqrt{6})^2$$

$$= 4 - 6$$

$$= -2$$

Answers

8. $\dfrac{2\sqrt{6}}{3}$ 9. $\dfrac{6\sqrt[3]{9y}}{y}$

Example 7 **Rationalizing the Denominator—Two Terms**

Simplify the expression by rationalizing the denominator. $\dfrac{-2}{2 + \sqrt{6}}$

Solution:

$$\dfrac{-2}{2 + \sqrt{6}}$$

$$= \dfrac{(-2)}{(2 + \sqrt{6})} \cdot \dfrac{(2 - \sqrt{6})}{(2 - \sqrt{6})}$$ Multiply the numerator and denominator by the conjugate of the denominator.

conjugates

$$= \dfrac{-2(2 - \sqrt{6})}{(2)^2 - (\sqrt{6})^2}$$ In the denominator, apply the formula $(a + b)(a - b) = a^2 - b^2$.

$$= \dfrac{-2(2 - \sqrt{6})}{4 - 6}$$ Simplify.

$$= \dfrac{-2(2 - \sqrt{6})}{-2}$$

$$= \dfrac{-2(2 - \sqrt{6})}{-2}$$

$$= 2 - \sqrt{6}$$

> **Avoiding Mistakes**
> When constructing the conjugate of an expression, change only the sign between the terms (not the sign of the leading term).

Skill Practice Simplify by rationalizing the denominator.

10. $\dfrac{8}{3 + \sqrt{5}}$

Example 8 **Rationalizing the Denominator—Two Terms**

Rationalize the denominator. $\dfrac{4 + \sqrt{x}}{\sqrt{x} - 7}$

Solution:

$$\dfrac{(4 + \sqrt{x})}{(\sqrt{x} - 7)} \cdot \dfrac{(\sqrt{x} + 7)}{(\sqrt{x} + 7)}$$ Multiply numerator and denominator by the conjugate of the denominator.

$$= \dfrac{(4)(\sqrt{x}) + (4)(7) + (\sqrt{x})(\sqrt{x}) + (\sqrt{x})(7)}{(\sqrt{x})^2 - (7)^2}$$ Apply the distributive property.

$$= \dfrac{4\sqrt{x} + 28 + x + 7\sqrt{x}}{x - 49}$$ Multiply the radicals.

$$= \dfrac{x + 11\sqrt{x} + 28}{x - 49}$$ Simplify.

Skill Practice Rationalize the denominator.

11. $\dfrac{\sqrt{y} - 4}{8 - \sqrt{y}}$

Answers
10. $6 - 2\sqrt{5}$
11. $\dfrac{y + 4\sqrt{y} - 32}{64 - y}$

Example 9	Rationalizing the Denominator—Two Terms

Rationalize the denominator. $\dfrac{3\sqrt{2} + 2\sqrt{5}}{\sqrt{2} - 4\sqrt{5}}$

Solution:

$\dfrac{(3\sqrt{2} + 2\sqrt{5})}{(\sqrt{2} - 4\sqrt{5})} \cdot \dfrac{(\sqrt{2} + 4\sqrt{5})}{(\sqrt{2} + 4\sqrt{5})}$ Multiply numerator and denominator by the conjugate of the denominator.

$= \dfrac{(3\sqrt{2}) \cdot (\sqrt{2}) + (3\sqrt{2})(4\sqrt{5}) + (2\sqrt{5})(\sqrt{2}) + (2\sqrt{5})(4\sqrt{5})}{(\sqrt{2})^2 - (4\sqrt{5})^2}$ Apply the distributive property.

$= \dfrac{6 + 12\sqrt{10} + 2\sqrt{10} + 40}{2 - 80}$ Multiply the radicals.

$= \dfrac{46 + 14\sqrt{10}}{-78}$ or $-\dfrac{46 + 14\sqrt{10}}{78}$

$= -\dfrac{\overset{1}{2}(23 + 7\sqrt{10})}{\underset{39}{78}}$ Factor and simplify.

$= -\dfrac{23 + 7\sqrt{10}}{39}$

Skill Practice Rationalize the denominator.

12. $\dfrac{5\sqrt{2} - 2\sqrt{5}}{\sqrt{2} - \sqrt{5}}$

Answer

12. $-\sqrt{10}$

Section 6.6   Practice Exercises

For the exercises in this set, assume that all variables represent positive real numbers unless otherwise stated.

Vocabulary and Key Concepts

1. **a.** In the simplified form of a radical expression, no _____ may appear in the denominator of a fraction.

   b. The division property of radicals indicates that if $\sqrt[n]{a}$ and $\sqrt[n]{b}$ are real numbers, then $\sqrt[n]{\dfrac{a}{b}} = \dfrac{\square}{\square}$ provided that $b \neq 0$.

   c. The simplified form of the expression $\sqrt[3]{\dfrac{64}{x^6}}$ is _____.

   d. The process of removing a radical from the denominator of a fraction is called _____ the denominator.

   e. The expression $\dfrac{\sqrt{3}}{3}$ (is/is not) in simplified form whereas $\dfrac{3}{\sqrt{3}}$ (is/is not) in simplified form.

   f. To rationalize the denominator for the expression $\dfrac{\sqrt{x} + 3}{\sqrt{x} - 2}$, multiply numerator and denominator by the conjugate of the _____.

Review Exercises

2. Simplify. $(4x^2y^4)^{1/2}(64y^{-3})^{1/3}$

For Exercises 3–10, perform the indicated operations.

3. $2y\sqrt{45} + 3\sqrt{20y^2}$ **4.** $3x\sqrt{72x} - 9\sqrt{50x^3}$ **5.** $(-6\sqrt{y} + 3)(3\sqrt{y} + 1)$ **6.** $(\sqrt{w} + 12)(2\sqrt{w} - 4)$

7. $(8 - \sqrt{t})^2$ **8.** $(\sqrt{p} + 4)^2$ **9.** $(\sqrt{2} + \sqrt{7})(\sqrt{2} - \sqrt{7})$ **10.** $(\sqrt{3} + 5)(\sqrt{3} - 5)$

Concept 2: Division Property of Radicals

For Exercises 11–22, simplify using the division property of radicals. Assume all variables represent positive real numbers. (See Examples 1–2.)

11. $\sqrt{\dfrac{49x^4}{y^6}}$ **12.** $\sqrt{\dfrac{100p^2}{q^8}}$ **13.** $\sqrt{\dfrac{8a^2}{x^6}}$ **14.** $\sqrt{\dfrac{4w^3}{25y^4}}$

15. $\sqrt[3]{\dfrac{-16j^3}{k^3}}$ **16.** $\sqrt[5]{\dfrac{32x}{y^{10}}}$ **17.** $\dfrac{\sqrt{72ab^5}}{\sqrt{8ab}}$ **18.** $\dfrac{\sqrt{6x^3}}{\sqrt{24x}}$

19. $\dfrac{\sqrt[4]{3b^3}}{\sqrt[4]{48b^{11}}}$ **20.** $\dfrac{\sqrt[3]{128wz^8}}{\sqrt[3]{2wz^2}}$ **21.** $\dfrac{\sqrt{3yz^2}}{\sqrt{w^4}}$ **22.** $\dfrac{\sqrt{50x^3z}}{\sqrt{9y^4}}$

Concept 3: Rationalizing the Denominator—One Term

The radical expressions in Exercises 23–30 have radicals in the denominator. Fill in the missing radicands to rationalize the denominators. (See Example 3.)

23. $\dfrac{x}{\sqrt{5}} = \dfrac{x}{\sqrt{5}} \cdot \dfrac{\sqrt{?}}{\sqrt{?}}$ **24.** $\dfrac{2}{\sqrt{x}} = \dfrac{2}{\sqrt{x}} \cdot \dfrac{\sqrt{?}}{\sqrt{?}}$ **25.** $\dfrac{7}{\sqrt[3]{x}} = \dfrac{7}{\sqrt[3]{x}} \cdot \dfrac{\sqrt[3]{?}}{\sqrt[3]{?}}$ **26.** $\dfrac{5}{\sqrt[4]{y}} = \dfrac{5}{\sqrt[4]{y}} \cdot \dfrac{\sqrt[4]{?}}{\sqrt[4]{?}}$

27. $\dfrac{8}{\sqrt{3z}} = \dfrac{8}{\sqrt{3z}} \cdot \dfrac{\sqrt{??}}{\sqrt{??}}$ **28.** $\dfrac{10}{\sqrt{7w}} = \dfrac{10}{\sqrt{7w}} \cdot \dfrac{\sqrt{??}}{\sqrt{??}}$ **29.** $\dfrac{1}{\sqrt[4]{8a^2}} = \dfrac{1}{\sqrt[4]{8a^2}} \cdot \dfrac{\sqrt[4]{??}}{\sqrt[4]{??}}$ **30.** $\dfrac{1}{\sqrt[3]{9b^2}} = \dfrac{1}{\sqrt[3]{9b^2}} \cdot \dfrac{\sqrt[3]{??}}{\sqrt[3]{??}}$

For Exercises 31–58, rationalize the denominator. (See Examples 4–6.)

31. $\dfrac{1}{\sqrt{3}}$ **32.** $\dfrac{1}{\sqrt{7}}$ **33.** $\sqrt{\dfrac{1}{x}}$ **34.** $\sqrt{\dfrac{1}{z}}$

35. $\dfrac{6}{\sqrt{2y}}$ **36.** $\dfrac{9}{\sqrt{3t}}$ **37.** $\sqrt{\dfrac{a^3}{2}}$ **38.** $\sqrt{\dfrac{b^3}{3}}$

39. $\dfrac{6}{\sqrt{8}}$ **40.** $\dfrac{2}{\sqrt{48}}$ **41.** $\dfrac{3}{\sqrt[3]{2}}$ **42.** $\dfrac{2}{\sqrt[3]{7}}$

43. $\dfrac{-6}{\sqrt[4]{x}}$ **44.** $\dfrac{-2}{\sqrt[5]{y}}$ **45.** $\dfrac{7}{\sqrt[3]{4}}$ **46.** $\dfrac{1}{\sqrt[3]{9}}$

47. $\sqrt[3]{\dfrac{4}{w^2}}$ **48.** $\sqrt[3]{\dfrac{5}{z^2}}$ **49.** $\sqrt[4]{\dfrac{16}{3}}$ **50.** $\sqrt[4]{\dfrac{81}{8}}$

51. $\dfrac{2}{\sqrt[3]{4x^2}}$ **52.** $\dfrac{6}{\sqrt[3]{3y^2}}$ **53.** $\dfrac{8}{7\sqrt{24}}$ **54.** $\dfrac{5}{3\sqrt{50}}$

55. $\dfrac{1}{\sqrt{x^7}}$ **56.** $\dfrac{1}{\sqrt{y^5}}$ **57.** $\dfrac{2}{\sqrt{8x^5}}$ **58.** $\dfrac{6}{\sqrt{27t^7}}$

Concept 4: Rationalizing the Denominator—Two Terms

59. What is the conjugate of $\sqrt{2} - \sqrt{6}$? **60.** What is the conjugate of $\sqrt{11} + \sqrt{5}$?

61. What is the conjugate of $\sqrt{x} + 23$? **62.** What is the conjugate of $17 - \sqrt{y}$?

For Exercises 63–82, rationalize the denominators. **(See Examples 7–9.)**

63. $\dfrac{4}{\sqrt{2} + 3}$ **64.** $\dfrac{6}{4 - \sqrt{3}}$ **65.** $\dfrac{8}{\sqrt{6} - 2}$ **66.** $\dfrac{-12}{\sqrt{5} - 3}$

67. $\dfrac{\sqrt{7}}{\sqrt{3} + 2}$ **68.** $\dfrac{\sqrt{8}}{\sqrt{3} + 1}$ **69.** $\dfrac{-1}{\sqrt{p} + \sqrt{q}}$ **70.** $\dfrac{6}{\sqrt{a} - \sqrt{b}}$

71. $\dfrac{x - 5}{\sqrt{x} + \sqrt{5}}$ **72.** $\dfrac{y - 2}{\sqrt{y} - \sqrt{2}}$ **73.** $\dfrac{\sqrt{w} + 2}{9 - \sqrt{w}}$ **74.** $\dfrac{10 - \sqrt{t}}{\sqrt{t} - 6}$

75. $\dfrac{3\sqrt{x} - \sqrt{y}}{\sqrt{y} + \sqrt{x}}$ **76.** $\dfrac{2\sqrt{a} + \sqrt{b}}{\sqrt{b} - \sqrt{a}}$ **77.** $\dfrac{3\sqrt{10}}{2 + \sqrt{10}}$ **78.** $\dfrac{4\sqrt{7}}{3 + \sqrt{7}}$

79. $\dfrac{2\sqrt{3} + \sqrt{7}}{3\sqrt{3} - \sqrt{7}}$ **80.** $\dfrac{5\sqrt{2} - \sqrt{5}}{5\sqrt{2} + \sqrt{5}}$ **81.** $\dfrac{\sqrt{5} + 4}{2 - \sqrt{5}}$ **82.** $\dfrac{3 + \sqrt{2}}{\sqrt{2} - 5}$

Mixed Exercises

For Exercises 83–86, write the English phrase as an algebraic expression. Then simplify the expression.

83. 16 divided by the cube root of 4

84. 21 divided by the principal fourth root of 27

85. 4 divided by the difference of x and the principal square root of 2

86. 8 divided by the sum of y and the principal square root of 3

87. The time $T(x)$ (in seconds) for a pendulum to make one complete swing back and forth is approximated by

$$T(x) = 2\pi\sqrt{\dfrac{x}{32}}$$

where x is the length of the pendulum in feet.

Determine the exact time required for one swing for a pendulum that is 1 ft long. Then approximate the time to the nearest hundredth of a second.

88. An object is dropped off a building x meters tall. The time $T(x)$ (in seconds) required for the object to hit the ground is given by

$$T(x) = \sqrt{\dfrac{10x}{49}}$$

Find the exact time required for the object to hit the ground if it is dropped off the First National Plaza Building in Chicago, a height of 230 m. Then round the time to the nearest hundredth of a second.

For Exercises 89–92, rationalize the denominator.

89. a. $\dfrac{1}{\sqrt{2}}$ **90. a.** $\dfrac{1}{\sqrt[3]{x}}$ **91. a.** $\dfrac{1}{\sqrt{5a}}$ **92. a.** $\dfrac{4}{\sqrt{2x}}$

b. $\dfrac{1}{\sqrt[3]{2}}$ **b.** $\dfrac{1}{\sqrt[3]{x^2}}$ **b.** $\dfrac{1}{\sqrt{5}+a}$ **b.** $\dfrac{4}{2-\sqrt{x}}$

Expanding Your Skills

For Exercises 93–98, simplify each term of the expression. Then add or subtract as indicated.

93. $\dfrac{\sqrt{6}}{2}+\dfrac{1}{\sqrt{6}}$ **94.** $\dfrac{1}{\sqrt{7}}+\sqrt{7}$ **95.** $\sqrt{15}-\sqrt{\dfrac{3}{5}}+\sqrt{\dfrac{5}{3}}$

96. $\sqrt{\dfrac{6}{2}}-\sqrt{12}+\sqrt{\dfrac{2}{6}}$ **97.** $\sqrt[3]{25}+\dfrac{3}{\sqrt[3]{5}}$ **98.** $\dfrac{1}{\sqrt[3]{4}}+\sqrt[3]{54}$

For Exercises 99–106, rationalize the numerator by multiplying both numerator and denominator by the conjugate of the numerator.

99. $\dfrac{\sqrt{3}+6}{2}$ **100.** $\dfrac{\sqrt{7}-2}{5}$ **101.** $\dfrac{\sqrt{a}-\sqrt{b}}{\sqrt{a}+\sqrt{b}}$ **102.** $\dfrac{\sqrt{p}+\sqrt{q}}{\sqrt{p}-\sqrt{q}}$

103. $\dfrac{\sqrt{5+3h}-\sqrt{5}}{h}$ **104.** $\dfrac{\sqrt{7+2h}-\sqrt{7}}{h}$ **105.** $\dfrac{\sqrt{4+5h}-2}{h}$ **106.** $\dfrac{\sqrt{9+4h}-3}{h}$

Section 6.7 Solving Radical Equations

Concepts

1. **Solutions to Radical Equations**
2. **Solving Radical Equations Involving One Radical**
3. **Solving Radical Equations Involving More than One Radical**
4. **Applications of Radical Equations and Functions**

1. Solutions to Radical Equations

An equation with one or more radicals containing a variable is called a **radical equation**. For example, $\sqrt[3]{x}=5$ is a radical equation. Recall that $(\sqrt[n]{a})^n=a$, provided $\sqrt[n]{a}$ is a real number. The basis of solving a radical equation is to eliminate the radical by raising both sides of the equation to a power equal to the index of the radical.

To solve the equation $\sqrt[3]{x}=5$, cube both sides of the equation.

$$\sqrt[3]{x}=5$$
$$(\sqrt[3]{x})^3=(5)^3$$
$$x=125$$

By raising each side of a radical equation to a power equal to the index of the radical, a new equation is produced. Note, however, that some of (or all) the solutions to the new equation may *not* be solutions to the original radical equation. For this reason, it is necessary to *check all potential solutions* in the original equation. For example, consider the equation $\sqrt{x}=-7$. This equation has no solution because by definition, the principal square root of x must be a nonnegative number. However, if we square both sides of the equation, it appears as though a solution exists.

$$(\sqrt{x})^2=(-7)^2$$
$$x=49 \qquad \text{The value 49 does not check in the original equation } \sqrt{x}=-7.$$
Therefore, 49 is an **extraneous solution**.

Solving a Radical Equation

Step 1 Isolate the radical. If an equation has more than one radical, choose one of the radicals to isolate.

Step 2 Raise each side of the equation to a power equal to the index of the radical.

Step 3 Solve the resulting equation. If the equation still has a radical, repeat steps 1 and 2.

*Step 4 Check the potential solutions in the original equation.

*In solving radical equations, extraneous solutions *potentially occur* only when each side of the equation is raised to an even power.

2. Solving Radical Equations Involving One Radical

Example 1 Solving an Equation Containing One Radical

Solve the equation. $\sqrt{p} + 5 = 9$

Solution:

$$\sqrt{p} + 5 = 9$$

$$\sqrt{p} = 4 \qquad \text{Isolate the radical.}$$

$$(\sqrt{p})^2 = 4^2 \qquad \text{Because the index is 2, square both sides.}$$

$$p = 16$$

$$\underline{\text{Check}:}\ p = 16 \qquad \text{Check } p = 16 \text{ as a potential solution.}$$

$$\sqrt{p} + 5 = 9$$

$$\sqrt{16} + 5 \stackrel{?}{=} 9$$

$$4 + 5 \stackrel{?}{=} 9 \ ✔ \qquad \text{True, 16 is a solution to the original equation.}$$

The solution set is $\{16\}$.

Skill Practice Solve.

1. $\sqrt{x} - 3 = 2$

Example 2 Solving an Equation Containing One Radical

Solve the equation. $\sqrt[3]{w - 1} - 2 = 2$

Solution:

$$\sqrt[3]{w - 1} - 2 = 2$$

$$\sqrt[3]{w - 1} = 4 \qquad \text{Isolate the radical.}$$

$$(\sqrt[3]{w - 1})^3 = (4)^3 \qquad \text{Because the index is 3, cube both sides.}$$

$$w - 1 = 64 \qquad \text{Simplify.}$$

$$w = 65$$

Answer

1. $\{25\}$

Check: $w = 65$

$\sqrt[3]{65 - 1} - 2 \stackrel{?}{=} 2$ Check $w = 65$ as a potential solution.

$\sqrt[3]{64} - 2 \stackrel{?}{=} 2$

$4 - 2 \stackrel{?}{=} 2$ ✔   True, 65 is a solution to the original equation.

The solution set is $\{65\}$.

Skill Practice Solve.

2. $\sqrt[3]{t + 2} + 5 = 3$

Example 3   **Solving an Equation Containing One Radical**

Solve the equation. $7 = (x + 3)^{1/4} + 9$

Solution:

$7 = (x + 3)^{1/4} + 9$

$7 = \sqrt[4]{x + 3} + 9$ Note that $(x + 3)^{1/4} = \sqrt[4]{x + 3}$.

$-2 = \sqrt[4]{x + 3}$ Isolate the radical.

$(-2)^4 = (\sqrt[4]{x + 3})^4$ Because the index is 4, raise both sides to the fourth power.

$16 = x + 3$

$x = 13$ Solve for x.

Check: $x = 13$

$7 = \sqrt[4]{x + 3} + 9$

$7 \stackrel{?}{=} \sqrt[4]{(13) + 3} + 9$

$7 \stackrel{?}{=} \sqrt[4]{16} + 9$

$7 \stackrel{?}{=} 2 + 9$ (false)   13 is *not* a solution to the original equation.

The equation $7 = \sqrt[4]{x + 3} + 9$ has no solution.

The solution set is the empty set, $\{\ \}$.

> **TIP:** After isolating the radical in Example 3, the equation shows a fourth root equated to a negative number:
>
> $-2 = \sqrt[4]{x + 3}$
>
> By definition, a principal fourth root of any real number must be nonnegative. Therefore, there can be no real solution to this equation.

Skill Practice Solve.

3. $3 = 6 + (x - 1)^{1/4}$

Example 4   **Solving an Equation Containing One Radical**

Solve the equation. $y + 2\sqrt{4y - 3} = 3$

Solution:

$y + 2\sqrt{4y - 3} = 3$

$2\sqrt{4y - 3} = 3 - y$ Isolate the radical term.

$(2\sqrt{4y - 3})^2 = (3 - y)^2$ Because the index is 2, square both sides.

$4(4y - 3) = 9 - 6y + y^2$ Note that $(2\sqrt{4y - 3})^2 = 2^2(\sqrt{4y - 3})^2$ and $(3 - y)^2 = (3 - y)(3 - y) = 9 - 6y + y^2$.

Answers

2. $\{-10\}$

3. $\{\ \}$ (The value 82 does not check.)

$$16y - 12 = 9 - 6y + y^2$$

$$0 = y^2 - 22y + 21 \quad \text{The equation is quadratic. Set one side equal to zero. Write the other side in descending order.}$$

$$0 = (y - 21)(y - 1) \quad \text{Factor.}$$

$$y - 21 = 0 \quad \text{or} \quad y - 1 = 0 \quad \text{Set each factor equal to zero.}$$

$$y = 21 \quad \text{or} \quad y = 1 \quad \text{Solve.}$$

Check: $y = 21$ Check: $y = 1$

$$y + 2\sqrt{4y - 3} = 3 \qquad\qquad y + 2\sqrt{4y - 3} = 3$$

$$21 + 2\sqrt{4(21) - 3} \overset{?}{=} 3 \qquad 1 + 2\sqrt{4(1) - 3} \overset{?}{=} 3$$

$$21 + 2\sqrt{81} \overset{?}{=} 3 \qquad\qquad 1 + 2\sqrt{1} \overset{?}{=} 3$$

$$21 + 18 \overset{?}{=} 3 \qquad\qquad\qquad 1 + 2 \overset{?}{=} 3$$

$$39 \overset{?}{=} 3 \text{ False} \qquad\qquad\qquad 3 \overset{?}{=} 3 \ \checkmark$$

The solution set is $\{1\}$. (The value 21 does not check.)

Skill Practice Solve.

4. $2\sqrt{m + 3} - m = 3$

3. Solving Radical Equations Involving More than One Radical

Example 5   Solving an Equation with Two Radicals

Solve the equation.

$$\sqrt[3]{2x - 4} = \sqrt[3]{1 - 8x}$$

Solution:

$$\sqrt[3]{2x - 4} = \sqrt[3]{1 - 8x}$$

$$(\sqrt[3]{2x - 4})^3 = (\sqrt[3]{1 - 8x})^3 \qquad \text{Because the index is 3, cube both sides.}$$

$$2x - 4 = 1 - 8x \qquad\qquad \text{Simplify.}$$

$$10x - 4 = 1 \qquad\qquad\qquad \text{Solve the resulting equation.}$$

$$10x = 5$$

$$x = \frac{1}{2} \qquad\qquad\qquad\qquad \text{Solve for } x.$$

Check: $x = \frac{1}{2}$

$$\sqrt[3]{2x - 4} = \sqrt[3]{1 - 8x}$$

$$\sqrt[3]{2\left(\frac{1}{2}\right) - 4} \overset{?}{=} \sqrt[3]{1 - 8\left(\frac{1}{2}\right)}$$

$$\sqrt[3]{1 - 4} \overset{?}{=} \sqrt[3]{1 - 4}$$

$$\sqrt[3]{-3} \overset{?}{=} \sqrt[3]{-3} \ \checkmark \ \text{(True)} \qquad \text{Therefore, } \frac{1}{2} \text{ is a solution to the original equation.}$$

The solution set is $\left\{\dfrac{1}{2}\right\}$.

Answer

4. $\{-3, 1\}$

Skill Practice Solve.

5. $\sqrt[5]{2y - 1} = \sqrt[5]{10y + 3}$

Example 6 Solving an Equation with Two Radicals

Solve the equation. $\sqrt{3m + 1} - \sqrt{m + 4} = 1$

Solution:

$\sqrt{3m + 1} - \sqrt{m + 4} = 1$

$\sqrt{3m + 1} = \sqrt{m + 4} + 1$ Isolate one of the radicals.

$(\sqrt{3m + 1})^2 = (\sqrt{m + 4} + 1)^2$ Square both sides.

$3m + 1 = m + 4 + 2\sqrt{m + 4} + 1$

Note: $(\sqrt{m + 4} + 1)^2$

$= (\sqrt{m + 4})^2 + 2(1)\sqrt{m + 4} + (1)^2$

$= m + 4 + 2\sqrt{m + 4} + 1$

$3m + 1 = m + 5 + 2\sqrt{m + 4}$ Combine *like* terms.

$2m - 4 = 2\sqrt{m + 4}$ Isolate the radical again.

$m - 2 = \sqrt{m + 4}$ Divide both sides by 2.

$(m - 2)^2 = (\sqrt{m + 4})^2$ Square both sides again.

$m^2 - 4m + 4 = m + 4$ The resulting equation is quadratic.

$m^2 - 5m = 0$ Set one side equal to zero.

$m(m - 5) = 0$ Factor.

$m = 0$ or $m = 5$

TIP: In Example 6, we divided the equation by 2 because all coefficients were divisible by 2. This made the coefficients smaller before we squared both sides of the equation a second time.

Check: $m = 0$

$\sqrt{3(0) + 1} - \sqrt{(0) + 4} \overset{?}{=} 1$

$\sqrt{1} - \sqrt{4} \overset{?}{=} 1$

$1 - 2 \overset{?}{=} 1$ (False)

Check: $m = 5$

$\sqrt{3(5) + 1} - \sqrt{(5) + 4} = 1$

$\sqrt{16} - \sqrt{9} \overset{?}{=} 1$

$4 - 3 \overset{?}{=} 1$ ✔

The solution set is $\{5\}$. (The value 0 does not check.)

Skill Practice Solve.

6. $\sqrt{3c + 1} - \sqrt{c - 1} = 2$

Answers

5. $\left\{ -\dfrac{1}{2} \right\}$ **6.** $\{1, 5\}$

Example 7   Solving an Equation with Two Radicals

Solve the equation. $\sqrt{5x + 4} = 1 + \sqrt{5x - 3}$

Solution:

$\sqrt{5x + 4} = 1 + \sqrt{5x - 3}$	One radical is already isolated.
$(\sqrt{5x + 4})^2 = (1 + \sqrt{5x - 3})^2$	Square both sides.
$5x + 4 = 1 + 2\sqrt{5x - 3} + 5x - 3$	*Note:* $(1 + \sqrt{5x - 3})^2 = 1^2 + 2 \cdot 1\sqrt{5x - 3} + (\sqrt{5x - 3})^2$
$5x + 4 = 2\sqrt{5x - 3} + 5x - 2$	Combine *like* terms.
$4 = 2\sqrt{5x - 3} - 2$	Subtract $5x$ from both sides.
$6 = 2\sqrt{5x - 3}$	Isolate the radical.
$3 = \sqrt{5x - 3}$	Divide both sides by 2.
$(3)^2 = (\sqrt{5x - 3})^2$	Square both sides again.
$9 = 5x - 3$	The resulting equation is linear.
$12 = 5x$	Solve for x.
$\dfrac{12}{5} = x$	Check: $x = \dfrac{12}{5}$

$$\sqrt{5\left(\frac{12}{5}\right) + 4} \overset{?}{=} 1 + \sqrt{5\left(\frac{12}{5}\right) - 3}$$

$$\sqrt{16} \overset{?}{=} 1 + \sqrt{9}$$

$$4 \overset{?}{=} 1 + 3 \quad \checkmark \text{ True}$$

The solution set is $\left\{\dfrac{12}{5}\right\}$.

Skill Practice Solve.

7. $\sqrt{4x - 3} = 2 - \sqrt{4x + 1}$

4. Applications of Radical Equations and Functions

Example 8   Applying a Radical Equation in Geometry

For a pyramid with a square base, the length of a side of the base b is given by

$$b = \sqrt{\frac{3V}{h}}$$

where V is the volume and h is the height.

The Pyramid of the Pharoah Khufu (known as the Great Pyramid) at Giza has a square base (Figure 6-3). If the distance around the bottom of the pyramid is 921.6 m and the height is 146.6 m, what is the volume of the pyramid?

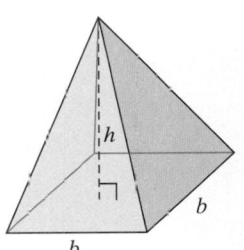

Figure 6-3

Answer

7. $\left\{\dfrac{3}{4}\right\}$

Solution:

$$b = \sqrt{\frac{3V}{h}}$$

$$b^2 = \left(\sqrt{\frac{3V}{h}}\right)^2 \qquad \text{Because the index is 2, square both sides.}$$

$$b^2 = \frac{3V}{h} \qquad \text{Simplify.}$$

$$b^2 \cdot h = \frac{3V}{h} \cdot h \qquad \text{Multiply both sides by } h.$$

$$b^2 h = 3V$$

$$\frac{b^2 h}{3} = \frac{3V}{3} \qquad \text{Divide both sides by 3.}$$

$$\frac{b^2 h}{3} = V$$

The length of a side b (in meters) is given by $\dfrac{921.6}{4} = 230.4$ m.

$$\frac{(230.4)^2(146.6)}{3} = V \qquad \text{Substitute } b = 230.4 \text{ and } h = 146.6.$$

$$2{,}594{,}046 \approx V$$

The volume of the Great Pyramid at Giza is approximately 2,594,046 m³.

Skill Practice

8. The length of the legs, s, of an isosceles right triangle is $s = \sqrt{2A}$, where A is the area. If the legs of the triangle are 9 in., find the area.

Example 9 **Applying a Radical Function**

On a certain surface, the speed $s(x)$ (in miles per hour) of a car before the brakes were applied can be approximated from the length of its skid marks x (in feet) by

$$s(x) = 3.8\sqrt{x} \qquad x \geq 0 \quad \text{See Figure 6-4.}$$

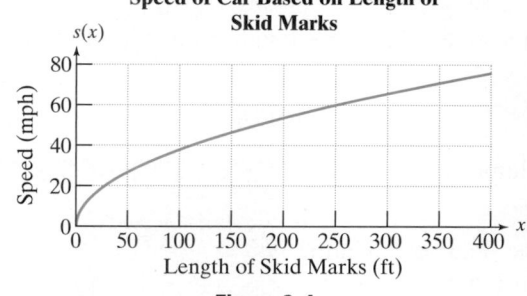

Figure 6-4

a. Find the speed of a car before the brakes were applied if its skid marks are 361 ft long.

b. How long would you expect the skid marks to be if the car had been traveling the speed limit of 50 mph? (Round to the nearest foot.)

Answer

8. 40.5 in.²

Solution:

a. $s(x) = 3.8\sqrt{x}$

$\quad s(361) = 3.8\sqrt{361}$ Substitute $x = 361$.

$\quad\quad\quad = 3.8(19)$

$\quad\quad\quad = 72.2$

If the skid marks are 361 ft, the car was traveling approximately 72.2 mph before the brakes were applied.

b. $s(x) = 3.8\sqrt{x}$

$\quad\quad 50 = 3.8\sqrt{x}$ Substitute $s(x) = 50$ and solve for x.

$\quad \dfrac{50}{3.8} = \sqrt{x}$ Isolate the radical.

$\quad \left(\dfrac{50}{3.8}\right)^2 = x$

$\quad\quad x \approx 173$

If the car had been going the speed limit (50 mph), then the length of the skid marks would have been approximately 173 ft.

Skill Practice When an object is dropped from a height of 64 ft, the time $t(x)$ (in seconds) it takes to reach a height x (in feet) is given by

$$t(x) = \frac{1}{4}\sqrt{64 - x}$$

9. Find the time to reach a height of 28 ft from the ground.

10. What is the height after 1 sec?

Answers

9. $\dfrac{3}{2}$ sec   **10.** 48 ft

Section 6.7   Practice Exercises

Vocabulary and Key Concepts

1. a. The equation $\sqrt{x + 5} + 7 = 11$ is an example of a _____ equation.

    b. The first step to solve the equation $\sqrt{x + 5} + 7 = 11$ is to _____ the radical by subtracting _____ from both sides of the equation.

    c. When solving a radical equation some potential solutions may not check in the original equation. These are called _____ solutions.

    d. To solve the equation $\sqrt[3]{w - 1} = 5$, raise both sides of the equation to the _____ power.

Review Exercises

2. Identify the equation as linear or quadratic. Then solve the equation.

    a. $2x + 3 = 23$ **b.** $2x^2 - 9x = 5$

For Exercises 3–6, simplify the radical expressions. Assume all variables represent positive real numbers.

3. $\sqrt{\dfrac{9w^3}{16}}$ **4.** $\sqrt{\dfrac{a^2}{3}}$ **5.** $\sqrt[3]{54c^4}$ **6.** $\sqrt{\dfrac{49}{5t^3}}$

For Exercises 7–10, simplify each expression. Assume all radicands represent positive real numbers.

7. $(\sqrt{4x-6})^2$ **8.** $(\sqrt{5y+2})^2$ **9.** $(\sqrt[3]{9p+7})^3$ **10.** $(\sqrt[3]{4t+13})^3$

Concept 2: Solving Radical Equations Involving One Radical

For Exercises 11–30, solve the equations. (See Examples 1–3.)

11. $\sqrt{x} = 10$ **12.** $\sqrt{y} = 7$ **13.** $\sqrt{x} + 4 = 6$ **14.** $\sqrt{x} + 2 = 8$

15. $\sqrt{5y+1} = 4$ **16.** $\sqrt{9z-5} - 2 = 9$ **17.** $6 = \sqrt{2z-3} - 3$ **18.** $4 = \sqrt{8+3a} - 1$

19. $\sqrt{x^2+5} = x+1$ **20.** $\sqrt{y^2-8} = y-2$ **21.** $\sqrt[3]{x-2} - 1 = 2$ **22.** $\sqrt[3]{2x-5} - 1 = 1$

23. $(15-w)^{1/3} + 7 = 2$ **24.** $(k+18)^{1/3} + 5 = 3$ **25.** $3 + \sqrt{x-16} = 0$ **26.** $12 + \sqrt{2x+1} = 0$

27. $2\sqrt{6a+7} - 2a = 0$ **28.** $2\sqrt{3-w} - w = 0$ **29.** $(2x-5)^{1/4} = -1$ **30.** $(x+16)^{1/4} = -4$

For Exercises 31–34, assume all variables represent positive real numbers.

31. Solve for V: $r = \sqrt[3]{\dfrac{3V}{4\pi}}$ **32.** Solve for V: $r = \sqrt{\dfrac{V}{h\pi}}$

33. Solve for h^2: $r = \pi\sqrt{r^2+h^2}$ **34.** Solve for d: $s = 1.3\sqrt{d}$

For Exercises 35–40, square the expression as indicated.

35. $(a+5)^2$ **36.** $(b+7)^2$ **37.** $(\sqrt{5a}-3)^2$

38. $(2+\sqrt{b})^2$ **39.** $(\sqrt{r-3}+5)^2$ **40.** $(2-\sqrt{2t-4})^2$

For Exercises 41–46, solve the radical equations, if possible. (See Example 4.)

41. $\sqrt{a^2+2a+1} = a+5$ **42.** $\sqrt{b^2-5b-8} = b+7$ **43.** $\sqrt{25w^2-2w-3} = 5w-4$

44. $\sqrt{4p^2-2p+1} = 2p-3$ **45.** $4\sqrt{p-2} - 2 = -p$ **46.** $x - 3\sqrt{x-5} = 5$

Concept 3: Solving Radical Equations Involving More than One Radical

For Exercises 47–70, solve the radical equations, if possible. (See Examples 5–7.)

47. $\sqrt[4]{h+4} = \sqrt[4]{2h-5}$ **48.** $\sqrt[4]{3b+6} = \sqrt[4]{7b-6}$ **49.** $\sqrt[3]{5a+3} - \sqrt[3]{a-13} = 0$

50. $\sqrt[3]{k-8} - \sqrt[3]{4k+1} = 0$ **51.** $\sqrt{5a-9} = \sqrt{5a} - 3$ **52.** $\sqrt{8+b} = 2 + \sqrt{b}$

53. $\sqrt{2h+5} - \sqrt{2h} = 1$ **54.** $\sqrt{3k-5} - \sqrt{3k} = -1$ **55.** $(t-9)^{1/2} - t^{1/2} = 3$

56. $(y-16)^{1/2} - y^{1/2} = 4$ **57.** $6 = \sqrt{x^2+3} - x$ **58.** $2 = \sqrt{y^2+5} - y$

59. $\sqrt{3t-7} = 2 - \sqrt{3t+1}$ **60.** $\sqrt{p-6} = \sqrt{p+2} - 4$ **61.** $\sqrt{z+1} + \sqrt{2z+3} = 1$

62. $\sqrt{2y+6} = \sqrt{7-2y} + 1$ **63.** $\sqrt{6m+7} - \sqrt{3m+3} = 1$ **64.** $\sqrt{5w+1} - \sqrt{3w} = 1$

65. $2 + 2\sqrt{2t + 3} + 2\sqrt{3t - 5} = 0$ **66.** $6 - 3\sqrt{3x + 1} + 3\sqrt{x - 1} = 0$ **67.** $3\sqrt{y - 3} = \sqrt{4y + 3}$

68. $\sqrt{5x - 8} = 2\sqrt{x - 1}$ **69.** $\sqrt{p + 7} = \sqrt{2p} + 1$ **70.** $\sqrt{t} = \sqrt{t - 12} + 2$

Concept 4: Applications of Radical Equations and Functions

71. If an object is dropped from an initial height h, its velocity at impact with the ground is given by

$$v = \sqrt{2gh}$$

where g is the acceleration due to gravity and h is the initial height. **(See Example 8.)**

  a. Find the initial height (in feet) of an object if its velocity at impact is 44 ft/sec. (Assume that the acceleration due to gravity is $g = 32$ ft/sec².)

  b. Find the initial height (in meters) of an object if its velocity at impact is 26 m/sec. (Assume that the acceleration due to gravity is $g = 9.8$ m/sec².) Round to the nearest tenth of a meter.

72. The time T (in seconds) required for a pendulum to make one complete swing back and forth is approximated by

$$T = 2\pi\sqrt{\frac{L}{g}}$$

where g is the acceleration due to gravity and L is the length of the pendulum (in feet).

  a. Find the length of a pendulum that requires 1.36 sec to make one complete swing back and forth. (Assume that the acceleration due to gravity is $g = 32$ ft/sec².) Round to the nearest tenth of a foot.

  b. Find the time required for a pendulum to complete one swing back and forth if the length of the pendulum is 4 ft. (Assume that the acceleration due to gravity is $g = 32$ ft/sec².) Round to the nearest tenth of a second.

73. The airline cost for x thousand passengers to travel round trip from New York to Atlanta is given by

$$C(x) = \sqrt{0.3x + 1}$$

where $C(x)$ is measured in millions of dollars and $x \geq 0$. **(See Example 9.)**

  a. Find the airline's cost for 10,000 passengers ($x = 10$) to travel from New York to Atlanta.

  b. If the airline charges $320 per passenger, find the profit made by the airline for flying 10,000 passengers from New York to Atlanta.

  c. Approximate the number of passengers who traveled from New York to Atlanta if the total cost for the airline was $4 million.

74. The time $t(d)$ in seconds it takes an object to drop d meters is given by

$$t(d) = \sqrt{\frac{d}{4.9}}$$

  a. Approximate the height of the Texas Commerce Tower in Houston if it takes an object 7.89 sec to drop from the top. Round to the nearest meter.

  b. Approximate the height of the Willis Tower in Chicago if it takes an object 9.51 sec to drop from the top. Round to the nearest meter.

Expanding Your Skills

 75. The number of hours needed to cook a turkey that weighs x pounds can be approximated by

$$t(x) = 0.90\sqrt[5]{x^3}$$

where $t(x)$ is the time in hours and x is the weight of the turkey in pounds.

a. Find the weight of a turkey that cooked for 4 hr. Round to the nearest pound.

b. Find $t(18)$ and interpret the result. Round to the nearest tenth of an hour.

For Exercises 76–79, use the Pythagorean theorem to find a, b, or c.

76. Find b when $a = 2$ and $c = y$.

77. Find b when $a = h$ and $c = 5$.

78. Find a when $b = x$ and $c = 8$.

79. Find a when $b = 14$ and $c = k$.

Graphing Calculator Exercises

80. Graph Y_1 and Y_2 on a viewing window defined by $-10 \le x \le 40$ and $-5 \le y \le 10$.

$$Y_1 = \sqrt{2x} \quad \text{and} \quad Y_2 = 8$$

Use an *Intersect* feature to approximate the x-coordinate of the point of intersection of the two graphs. How does the point of intersection relate to the solution to the equation $\sqrt{2x} = 8$?

81. Graph Y_1 and Y_2 on a viewing window defined by $-10 \le x \le 20$ and $-5 \le y \le 10$.

$$Y_1 = \sqrt{4x} \quad \text{and} \quad Y_2 = 6$$

Use an *Intersect* feature to approximate the x-coordinate of the point of intersection of the two graphs. How does the point of intersection relate to the solution to the equation $\sqrt{4x} = 6$?

82. Refer to Exercise 48. Graph Y_1 and Y_2 on a viewing window defined by $-5 \le x \le 20$ and $-1 \le y \le 4$.

$$Y_1 = \sqrt[4]{3x + 6} \quad \text{and} \quad Y_2 = \sqrt[4]{7x - 6}$$

Use an *Intersect* feature to approximate the x-coordinate of the point of intersection of the two graphs to support your solution to Exercise 48.

83. Refer to Exercise 47. Graph Y_1 and Y_2 on a viewing window defined by $-5 \le x \le 20$ and $-1 \le y \le 4$.

$$Y_1 = \sqrt[4]{x + 4} \quad \text{and} \quad Y_2 = \sqrt[4]{2x - 5}$$

Use an *Intersect* feature to approximate the x-coordinate of the point of intersection of the two graphs to support your solution to Exercise 47.

Section 6.8   Complex Numbers

Concepts

1. Definition of i
2. Powers of i
3. Definition of a Complex Number
4. Addition, Subtraction, and Multiplication of Complex Numbers
5. Division and Simplification of Complex Numbers

1. Definition of i

In Section 6.1, we learned that there are no real-valued square roots of a negative number. For example, $\sqrt{-9}$ is not a real number because no real number when squared equals -9. However, the square roots of a negative number are defined over another set of numbers called the **imaginary numbers**. The foundation of the set of imaginary numbers is the definition of the imaginary number i.

> **Definition of the Imaginary Number i**
>
> $$i = \sqrt{-1}$$
>
> *Note:* From the definition of i, it follows that $i^2 = -1$.

Using the imaginary number i, we can define the square root of any negative real number.

Definition of $\sqrt{-b}$ for $b > 0$

Let b be a positive real number. Then $\sqrt{-b} = i\sqrt{b}$.

Example 1 Simplifying Expressions in Terms of i

Simplify the expressions in terms of i.

a. $\sqrt{-64}$ **b.** $\sqrt{-50}$ **c.** $-\sqrt{-4}$ **d.** $\sqrt{-29}$

Solution:

a. $\sqrt{-64} = i\sqrt{64}$

$\qquad = 8i$

b. $\sqrt{-50} = i\sqrt{50}$

$\qquad = i\sqrt{5^2 \cdot 2}$

$\qquad = 5i\sqrt{2}$

c. $-\sqrt{-4} = -1 \cdot \sqrt{-4}$

$\qquad\quad = -1 \cdot 2i$

$\qquad\quad = -2i$

d. $\sqrt{-29} = i\sqrt{29}$

> **Avoiding Mistakes**
>
> In an expression such as $i\sqrt{29}$, the i is often written in front of the square root. The expression $\sqrt{29}\,i$ is also correct, but may be misinterpreted as $\sqrt{29i}$ (with i incorrectly placed under the radical).

Skill Practice Simplify the expressions in terms of i.

1. $\sqrt{-81}$ **2.** $\sqrt{-20}$ **3.** $-\sqrt{-36}$ **4.** $\sqrt{-7}$

The multiplication and division properties of radicals were presented in Sections 6.3 and 6.5 as follows:

If a and b represent real numbers such that $\sqrt[n]{a}$ and $\sqrt[n]{b}$ are both real, then

$$\sqrt[n]{ab} = \sqrt[n]{a} \cdot \sqrt[n]{b} \qquad \text{and} \qquad \sqrt[n]{\frac{a}{b}} = \frac{\sqrt[n]{a}}{\sqrt[n]{b}} \qquad b \neq 0$$

The conditions that $\sqrt[n]{a}$ and $\sqrt[n]{b}$ must both be real numbers prevent us from applying the multiplication and division properties of radicals for square roots with a negative radicand. Therefore, to multiply or divide radicals with a negative radicand, first write the radical in terms of the imaginary number i. This is demonstrated in Example 2.

Example 2 Simplifying a Product or Quotient in Terms of i

Simplify the expressions.

a. $\dfrac{\sqrt{-100}}{\sqrt{-25}}$ **b.** $\sqrt{-25} \cdot \sqrt{-9}$ **c.** $\sqrt{-5} \cdot \sqrt{-5}$

Answers

1. $9i$ **2.** $2i\sqrt{5}$ **3.** $-6i$ **4.** $i\sqrt{7}$

Solution:

a. $\dfrac{\sqrt{-100}}{\sqrt{-25}}$

$= \dfrac{10i}{5i}$ Simplify each radical in terms of i *before* dividing.

$= 2$

b. $\sqrt{-25} \cdot \sqrt{-9}$

$= 5i \cdot 3i$ Simplify each radical in terms of i first *before* multiplying.

$= 15i^2$ Multiply.

$= 15(-1)$ Recall that $i^2 = -1$.

$= -15$ Simplify.

c. $\sqrt{-5} \cdot \sqrt{-5}$

$= i\sqrt{5} \cdot i\sqrt{5}$

$= i^2 \cdot (\sqrt{5})^2$

$= -1 \cdot 5$

$= -5$

Skill Practice Simplify the expressions.

5. $\dfrac{\sqrt{-36}}{\sqrt{-9}}$ 6. $\sqrt{-16} \cdot \sqrt{-49}$ 7. $\sqrt{-2} \cdot \sqrt{-2}$

Avoiding Mistakes

In Example 2, we wrote the radical expressions in terms of i first, before multiplying or dividing. If we had mistakenly applied the multiplication or division property first, we would have obtained an incorrect answer.

Correct: $\sqrt{-25} \cdot \sqrt{-9}$

$= (5i)(3i) = 15i^2$

$= 15(-1) = -15$

↑ correct

Be careful: $\sqrt{-25} \cdot \sqrt{-9}$

$\neq \sqrt{225} = 15$

↑ (incorrect answer)

$\sqrt{-25}$ and $\sqrt{-9}$ are not real numbers. Therefore, the multiplication property of radicals cannot be applied.

---

2. Powers of i

From the definition of $i = \sqrt{-1}$, it follows that

$$i = i$$

$$i^2 = -1$$

$$i^3 = -i \qquad \text{because } i^3 = i^2 \cdot i = (-1)i = -i$$

$$i^4 = 1 \qquad \text{because } i^4 = i^2 \cdot i^2 = (-1)(-1) = 1$$

$$i^5 = i \qquad \text{because } i^5 = i^4 \cdot i = (1)i = i$$

$$i^6 = -1 \qquad \text{because } i^6 = i^4 \cdot i^2 = (1)(-1) = -1$$

Answers

5. 2   **6.** -28 **7.** -2

This pattern of values $i, -1, -i, 1, i, -1, -i, 1, \ldots$ continues for all subsequent powers of i. Table 6-1 lists several powers of i.

Table 6-1   Powers of i

$i^1 = i$	$i^5 = i$	$i^9 = i$
$i^2 = -1$	$i^6 = -1$	$i^{10} = -1$
$i^3 = -i$	$i^7 = -i$	$i^{11} = -i$
$i^4 = 1$	$i^8 = 1$	$i^{12} = 1$

To simplify higher powers of i, we can decompose the expression into multiples of i^4 ($i^4 = 1$) and write the remaining factors as i, i^2, or i^3.

Example 3   Simplifying Powers of i

Simplify the powers of i.

a. i^{13} **b.** i^{18} **c.** i^{107} **d.** i^{32}

Solution:

a. $i^{13} = (i^{12}) \cdot (i)$ Write the exponent as a multiple of 4 and a remainder.

$\quad = (i^4)^3 \cdot (i)$

$\quad = (1)^3(i)$ Recall that $i^4 = 1$.

$\quad = i$ Simplify.

b. $i^{18} = (i^{16}) \cdot (i^2)$ Write the exponent as a multiple of 4 and a remainder.

$\quad = (i^4)^4 \cdot (i^2)$

$\quad = (1)^4 \cdot (-1)$ $i^4 = 1$ and $i^2 = -1$

$\quad = -1$ Simplify.

c. $i^{107} = (i^{104}) \cdot (i^3)$ Write the exponent as a multiple of 4 and a remainder.

$\quad = (i^4)^{26}(i^3)$

$\quad = (1)^{26}(-i)$ $i^4 = 1$ and $i^3 = -i$

$\quad = -i$ Simplify.

d. $i^{32} = (i^4)^8$

$\quad = (1)^8$ $i^4 = 1$

$\quad = 1$ Simplify.

Skill Practice Simplify the powers of i.

8. i^{45} **9.** i^{22} **10.** i^{31} **11.** i^{80}

3. Definition of a Complex Number

We have already learned the definitions of the integers, rational numbers, irrational numbers, and real numbers. In this section, we define the complex numbers.

Answers

8. i **9.** -1

10. $-i$ **11.** 1

> ### Definition of a Complex Number
>
> A **complex number** is a number of the form $a + bi$, where a and b are real numbers and $i = \sqrt{-1}$.
>
> *Notes:*
>
> - If $b = 0$, then the complex number $a + bi$ is a real number.
> - If $b \neq 0$, then we say that $a + bi$ is an imaginary number.
> - The complex number $a + bi$ is said to be written in standard form. The quantities a and b are called the real and imaginary parts (respectively) of the complex number.
> - The complex numbers $a - bi$ and $a + bi$ are called **complex conjugates**.

From the definition of a complex number, it follows that all real numbers are complex numbers and all imaginary numbers are complex numbers. Figure 6-5 illustrates the relationship among the sets of numbers we have learned so far.

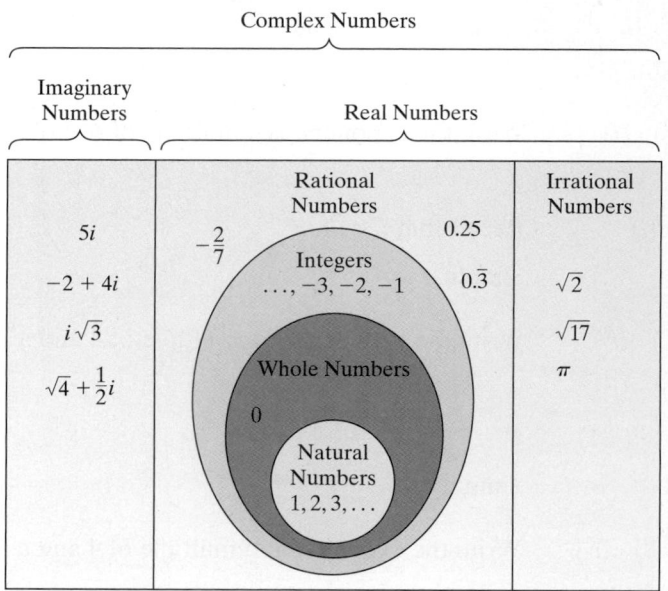

Figure 6-5

Example 4 **Identifying the Real and Imaginary Parts of a Complex Number**

Identify the real and imaginary parts of the complex numbers.

a. $-8 + 2i$ **b.** $\dfrac{3}{2}$ **c.** $-1.75i$

Solution:

a. $-8 + 2i$ -8 is the real part, and 2 is the imaginary part.

b. $\dfrac{3}{2} = \dfrac{3}{2} + 0i$ Rewrite $\frac{3}{2}$ in the form $a + bi$.
                      $\frac{3}{2}$ is the real part, and 0 is the imaginary part.

c. $-1.75i$

    $= 0 + -1.75i$ Rewrite $-1.75i$ in the form $a + bi$.
                       0 is the real part, and -1.75 is the imaginary part.

12. $22 - 14i$ **13.** -50 **14.** $15i$

> **TIP:** Example 4(b) illustrates that a real number is also a complex number.
>
> $$\frac{3}{2} = \frac{3}{2} + 0i$$
>
> Example 4(c) illustrates that an imaginary number is also a complex number.
>
> $$-1.75i = 0 + -1.75i$$

4. Addition, Subtraction, and Multiplication of Complex Numbers

The operations of addition, subtraction, and multiplication of real numbers also apply to imaginary numbers. To add or subtract complex numbers, combine the real parts and combine the imaginary parts. The commutative, associative, and distributive properties that apply to real numbers also apply to complex numbers.

Example 5 Adding and Subtracting Complex Numbers

Add or subtract as indicated. Write the answers in the form $a + bi$.

a. $(1 - 5i) + (-3 + 7i)$ **b.** $\left(-\frac{1}{4} + \frac{3}{5}i\right) - \left(\frac{1}{2} - \frac{1}{10}i\right)$

c. $\sqrt{-8} + \sqrt{-18}$

Solution:

real parts

a. $(1 - 5i) + (-3 + 7i) = (1 + -3) + (-5 + 7)i$ Add the real parts.
Add the imaginary parts.

imaginary parts

$$= -2 + 2i$$ Simplify.

b. $\left(-\frac{1}{4} + \frac{3}{5}i\right) - \left(\frac{1}{2} - \frac{1}{10}i\right) = -\frac{1}{4} + \frac{3}{5}i - \frac{1}{2} + \frac{1}{10}i$ Apply the distributive property.

$$= \left(-\frac{1}{4} - \frac{1}{2}\right) + \left(\frac{3}{5} + \frac{1}{10}\right)i$$ Add real parts.
Add imaginary parts.

$$= \left(-\frac{1}{4} - \frac{2}{4}\right) + \left(\frac{6}{10} + \frac{1}{10}\right)i$$ Get common denominators.

$$= -\frac{3}{4} + \frac{7}{10}i$$ Simplify.

c. $\sqrt{-8} + \sqrt{-18} = 2i\sqrt{2} + 3i\sqrt{2}$ Simplify each radical in terms of i.

$$= 5i\sqrt{2}$$ Combine *like* radicals.

15. $\left(\frac{1}{2} - \frac{1}{4}i\right) + \left(\frac{3}{5} + \frac{2}{3}i\right)$ **16.** $(-6 + 11i) - (-9 - 12i)$ **17.** $\sqrt{-45} - \sqrt{-20}$

Answers

12. real: 22; imaginary: -14
13. real: -50; imaginary 0
14. real: 0; imaginary: 15
15. $\frac{11}{10} + \frac{5}{12}i$ **16.** $3 + 23i$
17. $i\sqrt{5}$

Example 6 **Multiplying Complex Numbers**

Multiply the complex numbers. Write the answers in the form $a + bi$.

a. $(10 - 5i)(2 + 3i)$ **b.** $(1.2 + 0.5i)(1.2 - 0.5i)$

Solution:

a. $(10 - 5i)(2 + 3i)$

$= (10)(2) + (10)(3i) + (-5i)(2) + (-5i)(3i)$ Apply the distributive property.

$= 20 + 30i - 10i - 15i^2$

$= 20 + 20i - (15)(-1)$ Recall $i^2 = -1$.

$= 20 + 20i + 15$

$= 35 + 20i$ Write in the form $a + bi$.

b. $(1.2 + 0.5i)(1.2 - 0.5i)$

The expressions $(1.2 + 0.5i)$ and $(1.2 - 0.5i)$ are complex conjugates. The product is a difference of squares.

$$(a + b)(a - b) = a^2 - b^2$$

$(1.2 + 0.5i)(1.2 - 0.5i) = (1.2)^2 - (0.5i)^2$ Apply the formula, where $a = 1.2$ and $b = 0.5i$.

$= 1.44 - 0.25i^2$

$= 1.44 - 0.25(-1)$ Recall $i^2 = -1$.

$= 1.44 + 0.25$

$= 1.69 + 0i$

Skill Practice Multiply.

18. $(4 - 6i)(2 - 3i)$ **19.** $(1.5 + 0.8i)(1.5 - 0.8i)$

5. Division and Simplification of Complex Numbers

The product of a complex number and its complex conjugate is a real number. For example:

$$(5 + 3i)(5 - 3i) = 25 - 9i^2$$
$$= 25 - 9(-1)$$
$$= 25 + 9$$
$$= 34$$

To divide by a complex number, multiply the numerator and denominator by the complex conjugate of the denominator. This produces a real number in the denominator so that the resulting expression can be written in the form $a + bi$.

Answers

18. $-10 - 24i$ **19.** $2.89 + 0i$

Example 7   **Dividing by a Complex Number**

Divide the complex numbers and write the answer in the form $a + bi$.

$$\frac{4 - 3i}{5 + 2i}$$

Solution:

$\dfrac{4 - 3i}{5 + 2i}$ Multiply the numerator and denominator by the complex conjugate of the denominator:

$$\frac{(4 - 3i)}{(5 + 2i)} \cdot \frac{(5 - 2i)}{(5 - 2i)} = \frac{(4)(5) + (4)(-2i) + (-3i)(5) + (-3i)(-2i)}{(5)^2 - (2i)^2}$$

$$= \frac{20 - 8i - 15i + 6i^2}{25 - 4i^2}$$ Simplify numerator and denominator.

$$= \frac{20 - 23i + 6(-1)}{25 - 4(-1)}$$ Recall $i^2 = -1$.

$$= \frac{20 - 23i - 6}{25 + 4}$$

$$= \frac{14 - 23i}{29}$$ Simplify.

$$= \frac{14}{29} - \frac{23}{29}i$$ Write in the form $a + bi$.

TIP: In Example 7, we asked for the answer in the form $a + bi$. However, if the second term is negative, we often leave an answer in terms of subtraction: $\frac{14}{29} - \frac{23}{29}i$. This is the same as $\frac{14}{29} + (-\frac{23}{29}i)$.

Skill Practice Divide the complex numbers. Write the answer in the form $a + bi$.

20. $\dfrac{2 + i}{3 - 2i}$

Example 8   **Simplifying a Complex Number**

Simplify the complex number. $\dfrac{6 + \sqrt{-18}}{9}$

Solution:

$$\frac{6 + \sqrt{-18}}{9} = \frac{6 + i\sqrt{18}}{9}$$ Write the radical in terms of i.

$$= \frac{6 + 3i\sqrt{2}}{9}$$ Simplify $\sqrt{18} = 3\sqrt{2}$.

$$= \frac{3(2 + i\sqrt{2})}{9}$$ Factor the numerator.

$$= \frac{\overset{1}{3}(2 + i\sqrt{2})}{\underset{3}{9}}$$ Simplify.

$$= \frac{2 + i\sqrt{2}}{3} \quad \text{or} \quad \frac{2}{3} + \frac{\sqrt{2}}{3}i$$ Write in the form $a + bi$.

TIP: As an alternative approach in Example 8, the expression $\dfrac{6 + i\sqrt{18}}{9}$ can be written in the form $a + bi$ and then simplified.

$$\frac{6 + i\sqrt{18}}{9} = \frac{6}{9} + \frac{3\sqrt{2}}{9}i$$
$$= \frac{2}{3} + \frac{\sqrt{2}}{3}i$$

Answers

20. $\dfrac{4}{13} + \dfrac{7}{13}i$

21. $\dfrac{4 - i\sqrt{6}}{3}$ or $\dfrac{4}{3} - \dfrac{\sqrt{6}}{3}i$

Skill Practice Simplify the complex number.

21. $\dfrac{8 - \sqrt{-24}}{6}$

Section 6.8 Practice Exercises

Vocabulary and Key Concepts

1. a. A square root of a negative number is not a real number, but rather is an _____ number.

   b. $i = $ _____ , and $i^2 = $ _____ .

   c. For a positive number b, the value $\sqrt{-b} = $ _____ .

   d. A complex number is a number in the form _____ , where a and b are real numbers and $i = $ _____ .

   e. Given a complex number $a + bi$, the value a is called the _____ part, and _____ is called the imaginary part.

   f. The complex conjugate of $a - bi$ is _____ .

   g. Answer true or false. All real numbers are complex numbers.

   h. Answer true or false. All imaginary numbers are complex numbers.

Review Exercises

For Exercises 2–4, perform the indicated operations.

2. $-2\sqrt{5} - 3\sqrt{50} + \sqrt{125}$

3. $(3 - \sqrt{x})(3 + \sqrt{x})$

4. $(\sqrt{5} + \sqrt{2})^2$

For Exercises 5–7, solve the equations.

5. $\sqrt[3]{3p + 7} - \sqrt[3]{2p - 1} = 0$

6. $\sqrt{9t + 10} + 18 = 10$

7. $\sqrt{4a + 29} = 2\sqrt{a} + 5$

8. Rationalize the denominator. $\dfrac{2}{\sqrt{x} - 3}$

Concept 1: Definition of i

9. Simplify the expressions $\sqrt{-1}$ and $-\sqrt{1}$.

10. Simplify i^2.

For Exercises 11–30, simplify the expressions. **(See Examples 1–2.)**

11. $\sqrt{-144}$

12. $\sqrt{-81}$

13. $\sqrt{-3}$

14. $\sqrt{-17}$

15. $-\sqrt{-20}$

16. $-\sqrt{-75}$

17. $(2\sqrt{-25})(3\sqrt{-4})$

18. $(-4\sqrt{-9})(-3\sqrt{-1})$

19. $7\sqrt{-63} - 4\sqrt{-28}$

20. $7\sqrt{-3} - 4\sqrt{-27}$

21. $\sqrt{-7} \cdot \sqrt{-7}$

22. $\sqrt{-11} \cdot \sqrt{-11}$

23. $\sqrt{-9} \cdot \sqrt{-16}$

24. $\sqrt{-25} \cdot \sqrt{-36}$

25. $\sqrt{-15} \cdot \sqrt{-6}$

26. $\sqrt{-12} \cdot \sqrt{-50}$

27. $\dfrac{\sqrt{-50}}{\sqrt{25}}$

28. $\dfrac{\sqrt{-27}}{\sqrt{9}}$

29. $\dfrac{\sqrt{-90}}{\sqrt{-10}}$

30. $\dfrac{\sqrt{-125}}{\sqrt{-45}}$

Concept 2: Powers of i

For Exercises 31–42, simplify the powers of i. **(See Example 3.)**

31. i^7

32. i^{38}

33. i^{64}

34. i^{75}

35. i^{41}

36. i^{25}

37. i^{52}

38. i^0

39. i^{23}

40. i^{103}

41. i^6

42. i^{82}

Concept 3: Definition of a Complex Number

43. What is the complex conjugate of a complex number $a + bi$?

44. True or false?

   a. Every real number is a complex number.   **b.** Every complex number is a real number.

For Exercises 45–52, identify the real and imaginary parts of the complex number. **(See Example 4.)**

45. $-5 + 12i$ **46.** $22 - 16i$ **47.** $-6i$ **48.** $10i$

49. 35 **50.** -1 **51.** $\frac{3}{5} + i$ **52.** $-\frac{1}{2} - \frac{1}{4}i$

Concept 4: Addition, Subtraction, and Multiplication of Complex Numbers

For Exercises 53–76, perform the indicated operations. Write the answer in the form $a + bi$. **(See Examples 5–6.)**

53. $(2 - i) + (5 + 7i)$ **54.** $(5 - 2i) + (3 + 4i)$ **55.** $\left(\frac{1}{2} + \frac{2}{3}i\right) - \left(\frac{1}{5} - \frac{5}{6}i\right)$

56. $\left(\frac{11}{10} - \frac{7}{5}i\right) - \left(-\frac{2}{5} + \frac{3}{5}i\right)$ **57.** $\sqrt{-98} - \sqrt{-8}$ **58.** $\sqrt{-75} + \sqrt{-12}$

59. $(2 + 3i) - (1 - 4i) + (-2 + 3i)$ **60.** $(2 + 5i) - (7 - 2i) + (-3 + 4i)$

61. $(8i)(3i)$ **62.** $(2i)(4i)$ **63.** $6i(1 - 3i)$ **64.** $-i(3 + 4i)$

65. $(2 - 10i)(3 + 2i)$ **66.** $(4 + 7i)(2 - 3i)$ **67.** $(-5 + 2i)(5 + 2i)$ **68.** $(4 - 11i)(4 + 11i)$

69. $(4 + 5i)^2$ **70.** $(3 - 2i)^2$ **71.** $(2 + i)(3 - 2i)(4 + 3i)$ **72.** $(3 - i)(3 + i)(4 - i)$

73. $(-4 - 6i)^2$ **74.** $(-3 - 5i)^2$ **75.** $\left(-\frac{1}{2} - \frac{3}{4}i\right)\left(-\frac{1}{2} + \frac{3}{4}i\right)$ **76.** $\left(-\frac{2}{3} + \frac{1}{6}i\right)\left(-\frac{2}{3} - \frac{1}{6}i\right)$

Concept 5: Division and Simplification of Complex Numbers

For Exercises 77–90, divide the complex numbers. Write the answer in the form $a + bi$. **(See Example 7.)**

77. $\frac{2}{1 + 3i}$ **78.** $\frac{-2}{3 + i}$ **79.** $\frac{-i}{4 - 3i}$ **80.** $\frac{3 - 3i}{1 - i}$

81. $\frac{5 + 2i}{5 - 2i}$ **82.** $\frac{7 + 3i}{4 - 2i}$ **83.** $\frac{3 + 7i}{-2 - 4i}$ **84.** $\frac{-2 + 9i}{-1 - 4i}$

85. $\frac{13i}{-5 - i}$ **86.** $\frac{15i}{-2 - i}$ **87.** $\frac{2 + 3i}{6i}$ *(Hint:* Consider multiplying numerator and denominator by i or by $-i$. This will make the denominator a real number.)

88. $\frac{4 - i}{2i}$ **89.** $\frac{-10 + i}{i}$ **90.** $\frac{-6 - i}{-i}$

For Exercises 91–98, simplify the complex numbers. Write the answer in the form $a + bi$. **(See Example 8.)**

91. $\frac{2 + \sqrt{-16}}{8}$ **92.** $\frac{6 - \sqrt{-4}}{4}$ **93.** $\frac{-6 + \sqrt{-72}}{6}$ **94.** $\frac{-20 + \sqrt{-500}}{10}$

95. $\dfrac{-8 - \sqrt{-48}}{4}$ **96.** $\dfrac{-18 - \sqrt{-72}}{3}$ **97.** $\dfrac{-5 + \sqrt{-50}}{10}$ **98.** $\dfrac{14 + \sqrt{-98}}{7}$

Expanding Your Skills

For Exercises 99–102, determine if the complex number is a solution to the equation.

99. $x^2 - 4x + 5 = 0$; $2 + i$

100. $x^2 - 6x + 25 = 0$; $3 - 4i$

101. $x^2 + 12 = 0$; $-2i\sqrt{3}$

102. $x^2 + 18 = 0$; $3i\sqrt{2}$

Group Activity

Margin of Error of Survey Results

Materials: Calculator

Estimated time: 20 minutes

Group Size: 3

The members of the class will conduct a survey to estimate the percentage of the college population that answers "yes" to the several survey questions asked.

1. The members of the class will decide on three "Yes or No" questions that they will use to perform a survey. For example, here are some possible questions.

   "Do you study more than 10 hr a week?"

   "Do you work more than 25 hr a week?"

   "Are you taking more than 10 credit hours?"

   List the questions here:

   a. _____

   b. _____

   c. _____

2. Each student in the class will survey 20 people and record the number of "yes" responses in the table.

Question	Number of "Yes" Responses
1	
2	
3	

3. The instructor will then pool the results obtained from each student to get the total number of people surveyed by the whole class. Record the number of people surveyed (sample size) and the number of "yes" responses for each question in the table. Once the data are entered into the table, compute the percent, p, of

"yes" responses in the fourth column. Write this value in decimal form. Round the values of p to three decimal places. (*Note:* For now, leave the last column blank.)

Question	Sample Size	Number of "Yes" Responses	Percent "Yes" Responses, p	Margin of Error, E
1				
2				
3				

4. The value of p represents the percent (written in decimal form) of people in the *sample* who answered "yes." However, the percentage of people in the entire college population who would answer "yes" is unknown. The value of p is only an *estimate*. Statisticians often compute a margin of error associated with such an estimate by using the following formula.

$$E = 1.96 \sqrt{\frac{p(1-p)}{n}}$$ where E is the margin of error.
n is the sample size.
p is the percent (in decimal form) of "yes" responses.

Compute the margin of error for each of the three questions. Round the values of E to three decimal places. Record the results in the table.

5. The margin of error computed in step 4 is associated with what statisticians call a 95% level of confidence. To interpret the results, a statistician would say "With 95% confidence, the researcher estimates the percentage of 'yes' responses from the entire college population to be between $p \pm E$."

Interpret the results from each of the three questions.

Chapter 6   Summary

Section 6.1   Definition of an *n*th Root

Key Concepts	Examples

Key Concepts

b is an ***n*th root** of *a* if $b^n = a$.

The expression $\sqrt{a}$ represents the **principal square root** of *a*.

The expression $\sqrt[n]{a}$ represents the principal *n*th root of *a*.

$\sqrt[n]{a^n} = |a|$ if *n* is even.

$\sqrt[n]{a^n} = a$ if *n* is odd.

$\sqrt[n]{a}$ is not a real number if $a < 0$ and *n* is even.

$f(x) = \sqrt[n]{x}$ defines a **radical function**.

The Pythagorean Theorem

$a^2 + b^2 = c^2$

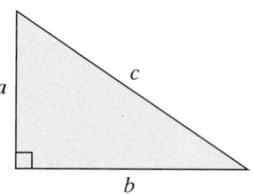

Examples

Example 1

2 is a square root of 4.

−2 is a square root of 4.

−3 is a cube root of −27.

Example 2

$\sqrt{36} = 6$ $\sqrt[3]{-64} = -4$

Example 3

$\sqrt[4]{(x+3)^4} = |x+3|$ $\sqrt[5]{(x+3)^5} = x+3$

Example 4

$\sqrt[4]{-16}$ is not a real number.

Example 5

For $g(x) = \sqrt{x}$ the domain is $[0, \infty)$.

For $h(x) = \sqrt[3]{x}$ the domain is $(-\infty, \infty)$.

Example 6

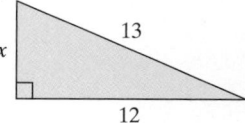

$x^2 + 12^2 = 13^2$

$x^2 + 144 = 169$

$x^2 = 25$

$x = \sqrt{25}$

$x = 5$

Section 6.2   Rational Exponents

Key Concepts

Let a be a real number and n be an integer such that $n > 1$. If $\sqrt[n]{a}$ exists, then

$$a^{1/n} = \sqrt[n]{a}$$

$$a^{m/n} = \left(\sqrt[n]{a}\right)^m = \sqrt[n]{a^m}$$

All properties of integer exponents hold for rational exponents, provided the roots are real-valued.

Examples

Example 1

$$121^{1/2} = \sqrt{121} = 11$$

Example 2

$$27^{2/3} = \left(\sqrt[3]{27}\right)^2 = (3)^2 = 9$$

Example 3

$$p^{1/3} \cdot p^{1/4} \qquad \frac{x^{4/3}}{x^{1/3}} \qquad (y^{-1/2})^6$$

$$= p^{1/3 + 1/4} \qquad = x^{4/3 - 1/3} \qquad = y^{(-1/2)(6)}$$

$$= p^{4/12 + 3/12} \qquad = x^{3/3} \qquad = y^{-3}$$

$$= p^{7/12} \qquad = x \qquad = \frac{1}{y^3}$$

Section 6.3   Simplifying Radical Expressions

Key Concepts

Let a and b represent real numbers such that $\sqrt[n]{a}$ and $\sqrt[n]{b}$ are both real. Then

$$\sqrt[n]{ab} = \sqrt[n]{a} \cdot \sqrt[n]{b} \qquad \text{Multiplication property}$$

A radical expression whose radicand is written as a product of prime factors is in simplified form if all the following conditions are met:

1. The radicand has no factor raised to a power greater than or equal to the index.
2. The radicand does not contain a fraction.
3. No radicals are in the denominator of a fraction.

Examples

Example 1

$$\sqrt{12} = \sqrt{4 \cdot 3}$$

$$= \sqrt{4} \cdot \sqrt{3}$$

$$= 2\sqrt{3}$$

Example 2

$$\sqrt[3]{16x^5 y^7}$$

$$= \sqrt[3]{2^4 x^5 y^7}$$

$$= \sqrt[3]{2^3 x^3 y^6 \cdot 2x^2 y}$$

$$= \sqrt[3]{2^3 x^3 y^6} \cdot \sqrt[3]{2x^2 y}$$

$$= 2xy^2 \sqrt[3]{2x^2 y}$$

Section 6.4 Addition and Subtraction of Radicals

Key Concepts

Like **radicals** have radical factors with the same index and the same radicand.

Use the distributive property to add and subtract *like* radicals.

Examples

Example 1

$2x\sqrt{7} - 5x\sqrt{7} + x\sqrt{7}$

$= (2 - 5 + 1)x\sqrt{7}$

$= -2x\sqrt{7}$

Example 2

$x\sqrt[4]{16x} - 3\sqrt[4]{x^5}$

$= 2x\sqrt[4]{x} - 3x\sqrt[4]{x}$

$= (2 - 3)x\sqrt[4]{x}$

$= -x\sqrt[4]{x}$

Section 6.5 Multiplication of Radicals

Key Concepts

The Multiplication Property of Radicals

If $\sqrt[n]{a}$ and $\sqrt[n]{b}$ are real numbers, then

$$\sqrt[n]{a} \cdot \sqrt[n]{b} = \sqrt[n]{ab}$$

To multiply or divide radicals with different indices, convert to rational exponents and use the properties of exponents.

Examples

Example 1

$3\sqrt{2}(\sqrt{2} + 5\sqrt{7} - \sqrt{6})$

$= 3\sqrt{4} + 15\sqrt{14} - 3\sqrt{12}$

$= 3 \cdot 2 + 15\sqrt{14} - 3 \cdot 2\sqrt{3}$

$= 6 + 15\sqrt{14} - 6\sqrt{3}$

Example 2

$\sqrt{p} \cdot \sqrt[5]{p^2}$

$= p^{1/2} \cdot p^{2/5}$

$= p^{5/10} \cdot p^{4/10}$

$= p^{9/10}$

$= \sqrt[10]{p^9}$

Section 6.6   Division of Radicals and Rationalization

Key Concepts

The Division Property of Radicals

If $\sqrt[n]{a}$ and $\sqrt[n]{b}$ are real numbers, then

$$\sqrt[n]{\dfrac{a}{b}} = \dfrac{\sqrt[n]{a}}{\sqrt[n]{b}} \qquad b \neq 0$$

The process of removing a radical from the denominator of an expression is called **rationalizing the denominator**.

- Rationalizing a denominator with one term

- Rationalizing a denominator with two terms involving square roots

Examples

Example 1

Simplify.

$$\sqrt{\dfrac{4x^5}{y^4}}$$

$$= \dfrac{\sqrt{4x^5}}{\sqrt{y^4}}$$

$$= \dfrac{2x^2\sqrt{x}}{y^2}$$

Example 2

Rationalize the denominator.

$$\dfrac{4}{\sqrt[3]{t}}$$

$$= \dfrac{4}{\sqrt[3]{t}} \cdot \dfrac{\sqrt[3]{t^2}}{\sqrt[3]{t^2}}$$

$$= \dfrac{4\sqrt[3]{t^2}}{\sqrt[3]{t^3}} = \dfrac{4\sqrt[3]{t^2}}{t}$$

Example 3

Rationalize the denominator.

$$\dfrac{\sqrt{2} - 3\sqrt{x}}{\sqrt{x} - \sqrt{2}}$$

$$= \dfrac{(\sqrt{2} - 3\sqrt{x})}{(\sqrt{x} - \sqrt{2})} \cdot \dfrac{(\sqrt{x} + \sqrt{2})}{(\sqrt{x} + \sqrt{2})}$$

$$= \dfrac{\sqrt{2} \cdot \sqrt{x} + \sqrt{2} \cdot \sqrt{2} - 3\sqrt{x} \cdot \sqrt{x} - 3\sqrt{x} \cdot \sqrt{2}}{(\sqrt{x})^2 - (\sqrt{2})^2}$$

$$= \dfrac{\sqrt{2x} + 2 - 3x - 3\sqrt{2x}}{x - 2}$$

$$= \dfrac{-3x - 2\sqrt{2x} + 2}{x - 2}$$

Section 6.7 Solving Radical Equations

Key Concepts

Steps to Solve a Radical Equation

1. Isolate the radical. If an equation has more than one radical, choose one of the radicals to isolate.
2. Raise each side of the equation to a power equal to the index of the radical.
3. Solve the resulting equation. If the equation still has a radical, repeat steps 1 and 2.
4. Check the potential solutions in the original equation.

Examples

Example 1

Solve:

$$\sqrt{b-5} - \sqrt{b+3} = 2$$
$$\sqrt{b-5} = \sqrt{b+3} + 2$$
$$(\sqrt{b-5})^2 = (\sqrt{b+3} + 2)^2$$
$$b - 5 = b + 3 + 4\sqrt{b+3} + 4$$
$$b - 5 = b + 7 + 4\sqrt{b+3}$$
$$-12 = 4\sqrt{b+3}$$
$$-3 = \sqrt{b+3}$$
$$(-3)^2 = (\sqrt{b+3})^2$$
$$9 = b + 3$$
$$6 = b$$

Check:

$$\sqrt{6-5} - \sqrt{6+3} \overset{?}{=} 2$$
$$\sqrt{1} - \sqrt{9} \overset{?}{=} 2$$
$$1 - 3 \overset{?}{=} 2 \qquad \text{(false)}$$

No solution, { }

Section 6.8   Complex Numbers

Key Concepts

$i = \sqrt{-1}$ and $i^2 = -1$

For a real number $b > 0$, $\sqrt{-b} = i\sqrt{b}$

A **complex number** is in the form $a + bi$, where a and b are real numbers. The value a is called the real part, and b is called the imaginary part.

To add or subtract complex numbers, combine the real parts and combine the imaginary parts.

Multiply complex numbers by using the distributive property.

Divide complex numbers by multiplying the numerator and denominator by the **complex conjugate** of the denominator.

Examples

Example 1

$\sqrt{-4} \cdot \sqrt{-9}$

$= (2i)(3i)$

$= 6i^2$

$= -6$

Example 2

$(3 - 5i) - (2 + i) + (3 - 2i)$

$= 3 - 5i - 2 - i + 3 - 2i$

$= 4 - 8i$

Example 3

$(1 + 6i)(2 + 4i)$

$= 2 + 4i + 12i + 24i^2$

$= 2 + 16i - 24(-1)$

$= -22 + 16i$

Example 4

$\dfrac{3}{2 - 5i}$

$= \dfrac{3}{(2 - 5i)} \cdot \dfrac{(2 + 5i)}{(2 + 5i)}$

$= \dfrac{6 + 15i}{4 - 25i^2}$

$= \dfrac{6 + 15i}{4 + 25}$

$= \dfrac{6 + 15i}{29}$

$= \dfrac{6}{29} + \dfrac{15}{29}i$

Chapter 6 Review Exercises

For the exercises in this set, assume that all variables represent positive real numbers unless otherwise stated.

Section 6.1

1. True or false?

   a. The principal nth root of an even-indexed root is always positive.

   b. The principal nth root of an odd-indexed root is always positive.

2. Explain why $\sqrt{(-3)^2} \neq -3$.

3. For $a > 0$ and $b > 0$, are the following statements true or false?

   a. $\sqrt{a^2 + b^2} = a + b$

   b. $\sqrt{(a+b)^2} = a + b$

For Exercises 4–6, simplify the radicals.

4. $\sqrt{\dfrac{50}{32}}$ 5. $\sqrt[4]{625}$ 6. $\sqrt{(-6)^2}$

7. Evaluate the function values for $f(x) = \sqrt{x-1}$.

   a. $f(10)$ **b.** $f(1)$ **c.** $f(8)$

   d. Write the domain of f in interval notation.

8. Evaluate the function values for $g(t) = \sqrt{5+t}$.

   a. $g(-5)$ **b.** $g(-4)$ **c.** $g(4)$

   d. Write the domain of g in interval notation.

9. Write the English expression as an algebraic expression: Four more than the quotient of the cube root of $2x$ and the principal fourth root of $2x$.

For Exercises 10–11, simplify the expression. Assume that x and y represent *any* real number.

10. **a.** $\sqrt{x^2}$ **b.** $\sqrt[3]{x^3}$

    c. $\sqrt[4]{x^4}$ **d.** $\sqrt[5]{(x+1)^5}$

11. **a.** $\sqrt{4y^2}$ **b.** $\sqrt[3]{27y^3}$

    c. $\sqrt[100]{y^{100}}$ **d.** $\sqrt[101]{y^{101}}$

12. Use the Pythagorean theorem to find the length of the third side of the triangle.

17 cm

15 cm

Section 6.2

13. Are the properties of exponents the same for rational exponents and integer exponents? Give an example. (Answers may vary.)

14. In the expression $x^{m/n}$ what does n represent?

15. Explain the process of eliminating a negative exponent from an algebraic expression.

For Exercises 16–21, simplify the expressions. Write the answer with positive exponents only.

16. $(-125)^{1/3}$ 17. $16^{-1/4}$

18. $\left(\dfrac{1}{16}\right)^{-3/4} - \left(\dfrac{1}{8}\right)^{-2/3}$ 19. $(b^{1/2} \cdot b^{1/3})^{12}$

20. $\left(\dfrac{x^{-1/4}y^{-1/3}z^{3/4}}{2^{1/3}x^{-1/3}y^{2/3}}\right)^{-12}$ 21. $\left(\dfrac{a^{12}b^{-4}c^{7}}{a^{3}b^{2}c^{4}}\right)^{1/3}$

For Exercises 22–23, rewrite the expressions by using rational exponents.

22. $\sqrt[4]{x^3}$ 23. $\sqrt[3]{2y^2}$

 For Exercises 24–26, use a calculator to approximate the expressions to four decimal places.

24. $10^{1/3}$ 25. $17.8^{2/3}$ 26. $\sqrt[5]{147^4}$

Section 6.3

27. List the criteria for a radical expression to be simplified.

For Exercises 28–31, simplify the radicals.

28. $\sqrt{108}$ 29. $\sqrt[4]{x^5yz^4}$

30. $-2\sqrt[3]{250a^3b^{10}}$ 31. $\sqrt[3]{\dfrac{-16a^4}{2ab^3}}$

32. Write an English phrase that describes the following mathematical expressions: (Answers may vary.)

a. $\sqrt{\dfrac{2}{x}}$

b. $(x + 1)^3$

33. An engineering firm made a mistake when building a $\frac{1}{4}$-mi bridge in the Florida Keys. The bridge was made without adequate expansion joints to prevent buckling during the heat of summer. During mid-June, the bridge expanded 1.5 ft, causing a vertical bulge in the middle. Calculate the height of the bulge h in feet. (*Note:* 1 mi = 5280 ft.) Round to the nearest foot.

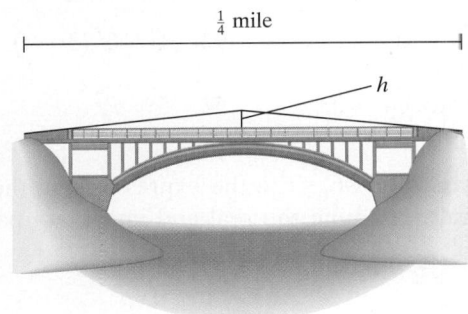

$\frac{1}{4}$ mile

Section 6.4

34. Complete the following statement: Radicals may be added or subtracted if . . .

For Exercises 35–38, determine whether the radicals may be combined, and explain your answer.

35. $\sqrt[3]{2x} - 2\sqrt{2x}$

36. $2 + \sqrt{x}$

37. $\sqrt[4]{3xy} + 2\sqrt[4]{3xy}$

38. $-4\sqrt{32} + 7\sqrt{50}$

For Exercises 39–42, add or subtract as indicated.

39. $4\sqrt{7} - 2\sqrt{7} + 3\sqrt{7}$

40. $2\sqrt[3]{64} + 3\sqrt[3]{54} - 16$

41. $\sqrt{50} + 7\sqrt{2} - \sqrt{8}$

42. $x\sqrt[3]{16x^2} - 4\sqrt[3]{2x^5} + 5x\sqrt[3]{54x^2}$

For Exercises 43–44, answer true or false. If an answer is false, explain why. Assume all variables represent positive real numbers.

43. $5 + 3\sqrt{x} = 8\sqrt{x}$

44. $\sqrt{y} + \sqrt{y} = \sqrt{2y}$

Section 6.5

For Exercises 45–56, multiply the radicals and simplify the answer.

45. $\sqrt{3} \cdot \sqrt{12}$

46. $\sqrt[4]{4} \cdot \sqrt[4]{8}$

47. $-2\sqrt{3}(\sqrt{7} - 3\sqrt{11})$

48. $-3\sqrt{5}(2\sqrt{3} - \sqrt{5})$

49. $(2\sqrt{x} - 3)(2\sqrt{x} + 3)$

50. $(\sqrt{y} + 4)(\sqrt{y} - 4)$

51. $(\sqrt{7y} - \sqrt{3x})^2$

52. $(2\sqrt{3w} + 5)^2$

53. $(-\sqrt{z} - \sqrt{6})(2\sqrt{z} + 7\sqrt{6})$

54. $(3\sqrt{a} - \sqrt{5})(\sqrt{a} + 2\sqrt{5})$

55. $\sqrt[3]{u} \cdot \sqrt[3]{u^5}$

56. $\sqrt{2} \cdot \sqrt[4]{w^3}$

Section 6.6

For Exercises 57–60, simplify the radicals.

57. $\sqrt{\dfrac{3y^5}{25x^6}}$

58. $\sqrt[3]{\dfrac{-16x^7y^6}{z^9}}$

59. $\dfrac{\sqrt{324w^7}}{\sqrt{4w^3}}$

60. $\dfrac{\sqrt[3]{3t^{14}}}{\sqrt[3]{192t^2}}$

For Exercises 61–68, rationalize the denominator.

61. $\sqrt{\dfrac{7}{2y}}$

62. $\sqrt{\dfrac{5}{3w}}$

63. $\dfrac{4}{\sqrt[3]{9p^2}}$

64. $\dfrac{-2}{\sqrt[3]{2x}}$

65. $\dfrac{-5}{\sqrt{15} + \sqrt{10}}$

66. $\dfrac{-6}{\sqrt{7} + \sqrt{5}}$

67. $\dfrac{t - 3}{\sqrt{t} - \sqrt{3}}$

68. $\dfrac{w - 7}{\sqrt{w} - \sqrt{7}}$

69. Write the mathematical expression as an English phrase. (Answers may vary.)

$$\dfrac{\sqrt{2}}{x^2}$$

Section 6.7

Solve the radical equations in Exercises 70–77, if possible.

70. $\sqrt{2y} = 7$

71. $\sqrt{a - 6} - 5 = 0$

72. $\sqrt[3]{2w - 3} + 5 = 2$

73. $\sqrt[4]{p + 12} - \sqrt[4]{5p - 16} = 0$

74. $\sqrt{t} + \sqrt{t - 5} = 5$

75. $\sqrt{8x + 1} = -\sqrt{x - 13}$

76. $\sqrt{2m^2 + 4} - \sqrt{9m} = 0$

77. $\sqrt{x + 2} = 1 - \sqrt{2x + 5}$

78. A tower is supported by stabilizing wires. Find the exact length of each wire, and then round to the nearest tenth of a meter.

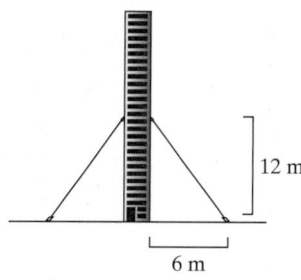

79. The velocity, $v(d)$, of an ocean wave depends on the water depth d as the wave approaches land.

$$v(d) = \sqrt{32d}$$

where $v(d)$ is in feet per second and d is in feet.

a. Find $v(20)$ and interpret its value. Round to one decimal place.

b. Find the depth of the water at a point where a wave is traveling at 16 ft/sec.

Section 6.8

80. Define a complex number.

81. Define an imaginary number.

82. Describe the first step in the process to simplify the expression.

$$\frac{3}{4 + 6i}$$

For Exercises 83–86, rewrite the expressions in terms of i.

83. $\sqrt{-16}$

84. $-\sqrt{-5}$

85. $\sqrt{-75} \cdot \sqrt{-3}$

86. $\dfrac{-\sqrt{-24}}{\sqrt{6}}$

For Exercises 87–90, simplify the powers of i.

87. i^{38}

88. i^{101}

89. i^{19}

90. $i^{1000} + i^{1002}$

For Exercises 91–94, perform the indicated operations. Write the final answer in the form $a + bi$.

91. $(-3 + i) - (2 - 4i)$

92. $(1 + 6i)(3 - i)$

93. $(4 - 3i)(4 + 3i)$

94. $(5 - i)^2$

For Exercises 95–96, write the expressions in the form $a + bi$, and determine the real and imaginary parts.

95. $\dfrac{17 - 4i}{-4}$

96. $\dfrac{-16 - 8i}{8}$

For Exercises 97–100, divide and simplify. Write the final answer in the form $a + bi$.

97. $\dfrac{2 - i}{3 + 2i}$

98. $\dfrac{10 + 5i}{2 - i}$

99. $\dfrac{5 + 3i}{-2i}$

100. $\dfrac{4i}{4 - i}$

For Exercises 101–102, simplify the expression.

101. $\dfrac{-8 + \sqrt{-40}}{12}$

102. $\dfrac{6 - \sqrt{-144}}{3}$

Chapter 6 Test

1. **a.** What is the principal square root of 36?

   b. What is the negative square root of 36?

2. Which of the following are real numbers?

   a. $-\sqrt{100}$ **b.** $\sqrt{-100}$

   c. $-\sqrt[3]{1000}$ **d.** $\sqrt[3]{-1000}$

3. Simplify.

   a. $\sqrt[3]{y^3}$ **b.** $\sqrt[4]{y^4}$

For Exercises 4–11, simplify the radicals. Assume that all variables represent positive numbers.

4. $\sqrt[4]{81}$

5. $\sqrt{\dfrac{16}{9}}$

6. $\sqrt[3]{32}$

7. $\sqrt{a^4 b^3 c^5}$

8. $\sqrt{18x^5 y^3 z^4}$

9. $\sqrt{\dfrac{32w^6}{2w}}$

10. $\sqrt[3]{\dfrac{x^6}{125y^3}}$

11. $\dfrac{2\sqrt{72}}{8}$

12. **a.** Evaluate the function values $f(-8)$, $f(-6)$, $f(-4)$, and $f(-2)$ for $f(x) = \sqrt{-2x-4}$.

   b. Write the domain of f in interval notation.

13. Use a calculator to evaluate $\dfrac{-3-\sqrt{5}}{17}$ to four decimal places.

For Exercises 14–15, simplify the expressions. Assume that all variables represent positive numbers.

14. $-27^{1/3}$

15. $8^{2/3} \cdot \left(\dfrac{25x^4 y^6}{z^2}\right)^{1/2}$

For Exercises 16–17, use rational exponents to multiply or divide. Write the final answer in radical form.

16. $\sqrt[6]{7} \cdot \sqrt{y}$

17. $\dfrac{\sqrt[3]{10}}{\sqrt[4]{10}}$

18. Add or subtract as indicated.
$$3\sqrt{5} + 4\sqrt{5} - 2\sqrt{20}$$

For Exercises 19–20, multiply the radicals.

19. $3\sqrt{x}(\sqrt{2} - \sqrt{5})$

20. $(2\sqrt{5} - 3\sqrt{x})(4\sqrt{5} + \sqrt{x})$

For Exercises 21–22, rationalize the denominator. Assume $x > 0$.

21. $\dfrac{-2}{\sqrt[5]{x}}$

22. $\dfrac{\sqrt{x} + 2}{3 - \sqrt{x}}$

23. Rewrite the expressions in terms of i.

   a. $\sqrt{-8}$ **b.** $2\sqrt{-16}$ **c.** $\dfrac{2 + \sqrt{-8}}{4}$

For Exercises 24–30, perform the indicated operations and simplify completely. Write the final answer in the form $a + bi$.

24. $(3 - 5i) - (2 + 6i)$

25. $(4 + i)(8 + 2i)$

26. $\sqrt{-16} \cdot \sqrt{-49}$

27. $(4 - 7i)^2$

28. $(2 - 10i)(2 + 10i)$

29. $\dfrac{3 - 2i}{3 - 4i}$

30. $\dfrac{5i}{3 - 5i}$

31. If the volume V of a sphere is known, the radius of the sphere can be computed by
$$r(V) = \sqrt[3]{\dfrac{3V}{4\pi}}.$$
Find $r(10)$ to two decimal places. Interpret the meaning in the context of the problem.

32. A patio 20 ft wide has a slanted roof, as shown in the figure. Find the length of the roof if there is an 8-in. overhang. Round the answer to the nearest foot.

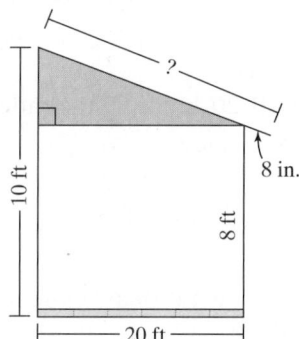

For Exercises 33–35, solve the radical equation.

33. $\sqrt[3]{2x + 5} = -3$

34. $\sqrt{5x + 8} = \sqrt{5x - 1} + 1$

35. $\sqrt{t + 7} - \sqrt{2t - 3} = 2$

Chapters 1–6   Cumulative Review Exercises

1. Simplify the expression.
$6^2 - 2[5 - 8(3 - 1) + 4 \div 2]$

2. Simplify the expression.
$3x - 3(-2x + 5) - 4y + 2(3x + 5) - y$

3. Solve. $9(2y + 8) = 20 - (y + 5)$

For Exercises 4–6, solve the inequalities. Write the answers in interval notation, if possible.

4. $2a - 4 < -14$ **5.** $11 \geq -x + 2 > 5$

6. $|9 - 4x| + 4 < 2$

7. Write an equation of the line that is parallel to the line $2x + y = 9$ and passes through the point $(3, -1)$. Write the answer in slope-intercept form.

8. Write an equation of a line parallel to the y-axis passing through the point $(-5, 8)$.

9. Solve the system of equations by using the addition method.

$$2x - 3y = 0$$

$$-4x + 3y = -1$$

10. Determine if $(2, -2, \frac{1}{2})$ is a solution to the system.

$$2x + y - 4z = 0$$

$$x - y + 2z = 5$$

$$3x + 2y + 2z = 4$$

11. Bennette and Pepe are landscapers and have to dig a ditch in which to put underground piping for a sprinkler system. Bennette can dig 40 yd of ditch in 3 hr and Pepe can dig 40 yd of ditch in 5 hr. How long will it take them to dig 40 yd together?

12. Given the function defined by $f(x) = 4x - 2$.

a. Find $f(-2), f(0), f(4)$, and $f(\frac{1}{2})$.

b. Write the ordered pairs that correspond to the function values in part (a).

c. Graph $y = f(x)$.

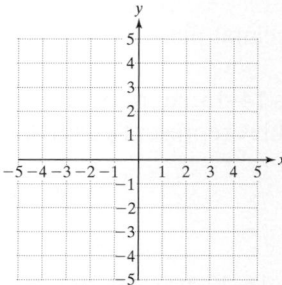

13. Simplify the expression. Write the final answer with positive exponents only.

$$\left(\frac{a^{3/2}b^{-1/4}c^{1/3}}{ab^{-5/4}c^0} \right)^{12}$$

14. Multiply or divide as indicated, and write the answer in scientific notation.

   a. $(3.5 \times 10^7)(4 \times 10^{-12})$

   b. $\dfrac{6.28 \times 10^5}{2 \times 10^{-4}}$

15. Multiply the polynomials $(2x + 5)(x - 3)$. What is the degree of the product?

16. Perform the indicated operations and simplify.
$\sqrt{3}(\sqrt{5} + \sqrt{6} + \sqrt{3})$

17. Divide. $(x^2 - x - 12) \div (x + 3)$

18. Simplify and subtract. $\sqrt[4]{\dfrac{1}{16}} - \sqrt[3]{\dfrac{8}{27}}$

19. Simplify. $\sqrt[3]{\dfrac{54c^4}{cd^3}}$

20. Add. $4\sqrt{45b^3} + 5b\sqrt{80b}$

21. Divide. $\dfrac{13i}{3 + 2i}$ Write the answer in the form $a + bi$.

22. Solve. $\dfrac{5}{y - 2} - \dfrac{3}{y - 4} = \dfrac{6}{y^2 - 6y + 8}$

23. Add. $\dfrac{3}{x^2 + 5x} + \dfrac{-2}{x^2 - 25}$

24. Divide. $\dfrac{a + 10}{2a^2 - 11a - 6} \div \dfrac{a^2 + 12a + 20}{6 - a}$

25. Perform the indicated operations.
$(-5x^2 - 4x + 8) - (3x - 5)^2$

26. Simplify. $\dfrac{-4}{\sqrt{3} - \sqrt{5}}$

27. Divide. $\dfrac{4}{3 - 5i}$

28. Solve. $12x^2 - 4x - 21 = 0$

29. Factor. $x^2 + 6x + 9 - y^2$

30. Factor. $x^3 + 8$

Quadratic Equations, Functions, and Inequalities

7

CHAPTER OUTLINE

7.1 Square Root Property and Completing the Square 582

7.2 Quadratic Formula 592

7.3 Equations in Quadratic Form 606

    Problem Recognition Exercises: Quadratic and Quadratic Type Equations 611

7.4 Graphs of Quadratic Functions 612

7.5 Vertex of a Parabola: Applications and Modeling 626

7.6 Polynomial and Rational Inequalities 636

    Problem Recognition Exercises: Recognizing Equations and Inequalities 648

    Group Activity: Creating a Quadratic Model of the Form $y = a(x - h)^2 + k$ 649

Chapter 7

In Chapter 7, we revisit quadratic equations. We have already learned how to solve quadratic equations by factoring and applying the zero product rule. Now we present two additional techniques that can be used to solve a quadratic equation even if the quadratic polynomial does not factor.

Are You Prepared?

To prepare for this chapter, we recommend that you review how to solve quadratic equations by using the zero product rule and how to simplify radical expressions. Match the answers to the exercises below to the letter on the right. Then write the letter in the appropriate space at the bottom of the page to define an important vocabulary word.

Solve.

1. $x^2 - 9 = 0$

2. $x^2 - 6x + 8 = 0$

3. $3x(x + 1) = 60$

Simplify.

4. $\dfrac{-5 \pm \sqrt{50}}{10}$ **5.** $\dfrac{7 \pm 21i}{7}$ **6.** $\dfrac{2 \pm \sqrt{-12}}{4}$

7. Determine the x-intercept(s). $f(x) = x^2 - 6x + 5$

8. Determine the y-intercept. $f(x) = x^2 - 6x + 5$

A function defined by $f(x) = ax^2 + bx + c$ $(a \neq 0)$ is called a

$\underline{}\ \underline{}\ \underline{}\ \underline{}\ \underline{}\ \underline{}\ \underline{}\ \underline{}\ \underline{}$ function.
$2\ 1\ 6\ 5\ 8\ 6\ 4\ 7\ 3$

Answers

D $1 \pm 3i$

A $\dfrac{1}{2} \pm \dfrac{\sqrt{3}}{2}i$

C $\{4, -5\}$

U $\{-3, 3\}$

T $\dfrac{-1 \pm \sqrt{2}}{2}$

Q $\{2, 4\}$

R $(0, 5)$

I $(1, 0)$ and $(5, 0)$

Section 7.1	Square Root Property and Completing the Square

Concepts

1. Solving Quadratic Equations by Using the Square Root Property
2. Solving Quadratic Equations by Completing the Square
3. Literal Equations

1. Solving Quadratic Equations by Using the Square Root Property

In Section 4.8 we learned how to solve a quadratic equation by factoring and applying the zero product rule. For example:

$$x^2 = 81$$

$$x^2 - 81 = 0 \qquad \text{Set one side equal to zero.}$$

$$(x - 9)(x + 9) = 0 \qquad \text{Factor.}$$

$$x - 9 = 0 \quad \text{or} \quad x + 9 = 0 \qquad \text{Set each factor equal to zero.}$$

$$x = 9 \quad \text{or} \qquad x = -9$$

The solution set is $\{9, -9\}$.

It is important to note that the zero product rule can only be used if the equation is factorable. In this section and Section 7.2, we will learn how to solve quadratic equations, factorable and nonfactorable.

Consider a quadratic equation of the form $x^2 = k$. The solutions are all numbers (real or imaginary) that when squared equal k, so for any nonzero constant k, there will be two solutions, $\sqrt{k}$ or $-\sqrt{k}$. For example:

$$x^2 = 25 \qquad \text{The solutions are 5 and } -5.$$

$$x^2 = -25 \qquad \text{The solutions are } 5i \text{ and } -5i.$$

This principle is stated formally as the **square root property**.

The Square Root Property

For any real number, k, if $x^2 = k$, then $x = \sqrt{k}$ or $x = -\sqrt{k}$.

Note: The solution may also be written as $\pm\sqrt{k}$, read "plus or minus the square root of k."

Example 1 Solving a Quadratic Equation by Using the Square Root Property

Use the square root property to solve the equation. $\quad 4p^2 = 9$

Solution:

$$4p^2 = 9$$

$$p^2 = \frac{9}{4} \qquad \text{Isolate } p^2 \text{ by dividing both sides by 4.}$$

$$p = \pm\sqrt{\frac{9}{4}} \qquad \text{Apply the square root property.}$$

$$p = \pm\frac{3}{2} \qquad \text{Simplify the radical.}$$

The solution set is $\left\{ \dfrac{3}{2}, -\dfrac{3}{2} \right\}$.

Skill Practice Solve using the square root property.

1. $25a^2 = 16$

For a quadratic equation, $ax^2 + bx + c = 0$, if $b = 0$, then the equation is easily solved by using the square root property. This is demonstrated in Example 2.

Example 2 Solving a Quadratic Equation by Using the Square Root Property

Use the square root property to solve the equation.

$$3x^2 + 75 = 0$$

Solution:

$\quad 3x^2 + 75 = 0 \qquad$ Rewrite the equation to fit the form $x^2 = k$.

$\quad\quad\quad 3x^2 = -75$

$\quad\quad\quad\quad x^2 = -25 \qquad$ The equation is now in the form $x^2 = k$.

$\quad\quad\quad\quad\ x = \pm\sqrt{-25} \qquad$ Apply the square root property.

$\quad\quad\quad\quad\quad = \pm 5i$

Check: $x = 5i$ Check: $x = -5i$

$3x^2 + 75 = 0$ $3x^2 + 75 = 0$

$3(5i)^2 + 75 \overset{?}{=} 0$ $3(-5i)^2 + 75 \overset{?}{=} 0$

$3(25i^2) + 75 \overset{?}{=} 0$ $3(25i^2) + 75 \overset{?}{=} 0$

$3(-25) + 75 \overset{?}{=} 0$ $3(-25) + 75 \overset{?}{=} 0$

$-75 + 75 \overset{?}{=} 0 \checkmark$ $-75 + 75 \overset{?}{=} 0 \checkmark$

The solution set is $\{\pm 5i\}$.

Avoiding Mistakes

A common mistake is to forget the $\pm$ symbol when solving the equation $x^2 = k$:

$$x = \pm\sqrt{k}$$

Skill Practice Solve using the square root property.

2. $8x^2 + 72 = 0$

Example 3 Solving a Quadratic Equation by Using the Square Root Property

Use the square root property to solve the equation.

$$(w + 3)^2 = 20$$

Solution:

$\quad (w + 3)^2 = 20 \qquad$ The equation is in the form $x^2 = k$, where $x = (w + 3)$.

$\quad\quad w + 3 = \pm\sqrt{20} \qquad$ Apply the square root property.

$\quad\quad w + 3 = \pm\sqrt{4 \cdot 5} \qquad$ Simplify the radical.

$\quad\quad w + 3 = \pm 2\sqrt{5}$

$\quad\quad\quad\quad\ w = -3 \pm 2\sqrt{5} \qquad$ Solve for w.

The solution set is $\{-3 \pm 2\sqrt{5}\}$.

TIP: Recall that $-3 \pm 2\sqrt{5}$ represents two solutions:

$-3 + 2\sqrt{5}$ and $-3 - 2\sqrt{5}$.

Skill Practice Solve using the square root property.

3. $(t - 5)^2 = 18$

Answers

1. $\left\{\dfrac{4}{5}, -\dfrac{4}{5}\right\}$ **2.** $\{\pm 3i\}$

3. $\{5 \pm 3\sqrt{2}\}$

2. Solving Quadratic Equations by Completing the Square

In Example 3 we used the square root property to solve an equation where the square of a binomial was equal to a constant.

$$\underbrace{(w + 3)^2}_{\substack{\text{Square of a} \\ \text{binomial}}} = \overset{\uparrow}{\underset{\text{Constant}}{20}}$$

The square of a binomial is the factored form of a perfect square trinomial. For example:

Perfect Square Trinomial	**Factored Form**
$x^2 + 10x + 25$ $\longrightarrow$	$(x + 5)^2$
$t^2 - 6t + 9$ $\longrightarrow$	$(t - 3)^2$
$p^2 - 14p + 49$ $\longrightarrow$	$(p - 7)^2$

For a perfect square trinomial with a leading coefficient of 1, the constant term is the square of one-half the linear term coefficient. For example:

$$x^2 + \overset{|}{\underset{[\frac{1}{2}(10)]^2}{10x}} + 25$$

In general an expression of the form $x^2 + bx + n$ is a perfect square trinomial if $n = (\frac{1}{2}b)^2$. The process to create a perfect square trinomial is called **completing the square**.

Example 4 **Completing the Square**

Determine the value of n that makes the polynomial a perfect square trinomial. Then factor the expression as the square of a binomial.

a. $x^2 + 12x + n$ **b.** $x^2 - 26x + n$

c. $x^2 + 11x + n$ **d.** $x^2 - \dfrac{4}{7}x + n$

Solution:

The expressions are in the form $x^2 + bx + n$. The value of n equals the square of one-half the linear term coefficient $(\frac{1}{2}b)^2$.

a. $x^2 + 12x + n$

$\quad x^2 + 12x + 36 \qquad n = [\frac{1}{2}(12)]^2 = (6)^2 = 36$

$\quad (x + 6)^2 \qquad\qquad$ Factored form

b. $x^2 - 26x + n$

$\quad x^2 - 26x + 169 \qquad n = [\frac{1}{2}(-26)]^2 = (-13)^2 = 169$

$\quad (x - 13)^2 \qquad\qquad$ Factored form

c. $x^2 + 11x + n$

$\quad x^2 + 11x + \dfrac{121}{4} \qquad n = [\frac{1}{2}(11)]^2 = (\frac{11}{2})^2 = \frac{121}{4}$

$\quad \left(x + \dfrac{11}{2}\right)^2 \qquad\qquad$ Factored form

d. $x^2 - \dfrac{4}{7}x + n$

$x^2 - \dfrac{4}{7}x + \dfrac{4}{49}$ $n = \left[\frac{1}{2}\left(-\frac{4}{7}\right)\right]^2 = \left(-\frac{2}{7}\right)^2 = \frac{4}{49}$

$\left(x - \dfrac{2}{7}\right)^2$ Factored form

Skill Practice Determine the value of n that makes the polynomial a perfect square trinomial. Then factor.

4. $x^2 + 20x + n$ **5.** $y^2 - 16y + n$

6. $a^2 - 5a + n$ **7.** $w^2 + \dfrac{7}{3}w + n$

The process of completing the square can be used to write a quadratic equation $ax^2 + bx + c = 0$ $(a \neq 0)$ in the form $(x - h)^2 = k$. Then the square root property can be used to solve the equation. The following steps outline the procedure.

Solving a Quadratic Equation $ax^2 + bx + c = 0$ by Completing the Square and Applying the Square Root Property

Step 1 Divide both sides by a to make the leading coefficient 1.

Step 2 Isolate the variable terms on one side of the equation.

Step 3 Complete the square.

- Add the square of one-half the linear term coefficient to both sides, $\left(\frac{1}{2}b\right)^2$.
- Factor the resulting perfect square trinomial.

Step 4 Apply the square root property and solve for x.

Example 5 **Solving a Quadratic Equation by Completing the Square and Applying the Square Root Property**

Solve by completing the square and applying the square root property.

$$x^2 - 6x + 13 = 0$$

Solution:

$x^2 - 6x + 13 = 0$ **Step 1:** Since the leading coefficient a is equal to 1, we do not have to divide by a. We can proceed to step 2.

$x^2 - 6x = -13$ **Step 2:** Isolate the variable terms on one side.

$x^2 - 6x + 9 = -13 + 9$ **Step 3:** To complete the square, add $\left[\frac{1}{2}(-6)\right]^2 = 9$ to both sides of the equation.

$(x - 3)^2 = -4$ Factor the perfect square trinomial.

Answers

4. $n = 100;\ (x + 10)^2$

5. $n = 64;\ (y - 8)^2$

6. $n = \dfrac{25}{4};\ \left(a - \dfrac{5}{2}\right)^2$

7. $n = \dfrac{49}{36};\ \left(w + \dfrac{7}{6}\right)^2$

$$x - 3 = \pm\sqrt{-4}$$ **Step 4:** Apply the square root property.

$$x - 3 = \pm 2i$$ Simplify the radical.

$$x = 3 \pm 2i$$ Solve for x.

The solutions are imaginary numbers and can be written as $3 + 2i$ and $3 - 2i$.

The solution set is $\{3 \pm 2i\}$.

Skill Practice Solve by completing the square and applying the square root property.

8. $z^2 - 4z + 26 = 0$

Example 6 **Solving a Quadratic Equation by Completing the Square and Applying the Square Root Property**

Solve by completing the square and applying the square root property.

$$2m^2 + 10m = 3$$

Solution:

$$2m^2 + 10m = 3$$ The variable terms are already isolated on one side of the equation.

$$\frac{2m^2}{2} + \frac{10m}{2} = \frac{3}{2}$$ Divide by the leading coefficient, 2.

$$m^2 + 5m = \frac{3}{2}$$

$$m^2 + 5m + \frac{25}{4} = \frac{3}{2} + \frac{25}{4}$$ Add $\left[\frac{1}{2}(5)\right]^2 = \left(\frac{5}{2}\right)^2 = \frac{25}{4}$ to both sides.

$$\left(m + \frac{5}{2}\right)^2 = \frac{6}{4} + \frac{25}{4}$$ Factor the left side and write the terms on the right with a common denominator.

$$\left(m + \frac{5}{2}\right)^2 = \frac{31}{4}$$

$$m + \frac{5}{2} = \pm\sqrt{\frac{31}{4}}$$ Apply the square root property.

$$m = -\frac{5}{2} \pm \frac{\sqrt{31}}{2}$$ Subtract $\frac{5}{2}$ from both sides and simplify the radical.

TIP: The solutions to Example 6 can also be written as:

$$\frac{-5 \pm \sqrt{31}}{2}$$

The solution set is $\left\{-\frac{5}{2} \pm \frac{\sqrt{31}}{2}\right\}$. The solutions are irrational numbers.

Skill Practice Solve by completing the square and applying the square root property.

9. $4x^2 + 12x = 5$

Answers

8. $\{2 \pm i\sqrt{22}\}$

9. $\left\{-\frac{3}{2} \pm \frac{\sqrt{14}}{2}\right\}$

Example 7   **Solving a Quadratic Equation by Completing the Square and Applying the Square Root Property**

Solve by completing the square and applying the square root property.

$$2x(2x - 10) = -30 + 6x$$

Solution:

$$2x(2x - 10) = -30 + 6x$$

$$4x^2 - 20x = -30 + 6x$$ Clear parentheses.

$$4x^2 - 26x + 30 = 0$$ Write the equation in the form $ax^2 + bx + c = 0$.

$$\frac{4x^2}{4} - \frac{26x}{4} + \frac{30}{4} = \frac{0}{4}$$ **Step 1:**   Divide both sides by the leading coefficient, 4.

$$x^2 - \frac{13}{2}x + \frac{15}{2} = 0$$

$$x^2 - \frac{13}{2}x = -\frac{15}{2}$$ **Step 2:**   Isolate the variable terms on one side.

$$x^2 - \frac{13}{2}x + \frac{169}{16} = -\frac{15}{2} + \frac{169}{16}$$ **Step 3:**   Add $[\frac{1}{2}(-\frac{13}{2})]^2 = (-\frac{13}{4})^2 = \frac{169}{16}$ to both sides.

$$\left(x - \frac{13}{4}\right)^2 = -\frac{120}{16} + \frac{169}{16}$$ Factor the perfect square trinomial. Rewrite the right-hand side with a common denominator.

$$\left(x - \frac{13}{4}\right)^2 = \frac{49}{16}$$

$$x - \frac{13}{4} = \pm\sqrt{\frac{49}{16}}$$ **Step 4:**   Apply the square root property.

$$x - \frac{13}{4} = \pm\frac{7}{4}$$ Simplify the radical.

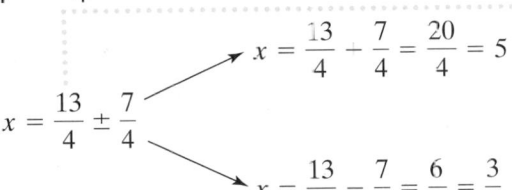

$$x = \frac{13}{4} + \frac{7}{4} = \frac{20}{4} = 5$$

$$x = \frac{13}{4} \pm \frac{7}{4}$$

$$x = \frac{13}{4} - \frac{7}{4} = \frac{6}{4} = \frac{3}{2}$$

The solution set is $\left\{5, \frac{3}{2}\right\}$.   The solutions are rational numbers.

TIP: In general, if the solutions to a quadratic equation are rational numbers, the equation can be solved by factoring and using the zero product rule. Consider the equation from Example 7.

$$2x(2x - 10) = -30 + 6x$$
$$4x^2 - 20x = -30 + 6x$$
$$4x^2 - 26x + 30 = 0$$
$$2(2x^2 - 13x + 15) = 0$$
$$2(x - 5)(2x - 3) = 0$$
$$x = 5 \quad \text{or} \quad x = \frac{3}{2}$$

Avoiding Mistakes

When the solutions are rational, combine the *like* terms. That is, do not leave the solution with the $\pm$ sign.

Skill Practice Solve by completing the square and applying the square root property.

10. $2y(y - 1) = 3 - y$

Answer

10. $\left\{\frac{3}{2}, -1\right\}$

3. Literal Equations

> **Example 8**   **Solving a Literal Equation**
>
> Ignoring air resistance, the distance d (in meters) that an object falls in t sec is given by the equation
>
> $$d = 4.9t^2 \qquad \text{where } t \geq 0$$
>
> **a.** Solve the equation for t. Do not rationalize the denominator.
>
> **b.** Using the equation from part (a), determine the amount of time required for an object to fall 500 m. Round to the nearest second.
>
> **Solution:**
>
> **a.** $d = 4.9t^2$
>
> $\dfrac{d}{4.9} = t^2$ Isolate the quadratic term. The equation is in the form $t^2 = k$.
>
> $t = \pm\sqrt{\dfrac{d}{4.9}}$ Apply the square root property.
>
> $= \sqrt{\dfrac{d}{4.9}}$ Because t represents time, $t \geq 0$. We reject the negative solution.
>
> **b.** $t = \sqrt{\dfrac{d}{4.9}}$
>
> $= \sqrt{\dfrac{500}{4.9}}$ Substitute $d = 500$.
>
> $t \approx 10.1$
>
> The object will require approximately 10.1 sec to fall 500 m.
>
> ---
>
> **Skill Practice** The formula for the area of a circle is $A = \pi r^2$, where r is the radius.
>
> **11.** Solve for r. (Do not rationalize the denominator.)
>
> **12.** Use the equation from the previous exercise to find the radius for an area of 15.7 cm². (Use 3.14 for π and round to two decimal places.)

Answers

11. $r = \sqrt{\dfrac{A}{\pi}}$

12. The radius is $\sqrt{5} \approx 2.24$ cm.

Section 7.1   Practice Exercises

Vocabulary and Key Concepts

1. a. The zero product rule states that if $ab = 0$, then $a = $ _____ or $b = $ _____ .

b. To apply the zero product rule, one side of the equation must be equal to _____ and the other side must be written in factored form.

c. The square root property states that for any real number k, if $x^2 = k$, then $x = $ _____ or $x = $ _____ .

d. To apply the square root property to the equation $t^2 + 2 = 11$, first subtract _____ from both sides. The solution set is _____ .

e. The process to create a perfect square trinomial is called _____ the square.

f. Fill in the blank to complete the square for the trinomial $x^2 + 20x +$ _____.

g. To use completing the square to solve the equation $4x^2 + 3x + 5 = 0$, the first step is to divide by _____ so that the coefficient of the x^2 term is _____.

h. Given the trinomial $y^2 + 8y + 16$, the coefficient of the linear term is _____.

Concept 1: Solving Quadratic Equations by Using the Square Root Property

For Exercises 2–21, solve the equations by using the square root property. Write imaginary solutions in the form $a + bi$. (See Examples 1–3.)

2. $x^2 = 100$

3. $y^2 = 4$

4. $a^2 = 5$

5. $k^2 - 7 = 0$

6. $4t^2 = 81$

7. $36u^2 = 121$

8. $3v^2 + 33 = 0$

9. $-2m^2 = 50$

10. $(p - 5)^2 = 9$

11. $(q + 3)^2 = 4$

12. $(3x - 2)^2 - 5 = 0$

13. $(2y + 3)^2 - 7 = 0$

14. $(h - 4)^2 = -8$

15. $(t + 5)^2 = -18$

16. $6p^2 - 3 = 2$

17. $15 = 4 + 3w^2$

18. $\left(x - \dfrac{3}{2}\right)^2 + \dfrac{7}{4} = 0$

19. $\left(m + \dfrac{4}{5}\right)^2 + \dfrac{3}{25} = 0$

20. $-x^2 + 4 = 13$

21. $-y^2 - 2 = 14$

22. Given the equation $x^2 = k$, match the following statements.

    a. If $k > 0$, then _____

    b. If $k < 0$, then _____

    c. If $k = 0$, then _____

    i. there will be one real solution.

    ii. there will be two real solutions.

    iii. there will be two imaginary solutions.

23. State two methods that can be used to solve the equation $x^2 - 36 = 0$. Then solve the equation by using both methods.

24. Explain the difference between solving the equations: $x = \sqrt{16}$ and $x^2 = 16$.

For Exercises 25–26, solve the equations.

25. a. $\sqrt{x} = 4$ **26. a.** $\sqrt{y} = 9$

    b. $x^2 = 4$ **b.** $y^2 = 9$

Concept 2: Solving Quadratic Equations by Completing the Square

For Exercises 27–38, find the value of n so that the expression is a perfect square trinomial. Then factor the trinomial. (See Example 4.)

27. $x^2 - 6x + n$

28. $x^2 + 12x + n$

29. $t^2 + 8t - n$

30. $v^2 - 18v + n$

31. $c^2 - c + n$

32. $x^2 + 9x + n$

33. $y^2 + 5y + n$

34. $a^2 - 7a + n$

35. $b^2 + \dfrac{2}{5}b + n$

36. $m^2 - \dfrac{2}{7}m + n$

37. $p^2 - \dfrac{2}{3}p + n$

38. $w^2 + \dfrac{3}{4}w + n$

39. Summarize the steps used in solving a quadratic equation by completing the square and applying the square root property.

40. What types of quadratic equations can be solved by completing the square and applying the square root property?

For Exercises 41–60, solve the quadratic equation by completing the square and applying the square root property. Write imaginary solutions in the form $a + bi$. **(See Examples 5–7.)**

41. $t^2 + 8t + 15 = 0$ **42.** $m^2 + 6m + 8 = 0$ **43.** $x^2 + 6x = -16$ **44.** $x^2 - 4x = -15$

45. $p^2 + 4p + 6 = 0$ **46.** $q^2 + 2q + 2 = 0$ **47.** $-3y - 10 = -y^2$ **48.** $-24 = -2y^2 + 2y$

49. $2a^2 + 4a + 5 = 0$ **50.** $3a^2 + 6a - 7 = 0$ **51.** $9x^2 - 36x + 40 = 0$ **52.** $9y^2 - 12y + 5 = 0$

53. $25p^2 - 10p = 2$ **54.** $9n^2 - 6n = 1$ **55.** $(2w + 5)(w - 1) = 2$ **56.** $(3p - 5)(p + 1) = -3$

57. $n(n - 4) = 7$ **58.** $m(m + 10) = 2$ **59.** $2x(x + 6) = 14$ **60.** $3x(x - 2) = 24$

Concept 3: Literal Equations

61. The distance (in feet) that an object falls in t sec is given by the equation $d = 16t^2$, where $t \geq 0$.

    a. Solve the equation for t. **(See Example 8.)**

    b. Using the equation from part (a), determine the amount of time required for an object to fall 1024 ft.

62. The volume of a can that is 4 in. tall is given by the equation $V = 4\pi r^2$, where r is the radius of the can, measured in inches.

    a. Solve the equation for r. Do not rationalize the denominator.

    b. Using the equation from part (a), determine the radius of a can with a volume of 12.56 in.³ Use 3.14 for π.

For Exercises 63–68, solve for the indicated variable.

63. $A = \pi r^2$ for r $(r > 0)$ **64.** $E = mc^2$ for c $(c > 0)$

65. $a^2 + b^2 + c^2 = d^2$ for a $(a > 0)$ **66.** $a^2 + b^2 = c^2$ for b $(b > 0)$

67. $V = \dfrac{1}{3}\pi r^2 h$ for r $(r > 0)$ **68.** $V = \dfrac{1}{3}s^2 h$ for s $(s > 0)$

69. A corner shelf is to be made from a triangular piece of plywood, as shown in the diagram. Find the distance x that the shelf will extend along the walls. Assume that the walls are at right angles. Round the answer to a tenth of a foot.

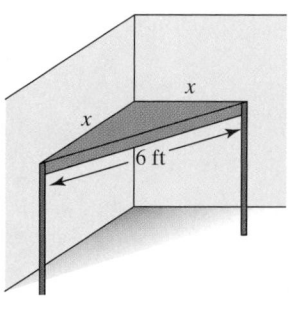

70. The volume of a box with a square bottom and a height of 4 in. is given by $V(x) = 4x^2$, where x is the length (in inches) of the sides of the bottom of the box.

    a. If the volume of the box is 289 in.³, find the dimensions of the box.

    b. Are there two possible answers to part (a)? Why or why not?

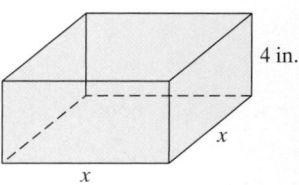

71. A square has an area of 50 in.^{2} What are the lengths of the sides? (Round to one decimal place.)

72. The amount of money A in an account with an interest rate r compounded annually is given by

$$A = P(1 + r)^t$$

where P is the initial principal and t is the number of years the money is invested.

a. If a \$10,000 investment grows to \$11,664 after 2 yr, find the interest rate.

b. If a \$6000 investment grows to \$7392.60 after 2 yr, find the interest rate.

c. Jamal wants to invest \$5000. He wants the money to grow to at least \$6500 in 2 yr to cover the cost of his son's first year at college. What interest rate does Jamal need for his investment to grow to \$6500 in 2 yr? Round to the nearest hundredth of a percent.

73. A textbook company has discovered that the profit for selling its books is given by

$$P(x) = -\frac{1}{8}x^2 + 5x$$

where x is the number of textbooks produced (in thousands) and $P(x)$ is the corresponding profit (in thousands of dollars). The graph of the function is shown at right.

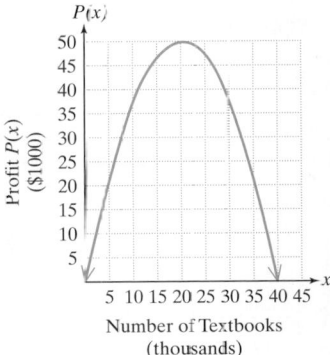

a. Approximate the number of books required to make a profit of \$20,000. [*Hint:* Let $P(x) = 20$. Then complete the square to solve for x.] Round to one decimal place.

b. Why are there two answers to part (a)?

74. If we ignore air resistance, the distance (in feet) that an object travels in free fall can be approximated by $d(t) = 16t^2$, where t is the time in seconds after the object was dropped.

a. If the CN Tower in Toronto is 1815 ft high, how long will it take an object to fall from the top of the building? Round to one decimal place.

b. If the Renaissance Tower in Dallas is 886 ft high, how long will it take an object to fall from the top of the building? Round to one decimal place.

Concepts

1. Derivation of the Quadratic Formula
2. Solving Quadratic Equations by Using the Quadratic Formula
3. Using the Quadratic Formula in Applications
4. Discriminant
5. Mixed Review: Methods to Solve a Quadratic Equation

1. Derivation of the Quadratic Formula

If we solve a quadratic equation in standard form $ax^2 + bx + c = 0, a > 0$, by completing the square and using the square root property, the result is a formula that gives the solutions for x in terms of a, b, and c.

$ax^2 + bx + c = 0$	Begin with a quadratic equation in standard form.
$\dfrac{ax^2}{a} + \dfrac{bx}{a} + \dfrac{c}{a} = \dfrac{0}{a}$	Divide by the leading coefficient.
$x^2 + \dfrac{b}{a}x + \dfrac{c}{a} = 0$	
$x^2 + \dfrac{b}{a}x = -\dfrac{c}{a}$	Isolate the terms containing x.
$x^2 + \dfrac{b}{a}x + \left(\dfrac{1}{2} \cdot \dfrac{b}{a}\right)^2 = \left(\dfrac{1}{2} \cdot \dfrac{b}{a}\right)^2 - \dfrac{c}{a}$	Add the square of $\frac{1}{2}$ the linear term coefficient to both sides of the equation.
$\left(x + \dfrac{b}{2a}\right)^2 = \dfrac{b^2}{4a^2} - \dfrac{c}{a}$	Factor the left side as a perfect square.
$\left(x + \dfrac{b}{2a}\right)^2 = \dfrac{b^2 - 4ac}{4a^2}$	Combine fractions on the right side by getting a common denominator.
$x + \dfrac{b}{2a} = \pm\sqrt{\dfrac{b^2 - 4ac}{4a^2}}$	Apply the square root property.
$x + \dfrac{b}{2a} = \dfrac{\pm\sqrt{b^2 - 4ac}}{2a}$	Simplify the denominator.
$x = -\dfrac{b}{2a} \pm \dfrac{\sqrt{b^2 - 4ac}}{2a}$	Subtract $\dfrac{b}{2a}$ from both sides.
$= \dfrac{-b \pm \sqrt{b^2 - 4ac}}{2a}$	Combine fractions.

TIP: When applying the quadratic formula, note that a, b, and c are constants. The variable is x.

The solutions to the equation $ax^2 + bx + c = 0$ for x in terms of the coefficients a, b, and c are given by the **quadratic formula**.

The Quadratic Formula

For a quadratic equation of the form $ax^2 + bx + c = 0$ $(a \neq 0)$ the solutions are

$$x = \frac{-b \pm \sqrt{b^2 - 4ac}}{2a}$$

The quadratic formula gives us another technique to solve a quadratic equation. This method will work regardless of whether the equation is factorable or not factorable.

2. Solving Quadratic Equations by Using the Quadratic Formula

Example 1 Solving a Quadratic Equation by Using the Quadratic Formula

Solve the quadratic equation by using the quadratic formula.

$$2x^2 - 3x = 5$$

Solution:

$2x^2 - 3x = 5$

$2x^2 - 3x - 5 = 0$ Write the equation in the form $ax^2 + bx + c = 0$.

$a = 2, \quad b = -3, \quad c = -5$ Identify a, b, and c.

$x = \dfrac{-b \pm \sqrt{b^2 - 4ac}}{2a}$ Apply the quadratic formula.

$= \dfrac{-(-3) \pm \sqrt{(-3)^2 - 4(2)(-5)}}{2(2)}$ Substitute $a = 2$, $b = -3$, and $c = -5$.

$= \dfrac{3 \pm \sqrt{9 + 40}}{4}$ Simplify.

$= \dfrac{3 \pm \sqrt{49}}{4}$

$= \dfrac{3 \pm 7}{4}$

$x = \dfrac{3 + 7}{4} = \dfrac{10}{4} = \dfrac{5}{2}$

$x = \dfrac{3 - 7}{4} = \dfrac{-4}{4} = -1$

> **Avoiding Mistakes**
> - The term, $-b$, represents the *opposite* of b.
> - Remember to write the *entire* numerator over $2a$.

The solution set is $\left\{ \dfrac{5}{2}, -1 \right\}$. Both solutions check in the original equation.

Skill Practice Solve the equation by using the quadratic formula.

1. $6x^2 - 5x = 4$

Example 2 Solving a Quadratic Equation by Using the Quadratic Formula

Solve the quadratic equation by using the quadratic formula. $-x(x - 6) = 11$

Solution:

$-x(x - 6) = 11$

$-x^2 + 6x - 11 = 0$ Write the equation in the form $ax^2 + bx + c = 0$.

$-1(-x^2 + 6x - 11) = -1(0)$ If the leading coefficient of the quadratic polynomial is negative, we suggest multiplying both sides of the equation by -1. Although this is not mandatory, it is generally easier to simplify the quadratic formula when the value of a is positive.

$x^2 - 6x + 11 = 0$

Answer

1. $\left\{ -\dfrac{1}{2}, \dfrac{4}{3} \right\}$

$$a = 1, b = -6, \text{ and } c = 11 \qquad \text{Identify } a, b, \text{ and } c.$$

$$x = \frac{-b \pm \sqrt{b^2 - 4ac}}{2a} \qquad \text{Apply the quadratic formula.}$$

$$= \frac{-(-6) \pm \sqrt{(-6)^2 - 4(1)(11)}}{2(1)} \qquad \text{Substitute } a = 1, b = -6, \text{ and } c = 11.$$

$$= \frac{6 \pm \sqrt{36 - 44}}{2} \qquad \text{Simplify.}$$

$$= \frac{6 \pm \sqrt{-8}}{2}$$

$$= \frac{6 \pm 2i\sqrt{2}}{2} \qquad \text{Simplify the radical.}$$

$$= \frac{2(3 \pm i\sqrt{2})}{2} \qquad \text{Factor the numerator.}$$

$$= \frac{2(3 \pm i\sqrt{2})}{2} \qquad \text{Simplify the fraction to lowest terms.}$$

$$= 3 \pm i\sqrt{2} \qquad \begin{matrix} x = 3 + i\sqrt{2} \\ \\ x = 3 - i\sqrt{2} \end{matrix} \qquad \begin{matrix} \text{The solutions are} \\ \text{imaginary numbers.} \end{matrix}$$

The solution set is $\{3 \pm i\sqrt{2}\}$.

Skill Practice Solve the equation by using the quadratic formula.

2. $-y(y + 4) = 12$

3. Using the Quadratic Formula in Applications

Example 3 Using the Quadratic Formula in an Application

A delivery truck travels south from Hartselle, Alabama, to Birmingham, Alabama, along Interstate 65. The truck then heads east to Atlanta, Georgia, along Interstate 20. The distance from Birmingham to Atlanta is 8 mi less than twice the distance from Hartselle to Birmingham. If the straight-line distance from Hartselle to Atlanta is 165 mi, find the distance from Hartselle to Birmingham and from Birmingham to Atlanta. (Round the answers to the nearest mile.)

Solution:

The motorist travels due south and then due east. Therefore, the three cities form the vertices of a right triangle (Figure 7-1).

Let x represent the distance between Hartselle and Birmingham.

Then $2x - 8$ represents the distance between Birmingham and Atlanta.

Use the Pythagorean theorem to establish a relationship among the three sides of the triangle.

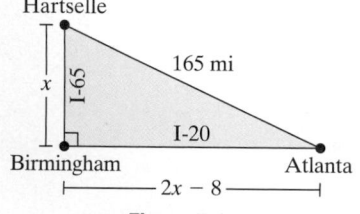

Figure 7-1

Answer

2. $\{-2 \pm 2i\sqrt{2}\}$

$$(x)^2 + (2x - 8)^2 = (165)^2$$

$$x^2 + 4x^2 - 32x + 64 = 27{,}225$$

$$5x^2 - 32x - 27{,}161 = 0$$ Write the equation in the form $ax^2 + bx - c = 0$.

$$a = 5 \qquad b = -32 \qquad c = -27{,}161$$ Identify a, b, and c.

$$x = \frac{-(-32) \pm \sqrt{(-32)^2 - 4(5)(-27{,}161)}}{2(5)}$$ Apply the quadratic formula.

$$= \frac{32 \pm \sqrt{1024 + 543{,}220}}{10}$$ Simplify.

$$= \frac{32 \pm \sqrt{544{,}244}}{10}$$

$$x = \frac{32 + \sqrt{544{,}244}}{10} \approx 76.97$$

$$x = \frac{32 - \sqrt{544{,}244}}{10} \approx -70.57$$

We reject the negative solution because distance cannot be negative. Rounding to the nearest whole unit, we have $x = 77$. Therefore, $2x - 8 = 2(77) - 8 = 146$.

The distance between Hartselle and Birmingham is 77 mi, and the distance between Birmingham and Atlanta is 146 mi.

Skill Practice

3. Steve and Tammy leave a campground, hiking on two different trails. Steve heads south and Tammy heads east. By lunchtime they are 9 mi apart. Steve walked 3 mi more than twice as many miles as Tammy. Find the distance each person hiked. (Round to the nearest tenth of a mile.)

Example 4 **Analyzing a Quadratic Function**

A model rocket is launched straight upward from the side of a 144-ft cliff (Figure 7-2). The initial velocity is 112 ft/sec. The height of the rocket $h(t)$ is given by

$$h(t) = -16t^2 + 112t + 144$$

where $h(t)$ is measured in feet and t is the time in seconds. Find the time(s) at which the rocket is 208 ft above the ground.

Solution:

$$h(t) = -16t^2 + 112t + 144$$

$$208 = -16t^2 + 112t + 144$$ Substitute 208 for $h(t)$.

$$16t^2 - 112t + 64 = 0$$ Write the equation in the form $at^2 + bt + c = 0$.

$$\frac{16t^2}{16} - \frac{112t}{16} + \frac{64}{16} = \frac{0}{16}$$ Divide by 16. This makes the coefficients smaller, and it is less cumbersome to solve.

$$t^2 - 7t + 4 = 0$$ The equation is not factorable. Apply the quadratic formula.

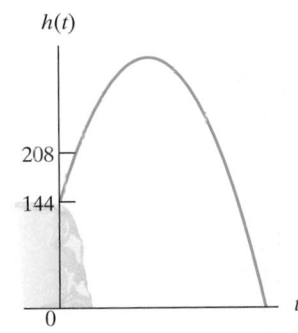

Figure 7-2

Answer

3. Tammy hiked 2.8 mi, and Steve hiked 8.6 mi.

$$t = \frac{-(-7) \pm \sqrt{(-7)^2 - 4(1)(4)}}{2(1)}$$ Let $a = 1, b = -7$, and $c = 4$.

$$= \frac{7 \pm \sqrt{33}}{2}$$

$$t = \frac{7 + \sqrt{33}}{2} \approx 6.37$$

$$t = \frac{7 - \sqrt{33}}{2} \approx 0.63$$

The rocket will reach a height of 208 ft after approximately 0.63 sec (on the way up) and after 6.37 sec (on the way down).

Skill Practice

4. A rocket is launched from the top of a 96-ft building with an initial velocity of 64 ft/sec. The height $h(t)$ of the rocket is given by $h(t) = -16t^2 + 64t + 96$. Find the time it takes for the rocket to hit the ground. [*Hint:* $h(t) = 0$ when the object hits the ground.]

4. Discriminant

The radicand within the quadratic formula is the expression $b^2 - 4ac$. This is called the **discriminant**. The discriminant can be used to determine the number of solutions to a quadratic equation as well as whether the solutions are rational, irrational, or imaginary numbers.

Using the Discriminant to Determine the Number and Type of Solutions to a Quadratic Equation

Consider the equation $ax^2 + bx + c = 0$, where a, b, and c are rational numbers and $a \neq 0$. The expression $b^2 - 4ac$ is called the *discriminant*. Furthermore,

- If $b^2 - 4ac > 0$, then there will be two real solutions.

  a. If $b^2 - 4ac$ is a perfect square, the solutions will be rational numbers.
  b. If $b^2 - 4ac$ is not a perfect square, the solutions will be irrational numbers.

- If $b^2 - 4ac < 0$, then there will be two imaginary solutions.
- If $b^2 - 4ac = 0$, then there will be one rational solution.

Example 5   **Using the Discriminant**

Use the discriminant to determine the type and number of solutions for each equation.

a. $2x^2 - 5x + 9 = 0$ **b.** $3x^2 = -x + 2$

c. $-2x(2x - 3) = -1$ **d.** $3.6x^2 = -1.2x - 0.1$

Solution:

For each equation, first write the equation in standard form $ax^2 + bx + c = 0$. Then determine the discriminant.

Equation	Discriminant	Solution Type and Number
a. $2x^2 - 5x + 9 = 0$	$b^2 - 4ac$ $= (-5)^2 - 4(2)(9)$ $= 25 - 72$ $= -47$	Because $-47 < 0$, there will be two imaginary solutions.

Answer

4. $2 + \sqrt{10} \approx 5.16$ sec

b. $3x^2 = -x + 2$

$3x^2 + x - 2 = 0$

$b^2 - 4ac$
$= (1)^2 - 4(3)(-2)$
$= 1 - (-24)$
$= 25$

$25 > 0$ and 25 is a perfect square. There will be two rational solutions.

c. $-2x(2x - 3) = -1$

$-4x^2 + 6x = -1$

$-4x^2 + 6x + 1 = 0$

$b^2 - 4ac$
$= (6)^2 - 4(-4)(1)$
$= 36 - (-16)$
$= 52$

$52 > 0$, but 52 is *not* a perfect square. There will be two irrational solutions.

d. $3.6x^2 = -1.2x - 0.1$

$3.6x^2 + 1.2x + 0.1 = 0$

$b^2 - 4ac$
$= (1.2)^2 - 4(3.6)(0.1)$
$= 1.44 - 1.44$
$= 0$

Because the discriminant equals 0, there will be only one rational solution.

Skill Practice Use the discriminant to determine the type and number of solutions for the equation.

5. $3y^2 + y + 3 = 0$

6. $4t^2 = 6t - 2$

7. $3t(t + 1) = 9$

8. $\dfrac{2}{3}x^2 - \dfrac{2}{3}x + \dfrac{1}{6} = 0$

With the discriminant we can determine the number of real-valued solutions to the equation $ax^2 + bx + c = 0$, and thus the number of x-intercepts to the function $f(x) = ax^2 + bx + c$. The following illustrations show the graphical interpretation of the three cases of the discriminant.

$f(x) = x^2 - 4x + 3$

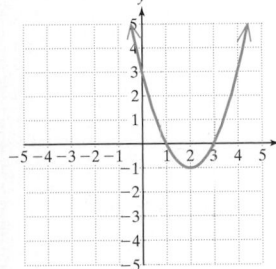

Use $x^2 - 4x + 3 = 0$ to find the value of the discriminant.

$$b^2 - 4ac = (-4)^2 - 4(1)(3)$$
$$= 4$$

Since the discriminant is positive, there are two real solutions to the quadratic equation. Therefore, there are two x-intercepts to the corresponding quadratic function, $(1, 0)$ and $(3, 0)$.

$f(x) = x^2 - x + 1$

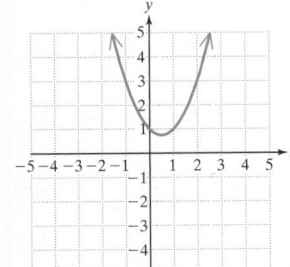

Use $x^2 - x + 1 = 0$ to find the value of the discriminant.

$$b^2 - 4ac = (-1)^2 - 4(1)(1)$$
$$= -3$$

Since the discriminant is negative, there are no real solutions to the quadratic equation. Therefore, there are no x-intercepts to the corresponding quadratic function.

Answers

5. -35; two imaginary solutions
6. 4; two rational solutions
7. 117; two irrational solutions
8. 0; one rational solution

$$f(x) = x^2 - 2x + 1$$

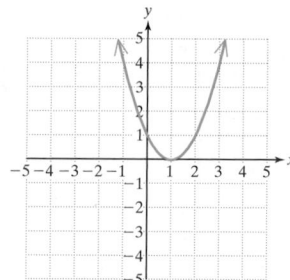

Use $x^2 - 2x + 1 = 0$ to find the value of the discriminant.

$$b^2 - 4ac = (-2)^2 - 4(1)(1)$$
$$= 0$$

Since the discriminant is zero, there is one real solution to the quadratic equation. Therefore, there is one x-intercept to the corresponding quadratic function, $(1, 0)$.

Example 6 Finding *x*- and *y*-Intercepts of a Quadratic Function

Given $f(x) = x^2 - 3x + 1$

a. Find the discriminant and use it to determine if there are any x-intercepts.

b. Find the x-intercept(s), if they exist.

c. Find the y-intercept.

Solution:

a. $a = 1$, $b = -3$, and $c = 1$.

The discriminant is $b^2 - 4ac = (-3)^2 - 4(1)(1)$
$$= 9 - 4$$
$$= 5$$

Since $5 > 0$, there are two x-intercepts.

b. The x-intercepts are given by the real solutions to the equation $f(x) = 0$. In this case, we have

$$f(x) = x^2 - 3x + 1 = 0$$

$$x^2 - 3x + 1 = 0 \qquad \text{The equation is in the form } ax^2 + bx + c = 0.$$

> **TIP:** Recall that an x-intercept is a point $(a, 0)$ where the graph of a function intersects the x-axis. A y-intercept is a point $(0, b)$ where the graph intersects the y-axis.

$$x = \frac{3 \pm \sqrt{(-3)^2 - (4)(1)(1)}}{2(1)} \qquad \text{Apply the quadratic formula.}$$

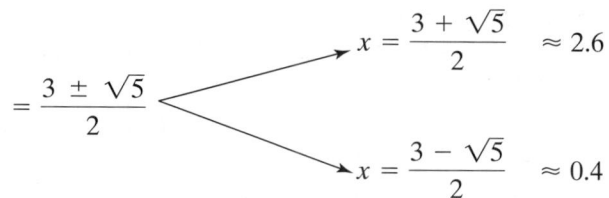

$$= \frac{3 \pm \sqrt{5}}{2}$$

$$x = \frac{3 + \sqrt{5}}{2} \approx 2.6$$

$$x = \frac{3 - \sqrt{5}}{2} \approx 0.4$$

The solutions are $\dfrac{3 + \sqrt{5}}{2}$ and $\dfrac{3 - \sqrt{5}}{2}$. Therefore, the x-intercepts are

$$\left(\frac{3 + \sqrt{5}}{2}, 0\right) \text{ and } \left(\frac{3 - \sqrt{5}}{2}, 0\right).$$

c. To find the y-intercept, evaluate $f(0)$.

$$f(0) = (0)^2 - 3(0) + 1 = 1$$

The y-intercept is located at $(0, 1)$.

The parabola is shown in the graph with the x- and y-intercepts labeled.

Answers

9. 17, there are two x-intercepts.

10. x-intercepts: $\left(\dfrac{-5 + \sqrt{17}}{2}, 0\right)$,

$\left(\dfrac{-5 - \sqrt{17}}{2}, 0\right)$

y-intercept: $(0, 2)$

Skill Practice Given $f(x) = x^2 + 5x + 2$,

9. Find the discriminant and use it to determine if there are any x-intercepts.

10. Find the x- and y-intercepts.

Example 7 Finding *x*- and *y*-Intercepts of a Quadratic Function

Given $f(x) = x^2 + x + 2$

a. Find the discriminant and use it to determine if there are any *x*-intercepts.

b. Find the *x*-intercept(s), if they exist.

c. Find the *y*-intercept.

Solution:

a. $a = 1, b = 1,$ and $c = 2$.

The discriminant is $b^2 - 4ac = (1)^2 - 4(1)(2)$

$$= 1 - 8$$

$$= -7$$

Since $-7 < 0$, there are no *x*-intercepts.

b. There are no *x*-intercepts.

c. $f(0) = (0)^2 + (0) + 2$

$$= 2$$

The *y*-intercept is located at $(0, 2)$.

The parabola is shown. *Note:* The graph does not intersect the *x*-axis.

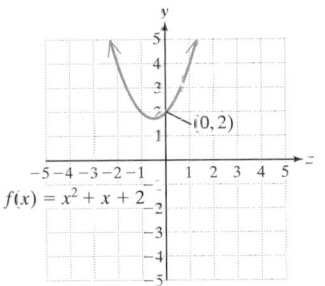

Skill Practice Given $f(x) = 2x^2 - 3x + 5$,

11. Find the discriminant and use it to determine if there are any *x*-intercepts.

12. Find the *y*-intercept.

5. Mixed Review: Methods to Solve a Quadratic Equation

Three methods have been presented to solve quadratic equations.

Methods to Solve a Quadratic Equation	
Factor and use the zero product rule (Section 4.8). • This method works well if you can factor the equation easily.	Example: $x^2 + 8x + 15 = 0$ factors as $(x + 3)(x + 5) = 0$
Use the square root property. Complete the square if necessary (Section 7.1). This method is particularly good if • the equation is of the form $ax^2 + c = 0$ or • the equation is of the form $x^2 + bx + c = 0$, where *b* is even.	Examples: $4x^2 + 9 = 0$ $x^2 + 10x - 3 = 0$
Quadratic Formula (Section 7.2) • This method works in all cases—just be sure to write the equation in the form $ax^2 + bx + c = 0$ before applying the formula.	Example: $7x^2 - 3x + 11 = 0$

Answers

11. -31, there are no *x*-intercepts.
12. $(0, 5)$

Before solving a quadratic equation, take a minute to analyze it first. Each problem must be evaluated individually before choosing the most efficient method to find its solutions.

Example 8 Solving a Quadratic Equation by Using Any Method

Solve the equation by using any method.

$$(x + 3)^2 + x^2 - 9x = 8$$

Solution:

$$(x + 3)^2 + x^2 - 9x = 8$$

$x^2 + 6x + 9 + x^2 - 9x - 8 = 0$	Clear parentheses and write the equation in the form $ax^2 + bx + c = 0$.
$2x^2 - 3x + 1 = 0$	This equation is factorable.
$(2x - 1)(x - 1) = 0$	Factor.
$2x - 1 = 0$ or $x - 1 = 0$	Apply the zero product rule.
$x = \frac{1}{2}$ or $x = 1$	Solve for x.

The solution set is $\{\frac{1}{2}, 1\}$.

This equation could have been solved by using any of the three methods, but factoring was the most efficient method.

Skill Practice Solve using any method.

13. $2t(t - 1) + t^2 = 5$

Example 9 Solving a Quadratic Equation by Using Any Method

Solve the equation by using any method.

$$x^2 + 5 = -2x$$

Solution:

$x^2 + 2x + 5 = 0$	The equation does not factor.
$x^2 + 2x = -5$	Because $a = 1$ and b is even, we can easily complete the square.
$x^2 + 2x + 1 = -5 + 1$	Add $[\frac{1}{2}(2)]^2 = 1^2 = 1$ to both sides.
$(x + 1)^2 = -4$	
$x + 1 = \pm\sqrt{-4}$	Apply the square root property.
$x = -1 \pm 2i$	Solve for x.

The solution set is $\{-1 \pm 2i\}$.

This equation could also have been solved by using the quadratic formula.

Answers

13. $\left\{-1, \dfrac{5}{3}\right\}$

14. $\{2 \pm i\sqrt{3}\}$

Skill Practice Solve using any method.

14. $x^2 - 4x = -7$

Example 10 Solving a Quadratic Equation by Using Any Method

Solve the equation by using any method. $\dfrac{1}{4}x^2 - \dfrac{1}{2}x + \dfrac{1}{3} = 0$

Solution:

$12 \cdot \left(\dfrac{1}{4}x^2 - \dfrac{1}{2}x + \dfrac{1}{3} \right) = 12 \cdot (0)$ Clear fractions.

$\qquad\qquad 3x^2 - 6x + 4 = 0$ The equation is in the form $ax^2 + bx + c = 0$.
The left-hand side does not factor.

$a = 3, b = -6,$ and $c = 4$

$x = \dfrac{-(-6) \pm \sqrt{(-6)^2 - 4(3)(4)}}{2(3)}$ Apply the quadratic formula.

$\quad = \dfrac{6 \pm \sqrt{-12}}{6}$ Simplify.

$\quad = \dfrac{6 \pm 2i\sqrt{3}}{6}$ Simplify the radical.

$\quad = \dfrac{\overset{1}{2}(3 \pm i\sqrt{3})}{\underset{3}{6}}$ Factor and simplify.

$\quad = \dfrac{3 \pm i\sqrt{3}}{3}$

$\quad = \dfrac{3}{3} \pm \dfrac{\sqrt{3}}{3}i$ Write in the form $a \pm bi$.

$\quad = 1 \pm \dfrac{\sqrt{3}}{3}i$ Simplify.

The solution set is $\left\{ 1 \pm \dfrac{\sqrt{3}}{3}i \right\}$.

Skill Practice Solve using any method.

15. $\dfrac{1}{5}x^2 - \dfrac{4}{5}x + \dfrac{1}{2} = 0$

Answer

15. $\left\{ \dfrac{4 \pm \sqrt{6}}{2} \right\}$

Example 11 Solving a Quadratic Equation by Using Any Method

Solve the equation by using any method. $9p^2 - 11 = 0$

Solution:

$$9p^2 - 11 = 0 \qquad \text{Because } b = 0, \text{ use the square root property.}$$

$$9p^2 = 11 \qquad \text{Isolate the variable term.}$$

$$p^2 = \frac{11}{9} \qquad \text{The equation is in the form } x^2 = k.$$

$$p = \pm\sqrt{\frac{11}{9}} \qquad \text{Apply the square root property.}$$

$$p = \pm\frac{\sqrt{11}}{3} \qquad \text{Simplify the radical.}$$

The solution set is $\left\{ \pm\dfrac{\sqrt{11}}{3} \right\}$.

Skill Practice Solve using any method.

16. $4y^2 - 13 = 0$

Answer

16. $\left\{ \pm\dfrac{\sqrt{13}}{2} \right\}$

Section 7.2 Practice Exercises

Vocabulary and Key Concepts

1. a. For the equation $ax^2 + bx + c = 0$ $(a \neq 0)$, the _____ formula gives the solutions as $x =$ _____.

b. To apply the quadratic formula, a quadratic equation must be written in the form _____ where $a \neq 0$.

c. To apply the quadratic formula to solve the equation $8x^2 - 42x - 27 = 0$, the value of a is _____, the value of b is _____, and the value of c is _____.

d. To apply the quadratic formula to solve the equation $3x^2 - 7x - 4 = 0$, the value of $-b$ is _____ and the value of the radicand is _____.

e. The radicand within the quadratic formula is _____ and is called the _____.

f. If the discriminant is negative, then the solutions to a quadratic equation will be (real/imaginary) numbers.

g. If the discriminant is positive, then the solutions to a quadratic equation will be (real/imaginary) numbers.

h. Given a quadratic function $f(x) = ax^2 + bx + c = 0$, the function will have no x-intercepts if the discriminant is (less than, greater than, equal to) zero.

Review Exercises

2. Use substitution to determine if $x = -3 + \sqrt{5}$ is a solution to $x^2 + 6x + 4 = 0$.

For Exercises 3–6, simplify the expression.

3. $\dfrac{16 - \sqrt{320}}{4}$

4. $\dfrac{18 + \sqrt{180}}{3}$

5. $\dfrac{14 - \sqrt{-147}}{7}$

6. $\dfrac{10 - \sqrt{-175}}{5}$

Concept 2: Solving Quadratic Equations by Using the Quadratic Formula

For Exercises 7–8, determine whether the equation is linear or quadratic.

7. $2(x - 5) + x^2 = 3x(x - 4) - 2x^2$

8. $5x(x + 3) - 9 = 4x^2 - 3(x + 1)$

For Exercises 9–34, solve the equation by using the quadratic formula. Write imaginary solutions in the form $a + bi$.
(See Examples 1–2.)

9. $x^2 + 11x - 12 = 0$

10. $5x^2 - 14x - 3 = 0$

11. $9y^2 - 2y + 5 = 0$

12. $2t^2 + 3t - 7 = 0$

13. $12p^2 - 4p + 5 = 0$

14. $-5n^2 + 4n - 6 = 0$

15. $-z^2 = -2z - 35$

16. $12x^2 - 5x = 2$

17. $y^2 + 3y = 8$

18. $k^2 + 4 = 6k$

19. $25x^2 - 20x + 4 = 0$

20. $9y^2 = -12y - 4$

21. $w(w - 6) = -14$

22. $m(m + 6) = -11$

23. $(x + 2)(x - 3) = 1$

24. $3y(y + 1) - 7y(y + 2) = 6$

25. $-4a^2 - 2a + 3 = 0$

26. $-2m^2 - 5m + 3 = 0$

27. $\frac{1}{2}y^2 + \frac{2}{3} = -\frac{2}{3}y$
(*Hint:* Clear fractions first.)

28. $\frac{2}{3}p^2 - \frac{1}{6}p + \frac{1}{2} = 0$

29. $\frac{1}{5}h^2 + h + \frac{3}{5} = 0$

30. $\frac{1}{4}w^2 + \frac{7}{4}w + 1 = 0$

31. $0.01x^2 + 0.06x + 0.08 = 0$
(*Hint:* Clear decimals first.)

32. $0.5y^2 - 0.7y + 0.2 = 0$

33. $0.3t^2 + 0.7t - 0.5 = 0$

34. $0.01x^2 + 0.04x - 0.07 = 0$

For Exercises 35–38, factor the expression. Then use the zero product rule and the quadratic formula to solve the equation. There should be three solutions to each equation. Write imaginary solutions in the form $a + bi$.

35. a. Factor. $x^3 - 27$

　　b. Solve. $x^3 - 27 = 0$

36. a. Factor. $64x^3 + 1$

　　b. Solve. $64x^3 + 1 = 0$

37. a. Factor. $3x^3 - 6x^2 + 6x$

　　b. Solve. $3x^3 - 6x^2 + 6x = 0$

38. a. Factor. $5x^3 + 5x^2 + 10x$

　　b. Solve. $5x^3 + 5x^2 + 10x = 0$

Concept 3: Using the Quadratic Formula in Applications

39. The volume of a cube is 27 ft³. Find the lengths of the sides.

40. The volume of a rectangular box is 64 ft³. If the width is 3 times longer than the height, and the length is 9 times longer than the height, find the dimensions of the box.

41. The hypotenuse of a right triangle measures 4 in. One leg of the triangle is 2 in. longer than the other leg. Find the lengths of the legs of the triangle. Round to one decimal place. **(See Example 3.)**

42. The length of one leg of a right triangle is 1 cm more than twice the length of the other leg. The hypotenuse measures 6 cm. Find the lengths of the legs. Round to one decimal place.

43. The hypotenuse of a right triangle is 10.2 m long. One leg is 2.1 m shorter than the other leg. Find the lengths of the legs. Round to one decimal place.

44. The hypotenuse of a right triangle is 17 ft long. One leg is 3.4 ft longer than the other leg. Find the lengths of the legs.

45. The fatality rate (in fatalities per 100 million vehicle miles driven) can be approximated for drivers x years old according to the function, $F(x) = 0.0036x^2 - 0.35x + 9.2$. (*Source:* U.S. Department of Transportation)

   a. Approximate the fatality rate for drivers 16 yr old.

   b. Approximate the fatality rate for drivers 40 yr old.

   c. Approximate the fatality rate for drivers 80 yr old.

   d. For what age(s) is the fatality rate approximately 2.5?

46. The braking distance (in feet) of a car going v mph is given by

$$d(v) = \frac{v^2}{20} + v \qquad v \geq 0$$

   a. Find the speed for a braking distance of 150 ft. Round to the nearest mile per hour.

   b. Find the speed for a braking distance of 100 ft. Round to the nearest mile per hour.

47. Mitch throws a baseball straight up in the air from a cliff that is 48 ft high. The initial velocity is 48 ft/sec. The height (in feet) of the object after t sec is given by $h(t) = -16t^2 + 48t + 48$. Find the time at which the height of the object is 64 ft. **(See Example 4.)**

48. An astronaut on the moon throws a rock into the air from the deck of a spacecraft that is 8 m high. The initial velocity of the rock is 2.4 m/sec. The height (in meters) of the rock after t sec is given by $h(t) = -0.8t^2 + 2.4t + 8$. Find the time at which the height of the rock is 6 m.

Concept 4: Discriminant

For Exercises 49–56,

   a. Write the equation in the form $ax^2 + bx + c = 0, a > 0$.

   b. Find the value of the discriminant.

   c. Use the discriminant to determine the number and type of solutions. **(See Example 5.)**

49. $x^2 + 2x = -1$

50. $12y - 9 = 4y^2$

51. $19m^2 = 8m$

52. $2n - 5n^2 = 0$

53. $5p^2 - 21 = 0$

54. $3k^2 = 7$

55. $4n(n - 2) - 5n(n - 1) = 4$

56. $(2x + 1)(x - 3) = -9$

For Exercises 57–62, determine the discriminant. Then use the discriminant to determine the number of x-intercepts for the function.

57. $f(x) = x^2 - 6x + 5$

58. $g(x) = -x^2 - 4x - 3$

59. $h(x) = 4x^2 + 12x + 9$

60. $k(x) = 25x^2 - 10x + 1$

61. $p(x) = 2x^2 + 3x + 6$

62. $m(x) = 3x^2 + 4x + 7$

For Exercises 63–68, find the x- and y-intercepts of the quadratic function. **(See Examples 6–7.)**

63. $f(x) = x^2 - 5x + 3$

64. $g(x) = 2x^2 + 7x + 2$

65. $g(x) = -x^2 + x - 1$

66. $f(x) = 2x^2 + x + 5$

67. $p(x) = 2x^2 + 5x - 2$

68. $h(x) = 3x^2 + 2x - 2$

Concept 5: Mixed Review: Methods to Solve a Quadratic Equation

For Exercises 69–86, solve the quadratic equation by using any method. Write imaginary solutions in the form $a + bi$. **(See Examples 8–11.)**

69. $a^2 + 2a + 10 = 0$

70. $4z^2 + 7z = 0$

71. $(x - 2)^2 + 2x^2 - 13x = 10$

72. $(x - 3)^2 + 3x^2 - 5x = 12$

73. $4y^2 - 20y + 43 = 0$

74. $k^2 + 18 = 4k$

75. $\left(x + \dfrac{1}{2}\right)^2 + 4 = 0$

76. $(2y + 3)^2 = 9$

77. $2y(y - 3) = -1$

78. $w(w - 5) = 4$

79. $(2t + 5)(t - 1) = (t - 3)(t + 8)$

80. $(b - 1)(b + 4) = (3b + 2)(b + 1)$

81. $\dfrac{1}{8}x^2 - \dfrac{1}{2}x + \dfrac{1}{4} = 0$

82. $\dfrac{1}{6}x^2 - \dfrac{1}{2}x + \dfrac{1}{4} = 0$

83. $32z^2 - 20z - 3 = 0$

84. $8k^2 - 14k + 3 = 0$

85. $4p^2 - 21 = 0$

86. $5h^2 - 120 = 0$

Sometimes students shy away from completing the square and using the square root property to solve a quadratic equation. However, sometimes this process leads to a simple solution. For Exercises 87–88, solve the equations two ways.

    a. Solve the equation by completing the square and applying the square root property.

    b. Solve the equation by applying the quadratic formula.

    c. Which technique was easier for you?

87. $x^2 + 6x = 5$

88. $x^2 - 10x = -27$

Graphing Calculator Exercises

89. Graph $Y_1 = x^3 - 27$. Compare the x-intercepts with the solutions to the equation $x^3 - 27 = 0$ found in Exercise 35.

90. Graph $Y_1 = 64x^3 + 1$. Compare the x-intercepts with the solutions to the equation $64x^3 + 1 = 0$ found in Exercise 36.

91. Graph $Y_1 = 3x^3 - 6x^2 + 6x$. Compare the x-intercepts with the solutions to the equation $3x^3 - 6x^2 + 6x = 0$ found in Exercise 37.

92. Graph $Y_1 = 5x^3 + 5x^2 + 10x$. Compare the x-intercepts with the solutions to the equation $5x^3 + 5x^2 + 10x = 0$ found in Exercise 38.

Section 7.3 Equations in Quadratic Form

Concepts

1. Solving Equations by Using Substitution
2. Solving Equations Reducible to a Quadratic

1. Solving Equations by Using Substitution

We have learned to solve a variety of different types of equations, including linear, radical, rational, and polynomial equations. Sometimes, however, it is necessary to use a quadratic equation as a tool to solve other types of equations.

In Example 1, we will solve the equation $(2x^2 - 5)^2 - 16(2x^2 - 5) + 39 = 0$. Notice that the terms in the equation are written in descending order by degree. Furthermore, the first two terms have the same base, $2x^2 - 5$, and the exponent on the first term is exactly double the exponent on the second term. The third term is a constant. An equation in this pattern is called **quadratic in form**.

exponent is double   third term is constant

$$(2x^2 - 5)^2 - 16(2x^2 - 5)^1 + 39 = 0.$$

To solve this equation we will use substitution as demonstrated in Example 1.

Example 1   Solving an Equation in Quadratic Form

Solve the equation.

$$(2x^2 - 5)^2 - 16(2x^2 - 5) + 39 = 0$$

Solution:

$$(2x^2 - 5)^2 - 16(2x^2 - 5) + 39 = 0$$

Substitute $u = (2x^2 - 5)$.

Once the substitution is made, the equation becomes quadratic in the variable u.

$$u^2 - 16u + 39 = 0$$

The equation is in the form $au^2 + bu + c = 0$.

$$(u - 13)(u - 3) = 0$$

Factor.

$$u = 13 \quad \text{or} \quad u = 3$$

Apply the zero product rule.

Reverse substitute.

$$2x^2 - 5 = 13 \quad \text{or} \quad 2x^2 - 5 = 3$$

Reverse substitute.

$$2x^2 = 18 \quad \text{or} \quad 2x^2 = 8$$

$$x^2 = 9 \quad \text{or} \quad x^2 = 4$$

Write the equations in the form $x^2 = k$.

$$x = \pm\sqrt{9} \quad \text{or} \quad x = \pm\sqrt{4}$$

Apply the square root property.

$$= \pm 3 \quad \text{or} \quad = \pm 2$$

The solution set is $\{3, -3, 2, -2\}$.

All solutions check in the original equation.

Skill Practice Solve the equation.

1. $(3t^2 - 10)^2 + 5(3t^2 - 10) - 14 = 0$

Answer

1. $\{1, -1, 2, -2\}$

For an equation written in descending order, notice that u was set equal to the variable factor on the middle term. This is generally the case.

Example 2   Solving an Equation in Quadratic Form

Solve the equation. $p^{2/3} - 2p^{1/3} = 8$

Solution:

$$p^{2/3} - 2p^{1/3} = 8$$

$$p^{2/3} - 2p^{1/3} - 8 = 0 \qquad \text{Set the equation equal to zero.}$$

$$(p^{1/3})^2 - 2(p^{1/3})^1 - 8 = 0 \qquad \text{Make the substitution } u = p^{1/3}.$$

Substitute $u = p^{1/3}$.

$$u^2 - 2u - 8 = 0 \qquad \begin{array}{l}\text{Then the equation is in the form}\\ au^2 + bu + c = 0.\end{array}$$

$$(u - 4)(u + 2) = 0 \qquad \text{Factor.}$$

$$u = 4 \qquad \text{or} \qquad u = -2 \qquad \text{Apply the zero product rule.}$$

Reverse substitute.

$$p^{1/3} = 4 \qquad \text{or} \qquad p^{1/3} = -2$$

$$\sqrt[3]{p} = 4 \qquad \text{or} \qquad \sqrt[3]{p} = -2 \qquad \begin{array}{l}\text{The equations are radical}\\ \text{equations.}\end{array}$$

$$(\sqrt[3]{p})^3 = (4)^3 \qquad \text{or} \qquad (\sqrt[3]{p})^3 = (-2)^3 \qquad \text{Cube both sides.}$$

$$p = 64 \qquad \text{or} \qquad p = -8$$

The solution set is $\{64, -8\}$.   All solutions check in the original equation.

Skill Practice Solve the equation.

2. $y^{2/3} - y^{1/3} = 12$

Example 3   Solving an Equation in Quadratic Form

Solve the equation. $x - \sqrt{x} - 12 = 0$

Solution:

The equation can be solved by first isolating the radical and then squaring both sides (this is left as an exercise—see Exercise 25). However, this equation is also quadratic in form. By writing $\sqrt{x}$ as $x^{1/2}$, we see that the exponent on the first term is exactly double the exponent on the middle term.

$$x^1 - x^{1/2} - 12 = 0$$

$$(x^{1/2})^2 - (x^{1/2})^1 - 12 = 0 \qquad \text{Let } u = x^{1/2}.$$

$$u^2 - u - 12 = 0$$

$$(u - 4)(u + 3) = 0 \qquad \text{Factor.}$$

$$u = 4 \text{ or } u = -3 \qquad \text{Solve for } u.$$

$$x^{1/2} = 4 \text{ or } x^{1/2} = -3 \qquad \text{Reverse substitute.}$$

Answer

2. $\{64, -27\}$

$\sqrt{x} = 4$ or $\sqrt{x} = -3$ Solve each equation for x. Recall that the principal square root of a number cannot

$x = 16$ equal -3.

The solution set is $\{16\}$.   The value 16 checks in the original equation.

Skill Practice Solve the equation.

3. $z - \sqrt{z} - 2 = 0$

2. Solving Equations Reducible to a Quadratic

Some equations are reducible to a quadratic equation. In Example 4, we solve a polynomial equation by factoring. The resulting factors are quadratic.

Example 4   Solving a Polynomial Equation

Solve the equation. $4x^4 + 7x^2 - 2 = 0$

Solution:

$$4x^4 + 7x^2 - 2 = 0$$ This is a polynomial equation.

$$(4x^2 - 1)(x^2 + 2) = 0$$ Factor.

$$4x^2 - 1 = 0 \text{ or } x^2 + 2 = 0$$ Set each factor equal to zero. Notice that the two equations are quadratic.

$$x^2 = \frac{1}{4} \quad \text{ or } \quad x^2 = -2$$ Each can be solved by the square root property.

$$x = \pm\sqrt{\frac{1}{4}} \quad \text{ or } \quad x = \pm\sqrt{-2}$$ Apply the square root property.

$$x = \pm\frac{1}{2} \quad \text{ or } \quad x = \pm i\sqrt{2}$$ Simplify the radicals.

The solution set is $\left\{\dfrac{1}{2}, -\dfrac{1}{2}, i\sqrt{2}, -i\sqrt{2}\right\}$.

Skill Practice Solve the equation.

4. $9x^4 + 35x^2 - 4 = 0$

Example 5   Solving a Rational Equation

Solve the equation. $\dfrac{3y}{y + 2} - \dfrac{2}{y - 1} = 1$

Solution:

$\dfrac{3y}{y + 2} - \dfrac{2}{y - 1} = 1$ This is a rational equation. The LCD is $(y + 2)(y - 1)$. Also note that the restrictions on y are $y \neq -2$ and $y \neq 1$.

$\left(\dfrac{3y}{y + 2} - \dfrac{2}{y - 1}\right) \cdot (y + 2)(y - 1) = 1 \cdot (y + 2)(y - 1)$ Multiply both sides by the LCD.

$\dfrac{3y}{\cancel{y + 2}} \cdot \cancel{(y + 2)}(y - 1) - \dfrac{2}{\cancel{y - 1}} \cdot (y + 2)\cancel{(y - 1)} = 1 \cdot (y + 2)(y - 1)$

$$3y(y - 1) - 2(y + 2) = (y + 2)(y - 1)$$ Clear fractions.

$$3y^2 - 3y - 2y - 4 = y^2 - y + 2y - 2$$ Apply the distributive property.

$$3y^2 - 5y - 4 = y^2 + y - 2$$ The equation is quadratic.

$$2y^2 - 6y - 2 = 0$$ Write the equation in descending order.

$$\frac{2y^2}{2} - \frac{6y}{2} - \frac{2}{2} = \frac{0}{2}$$ Each coefficient in the equation is divisible by 2. Therefore, if we divide both sides by 2, the coefficients in the equation are smaller. This will make it easier to apply the quadratic formula.

$$y^2 - 3y - 1 = 0$$

$$y = \frac{-(-3) \pm \sqrt{(-3)^2 - 4(1)(-1)}}{2(1)}$$ Apply the quadratic formula.

$$= \frac{3 \pm \sqrt{9 + 4}}{2}$$

$$= \frac{3 \pm \sqrt{13}}{2} \quad \begin{array}{l} \longrightarrow y = \dfrac{3 + \sqrt{13}}{2} \\[2mm] \longrightarrow y = \dfrac{3 - \sqrt{13}}{2} \end{array}$$

The solution set is $\left\{ \dfrac{3 + \sqrt{13}}{2}, \dfrac{3 - \sqrt{13}}{2} \right\}$.

Skill Practice Solve the equation.

5. $\dfrac{t}{2t - 1} - \dfrac{1}{t + 4} = 1$

Answer

5. $\left\{ \dfrac{-5 \pm 3\sqrt{5}}{2} \right\}$

Section 7.3   Practice Exercises

Vocabulary and Key Concepts

1. a. An equation that can be written in the form $au^2 + bu + c = 0$ where u represents an algebraic expression is said to be in _____ form.

   b. To use the method of substitution to solve the equation $(3x - 1)^2 + 2(3x - 1) - 8 = 0$, let $u =$ _____.

   c. To use the method of substitution to solve the equation $p^{2/3} - 2p^{1/3} - 15 = 0$, let $u =$ _____.

Review Exercises

For Exercises 2–7, solve the quadratic equations.

2. $16 = (2x - 3)^2$

3. $\left(x - \dfrac{3}{2} \right)^2 = \dfrac{7}{4}$

4. $n(n - 6) = -13$

5. $x(x + 8) = -16$

6. $6k^2 + 7k = 6$

7. $2x^2 - 8x - 44 = 0$

Concept 1: Solving Equations by Using Substitution

8. a. Solve the quadratic equation by factoring. $u^2 - 2u - 24 = 0$

   b. Solve the equation by using substitution. $(x - 5)^2 - 2(x - 5) - 24 = 0$

9. a. Solve the quadratic equation by factoring. $u^2 + 10u + 24 = 0$

   b. Solve the equation by using substitution. $(y^2 + 5y)^2 + 10(y^2 + 5y) + 24 = 0$

10. a. Solve the quadratic equation by factoring. $u^2 - 2u - 35 = 0$

   b. Solve the equation by using substitution. $(w^2 - 6w)^2 - 2(w^2 - 6w) - 35 = 0$

For Exercises 11–24, solve the equation by using substitution. **(See Examples 1–3.)**

11. $(x^2 - 2x)^2 + 2(x^2 - 2x) = 3$

12. $(x^2 + x)^2 - 8(x^2 + x) = -12$

13. $(y^2 - 4y)^2 - (y^2 - 4y) = 20$

14. $(w^2 - 2w)^2 - 11(w^2 - 2w) = -24$

15. $m^{2/3} - m^{1/3} - 6 = 0$

16. $2n^{2/3} + 7n^{1/3} - 15 = 0$

17. $2t^{2/5} + 7t^{1/5} + 3 = 0$

18. $p^{2/5} + p^{1/5} - 2 = 0$

19. $y + 6\sqrt{y} = 16$

20. $p - 8\sqrt{p} = -15$

21. $2x + 3\sqrt{x} - 2 = 0$

22. $3t + 5\sqrt{t} - 2 = 0$

23. $16\left(\dfrac{x+6}{4}\right)^2 + 8\left(\dfrac{x+6}{4}\right) + 1 = 0$

24. $9\left(\dfrac{x+3}{2}\right)^2 - 6\left(\dfrac{x+3}{2}\right) + 1 = 0$

25. In Example 3, we solved the equation $x - \sqrt{x} - 12 = 0$ by using substitution. Now solve this equation by first isolating the radical and then squaring both sides. Don't forget to check the potential solutions in the original equation. Do you obtain the same solution as in Example 3?

Concept 2: Solving Equations Reducible to a Quadratic

For Exercises 26–36, solve the equations. **(See Examples 4–5.)**

26. $t^4 + t^2 - 12 = 0$

27. $w^4 + 4w^2 - 45 = 0$

28. $x^2(9x^2 + 7) = 2$

29. $y^2(4y^2 + 17) = 15$

30. $\dfrac{y}{10} - 1 = -\dfrac{12}{5y}$

31. $1 + \dfrac{5}{x} = -\dfrac{3}{x^2}$

32. $\dfrac{x+5}{x} + \dfrac{x}{2} = \dfrac{x+19}{4x}$

33. $\dfrac{3x}{x+1} - \dfrac{2}{x-3} = 1$

34. $\dfrac{2t}{t-3} - \dfrac{1}{t+4} = 1$

35. $\dfrac{x}{2x-1} = \dfrac{1}{x-2}$

36. $\dfrac{z}{3z+2} = \dfrac{2}{z+1}$

Mixed Exercises

For Exercises 37–60, solve the equations.

37. $x^4 - 16 = 0$

38. $t^4 - 625 = 0$

39. $(4x + 5)^2 + 3(4x + 5) + 2 = 0$

40. $2(5x + 3)^2 - (5x + 3) - 28 = 0$

41. $4m^4 - 9m^2 + 2 = 0$

42. $x^4 - 7x^2 + 12 = 0$

43. $x^6 - 9x^3 + 8 = 0$

44. $x^6 - 26x^3 - 27 = 0$

45. $\sqrt{x^2 + 20} = 3\sqrt{x}$

46. $\sqrt{x^2 + 60} = 4\sqrt{x}$

47. $\sqrt{4t + 1} = t + 1$

48. $\sqrt{t + 10} = t + 4$

49. $2\left(\dfrac{t-4}{3}\right)^2 - \left(\dfrac{t-4}{3}\right) - 3 = 0$

50. $\left(\dfrac{x+1}{5}\right)^2 - 3\left(\dfrac{x+1}{5}\right) - 10 = 0$

51. $x^{2/3} + x^{1/3} = 20$

52. $x^{2/5} - 3x^{1/5} = -2$

53. $m^4 + 2m^2 - 8 = 0$

54. $2c^4 + c^2 - 1 = 0$

55. $a^3 + 16a - a^2 - 16 = 0$
(*Hint:* Factor by grouping first.)

56. $b^3 + 9b - b^2 - 9 = 0$

57. $x^3 + 5x - 4x^2 - 20 = 0$

58. $y^3 + 8y - 3y^2 - 24 = 0$

59. $\left(\dfrac{2}{x-3}\right)^2 - 8\left(\dfrac{2}{x-3}\right) + 12 = 0$

60. $\left(\dfrac{5}{x+1}\right)^2 - 6\left(\dfrac{5}{x+1}\right) - 16 = 0$

Graphing Calculator Exercises

61. a. Solve the equation $x^4 + 4x^2 + 4 = 0$.

  b. How many solutions are real and how many solutions are imaginary?

  c. How many x-intercepts do you anticipate for the function defined by $y = x^4 + 4x^2 + 4$?

  d. Graph $Y_1 = x^4 + 4x^2 + 4$ on a standard viewing window.

62. a. Solve the equation $x^4 - 2x^2 + 1 = 0$.

  b. How many solutions are real and how many solutions are imaginary?

  c. How many x-intercepts do you anticipate for the function defined by $y = x^4 - 2x^2 + 1$?

  d. Graph $Y_1 = x^4 - 2x^2 + 1$ on a standard viewing window.

63. a. Solve the equation $x^4 - x^3 - 6x^2 = 0$.

  b. How many solutions are real and how many solutions are imaginary?

  c. How many x-intercepts do you anticipate for the function defined by $y = x^4 - x^3 - 6x^2$?

  d. Graph $Y_1 = x^4 - x^3 - 6x^2$ on a standard viewing window.

64. a. Solve the equation $x^4 - 10x^2 + 9 = 0$.

  b. How many solutions are real and how many solutions are imaginary?

  c. How many x-intercepts do you anticipate for the function defined by $y = x^4 - 10x^2 + 9$?

  d. Graph $Y_1 = x^4 - 10x^2 + 9$ on a standard viewing window.

Problem Recognition Exercises

Quadratic and Quadratic Type Equations

For Exercises 1–4, solve each equation by

a. Completing the square and applying the square root property.

b. Using the quadratic formula.

1. $x^2 + 10x + 3 = 0$ **2.** $v^2 - 16v + 5 = 0$ **3.** $3t^2 + t + 4 = 0$ **4.** $4y^2 + 3y + 5 = 0$

In Exercises 5–24, we have presented all types of equations that you have learned up to this point. For each equation,

a. First determine the type of equation that is presented. Choose from: linear equation, quadratic equation, quadratic in form, rational equation, or radical equation.

b. Solve the equation by using a suitable method.

5. $t^2 + 5t - 14 = 0$

6. $a^2 - 9a + 20 = 0$

7. $a^4 - 10a^2 + 9 = 0$

8. $x^4 - 3x^2 - 4 = 0$

9. $x - 3x^{1/2} - 4 = 0$

10. $x^2 - 9x - 2 = 0$

11. $8b(b + 1) + 2(3b - 4) = 4b(2b + 3)$

12. $6x(x + 1) - 3(x + 4) = 3x(2x + 5)$

13. $5a(a + 6) = 10(3a - 1)$

14. $4x(x + 3) = 6(2x - 4)$

15. $\dfrac{t}{t + 5} + \dfrac{3}{t - 4} = \dfrac{17}{t^2 + t - 20}$

16. $\dfrac{v}{v + 4} + \dfrac{12}{v^2 + 7v + 12} = \dfrac{5}{v + 3}$

17. $c^2 - 20c - 1 = 0$

18. $d^2 + 18d + 4 = 0$

19. $2u(u - 3) = 4(2 - u)$

20. $3y(y + 2) = 9(y + 1)$

21. $\sqrt{2b + 3} = b$

22. $\sqrt{5t + 6} = t$

23. $x^{2/3} + 2x^{1/3} - 15 = 0$

24. $y^{2/3} + 5y^{1/3} + 4 = 0$

Section 7.4   Graphs of Quadratic Functions

Concepts

1. **Quadratic Functions of the Form** $f(x) = x^2 + k$
2. **Quadratic Functions of the Form** $f(x) = (x - h)^2$
3. **Quadratic Functions of the Form** $f(x) = ax^2$
4. **Quadratic Functions of the Form** $f(x) = a(x - h)^2 + k$

In Section 4.8, we defined a quadratic function as a function of the form $f(x) = ax^2 + bx + c\ (a \neq 0)$. We also learned that the graph of a quadratic function is a **parabola**. In this section, we will learn how to graph parabolas.

A parabola opens upward if $a > 0$ (Figure 7-3) and opens downward if $a < 0$ (Figure 7-4). If a parabola opens upward, the **vertex** is the lowest point on the graph. If a parabola opens downward, the **vertex** is the highest point on the graph. The **axis of symmetry** is the vertical line that passes through the vertex.

Figure 7-3

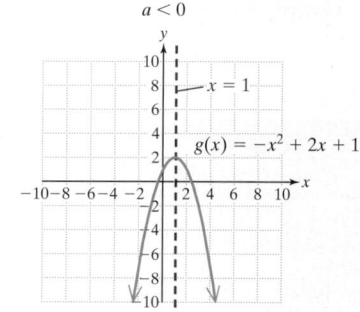

Figure 7-4

1. Quadratic Functions of the Form $f(x) = x^2 + k$

One technique for graphing a function is to plot a sufficient number of points on the graph until the general shape and defining characteristics can be determined. Then sketch a curve through the points.

Example 1 Graphing Quadratic Functions of the Form $f(x) = x^2 + k$

Graph the functions f, g, and h on the same coordinate system.

$$f(x) = x^2 \qquad g(x) = x^2 + 1 \qquad h(x) = x^2 - 2$$

Solution:

Several function values for f, g, and h are shown in Table 7-1 for selected values of x. The corresponding graphs are pictured in Figure 7-5.

Table 7-1

x	$f(x) = x^2$	$g(x) = x^2 + 1$	$h(x) = x^2 - 2$
-3	9	10	7
-2	4	5	2
-1	1	2	-1
0	0	1	-2
1	1	2	-1
2	4	5	2
3	9	10	7

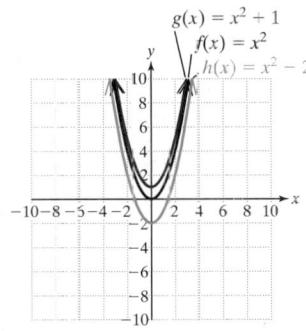

Figure 7-5

Skill Practice Refer to the graph of $f(x) = x^2 + k$ to determine the value of k.

1.

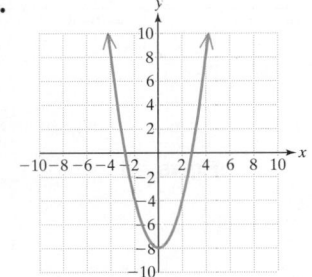

Notice that the graphs of $g(x) = x^2 + 1$ and $h(x) = x^2 - 2$ take on the same shape as $f(x) = x^2$. However, the y values of g are 1 greater than the y values of f. Hence, the graph of $g(x) = x^2 + 1$ is the same as the graph of $f(x) = x^2$ shifted *up* 1 unit. Likewise the y-values of h are 2 less than those of f. The graph of $h(x) = x^2 - 2$ is the same as the graph of $f(x) = x^2$ shifted *down* 2 units.

  A function of the type $f(x) = x^2 + k$ represents a vertical shift of the graph of $f(x) = x^2$. The functions in Example 1 illustrate the following properties of quadratic functions of the form $f(x) = x^2 + k$.

Answer

1. $k = -8$

Graphs of $f(x) = x^2 + k$

A function of the type $f(x) = x^2 + k$ represents a vertical shift of the graph of $f(x) = x^2$.

If $k > 0$, then the graph of $f(x) = x^2 + k$ is the same as the graph of $f(x) = x^2$ shifted *up* k units.

If $k < 0$, then the graph of $f(x) = x^2 + k$ is the same as the graph of $f(x) = x^2$ shifted *down* $|k|$ units.

Calculator Connections

Topic: Investigating Vertical Shifts of a Function

Try experimenting with a graphing calculator by graphing functions of the form $y = x^2 + k$ for several values of k.

Example 2 **Graphing Quadratic Functions of the Form $f(x) = x^2 + k$**

Sketch the functions defined by

a. $m(x) = x^2 - 4$ **b.** $n(x) = x^2 + \dfrac{7}{2}$

TIP: For more accuracy in the graph, plot one or two points near the vertex and use the symmetry of the curve to find additional points on the graph.

Solution:

a. $m(x) = x^2 - 4$

$m(x) = x^2 + (-4)$

Because $k = -4$, the graph is obtained by shifting the graph of $f(x) = x^2$ down $|-4|$ units (Figure 7-6).

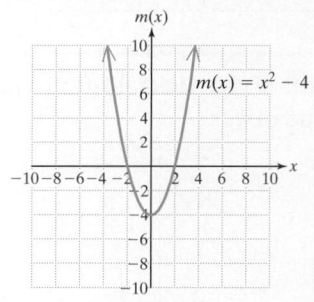

Figure 7-6

b. $n(x) = x^2 + \dfrac{7}{2}$

Because $k = \frac{7}{2}$, the graph is obtained by shifting the graph of $f(x) = x^2$ up $\frac{7}{2}$ units (Figure 7-7).

Figure 7-7

Answer

2.

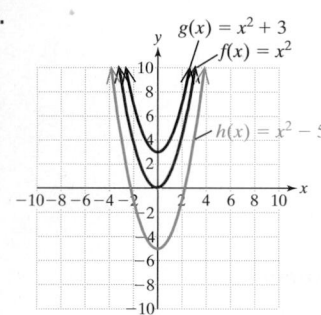

Skill Practice

2. Graph the functions f, g, and h on the same coordinate system.

$f(x) = x^2$ $g(x) = x^2 + 3$ $h(x) = x^2 - 5$

2. Quadratic Functions of the Form $f(x) = (x - h)^2$

The graph of $f(x) = x^2 + k$ represents a vertical shift (up or down) of the graph of $f(x) = x^2$. Example 3 shows that functions of the form $f(x) = (x - h)^2$ represent a horizontal shift (left or right) of the graph of $f(x) = x^2$.

Example 3   Graphing Quadratic Functions of the Form $f(x) = (x - h)^2$

Graph the functions $f, g,$ and h on the same coordinate system.

$$f(x) = x^2 \qquad g(x) = (x + 1)^2 \qquad h(x) = (x - 2)^2$$

Solution:

Several function values for $f, g,$ and h are shown in Table 7-2 for selected values of x. The corresponding graphs are pictured in Figure 7-8.

Table 7-2

x	$f(x) = x^2$	$g(x) = (x + 1)^2$	$h(x) = (x - 2)^2$
-4	16	9	36
-3	9	4	25
-2	4	1	16
-1	1	0	9
0	0	1	4
1	1	4	1
2	4	9	0
3	9	16	1
4	16	25	4
5	25	36	9

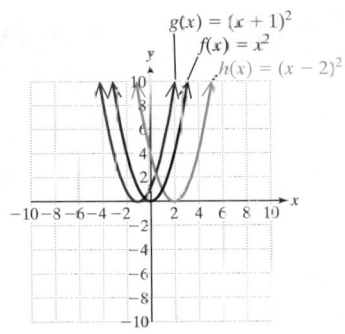

Figure 7-8

Skill Practice Refer to the graph of $f(x) = (x - h)^2$ to determine the value of h.

3.

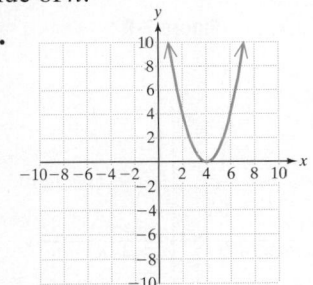

Example 3 illustrates the following properties of quadratic functions of the form $f(x) = (x - h)^2$.

Graphs of $f(x) = (x - h)^2$

A function of the type $f(x) = (x - h)^2$ represents a horizontal shift of the graph of $f(x) = x^2$.

If $h > 0$, then the graph of $f(x) = (x - h)^2$ is the same as the graph of $f(x) = x^2$ shifted h units to the *right*.

If $h < 0$, then the graph of $f(x) = (x - h)^2$ is the same as the graph of $f(x) = x^2$ shifted $|h|$ units to the *left*.

From Example 3 we have

$$h(x) = (x - 2)^2 \qquad \text{and} \qquad g(x) = (x + 1)^2$$
$$g(x) = [x - (-1)]^2$$

$y = x^2$ shifted 2 units to the right $\qquad$ $y = x^2$ shifted $|-1|$ unit to the left

Example 4   Graphing Functions of the Form $f(x) = (x - h)^2$

Sketch the functions p and q.

  a. $p(x) = (x - 7)^2$ **b.** $q(x) = (x + 1.6)^2$

Solution:

  a. $p(x) = (x - 7)^2$

     Because $h = 7 > 0$, shift the graph of $f(x) = x^2$ to the *right* 7 units (Figure 7-9).

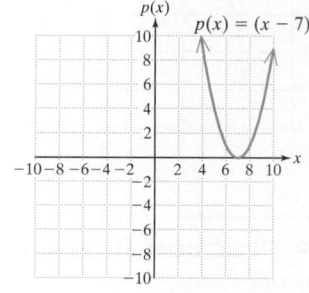

Figure 7-9

  b. $q(x) = (x + 1.6)^2$

    $q(x) = [x - (-1.6)]^2$

    Because $h = -1.6 < 0$, shift the graph of $f(x) = x^2$ to the *left* 1.6 units (Figure 7-10).

Figure 7-10

Answer

4.

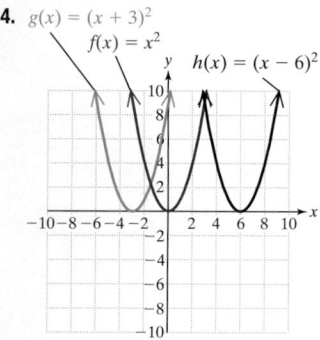

Skill Practice

  4. Graph the functions f, g, and h on the same coordinate system.
$$f(x) = x^2$$
$$g(x) = (x + 3)^2$$
$$h(x) = (x - 6)^2$$

3. Quadratic Functions of the Form $f(x) = ax^2$

Examples 5 and 6 investigate functions of the form $f(x) = ax^2$ $(a \neq 0)$.

Example 5 Graphing Functions of the Form $f(x) = ax^2$

Graph the functions f, g, and h on the same coordinate system.

$$f(x) = x^2 \qquad g(x) = 2x^2 \qquad h(x) = \frac{1}{2}x^2$$

Solution:

Several function values for f, g, and h are shown in Table 7-3 for selected values of x. The corresponding graphs are pictured in Figure 7-11.

Table 7-3

x	$f(x) = x^2$	$g(x) = 2x^2$	$h(x) = \frac{1}{2}x^2$
-3	9	18	$\frac{9}{2}$
-2	4	8	2
-1	1	2	$\frac{1}{2}$
0	0	0	0
1	1	2	$\frac{1}{2}$
2	4	8	2
3	9	18	$\frac{9}{2}$

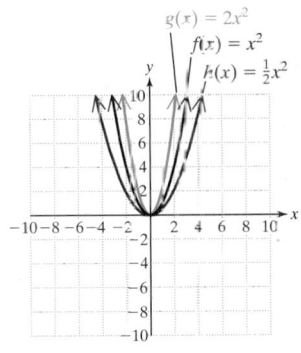

Figure 7-11

Skill Practice

5. Graph the functions f, g, and h on the same coordinate system.
   $f(x) = x^2$
   $g(x) = 3x^2$
   $h(x) = \frac{1}{3}x^2$

In Example 5, the function values defined by $g(x) = 2x^2$ are twice those of $f(x) = x^2$. The graph of $g(x) = 2x^2$ is the same as the graph of $f(x) = x^2$ *stretched vertically* by a factor of 2 (the graph appears narrower than $f(x) = x^2$).

In Example 5, the function values defined by $h(x) = \frac{1}{2}x^2$ are one-half those of $f(x) = x^2$. The graph of $h(x) = \frac{1}{2}x^2$ is the same as the graph of $f(x) = x^2$ *shrunk vertically* by a factor of $\frac{1}{2}$ (the graph appears wider than $f(x) = x^2$).

Answer

5.

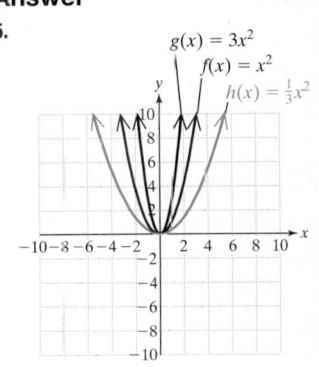

Example 6 Graphing Functions of the Form $f(x) = ax^2$

Graph the functions f, g, and h on the same coordinate system.

$$f(x) = -x^2 \qquad g(x) = -3x^2 \qquad h(x) = -\frac{1}{3}x^2$$

Solution:

Several function values for f, g, and h are shown in Table 7-4 for selected values of x. The corresponding graphs are pictured in Figure 7-12.

Table 7-4

x	$f(x) = -x^2$	$g(x) = -3x^2$	$h(x) = -\frac{1}{3}x^2$
-3	-9	-27	-3
-2	-4	-12	$-\frac{4}{3}$
-1	-1	-3	$-\frac{1}{3}$
0	0	0	0
1	-1	-3	$-\frac{1}{3}$
2	-4	-12	$-\frac{4}{3}$
3	-9	-27	-3

Figure 7-12

Skill Practice

6. Graph the functions f, g, and h on the same coordinate system.

$$f(x) = -x^2 \qquad g(x) = -2x^2 \qquad h(x) = -\frac{1}{2}x^2$$

Example 6 illustrates that if the coefficient of the squared term is negative, the parabola opens downward. The graph of $g(x) = -3x^2$ is the same as the graph of $f(x) = -x^2$ with a *vertical stretch* by a factor of $|-3|$. The graph of $h(x) = -\frac{1}{3}x^2$ is the same as the graph of $f(x) = -x^2$ with a *vertical shrink* by a factor of $|-\frac{1}{3}|$.

> ## Graphs of $f(x) = ax^2$
>
> **1.** If $a > 0$, the parabola opens upward. Furthermore,
> - If $0 < a < 1$, then the graph of $f(x) = ax^2$ is the same as the graph of $f(x) = x^2$ with a *vertical shrink* by a factor of a.
> - If $a > 1$, then the graph of $f(x) = ax^2$ is the same as the graph of $f(x) = x^2$ with a *vertical stretch* by a factor of a.
>
> **2.** If $a < 0$, the parabola opens downward. Furthermore,
> - If $0 < |a| < 1$, then the graph of $f(x) = ax^2$ is the same as the graph of $f(x) = -x^2$ with a *vertical shrink* by a factor of $|a|$.
> - If $|a| > 1$, then the graph of $f(x) = ax^2$ is the same as the graph of $f(x) = -x^2$ with a *vertical stretch* by a factor of $|a|$.

Answer

6.

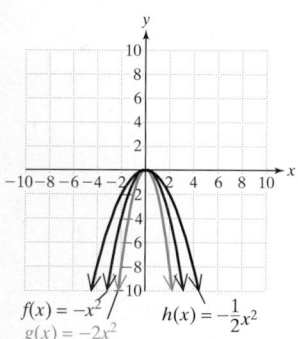

4. Quadratic Functions of the Form $f(x) = a(x - h)^2 + k$

We can summarize our findings from Examples 1–6 by graphing functions of the form $f(x) = a(x - h)^2 + k$ $(a \neq 0)$.

The graph of $f(x) = x^2$ has its vertex at the origin $(0, 0)$. The graph of $f(x) = a(x - h)^2 + k$ is the same as the graph of $f(x) = x^2$ shifted to the right or left h units and shifted up or down k units. Therefore, the vertex is shifted from $(0, 0)$ to (h, k). The axis of symmetry is the vertical line through the vertex. Therefore, the axis of symmetry must be the line $x = h$.

Graphs of $f(x) = a(x - h)^2 + k$

1. The vertex is located at (h, k).
2. The axis of symmetry is the line $x = h$.
3. If $a > 0$, the parabola opens upward, and k is the **minimum value** of the function.
4. If $a < 0$, the parabola opens downward, and k is the **maximum value** of the function.

Example 7 **Graphing a Function of the Form $f(x) = a(x - h)^2 + k$**

Given the function defined by

$$f(x) = 2(x - 3)^2 + 4$$

a. Identify the vertex.

b. Sketch the function.

c. Identify the axis of symmetry.

d. Identify the maximum or minimum value of the function.

e. Use the graph to determine the domain and range of f in interval notation.

TIP: We can find additional points to the right or left of the vertex and use the symmetry of the parabola to refine the sketch of the curve.

x	f(x)
4	6
5	12

Solution:

a. The function is in the form $f(x) = a(x - h)^2 + k$, where $a = 2$, $h = 3$, and $k = 4$. Therefore, the vertex is $(3, 4)$.

b. The graph of f is the same as the graph of $f(x) = x^2$ shifted to the right 3 units, shifted up 4 units, and stretched vertically by a factor of 2 (Figure 7-13).

c. The axis of symmetry is the line $x = 3$.

d. Because $a > 0$, the function opens upward. Therefore, the minimum function value is 4. Notice that the minimum value is the minimum y-value on the graph.

e. The domain is all real numbers: $(-\infty, \infty)$.

The vertex is the lowest point on the parabola, and its y-coordinate is 4. Therefore, the range is $[4, \infty)$.

$f(x) = 2(x - 3)^2 + 4$

$(3, 4)$

$x = 3$

Figure 7-13

Answers

7. a. Vertex: $(-1, -3)$

b.

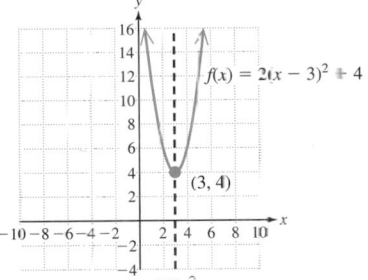

c. Axis of symmetry: $x = -1$

d. Minimum value: -3

e. Domain: $(-\infty, \infty)$; range: $[-3, \infty)$

Skill Practice

7. Given the function defined by $g(x) = 3(x + 1)^2 - 3$

a. Identify the vertex.

b. Sketch the graph.

c. Identify the axis of symmetry.

d. Identify the maximum or minimum value of the function.

e. Determine the domain and range in interval notation.

Topic: Using the Maximum and Minimum Features

Some graphing calculators have *Minimum* and *Maximum* features that enable the user to approximate the minimum and maximum values of a function.

Example 8   Graphing a Function of the Form $f(x) = a(x - h)^2 + k$

Given the function defined by

$$g(x) = -(x + 2)^2 - \frac{7}{4}$$

a. Identify the vertex.

b. Sketch the function.

c. Identify the axis of symmetry.

d. Identify the maximum or minimum value of the function.

e. Use the graph to determine the domain and range of f in interval notation.

Solution:

a. $g(x) = -(x + 2)^2 - \dfrac{7}{4}$

$= -1[x - (-2)]^2 + \left(-\dfrac{7}{4}\right)$

The function is in the form $g(x) = a(x - h)^2 + k$, where $a = -1$, $h = -2$, and $k = -\frac{7}{4}$. Therefore, the vertex is $(-2, -\frac{7}{4})$.

b. The graph of g is the same as the graph of $f(x) = x^2$ shifted to the left 2 units, shifted down $\frac{7}{4}$ units, and opening downward (Figure 7-14).

c. The axis of symmetry is the line $x = -2$.

d. The parabola opens downward, so the maximum function value is $-\frac{7}{4}$.

e. The domain is all real numbers: $(-\infty, \infty)$.

The vertex is the highest point on the parabola, and its y-coordinate is $-\frac{7}{4}$. Therefore, the range is $(-\infty, -\frac{7}{4}]$.

Figure 7-14

(graph, labeled $(-2, -\frac{7}{4})$, $g(x) = -(x + 2)^2 - \frac{7}{4}$, $x = -2$)

Skill Practice

8. Given the function defined by $h(x) = -\dfrac{1}{2}(x - 4)^2 + 2$

a. Identify the vertex.

b. Sketch the graph.

c. Identify the axis of symmetry.

d. Identify the maximum or minimum value of the function.

e. Determine the domain and range in interval notation.

Answers

8. a. Vertex: $(4, 2)$

b.

(graph)

c. Axis of symmetry: $x = 4$

d. Maximum value: 2

e. Domain: $(-\infty, \infty)$; range: $(-\infty, 2]$

Section 7.4 Practice Exercises

Vocabulary and Key Concepts

1. **a.** The graph of a quadratic function, $f(x) = ax^2 + bx + c$, is a _____.

    b. The parabola defined by $f(x) = ax^2 + bx + c$ ($a \neq 0$) will open upward if c _____ 0 and will open downward if a _____ 0.

    c. If a parabola opens upward, the vertex is the (highest/lowest) point on the graph. If a parabola opens downward, the vertex is the (highest/lowest) point on the graph.

    d. Given $f(x) = a(x - h)^2 + k$ ($a \neq 0$), the vertex of the parabola is given by the ordered pair _____. If the parabola opens _____, the value of k is the minimum value of the function. If the parabola opens _____, the value of k is the maximum value of the function.

    e. Given $f(x) = a(x - h)^2 + k$ ($a \neq 0$), the axis of symmetry of the parabola is the vertical line that passes through the _____. An equation of the axis of symmetry is _____.

Review Exercises

For Exercises 2–8, solve the equations.

2. $x^2 + x - 5 = 0$

3. $(y - 3)^2 = -4$

4. $\sqrt{2a + 2} = a + 1$

5. $5t(t - 2) = -3$

6. $2z^2 - 3z - 9 = 0$

7. $x^{2/3} + 5x^{1/3} + 6 = 0$

8. $m^2(m^2 + 6) = 27$

Concept 1: Quadratic Functions of the Form $f(x) = x^2 + k$

9. Describe how the value of k affects the graph of a function defined by $f(x) = x^2 + k$.

For Exercises 10–17, graph the functions. **(See Examples 1–2.)**

10. $g(x) = x^2 + 1$

11. $f(x) = x^2 + 2$

12. $p(x) = x^2 - 3$

 13. $q(x) = x^2 - 4$

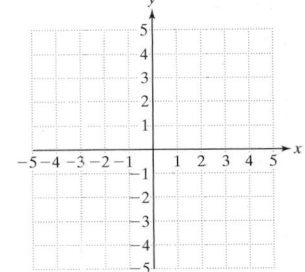

14. $T(x) = x^2 + \dfrac{3}{4}$

15. $S(x) = x^2 + \dfrac{3}{2}$

16. $M(x) = x^2 - \dfrac{5}{4}$

17. $n(x) = x^2 - \dfrac{1}{3}$

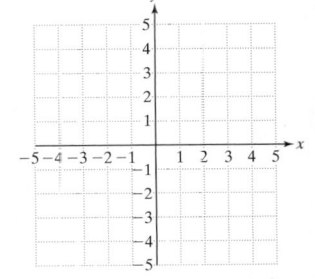

Concept 2: Quadratic Functions of the Form $f(x) = (x - h)^2$

18. Describe how the value of h affects the graph of a function defined by $f(x) = (x - h)^2$.

For Exercises 19–26, graph the functions. **(See Examples 3–4.)**

19. $r(x) = (x + 1)^2$ **20.** $h(x) = (x + 2)^2$ **21.** $k(x) = (x - 3)^2$ **22.** $L(x) = (x - 4)^2$

               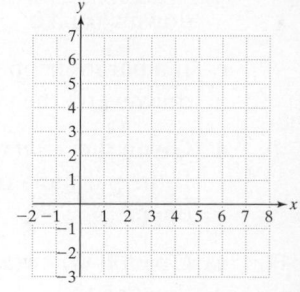

23. $A(x) = \left(x + \dfrac{3}{4}\right)^2$ **24.** $r(x) = \left(x + \dfrac{3}{2}\right)^2$ **25.** $W(x) = (x - 1.25)^2$ **26.** $V(x) = (x - 2.5)^2$

               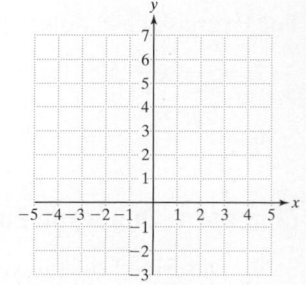

Concept 3: Quadratic Functions of the Form $f(x) = ax^2$

27. Describe how the value of a affects the graph of a function defined by $f(x) = ax^2$, where $a \neq 0$.

28. How do you determine whether the graph of a function defined by $h(x) = ax^2 + bx + c \ (a \neq 0)$ opens upward or downward?

For Exercises 29–36, graph the functions. **(See Examples 5–6.)**

29. $f(x) = 4x^2$ **30.** $g(x) = 3x^2$ **31.** $h(x) = \dfrac{1}{4}x^2$ **32.** $f(x) = \dfrac{1}{5}x^2$

33. $c(x) = -x^2$

34. $g(x) = -4x^2$

35. $v(x) = -\frac{1}{5}x^2$

36. $f(x) = -\frac{1}{4}x^2$

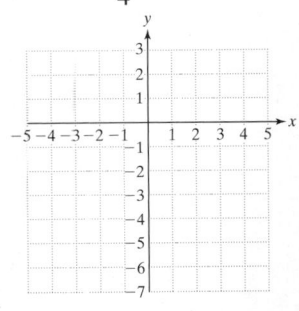

Concept 4: Quadratic Functions of the Form $f(x) = a(x - h)^2 + k$

For Exercises 37–44, match the function with its graph.

37. $f(x) = -\frac{1}{4}x^2$

38. $g(x) = (x + 3)^2$

39. $k(x) = (x - 3)^2$

40. $h(x) = \frac{1}{4}x^2$

41. $t(x) = x^2 + 2$

42. $m(x) = x^2 - 4$

43. $p(x) = (x + 1)^2 - 3$

44. $n(x) = -(x - 2)^2 + 3$

a.

b.

c.

d.

e.

f.

g.

h.

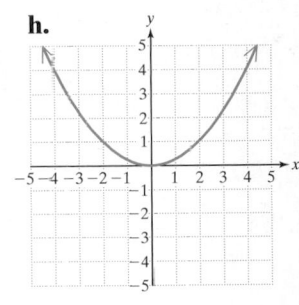

For Exercises 45–64, graph the parabola and the axis of symmetry. Label the coordinates of the vertex, and write the equation of the axis of symmetry. Use the graph to write the domain and range in interval notation. **(See Examples 7–8.)**

45. $f(x) = (x - 3)^2 + 2$

46. $f(x) = (x - 2)^2 + 3$

47. $f(x) = (x + 1)^2 - 3$

48. $f(x) = (x + 3)^2 - 1$

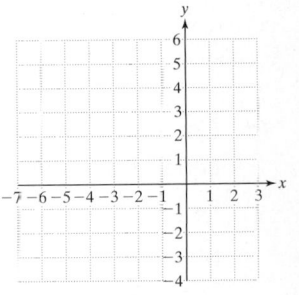

49. $f(x) = -(x - 4)^2 - 2$ **50.** $f(x) = -(x - 2)^2 - 4$ **51.** $f(x) = -(x + 3)^2 + 3$ **52.** $f(x) = -(x + 2)^2 + 2$

   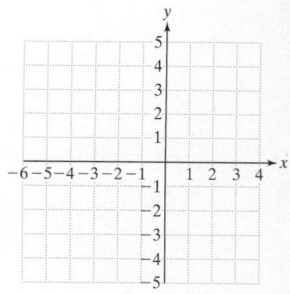

53. $f(x) = (x + 1)^2 + 1$ **54.** $f(x) = (x - 4)^2 - 4$ **55.** $f(x) = 3(x - 1)^2$ **56.** $f(x) = -3(x - 1)^2$

   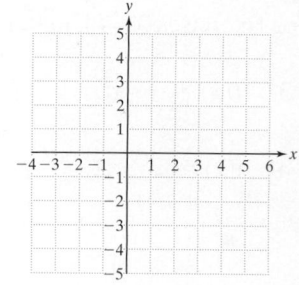

57. $f(x) = -4x^2 + 3$ **58.** $f(x) = 4x^2 + 3$ **59.** $f(x) = 2(x + 3)^2 - 1$ **60.** $f(x) = -2(x + 3)^2 - 1$

   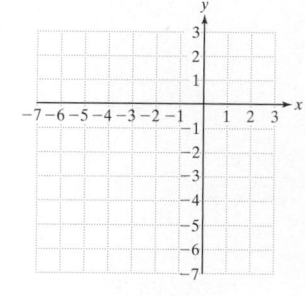

61. $f(x) = -\dfrac{1}{4}(x - 1)^2 + 2$ **62.** $f(x) = \dfrac{1}{4}(x - 1)^2 + 2$ **63.** $f(x) = \dfrac{1}{3}(x - 2)^2 + 1$ **64.** $f(x) = -\dfrac{1}{3}(x - 2)^2 + 1$

   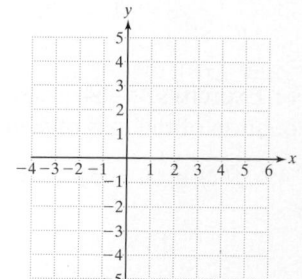

65. Compare the graphs of the following equations to the graph of $y = x^2$.

a. $y = x^2 + 3$

b. $y = (x + 3)^2$

c. $y = 3x^2$

66. Compare the graphs of the following equations to the graph of $y = x^2$.

a. $y = (x - 2)^2$

b. $y = 2x^2$

c. $y = x^2 - 2$

For Exercises 67–78, write the coordinates of the vertex and determine if the vertex is a maximum point or a minimum point. Then write the maximum or minimum value.

67. $f(x) = 4(x - 6)^2 - 9$

68. $g(x) = 3(x - 4)^2 - 7$

69. $p(x) = -\frac{2}{5}(x - 2)^2 + 5$

70. $h(x) = -\frac{3}{7}(x - 5)^2 + 10$

 71. $k(x) = \frac{1}{2}(x + 8)^2$

72. $m(x) = \frac{2}{9}(x + 11)^2$

73. $n(x) = -6x^2 + \frac{21}{4}$

74. $q(x) = -4x^2 + \frac{1}{6}$

75. $A(x) = 2(x - 7)^2 - \frac{3}{2}$

76. $B(x) = 5(x - 3)^2 - \frac{1}{4}$

77. $F(x) = 7x^2$

78. $G(x) = -7x^2$

79. True or false: The function defined by $g(x) = -5x^2$ has a maximum value but no minimum value.

80. True or false: The function defined by $f(x) = 2(x - 5)^2$ has a maximum value but no minimum value.

81. True or false: If the vertex $(-2, 8)$ represents a minimum point, then the minimum value is -2.

82. True or false: If the vertex $(-2, 8)$ represents a maximum point, then the maximum value is 8.

Expanding Your Skills

83. A suspension bridge is 120 ft long. Its supporting cable hangs in a shape that resembles a parabola. The function defined by $H(x) = \frac{1}{90}(x - 60)^2 + 30$ (where $0 \le x \le 120$) approximates the height of the supporting cable a distance of x ft from the end of the bridge (see figure).

a. What is the location of the vertex of the parabolic cable?

b. What is the minimum height of the cable?

c. How high are the towers at either end of the supporting cable?

84. A 50-m bridge over a crevasse is supported by a parabolic arch. The function defined by $f(x) = -0.16(x - 25)^2 + 100$ (where $0 \le x \le 50$) approximates the height of the supporting arch x meters from the end of the bridge (see figure).

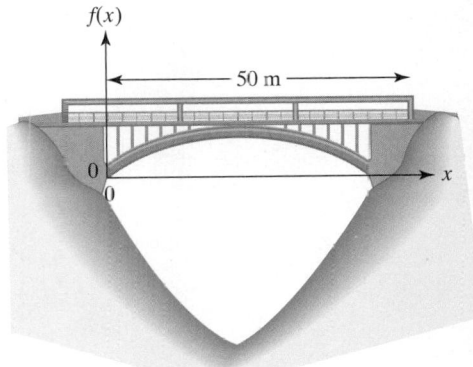

a. What is the location of the vertex of the arch?

b. What is the maximum height of the arch (relative to the origin)?

85. The staging platform for a fireworks display is 6 ft above ground, and the mortars leave the platform at 96 ft/sec. The height of the mortars, $h(t)$, can be modeled by

$h(t) = -16(t - 3)^2 + 150$, where t is the time in seconds after launch.

a. If the fuses are set for 3 sec after launch, at what height will the fireworks explode?

b. Will the fireworks explode at their maximum height? Explain.

Section 7.5 Vertex of a Parabola: Applications and Modeling

Concepts

1. Writing a Quadratic Function in the Form $f(x) = a(x - h)^2 + k$
2. Vertex Formula
3. Determining the Vertex and Intercepts of a Quadratic Function
4. Applications and Modeling of Quadratic Functions

1. Writing a Quadratic Function in the Form $f(x) = a(x - h)^2 + k$

The graph of a quadratic function is a parabola, and if the function is written in the form $f(x) = a(x - h)^2 + k$ $(a \neq 0)$, then the vertex is at (h, k). A quadratic function can be written in the form $f(x) = a(x - h)^2 + k$ $(a \neq 0)$ by completing the square. The process is similar to the steps outlined in Section 7.1 except that all algebraic manipulations are performed on the right-hand side of the equation.

Example 1 Writing a Quadratic Function in the Form $f(x) = a(x - h)^2 + k$ $(a \neq 0)$

Given: $f(x) = x^2 + 8x + 13$

a. Write the function in the form $f(x) = a(x - h)^2 + k$.

b. Identify the vertex, axis of symmetry, and minimum function value.

Solution:

a. $f(x) = x^2 + 8x + 13$

Rather than dividing by the leading coefficient on both sides, we will factor out the leading coefficient from the variable terms on the right-hand side.

$= 1(x^2 + 8x) + 13$

$= 1(x^2 + 8x \qquad) + 13$

Next, complete the square on the expression within the parentheses: $[\frac{1}{2}(8)]^2 = 16.$

$= 1(x^2 + 8x + 16 - 16) + 13$

Rather than add 16 to both sides of the function, we *add and subtract* 16 within the parentheses on the right-hand side. This has the effect of adding 0 to the right-hand side.

$= 1(x^2 + 8x + 16) - 16 + 13$

Use the associative property of addition to regroup terms and isolate the perfect square trinomial within the parentheses.

$= (x + 4)^2 - 3$

Factor and simplify.

b. $f(x) = (x + 4)^2 - 3$

The vertex is $(-4, -3)$.

The axis of symmetry is $x = -4$.

Because $a > 0$, the parabola opens upward.

The minimum value is -3 (Figure 7-15).

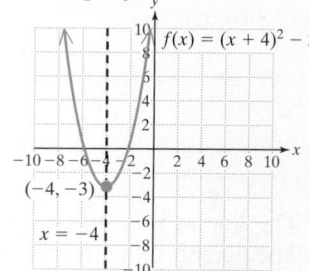

Figure 7-15

Answers

1. **a.** $f(x) = (x + 4)^2 - 17$
   b. Vertex: $(-4, -17)$; axis of symmetry: $x = -4$; minimum value: -17

Skill Practice

1. Given: $f(x) = x^2 + 8x - 1$
   a. Write the equation in the form $f(x) = a(x - h)^2 + k$.
   b. Identify the vertex, axis of symmetry, and minimum value of the function.

| **Example 2** | Analyzing a Quadratic Function |

Given: $f(x) = -2x^2 + 12x - 16$

a. Write the function in the form $f(x) = a(x - h)^2 + k$.

b. Find the vertex, axis of symmetry, and maximum function value.

c. Find the x- and y-intercepts.

d. Sketch the graph of the function.

Solution:

a. $f(x) = -2x^2 + 12x - 16$

To find the vertex, write the function in the form $f(x) = a(x - h)^2 + k$.

$$= -2(x^2 - 6x \quad) - 16$$

If the leading coefficient is not 1, factor the coefficient from the variable terms.

$$= -2(x^2 - 6x + 9 - 9) - 16$$

Add and subtract the quantity $[\frac{1}{2}(-6)]^2 = 9$ within the parentheses.

$$= -2(x^2 - 6x + 9) + (-2)(-9) - 16$$

To remove the term -9 from the parentheses, we must first apply the distributive property. When -9 is removed from the parentheses, it carries with it a factor of -2.

$$= -2(x - 3)^2 + 18 - 16$$

Factor and simplify.

$$= -2(x - 3)^2 + 2$$

b. $f(x) = -2(x - 3)^2 + 2$

The vertex is $(3, 2)$. The axis of symmetry is $x = 3$. Because $a < 0$, the parabola opens downward and the maximum value is 2.

c. The y-intercept is given by $f(0) = -2(0)^2 + 12(0) - 16 = -16$.

The y-intercept is $(0, -16)$.

To find the x-intercept(s), find the real solutions to the equation $f(x) = 0$.

$$f(x) = -2x^2 + 12x - 16$$

$0 = -2x^2 + 12x - 16$ Substitute $f(x) = 0$.

$0 = -2(x^2 - 6x + 8)$ Factor.

$0 = -2(x - 4)(x - 2)$

$x = 4 \quad$ or $\quad x = 2$

The x-intercepts are $(4, 0)$ and $(2, 0)$.

TIP: In Example 2(c), we could have used the form of the equation found in part (a) to find the intercepts.

d. Using the information from parts (a)–(c), sketch the graph (Figure 7-16).

Figure 7-16

Skill Practice

2. Given: $g(x) = -x^2 + 6x - 5$
   a. Write the equation in the form $g(x) = a(x - h)^2 + k$.
   b. Identify the vertex, axis of symmetry, and maximum value of the function.
   c. Determine the x- and y-intercepts.
   d. Graph the function.

2. Vertex Formula

Completing the square and writing a quadratic function in the form $f(x) = a(x - h)^2 + k$ $(a \neq 0)$ is one method to find the vertex of a parabola. Another method is to use the vertex formula. The **vertex formula** can be derived by completing the square on $f(x) = ax^2 + bx + c$ $(a \neq 0)$.

$f(x) = ax^2 + bx + c \qquad (a \neq 0)$

$= a\left(x^2 + \dfrac{b}{a}x \qquad\quad\right) + c$ Factor a from the variable terms.

$= a\left(x^2 + \dfrac{b}{a}x + \dfrac{b^2}{4a^2} - \dfrac{b^2}{4a^2}\right) + c$ Add and subtract $[\frac{1}{2}(b/a)]^2 = b^2/(4a^2)$ within the parentheses.

$= a\left(x^2 + \dfrac{b}{a}x + \dfrac{b^2}{4a^2}\right) + (a)\left(-\dfrac{b^2}{4a^2}\right) + c$ Apply the distributive property and remove the term $-b^2/(4a^2)$ from the parentheses.

$= a\left(x + \dfrac{b}{2a}\right)^2 - \dfrac{b^2}{4a} + c$ Factor the trinomial and simplify.

$= a\left(x + \dfrac{b}{2a}\right)^2 + c - \dfrac{b^2}{4a}$ Apply the commutative property of addition to reverse the last two terms.

$= a\left(x + \dfrac{b}{2a}\right)^2 + \dfrac{4ac}{4a} - \dfrac{b^2}{4a}$ Obtain a common denominator.

$= a\left(x + \dfrac{b}{2a}\right)^2 + \dfrac{4ac - b^2}{4a}$

$= a\left[x - \left(-\dfrac{b}{2a}\right)\right]^2 + \dfrac{4ac - b^2}{4a}$

$f(x) = a(x \quad - \quad h)^2 \quad + \quad k$

The function is in the form $f(x) = a(x - h)^2 + k$, where

$$h = \dfrac{-b}{2a} \qquad \text{and} \qquad k = \dfrac{4ac - b^2}{4a}$$

Therefore, the vertex is $\left(\dfrac{-b}{2a}, \dfrac{4ac - b^2}{4a}\right)$

Although the y-coordinate of the vertex is given by $\dfrac{4ac - b^2}{4a}$, it is usually easier to determine the x-coordinate of the vertex first and then find y by evaluating the function at $x = -b/(2a)$.

Answers

2. a. $g(x) = -(x - 3)^2 + 4$
   b. Vertex: $(3, 4)$; axis of symmetry: $x = 3$; maximum value: 4
   c. x-intercepts: $(5, 0)$ and $(1, 0)$; y-intercept: $(0, -5)$
   d.

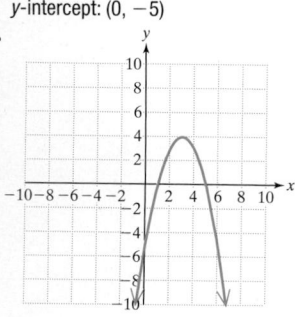

> **The Vertex Formula**
>
> For $f(x) = ax^2 + bx + c$ $(a \neq 0)$, the vertex is given by
>
> $$\left(\frac{-b}{2a}, \frac{4ac - b^2}{4a} \right) \quad \text{or} \quad \left(\frac{-b}{2a}, f\left(\frac{-b}{2a} \right) \right)$$

3. Determining the Vertex and Intercepts of a Quadratic Function

Example 3 Determining the Vertex and Intercepts of a Quadratic Function

Given: $h(x) = x^2 - 2x + 5$

a. Use the vertex formula to find the vertex.

b. Find the x- and y-intercepts.

c. Sketch the function.

Solution:

a. $h(x) = x^2 - 2x + 5$

$a = 1 \qquad b = -2 \qquad c = 5 \qquad$ Identify a, b, and c.

The x-coordinate of the vertex is $\dfrac{-b}{2a} = \dfrac{-(-2)}{2(1)} = 1$.

The y-coordinate of the vertex is $h(1) = (1)^2 - 2(1) + 5 = 4$.

The vertex is $(1, 4)$.

b. The y-intercept is given by $h(0) = (0)^2 - 2(0) + 5 = 5$.

The y-intercept is $(0, 5)$.

To find the x-intercept(s), find the real solutions to the equation $h(x) = 0$.

$h(x) = x^2 - 2x + 5$

$0 = x^2 - 2x + 5 \qquad$ This quadratic equation is not factorable. Apply the quadratic formula: $a = 1$, $b = -2$, $c = 5$.

$x = \dfrac{-(-2) \pm \sqrt{(-2)^2 - 4(1)(5)}}{2(1)}$

$= \dfrac{2 \pm \sqrt{4 - 20}}{2(1)}$

$= \dfrac{2 \pm \sqrt{-16}}{2}$

$= \dfrac{2 \pm 4i}{2}$

$= 1 \pm 2i$

The solutions to the equation $h(x) = 0$ are not real numbers. Therefore, there are no x-intercepts.

c.

Figure 7-17

> **TIP:** The location of the vertex and the direction that the parabola opens can be used to determine whether the function has any x-intercepts.
>
> Given $h(x) = x^2 - 2x + 5$, the vertex $(1, 4)$ is *above* the x-axis. Furthermore, because $a > 0$, the parabola opens upward. Therefore, it is not possible for the graph to cross the x-axis (Figure 7-17).

Skill Practice

3. Given: $f(x) = x^2 + 4x + 6$

    a. Use the vertex formula to find the vertex of the parabola.

    b. Determine the x- and y-intercepts.

    c. Sketch the graph.

4. Applications and Modeling of Quadratic Functions

Example 4   Applying a Quadratic Function

The crew from Extravaganza Entertainment launches fireworks at an angle of 60° from the horizontal. The height of one particular type of display can be approximated by:

$$h(t) = -16t^2 + 128\sqrt{3}\,t$$

where $h(t)$ is measured in feet and t is measured in seconds.

    a. How long will it take the fireworks to reach their maximum height? Round to the nearest second.

    b. Find the maximum height. Round to the nearest foot.

Solution:

$$h(t) = -16t^2 + 128\sqrt{3}\,t$$

This parabola opens downward; therefore, the maximum height of the fireworks will occur at the vertex of the parabola.

Answers

3. a. Vertex: $(-2, 2)$

    b. x-intercepts: none;

       y-intercepts: $(0, 6)$

    c.

$a = -16$ $b = 128\sqrt{3}$ $c = 0$

Identify a, b, and c, and apply the vertex formula.

The x-coordinate of the vertex is

$$\frac{-b}{2a} = \frac{-128\sqrt{3}}{2(-16)} = \frac{-128\sqrt{3}}{-32} \approx 6.9$$

The y-coordinate of the vertex is approximately

$$h(6.9) = -16(6.9)^2 + 128\sqrt{3}(6.9) \approx 768$$

The vertex is $(6.9, 768)$.

a. The fireworks will reach their maximum height in 6.9 sec.

b. The maximum height is 768 ft.

Skill Practice

4. An object is launched into the air with an initial velocity of 48 ft/sec from the top of a building 288 ft high. The height $h(t)$ of the object after t seconds is given by

$$h(t) = -16t^2 + 48t + 288$$

a. Find the time it takes for the object to reach its maximum height.

b. Find the maximum height.

In Section 2.3 we wrote a linear model given two points. Now we will learn how to write a quadratic model of a parabola given three points by using a system of equations and the standard form of a parabola: $y = ax^2 + bx + c$.

This process involves substituting the x- and y-coordinates of the three given points into the quadratic model. Then we solve the resulting system of three equations (Section 3.6).

Example 5 Writing a Quadratic Model

Write an equation of a parabola that passes through the points $(1, -1), (-1, -5)$, and $(2, 4)$.

Solution:

Substitute $(1, -1)$ into the equation
$y = ax^2 + bx + c$ $\longrightarrow$

$(-1) = a(1)^2 + b(1) + c$
$-1 = a + b + c$
$a + b + c = -1$

Substitute $(-1, -5)$ into the equation
$y = ax^2 + bx + c$ $\longrightarrow$

$(-5) = a(-1)^2 + b(-1) + c$
$-5 = a - b + c$
$a - b + c = -5$

Substitute $(2, 4)$ into the equation
$y = ax^2 + bx + c$ $\longrightarrow$

$(4) = a(2)^2 + b(2) + c$
$4 = 4a + 2b + c$
$4a + 2b + c = 4$

Solve the system:

$\boxed{A}$ $a + b + c = -1$

$\boxed{B}$ $a - b + c = -5$

$\boxed{C}$ $4a + 2b + c = 4$

Answers

4. a. 1.5 sec **b.** 324 ft

Notice that the c-variables all have a coefficient of 1. Therefore, we choose to eliminate the c-variable.

$$\boxed{A} \quad a + b + c = -1 \longrightarrow a + b + c = -1$$
$$\boxed{B} \quad a - b + c = -5 \xrightarrow{\text{Multiply by } -1.} \underline{-a + b - c = 5}$$
$$2b = 4$$
$$b = 2 \quad \boxed{D}$$

$$\boxed{B} \quad a - b + c = -5 \longrightarrow a - b + c = -5$$
$$\boxed{C} \quad 4a + 2b + c = 4 \xrightarrow{\text{Multiply by } -1.} \underline{-4a - 2b - c = -4}$$
$$-3a - 3b = -9 \quad \boxed{E}$$

Substitute b with 2 in equation $\boxed{E}$:
$$-3a - 3(2) = -9$$
$$-3a - 6 = -9$$
$$-3a = -3$$
$$a = 1$$

Substitute a and b in equation $\boxed{A}$ to solve for c:
$$(1) + (2) + c = -1$$
$$3 + c = -1$$
$$c = -4$$

Substitute $a = 1$, $b = 2$, and $c = -4$ in the standard form of the parabola for the final answer.

$$y = ax^2 + bx + c$$
$$y = (1)x^2 + (2)x + (-4) \longrightarrow y = x^2 + 2x - 4$$

A graph of $y = x^2 + 2x - 4$ is shown in Figure 7-18. Notice that the graph of the function passes through the points $(1, -1)$, $(-1, -5)$, and $(2, 4)$.

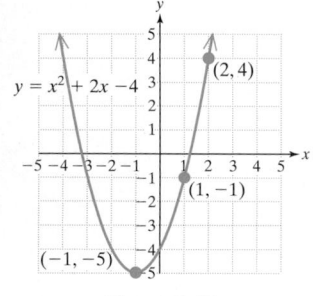

$y = x^2 + 2x - 4$

Figure 7-18

Answer

5. $a = -1$, $b = -1$, $c = 3$;
   $y = -x^2 - x + 3$

Skill Practice

5. Write an equation of the parabola that passes through the points $(1, 1)$, $(-2, 1)$, and $(3, -9)$.

Section 7.5 Practice Exercises

Vocabulary and Key Concepts

1. **a.** Given $f(x) = ax^2 + bx + c$ $(a \neq 0)$, the vertex formula gives the x-coordinate of the vertex as _____. The y-coordinate can be found by evaluating the function at $x =$ _____.

   b. Answer true or false. A parabola may have no x-intercepts.

   c. Answer true or false. A parabola may have one x-intercept.

   d. Answer true or false. A parabola may have two x-intercepts.

   e. Answer true or false. A parabola may have three x-intercepts.

Review Exercises

2. How does the graph of $f(x) = -2x^2$ compare with the graph of $y = x^2$?

3. How does the graph of $p(x) = \frac{1}{4}x^2$ compare with the graph of $y = x^2$?

4. How does the graph of $Q(x) = x^2 - \frac{8}{3}$ compare with the graph of $y = x^2$?

5. How does the graph of $r(x) = x^2 + 7$ compare with the graph of $y = x^2$?

6. How does the graph of $s(x) = (x - 4)^2$ compare with the graph of $y = x^2$?

7. How does the graph of $t(x) = (x + 10)^2$ compare with the graph of $y = x^2$?

8. Find the coordinates of the vertex of the parabola defined by $g(x) = 2(x + 3)^2 - 4$.

For Exercises 9–16, find the value of n to complete the square. Factor the resulting trinomial.

9. $x^2 - 8x + n$ **10.** $x^2 + 4x + n$ **11.** $y^2 + 7y + n$ **12.** $a^2 - a + n$

13. $b^2 + \dfrac{2}{9}b + n$ **14.** $m^2 - \dfrac{2}{7}m + n$ **15.** $t^2 - \dfrac{1}{3}t + n$ **16.** $p^2 + \dfrac{1}{4}p + n$

Concept 1: Writing a Quadratic Function in the Form $f(x) = a(x - h)^2 + k$

For Exercises 17–28, write the function in the form $f(x) = a(x - h)^2 + k$ by completing the square. Then identify the vertex. **(See Examples 1–2.)**

17. $g(x) = x^2 - 8x + 5$ **18.** $h(x) = x^2 + 4x + 5$ **19.** $n(x) = 2x^2 + 12x + 13$

20. $f(x) = 4x^2 + 16x + 19$ **21.** $p(x) = -3x^2 + 6x - 5$ **22.** $q(x) = -2x^2 + 12x - 11$

23. $k(x) = x^2 + 7x - 10$ **24.** $m(x) = x^2 - x - 8$ **25.** $F(x) = 5x^2 + 10x + 1$

26. $G(x) = 4x^2 + 4x - 7$ **27.** $P(x) = -2x^2 + x$ **28.** $Q(x) = -3x^2 + 12x$

Concept 2: Vertex Formula

For Exercises 29–40, find the vertex by using the vertex formula. **(See Example 3.)**

29. $Q(x) = x^2 - 4x + 7$ **30.** $T(x) = x^2 - 8x + 17$ **31.** $r(x) = -3x^2 - 6x - 5$

32. $s(x) = -2x^2 - 12x - 19$ **33.** $N(x) = x^2 + 8x + 1$ **34.** $M(x) = x^2 + 6x - 5$

35. $m(x) = \dfrac{1}{2}x^2 + x + \dfrac{5}{2}$ **36.** $n(x) = \dfrac{1}{2}x^2 + 2x + 3$ **37.** $k(x) = -x^2 + 2x + 2$

38. $h(x) = -x^2 + 4x - 3$ **39.** $A(x) = -\dfrac{1}{3}x^2 + x$ **40.** $B(x) = -\dfrac{2}{3}x^2 - 2x$

For Exercises 41–44, find the vertex two ways:

    a. by completing the square and writing in the form $f(x) = a(x - h)^2 + k$.

    b. by using the vertex formula.

41. $p(x) = x^2 + 8x + 1$ **42.** $F(x) = x^2 + 4x - 2$ **43.** $f(x) = 2x^2 + 4x + 6$ **44.** $g(x) = 3x^2 + 12x + 9$

Concept 3: Determining the Vertex and Intercepts of a Quadratic Function

For Exercises 45–52

    a. Find the vertex.

    b. Find the y-intercept.

    c. Find the x-intercept(s), if they exist.

    d. Use this information to graph the function. **(See Examples 2–3.)**

45. $f(x) = x^2 + 2x - 3$ **46.** $f(x) = x^2 + 4x + 3$ 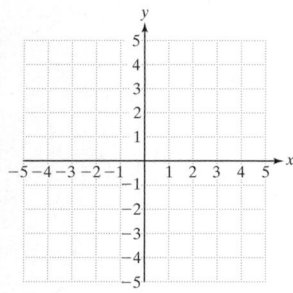 **47.** $f(x) = 2x^2 - 2x + 4$ **48.** $f(x) = 2x^2 - 12x + 19$

  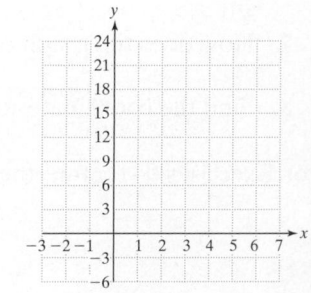

49. $f(x) = -x^2 + 3x - \dfrac{9}{4}$ **50.** $f(x) = -x^2 - \dfrac{3}{2}x - \dfrac{9}{16}$ **51.** $f(x) = -x^2 - 2x + 3$ 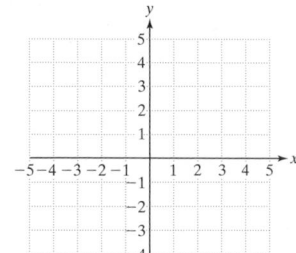 **52.** $f(x) = -x^2 - 4x$

  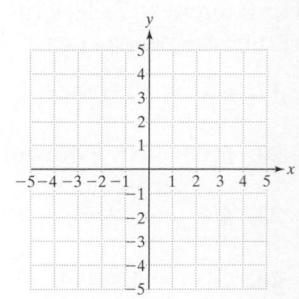

Concept 4: Applications and Modeling of Quadratic Functions

 53. Mia sells used MP3 players. The average cost to package MP3 players is given by the equation $C(x) = 2x^2 - 40x + 2200$, where x is the number of MP3 players she packages in a week. How many players must she package to minimize her average cost?

 54. Ben sells used iPods. The average cost to package iPods is given by the equation $C(x) = 3x^2 - 120x + 1300$, where x is the number of iPods packaged per month. Determine the number of iPods that Ben needs to package to minimize the average cost.

 55. A set of fireworks mortar shells is launched from the staging platform at 100 ft/sec from an initial height of 8 ft above the ground. The height of the fireworks, $h(t)$, can be modeled by

$h(t) = -16t^2 + 100t + 8$, where t is the time in seconds after launch. **(See Example 4.)**

a. What is the maximum height that the shells can reach before exploding?

b. For how many seconds after launch would the fuses need to be set so that the mortar shells would in fact explode at the maximum height?

 56. A baseball player throws a ball, and the height of the ball (in feet) can be approximated by

$$y(x) = -0.011x^2 + 0.577x + 5$$

where x is the horizontal position of the ball measured in feet from the origin.

a. For what value of x will the ball reach its highest point? Round to the nearest foot.

b. What is the maximum height of the ball? Round to the nearest tenth of a foot.

57. Gas mileage depends in part on the speed of the car. The gas mileage of a subcompact car is given by $m(x) = -0.04x^2 + 3.6x - 49$, where $20 \le x \le 70$ represents the speed in miles per hour and $m(x)$ is given in miles per gallon.

   a. At what speed will the car get the maximum gas mileage?

   b. What is the maximum gas mileage?

58. Gas mileage depends in part on the speed of the car. The gas mileage of a luxury car is given by $L(x) = -0.015x^2 + 1.44x - 21$, where $25 \le x \le 70$ represents the speed in miles per hour and $L(x)$ is given in miles per gallon.

   a. At what speed will the car get the maximum gas mileage?

   b. What is the maximum gas mileage? Round to the nearest whole unit.

59. Tetanus bacillus bacterium is cultured to produce tetanus toxin used in an inactive form for the tetanus vaccine. The amount of toxin produced per batch increases with time and then becomes unstable. The amount of toxin (in grams) as a function of time t (in hours) can be approximated by:

$$b(t) = -\frac{1}{1152}t^2 + \frac{1}{12}t$$

   a. How many hours will it take to produce the maximum yield?

   b. What is the maximum yield?

60. The bacterium *Pseudomonas aeruginosa* is cultured with an initial population of 10^4 active organisms. The population of active bacteria increases up to a point, and then due to a limited food supply and an increase of waste products, the population of living organisms decreases. Over the first 48 hr, the population can be approximated by:

$$P(t) = -1718.75t^2 + 82,500t + 10,000$$

   where $0 \le t \le 48$

   a. Find the time required for the population to reach its maximum value.

   b. What is the maximum population?

For Exercises 61–66, use the standard form of a parabola given by $y = ax^2 + bx + c$ to write an equation of a parabola that passes through the given points. **(See Example 5.)**

61. $(0, 4), (1, 0)$, and $(-1, -10)$ **62.** $(0, 3), (3, 0), (-1, 8)$ **63.** $(2, 1), (-2, 5)$, and $(1, -4)$

64. $(1, 2), (-1, -6)$, and $(2, -3)$ **65.** $(2, -4), (1, 1)$, and $(-1, -7)$ **66.** $(1, 4), (-1, 6)$, and $(2, 18)$

Expanding Your Skills

67. A farmer wants to fence a rectangular corral adjacent to the side of a barn; however, she has only 200 ft of fencing and wants to enclose the largest possible area. See the figure.

   a. If x represents the length of the corral and y represents the width, explain why the dimensions of the corral are subject to the constraint $2x + y = 200$.

   b. The area of the corral is given by $A = xy$. Use the constraint equation from part (a) to express A as a function of x, where $0 < x < 100$.

   c. Use the function from part (b) to find the dimensions of the corral that will yield the maximum area. [*Hint:* Find the vertex of the function from part (b).]

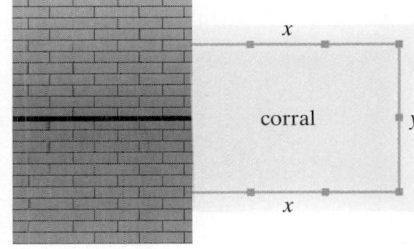

68. A veterinarian wants to construct two equal-sized pens of maximum area out of 240 ft of fencing. See the figure.

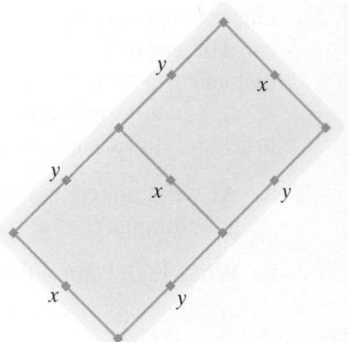

a. If x represents the length of each pen and y represents the width of each pen, explain why the dimensions of the pens are subject to the constraint $3x + 4y = 240$.

b. The area of each individual pen is given by $A = xy$. Use the constraint equation from part (a) to express A as a function of x, where $0 < x < 80$.

c. Use the function from part (b) to find the dimensions of an individual pen that will yield the maximum area. [*Hint:* Find the vertex of the function from part (b).]

Graphing Calculator Exercises

For Exercises 69–74, graph the functions in Exercises 45–50 on a graphing calculator. Use the *Max* or *Min* feature to approximate the vertex.

69. $Y_1 = x^2 + 2x - 3$ (Exercise 45)

70. $Y_1 = x^2 + 4x + 3$ (Exercise 46)

71. $Y_1 = 2x^2 - 2x + 4$ (Exercise 47)

72. $Y_1 = 2x^2 - 12x + 19$ (Exercise 48)

73. $Y_1 = -x^2 + 3x - \dfrac{9}{4}$ (Exercise 49)

74. $Y_1 = -x^2 - \dfrac{3}{2}x - \dfrac{9}{16}$ (Exercise 50)

Section 7.6 | Polynomial and Rational Inequalities

Concepts

1. Solving Quadratic and Polynomial Inequalities
2. Solving Rational Inequalities
3. Inequalities with "Special Case" Solution Sets

1. Solving Quadratic and Polynomial Inequalities

In this section, we will expand our study of solving inequalities.

   Quadratic inequalities are inequalities that can be written in one of the following forms:

$$ax^2 + bx + c \geq 0 \qquad ax^2 + bx + c \leq 0$$
$$ax^2 + bx + c > 0 \qquad ax^2 + bx + c < 0 \qquad \text{where } a \neq 0$$

Recall from Section 7.4 that the graph of a quadratic function defined by $f(x) = ax^2 + bx + c$ is a parabola that opens upward or downward.

- The inequality $ax^2 + bx + c > 0$ is asking "For what values of x is the function positive (above the x-axis)?"
- The inequality $ax^2 + bx + c < 0$ is asking "For what values of x is the function negative (below the x-axis)?"

The graph of a quadratic function can be used to answer these questions.

Example 1 Using a Graph to Solve a Quadratic Inequality

Use the graph of $f(x) = x^2 - 6x - 5$ in Figure 7-19 to solve the inequalities.

a. $x^2 - 6x + 5 < 0$ **b.** $x^2 - 6x + 5 > 0$

Solution:

From Figure 7-19, we see that the graph of $f(x) = x^2 - 6x + 5$ is a parabola opening upward. The function factors as $f(x) = (x - 1)(x - 5)$. The x-intercepts are $(1, 0)$ and $(5, 0)$, and the y-intercept is $(0, 5)$.

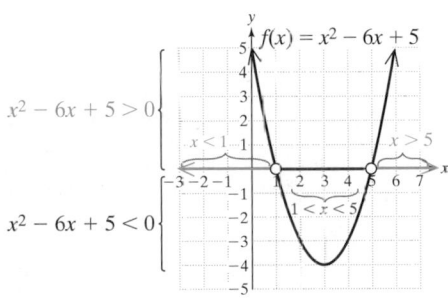

Figure 7-19

a. The solution to $x^2 - 6x + 5 < 0$ is the set of real numbers, x, for which $f(x) < 0$. Graphically, this is the set of all x values corresponding to the points where the parabola is below the x-axis (shown in red).

$$x^2 - 6x + 5 < 0 \quad \text{for } \{x \mid 1 < x < 5\} \text{ or equivalently, } (1, 5)$$

b. The solution to $x^2 - 6x + 5 > 0$ is the set of real numbers, x, for which $f(x) > 0$. This is the set of x values where the parabola is above the x-axis (shown in blue).

$$x^2 - 6x + 5 > 0 \quad \text{for } \{x \mid x < 1 \text{ or } x > 5\} \text{ or } (-\infty, 1) \cup (5, \infty)$$

Skill Practice Refer to the graph of $f(x) = x^2 + 3x - 4$ to solve the inequalities.

1. $x^2 + 3x - 4 > 0$
2. $x^2 + 3x - 4 < 0$

TIP: The inequalities in Example 1 are strict inequalities. Therefore, $x = 1$ and $x = 5$ (where $f(x) = 0$) are not included in the solution set. However, the corresponding inequalities using the symbols $\leq$ and $\geq$ do include the values where $f(x) = 0$.

The solution to $x^2 - 6x + 5 \leq 0$ is $\{x \mid 1 \leq x \leq 5\}$ or equivalently, $[1, 5]$.

The solution to $x^2 - 6x + 5 \geq 0$ is $\{x \mid x \leq 1 \text{ or } x \geq 5\}$ or $(-\infty, 1] \cup [5, \infty)$.

Answers

1. $\{x \mid x < -4 \text{ or } x > 1\}$; $(-\infty, -4) \cup (1, \infty)$
2. $\{x \mid -4 < x < 1\}$; $(-4, 1)$

Notice that $x = 1$ and $x = 5$ are the boundaries of the solution set to the inequalities in Example 1. These values are the solutions to the related equation $x^2 - 6x + 5 = 0$.

> ### Definition of Boundary Points
>
> The **boundary points** of an inequality consist of the real solutions to the related equation and the points where the inequality is undefined.

Recall that testing points in intervals bounded by these points is the basis of the **test point method** to solve inequalities.

TIP: In Section 1.7, we used the test point method to solve absolute value inequalities.

> ### Solving Inequalities by Using the Test Point Method
>
> **Step 1** Find the boundary points of the inequality.
>
> **Step 2** Plot the boundary points on the number line. This divides the number line into intervals.
>
> **Step 3** Select a test point from each interval and substitute it into the original inequality.
> - If a test point makes the original inequality true, then that interval is part of the solution set.
>
> **Step 4** Test the boundary points in the original inequality.
> - If the original inequality is strict ($<$ or $>$), do not include the boundary points in the solution set.
> - If the original inequality is defined using $\leq$ or $\geq$, then include the boundary points that are well defined within the inequality.
>
> *Note:* Any boundary point that makes an expression within the inequality undefined must *always* be excluded from the solution set.

Example 2 **Solving a Quadratic Inequality by Using the Test Point Method**

Solve the inequality by using the test point method.

$$2x^2 + 5x < 12$$

Solution:

$2x^2 + 5x < 12$ **Step 1:** Find the boundary points. Because polynomials are defined for all values of x, the only boundary points are the real solutions to the related equation.

$2x^2 + 5x = 12$ Solve the related equation.

$2x^2 + 5x - 12 = 0$

$(2x - 3)(x + 4) = 0$

$x = \frac{3}{2} \qquad x = -4$ The boundary points are $\frac{3}{2}$ and -4.

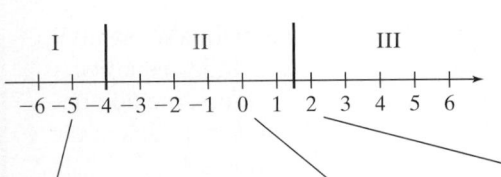

Step 2: Plot the boundary points.

Step 3: Select a test point from each interval.

Test $x = -5$

$2x^2 + 5x < 12$

$2(-5)^2 + 5(-5) \overset{?}{<} 12$

$50 - 25 \overset{?}{<} 12$

$25 \overset{?}{<} 12$ False

Test $x = 0$

$2x^2 + 5x < 12$

$2(0)^2 + 5(0) \overset{?}{<} 12$

$0 \overset{?}{<} 12$

$0 \overset{?}{<} 12$ True

Test $x = 2$

$2x^2 + 5x < 12$

$2(2)^2 + 5(2) \overset{?}{<} 12$

$8 + 10 \overset{?}{<} 12$

$18 \overset{?}{<} 12$ False

Step 4: Test the boundary points. The strict inequality excludes values of x for which $2x^2 + 5x = 12$. Therefore, the boundary points are *not* included in the solution set.

The solution is $\{x \mid -4 < x < \frac{3}{2}\}$ or equivalently in interval notation $\left(-4, \frac{3}{2}\right)$.

Skill Practice Solve the inequality.

3. $x^2 + x > 6$

Example 3   **Solving a Polynomial Inequality by Using the Test Point Method**

Solve the inequality by using the test point method.

$$x(x + 4)^2(x - 4) \geq 0$$

Solution:

$x(x + 4)^2(x - 4) \geq 0$

$x(x + 4)^2(x - 4) = 0$ **Step 1:** Find the boundary points.

$x = 0$ $x = -4$ $x = 4$

Step 2: Plot the boundary points.

Step 3: Select a test point from each interval.

Test $x = -5$: $-5(-5 + 4)^2(-5 - 4) \overset{?}{\geq} 0$ $45 \overset{?}{\geq} 0$ True

Test $x = -1$: $-1(-1 + 4)^2(-1 - 4) \overset{?}{\geq} 0$ $45 \overset{?}{\geq} 0$ True

Test $x = 1$: $1(1 + 4)^2(1 - 4) \overset{?}{\geq} 0$ $-75 \overset{?}{\geq} 0$ False

Test $x = 5$: $5(5 + 4)^2(5 - 4) \overset{?}{\geq} 0$ $405 \overset{?}{\geq} 0$ True

Answer

3. $(-\infty, -3) \cup (2, \infty)$

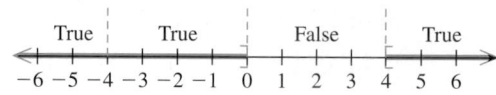

True   True   False   True

Step 4: The inequality symbol, $\geq$, includes equality. Therefore, include the boundary points in the solution set.

The solution is $\{x \mid x \leq 0 \text{ and } x \geq 4\}$, or equivalently in interval notation, $(-\infty, 0] \cup [4, \infty)$.

Skill Practice Solve the inequality.

4. $t(t - 5)(t + 2)^2 \geq 0$

TIP: In Example 3, one side of the inequality is factored, and the other side is zero. For inequalities written in this form, we can use a sign chart to determine the sign of each factor. Then the sign of the product (bottom row) is easily determined.

Sign of x	$-$	$-$	$+$	$+$
Sign of $(x + 4)^2$	$+$	$+$	$+$	$+$
Sign of $(x - 4)$	$-$	$-$	$-$	$+$
Sign of $x(x + 4)^2(x - 4)$	$+$	$+$	$-$	$+$

$-4 \qquad 0 \qquad 4$

The solution to the inequality $x(x + 4)^2(x - 4) \geq 0$ includes the intervals for which the product is positive (shown in blue).

The solution is $(-\infty, 0] \cup [4, \infty)$.

Example 4 **Solving a Polynomial Inequality by Using the Test Point Method**

Solve the inequality by using the test point method. $x^2 + x - 4 \geq 0$

Solution:

$x^2 + x - 4 \geq 0$

$x^2 + x - 4 = 0$

$x = \dfrac{-b \pm \sqrt{b^2 - 4ac}}{2a}$

$x = \dfrac{-1 \pm \sqrt{(1)^2 - 4(1)(-4)}}{2(1)}$

$x = \dfrac{-1 \pm \sqrt{17}}{2}$

$x = \dfrac{-1 - \sqrt{17}}{2} \approx -2.56$ and $x = \dfrac{-1 + \sqrt{17}}{2} \approx 1.56$

Step 1: Find the boundary points in the related equation. Since this equation is not factorable, use the quadratic formula to find the solutions.

Answer

4. $(-\infty, 0] \cup [5, \infty)$

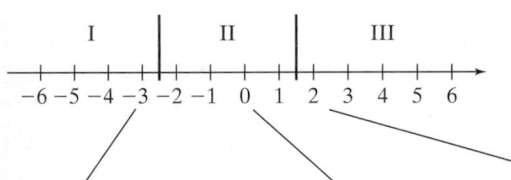

Step 2: Plot the boundary points.

Step 3: Select a test point from each interval.

Test $x = -3$

$x^2 + x - 4 \geq 0$

$(-3)^2 + (-3) - 4 \overset{?}{\geq} 0$

$2 \overset{?}{\geq} 0$ True

Test $x = 0$

$x^2 + x - 4 \geq 0$

$(0)^2 + (0) - 4 \overset{?}{\geq} 0$

$-4 \overset{?}{\geq} 0$ False

Test $x = 2$

$x^2 + x - 4 \geq 0$

$(2)^2 + (2) - 4 \overset{?}{\geq} 0$

$2 \overset{?}{\geq} 0$ True

Step 4: Test the boundary points. Both boundary points make the inequality true. Therefore, both boundary points are included in the solution set.

The solution is $\left\{ x \mid x \leq \dfrac{-1 - \sqrt{17}}{2} \text{ or } x \geq \dfrac{-1 + \sqrt{17}}{2} \right\}$ or equivalently

in interval notation: $\left(-\infty, \dfrac{-1 - \sqrt{17}}{2} \right] \cup \left[\dfrac{-1 + \sqrt{17}}{2}, \infty \right)$.

Skill Practice Solve the inequality.

5. $x^2 - 3x - 1 \leq 0$

2. Solving Rational Inequalities

The test point method can be used to solve rational inequalities. A **rational inequality** is an inequality in which one or more terms is a rational expression. The solution set to a rational inequality must exclude all values of the variable that make the inequality undefined. That is, exclude all values that make the denominator equal to zero for any rational expression in the inequality.

Example 5 **Solving a Rational Inequality by Using the Test Point Method**

Solve the inequality. $\dfrac{3}{x - 1} > 0$

Solution:

$\dfrac{3}{x - 1} > 0$

$\dfrac{3}{x - 1} = 0$

$(x - 1) \cdot \left(\dfrac{3}{x - 1} \right) = (x - 1) \cdot 0$

$3 = 0$

Step 1: Find the boundary points. Note that the inequality is undefined for $x = 1$, so $x = 1$ is a boundary point. To find any other boundary points, solve the related equation.

Clear fractions.

There is no solution to the related equation.

The only boundary point is $x = 1$.

Answer

5. $\left[\dfrac{3 - \sqrt{13}}{2}, \dfrac{3 + \sqrt{13}}{2} \right]$

Step 2: Plot boundary points.

Test $x = 0$ **Test $x = 2$**   **Step 3:** Select test points.

$$\frac{3}{(0) - 1} \overset{?}{>} 0$$ $$\frac{3}{(2) - 1} \overset{?}{>} 0$$

$$\frac{3}{-1} \overset{?}{>} 0 \text{ False}$$ $$\frac{3}{1} > 0 \text{ True}$$

Step 4: The boundary point $x = 1$ cannot be included in the solution set because it is undefined in the original inequality.

The solution is $\{x \mid x > 1\}$ or equivalently in interval notation, $(1, \infty)$.

Skill Practice Solve the inequality.

6. $\dfrac{-5}{y + 2} < 0$

TIP: Using a sign chart we see that the quotient of the factors 3 and $(x - 1)$ is positive on the interval $(1, \infty)$.

Therefore, the solution to the inequality $\dfrac{3}{x - 1} > 0$ is $(1, \infty)$.

The solution to the inequality $\dfrac{3}{x - 1} > 0$ can be confirmed from the graph of the related rational function, $f(x) = \dfrac{3}{x - 1}$ (see Figure 7-20).

- The graph is above the x-axis where $f(x) = \dfrac{3}{x - 1} > 0$ for $x > 1$ (shaded red).

- Also note that $x = 1$ cannot be included in the solution set because 1 is not in the domain of f.

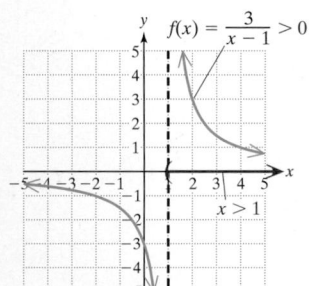

Figure 7-20

Example 6   Solving a Rational Inequality by Using the Test Point Method

Solve the inequality by using the test point method. $\dfrac{x + 2}{x - 4} \le 3$

Answer

6. $(-2, \infty)$

Solution:

$$\frac{x + 2}{x - 4} \leq 3$$

Step 1: Find the boundary points. Note that the inequality is undefined for $x = 4$. Therefore, $x = 4$ is automatically a boundary point. To find any other boundary points, solve the related equation.

$$\frac{x + 2}{x - 4} = 3$$

$$(x - 4)\left(\frac{x + 2}{x - 4}\right) = (x - 4)(3)$$　　Clear fractions.

$$x + 2 = 3(x - 4)$$　　Solve for x.

$$x + 2 = 3x - 12$$

$$-2x = -14$$

$$x = 7$$

The solution to the related equation is $x = 7$, and the inequality is undefined for $x = 4$. Therefore, the boundary points are $x = 4$ and $x = 7$.

I　　II　III

$$-3\ -2\ -1\ \ 0\ \ 1\ \ 2\ \ 3\ \ 4\ \ 5\ \ 6\ \ 7\ \ 8$$

Step 2: Plot boundary points.

Step 3: Select test points.

Test $x = 0$

$$\frac{x + 2}{x - 4} \leq 3$$

$$\frac{0 + 2}{0 - 4} \overset{?}{\leq} 3$$

$$-\frac{1}{2} \overset{?}{\leq} 3 \quad \text{True}$$

Test $x = 5$

$$\frac{x + 2}{x - 4} \leq 3$$

$$\frac{5 + 2}{5 - 4} \overset{?}{\leq} 3$$

$$\frac{7}{1} \overset{?}{\leq} 3 \quad \text{False}$$

Test $x = 8$

$$\frac{x + 2}{x - 4} \leq 3$$

$$\frac{8 + 2}{8 - 4} \overset{?}{\leq} 3$$

$$\frac{10}{4} \overset{?}{\leq} 3$$

$$\frac{5}{2} \overset{?}{\leq} 3 \quad \text{True}$$

Test $x = 4$

$$\frac{x + 2}{x - 4} \leq 3$$

$$\frac{4 + 2}{4 - 4} \overset{?}{\leq} 3$$

$$\frac{6}{0} \overset{?}{\leq} 3 \quad \text{Undefined}$$

Test $x = 7$

$$\frac{x + 2}{x - 4} \leq 3$$

$$\frac{7 + 2}{7 - 4} \overset{?}{\leq} 3$$

$$\frac{9}{3} \overset{?}{\leq} 3 \quad \text{True}$$

Step 4: Test the boundary points.

The boundary point $x = 4$ cannot be included in the solution set, because it is undefined in the inequality. The boundary point $x = 7$ makes the original inequality true and must be included in the solution set.

$$-3\ -2\ -1\ \ 0\ \ 1\ \ 2\ \ 3\ \ 4\ \ 5\ \ 6\ \ 7\ \ 8$$

The solution is $\{x \mid x < 4 \text{ or } x \geq 7\}$, or equivalently in interval notation, $(-\infty, 4) \cup [7, \infty)$.

Skill Practice Solve the inequality.

7. $\dfrac{x - 5}{x + 4} \leq -1$

Answer

7. $\left(-4, \dfrac{1}{2}\right]$

3. Inequalities with "Special Case" Solution Sets

The solution to an inequality is often one or more intervals on the real number line. Sometimes, however, the solution to an inequality may be a single point on the number line, the empty set, or the set of all real numbers.

Example 7   Solving Inequalities

Solve the inequalities.

     a. $x^2 + 6x + 9 \geq 0$ **b.** $x^2 + 6x + 9 > 0$

     c. $x^2 + 6x + 9 \leq 0$ **d.** $x^2 + 6x + 9 < 0$

Solution:

     a. $x^2 + 6x + 9 \geq 0$ Notice that $x^2 + 6x + 9$ is a perfect square trinomial.

         $(x + 3)^2 \geq 0$ Factor $x^2 + 6x + 9 = (x + 3)^2$.

The quantity $(x + 3)^2$ is a perfect square and is greater than or equal to zero for all real numbers, x. The solution is all real numbers, $(-\infty, \infty)$.

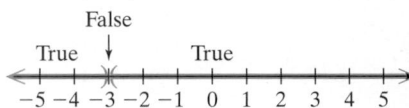

     b. $x^2 + 6x + 9 > 0$

         $(x + 3)^2 > 0$ This is the same inequality as in part (a) with the exception that the inequality is strict. The solution set does not include the point where $x^2 + 6x + 9 = 0$. Therefore, the boundary point $x = -3$ is *not* included in the solution set.

The solution set is $\{x \mid x \neq -3\}$ or equivalently $(-\infty, -3) \cup (-3, \infty)$.

     c. $x^2 + 6x + 9 \leq 0$

         $(x + 3)^2 \leq 0$

A perfect square cannot be less than zero. However, $(x + 3)^2$ is equal to zero at $x = -3$. Therefore, the solution set is $\{-3\}$.

           False   False

         $-6\ -5\ -4\ -3\ -2\ -1\ \ 0\ \ 1\ \ 2\ \ 3\ \ 4$

                      True

     d. $x^2 + 6x + 9 < 0$

         $(x + 3)^2 < 0$

A perfect square cannot be negative; therefore, there are no real numbers x such that $(x + 3)^2 < 0$. There is no solution, { }.

> **TIP:** The graph of $f(x) = x^2 + 6x + 9$, or equivalently $f(x) = (x + 3)^2$, is equal to zero at $x = -3$ and positive (above the x-axis) for all other values of x in its domain.
>
> *(graph of $f(x) = x^2 + 6x + 9$ with labeled axes)*

Answers

8. All real numbers; $(-\infty, \infty)$

9. $(-\infty, 2) \cup (2, \infty)$

10. $\{2\}$

11. { }

Skill Practice Solve the inequalities.

   8. $x^2 - 4x + 4 \geq 0$ **9.** $x^2 - 4x + 4 > 0$

   10. $x^2 - 4x + 4 \leq 0$ **11.** $x^2 - 4x + 4 < 0$

Section 7.6 Practice Exercises

Vocabulary and Key Concepts

1. a. An inequality of the form $ax^2 + bx + c > 0$ $(a \neq 0)$ is an example of a _____ inequality.

b. The boundary points of an inequality consist of the real _____ to the related equation and the points where the inequality is _____.

c. In solving an inequality by using the _____ _____ method, a point is selected from each interval formed by the boundary points and substituted into the original inequality.

d. If a test point makes the original inequality (true/false) then that interval is part of the solution set.

e. The inequality $\dfrac{4}{x + 7} > 0$ is an example of a _____ inequality.

f. The solution set to a rational inequality must exclude all values that make the _____ equal to zero for any rational expression in the original inequality.

g. The solution set to the inequality $(x + 3)^2 \leq -4$ is _____, whereas the solution set to the inequality $(x + 3)^2 \geq -4$ is _____.

h. The solution set to the inequality $(x + 3)^2 > 0$ (includes/excludes) -3, whereas the solution set to the inequality $(x + 3)^2 \leq 0$ (includes/excludes) -3.

Review Exercises

For Exercises 2–8, solve the equations.

2. $x^2 + 10x = -9$

3. $2t^2 - 3t + 4 = 0$

4. $3w^2 - w + 5 = 0$

5. $2m(m - 4)(m + 6) = 0$

6. $-n(3 - n)(n + 7) = 0$

7. $\dfrac{14}{x - 3} = 7$

8. $\dfrac{15}{x + 2} = 5$

Concept 1: Solving Quadratic and Polynomial Inequalities

For Exercises 9–12, estimate from the graph the intervals for which the inequality is true. **(See Example 1.)**

 9.

a. $p(x) > 0$ **b.** $p(x) < 0$

c. $p(x) \leq 0$ **d.** $p(x) \geq 0$

10.

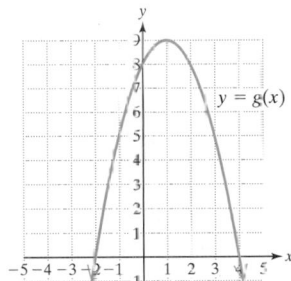

a. $g(x) > 0$ **b.** $g(x) < 0$

c. $g(x) \leq 0$ **d.** $g(x) \geq 0$

11.

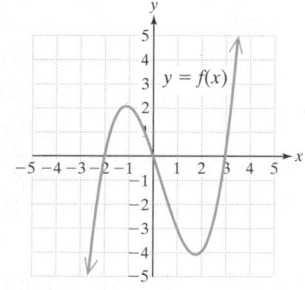

a. $f(x) > 0$ **b.** $f(x) < 0$

c. $f(x) \leq 0$ **d.** $f(x) \geq 0$

12.

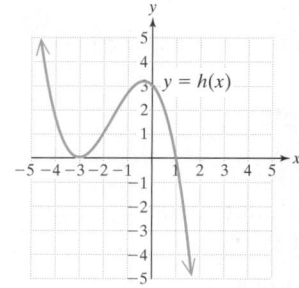

a. $h(x) > 0$ **b.** $h(x) < 0$

c. $h(x) \leq 0$ **d.** $h(x) \geq 0$

For Exercises 13–18, solve the equation and related inequalities. **(See Examples 2–4.)**

13. **a.** $3(4 - x)(2x + 1) = 0$

b. $3(4 - x)(2x + 1) < 0$

c. $3(4 - x)(2x + 1) > 0$

14. **a.** $5(y + 6)(3 - 5y) = 0$

b. $5(y + 6)(3 - 5y) < 0$

c. $5(y + 6)(3 - 5y) > 0$

15. **a.** $x^2 + 7x = 30$

b. $x^2 + 7x < 30$

c. $x^2 + 7x > 30$

16. **a.** $q^2 - 4q = 5$

b. $q^2 - 4q \leq 5$

c. $q^2 - 4q \geq 5$

17. **a.** $2p(p - 2) = p + 3$

b. $2p(p - 2) \leq p + 3$

c. $2p(p - 2) \geq p + 3$

18. **a.** $3w(w + 4) = 10 - w$

b. $3w(w + 4) < 10 - w$

c. $3w(w + 4) > 10 - w$

For Exercises 19–38, solve the polynomial inequality. Write the answer in interval notation. **(See Examples 2–4.)**

19. $(t - 7)(t - 1) < 0$

20. $(p - 4)(p - 2) > 0$

21. $-6(4 + 2x)(5 - x) > 0$

22. $-8(2t + 5)(6 - t) < 0$

23. $m(m + 1)^2(m + 5) \leq 0$

24. $w^2(3 - w)(w + 2) \geq 0$

25. $a^2 - 12a \leq -32$

26. $w^2 + 20w \geq -64$

27. $5x^2 - 2x - 1 > 0$

28. $3x^2 + 2x - 4 < 0$

29. $x^2 + 3x \leq 6$

30. $x^2 - 8 \leq 3x$

31. $b^2 - 121 < 0$

32. $c^2 - 25 < 0$

33. $3p(p - 2) - 3 \geq 2p$

34. $2t(t + 3) - t \leq 12$

35. $x^3 - x^2 \leq 12x$

36. $x^3 + 36 > 4x^2 + 9x$

37. $w^3 + w^2 > 4w + 4$

38. $2p^3 - 5p^2 \leq 3p$

Concept 2: Solving Rational Inequalities

For Exercises 39–42, solve the equation and related inequalities. **(See Examples 5–6.)**

39. **a.** $\dfrac{10}{x - 5} = 5$

b. $\dfrac{10}{x - 5} < 5$

c. $\dfrac{10}{x - 5} > 5$

40. **a.** $\dfrac{8}{a + 1} = 4$

b. $\dfrac{8}{a + 1} > 4$

c. $\dfrac{8}{a + 1} < 4$

41. **a.** $\dfrac{z + 2}{z - 6} = -3$

b. $\dfrac{z + 2}{z - 6} \leq -3$

c. $\dfrac{z + 2}{z - 6} \geq -3$

42. **a.** $\dfrac{w - 8}{w + 6} = 2$

b. $\dfrac{w - 8}{w + 6} \leq 2$

c. $\dfrac{w - 8}{w + 6} \geq 2$

For Exercises 43–54, solve the rational inequality. Write the answer in interval notation. **(See Examples 5–6.)**

43. $\dfrac{2}{x-1} \geq 0$

44. $\dfrac{-3}{x+2} \leq 0$

45. $\dfrac{b+4}{b-4} > 0$

46. $\dfrac{a+1}{a-3} < 0$

47. $\dfrac{3}{2x-7} < -1$

48. $\dfrac{8}{4x+9} > 1$

49. $\dfrac{x+1}{x-5} \geq 4$

50. $\dfrac{x-2}{x+6} \leq 5$

51. $\dfrac{1}{x} \leq 2$

52. $\dfrac{1}{x} \geq 3$

53. $\dfrac{(x+2)^2}{x} > 0$

54. $\dfrac{(x-3)^2}{x} < 0$

Concept 3: Inequalities with "Special Case" Solution Sets

For Exercises 55–70, solve the inequalities. **(See Example 7.)**

55. $x^2 + 10x + 25 \geq 0$

56. $x^2 + 6x + 9 < 0$

57. $x^2 + 2x + 1 < 0$

58. $x^2 + 8x + 16 \geq 0$

59. $x^4 + 3x^2 \leq 0$

60. $x^4 + 2x^2 \leq 0$

61. $x^2 + 12x - 36 < 0$

62. $x^2 + 12x + 36 \geq 0$

63. $x^2 + 3x + 5 < 0$

64. $2x^2 + 3x - 3 > 0$

65. $-5x^2 + x < 1$

66. $-3x^2 - x > 6$

67. $x^2 + 22x + 121 > 0$

68. $y^2 - 24y - 144 > 0$

69. $4t^2 - 12t \leq -9$

70. $4t^2 - 12t > -9$

Mixed Exercises

For Exercises 71–94, identify the inequality as one of the following types: linear, quadratic, rational, or polynomial (degree > 2). Then solve the inequality and write the answer in interval notation.

71. $2y^2 - 8 \leq 24$

72. $8p^2 - 18 > 0$

73. $(5x+2)^2 > -4$

74. $(3-7x)^2 < -1$

75. $4(x-2) < 6x - 3$

76. $-7(3-y) > 4 + 2y$

77. $\dfrac{2x+3}{x+1} \leq 2$

78. $\dfrac{5x-1}{x+3} \geq 5$

79. $4x^3 - 40x^2 + 100x > 0$

80. $2y^3 - 12y^2 + 18y < 0$

81. $2p^3 > 4p^2$

82. $w^3 \leq 5w^2$

83. $-3(x+4)^2(x-5) \geq 0$

84. $5x(x-2)(x-6)^2 \geq 0$

85. $x^2 - 2 < 0$

86. $y^2 - 3 > 0$

87. $x^2 + 5x - 2 \geq 0$

88. $t^2 + 7t + 3 \leq 0$

89. $\dfrac{a+2}{a-5} \geq 0$

90. $\dfrac{t+1}{t-2} \leq 0$

91. $2 \geq t - 3$

92. $-5p + 8 < p$

93. $4x^2 + 2 \geq -x$

94. $5x - 4 \geq 2x^2$

Expanding Your Skills

95. A fireworks display is planned for the 4th of July. The fire department wants to make sure that all fireworks explode at least 120 ft above the ground. The staging platform is 8 ft high, and all the mortar shells will be launched at 108 ft/sec. The height of the mortars, $h(t)$, can be modeled by

$$h(t) = -16t^2 + 108t + 8 \qquad \text{where } t \text{ is the time in seconds after launch.}$$

For what interval of time should the fuses be set? Round to the nearest tenth of a second.

Graphing Calculator Exercises

96. To solve the inequality $\dfrac{x}{x-2} > 0$ enter Y_1 as $x/(x-2)$ and determine where the graph is above the x-axis. Write the solution in interval notation.

97. To solve the inequality $\dfrac{x}{x-2} < 0$ enter Y_1 as $x/(x-2)$ and determine where the graph is below the x-axis. Write the solution in interval notation.

98. To solve the inequality $x^2 - 1 < 0$, enter Y_1 as $x^2 - 1$ and determine where the graph is below the x-axis. Write the solution in interval notation.

99. To solve the inequality $x^2 - 1 > 0$, enter Y_1 as $x^2 - 1$ and determine where the graph is above the x-axis. Write the solution in interval notation.

For Exercises 100–103, determine the solution by graphing the inequalities.

100. $x^2 + 10x + 25 \leq 0$

101. $-x^2 + 10x - 25 \geq 0$

102. $\dfrac{8}{x^2 + 2} < 0$

103. $\dfrac{-6}{x^2 + 3} > 0$

Problem Recognition Exercises

Recognizing Equations and Inequalities

At this point, you have learned how to solve a variety of equations and inequalities. Being able to distinguish the type of problem being posed is the first step in successfully solving it.

For Exercises 1–20,

a. Identify the problem type. Choose from

- linear equation
- quadratic equation
- rational equation
- absolute value equation
- radical equation
- equation quadratic in form

- polynomial equation
- linear inequality
- polynomial inequality
- rational inequality
- absolute value inequality
- compound inequality

b. Solve the equation or inequality. Write the solution to each inequality in interval notation if possible.

1. $(z^2 - 4)^2 - (z^2 - 4) - 12 = 0$

2. $3 + |4t - 1| < 6$

3. $2y(y - 4) \leq 5 + y$

4. $\sqrt[3]{11x - 3} + 4 = 6$

5. $-5 = -|w - 4|$

6. $\dfrac{5}{x-2} + \dfrac{3}{x+2} = 1$

7. $m^3 + 5m^2 - 4m - 20 \geq 0$

8. $-x - 4 > -5$ and $2x - 3 \leq 23$

9. $5 - 2[3 - (x - 4)] \leq 3x + 14$

10. $|2x - 6| = |x + 3|$

11. $\dfrac{3}{x-2} \leq 1$

12. $9 < |x + 4|$

13. $\sqrt{t + 8} - 6 = t$

14. $(4x - 3)^2 = -10$

15. $-4 - x > 2$ or $8 < 2x$

16. $\dfrac{1}{3}x - 2 = \dfrac{3}{4} + \dfrac{5}{6}x$

17. $x^2 - 10x \leq -25$

18. $\dfrac{10}{x^2 + 1} < 0$

19. $x - 13\sqrt{x} + 36 = 0$

20. $x^4 - 13x^2 + 36 = 0$

Group Activity

Creating a Quadratic Model of the Form $y = a(x - h)^2 + k$

Estimated time: 20 minutes

Group Size: 3

In Chapter 3, we modeled the path of a softball that was thrown from right field to third base. The data points are given in the table. The values of t represent the time in seconds after the ball was released, and y represents the height of the ball in feet.

Time (sec) t	0	0.2	0.4	0.6	0.8	1.0	1.2	1.4	1.6	1.8	2.0
Height (ft) y	5	11	16	19	21	22	21	19	16	12	6

1. Graph the points defined by (t, y) in the table.

2. Select a point that you think best represents the vertex of the parabola. Label this point (h, k).

$(h, k) =$ _____

3. Substitute the values of h and k into the formula $y = a(t - h)^2 + k$, and write the equation here.

$y = a(t -$ _____$)^2 +$ _____

Height of Softball vs. Time

4. Choose a different point (t, y) from the graph. Substitute these values into the equation in step 3 and then solve for a.

5. Substitute the values of h, k, and a that you found in steps 2 and 4 into the equation $y = a(t - h)^2 + k$.

6. Use the function found in step 5 to approximate the height of the ball after 0.7 sec.

7. Use the function found in step 5 to approximate the height of the ball after 1.8 sec. How well does this value match the observed value of 12 ft?

Chapter 7 Summary

| Section 7.1 | Square Root Property and Completing the Square |

Key Concepts

The **square root property** states that

If $x^2 = k$ then $x = \pm\sqrt{k}$

Follow these steps to solve a quadratic equation in the form $ax^2 + bx + c = 0$ $(a \neq 0)$ by completing the square and applying the square root property:

1. Divide both sides by a to make the leading coefficient 1.
2. Isolate the variable terms on one side of the equation.
3. Complete the square: Add the square of one-half the linear term coefficient to both sides of the equation. Then factor the resulting perfect square trinomial.
4. Apply the square root property and solve for x.

Examples

Example 1

$(x - 5)^2 = -13$

$x - 5 = \pm\sqrt{-13}$ (square root property)

$x = 5 \pm i\sqrt{13}$

The solution set is $\{5 \pm i\sqrt{13}\}$.

Example 2

$2x^2 - 12x - 16 = 0$

$\dfrac{2x^2}{2} - \dfrac{12x}{2} - \dfrac{16}{2} = \dfrac{0}{2}$

$x^2 - 6x - 8 = 0$

$x^2 - 6x = 8$

$x^2 - 6x + 9 = 8 + 9$ *Note:* $[\frac{1}{2}(-6)]^2 = 9$

$(x - 3)^2 = 17$

$x - 3 = \pm\sqrt{17}$

$x = 3 \pm \sqrt{17}$

The solution set is $\{3 \pm \sqrt{17}\}$.

Section 7.2   Quadratic Formula

Key Concepts

The solutions to a quadratic equation $ax^2 + bx + c = 0 \ (a \neq 0)$ are given by the **quadratic formula**

$$x = \frac{-b \pm \sqrt{b^2 - 4ac}}{2a}$$

The **discriminant** of a quadratic equation $ax^2 + bx + c = 0$ is $b^2 - 4ac$. If a, b, and c are rational numbers, then

1. If $b^2 - 4ac > 0$, then there will be two real solutions. Moreover,
   a. If $b^2 - 4ac$ is a perfect square, the solutions will be rational numbers.
   b. If $b^2 - 4ac$ is not a perfect square, the solutions will be irrational numbers.
2. If $b^2 - 4ac < 0$, then there will be two imaginary solutions.
3. If $b^2 - 4ac = 0$, then there will be one rational solution.

Three methods to solve a quadratic equation are

1. Factoring and applying the zero product rule.
2. Completing the square and applying the square root property.
3. Using the quadratic formula.

Example

Example 1

$$3x^2 - 2x + 4 = 0$$

$$a = 3 \qquad b = -2 \qquad c = 4$$

$$x = \frac{-(-2) \pm \sqrt{(-2)^2 - 4(3)(4)}}{2(3)}$$

$$= \frac{2 \pm \sqrt{4 - 48}}{6}$$

$$= \frac{2 \pm \sqrt{-44}}{6} \qquad \text{The discriminant is } -44. \text{ Therefore, there will be two imaginary solutions.}$$

$$= \frac{2 \pm 2i\sqrt{11}}{6}$$

$$= \frac{\overset{1}{2}(1 \pm i\sqrt{11})}{\underset{3}{6}}$$

$$= \frac{1 \pm i\sqrt{11}}{3}$$

$$= \frac{1}{3} \pm \frac{\sqrt{11}}{3}i$$

The solution set is $\left\{ \dfrac{1}{3} \pm \dfrac{\sqrt{11}}{3}i \right\}$.

Section 7.3   Equations in Quadratic Form

Key Concepts

Substitution can be used to solve equations that are in quadratic form.

Example

Example 1

$$x^{2/3} - x^{1/3} - 12 = 0 \quad \text{Let } u = x^{1/3}.$$

$$u^2 - u - 12 = 0$$

$$(u - 4)(u + 3) = 0$$

$$u = 4 \qquad \text{or} \qquad u = -3$$

$$x^{1/3} = 4 \qquad \text{or} \qquad x^{1/3} = -3$$

$$x = 64 \qquad \text{or} \qquad x = -27 \quad \text{Cube both sides.}$$

The solution set is $\{64, -27\}$.

Section 7.4 Graphs of Quadratic Functions

Key Concepts

A quadratic function of the form $f(x) = x^2 + k$ shifts the graph of $f(x) = x^2$ up k units if $k > 0$ and down $|k|$ units if $k < 0$.

Examples

Example 1

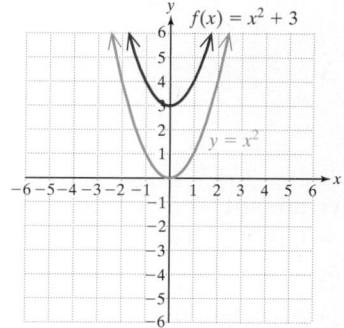

A quadratic function of the form $f(x) = (x - h)^2$ shifts the graph of $f(x) = x^2$ right h units if $h > 0$ and left $|h|$ units if $h < 0$.

Example 2

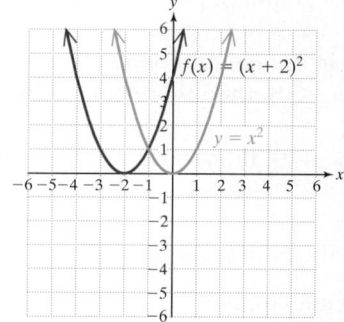

The graph of a quadratic function of the form $f(x) = ax^2$ is a parabola that opens upward when $a > 0$ and opens downward when $a < 0$. If $|a| > 1$, the graph of $f(x) = x^2$ is stretched vertically by a factor of $|a|$. If $0 < |a| < 1$, the graph of $f(x) = x^2$ is shrunk vertically by a factor of $|a|$.

Example 3

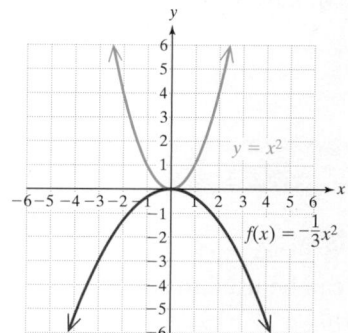

A quadratic function of the form $f(x) = a(x - h)^2 + k$ has vertex (h, k). If $a > 0$, the vertex represents the minimum point. If $a < 0$, the vertex represents the maximum point.

Example 4

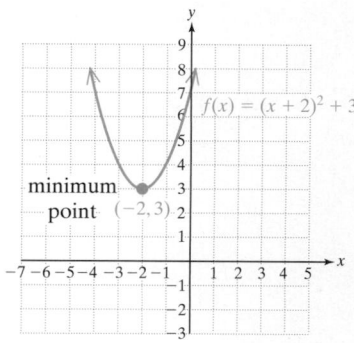

| Section 7.5 | Vertex of a Parabola: Applications and Modeling |

Key Concepts

Completing the square is a technique used to write a quadratic function $f(x) = ax^2 + bx + c$ $(a \neq 0)$ in the form $f(x) = a(x - h)^2 + k$ for the purpose of identifying the vertex (h, k).

The **vertex formula** finds the vertex of a quadratic function $f(x) = ax^2 + bx + c$ $(a \neq 0)$.

The vertex is

$$\left(\frac{-b}{2a}, \frac{4ac - b^2}{4a}\right) \quad \text{or} \quad \left(\frac{-b}{2a}, f\left(\frac{-b}{2a}\right)\right)$$

Examples

Example 1

$$f(x) = 3x^2 + 6x + 11$$
$$= 3(x^2 + 2x \qquad) + 11$$
$$= 3(x^2 + 2x + 1 - 1) + 11$$
$$= 3(x^2 + 2x - 1) - 3 + 11$$
$$= 3(x + 1)^2 + 8$$
$$= 3[x - (-1)]^2 + 8$$

The vertex is $(-1, 8)$. Because $a = 3 > 0$, the parabola opens upward and the vertex $(-1, 8)$ is a minimum point.

Example 2

$$f(x) = 3x^2 + 6x + 11$$
$$a = 3 \qquad b = 6 \qquad c = 11$$
$$x = \frac{-6}{2(3)} = -1$$
$$f(-1) = 3(-1)^2 + 6(-1) + 11 = 8$$

The vertex is $(-1, 8)$.

Section 7.6   Polynomial and Rational Inequalities

Key Concepts

The Test Point Method to Solve Polynomial and

Rational Inequalities

1. Find the boundary points of the inequality. (Boundary points are the real solutions to the related equation and points where the inequality is undefined.)
2. Plot the boundary points on the number line. This divides the number line into intervals.
3. Select a test point from each interval and substitute it into the original inequality.
   - If a test point makes the original inequality true, then that interval is part of the solution set.
4. Test the boundary points in the original inequality.
   - If the original inequality is strict ($<$ or $>$), do not include the boundary points in the solution set.
   - If the original inequality is defined using $\leq$ or $\geq$, then include the boundary points that are well defined within the inequality.

Note: Any boundary point that makes an expression within the inequality undefined must *always* be excluded from the solution set.

Example

Example 1

$$\frac{28}{2x - 3} \leq 4$$ The inequality is undefined for $x = \frac{3}{2}$. Find other possible boundary points by solving the related equation.

$$\frac{28}{2x - 3} = 4$$ Related equation

$$(2x - 3) \cdot \frac{28}{2x - 3} = (2x - 3) \cdot 4$$

$$28 = 8x - 12$$

$$40 = 8x$$

$$x = 5$$

The boundaries are $x = \dfrac{3}{2}$ and $x = 5$.

Interval I: Test $x = 1$: $\dfrac{28}{2(1) - 3} \overset{?}{\leq} 4$ True

Interval II: Test $x = 2$: $\dfrac{28}{2(2) - 3} \overset{?}{\leq} 4$ False

Interval III: Test $x = 6$: $\dfrac{28}{2(6) - 3} \overset{?}{\leq} 4$ True

The boundary point $x = \frac{3}{2}$ is not included because $\dfrac{28}{2x - 3}$ is undefined there. The boundary $x = 5$ does check in the original inequality.

The solution is $\left(-\infty, \frac{3}{2}\right) \cup [5, \infty)$.

Chapter 7   Review Exercises

Section 7.1

For Exercises 1–8, solve the equations by using the square root property.

1. $x^2 = 5$

2. $2y^2 = -8$

3. $a^2 = 81$

4. $3b^2 = -19$

5. $(x - 2)^2 = 72$

6. $(2x - 5)^2 = -9$

7. $(3y - 1)^2 = 3$

8. $3(m - 4)^2 = 15$

 9. The length of each side of an equilateral triangle is 10 in. Find the height of the triangle. Round the answer to the nearest tenth of an inch.

10 in.   10 in.

h

← 10 in. →

10. Use the square root property to find the length of the sides of a square whose area is 81 in².

 11. Use the square root property to find the length of the sides of a square whose area is 150 in². Round the answer to the nearest tenth of an inch.

For Exercises 12–15, find the value of n so that the expression is a perfect square trinomial. Then factor the trinomial.

12. $x^2 + 16x + n$

13. $x^2 - 9x + n$

14. $y^2 + \dfrac{1}{2}y + n$

15. $z^2 - \dfrac{2}{5}z + n$

For Exercises 16–21, solve the equation by completing the square and applying the square root property.

16. $w^2 + 4w + 13 = 0$

17. $4y^2 - 8y - 20 = 0$

18. $2x^2 = 12x + 6$

19. $-t^2 + 8t - 25 = 0$

20. $3x^2 + 2x = 1$

21. $b^2 + \dfrac{7}{2}b = 2$

22. Solve for r. $V = \pi r^2 h$ $(r > 0)$

23. Solve for s. $A = 6s^2$ $(s > 0)$

Section 7.2

24. Explain how the discriminant can determine the type and number of solutions to a quadratic equation with rational coefficients.

For Exercises 25–30, determine the type (rational, irrational, or imaginary) and number of solutions for the equations by using the discriminant.

25. $x^2 - 5x = -6$

26. $2y^2 = -3y$

27. $z^2 + 23 = 17z$

28. $a^2 + a + 1 = 0$

29. $10b - 1 = -25b^2$

30. $3x^2 + 15 = 0$

For Exercises 31–38, solve the equations by using the quadratic formula.

31. $y^2 - 4y + 1 = 0$

32. $m^2 - 5m + 25 = 0$

33. $6a(a - 1) = 10 + a$

34. $3x(x - 3) = x - 8$

35. $b^2 - \dfrac{4}{25} = \dfrac{3}{5}b$

36. $k^2 + 0.4k = 0.05$

37. $-32 + 4x - x^2 = 0$

38. $8y - y^2 = 10$

For Exercises 39–42, solve using any method.

39. $3x^2 - 4x = 6$

40. $\dfrac{w}{8} - \dfrac{2}{w} = \dfrac{3}{4}$

41. $y^2 + 14y = -46$

42. $(a + 1)^2 = 11$

 43. The landing distance that a certain plane will travel on a runway is determined by the initial landing speed at the instant the plane touches down. The function D relates landing distance in feet to initial landing speed s:

$$D(s) = \frac{1}{10}s^2 - 3s + 22 \text{ for } s \geq 50$$

where s is in feet per second.

a. Find the landing distance for a plane traveling 150 ft/sec at touchdown.

b. If the landing speed is too fast, the pilot may run out of runway. If the speed is too slow, the plane may stall. Find the maximum initial landing speed of a plane for a runway that is 1000 ft long. Round to one decimal place.

44. The recent population of Kenya (in thousands) can be approximated by $P(t) = 4.62t^2 + 564.6t + 13,128$, where t is the number of years since 1974.

a. If this trend continues, approximate the number of people in Kenya for the year 2025.

b. In what year after 1974 will the population of Kenya reach 50 million? (*Hint:* 50 million equals 50,000 thousand.)

45. A custom-built kitchen island is in the shape of a rectangle. The length is 1 ft more than twice the width. If the area is 22.32 ft^2, determine the dimensions of the island.

46. Lincoln, Nebraska, Kansas City, Missouri, and Omaha, Nebraska, form the vertices of a right triangle. The distance between Lincoln and Kansas City is 10 mi more than 3 times the distance between Lincoln and Omaha. If the distance from Omaha and Kansas City is 167 mi, find the distance between Lincoln and Omaha. Round to the nearest mile.

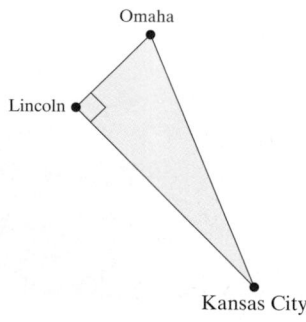

Section 7.3

For Exercises 47–56, solve the equations.

47. $x - 4\sqrt{x} - 21 = 0$

48. $n - 6\sqrt{n} + 8 = 0$

49. $y^4 - 11y^2 + 18 = 0$

50. $2m^4 - m^2 - 3 = 0$

51. $t^{2/5} + t^{1/5} - 6 = 0$

52. $p^{2/5} - 3p^{1/5} + 2 = 0$

53. $\dfrac{2t}{t+1} + \dfrac{-3}{t-2} = 1$

54. $\dfrac{1}{m-2} - \dfrac{m}{m+3} = 2$

55. $(x^2 + 5)^2 + 2(x^2 + 5) - 8 = 0$

56. $(x^2 - 3)^2 - 5(x^2 - 3) + 4 = 0$

Section 7.4

For Exercises 57–64, graph the functions and write the domain and range in interval notation.

57. $g(x) = x^2 - 5$

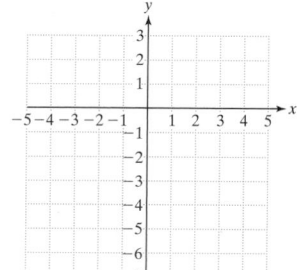

58. $f(x) = x^2 + 3$

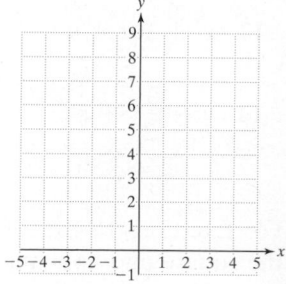

59. $h(x) = (x - 5)^2$

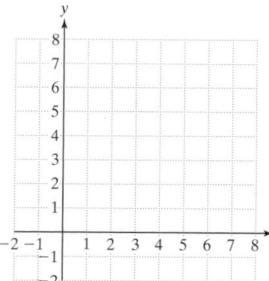

60. $k(x) = (x + 3)^2$

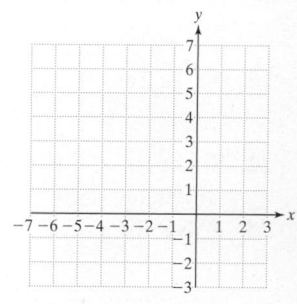

61. $m(x) = -2x^2$

62. $n(x) = -4x^2$

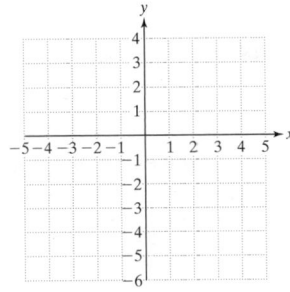

63. $p(x) = -2(x - 5)^2 - 5$

64. $q(x) = -4(x + 3)^2 + 3$

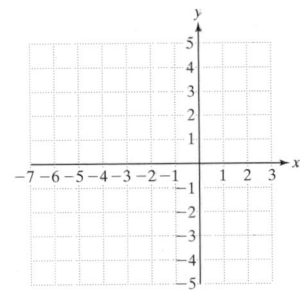

For Exercises 65–66, write the coordinates of the vertex of the parabola and determine if the vertex is a maximum point or a minimum point. Then write the maximum or the minimum value.

65. $t(x) = \dfrac{1}{3}(x - 4)^2 + \dfrac{5}{3}$

66. $s(x) = -\dfrac{5}{7}(x - 1)^2 - \dfrac{1}{7}$

For Exercises 67–68, write the equation of the axis of symmetry of the parabola.

67. $a(x) = -\dfrac{3}{2}\left(x + \dfrac{2}{11}\right)^2 - \dfrac{4}{13}$

68. $w(x) = -\dfrac{4}{3}\left(x - \dfrac{3}{16}\right)^2 + \dfrac{2}{9}$

Section 7.5

For Exercises 69–72, write the function in the form $f(x) = a(x - h)^2 + k$ by completing the square. Then write the coordinates of the vertex.

69. $z(x) = x^2 - 6x + 7$

70. $b(x) = x^2 - 4x - 44$

71. $p(x) = -5x^2 - 10x - 13$

72. $q(x) = -3x^2 - 24x - 54$

For Exercises 73–76, find the coordinates of the vertex of each parabola by using the vertex formula.

73. $f(x) = -2x^2 + 4x - 17$

74. $g(x) = -4x^2 - 8x + 3$

75. $m(x) = 3x^2 - 3x + 11$

76. $n(x) = 3x^2 + 2x - 7$

77. For the quadratic equation $y = \dfrac{3}{4}x^2 - 3x$,

   a. Write the coordinates of the vertex.

   b. Find the x- and y-intercepts.

   c. Use this information to sketch a graph of the parabola.

78. For the quadratic equation $y = -(x + 2)^2 + 4$,

   a. Write the coordinates of the vertex.

   b. Find the x- and y-intercepts.

   c. Use this information to sketch a graph of the parabola.

79. The height $h(t)$ (in feet) of a projectile fired vertically into the air from the ground is given by the equation $h(t) = -16t^2 + 96t$, where t represents the number of seconds that the projectile has been in the air.

   a. How long will it take the projectile to reach its maximum height?

   b. What is the maximum height?

80. The weekly profit, $P(x)$ (in dollars), for a catering service is given by $P(x) = -0.053x^2 + 15.9x - 7.5$. In this context, x is the number of meals prepared.

   a. Find the number of meals that should be prepared to obtain the maximum profit.

   b. What is the maximum profit?

81. Write an equation of a parabola that passes through the points $(-3, -4)$, $(-2, -5)$, and $(1, 4)$

82. Write an equation of a parabola that passes through the points $(4, 18)$, $(-2, 12)$, and $(-1, 8)$

Section 7.6

83. Solve the equation and inequalities. How do your answers to parts (a), (b), and (c) relate to the graph of $g(x) = x^2 - 4$?

a. $x^2 - 4 = 0$

b. $x^2 - 4 < 0$

c. $x^2 - 4 > 0$

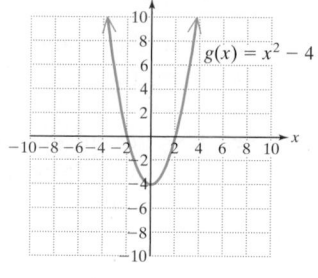

84. a. For what values of x is $\dfrac{4x}{x - 2}$ undefined?

b. Solve the equation. $\dfrac{4x}{x - 2} = 0$

c. Solve the inequality. $\dfrac{4x}{x - 2} \geq 0$

d. Solve the inequality. $\dfrac{4x}{x - 2} \leq 0$

For Exercises 85–96, solve the inequalities. Write the answers in interval notation.

85. $w^2 - 4w - 12 < 0$

86. $t^2 + 6t + 9 \geq 0$

87. $\dfrac{12}{x + 2} \leq 6$

88. $\dfrac{8}{p - 1} \geq -4$

89. $3y(y - 5)(y + 2) > 0$

90. $-3c(c + 2)(2c - 5) < 0$

91. $-x^2 - 4x \leq 1$

92. $y^2 + 4y > 5$

93. $\dfrac{w + 1}{w - 3} > 1$

94. $\dfrac{2a}{a + 3} \leq 2$

95. $t^2 + 10t + 25 \leq 0$

96. $-x^2 - 4x < 4$

Chapter 7   Test

For Exercises 1–3, solve the equation by using the square root property.

1. $(x + 3)^2 = 25$

2. $(p - 2)^2 = 12$

3. $(m + 1)^2 = -1$

4. Find the value of n so that the expression is a perfect square trinomial. Then factor the trinomial $d^2 + 11d + n$.

For Exercises 5–6, solve the equation by completing the square and applying the square root property.

5. $2x^2 + 12x - 36 = 0$

6. $2x^2 = 3x - 7$

For Exercises 7–8,

a. Write the equation in standard form $ax^2 + bx + c = 0$.

b. Identify a, b, and c.

c. Find the discriminant.

d. Determine the number and type (rational, irrational, or imaginary) of solutions.

7. $x^2 - 3x = -12$

8. $y(y - 2) = -1$

For Exercises 9–10, solve the equation by using the quadratic formula.

9. $3x^2 - 4x + 1 = 0$

10. $x(x + 6) = -11 - x$

11. The base of a triangle is 3 ft less than twice the height. The area of the triangle is 14 ft^2. Find the base and the height. Round the answers to the nearest tenth of a foot.

12. A circular garden has an area of approximately 450 ft^2. Find the radius. Round the answer to the nearest tenth of a foot.

For Exercises 13–21, solve the equation.

13. $x - \sqrt{x} - 6 = 0$ **14.** $y^{2/3} + 2y^{1/3} = 8$

15. $(3y - 8)^2 - 13(3y - 8) + 30 = 0$

16. $p^4 - 15p^2 = -54$ **17.** $3 = \dfrac{y}{2} - \dfrac{1}{y + 1}$

18. $2x^2 - 9x = 5$ **19.** $x^2 - 8x + 1 = 0$

20. $(x + 7)^2 = -24$ **21.** $x(x - 12) = -13$

22. Find the vertex of $y = x^2 - 6x - 8$ two ways,

    a. By completing the square.

    b. By using the vertex formula.

For Exercises 23–25, graph the functions. Use the graph to write the domain and range in interval notation.

23. $h(x) = x^2 - 4$

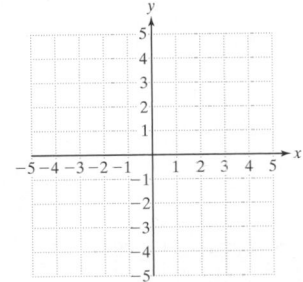

24. $f(x) = -(x - 4)^2$

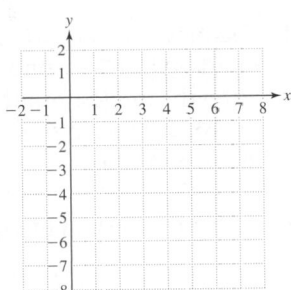

25. $g(x) = \dfrac{1}{2}(x - 2)^2 - 3$

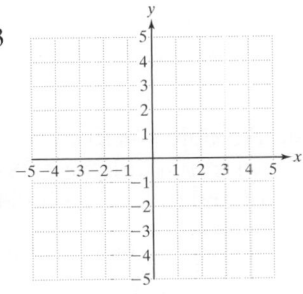

26. A child launches a toy rocket from the ground. The height of the rocket can be determined by its horizontal distance from the launch pad x by

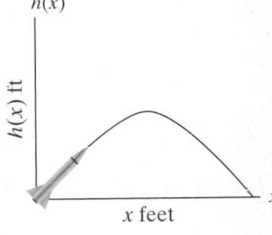

$$h(x) = -\dfrac{x^2}{256} + x$$

where x and $h(x)$ are in feet. How many feet from the launch pad will the rocket hit the ground?

27. The recent population of India (in millions) can be approximated by $P(t) = 0.135t^2 + 12.6t + 600$, where $t = 0$ corresponds to the year 1974.

    a. If this trend continues, approximate the number of people in India in the year 2014.

    b. Approximate the year in which the population of India reached 1 billion (1000 million). (Round to the nearest year.)

28. Explain the relationship between the graphs of $y = x^2$ and $y = x^2 - 2$.

29. Explain the relationship between the graphs of $y = x^2$ and $y = (x + 3)^2$.

30. Explain the relationship between the graphs of $y = 4x^2$ and $y = -4x^2$.

31. Given the function defined by

$$f(x) = -(x - 4)^2 + 2$$

    a. Identify the vertex of the parabola.

    b. Does this parabola open upward or downward?

    c. Does the vertex represent the maximum or minimum point of the function?

    d. What is the maximum or minimum value of the function f?

    e. Write an equation for the axis of symmetry.

32. For the function defined by $g(x) = 2x^2 - 20x + 51$, find the vertex by using two methods.

    a. Complete the square to write $g(x)$ in the form $g(x) = a(x - h)^2 + k$. Identify the vertex.

    b. Use the vertex formula to find the vertex.

33. Given: $f(x) = x^2 + 4x - 12$

    a. Write the equation for the function in the form $f(x) = a(x - h)^2 + k$.

    b. Determine the vertex.

    c. Find the x- and y-intercepts.

    d. Determine the maximum or minimum value.

    e. Write an equation for the axis of symmetry.

34. A farmer has 400 ft of fencing with which to enclose a rectangular field. The field is situated such that one of its sides is adjacent to a river and requires no fencing. The area of the field (in square feet) can be modeled by

$$A(x) = -\frac{x^2}{2} + 200x$$

where x is the length of the side parallel to the river (measured in feet).

a. Determine the value of x that maximizes the area of the field.

b. Determine the maximum area that can be enclosed.

For Exercises 35–40, solve the inequalities.

35. $\dfrac{2x - 1}{x - 6} \le 0$

36. $50 - 2a^2 > 0$

37. $y^3 + 3y^2 - 4y - 12 < 0$

38. $\dfrac{3}{w + 3} > 2$

39. $5x^2 - 2x + 2 < 0$

40. $t^2 + 22t + 121 \le 0$

Chapters 1–7 Cumulative Review Exercises

1. Given: $A = \{2, 4, 6, 8, 10\}$ and $B = \{2, 8, 12, 16\}$

    a. Find $A \cup B$.   **b.** Find $A \cap B$.

2. Perform the indicated operations and simplify.

$$(2x^2 - 5) - (x + 3)(5x - 2)$$

3. Simplify completely. $4^0 - \left(\dfrac{1}{2}\right)^{-3} - 81^{1/2}$

4. Perform the indicated operations. Write the answer in scientific notation:

$$(3 \times 10^{12})(6 \times 10^{-3})$$

5. a. Factor completely. $x^3 + 2x^2 - 9x - 18$

    b. Divide by using long division. Identify the quotient and remainder.

$$(x^3 + 2x^2 - 9x - 18) \div (x - 3)$$

6. Multiply. $(\sqrt{x} - \sqrt{2})(\sqrt{x} + \sqrt{2})$

7. Simplify. $\dfrac{4}{\sqrt{2x}}$

8. Jacques invests a total of $10,000 in two mutual funds. After 1 yr, one fund produced 12% growth, and the other lost 3%. Find the amount invested in each fund if the total investment grew by $900.

9. Solve the system of equations.

$$\frac{1}{9}x - \frac{1}{3}y = -\frac{13}{9}$$

$$x - \frac{1}{2}y = \frac{9}{2}$$

10. An object is fired straight up into the air from an initial height of 384 ft with an initial velocity of 160 ft/sec. The height in feet is given by

$$h(t) = -16t^2 + 160t + 384$$

where t is the time in seconds after launch.

    a. Find the height of the object after 3 sec.

    b. Find the height of the object after 7 sec.

    c. Find the time required for the object to hit the ground.

11. Solve. $(x - 3)^2 + 16 = 0$

12. Solve. $2x^2 + 5x - 1 = 0$

13. Find the value of n so that the expression is a perfect square trinomial. Then factor the trinomial $x^2 + 10x + n$.

14. Factor completely. $2x^3 + 250$

15. Graph the equation. $3x - 5y = 10$

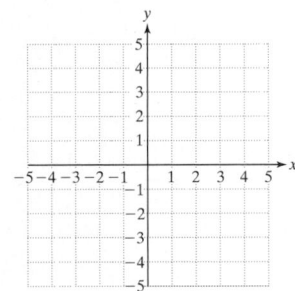

16. a. Find the x-intercepts of the function defined by $g(x) = 2x^2 - 9x + 10$.

   b. What is the y-intercept?

17. Explain why this relation is *not* a function.

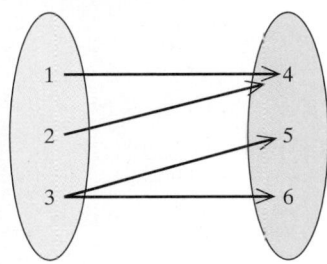

18. Graph the function defined by $f(x) = \dfrac{1}{x}$.

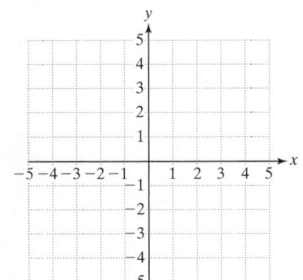

19. The quantity y varies directly as x and inversely as z. If $y = 15$ when $x = 50$ and $z = 10$, find y when $x = 65$ and $z = 5$.

20. The total number of flights (including passenger flights and cargo flights) at a large airport can be approximated by $F(x) = 300{,}000 + 0.008x$, where x is the number of passengers.

   a. Is this function linear, quadratic, constant, or other?

   b. Find the y-intercept and interpret its meaning in the context of this problem.

   c. What is the slope of the function and what does the slope mean in the context of this problem?

21. Let $m(x) = \sqrt{x + 4}$ and $n(x) = x^2 + 2$. Find

   a. The domain of m **b.** The domain of n

22. Consider the function $y = f(x)$ graphed here. Find

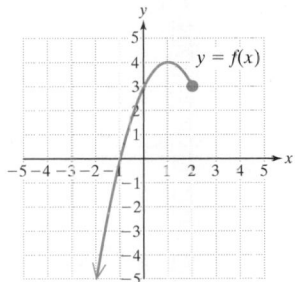

   a. The domain   **b.** The range

   c. $f(1)$ **d.** $f(0)$

   e. For what value(s) of x is $f(x) = 0$?

23. Solve. $\sqrt{8x + 5} = \sqrt{2x} + 2$

24. Solve for f. $\dfrac{1}{p} + \dfrac{1}{q} = \dfrac{1}{f}$

25. Solve. $\dfrac{15}{t^2 - 2t - 8} = \dfrac{1}{t - 4} + \dfrac{2}{t + 2}$

26. Simplify. $\dfrac{y - \dfrac{4}{y - 3}}{y - 4}$

27. Given the function defined by $f(x) = 2(x - 3)^2 + 1$

   a. Write the coordinates of the vertex.

   b. Does the graph of the function open upward or downward?

   c. Write the coordinates of the y-intercept.

   d. Find the x-intercepts, if possible.

   e. Sketch the function.

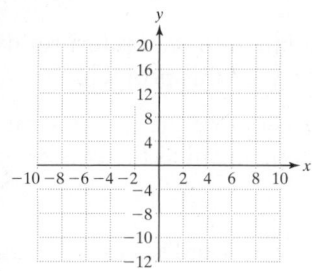

28. Find the vertex of the parabola.

$$f(x) = x^2 - 16x + 2$$

29. a. Solve the inequality. $2x^2 + x - 10 \geq 0$

   b. How does the answer in part (a) relate to the graph of the function $f(x) = 2x^2 + x - 10$?

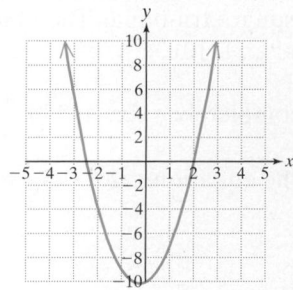

30. Solve the inequality. $\dfrac{8}{x - 3} \leq 4$

Exponential and Logarithmic Functions and Applications

8

CHAPTER OUTLINE

8.1 Algebra of Functions and Composition   664

8.2 Inverse Functions   671

8.3 Exponential Functions   680

8.4 Logarithmic Functions   690

    Problem Recognition Exercises: Identifying Graphs of Functions   703

8.5 Properties of Logarithms   704

8.6 The Irrational Number e and Change of Base   712

    Problem Recognition Exercises: Logarithmic and Exponential Forms   725

8.7 Logarithmic and Exponential Equations and Applications   726

    Group Activity: Creating a Population Model   738

Chapter 8

This chapter is devoted to the study of exponential and logarithmic functions. These functions are used to study many naturally occurring phenomena such as population growth, decay of radioactive matter, and growth of investments.

Are You Prepared?

To prepare yourself for this chapter, practice evaluating the expressions with exponents. Use the clues to fill in the boxes labeled A–H. Then fill in the remaining part of the grid so that every row, every column, and every 2 × 3 box contains the digits 1 through 6.

A. Value of x in the equation $5^x = 25$.

B. Value of x in the equation $3^x = 81$.

C. Value of x in the equation $8^{1/x} = 2$.

D. Real value of x in the equation $x^4 = 16$ (assume that $x > 0$).

E. Absolute value of x in the equation $2^x = \dfrac{1}{16}$.

F. Value of x in the equation $2^x = 64$.

G. Value of x in the equation $5^0 = x$.

H. Value of x in the equation $e^0 = x$.

Concepts

1. Algebra of Functions
2. Composition of Functions
3. Operations on Functions

1. Algebra of Functions

Addition, subtraction, multiplication, and division can be used to create a new function from two or more functions. The domain of the new function will be the intersection of the domains of the original functions.

Sum, Difference, Product, and Quotient of Functions

Given two functions f and g, the functions $f + g, f - g, f \cdot g$, and $\frac{f}{g}$ are defined as

$$(f + g)(x) = f(x) + g(x)$$

$$(f - g)(x) = f(x) - g(x)$$

$$(f \cdot g)(x) = f(x) \cdot g(x)$$

$$\left(\frac{f}{g}\right)(x) = \frac{f(x)}{g(x)} \qquad \text{provided } g(x) \neq 0$$

For example, suppose $f(x) = |x|$ and $g(x) = 3$. Taking the sum of the functions produces a new function denoted by $f + g$. In this case, $(f + g)(x) = |x| + 3$. Graphically, the y values of the function $f + g$ are given by the sum of the corresponding y values of f and g. This is depicted in Figure 8-1. The function $f + g$ appears in red. In particular, notice that $(f + g)(2) = f(2) + g(2) = 2 + 3 = 5$.

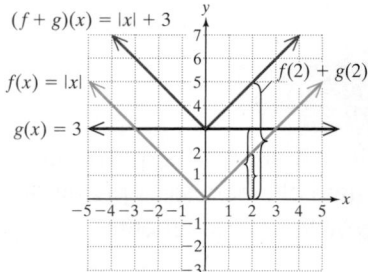

Figure 8-1

Example 1 **Adding, Subtracting, Multiplying, and Dividing Functions**

Given: $g(x) = 4x \qquad h(x) = x^2 - 3x \qquad k(x) = x - 2$

a. Find $(g + h)(x)$.   **b.** Find $(h - g)(x)$.

c. Find $(g \cdot k)(x)$.   **d.** Find $\left(\dfrac{k}{h}\right)(x)$.

Solution:

a. $(g + h)(x) = g(x) + h(x)$

$\qquad\qquad = (4x) + (x^2 - 3x)$

$\qquad\qquad = x^2 + x \qquad\qquad$ Simplify.

b. $(h - g)(x) = h(x) - g(x)$

$\qquad = (x^2 - 3x) - (4x)$

$\qquad = x^2 - 7x \qquad$ Simplify.

c. $(g \cdot k)(x) = g(x) \cdot k(x)$

$\qquad = (4x)(x - 2)$

$\qquad = 4x^2 - 8x \qquad$ Simplify.

d. $\left(\dfrac{k}{h}\right)(x) = \dfrac{k(x)}{h(x)}$

$\qquad = \dfrac{x - 2}{x^2 - 3x} \qquad$ From the denominator we have
$x^2 - 3x \neq 0$ or, equivalently,
$x(x - 3) \neq 0$. Hence, $x \neq 3$ and $x \neq 0$.

Skill Practice Given $f(x) = x - 1$, $g(x) = 5x^2 + x$, and $h(x) = x^2$, find

1. $(f + g)(x)$ **2.** $(g - f)(x)$ **3.** $(g \cdot h)(x)$ **4.** $\left(\dfrac{f}{h}\right)(x)$

2. Composition of Functions

Composition of Functions

The **composition** of f and g, denoted $f \circ g$, is defined by the rule

$$(f \circ g)(x) = f(g(x)) \qquad \text{provided that } g(x) \text{ is in the domain of } f.$$

Note: $f \circ g$ is also read as "f compose g."

For example, given $f(x) = 2x - 3$ and $g(x) = x + 5$, we have

$(f \circ g)(x) = f(g(x))$ ⋯⋯⋯⋯⋯⋯⋯⋯⋯⋯⋯⋯

$\qquad = f(x + 5) \qquad$ Substitute $g(x) = x + 5$ into the function f.

$\qquad = 2(x + 5) - 3$

$\qquad = 2x + 10 - 3$

$\qquad = 2x + 7$

Avoiding Mistakes

Be careful to note that the notation $f(g(x))$ represents function composition not multiplication.

In this composition, the function g is the innermost operation and acts on x first. Then the output value of function g becomes the domain element of the function f, as shown in the figure.

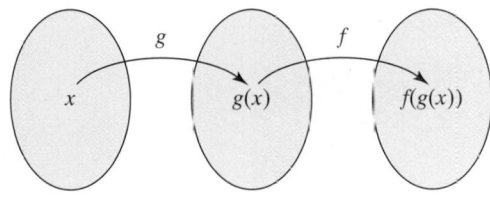

Answers

1. $5x^2 + 2x - 1$ **2.** $5x^2 + 1$

3. $5x^4 + x^3$ **4.** $\dfrac{x - 1}{x^2}$, $x \neq 0$

Example 2 Composing Functions

Given: $f(x) = x - 5$, $g(x) = x^2$, and $n(x) = \sqrt{x + 2}$, find

a. $(f \circ g)(x)$

b. $(g \circ f)(x)$

c. $(n \circ f)(x)$

Solution:

a. $(f \circ g)(x) = f(g(x))$

$= f(x^2)$ Evaluate the function f at x^2.

$= (x^2) - 5$ Replace x by x^2 in the function f.

$= x^2 - 5$

TIP: Examples 2(a) and 2(b) illustrate that the order in which two functions are composed may result in different functions. That is, $f \circ g$ does not necessarily equal $g \circ f$.

b. $(g \circ f)(x) = g(f(x))$

$= g(x - 5)$ Evaluate the function g at $(x - 5)$.

$= (x - 5)^2$ Replace x by $(x - 5)$ in function g.

$= x^2 - 10x + 25$ Simplify.

c. $(n \circ f)(x) = n(f(x))$

$= n(x - 5)$ Evaluate the function n at $x - 5$.

$= \sqrt{(x - 5) + 2}$ Replace x by the quantity $(x - 5)$ in function n.

$= \sqrt{x - 3}$

Skill Practice Given $f(x) = 2x^2$, $g(x) = x + 3$, and $h(x) = \sqrt{x - 1}$, find

5. $(f \circ g)(x)$ **6.** $(g \circ f)(x)$ **7.** $(h \circ g)(x)$

3. Operations on Functions

Example 3 Combining Functions

Given the functions defined by $f(x) = x - 7$ and $h(x) = 2x^3$, find the function values, if possible.

a. $(f \cdot h)(3)$ **b.** $\left(\dfrac{h}{f}\right)(7)$ **c.** $(h \circ f)(2)$

Solution:

a. $(f \cdot h)(3) = f(3) \cdot h(3)$ $(f \cdot h)(3)$ is a product (not a composition).

$= (3 - 7) \cdot 2(3)^3$

$= (-4) \cdot 2(27)$

$= -216$

Answers

5. $2x^2 + 12x + 18$ **6.** $2x^2 + 3$

7. $\sqrt{x + 2}$

b. $\left(\dfrac{h}{f}\right)(7) = \dfrac{h(7)}{f(7)}$

$\qquad = \dfrac{2(7)^3}{7-7}$ (undefined)   The value $x = 7$ makes the denominator zero.

Therefore, $\left(\dfrac{h}{f}\right)(7)$ is undefined.

c. $(h \circ f)(2) = h(f(2))$ Evaluate $f(2)$ first. $f(2) = 2 - 7 = -5$.

$\qquad = h(-5)$ Substitute the result into function h.

$\qquad = 2(-5)^3$

$\qquad = 2(-125)$

$\qquad = -250$

Skill Practice Given $h(x) = x + 4$ and $k(x) = x^2 - 3$, find

8. $(h \cdot k)(-2)$ **9.** $\left(\dfrac{k}{h}\right)(-4)$ **10.** $(k \circ h)(1)$

Example 4   **Finding Function Values from a Graph**

For the functions f and g pictured, find the function values if possible.

a. $g(2)$

b. $(f - g)(-3)$

c. $\left(\dfrac{g}{f}\right)(5)$

d. $(f \circ g)(-2)$

Solution:

a. $g(2) = -1$ The value $g(2)$ represents the y-value of $y = g(x)$ (the red graph) when $x = 2$. Because the point $(2, -1)$ lies on the graph, $g(2) = -1$.

b. $(f - g)(-3) = f(-3) - g(-3)$ Evaluate the difference of $f(-3)$ and $g(-3)$.

$\qquad = -2 - (3)$ Estimate function values from the graph.

$\qquad = -5$

c. $\left(\dfrac{g}{f}\right)(5) = \dfrac{g(5)}{f(5)}$ Evaluate the quotient of $g(5)$ and $f(5)$.

$\qquad = \dfrac{1}{0}$ (undefined)

The function $\dfrac{g}{f}$ is undefined at $x = 5$ because the denominator is zero.

TIP: For your reference, determining function values from a graph was first covered in Section 2.6.

Answers
8. 2   **9.** undefined   **10.** 22

d. $(f \circ g)(-2) = f(g(-2))$ — From the red graph, find $g(-2)$. We see that $g(-2) = 3$.

$= f(3)$ — From the blue graph, find the value of f at $x = 3$.

$= 1$

Skill Practice Find the values from the graph.

11. $g(3)$ **12.** $(f + g)(-4)$

13. $\left(\dfrac{f}{g}\right)(2)$ **14.** $(g \circ f)(-2)$

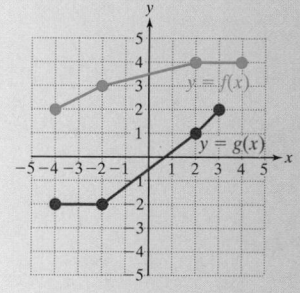

Answers

11. 2   **12.** 0   **13.** 4   **14.** 2

Section 8.1   Practice Exercises

Vocabulary and Key Concepts

1. a. Given the functions f and g, the function $(f + g)(x) = $ _____ + _____.

   b. Given the functions f and g, the function $\left(\dfrac{f}{g}\right)(x) = \dfrac{f(x)}{\square}$.

   c. The function f composed with g is defined by the rule $(f \circ g)(x) = $ _____.

Concept 1: Algebra of Functions

2. Given $f(x) = x^2$ and $g(x) = 2x - 3$, find

   a. $f(-2)$ **b.** $g(-2)$ **c.** $(f + g)(-2)$

For Exercises 3–14, refer to the functions defined below.

$$f(x) = x + 4 \qquad g(x) = 2x^2 + 4x \qquad h(x) = x^2 + 1 \qquad k(x) = \frac{1}{x}$$

Find the indicated functions. **(See Example 1.)**

3. $(f + g)(x)$ **4.** $(f - g)(x)$ **5.** $(g - f)(x)$

6. $(f + h)(x)$ **7.** $(f \cdot h)(x)$ **8.** $(h \cdot k)(x)$

9. $(g \cdot f)(x)$ **10.** $(f \cdot k)(x)$ **11.** $\left(\dfrac{h}{f}\right)(x)$

12. $\left(\dfrac{g}{f}\right)(x)$ **13.** $\left(\dfrac{f}{g}\right)(x)$ **14.** $\left(\dfrac{f}{h}\right)(x)$

Concept 2: Composition of Functions

For Exercises 15–22, find the indicated functions. Use f, g, h, and k as defined in Exercises 3–14. **(See Example 2.)**

15. $(f \circ g)(x)$ **16.** $(f \circ k)(x)$ **17.** $(g \circ f)(x)$

18. $(k \circ f)(x)$ **19.** $(k \circ h)(x)$ **20.** $(h \circ k)(x)$

21. $(k \circ g)(x)$ **22.** $(g \circ k)(x)$

23. Based on your answers to Exercises 15 and 17, is it true in general that $(f \circ g)(x) = (g \circ f)(x)$?

24. Based on your answers to Exercises 16 and 18, is it true in general that $(f \circ k)(x) = (k \circ f)(x)$?

For Exercises 25–28, find $(f \circ g)(x)$ and $(g \circ f)(x)$.

25. $f(x) = x^2 - 3x + 1$, $g(x) = 5x$ **26.** $f(x) = 3x^2 + 8$, $g(x) = 2x - 4$ **27.** $f(x) = |x|$, $g(x) = x^3 - 1$

28. $f(x) = \dfrac{1}{x + 2}$, $g(x) = |x + 2|$ **29.** For $h(x) = 5x - 4$, find $(h \circ h)(x)$.   **30.** For $k(x) = -x^2 + 1$, find $(k \circ k)(x)$.

Concept 3: Operations on Functions

For Exercises 31–46, refer to the functions defined below.

$$m(x) = x^3 \qquad n(x) = x - 3 \qquad r(x) = \sqrt{x + 4} \qquad p(x) = \dfrac{1}{x + 2}$$

Find each function value if possible. **(See Example 3.)**

31. $(m \cdot r)(0)$ **32.** $(n \cdot p)(0)$ **33.** $(m + r)(-4)$ **34.** $(n - m)(4)$

 35. $(r \circ n)(3)$ **36.** $(n \circ r)(5)$ **37.** $(p \circ m)(-1)$ **38.** $(m \circ n)(5)$

39. $(m \circ p)(2)$ **40.** $(r \circ m)(2)$ **41.** $(r + p)(-3)$ **42.** $(n + p)(-2)$

43. $(m \circ p)(-2)$ **44.** $(r \circ m)(-2)$ **45.** $\left(\dfrac{r}{n}\right)(12)$ **46.** $\left(\dfrac{n}{m}\right)(2)$

For Exercises 47–64, approximate each function value from the graph, if possible. **(See Example 4.)**

47. $f(-4)$ **48.** $f(1)$ **49.** $g(-2)$

50. $g(3)$ **51.** $(f + g)(2)$ **52.** $(g - f)(3)$

53. $(f \cdot g)(-1)$ **54.** $(g \cdot f)(-4)$ **55.** $\left(\dfrac{g}{f}\right)(0)$

56. $\left(\dfrac{f}{g}\right)(-2)$ **57.** $\left(\dfrac{f}{g}\right)(0)$ **58.** $\left(\dfrac{g}{f}\right)(-2)$

59. $(g \circ f)(-1)$ **60.** $(f \circ g)(0)$ **61.** $(f \circ g)(-4)$

62. $(g \circ f)(-4)$ **63.** $(g \circ g)(2)$ **64.** $(f \circ f)(-2)$

For Exercises 65–80, approximate each function value from the graph, if possible. **(See Example 4.)**

65. $a(-3)$ **66.** $a(1)$ **67.** $b(-1)$

68. $b(3)$ **69.** $(a - b)(-1)$ **70.** $(a + b)(0)$

71. $(b \cdot a)(1)$ **72.** $(a \cdot b)(2)$ **73.** $(b \circ a)(0)$

74. $(a \circ b)(-2)$ **75.** $(a \circ b)(-4)$ **76.** $(b \circ a)(-3)$

77. $\left(\dfrac{b}{a}\right)(3)$ **78.** $\left(\dfrac{a}{b}\right)(4)$ **79.** $(a \circ a)(-2)$ **80.** $(b \circ b)(1)$

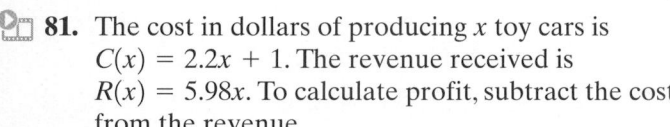

81. The cost in dollars of producing x toy cars is $C(x) = 2.2x + 1$. The revenue received is $R(x) = 5.98x$. To calculate profit, subtract the cost from the revenue.

   a. Write and simplify a function P that represents profit in terms of x.

   b. Find the profit of producing 50 toy cars.

82. The cost in dollars of producing x lawn chairs is $C(x) = 2.5x + 10.1$. The revenue for selling x lawn chairs is $R(x) = 6.99x$. To calculate profit, subtract the cost from the revenue.

   a. Write and simplify a function P that represents profit in terms of x.

   b. Find the profit in producing 100 lawn chairs.

83. The functions defined by $D(t) = 0.925t + 26.958$ and $R(t) = 0.725t + 20.558$ approximate the amount of child support (in billions of dollars) that was due and the amount of child support actually received in the United States since 2000. In each case, $t = 0$ corresponds to the year 2000.

   a. Find the function F defined by $F(t) = D(t) - R(t)$. What does F represent in the context of this problem?

   b. Find $F(4)$. What does this function value represent in the context of this problem?

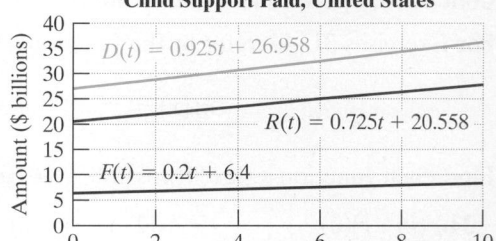

Difference Between Child Support Due and Child Support Paid, United States

Source: U.S. Bureau of the Census

84. The rural and urban populations in the South (in the United States) between the years 1900 and 1970 can be modeled by the following:

$$r(t) = -3.497t^2 + 266.2t + 20{,}220$$

$$u(t) = 0.0566t^3 + 0.952t^2 + 177.8t + 4593$$

The variable t represents the number of years since 1900, $r(t)$ represents the rural population in thousands, and $u(t)$ represents the urban population in thousands.

   a. Find the function T defined by $T(t) = r(t) + u(t)$. What does the function T represent in the context of this problem?

   b. Use the function T to approximate the total population in the South for the year 1940.

Rural and Urban Populations in the South, United States, 1900–1970

Source: Historical Abstract of the United States

85. Joe rides a bicycle and his wheels revolve at 80 revolutions per minute (rpm). Therefore, the total number of revolutions, r, is given by $r(t) = 80t$, where t is the time in minutes. For each revolution of the wheels of the bike, he travels approximately 7 ft. Therefore, the total distance he travels, D, depends on the total number of revolutions, r, according to the function $D(r) = 7r$.

   a. Find $(D \circ r)(t)$ and interpret its meaning in the context of this problem.

   b. Find Joe's total distance in feet after 10 min.

86. The area of a square is given by the function $a(x) = x^2$, where x is the length of the sides of the square. If carpeting costs $9.95 per square yard, then the cost, C, to carpet a square room is given by $C(a) = 9.95a$, where a is the area of the room in square yards.

   a. Find $(C \circ a)(x)$ and interpret its meaning in the context of this problem.

   b. Find the cost to carpet a square room if its floor dimensions are 15 yd by 15 yd.

Inverse Functions

1. Introduction to Inverse Functions

In Section 2.6, we defined a function as a set of ordered pairs (x, y) such that for every element x in the domain, there corresponds exactly one element y in the range. For example, the function f relates the weight of a package of deli meat, x, to its cost, y.

$$f = \{(1, 4), (2.5, 10), (4, 16)\}$$

That is, 1 lb of meat sells for $4, 2.5 lb sells for $10, and 4 lb sells for $16. Now suppose we create a new function in which the values of x and y are interchanged. The new function, called the **inverse of** f, denoted f^{-1}, relates the price of meat, x, to its weight, y.

$$f^{-1} = \{(4, 1), (10, 2.5), (16, 4)\}$$

Notice that interchanging the x- and y-values has the following outcome. The domain of f is the same as the range of f^{-1}, and the range of f is the domain of f^{-1}.

2. Definition of a One-to-One Function

A necessary condition for a function f to have an inverse function is that no two ordered pairs in f have different x-coordinates and the same y-coordinate. A function that satisfies this condition is called a **one-to-one function**. The function relating the weight of a package of meat to its price is a one-to-one function. However, consider the function g defined by

$g = \{(1, 4), (2, 3), (-2, 4)\}$ — same y, different x

Not a one-to-one function because two different x values map to the same y value.

This function is not one-to-one because the range element 4 has two different x-coordinates, 1 and -2. Interchanging the x- and y-values produces a relation that violates the definition of a function.

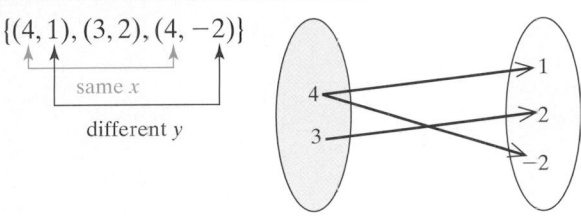

$\{(4, 1), (3, 2), (4, -2)\}$ — same x, different y

This relation is not a function because for $x = 4$, there are two different y-values.

Concepts

1. Introduction to Inverse Functions
2. Definition of a One-to-One Function
3. Definition of the Inverse of a Function
4. Finding an Equation of the Inverse of a Function

Avoiding Mistakes

f^{-1} denotes the inverse of a function. The -1 does not represent an exponent.

In Section 2.6, you learned the vertical line test to determine visually if a graph represents a function. Similarly, we use a **horizontal line test** to determine whether a function is one-to-one.

Using the Horizontal Line Test

Consider a function defined by a set of points (x, y) in a rectangular coordinate system. Then y is a *one-to-one* function of x if no horizontal line intersects the graph in more than one point.

To understand the horizontal line test, consider the functions f and g.

$$f = \{(1, 4), (2.5, 10), (4, 16)\} \qquad g = \{(1, 4), (2, 3), (-2, 4)\}$$

 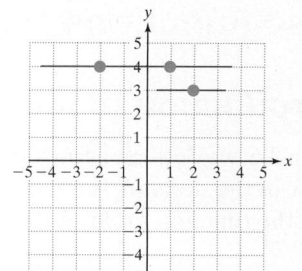

This function is one-to-one. No horizontal line intersects more than once.

This function is *not* one-to-one. A horizontal line intersects more than once.

Example 1 Identifying a One-to-One Function

For each function, determine if the function is one-to-one.

a.

b.

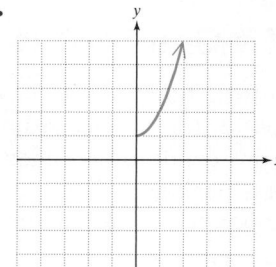

Solution:

a. Function is not one-to-one. A horizontal line intersects in more than one point.

b. Function is one-to-one. No horizontal line intersects more than once.

 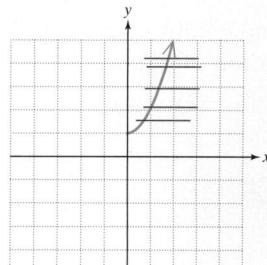

Skill Practice For each function determine if the function is one-to-one.

1.

2.

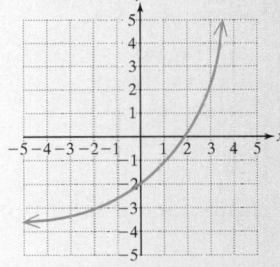

3. Definition of the Inverse of a Function

All one-to-one functions have an inverse function.

> ## Definition of an Inverse Function
> If f is a one-to-one function represented by ordered pairs of the form (x, y), then the inverse function, denoted by f^{-1}, is the set of ordered pairs given by (y, x).

Because the values of x and y are interchanged between a function, f, and its inverse, f^{-1}, we have the following important relationship (Figure 8-2).

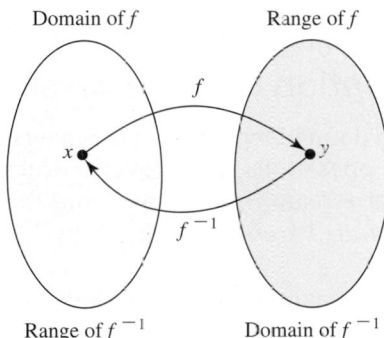

Figure 8-2

From Figure 8-2, we see that the operations performed by f are reversed by f^{-1}. This observation leads to the following important property.

> ## Inverse Function Property
> If f is a one-to-one function, then g is the inverse of f if and only if
> $$(f \circ g)(x) = x \quad \text{for all } x \text{ in the domain of } g$$
> and
> $$(g \circ f)(x) = x \quad \text{for all } x \text{ in the domain of } f.$$

Answers
1. Not one-to-one
2. One-to-one

Example 2 Composing a Function with Its Inverse

Show that the functions are inverses.

$$f(x) = 5x + 4 \qquad g(x) = \frac{x-4}{5}$$

Solution:

To show that the functions f and g are inverses, confirm that $(f \circ g)(x) = x$ and $(g \circ f)(x) = x$.

$$(f \circ g)(x) = f(g(x)) \qquad\qquad (g \circ f)(x) = g(f(x))$$

$$= f\left(\frac{x-4}{5}\right) \qquad\qquad = g(5x+4)$$

$$= 5\left(\frac{x-4}{5}\right) + 4 \qquad\qquad = \frac{(5x+4)-4}{5}$$

$$= x - 4 + 4 \qquad\qquad = \frac{5x}{5}$$

$$= x \checkmark \qquad\qquad = x \checkmark$$

Skill Practice Show that the functions are inverses.

3. $f(x) = 3x - 2$ and $g(x) = \frac{x+2}{3}$

4. Finding an Equation of the Inverse of a Function

For a one-to-one function defined by $y = f(x)$, the inverse is a function $y = f^{-1}(x)$ that performs the inverse operations in the reverse order. For example, the function defined by $f(x) = 2x + 1$ multiplies x by 2 and then adds 1. Therefore, the inverse function must *subtract* 1 from x and *divide* by 2. We have

$$f^{-1}(x) = \frac{x-1}{2} \qquad \text{The expression } f^{-1}(x) \text{ is read as "} f \text{ inverse of } x\text{."}$$

To facilitate the process of finding an equation of the inverse of a one-to-one function, we offer the following steps.

Finding an Equation of an Inverse of a Function

For a one-to-one function defined by $y = f(x)$, the equation of the inverse can be found as follows:

Step 1 Replace $f(x)$ by y.
Step 2 Interchange x and y.
Step 3 Solve for y.
Step 4 Replace y by $f^{-1}(x)$.

Answer

3. $(f \circ g)(x) = f(g(x))$
$\quad = 3\left(\frac{x+2}{3}\right) - 2 = x$
$(g \circ f)(x) = g(f(x))$
$\quad = \frac{3x-2+2}{3} = x$

Example 3 Finding an Equation of the Inverse of a Function

Find the inverse. $f(x) = 2x + 1$

Solution:

Foremost, we know the graph of f is a nonvertical line. Therefore, $f(x) = 2x + 1$ defines a one-to-one function. To find the inverse we have

$y = 2x + 1$ **Step 1:** Replace $f(x)$ by y.

$x = 2y + 1$ **Step 2:** Interchange x and y.

$x - 1 = 2y$ **Step 3:** Solve for y. Subtract 1 from both sides.

$\dfrac{x - 1}{2} = y$ Divide both sides by 2.

$f^{-1}(x) = \dfrac{x - 1}{2}$ **Step 4:** Replace y by $f^{-1}(x)$.

Skill Practice Find the inverse.

4. $f(x) = 4x + 6$

In Example 3, we can verify that $f(x) = 2x + 1$ and $f^{-1}(x) = \dfrac{x - 1}{2}$ are inverses by using function composition.

$$(f \circ f^{-1})(x) = f\big(f^{-1}(x)\big) = 2\left(\dfrac{x - 1}{2}\right) + 1 = x \quad \checkmark \quad \text{and}$$

$$(f^{-1} \circ f)(x) = f^{-1}(f(x)) = \dfrac{(2x + 1) - 1}{2} = x \quad \checkmark$$

The key step in determining the equation of the inverse of a function is to interchange x and y. By so doing, a point (a, b) on f corresponds to a point (b, a) on f^{-1}. For this reason, the graphs of f and f^{-1} are symmetric with respect to the line $y = x$ (Figure 8-3). Notice that the point $(-3, -5)$ of the function f corresponds to the point $(-5, -3)$ of f^{-1}. Likewise, $(1, 3)$ of f corresponds to $(3, 1)$ of f^{-1}.

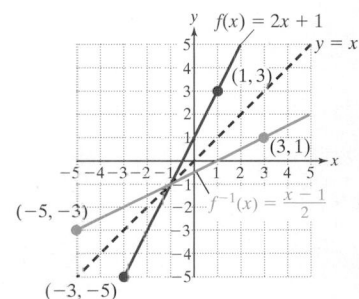

Figure 8-3

Example 4 Finding an Equation of the Inverse of a Function

Find the inverse of the one-to-one function. $g(x) = \sqrt[3]{5x} - 4$

Solution:

$y = \sqrt[3]{5x} - 4$ **Step 1:** Replace $g(x)$ by y.

$x = \sqrt[3]{5y} - 4$ **Step 2:** Interchange x and y.

$x + 4 = \sqrt[3]{5y}$ **Step 3:** Solve for y. Add 4 to both sides.

$(x + 4)^3 = (\sqrt[3]{5y})^3$ To eliminate the cube root, cube both sides.

$(x + 4)^3 = 5y$ Simplify the right side.

$\dfrac{(x + 4)^3}{5} = y$ Divide both sides by 5.

$g^{-1}(x) = \dfrac{(x + 4)^3}{5}$ **Step 4:** Replace y by $g^{-1}(x)$.

Answer

4. $f^{-1}(x) = \dfrac{x - 6}{4}$

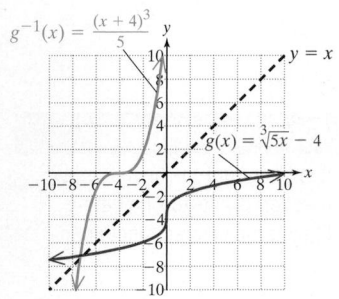

$g^{-1}(x) = \dfrac{(x+4)^3}{5}$

$g(x) = \sqrt[3]{5x} - 4$

Figure 8-4

5. $h(x) = \sqrt[3]{2x - 1}$

The graphs of g and g^{-1} from Example 4 are shown in Figure 8-4. Once again we see that the graphs of a function and its inverse are symmetric with respect to the line $y = x$.

For a function that is not one-to-one, sometimes we can restrict its domain to create a new function that is one-to-one. This is demonstrated in Example 5.

Example 5 **Finding the Equation of an Inverse of a Function with a Restricted Domain**

Given the function defined by $m(x) = x^2 + 4$ for $x \geq 0$, find an equation defining m^{-1}.

Solution:

From Section 7.4, we know that $y = x^2 + 4$ is a parabola with vertex at $(0, 4)$ (Figure 8-5). The graph represents a function that is not one-to-one. However, with the restriction on the domain $x \geq 0$, the graph of $m(x) = x^2 + 4$, $x \geq 0$, consists of only the "right" branch of the parabola (Figure 8-6). This *is* a one-to-one function.

Figure 8-5

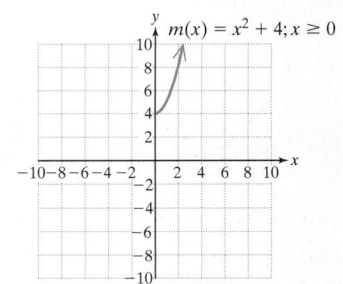

Figure 8-6

To find the inverse, we have

$y = x^2 + 4 \qquad x \geq 0$ **Step 1:** Replace $m(x)$ by y.

$x = y^2 + 4 \qquad y \geq 0$ **Step 2:** Interchange x and y. Notice that the restriction $x \geq 0$ becomes $y \geq 0$.

$x - 4 = y^2 \qquad y \geq 0$ **Step 3:** Solve for y. Subtract 4 from both sides.

$\sqrt{x - 4} = y \qquad y \geq 0$ Apply the square root property. Notice that we obtain the *positive* square root of $x - 4$ because of the restriction $y \geq 0$.

$m^{-1}(x) = \sqrt{x - 4}$ **Step 4:** Replace y by $m^{-1}(x)$. Notice that the domain of m^{-1} has the same values as the range of m.

$y \quad m(x) = x^2 + 4; x \geq 0$

$y = x$

$m^{-1}(x) = \sqrt{x - 4}$

Figure 8-7

Figure 8-7 shows the graphs of m and m^{-1}.

Answers

5. $h^{-1}(x) = \dfrac{x^3 + 1}{2}$

6. $g^{-1}(x) = \sqrt{x + 2}$

6. $g(x) = x^2 - 2 \qquad x \geq 0$

Section 8.2   Practice Exercises

Vocabulary and Key Concepts

1. **a.** Given the function $f = \{(1,2), (2,3), (3,4)\}$, write the set of ordered pairs representing f^{-1}.

   b. A necessary condition for a function f to be a _____-_____-_____ function is that no two ordered pairs in f have different x-coordinates and the same _____-coordinate.

   c. The function $f = \{(1,5), (-2,3), (-4,2), (2,5)\}$ (is/is not) a one-to-one function.

   d. A function defined by $y = f(x)$ (is/is not) a one-to-one function if no horizontal line intersects the graph of f in more than one point.

   e. The graph of a function and its inverse are symmetric with respect to the line _____.

   f. Let f be a one-to-one function and let g be the inverse of f. Then $(f \circ g)(x) =$ _____ and $(g \circ f)(x) =$ _____.

   g. The notation _____ is often used to represent the inverse of a function f and not the reciprocal of f.

   h. If (a, b) is a point on the graph of a one-to-one function f, then the corresponding ordered pair _____ is a point on the graph of f^{-1}.

Review Exercises

2. Write the domain and range of the relation $\{(3,4), (5, -2), (6,1), (3,0)\}$.

For Exercises 3–8, determine if the relation is a function by using the vertical line test. **(See Section 2.6.)**

3.

4.

5.

6.

7.

8.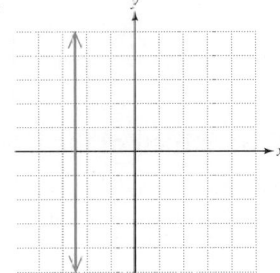

Concept 1: Introduction to Inverse Functions

For Exercises 9–12, write the inverse function for each function.

9. $g = \{(3,5), (8,1), (-3,9), (0,2)\}$

10. $f = \{(-6,2), (-9,0), (-2,-1), (3,4)\}$

11. $r = \{(a,3), (b,6), (c,9)\}$

12. $s = \{(-1,x), (-2,y), (-3,z)\}$

Concept 2: Definition of a One-to-One Function

13. The table relates a state, x, to the number of representatives in the House of Representatives, y, for a recent year. Does this relation define a one-to-one function? If so, write a function defining the inverse as a set of ordered pairs.

State x	Number of Representatives y
Colorado	7
California	53
Texas	32
Louisiana	7
Pennsylvania	19

14. The table relates a city, x, to its average January temperature, y. Does this relation define a one-to-one function? If so, write a function defining the inverse as a set of ordered pairs.

City x	Temperature y (°C)
Gainesville, Florida	13.6
Keene, New Hampshire	−6.0
Wooster, Ohio	−4.0
Rock Springs, Wyoming	−6.0
Lafayette, Louisiana	10.9

For Exercises 15–20, determine if the function is one-to-one by using the horizontal line test. **(See Example 1.)**

15.

16.

17.

18.

19.

20.
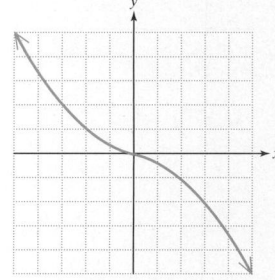

Concept 3: Definition of the Inverse of a Function

For Exercises 21–26, verify that f and g are inverse functions by showing that

    a. $(f \circ g)(x) = x$ **b.** $(g \circ f)(x) = x$ **(See Example 2.)**

21. $f(x) = 6x + 1$ and $g(x) = \dfrac{x - 1}{6}$

22. $f(x) = 5x - 2$ and $g(x) = \dfrac{x + 2}{5}$

23. $f(x) = \dfrac{\sqrt[3]{x}}{2}$ and $g(x) = 8x^3$

24. $f(x) = \dfrac{\sqrt[3]{x}}{3}$ and $g(x) = 27x^3$

25. $f(x) = x^2 + 1, x \geq 0$, and $g(x) = \sqrt{x - 1}, x \geq 1$

26. $f(x) = x^2 - 3, x \geq 0$, and $g(x) = \sqrt{x + 3}, x \geq -3$

Concept 4: Finding an Equation of the Inverse of a Function

For Exercises 27–42, write an equation of the inverse for each one-to-one function as defined. **(See Examples 3–5.)**

27. $h(x) = x + 4$

28. $k(x) = x - 3$

29. $m(x) = \frac{1}{3}x - 2$

30. $n(x) = 4x + 2$

31. $p(x) = -x + 10$

32. $q(x) = -x - \frac{2}{3}$

33. $n(x) = \frac{3x + 2}{5}$

34. $p(x) = \frac{2x - 7}{4}$

35. $h(x) = \frac{4x - 1}{3}$

36. $f(x) = \frac{6x + 3}{2}$

37. $f(x) = x^3 + 1$

38. $g(x) = \sqrt[3]{x} + 2$

39. $g(x) = \sqrt[3]{2x - 1}$

40. $f(x) = x^3 - 4$

41. $g(x) = x^2 + 9 \quad x \geq 0$

42. $m(x) = x^2 - 1 \quad x \geq 0$

43. The function defined by $f(x) = 0.3048x$ converts a length of x feet into $f(x)$ meters.

    a. Find the equivalent length in meters for a 4-ft board and a 50-ft wire.

    b. Find an equation defining $y = f^{-1}(x)$.

    c. Use the inverse function from part (b) to find the equivalent length in feet for a 1500-m race in track and field. Round to the nearest tenth of a foot.

44. The function defined by $s(x) = 1.47x$ converts a speed of x mph to $s(x)$ ft/sec.

    a. Find the equivalent speed in feet per second for a car traveling 60 mph.

    b. Find an equation defining $y = s^{-1}(x)$.

    c. Use the inverse function from part (b) to find the equivalent speed in miles per hour for a train traveling 132 ft/sec. Round to the nearest tenth.

For Exercises 45–51, answer true or false.

45. The function defined by $y = 2$ has an inverse function defined by $x = 2$.

46. The domain of any one-to-one function is the same as the domain of its inverse.

47. All linear functions with a nonzero slope have an inverse function.

48. The function defined by $g(x) = |x|$ is one-to-one.

49. The function defined by $k(x) = x^2$ is one-to-one.

50. The function defined by $h(x) = |x|$ for $x \geq 0$ is one-to-one.

51. The function defined by $L(x) = x^2$ for $x \geq 0$ is one-to-one.

52. Explain how the domain and range of a one-to-one function and its inverse are related.

53. If $(0, b)$ is the y-intercept of a one-to-one function, what is the x-intercept of its inverse?

54. If $(a, 0)$ is the x-intercept of a one-to-one function, what is the y-intercept of its inverse?

55. **a.** Find the domain and range of the function defined by $f(x) = \sqrt{x - 1}$.

    b. Find the domain and range of the function defined by $f^{-1}(x) = x^2 + 1, x \geq 0$.

56. **a.** Find the domain and range of the function defined by $g(x) = x^2 - 4, x \leq 0$.

    b. Find the domain and range of the function defined by $g^{-1}(x) = -\sqrt{x + 4}$.

For Exercises 57–60, the graph of $y = f(x)$ is given.

    a. State the domain of f.   **b.** State the range of f.

    c. State the domain of f^{-1}.   **d.** State the range of f^{-1}.

    e. Graph the function defined by $y = f^{-1}(x)$. The line $y = x$ is provided for your reference.

57.   **58.**   **59.**   **60.**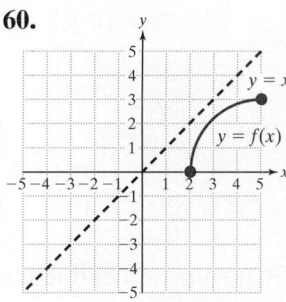

Expanding Your Skills

For Exercises 61–70, write an equation of the inverse of the one-to-one function.

61. $q(x) = \sqrt{x + 4}$ **62.** $v(x) = \sqrt{x + 16}$ **63.** $z(x) = -\sqrt{x + 4}$ **64.** $u(x) = -\sqrt{x + 16}$

65. $f(x) = \dfrac{x - 1}{x + 1}$ **66.** $p(x) = \dfrac{3 - x}{x + 3}$ **67.** $t(x) = \dfrac{2}{x - 1}$ **68.** $w(x) = \dfrac{4}{x + 2}$

69. $n(x) = x^2 + 9 \quad x \le 0$ **70.** $g(x) = x^2 - 1 \quad x \le 0$

Graphing Calculator Exercises

For Exercises 71–74, use a graphing calculator to graph each function on the standard viewing window. Use the graph of the function to determine if the function is one-to-one on the interval $-10 \le x \le 10$. If the function is one-to-one, find its inverse and graph both functions on the standard viewing window.

71. $f(x) = \sqrt[3]{x + 5}$ **72.** $k(x) = x^3 - 4$

73. $g(x) = 0.5x^3 - 2$ **74.** $m(x) = 3x - 4$

Section 8.3   Exponential Functions

Concepts

1. Definition of an Exponential Function
2. Approximating Exponential Expressions with a Calculator
3. Graphs of Exponential Functions
4. Applications of Exponential Functions

1. Definition of an Exponential Function

The concept of a function was first introduced in Section 2.6. Since then we have learned to recognize several categories of functions, including constant, linear, rational, and quadratic functions. In this section and the next, we will define two new types of functions called exponential and logarithmic functions.

    To introduce the concept of an exponential function, consider the following salary plans for a new job. Plan A pays $1 million for a month's work. Plan B starts with 2¢ on the first day, and every day thereafter the salary is doubled.

    At first glance, the million-dollar plan appears to be more favorable. Look, however, at Table 8-1, which shows the daily payments for 30 days under plan B.

Table 8-1

Day	Payment	Day	Payment	Day	Payment
1	2¢	11	$20.48	21	$20,971.52
2	4¢	12	$40.96	22	$41,943.04
3	8¢	13	$81.92	23	$83,886.08
4	16¢	14	$163.84	24	$167,772.16
5	32¢	15	$327.68	25	$335,544.32
6	64¢	16	$655.36	26	$671,088.64
7	$1.28	17	$1310.72	27	$1,342,177.28
8	$2.56	18	$2621.44	28	$2,684,354.56
9	$5.12	19	$5242.88	29	$5,368,709.12
10	$10.24	20	$10,485.76	30	$10,737,418.24

Notice that the salary on the 30th day for plan B is over $10 million. Taking the sum of the payments, we see the total salary for the 30-day period is $21,474,836.46.

The daily salary for plan B can be represented by the function $y = 2^x$, where x is the number of days on the job and y is the salary (in cents) for that day. An interesting feature of this function is that for every positive 1-unit change in x, the y-value doubles. The function $y = 2^x$ is called an exponential function.

Definition of an Exponential Function

Let b be any real number such that $b > 0$ and $b \neq 1$. Then for any real number x, a function of the form $f(x) = b^x$ is called an **exponential function**.

An exponential function is recognized as a function with a constant base and exponent, x. Notice that the base of an exponential function must be a positive real number not equal to 1.

2. Approximating Exponential Expressions with a Calculator

Up to this point, we have evaluated exponential expressions with integer exponents and rational exponents, for example, $4^3 = 64$ and $4^{1/2} = \sqrt{4} = 2$. However, how do we evaluate an exponential expression with an irrational exponent such as 4^π? In such a case, the exponent is a nonterminating and nonrepeating decimal. The value of 4^π can be thought of as the limiting value of a sequence of approximations using rational exponents:

$$4^{3.14} \approx 77.7084726$$

$$4^{3.141} \approx 77.81627412$$

$$4^{3.1415} \approx 77.87023095$$

$$\vdots$$

$$4^\pi \approx 77.88023365$$

An exponential expression can be evaluated at all rational numbers and at all irrational numbers. Therefore, the domain of an exponential function is all real numbers.

Example 1 **Approximating Exponential Expressions with a Calculator**

Approximate the expressions. Round the answers to four decimal places.

a. $8^{\sqrt{3}}$ **b.** $5^{-\sqrt{17}}$ **c.** $\sqrt{10}^{\sqrt{2}}$

Solution:

Calculator Connections

Topic: Investigating Exponential Expressions

On a calculator, use the $\boxed{\wedge}$, $\boxed{y^x}$, or $\boxed{x^y}$ key to approximate an expression with an irrational exponent.

```
8^√(3)
         36.66044576
5^-√(17)
         .00131242
(√(10))^√(2)
         5.09456117
```

a. $8^{\sqrt{3}} \approx 36.6604$ **b.** $5^{-\sqrt{17}} \approx 0.0013$

c. $\sqrt{10}^{\sqrt{2}} \approx 5.0946$

Skill Practice Approximate the value of the expressions. Round the answers to four decimal places.

1. 9^{π} **2.** $15^{\sqrt{5}}$ **3.** $\sqrt{7}^{\sqrt{3}}$

3. Graphs of Exponential Functions

The functions defined by $f(x) = 2^x$, $g(x) = (\frac{1}{2})^x$, $h(x) = 3^x$, and $k(x) = 5^x$ are all examples of exponential functions. Example 2 illustrates the two general shapes of exponential functions.

Example 2 **Graphing Exponential Functions**

Graph the functions f and g.

a. $f(x) = 2^x$ **b.** $g(x) = (\frac{1}{2})^x$

Solution:

Table 8-2 shows several function values for $f(x)$ and $g(x)$ for both positive and negative values of x. The graphs are shown in Figure 8-8.

Table 8-2

x	$f(x) = 2^x$	$g(x) = (\frac{1}{2})^x$
-4	$\frac{1}{16}$	16
-3	$\frac{1}{8}$	8
-2	$\frac{1}{4}$	4
-1	$\frac{1}{2}$	2
0	1	1
1	2	$\frac{1}{2}$
2	4	$\frac{1}{4}$
3	8	$\frac{1}{8}$
4	16	$\frac{1}{16}$

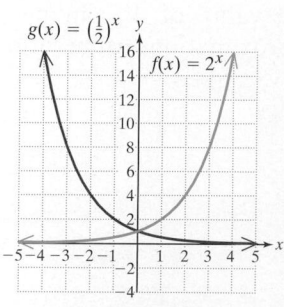

Figure 8-8

Answers

1. 995.0416   **2.** 426.4028

3. 5.3936

Skill Practice Graph the functions f and g on the same grid.

4. $f(x) = 3^x$ **5.** $g(x) = \left(\dfrac{1}{3}\right)^x$

The graphs in Figure 8-8 illustrate several important features of exponential functions.

Graphs of $f(x) = b^x$

The graph of an exponential function defined by $f(x) = b^x$ ($b > 0$ and $b \neq 1$) has the following properties.

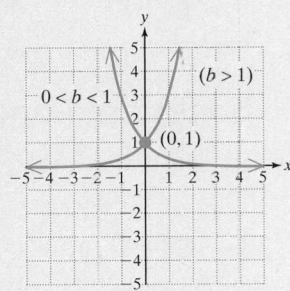

1. If $b > 1$, f is an *increasing* exponential function, sometimes called an **exponential growth function**.

   If $0 < b < 1$, f is a *decreasing* exponential function, sometimes called an **exponential decay function**.

2. The domain is the set of all real numbers, $(-\infty, \infty)$.

3. The range is $(0, \infty)$.

4. The line $y = 0$ (x-axis) is a horizontal asymptote.

5. The function passes through the point $(0, 1)$ because $f(0) = b^0 = 1$.

These properties indicate that the graph of an exponential function is an increasing function if the base is greater than 1. Furthermore, the base affects the rate of increase. Consider the graphs of $f(x) = 2^x$, $h(x) = 3^x$, and $k(x) = 5^x$ (Figure 8-9). For every positive 1-unit change in x, $f(x) = 2^x$ increases by 2 times, $h(x) = 3^x$ increases by 3 times, and $k(x) = 5^x$ increases by 5 times (Table 8-3).

Table 8-3

x	$f(x) = 2^x$	$h(x) = 3^x$	$k(x) = 5^x$
-3	$\frac{1}{8}$	$\frac{1}{27}$	$\frac{1}{125}$
-2	$\frac{1}{4}$	$\frac{1}{9}$	$\frac{1}{25}$
-1	$\frac{1}{2}$	$\frac{1}{3}$	$\frac{1}{5}$
0	1	1	1
1	2	3	5
2	4	9	25
3	8	27	125

Figure 8-9

Answers

4–5.

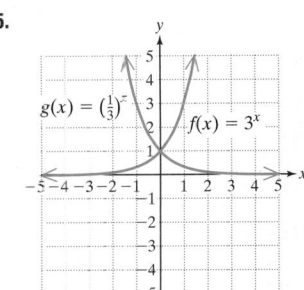

The graph of an exponential function is a *decreasing function* if the base is between 0 and 1. Consider the graphs of $g(x) = (\frac{1}{2})^x$, $m(x) = (\frac{1}{3})^x$, and $n(x) = (\frac{1}{5})^x$ (Table 8-4 and Figure 8-10).

Table 8-4

x	$g(x) = (\frac{1}{2})^x$	$m(x) = (\frac{1}{3})^x$	$n(x) = (\frac{1}{5})^x$
-3	8	27	125
-2	4	9	25
-1	2	3	5
0	1	1	1
1	$\frac{1}{2}$	$\frac{1}{3}$	$\frac{1}{5}$
2	$\frac{1}{4}$	$\frac{1}{9}$	$\frac{1}{25}$
3	$\frac{1}{8}$	$\frac{1}{27}$	$\frac{1}{125}$

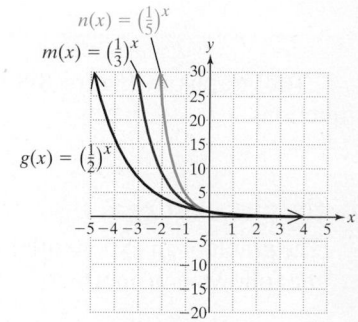

Figure 8-10

4. Applications of Exponential Functions

Exponential growth and decay can be found in a variety of real-world phenomena; for example,

- Population growth can often be modeled by an exponential function.
- The growth of an investment under compound interest increases exponentially.
- The mass of a radioactive substance decreases exponentially with time.
- The temperature of a cup of coffee decreases exponentially as it approaches room temperature.

A substance that undergoes radioactive decay is said to be radioactive. The *half-life* of a radioactive substance is the amount of time it takes for one-half of the original amount of the substance to change into something else. That is, after each half-life the amount of the original substance decreases by one-half.

In 1898, Marie Curie discovered the highly radioactive element radium. She shared the 1903 Nobel Prize in physics for her research on radio-activity and was awarded the 1911 Nobel Prize in chemistry for her discovery of radium and polonium. Radium 226 (an isotope of radium) has a half-life of 1620 years and decays into radon 222 (a radioactive gas).

Marie and Pierre Curie

Example 3 Applying an Exponential Decay Function

In a sample originally having 1 g of radium 226, the amount of radium 226 present after t years is given by

$$A(t) = (\tfrac{1}{2})^{t/1620}$$

where A is the amount of radium in grams and t is the time in years.

a. How much radium 226 will be present after 1620 yr?

b. How much radium 226 will be present after 3240 yr?

c. How much radium 226 will be present after 4860 yr?

Solution:

$$A(t) = \left(\frac{1}{2}\right)^{t/1620}$$

a. $A(1620) = \left(\frac{1}{2}\right)^{1620/1620}$ Substitute $t = 1620$.

$$= \left(\frac{1}{2}\right)^{1}$$

$$= 0.5$$

After 1620 yr (1 half-life), 0.5 g remains.

b. $A(3240) = \left(\frac{1}{2}\right)^{3240/1620}$ Substitute $t = 3240$.

$$= \left(\frac{1}{2}\right)^{2}$$

$$= 0.25$$

After 3240 yr (2 half-lives), the amount of the original substance is reduced by one-half, 2 times: 0.25 g remains.

c. $A(4860) = \left(\frac{1}{2}\right)^{4860/1620}$ Substitute $t = 4860$.

$$= \left(\frac{1}{2}\right)^{3}$$

$$= 0.125$$

After 4860 yr (3 half-lives), the amount of the original substance is reduced by one-half, 3 times: 0.125 g remains.

Skill Practice Cesium 137 is a radioactive metal with a short half-life of 30 years. In a sample originally having 1 g of cesium 137, the amount, $A(t)$ (in grams), of cesium 137 present after t years is given by

$$A(t) = \left(\frac{1}{2}\right)^{t/30}$$

6. How much cesium 137 will be present after 30 years?
7. How much cesium 137 will be present after 90 years?

Exponential functions are often used to model population growth. Suppose the initial value of a population at some time $t = 0$ is P_0. If the rate of increase of a population is r, then after t years, the population $P(t)$ is given by

$$P(t) = P_0(1 + r)^t$$

Answers
6. 0.5 g
7. 0.125 g

> Example 4 **Applying an Exponential Growth Function**
>
> The population of the Bahamas in 2008 was estimated at 321,000 with an annual rate of increase of 1.39%.
>
> **a.** Find a mathematical model that relates the population of the Bahamas as a function of the number of years since 2008.
>
> **b.** If the annual rate of increase remains the same, use this model to predict the population of the Bahamas in the year 2016. Round to the nearest thousand.
>
> **Solution:**
>
> **a.** The initial population is $P_0 = 321{,}000$, and the rate of increase is 1.39%.
>
> $P(t) = P_0(1 + r)^t$ Substitute $P_0 = 321{,}000$ and $r = 0.0139$.
>
> $ = 321{,}000(1 + 0.0139)^t$
>
> $ = 321{,}000(1.0139)^t$ Here $t = 0$ corresponds to the year 2008.
>
> **b.** Because the initial population ($t = 0$) corresponds to the year 2008, we use $t = 8$ to find the population in the year 2016.
>
> $P(8) = 321{,}000(1.0139)^8$
>
> $ \approx 358{,}000$ The population in the year 2016 will be approximately 358,000.

Skill Practice The population of Colorado in 2000 was approximately 3,700,000 with an annual increase of 2%.

8. Find a mathematical model that relates the population of Colorado as a function of the number of years since 2000.

9. Use this model to predict the population of Colorado in 2016. Round to the nearest thousand.

Answers

8. $P(t) = 3{,}700{,}000(1.02)^t$

9. Approximately 5,079,000

Section 8.3 Practice Exercises

Vocabulary and Key Concepts

1. a. Given a real number b, where $b > 0$ and $b \neq 1$, then a function of the form $f(x) = $ _____ is called an exponential function.

   b. The function defined by $f(x) = x^4$ (is/is not) an exponential function, whereas the function defined by $f(x) = 4^x$ (is/is not) an exponential function.

   c. The graph of $f(x) = \left(\dfrac{7}{2}\right)^x$ is (increasing/decreasing) over its domain.

   d. The graph of $f(x) = \left(\dfrac{2}{7}\right)^x$ is (increasing/decreasing) over its domain.

   e. In interval notation, the domain of an exponential function $f(x) = b^x$ is _____ and the range is _____.

   f. All exponential functions $f(x) = b^x$ pass through the point _____.

   g. The horizontal asymptote of an exponential function $f(x) = b^x$ is the line _____.

   h. The function defined by $f(x) = 1^x$ (is/is not) an exponential function.

Review Exercises

For Exercises 2–7, find the functions, using f and g as given. $f(x) = 2x^2 + x + 2$ $g(x) = 3x - 1$

2. $(f + g)(x)$ **3.** $(g - f)(x)$ **4.** $(f \cdot g)(x)$

5. $\left(\dfrac{g}{f}\right)(x)$ **6.** $(f \circ g)(x)$ **7.** $(g \circ f)(x)$

8. Find the inverse. $\{(2, 3), (0, 0), (-8, 4), (10, 12)\}$

For Exercises 9–16, evaluate the expression without the use of a calculator.

9. 5^2 **10.** 2^{-3} **11.** 10^{-3} **12.** 3^4

13. $36^{1/2}$ **14.** $27^{1/3}$ **15.** $16^{3/4}$ **16.** $8^{2/3}$

Concept 2: Approximating Exponential Expressions with a Calculator

For Exercises 17–24, evaluate the expression by using a calculator. Round to four decimal places. **(See Example 1.)**

17. $5^{1.1}$ **18.** $2^{\sqrt{3}}$ **19.** 10^{π} **20.** $3^{4.8}$

21. $36^{-\sqrt{2}}$ **22.** $27^{-0.5126}$ **23.** $16^{-0.04}$ **24.** $8^{-0.61}$

25. Solve for x.

    a. $3^x = 9$ **b.** $3^x = 27$ **c.** Between what two consecutive integers must the solution to $3^x = 11$ lie?

26. Solve for x.

    a. $5^x = 125$ **b.** $5^x = 625$ **c.** Between what two consecutive integers must the solution to $5^x = 130$ lie?

27. Solve for x.

    a. $2^x = 16$ **b.** $2^x = 32$ **c.** Between what two consecutive integers must the solution to $2^x = 30$ lie?

28. Solve for x.

    a. $4^x = 16$ **b.** $4^x = 64$ **c.** Between what two consecutive integers must the solution to $4^x = 20$ lie?

29. For $f(x) = \left(\frac{1}{5}\right)^x$ find $f(0), f(1), f(2), f(-1)$, and $f(-2)$.

30. For $g(x) = \left(\frac{2}{3}\right)^x$ find $g(0), g(1), g(2), g(-1)$, and $g(-2)$.

31. For $h(x) = 3^x$ use a calculator to find $h(0), h(1), h(-1), h(\sqrt{2})$, and $h(\pi)$. Round to two decimal places.

32. For $k(x) = 5^x$ use a calculator to find $k(0), k(1), k(-1), k(\pi)$, and $k(\sqrt{2})$. Round to two decimal places.

Concept 3: Graphs of Exponential Functions

33. How do you determine whether the graph of $f(x) = b^x$ ($b > 0, b \neq 1$) is increasing or decreasing?

34. For $f(x) = b^x$ ($b > 0, b \neq 1$), find $f(0)$.

Graph the functions defined in Exercises 35–42. Plot at least three points for each function. **(See Example 2.)**

35. $f(x) = 4^x$

36. $g(x) = 6^x$

37. $m(x) = \left(\dfrac{1}{8}\right)^x$

38. $n(x) = \left(\dfrac{1}{3}\right)^x$

39. $h(x) = 2^{x+1}$

40. $k(x) = 5^{x-1}$

41. $g(x) = 5^{-x}$

42. $f(x) = 2^{-x}$

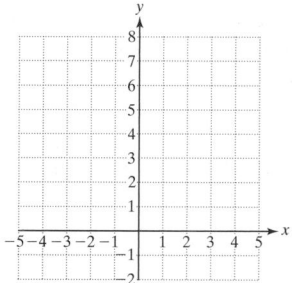

Concept 4: Applications of Exponential Functions

43. The half-life of the element radon (Rn 86) is 3.8 days. In a sample originally containing 1 g of radon, the amount left after t days is given by $A(t) = (0.5)^{t/3.8}$. (Round to two decimal places, if necessary.) **(See Example 3.)**

   a. How much radon will be present after 7.6 days?

   b. How much radon will be present after 10 days?

44. Nobelium, an element discovered in 1958, has a half-life of 10 min under certain conditions. In a sample containing 1 g of nobelium, the amount left after t min is given by $A(t) = (0.5)^{t/10}$. (Round to three decimal places.)

   a. How much nobelium is left after 5 min?

   b. How much nobelium is left after 1 hr?

45. Once an antibiotic is introduced to bacteria, the number of bacteria decreases exponentially. For example, beginning with 1 million bacteria, the amount present t days from the time penicillin is introduced is given by the function $A(t) = 1,000,000(2)^{-t/5}$. Rounding to the nearest thousand, determine how many bacteria are present after

a. 2 days   **b.** 1 week   **c.** 2 weeks

46. Once an antibiotic is introduced to bacteria, the number of bacteria decreases exponentially. For example, beginning with 1 million bacteria, the amount present t days from the time streptomycin is introduced is given by the function $A(t) = 1,000,000(2)^{-t/10}$. Rounding to the nearest thousand, determine how many bacteria are present after

a. 5 days   **b.** 1 week   **c.** 2 weeks

47. The population of Bangladesh was 153,000,000 in 2009 with an annual growth rate of 1.25%.
(See Example 4.)

a. Find a mathematical model that relates the population of Bangladesh as a function of the number of years after 2009.

b. If the annual rate of increase remains the same, use this model to predict the population of Bangladesh in the year 2050. Round to the nearest million.

48. The population of Fiji was 908,000 in 2009 with an annual growth rate of 0.07%.

a. Find a mathematical model that relates the population of Fiji as a function of the number of years after 2009.

b. If the annual rate of increase remains the same, use this model to predict the population of Fiji in the year 2050. Round to the nearest thousand.

49. Suppose $1000 is initially invested in an account and the value of the account grows exponentially. If the investment doubles in 7 yr, then the amount in the account t years after the initial investment is given by $A(t)$.

$$A(t) = 1000(2)^{t/7}$$

a. Find the amount in the account after 5 yr.

b. Find the amount in the account after 10 yr.

c. Find $A(0)$ and $A(7)$ and interpret the answers in the context of this problem.

50. Suppose $1500 is initially invested in an account and the value of the account grows exponentially. If the investment doubles in 8 yr, then the amount in the account t years after the initial investment is given by $A(t)$.

$$A(t) = 1500(2)^{t/8}$$

a. Find the amount in the account after 5 yr.

b. Find the amount in the account after 10 yr.

c. Find $A(0)$ and $A(8)$ and interpret the answers in the context of this problem.

Graphing Calculator Exercises

For Exercises 51–58, graph the functions on your calculator to support your solutions to the indicated exercises.

51. $f(x) = 4^x$
(see Exercise 35)

52. $g(x) = 6^x$
(see Exercise 36)

53. $m(x) = \left(\frac{5}{8}\right)^x$
(see Exercise 37)

54. $n(x) = \left(\frac{5}{3}\right)^x$
(see Exercise 38)

55. $h(x) = 2^{x+1}$
(see Exercise 39)

56. $k(x) = 5^{x-1}$
(see Exercise 40)

57. $g(x) = 5^{-x}$
(see Exercise 41)

58. $f(x) = 2^{-x}$
(see Exercise 42)

Concepts

1. Definition of a Logarithmic Function
2. Evaluating Logarithmic Expressions
3. The Common Logarithmic Function
4. Graphs of Logarithmic Functions
5. Applications of the Common Logarithmic Function

1. Definition of a Logarithmic Function

Consider the following equations in which the variable is located in the exponent of an expression. In some cases the solution can be found by inspection because the constant on the right-hand side of the equation is a perfect power of the base.

Equation	Solution
$5^x = 5$	$x = 1$
$5^x = 20$	$x = ?$
$5^x = 25$	$x = 2$
$5^x = 60$	$x = ?$
$5^x = 125$	$x = 3$

The equation $5^x = 20$ cannot be solved by inspection. However, we might suspect that x is between 1 and 2. Similarly, the solution to the equation $5^x = 60$ is between 2 and 3. To solve an exponential equation for an unknown exponent, we must use a new type of function called a logarithmic function.

Definition of a Logarithmic Function

If x and b are positive real numbers such that $b \neq 1$, then $y = \log_b x$ is called the **logarithmic function** with base b and

$$y = \log_b x \quad \text{is equivalent to} \quad b^y = x$$

Note: In the expression $y = \log_b x$, y is called the **logarithm**, b is called the **base**, and x is called the **argument**.

The expression $y = \log_b x$ is equivalent to $b^y = x$ and indicates that *the logarithm y is the exponent to which b must be raised to obtain x*. The expression $y = \log_b x$ is called the logarithmic form of the equation, and the expression $b^y = x$ is called the exponential form of the equation.

The definition of a logarithmic function suggests a close relationship with an exponential function of the same base. In fact, a logarithmic function is the inverse of the corresponding exponential function. For example, the following steps find the inverse of the exponential function defined by $f(x) = b^x$.

$f(x) = b^x$	
$y = b^x$	Replace $f(x)$ by y.
$x = b^y$	Interchange x and y.
$y = \log_b x$	Solve for y using the definition of a logarithmic function.
$f^{-1}(x) = \log_b x$	Replace y by $f^{-1}(x)$.

The concept of a logarithm is new and unfamiliar. Therefore, it is often advantageous to rewrite a logarithm in its exponential form.

Example 1 Converting from Logarithmic Form to Exponential Form

Rewrite the logarithmic equations in exponential form.

a. $\log_2 32 = 5$ **b.** $\log_{10}\left(\dfrac{1}{1000}\right) = -3$ **c.** $\log_5 1 = 0$

Solution:

Logarithmic Form		Exponential Form
a. $\log_2 32 = 5$	$\Leftrightarrow$	$2^5 = 32$
b. $\log_{10}\left(\dfrac{1}{1000}\right) = -3$	$\Leftrightarrow$	$10^{-3} = \dfrac{1}{1000}$
c. $\log_5 1 = 0$	$\Leftrightarrow$	$5^0 = 1$

Skill Practice Rewrite the logarithmic equations in exponential form.

1. $\log_3 9 = 2$ **2.** $\log_{10}\left(\dfrac{1}{100}\right) = -2$ **3.** $\log_8 1 = 0$

2. Evaluating Logarithmic Expressions

Example 2 Evaluating Logarithmic Expressions

Evaluate the logarithmic expressions.

a. $\log_8 64$ **b.** $\log_5\left(\dfrac{1}{125}\right)$ **c.** $\log_{1/2}\left(\dfrac{1}{8}\right)$

Solution:

a. $\log_8 64$ is the exponent to which 8 must be raised to obtain 64.

$y = \log_8 64$ Let y represent the value of the logarithm.

$8^y = 64$ Rewrite the expression in exponential form.

$8^y = 8^2$

$y = 2$ Therefore, $\log_8 64 = 2$.

b. $\log_5\left(\frac{1}{125}\right)$ is the exponent to which 5 must be raised to obtain $\frac{1}{125}$.

$y = \log_5\left(\dfrac{1}{125}\right)$ Let y represent the value of the logarithm.

$5^y = \dfrac{1}{125}$ Rewrite the expression in exponential form.

$5^y = \dfrac{1}{5^3} = 5^{-3}$

$y = -3$ Therefore, $\log_5\left(\frac{1}{125}\right) = -3$.

c. $\log_{1/2}\left(\frac{1}{8}\right)$ is the exponent to which $\frac{1}{2}$ must be raised to obtain $\frac{1}{8}$.

$y = \log_{1/2}\left(\dfrac{1}{8}\right)$ Let y represent the value of the logarithm.

$\left(\dfrac{1}{2}\right)^y = \dfrac{1}{8}$ Rewrite the expression in exponential form.

$\left(\dfrac{1}{2}\right)^y = \left(\dfrac{1}{2}\right)^3$

$y = 3$ Therefore, $\log_{1/2}\left(\frac{1}{8}\right) = 3$.

Skill Practice Evaluate the logarithmic expressions.

4. $\log_{10} 1000$ **5.** $\log_4\left(\dfrac{1}{16}\right)$ **6.** $\log_{1/3} 3$

Answers

1. $3^2 = 9$ **2.** $10^{-2} = \dfrac{1}{100}$
3. $8^0 = 1$ **4.** 3
5. -2 **6.** -1

Example 3 Evaluating Logarithmic Expressions

Evaluate the logarithmic expressions.

a. $\log_b b$ **b.** $\log_c c^7$ **c.** $\log_3 \sqrt[4]{3}$

Solution:

a. $\log_b b$ is the exponent to which b must be raised to obtain b.

$y = \log_b b$	Let y represent the value of the logarithm.
$b^y = b$	Rewrite the expression in exponential form.
$y = 1$	Therefore, $\log_b b = 1$.

b. $\log_c c^7$ is the exponent to which c must be raised to obtain c^7.

$y = \log_c c^7$	Let y represent the value of the logarithm.
$c^y = c^7$	Rewrite the expression in exponential form.
$y = 7$	Therefore, $\log_c c^7 = 7$.

c. $\log_3 \sqrt[4]{3} = \log_3 3^{1/4}$ is the exponent to which 3 must be raised to obtain $3^{1/4}$.

$y = \log_3 3^{1/4}$	Let y represent the value of the logarithm.
$3^y = 3^{1/4}$	Rewrite the expression in exponential form.
$y = \dfrac{1}{4}$	Therefore, $\log_3 \sqrt[4]{3} = \frac{1}{4}$.

Skill Practice Evaluate the logarithmic expressions.

7. $\log_x x$ **8.** $\log_b b^{10}$ **9.** $\log_5 \sqrt[3]{5}$

---

3. The Common Logarithmic Function

The logarithmic function with base 10 is called the **common logarithmic function** and is denoted by $y = \log x$. Notice that the base is not explicitly written but is understood to be 10. That is, $y = \log_{10} x$ is written simply as $y = \log x$.

Example 4 Evaluating Common Logarithms

Evaluate the logarithmic expressions.

a. $\log 100{,}000$ **b.** $\log 0.01$

Solution:

a. $\log 100{,}000$ is the exponent to which 10 must be raised to obtain 100,000.

$y = \log 100{,}000$	Let y represent the value of the logarithm.
$10^y = 100{,}000$	Rewrite the expression in exponential form.
$10^y = 10^5$	
$y = 5$	Therefore, $\log 100{,}000 = 5$.

Answers

7. 1   **8.** 10   **9.** $\dfrac{1}{3}$

b. log 0.01 is the exponent to which 10 must be raised to obtain 0.01 or $\frac{1}{100}$.

$y = \log 0.01$	Let y represent the value of the logarithm.
$10^y = 0.01$	Rewrite the expression in exponential form.
$10^y = 10^{-2}$	Note that $0.01 = \frac{1}{100}$ or 10^{-2}.
$y = -2$	Therefore, $\log 0.01 = -2$.

Skill Practice Evaluate the logarithmic expressions.

10. log 1,000,000   **11.** log 0.0001

On most calculators the ⃞LOG key is used to compute logarithms with base 10. This is demonstrated in Example 5.

Example 5 **Evaluating Common Logarithms on a Calculator**

Evaluate the common logarithms. Round the answers to four decimal places.

a. log 420   **b.** $\log\left(8.2 \times 10^9\right)$ **c.** log 0.0002

Solution:

a. $\log 420 \approx 2.6232$

b. $\log\left(8.2 \times 10^9\right) \approx 9.9138$

c. $\log 0.0002 \approx -3.6990$

Skill Practice Evaluate the common logarithms. Round answers to four decimal places.

12. log 1200   **13.** $\log\left(6.3 \times 10^5\right)$ **14.** log 0.00025

4. Graphs of Logarithmic Functions

In Section 8.3 we studied the graphs of exponential functions. In this section, we will graph logarithmic functions. First, recall that $f(x) = \log_b x$ is the inverse of $g(x) = b^x$. Therefore, the graph of $y = f(x)$ is symmetric to the graph of $y = g(x)$ about the line $y = x$, as shown in Figures 8-11 and 8-12.

Figure 8-11

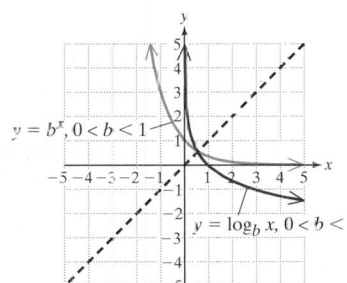

Figure 8-12

From Figures 8-11 and 8-12, we see that the range of $y = b^x$ is the set of positive real numbers. As expected, the domain of its inverse, the logarithmic function $y = \log_b x$, is also the set of positive real numbers. Therefore, the **domain of the logarithmic function** $y = \log_b x$ is the set of positive real numbers.

Answers
10. 6   **11.** −4
12. ≈ 3.0792   **13.** ≈ 5.7993
14. ≈ −3.6021

| **Example 6** | **Graphing Logarithmic Functions** |

Graph the logarithmic functions.

a. $y = \log_2 x$ **b.** $y = \log_{1/4} x$

Solution:

A logarithmic function is the inverse of an exponential function with the same base. For this reason, we can reverse the coordinates of the ordered pairs on an exponential function to obtain ordered pairs on the related logarithmic function. For example, to graph the function $y = \log_2 x$ reverse the coordinates of the ordered pairs from its inverse function, $y = 2^x$.

a. Exponential function   Logarithmic function

x	$y = 2^x$
-3	$\frac{1}{8}$
-2	$\frac{1}{4}$
-1	$\frac{1}{2}$
0	1
1	2
2	4
3	8

x	$y = \log_2 x$
$\frac{1}{8}$	-3
$\frac{1}{4}$	-2
$\frac{1}{2}$	-1
1	0
2	1
4	2
8	3

Reverse the order.

Figure 8-13

The graph of $y = \log_2 x$ is shown in Figure 8-13.

b. Exponential function   Logarithmic function

x	$y = \left(\frac{1}{4}\right)^x$
-3	64
-2	16
-1	4
0	1
1	$\frac{1}{4}$
2	$\frac{1}{16}$
3	$\frac{1}{64}$

x	$y = \log_{1/4} x$
64	-3
16	-2
4	-1
1	0
$\frac{1}{4}$	1
$\frac{1}{16}$	2
$\frac{1}{64}$	3

Reverse the order.

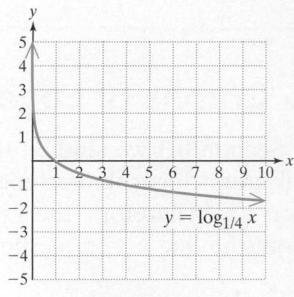

Figure 8-14

The graph of $y = \log_{1/4} x$ is shown in Figure 8-14.

Answers

15–16.

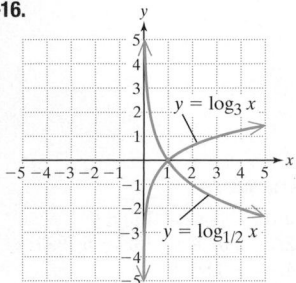

Skill Practice Graph.

15. $y = \log_3 x$ **16.** $y = \log_{1/2} x$

The general shape and important features of exponential and logarithmic functions are summarized as follows.

Graphs of Exponential and Logarithmic Functions

Exponential Functions   **Logarithmic Functions**

$f(x) = b^x$

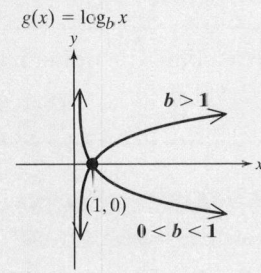

$g(x) = \log_b x$

Domain: $(-\infty, \infty)$
Range: $(0, \infty)$
Horizontal asymptote: $y = 0$
Passes through $(0, 1)$
If $b > 1$, the function is increasing.
If $0 < b < 1$, the function is decreasing.

Domain: $(0, \infty)$
Range: $(-\infty, \infty)$
Vertical asymptote: $x = 0$
Passes through $(1, 0)$
If $b > 1$, the function is increasing.
If $0 < b < 1$, the function is decreasing.

Notice that the roles of x and y are interchanged for the functions $y = b^x$ and $b^y = x$. Therefore, it is not surprising that the domain and range are reversed between exponential and logarithmic functions. Moreover, an exponential function passes through $(0, 1)$, whereas a logarithmic function passes through $(1, 0)$. An exponential function has a horizontal asymptote at $y = 0$, whereas a logarithmic function has a vertical asymptote at $x = 0$.

When graphing a logarithmic equation, it is helpful to know its domain.

Example 7 Identifying the Domain of a Logarithmic Function

Find the domain of the functions.

a. $f(x) = \log(4 - x)$ **b.** $g(x) = \log(2x + 6)$

Solution:

The domain of the function $y = \log_b x$ is the set of all positive real numbers. That is, the argument x must be greater than zero: $x > 0$.

a. $f(x) = \log(4 - x)$ The argument is $4 - x$.

     $4 - x > 0$ The argument of the logarithm must be greater than zero.

     $-x > -4$ Solve for x.

     $x < 4$ Divide by -1 and reverse the inequality sign.

The domain is $(-\infty, 4)$.

b. $g(x) = \log(2x + 6)$ The argument is $2x + 6$.

$2x + 6 > 0$ The argument of the logarithm must be greater than zero.

$2x > -6$

$x > -3$

The domain is $(-3, \infty)$.

Skill Practice Find the domain of the functions.

17. $f(x) = \log_3(x + 7)$ **18.** $g(x) = \log(4 - 8x)$

Calculator Connections

Topic: Graphing Common Logarithmic Functions

The graphs of $Y_1 = \log(4 - x)$ and $Y_2 = \log(2x + 6)$ are shown here and can be used to confirm the solutions to Example 7. Notice that each function has a vertical asymptote at the value of x where the argument equals zero.

---

5. Applications of the Common Logarithmic Function

Example 8   Applying a Common Logarithm to Compute pH

The pH (hydrogen potential) of a solution is defined as

$$\text{pH} = -\log[\text{H}^+]$$

where $[\text{H}^+]$ represents the concentration of hydrogen ions in moles per liter (mol/L). The pH scale ranges from 0 to 14. The midpoint of this range, 7, represents a neutral solution. Values below 7 are progressively more acidic, and values above 7 are progressively more alkaline. Based on the equation $\text{pH} = -\log[\text{H}^+]$, a 1-unit change in pH means a 10-fold change in hydrogen ion concentration.

a. Normal rain has a pH of 5.6. However, in some areas of the northeastern United States the rainwater is more acidic. What is the pH of a rain sample for which the concentration of hydrogen ions is 0.0002 mol/L?

b. Find the pH of household ammonia if the concentration of hydrogen ions is 1.0×10^{-11} mol/L.

Solution:

a. $\text{pH} = -\log[\text{H}^+]$

$= -\log(0.0002)$ Substitute $[\text{H}^+] = 0.0002$.

≈ 3.7 The pH of the rain sample is 3.7. (To compare this value with a familiar substance, note that the pH of orange juice is roughly 3.5.)

Answers

17. Domain: $(-7, \infty)$

18. Domain: $\left(-\infty, \dfrac{1}{2}\right)$

b. pH $= -\log[H^+]$

$\quad = -\log(1.0 \times 10^{-11})\qquad$ Substitute $[H^+] = 1.0 \times 10^{-11}$.

$\quad = -\log(10^{-11})$

$\quad = -(-11)$

$\quad = 11\qquad\qquad\qquad$ The pH of household ammonia is 11.

Skill Practice

19. A new all-natural shampoo on the market claims its hydrogen ion concentration is 5.88×10^{-7} mol/L. Use the formula pH $= -\log[H^+]$ to calculate the pH level of the shampoo.

Example 9 **Applying a Logarithmic Function to a Memory Model**

One method of measuring a student's retention of material after taking a course is to retest the student at specified time intervals after the course has been completed. A student's score on a calculus test t months after completing a course in calculus is approximated by

$$S(t) = 85 - 25\log(t+1)$$

where t is the time in months after completing the course and $S(t)$ is the student's score as a percent.

a. What was the student's score at $t = 0$?

b. What was the student's score after 2 months?

c. What was the student's score after 1 yr?

Solution:

a. $S(t) = 85 - 25\log(t+1)$

$S(0) = 85 - 25\log(0+1)\qquad$ Substitute $t = 0$.

$\quad = 85 - 25\log 1\qquad$ $\log 1 = 0$ because $10^0 = 1$.

$\quad = 85 - 25(0)$

$\quad = 85 - 0$

$\quad = 85\qquad\qquad$ The student's score at the time the course was completed was 85%.

b. $S(t) = 85 - 25\log(t+1)$

$S(2) = 85 - 25\log(2+1)$

$\quad = 85 - 25\log 3\qquad$ Use a calculator to approximate $\log 3$.

$\quad \approx 73.1\qquad\qquad$ The student's score dropped to 73.1%.

c. $S(t) = 85 - 25\log(t+1)$

$S(12) = 85 - 25\log(12+1)$

$\quad = 85 - 25\log 13\qquad$ Use a calculator to approximate $\log 13$.

$\quad \approx 57.2\qquad\qquad$ The student's score dropped to 57.2%.

Answer

19. pH ≈ 6.23

Skill Practice The memory model for a student's score on a statistics test t months after completion of the course in statistics is approximated by

$$S(t) = 92 - 28 \log (t + 1)$$

20. What was the student's score at the time the course was completed ($t = 0$)?

21. What was her score after 1 month?

22. What was the score after 2 months?

Answers

20. 92   **21.** 83.6   **22.** 78.6

Section 8.4 Practice Exercises

Vocabulary and Key Concepts

1. a. The function defined by $y = \log_b x$ is called the _____ function with base _____. The values of x and b are positive, and $b \neq 1$.

b. Given $y = \log_b x$, we call y the _____, we call b the _____, and we call x the _____.

c. In interval notation, the domain of $y = \log_b x$ is _____ and the range is _____.

d. A logarithmic function base b is the inverse of the _____ function base b.

e. The logarithmic function base 10 is called the _____ logarithmic function and is denoted by $y = \log x$.

f. The graph of a logarithmic function is (increasing/decreasing) if the base $b > 1$, whereas the graph is (increasing/decreasing) if the base is between 0 and 1.

g. Which values of x are *not* in the domain of $y = \log(x - 4)$? $x = 2, x = 3, x = 4, x = 5, x = 6$

h. The graphs of a logarithmic function base b and an exponential function base b are symmetric with respect to the line _____.

Review Exercises

2. For which graph of $y = b^x$ is $0 < b < 1$?

3. For which graph of $y = b^x$ is $b > 1$?

i.

ii.
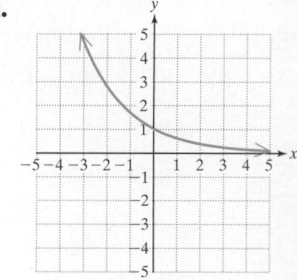

4. Let $f(x) = 6^x$.

a. Find $f(-2), f(-1), f(0), f(1)$, and $f(2)$.

b. Graph $y = f(x)$.

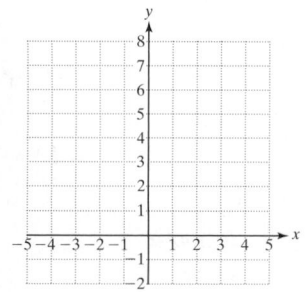

5. Let $s(x) = \left(\dfrac{2}{5}\right)^x$.

a. Find $s(-2), s(-1), s(0), s(1)$, and $s(2)$.

b. Graph $y = s(x)$.

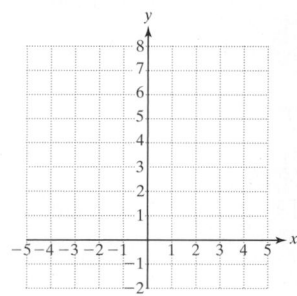

6. Let $r(x) = \left(\dfrac{3}{4}\right)^x$. Find $r(-2)$, $r(-1)$, $r(0)$, $r(1)$, and $r(2)$.

7. Let $g(x) = 3^x$. Find $g(-2)$, $g(-1)$, $g(0)$, $g(1)$, and $g(2)$.

Concept 1: Definition of a Logarithmic Function

8. For the equation $y = \log_b x$, identify the base, the argument, and the logarithm.

9. Rewrite the equation in exponential form. $\quad y = \log_b x$

10. Write the equation in logarithmic form. $\quad b^y = x$

For Exercises 11–22, write the equation in exponential form. **(See Example 1.)**

11. $\log_5 625 = 4$

12. $\log_{125} 25 = \dfrac{2}{3}$

13. $\log_{10}(0.0001) = -4$

14. $\log_{25}\left(\dfrac{1}{5}\right) = -\dfrac{1}{2}$

15. $\log_6 36 = 2$

16. $\log_2 128 = 7$

17. $\log_b 15 = x$

18. $\log_b 82 = y$

19. $\log_3 5 = x$

20. $\log_2 7 = x$

21. $\log_{14} x = 10$

22. $\log_{1/2} x = 6$

For Exercises 23–34, write the equation in logarithmic form.

23. $3^x = 81$

24. $10^3 = 1000$

25. $5^2 = 25$

26. $8^{1/3} = 2$

27. $7^{-1} = \dfrac{1}{7}$

28. $8^{-2} = \dfrac{1}{64}$

29. $b^x = y$

30. $b^y = x$

31. $e^x = y$

32. $e^y = x$

33. $\left(\dfrac{1}{3}\right)^{-2} = 9$

34. $\left(\dfrac{5}{2}\right)^{-1} = \dfrac{2}{5}$

Concept 2: Evaluating Logarithmic Expressions

For Exercises 35–50, evaluate the logarithms without the use of a calculator. **(See Examples 2–3.)**

35. $\log_7 49$

36. $\log_3 81$

37. $\log_{10} 0.1$

38. $\log_2\left(\dfrac{1}{16}\right)$

39. $\log_{16} 4$

40. $\log_8 2$

41. $\log_{7.2} 1$

42. $\log_{1/2} 2$

43. $\log_3 3^5$

44. $\log_9 9^3$

45. $\log_{10} 10$

46. $\log_7 1$

47. $\log_a a^3$

48. $\log_r r^4$

49. $\log_x \sqrt{x}$

50. $\log_y \sqrt[3]{y}$

Concept 3: The Common Logarithmic Function

For Exercises 51–58, evaluate the common logarithm without the use of a calculator. **(See Example 4.)**

51. $\log 10$

52. $\log 100$

53. $\log 1000$

54. $\log 10{,}000$

55. $\log(1.0 \times 10^6)$

56. $\log 0.1$

57. $\log 0.01$

58. $\log 0.001$

For Exercises 59–70, use a calculator to approximate the logarithms. Round to four decimal places. **(See Example 5.)**

59. $\log 6$

60. $\log 18$

61. $\log \pi$

62. $\log \left(\dfrac{1}{8} \right)$

63. $\log \left(\dfrac{1}{32} \right)$

64. $\log \sqrt{5}$

65. $\log 0.0054$

66. $\log 0.0000062$

67. $\log (3.4 \times 10^5)$

68. $\log (4.78 \times 10^9)$

69. $\log (3.8 \times 10^{-8})$

70. $\log (2.77 \times 10^{-4})$

71. Given: $\log 10 = 1$ and $\log 100 = 2$

    a. Estimate $\log 93$.

    b. Estimate $\log 12$.

   c. Evaluate the logarithms in parts (a) and (b) on a calculator and compare to your estimates.

72. Given: $\log \left(\dfrac{1}{10} \right) = -1$ and $\log 1 = 0$

    a. Estimate $\log \left(\dfrac{9}{10} \right)$.

    b. Estimate $\log \left(\dfrac{1}{5} \right)$.

   c. Evaluate the logarithms in parts (a) and (b) on a calculator and compare to your estimates.

Concept 4: Graphs of Logarithmic Functions

 73. Let $f(x) = \log_4 x$.

    a. Find the values of $f\left(\dfrac{1}{64} \right), f\left(\dfrac{1}{16} \right), f\left(\dfrac{1}{4} \right), f(1),$ $f(4), f(16),$ and $f(64)$.

    b. Graph $y = f(x)$. **(See Example 6.)**

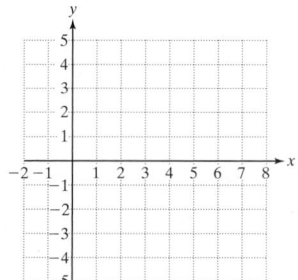

74. Let $g(x) = \log_2 x$.

    a. Find the values of $g\left(\dfrac{1}{8} \right), g\left(\dfrac{1}{4} \right), g\left(\dfrac{1}{2} \right), g(1), g(2),$ $g(4),$ and $g(8)$.

    b. Graph $y = g(x)$.

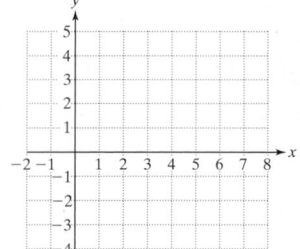

Graph the logarithmic functions in Exercises 75–78 by writing the function in exponential form and making a table of points.

75. $y = \log_3 x$

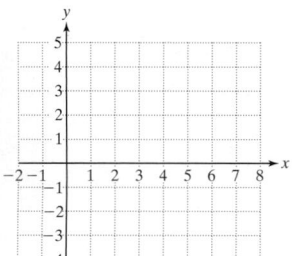

x	y

76. $y = \log_5 x$

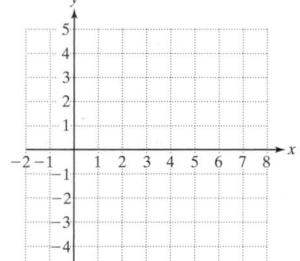

x	y

77. $y = \log_{1/2} x$ **(See Example 6.)**

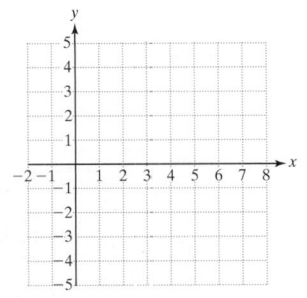

x	y

78. $y = \log_{1/e} x$

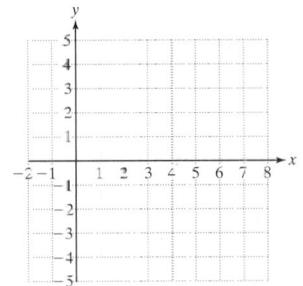

x	y

For Exercises 79–90, find the domain of the function and express the domain in interval notation. **(See Example 7.)**

79. $y = \log_7 (x - 5)$

80. $y = \log (x - 4)$

81. $g(x) = \log (2 - x)$

82. $f(x) = \log (1 - x)$

83. $y = \log (3x - 1)$

84. $y = \log_3 (2x + 1)$

85. $y = \log_3 (x + 1.2)$

86. $y = \log \left(x - \dfrac{1}{2} \right)$

87. $k(x) = \log (4 - 2x)$

88. $h(x) = \log (6 - 2x)$

89. $y = \log x^2$

90. $y = \log (x^2 + 1)$

Concept 5: Applications of the Common Logarithmic Function

For Exercises 91–92, use the formula $pH = -\log[H^+]$, where $[H^+]$ represents the concentration of hydrogen ions in moles per liter. Round to two decimal places. **(See Example 8.)**

91. Normally, the level of hydrogen ions in the blood is approximately 4.47×10^{-8} mol/L. Find the pH level of blood.

92. The normal pH level for streams and rivers is between 6.5 and 8.5. A high level of bacteria in a particular stream caused environmentalists to test the water. The level of hydrogen ions was found to be 0.006 mol/L. What is the pH level of the stream?

93. A graduate student in education is doing research to compare the effectiveness of two different teaching techniques designed to teach vocabulary to sixth-graders. The first group of students (group 1) was taught with method I, in which the students worked individually to complete the assignments in a workbook. The second group (group 2) was taught with method II, in which the students worked cooperatively in groups of four to complete the assignments in the same workbook.

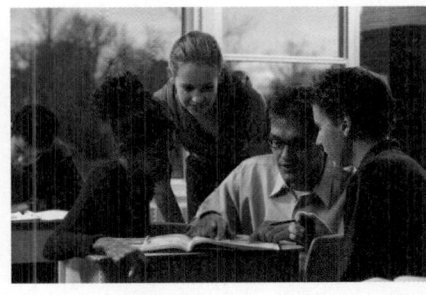

None of the students knew any of the vocabulary words before the study began. After completing the assignments in the workbook, the students were then tested on the vocabulary at 1-month intervals to assess how much material they had retained over time. The students' average score t months after completing the assignments are given by the following functions:

Method I: $S_1(t) = 91 - 30 \log(t + 1)$, where t is the time in months and $S_1(t)$ is the average score of students in group 1.

Method II: $S_2(t) = 88 - 15 \log(t + 1)$, where t is the time in months and $S_2(t)$ is the average score of students in group 2.

a. Complete the table to find the average scores for each group of students after the indicated number of months. Round to one decimal place. **(See Example 9.)**

b. Based on the table of values, what were the average scores for each group immediately after completion of the assigned material ($t = 0$)?

t (months)	0	1	2	6	12	24
$S_1(t)$						
$S_2(t)$						

c. Based on the table of values, which teaching method helped students retain the material better for a long time?

94. Generally, the more money a company spends on advertising, the higher the sales. Let a represent the amount of money spent on advertising (in \$100s). Then the amount of money in sales $S(a)$ (in \$1000s) is given by

$$S(a) = 10 + 20 \log(a + 1) \text{ where } a \geq 0.$$

a. The value of $S(1) \approx 16.0$ means that if \$100 is spent on advertising, \$16,000 is returned in sales. Find the values of $S(11)$, $S(21)$, and $S(31)$. Round to one decimal place. Interpret the meaning of each function value in the context of this problem.

b. The graph of $y = S(a)$ is shown here. Use the graph and your answers from part (a) to explain why the money spent in advertising becomes less "efficient" as it is used in larger amounts.

Graphing Calculator Exercises

For Exercises 95–100, graph the function on an appropriate viewing window. From the graph, identify the domain of the function and the location of the vertical asymptote.

95. $y = \log(x + 6)$

96. $y = \log(2x + 4)$

97. $y = \log(0.5x - 1)$

98. $y = \log(x + 8)$

99. $y = \log(2 - x)$

100. $y = \log(3 - x)$

Problem Recognition Exercises

Identifying Graphs of Functions

Match the function with the appropriate graph. Do not use a calculator.

1. $g(x) = 3^x$

2. $f(x) = \log x$

3. $h(x) = x^2$

4. $k(x) = -2x - 3$

5. $L(x) = |x|$

6. $m(x) = \sqrt{x}$

7. $B(x) = 3$

8. $A(x) = -x^2$

9. $n(x) = \sqrt[3]{x}$

10. $p(x) = x^3$

11. $q(x) = \dfrac{1}{x}$

12. $r(x) = x$

a.

b.

c.

d.

e.

f.

g.

h.

i.

j.

k.

l.

Section 8.5 Properties of Logarithms

Concepts

1. **Properties of Logarithms**
2. **Expanded Logarithmic Expressions**
3. **Single Logarithmic Expressions**

1. Properties of Logarithms

You have already been exposed to certain properties of logarithms that follow directly from the definition. Recall

$$y = \log_b x \quad \text{is equivalent to} \quad b^y = x \quad \text{for } x > 0, b > 0, \text{ and } b \neq 1$$

The following properties follow directly from the definition.

Property	Explanation
1. $\log_b 1 = 0$	Exponential form: $b^0 = 1$
2. $\log_b b = 1$	Exponential form: $b^1 = b$
3. $\log_b b^p = p$	Exponential form: $b^p = b^p$
4. $b^{\log_b x} = x$	Logarithmic form: $\log_b x = \log_b x$

Example 1 **Applying the Properties of Logarithms to Simplify Expressions**

Use the properties of logarithms to simplify the expressions. Assume that all variable expressions within the logarithms represent positive real numbers.

a. $\log_8 8 + \log_8 1$ **b.** $10^{\log (x+2)}$ **c.** $\log_{1/2} \left(\dfrac{1}{2}\right)^x$

Solution:

a. $\log_8 8 + \log_8 1 = 1 + 0 = 1$ Properties 2 and 1

b. $10^{\log (x+2)} = x + 2$ Property 4

c. $\log_{1/2} \left(\dfrac{1}{2}\right)^x = x$ Property 3

Skill Practice Use the properties of logarithms to simplify the expressions.

1. $\log_5 1 + \log_5 5$ **2.** $15^{\log_{15} 7}$ **3.** $\log_{1/3} \left(\dfrac{1}{3}\right)^c$

Three additional properties are useful when simplifying logarithmic expressions. The first is the product property for logarithms.

Product Property of Logarithms

Let b, x, and y be positive real numbers where $b \neq 1$. Then

$$\log_b (xy) = \log_b x + \log_b y$$

The logarithm of a product equals the sum of the logarithms of the factors.

Answers

1. 1   **2.** 7   **3.** c

Proof:

Let $M = \log_b x$, which implies $b^M = x$.

Let $N = \log_b y$, which implies $b^N = y$.

Then $xy = b^M b^N = b^{M+N}$.

Writing the expression $xy = b^{M+N}$ in logarithmic form, we have,

$$\log_b (xy) = M + N$$

$$\log_b (xy) = \log_b x + \log_b y \; \checkmark$$

To demonstrate the product property for logarithms, simplify the following expressions by using the order of operations.

$$\log_3 (3 \cdot 9) \stackrel{?}{=} \log_3 3 + \log_3 9$$

$$\log_3 27 \stackrel{?}{=} 1 + 2$$

$$3 \stackrel{?}{=} 3 \; \checkmark \text{ True}$$

Quotient Property of Logarithms

Let b, x, and y be positive real numbers where $b \neq 1$. Then

$$\log_b \left(\frac{x}{y} \right) = \log_b x - \log_b y$$

The logarithm of a quotient equals the difference of the logarithms of the numerator and denominator.

The proof of the quotient property for logarithms is similar to the proof of the product property and is omitted here. To demonstrate the quotient property for logarithms, simplify the following expressions by using the order of operations.

$$\log \left(\frac{1,000,000}{100} \right) \stackrel{?}{=} \log (1,000,000) - \log (100)$$

$$\log (10,000) \stackrel{?}{=} 6 - 2$$

$$4 \stackrel{?}{=} 4 \; \checkmark \text{ True}$$

Power Property of Logarithms

Let b and x be positive real numbers where $b \neq 1$. Let p be any real number. Then

$$\log_b x^p = p \log_b x$$

To demonstrate the power property for logarithms, simplify the following expressions by using the order of operations.

$$\log_4 4^2 \stackrel{?}{=} 2 \log_4 4$$

$$2 \stackrel{?}{=} 2 \cdot 1$$

$$2 \stackrel{?}{=} 2 \; \checkmark \text{ True}$$

The properties of logarithms are summarized in the box.

Summary of the Properties of Logarithms

Let b, x, and y be positive real numbers where $b \neq 1$, and let p be a real number. Then the following properties of logarithms are true.

1. $\log_b 1 = 0$	**5.** $\log_b(xy) = \log_b x + \log_b y$ **Product property for logarithms**
2. $\log_b b = 1$	**6.** $\log_b\left(\dfrac{x}{y}\right) = \log_b x - \log_b y$ **Quotient property for logarithms**
3. $\log_b b^p = p$	**7.** $\log_b x^p = p\log_b x$ **Power property for logarithms**
4. $b^{\log_b x} = x$	

2. Expanded Logarithmic Expressions

In many applications, it is advantageous to expand a logarithm into a sum or difference of simpler logarithms.

Example 2   **Writing a Logarithmic Expression in Expanded Form**

Write the expression as the sum or difference of logarithms of x, y, and z. Assume all variables represent positive real numbers.

$$\log_3\left(\frac{xy^3}{z^2}\right)$$

Solution:

$$\log_3\left(\frac{xy^3}{z^2}\right)$$

$$= \log_3(xy^3) - \log_3 z^2 \qquad \text{Quotient property for logarithms (property 6)}$$

$$= [\log_3 x + \log_3 y^3] - \log_3 z^2 \qquad \text{Product property for logarithms (property 5)}$$

$$= \log_3 x + 3\log_3 y - 2\log_3 z \qquad \text{Power property for logarithms (property 7)}$$

Skill Practice Write the expression as the sum or difference of logarithms of a, b, and c. Assume all variables represent positive real numbers.

4. $\log_5\left(\dfrac{a^2 b^3}{c}\right)$

Example 3   **Writing a Logarithmic Expression in Expanded Form**

Write the expression as the sum or difference of logarithms of x and y. Assume all variables represent positive real numbers.

$$\log\left(\frac{\sqrt{x}+y}{10}\right)$$

Answer

4. $2\log_5 a + 3\log_5 b - \log_5 c$

Solution:

$$\log\left(\frac{\sqrt{x+y}}{10}\right)$$

$= \log(\sqrt{x+y}) - \log 10$	Quotient property for logarithms (property 6)
$= \log(x+y)^{1/2} - 1$	Write $\sqrt{x+y}$ as $(x+y)^{1/2}$ and simplify $\log 10 = 1$.
$= \dfrac{1}{2}\log(x+y) - 1$	Power property for logarithms (property 7)

Skill Practice Write the expression as the sum or difference of logarithms of a and b. Assume $a > b$.

5. $\log\dfrac{(a-b)^2}{\sqrt{10}}$

Example 4 **Writing a Logarithmic Expression in Expanded Form**

Write the expression as the sum or difference of logarithms of x, y, and z. Assume all variables represent positive real numbers.

$$\log_b \sqrt[5]{\frac{x^4}{yz^3}}$$

Solution:

$$\log_b \sqrt[5]{\frac{x^4}{yz^3}}$$

$= \log_b\left(\dfrac{x^4}{yz^3}\right)^{1/5}$	Write $\sqrt[5]{\dfrac{x^4}{yz^3}}$ as $\left(\dfrac{x^4}{yz^3}\right)^{1/5}$.
$= \dfrac{1}{5}\log_b\left(\dfrac{x^4}{yz^3}\right)$	Power property for logarithms (property 7)
$= \dfrac{1}{5}[\log_b x^4 - \log_b(yz^3)]$	Quotient property for logarithms (property 6)
$= \dfrac{1}{5}[\log_b x^4 - (\log_b y + \log_b z^3)]$	Product property for logarithms (property 5)
$= \dfrac{1}{5}[\log_b x^4 - \log_b y - \log_b z^3]$	Distributive property
$= \dfrac{1}{5}[4\log_b x - \log_b y - 3\log_b z]$	Power property for logarithms (property 7)

or $\quad \dfrac{4}{5}\log_b x - \dfrac{1}{5}\log_b y - \dfrac{3}{5}\log_b z$

Skill Practice Write the expression as the sum or difference of logarithms of a, b, and c. Assume all variables represent positive real numbers.

6. $\log_3 \sqrt[4]{\dfrac{a}{bc^5}}$

Answers

5. $2\log(a-b) - \dfrac{1}{2}$

6. $\dfrac{1}{4}\log_3 a - \dfrac{1}{4}\log_3 b - \dfrac{5}{4}\log_3 c$

3. Single Logarithmic Expressions

In some applications, it is necessary to write a sum or difference of logarithms as a single logarithm.

Example 5 **Writing a Sum or Difference of Logarithms as a Single Logarithm**

Rewrite the expression as a single logarithm, and simplify the result, if possible.

$$\log_2 560 - \log_2 7 - \log_2 5$$

Solution:

$\log_2 560 - \log_2 7 - \log_2 5$

$= \log_2 560 - (\log_2 7 + \log_2 5)$	Factor out -1 from the last two terms.
$= \log_2 560 - \log_2 (7 \cdot 5)$	Product property for logarithms (property 5)
$= \log_2 \left(\dfrac{560}{7 \cdot 5} \right)$	Quotient property for logarithms (property 6)
$= \log_2 16$	Simplify inside parentheses.
$= \log_2 2^4$	Write 16 in base 2: 2^4
$= 4$	Property 3

Skill Practice Write the expression as a single logarithm, and simplify the result, if possible.

7. $\log_3 54 + \log_3 10 - \log_3 20$

Example 6 **Writing a Sum or Difference of Logarithms as a Single Logarithm**

Rewrite the expression as a single logarithm, and simplify the result, if possible. Assume all variable expressions within the logarithms represent positive real numbers.

$$2 \log x - \frac{1}{2} \log y + 3 \log z$$

Solution:

$2 \log x - \dfrac{1}{2} \log y + 3 \log z$

$= \log x^2 - \log y^{1/2} + \log z^3$	Power property for logarithms (property 7)
$= \log x^2 + \log z^3 - \log y^{1/2}$	Group terms with positive coefficients.
$= \log (x^2 z^3) - \log y^{1/2}$	Product property for logarithms (property 5)
$= \log \left(\dfrac{x^2 z^3}{y^{1/2}} \right)$ or $\log \left(\dfrac{x^2 z^3}{\sqrt{y}} \right)$	Quotient property for logarithms (property 6)

Skill Practice Write the expression as a single logarithm, and simplify, if possible.

8. $3 \log x + \dfrac{1}{3} \log y - 2 \log z$

Answers

7. 3   **8.** $\log \left(\dfrac{x^3 \sqrt[3]{y}}{z^2} \right)$

It is important to note that the properties of logarithms may be used to write a single logarithm as a sum or difference of logarithms. Furthermore, the properties may be used to write a sum or difference of logarithms as a single logarithm. In either case, these operations may change the domain.

For example, consider the function $y = \log_b x^2$. Using the power property for logarithms, we have $y = 2\log_b x$. Consider the domain of each function:

$y = \log_b x^2$ Domain: $(-\infty, 0) \cup (0, \infty)$

$y = 2\log_b x$ Domain: $(0, \infty)$

These two functions are equivalent only for values of x in the intersection of the two domains, that is, for $(0, \infty)$.

Section 8.5 Practice Exercises

Vocabulary and Key Concepts

1. a. Fill in the blanks to complete the basic properties of logarithms.
$\log_b b = $ _____ , $\log_b 1 = $ _____ , $\log_b b^x = $ _____ , and $b^{\log_b x} = $ _____ .

b. If $b, x,$ and y are positive real numbers and $b \neq 1$, then $\log_b(xy) = $ _____ and $\log_b\left(\dfrac{x}{y}\right) = $ _____ .

c. If b and x are positive real numbers and $b \neq 1$, then for any real number p, $\log_b x^p$ can be written as _____ .

d. Determine if the statement is true or false: $\log_b(xy) = (\log_b x)(\log_b y)$
Use the expression $\log_2 (4 \cdot 8)$ to help you answer.

e. Determine if the statement is true or false: $\log_b\left(\dfrac{x}{y}\right) = \dfrac{\log_b x}{\log_b y}$
Use the expression $\log_3\left(\dfrac{27}{9}\right)$ to help you answer.

f. Determine if the statement is true or false: $\log_b (x)^p = (\log_b x)^p$
Use the expression $\log(1000)^2$ to help you answer.

Review Exercises

For Exercises 2–5, find the values of the logarithmic and exponential expressions without using a calculator.

2. 8^{-2} **3.** $\log 10{,}000$ **4.** $\log_2 32$ **5.** 6^{-1}

For Exercises 6–9, approximate the values of the logarithmic and exponential expressions by using a calculator. Round to four decimal places.

6. $(\sqrt{2})^{\pi}$ **7.** $\log 8$ **8.** $\log 27$ **9.** $\pi^{\sqrt{2}}$

For Exercises 10–13, match the function with the appropriate graph.

10. $f(x) = 4^x$ **11.** $q(x) = \left(\dfrac{1}{5}\right)^x$ **12.** $h(x) = \log_5 x$ **13.** $k(x) = \log_{1/3} x$

a.

b.

c.

d.

Concept 1: Properties of Logarithms

14. Select the values that are equivalent to $\log_5 5^2$.

   a. $\log_5 25$

   b. $2 \log_5 5$

   c. $\log_5 5 + \log_5 5$

15. Select the values that are equivalent to $\log_2 2^3$.

   a. $3 \log_2 2$

   b. $\log_2 8$

   c. 3

16. Select the values that are equivalent to $\log 10^4$.

   a. 4

   b. $4 \log 10$

   c. $\log 10,000$

For Exercises 17–40, evaluate each expression. **(See Example 1.)**

17. $\log_3 3$ **18.** $\log 10$ **19.** $\log_5 5^4$ **20.** $\log_4 4^5$

21. $6^{\log_6 11}$ **22.** $7^{\log_7 2}$ **23.** $\log 10^3$ **24.** $\log_6 6^3$

25. $\log_3 1$ **26.** $\log_8 1$ **27.** $10^{\log 9}$ **28.** $8^{\log_8 5}$

29. $\log_{1/2} 1$ **30.** $\log_{1/3}\left(\dfrac{1}{3}\right)$ **31.** $\log_2 1 + \log_2 2^3$ **32.** $\log 10^4 + \log 10$

33. $\log_4 4 + \log_2 1$ **34.** $\log_7 7 + \log_4 4^2$ **35.** $\log_{1/4}\left(\dfrac{1}{4}\right)^{2x}$ **36.** $\log_{2/3}\left(\dfrac{2}{3}\right)^p$

37. $\log_a a^4$ **38.** $\log_y y^2$ **39.** $\log 10^2 - \log_3 3^2$ **40.** $\log_6 6^4 - \log 10^4$

41. Compare the expressions by approximating their values on a calculator. Which two expressions are equivalent?

   a. $\log(3 \cdot 5)$ **b.** $\log 3 \cdot \log 5$ **c.** $\log 3 + \log 5$

42. Compare the expressions by approximating their values on a calculator. Which two expressions are equivalent?

   a. $\log\left(\dfrac{6}{5}\right)$ **b.** $\dfrac{\log 6}{\log 5}$ **c.** $\log 6 - \log 5$

43. Compare the expressions by approximating their values on a calculator. Which two expressions are equivalent?

   a. $\log 20^2$ **b.** $[\log 20]^2$ **c.** $2 \log 20$

44. Compare the expressions by approximating their values on a calculator. Which two expressions are equivalent?

   a. $\log \sqrt{4}$ **b.** $\dfrac{1}{2} \log 4$ **c.** $\sqrt{\log 4}$

Concept 2: Expanded Logarithmic Expressions

For Exercises 45–62, expand into sums and/or differences of logarithms. Assume all variables represent positive real numbers. **(See Examples 2–4.)**

45. $\log_3\left(\dfrac{x}{5}\right)$

46. $\log_2\left(\dfrac{y}{z}\right)$

47. $\log(2x)$

48. $\log_6(xyz)$

49. $\log_5 x^4$

50. $\log_7 z^{1/3}$

51. $\log_4\left(\dfrac{ab}{c}\right)$

52. $\log_2\left(\dfrac{x}{yz}\right)$

53. $\log_b\left(\dfrac{\sqrt{xy}}{z^3 w}\right)$

54. $\log\left(\dfrac{a\sqrt[3]{b}}{cd^2}\right)$

55. $\log_2\left(\dfrac{x+1}{y^2\sqrt{z}}\right)$

56. $\log\left(\dfrac{a+1}{b\sqrt[3]{c}}\right)$

57. $\log\left(\sqrt[3]{\dfrac{ab^2}{c}}\right)$

58. $\log_5\left(\sqrt[4]{\dfrac{w^3 z}{x^2}}\right)$

59. $\log\left(\dfrac{1}{w^5}\right)$

60. $\log_3\left(\dfrac{1}{z^4}\right)$

61. $\log_b\left(\dfrac{\sqrt{a}}{b^3 c}\right)$

62. $\log_x\left(\dfrac{x}{y\sqrt{z}}\right)$

Concept 3: Single Logarithmic Expressions

For Exercises 63–78, write the expressions as a single logarithm and simplify if possible. Assume all variable expressions represent positive real numbers. **(See Examples 5–6.)**

63. $\log_3 270 - \log_3 2 - \log_3 5$

64. $\log_5 8 + \log_5 50 - \log_5 16$

65. $\log_7 98 - \log_7 2$

66. $\log_6 24 - \log_6 4$

67. $2\log_3 x - 3\log_3 y + \log_3 z$

68. $\log C + \log A + \log B + \log I + \log N$

69. $2\log_3 a - \dfrac{1}{4}\log_3 b + \log_3 c$

70. $\log_5 a - \dfrac{1}{2}\log_5 b - 3\log_5 c$

71. $\log_b x - 3\log_b x + 4\log_b x$

72. $2\log_3 z + \log_3 z - \dfrac{1}{2}\log_3 z$

73. $5\log_8 a - \log_8 1 + \log_8 8$

74. $\log_2 2 + 2\log_2 b - \log_2 1$

75. $2\log(x+6) + \dfrac{1}{3}\log y - 5\log z$

76. $\dfrac{1}{4}\log(a+1) - 2\log b - 4\log c$

77. $\log_b(x+1) - \log_b(x^2-1)$

78. $\log_x(p^2-4) - \log_x(p-2)$

Expanding Your Skills

For Exercises 79–90, find the values of the logarithms given that $\log_b 2 \approx 0.693$, $\log_b 3 \approx 1.099$, and $\log_b 5 \approx 1.609$.

79. $\log_b 6$

80. $\log_b 4$

81. $\log_b 12$

82. $\log_b 25$

83. $\log_b 81$

84. $\log_b 30$

85. $\log_b\left(\dfrac{5}{2}\right)$

86. $\log_b\left(\dfrac{25}{3}\right)$

87. $\log_b 10^6$

88. $\log_b 15^3$

89. $\log_b 5^{10}$

90. $\log_b 2^{12}$

91. The intensity of sound waves is measured in decibels and is calculated by the formula

$$B = 10 \log \left(\frac{I}{I_0} \right)$$

where I_0 is the minimum detectable decibel level.

a. Expand this formula by using the properties of logarithms.

b. Let $I_0 = 10^{-16} \, \text{W/cm}^2$ and simplify.

92. The Richter scale is used to measure the intensity of an earthquake and is calculated by the formula

$$R = \log \left(\frac{I}{I_0} \right)$$

where I_0 is the minimum level detectable by a seismograph.

a. Expand this formula by using the properties of logarithms.

b. Let $I_0 = 1$ and simplify.

Graphing Calculator Exercises

93. **a.** Graph $Y_1 = \log x^2$ and state its domain.

   b. Graph $Y_2 = 2 \log x$ and state its domain.

   c. For what values of x are the expressions $\log x^2$ and $2 \log x$ equivalent?

94. **a.** Graph $Y_1 = \log (x - 1)^2$ and state its domain.

   b. Graph $Y_2 = 2 \log (x - 1)$ and state its domain.

   c. For what values of x are the expressions $\log (x - 1)^2$ and $2 \log (x - 1)$ equivalent?

Section 8.6   The Irrational Number e and Change of Base

Concepts

1. **The Irrational Number** e
2. **Computing Compound Interest**
3. **The Natural Logarithmic Function**
4. **Change-of-Base Formula**
5. **Applications of the Natural Logarithmic Function**

1. The Irrational Number e

The exponential function base 10 is particularly easy to work with because integral powers of 10 represent different place positions in the base-10 numbering system. In this section, we introduce another important exponential function whose base is an irrational number called e.

Consider the expression $(1 + \frac{1}{x})^x$. The value of the expression for increasingly large values of x approaches a constant (Table 8-5).

Table 8-5

x	$\left(1 + \dfrac{1}{x}\right)^x$
100	2.70481382942
1000	2.71692393224
10,000	2.71814592683
100,000	2.71826823717
1,000,000	2.71828046932
1,000,000,000	2.71828182710

As x approaches infinity, the expression $(1 + \frac{1}{x})^x$ approaches a constant value that we call e. From Table 8-5, this value is approximately 2.718281828.

$$e \approx 2.718281828$$

The value of e is an irrational number (a nonterminating, nonrepeating decimal) and like the number π, it is a universal constant.

Example 1 Graphing $f(x) = e^x$

Graph the function defined by $f(x) = e^x$.

Solution:

Because the base of the function is greater than 1 ($e \approx 2.718281828$), the graph is an increasing exponential function. We can use a calculator to evaluate $f(x) = e^x$ at several values of x.

Calculator Connections

Topic: Evaluating Exponential Expressions Base *e*

Practice using your calculator by evaluating e^x for $x = 1$, $x = 2$, and $x = -1$.

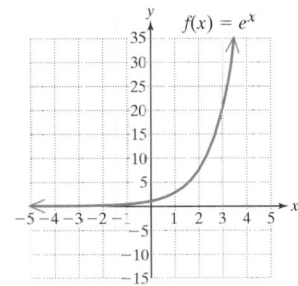

TIP: On a scientific calculator, you may need to enter the base and exponent of an exponential expression in reverse order. For example, enter e^2 as:

2   2ⁿᵈ   LN

If you are using your calculator correctly, your answers should match those found in Table 8-6. Values are rounded to three decimal places. The corresponding graph of $f(x) = e^x$ is shown in Figure 8-15.

Table 8-6

x	$f(x) = e^x$
-3	0.050
-2	0.135
-1	0.368
0	1.000
1	2.718
2	7.389
3	20.086

Figure 8-15

Skill Practice

1. Graph $f(x) = e^x + 1$.

---

2. Computing Compound Interest

One particularly interesting application of exponential functions is in computing compound interest.

1. If the number of compounding periods per year is finite, then the amount in an account is given by

$$A(t) = P\left(1 + \frac{r}{n}\right)^{nt}$$

where P is the initial principal, r is the annual interest rate, n is the number of times compounded per year, and t is the time in years that the money is invested.

Answer

1.

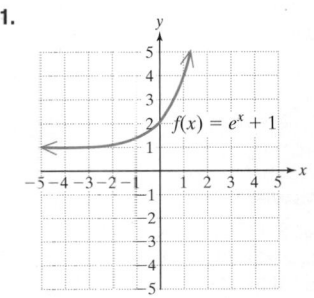

2. If the number of compound periods per year is infinite, then interest is said to be **compounded continuously**. In such a case, the amount in an account is given by

$$A(t) = Pe^{rt}$$

where P is the initial principal, r is the annual interest rate, and t is the time in years that the money is invested.

Example 2   Computing the Balance on an Account

Suppose $5000 is invested in an account earning 6.5% interest. Find the balance in the account after 10 yr under the following compounding options.

   a. Compounded annually   **b.** Compounded quarterly

   c. Compounded monthly   **d.** Compounded daily

   e. Compounded continuously

Solution:

Compounding Option	n Value	Formula	Result
Annually	$n = 1$	$A(10) = 5000\left(1 + \dfrac{0.065}{1}\right)^{(1)(10)}$	$9385.69
Quarterly	$n = 4$	$A(10) = 5000\left(1 + \dfrac{0.065}{4}\right)^{(4)(10)}$	$9527.79
Monthly	$n = 12$	$A(10) = 5000\left(1 + \dfrac{0.065}{12}\right)^{(12)(10)}$	$9560.92
Daily	$n = 365$	$A(10) = 5000\left(1 + \dfrac{0.065}{365}\right)^{(365)(10)}$	$9577.15
Continuously	Not applicable	$A(10) = 5000e^{(0.065)(10)}$	$9577.70

Notice that there is a $191.46 difference in the account balance between annual compounding and daily compounding. However, the difference between compounding daily and compounding continuously is small—$0.55. As n gets infinitely large, the function defined by

$$A(t) = P\left(1 + \frac{r}{n}\right)^{nt} \qquad \text{converges to } A(t) = Pe^{rt}.$$

Skill Practice

2. Suppose $1000 is invested at 5%. Find the balance after 8 yr under the following options.
   a. Compounded annually   **b.** Compounded quarterly
   c. Compounded monthly   **d.** Compounded daily
   e. Compounded continuously

3. The Natural Logarithmic Function

Recall that the common logarithmic function $y = \log x$ has a base of 10. Another important logarithmic function is called the **natural logarithmic function**. The natural logarithmic function has a base of e and is written as $y = \ln x$. That is,

$$y = \ln x = \log_e x$$

Answer
2. a. $1477.46   **b.** $1488.13
   c. $1490.59   **d.** $1491.78
   e. $1491.82

Example 3 Graphing $y = \ln x$

Graph $y = \ln x$.

Solution:

Because the base of the function $y = \ln x$ is e and $e > 1$, the graph is an increasing logarithmic function. We can use a calculator to find specific points on the graph of $y = \ln x$ by pressing the [LN] key.

  Practice using your calculator by evaluating $\ln x$ for the following values of x. If you are using your calculator correctly, your answers should match those found in Table 8-7. Values are rounded to three decimal places. The corresponding graph of $y = \ln x$ is shown in Figure 8-16.

TIP: On a scientific calculator, you may need to enter a logarithm and argument in reverse order. For example, enter ln 2 as

2 [LN]

Table 8-7

x	$\ln x$
0.25	−1.386
0.5	−0.693
1	0.000
2	0.693
3	1.099
4	1.386
5	1.609
6	1.792
7	1.946

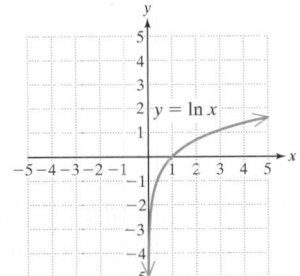

Figure 8-16

Calculator Connections

Topic: Evaluating Natural Logarithms

```
ln(1)
              0
ln(2)
    .6931471806
ln(3)
    1.098612289
```

Skill Practice

3. Graph $y = \ln x + 1$.

The properties of logarithms stated in Section 8.5 are also true for natural logarithms.

Properties of the Natural Logarithmic Function

Let x and y be positive real numbers, and let p be a real number. Then the following properties are true.

1. $\ln 1 = 0$ **5.** $\ln (xy) = \ln x + \ln y$ **Product property for logarithms**

2. $\ln e = 1$ **6.** $\ln \left(\dfrac{x}{y}\right) = \ln x - \ln y$ **Quotient property for logarithms**

3. $\ln e^p = p$ **7.** $\ln x^p = p \ln x$ **Power property for logarithms**

4. $e^{\ln x} = x$

Answer

3.

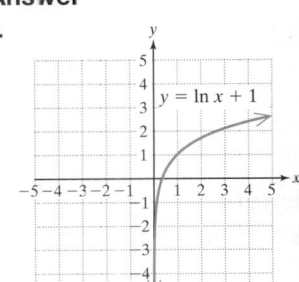

Example 4 **Simplifying Expressions with Natural Logarithms**

Simplify the expressions. Assume that all variable expressions within the logarithms represent positive real numbers.

a. $\ln e$ **b.** $\ln 1$ **c.** $\ln(e^{x+1})$ **d.** $e^{\ln(x+1)}$

Solution:

a. $\ln e = 1$ Property 2

b. $\ln 1 = 0$ Property 1

c. $\ln(e^{x+1}) = x + 1$ Property 3

d. $e^{\ln(x+1)} = x + 1$ Property 4

Skill Practice Simplify.

4. $\ln e^2$ **5.** $-3\ln 1$ **6.** $\ln e^{(x+y)}$ **7.** $e^{\ln(3x)}$

Example 5 **Writing a Sum or Difference of Natural Logarithms as a Single Logarithm**

Write the expression as a single logarithm. Assume that all variable expressions within the logarithms represent positive real numbers.

$$\ln x - \frac{1}{3}\ln y - \ln z$$

Solution:

$$\ln x - \frac{1}{3}\ln y - \ln z$$

$$= \ln x - \ln y^{1/3} - \ln z \qquad \text{Power property for logarithms (property 7)}$$

$$= \ln x - (\ln y^{1/3} + \ln z) \qquad \text{Factor out } -1 \text{ from the last two terms.}$$

$$= \ln x - \ln(y^{1/3} \cdot z) \qquad \text{Product property for logarithms (property 5)}$$

$$= \ln\left(\frac{x}{y^{1/3}z}\right) \qquad \text{Quotient property for logarithms (property 6)}$$

$$= \ln\left(\frac{x}{\sqrt[3]{y}\,z}\right) \qquad \text{Write the rational exponent as a radical.}$$

Skill Practice Write as a single logarithm.

8. $\frac{1}{4}\ln a - \ln b + \ln c$

Answers

4. 2 **5.** 0 **6.** $x + y$

7. $3x$ **8.** $\ln\left(\frac{\sqrt[4]{a}\,c}{b}\right)$

Example 6 **Writing a Logarithmic Expression in Expanded Form**

Write the expression as a sum or difference of logarithms of x and y. Assume all variable expressions within the logarithm represent positive real numbers.

$$\ln\left(\frac{e}{x^2\sqrt{y}}\right)$$

Solution:

$\ln\left(\dfrac{e}{x^2\sqrt{y}}\right)$

$= \ln e - \ln(x^2\sqrt{y})$ Quotient property for logarithms (property 6)

$= \ln e - (\ln x^2 + \ln \sqrt{y})$ Product property for logarithms (property 5)

$= 1 - \ln x^2 - \ln y^{1/2}$ Distributive property. Also simplify $\ln e = 1$ (property 2).

$= 1 - 2\ln x - \dfrac{1}{2}\ln y$ Power property for logarithms (property 7)

Skill Practice Write as a sum or difference of logarithms of x and y.

9. $\ln\left(\dfrac{x^3\sqrt{y}}{e^2}\right)$

4. Change-of-Base Formula

A calculator can be used to approximate the value of a logarithm with a base of 10 or a base of e by using the [LOG] key or the [LN] key, respectively. However, to use a calculator to evaluate a logarithmic expression with a base other than 10 or e, we must use the **change-of-base formula**.

> **Change-of-Base Formula**
>
> Let a and b be positive real numbers such that $a \neq 1$ and $b \neq 1$. Then for any positive real number x,
>
> $$\log_b x = \frac{\log_a x}{\log_a b}$$

Proof:

Let $M = \log_b x$, which implies that $b^M = x$.

Take the logarithm, base a, on both sides: $\log_a b^M = \log_a x$

Apply the power property for logarithms: $M \cdot \log_a b = \log_a x$

Divide both sides by $\log_a b$: $\dfrac{M \cdot \log_a b}{\log_a b} = \dfrac{\log_a x}{\log_a b}$

$$M = \frac{\log_a x}{\log_a b}$$

Because $M = \log_b x$, we have $\log_b x = \dfrac{\log_a x}{\log_a b}$ ✔

Answer

9. $3\ln x + \dfrac{1}{2}\ln y - 2$

The change-of-base formula converts a logarithm of one base to a ratio of logarithms of a different base. For the sake of using a calculator, we often apply the change-of-base formula with base 10 or base e.

Example 7   Using the Change-of-Base Formula

a. Use the change-of-base formula to evaluate $\log_4 80$ by using base 10. (Round to three decimal places.)

b. Use the change-of-base formula to evaluate $\log_4 80$ by using base e. (Round to three decimal places.)

Solution:

a. $\log_4 80 = \dfrac{\log_{10} 80}{\log_{10} 4} = \dfrac{\log 80}{\log 4} \approx \dfrac{1.903089987}{0.6020599913} \approx 3.161$

b. $\log_4 80 = \dfrac{\log_e 80}{\log_e 4} = \dfrac{\ln 80}{\ln 4} \approx \dfrac{4.382026635}{1.386294361} \approx 3.161$

To check the result, we see that $4^{3.161} \approx 80$.

Calculator Connections

Topic: Applying the Change-of-Base Formula

Use the change-of-base formula to evaluate $\log_4 80$.

```
log(80)/log(4)
          3.160964047
ln(80)/ln(4)
          3.160964047
```

Skill Practice

10. Use the change-of-base formula to evaluate $\log_5 95$ by using base 10. Round to three decimal places.

11. Use the change-of-base formula to evaluate $\log_5 95$ by using base e. Round to three decimal places.

Calculator Connections

Topic: Using the Change-of-Base Formula to Graph Logarithmic Functions

The change-of-base formula can be used to graph logarithmic functions with bases other than 10 or e. For example, to graph $Y_1 = \log_2 x$, we can enter the function as either

$$Y_1 = (\log x)/(\log 2) \quad \text{or} \quad Y_1 = (\ln x)/(\ln 2)$$

5. Applications of the Natural Logarithmic Function

Plant and animal tissue contains both carbon 12 and carbon 14. Carbon 12 is a stable form of carbon, whereas carbon 14 is a radioactive isotope with a half-life of approximately 5730 years. While a plant or animal is living, it takes in carbon from the atmosphere either through photosynthesis or through its food. The ratio of carbon 14 to carbon 12 in a living organism is constant and is the same as the ratio found in the atmosphere.

Answers
10. 2.829   **11.** 2.829

When a plant or animal dies, it no longer ingests carbon from the atmosphere. The amount of stable carbon 12 remains unchanged from the time of death, but the carbon 14 begins to decay. Because the rate of decay is constant, a tissue sample can be dated by comparing the percent of carbon 14 still present to the percentage of carbon 14 assumed to be in its original living state.

The age of a tissue sample is a function of the percent of carbon 14 still present in the organism according to the following model:

$$A(p) = \frac{\ln p}{-0.000121}$$

where $A(p)$ is the age in years and p is the percentage (in decimal form) of carbon 14 still present.

Example 8　Applying the Natural Logarithmic Function to Radioactive Decay

Using the formula

$$A(p) = \frac{\ln p}{-0.000121}$$

a. Find the age of a bone that has 72% of its initial carbon 14.

b. Find the age of the Iceman, a body uncovered in the mountains of northern Italy in 1991. Samples of his hair revealed that 52.7% of the original carbon 14 was present after his death.

Solution:

a. $\quad A(p) = \dfrac{\ln p}{-0.000121}$

$\quad A(0.72) = \dfrac{\ln 0.72}{-0.000121}$　　　Substitute 0.72 for p.

$\quad\quad\quad\quad \approx 2715$ years

b. $\quad A(p) = \dfrac{\ln p}{-0.000121}$

$\quad A(0.527) = \dfrac{\ln 0.527}{-0.000121}$　　　Substitute 0.527 for p.

$\quad\quad\quad\quad \approx 5300$ years　　　The body of the Iceman is approximately 5300 yr old.

Skill Practice

12. Use the formula

$$A(p) = \frac{\ln p}{-0.000121}$$

(where $A(p)$ is the age in years and p is the percent of carbon 14 still present) to determine the age of a human skull that has 90% of its initial carbon 14.

Answer

12. ≈ 871 yr

Section 8.6 Practice Exercises

Vocabulary and Key Concepts

1. **a.** As x becomes increasingly large, the value of $(1 + \frac{1}{x})^x$ approaches _____ ≈ 2.71828.

   b. The function $f(x) = e^x$ is the exponential function base _____.

   c. The logarithmic function base e is called the _____ logarithmic function and is denoted by $y =$ _____.

   d. If x is a positive real number, then $\ln 1 =$ _____, $\ln e =$ _____, $\ln e^p =$ _____, and $e^{\ln x} =$ _____.

   e. If x and y are positive real numbers, then $\ln(xy) =$ _____ and $\ln\left(\dfrac{x}{y}\right) =$ _____.

   f. If x is a positive real number, then for any real number p, $\ln x^p$ can be written as _____.

   g. The change-of-base formula states that $\log_b x = \dfrac{\log_a x}{\Box}$, where a is a positive real number and $a \neq 1$.

Review Exercises

For Exercises 2–3, write the expression as a single logarithm. Assume that all variables represent positive real numbers.

2. $\log_5 x + 3 \log_5 y - \log_5 z$

3. $\log_3 a - 5 \log_3 c + \dfrac{1}{2} \log_3 d$

For Exercises 4–5, write the expression as the sum or difference of logarithms of x, y, and z. Assume that all variables represent positive real numbers.

4. $\log\left(\dfrac{\sqrt{x}}{yz^4}\right)$

5. $\log_6 \sqrt[4]{\dfrac{xy^2}{z^3}}$

Concept 1: The Irrational Number e

6. From memory, write a decimal approximation of the number e, correct to three decimal places.

 For Exercises 7–10, graph the equation by completing the table and plotting points. Identify the domain. Round to two decimal places when necessary. **(See Example 1.)**

7. $y = e^{x+1}$

x	y
-4	
-3	
-2	
-1	
0	
1	

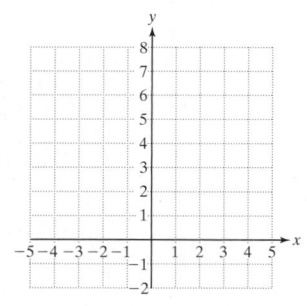

8. $y = e^{x+2}$

x	y
-5	
-4	
-3	
-2	
-1	
0	

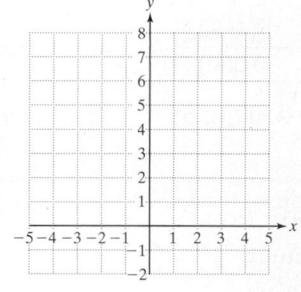

9. $y = e^x + 2$

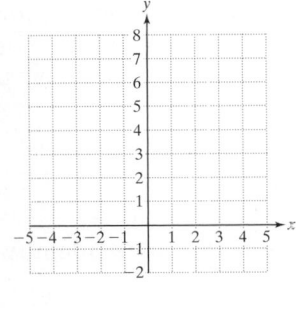

x	y
-2	
-1	
0	
1	
2	
3	

10. $y = e^x - 1$

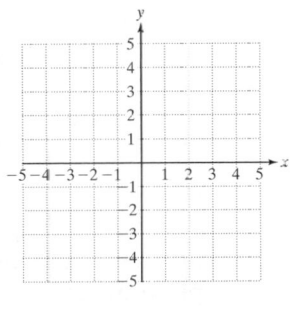

x	y
-4	
-3	
-2	
-1	
0	
1	

Concept 2: Computing Compound Interest

In Exercises 11–16, use the model

$$A(t) = P\left(1 + \frac{r}{n}\right)^{nt}$$

for interest compounded *n* times per year. Use the model $A(t) = Pe^{rt}$ for interest compounded continuously.

11. Suppose an investor deposits $10,000 in a certificate of deposit for 5 yr for which the interest is compounded monthly. Find the total amount of money in the account for the following interest rates. Compare your answers and comment on the effect of interest rate on an investment. **(See Example 2.)**

    a. $r = 4.0\%$ **b.** $r = 6.0\%$ **c.** $r = 8.0\%$ **d.** $r = 9.5\%$

12. Suppose an investor deposits $5000 in a certificate of deposit for 8 yr for which the interest is compounded quarterly. Find the total amount of money in the account for the following interest rates. Compare your answers and comment on the effect of interest rate on an investment.

    a. $r = 4.5\%$ **b.** $r = 5.5\%$ **c.** $r = 7.0\%$ **d.** $r = 9.0\%$

13. Suppose an investor deposits $8000 in a savings account for 10 yr at 4.5% interest. Find the total amount of money in the account for the following compounding options. Compare your answers. How does the number of compound periods per year affect the total investment?

    a. Compounded annually

    b. Compounded quarterly

    c. Compounded monthly

    d. Compounded daily

    e. Compounded continuously

14. Suppose an investor deposits $15,000 in a savings account for 8 yr at 5.0% interest. Find the total amount of money in the account for the following compounding options. Compare your answers. How does the number of compound periods per year affect the total investment?

    a. Compounded annually

    b. Compounded quarterly

    c. Compounded monthly

    d. Compounded daily

    e. Compounded continuously

15. Suppose an investor deposits $5000 in an account earning 6.5% interest compounded continuously. Find the total amount in the account for the following time periods. How does the length of time affect the amount of interest earned?

    a. 5 yr   **b.** 10 yr   **c.** 15 yr   **d.** 20 yr   **e.** 30 yr

16. Suppose an investor deposits $10,000 in an account earning 6.0% interest compounded continuously. Find the total amount in the account for the following time periods. How does the length of time affect the amount of interest earned?

   a. 5 yr   **b.** 10 yr   **c.** 15 yr   **d.** 20 yr   **e.** 30 yr

Concept 3: The Natural Logarithmic Function

For Exercises 17–20, graph the equation by completing the table and plotting the points. Identify the domain. Round to two decimal places when necessary. **(See Example 3.)**

17. $y = \ln(x - 2)$

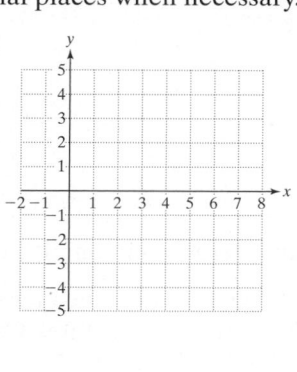

x	y
2.25	
2.50	
2.75	
3	
4	
5	
6	

18. $y = \ln(x - 1)$

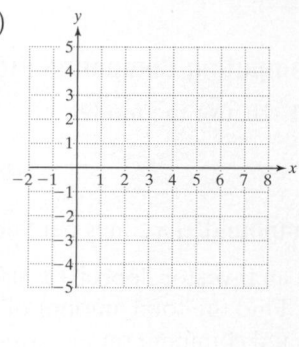

x	y
1.25	
1.50	
1.75	
2	
3	
4	
5	

19. $y = \ln x - 1$

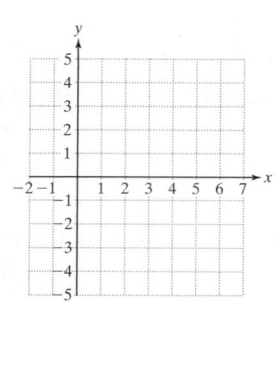

x	y
0.25	
0.5	
0.75	
1	
2	
3	
4	

20. $y = \ln x + 2$

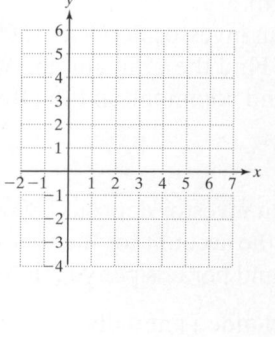

x	y
0.25	
0.5	
0.75	
1	
2	
3	
4	

21. **a.** Graph $f(x) = 10^x$ and $g(x) = \log x$.

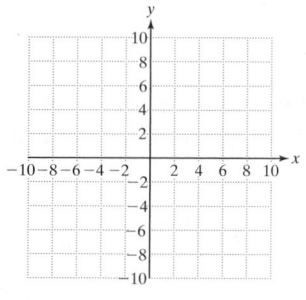

   b. Identify the domain and range of f.

   c. Identify the domain and range of g.

22. **a.** Graph $f(x) = e^x$ and $g(x) = \ln x$.

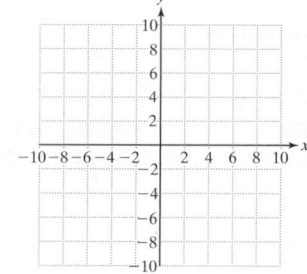

   b. Identify the domain and range of f.

   c. Identify the domain and range of g.

For Exercises 23–30, simplify the expressions. Assume all variables represent positive real numbers. **(See Example 4.)**

23. $\ln e$ **24.** $\ln e^2$ **25.** $\ln 1$ **26.** $e^{\ln x}$

27. $\ln e^{-6}$ **28.** $\ln e^{(5 - 3x)}$ **29.** $e^{\ln(2x+3)}$ **30.** $e^{\ln 4}$

For Exercises 31–38, write the expression as a single logarithm. Assume all variables represent positive real numbers. **(See Example 5.)**

31. $6 \ln p + \dfrac{1}{3} \ln q$

32. $2 \ln w + \ln z$

33. $\dfrac{1}{2} (\ln x - 3 \ln y)$

34. $\dfrac{1}{3} (4 \ln a - \ln b)$

35. $2 \ln a - \ln b - \dfrac{1}{3} \ln c$

36. $-\ln x + 3 \ln y - \ln z$

37. $4 \ln x - 3 \ln y - \ln z$

38. $\dfrac{1}{2} \ln c + \ln a - 2 \ln b$

For Exercises 39–46, write the expression as a sum and/or difference of $\ln a$, $\ln b$, and $\ln c$. Assume all variables represent positive real numbers. **(See Example 6.)**

39. $\ln \left(\dfrac{a}{b} \right)^2$

40. $\ln \sqrt[3]{\dfrac{a}{b}}$

41. $\ln (b^2 \cdot e)$

42. $\ln (\sqrt{c} \cdot e)$

43. $\ln \left(\dfrac{a^4 \sqrt{b}}{c} \right)$

44. $\ln \left(\dfrac{\sqrt{ab}}{c^3} \right)$

45. $\ln \left(\dfrac{ab}{c^2} \right)^{1/5}$

46. $\ln \sqrt{2ab}$

Concept 4: Change-of-Base Formula

47. a. Evaluate $\log_6 200$ by computing $\dfrac{\log 200}{\log 6}$ to four decimal places.

    b. Evaluate $\log_6 200$ by computing $\dfrac{\ln 200}{\ln 6}$ to four decimal places.

    c. How do your answers to parts (a) and (b) compare?

48. a. Evaluate $\log_8 120$ by computing $\dfrac{\log 120}{\log 8}$ to four decimal places.

    b. Evaluate $\log_8 120$ by computing $\dfrac{\ln 120}{\ln 8}$ to four decimal places.

    c. How do your answers to parts (a) and (b) compare?

For Exercises 49–60, use the change-of-base formula to approximate the logarithms to four decimal places. Check your answers by using the exponential key on your calculator. **(See Example 7.)**

49. $\log_2 7$

50. $\log_3 5$

51. $\log_8 24$

52. $\log_4 17$

53. $\log_8 0.012$

54. $\log_7 0.251$

55. $\log_9 1$

56. $\log_2 \left(\dfrac{1}{5} \right)$

57. $\log_4 \left(\dfrac{1}{100} \right)$

58. $\log_5 0.0025$

59. $\log_7 0.0006$

60. $\log_2 0.24$

Concept 5: Applications of the Natural Logarithmic Function

Under continuous compounding, the amount of time t in years required for an investment to double is a function of the interest rate r according to the formula:

$$t = \frac{\ln 2}{r}$$

 Use the formula for Exercises 61–63. **(See Example 8.)**

61. a. If you invest $5000, how long will it take the investment to reach $10,000 if the interest rate is 4.5%? Round to one decimal place.

   b. If you invest $5000, how long will it take the investment to reach $10,000 if the interest rate is 10%? Round to one decimal place.

   c. Using the doubling time found in part (b), how long would it take a $5000 investment to reach $20,000 if the interest rate is 10%?

62. a. If you invest $3000, how long will it take the investment to reach $6000 if the interest rate is 5.5%? Round to one decimal place.

   b. If you invest $3000, how long will it take the investment to reach $6000 if the interest rate is 8%? Round to one decimal place.

   c. Using the doubling time found in part (b), how long would it take a $3000 investment to reach $12,000 if the interest rate is 8%?

63. a. If you invest $4000, how long will it take the investment to reach $8000 if the interest rate is 3.5%? Round to one decimal place.

   b. If you invest $4000, how long will it take the investment to reach $8000 if the interest rate is 5%? Round to one decimal place.

   c. Using the doubling time found in part (b), how long would it take a $4000 investment to reach $16,000 if the interest rate is 5%?

64. On August 31, 1854, an epidemic of cholera was discovered in London, England, resulting from a contaminated community water pump at Broad Street. By the end of September more than 600 citizens who drank water from the pump had died.

   The cumulative number of deaths from cholera in the 1854 London epidemic can be approximated by

$$D(t) = 91 + 160 \ln (t + 1)$$

   where t is the number of days after the start of the epidemic ($t = 0$ corresponds to September 1, 1854).

   a. Approximate the total number of deaths as of September 1 ($t = 0$).

   b. Approximate the total number of deaths as of September 5, September 10, and September 20.

Graphing Calculator Exercises

65. a. Graph the function defined by $f(x) = \log_3 x$ by graphing $Y_1 = \dfrac{\log x}{\log 3}$.

   b. Graph the function defined by $f(x) = \log_3 x$ by graphing $Y_2 = \dfrac{\ln x}{\ln 3}$.

   c. Does it appear that $Y_1 = Y_2$ on the standard viewing window?

66. a. Graph the function defined by $f(x) = \log_7 x$ by graphing $Y_1 = \dfrac{\log x}{\log 7}$.

b. Graph the function defined by $f(x) = \log_7 x$ by graphing $Y_2 = \dfrac{\ln x}{\ln 7}$.

c. Does it appear that $Y_1 = Y_2$ on the standard viewing window?

For Exercises 67–69, graph the functions on your calculator.

67. Graph $s(x) = \log_{1/2} x$ **68.** Graph $y = e^{x-2}$ **69.** Graph $y = e^{x-1}$

Problem Recognition Exercises

Logarithmic and Exponential Forms

Fill out the table by writing the exponential expressions in logarithmic form, and the logarithmic expressions in exponential form. Use the fact that:

$$y = \log_b x \quad \text{is equivalent to} \quad b^y = x$$

	Exponential Form	**Logarithmic Form**
1.	$2^5 = 32$	
2.		$\log_3 81 = 4$
3.	$z^y = x$	
4.		$\log_b a = c$
5.	$10^3 = 1000$	
6.		$\log 10 = 1$
7.	$e^a = b$	
8.		$\ln p = q$
9.	$\left(\frac{1}{2}\right)^2 = \frac{1}{4}$	
10.		$\log_{1/3} 9 = -2$
11.	$10^{-2} = 0.01$	
12.		$\log 4 = x$
13.	$e^0 = 1$	
14.		$\ln e = 1$
15.	$25^{1/2} = 5$	
16.		$\log_{16} 2 = \frac{1}{4}$
17.	$e^t = s$	
18.		$\ln w = r$
19.	$15^{-2} = \frac{1}{225}$	
20.		$\log_3 p = -1$

Section 8.7   Logarithmic and Exponential Equations and Applications

Concepts

1. Solving Logarithmic Equations
2. Solving Exponential Equations
3. Applications

1. Solving Logarithmic Equations

Equations containing one or more logarithms are called **logarithmic equations**. For example:

$$\log_3(2x + 5) = 1 \qquad \text{and} \qquad \log_4 x = 1 - \log_4(x - 3)$$

To solve equations involving logarithms of first degree, we use the following guidelines.

Solving Logarithmic Equations

Step 1   Isolate the logarithms on one side of the equation.

Step 2   Write a sum or difference of logarithms as a single logarithm.

Step 3   Rewrite the equation in step 2 in exponential form.

Step 4   Solve the resulting equation from step 3.

Step 5   Check all solutions to verify that they are within the domain of the logarithmic expressions in the original equation.

Example 1   **Solving a Logarithmic Equation**

Solve the equation. $\log_3(2x + 5) = 1$

Solution:

$$\log_3(2x + 5) = 1$$ This equation has a single logarithm isolated on the left-hand side.

$$2x + 5 = 3^1$$ Write the logarithm in its equivalent exponential form.

$$2x + 5 = 3$$ Solve the resulting equation.

$$2x = -2$$

$$x = -1$$ <u>Check:</u> $x = -1$

$$\log_3(2x + 5) = 1$$

$$\log_3[2(-1) + 5] \overset{?}{=} 1$$

$$\log_3(3) \overset{?}{=} 1 \;\checkmark$$

The solution set is $\{-1\}$.

Skill Practice Solve the equation.

1. $\log_6(5x - 4) = 2$

Answer

1. $\{8\}$

Example 2   Solving a Logarithmic Equation

Solve the equation. $\log_4 x = 1 - \log_4(x - 3)$

Solution:

$$\log_4 x = 1 - \log_4(x - 3)$$

$\log_4 x + \log_4(x - 3) = 1$ Isolate the logarithms on one side of the equation.

$\log_4[x(x - 3)] = 1$ Write as a single logarithm.

$\log_4(x^2 - 3x) = 1$ Simplify inside the parentheses.

$x^2 - 3x = 4^1$ Write the equation in exponential form.

$x^2 - 3x - 4 = 0$ The resulting equation is quadratic.

$(x - 4)(x + 1) = 0$ Factor.

$x = 4$ or $x = -1$ Apply the zero product rule.

Notice that -1 is *not* a solution because $\log_4 x$ is not defined at $x = -1$. However, $x = 4$ *is* defined in both expressions $\log_4 x$ and $\log_4(x - 3)$. We can substitute $x = 4$ into the original equation to show that it checks.

Check: $x = 4$

$$\log_4 x = 1 - \log_4(x - 3)$$

$$\log_4 4 \overset{?}{=} 1 - \log_4(4 - 3)$$

$$1 \overset{?}{=} 1 - \log_4 1$$

$$1 \overset{?}{=} 1 - 0 \; \checkmark \; \text{True}$$

The solution set is $\{4\}$. (The value -1 does not check.)

Skill Practice Solve the equation.

2. $\log_3(x - 8) = 2 - \log_3 x$

Example 3   Solving a Logarithmic Equation

Solve the equation. $\log(x + 300) = 3.7$

Solution:

$\log(x + 300) = 3.7$ The equation has a single logarithm that is already isolated.

$10^{3.7} = x + 300$ Write the equation in exponential form.

$10^{3.7} - 300 = x$ Solve for x.

$x = 10^{3.7} - 300$

≈ 4711.87

Answer

2. $\{9\}$ (The value -1 does not check.)

Check: $x = 10^{3.7} - 300$ Check the exact value of x in the
                                              original equation.

$$\log(x + 300) = 3.7$$

$$\log[(10^{3.7} - 300) + 300] \overset{?}{=} 3.7$$

$$\log(10^{3.7} - 300 + 300) \overset{?}{=} 3.7$$

$$\log 10^{3.7} \overset{?}{=} 3.7 \qquad \text{Property 3 of logarithms:}$$
$$\qquad\qquad\qquad\qquad \log_b b^p = p$$

$$3.7 \overset{?}{=} 3.7 \checkmark \text{ True}$$

The solution $10^{3.7} - 300$ checks.

The solution set is $\{10^{3.7} - 300\}$.

Skill Practice Solve the equation.

3. $\log(p + 6) = 1.3$

The following property is another useful tool to solve a logarithmic equation. This is demonstrated in Example 4.

Equivalence of Logarithmic Expressions

Let x, y, and b represent real numbers such that $x > 0, y > 0, b > 0$, and $b \neq 1$. Then

$$\log_b x = \log_b y \quad \text{implies} \quad x = y$$

Example 4 **Solving a Logarithmic Equation**

Solve the equation.

$$\ln(x + 2) + \ln(x - 1) = \ln(9x - 17)$$

Solution:

$$\ln(x + 2) + \ln(x - 1) = \ln(9x - 17)$$

$$\ln[(x + 2)(x - 1)] = \ln(9x - 17) \qquad \text{Write the logarithms on the left-hand side as a single logarithm.}$$

$$\ln(x^2 + x - 2) = \ln(9x - 17) \qquad \text{Simplify.}$$

$$x^2 + x - 2 = 9x - 17 \qquad \text{Use the equivalence property of logarithms. That is, equate the arguments of each log.}$$

$$x^2 - 8x + 15 = 0 \qquad \text{Solve the resulting quadratic equation.}$$

$$(x - 5)(x - 3) = 0$$

$$x = 5 \quad \text{or} \quad x = 3$$

The solutions 5 and 3 are both within the domain of the logarithmic functions in the original equation. Both solutions check.

The solution set is $\{5, 3\}$.

Skill Practice Solve the equation.

4. $\ln(t - 3) + \ln(t - 1) = \ln(2t - 5)$

2. Solving Exponential Equations

An equation with one or more exponential expressions is called an **exponential equation**. The following property is often useful in solving exponential equations.

> **Equivalence of Exponential Expressions**
>
> Let x, y, and b be real numbers such that $b > 0$ and $b \neq 1$. Then
>
> $$b^x = b^y \quad \text{implies} \quad x = y$$

The equivalence of exponential expressions indicates that if two exponential expressions of the same base are equal, their exponents must be equal.

Example 5 Solving an Exponential Equation

Solve the equation. $\quad 4^{2x-9} = 64$

Solution:

$$4^{2x-9} = 64$$

$$4^{2x-9} = 4^3 \qquad \text{Write both sides with a common base.}$$

$$2x - 9 = 3 \qquad \text{If } b^x = b^y, \text{ then } x = y.$$

$$2x = 12 \qquad \text{Solve for } x.$$

$$x = 6$$

To check, substitute $x = 6$ into the original equation.

$$4^{2(6)-9} \stackrel{?}{=} 64$$

$$4^{12-9} \stackrel{?}{=} 64$$

$$4^3 \stackrel{?}{=} 64 \; ✔ \text{ True}$$

The solution set is $\{6\}$.

Skill Practice Solve the equation.

5. $2^{3x+1} = 8$

Answers

4. $\{4\}$ (The value 2 does not check.)

5. $\left\{\dfrac{2}{3}\right\}$

Example 6 Solving an Exponential Equation

Solve the equation. $(2^x)^{x+3} = \dfrac{1}{4}$

Solution:

$$(2^x)^{x+3} = \dfrac{1}{4}$$

$$2^{x^2+3x} = 2^{-2}$$ Apply the multiplication property of exponents. Write both sides of the equation with a common base.

$$x^2 + 3x = -2$$ If $b^x = b^y$, then $x = y$.

$$x^2 + 3x + 2 = 0$$ The resulting equation is quadratic.

$$(x + 2)(x + 1) = 0$$ Solve for x.

$$x = -2 \quad \text{or} \quad x = -1$$ Both solutions check.

The solution set is $\{-2, -1\}$.

Skill Practice Solve the equation.

6. $(3^x)^{x-5} = \dfrac{1}{81}$

Example 7 Solving an Exponential Equation

Solve the equation. $4^x = 25$

Solution:

Because 25 cannot be written as an integral power of 4, we cannot easily use the property that if $b^x = b^y$, then $x = y$. Instead, we can take a logarithm of any base on both sides of the equation. Then by applying the power property of logarithms, the unknown exponent can be written as a factor.

$$4^x = 25$$

$$\log 4^x = \log 25$$ Take the common logarithm of both sides.

$$x \log 4 = \log 25$$ Apply the power property of logarithms to express the exponent as a factor. This is now a linear equation in x.

$$\dfrac{x \log 4}{\log 4} = \dfrac{\log 25}{\log 4}$$ Solve for x.

$$x = \dfrac{\log 25}{\log 4} \text{ or approximately 2.322.}$$

The solution set is $\left\{\dfrac{\log 25}{\log 4}\right\}$ or equivalently $\left\{\dfrac{\ln 25}{\ln 4}\right\}$.

Skill Practice Solve the equation.

7. $5^x = 32$

Answers

6. $\{1, 4\}$

7. $\left\{\dfrac{\log 32}{\log 5}\right\}$ or $\left\{\dfrac{\ln 32}{\ln 5}\right\}$

TIP: The equation from Example 7 is written in the form $b^x =$ constant. In this special case, we also have the option of rewriting the equation in its corresponding logarithmic form to solve for x.

$$4^x = 25$$

$$x = \log_4 25$$

$$= \frac{\ln 25}{\ln 4} \quad \text{(Change-of-base formula)}$$

Solving Exponential Equations

Step 1 Isolate one of the exponential expressions in the equation.

Step 2 Take a logarithm on both sides of the equation. (The natural logarithmic function or the common logarithmic function is often used so that the final answer can be approximated with a calculator.)

Step 3 Use the power property of logarithms (property 7) to write exponents as factors. Recall: $\log_b M^p = p \log_b M$.

Step 4 Solve the resulting equation from step 3.

Example 8 **Solving an Exponential Equation by Taking a Logarithm on Both Sides**

Solve the equation. $e^{-3.6x} - 2 = 7.74$

Solution:

$e^{-3.6x} = 9.74$ Add 2 to both sides to isolate the exponential expression.

$\ln e^{-3.6x} = \ln 9.74$ The exponential expression has a base of e, so it is convenient to take the natural logarithm of both sides.

$(-3.6x) \ln e = \ln 9.74$ Use the power property of logarithms.

$-3.6x = \ln 9.74$ Simplify (recall that $\ln e = 1$).

$$x = \frac{\ln 9.74}{-3.6} \approx -0.632$$

The solution set is $\left\{ -\dfrac{\ln 9.74}{3.6} \right\}$.

Skill Practice Solve the equation.

8. $e^{-0.2t} + 1 = 8.52$

Answer

8. $\left\{ \dfrac{\ln 7.52}{-0.2} \right\}$

Example 9   Solving an Exponential Equation by Taking a Logarithm on Both Sides

Solve the equation. $2^{x+3} = 7^x$

Solution:

$$2^{x+3} = 7^x$$

$$\ln 2^{(x+3)} = \ln 7^x$$ Take the natural logarithm of both sides.

$$(x + 3) \ln 2 = x \ln 7$$ Use the power property of logarithms.

$$x(\ln 2) + 3(\ln 2) = x \ln 7$$ Apply the distributive property.

$$x(\ln 2) - x(\ln 7) = -3 \ln 2$$ Collect x-terms on one side.

$$x(\ln 2 - \ln 7) = -3 \ln 2$$ Factor out x.

$$\frac{x(\ln 2 - \ln 7)}{(\ln 2 - \ln 7)} = \frac{-3 \ln 2}{\ln 2 - \ln 7}$$ Solve for x.

$$x = \frac{-3 \ln 2}{\ln 2 - \ln 7} \approx 1.66$$

The solution set is $\left\{ -\dfrac{3 \ln 2}{\ln 2 - \ln 7} \right\}$.

TIP: The exponential equation $2^{x+3} = 7^x$ could have been solved by taking a logarithm of *any* base on both sides of the equation.

Skill Practice Solve the equation.

9. $3^x = 8^{x+2}$

3. Applications

Example 10   Applying an Exponential Function to World Population

The population of the world was estimated to have reached 6.5 billion in April 2006. The population growth rate for the world is estimated to be 1.4%. (*Source: U.S. Census Bureau*)

$$P(t) = 6.5(1.014)^t$$

represents the world population in billions as a function of the number of years after April 2006 ($t = 0$ represents April 2006).

a. Use the function to estimate the world population in April 2010.

b. Use the function to estimate the amount of time after April 2006 required for the world population to reach 13 billion.

Solution:

a. $P(t) = 6.5(1.014)^t$

$$P(4) = 6.5(1.014)^4$$ The year 2010 corresponds to $t = 4$.

$$\approx 6.87$$ In 2010, the world's population was approximately 6.87 billion.

Answer

9. $\left\{ \dfrac{2 \ln 8}{\ln 3 - \ln 8} \right\}$

b. $\quad P(t) = 6.5(1.014)^t$

$\quad\quad 13 = 6.5(1.014)^t \quad\quad\quad$ Substitute $P(t) = 13$ and solve for t.

$\quad\quad \dfrac{13}{6.5} = \dfrac{6.5(1.014)^t}{6.5} \quad\quad$ Isolate the exponential expression on one side of the equation.

$\quad\quad\quad 2 = 1.014^t$

$\quad\quad \ln 2 = \ln 1.014^t \quad\quad$ Take the natural logarithm of both sides.

$\quad\quad \ln 2 = t \ln 1.014 \quad\quad$ Use the power property of logarithms.

$\quad \dfrac{\ln 2}{\ln 1.014} = \dfrac{t \ln 1.014}{\ln 1.014} \quad\quad$ Solve for t.

$\quad\quad\quad t = \dfrac{\ln 2}{\ln 1.014} \approx 50 \quad\quad$ The population will reach 13 billion (double the April 2006 value) approximately 50 yr after 2006.

Note: It has taken thousands of years for the world's population to reach 6.5 billion. However, with a growth rate of 1.4%, it will take only 50 years to gain an additional 6.5 billion, if this trend continues.

Skill Practice Use the population function from Example 10.

10. Estimate the world population in April 2020.

11. Estimate the year in which the world population will reach 9 billion.

On Friday, April 25, 1986, a nuclear accident occurred at the Chernobyl nuclear reactor, resulting in radioactive contaminates being released into the atmosphere. The most hazardous isotopes released in this accident were ^{137}Cs (cesium 137), ^{131}I (iodine 131), and ^{90}Sr (strontium 90). People living close to Chernobyl (in Ukraine) were at risk of radiation exposure from inhalation, from absorption through the skin, and from food contamination. Years after the incident, scientists have seen an increase in the incidence of thyroid disease among children living in the contaminated areas. Because iodine is readily absorbed in the thyroid gland, scientists suspect that radiation from iodine 131 is the cause.

Example 11 Applying an Exponential Equation to Radioactive Decay

The half-life of radioactive iodine, ^{131}I, is 8.04 days. If 10 g of iodine 131 is initially present, then the amount of radioactive iodine still present after t days is approximated by

$$A(t) = 10e^{-0.0862t}$$

where t is the time in days.

a. Use the model to approximate the amount of ^{131}I still present after 2 weeks. Round to the nearest 0.1 g.

b. How long will it take for the amount of ^{131}I to decay to 0.5 g? Round to the nearest 0.1 day.

Answers

10. Approximately 7.9 billion

11. The year 2029

Solution:

a. $A(t) = 10e^{-0.0862t}$

$A(14) = 10e^{-0.0862(14)}$ Substitute $t = 14$ (2 weeks).

$\approx 3.0 \text{ g}$ There was 3.0 g still present after 2 weeks.

b. $A(t) = 10e^{-0.0862t}$

$0.5 = 10e^{-0.0862t}$ Substitute $A = 0.5$.

$\dfrac{0.5}{10} = \dfrac{\cancel{10}e^{-0.0862t}}{\cancel{10}}$ Isolate the exponential expression.

$0.05 = e^{-0.0862t}$

$\ln 0.05 = \ln e^{-0.0862t}$ Take the natural logarithm of both sides.

$\ln 0.05 = -0.0862t$ The resulting equation is linear.

$\dfrac{\ln 0.05}{-0.0862} = \dfrac{\cancel{-0.0862}t}{\cancel{-0.0862}}$ Solve for t.

$t = \dfrac{\ln 0.05}{-0.0862} \approx 34.8 \text{ days}$ It will take 34.8 days for 10 g of ^{131}I to decay to 0.5 g.

Skill Practice Radioactive strontium 90 (^{90}Sr) has a half-life of 28 yr. If 100 g of strontium 90 is initially present, the amount left after t years is approximated by

$$A(t) = 100e^{-0.0248t}$$

12. Find the amount of ^{90}Sr present after 85 yr.

13. How long will it take for the amount of ^{90}Sr to decay to 40 g?

Answers

12. 12.1 g   **13.** 36.9 years

Section 8.7 Practice Exercises

Vocabulary and Key Concepts

1. a. The equivalence property of logarithmic expressions states that if $\log_b x = \log_b y$, then _____.

b. The equivalence property of exponential expressions states that if $b^x = b^y$, then _____.

Review Exercises

For Exercises 2–6, write the expression as a single logarithm. Assume all variables represent positive real numbers.

2. $\log_b (x - 1) + \log_b (x + 2)$

3. $\log_b x + \log_b (2x + 3)$

4. $\log_b x - \log_b (1 - x)$

5. $\log_b (x + 2) - \log_b (3x - 5)$

6. $2 \log_b x - \log_b (x + 1)$

Concept 1: Solving Logarithmic Equations

For Exercises 7–38, solve the logarithmic equations. **(See Examples 1–4.)**

7. $\log_3 x = 2$

8. $\log_4 x = 9$

9. $\log p = 42$

10. $\log q = \dfrac{1}{2}$

11. $\ln x = 0.08$

12. $\ln x = 19$

13. $\log (x + 40) = -9.2$

14. $\log (z - 3) = -6.7$

15. $\log_x 25 = 2 \qquad (x > 0)$

16. $\log_x 100 = 2 \qquad (x > 0)$

17. $\log_b 10{,}000 = 4 \qquad (b > 0)$

18. $\log_b e^3 = 3 \qquad (b > 0)$

19. $\log_y 5 = \dfrac{1}{2} \qquad (y > 0)$

20. $\log_b 8 = \dfrac{1}{2} \qquad (b > 0)$

21. $\log_4 (c + 5) = 3$

22. $\log_5 (a - 4) = 2$

23. $\log_5 (4y + 1) = 1$

24. $\log_6 (5t - 2) = 1$

25. $\ln (1 - x) = 0$

26. $\log_4 (2 - x) = 1$

27. $\log_3 8 - \log_3 (x + 5) = 2$

28. $\log_2 (x + 3) - \log_2 (x + 2) = 1$

29. $\log_2 (h - 1) + \log_2 (h + 1) = 3$

30. $\log_3 k + \log_3 (2k + 3) = 2$

31. $\log (x + 2) = \log (3x - 6)$

32. $\log x = \log (1 - x)$

33. $\ln x - \ln (4x - 9) = 0$

34. $\ln (x + 5) - \ln x = \ln (4x)$

35. $\log_5 (3t + 2) - \log_5 t = \log_5 4$

36. $\log (6y - 7) + \log y = \log 5$

37. $\log (4m) = \log 2 + \log (m - 3)$

38. $\log (-h) + \log 3 = \log (2h - 15)$

Concept 2: Solving Exponential Equations

For Exercises 39–54, solve the exponential equation by using the property that $b^x = b^y$ implies $x = y$, for $b > 0$ and $b \neq 1$. **(See Examples 5–6.)**

39. $5^x = 625$

40. $3^x = 81$

41. $2^{-x} = 64$

42. $6^{-x} = 216$

43. $36^x = 6$

44. $343^x = 7$

45. $4^{2x-1} = 64$

46. $5^{3x-1} = 125$

47. $81^{3x-4} = \dfrac{1}{243}$

48. $4^{2x-7} = \dfrac{1}{128}$

49. $\left(\dfrac{2}{3}\right)^{-x+4} = \dfrac{8}{27}$

50. $\left(\dfrac{1}{4}\right)^{3x+2} = \dfrac{1}{64}$

51. $16^{-x+1} = 8^{5x}$

52. $27^{(1/3)x} = \left(\dfrac{1}{9}\right)^{2x-1}$

53. $(4^x)^{x+1} = 15$

54. $(3^x)^{x+2} = \dfrac{1}{3}$

For Exercises 55–74, solve the exponential equation by taking the logarithm of both sides. **(See Examples 7–9.)**

55. $8^a = 21$

56. $6^y = 39$

57. $e^x = 8.1254$

58. $e^x = 0.3151$

59. $10^t = 0.0138$

60. $10^p = 16.8125$

61. $e^{0.07h} - 6 = 9$

62. $e^{0.03k} + 7 = 11$

63. $e^{1.2t} = 3$

64. $e^{1.5t} = 7$

65. $3^{x+1} = 5^x$

66. $2^{x-1} = 7^x$

67. $2^{x+2} = 6^x$

68. $5^{x-2} = 3^x$

69. $32e^{0.04m} = 128$

70. $8e^{0.05n} = 160$

71. $6e^{x/3} = 125$

72. $8e^{x/5} = 155$

73. $5^{x-2} - 4 = 16$

74. $2^{x+4} + 17 = 50$

Concept 3: Applications

75. The population of China can be modeled by the function

$$P(t) = 1237(1.0095)^t$$

where $P(t)$ is in millions and t is the number of years since 1998. **(See Example 10.)**

   a. Using this model, what was the population in the year 2002?

   b. Predict the population in the year 2016.

   c. If this growth rate continues, in what year will the population reach 2 billion people (2 billion is 2000 million)?

76. The population of Delhi, India, can be modeled by the function

$$P(t) = 9817(1.031)^t$$

where $P(t)$ is in thousands and t is the number of years since 2001.

   a. Using this model, predict the population in the year 2018.

   b. If this growth rate continues, in what year will the population reach 15 million (15 million is 15,000 thousand)?

77. The growth of a certain bacteria in a culture is given by the model

$$A(t) = 500e^{0.0277t}$$

where $A(t)$ is the number of bacteria and t is time in minutes.

   a. What is the initial number of bacteria?

   b. What is the population after 10 min?

   c. How long will it take for the population to double (that is, reach 1000)?

78. The population of the bacteria *Salmonella typhimurium* is given by the model

$$A(t) = 300e^{0.01733t}$$

where $A(t)$ is the number of bacteria and t is time in minutes.

   a. What is the initial number of bacteria?

   b. What is the population after 10 min?

   c. How long will it take for the population to double?

79. Suppose $5000 is invested at 7% interest compounded continuously. How long will it take for the investment to grow to $10,000? Use the model $A(t) = Pe^{rt}$ and round to the nearest tenth of a year.

80. Suppose $2000 is invested at 10% interest compounded continuously. How long will it take for the investment to triple? Use the model $A(t) = Pe^{rt}$ and round to the nearest year.

81. Suppose that $8000 is invested at 4.5% interest compounded monthly. How long will it take the investment to reach $10,000? Use the model $A(t) = P\left(1 + \dfrac{r}{n}\right)^{nt}$ and round to the nearest year.

82. Suppose that a couple invests $30,000 in a bond fund that pays 6.5% interest compounded quarterly. How long will it take the investment to reach $250,000? Use the model $A(t) = P\left(1 + \dfrac{r}{n}\right)^{nt}$ and round to the nearest year.

83. Phosphorus 32 (^{32}P) has a half-life of approximately 14 days. If 10 g of ^{32}P is present initially, then the amount of phosphorus 32 still present after t days is given by $A(t) = 10(0.5)^{t/14}$. **(See Example 11.)**

   a. Find the amount of phosphorus 32 still present after 5 days. Round to the nearest tenth of a gram.

   b. Find the amount of time necessary for the amount of ^{32}P to decay to 4 g. Round to the nearest tenth of a day.

84. Polonium 210 (^{210}Po) has a half-life of approximately 138.6 days. If 4 g of ^{210}Po is present initially, then the amount of polonium 210 still present after t days is given by $A(t) = 4e^{-0.005t}$.

   a. Find the amount of polonium 210 still present after 50 days. Round to the nearest tenth of a gram.

   b. Find the amount of time necessary for the amount of ^{210}P to decay to 0.5 g. Round to the nearest tenth of a day.

The decibel level of sound can be found by the equation $D = 10 \log\left(\dfrac{I}{I_0}\right)$, where I is the intensity of the sound and I_0 is the intensity of the least audible sound that an average person can hear. Generally I_0 is found as 10^{-12} watt per square meter (W/m^2). Use this information to answer Exercises 85–86.

85. Given that heavy traffic has a decibel level of 89.3 and that $I_0 = 10^{-12}$, find the intensity of the sound of heavy traffic.

86. Given that normal conversation has a decibel level of 65 and $I_0 = 10^{-12}$, find the intensity of the sound of normal conversation.

87. Suppose you save $10,000 from working an extra job. Rather than spending the money, you decide to save the money for retirement by investing in a mutual fund that averages 12% per year. How long will it take for this money to grow to $1,000,000? Use the model $A(t) = Pe^{rt}$ and round to the nearest tenth of a year.

88. The model $A = Pe^{rt}$ is used to compute the total amount of money in an account after t years at an interest rate r, compounded continuously. The value P is the initial principal. Find the amount of time required for the investment to double as a function of the interest rate. (*Hint:* Substitute $A = 2P$ and solve for t.)

Expanding Your Skills

89. The isotope of plutonium of mass 238 (written ^{238}Pu) is used to make thermoelectric power sources for spacecraft. The heat and electric power derived from such units have made the Voyager, Gallileo, and Cassini missions to the outer reaches of our solar system possible. The half-life of ^{238}Pu is 87.7 yr.

   Suppose a space probe was launched in the year 2002 with 2.0 kg of ^{238}Pu. Then the amount of ^{238}Pu available to power the spacecraft decays over time according to

$$P(t) = 2e^{-0.0079t}$$

where $t \geq 0$ is the number of years since 2002 and $P(t)$ is the amount of plutonium still present (in kilograms).

   a. Suppose the space probe is due to arrive at Pluto in the year 2045. How much plutonium will remain when the spacecraft reaches Pluto? Round to two decimal places.

   b. If 1.5 kg of ^{238}Pu is required to power the spacecraft's data transmitter, will there be enough power in the year 2045 for us to receive close-up images of Pluto?

90. ^{99m}Tc is a radionuclide of technetium that is widely used in nuclear medicine. Although its half-life is only 6 hr, the isotope is continuously produced via the decay of its longer-lived parent ^{99}Mo (molybdenum 99) whose half-life is approximately 3 days. The ^{99}Mo generators (or "cows") are sold to hospitals in which the ^{99m}Tc can be "milked" as needed over a period of a few weeks. Once separated from its parent, the ^{99m}Tc may be chemically incorporated into a variety of imaging agents, each of which is designed to be taken up by a specific target organ within the body. Special cameras, sensitive to the gamma rays emitted by the technetium, are then used to record a "picture" (similar in appearance to an X-ray film) of the selected organ.

Suppose a technician prepares a sample of ^{99m}Tc-pyrophosphate to image the heart of a patient suspected of having had a mild heart attack. If the injection contains 10 millicuries (mCi) of ^{99m}Tc at 1:00 P.M., then the amount of technetium still present is given by

$$T(t) = 10e^{-0.1155t}$$

where $t > 0$ represents the time in hours after 1:00 P.M. and $T(t)$ represents the amount of ^{99m}Tc (in millicuries) still present.

a. How many millicuries of ^{99m}Tc will remain at 4:20 P.M. when the image is recorded? Round to the nearest tenth of a millicurie.

b. How long will it take for the radioactive level of the ^{99m}Tc to reach 2 mCi? Round to the nearest tenth of an hour.

For Exercises 91–94, solve the equations.

91. $(\log x)^2 - 2 \log x - 15 = 0$
(*Hint:* Let $u = \log x$.)

92. $(\log_2 z)^2 - 3 \log_2 z - 4 = 0$

93. $(\log_3 w)^2 + 5 \log_3 w + 6 = 0$

94. $(\ln x)^2 - 2 \ln x = 0$

Graphing Calculator Exercises

95. Graph $Y_1 = 8^x$ and $Y_2 = 21$ on a window where $0 \leq x \leq 5$ and $0 \leq y \leq 40$. Use the graph and an *Intersect* feature to support your answer to Exercise 55.

96. Graph $Y_1 = 6^x$ and $Y_2 = 39$ on a window where $0 \leq x \leq 5$ and $0 \leq y \leq 50$. Use the graph and an *Intersect* feature to support your answer to Exercise 56.

Group Activity

Creating a Population Model

Materials: A computer with Internet access and a calculator

Estimated time: 15–20 minutes

Group Size: 3

1. Go to a website such as http://www.census.gov/ and determine the population of your state for the year 2000, and then for another year after that time. For example, the population of Florida in the year 2000 was approximately 16,050,000. Florida's population in 2006 was approximately 18,090,000.

2. Let the variable t represent the number of years since the year 2000. Therefore, $t = 0$ represents the year 2000, $t = 1$ represents the year 2001, and so on. Write your data from step 1 as ordered pairs. For example: (0, 16050000) and (6, 18090000). (_____ , _____) (_____ , _____)

3. Population growth can be modeled by an exponential function of the form:

   $P(t) = P_0 e^{kt}$ where P_0 is the initial population for $t = 0$

   $P(t)$ is the population at time, t

   k is the growth constant

   Substitute the population for the year 2000 into the model for P_0, and write the function here:

   $P(t) = $ _____ e^{kt} [For example: $P(t) = 16,050,000e^{kt}$]

4. Substitute the population value and t value from the second ordered pair into the function for $P(t)$ and t, respectively.   [For example: $18{,}090{,}000 = 16{,}050{,}000e^{k(6)}$]

5. Solve the equation from step 4 for the value of k. Round k to five decimal places.

6. Substitute the values of P_0 and k in the original model, $P(t) = P_0e^{kt}$, and write the function here:

$P(t) = $ _____

You should now have a function that models the population of your state as a function of the time in years since 2000.

7. Assuming that the population follows this trend, predict the population of your state in the year 2015.

8. Assuming that the population follows this trend, predict the population of your state in the year 2050.

Chapter 8   Summary

Section 8.1   Algebra of Functions and Composition

Key Concepts

Algebra of Functions

Given two functions f and g, the functions $f - g, f - g, f \cdot g$, and $\dfrac{f}{g}$ are defined as

$(f + g)(x) = f(x) + g(x)$

$(f - g)(x) = f(x) - g(x)$

$(f \cdot g)(x) = f(x) \cdot g(x)$

$\left(\dfrac{f}{g}\right)(x) = \dfrac{f(x)}{g(x)}$ provided $g(x) \neq 0$

Examples

Example 1

Let $g(x) = 5x + 1$ and $h(x) = x^3$. Find:

1. $(g + h)(3) = g(3) + h(3) = 16 + 27 = 43$

2. $(g \cdot h)(-1) = g(-1) \cdot h(-1) = (-4) \cdot (-1) = 4$

3. $(g - h)(x) = 5x + 1 - x^3$

4. $\left(\dfrac{g}{h}\right)(x) = \dfrac{5x + 1}{x^3}, x \neq 0$

Composition of Functions

The composition of f and g, denoted $f \circ g$, is defined by the rule

$(f \circ g)(x) = f(g(x))$ provided that $g(x)$ is in the domain of f

Example 2

Find $(f \circ g)(x)$ given the functions defined by $f(x) = 4x + 3$ and $g(x) = 7x$.

$(f \circ g)(x) = f(g(x))$

$= f(7x)$

$= 4(7x) + 3$

$= 28x + 3$

Section 8.2 | Inverse Functions

Key Concepts

Horizontal Line Test

Consider a function defined by a set of points (x, y) in a rectangular coordinate system. Then y is a one-to-one function of x if no horizontal line intersects the graph in more than one point.

Finding an Equation of the Inverse of a Function

For a one-to-one function defined by $y = f(x)$, the equation of the inverse can be found as follows:

1. Replace $f(x)$ by y.
2. Interchange x and y.
3. Solve for y.
4. Replace y by $f^{-1}(x)$.

The graphs defined by $y = f(x)$ and $y = f^{-1}(x)$ are symmetric with respect to the line $y = x$.

Definition of an Inverse Function

If f is a one-to-one function represented by ordered pairs of the form (x, y), then the inverse function, denoted f^{-1}, is the set of ordered pairs (y, x).

Inverse Function Property

If f is a one-to-one function, then g is the inverse of f if and only if $(f \circ g)(x) = x$ for all x in the domain of g, and $(g \circ f)(x) = x$ for all x in the domain of f.

Examples

Example 1

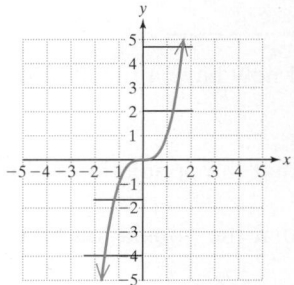

The function is one-to-one because it passes the horizontal line test.

Example 2

Find the inverse of the one-to-one function defined by $f(x) = 3 - x^3$.

1. $y = 3 - x^3$

2. $x = 3 - y^3$

3. $\quad x - 3 = -y^3$

$\quad -x + 3 = y^3$

$\quad \sqrt[3]{-x + 3} = y$

4. $f^{-1}(x) = \sqrt[3]{-x + 3}$

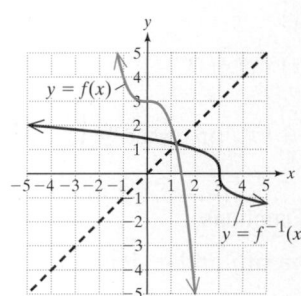

Example 3

Verify that the functions defined by $f(x) = x - 1$ and $g(x) = x + 1$ are inverses.

$(f \circ g)(x) = f(x + 1) = (x + 1) - 1 = x$

$(g \circ f)(x) = g(x - 1) = (x - 1) + 1 = x$

Section 8.3   Exponential Functions

Key Concepts

A function $f(x) = b^x$ $(b > 0, b \neq 1)$ is an **exponential function**.

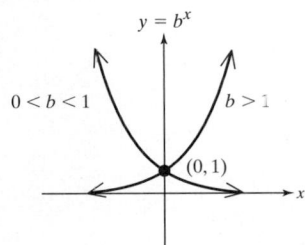

The domain is $(-\infty, \infty)$.

The range is $(0, \infty)$.

The line $y = 0$ (x-axis) is a horizontal asymptote.

The y-intercept is $(0, 1)$.

Example

Example 1

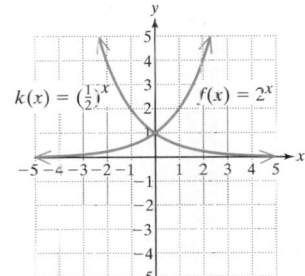

Section 8.4   Logarithmic Functions

Key Concepts

The function $y = \log_b x$ is a **logarithmic function**.

$y = \log_b x \Leftrightarrow b^y = x$ $(x > 0, b > 0, b \neq 1)$

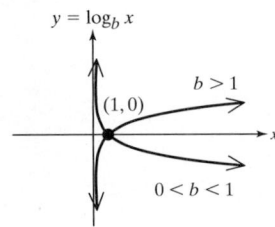

For $y = \log_b x$, the domain is $(0, \infty)$.

The range is $(-\infty, \infty)$.

The line $x = 0$ (y-axis) is a vertical asymptote.

The x-intercept is $(1, 0)$.

The function $f(x) = \log x$ is the **common logarithmic function** (base 10).

Examples

Example 1

$\log_4 64 = 3$ because $4^3 = 64$

Example 2

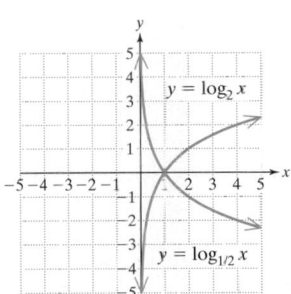

Example 3

$\log 10,000 = 4$ because $10^4 = 10,000$

Section 8.5　Properties of Logarithms

Key Concepts

Let b, x, and y be positive real numbers where $b \neq 1$, and let p be a real number. Then the following properties are true.

1. $\log_b 1 = 0$

2. $\log_b b = 1$

3. $\log_b b^p = p$

4. $b^{\log_b x} = x$

5. $\log_b (xy) = \log_b x + \log_b y$

6. $\log_b \left(\dfrac{x}{y}\right) = \log_b x - \log_b y$

7. $\log_b x^p = p \log_b x$

The **properties of logarithms** can be used to write multiple logarithms as a single logarithm.

The **properties of logarithms** can be used to write a single logarithm as a sum or difference of logarithms.

Examples

Example 1

1. $\log_5 1 = 0$

2. $\log_6 6 = 1$

3. $\log_4 4^7 = 7$

4. $2^{\log_2 5} = 5$

5. $\log (5x) = \log 5 + \log x$

6. $\log_7 \left(\dfrac{z}{10}\right) = \log_7 z - \log_7 10$

7. $\log x^5 = 5 \log x$

Example 2

$$\log x - \frac{1}{2} \log y - 3 \log z$$

$$= \log x - (\log y^{1/2} + \log z^3)$$

$$= \log x - \log (\sqrt{y}z^3)$$

$$= \log \left(\frac{x}{\sqrt{y}z^3}\right)$$

Example 3

$$\log \sqrt[3]{\frac{x}{y^2}}$$

$$= \frac{1}{3} \log \left(\frac{x}{y^2}\right)$$

$$= \frac{1}{3}(\log x - \log y^2)$$

$$= \frac{1}{3}(\log x - 2 \log y)$$

$$= \frac{1}{3} \log x - \frac{2}{3} \log y$$

Section 8.6 — The Irrational Number e and Change of Base

Key Concepts

The function $y = e^x$ is the exponential function with base e.

The **natural logarithm function** $y = \ln x$ is the logarithm function with base e.

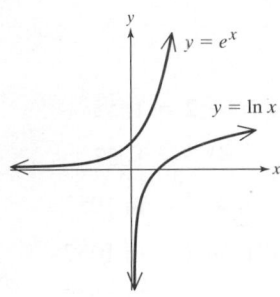

Change-of-Base Formula

$$\log_b x = \frac{\log_a x}{\log_a b} \qquad a > 0, a \neq 1, b > 0, b \neq 1$$

Examples

Example 1

Use a calculator to approximate the value of the expressions.

$$e^{7.5} \approx 1808.0424$$

$$e^{-\pi} \approx 0.0432$$

$$\ln 107 \approx 4.6728$$

$$\ln\left(\frac{1}{\sqrt{2}}\right) \approx -0.3466$$

Example 2

$$\log_3 59 = \frac{\log 59}{\log 3} \approx 3.7115$$

Section 8.7 — Logarithmic and Exponential Equations and Applications

Key Concepts

Guidelines to Solve Logarithmic Equations

1. Isolate the logarithms on one side of the equation.
2. Write a sum or difference of logarithms as a single logarithm.
3. Rewrite the equation in step 2 in exponential form.
4. Solve the resulting equation from step 3.
5. Check all solutions to verify that they are within the domain of the logarithmic expressions in the equation.

Examples

Example 1

$$\log (3x - 1) + 1 = \log (2x + 1)$$

Step 1: $\log (3x - 1) - \log (2x + 1) = -1$

Step 2: $\log\left(\dfrac{3x - 1}{2x + 1}\right) = -1$

Step 3: $10^{-1} = \dfrac{3x - 1}{2x + 1}$

Step 4: $\dfrac{1}{10} = \dfrac{3x - 1}{2x + 1}$

$$2x + 1 = 10(3x - 1)$$

$$2x + 1 = 30x - 10$$

$$-28x = -11$$

$$x = \frac{11}{28} \qquad \text{The solution checks.}$$

Step 5: The solution set is $\left\{\dfrac{11}{28}\right\}$.

The equivalence of exponential expressions can be used to solve **exponential equations**.

If $b^x = b^y$, then $x = y$.

Guidelines to Solve Exponential Equations

1. Isolate one of the exponential expressions in the equation.
2. Take a logarithm of both sides of the equation.
3. Use the power property of logarithms to write exponents as factors.
4. Solve the resulting equation from step 3.

Example 2

$5^{2x} = 125$

$5^{2x} = 5^3$ implies that $2x = 3$

$$x = \frac{3}{2}$$

The solution set is $\left\{ \frac{3}{2} \right\}$.

Example 3

$$4^{x+1} - 2 = 1055$$

Step 1: $4^{x+1} = 1057$

Step 2: $\ln 4^{x+1} = \ln 1057$

Step 3: $(x + 1) \ln 4 = \ln 1057$

Step 4: $x + 1 = \dfrac{\ln 1057}{\ln 4}$

$$x = \frac{\ln 1057}{\ln 4} - 1 \approx 4.023$$

The solution set is $\left\{ \dfrac{\ln 1057}{\ln 4} - 1 \right\}$.

Chapter 8   Review Exercises

Section 8.1

For Exercises 1–8, refer to the functions defined here.

$$f(x) = x - 7 \qquad g(x) = -2x^3 - 8x$$

$$m(x) = x^2 \qquad n(x) = \frac{1}{x - 2}$$

Find the indicated function values.

1. $(f - g)(x)$

2. $(f + g)(x)$

3. $(f \cdot n)(x)$

4. $(f \cdot m)(x)$

5. $\left(\dfrac{f}{g} \right)(x)$

6. $\left(\dfrac{g}{f} \right)(x)$

7. $(m \circ f)(x)$

8. $(n \circ f)(x)$

For Exercises 9–12, refer to the functions defined for Exercises 1–8. Find the function values, if possible.

9. $(m \circ g)(-1)$

10. $(n \circ g)(-1)$

11. $(f \circ g)(4)$

12. $(g \circ f)(8)$

13. Given $f(x) = 2x + 1$ and $g(x) = x^2$.

 a. Find $(g \circ f)(x)$.

 b. Find $(f \circ g)(x)$.

 c. Based on your answers to part (a), is $f \circ g$ equal to $g \circ f$?

For Exercises 14–19, refer to the graph. Approximate the function values, if possible.

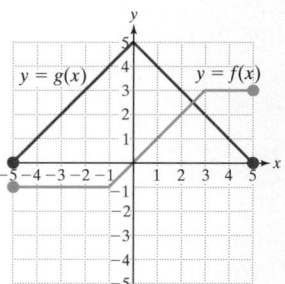

14. $\left(\dfrac{f}{g} \right)(1)$

15. $(f \cdot g)(-2)$

16. $(f + g)(-4)$ **17.** $(f - g)(2)$

18. $(g \circ f)(-3)$ **19.** $(f \circ g)(4)$

Section 8.2

For Exercises 20–21, determine if the function is one-to-one by using the horizontal line test.

20.   **21.**

        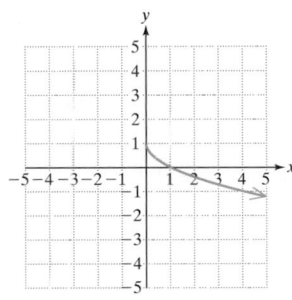

For Exercises 22–26, write the inverse for each one-to-one function.

22. $\{(3, 5), (2, 9), (0, -1), (4, 1)\}$

23. $q(x) = \dfrac{3}{4}x - 2$ **24.** $g(x) = \sqrt[5]{x} + 3$

25. $f(x) = (x - 1)^3$ **26.** $n(x) = \dfrac{4}{x - 2}$

27. Verify that the functions defined by $f(x) = 5x - 2$ and $g(x) = \frac{1}{5}x + \frac{2}{5}$ are inverses by showing that $(f \circ g)(x) = x$ and $(g \circ f)(x) = x$.

28. Graph the functions q and q^{-1} from Exercise 23 on the same grid. What can you say about the relationship between these two graphs?

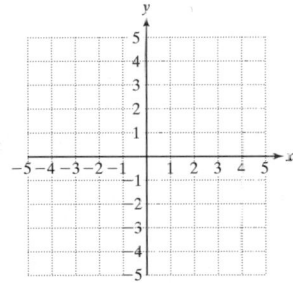

29. a. Find the domain and range of the function defined by $h(x) = \sqrt{x + 1}$.

b. Find the domain and range of the function defined by $k(x) = x^2 - 1, x \geq 0$.

30. Determine the inverse of the function $p(x) = \sqrt{x} + 2$.

Section 8.3

 For Exercises 31–38, evaluate the exponential expressions. Use a calculator and round to three decimal places, if necessary.

31. 4^5 **32.** 6^{-2}

33. $8^{1/3}$ **34.** $\left(\dfrac{1}{100}\right)^{-1/2}$

35. 2^{π} **36.** $5^{\sqrt{3}}$

37. $(\sqrt{7})^{1/2}$ **38.** $\left(\dfrac{3}{4}\right)^{4/3}$

For Exercises 39–42, graph the functions.

39. $f(x) = 3^x$ **40.** $g(x) = \left(\dfrac{1}{4}\right)^x$

        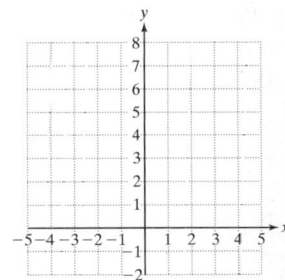

41. $h(x) = 5^{-x}$ **42.** $k(x) = \left(\dfrac{2}{5}\right)^{-x}$

        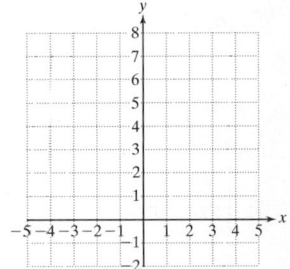

43. a. Does the graph of $y = b^x, b > 0, b \neq 1$, have a vertical or a horizontal asymptote?

   b. Write an equation of the asymptote.

44. Background radiation is radiation that we are exposed to from naturally occurring sources including the soil, the foods we eat, and the Sun. Background radiation varies depending on where we live. A typical background radiation level is 150 millirems (mrem) per year. (A rem is a measure of energy produced from radiation.) Suppose a substance emits 30,000 mrem per year and has a half-life of 5 yr. The function defined by

$$A(t) = 30,000\left(\frac{1}{2}\right)^{t/5}$$

gives the radiation level (in millirems) of this substance after t years.

   a. What is the radiation level after 5 yr?

   b. What is the radiation level after 15 yr?

   c. Will the radiation level of this substance be below the background level of 150 mrem after 50 yr?

Section 8.4

For Exercises 45–52, evaluate the logarithms without using a calculator.

45. $\log_3\left(\dfrac{1}{27}\right)$ **46.** $\log_5 1$

47. $\log_7 7$ **48.** $\log_2 2^8$

49. $\log_2 16$ **50.** $\log_3 81$

51. $\log 100{,}000$ **52.** $\log_8\left(\dfrac{1}{8}\right)$

For Exercises 53–54, graph the logarithmic functions.

53. $q(x) = \log_3 x$ **54.** $r(x) = \log_{1/2} x$

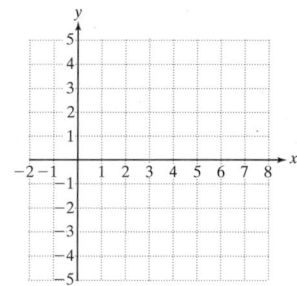

55. a. Does the graph of $y = \log_b x$ have a vertical or a horizontal asymptote?

   b. Write an equation of the asymptote.

56. Acidity of a substance is measured by its pH. The pH can be calculated by the formula $pH = -\log[H^+]$, where $[H^+]$ is the hydrogen ion concentration.

   a. What is the pH of a fruit with a hydrogen ion concentration of 0.00316 mol/L? Round to one decimal place.

   b. What is the pH of an antacid tablet with $[H^+] = 3.16 \times 10^{-10}$? Round to one decimal place.

Section 8.5

For Exercises 57–60, evaluate the logarithms without using a calculator.

57. $\log_8 8$ **58.** $\log_{11} 11^6$

59. $\log_{1/2} 1$ **60.** $12^{\log_{12} 7}$

61. Complete the properties. Assume x, y, and b are positive real numbers such that $b \neq 1$.

   a. $\log_b(xy) =$

   b. $\log_b x - \log_b y =$

   c. $\log_b x^p =$

For Exercises 62–65, write the logarithmic expressions as a single logarithm and simplify if possible.

62. $\dfrac{1}{4}(\log_b y - 4\log_b z + 3\log_b x)$

63. $\dfrac{1}{2}\log_3 a + \dfrac{1}{2}\log_3 b - 2\log_3 c - 4\log_3 d$

64. $\log 540 - 3\log 3 - 2\log 2$

65. $-\log_4 18 + \log_4 6 + \log_4 3 - \log_4 1$

66. Which of the following is equivalent to $\dfrac{2\log 7}{\log 7 + \log 6}$?

   a. $\dfrac{\log 7}{\log 6}$ **b.** $\dfrac{\log 49}{\log 42}$ **c.** $\log\left(\dfrac{7}{6}\right)$

67. Which of the following is equivalent to $\dfrac{\log 8^{-3}}{\log 2 + \log 4}$?

a. -3 **b.** $-3 \log \left(\dfrac{4}{3}\right)$ **c.** $\dfrac{-3 \log 4}{\log 3}$

Section 8.6

For Exercises 68–75, use a calculator to approximate the expressions to four decimal places.

68. e^5

69. $e^{\sqrt{7}}$

70. $32e^{0.008}$

71. $58e^{-0.0125}$

72. $\ln 6$

73. $\ln \left(\dfrac{1}{9}\right)$

74. $\log 22$

75. $\log e^3$

For Exercises 76–79, use the change-of-base formula to approximate the logarithms to four decimal places.

76. $\log_2 10$

77. $\log_9 80$

78. $\log_5 0.26$

79. $\log_4 0.0062$

80. An investor wants to deposit \$20,000 in an account for 10 yr at 5.25% interest. Compare the amount she would have if her money were invested with the following different compounding options. Use

$$A(t) = P\left(1 + \frac{r}{n}\right)^{nt}$$

for interest compounded n times per year and $A(t) = Pe^{rt}$ for interest compounded continuously.

a. Compounded annually

b. Compounded quarterly

c. Compounded monthly

d. Compounded continuously

81. To measure a student's retention of material at the end of a course, researchers give the student a test on the material every month for 24 months after the course is over. The student's average score t months after completing the course is given by

$$S(t) = 75e^{-0.5t} + 20$$

where $S(t)$ is the test score.

a. Find $S(0)$ and interpret the result.

b. Find $S(6)$ and interpret the result.

c. Find $S(12)$ and interpret the result.

For Exercises 82–89, identify the domain. Write the answer in interval notation.

82. $f(x) = e^x$

83. $g(x) = e^{x+6}$

84. $h(x) = e^{x-3}$

85. $k(x) = \ln x$

86. $q(x) = \ln (x + 5)$

87. $p(x) = \ln (x - 7)$

88. $r(x) = \ln (3x - 4)$

89. $w(x) = \ln (5 - x)$

Section 8.7

For Exercises 90–107, solve the equations.

90. $\log_5 x = 3$

91. $\log_7 x = -2$

92. $\log_6 y = 3$

93. $\log_3 y = \dfrac{1}{12}$

94. $\log (2w - 1) = 3$

95. $\log_2 (3w + 5) = 5$

96. $-1 + \log p = -\log (p - 3)$

97. $\log_4 (2 + t) - 3 = \log_4 (3 - 5t)$

98. $4^{3x+5} = 16$

99. $5^{7x} = 625$

100. $4^a = 21$

101. $5^a = 18$

102. $e^{-x} = 0.1$

103. $e^{-2x} = 0.06$

104. $10^{2n} = 1512$

105. $10^{-3m} = \dfrac{1}{821}$

106. $2^{x+3} = 7^x$

107. $14^{x-5} = 6^x$

108. Radioactive iodine (^{131}I) is used to treat patients with a hyperactive (overactive) thyroid. The half-life of radioactive iodine is 8.04 days. If a patient is given an initial dose of 2 μg. then the amount of iodine in the body after t days is approximated by

$$A(t) = 2e^{-0.0862t}$$

where t is the time in days and $A(t)$ is the amount (in micrograms) of ^{131}I remaining.

a. How much radioactive iodine is present after a week? Round to two decimal places.

b. How much radioactive iodine is present after 30 days? Round to two decimal places.

c. How long will it take for the level of radioactive iodine to reach 0.5 μg?

109. The growth of a certain bacteria in a culture is given by the model $A(t) = 150e^{0.007t}$, where $A(t)$ is the number of bacteria and t is time in minutes.

a. What is the initial number of bacteria?

b. What is the population after $\frac{1}{2}$ hr?

c. How long will it take for the population to double?

110. The value of a car is depreciated with time according to

$$V(t) = 15{,}000e^{-0.15t}$$

where $V(t)$ is the value in dollars and t is the time in years.

a. Find $V(0)$ and interpret the result in the context of this problem.

b. Find $V(10)$ and interpret the result in the context of this problem. Round to the nearest dollar.

c. Find the time required for the value of the car to drop to $5000. Round to the nearest tenth of a year.

Chapter 8   Test

For Exercises 1–9, refer to these functions.

$$f(x) = x - 4 \qquad g(x) = x^2 + 2 \qquad h(x) = \frac{1}{x}$$

Find the function values if possible.

1. $\left(\dfrac{f}{g}\right)(x)$

2. $(h \cdot g)(x)$

3. $(g \circ f)(x)$

4. $(h \circ f)(x)$

5. $(f - g)(7)$

6. $(h + f)(2)$

7. $(h \circ g)(4)$

8. $(g \circ f)(0)$

9. $\left(\dfrac{g}{f}\right)(x)$

10. Explain how to determine graphically if a function is one-to-one.

11. Which of the functions is one-to-one?

a.

b.

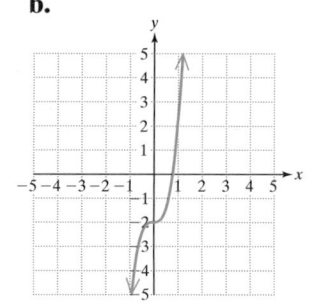

12. Write an equation of the inverse of the one-to-one function defined by $f(x) = \frac{1}{4}x + 3$.

13. Write an equation of the inverse of the function defined by $g(x) = (x - 1)^2$, $x \geq 1$.

14. Given the graph of the function $y = p(x)$, graph its inverse $p^{-1}(x)$.

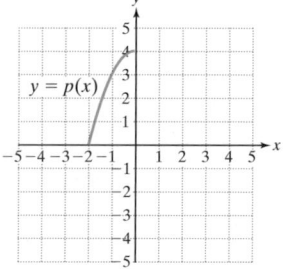

15. Use a calculator to approximate the expression to four decimal places.

a. $10^{2/3}$ **b.** $3^{\sqrt{10}}$ **c.** 8^{π}

16. Graph $f(x) = 4^{x-1}$.

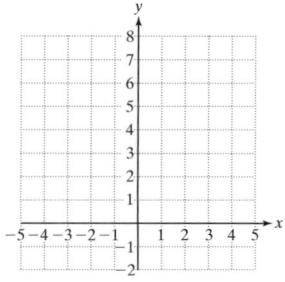

17. a. Write in logarithmic form. $16^{3/4} = 8$

   b. Write in exponential form. $\log_x 31 = 5$

18. Graph $g(x) = \log_3 x$.

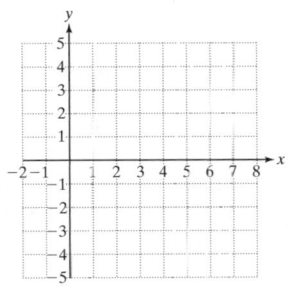

19. Complete the change-of-base formula:
   $\log_b n = \underline{\hspace{1cm}}$

20. Use a calculator to approximate the expression to four decimal places:

   a. $\log 21$ **b.** $\log_4 13$ **c.** $\log_{1/2} 6$

21. Using the properties of logarithms, expand and simplify. Assume all variables represent positive real numbers.

   a. $-\log_3\left(\dfrac{3}{9x}\right)$ **b.** $\log\left(\dfrac{1}{10^5}\right)$

22. Write as a single logarithm. Assume all variables represent positive real numbers.

   a. $\dfrac{1}{2}\log_b x + 3\log_b y$ **b.** $\log a - 4\log a$

23. Use a calculator to approximate the expression to four decimal places, if necessary.

   a. $e^{1/2}$ **b.** e^{-3}

   c. $\ln\left(\dfrac{1}{3}\right)$ **d.** $\ln e$

24. Identify the graphs as $y = e^x$ or $y = \ln x$.

   a.

   b.

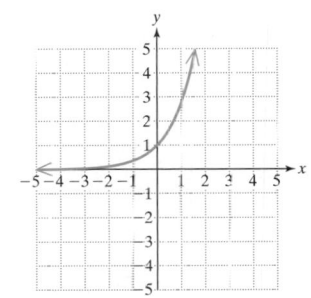

25. Researchers found that t months after taking a course, students remembered $p\%$ of the material according to

$$p(t) = 92 - 20\ln(t+1)$$

where $0 \le t \le 24$ is the time in months.

   a. Find $p(4)$ and interpret the results.

   b. Find $p(12)$ and interpret the results.

   c. Find $p(0)$ and interpret the results.

26. The population of New York City has a 2% growth rate and can be modeled by the function $P(t) = 8008(1.02)^t$, where $P(t)$ is in thousands and t is in years ($t = 0$ corresponds to the year 2000).

   a. Using this model, predict the population in the year 2010.

   b. If this growth rate continues, in what year will the population reach 12 million (12 million is 12,000 thousand)?

27. A certain bacterial culture grows according to

$$P(t) = \frac{1{,}500{,}000}{1 + 5000e^{-0.8t}}$$

where P is the population of the bacteria and t is the time in hours.

   a. Find $P(0)$ and interpret the result. Round to the nearest whole number.

   b. How many bacteria will be present after 6 hr?

   c. How many bacteria will be present after 12 hr?

   d. How many bacteria will be present after 18 hr?

For Exercises 28–35, solve the exponential and logarithmic equations.

28. $\log x + \log(x - 21) = 2$

29. $\log_{1/2} x = -5$

30. $\ln(x + 7) = 2.4$ **31.** $3^{x+4} = \dfrac{1}{27}$

32. $4^x = 50$ **33.** $e^{2.4x} = 250$

34. $68 = 10 - 2^{x-3}$ **35.** $4^{x+7} = 5^x$

36. Atmospheric pressure P decreases exponentially with altitude x according to

$$P(x) = 760e^{-0.000122x}$$

where $P(x)$ is the pressure measured in millimeters of mercury (mm Hg) and x is the altitude measured in meters.

a. Find $P(2500)$ and interpret the result. Round to one decimal place.

b. Find the pressure at sea level.

c. Find the altitude at which the pressure is 633 mm Hg.

 37. Use the formula $A(t) = Pe^{rt}$ to compute the value of an investment under continuous compounding.

a. If $2000 is invested at 7.5% compounded continuously, find the value of the investment after 5 yr.

b. How long will it take the investment to double? Round to two decimal places.

Chapters 1–8 Cumulative Review Exercises

1. Simplify completely.

$$\frac{8 - 4 \cdot 2^2 + 15 \div 5}{|-3 + 7|}$$

2. Divide. $\dfrac{-8p^2 + 4p^3 + 6p^5}{8p^2}$

3. Divide $(t^4 - 13t^2 + 36) \div (t - 2)$. Identify the quotient and remainder.

4. Simplify. $\sqrt{x^2 - 6x + 9}$

5. Simplify. $\dfrac{4}{\sqrt[3]{40}}$

6. Find the length of the missing side.

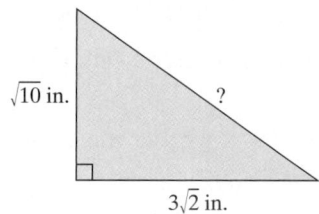
$\sqrt{10}$ in.   ?
$3\sqrt{2}$ in.

7. Simplify. Write the answer with positive exponents only.

$$\frac{2^{2/5}c^{-1/4}d^{1/5}}{2^{-8/5}c^{3/4}d^{1/10}}$$

8. Find the area of the rectangle.

$(\sqrt{5} - \sqrt{2})$ m
$(\sqrt{3} + \sqrt{6})$ m

9. Perform the indicated operation.

$$\frac{4 - 3i}{2 + 5i}$$

10. Solve. $2x(x - 7) = x - 18$

11. How many liters of pure alcohol must be mixed with 8 L of 20% alcohol to bring the concentration up to 50% alcohol?

12. Bank robbers leave the scene of a crime and head north through winding mountain roads to their hideaway. Their average rate of speed is 40 mph. The police leave 6 min later in pursuit. If the police car averages 50 mph traveling the same route, how long will it take the police to catch the bank robbers?

13. Solve the system by using the Gauss-Jordan method.

$$5x + 10y = 25$$
$$-2x + 6y = -20$$

14. Solve. $-2[w - 3(w + 1)] = 4 - 7(w + 3)$

15. Solve for x. $ax - c = bx + d$

16. Solve for t. $s = \dfrac{1}{2}gt^2$ $\quad t \geq 0$

17. Solve. $3w(w - 2) = 6 + 2w$

18. Find the x-intercepts of the function defined by $f(x) = |x - 5| - 2$.

19. Let $f(t) = 6$, $g(t) = -5t$, and $h(t) = 2t^2$. Find

a. $(f \cdot g)(t)$ **b.** $(g \circ h)(t)$ **c.** $(h - g)(t)$

20. Solve. $|2q - 5| = |2q + 5|$

21. a. Find an equation of the line parallel to the y-axis and passing through the point $(2, 6)$.

b. Find an equation of the line perpendicular to the y-axis and passing through the point $(2, 6)$.

c. Find an equation of the line perpendicular to the line $2x + y = 4$ and passing through the point $(2, 6)$. Write the answer in slope-intercept form.

22. The smallest angle in a triangle measures one-half the largest angle. The smallest angle measures $20°$ less than the middle angle. Find the measures of all three angles.

23. Solve the system.

$$\frac{1}{2}x - \frac{1}{4}y = 1$$

$$-2x + y = -4$$

24. Find the domain. $f(x) = \sqrt{2x - 1}$.

25. Find $f^{-1}(x)$, given $f(x) = 5x - \frac{2}{3}$.

26. The volume of a gas varies directly as its temperature and inversely with pressure. At a temperature of 100 kelvins (K) and a pressure of 10 newtons per square meter (N/m^2), the gas occupies a volume of $30 \ m^3$. Find the volume at a temperature of 200 K and pressure of $15 \ N/m^2$.

27. Perform the indicated operations.

$$\frac{5x - 10}{x^2 - 4x + 4} \div \frac{5x^2 - 125}{25 - 5x} \cdot \frac{x^3 + 125}{10x + 5}$$

28. Perform the indicated operations.

$$\frac{x}{x - y} + \frac{y}{y - x} + x$$

29. Given the equation

$$\frac{2}{x - 4} = \frac{5}{x + 2}$$

a. Are there any restrictions on x for which the rational expressions are undefined?

b. Solve the equation.

c. Solve the related inequality.

$$\frac{2}{x - 4} \geq \frac{5}{x + 2}$$

Write the answer in interval notation.

30. Two more than 3 times the reciprocal of a number is $\frac{5}{4}$ less than the number. Find all such numbers.

31. Solve. $\sqrt{-x} = x + 6$

32. Solve the inequality $2|x - 3| + 1 > -7$. Write the answer in interval notation.

33. Four million *Escherichia coli (E. coli)* bacteria are present in a laboratory culture. An antibacterial agent is introduced and the population of bacteria decreases by one-half every 6 hr according to

$$P(t) = 4{,}000{,}000\left(\frac{1}{2}\right)^{t/6} \qquad t \geq 0$$

where t is the time in hours.

a. Find the population of bacteria after 6, 12, 18, 24, and 30 hr.

b. Predict the time required for the population to decrease to 15,625 bacteria.

34. Evaluate the expressions without a calculator.

a. $\log_7 49$ **b.** $\log_4\left(\frac{1}{64}\right)$

c. $\log 1{,}000{,}000$ **d.** $\ln e^3$

35. Use a calculator to approximate the expressions to four decimal places.

a. $\pi^{4.7}$ **b.** e^{π}

c. $(\sqrt{2})^{-5}$ **d.** $\log 5362$

e. $\ln 0.67$ **f.** $\log_4 37$

36. Solve. $5^2 = 125^x$

37. Solve. $e^x = 100$

38. Solve. $\log_3(x + 6) - 3 = -\log_3 x$

39. Write the following expression as a single logarithm.

$$\frac{1}{2}\log z - 2\log x - 3\log y$$

40. Write the following expression as a sum or difference of logarithms

$$\ln \sqrt[3]{\frac{x^2}{y}}$$

Conic Sections

CHAPTER OUTLINE

9.1 Distance Formula, Midpoint Formula, and Circles 754

9.2 More on the Parabola 765

9.3 The Ellipse and Hyperbola 774

Problem Recognition Exercises: Formulas and Conic Sections 783

9.4 Nonlinear Systems of Equations in Two Variables 784

9.5 Nonlinear Inequalities and Systems of Inequalities 791

Group Activity: Investigating the Graphs of Conic Sections on a Calculator 800

Chapter 9

In this chapter, we present several new types of graphs, called conic sections. These include the parabola, the circle, the ellipse, and the hyperbola. These shapes are found in a variety of applications. For example, a reflecting telescope has a mirror whose cross section is in the shape of a parabola, and planetary orbits are modeled by ellipses.

Are You Prepared?

To prepare yourself for this chapter, review the process of completing the square. For Exercises 1–6 below, find the value of k that makes the trinomial a perfect square trinomial. Then factor the trinomial. Find the answer at the right and then record the corresponding letter in the space at the bottom of the page.

Answers

1. $x^2 + 6x + k$

2. $x^2 - 8x + k$

3. $x^2 - 7x + k$

4. $x^2 + 3x + k$

5. $x^2 - \dfrac{2}{5}x + k$

6. $x^2 - \dfrac{1}{4}x + k$

I $\dfrac{49}{4}; \left(x - \dfrac{7}{2}\right)^2$ **O** $\dfrac{9}{4}; \left(x + \dfrac{3}{2}\right)^2$

C $\dfrac{1}{25}; \left(x - \dfrac{1}{5}\right)^2$ **N** $\dfrac{1}{64}; \left(x - \dfrac{1}{8}\right)^2$

S $9; (x + 3)^2$ **T** $16; (x - 4)^2$

7. Plot the points (3, 1), (7, 1), and (7, 6). Use the Pythagorean theorem to determine the distance between the points (3, 1) and (7, 6) and choose the correct answer.

  A. 4   **B.** 5   **C.** $\sqrt{29}$ **D.** 3   **E.** $\sqrt{41}$

The parabola, ellipse, hyperbola, are all

$\overline{}\ \overline{}\ \overline{}\ \overline{}\ \overline{}\quad \overline{}\ \overline{}\ \overline{}\ \overline{}\ \overline{}\ \overline{}\ \overline{}\ \overline{}$
 5 4 6 3 5   1 7 5 2 3 4 6 1

Concepts

1. Distance Formula
2. Circles
3. Writing an Equation of a Circle
4. The Midpoint Formula

1. Distance Formula

Suppose we are given two points (x_1, y_1) and (x_2, y_2) in a rectangular coordinate system. The distance between the two points can be found by using the Pythagorean theorem (Figure 9-1).

First draw a right triangle with the distance d as the hypotenuse. The length of the horizontal leg a is $|x_2 - x_1|$, and the length of the vertical leg b is $|y_2 - y_1|$. From the Pythagorean theorem we have

Figure 9-1

$$d^2 = a^2 + b^2$$
$$= (x_2 - x_1)^2 + (y_2 - y_1)^2$$
$$d = \pm\sqrt{(x_2 - x_1)^2 + (y_2 - y_1)^2}$$
$$= \sqrt{(x_2 - x_1)^2 + (y_2 - y_1)^2}$$

Because distance is positive, reject the negative value.

The Distance Formula

The distance d between the points (x_1, y_1) and (x_2, y_2) is

$$d = \sqrt{(x_2 - x_1)^2 + (y_2 - y_1)^2}$$

Example 1 Finding the Distance Between Two Points

Find the distance between the points $(-2, 3)$ and $(4, -1)$ (Figure 9-2).

Solution:

$$
\begin{array}{ll}
(-2, 3) \quad \text{and} \quad (4, -1) & \\
(x_1, y_1) \quad\quad\quad (x_2, y_2) & \text{Label the points.}
\end{array}
$$

$$d = \sqrt{(x_2 - x_1)^2 + (y_2 - y_1)^2}$$
$$= \sqrt{[(4) - (-2)]^2 + [(-1) - (3)]^2} \quad \text{Apply the distance formula.}$$
$$= \sqrt{(6)^2 + (-4)^2}$$
$$= \sqrt{36 + 16}$$
$$= \sqrt{52}$$
$$= \sqrt{4 \cdot 13}$$
$$= 2\sqrt{13}$$

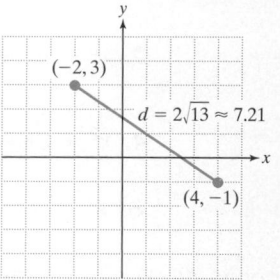

Figure 9-2

Skill Practice

1. Find the distance between the points $(-4, -2)$ and $(2, -5)$.

Answer

1. $3\sqrt{5}$

TIP: The order in which the points are labeled does not affect the result of the distance formula. For example, if the points in Example 1 had been labeled in reverse, the distance formula would still yield the same result:

$(-2, 3)$ and $(4, -1)$ $d = \sqrt{(x_2 - x_1)^2 + (y_2 - y_1)^2}$
(x_2, y_2) (x_1, y_1) $= \sqrt{[(-2) - (4)]^2 + [3 - (-1)]^2}$
$= \sqrt{(-6)^2 + (4)^2}$
$= \sqrt{36 + 16}$
$= \sqrt{52}$
$= 2\sqrt{13}$

2. Circles

A **circle** is defined as the set of all points in a plane that are equidistant from a fixed point called the **center**. The fixed distance from the center is called the **radius** and is denoted by r, where $r > 0$.

Suppose a circle is centered at the point (h, k) and has radius, r (Figure 9-3). The distance formula can be used to derive an equation of the circle.

Let (x, y) be any arbitrary point on the circle. Then, by definition, the distance between (h, k) and (x, y) must be r.

$$\sqrt{(x - h)^2 + (y - k)^2} = r$$
$$(x - h)^2 + (y - k)^2 = r^2 \qquad \text{Square both sides.}$$

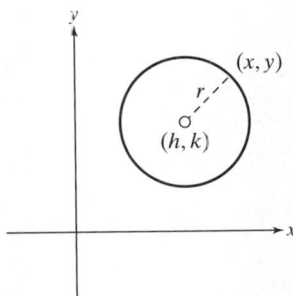

Figure 9-3

Standard Equation of a Circle

The **standard equation of a circle**, centered at (h, k) with radius r, is given by

$$(x - h)^2 + (y - k)^2 = r^2 \qquad \text{where } r > 0.$$

Note: If a circle is centered at the origin $(0, 0)$, then $h = 0$ and $k = 0$, and the equation simplifies to $x^2 + y^2 = r^2$.

Example 2   **Graphing a Circle**

Find the center and radius of the circle. Then graph the circle.

$$(x - 3)^2 + (y + 4)^2 = 36$$

Solution:

$$(x - 3)^2 + (y + 4)^2 = 36$$
$$(x - 3)^2 + [y - (-4)]^2 = (6)^2$$
$$h = 3, k = -4, \text{ and } r = 6$$

Figure 9-4

The equation is in standard form $(x - h)^2 + (y - k)^2 = r^2$.
The center is $(3, -4)$, and the radius is given by $r = 6$.
To graph the circle, first locate the center. From the center, mark points 6 units to the right, left, above, and below the center.
Then sketch the circle through these four points (Figure 9-4).

Note that the center of the circle is not actually part of the circle. It is drawn as an open dot for reference.

Answer

2. Center: $(-1, 2)$; radius: 3

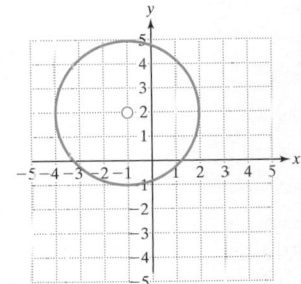

Skill Practice Find the center and radius of the circle. Then graph the circle.

2. $(x + 1)^2 + (y - 2)^2 = 9$

Calculator Connections

Topic: Graphing a Circle

Graphing calculators are designed to graph *functions*, in which *y* is written in terms of *x*. A circle is not a function. However, it can be graphed as the union of two functions—one representing the top semicircle and the other representing the bottom semicircle.

Solving for *y* in Example 2, we have

$$(x - 3)^2 + (y + 4)^2 = 36$$

Graph these functions as Y_1 and Y_2, using a square viewing window.

$$(y + 4)^2 = 36 - (x - 3)^2$$

$$y + 4 = \pm\sqrt{36 - (x - 3)^2}$$

$$y = -4 \pm \sqrt{36 - (x - 3)^2}$$

$$Y_1 = -4 + \sqrt{36 - (x - 3)^2}$$

$$Y_2 = -4 - \sqrt{36 - (x - 3)^2}$$

Notice that the image from the calculator does not show the upper and lower semicircles connecting at their endpoints, when in fact the semicircles should "hook up." This is due to the calculator's limited resolution.

Example 3   **Graphing a Circle**

Find the center and radius of each circle. Then graph the circle.

a. $x^2 + \left(y - \dfrac{10}{3}\right)^2 = \dfrac{25}{9}$ **b.** $x^2 + y^2 = 10$

Solution:

a. $x^2 + \left(y - \dfrac{10}{3}\right)^2 = \dfrac{25}{9}$

$$(x - 0)^2 + \left(y - \dfrac{10}{3}\right)^2 = \left(\dfrac{5}{3}\right)^2$$

The equation is now in standard form $(x - h)^2 + (y - k)^2 = r^2$.
The center is $\left(0, \dfrac{10}{3}\right)$ and the radius is $\dfrac{5}{3}$ (Figure 9-5).

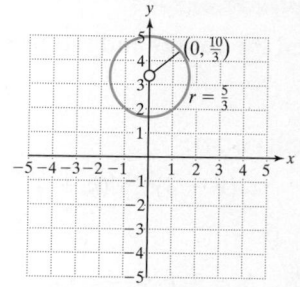

Figure 9-5

b. $x^2 + y^2 = 10$

$(x - 0)^2 + (y - 0)^2 = (\sqrt{10})^2$

$(x - h)^2 + (y - k)^2 = r^2$

The center is $(0, 0)$ and the radius is $\sqrt{10} \approx 3.16$ (Figure 9-6).

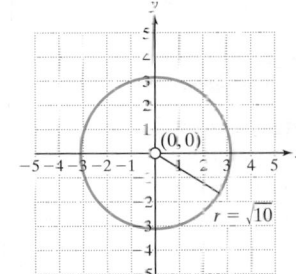

Figure 9-6

Skill Practice Find the center and radius of the circle. Then graph the circle.

3. $\left(x + \dfrac{7}{2}\right)^2 + y^2 = \dfrac{9}{4}$

Sometimes it is necessary to complete the square to write an equation of a circle in standard form.

Example 4 Writing the Equation of a Circle in the Form $(x - h)^2 + (y - k)^2 = r^2$

Identify the center and radius of the circle given by the equation $x^2 + y^2 + 2x - 16y + 61 = 0$.

Solution:

$x^2 + y^2 + 2x - 16y + 61 = 0$ — To identify the center and radius, write the equation in the form $(x - h)^2 + (y - k)^2 = r^2$.

$(x^2 + 2x \quad) + (y^2 - 16y \quad) = -61$ — Group the x terms and group the y terms. Move the constant to the right-hand side.

$(x^2 + 2x + 1) + (y^2 - 16y + 64) = -61 + 1 + 64$ —
- Complete the square on x. Add $\left[\frac{1}{2}(2)\right]^2 = 1$ to both sides of the equation.
- Complete the square on y. Add $\left[\frac{1}{2}(-16)\right]^2 = 64$ to both sides of the equation.

$(x + 1)^2 + (y - 8)^2 = 4$ — Factor and simplify.

$[x - (-1)]^2 + (y - 8)^2 = 2^2$ — Standard form: $(x - h)^2 + (y - k)^2 = r^2$.

The center is $(-1, 8)$ and the radius is 2.

Skill Practice

4. Identify the center and radius of the circle given by the equation
$$x^2 + y^2 - 10x + 4y - 7 = 0$$

Answers

3. Center: $\left(-\dfrac{7}{2}, 0\right)$; radius: $\dfrac{3}{2}$

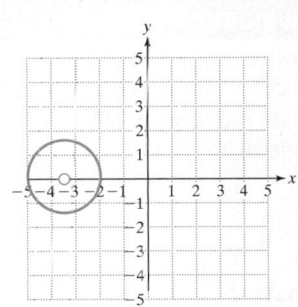

4. Center: $(5, -2)$; radius: 6

3. Writing an Equation of a Circle

Example 5   Writing an Equation of a Circle

Write an equation of the circle shown
in Figure 9-7.

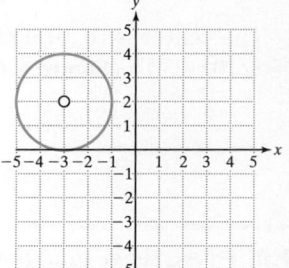

Solution:

The center is at $(-3, 2)$; therefore, $h = -3$
and $k = 2$. From the graph, $r = 2$.

$$(x - h)^2 + (y - k)^2 = r^2$$

$$[x - (-3)]^2 + (y - 2)^2 = (2)^2$$

$$(x + 3)^2 + (y - 2)^2 = 4$$

Figure 9-7

Skill Practice

5. Write an equation for a circle whose center is $(6, -1)$ and whose radius is 8.

4. The Midpoint Formula

Consider two points in the coordinate plane and the line segment determined by the
points. It is sometimes necessary to determine the point that is halfway between
the endpoints of the segment. This point is
called the *midpoint*. If the coordinates of the
endpoints are represented by (x_1, y_1) and
(x_2, y_2), then the midpoint of the segment is
given by the following formula.

Midpoint formula: $\left(\dfrac{x_1 + x_2}{2}, \dfrac{y_1 + y_2}{2} \right)$

Example 6   Finding the Midpoint of a Segment

Find the midpoint of the line segment with the given endpoints.

a. $(-4, 6)$ and $(8, 1)$ **b.** $(-1.2, -3.1)$ and $(-6.6, 1.2)$

Solution:

a. $(-4, 6)$ and $(8, 1)$
     (x_1, y_1) (x_2, y_2)

$$\left(\frac{-4 + 8}{2}, \frac{6 + 1}{2} \right)$$ Apply the midpoint formula: $\left(\dfrac{x_1 + x_2}{2}, \dfrac{y_1 + y_2}{2} \right)$.

$$\left(2, \frac{7}{2} \right)$$ Simplify.

The midpoint of the segment is $\left(2, \frac{7}{2} \right)$.

Answer

5. $(x - 6)^2 + (y + 1)^2 = 64$

b. $(-1.2, -3.1)$ and $(-6.6, 1.2)$
  (x_1, y_1) (x_2, y_2)

$$\left(\frac{-1.2 + (-6.6)}{2}, \frac{-3.1 + 1.2}{2}\right)$$ Apply the midpoint formula.

$(-3.9, -0.95)$ Simplify.

Skill Practice Find the midpoint of the line segment with the given endpoints.

6. $(5, 6)$ and $(-10, 4)$ **7.** $(-2.6, -6.3)$ and $(1.2, 4.1)$

Example 7 **Applying the Midpoint Formula**

Suppose that $(-2, 3)$ and $(4, 1)$ are endpoints of a diameter of a circle.

 a. Find the center of the circle. **b.** Write an equation of the circle.

Solution:

a. Because the midpoint of a diameter of a circle is the center of the circle, apply the midpoint formula. See Figure 9-8.

$(-2, 3)$ and $(4, 1)$
(x_1, y_1) (x_2, y_2)

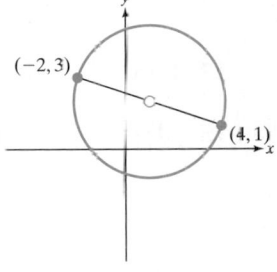

$$\left(\frac{x_1 + x_2}{2}, \frac{y_1 + y_2}{2}\right)$$

$$\left(\frac{-2 + 4}{2}, \frac{3 + 1}{2}\right)$$ Apply the midpoint formula.

$(1, 2)$ Simplify.

Figure 9-8

The center of the circle is $(1, 2)$.

b. The radius can be determined by finding the distance between the center and either endpoint of the diameter.

$(1, 2)$ and $(4, 1)$
(x_1, y_1) (x_2, y_2)

$d = \sqrt{(x_2 - x_1)^2 + (y_2 - y_1)^2}$ Apply the distance formula.

$d = \sqrt{(4 - 1)^2 + (1 - 2)^2}$ Substitute $(1, 2)$ and $(4, 1)$ for (x_1, y_1) and (x_2, y_2).

$= \sqrt{(3)^2 + (-1)^2}$ Simplify.

$= \sqrt{10}$

The circle is represented by $(x - 1)^2 + (y - 2)^2 = (\sqrt{10})^2$

   or $(x - 1)^2 + (y - 2)^2 = 10$.

Skill Practice

 8. Suppose that $(3, 2)$ and $(7, 10)$ are endpoints of a diameter of a circle.
  a. Find the center of the circle.
  b. Write an equation of the circle.

Answers

6. $\left(-\dfrac{5}{2}, 5\right)$ **7.** $(-0.7, -1.1)$

8. a. $(5, 6)$
 b. $(x - 5)^2 + (y - 6)^2 = 20$

Section 9.1 Practice Exercises

Vocabulary and Key Concepts

1. a. The distance between two distinct points (x_1, y_1) and (x_2, y_2) is given by $d = $ _____.

   b. A _____ is the set of all points equidistant from a fixed point called the _____.

   c. The distance from the center of a circle to any point on the circle is called the _____ and is often denoted by r.

   d. The standard equation of a circle with center (h, k) and radius r is given by _____.

   e. The midpoint of the line segment with endpoints (x_1, y_1) and (x_2, y_2) is given by _____.

Concept 1: Distance Formula

For Exercises 2–16, use the distance formula to find the distance between the two points. **(See Example 1.)**

2. $(-2, 7)$ and $(4, -5)$

3. $(1, 10)$ and $(-2, 4)$

4. $(0, 5)$ and $(-3, 8)$

5. $(6, 7)$ and $(3, 2)$

6. $\left(\dfrac{2}{3}, \dfrac{1}{5}\right)$ and $\left(-\dfrac{5}{6}, \dfrac{3}{10}\right)$

7. $\left(-\dfrac{1}{2}, \dfrac{5}{8}\right)$ and $\left(-\dfrac{3}{2}, \dfrac{1}{4}\right)$

8. $(4, 13)$ and $(4, -6)$

9. $(-2, 5)$ and $(-2, 9)$

10. $(8, -6)$ and $(-2, -6)$

11. $(7, 2)$ and $(15, 2)$

12. $(-6, -2)$ and $(-3, -5)$

13. $(-1, -5)$ and $(-5, -9)$

14. $(3\sqrt{5}, 2\sqrt{7})$ and $(-\sqrt{5}, -3\sqrt{7})$

15. $(4\sqrt{6}, -2\sqrt{2})$ and $(2\sqrt{6}, \sqrt{2})$

16. $(6, 0)$ and $(0, -1)$

17. Explain how to find the distance between 5 and -7 on the y-axis.

18. Explain how to find the distance between 15 and -37 on the x-axis.

19. Find the values of y such that the distance between the points $(4, 7)$ and $(-4, y)$ is 10 units.

20. Find the values of x such that the distance between the points $(-4, -2)$ and $(x, 3)$ is 13 units.

21. Find the values of x such that the distance between the points $(x, 2)$ and $(4, -1)$ is 5 units.

22. Find the values of y such that the distance between the points $(-5, 2)$ and $(3, y)$ is 10 units.

For Exercises 23–26, determine if the three points define the vertices of a right triangle.

23. $(-3, 2), (-2, -4)$, and $(3, 3)$

24. $(1, -2), (-2, 4)$, and $(7, 1)$

25. $(-3, -2), (4, -3)$, and $(1, 5)$

26. $(1, 4), (5, 3)$, and $(2, 0)$

Concept 2: Circles

For Exercises 27–47, identify the center and radius of the circle and then graph the circle. Complete the square, if necessary. **(See Examples 2–4.)**

27. $(x - 4)^2 + (y + 2)^2 = 9$

28. $(x - 3)^2 + (y + 1)^2 = 16$

29. $(x + 1)^2 + (y + 1)^2 = 1$

Animation

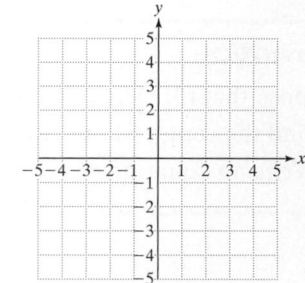

30. $(x - 4)^2 + (y - 4)^2 = 4$

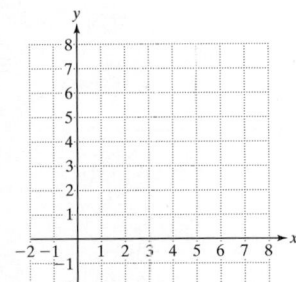

31. $x^2 + (y - 2)^2 = 4$

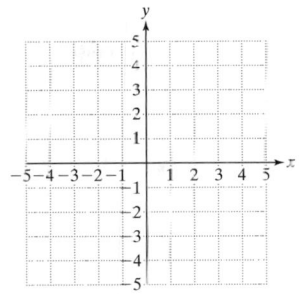

32. $(x + 1)^2 + y^2 = 1$

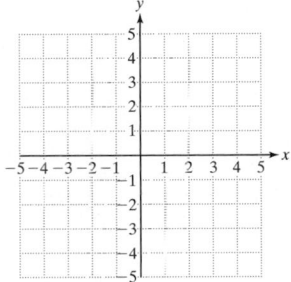

33. $(x - 3)^2 + y^2 = 8$

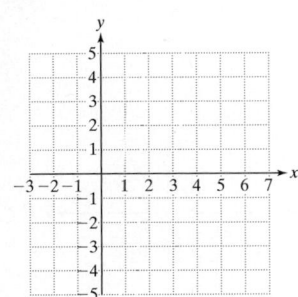

34. $x^2 + (y + 2)^2 = 20$

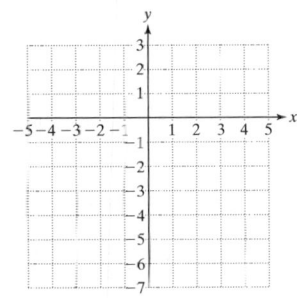

35. $x^2 + y^2 = 6$

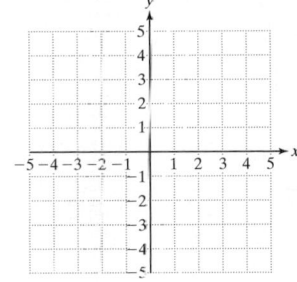

36. $x^2 + y^2 = 15$

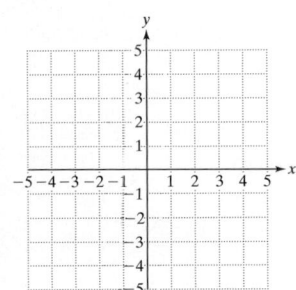

37. $\left(x + \dfrac{4}{5}\right)^2 + y^2 = \dfrac{64}{25}$

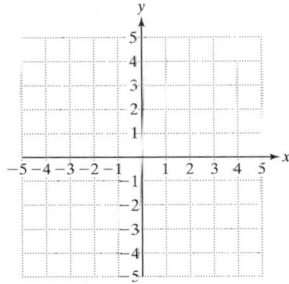

38. $x^2 + \left(y - \dfrac{5}{2}\right)^2 = \dfrac{9}{4}$

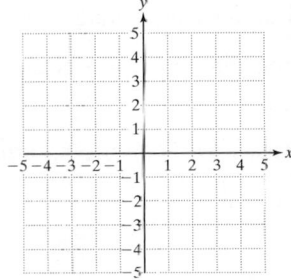

39. $x^2 + y^2 - 2x - 6y - 26 = 0$

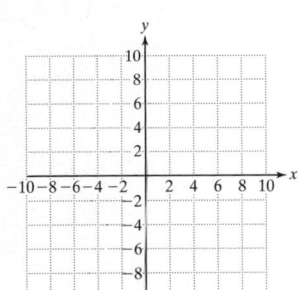

40. $x^2 + y^2 + 4x - 8y + 16 = 0$

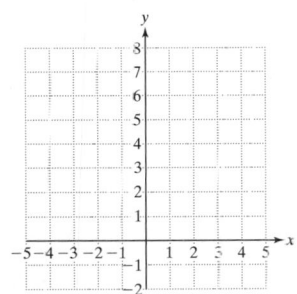

41. $x^2 + y^2 - 6y + 5 = 0$

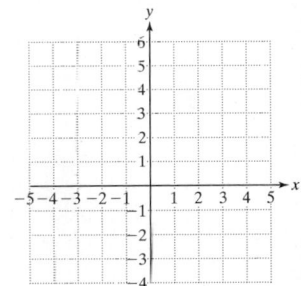

42. $x^2 + 2x + y^2 - 15 = 0$

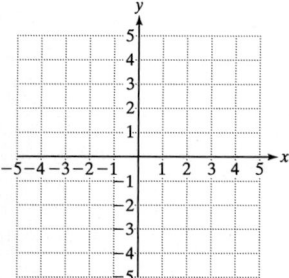

43. $x^2 + y^2 + 6y + \dfrac{65}{9} = 0$

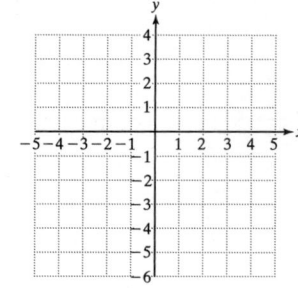

44. $x^2 + y^2 - 12x + 12y + 71 = 0$

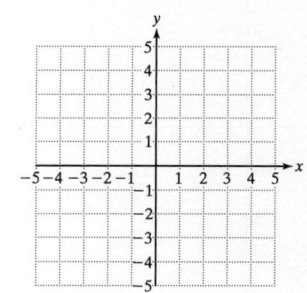

45. $x^2 + y^2 + 2x + 4y - 4 = 0$

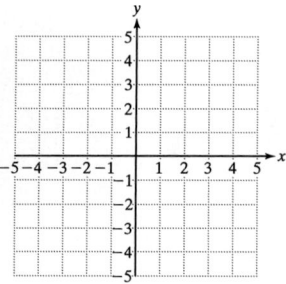

46. $2x^2 + 2y^2 = 32$

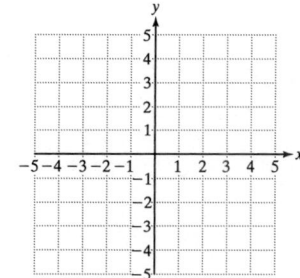

47. $3x^2 + 3y^2 = 3$

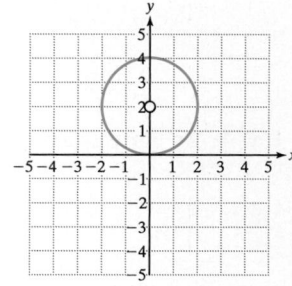

Concept 3: Writing an Equation of a Circle

48. If the diameter of a circle is 10 ft, what is the radius?

For Exercises 49–54, write an equation that represents the graph of the circle. **(See Example 5.)**

49.

50.

51.

52.

53.

54.

55. Write an equation of a circle centered at the origin with a radius of 7.

56. Write an equation of a circle centered at the origin with a radius of 12.

57. Write an equation of a circle centered at $(-3, -4)$ with a diameter of 12.

58. Write an equation of a circle centered at $(5, -1)$ with a diameter of 8.

59. A cell tower is a site where antennas, transmitters, and receivers are placed to create a cellular network. Suppose that a cell tower is located at a point $A(5, 3)$ on a map and its range is 1.5 mi. Write an equation that represents the boundary of the area that can receive a signal from the tower. Assume that all distances are in miles.

60. A radar transmitter on a ship has a range of 20 nautical miles. If the ship is located at a point $(-28, 32)$ on a map, write an equation for the boundary of the area within the range of the ship's radar. Assume that all distances on the map are represented in nautical miles.

Concept 4: The Midpoint Formula

For Exercises 61–64, find the midpoint of the line segment. Check your answers by plotting the midpoint on the graph.

61.

62.

63.

64.

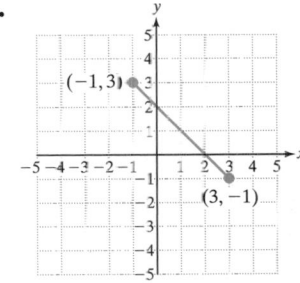

For Exercises 65–72, find the midpoint of the line segment between the two given points. **(See Example 6.)**

65. $(4, 0)$ and $(-6, 12)$

66. $(-7, 2)$ and $(-3, -2)$

67. $(-3, 8)$ and $(3, -2)$

68. $(0, 5)$ and $(4, -5)$

69. $(5, 2)$ and $(-6, 1)$

70. $(-9, 3)$ and $(0, -4)$

71. $(-2.4, -3.1)$ and $(1.6, 1.1)$

72. $(0.8, 5.3)$ and $(-4.2, 7.1)$

73. Two courier trucks leave the warehouse to make deliveries. One travels 20 mi north and 30 mi east. The other truck travels 5 mi south and 50 mi east. If the two drivers want to meet for lunch at a restaurant at a point halfway between them, where should they meet relative to the warehouse? (*Hint:* Label the warehouse as the origin, and find the coordinates of the restaurant. See the figure.)

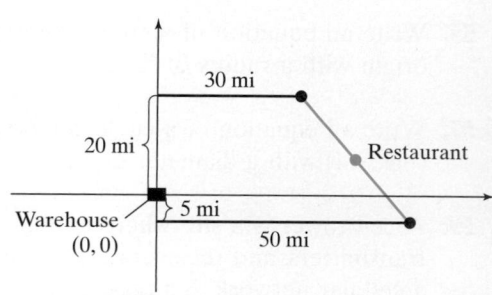

74. A map of a hiking area is drawn so that the Visitor Center is at the origin of a rectangular grid. Two hikers are located at positions $(-1, 1)$ and $(-3, -2)$ with respect to the Visitor Center where all units are in miles. A campground is located exactly halfway between the hikers. What are the coordinates of the campground?

For Exercises 75–78, the two given points are endpoints of a diameter of a circle.

   a. Find the center of the circle.

   b. Write an equation of the circle. **(See Example 7.)**

75. $(-1, 2)$ and $(3, 4)$ **76.** $(-3, 3)$ and $(7, -1)$ **77.** $(-2, 3)$ and $(2, 3)$ **78.** $(-1, 3)$ and $(-1, -3)$

Expanding Your Skills

79. Write an equation of a circle whose center is $(4, 4)$ and is tangent to the x- and y-axes. (*Hint:* Sketch the circle first.)

80. Write an equation of a circle whose center is $(-3, 3)$ and is tangent to the x- and y-axes. (*Hint:* Sketch the circle first.)

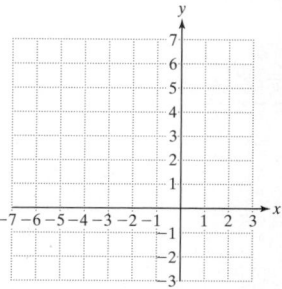

81. Write an equation of a circle whose center is $(1, 1)$ and that passes through the point $(-4, 3)$.

82. Write an equation of a circle whose center is $(-3, -1)$ and that passes through the point $(5, -2)$.

Graphing Calculator Exercises

For Exercises 83–88, graph the circles from the indicated exercise on a square viewing window, and approximate the center and the radius from the graph.

83. $(x - 4)^2 + (y + 2)^2 = 9$ (Exercise 27)

84. $(x - 3)^2 + (y + 1)^2 = 16$ (Exercise 28)

85. $x^2 + (y - 2)^2 = 4$ (Exercise 31)

86. $(x + 1)^2 + y^2 = 1$ (Exercise 32)

87. $x^2 + y^2 = 6$ (Exercise 35)

88. $x^2 + y^2 = 15$ (Exercise 36)

More on the Parabola

1. Introduction to Conic Sections

Recall that the graph of a second-degree equation of the form $y = ax^2 + bx + c$ ($a \neq 0$) is a parabola. In Section 9.1 we learned that the graph of $(x - h)^2 + (y - k)^2 = r^2$ is a circle. These and two other types of figures called ellipses and hyperbolas are called **conic sections**. Conic sections derive their names because each is the intersection of a plane and a double-napped cone.

Concepts

1. Introduction to Conic Sections
2. Parabola—Vertical Axis of Symmetry
3. Parabola—Horizontal Axis of Symmetry
4. Vertex Formula

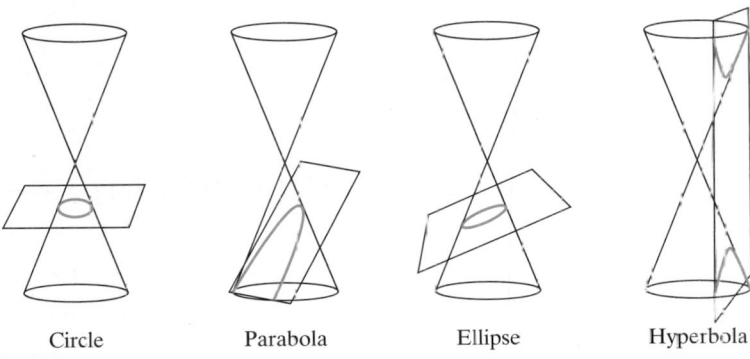

Circle   Parabola   Ellipse   Hyperbola

2. Parabola—Vertical Axis of Symmetry

A **parabola** is defined by a set of points in a plane that are equidistant from a fixed line (called the directrix) and a fixed point (called the focus) not on the directrix (Figure 9-9). Parabolas have numerous real-world applications. For example, a flashlight has a mirror with the cross section in the shape of a parabola. The bulb is located at the focus. The light hits the mirror and is reflected outward parallel to the sides of the cylinder. This forms a beam of light (Figure 9-10).

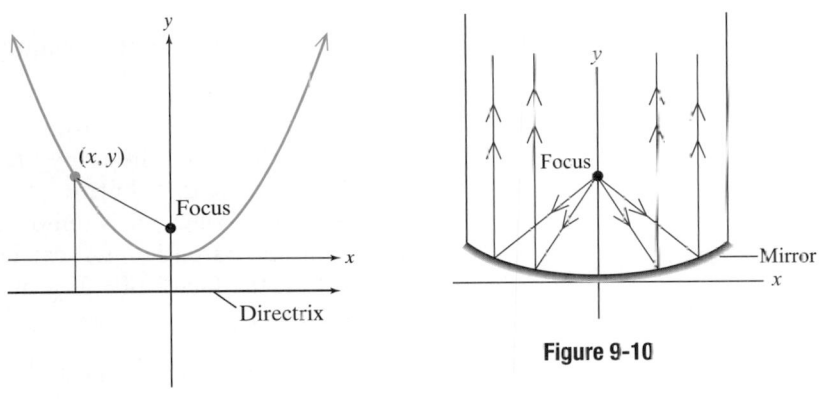

Figure 9-9

Figure 9-10

The graph of the solutions to the quadratic equation $y = ax^2 + bx + c$ is a parabola. We graphed parabolas of this type in Section 7.4. Recall that the **vertex** is the highest or the lowest point of a parabola. The **axis of symmetry** of the parabola is a line that passes through the vertex and is perpendicular to the directrix (Figure 9-11).

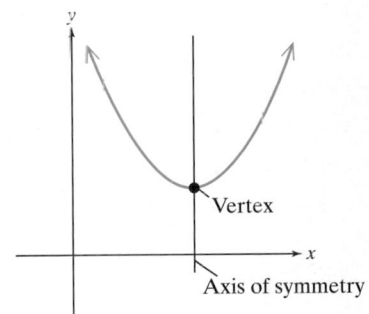

Figure 9-11

> ### Standard Form of the Equation of a Parabola—Vertical Axis of Symmetry
>
> The standard form of the equation of a parabola with vertex (h, k) and vertical axis of symmetry is
>
> $$y = a(x - h)^2 + k \qquad \text{where } a \neq 0$$
>
> If $a > 0$, the parabola opens upward; and if $a < 0$, the parabola opens downward.
>
> The axis of symmetry is given by $x = h$.

In Section 7.5, we learned that it is sometimes necessary to complete the square to write the equation of a parabola in standard form.

Example 1   Graphing a Parabola by First Completing the Square

Given the equation of the parabola $y = -2x^2 + 4x + 1$:

a. Write the equation in standard form $y = a(x - h)^2 + k$.

b. Identify the vertex and axis of symmetry. Determine if the parabola opens upward or downward.

c. Graph the parabola.

Solution:

a. Complete the square to write the equation in the form $y = a(x - h)^2 + k$.

$y = -2x^2 + 4x + 1$

$y = -2(x^2 - 2x) + 1$ Factor out -2 from the variable terms.

$y = -2(x^2 - 2x + 1 - 1) + 1$ Add and subtract the quantity $\left[\frac{1}{2}(-2)\right]^2 = 1$.

$y = -2(x^2 - 2x + 1) + (-2)(-1) + 1$ Remove the -1 term from within the parentheses by first applying the distributive property. When -1 is removed

$y = -2(x - 1)^2 + 2 + 1$ from the parentheses, it carries with it the factor of -2 from

$y = -2(x - 1)^2 + 3$ outside the parentheses.

The equation is in the form $y = a(x - h)^2 + k$ where $a = -2$, $h = 1$, and $k = 3$.

b. The vertex is $(1, 3)$. Because a is negative, the parabola opens downward. The axis of symmetry is $x = 1$.

c. To graph the parabola, we know that its orientation is downward because the leading term is negative. Furthermore, we know the vertex is $(1, 3)$. To find other points on the parabola, select several values of x and solve for y. Recall that the y-intercept is found by substituting $x = 0$ and solving for y.

x	y	
1	3	← Vertex
0	1	← y-intercept
−1	−5	
2	1	
3	−5	

Choose x.   Solve for y.

Notice that the points $(2, 1)$ and $(3, -5)$ are the mirror images of the points $(0, 1)$ and $(-1, -5)$ around the axis of symmetry, $x = 1$. For this reason, once we plot several points on one side of the axis of symmetry, the points on the other side automatically follow as a mirror image.

Skill Practice Given the equation of the parabola $y = 3x^2 + 6x + 4$:

1. Write the equation in standard form.
2. Identify the vertex and axis of symmetry.
3. Graph the parabola.

3. Parabola—Horizontal Axis of Symmetry

We have seen that the graph of a parabola $y = ax^2 + bx + c$ opens upward if $a > 0$ and downward if $a < 0$. A parabola can also open to the left or right. In such a case, the "roles" of x and y are essentially interchanged in the equation. Thus, the graph of $x = ay^2 + by + c$ opens to the right if $a > 0$ (Figure 9-12) and to the left if $a < 0$ (Figure 9-13).

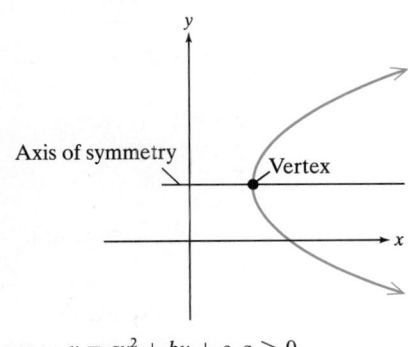

$x = ay^2 + by + c, a > 0$

Figure 9-12

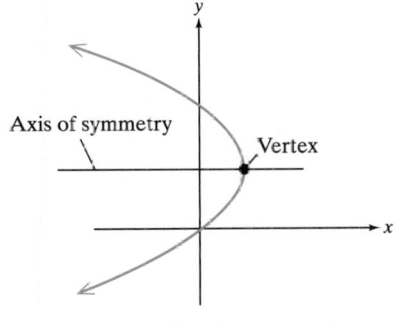

$x = ay^2 + by + c, a < 0$

Figure 9-13

Answers

1. $y = 3(x + 1)^2 + 1$
2. Vertex: $(-1, 1)$; axis of symmetry: $x = -1$
3.

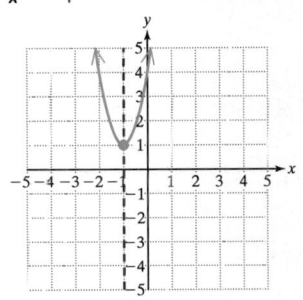

Standard Form of the Equation of a Parabola—Horizontal Axis of Symmetry

The standard form of the equation of a parabola with vertex (h, k) and horizontal axis of symmetry is

$$x = a(y - k)^2 + h \qquad \text{where } a \neq 0$$

If $a > 0$, the parabola opens to the right and if $a < 0$, the parabola opens to the left.

The axis of symmetry is given by $y = k$.

Example 2 **Graphing a Parabola with a Horizontal Axis of Symmetry**

Given the equation of the parabola $x = 4y^2$:

a. Determine the coordinates of the vertex and the equation of the axis of symmetry.

b. Use the value of a to determine if the parabola opens to the right or left.

c. Plot several points and graph the parabola.

Solution:

a. The equation can be written in the form $x = a(y - k)^2 + h$:
$$x = 4(y - 0)^2 + 0$$

Therefore, $h = 0$ and $k = 0$.

The vertex is $(0, 0)$. The axis of symmetry is $y = 0$ (the x-axis).

b. Because a is positive, the parabola opens to the right.

c. The vertex of the parabola is $(0, 0)$. To find other points on the graph, select values for y and solve for x.

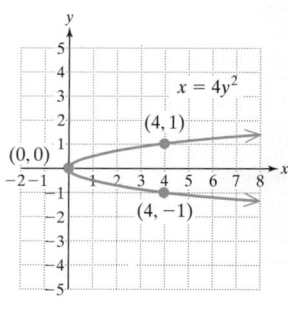

x	y
0	0	←——Vertex
4	1
4	−1
16	2
16	−2

Solve for x.   Choose y.

TIP: For Example 2, the x-axis is the axis of symmetry. Therefore, once we find points on the parabola above the axis of symmetry, their mirror image will be points on the parabola below the axis of symmetry.

Answers

4. Vertex: $(1, 0)$; axis of symmetry: $y = 0$
5. Parabola opens left.
6.

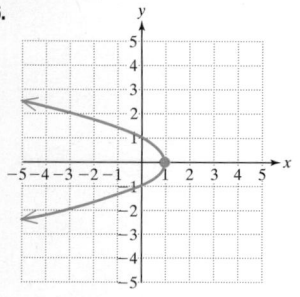

Skill Practice Given the equation $x = -y^2 + 1$:

4. Identify the vertex and the axis of symmetry.

5. Determine if the parabola opens to the right or to the left.

6. Graph the parabola.

Calculator Connections

Topic: Graphing a Parabola Opening Left or Right

Graphing calculators are designed to graph functions, where y is written as a function of x. The parabola from Example 2 is not a function. It can be graphed, however, as the union of two functions, one representing the top branch and the other representing the bottom branch.

Solving for y in Example 2, we have:

$$x = 4y^2$$

$$y^2 = \frac{x}{4}$$

$$y = \pm\sqrt{\frac{x}{4}}$$

$$y = \pm\frac{\sqrt{x}}{2}$$

Graph these functions as Y_1 and Y_2.

$$Y_1 = \frac{\sqrt{x}}{2} \quad \text{and} \quad Y_2 = -\frac{\sqrt{x}}{2}$$

Example 3 **Graphing a Parabola by First Completing the Square**

Given the equation of the parabola $x = -y^2 + 8y - 14$:

a. Write the equation in standard form $x = a(y - k)^2 + h$.

b. Identify the vertex and axis of symmetry. Determine if the parabola opens to the right or left.

c. Graph the parabola.

Solution:

a. Complete the square to write the equation in the form $x = a(y - k)^2 + h$.

$x = -y^2 + 8y - 14$

$x = -1(y^2 - 8y) - 14$ Factor out -1 from the variable terms.

$x = -1(y^2 - 8y + 16 - 16) - 14$ Add and subtract the quantity $[\frac{1}{2}(-8)]^2 = 16$.

$x = -1(y^2 - 8y + 16) + (-1)(-16) - 14$ Remove the -16 term from within the parentheses by first applying the distributive property. When -16 is removed from the parentheses, it carries with it the factor of -1 from outside the parentheses.

$x = -1(y - 4)^2 + 16 - 14$

$x = -1(y - 4)^2 + 2$

The equation is in the form $x = a(y - k)^2 + h$, where $a = -1, h = 2$, and $k = 4$.

b. The vertex is $(2, 4)$. Because a is negative, the parabola opens to the left. The axis of symmetry is $y = 4$.

c.

x	y
2	4	← Vertex
1	5
1	3
−2	6
−2	2

Solve   Choose
for x. y.

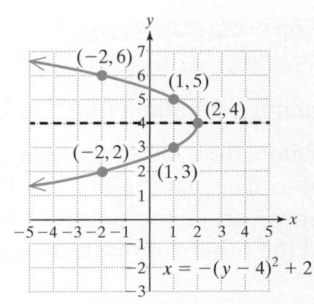

$x = -(y - 4)^2 + 2$

Skill Practice Given the equation of the parabola $x = y^2 + 4y + 7$:

7. Determine the coordinates of the vertex.

8. Determine if the parabola opens to the right or to the left.

9. Graph the parabola.

4. Vertex Formula

From Section 7.5, we learned that the vertex formula can also be used to find the vertex of a parabola.

For a parabola defined by $y = ax^2 + bx + c$,

- The x-coordinate of the vertex is given by $x = \frac{-b}{2a}$.
- The y-coordinate of the vertex is found by substituting this value for x into the original equation and solving for y.

For a parabola defined by $x = ay^2 + by + c$,

- The y-coordinate of the vertex is given by $y = \frac{-b}{2a}$.
- The x-coordinate of the vertex is found by substituting this value for y into the original equation and solving for x.

Example 4 **Finding the Vertex of a Parabola by Using the Vertex Formula**

Find the vertex by using the vertex formula. $x = y^2 + 4y + 5$

Solution:

$x = y^2 + 4y + 5$ The parabola is in the form $x = ay^2 + by + c$.

$a = 1$ $b = 4$ $c = 5$ Identify a, b, and c.

The y-coordinate of the vertex is given by $y = \dfrac{-b}{2a} = \dfrac{-(4)}{2(1)} = -2$.

The x-coordinate of the vertex is found by substitution:
$$x = (-2)^2 + 4(-2) + 5 = 1.$$

The vertex is $(1, -2)$.

Skill Practice Find the vertex by using the vertex formula.

10. $x = -4y^2 + 12y$

Answers

7. Vertex: $(3, -2)$

8. Parabola opens right.

9.

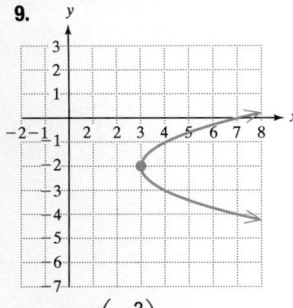

10. Vertex: $\left(9, \dfrac{3}{2}\right)$

| Example 5 | Find the Vertex of a Parabola by Using the Vertex Formula |

Find the vertex by using the vertex formula. $y = \dfrac{1}{2}x^2 - 3x + \dfrac{5}{2}$

Solution:

$y = \dfrac{1}{2}x^2 - 3x + \dfrac{5}{2}$ The parabola is in the form $y = ax^2 + bx - c$.

$a = \dfrac{1}{2} \quad b = -3 \quad c = \dfrac{5}{2}$ Identify a, b, and c.

The x-coordinate of the vertex is given by $x = \dfrac{-b}{2a} = \dfrac{-(-3)}{2\left(\dfrac{1}{2}\right)} = 3$.

The y-coordinate of the vertex is found by substitution.

$$y = \dfrac{1}{2}(3)^2 - 3(3) + \dfrac{5}{2}$$

$$= \dfrac{9}{2} - 9 + \dfrac{5}{2}$$

$$= \dfrac{14}{2} - 9$$

$$= 7 - 9$$

$$= -2$$

The vertex is $(3, -2)$.

Skill Practice Find the vertex by using the vertex formula.

11. $y = \dfrac{3}{4}x^2 + 3x + 5$

Answer

11. $(-2, 2)$

Section 9.2   Practice Exercises

Vocabulary and Key Concepts

1. a. A circle, a parabola, an ellipse, and a hyperbola are curves that collectively are called _____ sections. These types of curves derive their names because each is the intersection of a _____ and a double-napped cone.

  b. A _____ is defined by a set of points in a plane that are equidistant from a fixed line (called the _____) and a fixed point (called the _____).

  c. Given a parabola defined by $y = a(x - h)^2 + k$ ($a \neq 0$), the _____ is (h, k) and is the highest or lowest point on the graph. The axis of _____ is the vertical line through the vertex whose equation is $x = h$.

  d. Given a parabola defined by $x = a(y - k)^2 + h$ ($a \neq 0$), the axis of symmetry is the horizontal line whose equation is _____.

Review Exercises

2. Determine the distance between the points $(1, 1)$ and $(2, -2)$.

3. Determine the distance from the origin to the point $(4, -3)$.

4. Determine the distance between the points $(11, 3)$ and $(5, 3)$.

For Exercises 5–6, identify the center and radius of the circle and then graph the circle.

5. $x^2 + (y + 1)^2 = 16$

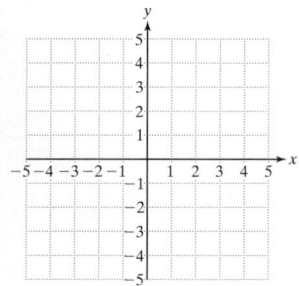

6. $(x - 3)^2 + y^2 = 4$

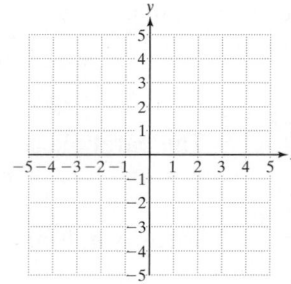

7. Find the midpoint of the line segment between the points $(7, -3)$ and $(4, 5)$.

8. Find the midpoint of the line segment between the points $(-3, 0)$ and $(-1, 9)$.

Concept 2: Parabola—Vertical Axis of Symmetry

For Exercises 9–16, use the equation of the parabola in standard form $y = a(x - h)^2 + k$ to determine the coordinates of the vertex and the equation of the axis of symmetry (complete the square if necessary). Then graph the parabola.
(See Example 1.)

9. $y = (x + 2)^2 + 1$

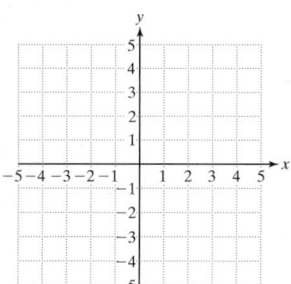

10. $y = (x - 1)^2 + 1$

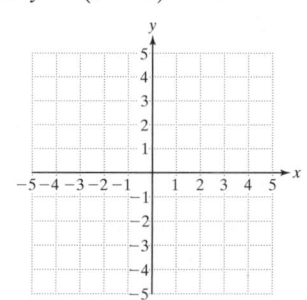

11. $y = x^2 - 4x + 3$

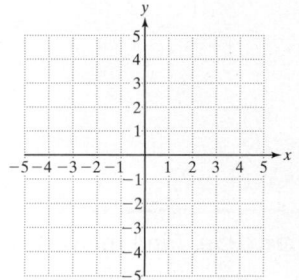

12. $y = x^2 + 6x + 5$

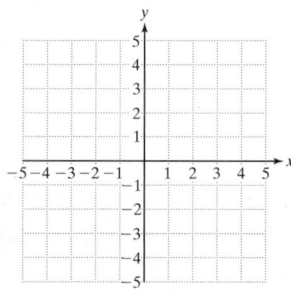

13. $y = -2x^2 + 8x$

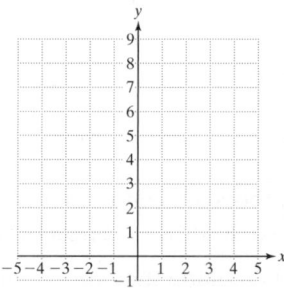

14. $y = -3x^2 - 6x$

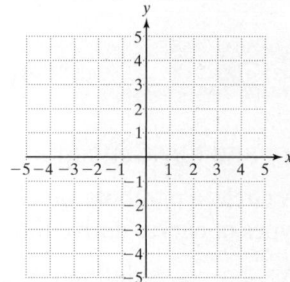

15. $y = -x^2 - 3x + 2$

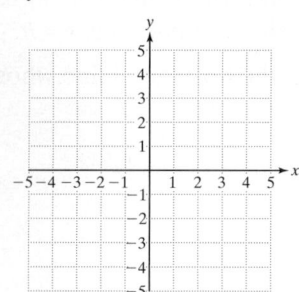

16. $y = -x^2 + x - 4$

Animation

Concept 3: Parabola—Horizontal Axis of Symmetry

For Exercises 17–24, use the equation of the parabola in standard form $x = a(y - k)^2 + h$ to determine the coordinates of the vertex and the axis of symmetry (complete the square if necessary). Then graph the parabola. (See Examples 2–3.)

17. $x = y^2 - 3$

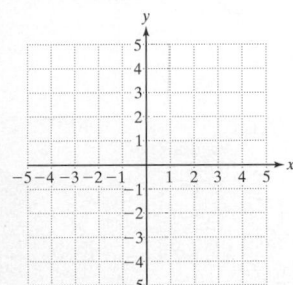

18. $x = y^2 + 1$

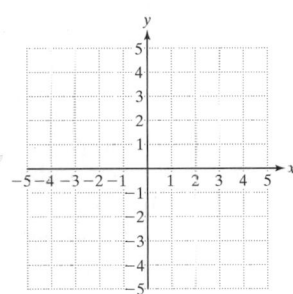

19. $x = -(y - 3)^2 - 3$

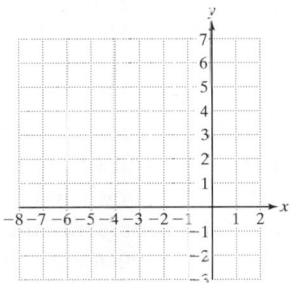

20. $x = -2(y + 2)^2 + 1$

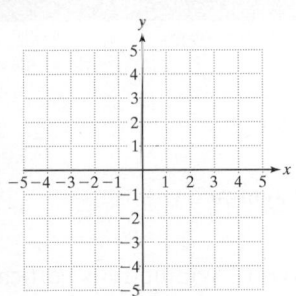

21. $x = -y^2 + 4y - 4$

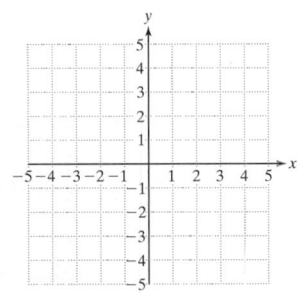

22. $x = -4y^2 - 4y - 2$

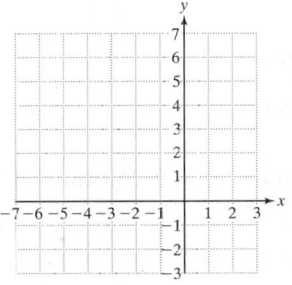

23. $x = y^2 - 2y + 2$

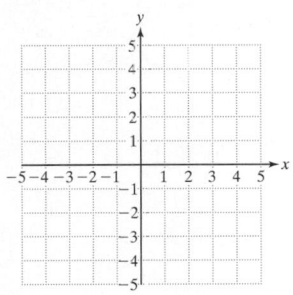

24. $x = y^2 + 4y + 1$

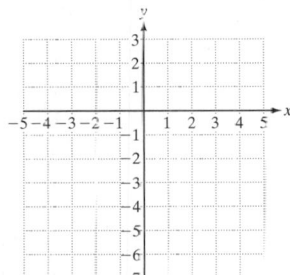

Concept 4: Vertex Formula

For Exercises 25–33, determine the vertex by using the vertex formula. (See Examples 4–5.)

25. $y = x^2 - 4x + 3$

26. $y = x^2 + 6x - 2$

27. $x = y^2 + 2y + 6$

28. $x = y^2 - 8y + 3$

29. $y = -\frac{1}{4}x^2 + x + \frac{3}{4}$

30. $y = -\frac{1}{2}x^2 - x + \frac{1}{2}$

31. $y = x^2 - 3x + 2$

32. $x = -2y^2 + 16y + 1$

33. $x = -3y^2 - 6y + 7$

34. Ricardo has a satellite dish for his television. The cross sections of the satellite dish are parabolic in shape. A cross section through the vertex can be described by the equation

$$d(x) = \frac{2}{25}x^2 - \frac{2}{5}x$$

where $d(x)$ is the depth of the dish (in feet) at a distance of x ft from the edge of the dish. How deep is the satellite dish?

35. Water from an outdoor fountain is projected outward from a jet on the side wall of the fountain. The water follows a parabolic path, and can be modeled by $h(x) = -x^2 + 10x - 3$. In this function, $h(x)$ represents the height of the water, x ft from the jet. Find the maximum height of the water.

Mixed Exercises

36. Explain how to determine whether a parabola opens upward, downward, left, or right.

37. Explain how to determine whether a parabola has a vertical or horizontal axis of symmetry.

For Exercises 38–49, use the equation of the parabola first to determine whether the axis of symmetry is vertical or horizontal. Then determine if the parabola opens upward, downward, left, or right.

38. $y = (x - 2)^2 + 3$ **39.** $y = (x - 4)^2 + 2$ **40.** $y = -2(x + 1)^2 - 4$

41. $y = -3(x + 2)^2 - 1$ **42.** $x = y^2 + 4$ **43.** $x = y^2 - 2$

44. $x = -(y + 3)^2 + 2$ **45.** $x = -2(y - 1)^2 - 3$ **46.** $y = -2x^2 - 5$

47. $y = -x^2 + 3$ **48.** $x = 2y^2 + 3y - 2$ **49.** $x = y^2 - 5y + 1$

Section 9.3   The Ellipse and Hyperbola

Concepts

1. Standard Form of an Equation of an Ellipse
2. Standard Form of an Equation of a Hyperbola

1. Standard Form of an Equation of an Ellipse

In this section, we will study the two remaining conic sections: the ellipse and the hyperbola. An **ellipse** is the set of all points (x, y) such that the sum of the distances between (x, y) and two distinct points is a constant. The fixed points are called the foci (plural of *focus*) of the ellipse.

To visualize an ellipse, consider the following application. Suppose Sonya wants to cut an elliptical rug from a rectangular rug to avoid a stain made by the family dog. She places two tacks along the center horizontal line. Then she ties the ends of a slack piece of rope to each tack. With the rope pulled tight, she traces out a curve. This curve is an ellipse, and the tacks are located at the foci of the ellipse (Figure 9-14).

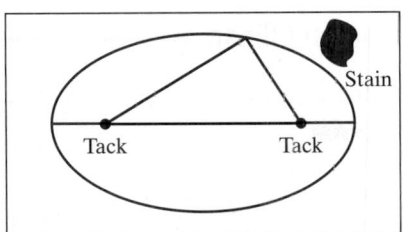

Figure 9-14

We will first graph ellipses that are centered at the origin.

Standard Form of an Equation of an Ellipse Centered at the Origin

An ellipse with the center at the origin has the equation $\dfrac{x^2}{a^2} + \dfrac{y^2}{b^2} = 1$, where a and b are positive real numbers. In the standard form of the equation, the right side must equal 1.

To graph an ellipse centered at the origin, find the x- and y-intercepts.

To find the x-intercepts, let $y = 0$.

$$\frac{x^2}{a^2} + \frac{0}{b^2} = 1$$

$$\frac{x^2}{a^2} = 1$$

$$x^2 = a^2$$

$$x = \pm\sqrt{a^2}$$

$$x = \pm a$$

The x-intercepts are $(a, 0)$ and $(-a, 0)$. See Figure 9-15.

To find the y-intercepts, let $x = 0$.

$$\frac{0}{a^2} + \frac{y^2}{b^2} = 1$$

$$\frac{y^2}{b^2} = 1$$

$$y^2 = b^2$$

$$y = \pm\sqrt{b^2}$$

$$y = \pm b$$

The y-intercepts are $(0, b)$ and $(0, -b)$. See Figure 9-15.

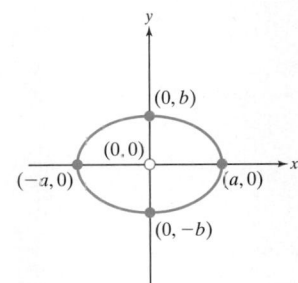

Figure 9-15

Example 1   Graphing an Ellipse

Graph the ellipse given by the equation. $\dfrac{x^2}{9} + \dfrac{y^2}{4} = 1$

Solution:

The equation can be written as $\dfrac{x^2}{3^2} + \dfrac{y^2}{2^2} = 1$.

The equation is in standard form $\dfrac{x^2}{a^2} + \dfrac{y^2}{b^2} = 1$, therefore, $a = 3$ and $b = 2$.

Graph the intercepts $(3, 0), (-3, 0), (0, 2), (0, -2)$ and sketch the ellipse.

Answer

1.

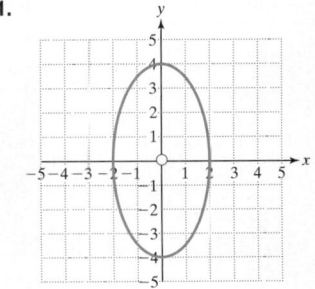

Skill Practice

1. Graph the ellipse given by the equation $\dfrac{x^2}{4} + \dfrac{y^2}{16} = 1$.

Example 2 Graphing an Ellipse

Graph the ellipse given by the equation. $25x^2 + y^2 = 25$

Solution:

$25x^2 + y^2 = 25$ Divide both sides by 25.

$$\frac{25x^2}{25} + \frac{y^2}{25} = \frac{25}{25}$$

$$x^2 + \frac{y^2}{25} = 1$$

The equation can then be written as $\dfrac{x^2}{1^2} + \dfrac{y^2}{5^2} = 1$.

The equation is in standard form $\dfrac{x^2}{a^2} + \dfrac{y^2}{b^2} = 1$,

therefore, $a = 1$ and $b = 5$.

Graph the intercepts $(1, 0), (-1, 0), (0, 5)$, and $(0, -5)$ and sketch the ellipse.

Skill Practice

2. Graph the ellipse given by the equation $x^2 + 9y^2 = 9$.

A circle is a special case of an ellipse where $a = b$. Therefore, it is not surprising that we graph an ellipse centered at the point (h, k) in much the same way we graph a circle.

Example 3 Graphing an Ellipse Whose Center Is Not at the Origin

Graph the ellipse given by the equation. $\dfrac{(x-1)^2}{16} + \dfrac{(y+3)^2}{4} = 1$

Solution:

Just as we would find the center of a circle, we see that the center of the ellipse is $(1, -3)$. Now we can use the values of a and b to help us plot four strategic points to define the curve.

The equation can be written as
$$\frac{(x-1)^2}{(4)^2} + \frac{(y+3)^2}{(2)^2} = 1.$$

From this, we have $a = 4$ and $b = 2$. To sketch the curve, locate the center at $(1, -3)$. Then move $a = 4$ units to the left and right of the center and plot two points. Similarly, move $b = 2$ units up and down from the center and plot two points.

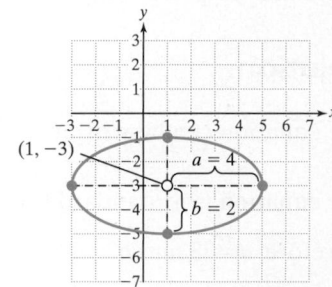

Skill Practice

3. Graph the ellipse $\dfrac{(x+1)^2}{4} + \dfrac{(y-4)^2}{9} = 1$.

Answers

2.

3.

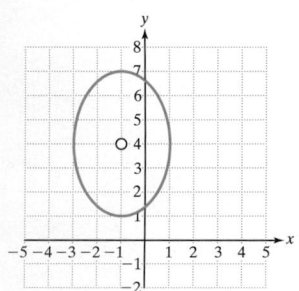

2. Standard Form of an Equation of a Hyperbola

A **hyperbola** is the set of all points (x, y) such that the absolute value of the *difference* of the distances between (x, y) and two distinct points is a constant. The fixed points are called the foci of the hyperbola. The graph of a hyperbola has two parts, called branches. A hyperbola has two vertices that lie on an axis of symmetry called the **transverse axis**. For the hyperbolas studied here, the transverse axis is either horizontal or vertical.

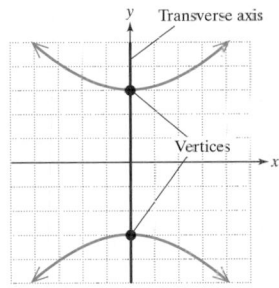

Standard Forms of an Equation of a Hyperbola with Center at the Origin

Let a and b represent positive real numbers.

Horizontal transverse axis:

The standard form of an equation of a hyperbola with a *horizontal transverse axis* and center at the origin is given by $\dfrac{x^2}{a^2} - \dfrac{y^2}{b^2} = 1$.

Note: The x term is positive. The branches of the hyperbola open left and right.

Vertical transverse axis:

The standard form of an equation of a hyperbola with a *vertical transverse axis* and center at the origin is given by $\dfrac{y^2}{b^2} - \dfrac{x^2}{a^2} = 1$.

Note: The y term is positive. The branches of the hyperbola open up and down.

In the standard forms of an equation of a hyperbola, the right side must equal 1.

To graph a hyperbola centered at the origin, first construct a reference rectangle. Draw this rectangle by using the points (a, b), $(-a, b)$, $(a, -b)$, and $(-a, -b)$. Asymptotes lie on the diagonals of the rectangle. The branches of the hyperbola are drawn to approach the asymptotes.

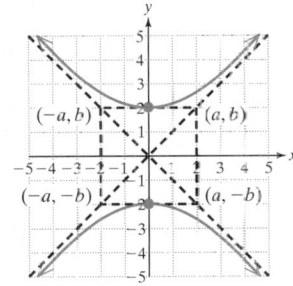

Example 4 Graphing a Hyperbola

Graph the hyperbola given by the equation. $\dfrac{x^2}{36} - \dfrac{y^2}{9} = 1$

a. Determine whether the transverse axis is horizontal or vertical.

b. Draw the reference rectangle and asymptotes.

c. Graph the hyperbola and label the vertices.

Solution:

a. Since the x-term is positive, the transverse axis is horizontal.

b. The equation can be written $\dfrac{x^2}{6^2} - \dfrac{y^2}{3^2} = 1$;

 therefore, $a = 6$ and $b = 3$. Graph the reference rectangle from the points $(6, 3), (6, -3), (-6, 3), (-6, -3)$.

c. The vertices are $(-6, 0)$ and $(6, 0)$.

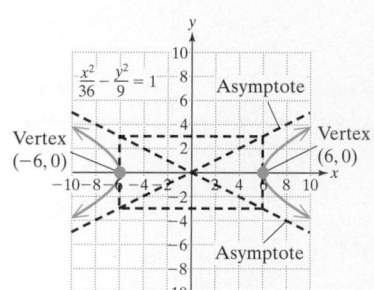

Skill Practice

4. Graph the hyperbola $x^2 - \dfrac{y^2}{9} = 1$.

Example 5 Graphing a Hyperbola

Graph the hyperbola given by the equation. $y^2 - 4x^2 - 16 = 0$

a. Write the equation in standard form to determine whether the transverse axis is horizontal or vertical.

b. Draw the reference rectangle and asymptotes.

c. Graph the hyperbola and label the vertices.

Solution:

a. $y^2 - 4x^2 - 16 = 0$

 $y^2 - 4x^2 = 16$ Isolate the constant term on the right.

 $\dfrac{y^2}{16} - \dfrac{4x^2}{16} = \dfrac{16}{16}$ Divide by 16 to make the constant term 1.

 $\dfrac{y^2}{16} - \dfrac{x^2}{4} = 1$

 Since the y term is positive, the transverse axis is vertical.

b. The equation can be written $\dfrac{y^2}{4^2} - \dfrac{x^2}{2^2} = 1$;

 therefore, $a = 2$ and $b = 4$. Graph the reference rectangle from the points $(2, 4)$, $(2, -4), (-2, 4), (-2, -4)$.

c. The vertices are $(0, 4)$ and $(0, -4)$.

Answers

4.

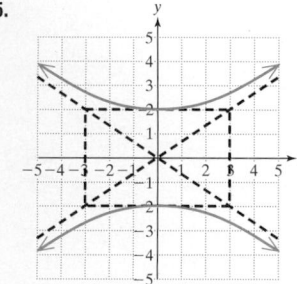

5.

Skill Practice

5. Graph the hyperbola $9y^2 - 4x^2 = 36$.

The following summarizes the steps for graphing ellipses and hyperbolas.

Graphing an Ellipse with Center at the Origin

1. Write the equation in standard form:
$$\frac{x^2}{a^2} + \frac{y^2}{b^2} = 1$$

2. Plot the intercepts.
   x-intercepts: $(a, 0)$ and $(-a, 0)$
   y-intercepts: $(0, b)$ and $(0, -b)$

3. Sketch the ellipse through the intercepts.

Graphing a Hyperbola with Center at the Origin

1. Write the equation in standard form:
   - Horizontal transverse axis:
   $$\frac{x^2}{a^2} - \frac{y^2}{b^2} = 1$$
   - Vertical transverse axis:
   $$\frac{y^2}{b^2} - \frac{x^2}{a^2} = 1$$

2. Draw the reference rectangle through the points $(a, b), (a, -b), (-a, b),$ and $(-a, -b)$.

3. Draw the asymptotes.

4. Plot the vertices and sketch the hyperbola.

Section 9.3 Practice Exercises

Vocabulary and Key Concepts

1. **a.** A(n) _____ is the set of all points (x, y) such that the sum of the distances between (x, y) and two distinct points (called _____) is constant.

   b. The standard form of an equation of an _____ with center at the origin is $\frac{x^2}{a^2} + \frac{y^2}{b^2} = 1$.

   c. A(n) _____ is the set of all points (x, y) such that the absolute value of the difference in distances between (x, y) and two distinct points (called _____) is a constant.

   d. A hyperbola has two vertices that lie on an axis of symmetry called the _____ axis.

   e. The standard form of an equation of a hyperbola with a *horizontal* transverse axis and center at $(0, 0)$, is given by _____.

   f. The standard form of an equation of a hyperbola with a *vertical* transverse axis and center at $(0, 0)$, is given by _____.

Review Exercises

2. Write the standard form of a circle with center at (h, k) and radius, r.

For Exercises 3–4, identify the center and radius of the circle.

3. $x^2 + y^2 - 16x + 12y = 0$

4. $x^2 + y^2 - 4x + 4y - 1 = 0$

For Exercises 5–6, identify the vertex and the axis of symmetry.

5. $y = 3(x + 3)^2 - 1$

6. $x = -\frac{1}{4}(y - 1)^2 - 6$

7. Write an equation for a circle whose center has coordinates $\left(\frac{1}{2}, \frac{5}{2}\right)$ with radius equal to $\frac{1}{2}$.

8. Write the equation for the circle centered at the origin and with radius equal to $\frac{1}{8}$.

Concept 1: Standard Form of an Equation of an Ellipse

For Exercises 9–16, find the x- and y-intercepts and graph the ellipse. **(See Examples 1–2.)**

9. $\dfrac{x^2}{4} + \dfrac{y^2}{9} = 1$

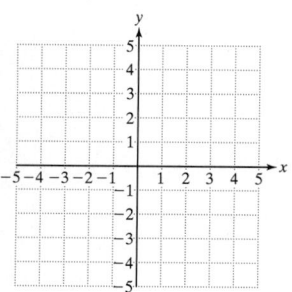

10. $\dfrac{x^2}{16} + \dfrac{y^2}{25} = 1$

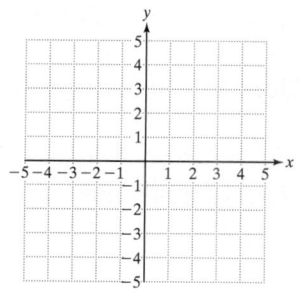

11. $\dfrac{x^2}{16} + \dfrac{y^2}{9} = 1$

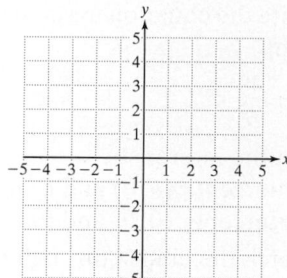

12. $\dfrac{x^2}{36} + \dfrac{y^2}{4} = 1$

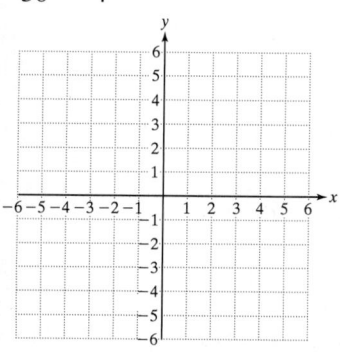

13. $4x^2 + y^2 = 4$

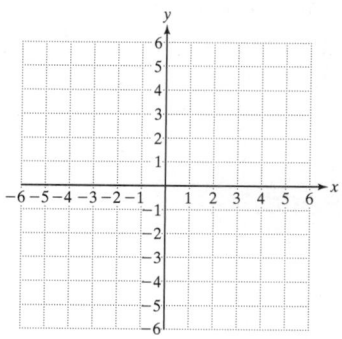

14. $9x^2 + y^2 = 36$

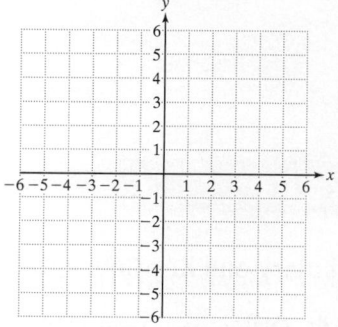

15. $x^2 + 25y^2 - 25 = 0$

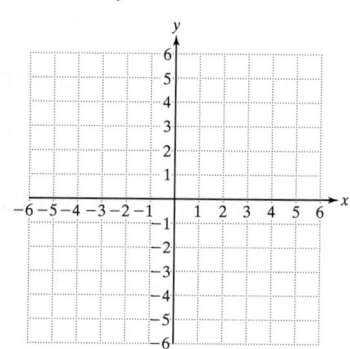

16. $4x^2 + 9y^2 = 144$

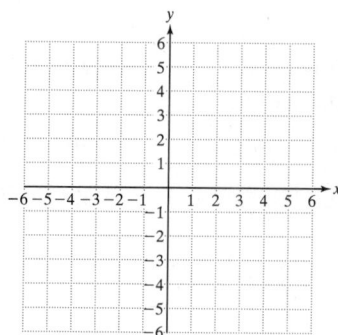

For Exercises 17–24, identify the center (h, k) of the ellipse. Then graph the ellipse. **(See Example 3.)**

17. $\dfrac{(x - 4)^2}{4} + \dfrac{(y - 5)^2}{9} = 1$

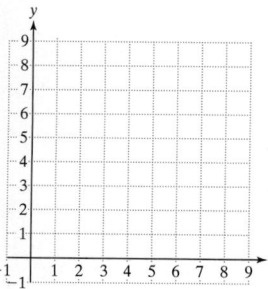

18. $\dfrac{(x + 2)^2}{9} + \dfrac{(y - 1)^2}{16} = 1$

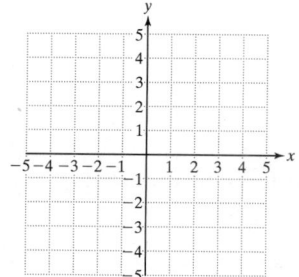

19. $\dfrac{(x + 1)^2}{25} + \dfrac{(y - 2)^2}{9} = 1$

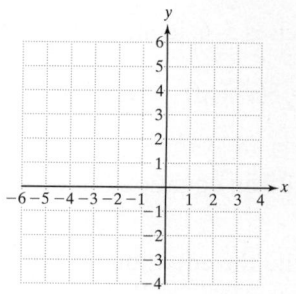

20. $\dfrac{(x-4)^2}{25} + \dfrac{(y+1)^2}{16} = 1$

21. $\dfrac{(x-2)^2}{9} + (y+3)^2 = 1$

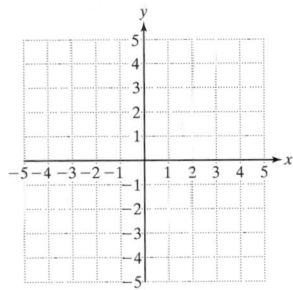

22. $(x-5)^2 + \dfrac{(y-3)^2}{4} = 1$

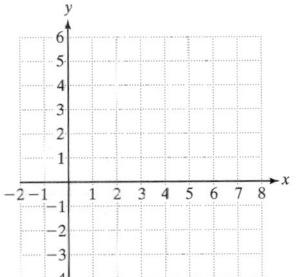

23. $\dfrac{x^2}{36} + \dfrac{(y-1)^2}{25} = 1$

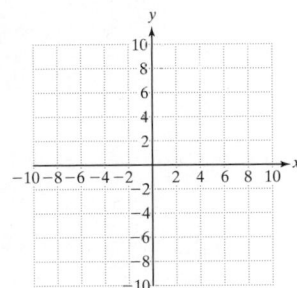

24. $\dfrac{(x+5)^2}{4} + \dfrac{y^2}{16} = 1$

Concept 2: Standard Form of an Equation of a Hyperbola

For Exercises 25–32, determine whether the transverse axis is horizontal or vertical.

25. $\dfrac{y^2}{6} - \dfrac{x^2}{18} = 1$

26. $\dfrac{y^2}{10} - \dfrac{x^2}{4} = 1$

27. $\dfrac{x^2}{20} - \dfrac{y^2}{15} = 1$

28. $\dfrac{x^2}{12} - \dfrac{y^2}{9} = 1$

29. $x^2 - y^2 = 12$

30. $x^2 - y^2 = 15$

31. $x^2 - 3y^2 = -9$

32. $2x^2 - y^2 = -10$

For Exercises 33–40, use the equation in standard form to graph the hyperbola. Label the vertices of the hyperbola.
(See Examples 4–5.)

33. $\dfrac{x^2}{25} - \dfrac{y^2}{16} = 1$

34. $\dfrac{x^2}{9} - \dfrac{y^2}{36} = 1$

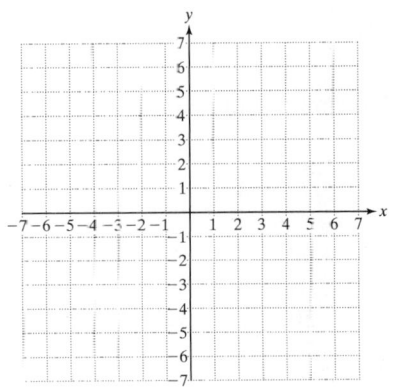

35. $\dfrac{y^2}{4} - \dfrac{x^2}{4} = 1$

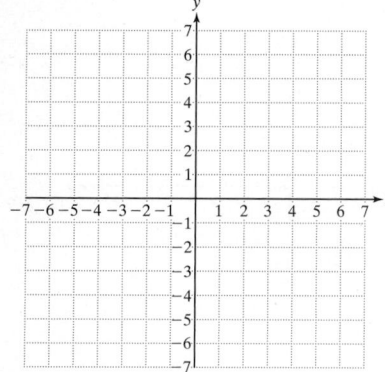

36. $\dfrac{y^2}{9} - \dfrac{x^2}{9} = 1$

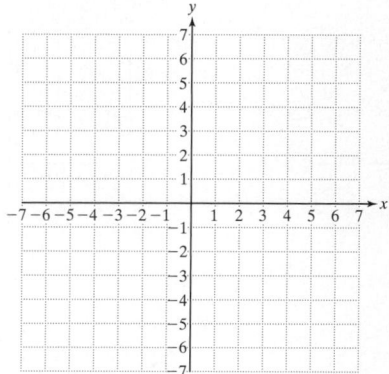

37. $36x^2 - y^2 = 36$

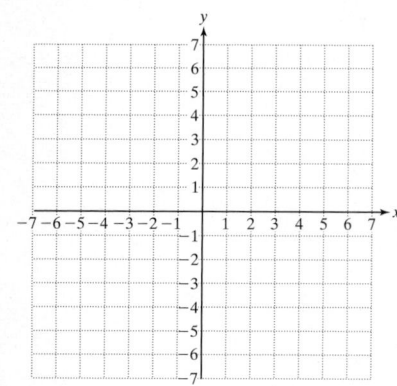

38. $x^2 - 25y^2 = 25$

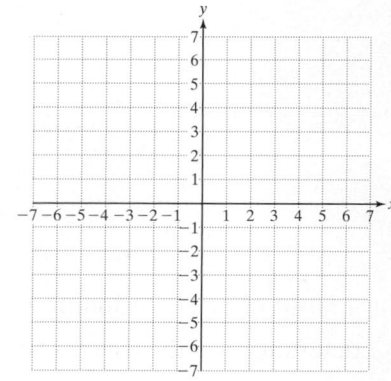

39. $y^2 - 4x^2 - 16 = 0$

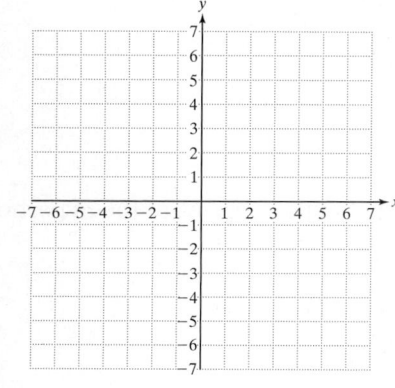

40. $y^2 - 4x^2 - 36 = 0$

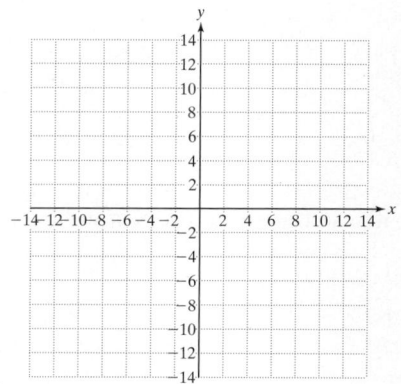

Mixed Exercises

For Exercises 41–52, determine if the equation represents an ellipse or a hyperbola.

41. $\dfrac{x^2}{6} - \dfrac{y^2}{10} = 1$

42. $\dfrac{x^2}{14} + \dfrac{y^2}{2} = 1$

43. $\dfrac{y^2}{4} + \dfrac{x^2}{16} = 1$

44. $\dfrac{x^2}{5} + \dfrac{y^2}{10} = 1$

45. $4x^2 + y^2 = 16$

46. $-3x^2 - 4y^2 = -36$

47. $-y^2 + 2x^2 = -10$

48. $x^2 - y^2 = -1$

49. $5x^2 + y^2 - 10 = 0$

50. $5x^2 - 3y^2 = 15$

51. $y^2 - 6x^2 = 6$

52. $3x^2 + 5y^2 = 15$

Expanding Your Skills

 53. An arch for a tunnel is in the shape of a semiellipse. The distance between vertices is 120 ft, and the height to the top of the arch is 50 ft. Find the height of the arch 10 ft from the center. Round to the nearest foot.

 54. A bridge over a gorge is supported by an arch in the shape of a semiellipse. The length of the bridge is 400 ft, and the height is 100 ft. Find the height of the arch 50 ft from the center. Round to the nearest foot.

In Exercises 55–58, graph the hyperbola centered at (h, k) by following these guidelines. First locate the center. Then draw the reference rectangle relative to the center (h, k). Using the reference rectangle as a guide, sketch the hyperbola. Label the center and vertices.

55. $\dfrac{(x-1)^2}{9} - \dfrac{(y+2)^2}{4} = 1$

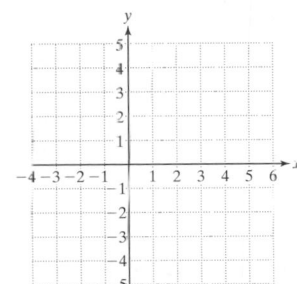

56. $\dfrac{(x+2)^2}{16} - \dfrac{(y-3)^2}{4} = 1$

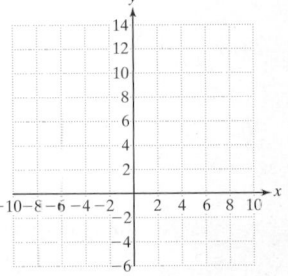

57. $\dfrac{(y-1)^2}{4} - (x+3)^2 = 1$

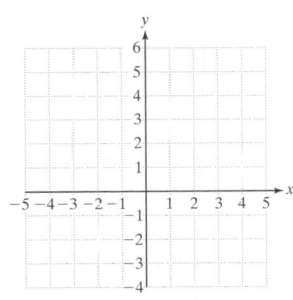

58. $(y+2)^2 - \dfrac{(x+3)^2}{4} = 1$

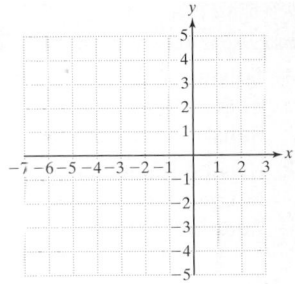

Problem Recognition Exercises

Formulas and Conic Sections

For Exercises 1–8, identify the formula.

1. $(x - h)^2 + (y - k)^2 = r^2$

2. $\dfrac{x^2}{a^2} + \dfrac{y^2}{b^2} = 1$

3. $\sqrt{(x_2 - x_1)^2 + (y_2 - y_1)^2}$

4. $\left(\dfrac{x_1 + x_2}{2}, \dfrac{y_1 + y_2}{2}\right)$

5. $y = a(x - h)^2 + k$

6. $\dfrac{x^2}{a^2} - \dfrac{y^2}{b^2} = 1$

7. $\dfrac{y^2}{b^2} - \dfrac{x^2}{a^2} = 1$

8. $x = a(y - k)^2 + h$

For Exercises 9–30, identify the equation as representing a circle, parabola, ellipse, hyperbola, or none of the above.

9. $y = -2(x - 3)^2 + 4$

10. $\dfrac{x^2}{4} - \dfrac{y^2}{1} = 1$

11. $(x + 3)^2 + (y + 2)^2 = 4$

12. $(x - 2)^2 + (y - 4)^2 = 9$

13. $\dfrac{x^2}{9} - \dfrac{y^2}{9} = 1$

14. $\dfrac{x^2}{9} + \dfrac{y^2}{16} = 1$

15. $\dfrac{x^2}{16} + \dfrac{y^2}{4} = 0$

16. $x^2 + y^2 - 2x + 4y - 4 = 0$

17. $y = \dfrac{1}{2}(x + 2)^2 - 3$

18. $\dfrac{x^2}{4} - \dfrac{y^2}{2} = 1$

19. $x = (y + 2)^2 - 4$

20. $x^2 + y^2 + 6x + 8 = 0$

21. $(x - 1)^2 + (y + 1)^2 = 0$

22. $x = -(y - 2)^2 - 1$

23. $\dfrac{x^2}{25} + \dfrac{y^2}{4} = 1$

24. $x^2 + y^2 = 15$

25. $y = (x - 6)^2 + 4$

26. $\dfrac{(x + 1)^2}{2} + \dfrac{(y + 1)^2}{5} = 1$

27. $\dfrac{y^2}{3} - \dfrac{x^2}{3} = 1$

28. $3x^2 + 3y^2 = 1$

29. $\dfrac{x^2}{9} + \dfrac{y^2}{12} = 1$

30. $x = (y + 2)^2 - 5$

Section 9.4   Nonlinear Systems of Equations in Two Variables

Concepts

1. Solving Nonlinear Systems of Equations by the Substitution Method
2. Solving Nonlinear Systems of Equations by the Addition Method

1. Solving Nonlinear Systems of Equations by the Substitution Method

Recall that a linear equation in two variables x and y is an equation that can be written in the form $Ax + By = C$, where A and B are not both zero. In Sections 3.1–3.3, we solved systems of linear equations in two variables by using the graphing method, the substitution method, and the addition method. In this section, we will solve *nonlinear* systems of equations by using the same methods. A **nonlinear system of equations** is a system in which at least one of the equations is nonlinear.

Graphing the equations in a nonlinear system helps to determine the number of solutions and to approximate the coordinates of the solutions. The substitution method is used most often to solve a nonlinear system of equations analytically.

Example 1   Solving a Nonlinear System of Equations

Given the system
$$x - 7y = -25$$
$$x^2 + y^2 = 25$$

a. Solve the system by graphing.

b. Solve the system by the substitution method.

Solution:

a. $x - 7y = -25$ is a line $\left(\text{the slope-intercept form is } y = \frac{1}{7}x + \frac{25}{7}\right)$.

$x^2 + y^2 = 25$ is a circle centered at the origin with radius 5.

From Figure 9-16, we appear to have two solutions $(-4, 3)$ and $(3, 4)$.

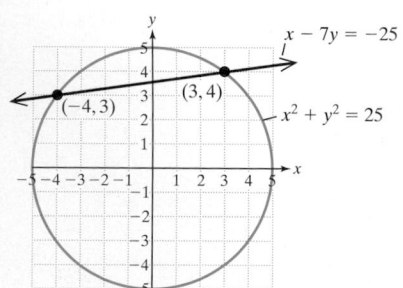

Figure 9-16

b. To use the substitution method, isolate one of the variables from one of the equations. We will solve for x in the first equation.

$$\boxed{A} \qquad x - 7y = -25 \xrightarrow{\text{Solve for } x.} x = 7y - 25$$

$$\boxed{B} \qquad x^2 + y^2 = 25$$

$$\boxed{B} \quad (7y - 25)^2 + y^2 = 25 \qquad \text{Substitute } (7y - 25) \text{ for } x \text{ in the second equation.}$$

$$49y^2 - 350y + 625 + y^2 = 25 \qquad \text{The resulting equation is quadratic in } y.$$

$$50y^2 - 350y + 600 = 0 \qquad \text{Set the equation equal to zero.}$$

$$50(y^2 - 7y + 12) = 0 \qquad \text{Factor.}$$

$$50(y - 3)(y - 4) = 0$$

$$y = 3 \quad \text{or} \quad y = 4$$

For each value of y, find the corresponding x value from the equation $x = 7y - 25$.

$y = 3$: $x = 7(3) - 25 = -4$ The solution point is $(-4, 3)$.

$y = 4$: $x = 7(4) - 25 = 3$ The solution point is $(3, 4)$.
(See Figure 9-16.)

The solution set is $\{(-4, 3), (3, 4)\}$.

Skill Practice Given the system

$$2x + y = 5$$
$$x^2 + y^2 = 50$$

1. Solve the system by graphing.

2. Solve the system by the substitution method.

Example 2   **Solving a Nonlinear System by the Substitution Method**

Given the system $y = \sqrt{x}$
$$x^2 + y^2 = 20$$

a. Sketch the graphs.

b. Solve the system by the substitution method.

Solution:

a. $y = \sqrt{x}$ is one of the six basic functions graphed in Section 2.7.

$x^2 + y^2 = 20$ is a circle centered at the origin with radius $\sqrt{20} \approx 4.5$.

From Figure 9-17, we see that there is one solution.

Figure 9-17

Answers

1.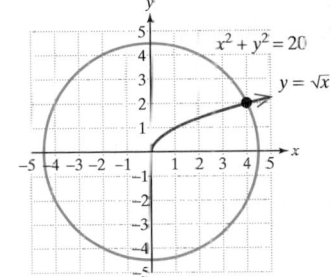

The solutions appear to be $(-1, 7)$ and $(5, -5)$.

2. $\{(-1, 7), (5, -5)\}$

b. To use the substitution method, we will substitute $y = \sqrt{x}$ into equation $\boxed{B}$.

$\boxed{A}$ $\qquad\qquad y = \sqrt{x}$ $\qquad\qquad$ Substitute $y = \sqrt{x}$ into the second equation.

$\boxed{B}$ $\qquad x^2 + y^2 = 20$

$\boxed{B}$ $\quad x^2 + (\sqrt{x})^2 = 20$

$\qquad\qquad x^2 + x = 20$

$\qquad x^2 + x - 20 = 0$ $\qquad$ Set the second equation equal to zero.

$\quad (x + 5)(x - 4) = 0$ $\qquad$ Factor.

$x \cancel{=} -5 \quad$ or $\quad x = 4$ $\qquad$ Reject $x = -5$ because it is not in the domain of $y = \sqrt{x}$.

Substitute $x = 4$ into the equation $y = \sqrt{x}$.

If $x = 4$, then $y = \sqrt{4} = 2$. $\qquad$ The solution set is $\{(4, 2)\}$.

Skill Practice Given the system

$$y = \sqrt{2x}$$
$$x^2 + y^2 = 8$$

3. Sketch the graphs.

4. Solve the system by using the substitution method.

Calculator Connections

Topic: Graphing a Nonlinear System of Equations

Graph the equations from Example 2 to confirm your solution to the system of equations. Use an *Intersect* feature to approximate the point of intersection. Recall that the circle must be entered into the calculator as two functions:

$$Y_1 = \sqrt{20 - x^2}$$
$$Y_2 = -\sqrt{20 - x^2}$$
$$Y_3 = \sqrt{x}$$

Example 3 Solving a Nonlinear System by the Substitution Method

Answers

3.

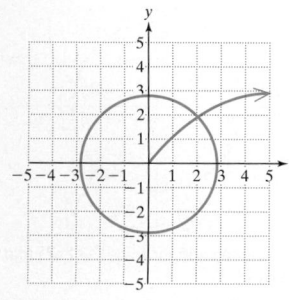

4. $\{(2, 2)\}$

Solve the system by using the substitution method. $\qquad y = \sqrt[3]{x}$

$\qquad\qquad\qquad\qquad\qquad\qquad\qquad\qquad\qquad\qquad\quad y = x$

Solution:

$\boxed{A}$ $\quad y = \sqrt[3]{x}$

$\boxed{B}$ $\quad y = x$

$\qquad \sqrt[3]{x} = x$ $\qquad\qquad\qquad$ Because y is isolated in both equations, we can equate the expressions for y.

$\qquad (\sqrt[3]{x})^3 = (x)^3$ $\qquad\qquad$ To solve the radical equation, raise both sides to the third power.

$x = x^3$ This is a third-degree polynomial equation.

$0 = x^3 - x$ Set the equation equal to zero.

$0 = x(x^2 - 1)$ Factor out the GCF.

$0 = x(x + 1)(x - 1)$ Factor completely.

$x = 0$ or $x = -1$ or $x = 1$

For each value of x, find the corresponding y-value from either original equation. We will use equation $\boxed{\text{B}}$: $y = x$.

    If $x = 0$, then $y = 0$.   The solution point is $(0, 0)$.

    If $x = -1$, then $y = -1$.   The solution point is $(-1, -1)$.

    If $x = 1$, then $y = 1$.   The solution point is $(1, 1)$.

The solution set is $\{(0, 0), (-1, -1), (1, 1)\}$. See Figure 9-18.

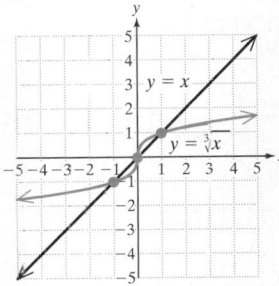

Figure 9-18

Skill Practice Solve the system by using the substitution method.

5. $y = \sqrt[3]{9x}$

   $y = x$

2. Solving Nonlinear Systems of Equations by the Addition Method

The substitution method is used most often to solve a system of nonlinear equations. In some situations, however, the addition method offers an efficient means of finding a solution. Example 4 demonstrates that we can eliminate a variable from both equations provided the terms containing the corresponding variables are *like* terms.

> **Example 4** **Solving a Nonlinear System of Equations by the Addition Method**

Solve the system. $2x^2 + y^2 = 17$

                    $x^2 + 2y^2 = 22$

Solution:

$\boxed{\text{A}}$ $2x^2 + y^2 = 17$ Notice that the y^2 terms are *like* in each equation.

$\boxed{\text{B}}$ $x^2 + 2y^2 = 22$ To eliminate the y^2 terms, multiply the first equation by -2.

$\boxed{\text{A}}$ $2x^2 + y^2 = 17$ $\xrightarrow{\text{Multiply by } -2.}$ $-4x^2 - 2y^2 = -34$

$\boxed{\text{B}}$ $x^2 + 2y^2 = 22$ $\xrightarrow{\hspace{2cm}}$ $\underline{ x^2 + 2y^2 = 22}$

                                         $-3x^2 = -12$ Eliminate the y^2 term.

$$\frac{-3x^2}{-3} = \frac{-12}{-3}$$

$$x^2 = 4$$

$$x = \pm 2 \qquad \text{Use the square root property.}$$

> **TIP:** In Example 4, the x^2 terms are also *like* in both equations. We could have eliminated the x^2 terms by multiplying equation $\boxed{\text{B}}$ by -2.

Answer

5. $\{(0, 0), (3, 3), (-3, -3)\}$

Substitute each value of x into one of the original equations to solve for y. We will use equation $\boxed{A}$: $2x^2 + y^2 = 17$.

$x = 2$: $\boxed{A}$ $\quad 2(2)^2 + y^2 = 17$

$$8 + y^2 = 17$$

$$y^2 = 9$$

$$y = \pm 3 \quad \text{The solution points are } (2, 3) \text{ and } (2, -3).$$

$x = -2$: $\boxed{A}$ $\quad 2(-2)^2 + y^2 = 17$

$$8 + y^2 = 17$$

$$y^2 = 9$$

$$y = \pm 3 \quad \text{The solution points are } (-2, 3) \text{ and } (-2, -3).$$

The solution set is $\{(2, 3), (2, -3), (-2, 3), (-2, -3)\}$.

TIP: The two equations in Example 4 are ellipses but an exact graph is difficult to render by hand.

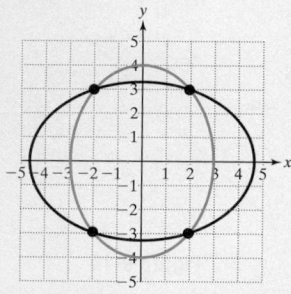

To check the solutions to the system, verify that each ordered pair satisfies both equations.

Skill Practice Solve the system by using the addition method.

6. $\quad x^2 - y^2 = 24$
$\quad\quad 3x^2 + y^2 = 76$

TIP: It is important to note that the addition method can be used only if two equations share a pair of *like* terms. The substitution method is effective in solving a wider range of systems of equations. The system in Example 4 could also have been solved by using substitution.

$\boxed{A}$ $\quad\quad 2x^2 + y^2 = 17 \xrightarrow{\text{Solve for } y^2.} y^2 = 17 - 2x^2$

$\boxed{B}$ $\quad\quad x^2 + 2y^2 = 22$

$\boxed{B}$ $\quad x^2 + 2(17 - 2x^2) = 22 \quad\quad\quad x = 2: \quad y^2 = 17 - 2(2)^2$

$\quad\quad\quad x^2 + 34 - 4x^2 = 22 \quad\quad\quad\quad\quad\quad\quad y^2 = 9$

$\quad\quad\quad\quad\quad\quad -3x^2 = -12 \quad\quad\quad\quad\quad\quad\quad\quad y = \pm 3 \quad \text{The solutions are}$
$\quad\quad\quad\quad\quad\quad\quad\quad\quad\quad\quad\quad\quad\quad\quad\quad\quad\quad\quad\quad (2, 3) \text{ and } (2, -3).$

$\quad\quad\quad\quad\quad\quad\quad\quad x^2 = 4$

$\quad\quad\quad\quad\quad\quad x = \pm 2 \longrightarrow x = -2: \quad y^2 = 17 - 2(-2)^2$

$\quad\quad\quad\quad\quad\quad\quad\quad\quad\quad\quad\quad\quad\quad\quad\quad\quad y^2 = 9$

Answer

6. $\{(5, 1), (5, -1), (-5, 1), (-5, -1)\}$

$\quad\quad\quad\quad\quad\quad\quad\quad\quad\quad\quad\quad\quad\quad\quad\quad y = \pm 3 \quad \text{The solutions are}$
$\quad\quad\quad\quad\quad\quad\quad\quad\quad\quad\quad\quad\quad\quad\quad\quad\quad\quad\quad\quad (-2, 3) \text{ and } (-2, -3).$

Section 9.4 Practice Exercises

Vocabulary and Key Concepts

1. a. A _____ system of equations in two variables is a system in which at least one of the equations is nonlinear.

b. Graphically, the solution set to a nonlinear system of equations is the set of ordered pairs representing the points of _____ of the graphs of the two equations.

Review Exercises

2. Write the distance formula between two points (x_1, y_1) and (x_2, y_2) from memory.

3. Find the distance between the two points $(8, -1)$ and $(1, -8)$.

4. Write an equation representing the set of all points 2 units from the point $(-1, 1)$.

5. Write an equation representing the set of all points 8 units from the point $(-5, 3)$.

6. Write the formula for the midpoint between the points (x_1, y_1) and (x_2, y_2).

7. Find the midpoint of the segment with the endpoints $(3, -9)$ and $(-4, 2)$.

8. Find the midpoint of the segment with the endpoints $(2.4, 1.8)$ and $(-1.6, -5.2)$.

Concept 1: Solving Nonlinear Systems of Equations by the Substitution Method

For Exercises 9–16, use sketches to explain.

9. How many points of intersection are possible between a line and a parabola?

10. How many points of intersection are possible between a line and a circle?

11. How many points of intersection are possible between two distinct circles?

12. How many points of intersection are possible between two distinct parabolas of the form $y = ax^2 + bx + c$, $a \neq 0$?

13. How many points of intersection are possible between a circle and a parabola?

14. How many points of intersection are possible between two distinct lines?

15. How many points of intersection are possible with an ellipse and a hyperbola?

16. How many points of intersection are possible with an ellipse and a parabola?

For Exercises 17–22, sketch each system of equations. Then solve the system by the substitution method. **(See Example 1.)**

17. $y = x + 3$
$x^2 + y = 9$

18. $y = x - 2$
$x^2 + y = 4$

19. $x^2 + y^2 = 1$
$y = x + 1$

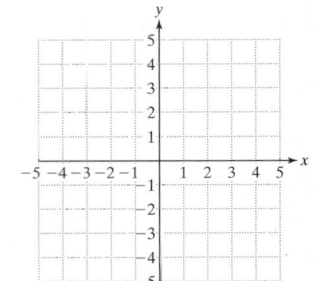

20. $x^2 + y^2 = 25$

$y = 2x$

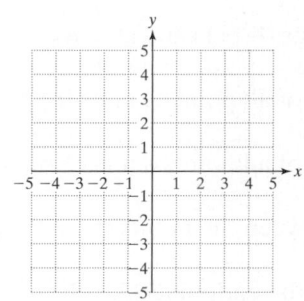

21. $x^2 + y^2 = 6$

$y = x^2$

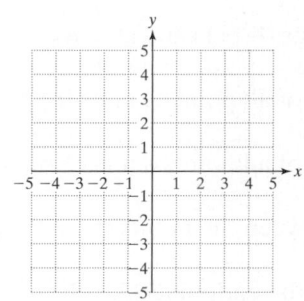

22. $x^2 + y^2 = 12$

$y = x^2$

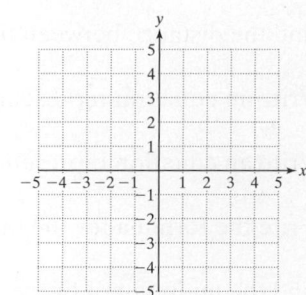

For Exercises 23–31, solve the system by the substitution method. **(See Examples 2–3.)**

23. $y = \sqrt{x}$

$2x^2 - y^2 = 1$

24. $x^2 + y^2 = 30$

$y = \sqrt{x}$

25. $y = x^2$

$y = -\sqrt{x}$

26. $y = -x^2$

$y = -\sqrt{x}$

27. $y = x^2$

$y = (x - 3)^2$

28. $y = (x + 4)^2$

$y = x^2$

29. $y = x^2 + 6x$

$y = 4x$

30. $y = 3x^2 - 6x$

$y = 3x$

31. $x^2 - 5x + y = 0$

$y = 3x + 1$

Concept 2: Solving Nonlinear Systems of Equations by the Addition Method

For Exercises 32–48, solve the system of nonlinear equations by the addition method if appropriate. **(See Example 4.)**

32. $x^2 + y^2 = 13$

$x^2 - y^2 = 5$

33. $4x^2 - y^2 = 4$

$4x^2 + y^2 = 4$

34. $9x^2 + 4y^2 = 36$

$x^2 + y^2 = 9$

35. $x^2 + y^2 = 4$

$2x^2 + y^2 = 8$

36. $3x^2 + 4y^2 = 16$

$2x^2 - 3y^2 = 5$

37. $2x^2 - 5y^2 = -2$

$3x^2 + 2y^2 = 35$

38. $x^2 + y^2 = 169$

$12x - 5y = 0$

39. $x^2 + y^2 = 100$

$4y - 3x = 0$

40. $\dfrac{x^2}{4} + \dfrac{y^2}{9} = 1$

$x^2 + y^2 = 4$

41. $\dfrac{x^2}{16} + \dfrac{y^2}{4} = 1$

$x^2 + y^2 = 4$

42. $\dfrac{x^2}{10} + \dfrac{y^2}{10} = 1$

$2x^2 + y^2 = 11$

43. $2y = -x + 2$

$x^2 + 4y^2 = 4$

44. $2y = -3x - 6$

$9x^2 + 4y^2 = 36$

45. $x^2 - xy = -4$

$2x^2 - xy = 12$

46. $x^2 - xy = 3$

$2x^2 + xy = 6$

47. $3x^2 + 2xy = 4$

$x^2 - xy = 3$

48. $x^2 - 3xy = 8$

$x^2 + xy = 4$

Expanding Your Skills

49. The sum of two numbers is 7. The sum of the squares of the numbers is 25. Find the numbers.

50. The sum of the squares of two numbers is 100. The sum of the numbers is 2. Find the numbers.

51. The sum of the squares of two numbers is 32. The difference of the squares of the numbers is 18. Find the numbers.

52. The sum of the squares of two numbers is 24. The difference of the squares of the numbers is 8. Find the numbers.

Graphing Calculator Exercises

For Exercises 53–56, use the *Intersect* feature to approximate the solutions to the system.

53. $y = x + 3$ (Exercise 17)

$x^2 + y = 9$

54. $y = x - 2$ (Exercise 18)

$x^2 + y = 4$

55. $y = x^2$ (Exercise 25)

$y = -\sqrt{x}$

56. $y = -x^2$ (Exercise 26)

$y = -\sqrt{x}$

For Exercises 57–58, graph the system on a square viewing window. What can be said about the solution to the system?

57. $x^2 + y^2 = 4$

$y = x^2 + 3$

58. $x^2 + y^2 = 16$

$y = -x^2 - 5$

Nonlinear Inequalities and Systems of Inequalities

1. Nonlinear Inequalities in Two Variables

In Section 3.5 we graphed the solution sets to linear inequalities in two variables. For example, to graph the solution set to $y \leq 2x + 1$, first graph the related equation $y = 2x + 1$. This is the line shown in Figure 9-19. Then using test points, we see that points on and below the line make up the solution set to the inequality.

Concepts

1. Nonlinear Inequalities in Two Variables
2. Systems of Nonlinear Inequalities in Two Variables

Test Point Above: (–2, 2)

$y \leq 2x + 1$

$2 \overset{?}{\leq} 2(-2) + 1$

$2 \overset{?}{\leq} -3$ False

Test Point Below: (0, 0)

$y \leq 2x + 1$

$0 \overset{?}{\leq} 2(0) + 1$

$0 \overset{?}{\leq} 1$ ✓True

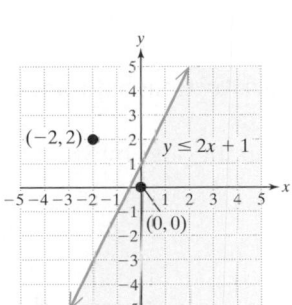

Figure 9-19

Example 1   Graphing a Nonlinear Inequality in Two Variables

Graph the solution set of the inequality $x^2 + y^2 < 16$.

Avoiding Mistakes

The first step in solving an inequality in two variables is to graph the related equation.

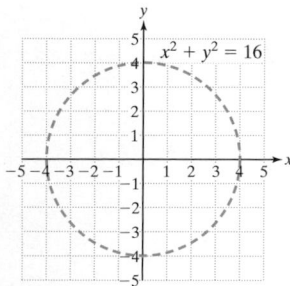

Figure 9-20

Solution:

The related equation $x^2 + y^2 = 16$ is a circle of radius 4, centered at the origin. Graph the related equation by using a dashed curve because the points satisfying the equation $x^2 + y^2 = 16$ are not part of the solution to the strict inequality $x^2 + y^2 < 16$. See Figure 9-20.

Notice that the dashed curve separates the xy-plane into two regions, one "inside" the circle, the other "outside" the circle. Select a test point from each region and test the point in the original inequality.

Test Point "Inside": (0, 0)	**Test Point "Outside": (4, 4)**
$x^2 + y^2 < 16$	$x^2 + y^2 < 16$
$(0)^2 + (0)^2 \overset{?}{<} 16$	$(4)^2 + (4)^2 \overset{?}{<} 16$
$0 \overset{?}{<} 16$ True	$32 \overset{?}{<} 16$ False

The inequality $x^2 + y^2 < 16$ is true at the test point $(0, 0)$. Therefore, the solution set is the region "inside" the circle. See Figure 9-21.

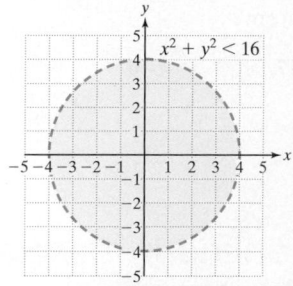

Figure 9-21

Skill Practice

1. Graph the solution set of the inequality $x^2 + y^2 \geq 9$.

Example 2   Graphing a Nonlinear Inequality in Two Variables

Graph the solution set of the inequality $9y^2 \geq 36 + 4x^2$.

Solution:

First graph the related equation $9y^2 = 36 + 4x^2$. Notice that the equation can be written in the standard form of a hyperbola.

Answer

1.

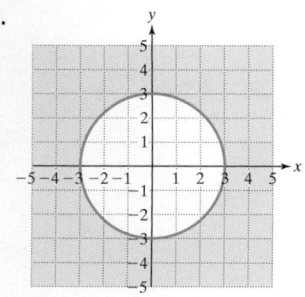

$$9y^2 = 36 + 4x^2$$

$$9y^2 - 4x^2 = 36 \qquad \text{Subtract } 4x^2 \text{ from both sides.}$$

$$\frac{9y^2}{36} - \frac{4x^2}{36} = \frac{36}{36} \qquad \text{Divide both sides by 36.}$$

$$\frac{y^2}{4} - \frac{x^2}{9} = 1$$

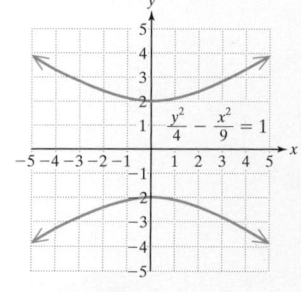

Figure 9-22

Graph the hyperbola as a solid curve, because the original inequality includes equality. See Figure 9-22.

The hyperbola divides the xy-plane into three regions: a region above the upper branch, a region between the branches, and a region below the lower branch. Select a test point from each region.

$$9y^2 \geq 36 + 4x^2$$

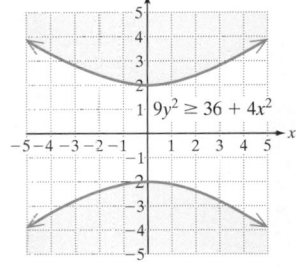

Figure 9-23

Test: (0, 3)

$9(3)^2 \overset{?}{\geq} 36 - 4(0)^2$

$81 \overset{?}{\geq} 36$ True

Test: (0, 0)

$9(0)^2 \overset{?}{\geq} 36 + 4(0)^2$

$0 \overset{?}{\geq} 36$ False

Test: (0, −3)

$9(-3)^2 \overset{?}{\geq} 36 + 4(0)^2$

$81 \overset{?}{\geq} 36$ True

Shade the regions above the top branch and below the bottom branch of the hyperbola. See Figure 9-23.

Skill Practice

2. Graph the solution set of the inequality $9x^2 < 144 - 16y^2$.

2. Systems of Nonlinear Inequalities in Two Variables

In Section 9.4 we solved systems of nonlinear equations in two variables. The solution set for such a system is the set of ordered pairs that satisfy both equations simultaneously. We will now solve systems of nonlinear inequalities in two variables. Similarly, the solution set is the set of all ordered pairs that simultaneously satisfy each inequality. To solve a system of inequalities, we will graph the solution to each individual inequality and then take the intersection of the solution sets.

Example 3   **Graphing a System of Nonlinear Inequalities in Two Variables**

Graph the solution set.

$$y > \frac{1}{3}x^2$$

$$y < -x^2 + 2$$

Solution:

The solution to $y > \frac{1}{3}x^2$ is the set of points above the parabola $y = \frac{1}{3}x^2$. See Figure 9-24. The solution to $y < -x^2 + 2$ is the set of points below the parabola $y = -x^2 + 2$. See Figure 9-25.

Figure 9-24

Figure 9-25

Figure 9-26

Answer

2.

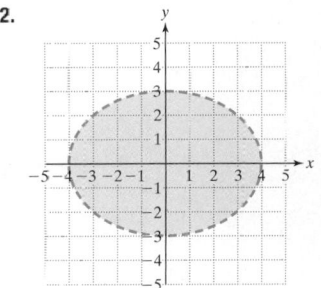

The solution to the system of inequalities is the intersection of the solution sets of the individual inequalities. See Figure 9-26.

Example 4 **Graphing a System of Nonlinear Inequalities in Two Variables**

Graph the solution set.
$$y > e^x$$
$$y < -x^2 + 4$$

Solution:

The solution to $y > e^x$ is the set of points above the curve $y = e^x$. See Figure 9-27. The solution to $y < -x^2 + 4$ is the set of points below the parabola $y = -x^2 + 4$. See Figure 9-28.

Figure 9-27

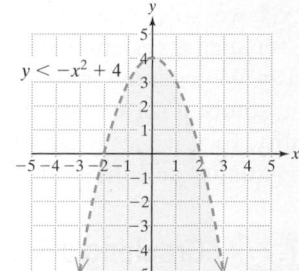

Figure 9-28

The solution to the system of inequalities is the intersection of the solution sets of the individual inequalities. See Figure 9-29.

Answers

3.

4.

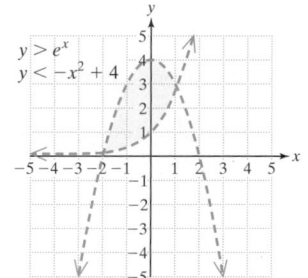

Figure 9-29

Section 9.5   Practice Exercises

Review Exercises

For Exercises 1–12, match the equation with its graph.

1. $y = \left(\dfrac{1}{3}\right)^x$ **2.** $y = 4x - 1$ **3.** $y = -4x^2$ **4.** $y = e^x$

5. $y = x^3$ **6.** $\dfrac{x^2}{4} + \dfrac{y^2}{9} = 1$ **7.** $\dfrac{x^2}{4} - \dfrac{y^2}{9} = 1$ **8.** $y = \dfrac{1}{x}$

9. $y = \log_2(x)$ **10.** $y = \sqrt{x}$ **11.** $(x + 2)^2 - (y - 1)^2 = 4$ **12.** $x = -y^2$

a.   **b.**   **c.**   **d.**

e.   **f.**   **g.**   **h.**

i.   **j.**   **k.**   **l.**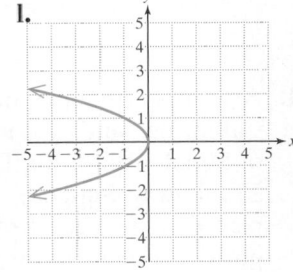

Concept 1: Nonlinear Inequalities in Two Variables

13. True or false? The point $(2, 3)$ satisfies the inequality $-x^2 + y^3 > 1$.

14. True or false? The point $(4, -2)$ satisfies the inequality $4x^2 - 2x + 1 + y^2 < 3$.

15. True or false? The point $(5, 4)$ satisfies the system of inequalities.

$$\frac{x^2}{36} + \frac{y^2}{25} < 1$$

$$x^2 + y^2 \geq 4$$

16. True or false? The point $(1, -2)$ satisfies the system of inequalities.

$$y < x^2$$

$$y > x^2 - 4$$

17. a. Graph the solution set for $x^2 + y^2 \leq 9$.

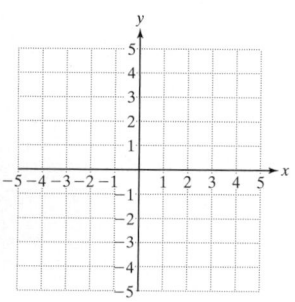

b. Describe the solution set for the inequality $x^2 + y^2 \geq 9$.

c. Describe the solution set for the equation $x^2 + y^2 = 9$

18. a. Graph the solution set for $\dfrac{x^2}{4} + \dfrac{y^2}{9} \geq 1$.

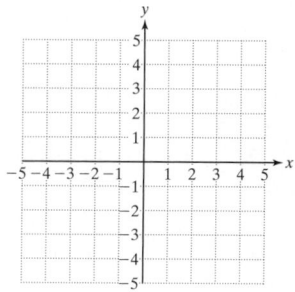

b. Describe the solution set for the inequality $\dfrac{x^2}{4} + \dfrac{y^2}{9} \leq 1$.

c. Describe the solution set for the equation $\dfrac{x^2}{4} + \dfrac{y^2}{9} = 1$

19. a. Graph the solution set for $y \geq x^2 + 1$.

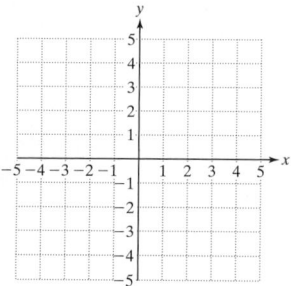

b. How would the solution change for the strict inequality $y > x^2 + 1$?

20. a. Graph the solution set for $\dfrac{x^2}{4} - \dfrac{y^2}{9} \leq 1$.

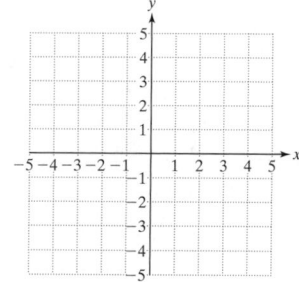

b. How would the solution change for the strict inequality $\dfrac{x^2}{4} - \dfrac{y^2}{9} < 1$?

21. A weak earthquake occurred in northern California roughly 4 mi south and 3 mi east of Sunol, California. The quake could be felt 25 mi away. Suppose the origin of a map is placed at the center of Sunol with the positive x-axis pointing east and the positive y-axis pointing north. Find an inequality that describes the points on the map for which the earthquake could be felt.

22. A coordinate system is placed at the center of a town with the positive x-axis pointing east, and the positive y-axis pointing north. A cell tower is located 2 mi west and 4 mi north of the center of town. If the tower has a 30-mi range, write an inequality that represents the points on the map serviced by this tower.

For Exercises 23–37, graph the solution set. **(See Examples 1–2.)**

23. $2x + y \geq 1$

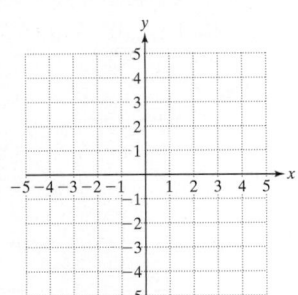

24. $3x + 2y \geq 6$

25. $x \leq y^2$

26. $y \leq -x^2$

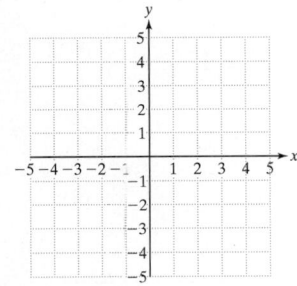

27. $(x - 1)^2 + (y + 2)^2 > 9$

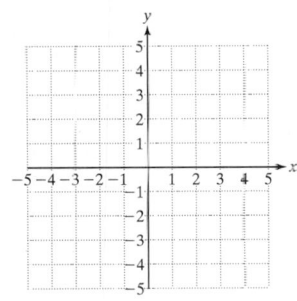

28. $(x + 1)^2 + (y - 4)^2 > 1$

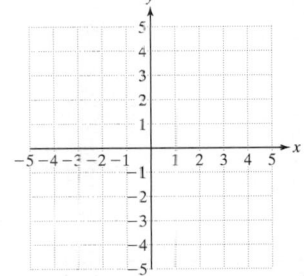

29. $x + y^2 \geq 4$

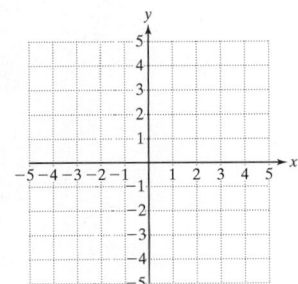

30. $x^2 + 2x + y - 1 \leq 0$

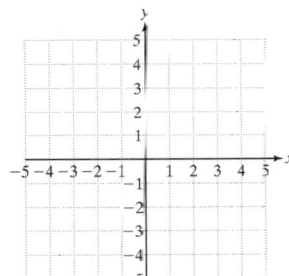

31. $9x^2 - y^2 > 9$

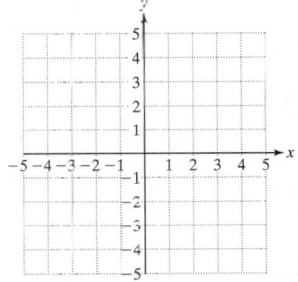

32. $y^2 - 4x^2 \leq 4$

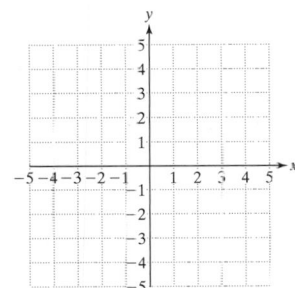

33. $x^2 + 16y^2 \leq 16$

34. $4x^2 - y^2 \leq 4$

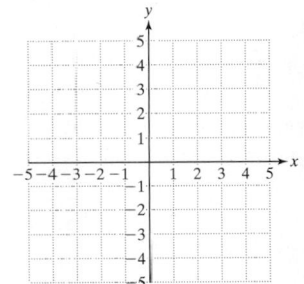

35. $y \leq \ln x$

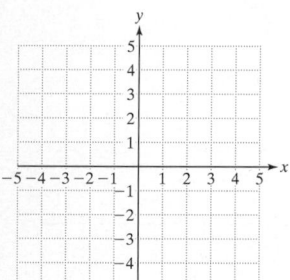

36. $y \leq \log x$

37. $y > 5^x$

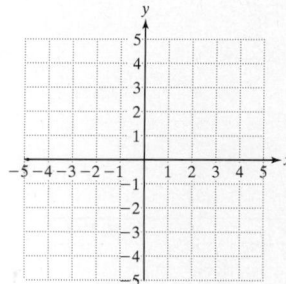

Concept 2: Systems of Nonlinear Inequalities in Two Variables

For Exercises 38–51, graph the solution set to the system of inequalities. **(See Examples 3–4.)**

38. $y < \sqrt{x}$
$x > 1$

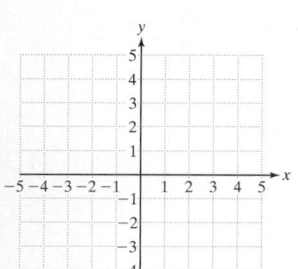

39. $y \geq \sqrt{x}$
$x \geq 0$

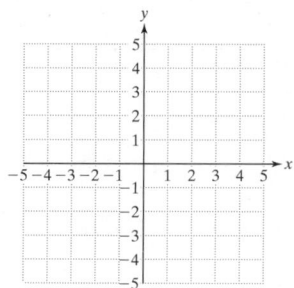

40. $\dfrac{x^2}{36} + \dfrac{y^2}{25} < 1$
$x^2 + y^2 \geq 4$

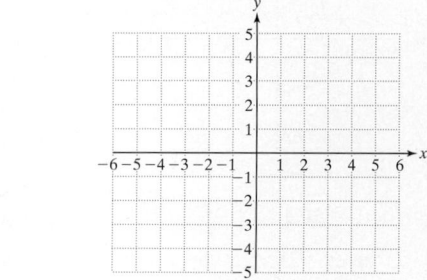

41. $x^2 - y^2 \geq 1$
$x \leq 0$

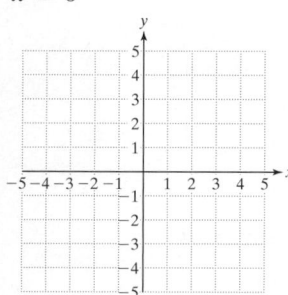

42. $y < x^2$
$y > x^2 - 4$

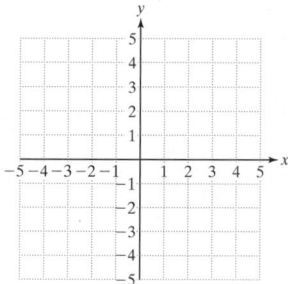

43. $y^2 - x^2 \geq 1$
$y \geq 0$

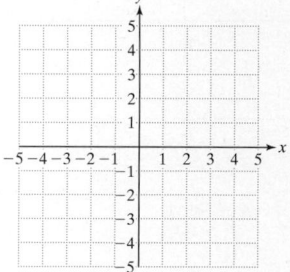

44. $y < \dfrac{1}{x}$
$y > 0$
$y < x$

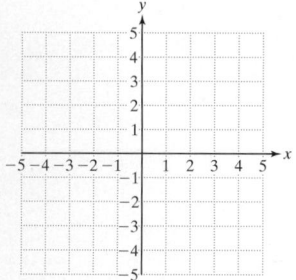

45. $y > x^3$
$y < 8$
$x > 0$

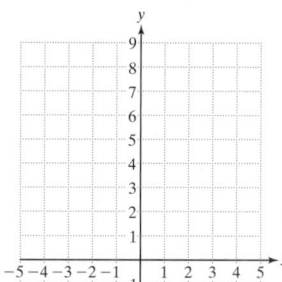

46. $x^2 + y^2 \geq 25$
$x^2 + y^2 \leq 9$

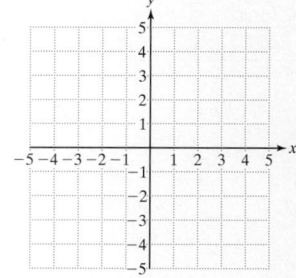

47. $\dfrac{x^2}{4} + \dfrac{y^2}{25} \geq 1$

$x^2 + \dfrac{y^2}{4} \leq 1$

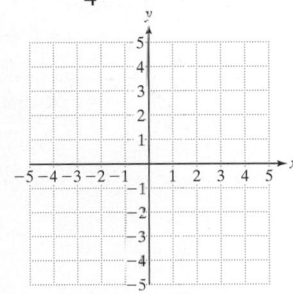

48. $x < -(y - 1)^2 + 3$

$x + y > 2$

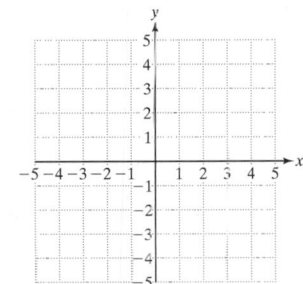

49. $x > (y - 2)^2 + 1$

$x - y < 1$

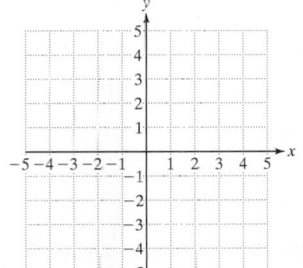

50. $x^2 + y^2 < 25$

$y < \dfrac{4}{3}x$

$y > -\dfrac{4}{3}x$

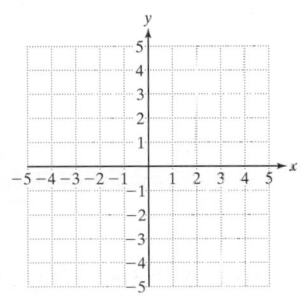

51. $y < e^x$

$y > 1$

$x < 2$

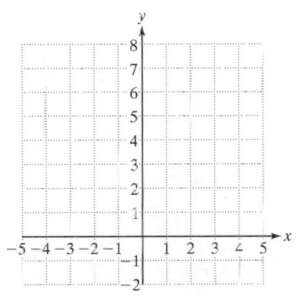

Expanding Your Skills

For Exercises 52–55, graph the compound inequalities.

52. $x^2 + y^2 \leq 36$ or $x + y \geq 0$

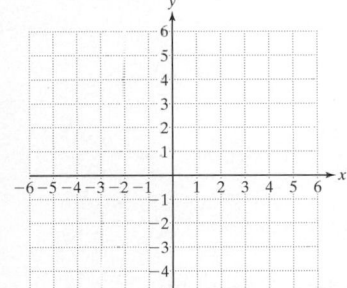

53. $y \leq -x^2 + 4$ or $y \geq x^2 - 4$

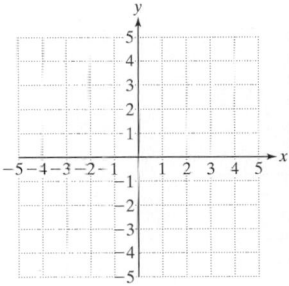

54. $y + 1 \geq x^2$ or $y + 1 \leq -x^2$

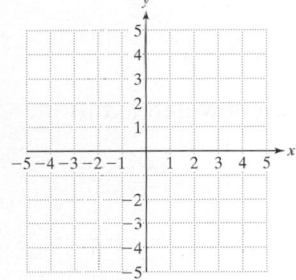

55. $(x + 2)^2 + (y + 3)^2 \leq 4$ or $x \geq y^2$

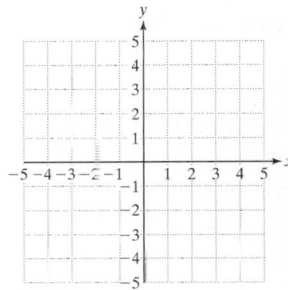

Group Activity

Investigating the Graphs of Conic Sections on a Calculator

Materials: Graphing calculator

Estimated time: 15 minutes

Group Size: 3

1. a. Given $x^2 + y^2 = 25$, use analytical methods to determine the x- and y-intercepts.

    x-intercepts: _____ y-intercepts: _____

  b. Solve the equation for y: $x^2 + y^2 = 25$. (*Hint:* When taking the square root of both sides of the equation, don't forget the $\pm$.)

    $Y_1 = $ _____

    $Y_2 = $ _____

  c. Graph the two functions from part (b) on a graphing calculator. Use a window defined by $-7.6 \le x \le 7.6$ and $-5 \le y \le 5$.

  d. Does it look like the functions graphed on the calculator intersect with the x-axis? Why do the x-intercepts not show up in the graph? If you relied solely on your calculator to find the x-intercepts, do you think that you would always get reliable results?

2. a. Given $(x + 1)^2 - \dfrac{y^2}{9} = 1$, use analytical methods to find the x- and y-intercepts.

    x-intercepts: _____ y-intercepts: _____

  b. Solve the equation for y. $(x + 1)^2 - \dfrac{y^2}{9} = 1$

    $Y_1 = $ _____

    $Y_2 = $ _____

  c. Graph the two functions from part (b) on a graphing calculator. Use a window defined by $-7.6 \le x \le 7.6$ and $-5 \le y \le 5$.

  d. Does it look like the functions graphed on the calculator intersect with the x-axis? Why do the x-intercepts not show up in the graph? If you relied solely on your calculator to find the x-intercepts, do you think that you would always get reliable results?

3. Solve the equation for y and graph the functions on a graphing calculator. Use a window defined by $-7.6 \le x \le 7.6$ and $-5 \le y \le 5$.

  a. $\dfrac{x^2}{4} + \dfrac{y^2}{16} = 1$ $Y_1 = $ _____

                             $Y_2 = $ _____

  b. What are the x-intercepts for the equation $\dfrac{x^2}{4} + \dfrac{y^2}{16} = 1$?

    x-intercepts: _____

Chapter 9   Summary

Section 9.1   Distance Formula, Midpoint Formula, and Circles

Key Concepts

The distance between two points (x_1, y_1) and (x_2, y_2) is

$$d = \sqrt{(x_2 - x_1)^2 + (y_2 - y_1)^2}$$

The standard equation of a **circle** with center (h, k) and radius r is

$$(x - h)^2 + (y - k)^2 = r^2$$

The midpoint between two points is found by using the formula:

$$\left(\frac{x_1 + x_2}{2}, \frac{y_1 + y_2}{2}\right)$$

Examples

Example 1

Find the distance between two points.

$(5, -2)$ and $(-1, -6)$

$$\begin{aligned}
d &= \sqrt{(-1 - 5)^2 + [-6 - (-2)]^2} \\
&= \sqrt{(-6)^2 + (-4)^2} \\
&= \sqrt{36 + 16} \\
&= \sqrt{52} = 2\sqrt{13}
\end{aligned}$$

Example 2

Find the center and radius of the circle.

$$x^2 + y^2 - 8x + 6y = 0$$
$$(x^2 - 8x + 16) + (y^2 + 6y + 9) = 16 + 9$$
$$(x - 4)^2 + (y + 3)^2 = 25$$

The center is $(4, -3)$ and the radius is 5.

Example 3

Find the midpoint between $(-3, 1)$ and $(5, 7)$.

$$\left(\frac{-3 + 5}{2}, \frac{1 + 7}{2}\right) = (1, 4)$$

Section 9.2   More on the Parabola

Key Concepts

A **parabola** is the set of points in a plane that are equidistant from a fixed line (called the directrix) and a fixed point (called the focus) not on the directrix.

The standard form of an equation of a parabola with **vertex** (h, k) and vertical **axis of symmetry** is

$$y = a(x - h)^2 + k \qquad \text{where } a \neq 0$$

- The equation of the axis of symmetry is $x = h$.
- If $a > 0$, the parabola opens upward.
- If $a < 0$, the parabola opens downward.

The standard form of an equation of a parabola with vertex (h, k) and horizontal axis of symmetry is

$$x = a(y - k)^2 + h \qquad \text{where } a \neq 0$$

- The equation of the axis of symmetry is $y = k$.
- If $a > 0$, the parabola opens to the right.
- If $a < 0$, the parabola opens to the left.

Examples

Example 1

Given the parabola $y = (x - 2)^2 + 1$:

The vertex is $(2, 1)$.

The axis of symmetry is $x = 2$.

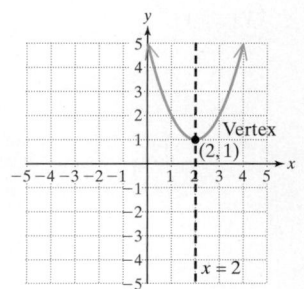

Example 2

Given the parabola $x = -\dfrac{1}{4}y^2 + 1$,

determine the coordinates of the vertex and the equation of the axis of symmetry.

$$x = -\frac{1}{4}(y - 0)^2 + 1$$

The vertex is $(1, 0)$.

The axis of symmetry is $y = 0$.

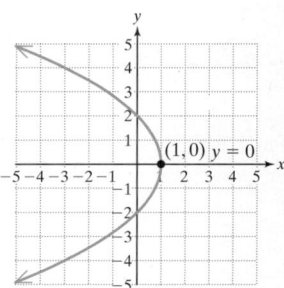

Section 9.3   The Ellipse and Hyperbola

Key Concepts

An **ellipse** is the set of all points (x, y) such that the sum of the distances between (x, y) and two distinct points (called foci) is constant.

Standard Form of an Ellipse with Center at the Origin

An ellipse with the center at the origin has the equation

$$\frac{x^2}{a^2} + \frac{y^2}{b^2} = 1$$

where a and b are positive real numbers.

For an ellipse centered at the origin, the x-intercepts are given by $(a, 0)$ and $(-a, 0)$, and the y-intercepts are given by $(0, b)$ and $(0, -b)$.

A **hyperbola** is the set of all points (x, y) such that the absolute value of the difference of the distances between (x, y) and two distinct points is a constant. The fixed points are called the foci of the hyperbola.

Standard Forms of an Equation of a Hyperbola

Let a and b represent positive real numbers.

Horizontal Transverse Axis. The standard form of an equation of a hyperbola with a horizontal transverse axis and center at the origin is given by

$$\frac{x^2}{a^2} - \frac{y^2}{b^2} = 1$$

Vertical Transverse Axis. The standard form of an equation of a hyperbola with a vertical transverse axis and center at the origin is given by

$$\frac{y^2}{b^2} - \frac{x^2}{a^2} = 1$$

Examples

Example 1

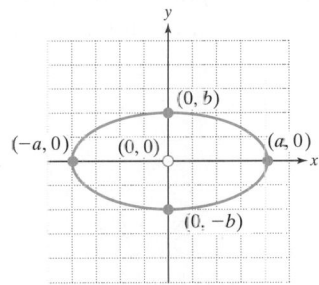

Example 2

$$\frac{x^2}{25} + \frac{y^2}{9} = 1$$

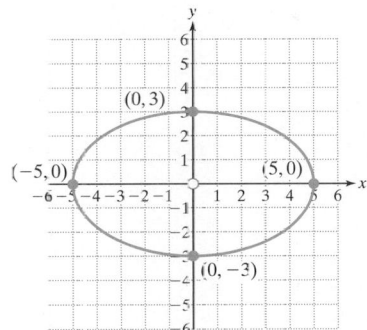

Example 3

$$\frac{x^2}{4} - \frac{y^2}{16} = 1$$

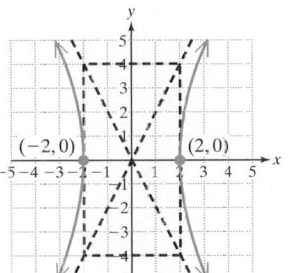

Example 4

$$\frac{y^2}{4} - \frac{x^2}{16} = 1$$

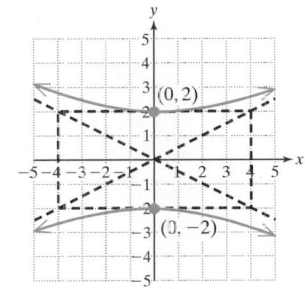

Section 9.4 Nonlinear Systems of Equations in Two Variables

Key Concepts

A **nonlinear system of equations** can be solved by graphing or by the substitution method.

$$2x^2 + y^2 = 15$$
$$x^2 - y = 0$$

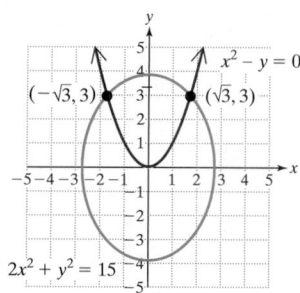

Examples

Example 1

$\boxed{A}$ $2x^2 + y^2 = 15$

$\boxed{B}$ $x^2 - y = 0$ Solve for y: $y = x^2$

$\boxed{A}$ $2x^2 + (x^2)^2 = 15$ Substitute in first equation.

$$2x^2 + x^4 = 15$$

$$x^4 + 2x^2 - 15 = 0$$

$$(x^2 + 5)(x^2 - 3) = 0$$

$x^2 + 5 = 0$ or $x^2 - 3 = 0$

$\cancel{x^2 = -5}$ or $x^2 = 3$

$x = \pm\sqrt{3}$

If $x = \sqrt{3}$ then $y = (\sqrt{3})^2 = 3$.

If $x = -\sqrt{3}$ then $y = (-\sqrt{3})^2 = 3$.

Points of intersection are $(\sqrt{3}, 3)$ and $(-\sqrt{3}, 3)$.

The solution set is $\{(\sqrt{3}, 3), (-\sqrt{3}, 3)\}$.

Section 9.5 Nonlinear Inequalities and Systems of Inequalities

Key Concepts

Graph a nonlinear inequality by using the test point method. That is, graph the related equation. Then choose test points in each region to determine where the inequality is true.

Example 1

$y \geq x^2$

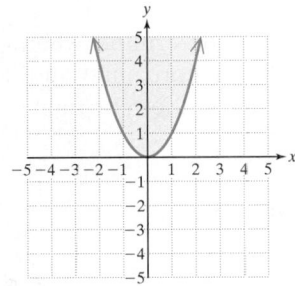

Graph a system of nonlinear inequalities by finding the intersection of the solution sets. That is, graph the solution set for each individual inequality, then take the intersection.

Examples

Example 2

$x^2 + y^2 < 4$

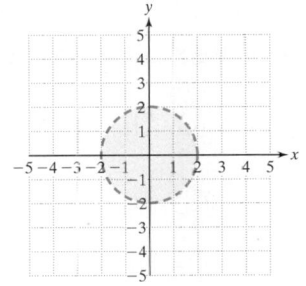

Example 3

$x \geq 0, \qquad y > x^2, \qquad$ and $\qquad x^2 + y^2 < 4$

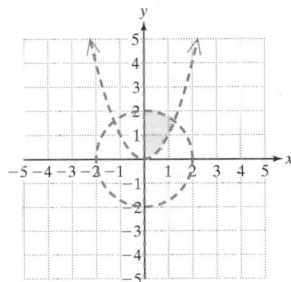

Chapter 9 Review Exercises

Section 9.1

For Exercises 1–4, find the distance between the two points by using the distance formula.

1. $(-6, 3)$ and $(0, 1)$ **2.** $(4, 13)$ and $(-1, 5)$

3. Find x such that $(x, 5)$ is 5 units from $(2, 9)$.

4. Find x such that $(-3, 4)$ is 3 units from $(x, 1)$.

For Exercises 5–8, find the center and the radius of the circle.

5. $(x - 12)^2 + (y - 3)^2 = 16$

6. $(x + 7)^2 + (y - 5)^2 = 81$

7. $(x + 3)^2 + (y + 8)^2 = 20$

8. $(x - 1)^2 + (y + 6)^2 = 32$

9. A stained glass window is in the shape of a circle with a 16-in. diameter. Find an equation of the circle relative to the origin for each of the following graphs.

a.

b.
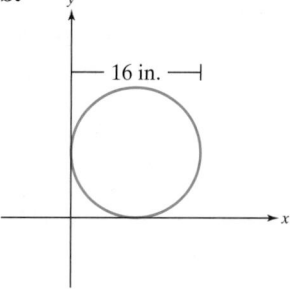

For Exercises 10–13, write the equation of the circle in standard form by completing the square.

10. $x^2 + y^2 + 12x - 10y + 51 = 0$

11. $x^2 + y^2 + 4x + 16y + 60 = 0$

12. $x^2 + y^2 - x - 4y + \dfrac{1}{4} = 0$

13. $x^2 + y^2 - 6x - \dfrac{2}{3}y + \dfrac{1}{9} = 0$

14. Write an equation of a circle with center at the origin and a diameter of 7 m.

15. Write an equation of a circle with center $(0, 2)$ and radius 3 m.

For Exercises 16–17, find the midpoint of the segment with the given endpoints.

16. $(-3, 1)$ and $(-5, -5)$ **17.** $(0, 9)$ and $(-2, 7)$

Section 9.2

For Exercises 18–21, determine whether the axis of symmetry is vertical or horizontal and if the parabola opens upward, downward, left, or right.

18. $y = -2(x - 3)^2 + 2$

19. $x = 3(y - 9)^2 + 1$

20. $x = -(y + 4)^2 - 8$

21. $y = (x + 3)^2 - 10$

For Exercises 22–25, determine the coordinates of the vertex and the equation of the axis of symmetry. Then use this information to graph the parabola.

22. $x = -(y - 1)^2$
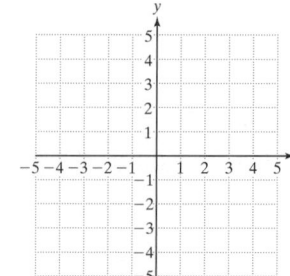

23. $y = (x + 2)^2$

24. $y = -\dfrac{1}{4}x^2$
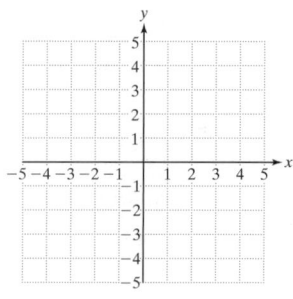

25. $x = 2y^2 - 1$
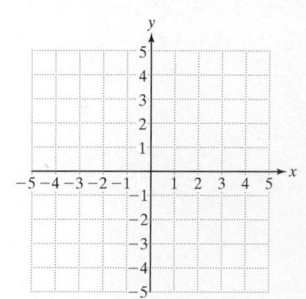

For Exercises 26–29, write the equation in standard form $y = a(x - h)^2 + k$ or $x = a(y - k)^2 + h$. Then identify the vertex and axis of symmetry.

26. $y = x^2 - 6x + 5$

27. $x = y^2 + 4y + 2$

28. $x = -4y^2 + 4y$

29. $y = -2x^2 - 2x$

Section 9.3

For Exercises 30–31, identify the x- and y-intercepts. Then graph the ellipse.

30. $\dfrac{x^2}{9} + \dfrac{y^2}{25} = 1$

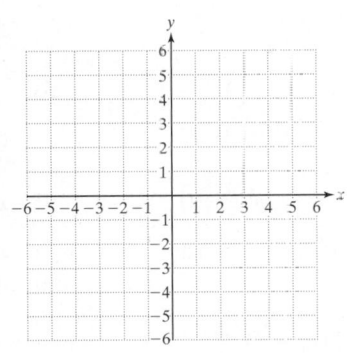

31. $x^2 + 4y^2 = 36$

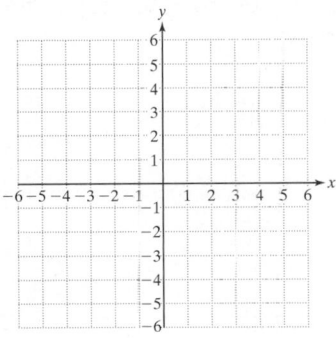

For Exercises 32–33, identify the center of the ellipse and graph the ellipse.

32. $\dfrac{(x - 5)^2}{4} + \dfrac{(y + 3)^2}{16} = 1$

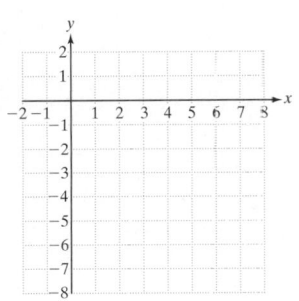

33. $\dfrac{x^2}{25} + \dfrac{(y - 2)^2}{9} = 1$

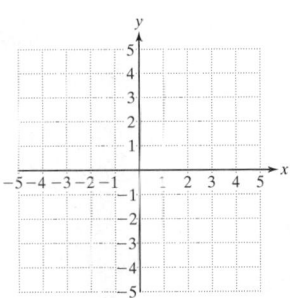

For Exercises 34–37, determine whether the transverse axis is horizontal or vertical.

34. $\dfrac{x^2}{12} - \dfrac{y^2}{16} = 1$

35. $\dfrac{y^2}{9} - \dfrac{x^2}{9} = 1$

36. $y^2 - 8x^2 = 16$

37. $3x^2 - y^2 = 18$

For Exercises 38–39, graph the hyperbola by first drawing the reference rectangle and the asymptotes. Label the vertices.

38. $\dfrac{x^2}{4} - y^2 = 1$

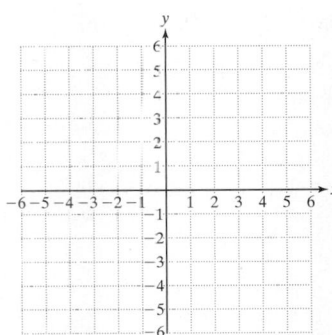

39. $y^2 - x^2 = 16$

For Exercises 40–43, identify the equations as representing an ellipse or a hyperbola.

40. $\dfrac{x^2}{4} - \dfrac{y^2}{9} = 1$

41. $\dfrac{x^2}{16} + \dfrac{y^2}{9} = 1$

42. $\dfrac{x^2}{4} + \dfrac{y^2}{1} = 1$

43. $\dfrac{y^2}{1} - \dfrac{x^2}{16} = 1$

Section 9.4

For Exercises 44–47,

    a. Identify each equation as representing a line, a parabola, a circle, an ellipse, or a hyperbola.

    b. Graph both equations on the same coordinate system.

    c. Solve the system analytically and verify the answers from the graph.

44. $3x + 2y = 10$

    $y = x^2 - 5$

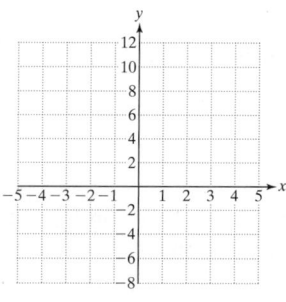

45. $4x + 2y = 10$

    $y = x^2 - 10$

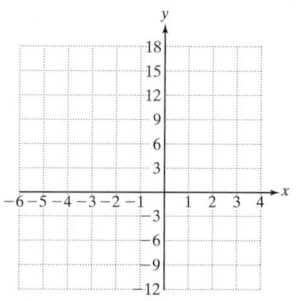

46. $x^2 + y^2 = 9$

    $2x + y = 3$

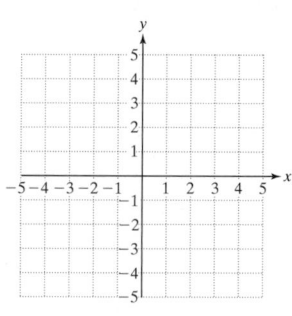

47. $x^2 + y^2 = 16$

    $x - 2y = 8$

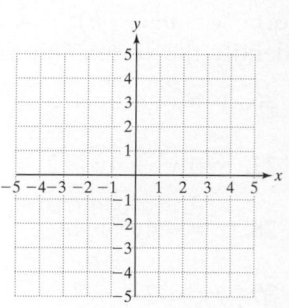

For Exercises 48–53, solve the system of nonlinear equations by using either the substitution method or the addition method.

48. $x^2 + 2y^2 = 8$

    $2x - y = 2$

49. $x^2 + 4y^2 = 29$

    $x - y = -4$

50. $x - y = 4$

    $y^2 = 2x$

51. $y = x^2$

    $6x^2 - y^2 = 8$

52. $x^2 + y^2 = 10$

    $x^2 + 9y^2 = 18$

53. $x^2 + y^2 = 61$

    $x^2 - y^2 = 11$

Section 9.5

For Exercises 54–59, graph the solution set to the inequality.

54. $\dfrac{x^2}{16} + \dfrac{y^2}{81} < 1$

55. $\dfrac{x^2}{25} + \dfrac{y^2}{4} > 1$

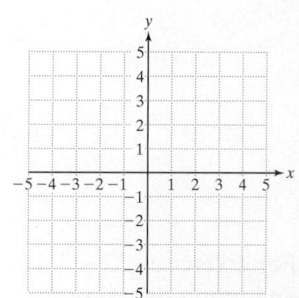

56. $(x - 3)^2 + (y + 1)^2 \geq 9$

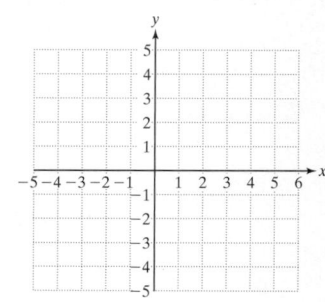

57. $(x + 2)^2 + (y + 1)^2 \leq 4$

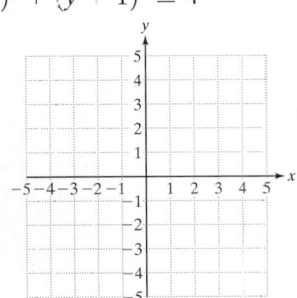

For Exercises 60–61, graph the solution set to the system of nonlinear inequalities.

60. $y > 2^x$

$x^2 + y^2 < 4$

61. $y < x^2$

$x^2 + y^2 < 9$

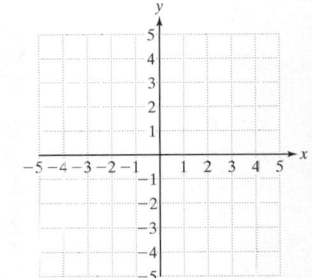

58. $y > (x - 1)^2$

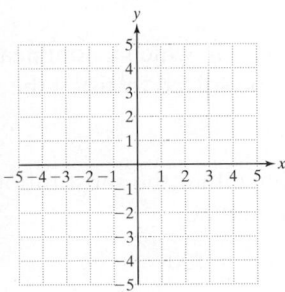

59. $x^2 - \dfrac{y^2}{4} \leq 1$

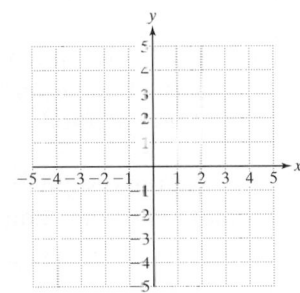

Chapter 9   Test

1. Use the distance formula to find the distance between the two points $(-5, 2)$ and $(-1, 6)$.

2. Determine the center and radius of the circle.

$$\left(x - \frac{5}{6}\right)^2 + \left(y + \frac{1}{3}\right)^2 = \frac{25}{49}$$

3. Determine the center and radius of the circle.

$$x^2 + y^2 - 4y - 5 = 0$$

4. Let $(0, 4)$ be the center of a circle that passes through the point $(-2, 5)$.

  a. What is the radius of the circle?

  b. Write the equation of the circle in standard form.

5. Find the center of the circle that has a diameter with endpoints $(7.3, -1.2)$ and $(0.3, 5.1)$.

6. Determine the vertex and the equation of the axis of symmetry. Then graph the parabola.

$$x = -(y - 2)^2 + 3$$

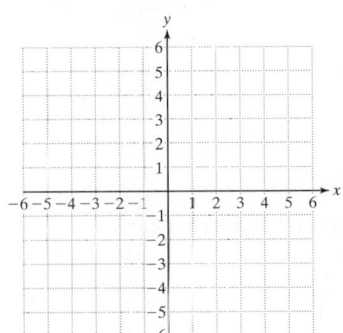

7. Write the equation in standard form $y = a(x - h)^2 + k$, and graph the parabola.

$y = x^2 + 4x + 5$

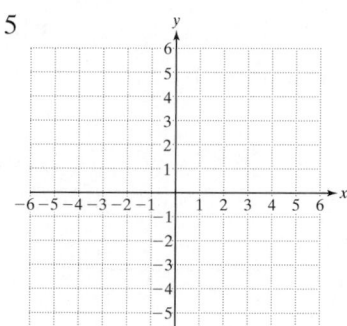

8. Graph the ellipse.

$\dfrac{x^2}{16} + \dfrac{y^2}{49} = 1$

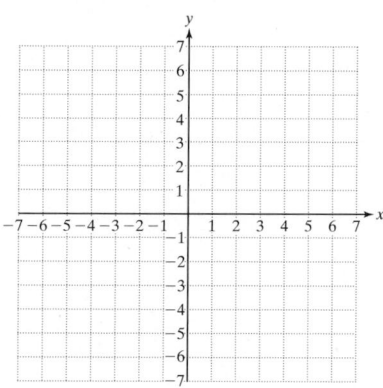

9. Graph the ellipse.

$\dfrac{(x + 4)^2}{25} + (y - 3)^2 = 1$

10. Graph the hyperbola.

$y^2 - \dfrac{x^2}{4} = 1$

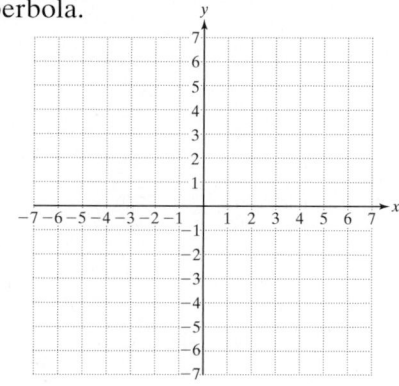

For Exercises 11–12, solve the system and identify the correct graph.

11. $16x^2 + 9y^2 = 144$

$4x - 3y = -12$

12. $x^2 + 4y^2 = 4$

$4x - 3y = -12$

a.

b.

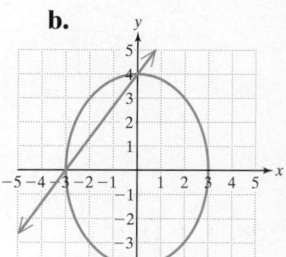

13. Describe the circumstances in which a nonlinear system of equations can be solved by using the addition method.

14. Solve the system by using either the substitution method or the addition method.

$$25x^2 + 4y^2 = 100$$
$$25x^2 - 4y^2 = 100$$

For Exercises 15–18, graph the solution set.

15. $x \leq y^2 + 1$

16. $y \geq -\dfrac{1}{3}x + 1$

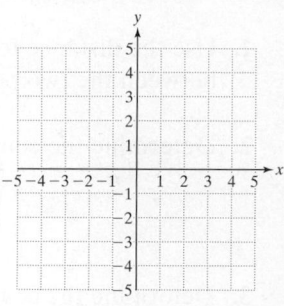

17. $x < y^2 + 1$

$y > -\dfrac{1}{3}x + 1$

18. $y < \sqrt{x}$

$y > x - 2$

$x > 0$

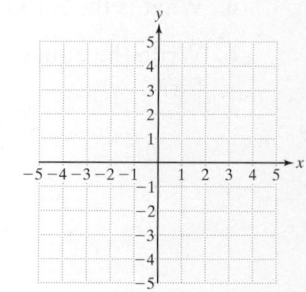

Chapters 1–9　Cumulative Review Exercises

1. Solve.　$5(2y - 1) = 2y - 4 + 8y - 1$

2. Solve the inequality. Graph the solution and write the solution in interval notation.

$$4(x - 1) + 2 > 3x + 8 - 2x$$

3. The product of two integers is 150. If one integer is 5 less than twice the other, find the integers.

4. For $5y - 3x - 15 = 0$:

　a. Find the x- and y-intercepts.

　b. Find the slope.

　c. Graph the line.

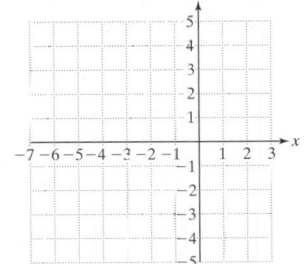

5. Find the slope and y-intercept of $3x - 4y = 6$.

6. A collection of dimes and quarters has a total value of \$2.45. If there are 17 coins, how many of each type are there?

7. Solve the system.

$$\begin{aligned} x + y &= -1 \\ 2x \quad - z &= 3 \\ y + 2z &= -1 \end{aligned}$$

8. Solve the system.

$$\begin{aligned} 4x - 2y &= 7 \\ -3x + 5y &= 0 \end{aligned}$$

9. Solve by using the Gauss-Jordan method.

$$\begin{aligned} 3x - 4y &= 6 \\ x + 2y &= 12 \end{aligned}$$

10. For $f(x) = 3x - x^2 - 12$, find the function values $f(0), f(-1), f(2),$ and $f(4)$.

11. Solve the inequality.　$x^2 + 6x < -8$

12. The quantity z varies jointly as y and as the square of x. If z is 80 when x is 5 and y is 2, find z when $x = 2$ and $y = 5$.

13. For $f(x) = \sqrt{x + 1}$ and $g(x) = x^2 + 6$, find $(g \circ f)(x)$.

14. **a.** Find the value of the expression $x^3 + x^2 + x + 1$ for $x = -2$.

　b. Factor the expression $x^3 + x^2 + x + 1$ and find the value when x is -2.

　c. Compare the values for parts (a) and (b).

15. Factor completely.　$x^2 - y^2 - 6x - 6y$

16. Multiply.　$(x - 4)(x^2 + 2x + 1)$

17. Solve.　$2x(x - 7) = x - 18$

18. Simplify.　$\dfrac{3a^2 - a - 2}{3a^2 + 8a + 4}$

19. Subtract.　$\dfrac{2}{x + 3} - \dfrac{x}{x - 2}$

20. Solve.　$\dfrac{2}{x + 3} - \dfrac{x}{x - 2} = \dfrac{-4}{x^2 + x - 6}$

21. Solve the radical equations.

　a. $\sqrt{2x - 5} = -3$

　b. $\sqrt[3]{2x - 5} = -3$

For Exercises 22–23, perform the indicated operations with complex numbers.

22. $6i(4 + 5i)$　　　**23.** $\dfrac{3}{4 - 5i}$

24. Find the length of the missing side.

25. An automobile starts from rest and accelerates at a constant rate for 10 sec. The distance, $d(t)$, in feet traveled by the car is given by

$$d(t) = 4.4t^2$$

where $0 \leq t \leq 10$ is the time in seconds.

a. How far has the car traveled after 2, 3, and 4 sec, respectively?

b. How long will it take for the car to travel 281.6 ft?

26. Solve the equation $125w^3 + 1 = 0$ by factoring and using the quadratic formula. (*Hint:* You will find one real solution and two imaginary solutions.)

27. Solve. $\dfrac{x}{x + 2} - \dfrac{3}{x - 1} = \dfrac{1}{x^2 + x - 2}$

28. Find the coordinates of the vertex of the parabola defined by $f(x) = x^2 + 10x - 11$ by completing the square.

29. Graph the quadratic function defined by $g(x) = -x^2 - 2x + 3$.

a. Label the x-intercepts.

b. Label the y-intercept.

c. Label the vertex.

30. Solve the inequality and write the answer in interval notation.

$$|x - 9| - 3 < 7$$

31. Solve the inequality: $|2x - 5| \geq 4$

32. Write the expression in logarithmic form. $8^{5/3} = 32$

33. Solve. $5^2 = 125^x$

34. For $h(x) = x^3 - 1$, find $h^{-1}(x)$.

35. Write an equation representing the set of all points 4 units from the point $(0, 5)$.

36. Graph the ellipse. $\dfrac{x^2}{4} + \dfrac{y^2}{25} = 1$

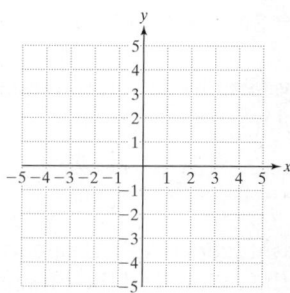

37. Determine the center of the circle, given the endpoints of a diameter.

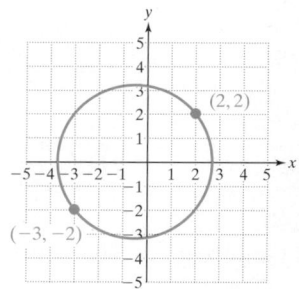

38. Solve the system of nonlinear equations.

$$x^2 + y^2 = 16$$
$$y = -x^2 - 4$$

39. Graph the solution set.

$$y^2 - x^2 < 1$$

40. Graph the solution set to this system.

$$y > \left(\dfrac{1}{2}\right)^x$$
$$x < 0$$

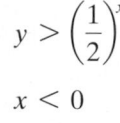

Binomial Expansions, Sequences, and Series

10

CHAPTER OUTLINE

10.1 Binomial Expansions   814

10.2 Sequences and Series   820

10.3 Arithmetic Sequences and Series   829

10.4 Geometric Sequences and Series   835

  Problem Recognition Exercises: Identifying Arithmetic and Geometric Sequences   843

  Group Activity: Investigating Mean and Standard Deviation   844

Chapter 10

We begin this chapter by learning how to expand expressions of the form $(a + b)^n$. Then we study a new type of function called a sequence. A sequence is a progression of terms, and the sum of the terms in a sequence is called a series. We will study arithmetic and geometric sequences. These are sequences in which successive terms differ by a fixed constant or by a fixed ratio.

Are You Prepared?

To practice working with sequences, try Exercises 1–8. In each case, fill in the blanks with the next three terms that follow the given pattern.

1. 1, 4, 7, 10, _____, _____, _____

2. 6, −2, −10, −18, _____, _____, _____

3. 3, 6, 12, 24, _____, _____, _____

4. 9, 3, 1, $\frac{1}{3}$, _____, _____, _____

5. 1, 4, 9, 16, _____, _____, _____

6. −4, 12, −36, 108, _____, _____, _____

7. $x^7, x^6y, x^5y^2, x^4y^3$, _____, _____, _____

8. $a^6b^3, a^5b^4, a^4b^5, a^3b^6$, _____, _____, _____

9. Expand the expression. $(a + b)^2 =$ _____

10. Expand the expression. $(a + b)^3 =$ _____

11. Based on your answers to Exercises 9 and 10, how many terms are in the expansion of $(a + b)^n$? _____

Concepts

1. **Binomial Expansions and Pascal's Triangle**
2. **Factorial Notation**
3. **The Binomial Theorem**

| Section 10.1 | Binomial Expansions |

1. Binomial Expansions and Pascal's Triangle

In Section 4.3 we learned how to square a binomial.

$$(a + b)^2 = a^2 + 2ab + b^2$$

The expression $a^2 + 2ab + b^2$ is called the **binomial expansion** of $(a + b)^2$. To expand $(a + b)^3$, we can find the product $(a + b)(a + b)^2$.

$$(a + b)(a + b)^2 = (a + b)(a^2 + 2ab + b^2)$$

$$= a^3 + 2a^2b + ab^2 + a^2b + 2ab^2 + b^3$$
$$= a^3 + 3a^2b + 3ab^2 + b^3$$

Similarly, to expand $(a + b)^4$, we can multiply $(a + b)$ by $(a + b)^3$. Using this method, we can expand several powers of $(a + b)$ to find the following pattern:

$$(a + b)^0 = 1$$
$$(a + b)^1 = a + b$$
$$(a + b)^2 = a^2 + 2ab + b^2$$
$$(a + b)^3 = a^3 + 3a^2b + 3ab^2 + b^3$$
$$(a + b)^4 = a^4 + 4a^3b + 6a^2b^2 + 4ab^3 + b^4$$
$$(a + b)^5 = a^5 + 5a^4b + 10a^3b^2 + 10a^2b^3 + 5ab^4 + b^5$$

Notice that the exponents on a decrease from left to right, while the exponents on b increase from left to right. Also observe that for each term, the sum of the exponents is equal to the exponent to which $(a + b)$ is raised. Finally, notice that the number of terms in the expansion is exactly 1 more than the power to which $(a + b)$ is raised. For example, the expansion of $(a + b)^4$ has five terms, and the expansion of $(a + b)^5$ has six terms.

With these guidelines in mind, we know that $(a + b)^6$ will contain seven terms involving

$$a^6 \qquad a^5b \qquad a^4b^2 \qquad a^3b^3 \qquad a^2b^4 \qquad ab^5 \qquad b^6$$

We can complete the expansion of $(a + b)^6$ if we can determine the correct coefficients of each term.

If we write the coefficients for several expansions of $(a + b)^n$, where $n \geq 0$, we have a triangular array of numbers.

$(a + b)^0 = 1$ 1

$(a + b)^1 = 1a + 1b$ 1   1

$(a + b)^2 = 1a^2 + 2ab + 1b^2$ 1   2   1

$(a + b)^3 = 1a^3 + 3a^2b + 3ab^2 + 1b^3$ 1   3   3   1

$(a + b)^4 = 1a^4 + 4a^3b + 6a^2b^2 + 4ab^3 + 1b^4$ 1   4   6   4   1

$(a + b)^5 = 1a^5 + 5a^4b + 10a^3b^2 + 10a^2b^3 + 5ab^4 + 1b^5$ 1   5   10   10   5   1

Each row begins and ends with a 1, and each entry in between is the sum of the two entries from the line above. For example, in the last row, $5 = 1 + 4$, $10 = 4 + 6$, and so on. This triangular array of coefficients for a binomial expansion is called **Pascal's triangle**, named after French mathematician Blaise Pascal (1623–1662).

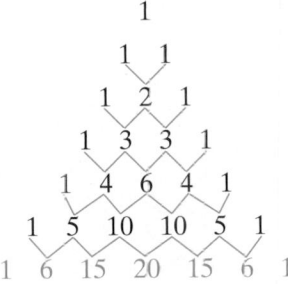

By using the pattern shown in Pascal's triangle, the coefficients corresponding to $(a + b)^6$ would be 1, 6, 15, 20, 15, 6, 1. By inserting the coefficients, the sum becomes

$$(a + b)^6 = 1a^6 + 6a^5b + 15a^4b^2 + 20a^3b^3 + 15a^2b^4 + 6ab^5 - 1b^6$$

2. Factorial Notation

Although Pascal's triangle provides an easy method to find the coefficients of $(a + b)^n$, it is impractical for large values of n. A more efficient method to find the coefficients of a binomial expansion involves **factorial notation**.

> ### Definition of $n!$
>
> Let n be a positive integer. Then $n!$ (read as "n factorial") is defined as the product of integers from 1 through n. That is,
>
> $$n! = n(n - 1)(n - 2) \cdots (2)(1)$$
>
> *Note:* We define $0! = 1$.

Example 1 Evaluating Factorial Notation

Evaluate the expressions.

  a. 4!   **b.** 10!   **c.** 0!

Solution:

  a. $4! = 4 \cdot 3 \cdot 2 \cdot 1 = 24$

  b. $10! = 10 \cdot 9 \cdot 8 \cdot 7 \cdot 6 \cdot 5 \cdot 4 \cdot 3 \cdot 2 \cdot 1 = 3,628,800$

  c. $0! = 1$ by definition

Skill Practice Evaluate the expressions.

  1. 3!   **2.** 8!   **3.** 1!

Calculator Connections

Topic: Evaluating Expressions with Factorial Notation

Most calculators have a $\boxed{!}$ function. Try evaluating the expressions from Example 1 on a calculator:

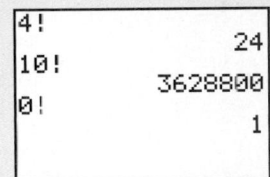

Answers

1. 6   **2.** 40,320   **3.** 1

Sometimes factorial notation is used with other operations such as multiplication and division.

Example 2 Evaluating Expressions with Factorials

Evaluate the expressions.

a. $\dfrac{4!}{4! \cdot 0!}$ **b.** $\dfrac{4!}{3! \cdot 1!}$ **c.** $\dfrac{4!}{2! \cdot 2!}$ **d.** $\dfrac{4!}{1! \cdot 3!}$ **e.** $\dfrac{4!}{0! \cdot 4!}$

Solution:

a. $\dfrac{4!}{4! \cdot 0!} = \dfrac{4!}{4! \cdot 1} = 1$

b. $\dfrac{4!}{3! \cdot 1!} = \dfrac{4 \cdot 3!}{3! \cdot 1} = 4$

c. $\dfrac{4!}{2! \cdot 2!} = \dfrac{4 \cdot 3 \cdot 2!}{2! \cdot 2 \cdot 1} = \dfrac{12}{2} = 6$

d. $\dfrac{4!}{1! \cdot 3!} = \dfrac{4 \cdot 3!}{1 \cdot 3!} = 4$

e. $\dfrac{4!}{0! \cdot 4!} = \dfrac{4!}{1 \cdot 4!} = 1$

Skill Practice Evaluate.

4. $\dfrac{5!}{5!0!}$ **5.** $\dfrac{5!}{4!1!}$ **6.** $\dfrac{5!}{3!2!}$

3. The Binomial Theorem

Notice from Example 2 that the values of

$$\dfrac{4!}{4! \cdot 0!} \qquad \dfrac{4!}{3! \cdot 1!} \qquad \dfrac{4!}{2! \cdot 2!} \qquad \dfrac{4!}{1! \cdot 3!} \quad \text{and} \quad \dfrac{4!}{0! \cdot 4!}$$

correspond to the values $1, 4, 6, 4, 1$, which are the coefficients for the expansion of $(a + b)^4$. Generalizing this pattern, we see the coefficients for the terms in the expansion of $(a + b)^n$ are given by

$$\dfrac{n!}{r! \cdot (n - r)!}$$

where r corresponds to the exponent on the factor of a and $(n - r)$ corresponds to the exponent on the factor of b. Using this formula for the coefficients in a binomial expansion results in the **binomial theorem**.

Answers
4. 1   **5.** 5   **6.** 10

The Binomial Theorem

For any positive integer n,

$$(a + b)^n = \frac{n!}{n! \cdot 0!} a^n + \frac{n!}{(n-1)! \cdot 1!} a^{(n-1)}b + \frac{n!}{(n-2)! \cdot 2!} a^{(n-2)}b^2$$
$$+ \cdots + \frac{n!}{0! \cdot n!} b^n$$

Example 3   Applying the Binomial Theorem

Write out the first three terms of the expansion of $(a + b)^{10}$.

Solution:

The first three terms of $(a + b)^{10}$ are

$$\frac{10!}{10! \cdot 0!} a^{10} + \frac{10!}{9! \cdot 1!} a^9 b + \frac{10!}{8! \cdot 2!} a^8 b^2$$

$$= \frac{10!}{10! \cdot 1} a^{10} + \frac{10 \cdot 9!}{9! \cdot 1!} a^9 b + \frac{10 \cdot 9 \cdot 8!}{8! \cdot 2 \cdot 1} a^8 b^2$$

$$= a^{10} + 10a^9 b + 45a^8 b^2$$

> **TIP:** When applying the binomial theorem, exponents on a and b must add up to n.

Skill Practice

7. Write out the first three terms of $(x + y)^5$.

Example 4   Applying the Binomial Theorem

Use the binomial theorem to expand $(2m + 7)^3$.

Solution:

This expression is of the form $(a + b)^3$, where $a = 2m$ and $b = 7$. Applying the binomial theorem we have

$$(a + b)^3 = \frac{3!}{3! \cdot 0!} a^3 + \frac{3!}{2! \cdot 1!} a^2 b + \frac{3!}{1! \cdot 2!} ab^2 + \frac{3!}{0! \cdot 3!} b^3$$

$$= \frac{3!}{3! \cdot 0!}(2m)^3 + \frac{3!}{2! \cdot 1!}(2m)^2(7) + \frac{3!}{1! \cdot 2!}(2m)(7)^2 + \frac{3!}{0! \cdot 3!}(7)^3$$

$$= 1 \cdot (8m^3) + 3 \cdot (4m^2)(7) + 3 \cdot (2m)(49) + 1 \cdot (343)$$

$$= 8m^3 + 84m^2 + 294m + 343$$

> **TIP:** The values of $\dfrac{n!}{r! \cdot (n-r)!}$ can also be found by using Pascal's triangle.
>
> $$\begin{matrix} & & & 1 & & & \\ & & 1 & & 1 & & \\ & 1 & & 2 & & 1 & \\ 1 & & 3 & & 3 & & 1 \end{matrix}$$

Skill Practice

8. Use the binomial theorem to expand $(3y + 4)^3$.

Use the binomial theorem to expand $(3x^2 - 5y)^4$.

Solution:

Write $(3x^2 - 5y)^4$ as $[(3x^2) + (-5y)]^4$. In this case, the expressions $3x^2$ and $-5y$ may be substituted for a and b in the expansion of $(a + b)^4$.

$$(a + b)^4 = \frac{4!}{4! \cdot 0!}a^4 + \frac{4!}{3! \cdot 1!}a^3b + \frac{4!}{2! \cdot 2!}a^2b^2 + \frac{4!}{1! \cdot 3!}ab^3 + \frac{4!}{0! \cdot 4!}b^4$$

$$= \frac{4!}{4! \cdot 0!}(3x^2)^4 + \frac{4!}{3! \cdot 1!}(3x^2)^3(-5y) + \frac{4!}{2! \cdot 2!}(3x^2)^2(-5y)^2$$

$$+ \frac{4!}{1! \cdot 3!}(3x^2)(-5y)^3 + \frac{4!}{0! \cdot 4!}(-5y)^4$$

$$= 1 \cdot (81x^8) + 4 \cdot (27x^6)(-5y) + 6 \cdot (9x^4)(25y^2) + 4 \cdot (3x^2)(-125y^3)$$
$$+ 1 \cdot (625y^4)$$

$$= 81x^8 - 540x^6y + 1350x^4y^2 - 1500x^2y^3 + 625y^4$$

Skill Practice

9. Use the binomial theorem to expand $(2a - 3b^2)^4$.

The binomial theorem may also be used to find a specific term in a binomial expansion.

Example 6 Finding a Specific Term in a Binomial Expansion

Find the fourth term of the expansion $(a + b)^{13}$.

Solution:

There are 14 terms in the expansion of $(a + b)^{13}$. The first term will have variable factors $a^{13}b^0$. The second term will have variable factors $a^{12}b^1$. The third term will have $a^{11}b^2$, and the fourth term will have $a^{10}b^3$. Hence, the fourth term is

$$\frac{13!}{10! \cdot 3!}a^{10}b^3 = 286a^{10}b^3$$

Skill Practice

10. Find the fourth term of $(x + y)^8$.

From Example 6, we see that for the kth term in the expansion $(a + b)^n$, where k is an integer greater than zero, the variable factors are $a^{n-(k-1)}$ and b^{k-1}. Therefore, to find the kth term of $(a + b)^n$, we can make the following generalization.

Example 7   **Finding a Specific Term in a Binomial Expansion**

Find the sixth term of $(p^3 + 2w)^8$.

Solution:

Apply the formula.

$$\frac{n!}{[n - (k - 1)]! \cdot (k - 1)!} \cdot a^{n-(k-1)} \cdot b^{k-1} \quad \text{with } n = 8, k = 6, a = p^3, \text{ and } b = 2w$$

$$\frac{8!}{[8 - (6 - 1)]! \cdot (6 - 1)!} \cdot (p^3)^{8-(6-1)} \cdot (2w)^{6-1}$$

$$= \frac{8!}{(3)! \cdot (5)!} \cdot (p^3)^3 \cdot (2w)^5$$

$$= 56 \cdot (p^9) \cdot (32w^5)$$

$$= 1792 p^9 w^5$$

Skill Practice

11. Find the fifth term of $(t - m^2)^8$.

Answer

11. $70t^4 m^8$

Section 10.1 Practice Exercises

Vocabulary and Key Concepts

1. **a.** The expanded form of $(a + b)^2 = $ _____ . The expanded form of $(a + b)^3 = $ _____ .
   These are both called _____ expansions.

   b. Given a positive integer n, the value of $n!$ is _____ . Furthermore, $n!$ is read
   as "n _____ ."

   c. $3! = $ _____ , $2! = $ _____ , $1! = $ _____ , and $0! = $ _____ .

   d. The _____ theorem states that for any positive integer n,

   $$(a + b)^n = \frac{n!}{n! \cdot 0!} a^n + \frac{n!}{(n - 1)! \cdot 1!} a^{(n-1)}b + \frac{n!}{(n - 2)! \cdot 2!} a^{(n-2)}b^2 + \cdots + \frac{n!}{0! \cdot n!} b^n.$$

   e. The coefficients of the binomial expansion of $(a + b)^n$ can also be found by using _____ triangle.

Concept 1: Binomial Expansions and Pascal's Triangle

For Exercises 2–7, expand the binomials. Use Pascal's triangle to find the coefficients.

2. $(x + y)^4$ 3. $(a + b)^3$ 4. $(4 + p)^3$

5. $(1 + g)^4$ 6. $(a^2 + b)^6$ 7. $(p + q^2)^7$

For Exercises 8–13, rewrite each binomial of the form $(a - b)^n$ as $[a + (-b)]^n$. Then expand the binomials. Use Pascal's triangle to find the coefficients.

8. $(t - 2)^5$ 9. $(s - t)^5$

10. $(p^2 - w)^3$ 11. $(5 - u^3)^4$

12. $(a - b^2)^4$ 13. $(x^2 - 4)^3$

14. For $a > 0$ and $b > 0$, what happens to the signs of the terms when expanding the binomial $(a - b)^n$ compared with $(a + b)^n$?

Concept 2: Factorial Notation

For Exercises 15–18, evaluate the expression. **(See Example 1.)**

15. $5!$ **16.** $3!$ **17.** $0!$ **18.** $1!$

19. True or false: $0! \neq 1!$ **20.** True or false: $n!$ is defined for negative integers.

21. True or false: $n! = n$ for $n = 1$ and 2.   **22.** Show that $9! = 9 \cdot 8!$

23. Show that $6! = 6 \cdot 5!$ **24.** Show that $8! = 8 \cdot 7!$

For Exercises 25–32, evaluate the expression. **(See Example 2.)**

25. $\dfrac{8!}{4!}$ **26.** $\dfrac{7!}{5!}$ **27.** $\dfrac{3!}{0!}$ **28.** $\dfrac{4!}{0!}$

29. $\dfrac{8!}{3! \cdot 5!}$ **30.** $\dfrac{6!}{2! \cdot 4!}$ **31.** $\dfrac{4!}{0! \cdot 4!}$ **32.** $\dfrac{6!}{0! \cdot 6!}$

Concept 3: The Binomial Theorem

For Exercises 33–36, find the first three terms of the expansion. **(See Example 3.)**

33. $(m + n)^{11}$ **34.** $(p + q)^9$ **35.** $(u^2 - v)^{12}$ **36.** $(r - s^2)^8$

37. How many terms are in the expansion of $(a + b)^8$?

38. How many terms are in the expansion of $(x + y)^{13}$?

For Exercises 39–50, use the binomial theorem to expand the binomials. **(See Examples 4–5.)**

39. $(s + t)^6$ **40.** $(h + k)^4$ **41.** $(b - 3)^3$ **42.** $(c - 2)^5$

43. $(2x + y)^4$ **44.** $(p + 3q)^3$ **45.** $(c^2 - d)^7$ **46.** $(u - v^3)^6$

47. $\left(\dfrac{a}{2} - b\right)^5$ **48.** $\left(\dfrac{s}{3} + t\right)^5$ **49.** $(x + 4y)^4$ **50.** $(3y - w)^3$

For Exercises 51–56, find the indicated term of the binomial expansion. **(See Examples 6–7.)**

51. $(m - n)^{11}$; sixth term   **52.** $(p - q)^9$; fourth term   **53.** $(u^2 - v)^{12}$; fifth term

54. $(2r - s^2)^8$; sixth term   **55.** $(5f + g)^9$; 10th term   **56.** $(4m + n)^{10}$; 11th term

Section 10.2 Sequences and Series

Concepts

1. Finite and Infinite Sequences
2. Series

1. Finite and Infinite Sequences

In day-to-day life, we think of a sequence as a set of items with some order or pattern. In mathematics, a sequence is a list of terms that correspond to the set of positive integers.

    For example, suppose Justine is allowed to choose between two salary plans. Plan A offers $48,000 as a starting salary with a 5% raise each year. Plan B offers

$50,000 with a $2000 raise each year. Then the following sequences represent the salaries for each plan for the first 5 yr:

Plan A: $48,000 $50,400 $52,920 $55,566 $58,344.30

Plan B: $50,000 $52,000 $54,000 $56,000 $58,000

By studying sequences we can see that Plan A yields a better salary if Justine plans to stay in the job for at least 5 yr.

These sequences are called *finite sequences* because they have a finite number of terms. The sequence 1, 4, 9, 16, 25, . . . is called an *infinite sequence* because it continues indefinitely.

Because the terms in a sequence are related to the set of positive integers, we give a formal definition of finite and infinite sequences, using the language of functions.

> ### Definition of Finite and Infinite Sequences
>
> An **infinite sequence** is a function whose domain is the set of positive integers. A **finite sequence** is a function whose domain is the set of the first n positive integers.

For any positive integer n, the value of the sequence is denoted by a_n (read as "a sub n"). The values $a_1, a_2, a_3, \ldots$ are called the **terms of the sequence**. The expression a_n defines the **nth term** (or general term) **of the sequence**.

Example 1   Listing the Terms of a Sequence

List the terms of the following sequences.

a. $a_n = 3n^2 - 4, \quad 1 \le n \le 4$ **b.** $a_n = 3 \cdot 2^n$

Solution:

a. The domain is restricted to the first four positive integers, indicating that the sequence is finite.

n	a_n
1	$3(1)^2 - 4 = -1$
2	$3(2)^2 - 4 = 8$
3	$3(3)^2 - 4 = 23$
4	$3(4)^2 - 4 = 44$

The sequence is $-1, 8, 23, 44$.

b. The sequence $a_n = 3 \cdot 2^n$ has no restrictions on its domain; therefore, it is an infinite sequence.

n	a_n
1	$3 \cdot 2^1 = 6$
2	$3 \cdot 2^2 = 12$
3	$3 \cdot 2^3 = 24$
4	$3 \cdot 2^4 = 48$

The sequence is $6, 12, 24, 48, \ldots$

. . .

Skill Practice List the terms of the sequences.

1. $a_n = n^3 - 2, \quad 1 \le n \le 3$ **2.** $a_n = \left(\dfrac{2}{3}\right)^n$

Answers

1. $-1, 6, 25$ **2.** $\dfrac{2}{3}, \dfrac{4}{9}, \dfrac{8}{27}, \ldots$

Calculator Connections

Topic: Displaying Terms of a Sequence

If the nth term of a sequence is known, a seq(function on a graphing utility can display a list of terms. Finding the first four terms of the sequence from Example 1(b) is outlined here: $a_n = 3 \cdot 2^n$.

- Select the MODE key and highlight SEQ.
- Then access the seq(function from the LIST menu and enter the parameters as prompted.
- Paste the seq(command on the home screen and select ENTER.

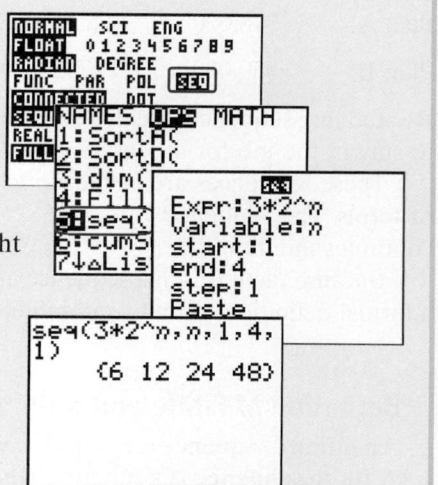

Sometimes the terms of a sequence may have alternating signs. Such a sequence is called an **alternating sequence**.

Example 2   **Listing the Terms of an Alternating Sequence**

List the first four terms of each alternating sequence.

a. $a_n = (-1)^n \cdot \dfrac{1}{n}$ **b.** $a_n = (-1)^{n+1} \cdot \left(\dfrac{2}{3}\right)^n$

Solution:

a.

n	a_n
1	$(-1)^1 \cdot \dfrac{1}{1} = -1$
2	$(-1)^2 \cdot \dfrac{1}{2} = \dfrac{1}{2}$
3	$(-1)^3 \cdot \dfrac{1}{3} = -\dfrac{1}{3}$
4	$(-1)^4 \cdot \dfrac{1}{4} = \dfrac{1}{4}$

The first four terms are $-1, \frac{1}{2}, -\frac{1}{3}, \frac{1}{4}$.

TIP: Notice that the factor $(-1)^n$ makes the even-numbered terms positive and the odd-numbered terms negative.

b.

n	a_n
1	$(-1)^{1+1} \cdot \left(\dfrac{2}{3}\right)^1 = (-1)^2 \cdot \left(\dfrac{2}{3}\right) = \dfrac{2}{3}$
2	$(-1)^{2+1} \cdot \left(\dfrac{2}{3}\right)^2 = (-1)^3 \cdot \left(\dfrac{4}{9}\right) = -\dfrac{4}{9}$
3	$(-1)^{3+1} \cdot \left(\dfrac{2}{3}\right)^3 = (-1)^4 \cdot \left(\dfrac{8}{27}\right) = \dfrac{8}{27}$
4	$(-1)^{4+1} \cdot \left(\dfrac{2}{3}\right)^4 = (-1)^5 \cdot \left(\dfrac{16}{81}\right) = -\dfrac{16}{81}$

TIP: Notice that the factor $(-1)^{n+1}$ makes the odd-numbered terms positive and the even-numbered terms negative.

The first four terms are $\frac{2}{3}, -\frac{4}{9}, \frac{8}{27}, -\frac{16}{81}$.

Skill Practice List the first four terms of each alternating sequence.

3. $a_n = (-1)^n \left(\dfrac{1}{2}\right)^n$ **4.** $a_n = (-1)^{n-1} \left(\dfrac{2}{n}\right)$

In Examples 1 and 2, we were given the formula for the nth term of a sequence and asked to list several terms of the sequence. We now consider the reverse process. Given several terms of the sequence, we will find a formula for the nth term. To do so, look for a pattern that establishes each term as a function of the term number.

Example 3 Finding the nth Term of a Sequence

Find a formula for the nth term of the sequence.

a. $\dfrac{1}{2}, \dfrac{2}{3}, \dfrac{3}{4}, \dfrac{4}{5}, \dots$ **b.** $-2, 4, -6, 8, -10, \dots$ **c.** $\dfrac{1}{2}, \dfrac{1}{4}, \dfrac{1}{8}, \dfrac{1}{16}, \dots$

Solution:

a. For each term in the sequence, the numerator is equal to the term number, and the denominator is equal to 1 more than the term number. Therefore, the nth term may be given by

$$a_n = \frac{n}{n+1}$$

b. The odd-numbered terms are negative, and the even-numbered terms are positive. The factor $(-1)^n$ will produce the required alternation of signs. The numbers 2, 4, 6, 8, 10, ... are equal to $2(1), 2(2), 2(3), 2(4), 2(5), \dots$. Therefore, the nth term may be given by

$$a_n = (-1)^n \cdot 2n$$

c. The denominators are consecutive powers of 2. The sequence can be written as

$$\frac{1}{2^1}, \frac{1}{2^2}, \frac{1}{2^3}, \frac{1}{2^4}, \dots$$

Therefore, the nth term may be given by

$$a_n = \frac{1}{2^n}$$

Skill Practice Find a formula for the nth term of the sequence.

5. $\dfrac{1}{3}, \dfrac{1}{9}, \dfrac{1}{27}, \dfrac{1}{81}, \dots$ **6.** $-1, 4, -9, 16, -25, \dots$ **7.** $\dfrac{3}{2}, \dfrac{4}{3}, \dfrac{5}{4}, \dfrac{6}{5}, \dfrac{7}{6}, \dots$

Answers

3. $-\dfrac{1}{2}, \dfrac{1}{4}, -\dfrac{1}{8}, \dfrac{1}{16}$

4. $2, -1, \dfrac{2}{3}, -\dfrac{1}{2}$

5. $a_n = \dfrac{1}{3^n}$ **6.** $a_n = (-1)^n n^2$

7. $a_n = \dfrac{n+2}{n+1}$

> **TIP:** The first few terms of a sequence are not sufficient to define the sequence uniquely. For example, consider the sequence
>
> $$\frac{1}{2}, \frac{1}{4}, \frac{1}{8}, \cdots$$
>
> The following formulas both produce the first three terms, but differ at the fourth term:
>
> $$a_n = \frac{1}{2^n} \quad \text{and} \quad b_n = \frac{1}{n^2 - n + 2}$$
>
> $$a_4 = \frac{1}{16} \quad \text{whereas} \quad b_4 = \frac{1}{14}$$
>
> To define a sequence uniquely, the nth term must be provided.

Example 4 Using a Sequence in an Application

A child drops a ball from a height of 4 ft. With each bounce, the ball rebounds to 50% of its height. Write a sequence whose terms represent the heights from which the ball falls (begin with the initial height from which the ball was dropped).

Solution:

The ball first drops 4 ft and then rebounds to a new height of 0.50(4 ft) = 2 ft. Similarly, it falls from 2 ft and rebounds 0.50(2 ft) = 1 ft. Repeating this process, we have

$$4, 2, 1, \frac{1}{2}, \frac{1}{4}, \frac{1}{8}, \cdots$$

The nth term can be represented by $a_n = 4 \cdot (0.50)^{n-1}$.

Skill Practice

8. The first swing of a pendulum measures $30°$. If each swing that follows is 25% less, list the first three terms of this sequence.

2. Series

In many mathematical applications it is important to find the sum of the terms of a sequence. For example, suppose that the yearly interest earned in an account over a 4-yr period is given by the sequence

$$\$250, \quad \$265, \quad \$278.25, \quad \$292.16$$

The sum of the terms gives the total interest earned

$$\$250 + \$265 + \$278.25 + \$292.16 = \$1085.41$$

By adding the terms of a sequence, we obtain a series.

Definition of a Series

The indicated sum of the terms of a sequence is called a **series**.

Answer

8. $30°$, $22.5°$, $16.875°$

As with a sequence, a series may be a finite or an infinite sum of terms.

A convenient notation used to denote the sum of a set of terms is called **summation notation**, or sigma notation. The Greek letter Σ (sigma) is used to indicate sums. For example, the sum of the first four terms of the sequence defined by $a_n = n^3$ are denoted by

$$\sum_{n=1}^{4} n^3$$

This is read as "the sum from n equals 1 to 4 of n^3" and is simplified as

$$\sum_{n=1}^{4} n^3 = (1)^3 + (2)^3 - (3)^3 + (4)^3$$

$$= 1 + 8 + 27 + 64$$

$$= 100$$

In this example, the letter n is called the **index of summation**. Many times, the letters $i, j,$ and k are also used for the index of summation.

Example 5 Finding a Sum from Summation Notation

Find the sum. $\displaystyle\sum_{i=1}^{4} 2^i$

Solution:

$$\sum_{i=1}^{4} 2^i = 2^1 + 2^2 + 2^3 + 2^4$$

$$= 2 + 4 + 8 + 16$$

$$= 30$$

Skill Practice Find the sum.

9. $\displaystyle\sum_{i=1}^{4} i(i + 4)$

Example 6 Finding a Sum from Summation Notation

Find the sum. $\displaystyle\sum_{i=1}^{3} (-1)^{i+1} \cdot (3i + 4)$

Solution:

$$\sum_{i=1}^{3} (-1)^{i+1} \cdot (3i + 4) = (-1)^{1+1} \cdot [3(1) + 4] + (-1)^{2+1} \cdot [3(2) + 4]$$

$$+ (-1)^{3+1} \cdot [3(3) + 4]$$

$$= (-1)^2 \cdot (7) + (-1)^3 \cdot (10) + (-1)^4 \cdot (13)$$

$$= 7 - 10 + 13$$

$$= 10$$

Avoiding Mistakes

The letter i is used as a variable for the index of summation. In this context, it does not represent an imaginary number.

Skill Practice Find the sum.

10. $\displaystyle\sum_{i=1}^{4} (-1)^i (4i - 3)$

Answers

9. 70   **10.** 8

Example 7 **Converting to Summation Notation**

Write the series in summation notation.

a. $\dfrac{2}{3} + \dfrac{4}{9} + \dfrac{8}{27} + \dfrac{16}{81}$ Use n as the index of summation.

b. $1 - \dfrac{1}{2} + \dfrac{1}{3} - \dfrac{1}{4} + \dfrac{1}{5} - \dfrac{1}{6}$ Use j as the index of summation.

Solution:

a. The sum can be written as

$$\left(\dfrac{2}{3}\right)^1 + \left(\dfrac{2}{3}\right)^2 + \left(\dfrac{2}{3}\right)^3 + \left(\dfrac{2}{3}\right)^4$$

Taking n from 1 to 4, we have

$$\sum_{n=1}^{4} \left(\dfrac{2}{3}\right)^n$$

b. The even-numbered terms are negative. The factor $(-1)^{j+1}$ is negative for even values of j. Therefore, the series can be written as

$$\sum_{j=1}^{6} (-1)^{j+1} \cdot \dfrac{1}{j}$$

Answers

11. $\displaystyle\sum_{i=1}^{4} \dfrac{1}{4^i}$

12. $\displaystyle\sum_{i=1}^{5} (-1)^{i+1} \dfrac{1}{2i}$

Skill Practice Write the series in summation notation.

11. $\dfrac{1}{4} + \dfrac{1}{16} + \dfrac{1}{64} + \dfrac{1}{256}$ **12.** $\dfrac{1}{2} - \dfrac{1}{4} + \dfrac{1}{6} - \dfrac{1}{8} + \dfrac{1}{10}$

Section 10.2   Practice Exercises

Vocabulary and Key Concepts

1. a. A(n) (finite/infinite) sequence is a function whose domain is the set of positive integers. A(n) (finite/infinite) sequence is a function whose domain is the set of the first n positive integers.

   b. Given a sequence, the values a_1, a_2, a_3, and so on are called the _____ of the sequence. The expression a_n defines the _____ term (or general term) of the sequence.

   c. In an _____ sequence, consecutive terms alternate in sign.

   d. A sum of a sequence is called a _____.

   e. To represent a sum of terms, _____ notation is often used and is represented by the Greek letter $\sum$.

   f. Given the sum $\displaystyle\sum_{n=1}^{5} n^2$, the letter n is called the _____ of summation. In expanded form, this series is $(1)^2 + (2)^2 + $ _____ $= 55$.

Review Exercises

For Exercises 2–4, evaluate the expression.

2. $9!$ **3.** $\dfrac{8!}{2! \cdot 6!}$ **4.** $\dfrac{7!}{7! \cdot 0!}$

For Exercises 5–6, expand the binomial.

5. $(2x + z)^4$

6. $(4m - 3t)^3$

Concept 1: Finite and Infinite Sequences

For Exercises 7–18, list the terms of each sequence. **(See Examples 1–2.)**

7. $a_n = 3n + 1, \quad 1 \le n \le 5$

8. $a_n = -2n + 3, \quad 1 \le n \le 5$

9. $a_n = \sqrt{n + 2}, \quad 1 \le n \le 4$

10. $a_n = \sqrt{n - 1}, \quad 1 \le n \le 4$

11. $a_n = (-1)^n \dfrac{n + 1}{n + 2}, \quad 1 \le n \le 4$

12. $a_n = (-1)^n \dfrac{n - 1}{n + 2}, \quad 1 \le n \le 4$

13. $a_n = (-1)^{n+1} (n^2 - 1), \quad 1 \le n \le 3$

14. $a_n = (-1)^{n+1} (n^2), \quad 1 \le n \le 3$

15. $a_n = n^2 - n, \quad 1 \le n \le 6$

16. $a_n = n(n^2 - 1), \quad 1 \le n \le 6$

17. $a_n = (-1)^n 3^n, \quad 1 \le n \le 4$

18. $a_n = (-1)^n n, \quad 1 \le n \le 4$

19. If the nth term of a sequence is $a_n = (-1)^n \cdot n^2$, which terms are positive and which are negative?

20. If the nth term of a sequence is $a_n = (-1)^{n-1} \cdot \frac{1}{n}$, which terms are positive and which are negative?

For Exercises 21–32, find a formula for the nth term of the sequence. Answers may vary. **(See Example 3.)**

21. $2, 4, 6, 8, \ldots$

22. $3, 6, 9, 12, \ldots$

23. $1, 3, 5, 7, \ldots$

24. $3, 5, 7, 9, \ldots$

25. $1, \dfrac{1}{4}, \dfrac{1}{9}, \dfrac{1}{16}, \ldots$

26. $\dfrac{1}{2}, \dfrac{2}{3}, \dfrac{3}{4}, \dfrac{4}{5}, \ldots$

27. $1, -1, 1, -1, \ldots$

28. $-1, 1, -1, 1, \ldots$

29. $-2, 4, -8, 16, \ldots$

30. $3, -9, 27, -81, \ldots$

31. $\dfrac{3}{5}, \dfrac{3}{25}, \dfrac{3}{125}, \dfrac{3}{625}, \ldots$

32. $\dfrac{1}{4}, \dfrac{1}{16}, \dfrac{1}{64}, \dfrac{1}{256}, \ldots$

33. Edmond borrowed $500. To pay off the loan, he agreed to pay 2% of the balance plus $50 each month. Write a sequence representing the amount Edmond will pay each month for the next 4 months. Round each term to the nearest cent.

34. Janice deposited $1000 in a savings account that pays 3% interest compounded annually. Write a sequence representing the amount Janice receives in interest each year for the first 4 yr. Round each term to the nearest cent.

35. A certain bacteria culture doubles its size each day. If there are 25,000 bacteria on the first day, write a sequence representing the population each day for the first week (7 days). **(See Example 4.)**

36. A radioactive chemical decays by one-half of its amount each week. If there is 16 g of the chemical in week 1, write a sequence representing the amount present each week for 2 months (8 weeks).

Concept 2: Series

37. What is the difference between a sequence and a series?

38. a. Identify the index of summation for the series.

$$\sum_{k=1}^{7} (k^2 + 1)$$

b. How many terms does this series have?

For Exercises 39–54, find the sums. (See Examples 5–6.)

39. $\displaystyle\sum_{i=1}^{4} (3i^2)$ **40.** $\displaystyle\sum_{i=1}^{4} (2i^2)$ **41.** $\displaystyle\sum_{j=0}^{4} \left(\frac{1}{2}\right)^j$ **42.** $\displaystyle\sum_{j=0}^{4} \left(\frac{1}{3}\right)^j$

43. $\displaystyle\sum_{i=1}^{6} 5$ **44.** $\displaystyle\sum_{i=1}^{7} 3$ **45.** $\displaystyle\sum_{j=1}^{4} (-1)^j (5j)$ **46.** $\displaystyle\sum_{j=1}^{4} (-1)^j (4j)$

47. $\displaystyle\sum_{i=1}^{4} \frac{i+1}{i}$ **48.** $\displaystyle\sum_{i=2}^{5} \frac{i-1}{i}$ **49.** $\displaystyle\sum_{j=1}^{3} (j+1)(j+2)$ **50.** $\displaystyle\sum_{j=1}^{3} j(j+2)$

51. $\displaystyle\sum_{k=1}^{7} (-1)^k$ **52.** $\displaystyle\sum_{k=0}^{5} (-1)^{k+1}$ **53.** $\displaystyle\sum_{k=1}^{5} k^2$ **54.** $\displaystyle\sum_{k=1}^{5} 2^k$

For Exercises 55–66, write the series in summation notation. (See Example 7.)

55. $1 + 2 + 3 + 4 + 5 + 6$

56. $1 - 2 + 3 - 4 + 5 - 6$

57. $4 + 4 + 4 + 4 + 4$

58. $8 + 8 + 8 + 8 + 8$

59. $4 + 8 + 12 + 16 + 20$

60. $3 + 6 + 9 + 12 + 15$

61. $\dfrac{1}{3} - \dfrac{1}{9} + \dfrac{1}{27} - \dfrac{1}{81}$

62. $\dfrac{1}{2} - \dfrac{1}{4} + \dfrac{1}{8} - \dfrac{1}{16}$

63. $\dfrac{5}{11} + \dfrac{6}{22} + \dfrac{7}{33} + \dfrac{8}{44} + \dfrac{9}{55} + \dfrac{10}{66}$

64. $-\dfrac{5}{7} - \dfrac{10}{8} - \dfrac{15}{9} - \dfrac{20}{10} - \dfrac{25}{11} - \dfrac{30}{12} - \dfrac{35}{13}$

65. $x + x^2 + x^3 + x^4 + x^5$

66. $y^2 + y^4 + y^6 + y^8 + y^{10}$

Expanding Your Skills

Some sequences are defined by a recursion formula, which defines each term of a sequence in terms of one or more of its preceding terms. For example, if $a_1 = 5$ and $a_n = 2a_{n-1} + 1$ for $n > 1$, then the terms of the sequence are 5, 11, 23, 47, …. In this case, each term after the first is one more than twice the term before it.

For Exercises 67–70, list the first five terms of the sequence.

67. $a_1 = -3$, $a_n = a_{n-1} + 5$ for $n > 1$

68. $a_1 = 10$, $a_n = a_{n-1} - 3$ for $n > 1$

69. $a_1 = 5$, $a_n = 4a_{n-1} + 1$ for $n > 1$

70. $a_1 = -2$, $a_n = -3a_{n-1} + 4$ for $n > 1$

71. A famous sequence in mathematics is called the Fibonacci sequence, named after the Italian mathematician Leonardo Fibonacci of the thirteenth century. The Fibonacci sequence is defined by

$$a_1 = 1$$
$$a_2 = 1$$
$$a_n = a_{n-1} + a_{n-2} \quad \text{for } n > 2$$

This definition implies that beginning with the third term, each term is the sum of the preceding two terms. Write out the first 10 terms of the Fibonacci sequence.

| Arithmetic Sequences and Series | Section 10.3 |

1. Arithmetic Sequences

In this section and the next, we will study two special types of sequences and series. The first is called an arithmetic sequence, for example:

$$4, 7, 10, 13, 16, \ldots$$

This sequence is an arithmetic sequence. Note the characteristic that each successive term after the first is a fixed value more than the previous term (in this case the terms differ by 3).

Concepts

1. Arithmetic Sequences
2. Arithmetic Series

> **Definition of an Arithmetic Sequence**
>
> An **arithmetic sequence** is a sequence in which the difference between consecutive terms is constant.

The fixed difference between a term and its predecessor is called the **common difference** and is denoted by the letter d. The common difference is the difference between a term and its predecessor. That is,

$$d = a_{n+1} - a_n$$

Furthermore, if a_1 is the first term, then

$$a_1 = a_1 + 0d$$
$$a_2 = a_1 + 1d \quad \text{is the second term,}$$
$$a_3 = a_1 + 2d \quad \text{is the third term,}$$
$$a_4 = a_1 + 3d \quad \text{is the fourth term, and so on.}$$
$$\cdots$$

In general, $a_n = a_1 + (n - 1)d$.

> **nth Term of an Arithmetic Sequence**
>
> The nth term of an arithmetic sequence is given by
>
> $$a_n = a_1 + (n - 1)d$$
>
> where a_1 is the first term of the sequence and d is the common difference.

Example 1 Writing the nth Term of an Arithmetic Sequence

Write the nth term of the sequence. $\quad 9, 2, -5, -12, \ldots$

Solution:

$$9, \quad 2, \quad -5, \quad -12, \ldots$$
$$\quad -7 \quad -7 \quad -7$$

The common difference can be found by subtracting a term from its predecessor: $2 - 9 = -7$

TIP: Notice that the terms of an arithmetic sequence increase or decrease linearly. In Example 1, the terms decrease by 7 for each subsequent term. This is similar to a line with a slope of -7.

With $a_1 = 9$ and $d = -7$, we have

$$a_n = 9 + (n - 1)(-7)$$
$$= 9 - 7n + 7$$
$$= -7n + 16$$

Skill Practice Write the nth term of the sequence.

1. $-3, -1, 1, 3, 5, \ldots$

In Example 1, the common difference between terms is -7. Accordingly, each term of the sequence *decreases* by 7.

The formula $a_n = a_1 + (n - 1)d$ contains four variables: $a_n, a_1, n,$ and d. Consequently, if we know the value of three of the four variables, we can solve for the fourth.

Example 2 Finding a Specified Term of an Arithmetic Sequence

Find the ninth term of the arithmetic sequence in which $a_1 = -4$ and $a_{22} = 164$.

Solution:

To find the value of the ninth term a_9, we need to determine the value of d. To find d, substitute $a_1 = -4$, $n = 22$, and $a_{22} = 164$ into the formula $a_n = a_1 + (n - 1)d$.

$$a_n = a_1 + (n - 1)d$$
$$164 = -4 + (22 - 1)d$$
$$164 = -4 + 21d$$
$$168 = 21d$$
$$d = 8$$

Therefore, $a_n = -4 + (n - 1)(8)$

$$a_9 = -4 + (9 - 1)(8)$$
$$= -4 + (8)(8)$$
$$= 60$$

Skill Practice

2. Find the tenth term of the arithmetic sequence in which $a_1 = 14$ and $a_{15} = -28$. (*Hint:* First find d.)

Example 3 Finding the Number of Terms in an Arithmetic Sequence

Find the number of terms of the sequence $7, 3, -1, -5, \ldots, -113$.

Answers

1. $2n - 5$ 2. $d = -3, a_{10} = -13$

Solution:

To find the number of terms n, we can substitute $a_1 = 7$, $d = -4$, and $a_n = -113$ into the formula for the nth term.

$$a_n = a_1 + (n - 1)d$$

$$-113 = 7 + (n - 1)(-4)$$

$$-113 = 7 - 4n + 4$$

$$-113 = 11 - 4n$$

$$-124 = -4n$$

$$n = 31$$

Skill Practice Find the number of terms of the sequence.

3. $-15, -11, -7, -3, \ldots, 81$

2. Arithmetic Series

The indicated sum of an arithmetic sequence is called an **arithmetic series**. For example, the series

$$3 + 7 + 11 + 15 + 19 + 23$$

is an arithmetic series because the common difference between terms is constant (4). Adding the terms in a lengthy sum is cumbersome, so we offer the following "shortcut," which is developed here. Let S represent the sum of the terms in the series.

$S = 3 + 7 + 11 + 15 + 19 + 23$	Add the terms in ascending order.
$S = 23 + 19 + 15 + 11 + 7 + 3$	Add the terms in descending order.
$2S = 26 + 26 + 26 + 26 + 26 - 26$	Adding the two series produces six terms of 26.

$$2S = 6 \cdot 26$$

$$S = \frac{6 \cdot 26}{2}$$

$$= 78$$

By adding the terms in ascending and descending order, we double the sum but create a pattern that is easily added. This is true in general. To find the sum, S_n, of the first n terms of the arithmetic series $a_1 + a_2 + a_3 + \cdots + a_n$, we have

$S_n = a_1 + (a_1 + d) + (a_1 + 2d) + \cdots + a_n$	Ascending order
$S_n = a_n + (a_n - d) + (a_n - 2d) + \cdots + a_1$	Descending order

$$2S_n = (a_1 + a_n) + (a_1 + a_n) + (a_1 + a_n) + \cdots + (a_1 + a_n)$$

$$2S_n = n(a_1 + a_n)$$

$$S_n = \frac{n}{2}(a_1 + a_n)$$

Answer

3. 25

Sum of an Arithmetic Series

The sum, S_n, of the first n terms of an arithmetic series is given by

$$S_n = \frac{n}{2}(a_1 + a_n)$$

where a_1 is the first term of the series and a_n is the nth term of the series.

Example 4   Finding the Sum of an Arithmetic Series

Find the sum of the series. $\displaystyle\sum_{i=1}^{25} (2i + 3)$

Avoiding Mistakes

When we apply the sum formula $S_n = \dfrac{n}{2}(a_1 + a_n)$ to find the sum $\displaystyle\sum_{i=1}^{n} a_n$, the index of summation, i, must begin at 1.

Solution:

In this series, $n = 25$. Furthermore, $a_1 = 2(1) + 3 = 5$ and $a_{25} = 2(25) + 3 = 53$. Therefore,

$$S_{25} = \frac{25}{2}(5 + 53) \qquad \text{Apply the formula } S_n = \frac{n}{2}(a_1 + a_n).$$

$$= \frac{25}{2}(58) \qquad\qquad \text{Simplify.}$$

$$= 725$$

Skill Practice   Find the sum of the series.

4. $\displaystyle\sum_{i=1}^{20} (3i - 2)$

Example 5   Finding the Sum of an Arithmetic Series

Find the sum of the series.

$$-3 + (-5) + (-7) + \cdots + (-127)$$

Solution:

For this series, $a_1 = -3$ and $a_n = -127$. However, to determine the sum, we also need to find the value of n. The difference between the second term and its predecessor is $-5 - (-3) = -2$. Thus, $d = -2$. We have

$$-127 = -3 + (n - 1)(-2) \qquad \text{Apply the formula } a_n = a_1 + (n - 1)d.$$

$$-127 = -3 - 2n + 2 \qquad\qquad \text{Apply the distributive property.}$$

$$-127 = -1 - 2n \qquad\qquad\quad \text{Combine } like \text{ terms.}$$

$$-126 = -2n \qquad\qquad\qquad \text{Solve for } n.$$

$$n = 63$$

Using $n = 63$, $a_1 = -3$, and $a_{63} = -127$, we have

$$S_n = \frac{n}{2}(a_1 + a_n) = \frac{63}{2}[-3 + (-127)] = \frac{63}{2}(-130) = -4095$$

Skill Practice   Find the sum of the series.

5. $8 + 3 + (-2) + \cdots + (-137)$

Answers

4. 590   **5.** −1935

Section 10.3 Practice Exercises

Vocabulary and Key Concepts

1. **a.** An _____ sequence is a sequence in which the difference between consecutive terms is constant.

   b. The common _____ between a term and its predecessor in an arithmetic sequence is often denoted by the letter d.

   c. The nth term of an arithmetic sequence is given by $a_n =$ _____ , where a_n is the nth term, a_1 is the first term, n is the number of terms, and _____ is the common difference between terms.

   d. An indicated sum of an arithmetic sequence is called an arithmetic _____ .

   e. The sum S_n of the first n terms of an arithmetic series is given by $S_n =$ _____ , where a_1 is the first term and a_n is the nth term.

Review Exercises

2. Write the first four terms of the sequence. $c_n = \dfrac{n^2}{n+1}$

3. Write the first four terms of the sequence. $d_n = (-1)^n$

4. Find the sum. $\displaystyle\sum_{n=1}^{10} (-1)^{n+1}$

5. Find the sum. $\displaystyle\sum_{n=1}^{5} (1 - 4n)$

Concept 1: Arithmetic Sequences

6. Explain how to determine if a sequence is arithmetic.

For Exercises 7–12, the first term of an arithmetic sequence is given along with its common difference. Write the first five terms of the sequence.

7. $a_1 = 3, d = 8$

8. $a_1 = -5, d = 2$

9. $a_1 = 80, d = -20$

10. $a_1 = -4, d = -5$

11. $a_1 = 3, d = \dfrac{3}{4}$

12. $a_1 = 1, d = \dfrac{2}{3}$

For Exercises 13–18, find the common difference d for each arithmetic sequence.

13. $1, 3, 5, 7, 9, \ldots$

14. $2, 8, 14, 20, 26, \ldots$

15. $6, 3, 0, -3, -6, \ldots$

16. $8, 3, -2, -7, -12, \ldots$

17. $-7, -9, -11, -13, -15, \ldots$

18. $-15, -11, -7, -3, 1, \ldots$

For Exercises 19–24, write the first five terms of the arithmetic sequence.

19. $a_1 = 3, d = 5$

20. $a_1 = -3, d = 2$

21. $a_1 = 2, d = \dfrac{1}{2}$

22. $a_1 = 0, d = \dfrac{1}{3}$

23. $a_1 = 2, d = -4$

24. $a_1 = 10, d = -6$

For Exercises 25–33, write the nth term of the sequence. **(See Example 1.)**

25. $0, 5, 10, 15, 20, \ldots$

26. $7, 12, 17, 22, 27, \ldots$

27. $-2, -4, -6, -8, -10, \ldots$

28. $1, -3, -7, -11, -15, \ldots$

29. $2, \dfrac{5}{2}, 3, \dfrac{7}{2}, 4, \ldots$

30. $1, \dfrac{4}{3}, \dfrac{5}{3}, 2, \dfrac{7}{3}, \ldots$

31. $21, 17, 13, 9, 5, \ldots$

32. $9, 6, 3, 0, -3, \ldots$

33. $-8, -2, 4, 10, 16, \ldots$

For Exercises 34–41, find the indicated term of each arithmetic sequence. **(See Example 2.)**

34. Find the eighth term given $a_1 = -6$ and $d = -3$.

35. Find the sixth term given $a_1 = -3$ and $d = 4$.

36. Find the 12th term given $a_1 = -8$ and $d = -2$.

37. Find the ninth term given $a_1 = -1$ and $d = 6$.

38. Find the tenth term given $a_1 = 1$ and $a_7 = 31$.

39. Find the seventh term given $a_1 = 0$ and $a_{10} = -45$.

40. Find the sixth term given $a_1 = 3$ and $a_{13} = 39$.

41. Find the 11th term given $a_1 = 12$ and $a_6 = -18$.

For Exercises 42–49, find the number of terms, n, of each arithmetic sequence. **(See Example 3.)**

42. $2, 0, -2, \ldots, -56$

43. $8, 13, 18, \ldots, 98$

44. $1, -3, -7, \ldots, -67$

45. $1, 5, 9, \ldots, 85$

46. $1, \dfrac{3}{4}, \dfrac{1}{2}, \ldots, -4$

47. $2, \dfrac{5}{2}, 3, \ldots, 13$

48. $-\dfrac{5}{3}, -1, -\dfrac{1}{3}, \ldots, \dfrac{37}{3}$

49. $\dfrac{13}{3}, \dfrac{19}{3}, \dfrac{25}{3}, \ldots, \dfrac{73}{3}$

50. If the third and fourth terms of an arithmetic sequence are 18 and 21, what are the first and second terms?

51. If the third and fourth terms of an arithmetic sequence are -8 and -11, what are the first and second terms?

Concept 2: Arithmetic Series

52. Explain the difference between an arithmetic sequence and an arithmetic series.

For Exercises 53–66, find the sum of the arithmetic series. **(See Examples 4–5.)**

53. $\displaystyle\sum_{i=1}^{20} (3i + 2)$

54. $\displaystyle\sum_{i=1}^{15} (2i - 3)$

55. $\displaystyle\sum_{i=1}^{20} (i + 4)$

56. $\displaystyle\sum_{i=1}^{25} (i - 3)$

57. $\displaystyle\sum_{j=1}^{10} (4 - j)$

58. $\displaystyle\sum_{j=1}^{10} (6 - j)$

59. $\displaystyle\sum_{j=1}^{15} \left(\dfrac{2}{3}j + 1\right)$

60. $\displaystyle\sum_{j=1}^{15} \left(\dfrac{1}{2}j - 2\right)$

61. $4 + 8 + 12 + \cdots + 84$

62. $4 + 9 + 14 + \cdots + 49$

63. $6 + 8 + 10 + \cdots + 34$

64. $-4 + (-3) + (-2) + \cdots + 12$

65. $-3 + (-7) + (-11) + \cdots + (-39)$

66. $2 + 5 + 8 + \cdots + 53$

67. Find the sum of the first 100 positive integers.

68. Find the sum of the first 50 positive even integers.

69. The seating in a certain theater is arranged so that there are 30 seats in row 1, 32 in row 2, 34 in row 3, and so on. If there are 20 rows, how many total seats are there? What is the total revenue if the average ticket price is $15 per seat and the theater is sold out?

70. A triangular array of dominoes has one domino in the first row, two dominoes in the second row, three dominoes in the third row, and so on. If there are 15 rows, how many dominoes are in the array?

Geometric Sequences and Series

1. Geometric Sequences

The sequence 2, 4, 8, 16, 32, . . . is not an arithmetic sequence because the difference between terms is not constant. However, a different pattern exists. Notice that each term after the first is 2 times the preceding term. This sequence is called a geometric sequence.

Definition of a Geometric Sequence

A **geometric sequence** is a sequence in which the ratio between each term and its predecessor is constant.

The constant quotient of a term and its predecessor is called the **common ratio** and is denoted by r. The common ratio is found by dividing a term by the preceding term. That is,

$$r = \frac{a_{n-1}}{a_n}$$

For the sequence 2, 4, 8, 16, 32, . . . we have $r = \frac{4}{2} = \frac{8}{4} = \frac{16}{8} = \frac{32}{16} = 2$.

If a_1 denotes the first term of a geometric sequence, then

$a_2 = a_1 r$ is the second term,

$a_3 = a_1 r^2$ is the third term,

$a_4 = a_1 r^3$ is the fourth term, and so on.

. . .

This pattern gives $a_n = a_1 r^{n-1}$.

nth Term of a Geometric Sequence

The nth term of a geometric sequence is given by

$$a_n = a_1 r^{n-1}$$

where a_1 is the first term and r is the common ratio.

Example 1 Finding the nth Term of a Geometric Sequence

Find the nth term of the sequence.   **a.** $-1, 4, -16, 64, \ldots$ **b.** $12, 8, \frac{16}{3}, \frac{32}{9}, \ldots$

Solution:

a. The common ratio is found by dividing any term (after the first) by its predecessor.

$$r = \frac{4}{-1} = -4$$

With $r = -4$ and $a_1 = -1$, we have $a_n = -1(-4)^{n-1}$.

b. The common ratio is $r = \frac{8}{12} = \frac{2}{3}$. With $a_1 = 12$ and $r = \frac{2}{3}$, we have

$$a_n = 12\left(\frac{2}{3}\right)^{n-1}$$

Skill Practice Find the nth term of the geometric sequence.

1. $6, 12, 24, 48, \ldots$ **2.** $6, 2, \dfrac{2}{3}, \dfrac{2}{9}, \ldots$

The formula $a_n = a_1 r^{n-1}$ contains the variables $a_n, a_1, n,$ and r. If we know the value of three of the four variables, we can find the fourth.

Example 2   **Finding a Specified Term of a Geometric Sequence**

Given $a_n = 6\left(\dfrac{1}{2}\right)^{n-1}$, find a_5.

Solution:

$$a_n = 6\left(\frac{1}{2}\right)^{n-1}$$

$$a_5 = 6\left(\frac{1}{2}\right)^{5-1} = 6\left(\frac{1}{2}\right)^4 = 6\left(\frac{1}{16}\right) = \frac{3}{8}$$

Skill Practice

3. Given $a_n = 3(-2)^{n-1}$, find a_6.

Example 3   **Finding a Specified Term of a Geometric Sequence**

Find the first term of the geometric sequence where $a_5 = -162$ and $r = 3$.

Solution:

$-162 = a_1(3)^{5-1}$ Substitute $a_5 = -162, n = 5,$ and $r = 3$ into the formula $a_n = a_1 r^{n-1}$.

$-162 = a_1(3)^4$ Simplify and solve for a_1.

$-162 = a_1(81)$

$a_1 = -2$

Skill Practice

4. Find the first term of the geometric sequence where $a_4 = \frac{25}{32}$ and $r = \frac{1}{4}$.

2. Geometric Series

The indicated sum of a geometric sequence is called a **geometric series**. For example,

$$1 + 3 + 9 + 27 + 81 + 243$$

is a geometric series. To find the sum, consider the following procedure. Let S represent the sum

$$S = 1 + 3 + 9 + 27 + 81 + 243$$

Answers

1. $a_n = 6(2)^{n-1}$ **2.** $a_n = 6\left(\dfrac{1}{3}\right)^{n-1}$

3. -96 **4.** 50

Now multiply S by the common ratio, which in this case is 3.

$$3S = 3 + 9 + 27 + 81 + 243 + 729$$

Then

$$3S - S = (3 + 9 + 27 + 81 + 243 + 729) - (1 + 3 + 9 + 27 + 81 + 243)$$

$$2S = 3 + 9 + 27 + 81 + 243 + 729 - 1 - 3 - 9 - 27 - 81 - 243$$

$$2S = 729 - 1 \qquad \text{The terms in red form a sum of zero.}$$

$$2S = 728$$

$$S = 364$$

A similar procedure can be used to find the sum S_n of the first n terms of any geometric series. Subtract rS_n from S_n.

$$S_n = a_1 \qquad + a_1 r \qquad + a_1 r^2 \qquad + \cdots + a_1 r^{n-1}$$

$$\underline{rS_n = a_1 r \qquad + a_1 r^2 \qquad + a_1 r^3 \qquad + \cdots + a_1 r^n}$$

$$S_n - rS_n = (a_1 - a_1 r) + (a_1 r - a_1 r^2) + (a_1 r^2 - a_1 r^3) + \cdots + (a_1 r^{n-1} - a_1 r^n)$$

$$S_n - rS_n = a_1 - a_1 r^n \qquad \text{The terms in red form a sum of zero.}$$

$$S_n(1 - r) = a_1(1 - r^n) \qquad \text{Factor each side of the equation.}$$

$$S_n = \frac{a_1(1 - r^n)}{1 - r} \qquad \text{Divide by } (1 - r).$$

Sum of a Geometric Series

The sum, S_n, of the first n terms of a geometric series $\displaystyle\sum_{i=1}^{n} a_1 r^{i-1}$ is given by

$$S_n = \frac{a_1(1 - r^n)}{1 - r}$$

where a_1 is the first term of the series, r is the common ratio, and $r \neq 1$.

Example 4 Finding the Sum of a Geometric Series

Find the sum of the series. $\displaystyle\sum_{i=1}^{6} 4\left(\frac{1}{2}\right)^{i-1}$

Solution:

By expanding the terms of this series, we see that the series is geometric.

$$\sum_{i=1}^{6} 4\left(\frac{1}{2}\right)^{i-1} = 4 + 2 + 1 + \frac{1}{2} + \frac{1}{4} + \frac{1}{8}$$

We have: $a_1 = 4,$ $\quad r = \dfrac{1}{2},$ $\quad$ and $\quad n = 6,$

$$S_n = \frac{a_1(1 - r^n)}{1 - r} = \frac{4\left[1 - \left(\frac{1}{2}\right)^6\right]}{1 - \frac{1}{2}} = \frac{4\left(1 - \frac{1}{64}\right)}{\frac{1}{2}} = 8\left(\frac{63}{64}\right) = \frac{63}{8}$$

Skill Practice Find the sum of the geometric series.

5. $\dfrac{1}{9} + \dfrac{1}{3} + 1 + 3 + 9 + 27$

Answer

5. $\dfrac{364}{9}$

Example 5 Finding the Sum of a Geometric Series

Find the sum of the series $5 + 10 + 20 + \cdots + 5120$.

Solution:

The common ratio is 2 and $a_1 = 5$. The nth term of the sequence can be written as $a_n = 5(2)^{n-1}$. To find the value of n, substitute 5120 for a_n.

$$5120 = 5(2)^{n-1}$$

$$\frac{5120}{5} = \frac{5(2)^{n-1}}{5} \qquad \text{Divide both sides by 5.}$$

$$1024 = 2^{n-1} \qquad \text{To solve the exponential equation, write each side as}$$
$$2^{10} = 2^{n-1} \qquad \text{a power of 2.}$$

$$10 = n - 1 \qquad \text{From Section 8.7, we know that if } b^x = b^y, \text{then } x = y.$$

$$n = 11$$

With $a_1 = 5$, $r = 2$, and $n = 11$, we have

$$S_n = \frac{a_1(1 - r^n)}{1 - r} = \frac{5(1 - 2^{11})}{1 - 2} = \frac{5(1 - 2048)}{-1} = \frac{5(-2047)}{-1} = 10{,}235$$

Skill Practice Find the sum of the geometric series.

6. $3 + 6 + 12 + \cdots + 768$

Consider a geometric series where $|r| < 1$. For increasing values of n, r^n decreases. For example,

$$\left(\frac{1}{2}\right)^5 = 0.03125 \qquad \left(\frac{1}{2}\right)^{10} \approx 0.00097656 \qquad \left(\frac{1}{2}\right)^{15} \approx 0.00003052$$

For $|r| < 1$, r^n approaches 0 as n gets larger and larger. As n approaches infinity, the sum

$$S = \frac{a_1(1 - r^n)}{1 - r} \qquad \text{approaches} \qquad \frac{a_1(1 - 0)}{1 - r} = \frac{a_1}{1 - r}$$

Sum of an Infinite Geometric Series

Given an infinite geometric series $a_1 + a_1 r + a_1 r^2 + \cdots$, with $|r| < 1$, the sum S of all terms in the series is given by

$$S = \frac{a_1}{1 - r}$$

Note: If $|r| \geq 1$, then the sum does not exist.

Example 6 Finding the Sum of an Infinite Geometric Series

Find the sum of the series. $\displaystyle\sum_{i=1}^{\infty}\left(\frac{1}{3}\right)^{i-1}$

Solution:

In this example, the upper limit of summation is ∞. This indicates that we have an infinite series where the pattern for the sum of terms continues indefinitely.

$$\sum_{i=1}^{\infty}\left(\frac{1}{3}\right)^{i-1} = 1 + \frac{1}{3} + \frac{1}{9} + \frac{1}{27} + \frac{1}{81} + \cdots$$

Answer

6. 1533

The series is geometric with $a_1 = 1$ and $r = \frac{1}{3}$. Because $|r| = |\frac{1}{3}| < 1$, we have

$$S = \frac{a_1}{1-r} = \frac{1}{1-\frac{1}{3}} = \frac{1}{\frac{2}{3}} = \frac{3}{2}$$

The sum is $\frac{3}{2}$.

Skill Practice Find the sum of the series.

7. $\displaystyle\sum_{i=1}^{\infty} 4\left(\frac{3}{4}\right)^{i-1}$

Example 7 **Using a Geometric Series in a Physics Application**

A child drops a ball from a height of 4 ft. With each bounce, the ball rebounds to 50% of its original height. Determine the total distance traveled by the ball.

Solution:

The heights (in ft) from which the ball drops are given by the sequence

$$4, 2, 1, \frac{1}{2}, \frac{1}{4}, \ldots$$

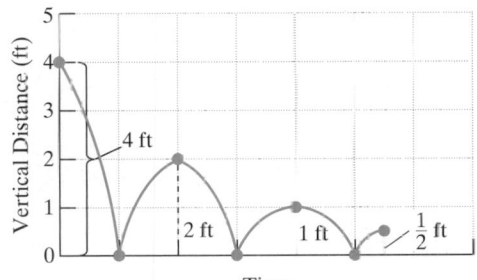

After the ball falls from its initial height of 4 ft, the distance traveled for every bounce thereafter is doubled (the ball travels up and down). Therefore, the total distance traveled is given by the series

$$4 + 2 \cdot 2 + 2 \cdot 1 + 2 \cdot \frac{1}{2} + 2 \cdot \frac{1}{4} + \cdots$$

or equivalently

$$4 + 4 + 2 + 1 + \frac{1}{2} + \cdots$$

The series $4 + 2 + 1 + \frac{1}{2} + \cdots$ is an infinite geometric series with $a_1 = 4$ and $r = \frac{1}{2}$.

Initial height from which the ball was dropped   Infinite geometric series

$$4 \quad + \quad 4 + 2 + 1 + \frac{1}{2} + \cdots$$

$$= 4 + \frac{4}{1-\frac{1}{2}}$$

$$= 4 + \frac{4}{\frac{1}{2}}$$

$$= 4 + 8 = 12$$

The ball traveled a total of 12 ft.

Skill Practice

8. A child drops a ball from a height of 3 ft. With each bounce, the ball rebounds to $\frac{1}{3}$ of its original height. Determine the total distance traveled by the ball.

Answers

7. 16   **8.** 6 ft

Section 10.4   Practice Exercises

Vocabulary and Key Concepts

1. **a.** A _____ sequence is a sequence in which the ratio between each term and its predecessor is constant.

   b. The common _____ between a term and its predecessor in a geometric sequence is often denoted by r.

   c. The nth term of a geometric sequence is given by $a_n =$ _____, where a_n is the nth term in the sequence, a_1 is the first term, and r is the common ratio.

   d. The sum S_n of the first n terms of a geometric series $\sum_{i=1}^{n} a_1 r^{i-1}$ is given by $S_n =$ _____, where a_1 is the first term and r is the common ratio.

   e. Given an infinite geometric series $a_1 + a_2 r + a_1 r^2 + \cdots$ with $|r| < 1$, the sum of all terms in the sequence is given by $S =$ _____. If $|r| \geq$ _____, then the sum does not exist.

Review Exercises

2. Determine the common difference. $7, 1, -5, -11, \ldots$

3. Write an expression for the nth term of the sequence. $-4, 0, 4, 8, 12, \ldots$

4. Find the sum of the series. $\displaystyle\sum_{i=1}^{6} (3i^2 - 4)$

5. Find the fifth term of the arithmetic sequence with $a_1 = -15$ and $d = 10$.

Concept 1: Geometric Sequences

6. Explain how to determine if a sequence is geometric.

For Exercises 7–12, the first term of a geometric sequence is given along with the common ratio. Write the first four terms of the sequence.

7. $a_1 = 1; r = 10$ 8. $a_1 = 2; r = 5$ 9. $a_1 = 64; r = \dfrac{1}{2}$

10. $a_1 = 3; r = \dfrac{2}{3}$ 11. $a_1 = 8; r = -\dfrac{1}{4}$ 12. $a_1 = 9; r = -\dfrac{1}{3}$

For Exercises 13–18, determine the common ratio, r, for the geometric sequence.

13. $5, 10, 20, 40, \ldots$ 14. $-2, -1, -\dfrac{1}{2}, -\dfrac{1}{4}, \ldots$ 15. $8, -2, \dfrac{1}{2}, -\dfrac{1}{8}, \ldots$

16. $4, -12, 36, -108, \ldots$ 17. $3, -6, 12, -24, \ldots$ 18. $1, 4, 16, 64, \ldots$

For Exercises 19–24, write the first five terms of the geometric sequence.

19. $a_1 = -3, r = -2$ 20. $a_1 = -4, r = -1$ 21. $a_1 = 6, r = \dfrac{1}{2}$

22. $a_1 = 8, r = \dfrac{1}{4}$ 23. $a_1 = -1, r = 6$ 24. $a_1 = 2, r = -3$

For Exercises 25–30, find the nth term of each geometric sequence. **(See Example 1.)**

25. $3, 12, 48, 192, \ldots$

26. $2, 6, 18, 54, \ldots$

27. $-5, 15, -45, 135, \ldots$

28. $-6, 12, -24, 48, \ldots$

29. $\dfrac{1}{2}, 2, 8, 32, \ldots$

30. $\dfrac{16}{3}, 4, 3, \dfrac{9}{4}, \ldots$

For Exercises 31–40, find the indicated term of each geometric sequence. **(See Examples 2–3.)**

31. Given $a_n = 2\left(\frac{1}{2}\right)^{n-1}$, find a_8.

32. Given $a_n = -3\left(\frac{1}{2}\right)^{n-1}$, find a_8.

33. Given $a_n = 4\left(-\frac{3}{2}\right)^{n-1}$, find a_6.

34. Given $a_n = 6\left(-\frac{1}{3}\right)^{n-1}$, find a_6.

35. Given $a_n = -3(2)^{n-1}$, find a_5.

36. Given $a_n = 5(3)^{n-1}$, find a_4.

37. Given $a_5 = -\frac{16}{9}$ and $r = -\frac{2}{3}$, find a_1.

38. Given $a_6 = \frac{5}{16}$ and $r = -\frac{1}{2}$, find a_1.

39. Given $a_7 = 8$ and $r = 2$, find a_1.

40. Given $a_5 = 27$ and $r = 3$, find a_1.

41. If the second and third terms of a geometric sequence are 16 and 64, what is the first term?

42. If the second and third terms of a geometric sequence are $\frac{1}{3}$ and $\frac{1}{9}$, what is the first term?

Concept 2: Geometric Series

43. Explain the difference between a geometric sequence and a geometric series.

44. a. Write the series in expanded form. $\displaystyle\sum_{i=1}^{5} 4(2)^{i-1}$

    b. Find the sum. $\displaystyle\sum_{i=1}^{5} 4(2)^{i-1}$

45. a. Write the series in expanded form. $\displaystyle\sum_{i=1}^{4} 3(4)^{i-1}$

    b. Find the sum. $\displaystyle\sum_{i=1}^{4} 3(4)^{i-1}$

46. a. Write the series in expanded form. $\displaystyle\sum_{i=1}^{4} 6\left(\frac{1}{2}\right)^{i-1}$

    b. Find the sum. $\displaystyle\sum_{i=1}^{4} 6\left(\frac{1}{2}\right)^{i-1}$

For Exercises 47–56, find the sum of the geometric series. **(See Examples 4–5.)**

47. $10 + 2 + \dfrac{2}{5} + \dfrac{2}{25} + \dfrac{2}{125}$

48. $1 + 3 + 9 + 27 + 81 + 243$

49. $-2 + 1 + \left(-\dfrac{1}{2}\right) + \dfrac{1}{4} + \left(-\dfrac{1}{8}\right)$

50. $\dfrac{1}{4} + (-1) + 4 + (-16) + 64$

51. $12 + 16 + \dfrac{64}{3} + \dfrac{256}{9} + \dfrac{1024}{27}$

52. $9 + 6 + 4 + \dfrac{8}{3} + \dfrac{16}{9}$

53. $1 + \dfrac{2}{3} + \dfrac{4}{9} + \cdots + \dfrac{32}{243}$

54. $\dfrac{8}{3} + 2 + \dfrac{3}{2} + \cdots + \dfrac{243}{512}$

55. $-4 + 8 + (-16) + \cdots + (-256)$

56. $-\dfrac{7}{3} + 7 + (-21) + \cdots + 5103$

57. A deposit of $1000 is made in an account that earns 5% interest compounded annually. The balance in the account after n years is given by

$$a_n = 1000(1.05)^n \quad \text{for } n \geq 1$$

a. List the first four terms of the sequence. Round to the nearest cent.

b. Find the balance after 10 yr, 20 yr, and 40 yr by computing a_{10}, a_{20}, and a_{40}. Round to the nearest cent.

58. A home purchased for $125,000 increases by 4% of its value each year. The value of the home after n years is given by

$$a_n = 125,000(1.04)^n \quad \text{for } n \geq 1$$

a. List the first four terms of the sequence. Round to the nearest dollar.

b. Find the value of the home after 5 yr, 10 yr, and 20 yr by computing a_5, a_{10}, and a_{20}. Round to the nearest dollar.

For Exercises 59–64, first find the common ratio r. Then determine the sum of the infinite series, if it exists.
(See Example 6.)

59. $1 + \dfrac{1}{6} + \dfrac{1}{36} + \dfrac{1}{216} + \cdots$

60. $-2 + \left(-\dfrac{1}{2}\right) + \left(-\dfrac{1}{8}\right) + \left(-\dfrac{1}{32}\right) + \cdots$

61. $\displaystyle\sum_{i=1}^{\infty} \left(-\dfrac{1}{4}\right)^{i-1}$

62. $\displaystyle\sum_{i=1}^{\infty} \left(-\dfrac{1}{5}\right)^{i-1}$

63. $\dfrac{2}{3} + (-1) + \dfrac{3}{2} + \left(-\dfrac{9}{4}\right) + \cdots$

64. $3 + 5 + \dfrac{25}{3} + \dfrac{125}{9} + \cdots$

65. Suppose $200 million is spent by tourists at a certain resort town. Further suppose that 75% of the revenue is respent in the community and then respent over and over, each time at a rate of 75%. The series

$$200 + 200(0.75) + 200(0.75)^2 + 200(0.75)^3 + \cdots$$

gives the total amount spent (and respent) in the community. Find the sum of the infinite series. **(See Example 7.)**

66. A bungee jumper jumps off a platform and stretches the cord 80 ft before rebounding upward. Each successive bounce stretches the cord 60% of its previous length. The total vertical distance traveled is given by

$$80 + 2(0.60)(80) + 2(0.60)^2(80) + 2(0.60)^3(80) + \cdots$$

After the first term, the series is an infinite geometric series. Compute the total vertical distance traveled.

67. A ball drops from a height of 4 ft. With each bounce, the ball rebounds to $\frac{3}{4}$ of its height. The total vertical distance traveled is given by

$$4 + 2\left(\dfrac{3}{4}\right)(4) + 2\left(\dfrac{3}{4}\right)^2(4) + 2\left(\dfrac{3}{4}\right)^3(4) + \cdots$$

After the first term, the series is an infinite geometric series. Compute the total vertical distance traveled.

68. The repeating decimal number $0.\overline{2}$ can be written as an infinite geometric series by

$$\frac{2}{10} + \frac{2}{100} + \frac{2}{1000} + \cdots$$

   a. What is a_1?   **b.** What is r?   **c.** Find the sum of the series.

69. The repeating decimal number $0.\overline{7}$ can be written as an infinite geometric series by

$$\frac{7}{10} + \frac{7}{100} + \frac{7}{1000} + \cdots$$

   a. What is a_1?   **b.** What is r?   **c.** Find the sum of the series.

Expanding Your Skills

70. The yearly salary for job A is $40,000 initially with an annual raise of $3000 per year. The yearly salary for job B is $38,000 initially with an annual raise of 6% per year.

   a. Find the total earnings for job A over 20 yr. Is this an arithmetic or geometric series?

   b. For job B, what is the amount of the raise after 1 yr?

   c. Find the total earnings for job B over 20 yr. Round to the nearest dollar. Is this an arithmetic or geometric series?

   d. What is the difference in total salary earned over 20 yr between job A and job B?

71. **a.** Brook has a job that pays $48,000 the first year. She receives a 4% raise each year. Find the sum of her yearly salaries over a 20-yr period. Round to the nearest dollar.

   b. Chamille has a job that pays $48,000 the first year. She receives a 4.5% raise each year. Find the sum of her yearly salaries over a 20-yr period. Round to the nearest dollar.

   c. Chamille's raise each year was 0.5% higher than Brook's raise. How much more total income did Chamille receive than Brook over 20 yr?

Problem Recognition Exercises

Identifying Arithmetic and Geometric Sequences

For Exercises 1–18, determine if the sequence is arithmetic, geometric, or neither. If the sequence is arithmetic, find d. If the sequence is geometric, find r.

1. $5, -\dfrac{5}{2}, \dfrac{5}{4}, -\dfrac{5}{8}, \ldots$

2. $\dfrac{1}{2}, 1, \dfrac{3}{2}, 2, \dfrac{5}{2}, \ldots$

3. $-2, -4, -8, -16, -32, \ldots$

4. $-2, -4, -7, -11, -16, \ldots$

5. $-\dfrac{1}{3}, \dfrac{1}{3}, 1, \dfrac{5}{3}, \ldots$

6. $1, -\dfrac{3}{2}, \dfrac{9}{4}, -\dfrac{27}{8}, \ldots$

7. $2, 6, 11, 17, 24, \ldots$

8. $2, -2, 2, -2, 2, \ldots$

9. $-2, -4, -6, -8, -10, \ldots$

10. $1, 3, 1, 3, 1, \ldots$

11. $0, 1, 0, 1, 0, 1, \ldots$

12. $2, 6, 10, 14, 18, 22, \ldots$

13. $\dfrac{1}{4}\pi, \dfrac{1}{2}\pi, \dfrac{3}{4}\pi, \pi, \dfrac{5}{4}\pi, \cdots$

14. $-1, 1, -1, 1, -1, \ldots$

15. $1, 4, 9, 16, 25, \ldots$

16. $\dfrac{1}{3}e, \dfrac{2}{3}e, e, \dfrac{4}{3}e, \dfrac{5}{3}e, \cdots$

17. $\dfrac{\pi}{2}, \dfrac{\pi}{3}, \dfrac{\pi}{4}, \dfrac{\pi}{5}, \dfrac{\pi}{6}, \cdots$

18. $2, 6, 18, 54, 162, \ldots$

Group Activity

Investigating Mean and Standard Deviation

Materials: A computer with Internet access and a calculator

Estimated time: 15–20 minutes

Group Size: 2

To study the characteristics of a population, statisticians often use a sample or part of the population. A sample of data values is smaller and less cumbersome than data from the entire population. Furthermore, if the sample is taken randomly, the sample is generally a good representation of the population as a whole.

Once a collection of data is obtained, statisticians may compute the mean and standard deviation of the sample. The mean is the average of the data values. The standard deviation is a measure of variability and is based on finding the deviation between the given data values and the mean.

The formulas to compute sample mean (denoted by $\bar{x}$) and sample standard deviation (denoted by s) are as follows. In each case, the sample size is n, and the individual data values are denoted by $x_1, x_2, x_3, \ldots, x_n$. Notice that sometimes the summations are given without the index of summation. This is common practice in statistics. The index of summation is understood to be from 1 to n.

$$\bar{x} = \frac{\sum_{i=1}^{n} x_i}{n}$$

- Take the sum of the data values $(x_1 + x_2 + x_3 + \cdots + x_n)$ and divide by n.

$$s = \sqrt{\frac{n \cdot \Sigma(x^2) - (\Sigma x)^2}{n(n-1)}}$$

- For the first term in the numerator, square each value of x first, then take the sum and multiply by the sample size, n.

- For the second term in the numerator, add the values of x first and then square the result.

1. The following data represent a sample of eight breakfast cereals and the corresponding number of calories for a 1-cup serving. Find the mean number of calories. Round to one decimal place.

160   120   140   240   180   210   380   320

Brand	x	x^2
1	160	
2	120	
3	140	
4	240	
5	180	
6	210	
7	380	
8	320	
Sum:		

2. Use the data from Question 1 to compute the standard deviation. Use the given chart to help you organize the calculations. Round to one decimal place.

3. Select two cities in different parts of the country that have roughly the same latitude (same distance north from the equator) such as Phoenix, Arizona, and Charleston, South Carolina. Visit a website that gives the average monthly temperatures for each city for a 1-yr period. Record the data here.

	Jan.	Feb.	Mar.	Apr.	May	June	July	Aug.	Sept.	Oct.	Nov.	Dec.
City 1 temp.												
City 2 temp.												

4. Compute the mean and standard deviation for the average monthly temperature for each city.

5. Based on the mean and standard deviation, what conclusions can you make about the differences between the monthly temperatures of each city?

Chapter 10 Summary

Section 10.1 Binomial Expansions

Key Concepts

For a whole number, n, the expression $n!$ (read as "n factorial") is defined as the product of integers from 1 to n, for $n \geq 1$.

$$n! = n(n-1)(n-2)\ldots(2)(1)$$

$0! = 1$ by definition

For a positive integer, n, the expression $(a + b)^n$ can be expanded using the **binomial theorem**:

$$(a + b)^n = \frac{n!}{n! \cdot 0!}a^n + \frac{n!}{(n-1)! \cdot 1!}a^{(n-1)}b$$

$$+ \frac{n!}{(n-2)! \cdot 2!}a^{(n-2)}b^2 + \cdots + \frac{n!}{0! \cdot n!}b^n$$

The coefficients of the expansion may also be found by using **Pascal's triangle**.

$$
\begin{array}{ccccccccc}
 & & & & 1 & & & & \\
 & & & 1 & & 1 & & & \\
 & & 1 & & 2 & & 1 & & \\
 & 1 & & 3 & & 3 & & 1 & \\
1 & & 4 & & 6 & & 4 & & 1
\end{array}
$$

Let n and k be positive integers such that $k \leq n$. Then the kth term in the expansion of $(a + b)^n$ is

$$\frac{n!}{[n - (k - 1)]! \cdot (k - 1)!} \cdot a^{n-(k-1)} \cdot b^{k-1}$$

Examples

Example 1

$6! = (6)(5)(4)(3)(2)(1) = 720$

Example 2

$$\frac{7!}{2! \cdot 5!} = \frac{7 \cdot \overset{3}{6} \cdot 5 \cdot 4 \cdot 3 \cdot 2 \cdot 1}{(2 \cdot 1)(5 \cdot 4 \cdot 3 \cdot 2 \cdot 1)} = 21$$

Example 3

$$(2x - 4y)^4 = [2x + (-4y)]^4$$

$$= \frac{4!}{4! \cdot 0!}(2x)^4 + \frac{4!}{3! \cdot 1!}(2x)^3(-4y) + \frac{4!}{2! \cdot 2!}(2x)^2(-4y)^2 +$$

$$\frac{4!}{1! \cdot 3!}(2x)(-4y)^3 + \frac{4!}{0! \cdot 4!}(-4y)^4$$

$$= 1(2x)^4 + 4(2x)^3(-4y) + 6(2x)^2(-4y)^2$$

$$+ 4(2x)(-4y)^3 + 1(-4y)^4$$

$$= 16x^4 - 128x^3y + 384x^2y^2 - 512xy^3 + 256y^4$$

Example 4

The third term of $(3x + y^2)^7$ is

$$\frac{7!}{[7 - (3 - 1)]! \cdot (3 - 1)!} \cdot (3x)^{7-(3-1)} \cdot (y^2)^{3-1}$$

$$= \frac{7!}{5! \cdot 2!}(3x)^5(y^2)^2$$

$$= 5103x^5y^4$$

Section 10.2 Sequences and Series

Key Concepts

An **infinite sequence** is a function whose domain is the set of positive integers.

A **finite sequence** is a function whose domain is the set of the first n positive integers.

The expression a_n defines the nth term (or general term) of a sequence.

A **series** is the sum of a sequence.

Examples

Example 1

$a_n = 8 - 3n$

$\quad = 5, 2, -1, -4, -7, -10, \ldots$

Example 2

$a_n = (-1)^n \dfrac{n}{n+1} \qquad$ for $1 \le n \le 5$

$\quad = -\dfrac{1}{2}, \dfrac{2}{3}, -\dfrac{3}{4}, \dfrac{4}{5}, -\dfrac{5}{6}$

Example 3

$\displaystyle\sum_{i=1}^{4} 5i = 5 + 10 + 15 + 20 = 50$

Section 10.3 Arithmetic Sequences and Series

Key Concepts

An **arithmetic sequence** is a sequence in which the **common difference**, d, between two consecutive terms is a constant.

The nth term of an arithmetic sequence is given by

$a_n = a_1 + (n-1)d$

This formula has four variables, a_n, a_1, n, and d. Sometimes it is necessary to solve for one or more of the variables before the nth term can be identified.

Examples

Example 1

Write the nth term of the sequence:
$10, 8, 6, 4, \ldots$

$a_1 = 10 \quad$ (the first term)

$\ d = -2 \quad$ (the common difference)

$a_n = 10 + (n-1) \cdot (-2)$

$a_n = -2n + 12$

Example 2

Find the 12th term of the arithmetic sequence in which $a_1 = -6$ and $a_{20} = 51$.

$a_n = a_1 + (n-1)d \qquad$ Find d first.

$51 = -6 + (20-1)d \qquad$ Substitute known values for a_1 and a_{20}.

$57 = 19d$

$\ 3 = d$

$a_n = -6 + (n-1) \cdot 3 \longrightarrow a_{12} = -6 + (12-1) \cdot 3$

$\qquad\qquad\qquad\qquad\qquad\qquad a_{12} = 27$

The sum of an arithmetic sequence is an **arithmetic series**. The sum of the first n terms of an arithmetic series is given by

$$S_n = \frac{n}{2}(a_1 + a_n)$$

Example 3

Find the sum. $\displaystyle\sum_{i=1}^{50}(3i - 1)$

$n = 50,\ a_1 = 2,\ a_{50} = 149$

$$S_n = \frac{n}{2}(a_1 + a_n) = \frac{50}{2}(2 + 149)$$

$$= 3775$$

Section 10.4 Geometric Sequences and Series

Key Concepts

A **geometric sequence** is a sequence in which the ratio between each term and its predecessor is constant. This constant is called the **common ratio**, r.

The nth term of a geometric sequence is given by

$$a_n = a_1 r^{n-1}$$

The sum of a geometric sequence is called a **geometric series**. The sum of the first n terms of a geometric series is given by

$$S_n = \frac{a_1(1 - r^n)}{1 - r}$$

The sum of an infinite geometric series is given by

$$S_n = \frac{a_1}{1 - r} \quad \text{provided } |r| < 1$$

Examples

Example 1

Find the common ratio: $\dfrac{8}{5}, \dfrac{6}{5}, \dfrac{9}{10}, \dfrac{27}{40}, \cdots$

$$r = \frac{6}{5} \div \frac{8}{5} = \frac{6}{5} \cdot \frac{5}{8} = \frac{30}{40} = \frac{3}{4}$$

Example 2

Find the fourth term.

$$a_n = 3\left(\frac{2}{3}\right)^{a-1} \longrightarrow a_4 = 3\left(\frac{2}{3}\right)^{4-1} = 3\left(\frac{2}{3}\right)^3 = \frac{8}{9}$$

Example 3

Find the sum. $48, 24, 12, 6, 3, \frac{3}{2}$

$a_1 = 48,\ r = \frac{1}{2},\ n = 6$

$$S_n = \frac{a_1(1 - r^n)}{1 - r} = \frac{48\left[1 - \left(\frac{1}{2}\right)^6\right]}{1 - \frac{1}{2}} = \frac{189}{2}$$

Example 4

Find the sum of the series: $\ 4 - 2 + 1 - \frac{1}{2} + \cdots$

$a_1 = 4$ and the common ratio is $-2 \div 4 = -\frac{1}{2}$.

Since $\left|-\frac{1}{2}\right| < 1$,

$$S_n = \frac{a_1}{1 - r} = \frac{4}{1 - \left(-\frac{1}{2}\right)} = \frac{4}{\frac{3}{2}} = \frac{8}{3}$$

Chapter 10   Review Exercises

Section 10.1

For Exercises 1–4, simplify the expression.

1. $8!$ **2.** $5!$ **3.** $\dfrac{12!}{10! \cdot 2!}$ **4.** $\dfrac{9!}{6! \cdot 3!}$

For Exercises 5–6, expand the binomial.

5. $(x^2 + 4)^5$

6. $(c - 3d)^4$

7. Find the first three terms of the binomial expansion. $(a - 2b)^{11}$

8. Find the fourth term of the binomial expansion. $(x^2 + 3y)^8$

9. Find the eighth term of the binomial expansion. $(5x - y^3)^{10}$

10. Find the middle term of the binomial expansion. $(a + 2b)^6$

Section 10.2

For Exercises 11–14, write the terms of the sequence.

11. $a_n = -3n + 4; \ 1 \le n \le 5$

12. $a_n = -2n^3; \ 1 \le n \le 3$

13. $a_n = (-1)^{n+1} \cdot \dfrac{n}{n+2}; \ 1 \le n \le 4$

14. $a_n = \left(-\dfrac{2}{3}\right)^n; \ 1 \le n \le 4$

For Exercises 15–16, write a formula for the nth term of the sequence.

15. $\dfrac{1}{2}, \dfrac{2}{3}, \dfrac{3}{4}, \dfrac{4}{5}, \cdots$

16. $-\dfrac{3}{2}, \dfrac{3}{4}, -\dfrac{3}{8}, \dfrac{3}{16}, \cdots$

17. What is the index of summation for the series?

$$\sum_{k=1}^{7} (2n + 3)$$

18. How many terms are in the series? $\displaystyle\sum_{j=1}^{5} j^3$

For Exercises 19–20, find the sum of the series.

19. $\displaystyle\sum_{i=1}^{7} 5(-1)^{i-1}$ **20.** $\displaystyle\sum_{n=1}^{5} (-2)^n$

21. Write the sum in summation notation.

$$\dfrac{4}{1} + \dfrac{5}{2} + \dfrac{6}{3} + \dfrac{7}{4} + \dfrac{8}{5} + \dfrac{9}{6} + \dfrac{10}{7}$$

22. Study the figures and the corresponding expression for the area of each region. Write the nth term of a sequence representing the area of a region with base n.

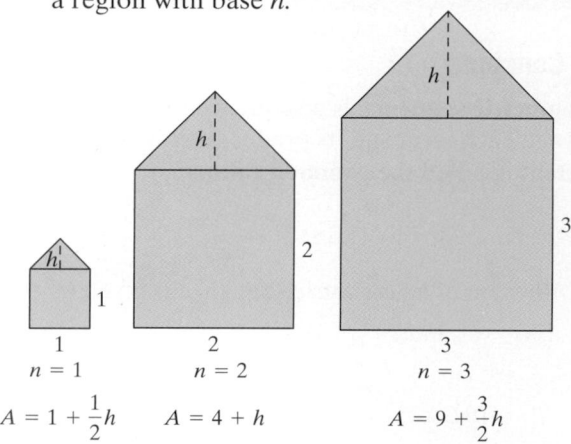

$A = 1 + \dfrac{1}{2}h \qquad A = 4 + h \qquad A = 9 + \dfrac{3}{2}h$

Section 10.3

For Exercises 23–24, write the first five terms of the arithmetic sequence.

23. $a_1 = -12, \ d = 1.5$ **24.** $a_1 = \dfrac{2}{3}, \ d = \dfrac{1}{3}$

For Exercises 25–26, write an expression for the nth term of the sequence.

25. $1, 11, 21, 31, \ldots$ **26.** $6, -8, -22, -36, \ldots$

For Exercises 27–28, find the indicated term of the arithmetic sequence.

27. $a_n = \dfrac{1}{2} + (n-1) \cdot \dfrac{1}{4};$ 17th term

28. $a_1 = -11, d = -6;$ 25th term

For Exercises 29–30, find the number of terms.

29. $3, 8, 13, 18, \ldots, 118$

30. $4, 1, -2, -5, \ldots, -50$

31. Find the common difference for an arithmetic sequence, given $a_1 = -40$ and $a_{19} = 140$.

32. Write an expression for the nth term of the arithmetic sequence, $14, 9, 4, -1, -6, \ldots$.

For Exercises 33–36, find the sum of the arithmetic series.

33. $\displaystyle\sum_{k=1}^{40} (5 - k)$ **34.** $\displaystyle\sum_{i=1}^{20} (0.25i + 2)$

35. $-6 + (-4) + (-2) + \cdots + 34$

36. $4 + 5 + 6 + \cdots + 109$

Section 10.4

For Exercises 37–38, find the common ratio.

37. $5, 15, 45, 135, \ldots$

38. $2, 2.4, 2.88, 3.456$

For Exercises 39–40, write the first four terms of the geometric series.

39. $a_1 = -1$ and $r = -\dfrac{1}{4}$

40. $a_n = 10(2)^{n-1}$

For Exercises 41–42, write an expression for the nth term of the geometric series.

41. $-4, -8, -16, -32, \ldots$

42. $6, -2, \dfrac{2}{3}, -\dfrac{2}{9}, \ldots$

43. Given $a_n = 4\left(\dfrac{2}{3}\right)^{n-1}$, find a_6.

44. Given $a_n = 10(3)^{n-1}$, find a_4.

45. Given $a_7 = \dfrac{1}{16}$ and $r = \dfrac{1}{2}$, find a_1.

46. Given $a_5 = 81$ and $r = 3$, find a_1.

47. Find the sum of the series. $\displaystyle\sum_{n=1}^{8} 5(2)^{n-1}$

48. Find the sum of the series.
$\dfrac{1}{3} + (-1) + 3 + (-9) + \cdots + (-81)$

49. Find the sum of the series.
$6 + 4 + \dfrac{8}{3} + \dfrac{16}{9} + \cdots$

50. Find the sum of the series.
$10 + 2 + \dfrac{2}{5} + \dfrac{2}{25} + \cdots$

51. An investment of \$10,000 is originally invested in a mutual fund. After 10 yr, the yearly return averaged approximately 7%. The balance on the account is given by $b_n = 10,000(1.07)^n$.

   a. Find the first three terms of the sequence.

   b. Interpret the value of b_3.

   c. Find the amount in the account after 10 yr.

Chapter 10 Test

For Exercises 1–2, simplify.

1. $0!$ **2.** $\dfrac{10!}{6! \cdot 4!}$

3. Write the expansion of $(a + b)^4$. Use Pascal's triangle to determine the coefficients of the expansion.

4. Expand. $(3y - 2x^2)^4$

5. Find the sixth term. $(a - c^3)^8$

6. Write the terms of the sequence.
$a_n = -\dfrac{3}{n + 2}; 1 \le n \le 4$

7. Find the sum. $\displaystyle\sum_{i=1}^{5} i^2 + 2$

8. a. An 8-in. tomato seedling is planted on Sunday. Each day, it grows an average of 1.5 in. Write five terms representing the height of the plant each day, beginning with Sunday.

   b. Write the nth term of the sequence.

9. Write the series in summation notation.
$x^3 + x^6 + x^9 + x^{12}$

10. Find the common difference. $\quad 3, \dfrac{13}{4}, \dfrac{7}{2}, \dfrac{15}{4}, \dots$

11. Find the common ratio. $\quad 9, 3, 1, \dfrac{1}{3}, \dots$

12. Given an arithmetic sequence with $a_1 = -15$ and $d = -3$,

    a. Write the first four terms of the sequence.

    b. Write an expression for the nth term of the sequence.

    c. Find the 40th term of the sequence.

13. Given a geometric sequence with $a_1 = 4$ and $r = -2$,

    a. Write the first four terms of the sequence.

    b. Write an expression for the nth term of the sequence.

    c. Find the 10th term of the sequence.

14. Write an expression for the nth term of the sequence. $\quad \dfrac{3}{5}, \dfrac{3}{10}, \dfrac{3}{20}, \dfrac{3}{40}, \dots$

15. Write an expression for the nth term of the sequence. $\quad 20, 18, 16, 14, \dots$

16. Find the number of terms in the sequence. $3, 10, 17, 24, \dots, 213$

17. Find the number of terms in the sequence. $1, \dfrac{2}{3}, \dfrac{4}{9}, \dots, \dfrac{64}{729}$

18. Find the sum of the arithmetic series.
$$\sum_{i=1}^{100} (8 + 2i)$$

19. Find the sum of the geometric series.
$$\sum_{n=1}^{10} 8 \left(\dfrac{1}{2} \right)^{n-1}$$

20. Find the sum of the infinite geometric series.
$4 + 2 + 1 + \frac{1}{2} + \cdots$

21. Given a geometric series with $a_6 = 9$ and $r = 3$, find a_1.

22. Find the 18th term of the arithmetic sequence in which $a_1 = -66$ and $a_{40} = -378$.

23. Suppose a student smokes one pack of cigarettes per day at a cost of $3.00 per pack.

    a. If the student quits smoking and puts the $3.00 saved per day in a jar, how much money will the student have saved at the end of 1 yr?

    b. If the student takes the savings from part (a) and invests the money in a bond fund paying 6% annual interest, how much will this money be worth a year later?

    c. Suppose that the student reinvests the principal and interest each year so that the money is compounded annually. The student will have:

        end of 1 yr $1095
        end of 2 yr ($1095)(1.06) = $1160.70
        end of 3 yr ($1095)(1.06)^2 = $1230.34
        $\cdots$

    The sequence $a_n = 1095(1.06)^{n-1}$ represents the value of the investment at the end of n years. Find the value of the investment after 30 yr. (Round to two decimal places.)

    d. Now suppose that the student saves $1095 per year each year for 30 yr. Further suppose that the student invests each year's savings at 6% interest compounded annually. The following series represents the total savings plus interest over 30 yr. Find the sum.

$$\sum_{n=1}^{30} 1095(1.06)^{n-1}$$

Chapter 10   Cumulative Review Exercises

For Exercises 1–8, simplify the expressions.

1. $(2x - 3)(x - 4) - (x - 5)^2$

2. $\dfrac{5a^2b^{-3}}{25a^{-4}b} \cdot \left(\dfrac{10c^0}{b^3}\right)^2$

3. $64^{-2/3}$

4. $\dfrac{5}{\sqrt{10}}$

5. $\dfrac{2c^2 - 3c - 20}{2c^2 - 32}$

6. $\dfrac{x^{-1} - y^{-1}}{y^{-2} - x^{-2}}$

7. $\log_3 81$

8. $\sqrt[3]{54x^5y^{12}z^2}$

For Exercises 9–12, perform the indicated operations.

9. $\dfrac{a^3 + 64}{16 - a^2} \div \dfrac{a^3 - 4a^2 + 16a}{a^2 - 3a - 4}$

10. $\dfrac{1}{x^2 - 7x + 10} + \dfrac{1}{x^2 + 8x - 20}$

11. $\sqrt{50x^5} + 4x\sqrt{2x^3} - x^2\sqrt{8x}$

12. $(\sqrt{x} + 3\sqrt{y})(\sqrt{x} - 3\sqrt{y})$

13. Divide using long division or synthetic division.
$$\dfrac{2x^4 - x^3 + 5x - 7}{x - 2}$$

For Exercises 14–17, factor completely.

14. $6a^2 - 17a - 14$

15. $75c^3 - 12c$

16. $27x^3 + 64y^6$

17. $w^3 + 9w^2 - 4w - 36$

For Exercises 18–25, solve the equation.

18. $\dfrac{1}{2}t - \dfrac{3}{5} = \dfrac{1}{10}t + 1$

19. $\dfrac{9}{x - 4} + \dfrac{3}{x + 2} = \dfrac{6}{x^2 - 2x - 8}$

20. $\sqrt[3]{4x + 3} - 9 = -7$

21. $(5y - 2)^2 + 7 = 4$

22. $c^4 - 17c^2 + 16 = 0$

23. $|x + 5| = |3x - 2|$

24. $5^{2x - 1} = 125$

25. $\log x + \log(x + 3) = 1$

26. Write the expression as a single logarithm.
$$\ln x - 5 \ln y + \dfrac{1}{2} \ln z$$

27. Use a calculator to approximate the value of the expression. Round to four decimal places.
$$\log_7 895$$

For Exercises 28–32, solve the inequality. Write the answer in interval notation if possible.

28. $2x - 4 < 6$ and $-3x \le 6$

29. $3x(x + 5)(x - 1) \ge 0$

30. $x^2 - 10x \le -25$

31. $\dfrac{3}{x + 4} \le 1$

32. $|2x - 4| + 5 > 7$

For Exercises 33–35, graph the equation.

33. $3x - 2y = 6$

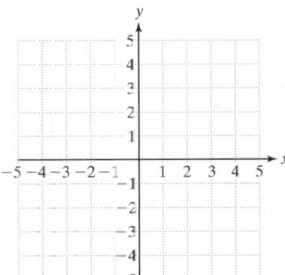

34. $(x - 2)^2 + y^2 = 9$

35. $y = 2^x$

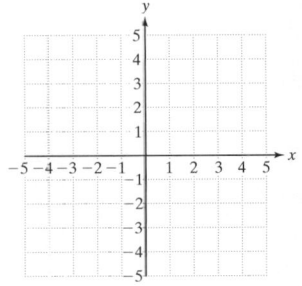

36. Graph the solution set. $\dfrac{x^2}{9} + \dfrac{y^2}{25} \le 1$

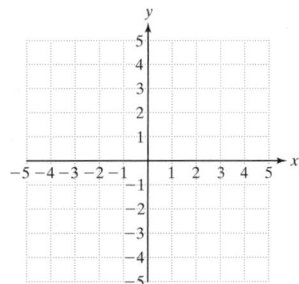

37. Given $f(x) = x^2 + 6x + 5$,

  a. Determine the x-intercepts.

  b. Determine the y-intercept.

  c. Determine the vertex of the parabola.

38. Determine the slope and y-intercept of the line defined by $4x + y = 6$

39. Write an equation of the line passing through the point $(4, -2)$ and perpendicular to the line defined by $x - 5y = 2$.

40. Refer to the graph of $y = f(x)$.

  a. Find $f(-1)$.

  b. Find $f(4)$.

  c. For what value(s) of x is $f(x) = 3$?

  d. Write the domain in interval notation.

  e. Write the range in interval notation.

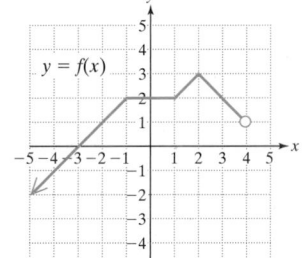

41. Given $f(x) = \sqrt{5 - x}$, write the domain of f.

For Exercises 42–43, solve the system.

42. $x = 1 - 2y$
     $7y = 3x - 16$

43. $x \qquad + 2z = 7$
     $2x + 4y \qquad = 10$
     $\qquad -3y + 5z = 9$

44. Given the points $(9, -4)$ and $(3, 0)$:

  a. Find the distance between the points.

  b. Find the midpoint of the line segment between the two points.

45. Given $A = \frac{1}{2}(b_1 + b_2)h$, solve for b_1.

46. The time, $t(n)$ (in minutes), required for a rat to run through a maze depends on the number of trials, n, that the rat has practiced. Find $t(9)$ and interpret its meaning in the context of this problem.

$$t(n) = \dfrac{3n + 15}{n + 1} \text{ for } n \ge 1$$

47. The speed of a car varies inversely as the time to travel a fixed distance. A car traveling the speed limit of 60 mph travels between two points in 10 sec. How fast is a car moving if it takes only 8 sec to cover the same distance?

48. Two angles are supplementary. One angle measures $9°$ more than twice the other angle. Find the measures of the angles.

49. Chemique invests $3000 less in an account earning 5% simple interest than she does in an account earning 6.5% simple interest. At the end of 1 yr, she earns a total of $770 in interest. Find the amount invested in each account.

50. Against a headwind, a plane can travel 4950 mi in 11 hr. On the return trip flying with the same wind, the plane can fly the same distance in 9 hr. Find the wind speed and the speed of the plane in still air.

Additional Topics Appendix

Determinants and Cramer's Rule

1. Introduction to Determinants

Associated with every square matrix is a real number called the **determinant** of the matrix. A determinant of a square matrix **A**, denoted det **A**, is written by enclosing the elements of the matrix within two vertical bars. For example,

Concepts

1. Introduction to Determinants
2. Determinant of a 3 × 3 Matrix
3. Cramer's Rule

$$\text{If} \quad \mathbf{A} = \begin{bmatrix} 2 & -1 \\ 6 & 0 \end{bmatrix} \quad \text{then} \quad \det \mathbf{A} = \begin{vmatrix} 2 & -1 \\ 6 & 0 \end{vmatrix}$$

$$\text{If} \quad \mathbf{B} = \begin{bmatrix} 0 & -5 & 1 \\ 4 & 0 & \frac{1}{2} \\ -2 & 10 & 1 \end{bmatrix} \quad \text{then} \quad \det \mathbf{B} = \begin{vmatrix} 0 & -5 & 1 \\ 4 & 0 & \frac{1}{2} \\ -2 & 10 & 1 \end{vmatrix}$$

Determinants have many applications in mathematics, including solving systems of linear equations, finding the area of a triangle, determining whether three points are collinear, and finding an equation of a line between two points.

The determinant of a 2 × 2 matrix is defined as follows:

Definition of a Determinant of a 2 × 2 Matrix

The determinant of the matrix $\begin{bmatrix} a & b \\ c & d \end{bmatrix}$ is the real number $ad - bc$. It is written as

$$\begin{vmatrix} a & b \\ c & d \end{vmatrix} = ad - bc$$

Example 1   Evaluating a 2 × 2 Determinant

Evaluate the determinants.

a. $\begin{vmatrix} 6 & -2 \\ 5 & \frac{1}{3} \end{vmatrix}$ **b.** $\begin{vmatrix} 2 & -11 \\ 0 & 0 \end{vmatrix}$

Solution:

a. $\begin{vmatrix} 6 & -2 \\ 5 & \frac{1}{3} \end{vmatrix}$ For this determinant, $a = 6, b = -2, c = 5,$ and $d = \frac{1}{3}$.

$$ad - bc = (6)\left(\frac{1}{3}\right) - (-2)(5)$$

$$= 2 + 10$$

$$= 12$$

TIP: Example 1(b) illustrates that the value of a determinant having a row of all zeros is 0. The same is true for a determinant having a column of all zeros.

b. $\begin{vmatrix} 2 & -11 \\ 0 & 0 \end{vmatrix}$ For this determinant, $a = 2, b = -11, c = 0, d = 0$.

$$ad - bc = (2)(0) - (-11)(0)$$
$$= 0 - 0$$
$$= 0$$

Skill Practice Evaluate the determinants.

1. $\begin{vmatrix} 2 & 8 \\ -1 & 5 \end{vmatrix}$ **2.** $\begin{vmatrix} -6 & 0 \\ 4 & 0 \end{vmatrix}$

2. Determinant of a 3 × 3 Matrix

To find the determinant of a 3×3 matrix, we first need to define the **minor** of an element of the matrix. For any element of a 3×3 matrix, the minor of that element is the determinant of the 2×2 matrix obtained by deleting the row and column in which the element resides. For example, consider the matrix

$$\begin{bmatrix} 5 & -1 & 6 \\ 0 & -7 & 1 \\ 4 & 2 & 6 \end{bmatrix}$$

The minor of the element 5 is found by deleting the first row and first column and then evaluating the determinant of the remaining 2×2 matrix:

$$\begin{bmatrix} 5 & -1 & 6 \\ 0 & -7 & 1 \\ 4 & 2 & 6 \end{bmatrix}$$ Now evaluate the determinant: $\begin{vmatrix} -7 & 1 \\ 2 & 6 \end{vmatrix} = (-7)(6) - (1)(2)$
$$= -44$$

For this matrix, the minor for the element 5 is -44.

To find the minor of the element -7, delete the second row and second column, and then evaluate the determinant of the remaining 2×2 matrix.

$$\begin{bmatrix} 5 & -1 & 6 \\ 0 & -7 & 1 \\ 4 & 2 & 6 \end{bmatrix}$$ Now evaluate the determinant: $\begin{vmatrix} 5 & 6 \\ 4 & 6 \end{vmatrix} = (5)(6) - (6)(4) = 6$

For this matrix, the minor for the element -7 is 6.

Example 2   Determining the Minor for Elements in a 3 × 3 Matrix

Find the minor for each element in the first column of the matrix.

$$\begin{bmatrix} 3 & 4 & -1 \\ 2 & -4 & 5 \\ 0 & 1 & -6 \end{bmatrix}$$

Solution:

For 3: $\begin{bmatrix} 3 & 4 & -1 \\ 2 & -4 & 5 \\ 0 & 1 & -6 \end{bmatrix}$ The minor is: $\begin{vmatrix} -4 & 5 \\ 1 & -6 \end{vmatrix} = (-4)(-6) - (5)(1) = 19$

Answers

1. 18   **2.** 0

For 2: $\begin{bmatrix} 3 & 4 & -1 \\ 2 & -4 & 5 \\ 0 & 1 & -6 \end{bmatrix}$ The minor is: $\begin{vmatrix} 4 & -1 \\ 1 & -6 \end{vmatrix} = (4)(-6) - (-1)(1) = -23$

For 0: $\begin{bmatrix} 3 & 4 & -1 \\ 2 & -4 & 5 \\ 0 & 1 & -6 \end{bmatrix}$ The minor is: $\begin{vmatrix} 4 & -1 \\ -4 & 5 \end{vmatrix} = (4)(5) - (-1)(-4) = 16$

Skill Practice Find the minor for the element 3.

3. $\begin{bmatrix} -1 & 8 & -6 \\ \frac{1}{2} & 3 & 2 \\ 5 & 7 & 4 \end{bmatrix}$

The determinant of a 3 × 3 matrix is defined as follows.

Definition of a Determinant of a 3 × 3 Matrix

$$\begin{vmatrix} a_1 & b_1 & c_1 \\ a_2 & b_2 & c_2 \\ a_3 & b_3 & c_3 \end{vmatrix} = a_1 \cdot \begin{vmatrix} b_2 & c_2 \\ b_3 & c_3 \end{vmatrix} - a_2 \cdot \begin{vmatrix} b_1 & c_1 \\ b_3 & c_3 \end{vmatrix} + a_3 \cdot \begin{vmatrix} b_1 & c_1 \\ b_2 & c_2 \end{vmatrix}$$

From this definition, we see that the determinant of a 3 × 3 matrix can be written as

$$a_1 \cdot (\text{minor of } a_1) - a_2 \cdot (\text{minor of } a_2) + a_3 \cdot (\text{minor of } a_3)$$

Evaluating determinants in this way is called *expanding minors*.

Example 3   Evaluating a 3 × 3 Determinant

Evaluate the determinant. $\begin{vmatrix} 2 & 4 & 2 \\ 1 & -3 & 0 \\ -5 & 5 & -1 \end{vmatrix}$

Solution:

$$\begin{vmatrix} 2 & 4 & 2 \\ 1 & -3 & 0 \\ -5 & 5 & -1 \end{vmatrix} = 2 \cdot \begin{vmatrix} -3 & 0 \\ 5 & -1 \end{vmatrix} - (1) \cdot \begin{vmatrix} 4 & 2 \\ 5 & -1 \end{vmatrix} + (-5) \cdot \begin{vmatrix} 4 & 2 \\ -3 & 0 \end{vmatrix}$$

$$= 2[(-3)(-1) - (0)(5)] - 1[(4)(-1) - (2)(5)] - 5[(4)(0) - (2)(-3)]$$

$$= 2(3) - 1(-14) - 5(6)$$

$$= 6 + 14 - 30$$

$$= -10$$

Skill Practice Evaluate the determinant.

4. $\begin{vmatrix} -2 & 4 & 9 \\ 5 & -1 & 2 \\ 1 & 1 & 6 \end{vmatrix}$

Answers

3. $\begin{vmatrix} -1 & -6 \\ 5 & 4 \end{vmatrix} = 26$ 4. -42

TIP: There is another method to determine the signs for each term of the expansion. For the a_{ij} element, multiply the term by $(-1)^{i+j}$.

Although we defined the determinant of a matrix by expanding the minors of the elements in the first column, *any row or column can be used*. However, we must choose the correct sign to apply to each term in the expansion. The following array of signs is helpful.

$$
\begin{array}{ccc}
+ & - & + \\
- & + & - \\
+ & - & +
\end{array}
$$

The signs alternate for each row and column, beginning with $+$ in the first row, first column.

Example 4   Evaluating a 3 × 3 Determinant

Evaluate the determinant by expanding minors about the elements in the second row.

$$
\begin{vmatrix}
2 & 4 & 2 \\
1 & -3 & 0 \\
-5 & 5 & -1
\end{vmatrix}
$$

Solution:

Signs obtained from the array of signs

$$
\begin{vmatrix}
2 & 4 & 2 \\
1 & -3 & 0 \\
-5 & 5 & -1
\end{vmatrix}
= -(1) \cdot \begin{vmatrix} 4 & 2 \\ 5 & -1 \end{vmatrix} + (-3) \cdot \begin{vmatrix} 2 & 2 \\ -5 & -1 \end{vmatrix} - (0) \cdot \begin{vmatrix} 2 & 4 \\ -5 & 5 \end{vmatrix}
$$

$$
= -1[(4)(-1) - (2)(5)] - 3[(2)(-1) - (2)(-5)] - 0
$$

$$
= -1(-14) - 3(8)
$$

$$
= 14 - 24
$$

$$
= -10
$$

Skill Practice Evaluate the determinant.

5. $\begin{vmatrix} 4 & -1 & 2 \\ 3 & 6 & -8 \\ 0 & \frac{1}{2} & 5 \end{vmatrix}$

Notice that the value of the determinant obtained in Examples 3 and 4 is the same.

Calculator Connections

Topic: Evaluating a Determinant

The determinant of a matrix can be evaluated on a graphing calculator. First use the matrix editor to enter the elements of the matrix. Then use a *det* function to evaluate the determinant. The determinant from Examples 3 and 4 is evaluated below.

```
MATRIX[A] 3 ×3
[ 2    4    2  ]
[ 1   -3    0  ]
[ -5   5   -1  ]
```

```
det([A])
           -10
```

Answer

5. 154

In Example 4, the third term in the expansion of minors was zero because the element 0 when multiplied by its minor is zero. To simplify the arithmetic in evaluating a determinant of a 3×3 matrix, expand about the row or column that has the most 0 elements.

3. Cramer's Rule

In Sections 3.2, 3.3, and 3.7, we learned three methods to solve a system of linear equations: the substitution method, the addition method, and the Gauss-Jordan method. In this section, we will learn another method called **Cramer's rule** to solve a system of linear equations.

Cramer's Rule for a 2 × 2 System of Linear Equations

The solution to the system
$$a_1 x + b_1 y = c_1$$
$$a_2 x + b_2 y = c_2$$

is given by
$$x = \frac{\mathbf{D}_x}{\mathbf{D}} \quad \text{and} \quad y = \frac{\mathbf{D}_y}{\mathbf{D}}$$

where $\mathbf{D} = \begin{vmatrix} a_1 & b_1 \\ a_2 & b_2 \end{vmatrix}$ (and $\mathbf{D} \neq 0$) $\quad \mathbf{D}_x = \begin{vmatrix} c_1 & b_1 \\ c_2 & b_2 \end{vmatrix} \quad \mathbf{D}_y = \begin{vmatrix} a_1 & c_1 \\ a_2 & c_2 \end{vmatrix}$

Example 5 **Using Cramer's Rule to Solve a 2 × 2 System of Linear Equations**

Solve the system by using Cramer's rule.
$$3x - 5y = 11$$
$$-x + 3y = -5$$

Solution:

For this system:
$$a_1 = 3 \qquad b_1 = -5 \qquad c_1 = 11$$
$$a_2 = -1 \qquad b_2 = 3 \qquad c_2 = -5$$

$$\mathbf{D} = \begin{vmatrix} 3 & -5 \\ -1 & 3 \end{vmatrix} = (3)(3) - (-5)(-1) = 9 - 5 = 4$$

$$\mathbf{D}_x = \begin{vmatrix} 11 & -5 \\ -5 & 3 \end{vmatrix} = (11)(3) - (-5)(-5) = 33 - 25 = 8$$

$$\mathbf{D}_y = \begin{vmatrix} 3 & 11 \\ -1 & -5 \end{vmatrix} = (3)(-5) - (11)(-1) = -15 + 11 = -4$$

Therefore, $\quad x = \dfrac{\mathbf{D}_x}{\mathbf{D}} = \dfrac{8}{4} = 2 \qquad y = \dfrac{\mathbf{D}_y}{\mathbf{D}} = \dfrac{-4}{4} = -1$

Check the ordered pair $(2, -1)$ in both original equations.

Check: $3x - 5y = 11 \longrightarrow 3(2) - 5(-1) \overset{?}{=} 11$ ✔

$\qquad -x + 3y = -5 \longrightarrow -(2) + 3(-1) \overset{?}{=} -5$ ✔

The solution set is $\{(2, -1)\}$.

Skill Practice Solve using Cramer's rule.

6. $2x + y = 5$
$\quad -x - 3y = 5$

Answer

6. $\{(4, -3)\}$

TIP: Here are some memory tips to help you remember Cramer's rule.

Coefficients of
x-terms   y-terms

1. The determinant **D** is the determinant of the coefficients of x and y.

$$\mathbf{D} = \begin{vmatrix} a_1 & b_1 \\ a_2 & b_2 \end{vmatrix}$$

x-coefficients
replaced by c_1 and c_2

2. The determinant $\mathbf{D}_x$ has the column of x-term coefficients replaced by c_1 and c_2.

$$\mathbf{D}_x = \begin{vmatrix} c_1 & b_1 \\ c_2 & b_2 \end{vmatrix}$$

y-coefficients
replaced by c_1 and c_2

3. The determinant $\mathbf{D}_y$ has the column of y-term coefficients replaced by c_1 and c_2.

$$\mathbf{D}_y = \begin{vmatrix} a_1 & c_1 \\ a_2 & c_2 \end{vmatrix}$$

It is important to note that the linear equations must be written in standard form to apply Cramer's rule.

Example 6   Using Cramer's Rule to Solve a 2 × 2 System of Linear Equations

Solve the system by using Cramer's rule.

$$-16y = -40x - 7$$
$$40y = 24x + 27$$

Solution:

$$\begin{array}{lll} -16y = -40x - 7 & \longrightarrow & 40x - 16y = -7 \\ 40y = 24x + 27 & \longrightarrow & -24x + 40y = 27 \end{array} \quad \begin{array}{l} \text{Rewrite each equation} \\ \text{in standard form.} \end{array}$$

For this system:

$$\begin{array}{lll} a_1 = 40 & b_1 = -16 & c_1 = -7 \\ a_2 = -24 & b_2 = 40 & c_2 = 27 \end{array}$$

$$\mathbf{D} = \begin{vmatrix} 40 & -16 \\ -24 & 40 \end{vmatrix} = (40)(40) - (-16)(-24) = 1216$$

$$\mathbf{D}_x = \begin{vmatrix} -7 & -16 \\ 27 & 40 \end{vmatrix} = (-7)(40) - (-16)(27) = 152$$

$$\mathbf{D}_y = \begin{vmatrix} 40 & -7 \\ -24 & 27 \end{vmatrix} = (40)(27) - (-7)(-24) = 912$$

Therefore, $\quad x = \dfrac{\mathbf{D}_x}{\mathbf{D}} = \dfrac{152}{1216} = \dfrac{1}{8} \qquad y = \dfrac{\mathbf{D}_y}{\mathbf{D}} = \dfrac{912}{1216} = \dfrac{3}{4}$

The ordered pair $\left(\dfrac{1}{8}, \dfrac{3}{4} \right)$ checks in both original equations.

The solution set is $\left\{ \left(\dfrac{1}{8}, \dfrac{3}{4} \right) \right\}$.

Skill Practice Solve by using Cramer's rule.

7. $9x = 12y - 8$
    $30y = -18x - 7$

Answer

7. $\left\{ \left(-\dfrac{2}{3}, \dfrac{1}{6} \right) \right\}$

Cramer's rule can be used to solve a 3×3 system of linear equations by using a similar pattern of determinants.

Cramer's Rule for a 3 × 3 System of Linear Equations

The solution to the system

$$a_1 x + b_1 y + c_1 z = d_1$$
$$a_2 x + b_2 y + c_2 z = d_2$$
$$a_3 x + b_3 y + c_3 z = d_3$$

is given by $\qquad x = \dfrac{\mathbf{D}_x}{\mathbf{D}} \qquad y = \dfrac{\mathbf{D}_y}{\mathbf{D}} \qquad$ and $\qquad z = \dfrac{\mathbf{D}_z}{\mathbf{D}}$

where $\mathbf{D} = \begin{vmatrix} a_1 & b_1 & c_1 \\ a_2 & b_2 & c_2 \\ a_3 & b_3 & c_3 \end{vmatrix}$ (and $\mathbf{D} \neq 0$) $\qquad \mathbf{D}_x = \begin{vmatrix} d_1 & b_1 & c_1 \\ d_2 & b_2 & c_2 \\ d_3 & b_3 & c_3 \end{vmatrix}$

$$\mathbf{D}_y = \begin{vmatrix} a_1 & d_1 & c_1 \\ a_2 & d_2 & c_2 \\ a_3 & d_3 & c_3 \end{vmatrix} \qquad\qquad\qquad \mathbf{D}_z = \begin{vmatrix} a_1 & b_1 & d_1 \\ a_2 & b_2 & d_2 \\ a_3 & b_3 & d_3 \end{vmatrix}$$

Example 7 **Using Cramer's Rule to Solve a 3 × 3 System of Linear Equations**

Solve the system by using Cramer's rule.

$$2x - 3y + 5z = 11$$
$$-5x + 7y - 2z = -6$$
$$9x - 2y + 3z = 4$$

Solution:

$$\mathbf{D} = \begin{vmatrix} 2 & -3 & 5 \\ -5 & 7 & -2 \\ 9 & -2 & 3 \end{vmatrix} = 2 \cdot \begin{vmatrix} 7 & -2 \\ -2 & 3 \end{vmatrix} - (-5) \cdot \begin{vmatrix} -3 & 5 \\ -2 & 3 \end{vmatrix} + 9 \cdot \begin{vmatrix} -3 & 5 \\ 7 & -2 \end{vmatrix}$$

$$= 2(17) + 5(1) + 9(-29)$$

$$= -222$$

> **TIP:** In Example 7, we expanded the determinants about the first column.

$$\mathbf{D}_x = \begin{vmatrix} 11 & -3 & 5 \\ -6 & 7 & -2 \\ 4 & -2 & 3 \end{vmatrix} = 11 \cdot \begin{vmatrix} 7 & -2 \\ -2 & 3 \end{vmatrix} - (-6) \cdot \begin{vmatrix} -3 & 5 \\ -2 & 3 \end{vmatrix} + 4 \cdot \begin{vmatrix} -3 & 5 \\ 7 & -2 \end{vmatrix}$$

$$= 11(17) + 6(1) + 4(-29)$$

$$= 77$$

$$\mathbf{D}_y = \begin{vmatrix} 2 & 11 & 5 \\ -5 & -6 & -2 \\ 9 & 4 & 3 \end{vmatrix} = 2 \cdot \begin{vmatrix} -6 & -2 \\ 4 & 3 \end{vmatrix} - (-5) \cdot \begin{vmatrix} 11 & 5 \\ 4 & 3 \end{vmatrix} + 9 \cdot \begin{vmatrix} 11 & 5 \\ -6 & -2 \end{vmatrix}$$

$$= 2(-10) + 5(13) + 9(8)$$

$$= 117$$

$$\mathbf{D}_z = \begin{vmatrix} 2 & -3 & 11 \\ -5 & 7 & -6 \\ 9 & -2 & 4 \end{vmatrix} = 2 \cdot \begin{vmatrix} 7 & -6 \\ -2 & 4 \end{vmatrix} - (-5) \cdot \begin{vmatrix} -3 & 11 \\ -2 & 4 \end{vmatrix} + 9 \cdot \begin{vmatrix} -3 & 11 \\ 7 & -6 \end{vmatrix}$$

$$= 2(16) + 5(10) + 9(-59)$$

$$= -449$$

$$x = \frac{\mathbf{D}_x}{\mathbf{D}} = \frac{77}{-222} = -\frac{77}{222}$$

$$y = \frac{\mathbf{D}_y}{\mathbf{D}} = \frac{117}{-222} = -\frac{39}{74}$$

$$z = \frac{\mathbf{D}_z}{\mathbf{D}} = \frac{-449}{-222} = \frac{449}{222}$$

The solution $\left(-\dfrac{77}{222}, -\dfrac{39}{74}, \dfrac{449}{222}\right)$ checks in each of the original equations.

The solution set is $\left\{\left(-\dfrac{77}{222}, -\dfrac{39}{74}, \dfrac{449}{222}\right)\right\}$.

Skill Practice Solve by using Cramer's rule.

8. $x + 3y - 3z = -14$
   $x - 4y + z = 2$
   $x + y + 2z = 6$

Cramer's rule may seem cumbersome for solving a 3 × 3 system of linear equations. However, it provides convenient formulas that can be programmed into a computer or calculator to solve for x, y, and z. Cramer's rule can also be extended to solve a 4 × 4 system of linear equations, a 5 × 5 system of linear equations, and in general an $n \times n$ system of linear equations.

It is important to remember that Cramer's rule does not apply if $\mathbf{D} = 0$. In such a case, either the equations are dependent or the system is inconsistent, and another method may be needed to analyze the system.

Example 8   Analyzing a Dependent System of Equations

Solve the system. Use Cramer's rule if possible. $2x - 3y = 6$
$-6x + 9y = -18$

Solution:

$$\mathbf{D} = \begin{vmatrix} 2 & -3 \\ -6 & 9 \end{vmatrix} = (2)(9) - (-3)(-6) = 18 - 18 = 0$$

TIP: When Cramer's rule does not apply, that is, when $\mathbf{D} = 0$, you may also use the substitution method or the Gauss-Jordan method to get a solution.

Because $\mathbf{D} = 0$, Cramer's rule does not apply. Using the addition method to solve the system, we have

Multiply by 3.
$2x - 3y = 6 \quad \longrightarrow \quad 6x - 9y = 18$

$-6x + 9y = -18 \quad \longrightarrow \quad \underline{-6x + 9y = -18}$

$0 = 0$ The equations are dependent.

Answer

8. $\{(-2, 0, 4)\}$

The solution set is $\{(x, y) \mid 2x - 3y = 6\}$.

Skill Practice Solve. Use Cramer's rule if possible.

9. $x - 6y = 2$
$2x - 12y = -2$

TIP: In a 2 × 2 system of equations, if **D** = 0, then the equations are dependent or the system is inconsistent.

- If **D** = 0 and both $\mathbf{D}_x = 0$ and $\mathbf{D}_y = 0$, then the equations are dependent and the system has infinitely many solutions.
- If **D** = 0 and either $\mathbf{D}_x \neq 0$ or $\mathbf{D}_y \neq 0$, then the system is inconsistent and has no solution.

Answer

9. { }; inconsistent system

Section A.1 Practice Exercises

Vocabulary and Key Concepts

1. a. Given the matrix $\mathbf{A} = \begin{bmatrix} a & b \\ c & d \end{bmatrix}$, the _____ of **A** is denoted det **A** and is written as $\begin{vmatrix} a & b \\ c & d \end{vmatrix}$. The value of det **A** is the real number equal to _____.

   b. Given a 3 × 3 matrix, the _____ of an element in the matrix is the determinant of the 2 × 2 matrix formed by deleting the row and column in which the element resides.

   c. Complete the expression on the right to represent the value of the 3 × 3 determinant shown here.

$$\begin{vmatrix} a_1 & b_1 & c_1 \\ a_2 & b_2 & c_2 \\ a_3 & b_3 & c_3 \end{vmatrix} = a_1 \begin{vmatrix} b_2 & c_2 \\ b_3 & c_3 \end{vmatrix} - \square \begin{vmatrix} b_1 & c_1 \\ b_3 & c_3 \end{vmatrix} + a_3 \begin{vmatrix} \square & \square \\ \square & \square \end{vmatrix}$$

Concept 1: Introduction to Determinants

For Exercises 2–7, evaluate the determinant of the 2 × 2 matrix. **(See Example 1.)**

2. $\begin{vmatrix} -3 & 1 \\ 5 & 2 \end{vmatrix}$

3. $\begin{vmatrix} 5 & 6 \\ 4 & 8 \end{vmatrix}$

4. $\begin{vmatrix} -2 & 2 \\ -3 & -5 \end{vmatrix}$

5. $\begin{vmatrix} 5 & -1 \\ 1 & 0 \end{vmatrix}$

6. $\begin{vmatrix} \frac{1}{2} & 3 \\ -2 & 4 \end{vmatrix}$

7. $\begin{vmatrix} -3 & \frac{1}{4} \\ 8 & -2 \end{vmatrix}$

Concept 2: Determinant of a 3 × 3 Matrix

For Exercises 8–11, evaluate the minor corresponding to the given element from matrix **A**. **(See Example 2.)**

$$\mathbf{A} = \begin{bmatrix} 4 & -1 & 8 \\ 2 & 6 & 0 \\ -7 & 5 & 3 \end{bmatrix}$$

8. 4

9. −1

10. 2

11. 3

For Exercises 12–15, evaluate the minor corresponding to the given element from matrix **B**.

$$\mathbf{B} = \begin{bmatrix} -2 & 6 & 0 \\ 4 & -2 & 1 \\ 5 & 9 & -1 \end{bmatrix}$$

12. 6   **13.** 5   **14.** 1   **15.** 0

16. Construct the sign array for a 3×3 matrix.

17. Evaluate the determinant of matrix **B**, using expansion by minors. **(See Exercises 3–4.)**

$$\mathbf{B} = \begin{bmatrix} 0 & 1 & 2 \\ 3 & -1 & 2 \\ 3 & 2 & -2 \end{bmatrix}$$

   a. About the first column

   b. About the second row

18. Evaluate the determinant of matrix **C**, using expansion by minors.

$$\mathbf{C} = \begin{bmatrix} 4 & 1 & 3 \\ 2 & -2 & 1 \\ 3 & 1 & 2 \end{bmatrix}$$

   a. About the first row

   b. About the second column

19. When evaluating the determinant of a 3×3 matrix, explain the advantage of being able to choose any row or column about which to expand minors.

For Exercises 20–25, evaluate the determinant. **(See Examples 3–4.)**

20. $\begin{vmatrix} 8 & 2 & -4 \\ 4 & 0 & 2 \\ 3 & 0 & -1 \end{vmatrix}$

21. $\begin{vmatrix} 5 & 2 & 1 \\ 3 & -6 & 0 \\ -2 & 8 & 0 \end{vmatrix}$

22. $\begin{vmatrix} -2 & 1 & 3 \\ 1 & 4 & 4 \\ 1 & 0 & 2 \end{vmatrix}$

23. $\begin{vmatrix} 3 & 2 & 1 \\ 1 & -1 & 2 \\ 1 & 0 & 4 \end{vmatrix}$

24. $\begin{vmatrix} -5 & 4 & 2 \\ 0 & 0 & 0 \\ 3 & -1 & 5 \end{vmatrix}$

25. $\begin{vmatrix} 0 & 5 & -8 \\ 0 & -4 & 1 \\ 0 & 3 & 6 \end{vmatrix}$

For Exercises 26–31, evaluate the determinant.

26. $\begin{vmatrix} x & 3 \\ y & -2 \end{vmatrix}$

27. $\begin{vmatrix} a & 2 \\ b & 8 \end{vmatrix}$

28. $\begin{vmatrix} a & 5 & -1 \\ b & -3 & 0 \\ c & 3 & 4 \end{vmatrix}$

29. $\begin{vmatrix} x & 0 & 3 \\ y & -2 & 6 \\ z & -1 & 1 \end{vmatrix}$

30. $\begin{vmatrix} p & 0 & q \\ r & 0 & s \\ t & 0 & u \end{vmatrix}$

31. $\begin{vmatrix} f & e & 0 \\ d & c & 0 \\ b & a & 0 \end{vmatrix}$

Concept 3: Cramer's Rule

For Exercises 32–34, evaluate the determinants represented by $\mathbf{D}$, $\mathbf{D}_x$, and $\mathbf{D}_y$.

32. $x - 4y = 2$
$3x + 2y = 1$

33. $4x + 6y = 9$
$-2x + y = 12$

34. $-3x + 8y = -10$
$5x + 5y = -13$

For Exercises 35–40, solve the system by using Cramer's rule. **(See Examples 5–6.)**

35. $2x + y = 3$
$x - 4y = 6$

36. $2x - y = -1$
$3x + y = 6$

37. $4y = x - 8$
$3x = -7y + 5$

38. $7x - 4 = -3y$
$5x = 4y + 9$

39. $4x - 3y = 5$
$2x + 5y = 7$

40. $2x + 3y = 4$
$6x - 12y = -5$

For Exercises 41–46, solve for the indicated variable by using Cramer's rule. **(See Example 7.)**

41. $2x - y + 3z = 9$
$x + 4y + 4z = 5$ for x
$3x + 2y + 2z = 5$

42. $x + 2y + 3z = 8$
$2x - 3y + z = 5$ for y
$3x - 4y + 2z = 9$

43. $3x - 2y + 2z = 5$
$6x + 3y - 4z = -1$ for z
$3x - y + 2z = 4$

44. $4x + 4y - 3z = 3$
$8x + 2y + 3y = 0$ for x
$4x - 4y + 6z = -3$

45. $5x + 6z = 5$
$-2x + y = -6$ for y
$3y - z = 3$

46. $8x + y = 1$
$7y + z = 0$ for y
$x - 3z = -2$

47. When does Cramer's rule not apply in solving a system of equations?

48. How can a system be solved if Cramer's rule does not apply?

For Exercises 49–58, solve the system by using Cramer's rule, if possible. Otherwise, use another method.
(See Example 8.)

49. $4x - 2y = 3$
$-2x + y = 1$

50. $6x - 6y = 5$
$x - y = 8$

51. $4x + y = 0$
$x - 7y = 0$

52. $-3x - 2y = 0$
$-x + 5y = 0$

53. $x + 5y = 3$
$2x + 10y = 6$

54. $-2x - 10y = -4$
$x + 5y = 2$

55. $x = 3$
$-x + 3y = 3$
$y + 2z = 4$

56. $4x + z = 7$
$y = 2$
$x + z = 4$

57. $x + y + 8z = 3$
$2x + y + 11z = 4$
$x + 3z = 0$

58. $-8x + y + z = 6$
$2x - y + z = 3$
$3x - z = 0$

Expanding Your Skills

For Exercises 59–62, solve the equation.

59. $\begin{vmatrix} 6 & x \\ 2 & -4 \end{vmatrix} = 14$

60. $\begin{vmatrix} y & -2 \\ 8 & 7 \end{vmatrix} = 30$

61. $\begin{vmatrix} 3 & 1 & 0 \\ 0 & 4 & -2 \\ 1 & 0 & w \end{vmatrix} = 10$

62. $\begin{vmatrix} -1 & 0 & 2 \\ 4 & t & 0 \\ 0 & -5 & 3 \end{vmatrix} = -4$

For Exercises 63–64, evaluate the determinant by using expansion by minors about the first column.

63. $\begin{vmatrix} 1 & 0 & 3 & 0 \\ 0 & 1 & 2 & 4 \\ -2 & 0 & 0 & 1 \\ 4 & -1 & -2 & 0 \end{vmatrix}$

64. $\begin{vmatrix} 5 & 2 & 0 & 0 \\ 0 & 4 & -1 & 1 \\ -1 & 0 & 3 & 0 \\ 0 & -2 & 1 & 0 \end{vmatrix}$

For Exercises 65–66, refer to the following system of four variables.

$$x + y + z + w = 0$$
$$2x \quad\;\; - z + w = 5$$
$$2x + y \quad\;\; - w = 0$$
$$y + z \quad\quad = -1$$

65. a. Evaluate the determinant **D**.

    b. Evaluate the determinant $\mathbf{D}_x$.

    c. Solve for x by computing $\dfrac{\mathbf{D}_x}{\mathbf{D}}$.

66. a. Evaluate the determinant $\mathbf{D}_y$.

    b. Solve for y by computing $\dfrac{\mathbf{D}_y}{\mathbf{D}}$.

67. Two angles are complementary. The measure of one angle is $\frac{5}{7}$ the measure of the other. Find the measures of the two angles.

68. Two angles are supplementary. The measure of the larger angle is $61°$ more than $\frac{3}{4}$ the measure of the smaller angle. Find the measures of the angles.

69. An electronics store sells iPods, iPads, and iPhones. In a recent month, twice as many iPods as iPhones were sold. The number of iPods sold was five more than the sum of iPads and iPhones. The total number of devices sold was 75. Find the number iPods, iPads, and iPhones sold.

70. A group was surveyed regarding e-readers. The group represented a total of 100 e-reader users. The number of Kindles used was 44 more than the sum of the number Kobos and iPads. The number of Kobos was 50 less than the number of Kindles. Find the number of each type of e-reader.

71. Suppose 1000 people were surveyed in southern California, and 445 said that they worked-out at least three times a week. If $\frac{1}{2}$ of the women and $\frac{3}{8}$ of the men said that they worked-out at least three times a week, how many men and how many women were in the survey?

72. During a 1-hr television program, there were 22 commercials. Some commercials were 15 sec and some were 30 sec long. Find the number of 15-sec commercials and the number of 30-sec commercials if the total playing time for commercials was 9.5 min.

Student Answer Appendix

Chapter R

Chapter Opener Puzzle

$$\frac{D}{4}\,\frac{I}{3}\,\frac{S}{10}\,\frac{T}{1}\,\frac{R}{5}\,\frac{I}{3}\,\frac{B}{2}\,\frac{U}{12}\,\frac{T}{1}\,\frac{I}{3}\,\frac{V}{11}\,\frac{E}{8}\quad\frac{P}{7}\,\frac{R}{5}\,\frac{O}{9}\,\frac{P}{7}\,\frac{E}{8}\,\frac{R}{5}\,\frac{T}{1}\,\frac{Y}{6}$$

Section R.2 Practice Exercises, pp. 13–16

1. a. set **b.** inequalities **c.** a is less than b **d.** c is greater than or equal to d **e.** 5 is not equal to 6 **f.** infinity; negative infinity **g.** $\{x \mid x > 5\}$; interval **h.** excludes; includes **i.** parenthesis

3.

5. $\dfrac{-10}{1}$ **7.** $\dfrac{-3}{5}$ **9.** $\dfrac{0}{1}$

11.

	Real Numbers	Irrational Numbers	Rational Numbers	Integers	Whole Numbers	Natural Numbers
5	✓		✓	✓	✓	✓
$-\sqrt{9}$	✓		✓	✓		
-1.7	✓		✓			
$\dfrac{1}{2}$	✓		✓			
$\sqrt{7}$	✓	✓				
$\dfrac{0}{4}$	✓		✓	✓	✓	
$0.\overline{2}$	✓		✓			

13. $<$ **15.** $>$ **17.** $>$ **19.** $<$ **21.** $(2, \infty)$
23. $(-\infty, 0]$ **25.** $(-5, 0]$ **27.** $[-4.7, \infty)$
29. $(-1, \infty)$ **31.** $(-\infty, -2]$
33. $(-\infty, \frac{9}{2})$ **35.** $(-2.5, 4.5]$
37. $(-\infty, -3)$ **39.** $(\frac{5}{2}, \infty)$
41. $[2, \infty)$ **43.** $(-4, 4)$
45. $[-3, 0]$

47. All real numbers less than -4 **49.** All real numbers greater than -2 and less than or equal to 7 **51.** All real numbers between -180 and 90, inclusive **53.** All real numbers
55. $a \geq 18$ **57.** $c \leq 25$ **59.** $s \geq 261$
61. $r \leq 4.5$ **63.** $18 \leq a \leq 25$ **65.** $p < 130$
67. $130 \leq p \leq 139$ **69.** $2.2 \leq \mathrm{pH} \leq 2.4$ acidic
71. $3.0 \leq \mathrm{pH} \leq 3.5$ acidic

Section R.3 Practice Exercises, pp. 27–30

1. a. opposites **b.** $|a|$; 0 **c.** base; n **d.** radical; square **e.** negative; positive **f.** positive; negative **g.** $-b$; positive **h.** $\dfrac{1}{a}$, 1 **i.** 0 **j.** 0; undefined **3.** $(-\infty, 4)$ **5.** $[-3, -1)$
7. Distance can never be negative. **9.** Negative
11.

Number	Opposite	Reciprocal	Absolute Value
6	-6	$\dfrac{1}{6}$	6
$\dfrac{1}{11}$	$-\dfrac{1}{11}$	11	$\dfrac{1}{11}$
-8	8	$-\dfrac{1}{8}$	8
$-\dfrac{13}{10}$	$\dfrac{13}{10}$	$-\dfrac{10}{13}$	$\dfrac{13}{10}$
0	0	Undefined	0
-3	3	$-0.\overline{3}$	3

13. $<$ **15.** $=$ **17.** $<$ **19.** $<$ **21.** -4 **23.** -19
25. -7 **27.** 14 **29.** -22.1 **31.** -8.1 **33.** $-\dfrac{5}{3}$ or $-1\dfrac{2}{3}$
35. $-\dfrac{67}{45}$ **37.** -32 **39.** $\dfrac{8}{21}$ **41.** $\dfrac{3}{5}$ **43.** $-\dfrac{18}{5}$
45. Undefined **47.** 0 **49.** 3.72 **51.** $\dfrac{5}{11}$ **53.** 64 **55.** -49
57. 49 **59.** $\dfrac{125}{27}$ **61.** 3 **63.** Not a real number **65.** $\dfrac{1}{2}$
67. -7 **69.** 32 **71.** 40 **73.** 25 **75.** 13 **77.** 17
79. -11 **81.** -603 **83.** $\dfrac{109}{150}$ **85.** 5.4375 **87.** $\dfrac{2}{3}$
89. -1 **91.** 21 **93.** Undefined **95.** $\dfrac{9}{10}$ **97.** $-10.1°C$
99. a. $25°C$ **b.** $100°C$ **c.** $0°C$ **d.** $-40°C$ **101.** $4\dfrac{1}{6}$ gal
103. 9 in.^{2} **105.** 8.06 cm^2 **107.** 14.1 ft^3 **109.** 26.8 ft^3
111. 141.4 in.^{3}

Section R.3 Graphing Calculator Exercises, p. 30

113. $12/(6-2)$ **114.** $(24-6)/3$

115. $\dfrac{2}{3}$ or 0.6666667 **116.** 6

Section R.4 Practice Exercises, pp. 36–38

1. a. constant **b.** coefficient **c.** 1; 1 **d.** like **3.** $-\dfrac{1}{2}$
5. $(3, \infty)$ **7.** $\left(-\dfrac{5}{2}, 3\right]$ **9. a.** 3 terms **b.** 6 **c.** 2, -5, 6

SA-1

11. a. 5 terms **b.** -7 **c.** $1, -7, 1, -4, 1$ **13.** a   **15.** f
17. e   **19.** i   **21.** b   **23.** g   **25.** d   **27.** h   **29.** a
31. $2x - 6y + 16$ **33.** $-40s + 90t + 30$ **35.** $7w - 5z$
37. $\frac{1}{2}a - 2b + \frac{8}{5}$ **39.** $7.8x - 12.3$ **41.** $14c - 16 - 30d + 5f$
43. $14y - 2x$ **45.** $6p^2 + p - 6$ **47.** $-p^2 - 3p$
49. $n^3 + m - 6$ **51.** $7ab + 8a$ **53.** $16xy^2 - 5y^2$ **55.** $8x - 23$
57. $-4c - 6$ **59.** $-9w + 10$ **61.** $4z - 16$ **63.** $7s - 26$
65. $-12w + 13$ **67.** 0 **69.** $4c + 2$ **71.** $1.4x + 10.2$
73. $-2a^2 + 3a + 38$ **75.** $2y^2 - 3y - 5$ **77.** $-62.7x + 220$
79. $-4m + 15n + 2$ **81.** $0,$ for example: $3 + 0 = 3$
83. Reciprocal   **85.** No, for example: $6 - 5 \neq 5 - 6$ $1 \neq -1$
87. a. $x(y + z)$ **b.** xy **c.** xz **d.** $xy + xz$
e. $x(y + z) = xy + xz$ the distributive property of multiplication
over addition

Chapter R Review Exercises, pp. 41–42

1. 0 **2.** For example: $-\frac{1}{2}, -\frac{3}{4}, \frac{5}{2}$ **3.** For example: $-2, -1, 0, 1, 2$
4. All real numbers between 7 and 16   **5.** All real numbers
greater than 0 but less than or equal to 2.6   **6.** All real numbers
between -6 and -3, inclusive   **7.** All real numbers greater than 8
8. All real numbers less than or equal to 13   **9.** All real numbers
10. $(-\infty, 2)$

11. $[0, \infty)$

12. $(-1, 5)$

13. True   **14.** $8, -\frac{1}{8}, 8$ **15.** $-\frac{4}{9}, \frac{9}{4}, \frac{4}{9}$ **16.** $16, 2$
17. $625, 5$ **18.** -2 **19.** 3 **20.** -21.6 **21.** -8.151
22. $-\frac{25}{13}$ or $-1\frac{12}{13}$ **23.** $\frac{4}{11}$ **24.** $\frac{119}{100}$ or $1\frac{19}{100}$ **25.** $\frac{7}{6}$ or $1\frac{1}{6}$
26. 3 **27.** -8 **28.** 6 **29.** 48 **30.** 11 **31.** 37
32. 75 **33.** 4 **34.** 256 **35.** 756 in.^{2} **36.** $3x + 15y$
37. $\frac{1}{2}x + 4y - \frac{5}{2}$ **38.** $4x - 10y + z$ **39.** $-13a + b + 5c$
40. $7q - 14$ **41.** $9p + 3$ **42.** $-6y - 5$ **43.** $9x - 1$
44. For example: $3 + x = x + 3$ **45.** For example: $5(2y) = (5 \cdot 2)y$

Chapter R Test, p. 42

1. a. $-5, -4, -3, -2, -1, 0, 1, 2$ **b.** For example: $\frac{3}{2}, \frac{5}{4}, \frac{8}{5}$
2. a. $\frac{1}{2}, -2, \frac{1}{2}$ **b.** $-4, \frac{1}{4}, 4$ **c.** $0,$ no reciprocal exists, 0
3. The interval $[4, \infty)$ includes all real numbers 4 and greater,
whereas the interval $(4, \infty)$ does not include the endpoint, 4.
4. True
5. $\left(-\infty, -\frac{4}{3}\right)$ **6.** $[12, \infty)$

7. $x \leq 5$ **8.** $p \geq 7$ **9.** 6 **10.** -17 **11.** $\frac{1}{4}$ **12.** $-\frac{16}{3}$
13. $z = 1.1$ **14.** $-2b - 6$ **15.** $2x - 1$ **16.** $-2x + 1$
17. False   **18.** True   **19.** True   **20.** True

Chapter 1
Chapter Opener Puzzle

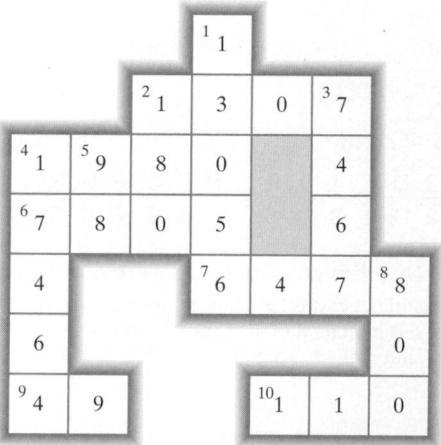

Section 1.1 Practice Exercises, pp. 53–56

1. a. equation **b.** solution **c.** linear **d.** first **e.** solution
set **f.** solution **g.** conditional **h.** contradiction **i.** empty
set; $\{\ \}$ or $\varnothing$ **j.** identity   **3.** $3x - 3y + 14xy$ **5.** $5z - 20$
7. Linear   **9.** Nonlinear   **11.** Linear   **13.** b   **15.** $\{12\}$
17. $\{-2\}$ **19.** $\left\{\frac{21}{20}\right\}$ **21.** $\{-40\}$ **23.** $\{-1.1\}$ **25.** $\{6.7\}$
27. $\{11\}$ **29.** $\{13\}$ **31.** $\{-7\}$ **33.** $\{13\}$ **35.** $\left\{\frac{3}{2}\right\}$
37. $\{0\}$ **39.** $\{3\}$ **41.** $\left\{\frac{7}{4}\right\}$ **43.** $\{0\}$ **45.** $\left\{\frac{4}{3}\right\}$
47. $\{-4\}$ **49.** $\{3\}$ **51.** $\{-6\}$ **53.** $\{1\}$ **55.** $\{2\}$
57. It is an equation that is true for some values of the variable but
false for other values.   **59.** Identity; $\{x \mid x$ is a real number$\}$
61. Conditional equation; $\{0\}$ **63.** Contradiction; $\{\ \}$ **65.** $\{16\}$
67. $\{2\}$ **69.** $\{3\}$ **71.** $\{b \mid b$ is a real number$\}$ **73.** $\{\ \}$
75. $\left\{\frac{8}{5}\right\}$ **77.** $\left\{-\frac{3}{2}\right\}$ **79.** $\{p \mid p$ is a real number$\}$
81. $\left\{\frac{33}{5}\right\}$ **83.** $\{\ \}$ **85.** $\left\{\frac{3}{7}\right\}$ **87.** $\left\{\frac{3}{4}\right\}$ **89.** $\{60\}$
91. $\{6\}$ **93.** $\{\ \}$ **95.** $\left\{-\frac{9}{2}\right\}$ **97.** The family used 1024 kWh.
99. a. $y + 8$ **b.** $\{-8\}$ **c.** To simplify an equation, clear
parentheses and combine *like* terms. To solve an equation, isolate
the variable to find a solution.

Chapter 1 Problem Recognition Exercises, p. 56

1. Expression; $-4x + 4$ **2.** Expression; $-7y + 5$
3. Equation; $\{1\}$ **4.** Equation; $\{0\}$
5. Expression; $-10a - 39$ **6.** Expression; $28x - 10$
7. Equation; $\left\{\frac{1}{2}\right\}$ **8.** Equation; $\left\{\frac{3}{4}\right\}$
9. Equation; $\left\{-\frac{19}{6}\right\}$ **10.** Equation; $\left\{\frac{25}{22}\right\}$
11. Expression; $-\frac{1}{6}v - \frac{1}{10}$ **12.** Expression; $-\frac{17}{8}t + \frac{1}{2}u$
13. Equation; $\{\ \}$ **14.** Equation; $\{\ \}$ **15.** Equation; $\left\{\frac{39}{8}\right\}$
16. Equation; $\left\{\frac{2}{25}\right\}$ **17.** Expression; $0.17c + 4.495$
18. Expression; $-1.006k - 0.78$ **19.** Equation;
$\{p \mid p$ is a real number$\}$ **20.** Equation; $\{u \mid u$ is a real number$\}$

Section 1.2 Practice Exercises, pp. 65–68

1. a. consecutive **b.** even; odd **c.** 1; 2; 2 **d.** $x + 1$
e. $x + 2$ **f.** $x + 2; x + 4$ **g.** Prt; interest **h.** \$1300 **i.** 0.48 L;
$0.12(x + 8)$ **j.** $\dfrac{d}{t}; \dfrac{d}{r}$ **3.** $\left\{\dfrac{13}{7}\right\}$ **5.** $\{6\}$ **7.** $\left\{\dfrac{18}{5}\right\}$
9. $x + 5$ **11.** $2t - 7$ **13.** The numbers are 5 and 13.
15. The number is -3. **17.** The page numbers are 111 and 112.
19. The integers are -75 and -73. **21.** The integers are -22
and -20. **23.** The integers are 31, 33, and 35. **25.** She would
pay \$5100 for 4 yr at 8.5% and \$5812.50 for 5 yr at 7.75%; the 8.5%
option for 4 yr requires less interest. **27.** She must sell \$60,000.
29. The total for merchandise was \$1197.02 and the sales tax
was \$96.36. **31.** The price before markup was \$35.90. **33.** He
invested \$8500 in the 2% account and \$4000 in the 5% account.
35. \$12,000 was borrowed at 6% and \$6000 was borrowed at 11%.
37. She invested \$12,000 in the 4% account and \$8000 in the 3%
account. **39.** 8 oz should be used. **41.** 2 L should be used.
43. 12.5 L of 18% solution should be added to 7.5 L of 10%
solution. **45.** 2 oz should be added. **47.** The plane flies 300 mph
from Atlanta to Fort Lauderdale and 240 mph on the return trip.
49. The speeds are 46 mph and 50 mph. **51.** The integers are
10 and 20. **53.** The original price was \$199. **55.** The speeds
are 20 mph and 40 mph. **57.** She deposited \$4500 at 5% and
\$9000 at 6%. **59.** Mix 2.5 lb of black tea and 1.5 lb of green tea.
61. The median price the previous year was \$215,000.

Section 1.3 Practice Exercises, pp. 73–76

1. a. $2l + 2w$ **b.** 90° **c.** supplementary **d.** 180°
3. $\left\{\dfrac{40}{13}\right\}$ **5.** $\{\ \}$ **7.** The court is 29.5 ft wide and 59 ft long.
9. The sides are 6 m, 8 m, and 10 m. **11. a.** The dimensions
are $12\frac{1}{2}$ yd by 8 yd. **b.** Raoul needs 41 yd. **13.** The width is 9 ft
and the length is 11 ft. **15.** The angles are 30°, 30°, and 120°.
17. The angles are 15° and 75°. **19.** $x = 20$; 139°, 41°
21. $x = 27.5$; 60°, 30° **23.** $x = 18$; 36°, 91°, 53°
25. $x = 23$; 42°, 48° **27. a.** $r = \dfrac{d}{t}$
b. Franchitti's average speed was 161.6 mph.
29. a. $t = \dfrac{I}{Pr}$ **b.** 7 yr **31.** $l = \dfrac{A}{w}$ **33.** $P = \dfrac{I}{rt}$
35. $K_1 = K_2 - W$ **37.** $C = \frac{5}{9}(F - 32)$ or $C = \dfrac{5F - 160}{9}$
39. $v^2 = \dfrac{2K}{m}$ **41.** $a = \dfrac{v - v_0}{t}$
43. $v_2 = \dfrac{w}{p} + v_1$ or $v_2 = \dfrac{w + pv_1}{p}$ **45.** $y = \dfrac{c - ax}{b}$
47. $B = \dfrac{3V}{h}$ **49.** $y = -3x + 6$ **51.** $y = \frac{5}{4}x - 5$
53. $y = -3x - \dfrac{13}{2}$ **55.** $y = x - 2$
57. $y = -\dfrac{27}{4}x + \dfrac{15}{4}$ **59.** $y = \frac{3}{2}x$ **61. a.** $x = z\sigma + \mu$
b. $x = 130$ **63.** a, b, c **65.** a, b **67.** $t = \dfrac{12}{6 - r}$
69. $x = \dfrac{-2}{a - 6}$ or $x = \dfrac{2}{6 - a}$ **71.** $P = \dfrac{A}{1 + rt}$
73. $m = \dfrac{T}{g - f}$ **75.** $x = \dfrac{z - by}{a - c}$ or $x = \dfrac{by - z}{c - a}$

Section 1.4 Practice Exercises, pp. 82–85

1. a. linear inequality **b.** negative **c.** Both statements are
correct.

	Set-Builder Notation	Interval Notation	Graph
3.	$\{x \mid x > 5\}$	$(5, \infty)$	
5.	$\{x \mid -3 < x \le 6\}$	$(-3, 6]$	
7.	$\{x \mid x \ge 4\}$	$[4, \infty)$	

	Set Notation	Interval Notation	Graph
9a.	$\{-3\}$	n/a	
b.	$\{x \mid x > -3\}$	$(-3, \infty)$	
c.	$\{x \mid x < -3\}$	$(-\infty, -3)$	

11. a. $\{y \mid y \le -1\}$ **b.** $(-\infty, -1]$
13. a. $\{x \mid x \ge 10\}$ **b.** $[10, \infty)$
15. a. $\{z \mid z > \frac{13}{6}\}$ **b.** $\left(\dfrac{13}{6}, \infty\right)$
17. a. $\{t \mid t < -12\}$ **b.** $(-\infty, -12)$
19. a. $\{y \mid y < \frac{13}{8}\}$ **b.** $\left(-\infty, \dfrac{13}{8}\right)$
21. a. $\{a \mid a \le -21\}$ **b.** $(-\infty, -21]$
23. a. $\{x \mid x > -3\}$ **b.** $(-3, \infty)$
25. a. $\{x \mid x \ge \frac{9}{10}\}$ **b.** $\left[\dfrac{9}{10}, \infty\right)$
27. a. $\{p \mid p < -3\}$ **b.** $(-\infty, -3)$
29. a. $\{t \mid t < 5\}$ **b.** $(-\infty, 5)$
31. a. $\{y \mid y \ge -\frac{5}{4}\}$ **b.** $\left[-\dfrac{5}{4}, \infty\right)$
33. a. $\{k \mid k \ge 1.5\}$ **b.** $[1.5, \infty)$
35. a. $\{x \mid x \le 12\}$ **b.** $(-\infty, 12]$
37. a. $\{b \mid b \le -0.5\}$ **b.** $(-\infty, -0.5]$
39. a. $\{c \mid c \le -\frac{5}{11}\}$ **b.** $\left(-\infty, -\dfrac{5}{11}\right]$
41. a. $\{y \mid y < -6\}$ **b.** $(-\infty, -6)$
43. a. $\{x \mid x > -3\}$ **b.** $(-3, \infty)$
45. a. $\{a \mid a \le 0\}$ **b.** $(-\infty, 0]$

47. a. Nadia needs to score at least a 70 but less than 120 to get
a B average. **b.** It would be impossible for Nadia to get an A
because she would have to earn 120 on her last quiz and it is
impossible to earn more than 100. **49.** Boys 8 yr old or older will
be on average at least 51 in. tall. **51.** Boys 6 yr old or younger will
be on average no more than 46 in. tall. **53. a.** She needs to sell in
excess of \$375,000. **b.** She needs to sell in excess of \$1,375,000.

c. The base salary is still the same. The increase comes solely from commission.   **55.** More than 73 jackets must be sold.   **57.** $>$ **59.** $<$

Section 1.5 Practice Exercises, pp. 93–97

1. a. union; $A \cup B$ **b.** intersection; $A \cap B$ **c.** intersection **d.** $a < x < b$ **e.** union   **3.** $(-\infty, 1)$ **5.** $[-16, \infty)$

7. a. $\{-3, -1\}$ **b.** $\{-4, -3, -2, -1, 0, 1, 3, 5\}$ **9.** $[-7, -4)$

11. $(-\infty, -4) \cup (2, \infty)$ **13.** $\{\ \}$ **15.** $[-7, \infty)$

17. $[0, 5)$ **19.** $[-7, \infty)$ **21. a.** $[-1, 5)$ **b.** $(-2, \infty)$

23. a. $(-1, 3)$ **b.** $\left(-\frac{5}{2}, \frac{9}{2}\right)$ **25. a.** $(0, 2]$ **b.** $(-4, 5]$

27. $[-2, 3]$

29. $(3, 6)$

31. $(-\infty, 2]$

33. $[8, \infty)$

35. $\{\ \}$ **37.** $-4 \le t$ and $t < \frac{3}{4}$ **39.** The statement $6 < x < 2$ is equivalent to $6 < x$ and $x < 2$. However, no real number is greater than 6 and also less than 2.

41. The statement $-5 > y > -2$ is equivalent to $-5 > y$ and $y > -2$. However, no real number is less than -5 and also greater than -2.

43. $\left[\frac{5}{2}, 7\right)$

45. $(-6, 6]$

47. $(2, 8)$

49. $\left[-\frac{10}{3}, -\frac{7}{3}\right]$

51. $\left[-\frac{1}{2}, \frac{3}{2}\right)$

53. $(-0.4, 0.2)$

55. $(-\infty, -2) \cup [2, \infty)$

57. $(-\infty, 2]$

59. $[-6, \infty)$

61. $(-\infty, \infty)$

63. $(-\infty, -4] \cup (4.5, \infty)$

65. a. $(-10, 8)$ **b.** $(-\infty, \infty)$ **67. a.** $(-\infty, -8] \cup [1.3, \infty)$ **b.** $\{\ \}$

69. $\left(-\frac{11}{2}, \frac{7}{2}\right]$ **71.** $\left(-\frac{13}{2}, \infty\right)$ **73.** $(-\infty, 1) \cup (5, \infty)$

75. a. $4800 \le x \le 10{,}800$ **b.** $x < 4800$ or $x > 10{,}800$

77. a. $44\% < x < 48\%$ **b.** $x \le 44\%$ or $x \ge 48\%$

79. All real numbers between $-\frac{3}{2}$ and 6

81. All real numbers greater than 2 or less than -1

83. a. Amy would need 82% or better on her final exam. **b.** If Amy scores at least 32% and less than 82% on her final exam she will receive a "B" in the class.   **85.** $32° \le F \le 42.08°$

Section 1.6 Practice Exercises, pp. 102–103

1. a. absolute; $\{a, -a\}$ **b.** Subtract 5 from both sides. **c.** $y; -y$ **d.** $\{\ \}; \{-4\}$ **3.** $(-\infty, -2] \cup [2, \infty)$ **5.** $[-6, 10)$

7. $\{7, -7\}$ **9.** $\{6, -6\}$ **11.** $\{\ \}$ **13.** $\{-2, 2\}$

15. $\{\sqrt{2}, -\sqrt{2}\}$ **17.** $\{\ \}$ **19.** $\{0\}$ **21.** $\left\{4, -\frac{4}{3}\right\}$

23. $\left\{\frac{9}{2}, -\frac{1}{2}\right\}$ **25.** $\left\{\frac{10}{7}, -\frac{8}{7}\right\}$ **27.** $\{\ \}$ **29.** $\{-12, 28\}$

31. $\left\{\frac{5}{2}, -\frac{7}{2}\right\}$ **33.** $\left\{\frac{7}{3}\right\}$ **35.** $\{\ \}$ **37.** $\left\{\frac{3}{2}\right\}$

39. $\left\{-4, \frac{16}{5}\right\}$ **41.** $\left\{-\frac{1}{2}, \frac{7}{6}\right\}$ **43.** $\left\{\frac{5}{2}, -\frac{3}{2}\right\}$ **45.** $\left\{-4, \frac{1}{3}\right\}$

47. $\left\{\frac{1}{2}\right\}$ **49.** $\left\{-\frac{1}{16}\right\}$ **51.** $\{x \mid x \text{ is a real number}\}$

53. $\{-1.44, -0.4\}$ **55.** $\{\ \}$ **57.** $\{w \mid w \text{ is a real number}\}$

59. $\{\ \}$ **61.** $|x| = 6$ **63.** $|x| = \frac{4}{3}$

Section 1.7 Practice Exercises, pp. 110–113

1. a. $-a; a$ **b.** $-a; >$ **c.** $\{\ \}; (-\infty, \infty)$ **d.** includes; excludes

3. $\left\{\frac{4}{7}, \frac{6}{7}\right\}$ **5.** $(-3, -1]$

7. $(-\infty, \infty)$ **9. a.** $\{-5, 5\}$

b. $(-\infty, -5) \cup (5, \infty)$

c. $(-5, 5)$

11. a. $\{10, -4\}$

b. $(-\infty, -4) \cup (10, \infty)$

c. $(-4, 10)$

13. a. $\{\ \}$ **b.** $(-\infty, \infty)$ **c.** $\{\ \}$
15. a. $\{\ \}$ **b.** $(-\infty, \infty)$ **c.** $\{\ \}$
17. a. $\{0\}$ **b.** $(-\infty, 0) \cup (0, \infty)$

c. $\{\ \}$ **19. a.** $\{7\}$ **b.** $(-\infty, 7) \cup (7, \infty)$ **c.** $\{\ \}$

21. $(-\infty, -6) \cup (6, \infty)$

23. $[-3, 3]$

25. $(-\infty, \infty)$
27. $(-\infty, -2] \cup [3, \infty)$

29. $\{\ \}$ **31.** $[-10, 14]$

33. $\left(-\infty, -\frac{5}{4}\right] \cup \left[\frac{23}{4}, \infty\right)$

35. $\left(-\frac{21}{2}, \frac{19}{2}\right)$

37. $(-\infty, \infty)$ **39.** $\{\ \}$

41. $\{-5\}$

43. $(-\infty, 6) \cup (6, \infty)$

45. $(-\infty, 2] \cup [4, \infty)$

47. $(4, 8)$

49. $(-10.5, 4.5)$

51. $|x| > 7$ **53.** $|x - 2| \le 13$ **55.** $|x - 32| \le 0.05$

57. $\left|x - 6\frac{3}{4}\right| \le \frac{1}{8}$ **59.** The solution set is $\{w \mid 1.99 \le w \le 2.01\}$ or, equivalently in interval notation, $[1.99, 2.01]$. This means that the actual width of the bolt could be between 1.99 cm and 2.01 cm inclusive.   **61.** b   **63.** a

Chapter 1 Problem Recognition Exercises, pp. 113–114

1. a. $\{9\}$ **b.** $\{9, -3\}$ **c.** $(-3, 9)$ **d.** $(-\infty, -3] \cup [9, \infty)$

2. a. $\left\{-\dfrac{22}{5}\right\}$ **b.** { } **c.** { } **d.** $(-\infty, \infty)$

3. a. $\{-7\}$ **b.** $(-\infty, -7)$ **c.** $[-7, \infty)$
4. a. $\{29\}$ **b.** $[29, \infty)$ **c.** $(-\infty, 29)$

5. a. $\left\{\dfrac{1}{2}, -\dfrac{1}{6}\right\}$ **b.** $\left\{\dfrac{1}{2}\right\}$ **6. a.** $(-7, 6]$ **b.** $(-7, 6]$

7. a. $(-5, \infty)$ **b.** $[1, \infty)$ **8. a.** $(-\infty, -7) \cup (2, \infty)$ **b.** { }
9. a. Linear equation **b.** $\{-6\}$ **10. a.** Linear equation
b. $\{9\}$ **11. a.** Absolute value inequality **b.** $[-6, -2]$
12. a. Absolute value inequality **b.** { }
13. a. Compound inequality **b.** $(-6, 9)$
14. a. Compound inequality **b.** $(-\infty, 2) \cup [7, \infty)$
15. a. Absolute value equation **b.** $\{4, -16\}$
16. a. Absolute value equation **b.** $\{1, -6\}$
17. a. Linear inequality **b.** $(-\infty, 12]$ **18. a.** Linear
inequality   **b.** $[-1, \infty)$ **19. a.** Absolute value inequality
b. $(-\infty, -3] \cup [12, \infty)$ **20. a.** Absolute value inequality
b. $(-5, 25)$ **21. a.** Absolute value equation **b.** { }

22. a. Absolute value equation **b.** $\left\{-\dfrac{1}{5}\right\}$

23. a. Linear equation **b.** $\{-16\}$ **24. a.** Linear equation
b. $\{-11\}$ **25. a.** Compound inequality **b.** $(8, 12]$
26. a. Compound inequality **b.** $[-6, -3)$ **27. a.** Linear
equation **b.** $(-\infty, \infty)$ **28. a.** Linear equation **b.** { }

Chapter 1 Review Exercises, pp. 122–125

1. The empty set; no solution   **2.** All real numbers

3. $\{-5\}$; a conditional equation   **4.** $\left\{\dfrac{1}{8}\right\}$; a conditional

equation   **5.** $\{-18.075\}$; a conditional equation

6. $\{0.22\}$; a conditional equation   **7.** $\left\{\dfrac{31}{6}\right\}$;
a conditional equation   **8.** $\{3\}$; a conditional equation

9. $\left\{-\dfrac{21}{8}\right\}$; a conditional equation   **10.** { }; contradiction

11. { }; contradiction   **12.** $\{m \mid m \text{ is a real number}\}$; identity
13. $x, x+1, x+2$ **14.** $x, x+2$ **15.** Distance equals rate
times time.   **16.** Simple interest equals principal times rate times
time.   **17. a.** $23,856 **b.** $61,344   **18.** There were 6.7 million
men.   **19.** There were 16,600 deaths due to alcohol-related
accidents.   **20.** The integers are $-4, -2,$ and 0.   **21.** The pieces
are 2 ft and $\dfrac{2}{3}$ ft.   **22.** She invested $1500 in 6% account and
$3500 in 9% account.   **23.** Mix 2 L of 10% solution.
24. Lynn drives 45 mph and Linda drives 30 mph.   **25.** The
width is 9 ft, and the length is 11 ft.   **26.** $x = 52; 27°$

27. $x = 30; 29°, 61°$ **28.** $y = \dfrac{3x-4}{2}$ or $y = \dfrac{3}{2}x - 2$

29. $y = 6x + 12$ **30.** $h = \dfrac{S - 2\pi r}{\pi r^2}$ or $h = \dfrac{S}{\pi r^2} - \dfrac{2}{r}$

31. $b = \dfrac{2A}{h}$ **32. a.** $\pi = \dfrac{C}{2r}$ **b.** $\pi \approx 3.14$

33. ⟵————→ $-\dfrac{4}{3}$ **a.** $\{x \mid x < -\frac{4}{3}\}$ **b.** $(-\infty, -\frac{4}{3})$

34. ⟵—[————→ $-\dfrac{3}{2}$ **a.** $\{x \mid x \geq -\frac{3}{2}\}$ **b.** $[-\frac{3}{2}, \infty)$

35. ⟵————→ $-\dfrac{8}{13}$ **a.** $\{x \mid x < -\frac{8}{13}\}$ **b.** $(-\infty, -\frac{8}{13})$

36. ⟵——[———→ $\dfrac{46}{7}$ **a.** $\{x \mid x \geq \frac{46}{7}\}$ **b.** $[\frac{46}{7}, \infty)$

37. ⟵]————→ $-\dfrac{67}{4}$ **a.** $\{x \mid x \leq -\frac{67}{4}\}$ **b.** $(-\infty, -\frac{67}{4}]$

38. ⟵————]→ $\dfrac{29}{2}$ **a.** $\{x \mid x \leq \frac{29}{2}\}$ **b.** $(-\infty, \frac{29}{2}]$

39. Dave must earn at least 95 on his fifth test.
40. $C \cap D$ is the set of $C \cup D$ is the set of
elements common to   elements in either set C,
both sets C and D.   set D, or both sets.

41. $[-10, 1)$ **42.** $(-\infty, \infty)$ **43.** $(-\infty, \infty)$ **44.** $(-1, 1)$

45. $(-\infty, -3] \cup (-1, \infty)$ **46.** { } **47.** $\left(-\dfrac{11}{4}, 4\right]$ **48.** $[-5, 2)$

49. { }   **50.** { }   **51.** $(-\infty, 5] \cup (12, \infty)$

52. $(-\infty, -2) \cup [2, \infty)$ **53.** $\left(-\infty, \dfrac{1}{2}\right)$ **54.** $(-5, \infty)$

55. $\left[0, \dfrac{4}{3}\right]$ **56.** $[-7, -2)$

57. All real numbers between -6 and 12
58. a. $140 \leq x \leq 225$ **b.** $x < 140$ or $x > 225$
59. a. $125 \leq x \leq 200$ **b.** $x < 125$ or $x > 200$
60. a. For example, for a 47-year-old person, the maximum
recommended heart rate is 173.   **b.** Given a maximum
recommended heart rate of 173, the interval is $(104, 130)$.
61. $\{10, -10\}$ **62.** $\{17, -17\}$ **63.** $\{1.3, 7.4\}$
64. $\{-0.44, 2.54\}$ **65.** $\{5, -9\}$ **66.** $\{3, 1\}$

67. { }   **68.** { }   **69.** $\left\{\dfrac{3}{7}\right\}$ **70.** $\left\{-\dfrac{5}{4}\right\}$

71. $\left\{6, \dfrac{4}{5}\right\}$ **72.** $\left\{-\dfrac{1}{2}\right\}$ **73.** $\{d \mid d \text{ is a real number}\}$

74. { }   **75.** Both expressions give the distance between 3 and -2.

76. $|x| > 5$ **77.** $|x| < 4$ **78.** $|x| < 6$ **79.** $|x| > \dfrac{2}{3}$

80. $(-\infty, -14] \cup [2, \infty)$ ⟵]———[→ -14 ... 2

81. $[-11, -5]$ ⟵[———]→ -11 ... -5

82. $\left(-\infty, \dfrac{1}{7}\right) \cup \left(\dfrac{1}{7}, \infty\right)$ ⟵——✕——→ $\dfrac{1}{7}$

83. $\left(-\infty, -\dfrac{1}{5}\right) \cup \left(-\dfrac{1}{5}, \infty\right)$ ⟵—✕———→ $-\dfrac{1}{5}$

84. $\left[-2, -\dfrac{2}{3}\right]$ ⟵[———]→ -2 ... $-\dfrac{2}{3}$

85. $\left[0, \dfrac{6}{5}\right]$ ⟵[———]→ 0 ... $\dfrac{6}{5}$

86. $(2, 22)$ ⟵(———)→ 2 ... 22

87. $(-12, 0)$ ⟵(———)→ -12 ... 0

88. $(-\infty, \infty)$ ⟵————→

89. $(-\infty, \infty)$ ⟵————→

90. $(-\infty, \infty)$ ⟵————→

91. $(-\infty, \infty)$ ⟵————→

92. { }   **93.** { }   **94.** If an absolute value is less than a
negative number, there will be no solution.   **95.** If an absolute
value is greater than a negative number, then all real numbers are
solutions.   **96.** $0.17 \leq p \leq 0.23$ or, equivalently in interval
notation, $[0.17, 0.23]$. This means that the actual percentage of
viewers is estimated to be between 17% and 23% inclusive.
97. $3\frac{1}{8} \leq L \leq 3\frac{5}{8}$ or, equivalently in interval notation, $[3\frac{1}{8}, 3\frac{5}{8}]$. This
means that the actual length of the screw may be between $3\frac{1}{8}$ in. and
$3\frac{5}{8}$ in., inclusive.

Chapter 1 Test, pp. 125–126

1. $\{133\}$ **2.** $\left\{\dfrac{4}{5}\right\}$ **3.** $\{142,500\}$ **4.** $\left\{\dfrac{14}{9}\right\}$

5. $\{10, -22\}$ **6.** $\{-8, 2\}$ **7.** $\{3, 0\}$ **8.** $\{\ \}$ **9.** $\{-1\}$

10. a. Identity   **b.** Contradiction   **c.** Conditional equation

11. The numbers are 18 and 90.   **12. a.** It took 0.4 hr (24 min).

b. The distance is 1.8 mi.   **13.** Shawnna invested \$1000 in the CD.

14. Each side is 27 in.   **15.** The numbers are 31, 33, and 35.

16. 24 gal of 20% solution must be used.   **17.** $y = -2x + 3$

18. $z = \dfrac{x - \mu}{\sigma}$

19. $(34, \infty)$

20. $\left(-\infty, \dfrac{18}{5}\right]$

21. $\left(-\dfrac{1}{3}, 2\right]$

22. $\left[-\dfrac{1}{3}, 2\right]$ **23.** $(-2, 13]$ **24.** $(-\infty, -24] \cup [-15, \infty)$

25. $[-3, 0)$ **26.** $(-\infty, \infty)$ **27.** $\{\ \}$ **28.** $\{\ \}$

29. $\left(-\infty, -\dfrac{1}{3}\right] \cup \left[\dfrac{17}{3}, \infty\right)$ **30.** $(-18.75, 17.25)$ **31.** $(-\infty, \infty)$

32. $[-3, 8]$ **33.** It can carry at most seven more passengers.

34. a. $9 \le x \le 33$ **b.** $x < 9$ or $x > 33$

35. $|x - 15.41| \le 0.01$

Chapter 2

Chapter Opener Puzzle

A graph intersects the x-axis at an

$\underset{1}{X} \text{-} \underset{2}{I} \ \underset{4}{N} \ \underset{3}{T} \ \underset{5}{E} \ \underset{3}{R} \underset{}{C} \ \underset{}{E} \ \underset{4}{P} \ \underset{}{T}$.

Section 2.1 Practice Exercises, pp. 139–145

1. a. x; y-axis   **b.** ordered   **c.** origin; $(0, 0)$
d. quadrants   **e.** negative   **f.** III   **g.** $Ax + By = C$
h. x-intercept   **i.** y-intercept   **j.** vertical   **k.** horizontal
3. For (x, y), if $x > 0$, $y > 0$ the point is in quadrant I. If $x < 0$, $y > 0$ the point is in quadrant II. If $x < 0$, $y < 0$ the point is in quadrant III. If $x > 0$, $y < 0$ the point is in quadrant IV.

5.

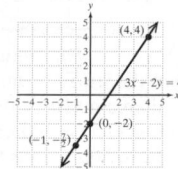

7. 0   **9.** $A(-4, 5)$, II; $B(-2, 0)$, x-axis; $C(1, 1)$, I; $D(4, -2)$, IV; $E(-5, -3)$, III   **11. a.** Yes   **b.** No   **c.** Yes
13. a. No   **b.** Yes   **c.** No

15.

x	y
0	-2
4	4
-1	$-\dfrac{7}{2}$

17.

x	y
0	0
5	-1
-5	1

19.

21.

23.

25.

27.

29.

31. To find an x-intercept, substitute $y = 0$ and solve for x. To find a y-intercept, substitute $x = 0$ and solve for y.

33. a. $(9, 0)$ **b.** $(0, 6)$ **35. a.** $(4, 0)$ **b.** $(0, -2)$
c.   **c.**

37. a. $(0, 0)$ **b.** $(0, 0)$ **39. a.** $(-2, 0)$ **b.** $(0, 4)$
c.   **c.**

41. a. $\left(\dfrac{3}{2}, 0\right)$ **b.** $(0, 2)$ **43. a.** $(0, 0)$ **b.** $(0, 0)$
c.   **c.**

45. a. \$55,000   **b.** \$39,000   **c.** The y-intercept is $(0, 15,000)$. For \$0 in sales, the salary will be \$15,000.
d. Total sales cannot be negative.   **47. a.** \$1200   **b.** 4 yr
c. $(0, 1500)$; The y-intercept represents the initial value.
d. $(5, 0)$; The x-intercept indicates that once the computer is 5 yr old, its value is \$0.

49. Horizontal; no x-intercept; y-intercept $(0, -1)$

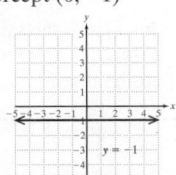

51. Vertical; x-intercept $(2, 0)$; no y-intercept

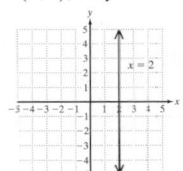

53. Vertical; x-intercept $\left(-\frac{1}{2}, 0\right)$; no y-intercept

55. Horizontal; no x-intercept; y-intercept $(0, -4)$

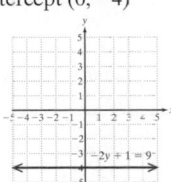

57. A horizontal line parallel to the x-axis will not have an x-intercept. A vertical line parallel to the y-axis will not have a y-intercept.　　**59.** b, c, d　　**61.** x-intercept $(2, 0)$; y-intercept $(0, 3)$　　**63.** x-intercept $(a, 0)$; y-intercept $(0, b)$

Section 2.1 Graphing Calculator Exercises, p. 145

65. $y = \frac{2}{3}x - \frac{7}{3}$

66. $y = -2x - 1$

67. $y = 3$

68. $y = -5$

69.

70.

71.

72.

73. The lines look nearly indistinguishable. However, the linear equations are different so the lines are different.

74. The lines look nearly indistinguishable. However, the linear equations are different so the lines are different.

Section 2.2 Practice Exercises, pp. 152–157

1. a. slope; $\dfrac{y_2 - y_1}{x_2 - x_1}$　　**b.** parallel; same　　**c.** right　　**d.** -1

3. a. 4　**b.** 8　**c.** 6

5. No x-intercept; y-intercept: $(0, 2)$

7. $m = \dfrac{24}{7}$　　**9.** $m = \dfrac{1}{9}$　　**11.** $m = \dfrac{4}{100}$ or $\dfrac{1}{25}$

13. $m = \dfrac{1}{2}$　　**15.** $m = -\dfrac{5}{3}$　　**17.** $m = -2$

19. $m = -\dfrac{3}{2}$　　**21.** m is undefined.　　**23.** $m = 0$

25. $m = \dfrac{1}{2}$　　**27.** $m = -\dfrac{1}{6}$　　**29.** $m = 0$

31. Line rising to the right: positive slope　Line falling to the right: negative slope　Horizontal line: zero slope　Vertical line: undefined slope　　**33.** $m = 0$

35. $m = \dfrac{1}{10}$　　**37.** $m = -1$　　**39. a.** $m = 5$　**b.** $m = -\dfrac{1}{5}$

41. a. $m = -\dfrac{4}{7}$　**b.** $m = \dfrac{7}{4}$　　**43. a.** $m = 0$

b. m is undefined.　　**45.** No, because the product of slopes of perpendicular lines must be -1. The product of two positive numbers is not negative.　　**47.** Undefined　　**49.** $m = 0$

51. Undefined　　**53.** $m_1 = 2$; $m_2 = -\dfrac{1}{2}$; perpendicular

55. $m_1 = -1$; $m_2 = 3$; neither　　**57.** m_1 is undefined; $m_2 = 0$; perpendicular. One line is horizontal and one is vertical.

59. $m_1 = 1$; $m_2 = 1$; parallel　　**61. a.** $m = 20.25$　**b.** The number of cell phone subscriptions increased at a rate of 20.25 million per year during this period.

63. a. $m = 6$　**b.** The weight of boys tends to increase at a rate of 6 lb/yr during this period of growth.

65. a. $m = 2$　**b.** $m = 2$　**c.** $m = 2$

67. a. $m = -2$　**b.** $m = 2$　**c.** $m = 5$

69. For example: $(1, 2)$ **71.** For example: $(2, 0)$

 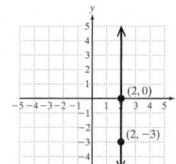

73. For example: $(2, 0)$

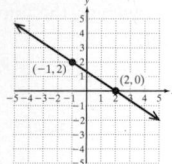

75. $y = 15$ **77. a.** Pitch: $\dfrac{1}{6}$ **b.** Slope: $\dfrac{1}{3}$

Section 2.3 Practice Exercises, pp. 166–170

1. a. $y = mx + b$ **b.** standard   **c.** horizontal   **d.** vertical
e. slope; y-intercept   **f.** $y - y_1 = m(x - x_1)$
3. a. 0   **b.** undefined
5. If the slope of one line is the opposite of the reciprocal of
the other, then the lines are perpendicular.   **7.** Slope: $-\dfrac{2}{3}$;
y-intercept: $(0, -4)$ **9.** Slope: 3; y-intercept: $(0, 2)$
11. Slope: -17; y-intercept: $(0, 0)$ **13.** Slope: 0;
y-intercept: $(0, 9)$ **15.** Slope: $-\dfrac{2}{3}$; y-intercept: $\left(0, \dfrac{3}{4}\right)$
17. Slope: 0.625; y-intercept: $(0, -1.2)$
19. d   **21.** f   **23.** b
25.   **27.**   **29.**

  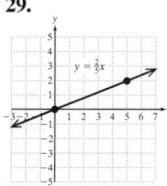

31. $y = -\dfrac{A}{B}x + \dfrac{C}{B}$. The slope is given by $m = -\dfrac{A}{B}$.
The y-intercept is $\left(0, \dfrac{C}{B}\right)$.   **33.** Perpendicular
35. Parallel   **37.** Neither   **39.** $y = 3x + 5$
41. $y = 2x - 11$ **43.** $y = -\dfrac{4}{5}x + 8$ **45.** $y = 3x - 2$
or $3x - y = 2$ **47.** $y = 2x + 3$ or $2x - y = -3$
49. $y = -3x - 11$ or $3x + y = -11$ **51.** $y = -\dfrac{4}{5}x + \dfrac{9}{5}$ or
$4x + 5y = 9$ **53.** $y = -\dfrac{4}{3}x + 4$ or $4x + 3y = 12$
55. $y = x + 6$ or $x - y = -6$ **57.** $y = 2$
59. $y = -\dfrac{3}{4}x + \dfrac{17}{4}$ or $3x + 4y = 17$ **61.** $y = \dfrac{4}{3}x - 2$ or
$4x - 3y = 6$ **63.** $y = \dfrac{3}{4}x - \dfrac{13}{2}$ or $3x - 4y = 26$
65. $y = -\dfrac{1}{5}x - \dfrac{13}{5}$ or $x + 5y = -13$ **67.** $y = \dfrac{3}{2}x - 6$ or
$3x - 2y = 12$ **69.** $y = \dfrac{3}{2}x + 1$ or $3x - 2y = -2$
71. $y = -\dfrac{1}{2}x + \dfrac{7}{2}$ or $x + 2y = 7$ **73.** $y = -3x$ or
$3x + y = 0$ **75.** $y = -3$ **77.** $x = 2$ **79.** $y = 5$
81. $x = 5$ **83.** $x = -2$ is not in slope-intercept form. No
y-intercept, undefined slope.   **85.** $y = 3$ is in slope-intercept
form, $y = 0x + 3$. Slope is 0 and y-intercept is $(0, 3)$.
87. $y = -2x + 3$ **89.** $y = 2$

Section 2.3 Graphing Calculator Exercises, p. 170

91. The lines have the same slope but different y-intercepts; they are parallel lines.

92. The lines have the same slope but different y-intercepts; they are parallel lines.

93. The lines have different slopes but the same y-intercept.

94. The lines have different slopes but the same y-intercept.

95. The lines are perpendicular.

96. The lines are perpendicular.

97.

98.

Chapter 2 Problem Recognition Exercises, p. 171

1. b, f   **2.** a, c, d, h   **3.** a   **4.** b, g   **5.** c, e
6. a   **7.** c, h   **8.** b   **9.** e   **10.** g   **11.** c, h
12. a   **13.** g   **14.** e   **15.** h   **16.** f   **17.** e
18. g   **19.** d, h   **20.** b, f

Section 2.4 Practice Exercises, pp. 176–182

1. model
3. a. $m = -\dfrac{1}{3}$ **b.** $y = -\dfrac{1}{3}x - 1$ or $x + 3y = -3$
c.

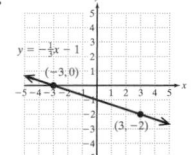

5. a. $m = 0$ **b.** $y = 3$
c.

7. a. $y = 120x + 65$

b.

Cost of Rental Car by Number of Days

$y = 120x + 65$

c. $(0, 65)$; The base cost to rent the car is $65. **d.** A 2-day rental costs $305. A 5-day rental costs $665, and a 7-day rental costs $905. **e.** Yes; $799 is less expensive than $905. **f.** $577.70 **g.** No, the car cannot be driven for a negative number of days.

9. a. $y = 52x + 2742$

b.

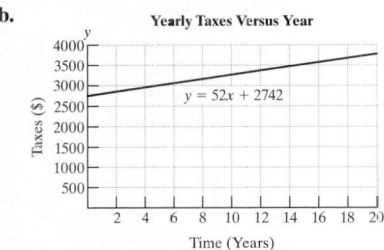

Yearly Taxes Versus Year

$y = 52x + 2742$

c. $m = 52$. Taxes increase at a rate of $52 per year.
d. $(0, 2742)$. Initial year $(x = 0)$ taxes were $2742.
e. 10 years: $3262; 15 years: $3522

11. a. 0.8 mi, 2.4 mi, and 3.2 mi **b.** 21 sec

13. a. $73.30 **b.** $54.50; The approximate value differs from the actual value by $1.30. **c.** $m = 9.4$; The amount spent per person on video games increased by an average rate of $9.40 per year.
d. $(0, 35.7)$; The y-intercept means that the average amount spent on video games per person was $35.70 in the year 2006.

15. a.

Windchill Versus Wind Speed for Fixed Temperature of 5° Fahrenheit

$(20, -15)$ $(50, -24)$

b. $y = -0.3x - 9$ **c.** $-21°F$ **d.** $-22.8°F$ **e.** $m = -0.3$; This means that windchill decreases at a rate of 0.3°F for every 1 mph increase in wind speed. **17. a.** $y = 15x + 155$ **b.** The slope is 15 and means that the number of associate degrees awarded in the United States increased by 15 thousand per year. **c.** 830 thousand

19. a.

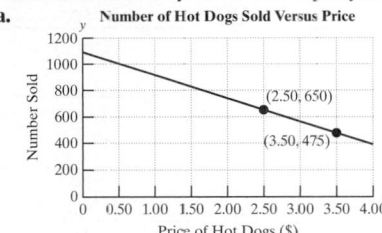

Number of Hot Dogs Sold Versus Price

$(2.50, 650)$

$(3.50, 475)$

b. $y = -175x + 1087.5$ **c.** Approximately 388 hot dogs would be sold.

21. a.

Grade Point Average Versus Weekly Study Time

b. yes **c.** $y = 0.05x + 1.7$ **d.** 3.2 **e.** This model is not reasonable for study times greater than 46 hr per week, because the GPA would exceed 4.0. **23.** Collinear
25. Not collinear

Section 2.4 Graphing Calculator Exercises, p. 182

27.

X	Y1
0	0
4	.8
8	1.6
12	2.4
16	3.2
20	4
24	4.8
X=0	

28.

X	Y1
0	0
6	15
12	30
18	45
24	60
30	75
36	90
X=0	

29. Y1=-175X+1087.5   Y1=-175X+1087.5

X=2.5 Y=650   X=3.5 Y=475

30. Y1=-40=X+1420   Y1=-400X+1420

X=1 Y=1020   X=1.5 Y=820

Section 2.5 Practice Exercises, pp. 188–191

1. a. relation **b.** domain **c.** range

3. a. Vertical **b.** Undefined **c.** $\left(\frac{7}{2}, 0\right)$ **d.** None

5. a. Slanted **b.** $m = \frac{3}{2}$ **c.** $\left(\frac{4}{3}, 0\right)$ **d.** $(0, -2)$

7. a. {(Northeast, 54.1), (Midwest, 65.6), (South, 110.7), (West, 70.7)} **b.** Domain: {Northeast, Midwest, South, West}; range: {54.1, 65.6, 110.7, 70.7} **9. a.** {(USSR, 1961), (USA, 1962), (Poland, 1978), (Vietnam, 1980), (Cuba, 1980)} **b.** Domain: {USSR, USA, Poland, Vietnam, Cuba}; range: {1961, 1962, 1978, 1980}
11. a. {(A, 1), (A, 2), (B, 2), (C, 3), (D, 5), (E, 4)}
b. Domain: {A, B, C, D, E}; range: {1, 2, 3, 4, 5}
13. a. {(−4, 4), (1, 1), (2, 1), (3, 1), (4, −2)}
b. Domain: {−4, 1, 2, 3, 4}; range: {−2, 1, 4}
15. Domain: [0, 4]; range: [−2, 2]   **17.** Domain: [−5, 3]; range: [−2.1, 2.8] **19.** Domain: (−∞, 2]; range: (−∞, ∞)
21. Domain: (−∞, ∞); range: (−∞, ∞)   **23.** Domain: {−3}; range: (−∞, ∞)   **25.** Domain: (−∞, 2); range: [−1.3, ∞)
27. Domain: {−3, −1, 1, 3}; range: {0, 1, 2, 3}
29. Domain: [−4, 5); range: {−2, 1, 3}   **31. a.** 2.85
b. 9.33 **c.** Dec. **d.** Nov. **e.** 7.63 **f.** {Jan., Feb., Mar., Apr., May, June, July, Aug., Sept., Oct., Nov., Dec.}
33. a. 31.876 million or 31,876,000 **b.** The year 2012
35. a. For example: {(Julie, New York), (Peggy, Florida), (Stephen, Kansas), (Pat, New York)} **b.** Domain: {Julie, Peggy, Stephen, Pat}; range: {New York, Florida, Kansas}
37. $y = 2x - 1$ **39.** $y = x^2$

Section 2.6 Practice Exercises, pp. 199–204

1. a. function **b.** vertical **c.** $2x + 1$ **d.** domain; range
e. denominator; negative **f.** -2 **g.** 3 **h.** $(1, 6)$
3. a. {(Kevin, Kayla), (Kevin, Kira), (Kathleen, Katie), (Kathleen, Kira)} **b.** {Kevin, Kathleen} **c.** {Kayla, Katie, Kira}
5. Function   **7.** Not a function   **9.** Function
11. Not a function   **13.** Function   **15.** Not a function
17. When x is 2, the function value y is 5.   **19.** $(0, -2)$
21. -11 **23.** 1   **25.** 2   **27.** $6t - 2$ **29.** 7
31. 4   **33.** 4   **35.** $6x + 4$ **37.** $-4x^2 - 8x + 1$
39. $-\pi^2 + 4\pi + 1$ **41.** 7   **43.** $-6a - 2$
45. $|-c - 2|$ **47.** 1   **49.** 7   **51.** -18.8 **53.** -7

55. 2π **57.** -5 **59.** 4   **61. a.** 3   **b.** 1   **c.** 1
d. $x = -3$ **e.** $x = 0, x = 2$ **f.** $(-\infty, 3]$ **g.** $(-\infty, 5]$
  63. a. 3   **b.** $H(4)$ is not defined because 4 is not in the domain
of H.   **c.** 4   **d.** $x = -3$ and $x = 2$ **e.** All x on the interval $[-2, 1]$
f. $[-4, 4)$ **g.** $[2, 5)$ **65. a.** -4 **b.** 0   **c.** -3 **d.** $x = -1$
e. There are no such values of x.   **f.** $(-\infty, \infty)$
g. $(-\infty, -3] \cup (-2, \infty)$ **67.** $\{-3, -7, -\frac{3}{2}, 1.2\}$ **69.** $\{6, 0\}$
71. -3 and 1.2   **73.** 6 and 1   **75.** -3
77. The domain is the set of all real numbers for which the
denominator is not zero. Set the denominator equal to zero, and
solve the resulting equation. The solution(s) to the equation must be
excluded from the domain. In this case, setting $x - 2 = 0$ indicates
that $x = 2$ must be excluded from the domain. The domain is
$(-\infty, 2) \cup (2, \infty)$.   **79.** $(-\infty, -6) \cup (-6, \infty)$
81. $(-\infty, 0) \cup (0, \infty)$ **83.** $(-\infty, \infty)$ **85.** $[-7, \infty)$
87. $[3, \infty)$ **89.** $(-\infty, \frac{1}{2}]$ **91.** $(-\infty, \infty)$ **93.** $(-\infty, \infty)$
95. a. $h(1) = 64$ and $h(1.5) = 44$ **b.** $h(1) = 64$ means that after
1 sec, the height of the ball is 64 ft. $h(1.5) = 44$ means that after
1.5 sec, the height of the ball is 44 ft.   **97. a.** $d(1) = 11.5$ and
$d(1.5) = 17.25$ **b.** $d(1) = 11.5$ means that after 1 hr, the distance
traveled is 11.5 mi. $d(1.5) = 17.25$ means that after 1.5 hr, the
distance traveled is 17.25 mi.   **99.** $f(x) = 2x + 3$
101. $f(x) = |x| - 10$ **103.** $(-2, \infty)$

Section 2.6 Graphing Calculator Exercises, p. 204

105.

106.

Section 2.7 Practice Exercises, pp. 211–216

1. a. linear   **b.** constant   **c.** quadratic   **d.** parabola
e. 0   **f.** y **3. a.** Yes   **b.** $\{7, 2, -5\}$ **c.** $\{3\}$
5. a. $f(0) = 2, f(-3) = 1, f(-4) = 0$, and $f(-5)$ cannot be
evaluated because $x = -5$ is not in the domain of f.
b. $[-4, \infty)$ **7.** horizontal
9. Domain $(-\infty, \infty)$; range $\{2\}$

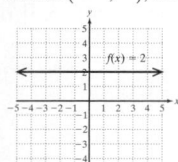

11.

x	$f(x)$	x	$f(x)$
-2	$-\frac{1}{2}$	$\frac{1}{4}$	4
-1	-1	$\frac{1}{2}$	2
$-\frac{1}{2}$	-2	1	1
$-\frac{1}{4}$	-4	2	$\frac{1}{2}$

13.

x	$h(x)$
-2	-8
-1	-1
0	0
1	1
2	8

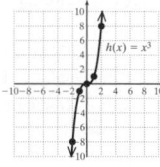

15.

x	$q(x)$
-2	4
-1	1
0	0
1	1
2	4

17. Quadratic   **19.** Linear   **21.** Constant
23. None of these   **25.** Linear   **27.** None of these

29. x-intercept: $(2, 0)$;   **31.** x-intercept: $\left(\frac{5}{6}, 0\right)$;
   y-intercept: $(0, -10)$ y-intercept: $(0, 5)$

33. x-intercept: none;   **35.** x-intercept: $(-3, 0)$;
   y-intercept: $(0, 18)$ y-intercept: $(0, 2)$

37. a. $x = -1$ **b.** $f(0) = 1$ **39. a.** $x = -2$ and $x = 2$
b. $f(0) = -2$ **41. a.** None   **b.** $f(0) = 2$
43. a. $(-\infty, \infty)$ **b.** $(0, 0)$ **c.** vi   **45. a.** $(-\infty, \infty)$
b. $(0, 1)$ **c.** viii   **47. a.** $[-1, \infty)$ **b.** $(0, 1)$ **c.** vii
49. a. $(-\infty, 3) \cup (3, \infty)$ **b.** $\left(0, -\frac{1}{3}\right)$ **c.** ii
51. a. $(-\infty, \infty)$ **b.** $(0, 2)$ **c.** iv   **53. a.** Linear
b. $G(90) = 82.5$. This means that if the student gets a 90% score
on her final exam, then her overall course average is 82.5%.
c. $G(50) = 72.5$. This means that if the student gets a 50% score
on her final exam, then her overall course average is 72.5%.

55. $f(x) = x^2 + 2$ **57.** $f(x) = 3$ **59.** $f(x) = \frac{1}{2}x - 2$

Section 2.7 Graphing Calculator Exercises, p. 216

61.   **62.**

63.   **64.**

65.   **66.**

67. x-intercept: $(8,0)$; y-intercept: $(0,1)$ **68.** x-intercept: $(-6,0)$; y-intercept: $(0,-3)$ **69.** x-intercept: $(-5,0)$; y-intercept: $(0,4)$ **70.** x-intercept: $(2,0)$; y-intercept: $(0,-7)$

Chapter 2 Problem Recognition Exercises, p. 217

1. a, c, d, f, g   **2.** a, b, d, f, h   **3.** 4   **4.** 2
5. $\{0, 1, \frac{1}{2}, -3, 2\}$ **6.** $\{4, 3, 6, 1, 10\}$ **7.** $[-2, 4]$
8. $[0, \infty)$ **9.** 0   **10.** $\left(\frac{9}{5}, 0\right)$ **11.** $(0, -1), (0, 1)$
12. 1   **13.** c   **14.** d   **15.** 3

Chapter 2 Review Exercises, pp. 225–230

1.   **2.** No   **3.** Yes

4. $A(0,0)$; $B(2,1)$; $C(0,-4)$; $D(-2,-4)$; $E(-2,0)$; $F(-5,1)$; $G(4,-3)$
5.

x	y
0	3
-2	0
1	$\frac{9}{2}$

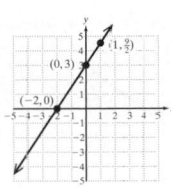

6.

x	y
0	$\frac{13}{2}$
5	$\frac{13}{2}$
-4	$\frac{13}{2}$

7.

x	y
4	0
4	1
4	-2

8. x-intercept $(-3,0)$; y-intercept $(0,2)$ **9.** x- and y-intercept $(0,0)$

    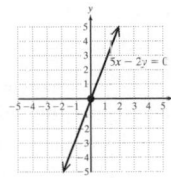

10. No x-intercept; y-intercept $(0,3)$ **11.** x-intercept $(-2,0)$; no y-intercept

12. a. $m = \frac{1}{2}$ **b.** $m = -3$ **c.** $m = 0$

13. For example: $y = 2x$ **14.** For example: $y = -\frac{3}{4}x$

    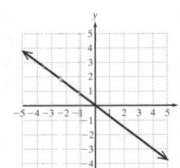

15. $m = 2$ **16.** $m = \frac{7}{10}$ **17.** $m = 0$
18. Undefined slope   **19.** Perpendicular
20. Perpendicular   **21.** Neither   **22.** Parallel
23. a. $m = 53$ **b.** The enrollment increased at a rate of 53 students per year.
24. $m = \frac{3}{4}$ **25. a.** $y = k$ **b.** $y - y_1 = m(x - x_1)$
c. $Ax + By = C$ **d.** $x = k$ **e.** $y = mx - b$
26. $y = \frac{1}{9}x + 6$ or $x - 9y = -54$
27. $y = -\frac{2}{3}x + 2$ or $2x + 3y = 6$
28. $y = \frac{10}{3}x + \frac{77}{3}$ or $10x - 3y = -77$
29. $y = 3x - 20$ or $3x - y = 20$
30. $y = -\frac{4}{3}x - 3$ or $4x + 3y = -9$
31. a. $y = -2$ **b.** $x = -3$ **c.** $x = -3$ **d.** $y = -2$
32. Yes, (a) and (d) are the same; (b) and (c) are the same.
33. a. $y = 0.75x + 50$

b.

c. The daily fixed cost is $50 if no ice cream is sold.
d. \$387.50   **e.** 0.75   **f.** The cost increases at a rate of \$0.75 per ice cream product.
34. a.

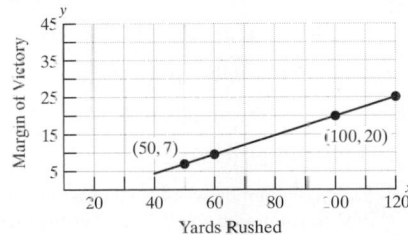

b. $y = 0.26x - 6$ **c.** The margin of victory would be -6 points, which means they would lose by 6 points.
35. Domain $\left\{\frac{1}{3}, 6, \frac{1}{4}, 7\right\}$; range $\left\{10, -\frac{1}{2}, 4, \frac{2}{5}\right\}$
36. Domain $\{-3, -1, 0, 2, 3\}$; range $\left\{-2, 0, 1, \frac{5}{2}\right\}$
37. Domain $[-3, 9]$; range $[0, 60]$ **38.** Domain $[-4, 4]$; range $[1, 3]$
39.   **40.**

    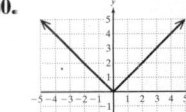

41. a. Not a function   **b.** $[1, 3]$ **c.** $[-4, 4]$
42. a. Function   **b.** $(-\infty, \infty)$ **c.** $(-\infty, 0.35]$
43. a. Function   **b.** $\{1, 2, 3, 4\}$ **c.** $\{3\}$
44. a. Not a function   **b.** $\{0, 4\}$ **c.** $\{2, 3, 4, 5\}$
45. a. Not a function   **b.** $\{4, 9\}$ **c.** $\{2, -2, 3, -3\}$
46. a. Function   **b.** $\{6, 7, 8, 9\}$ **c.** $\{9, 10, 11, 12\}$
47. -4 **48.** 2 **49.** 2 **50.** $6t^2 - 4$ **51.** $6b^2 - 4$
52. $6\pi^2 - 4$ **53.** $6a^2 - 4$ **54.** 20 **55.** $(-\infty, \infty)$
56. $(-\infty, 11) \cup (11, \infty)$ **57.** $[8, \infty)$ **58.** $[-2, \infty)$
59. a. $98   **b.** $123   **c.** $148
60.

61.

62.

63.

64.

65.

66.

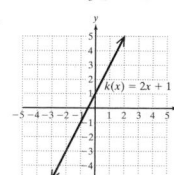

67.

68. a. $s(4) = 4, s(-3) = 25, s(2) = 0, s(1) = 1, s(0) = 4$
b. $(-\infty, \infty)$ **69. a.** $r(2)$ is not a real number,
$r(4) = 0, r(5) = 2, r(8) = 4$ **b.** $[4, \infty)$

70. a. $h(-3) = -\dfrac{1}{2}, h(0) = -1, h(2) = -3, h(5) = \dfrac{3}{2}$

b. $(-\infty, 3) \cup (3, \infty)$
71. a. $k(-5) = -2, k(-4) = -1, k(-3) = 0,$

$k(2) = -5$ **b.** $(-\infty, \infty)$ **72.** x-intercept: $\left(\dfrac{7}{4}, 0\right)$;

y-intercept: $(0, -7)$

73. x-intercept: $\left(\dfrac{9}{2}, 0\right)$; y-intercept: $(0, 9)$

74. a. $b(0) = 4.5$. In 1985 consumption was 4.5 gal of bottled water per
capita. $b(7) = 9.4$. In 1992 consumption was 9.4 gal of bottled water per
capita.   **b.** $m = 0.7$. Consumption increased at a rate of 0.7 gal/year.
75. -1 **76.** 0 **77.** $x = 0$ and $x = 4$
78. There are no values of x for which $g(x) = 4$.
79. $(-4, \infty)$ **80.** $(-\infty, 1]$

Chapter 2 Test, pp. 230–233

1. $(0, -9)$ $(6, 0)$ $(4, -3)$

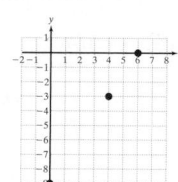

2. a. False, the product is positive in quadrant III also.
b. False, the quotient is negative in quadrant II also.

c. True   **d.** True   **3.** Yes   **4.** To find the x-intercept, let $y = 0$
and solve for x. To find the y-intercept, let $x = 0$ and solve for y.
   5. x-intercept: $(4, 0)$;   **6.** x-intercept: $(-4, 0)$,
y-intercept: $(0, -3)$ no y-intercept

7. x- and y-intercept: $(0, 0)$ **8.** no x-intercept;
                                              y-intercept: $(0, -3)$

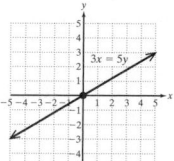

9. a. $m = \dfrac{5}{8}$ **b.** $m = \dfrac{6}{5}$

10. a. $y = \frac{3}{4}x + 1$ **b.** Slope: $\frac{3}{4}$; y-intercept: $(0, 1)$

c.

11. a. -7 **b.** $\dfrac{1}{7}$ **12. a.** The slopes are the same.

b. The slope of one line is the opposite of the reciprocal of
the slope of the other.   **13.** Neither
14. a. Perpendicular   **b.** Parallel   **c.** Perpendicular
d. Neither   **15. a.** For example: $y = 3x + 2$
b. For example: $x = 2$ **c.** For example: $y = 3; m = 0$
d. For example: $y = -2x$ **16.** $y = \frac{3}{2}x - 6$ or $3x - 2y = 12$

17. $y = 2x - 11$ or $2x - y = 11$ **18.** $y = -2x + \dfrac{31}{2}$

19. $y = \dfrac{1}{3}x + \dfrac{1}{3}$

20. a. $y = 300x + 800$

b.

c. Jack earns a base monthly salary of $800.   **d.** $5900
21. a. $(0, 66)$. For a woman born in 1940, life expectancy was

about 66 yr.   **b.** $m = \dfrac{3}{10}$; Life expectancy rises at a rate

of 3 yr for every 10 yr.   **c.** $y = \dfrac{3}{10}x + 66$

d. 82.2 yr; It is greater than the reported value.
22.   **23.**

24.

25.

26. a. Not a function   **b.** $\{-3, -1, 1, 3\}$
c. $\{-2, -1, 1, 3\}$ **27. a.** Function   **b.** $(-\infty, \infty)$
c. $(-\infty, 0]$ **28.** $(-\infty, -7) \cup (-7, \infty)$ **29.** $[-7, \infty)$
30. $(-\infty, \infty)$ **31. a.** $r(-2) = 9, r(0) = 1, r(3) = 4$
b. $(-\infty, \infty)$ **32. a.** $s(0) = 36$. In 1985 the per capita consumption
was 36 gal. $s(7) = 47.2$. In 1992 the per capita consumption was 47.2
gal.   **b.** $m = 1.6$. Consumption increases at a rate of 1.6 gal/year.
33. Quadratic   **34.** Linear   **35.** Constant
36. None of these   **37.** To find the x-intercept(s), solve for
the real solutions of the equation $f(x) = 0$. To find the y-intercept,
find $f(0)$.   **38.** x-intercept: $(-12, 0)$; y-intercept: $(0, 9)$
39. 1   **40.** 2   **41.** $(-1, 7]$ **42.** $[-1, 4)$ **43.** False
44. $(6, 0)$ **45.** $x = 6$ **46.** All x in the interval $[1, 3]$ and $x = 5$

Chapters 1–2 Cumulative Review Exercises, pp. 233–234

1. -1 **2.** 4   **3.** $21x - 62y + 12$ **4.** $\{-15\}$ **5.** $\{\ \}$
6. $\{8, 2\}$ **7.** $[-7, 7)$ **8.** $(-2, 3)$ **9.** $(-\infty, \infty)$
10. $(-4, 5)$ **11.** $(-\infty, -7] \cup [13, \infty)$
12. a. 1   **b.** 33   **13.** $-\dfrac{1}{5}$ **14.** No
15. a. x-intercept: $\left(\dfrac{10}{3}, 0\right)$; y-intercept: $(0, -2)$ **b.** $m = \dfrac{3}{5}$
c.

16. a. no x-intercept; y-intercept: $(0, 3)$ **b.** $m = 0$
c.

17. $x = 7$ **18.** $m = \dfrac{5}{2}$ **19.** Domain: $\{3, 4\}$; range:
$\{-1, -5, -8\}$; not a function   **20.** $y = -2x - 2$
21. $y = -4x$ **22.** She spent $5.98 on drinks and $11.96 on
popcorn.   **23.** $(-\infty, 15) \cup (15, \infty)$ **24.** She used 10 L of the
20% solution.   **25.** Los Angeles gets 16.9 in. per year and Seattle
gets 33.1 in. per year.

Chapter 3
Chapter Opener Puzzle

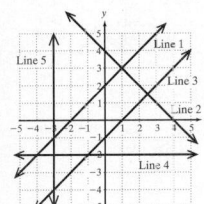

S	O	L	U	T	I	O	N
4	1	5	6	2	3	1	7

1. a. system   **b.** solution   **c.** intersect   **d.** consistent
e. the empty set, $\{\ \}$ **f.** dependent   **g.** independent
3. $(2, 11)$ is a solution.   **5.** $(-1, 4)$ is a solution.
7. None   **9. a.** Consistent   **b.** Independent   **c.** One solution
11. a. Inconsistent   **b.** Independent   **c.** Zero solutions
13. a. Consistent   **b.** Dependent   **c.** Infinitely many solutions
15. $\{(-2, 1)\}$ **17.** $\{(1, 1)\}$

        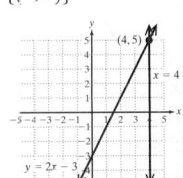

19. $\{(3, -4)\}$ **21.** $\{(4, 5)\}$

23. No solution; $\{\ \}$;   **25.** Infinitely many solutions;
inconsistent system $\left\{(x, y) \mid y = \dfrac{2}{3}x - 1\right\}$;
dependent equations

27. $\{(2, -2)\}$ **29.** Infinitely many solutions;
$\{(x, y) \mid -x + 3y = 6\}$;
dependent equations

31. No solution: $\{\ \}$; inconsistent system

33. False   **35.** True
37. For example: $x + y = 9$ **39.** $C = 5, D = 4$
$2x + y = 13$

41. $\{(-3.5, -4.21)\}$ **42.** $\{(-4.6, 5.1)\}$

43. $\{(2.14, 3.02)\}$ **44.** $\{(4.5, 2.2)\}$

  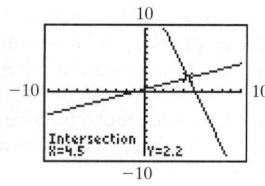

Section 3.2 Practice Exercises, pp. 251–253

1. One solution **3.** No solution **5.** No
7. $\{(0, 3)\}$

9. $\{(-1, 2)\}$ **11.** $\{(3, 1)\}$ **13.** $\{(12, 0)\}$ **15.** $\{(5, -3)\}$
17. $\left\{\left(-\frac{1}{4}, \frac{1}{2}\right)\right\}$ **19.** $\{(2, 0)\}$ **21.** $\{(1, -2)\}$
23. Infinitely many solutions; $\{(x, y) \mid x = 3y - 1\}$; dependent
equations **25.** No solution; $\{\ \}$; inconsistent system
27. No solution; $\{\ \}$; inconsistent system
29. Infinitely many solutions; $\{(x, y) \mid 3x - y = 7\}$; dependent
equations **31.** When solving a system, if you get an identity, such
as $0 = 0$ or $5 = 5$, then the equations are dependent. **33.** $\{(8, 5)\}$
35. $\{(5, 3)\}$ **37.** No solution; $\{\ \}$; inconsistent system
39. $\{(0, 3)\}$ **41.** $\{(-7, 6)\}$ **43.** Infinitely many solutions;
$\{(x, y) \mid 2x - y = 6\}$; dependent equations **45.** $\{(10, -32.1)\}$
47. $\left\{\left(\frac{1}{2}, \frac{3}{4}\right)\right\}$ **49.** $\{(-2, 4)\}$ **51.** $\{(1, 1)\}$ **53.** $\{(6, 1)\}$
55. $\{(5, -1)\}$ **57. a.** $y = \frac{4}{9}x + \frac{25}{9}$ **b.** $y = -\frac{4}{7}x + \frac{23}{7}$ **c.** $\left(\frac{1}{2}, 3\right)$
59. a. At Glendale Lakes: $y = 800x + 250$
At the Breakers: $y = 750x + 500$ **b.** 5 months

Section 3.3 Practice Exercises, pp. 258–260

1. a. -3 **b.** 5 **3.** One solution
5. $\{(-2, -5)\}$ **7.** $\{(3, -1)\}$ **9.** $\{(12, -8)\}$
11. $\{(1, -1)\}$ **13.** $\{(2, -1)\}$ **15.** $\{(0, 0)\}$
17. Infinitely many solutions; $\{(x, y) \mid 3x - 2y = 1\}$; dependent
equations **19.** No solution; $\{\ \}$; inconsistent system
21. Infinitely many solutions; $\{(x, y) \mid 12x - 4y = 2\}$; dependent
equations **23.** No solution; $\{\ \}$; inconsistent system
25. Use the substitution method if one equation has x or y
already isolated. **27.** False **29.** True **31.** True
33. $\{(-2, -3)\}$ **35.** $\{(-\frac{1}{2}, 2)\}$ **37.** $\{(4, 0)\}$
39. No solution; $\{\ \}$; inconsistent system **41.** $\{(12, 30)\}$
43. Infinitely many solutions; $\{(x, y) \mid x = \frac{3}{2}y\}$; dependent equations
45. $\{(0, 4)\}$ **47.** $\{(-\frac{1}{2}, -\frac{5}{2})\}$ **49.** $\{(30, 10)\}$ **51.** $\{(\frac{4}{3}, -1)\}$
53. $\{(1, 5)\}$ **55.** No solution; $\{\ \}$; inconsistent system
57. 112 mi in the city and 120 mi on the highway.
59. $\{(-\frac{56}{41}, \frac{221}{41})\}$ **61.** $\{(-\frac{13}{7}, -\frac{37}{14})\}$

Chapter 3 Problem Recognition Exercises, p. 260

1. $\{(2, 4)\}$ **2.** Infinitely many solutions; $\{(x, y) \mid 3x - 2y = 4\}$;
dependent equations **3.** No solution; $\{\ \}$; inconsistent system
4. $\{(1, 2)\}$ **5.** $\left\{\left(\frac{1}{2}, -11\right)\right\}$ **6.** $\{(-3, 1)\}$
7. $\{(-2, -4)\}$ **8.** $\{(2, 2)\}$

Section 3.4 Practice Exercises, pp. 266–269

1. a. \$60; $12x$ **b.** 2 L; $0.10x$ **c.** \$200; $0.04y$ **d.** $b - c$; $b + c$
e. 180° **f.** 180° **g.** 90° **h.** 90° **3.** $\{(5, -1)\}$ **5.** 552 tickets
were sold at \$20 each, and 638 were sold at \$30 each.

7. Hamburgers cost \$4.60 and fish sandwiches cost \$5.20.
9. Vanilla has 12 g of fat per scoop and mud pie has 16 g of fat
per scoop. **11.** Mix 4 oz of 18% moisturizer and 8 oz of 24%
moisturizer. **13.** Mix 2 L of 8% nitrogen fertilizer and 6 L of 12%
nitrogen fertilizer. **15.** Mix 1 oz of pure bleach and 11 oz of 4%
solution. **17.** He invested \$4500 in the stock fund and \$1500 in
the bond fund. **19.** He borrowed \$2800 at 5.5% and \$2600
at 3.5%. **21.** Alina borrowed \$10,000 from the bank charging
6% interest and \$5000 from the bank charging 7% interest.
23. The boat travels 6 mph in still water, and the current is 2 mph.
25. The plane's still air speed is 720 km/hr, and the speed of the
wind is 80 km/hr. **27.** Stephen's walking speed on nonmoving
ground is 3.5 ft/sec. The sidewalk moves at 1.5 ft/sec. **29.** The
angles are 69° and 21°. **31.** The angles are 134.5° and 45.5°.
33. The angles are 28° and 62°. **35.** 7.5 g of pure gold is
needed. **37.** The boat's speed in still water is 5.2 mph and the
current speed is 1.2 mph. **39.** Grandstand tickets cost \$330 each
and general admission tickets cost \$175 each. **41.** Svetlana invested
\$1750 at 2% and \$1250 at 1.3%. **43.** The length is 11 m and the
width is 10 m. **45.** There are eleven 50¢ pieces and ten \$1 coins.
47. a. $f(x) = 0.15x$ **b.** $g(x) = 0.10x + 4.95$ **c.** 99 text messages

Section 3.5 Practice Exercises, pp. 277–282

1. a. linear **b.** is not; is **c.** dashed; is not **d.** solid; is
3. $[1, 3)$ **5.** $(-\infty, -4] \cup (-2, \infty)$ **7. a.** No **b.** No
c. Yes **d.** Yes **9. a.** No **b.** Yes **c.** Yes **d.** No
11. $>$ **13.** $\geq$ **15.** $\geq$, $\leq$
17. **19.** **21.**

23. **25.** **27.**

29. **31.** **33.**

35. **37.** **39.**

41. **43.** **45.**

47.

49.

51.

53.

55.

57.

59.

61. a. $2x + 2y \le 40$
b.
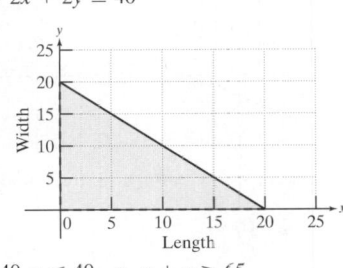

63. a. $x \ge 0, y \ge 0$ **b.** $x \le 40, y \le 40$ **c.** $x + y \ge 65$
d.

e. Yes; The point $(35, 40)$ means that Karen works 35 hr and Todd works 40 hr. **f.** No; The point $(20, 40)$ means that Karen works 20 hr and Todd works 40 hr. This does not satisfy the constraint that there must be at least 65 hr total.

Section 3.6 Practice Exercises, pp. 290–292

1. a. linear **b.** ordered triples **3.** $\{(1, 1)\}$
5. Marge's speed in still air is 15 mph and the wind speed is 3 mph.
7. $(4, 0, 2)$ is a solution. **9.** $(1, 1, 1)$ is a solution.
11. $\{(1, -2, 4)\}$ **13.** $\{(-1, 2, 0)\}$ **15.** $\{(1, -4, -2)\}$
17. $\{(1, 2, 3)\}$ **19.** $\{(-6, 1, 7)\}$ **21.** $\left\{\left(\frac{1}{2}, \frac{2}{3}, -\frac{5}{6}\right)\right\}$
23. The angles are $67°, 82°,$ and $31°$. **25.** The sides are 10 cm, 18 cm, and 27 cm. **27.** The fiber supplement has 3 g; the oatmeal has 4 g; and the cereal has 1.5 g. **29.** There are three par 3s, nine par 4s, and six par 5s. **31.** Hats cost $15, T-shirts cost $25, and jackets cost $45. **33.** Walter invested $7000 in small caps, $6000 in the balanced fund, and $12,000 in global markets.
35. Dependent equations **37.** Inconsistent system
39. $\{(1, 3, 1)\}$ **41.** $\{(-2, -1, -3)\}$
43. Dependent equations **45.** Inconsistent system
47. $(0, 0, 0)$ is the only solution. **49.** Dependent equations

Section 3.7 Practice Exercises, pp. 299–301

1. a. matrix; rows; columns **b.** column; one; square
c. coefficient; augmented **d.** row echelon
3. $\left\{\left(12, \frac{1}{2}\right)\right\}$ **5.** $\{(1, 3, -3)\}$ **7. a.** 3×1 **b.** column matrix

9. a. 2×2 **b.** square matrix **11. a.** 1×4 **b.** row matrix
13. a. 2×3 **b.** none of these
15. $\begin{bmatrix} 1 & -2 & | & -1 \\ 2 & 1 & | & -7 \end{bmatrix}$ **17.** $\begin{bmatrix} 1 & -2 & 1 & | & 5 \\ 2 & 6 & 3 & | & -2 \\ 3 & -1 & -2 & | & 1 \end{bmatrix}$
19. $4x + 3y = 6$ **21.** $x = 4, y = -1, z = 7$
    $12x + 5y = -6$
23. a. 7 **b.** -2 **25.** $\begin{bmatrix} 1 & \frac{1}{2} & | & \frac{11}{2} \\ 2 & -1 & | & 1 \end{bmatrix}$
27. $\begin{bmatrix} 1 & -4 & | & 3 \\ 5 & 2 & | & 1 \end{bmatrix}$ **29.** $\begin{bmatrix} 1 & 5 & | & 2 \\ 0 & 11 & | & 5 \end{bmatrix}$
31. a. $\begin{bmatrix} 1 & 3 & 0 & | & -1 \\ 0 & -11 & -5 & | & 10 \\ -2 & 0 & -3 & | & 10 \end{bmatrix}$ **b.** $\begin{bmatrix} 1 & 3 & 0 & | & -1 \\ 0 & -11 & -5 & | & 10 \\ 0 & 6 & -3 & | & 8 \end{bmatrix}$
33. True **35.** True **37.** Interchange rows 1 and 2.
39. Multiply row 1 by -3 and add to row 2. Replace row 2 with the result. **41.** $\{(-3, -1)\}$ **43.** $\{(-21, 9)\}$ **45.** Infinitely many solutions; $\{(x, y) \mid x + 3y = 3\}$; dependent equations
47. $\{(3, -1)\}$ **49.** $\{(-10, 3)\}$ **51.** No solution; $\{\ \}$; inconsistent system **53.** $\{(1, 2, 3)\}$ **55.** $\{(1, -2, 0)\}$

Section 3.7 Graphing Calculator Exercises, p. 302

57. **58.** **59.**

60. **61.** **62.**

Chapter 3 Review Exercises, pp. 310–314

1. a. No **b.** Yes **2.** False **3.** True **4.** True
5. $\{(3, 2)\}$ **6.** $\{(-4, -1)\}$

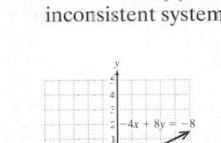

7. Infinitely many solutions; $\{(x, y) \mid 6x + 2y = 4\}$; dependent equations **8.** No solution; $\{\ \}$; inconsistent system

9. $\{(4, -1)\}$ **10.** $\{(2, -7)\}$ **11.** No solution; $\{\ \}$; inconsistent system **12.** No solution; $\{\ \}$; inconsistent system **13.** Infinitely many solutions; $\{(x, y) \mid 10x - 2y = 3\}$; dependent equations **14.** Infinitely many solutions; $\{(x, y) \mid 4x + y = 7\}$; dependent equations **15.** 30 months **16.** 8 days
17. $\{(1, 1)\}$ **18.** $\{(2, -1)\}$ **19.** $\{(2, -1)\}$
20. Infinitely many solutions; $\{(x, y) \mid 3x + y = 1\}$; dependent equations **21.** No solution; $\{\ \}$; inconsistent system
22. $\{(6, -2)\}$ **23.** $\{(2, -3)\}$ **24.** $\{(-10, 25)\}$

25. $\{(0,6)\}$ **26.** $\{(-2,7)\}$ **27.** She invested $4500 at 5%.
28. There were 21 student tickets and 33 adult tickets.
29. 10 L of 20% solution must be mixed with 6 L of 50% solution.
30. The plane travels 150 mph in still air, and the wind speed is
10 mph.   **31. a.** $f(x) = 0.1x + 9.95$ **b.** $g(x) = 0.08x + 12.95$
c. 150 min   **32.** The angles are 76° and 14°.

33.   **34.**   **35.**

36.   **37.**   **38.**

39.   **40.**   **41.**

42.

43. a. $x \geq 0, y \geq 0$ **b.** $x + y \leq 100$
c. $x \geq 4y$
d.

44. $\{(2,-1,5)\}$ **45.** Inconsistent system
46. Dependent equations   **47.** $\{(-1,-2,2)\}$
48. The sides are 5 ft, 12 ft, and 13 ft.
49. The pumps can drain 250, 300, and 400 gal/hr.
50. The angles measure 113°, 38°, and 29°.
51. 3×3 **52.** 3×2 **53.** 1×4 **54.** 3×1
55. $\begin{bmatrix} 1 & 1 & | & 3 \\ 1 & -1 & | & -1 \end{bmatrix}$ **56.** $\begin{bmatrix} 1 & -1 & 1 & | & 4 \\ 2 & -1 & 3 & | & 8 \\ -2 & 2 & -1 & | & -9 \end{bmatrix}$
57. $x = 9, y = -3$ **58.** $x = -5, y = 2, z = -8$ **59. a.** 4
b. $\begin{bmatrix} 1 & 3 & | & 1 \\ 0 & -13 & | & 2 \end{bmatrix}$ **60. a.** $\begin{bmatrix} 1 & 2 & 0 & | & -3 \\ 0 & -9 & 1 & | & 12 \\ -3 & 2 & 2 & | & 5 \end{bmatrix}$
b. $\begin{bmatrix} 1 & 2 & 0 & | & -3 \\ 0 & -9 & 1 & | & 12 \\ 0 & 8 & 2 & | & -4 \end{bmatrix}$ **61.** $\{(1,2)\}$ **62.** $\{(-3,6)\}$
63. $\{(1,3,-2)\}$ **64.** $\{(6,1,-1)\}$

Chapter 3 Test, pp. 314–316

1. Yes   **2.** b   **3.** c   **4.** a
5. $\{(1,4)\}$ **6.** $\{(-2,1)\}$

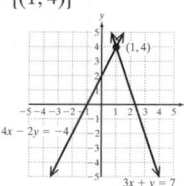

7. $\{(-4,5)\}$ **8.** $\left\{\left(\frac{1}{2},\frac{1}{4}\right)\right\}$ **9.** $\{(0,-3)\}$
10. Infinitely many solutions; $\{(x,y) \mid 3x - 5y = -7\}$;
dependent equations   **11.** $\{(2,-6)\}$ **12.** No solution; $\{\ \}$;
inconsistent system   **13.** $\{(5,0)\}$

14.   **15.**   **16.**

17. a. $x \geq 0, y \geq 0$ **b.** $300x + 400y \geq 1200$
c.

18. $\{(16,-37,9)\}$ **19.** $\{(2,-2,5)\}$ **20.** She borrowed $1200
at 6.5% and $3800 at 5%.   **21.** Mix 80 L of 20% solution with
120 L of 60% solution.   **22.** The angles are 80° and 10°.
23. Joanne can process 142 orders, Kent can process 162 orders,
and Geoff can process 200 orders.

24. For example: $\begin{bmatrix} 2 & 1 \\ 0 & -4 \\ 2.6 & 7 \end{bmatrix}$

25. a. $\begin{bmatrix} 1 & 2 & 1 & | & -3 \\ 0 & -8 & -3 & | & 10 \\ -5 & -6 & 3 & | & 0 \end{bmatrix}$ **b.** $\begin{bmatrix} 1 & 2 & 1 & | & -3 \\ 0 & -8 & -3 & | & 10 \\ 0 & 4 & 8 & | & -15 \end{bmatrix}$
26. $\{(6,-1)\}$ **27.** $\{(2,-4,3)\}$

Chapters 1–3 Cumulative Review Exercises, pp. 316–317

1. -14 **2.** $-56w + 97$ **3.** $\{\ \}$ **4.** $\left\{-\frac{19}{12}\right\}$
5. $\{4,-1\}$ **6.** $(-2,\infty)$ **7.** $(-\infty,-2] \cup (4,\infty)$
8. $\{\ \}$ **9.** $[3,13]$ **10.** $(-6,14)$ **11.** $(-\infty,-4) \cup (0,\infty)$
12.
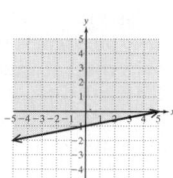

13. Slope $\frac{5}{2}$; x-intercept $(3,0)$; y-intercept $\left(0,-\frac{15}{2}\right)$
14.   **15.**

16. 0   **17.** $y = -4x + 4$ **18.** No solution; { }; inconsistent system   **19.** $\{(1,2)\}$ **20.** The lengths of the sides are $1\frac{1}{2}$ m, $2\frac{1}{2}$ m, and 3 m.   **21. a.** $f(x) = 2.50x + 25$
b. $g(x) = 3x + 10$ **c.** 30 DVDs   **22.** $\{(2, 0, -1)\}$ **23.** 2×3
24. For example: $\begin{bmatrix} 2 & 3 & 1 & 7 \\ -1 & 4 & 2 & 6 \end{bmatrix}$ **25.** $\{(1,1)\}$

Chapter 4

Chapter Opener Puzzle

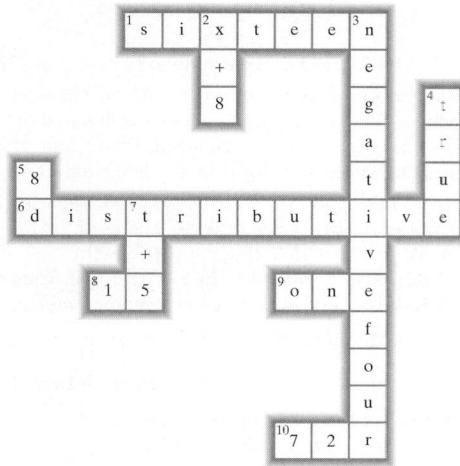

Section 4.1 Practice Exercises, pp. 326–329

1. a. exponent   **b.** 1   **c.** $\left(\frac{1}{b}\right)^n$ or $\frac{1}{b^n}$ **d.** scientific notation
3. $ab^3 = a \cdot b \cdot b \cdot b$
$(ab)^3 = (ab) \cdot (ab) \cdot (ab) = a \cdot a \cdot a \cdot b \cdot b \cdot b = a^3 b^3$
5. For example: $(xy)^3 = x^3 y^3$ **7.** For example: $\dfrac{x^5}{x^2} = x^3$
9. For example: $x^0 = 1$ $(x \neq 0)$ **11.** 3   **13.** $\frac{1}{25}$ **15.** $-\frac{1}{25}$
17. $\frac{1}{25}$ **19.** -64 **21.** $\frac{16}{81}$ **23.** $-\frac{125}{8}$ **25.** 1   **27.** $10a$
29. y^8 **31.** 13^2 or 169   **33.** y^8 **35.** $3^4 x^8$ or $81x^8$ **37.** $\frac{1}{p^3}$
39. $\frac{1}{7^3}$ or $\frac{1}{343}$ **41.** $\frac{1}{w^2}$ **43.** $\frac{1}{a^7}$ **45.** r^2 **47.** $\frac{1}{z^c}$ **49.** $a^2 b^2$
51. 1   **53.** $\frac{65}{4}$ **55.** $\frac{26}{25}$ **57.** 3   **59.** $\frac{5}{2}$ **61.** $\frac{q^2}{p^3}$
63. $-\frac{3b^7}{2a^3}$ **65.** $\frac{x^{16}}{81y^{20}z^8}$ **67.** $-\frac{4m^4}{n^2}$ **69.** $\frac{4q^{11}}{p^4}$ **71.** $\frac{5x^8}{y^2}$
73. $\frac{16a^2}{b^6}$ **75.** $-\frac{27y^{27}}{8x^{24}}$ **77.** $\frac{4x^{18}}{y^{10}}$ **79.** $27x^3y^9$
81. a. $\$8 \times 10^9$ **b.** 3×10^6 DVDs   **c.** 1.4×10^{13} eV
d. 1.602×10^{-19} J   **83. a.** 200,000,000,000   **b.** 0.000004 m
c. 108,200,000,000 m   **85.** 3.5×10^5 **87.** Proper   **89.** Proper
91. 3.38×10^{-4} **93.** 1.608×10^4 **95.** 3.4×10^{13}
97. 2.5×10^3 **99.** 1.204×10^{24} hydrogen atoms and 6.02×10^{23} oxygen atoms   **101.** There are 2×10^4 or 20,000 people per square mile.   **103.** $\$5.25 \times 10^{10}$ **105. a.** 540 months
b. \$10,800   **c.** \$55,395.45   **107.** y^{2a+2} **109.** x^{2a-4} **111.** $x^{a-6}y^6$

Section 4.2 Practice Exercises, pp. 335–339

1. a. polynomial   **b.** coefficient; n **c.** $1; 1$ **d.** one   **e.** binomial
f. trinomial   **g.** leading; leading coefficient   **h.** greatest   **i.** zero

j. exponents   **k.** polynomial   **3.** $10c^2$ **5.** 1.7×10^4
7. $-6a^3 + a^2 - a$; leading coefficient -6; degree 3
9. $3x^4 - 6x^2 - x - 1$; leading coefficient 3; degree 4
11. $-t^2 + 100$; leading coefficient -1; degree 2
13. For example: $3x^5$ **15.** For example: $x^2 + 2x + 1$
17. For example: $6x^4 - x^2$ **19.** $m^2 + 10m$
21. $3x^4 - 2x^3 - 8x^2 + 2x$ **23.** $2w^3 + \frac{1}{9}w^2 + 0.9w$
25. $17x^2y - 4xy - 14$ **27.** $4a^2 - 11a - 7$
29. $9x^3 - 5x^2 + 2x - 8$ **31.** $30y^3$ **33.** $-4p^3 - 2p + 12$
35. $11ab^2 - a^2b$ **37.** $6z^5 - 6z^2$ **39.** $-2x^3 + 4x^2 + 5$
41. $-2xy^3 + 3x^2y + xy + 5$ **43.** $t^3 - 13t^2 - 9t - 13$
45. $\frac{1}{2}a^2 - \frac{9}{10}ab + \frac{3}{5}b^2 + 8$ **47.** $-x^2 + 6x - 16$
49. $3x^5 - x^4 - 4x^3 + 11$ **51.** $4y^3 + 5y^2$ **53.** $-7r^4 - 11r$
55. $9x^2 + 5xy + 11y$ **57.** $18ab - 42b^2$ **59.** $3p - 9$
61. $2m^2 + 6$ **63.** $7x^3 + 4x - 5$ **65.** $12a^2b + ab - 5ab^2$
67. $12x^5 - 6x^2 + 1$ **69.** $8a^2b - 5ab^2 + 4ab$
71. $-3x^3 - 16x^2 + 10x + 1$
73. $-2.2p^5 - 15.5p^4 - 8.5p^3 - 5p^2 - 7.9p$ **75.** $12x^3 + 2x$

77. Yes; degree 2   **79.** No; the term $-\dfrac{3}{x} = 3x^{-1}$ and -1 is not a whole number.   **81.** Yes; degree 0   **83.** No; the term $|x|$ is not of the form ax^n.   **85. a.** -17 **b.** -8 **c.** -5 **d.** -4
87. a. $\frac{1}{4}$ **b.** $\frac{9}{4}$ **c.** $-\frac{7}{4}$ **d.** $\frac{3}{4}$ **89.** $P(x) = 4x + 6$
91. a. $P(z) = 6.6x - 99$ **b.** \$231   **93. a.** $D(0) = 1636$ means that in the year 1990, the annual dormitory charge was \$1636. $D(18) = 4048$ means that in the year 2008, the annual dormitory charge was \$4048.   **b.** \$5896   **95. a.** $W(0) = 6580; W(5) = 7295$; $W(10) = 8010$ **b.** $W(10) = 8010$ means that in the year 2010, 8010 thousand women (8,010,000) were due child support.
97. a. $(0, 0)$; at $t = 0$ sec, the position of the rocket is at the origin.
b. $(25, 27.3)$; after 1 sec, the position of the rocket is $(25, 27.3)$.
c. $(50, 22.6)$

Section 4.3 Practice Exercises, pp. 346–349

1. a. distributive   **b.** $4x - 7$ **c.** squares; $a^2 - b^2$ **d.** perfect; $a^2 - 2ab + b^2$ **3.** $6x^2 + x - 6$ **5. a.** -3 **b.** 9   **c.** -3
7. $-42x^5y^6$ **9.** $11a^7b^8c^{10}$ **11.** $\frac{2}{5}a - \frac{3}{5}$
13. $2m^5n^5 - 6m^4n^4 + 8m^3n^3$ **15.** $3x^2y^2 - 4x^2y^3$
17. $x^2 - zy - 2y^2$ **19.** $12x^2 + 28x - 5$ **21.** $2y^4 - 21y^2 - 36$
23. $25s^2 + 5st - 6t^2$ **25.** $5n^3 + 3n^2 + 50n + 30$
27. $3.25a^2 - 0.9ab - 28b^2$ **29.** $6x^3 + 7x^2y + 4xy^2 + y^3$
31. $x^3 - 343$ **33.** $4a^4 - 17a^3b + 8a^2b^2 - 5ab^3 + b^4$
35. $\frac{1}{2}a^2 - ab + \frac{1}{2}ac - 12b^2 + 8bc - c^2$
37. $-3x^3 + 11x^2 - 7x - 5$ **39.** $\frac{1}{10}y^2 - 8y + 150$ **41.** $a^2 - 64$
43. $9p^2 - 1$ **45.** $x^2 - \frac{1}{9}$ **47.** $9h^2 - k^2$ **49.** $9h^2 - 6hk + k^2$
51. $t^2 - 14t + 49$ **53.** $u^2 + 6uv + 9v^2$ **55.** $h^2 + \frac{1}{3}hk + \frac{1}{36}k^2$
57. $4z^4 - w^6$ **59.** $25x^4 - 30x^2y + 9y^2$ **61. a.** Yes   **b.** No
63. a. $A^2 - B^2$ **b.** $x^2 + 2xy + y^2 - B^2$; Both are examples of multiplying conjugates to get a difference of squares.
65. $w^2 + 2wv + v^2 - 4$ **67.** $4 - x^2 - 2xy - y^2$
69. $9a^2 - 24a + 16 - b^2$ **71.** Write $(x + y)^3$ as $(x + y)^2(x + y)$. Square the binomial and then use the distributive property to multiply the resulting trinomial by the remaining factor of $(x + y)$.
73. $8x^3 + 12x^2y + 6xy^2 + y^3$ **75.** $64a^3 - 48a^2b + 12ab^2 - b^3$
77. Multiply and simplify the first two binomials. Then multiply the resulting trinomial by the third binomial, using the distributive property.   **79.** $6a^4 + 32a^3 + 10a^2$ **81.** $x^3 + 5x^2 - 9x - 45$
83. $-16x^2 + 2x - 22$ **85.** $-3y^2 - 10y - 8$ **87.** $(r + t)^2$

89. $x^2 - y^3$ **91.** The sum of p cubed and q squared   **93.** The product of x and the square of y **95.** $A(x) = 4x^2 + 70x + 300$
97. a. $V(x) = 4x^3 - 32x^2 + 64x$ **b.** 36 in.^{3} **99.** $x^2 - 4x + 4$
101. $x^2 - 4$ **103.** $x^2 - 9$ **105.** $9x^3 + 30x^2$ **107.** $2x + h - 3$
109. Multiply $(x + 2)^2(x + 2)^2$ by squaring the binomials. Then multiply the resulting trinomials, using the distributive property.
111. $(5x - 6)$ **113.** $(2y - 1)$

Section 4.4 Practice Exercises, pp. 356–359

1. a. division; quotient; remainder   **b.** Synthetic
3. a. $-4a - 11b$ **b.** $5a^2 - 49ab - 10b^2$
5. a. $7x^2 + 2$ **b.** $6x^4 - 5x^3 + x^2 - 2x$
7. For example: $(5y + 1)^2 = 25y^2 + 10y + 1$ **9.** $-4t^3 + t - 5$
11. $12 + 8y + 2y^2$ **13.** $x^2 + 3xy - y^2$ **15.** $2y^2 + 3y - 8$
17. $-\dfrac{1}{2}p^3 + p^2 - \dfrac{1}{3}p + \dfrac{1}{6}$ **19.** $a^2 + 5a + 1 - \dfrac{5}{a}$
21. $-3s^2t + 4s - \dfrac{5}{t^2}$ **23.** $8p^2q^6 - 9p^3q^5 - 11p - \dfrac{4}{p^2q}$
25. a. Divisor $(x - 2)$; quotient $(2x^2 - 3x - 1)$; remainder (-3)
b. Multiply the quotient and the divisor, then add the remainder. The result should equal the dividend.
27. $x + 7 + \dfrac{-9}{x + 4}$ **29.** $3y^2 + 2y + 2 + \dfrac{9}{y - 3}$ **31.** $-4a + 11$
33. $6y - 5$ **35.** $6x^2 + 4x + 5 + \dfrac{22}{3x - 2}$ **37.** $4a^2 - 2a + 1$
39. $x^2 - 2x + 2$ **41.** $x^2 + 2x + 5$ **43.** $x^2 - 1 + \dfrac{8}{x^2 - 2}$
45. $n^3 + 2n^2 + 4n + 8$ **47.** The divisor must be of the form $x - r$.
49. No. The divisor is not of the form $x - r$.   **51. a.** $x - 5$
b. $x^2 + 3x + 11$ **c.** 58   **53.** $x + 6$ **55.** $t - 4$
57. $5y + 10 + \dfrac{11}{y - 1}$ **59.** $3y^2 - 2y + 2 + \dfrac{-3}{y + 3}$
61. $x^2 - x - 2$ **63.** $a^4 + 2a^3 + 4a^2 + 8a + 16$
65. $x^2 + 6x + 36$ **67.** $6t^2 + 3t - 3 + \dfrac{5}{t + \frac{2}{3}}$ **69.** $4w^3 + 2w^2 + 6$
71. $-x^2 - 4x + 13 + \dfrac{-54}{x + 4}$ **73.** $2x - 1 + \dfrac{3}{x}$
75. $4y - 3 + \dfrac{4y + 5}{3y^2 - 2y + 5}$ **77.** $2x^2 + 3x - 1$
79. $2k^3 - 4k^2 + 1 - \dfrac{5}{k^4}$ **81.** $x + \dfrac{9}{5} + \dfrac{2}{x}$ **83. a.** -84
b. $4x^2 - 6x + 16 + \dfrac{-84}{x + 4}$ **c.** The values are the same.
85. $P(r)$ equals the remainder of $P(x) \div (x - r)$.
87. a. $8x + 5$ **b.** Yes

Chapter 4 Problem Recognition Exercises, pp. 359–360

1. a. $9x^2 + 6x + 1$ **b.** $9x^2 - 1$ **c.** 2   **2. a.** -10 **b.** $81m^2 - 25$
c. $81m^2 - 90m + 25$ **3. a.** $2x + 4 - \dfrac{5}{x}$ **b.** $2x + 5 + \dfrac{-5}{2x - 1}$
c. $4x + 12 + \dfrac{2}{x - 1}$ **4. a.** $y - 5 + \dfrac{4}{3y}$ **b.** $y - 7 + \dfrac{46}{3y + 6}$
c. $3y - 33 + \dfrac{202}{y + 6}$ **5. a.** -30 **b.** $-10p - 50$ **c.** 0
6. a. $-8x - 32$ **b.** -20 **c.** 0   **7.** $2t^2 + t - 1$
8. $-15x^4 - 5x^3 + 10x^2$ **9.** $36z^2 - 25$ **10.** $9y^3 + 2y^2 - 3y + 1$
11. $6b^2 - 11b + 4$ **12.** $10a^3 + 19a^2 + 11a + 2$
13. $t^3 - 6t^2 + 8t + 3$ **14.** $2b^2 + 4b + 5$ **15.** $k^2 + 4k + 25$
16. $3x^3 + x^2 - 5$ **17.** $7t^3 - 12t^2 + 2t$ **18.** $-xy^2 + 2y + \dfrac{x}{7y}$
19. $\dfrac{11}{12}p^3 - \dfrac{1}{2}p^2 + \dfrac{1}{5}p + 5$ **20.** $-30.6w^6 + 15.6w^5 - 7.2w^4$
21. $36a^4 - 48a^2b + 16b^2$ **22.** $\dfrac{1}{4}z^4 - \dfrac{1}{9}$ **23.** $m^2 - 8m - 7$
24. $x^2 + 3x - 14$ **25.** $2m^4 - 8m^3 - 13m^2 + 46m - 21$
26. $x^2 + 4x + 16$ **27.** $25 - 10a - 10b + a^2 + 2ab + b^2$

28. $a^2 - x^2 + 2xy - y^2$ **29.** $4xy$ **30.** $a^3 - 12a^2 + 48a - 64$
31. $-\dfrac{1}{8}x^2 + \dfrac{1}{3}x - \dfrac{1}{6}$ **32.** $-\dfrac{1}{2}x^6y^4z^5w^3$

Section 4.5 Practice Exercises, pp. 366–368

1. a. product   **b.** greatest common factor   **c.** greatest common factor   **d.** grouping   **3.** $9t^4 + 5t^3 - 6t^2 - 6t$
5. $5y^4 + 5y^3 + 7y^2 - 3y - 6$ **7.** $-3v^2 + 6v - 1$
9. $3(x + 4)$ **11.** $2z(3z + 2)$ **13.** $4p(p^5 - 1)$
15. $12x^2(x^2 - 3)$ **17.** $9t(st + 3)$ **19.** $9a^2b^3(a^2 + 3ab - 2b^2)$
21. $5xy(2x + 3y - 1)$ **23.** $b(13b - 11a^2 - 12a)$
25. $-1(x^2 + 10x - 7)$ **27.** $-3xy(4x^2 + 2x + 1)$
29. $-t(2t^2 - 11t + 3)$ **31.** $(3z - 2b)(2a - 5)$
33. $(2x - 3)(2x^2 + 1)$ **35.** $(2x + 1)^2(y - 3)$
37. $3(x - 2)^2(y + 2)$ **39.** For example: $3x^3 + 6x^2 + 12x^4$
41. For example: $6(c + d) + y(c + d)$ **43. a.** $(2x - y)(a + 3b)$
b. $(2w - 1)(5w - 3b)$ **c.** In part (b), $-3b$ was factored out so that the signs in the last two terms were changed. The resulting binomial factor matches the binomial factor from the first two terms.
45. $(y + 4)(y^2 + 3)$ **47.** $(p - 7)(6 + q)$ **49.** $(m + n)(2x + 3y)$
51. $(2x - 3y)(5a - 4b)$ **53.** $(x^2 - 3)(x - 1)$
55. $6p(p + 3)(q - 5)$ **57.** $100(x - 3)(x^2 + 2)$
59. $(3a + b)(2x - y)$ **61.** $(4 - b)(a + 3)$ **63.** Cannot be factored   **65.** It is not possible to get a common binomial factor regardless of the order of terms.   **67.** $A = \dfrac{U}{v + cw}$
69. $y = \dfrac{bx}{c - a}$ or $y = \dfrac{-bx}{a - c}$ **71.** Length $= 2w + 1$
73. $(a + 3)^4(6a + 19)$ **75.** $18(3x + 5)^2(4x + 5)$
77. $(t + 4)(t + 3)$ **79.** $5w^2(2w - 1)^2(7w - 3)$

Section 4.6 Practice Exercises, pp. 379–381

1. a. positive   **b.** opposite   **c.** Both are correct.
d. $2(3x - 5)(x + 1)$ **e.** $(a + b)^2$; $(a - b)^2$
3. $6c^2d^4e^7(6d^3e^4 + 2cde^8 - 1)$ **5.** $(3a - b)(2x - 1)$
7. $(z + 2)(wz - 33a)$ **9.** $(b - 8)(b - 4)$
11. $(y + 12)(y - 2)$ **13.** $(x + 10)(x + 3)$ **15.** $(c - 8)(c + 2)$
17. $(2x + 3)(x - 5)$ **19.** $(6a - 5)(a + 1)$ **21.** $(s + 3t)(s - 2t)$
23. $3(x - 18)(x - 2)$ **25.** $2(c - 4)(c + 3)$ **27.** $2(x - y)(x + 5y)$
29. Prime   **31.** $(3x + 5y)(x + 3y)$ **33.** $5uv(u - 3v)^2$
35. $x(x - 7)(x + 2)$ **37.** $(2z - 5)(5z + 1)$ **39.** Prime
41. $-2(t + 4)(t - 10)$ **43.** $(7a - 4)(2a + 3)$
45. $2b(3a + 2)(a + 3)$ **47. a.** $x^2 + 10x + 25$ **b.** $(x + 5)^2$
49. a. $9x^2 - 12xy + 4y^2$ **b.** $(3x - 2y)^2$ **51.** $30x$
53. $16z^2t$ **55.** $(y - 4)^2$ **57.** $(8m + 5)^2$
59. Not a perfect square trinomial   **61.** $(3a - 5b)^2$
63. $4(4t^2 - 20tv + 5v^2)$; Not a perfect square trinomial
65. $5(b^2 - 2)^2$ **67. a.** $(u - 5)^2$ **b.** $(x^2 - 5)^2$ **c.** $(a - 4)^2$
69. a. $(u + 13)(u - 2)$ **b.** $(w^3 + 13)(w^3 - 2)$ **c.** $(y + 9)(y - 6)$
71. $(3x - 4)(3x + 1)$ **73.** $(2x - 9)(x - 1)$ **75.** $(3y + 11)(y + 6)$
77. $(3y^3 + 2)(y^3 + 3)$ **79.** $(4p^2 + 1)(p^2 + 1)$
81. $(x^2 + 12)(x^2 + 3)$ **83.** The factorization $(2y - 1)(2y - 4)$ is not factored completely because the factor $2y - 4$ has a GCF of 2.
85. $(w^2 + 6)^2$ **87.** $(9w + 5)^2$ **89.** $3(a + b)(x - 2)$
91. $2abc^2(6a + 2b - 3c)$ **93.** $-2x(5x - 6)(2x - 5)$
95. Prime   **97.** $(2w^2 - 15)(w^2 - 2)$
99. $(1 - d)(1 - 3d)$ or $(3d - 1)(d - 1)$
101. $(a + 2b)(x - 5a)$ **103.** $8(z - 4w)(z + 7w)$
105. $(y + x)(a - 5c)$ **107.** $g(x) = (3x + 2)(x + 4)$
109. $n(t) = (t + 10)^2$ **111.** $Q(x) = x^2(x + 4)(x + 2)$
113. $k(a) = (a - 4)(a^2 + 2)$

Section 4.7 Practice Exercises, pp. 389–391

1. a. difference; $(a + b)(a - b)$ **b.** sum   **c.** is not   **d.** square
e. sum; cubes   **f.** difference; cubes   **g.** $(a - b)(a^2 + ab + b^2)$
h. $(a + b)(a^2 - ab + b^2)$ **3.** $(2x - 5)^2$ **5.** $(5 + 3y)(2x + 1)$
7. $4(8p + 1)(p - 1)$ **9.** Look for a binomial of the form $a^2 - b^2$. This factors as $(a + b)(a - b)$.   **11.** $(x - 3)(x + 3)$
13. $(4 - 7w)(4 + 7w)$ **15.** $2(2a - 9b)(2a + 9b)$ **17.** Prime

19. $2(a^2 + 4)(a - 2)(a + 2)$ **21.** $(7 - k^3)(7 + k^3)$
23. $(x - 4)(x + 4)(x - 1)$ **25.** $(2x + 1)(2x - 1)(x + 3)$
27. $(9y + 7)(y - 2)(y + 2)$ **29.** $(7x + 2 - y)(7x + 2 + y)$
31. $(w - 3n + 1)(w + 3n - 1)$ **33.** $(p^2 - 5 - t^2)(p^2 - 5 + t^2)$
35. $(3u^2 - 2v^2 + 5)(3u^2 + 2v^2 - 5)$ **37.** Look for a binomial of the form $a^3 + b^3$. This factors as $(a + b)(a^2 - ab + b^2)$.
39. $(2x - 1)(4x^2 + 2x + 1)$ **41.** $(5c + 3)(25c^2 - 15c + 9)$
43. $(x - 10)(x^2 + 10x + 100)$ **45.** $(4t^2 + 1)(16t^4 - 4t^2 + 1)$
47. $2(10y^2 + x)(100y^4 - 10y^2x + x^2)$
49. $2z(2z - 3)(4z^2 + 6z + 9)$ **51.** $(p^4 - 5)(p^8 + 5p^4 + 25)$
53. $\left(6y - \dfrac{1}{5}\right)\left(6y + \dfrac{1}{5}\right)$ **55.** $2(3d^6 - 4)(3d^6 + 4)$
57. $2(121v^2 + 16)$ **59.** $4(x - 2)(x + 2)$ **61.** $(5 - 7q)(5 + 7q)$
63. $(t + 2s - 6)(t + 2s + 6)$ **65.** $(3 - t)(9 + 3t + t^2)$
67. $\left(3a + \dfrac{1}{2}\right)\left(9a^2 - \dfrac{3}{2}a + \dfrac{1}{4}\right)$ **69.** $2(m + 2)(m^2 - 2m + 4)$
71. $(x - y)(x + y)(x^2 + y^2)$
73. $(a + b)(a^2 - ab + b^2)(a^6 - a^3b^3 + b^6)$
75. $\left(\dfrac{1}{2}p - \dfrac{1}{5}\right)\left(\dfrac{1}{4}p^2 + \dfrac{1}{10}p + \dfrac{1}{25}\right)$ **77.** Prime
79. $\left(\dfrac{1}{5}x - \dfrac{1}{2}y\right)\left(\dfrac{1}{5}x + \dfrac{1}{2}y\right)$
81. $(a + b)(a^2 - ab + b^2)(a - b)(a^2 + ab + b^2)$
83. $(2 + y)(4 - 2y + y^2)(2 - y)(4 + 2y + y^2)$
85. $(h^2 + k^2)(h^4 - h^2k^2 + k^4)$ **87.** $(2x^2 + 5)(4x^4 - 10x^2 + 25)$
89. $4x^2 - 9$ **91.** $8a^3 - 27$ **93.** $64x^6 + y^3$ **95. a.** $x^2 - y^2$
b. $(x + y)(x - y)$ **c.** 20 in.^{2} **97.** $(x + y)(x - y + 1)$
99. $(x + y)(x^2 - xy + y^2 + 1)$
101. $(3a - c)(3a + c)(4a - 1)(16a^2 + 4a + 1)$

Chapter 4 Problem Recognition Exercises, pp. 392–394

1. A polynomial whose only factors are 1 and itself
2. Factor out the GCF.   **3.** Difference of squares $a^2 - b^2$, difference of cubes $a^3 - b^3$, or sum of cubes $a^3 + b^3$
4. Look for a perfect square trinomial, $a^2 + 2ab + b^2$ or $a^2 - 2ab + b^2$.
5. Try factoring by grouping 2 terms by 2 terms or by grouping 3 terms by 1 term.   **6.** Let $u = (4x^2 + 1)$. The polynomial becomes $3u^2 + 20u + 12$. Factor this simpler expression and then back substitute.   **7. a.** Trinomial   **b.** $3(2x + 3)(x - 5)$
8. a. Trinomial   **b.** $m(4m + 1)(2m - 3)$
9. a. Difference of squares   **b.** $2(2a - 5)(2a + 5)$
10. a. Grouping   **b.** $(b + y)(a - b)$
11. a. Trinomial   **b.** $(2u - v)(7u - 2v)$
12. a. Perfect square trinomial   **b.** $(3p - 2q)^2$
13. a. Difference of cubes   **b.** $2(2x - 1)(4x^2 + 2x + 1)$
14. a. Sum of squares **b.** Prime
15. a. Sum of cubes   **b.** $(3y + 5)(9y^2 - 15y + 25)$
16. a. None of these **b.** Prime
17. a. Sum of cubes   **b.** $2(4p^2 + 3q)(16p^4 - 12p^2q + 9q^2)$
18. a. Perfect square trinomial   **b.** $5(b - 3)^2$
19. a. Difference of squares   **b.** $(2a - 1)(2a + 1)(4a^2 + 1)$
20. a. Perfect square trinomial   **b.** $(9u - 5v)^2$
21. a. Grouping   **b.** $(p - 6 - c)(p - 6 + c)$
22. a. Sum of squares   **b.** $4(x^2 + 4)$
23. a. Grouping   **b.** $2(2x - y)(3a + b)$
24. a. Difference of cubes   **b.** $(5y - 2)(25y^2 + 10y + 4)$
25. a. Trinomial   **b.** $(5y - 1)(y + 3)$
26. a. Difference of squares   **b.** $2(m^2 - 8)(m^2 + 8)$
27. a. Difference of squares   **b.** $(t - 10)(t + 10)$
28. a. Difference of squares   **b.** $(2m - 7n)(2m + 7n)$
29. a. Sum of cubes   **b.** $(y + 3)(y^2 - 3y + 9)$
30. a. Sum of cubes   **b.** $(x + 1)(x^2 - x + 1)$
31. a. Trinomial   **b.** $(d - 4)(d + 7)$
32. a. Trinomial   **b.** $(c + 8)(c - 3)$
33. a. Perfect square trinomial   **b.** $(x - 6)^2$
34. a. Perfect square trinomial   **b.** $(p + 8)^2$
35. a. Grouping   **b.** $(ax + b)(2x - 5)$
36. a. Grouping   **b.** $(4x + a)(2x - b)$

37. a. Trinomial   **b.** $(2y - 1)(5y + 4)$
38. a. Trinomial   **b.** $(3z + 2)(4z + 1)$
39. a. Difference of squares   **b.** $10(p - 8)(p + 8)$
40. a. Difference of squares **b.** $2(5a - 6)(5a + 6)$
41. a. Difference of cubes   **b.** $z(z - 4)(z^2 + 4z + 16)$
42. a. Difference of cubes **b.** $t(t - 2)(t^2 + 2t + 4)$
43. a. Trinomial   **b.** $b(b + 5)(b - 9)$
44. a. Trinomial **b.** $y(y - 4)(y - 10)$
45. a. Perfect square trinomial   **b.** $(3w + 4x)^2$
46. a. Perfect square trinomial   **b.** $(2k - 5p)^2$
47. a. Grouping   **b.** $10(2x + a)(3x - 1)$
48. a. Grouping   **b.** $10(5x + c)(x - 4)$
49. a. Difference of squares   **b.** $(w^2 + 4)(w - 2)(w + 2)$
50. a. Difference of squares **b.** $(k^2 + 9)(k - 3)(k + 3)$
51. a. Difference of cubes   **b.** $(t^2 - 2)(t^4 + 2t^2 + 4)$
52. a. Sum of cubes **b.** $(p^2 + 3)(p^4 - 3p^2 + 9)$
53. a. Trinomial   **b.** $(4p - 1)(2p - 5)$
54. a. Trinomial **b.** $(3m + 4)(3m - 5)$
55. a. Perfect square trinomial   **b.** $(6y - 1)^2$
56. a. Perfect square trinomial   **b.** $(3a + 7)^2$
57. a. Sum of squares   **b.** $2(x^2 + 25)$
58. a. Sum of squares   **b.** $4(y^2 + 16)$
59. a. Trinomial   **b.** $s^2(3r - 2)(4r + 5)$
60. a. Trinomial   **b.** $w^2(7z + 4)(z - 2)$
61. a. Trinomial   **b.** $(x - 3y)(x + 11y)$
62. a. Trinomial   **b.** $(s + 3t)(s - 12t)$
63. a. Sum of cubes   **b.** $(m^2 + n)(m^4 - m^2n + n^2)$
64. a. Difference of cubes **b.** $(a - b^2)(a^2 + ab^2 + b^4)$
65. a. None of these   **b.** $x(x - 4)$
66. a. None of these   **b.** $y(y - 9)$ **67.** $(x - y)(x + y)^2$
68. $(u - v)^2(u + v)$ **69.** $(a + 3)^4(6a + 19)$
70. $(4 - b)^3(2 - b)$ **71.** $18(3x + 5)^2(4x + 5)$
72. $5(2y - 3)^2(6y + 11)$ **73.** $\left(\dfrac{1}{10}x + \dfrac{1}{7}\right)^2$ **74.** $\left(\dfrac{1}{5}a + \dfrac{1}{6}\right)^2$
75. $5x^2(5x^2 - 6)$ **76.** $x^3(x^3 - 2)$ **77.** $(4p^2 + q^2)(2p - q)(2p + q)$
78. $(s^2t^2 + 9)(st - 3)(st + 3)$ **79.** $\left(y + \dfrac{1}{4}\right)\left(y^2 - \dfrac{1}{4}y + \dfrac{1}{16}\right)$
80. $\left(z + \dfrac{1}{5}\right)\left(z^2 - \dfrac{1}{5}z + \dfrac{1}{25}\right)$ **81.** $(a + b)(a - b)(6a + b)$
82. $(2p - q)(2p + q)(p + 3q)$ **83.** $\left(\dfrac{1}{3}t + \dfrac{1}{4}\right)^2$ **84.** $\left(\dfrac{1}{5}y + \dfrac{1}{2}\right)^2$
85. $(x + 6 - a)(x + 6 + a)$ **86.** $(a + 5 + b)(a + 5 - b)$
87. $(p + q - 9)(p + q + 9)$ **88.** $(m - n - 3)(m - n + 3)$
89. $(b - x - 2)(b + x + 2)$ **90.** $(p - y + 3)(p + y - 3)$
91. $(2 + u - v)(2 - u + v)$ **92.** $(5 - a - b)(5 + a + b)$
93. $(3a + b)(2x - y)$ **94.** $(q + 3)(5p - 4)$
95. $(u - 2)(u + 2)(u^2 + 2u - 4)(u^2 - 2u + 4)$
96. $(1 - v)(1 + v)(1 + v + v^2)(1 - v + v^2)$
97. $(x^4 + 1)(x^2 + 1)(x + 1)(x - 1)$
98. $(y^4 + 15)(y^2 + 4)(y + 2)(y - 2)$
99. $(a + b)(a - b + 1)$ **100.** $(5c - 3d)(5c + 3d + 1)$
101. $(x + y)(x^2 - xy + y^2)(5w - 2z)$

Section 4.8 Practice Exercises, pp. 403–407

1. a. quadratic   **b.** 0;0   **c.** Pythagorean; c^2 **d.** quadratic
e. $f(x) = 0; y$ **f.** $x + 1; x + 2; x + 2$ **g.** lw **h.** $\dfrac{1}{2}bh$
3. $x(10x + 3)$ **5.** $(p - 5)(2p + 1)$ **7.** $(t - 1)(t^2 + t + 1)$
9. The equation must be set equal to 0, and the polynomial must be factored.   **11.** Correct form   **13.** Incorrect form. Polynomial is not factored.   **15.** Incorrect form. Equation is not set equal to 0.
17. a. $(w + 9)(w - 9)$ **b.** $\{-9, 9\}$ **19. a.** $(3x - 1)(x + 5)$
b. $\left\{\dfrac{1}{3}, -5\right\}$ **21.** $\{-3, -5\}$ **23.** $\left\{-\dfrac{9}{2}, \dfrac{1}{5}\right\}$ **25.** $\left\{0, -4, \dfrac{3}{10}\right\}$
27. $\{0.4, -2.1\}$ **29.** $\{-9, 3\}$ **31.** $\left\{-3, \dfrac{1}{2}\right\}$ **33.** $\left\{0, \dfrac{3}{2}\right\}$
35. $\left\{\dfrac{23}{3}\right\}$ **37.** $\{-3\}$ **39.** $\left\{-\dfrac{1}{3}, 2\right\}$ **41.** $\left\{-1, \dfrac{1}{2}, 3\right\}$

43. $\{-5, 4\}$ **45.** $\left\{\dfrac{5}{2}, -1\right\}$ **47.** $\{-12, 5\}$ **49.** $\left\{-\dfrac{1}{11}\right\}$

51. $\{0, 6, -2\}$ **53.** $\{0, 4, -4\}$ **55.** $\left\{3, -3, -\dfrac{5}{2}\right\}$ **57.** $5, -5$

59. $4, -3$ **61.** $-7, -6$ or $6, 7$ **63.** $-9, -7$ or $7, 9$
65. The length is 7 ft, and the width is 5 ft.
67. The length is 20 yd, and the width is 15 yd.
69. a. The base is 5 in., and the height is 6 in.   **b.** The area is 15 in.^{2}
71. The base is 10 ft, and the height is 5 ft.   **73.** The integers are 4
and 5.   **75. a.** 14 mi   **b.** The alternative route using superhighways
77. The lengths are 6 m, 8 m, and 10 m.   **79.** The radius is 2 units.
81. a. $0, 3$ **b.** $f(0) = 0$ **83. a.** $7, -1$ **b.** $f(0) = -7$
85. x-intercepts: $(2, 0), (-1, 0), (0, 0)$; y-intercept: $(0, 0)$
87. x-intercept: $(1, 0)$; y-intercept: $(0, 1)$ **89.** $(3, 0)(-3, 0)$; d
91. $(-1, 0)$; a   **93. a.** The function is in the form
$s(t) = at^2 + bt + c$.   **b.** $(0, 0)$ and $(100, 0)$ **c.** At 0 sec and 100 sec,
the rocket is at ground level (height $= 0$).   **d.** At 1 sec and 99 sec
95. $f(x) = (x - 5)(x - 2)$; $x = 5$ and $x = 2$ represent x-intercepts.
97. $f(x) = (x + 1)^2$; $x = -1$ represents the x-intercept.
99. $f(x) = -(x + 1)(x + 5)$; $x = -1$ and $x = -5$ represent the x-
intercepts.   **101.** The radius is 6 ft.   **103.** The length is 8 ft and
the width is 6 ft.   **105.** $(x - 2)(x + 2) = 0$ or $x^2 - 4 = 0$
107. $(x - 0)(x + 3) = 0$ or $x^2 + 3x = 0$

Section 4.8 Graphing Calculator Exercises, p. 407

108. $(2, 0)(-1, 0)$

109. $(-5, 0)(4, 0)$

110. $(3, 0)$ **111.** $(-2, 0)$

Chapter 4 Review Exercises, pp. 414–417

1. $3^5 x^5$ or $243x^5$ **2.** $-\dfrac{18}{x^{12}}$ **3.** $-3xy^2$ **4.** $\dfrac{3x^3}{2y^2}$ **5.** $-\dfrac{b^{15}}{8a^6}$

6. $\dfrac{a^4}{16b^6}$ **7.** $\dfrac{5^4 y^{24}}{4^4 x^{20}}$ **8.** $\dfrac{x^{10} y^5}{5^5}$ **9. a.** 3.6866×10^9

b. 1×10^{-6} **10. a.** 1×10^{-3} **b.** 5.1557×10^9
11. a. 0.001 **b.** 0.000000001 **12. a.** $5,230,000,000$ ft^2
b. $\$1,091,000,000,000$ **13.** 6.25×10^9 **14.** 2×10^{-8}
15. 3.24×10^7 **16.** 3.64×10^{-8} **17.** Trinomial; degree 4
18. Monomial; degree 0   **19. a.** -7 **b.** -23 **c.** 5
20. a. 12   **b.** 10   **c.** -2 **21. a.** $A(5) \approx 25$ means that in the
year 2005, Americans on average consumed approximately 25 gal of
bottled water each.   **b.** $A(15) \approx 49$ means that in the year 2015,
Americans will consume approximately 49 gal of bottled water each
if this trend continues.   **22.** $-2x^2 - 3x - xy - 1$
23. $20xy - 18xz - 3yz$ **24.** $3a^3 + 5a^2 + 6a$ **25.** $-2a^2 + 6a$
26. $x^4 - x^2 - 1$ **27.** $x^4 + \dfrac{3}{4}x^2$ **28.** $6x - 6y$ **29.** $-5x + 9y$

30. $-11x + 1$ **31.** $4x^2 - 11x$ **32.** $-3x - 11$ **33.** $-3x$
34. $2x^3 - 14x^2 - 8x$ **35.** $-18x^3 + 15x^2 - 12x$
36. $x^2 - x - 42$ **37.** $x^2 - 11x + 18$ **38.** $\dfrac{1}{4}x^2 - 2x - 5$

39. $2y^2 + \dfrac{1}{5}y - \dfrac{1}{25}$ **40.** $27x^3 + 125$ **41.** $x^3 - y^3$

42. $4x^2 - 20x + 25$ **43.** $\dfrac{1}{4}x^2 + 4x + 16$ **44.** $9y^2 - 121$

45. $36w^2 - 1$ **46.** $\dfrac{4}{9}t^2 - 16$ **47.** $z^2 - \dfrac{1}{16}$

48. $x^2 + 4x + 4 - b^2$ **49.** $c^2 - w^2 - 6w - 9$
50. $8x^3 + 12x^2 + 6x + 1$ **51.** $y^6 - 9y^4 + 27y^2 - 27$
52. a. $4x^2 + 12x + 9$ **b.** $16x^2 + 24x + 9$ **c.** $12x^2 + 12x$
53. a. $P(x) = 8x + 4$ **b.** $A(x) = 3x^2 + 2x$
54. a. False; cannot add unlike terms   True   **b.** True   False;
when subtracting *like* terms, keep the variable the same.
55. $2x^2 + 4xy - 3y^2$ **56.** $-2x^2 - 3x + 4$
57. a. $3y^3 - 2y^2 + 6y - 4$ **b.** Quotient $3y^3 - 2y^2 + 6y - 4$; no
remainder   **c.** Multiply the quotient and the divisor.
58. $x + 2$ **59.** $x + 4 + \dfrac{-32}{x + 4}$

60. $2x^3 + 2x^2 + 8x + 24 + \dfrac{72x - 4}{x^2 - 3x}$

61. $2x^3 - 2x^2 + 5x - 4 + \dfrac{4x - 4}{x^2 + x}$
62. Synthetic division can be used with a divisor in the form $x - r$.
63. a. $x - 3$ **b.** $2x^3 + 11x^2 + 31x + 99$ **c.** 298
64. $t^2 - t + 6$ **65.** $x + 2 + \dfrac{4}{x + 5}$ **66.** $x + 4 + \dfrac{4}{x + 4}$

67. $w^2 - 3w - 9 + \dfrac{-19}{w - 3}$ **68.** $p^3 + 2p^2 + 4p + 8$

69. $x(-x^2 - 4x + 11)$ or $-x(x^2 + 4x - 11)$ **70.** $7(3w^3 - w + 2)$
71. $(x - 7)(5x - 2)$ **72.** $(t + 4)(3t + 5)$
73. $(m - 8)(m^2 + 1)$ **74.** $12(2x - 3)(x^2 + 3)$
75. $x(2a + b)(2x - 3)$ **76.** $(y - 6)(y^2 + 1)$ **77.** The
trinomial must be of the form $a^2 + 2ab + b^2$ or $a^2 - 2ab + b^2$.
78. $(3x + 2y)(6x + 5y)$ **79.** $(3m - 5t)(m + 2t)$
80. $5a^2(3 + 4a)(4 - a)$ or $-5a^2(4a + 3)(a - 4)$
81. $k^2(3k + 2)(2k + 1)$ **82.** $(7x - 6)^2$ **83.** $2(5z + 4)^2$
84. $(9w + 7)(9w + 1)$ **85.** $(4x - 3)^2$
86. $3(3a^2 - 1)(2a^2 + 5)$ **87.** $(w^2 + 1)(3w^2 - 5)$
88. $(5 - y)(5 + y)$ **89.** $\left(x - \dfrac{1}{3}\right)\left(x^2 + \dfrac{1}{3}x + \dfrac{1}{9}\right)$

90. Prime   **91.** $h(h^2 + 9)$ **92.** $(a + 4)(a^2 - 4a + 16)$
93. $(k^2 + 4)(k - 2)(k + 2)$ **94.** $y(3y - 2)(3y + 2)$
95. $(x - 4y - 3)(x - 4y + 3)$ **96.** $(a + 6 - b)(a + 6 + b)$
97. $(t + 8 - 5c)(t + 8 + 5c)$ **98.** $(y - 3 - 4x)(y - 3 + 4x)$
99. It can be written in the form $ax^2 + bx + c = 0$ ($a \neq 0$).
100. It is a parabola.   **101.** Quadratic   **102.** Quadratic
103. Linear   **104.** Quadratic   **105. a.** $(5x - 4)(x + 2)$

b. $\left\{\dfrac{4}{5}, -2\right\}$ **106. a.** $(3x - 7)(x - 4)$ **b.** $\left\{\dfrac{7}{3}, 4\right\}$ **107.** $\{-3, 5\}$

108. $\left\{\dfrac{3}{8}, 7\right\}$ **109.** $\left\{-\dfrac{4}{3}, -1\right\}$ **110.** $\left\{1, -5, \dfrac{9}{2}\right\}$

111. $(1, 0)(-1, 0)(0, 4)$; b   **112.** $(1, 0)(-1, 0)(0, -2)$; d
113. $(2, 0)(-2, 0)(0, 40)$; c   **114.** $(2, 0)(-2, 0)\left(0, \dfrac{1}{2}\right)$; a
115. Length 15 ft; width 8 ft; height 10 ft
116. a.

Time t (sec)	Height $h(t)$ (ft)
0	-1280
1	-624
3	592
10	3840
20	5760
30	4480
42	-1280

b. The position of the missile is below sea level.   **c.** The missile will be at sea level after 2 sec and again after 40 sec.

Chapter 4 Test, pp. 418–419

1. $5a^{13}$ **2.** x^{11} **3.** $\dfrac{9x^{12}}{25y^{14}}$ **4.** $\dfrac{8y^{12}}{x^7}$ **5.** 5.68

6. 2.3×10^9 **7.** $F(-1) = 1$ $F(2) = 40$ $F(0) = 8$
8. $8x^2 - 8x + 8$ **9.** $2a^3 - 13a^2 + 2a + 45$
10. $2x^2 - \dfrac{23}{3}x - 6$ **11.** $25x^2 - 16y^2$ **12.** The expression $25x^2 + 49$ does not include the middle term $70x$.
13. $49x^2 - 56x + 16$ **14.** $x^2y^3 + \dfrac{5}{2}xy - 3y^2 - \dfrac{1}{2}$
15. $5p^2 - p + 1$ **16.** $y^3 + 2y^2 + 4y + 6 + \dfrac{17}{y - 2}$
17. See page 392.   **18.** See page 395.
19. $3(a + 6b)(a + 3b)$ **20.** $(c - 1)(c + 1)(c^2 + 1)$
21. $(y - 7)(x + 3)$ **22.** Prime   **23.** $-10(u - 2)(u - 1)$
24. $3(2t - 5)(2t + 5)$ **25.** $5(y - 5)^2$ **26.** $7q(3q + 2)$
27. $(2x + 1)(x - 2)(x + 2)$ **28.** $(y - 5)(y^2 + 5y + 25)$
29. $(x + 4 - y)(x + 4 + y)$ **30.** $r^2(r^2 + 16)(r - 4)(r + 4)$
31. $(x^2 + 3)(x^2 + 2)$ **32.** $(2 - c)(6a + b)$ **33.** $\left\{\dfrac{3}{2}, -5\right\}$
34. $\{0, 7\}$ **35.** $\{8, -2\}$ **36.** $\left\{\dfrac{1}{5}, -1\right\}$ **37.** $\left\{0, \dfrac{1}{4}, -\dfrac{1}{4}\right\}$
38. $\left\{-\dfrac{1}{4}\right\}$ **39.** 256 ft   **40. a.** $P(4) = 127.292$. In the year 2004, the population of Japan was approximately 127.292 million.
b. 126.968 million   **c.** 124.52 million   **41.** $(4, 0)(2, 0)(0, 8)$; c
42. $(-4, 0)(3, 0)(-3, 0)(0, -36)$; b
43. $(-3, 0)(-1, 0)(0, -6)$; d   **44.** $(4, 0)(-3, 0)(0, 0)$; a

Chapters 1–4 Cumulative Review Exercises, pp. 419–420

1. $[5, 12]$

2. $-x^2 - 5x - 10$
3. a. 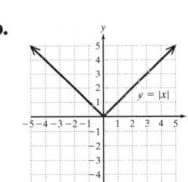   **b.**

4. $\dfrac{71}{8}$ **5.** 1.97×10^8 people   **6.** Penn State scored 26 points and Florida State scored 23 points.   **7.** $37°, 74°, 69°$

8. $x^2 - 4x + 16$
9. $m = \dfrac{4}{3}$; y-intercept $(0, 3)$

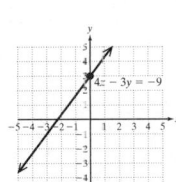

10. $y = -5$ **11.** $\dfrac{a^6}{8b^{30}}$ **12.** $\{(0, 3, 2)\}$ **13. a.** Function
b. Not a function   **14.** $m = 7$ **15.** 7   **16.** $\{8\}$
17. $y = \dfrac{3}{2}x - \dfrac{5}{2}$ **18. a.** The fourth test would have to be 107; therefore, it is not possible.   **b.** Student can score between 67 and 100, inclusive.   **19.** 15 L of 40% solution; 10 L of 15% solution
20. $8b^3 - 6b^2 + 4b - 3$ **21.** $3a^3 + 5a^2 - 2a + 5$
22. $3w - \dfrac{5}{2} - \dfrac{1}{w}$ **23.** $\{7, -2\}$ **24.** $\{0, -4, -5\}$
25. $(\infty, 1) \cup (4, \infty)$

Chapter 5

Chapter Opener Puzzle

To $\underset{2}{\underline{\text{add}}}$ or $\underset{4}{\underline{\text{subtract}}}$ $\underset{8}{\underline{\text{rational}}}$ $\underset{3}{\underline{\text{expressions,}}}$ $\underset{1}{\underline{\text{we}}}$ $\underset{6}{\underline{\text{need}}}$ a $\underset{5}{\underline{\text{common}}}$ $\underset{7}{\underline{\text{denominator.}}}$

Section 5.1 Practice Exercises, pp. 428–431

1. a. rational   **b.** denominator   **c.** $\dfrac{p}{q}$ **d.** $1; -1$
3. $k(0) = -\dfrac{3}{4}, k(-1) = -1, k(2) = -\dfrac{1}{2}, k(-4)$ is undefined
5. $n(1) = 2, n(0) = 1, n\left(-\dfrac{1}{3}\right) = 0, n(-1) = -1$
7. a. $\{x \mid x$ is a real number, $x \neq 0\}$ **b.** $(-\infty, 0) \cup (0, \infty)$
9. a. $\{v \mid v$ is a real number, $v \neq 7\}$ **b.** $(-\infty, 7) \cup (7, \infty)$
11. a. $\{x \mid x$ is a real number, $x \neq \frac{5}{2}\}$ **b.** $(-\infty, \frac{5}{2}) \cup (\frac{5}{2}, \infty)$
13. a. $\{q \mid q$ is a real number, $q \neq -9, q \neq 3\}$
b. $(-\infty, -9) \cup (-9, 3) \cup (3, \infty)$ **15. a.** $\{c \mid c$ is a real number$\}$
b. $(-\infty, \infty)$ **17. a.** $\{x \mid x$ is a real number, $x \neq -5, x \neq 5\}$
b. $(-\infty, -5) \cup (-5, 5) \cup (5, \infty)$ **19. a.** $\{x \mid x$ is a real number$\}$
b. $(-\infty, \infty)$ **21.** $(-\infty, -4) \cup (-4, \infty)$; b   **23.** $(-\infty, 4) \cup (4, \infty)$; d
25. a. $\dfrac{2x}{y}$ **b.** Cannot be simplified   **27. a.** $\dfrac{(x + 4)(x + 2)}{(x + 4)(x - 1)}$
b. $x \neq -4, x \neq 1$ **c.** $\dfrac{x + 2}{x - 1}$ **29. a.** $\dfrac{(x - 9)^2}{(x - 9)(x + 9)}$
b. $x \neq 9, x \neq -9$ **c.** $\dfrac{x - 9}{x + 9}$ **31.** $\dfrac{25x^2}{9y^3}$ **33.** $\dfrac{w^8z^3}{2}$ **35.** $-\dfrac{1}{4m^2n^3}$
37. $\dfrac{2}{3}$ **39.** $\dfrac{1}{x + 5}$ **41.** $-\dfrac{1}{3c - 5}$ **43.** $-\dfrac{t + 4}{t + 3}$
45. $\dfrac{(2p - 1)^2}{p + 1}$ **47.** $\dfrac{3 - z}{2z - 5}$ **49.** $\dfrac{2(z^2 - 4z + 16)}{z + 4}$
51. $-\dfrac{2x - 5}{2}$ **53.** 1   **55.** -1 **57.** -1 **59.** -2
61. $\dfrac{c + 4}{c - 4}$; cannot be simplified   **63.** $-\dfrac{1}{12(x + y)}$
65. $\dfrac{t - 1}{t + 6}$ **67.** $\dfrac{4}{p - 3}$ **69.** $-\dfrac{2a}{b}$ **71.** $-\dfrac{x + y}{8}$
73. $\dfrac{1}{2b - 3}$ **75.** $\dfrac{x - 17}{2x - 3}$ **77.** $\dfrac{(a - 5)^2}{a - 2}$ **79.** $-\dfrac{2x}{5}$
81. $\dfrac{x - 5}{x + 2}$ **83.** $\dfrac{t^2 - 2t + 4}{3t - 5}$ **85.** For example: $\dfrac{1}{x - 2}$
87. For example: $f(x) = \dfrac{1}{x + 5}$

Section 5.2 Practice Exercises, pp. 434–436

1. a. $\dfrac{pr}{qs}$ **b.** $\dfrac{ps}{qr}$ **3.** $\dfrac{t + 1}{t - 1}$ **5.** $-\dfrac{1}{p + 1}$ **7.** $\dfrac{1}{x - 3}$
9. $\dfrac{bc^3}{2}$ **11.** $\dfrac{2}{z^3}$ **13.** $12r^3$ **15.** $\dfrac{x(2x - 3)}{y^2(x + 1)}$ **17.** $\dfrac{2(3w - 7)}{5w + 4}$
19. 15   **21.** $y^2(y - 2)$ **23.** $\dfrac{11}{5a^4b^3}$ **25.** $\dfrac{r + 3}{4r^2}$
27. $\dfrac{1}{(p + 2)(6p - 7)}$ **29.** $\dfrac{4}{(b + 2)(b + 3)}$ **31.** $\dfrac{2s + 5t}{s + 5t}$
33. $\dfrac{2a + b^2}{a^2 + b^2}$ **35.** $-\dfrac{4x}{3(x + 2)}$ **37.** $\dfrac{9}{4y^2}$ **39.** $\dfrac{24}{y^2}$
41. $\dfrac{2a}{5}$ **43.** $\dfrac{2(x + 3)}{x + 2}$ **45.** $m - n$ **47.** $-\dfrac{x - 3y}{x + 3y}$ or $\dfrac{3y - x}{x + 3y}$
49. 1   **51.** $\dfrac{a - 3}{a + 4}$ **53.** $\dfrac{40(2x + 1)}{3}$ **55.** $\dfrac{(x - 6)(3x + 2)}{3x - 2}$
57. $\dfrac{2k}{h^3}$ cm^2 **59.** $\dfrac{5x}{4}$ ft^2

Section 5.3 Practice Exercises, pp. 444–447

1. a. $\dfrac{p+r}{q}$; $\dfrac{p-r}{q}$ **b.** least common denominator

3. $\dfrac{3(b+1)}{2(b-1)}$ **5.** $\dfrac{5a+1}{2}$ **7.** $\dfrac{2}{x}$ **9.** $\dfrac{1}{x+1}$

11. $\dfrac{2x+5}{(2x+9)(x-6)}$ **13.** 2 **15.** $40x$ **17.** $30m^4n^7$

19. $(x-4)(x+2)(x-6)$ **21.** $x^2(x-1)(x+7)^2$

23. $(x-6)(x-2)$ **25.** $a-4$ or $4-a$ **27.** $15xy$

29. $2x^3+4x^2$ **31.** y^2-y **33.** $\dfrac{8p-15}{6p^2}$ **35.** $\dfrac{-t-s}{st}$

37. $\dfrac{1}{3}$ **39.** $\dfrac{2b+20}{b(b+5)}$ **41.** $\dfrac{2x}{x-6}$ or $\dfrac{-2x}{6-x}$

43. $\dfrac{6b^2+5b+4}{(b-4)(b+1)}$ **45.** $\dfrac{10x}{(2x+1)(x-2)}$ **47.** $\dfrac{3y-1}{y+4}$

49. $\dfrac{11x+6}{(x-6)(x+6)(x+3)}$ **51.** $-\dfrac{3}{w}$ **53.** 1

55. $\dfrac{10-x}{3(x-5)(x+5)}$ or $\dfrac{x-10}{3(5-x)(5+x)}$

57. $\dfrac{m-5}{(m+5)(m+3)}$ **59.** $\dfrac{x^2+7x+6}{2x^2}$ **61.** $\dfrac{w^2-3}{w-2}$

63. $\dfrac{-x^2+13x-3}{x(x-1)^2}$ **65.** $-\dfrac{3}{t+3}$ **67.** $\dfrac{x^2+x+6}{2(x+1)}$

69. $\dfrac{2z^2+15z}{(z-3)(z+4)}$ **71.** $\dfrac{-2y}{(x-y)(x+y)}$ or $\dfrac{2y}{(y-x)(y+x)}$

73. $\dfrac{2p^2+2p-3}{3(p+4)}$ **75.** $\dfrac{x^2+13x+38}{(x+5)^3}$ **77.** $\dfrac{5}{(z+1)(z-1)}$

79. $\dfrac{5x^2+12x+24}{(x-2)(x+2)(x^2+2x+4)}$ **81.** $\dfrac{3x^2+5x+18}{3x^2}$ cm

83. $\dfrac{4x^2-2x+50}{(x-3)(x+5)}$ m

Section 5.4 Practice Exercises, pp. 452–453

1. complex   **3.** $\dfrac{5b^2}{3a}$ **5.** $\dfrac{10+3x}{2x^2}$ **7.** $\dfrac{2a+8}{(a-5)(a+1)}$

9. $\dfrac{5x^2}{27}$ **11.** $\dfrac{1}{x}$ **13.** $\dfrac{10}{3}$ **15.** -4 **17.** $28y$ **19.** -8

21. $\dfrac{3-p}{p-1}$ **23.** $\dfrac{2a+3}{4-9a}$ **25.** $-\dfrac{t}{t+1}$ **27.** $-\dfrac{4(w-1)}{w+2}$

29. $\dfrac{1}{y+4}$ **31.** $\dfrac{x+2}{x-1}$ **33.** $\dfrac{t^2}{(t+1)^2}$ **35.** $\dfrac{-a+2}{-a-3}$

37. $-\dfrac{y+1}{y-5}$ **39.** $-\dfrac{2}{x(x+h)}$ **41.** $\dfrac{1}{x(x^2+3)}$ **43.** $\dfrac{2b^2+3a}{b(b-a)}$

45. $\dfrac{-1}{4(4+h)}$ **47.** $\dfrac{-6}{x(x+h)}$ **49.** $m=\dfrac{y_2-y_1}{x_2-x_1}$ **51.** $\dfrac{63}{10}$

53. $\dfrac{4}{3}$ **55.** $(x^{-1}+y^{-1})^{-1}=\dfrac{1}{x^{-1}+y^{-1}}=\dfrac{1}{(1/x)+(1/y)}$

simplifies to $\dfrac{xy}{y+x}$ **57.** $-x(x-1)$

Chapter 5 Problem Recognition Exercises, p. 454

1. $\dfrac{4y^2-8y+9}{2y(2y-3)}$ **2.** $\dfrac{x^2+x-13}{x-4}$ **3.** $\dfrac{4x-1}{4x-3}$ **4.** $\dfrac{a-5}{3a}$

5. $\dfrac{2y-5}{(y-1)(y+1)}$ **6.** $\dfrac{4}{w+4}$ **7.** $\dfrac{a+4}{2}$ **8.** $\dfrac{(t-3)(t+2)}{t}$

9. $\dfrac{a}{2a-1}$ **10.** $\dfrac{2xy^2}{5}$ **11.** $\dfrac{-x^2+4xy-y^2}{(x-y)(x+y)}$ or $\dfrac{x^2-4xy+y^2}{(y-x)(y+x)}$

12. $3(x-4)$ **13.** $\dfrac{-x^2+4x+14}{6(x-2)}$ **14.** $\dfrac{x^2+14x+99}{10(x+7)}$

15. $-\dfrac{1}{3}$ **16.** $\dfrac{1}{9}$ **17.** $y-1$ **18.** $\dfrac{5t^2-6t-17}{(t+2)(t-3)}$

19. $\dfrac{3(x+4)}{2x-3}$ **20.** $\dfrac{3z^3}{2x^2y^2}$ **21.** $\dfrac{9x^2+6x+15}{4(3x-1)}$ **22.** $\dfrac{2x+3}{1-5x}$

23. $\dfrac{y^2+11y-1}{(y+3)(y-2)}$ **24.** $a-10$

Section 5.5 Practice Exercises, pp. 460–462

1. a. rational   **b.** denominator   **c.** No

3. $\dfrac{y-5}{(y+1)(y-1)^2}$ **5.** $\dfrac{1}{2t-1}$ **7.** xy **9.** $\{-14\}$

11. $\{-24\}$ **13.** $\left\{-\dfrac{15}{22}\right\}$ **15.** $\{6\}$ **17.** $\{5\}$ **19.** $\{6\}$

21. $\{-25\}$ **23.** $\{3,1\}$ **25.** $\{4,-4\}$ **27.** $\{8,-2\}$

29. $\{-2\}$ (The value 2 does not check.)   **31.** $\{60\}$

33. $\{\ \}$ (The value 5 does not check.)

35. $\left\{\dfrac{31}{5}\right\}$ **37.** $\{5\}$ (The value -2 does not check.)

39. $\{\ \}$ (The value 4 does not check.)

41. $\left\{-\dfrac{11}{4}\right\}$ **43.** $m=\dfrac{FK}{a}$ **45.** $E=\dfrac{IR}{K}$

47. $R=\dfrac{E-Ir}{I}$ or $R=\dfrac{E}{I}-r$

49. $B=\dfrac{2A-hb}{h}$ or $B=\dfrac{2A}{h}-b$

51. $t=\dfrac{b}{x-a}$ or $t=\dfrac{-b}{a-x}$

53. $x=\dfrac{y}{1-yz}$ or $x=\dfrac{-y}{yz-1}$

55. $h=\dfrac{2A}{a+b}$ **57.** $R=\dfrac{R_1R_2}{R_2+R_1}$

59. $t_2=\dfrac{s_2-s_1+vt_1}{v}$ or $t_2=\dfrac{s_2-s_1}{v}+t_1$ **61.** $\left\{-\dfrac{8}{5}\right\}$

63. $\{-6,2\}$ **65.** $\left\{\dfrac{11}{2}\right\}$ **67.** $\left\{\dfrac{11}{2}\right\}$ **69.** $\{-5,5\}$

71. $\{d|d$ is a real number$\}$ **73.** $\left\{-\dfrac{5}{2}\right\}$ **75.** $y=5$ **77.** $x=5$

Chapter 5 Problem Recognition Exercises, pp. 462–463

1. a. $\dfrac{2w+30}{(w-5)(w+5)}$ **b.** $\{-15\}$ **c.** The problem in part (a) is an expression, and the problem in part (b) is an equation.

2. a. $\dfrac{-3x-8}{6(x+2)}$ **b.** $\left\{-\dfrac{8}{3}\right\}$ **c.** The problem in part (a) is an expression, and the problem in part (b) is an equation.

3. $\dfrac{1}{a+1}$ **4.** $\dfrac{1}{c+2}$ **5.** $\{1\}$ **6.** $\{5\}$ **7.** $\dfrac{x^2-12}{x(x-1)}$

8. $\dfrac{23}{5(t-4)}$ **9.** $\{-1\}$ **10.** $\left\{\dfrac{3}{2},-1\right\}$ **11.** $\dfrac{8p-11}{4(2p-3)}$

12. $\dfrac{-(2x+1)(x-1)}{2(x+2)}$ **13.** $\dfrac{x+3}{6x^2}$ **14.** $\dfrac{15a+2}{12a^2}$ **15.** $\left\{-\dfrac{4}{15}\right\}$

16. $\{-10\}$ **17.** $\dfrac{c+7}{(c+3)^2(c+1)}$ **18.** $\dfrac{-2y+13}{(y-5)^2(y-2)}$

19. $\{\ \}$ (The value 4 does not check.)

20. $\{\ \}$ (The value -2 does not check.)

Section 5.6 Practice Exercises, pp. 470–473

1. a. proportion   **b.** proportional   **3.** $\{-6,1\}$

5. $\dfrac{9}{2(5t-1)}$ **7.** $\left\{-\dfrac{6}{7},3\right\}$ **9.** $\dfrac{12(p+1)}{p+8}$ **11.** $\{8\}$

13. $\{6\}$ **15.** $\{-15\}$ **17.** $\left\{\dfrac{1}{3}\right\}$ **19.** $\left\{\dfrac{20}{9}\right\}$ **21.** $\{5,-1\}$

23. $\left\{\dfrac{3}{7},-\dfrac{3}{7}\right\}$ **25.** $\{8,-5\}$ **27.** There must be 6 adults.

29. A 14-oz box contains 84 g of fat.   **31.** 1000 swordfish were caught.   **33.** Pam needs 11.5 gal of gas.   **35.** There are approximately 4000 bison in the park.   **37.** There are 31 men.

39. 595 are men and 500 are women.   **41.** $a = 8$ ft, $z = 8.4$ ft
43. $x = 12$ in., $y = 13$ in.. $a = 4.2$ in.   **45.** The number is 3.
47. The number is $\frac{2}{5}$.   **49. a.** $x + 7$ **b.** $\frac{48}{x}$ **c.** $\frac{83}{x + 7}$
51. The motorist drives 40 mph in the rain and 60 mph in sunny weather.   **53.** The Broadmoor truck travels 70.4 mph and the Wescott truck travels 76.8 mph.   **55.** The cyclist rode 10 mph against the wind.   **57.** Celeste walks 5 ft/sec on the ground and 7 ft/sec on the moving walkway.   **59.** Joe runs 6 mph and Beatrice runs 8 mph.
61. It will take them $\frac{24}{7}$ hr.   **63.** It will take them $\frac{20}{3}$ hr.
65. a. The new pump will take 20 hr. **b.** The technician should return at noon on Friday.   **67.** Gus would take 6 hr; Sid would take 12 hr.

Section 5.7 Practice Exercises, pp. 479–482

1. a. kx **b.** $\frac{k}{x}$ **c.** kxw **3.** $\frac{44}{3}$ **5.** $\frac{16}{3}$ **7.** $\frac{1}{8}$
9. Inversely   **11.** $T = kq$ **13.** $b = \frac{k}{c}$ **15.** $Q = \frac{kx}{y}$
17. $c = kst$ **19.** $L = kw\sqrt{v}$ **21.** $x = \frac{ky^2}{z}$ **23.** $k = \frac{9}{2}$
25. $k = 512$ **27.** $k = 1.75$ **29.** $x = 70$ **31.** $b = 6$
33. $Z = 56$ **35.** $Q = 9$ **37.** $L = 9$ **39.** $B = \frac{15}{2}$
41. a. 3.6 g **b.** 4.5 g **c.** 2.4 g   **43. a.** $0.40 **b.** $0.30 **c.** $1.00
45. 355,000 tons   **47.** 156.25 lumen/m^2 **49.** 18.5 A
51. $3500   **53.** 42.6 ft   **55.** 150 ft^2 **57.** 300 W
59. a. $A = kl^2$ **b.** The area is 4 times the original.
c. The area is 9 times the original.

Chapter 5 Review Exercises, pp. 489–492

1. a. $k(2) = \frac{2}{3}$, $k(0) = 0$, $k(1)$ undefined, $k(-1)$ undefined,
$k\left(\frac{1}{2}\right) = -\frac{2}{3}$ **b.** $(-\infty, -1) \cup (-1, 1) \cup (1, \infty)$
2. a. $h(1) = \frac{1}{2}$, $h(0) = 0$, $h(-1) = -\frac{1}{2}$, $h(-3) = -\frac{3}{10}$,
$h\left(\frac{1}{2}\right) = \frac{2}{5}$ **b.** $(-\infty, \infty)$ **3.** $2a$ **4.** $\frac{xz^2}{5}$ **5.** $x - 1$
6. $\frac{k + 5}{k - 3}$ **7.** $-\frac{x^2 + 3x + 9}{3 + x}$ **8.** $-(a^2 + 9)(a + 3)$
9. $\frac{-2t + 5}{t + 7}$ **10.** $\frac{y(y + 2)}{y - 3}$ **11.** c; $(-\infty, 3) \cup (3, \infty)$
12. a; $(-\infty, -2) \cup (-2, \infty)$ **13.** b; $(-\infty, 0) \cup (0, 3) \cup (3, \infty)$
14. d; $(-\infty, \infty)$ **15.** $\frac{a}{2}$ **16.** $-\frac{3}{2}$ **17.** $\frac{x - y}{5x}$ or $\frac{y - x}{5x}$
18. $(x + 8)^2$ **19.** $\frac{7(k - 4)}{2(k - 2)}$ **20.** $\frac{a + 1}{a - 1}$ **21.** $\frac{x - 5}{x - 4}$
22. $-\frac{1}{b}$ **23.** $\frac{8}{9w^2}$ **24.** $\frac{5(y^2 + 1)}{7(y^2 - 2y + 4)}$ **25.** $\frac{5}{2}$
26. $(3k + 5)(k + 5)$ **27.** $\frac{x^2 + x - 1}{x^3}$ **28.** $\frac{2(3x + 4)}{(x + 2)(x - 2)}$
29. $\frac{y - 3}{2y - 1}$ or $\frac{3 - y}{1 - 2y}$ **30.** $\frac{a - 4}{2(a + 3)}$ **31.** $\frac{4k^2 - k + 3}{(k + 1)^2(k - 1)}$
32. $\frac{4x^2 + 17x + 11}{x + 4}$ **33.** $\frac{2(a^2 - 5)}{(a - 5)(a + 3)}$
34. $\frac{13x + 19}{(x + 1)(x + 3)(x + 2)}$ **35.** $\frac{6(7 - 4x)}{3x - 5}$ or $\frac{-6(4x - 7)}{3x - 5}$
36. $\frac{2(16k - 9)}{(4k + 3)(k - 1)(4k - 3)}$ **37.** $\frac{9a^2 + a + 4}{(3a - 1)(a - 2)}$
38. $\frac{5y^2 - 17y - 9}{(y - 3)(y + 2)}$ **39.** $\frac{1}{x + 1}$ **40.** $\frac{(k - 2)(k + 2)}{15}$
41. $\frac{x(2y + 1)}{4y}$ **42.** $-y$ **43.** $\frac{1 + a}{1 - a}$ or $-\frac{a + 1}{a - 1}$

44. $2x$ **45.** $\frac{y}{x - y}$ **46.** $\frac{a(5b + 1)}{3b}$ **47.** $m = \frac{1}{18}$
48. $m = \frac{64}{23}$ **49.** {3} (The value 1 does not check.)
50. { } (The values 3 and -3 do not check.)   **51.** {0, 17}
52. $\left\{-\frac{1}{3}\right\}$ **53.** {5, 1}   **54.** {1}
55. $x = \frac{b}{c - a}$ or $x = \frac{-b}{a - c}$ **56.** $P = \frac{A}{rt + 1}$
57. $\left\{\frac{15}{2}\right\}$ **58.** $\left\{\frac{216}{7}\right\}$ **59.** $\left\{-\frac{7}{11}\right\}$ **60.** $\left\{-\frac{3}{4}\right\}$
61. Manning would gain 231 yd.   **62.** Erik can buy $253.80 Canadian.   **63.** Tony averaged 20 mph the first day and 15 mph the second day.   **64.** His speed driving was 60 mph.   **65.** It would take $\frac{40}{9}$ hr.   **66.** The larger pipe would take 9 hr and the smaller would take 18 hr.   **67. a.** $F = kd$ **b.** $k = 3$ **c.** 12.6 lb
68. $y = 16$ **69.** $y = 12$ **70.** 48 km

Chapter 5 Test, pp. 492–493

1. a. $h(0) = \frac{2}{7}$, $h(5) = \frac{1}{6}$, $h(7)$ is undefined, $h(-7)$ is undefined
b. $(-\infty, -7) \cup (-7, 7) \cup (7, \infty)$
2. $(-\infty, \infty)$ **3. a.** $\{x \mid x$ is a real number and $x \neq 4, x \neq -3\}$
b. $f(x) = \frac{2}{x - 4}$ **4.** $\frac{2m^2}{3n}$ **5.** $\frac{9(x + 1)}{3x + 5}$ **6.** $m = -\frac{23}{9}$
7. $-(x - 3)$ or $-x + 3$ **8.** $x - 4$ **9.** $\frac{x^2 + 5x + 2}{x + 1}$
10. $\frac{3}{4}$ **11.** $\frac{u^2v^2}{2(v^2 - uv + u^2)}$ **12.** $-a$ **13.** $\frac{1}{(x + 5)(x + 3)}$
14. {3}   **15.** {0, 4}   **16.** { } (The value 4 does not check.)
17. $T = \frac{1}{p - v}$ or $T = \frac{-1}{v - p}$ **18.** $m_1 = \frac{Fr^2}{Gm_2}$
19. $\frac{1}{6}$ or 2   **20.** $a = 14$ m. $y = 15$ m
21. The cities are 1960 mi apart.   **22.** Lance rides 16 mph against the wind and 20 mph with the wind.
23. It would take $\frac{20}{7}$ hr working together.
24. $x = \frac{ky}{t^2}$ **25.** 3.3 sec

Chapters 1–5 Cumulative Review Exercises, pp. 493–494

1.

Set \ Number	−22	π	6	$-\sqrt{2}$
Real numbers	✔	✔	✔	✔
Irrational numbers		✔		✔
Rational numbers	✔		✔	
Integers	✔		✔	
Whole numbers			✔	
Natural numbers			✔	

2. $\left\{\left(-2, \frac{1}{2}\right)\right\}$ **3.** $x^2 - x - 13$
4. a. $b_1 = \frac{2A - hb_2}{h}$ or $b_1 = \frac{2A}{h} - b_2$ **b.** 10 cm
5. The dimensions are 50 m by 30 m.   **6.** $\{(-1, 2, -4)\}$
7. $y = -\frac{1}{3}x + 4$
8.   **9.** 75 mph

10. $(0,0), \left(\frac{3}{4},0\right), \left(\frac{2}{3},0\right)$ **11.** $8(2y - z^2)(4y^2 + 2yz^2 + z^4)$

12. $(5x + 2)(2x - 1)$ **13.** $\dfrac{x + a}{4(x - 7)}$ **14.** $\dfrac{x + 3}{(x + 10)(x - 2)}$

15. $\dfrac{c - 7}{c}$ **16.** $\{1, -1, 5\}$ **17.** $\{3, -1\}$ **18.** The distance is 278 mi.   **19. a.** Vertical line; slope is undefined   **b.** Horizontal line; $m = 0$ **20.** \$3500

Chapter 6

Chapter Opener Puzzle

I	M	A	G	I	N	A	R	Y		N	U	M	B	E	R
4	8	7	1	4	3	7	6	9		3	5	8	10	2	6

Section 6.1 Practice Exercises, pp. 504–507

1. a. $b; a$ **b.** principal   **c.** $b^n; a$ **d.** index; radicand   **e.** cube   **f.** is not; is   **g.** even; odd   **h.** Pythagorean; c^2 **i.** $[0, \infty); (-\infty, \infty)$ **j.** -5 and -4
3. a. $8, -8$ **b.** 8 **c.** There are two square roots for every positive number. $\sqrt{64}$ identifies the positive square root.
5. a. 9 **b.** -9 **7.** There is no real number b such that $b^2 = -36$.   **9.** 7 **11.** -7 **13.** Not a real number
15. $\dfrac{8}{3}$ **17.** 0.9 **19.** -0.4 **21. a.** 8 **b.** 4
c. -8 **d.** -4 **e.** Not a real number   **f.** -4
23. -3 **25.** $\dfrac{1}{2}$ **27.** 2 **29.** $-\dfrac{5}{4}$ **31.** Not a real number   **33.** 10 **35.** -0.2 **37.** 0.5 **39.** $|a|$
41. a **43.** $|a|$ **45.** $|x + 1|$ **47.** $|x| y^2$ **49.** $-\dfrac{x}{y}$
51. $\dfrac{2}{|x|}$ **53.** -92 **55.** 2 **57.** -923 **59.** y^4
61. $\dfrac{a^3}{b}$ **63.** $-\dfrac{5}{q}$ **65.** $3xy^2z$ **67.** $\dfrac{hk^2}{4}$ **69.** $-\dfrac{t}{3}$
71. $2y^2$ **73.** $2p^2q^3$ **75.** 9 cm   **77.** 13 ft
79. They were 5 mi apart.   **81.** They are 25 mi apart.
83. a. Not a real number   **b.** Not a real number
c. 0 **d.** 1 **e.** 2 Domain: $[2, \infty)$ **85. a.** -2 **b.** -1
c. 0 **d.** 1 Domain: $(-\infty, \infty)$ **87.** $\left(-\infty, \dfrac{5}{2}\right]$
89. $(-\infty, \infty)$ **91.** $[5, \infty)$ **93.** $(-\infty, \infty)$
95. a. $(-\infty, 1]$ **97. a.** $[-3, \infty)$
b.   **b.**

99. a. $[0, \infty)$ **101. a.** $(-\infty, \infty)$
b.   **b.**

103. $q + p^2$ **105.** $\dfrac{6}{\sqrt[3]{x}}$ **107.** 8 in.

Section 6.1 Graphing Calculator Exercises, p. 507

109. 8.3066 **110.** 76.1446 **111.** 3.7100
112. -0.5566 **113.** 15.6525 **114.** -6.2403
115. -0.1235 **116.** 1.0622

117.   **118.**

119.   **120.**

Section 6.2 Practice Exercises, pp. 512–515

1. a. $\sqrt[n]{a}$ **b.** $(\sqrt[n]{a})^m$ or $\sqrt[n]{a^m}$ **c.** $\dfrac{1}{\sqrt{x}}$ **d.** $2; \dfrac{1}{2}$
3. 5 **5.** 3 **7.** 12 **9.** -12 **11.** Not a real number
13. -4 **15.** $\dfrac{1}{5}$ **17.** $-\dfrac{1}{7}$
19. $a^{m/n} = \sqrt[n]{a^m}$; The numerator of the exponent represents the power of the base. The denominator of the exponent represents the index of the radical.   **21. a.** 8 **b.** -8 **c.** Not a real number
d. $\dfrac{1}{8}$ **e.** $-\dfrac{1}{8}$ **f.** Not a real number
23. a. 125 **b.** -125 **c.** Not a real number   **d.** $\dfrac{1}{125}$
e. $-\dfrac{1}{125}$ **f.** Not a real number   **25.** $\dfrac{1}{512}$ **27.** 27
29. $-\dfrac{1}{81}$ **31.** $\dfrac{27}{1000}$ **33.** Not a real number   **35.** -2
37. -2 **39.** 6 **41.** 10 **43.** $\dfrac{3}{4}$ **45.** 1 **47.** $\dfrac{9}{2}$
49. $\sqrt[3]{q^2}$ **51.** $6\sqrt[4]{y^3}$ **53.** $\sqrt[3]{x^2y}$ **55.** $\dfrac{6}{\sqrt[5]{r^2}}$
57. $x^{1/3}$ **59.** $10b^{1/2}$ **61.** $y^{2/3}$ **63.** $(a^2b^3)^{1/4}$
65. $\dfrac{1}{x}$ **67.** p **69.** y^2 **71.** $6^{2/5}$ **73.** $\dfrac{4}{t^{5/3}}$
75. a^7 **77.** $\dfrac{25a^4d}{c}$ **79.** $\dfrac{y^9}{x^8}$ **81.** $\dfrac{2z^3}{w}$ **83.** $5xy^2z^3$
85. $x^{13}z^{4/3}$ **87.** $\dfrac{x^3y^2}{z^5}$ **89. a.** 10.9% **b.** 8.8%
c. The account in part (a).   **91.** 2.7 in.   **93.** $\sqrt{y}$
95. $\sqrt[4]{z}$ **97.** $\sqrt[3]{x^2}$ **99.** $y\sqrt{x}$ **101.** $4x^4y^3$
103. $2x\sqrt[3]{y^2z}$ **105.** $\sqrt[6]{x}$ **107.** $\sqrt[15]{w}$ **109.** 3
111. 0.3761 **113.** 2.9240 **115.** 31.6228

Section 6.3 Practice Exercises, pp. 520–522

1. a. $\sqrt[n]{a}; \sqrt[n]{b}$ **b.** The exponent within the radicand is greater than the index.   **c.** is not   **d.** 3 **e.** t^{12} **f.** No. $\sqrt{2}$ is an irrational number and the decimal form is a nonterminating, nonrepeating decimal.
3. $\dfrac{p}{q^5}$ **5.** $\sqrt[7]{x^4}$ **7.** $y^{9/2}$ **9.** $x^2\sqrt{x}$ **11.** $q^2\sqrt[3]{q}$
13. $a^2b^2\sqrt{a}$ **15.** $-x^2y^3\sqrt[4]{y}$ **17.** $2\sqrt{7}$ **19.** $2\sqrt{5}$
21. $15\sqrt{2}$ **23.** $3\sqrt[3]{2}$ **25.** $5b\sqrt{ab}$ **27.** $2x^2\sqrt[3]{5x}$
29. $-2x^2z\sqrt[3]{2y}$ **31.** $2wz\sqrt[4]{5z^3}$ **33.** x **35.** p^2
37. 5 **39.** $\dfrac{1}{2}$ **41.** $\dfrac{5\sqrt[3]{2}}{3}$ **43.** $\dfrac{5\sqrt[3]{9}}{6}$ **45.** $4\sqrt{5}$

47. $-30\sqrt{3}$ **49.** $5x^2y\sqrt{y}$ **51.** $3yz\sqrt[3]{x^2z}$ **53.** $2w^2$

55. $\dfrac{1}{10y^6}$ **57.** $\dfrac{2}{b}$ **59.** $2a^7b^4c^{15}d^{11}\sqrt{2c}$

61. $3a^2b\sqrt[3]{2b}$ **63.** $-10a^2b^2\sqrt{3ac}$ **65.** $x\sqrt[4]{7xy}$

67. $3a^2b\sqrt{6}$ **69.** $2\sqrt{3}$ **71.** $\dfrac{3\sqrt{5}}{4}$

73. $\dfrac{1}{\sqrt[3]{w^6}}$ simplifies to $\dfrac{1}{w^2}$ **75.** $\sqrt{k^3}$ simplifies to $k\sqrt{k}$

77. $2\sqrt{41}$ ft **79.** $6\sqrt{5}$ m **81.** The distance is $90\sqrt{2}$ ft or approximately 127.3 ft. **83.** The path from A to B and B to C is faster.

Section 6.4 Practice Exercises, pp. 525–527

1. a. index; radicand **b.** $2\sqrt{3x}$ **c.** cannot; can **d.** $4\sqrt{2}$

3. $-xy\sqrt[4]{x^3}$ **5.** $\dfrac{b^2}{2}$ **7.** $(x^3y)^{1/4}$ **9.** $y^{11/12}$

11. a. Not *like* radicals **b.** *Like* radicals
c. Not *like* radicals **13. a.** Both expressions can be simplified by using the distributive property. **b.** Neither expression can be simplified because the terms do not contain *like* radicals or *like* terms. **15.** $9\sqrt{5}$ **17.** $2\sqrt[3]{tw}$
19. $5\sqrt{10}$ **21.** $8\sqrt[4]{3}-\sqrt[4]{14}$ **23.** $2\sqrt{x}+2\sqrt{y}$
25. Cannot be simplified further
27. Cannot be simplified further **29.** $\dfrac{29}{18}z\sqrt[3]{6}$
31. $0.7x\sqrt{y}$ **33.** Simplify each radical: $3\sqrt{2}+35\sqrt{2}$. Then add *like* radicals: $38\sqrt{2}$ **35.** 15 **37.** $8\sqrt{3}$
39. $3\sqrt{7}$ **41.** $-5\sqrt{2}+3\sqrt{3}$ **43.** $\sqrt[3]{3}$ **45.** $-5\sqrt{2a}$
47. $8s^2t^2\sqrt[3]{s^2}$ **49.** $6x\sqrt[3]{x}$ **51.** $14p^2\sqrt{5}$
53. $-\sqrt[3]{a^2b}$ **55.** $(5x+6)\sqrt{x}$ **57.** $(5x-6)\sqrt{2}$
59. $33d\sqrt[3]{2c}$ **61.** $2a^2b\sqrt{6a}$ **63.** $5x\sqrt[3]{2}-6\sqrt[3]{x}$
65. False; $\sqrt{9}+\sqrt{16}\neq\sqrt{9+16}$; $7\neq5$ **67.** True
69. False; $\sqrt{y}+\sqrt{y}=2\sqrt{y}\neq\sqrt{2y}$
71. False; $2w\sqrt{5}+4w\sqrt{5}=6w\sqrt{5}\neq6w^2\sqrt{5}$
73. $\sqrt{48}+\sqrt{12}$ simplifies to $6\sqrt{3}$
75. $5\sqrt[3]{x^6}-x^2$ simplifies to $4x^2$ **77.** The difference of the principal square root of 18 and the square of 5
79. The sum of the principal fourth root of x and the cube of y
81. $9\sqrt{6}$ cm ≈ 22.0 cm **83.** $x=2\sqrt{2}$ ft
85. a. $10\sqrt{5}$ yd **b.** 22.36 yd **c.** \$105.95

Section 6.5 Practice Exercises, pp. 533–535

1. a. $\sqrt[n]{ab}$ **b.** x **c.** a **d.** conjugates **e.** $m-n$
f. $c+8\sqrt{c}+16$ **3.** $-2ab\sqrt{5bc}$ **5.** $x^{1/6}y^{7/12}$ **7.** $2\sqrt[3]{7}$
9. $\sqrt[3]{21}$ **11.** $2\sqrt{5}$ **13.** $4\sqrt[4]{4}$ **15.** $8\sqrt[3]{20}$
17. $-24ab\sqrt{a}$ **19.** $6\sqrt{10}$ **21.** $6x\sqrt{2}$ **23.** $10a^3b^2\sqrt{b}$
25. $-24x^2y^2\sqrt{2y}$ **27.** $6ab\sqrt[3]{2a^2b^2}$ **29.** $12-6\sqrt{3}$
31. $2\sqrt{3}-\sqrt{6}$ **33.** $-2x-\dfrac{7}{3}\sqrt{x}$ **35.** $-8+7\sqrt{30}$
37. $x-5\sqrt{x}-36$ **39.** $\sqrt[3]{y^2}-\sqrt[3]{y}-6$
41. $9a-28\sqrt{ab}+3b$ **43.** $8\sqrt{p}+3p+5\sqrt{pq}+16\sqrt{q}-2q$
45. 15 **47.** $3y$ **49.** 6 **51.** 709 **53. a.** x^2-y^2
b. x^2-25 **55.** $29+8\sqrt{13}$ **57.** $p-2\sqrt{7p}+7$
59. $2a-6\sqrt{2ab}+9b$ **61.** $3-x^2$ **63.** 4
65. $\dfrac{4}{9}x-\dfrac{1}{4}y$ **67. a.** $3-x$ **b.** $3+2\sqrt{3x}+x$
c. $3-2\sqrt{3x}+x$ **69.** True **71.** False;
$(x-\sqrt{5})^2=x^2-2x\sqrt{5}+5$ **73.** False; 5 is multiplied by 3 only.
75. True **77.** $6x$ **79.** $3x+1$
81. $x+19-8\sqrt{x+3}$ **83.** $2t+10\sqrt{2t-3}+22$
85. $12\sqrt{5}$ ft² **87.** $18\sqrt{15}$ in.² **89.** $\sqrt[4]{x^3}$
91. $\sqrt[15]{(2z)^8}$ **93.** $p^2\sqrt[6]{p}$ **95.** $u\sqrt[6]{u}$
97. $\sqrt[6]{x^2y}$ **99.** $\sqrt[4]{2^3\cdot3^2}$ or $\sqrt[4]{72}$ **101.** $\sqrt[12]{2^4 5^3 x^7 y^7}$
103. $2m\sqrt[6]{3^3 2mn^5}$ **105.** $a+b$

Chapter 6 Problem Recognition Exercises, p. 536

1. a. $2\sqrt{6}$ **b.** $2\sqrt[3]{3}$ **2. a.** $3\sqrt{6}$ **b.** $3\sqrt[3]{2}$
3. a. $10y^3\sqrt{2}$ **b.** $2y^2\sqrt[3]{25}$ **4. a.** $4z^7\sqrt{2z}$ **b.** $2z^5\sqrt[3]{4}$
5. a. $4\sqrt{5}$ **b.** $2\sqrt[3]{10}$ **c.** $2\sqrt{5}$ **6. a.** $4\sqrt{3}$ **b.** $2\sqrt[3]{6}$ **c.** $2\sqrt[4]{3}$
7. a. $x^2y^3\sqrt{x}$ **b.** $xy^2\sqrt[3]{x^2}$ **c.** $xy\sqrt[6]{xy^2}$
8. a. $a^5b^4\sqrt{b}$ **b.** $a^3b^3\sqrt[3]{a}$ **c.** $a^2b^2\sqrt[4]{a^2b}$ **9. a.** $2st^2\sqrt[3]{4s^2}$
b. $2st\sqrt[4]{2st^2}$ **c.** $2st\sqrt[6]{t}$ **10. a.** $2v^2w^6\sqrt[3]{12vw^2}$
b. $2vw^5\sqrt[4]{6v^3}$ **c.** $2vw^4\sqrt[5]{3v^2}$ **11. a.** $2\sqrt{5}$ **b.** 5
12. a. $2\sqrt{10}$ **b.** 10 **13. a.** $-3\sqrt{6}$ **b.** 60
14. a. $-7\sqrt{7}$ **b.** 210 **15. a.** $3\sqrt{2}$ **b.** 4
16. a. $3\sqrt{3}$ **b.** 6 **17. a.** $7\sqrt{2}$ **b.** 240
18. a. $-\sqrt{2}$ **b.** 60 **19. a.** $14\sqrt[3]{3}$ **b.** $48\sqrt[3]{9}$
20. a. $\sqrt[3]{2}$ **b.** $30\sqrt[3]{4}$

Section 6.6 Practice Exercises, pp. 543–546

1. a. radical **b.** $\dfrac{\sqrt[n]{a}}{\sqrt[n]{b}}$ **c.** $\dfrac{4}{x^2}$ **d.** rationalizing **e.** is; is not
f. denominator **3.** $12y\sqrt{5}$ **5.** $-18y+3\sqrt{y}+3$
7. $64-16\sqrt{t}+t$ **9.** -5 **11.** $\dfrac{7x^2}{y^3}$ **13.** $\dfrac{2a\sqrt{2}}{x^3}$
15. $\dfrac{-2j\sqrt[3]{2}}{k}$ **17.** $3b^2$ **19.** $\dfrac{1}{2b^2}$ **21.** $\dfrac{z\sqrt{3y}}{w^2}$ **23.** $\dfrac{\sqrt{5}}{\sqrt{5}}$
25. $\dfrac{\sqrt[3]{x^2}}{\sqrt[3]{x^2}}$ **27.** $\dfrac{\sqrt{3z}}{\sqrt{3z}}$ **29.** $\dfrac{\sqrt[4]{2a^2}}{\sqrt[4]{2a^2}}$ **31.** $\dfrac{\sqrt{3}}{3}$
33. $\dfrac{\sqrt{x}}{x}$ **35.** $\dfrac{3\sqrt{2y}}{y}$ **37.** $\dfrac{a\sqrt{2a}}{2}$ **39.** $\dfrac{3\sqrt{2}}{2}$
41. $\dfrac{3\sqrt[3]{4}}{2}$ **43.** $\dfrac{-6\sqrt[4]{x^3}}{x}$ **45.** $\dfrac{7\sqrt[3]{2}}{2}$ **47.** $\dfrac{\sqrt[3]{4w}}{w}$
49. $\dfrac{2\sqrt[3]{27}}{3}$ **51.** $\dfrac{\sqrt[3]{2x}}{x}$ **53.** $\dfrac{2\sqrt{6}}{21}$ **55.** $\dfrac{\sqrt{x}}{x^4}$
57. $\dfrac{\sqrt{2x}}{2x^3}$ **59.** $\sqrt{2}+\sqrt{6}$ **61.** $\sqrt{x}-23$
63. $\dfrac{4\sqrt{2}-12}{-7}$ or $\dfrac{-4\sqrt{2}+12}{7}$ **65.** $4\sqrt{6}+8$
67. $-\sqrt{21}+2\sqrt{7}$ **69.** $\dfrac{-\sqrt{p}-\sqrt{q}}{p-q}$ **71.** $\sqrt{x}-\sqrt{5}$
73. $\dfrac{w+11\sqrt{w}+18}{81-w}$ **75.** $\dfrac{4\sqrt{xy}-3x-y}{y-x}$
77. $5-\sqrt{10}$ **79.** $\dfrac{5+\sqrt{21}}{4}$ **81.** $-6\sqrt{5}-13$
83. $\dfrac{16}{\sqrt[3]{2}}$ simplifies to $8\sqrt[3]{2}$
85. $\dfrac{4}{x-\sqrt{2}}$ simplifies to $\dfrac{4x+4\sqrt{2}}{x^2-2}$
87. $\dfrac{\pi\sqrt{2}}{4}$ sec ≈ 1.11 sec **89. a.** $\dfrac{\sqrt{2}}{2}$ **b.** $\dfrac{\sqrt[3]{4}}{2}$
91. a. $\dfrac{\sqrt{5a}}{5a}$ **b.** $\dfrac{\sqrt{5}-a}{5-a^2}$ **93.** $\dfrac{2\sqrt{6}}{3}$ **95.** $\dfrac{17\sqrt{15}}{15}$
97. $\dfrac{8\sqrt[3]{25}}{5}$ **99.** $\dfrac{-33}{2\sqrt{3}-12}$ **101.** $\dfrac{a-b}{a+2\sqrt{ab}+b}$
103. $\dfrac{3}{\sqrt{5+3h}+\sqrt{5}}$ **105.** $\dfrac{5}{\sqrt{4+5h}+2}$

Section 6.7 Practice Exercises, pp. 553–556

1. a. radical **b.** isolate; 7 **c.** extraneous **d.** third
3. $\dfrac{3w\sqrt{w}}{4}$ **5.** $3c\sqrt[3]{2c}$ **7.** $4x-6$ **9.** $9p+7$
11. {100} **13.** {4} **15.** {3} **17.** {42} **19.** {2}
21. {29} **23.** {140} **25.** { } (The value 25 does not check.)
27. -7} (The value -1 does not check.)
29. { } (The value 3 does not check.)

31. $V = \dfrac{4\pi r^3}{3}$ **33.** $h^2 = \dfrac{r^2 - \pi^2 r^2}{\pi^2}$ or $h^2 = \dfrac{r^2}{\pi^2} - r^2$

35. $a^2 + 10a + 25$ **37.** $5a - 6\sqrt{5a} + 9$

39. $r + 22 + 10\sqrt{r-3}$ **41.** $\{-3\}$

43. $\{\ \}$ (The value $\frac{1}{2}$ does not check.)

45. $\{2\}$ (The value 18 does not check.)   **47.** $\{9\}$

49. $\{-4\}$ **51.** $\left\{\dfrac{9}{5}\right\}$ **53.** $\{2\}$

55. $\{\ \}$ (The value 9 does not check.)

57. $\left\{-\dfrac{11}{4}\right\}$ **59.** $\{\ \}$ (The value $\frac{8}{3}$ does not check.)

61. $\{-1\}$ (The value 3 does not check.)   **63.** $\left\{\dfrac{1}{3}, -1\right\}$

65. $\{\ \}$ (The values 3 and 23 do not check.)

67. $\{6\}$ **69.** $\{2\}$ (The value 18 does not check.)

71. a. 30.25 ft   **b.** 34.5 m   **73. a.** $2 million

b. $1.2 million   **c.** 50,000 passengers   **75. a.** 12 lb

b. $t(18) = 5.1$. An 18-lb turkey will take about 5.1 hr to cook.

77. $b = \sqrt{25 - h^2}$ **79.** $a = \sqrt{k^2 - 196}$

Section 6.7 Graphing Calculator Exercises, p. 556

80. The x-coordinate of the point of intersection is the solution to the equation.

81. The x-coordinate of the point of intersection is the solution to the equation.

82.

83.

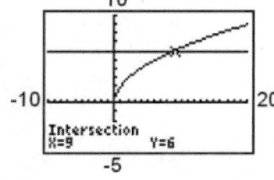

Section 6.8 Practice Exercises, pp. 564–566

1. a. imaginary   **b.** $\sqrt{-1}; -1$ **c.** $i\sqrt{b}$ **d.** $a + bi; \sqrt{-1}$

e. real; b **f.** $a + bi$ **g.** True   **h.** True

3. $9 - x$ **5.** $\{-8\}$ **7.** $\left\{\dfrac{1}{25}\right\}$ **9.** $\sqrt{-1} = i$ and

$-\sqrt{1} = -1$ **11.** $12i$ **13.** $i\sqrt{3}$ **15.** $-2i\sqrt{5}$

17. -60 **19.** $13i\sqrt{7}$ **21.** -7 **23.** -12

25. $-3\sqrt{10}$ **27.** $i\sqrt{2}$ **29.** 3   **31.** $-i$ **33.** 1

35. i **37.** 1   **39.** $-i$ **41.** -1 **43.** $a - bi$

45. Real: -5; imaginary: 12   **47.** Real: 0; imaginary: -6

49. Real: 35; imaginary: 0   **51.** Real: $\dfrac{3}{5}$; imaginary: 1

53. $7 + 6i$ **55.** $\dfrac{3}{10} + \dfrac{3}{2}i$ **57.** $5i\sqrt{2}$

59. $-1 + 10i$ **61.** $-24 + 0i$ **63.** $18 + 6i$

65. $26 - 26i$ **67.** $-29 + 0i$ **69.** $-9 + 40i$

71. $35 + 20i$ **73.** $-20 + 48i$ **75.** $\dfrac{13}{16} + 0i$

77. $\dfrac{1}{5} - \dfrac{3}{5}i$ **79.** $\dfrac{3}{25} - \dfrac{4}{25}i$ **81.** $\dfrac{21}{29} + \dfrac{20}{29}i$

83. $-\dfrac{17}{10} - \dfrac{1}{10}i$ **85.** $-\dfrac{1}{2} - \dfrac{5}{2}i$ **87.** $\dfrac{1}{2} - \dfrac{1}{3}i$

89. $1 + 10i$ **91.** $\dfrac{1}{4} + \dfrac{1}{2}i$ **93.** $-1 + i\sqrt{2}$

95. $-2 - i\sqrt{3}$ **97.** $-\dfrac{1}{2} + \dfrac{\sqrt{2}}{2}i$ **99.** Yes   **101.** Yes

Chapter 6 Review Exercises, pp. 574–576

1. a. False; $\sqrt{0} = 0$ is not positive.   **b.** False; $\sqrt[3]{-8} = -2$

2. $\sqrt{(-3)^2} = \sqrt{9} = 3$ **3. a.** False   **b.** True

4. $\dfrac{5}{4}$ **5.** 5   **6.** 6   **7. a.** 3   **b.** 0   **c.** $\sqrt{7}$

d. $[1, \infty)$ **8. a.** 0   **b.** 1   **c.** 3   **d.** $[-5, \infty)$

9. $\dfrac{\sqrt[3]{2x}}{\sqrt[4]{2x}} + 4$ **10. a.** $|x|$ **b.** x **c.** $|x|$ **d.** $x + 1$

11. a. $2|y|$ **b.** $3y$ **c.** $|y|$ **d.** y **12.** 8 cm

13. Yes, provided the expressions are defined. For example: $x^5 \cdot x^3 = x^8$ and $x^{1/5} \cdot x^{2/5} = x^{3/5}$ **14.** n represents the root.

15. Take the reciprocal of the base and change the exponent to positive.   **16.** -5 **17.** $\dfrac{1}{2}$ **18.** 4

19. b^{10} **20.** $\dfrac{16y^{12}}{xz^9}$ **21.** $\dfrac{a^3 c}{b^2}$ **22.** $x^{3/4}$

23. $(2y^2)^{1/3}$ **24.** 2.1544   **25.** 6.8173   **26.** 54.1819

27. See page 516.   **28.** $6\sqrt{3}$ **29.** $xz\sqrt[4]{xy}$

30. $-10ab^3\sqrt[3]{2b}$ **31.** $-\dfrac{2a}{b}$

32. a. The principal square root of the quotient of 2 and x.

b. The cube of the sum of x and 1   **33.** 31 ft

34. they are *like* radicals.   **35.** Cannot be combined; the indices are different.   **36.** Cannot be combined; one term has a radical, but the other does not.   **37.** Can be combined: $3\sqrt[4]{3xy}$ **38.** Can be added after simplifying: $19\sqrt{2}$

39. $5\sqrt{7}$ **40.** $9\sqrt[3]{2} - 8$ **41.** $10\sqrt{2}$ **42.** $13x\sqrt[3]{2x^2}$

43. False; 5 and $3\sqrt{x}$ are not *like* radicals.

44. False; $\sqrt{y} + \sqrt{y} = 2\sqrt{y}$ (add the coefficients).

45. 6   **46.** $2\sqrt[4]{2}$ **47.** $-2\sqrt{21} + 6\sqrt{33}$

48. $-6\sqrt{15} + 15$ **49.** $4x - 9$ **50.** $y - 16$

51. $7y - 2\sqrt{21xy} + 3x$ **52.** $12w + 20\sqrt{3w} + 25$

53. $-2z - 9\sqrt{6z} - 42$ **54.** $3a + 5\sqrt{5a} - 10$

55. $u^2\sqrt[6]{u^5}$ **56.** $\sqrt[4]{4w^3}$ **57.** $\dfrac{y^2\sqrt{3y}}{5x^3}$

58. $\dfrac{-2x^2y^2\sqrt[3]{2x}}{z^3}$ **59.** $9w^2$ **60.** $\dfrac{t^4}{4}$ **61.** $\dfrac{\sqrt{14y}}{2y}$

62. $\dfrac{\sqrt{15w}}{3w}$ **63.** $\dfrac{4\sqrt[3]{3p}}{3p}$ **64.** $-\dfrac{\sqrt[3]{4x^2}}{x}$

65. $\sqrt{10} - \sqrt{15}$ **66.** $3\sqrt{5} - 3\sqrt{7}$ **67.** $\sqrt{t} + \sqrt{3}$

68. $\sqrt{w} + \sqrt{7}$ **69.** The quotient of the principal square root of 2 and the square of x

70. $\left\{\dfrac{49}{2}\right\}$ **71.** $\{31\}$ **72.** $\{-12\}$ **73.** $\{7\}$

74. $\{9\}$ **75.** $\{\ \}$ (The value -2 does not check.)

76. $\left\{\dfrac{1}{2}, 4\right\}$ **77.** $\{-2\}$ (The value 2 does not check.)

78. $6\sqrt{5}$ m ≈ 13.4 m   **79. a.** $v(20) \approx 25.3$ ft/sec. When the water depth is 20 ft, a wave travels about 25.3 ft/sec.

b. 8 ft   **80.** $a + bi$, where a and b are real numbers and $i = \sqrt{-1}$

81. $a + bi$, where $b \neq 0$

82. Multiply the numerator and denominator by $4 - 6i$, which is the complex conjugate of the denominator.

83. $4i$ **84.** $-i\sqrt{5}$ **85.** -15 **86.** $-2i$ **87.** -1

88. i **89.** $-i$ **90.** 0   **91.** $-5 + 5i$ **92.** $9 + 17i$

93. $25 + 0i$ **94.** $24 - 10i$

95. $-\dfrac{17}{4} + i$; real part: $-\dfrac{17}{4}$; imaginary part: 1

96. $-2 - i$; real part: -2; imaginary part: -1

97. $\dfrac{4}{13} - \dfrac{7}{13}i$ **98.** $3 - 4i$ **99.** $-\dfrac{3}{2} + \dfrac{5}{2}i$

100. $-\dfrac{4}{17} + \dfrac{16}{17}i$ **101.** $-\dfrac{2}{3} + \dfrac{\sqrt{10}}{6}i$ **102.** $2 - 4i$

Chapter 6 Test, pp. 577–578

1. a. 6 **b.** -6 **2. a.** Real **b.** Not real **c.** Real

d. Real   **3. a.** y **b.** $|y|$ **4.** 3   **5.** $\dfrac{4}{3}$ **6.** $2\sqrt[3]{4}$

7. $a^2bc^2\sqrt{bc}$ **8.** $3x^2yz^2\sqrt{2xy}$ **9.** $4w^2\sqrt{w}$ **10.** $\dfrac{x^2}{5y}$

11. $\dfrac{3\sqrt{2}}{2}$ **12. a.** $f(-8) = 2\sqrt{3};\ f(-6) = 2\sqrt{2};$
$f(-4) = 2;\ f(-2) = 0$ **b.** $(-\infty, -2]$ **13.** -0.3080

14. -3 **15.** $\dfrac{20x^2y^3}{z}$ **16.** $\sqrt[6]{7y^3}$ **17.** $\sqrt[12]{10}$

18. $3\sqrt{5}$ **19.** $3\sqrt{2x} - 3\sqrt{5x}$ **20.** $40 - 10\sqrt{5x} - 3x$
21. $\dfrac{-2\sqrt[3]{x^2}}{x}$ **22.** $\dfrac{x + 6 + 5\sqrt{x}}{9 - x}$ **23. a.** $2i\sqrt{2}$ **b.** $8i$

c. $\dfrac{1}{2} + \dfrac{\sqrt{2}}{2}i$ **24.** $1 - 11i$ **25.** $30 + 16i$ **26.** -28

27. $-33 - 56i$ **28.** 104   **29.** $\dfrac{17}{25} + \dfrac{6}{25}i$

30. $-\dfrac{15}{17} + \dfrac{9}{17}i$ **31.** $r(10) \approx 1.34$; the radius of a sphere
of volume 10 cubic units is approximately 1.34 units.

32. 21 ft   **33.** $\{-16\}$ **34.** $\left\{\dfrac{17}{5}\right\}$

35. $\{2\}$ (The value 42 does not check.)

Chapters 1–6 Cumulative Review Exercises, pp. 578–579

1. 54   **2.** $15x - 5y - 5$ **3.** $\{-3\}$ **4.** $(-\infty, -5)$
5. $[-9, -3)$ **6.** $\{\ \}$ **7.** $y = -2x + 5$ **8.** $x = -5$
9. $\left\{\left(\dfrac{1}{2}, \dfrac{1}{3}\right)\right\}$ **10.** $\left(2, -2, \dfrac{1}{2}\right)$ is not a solution.
11. Together it will take them $1\dfrac{7}{8}$ hr.

12. a. $f(-2) = -10; f(0) = -2; f(4) = 14; f\left(\dfrac{1}{2}\right) = 0$

b. $(-2, -10), (0, -2), (4, 14), \left(\dfrac{1}{2}, 0\right)$

c.

13. $a^6b^{12}c^4$ **14. a.** 1.4×10^{-4} **b.** 3.14×10^9
15. $2x^2 - x - 15$; 2nd degree   **16.** $\sqrt{15} + 3\sqrt{2} + 3$

17. $x - 4$ **18.** $-\dfrac{1}{6}$ **19.** $\dfrac{3c\sqrt[3]{2}}{d}$ **20.** $32b\sqrt{5b}$

21. $2 + 3i$ **22.** $\{10\}$ **23.** $\dfrac{x - 15}{x(x + 5)(x - 5)}$

24. $\dfrac{-1}{(2a + 1)(a + 2)}$ **25.** $-14x^2 + 26x - 17$

26. $2\sqrt{3} + 2\sqrt{5}$ **27.** $\dfrac{6}{17} + \dfrac{10}{17}i$ **28.** $\left\{\dfrac{7}{6}, -\dfrac{3}{2}\right\}$
29. $(x + 3 - y)(x + 3 + y)$ **30.** $(x^2 + 2)(x^4 - 2x^2 + 4)$

Chapter 7

Chapter Opener Puzzle

Q	U	A	D	R	A	T	I	C
2	1	6	5	8	6	4	7	3

Section 7.1 Practice Exercises, pp. 588–591

1. a. $0; 0$ **b.** 0 **c.** $\sqrt{k};\ -\sqrt{k}$ **d.** $2; \{3, -3\}$ **e.** completing
f. 100 **g.** $4; 1$ **h.** 8

3. $\{\pm 2\}$ **5.** $\{\pm\sqrt{7}\}$ **7.** $\left\{\dfrac{11}{6}, -\dfrac{11}{6}\right\}$ **9.** $\{\pm 5i\}$

11. $\{-1, -5\}$ **13.** $\left\{\dfrac{-3 \pm \sqrt{7}}{2}\right\}$ **15.** $\{-5 \pm 3i\sqrt{2}\}$

17. $\left\{\pm\dfrac{\sqrt{33}}{3}\right\}$ **19.** $\left\{-\dfrac{4}{5} \pm \dfrac{\sqrt{3}}{5}i\right\}$ **21.** $\{4i, -4i\}$

23. 1. Factoring and applying the zero product rule. 2. Applying
the square root property. $\{\pm 6\}$ **25. a.** $\{16\}$ **b.** $\{2, -2\}$
27. $n = 9; (x - 3)^2$ **29.** $n = 16; (t + 4)^2$

31. $n = \dfrac{1}{4}; \left(c - \dfrac{1}{2}\right)^2$ **33.** $n = \dfrac{25}{4}; \left(y + \dfrac{5}{2}\right)^2$

35. $n = \dfrac{1}{25}; \left(b + \dfrac{1}{5}\right)^2$ **37.** $n = \dfrac{1}{9}; \left(p - \dfrac{1}{3}\right)^2$

39. See page 585.   **41.** $\{-3, -5\}$ **43.** $\{-3 \pm i\sqrt{7}\}$

45. $\{-2 \pm i\sqrt{2}\}$ **47.** $\{5, -2\}$ **49.** $\left\{-1 \pm \dfrac{\sqrt{6}}{2}i\right\}$

51. $\left\{z = \dfrac{2}{3}i\right\}$ **53.** $\left\{\dfrac{1}{5} \pm \dfrac{\sqrt{3}}{5}\right\}$ **55.** $\left\{-\dfrac{3}{4} \pm \dfrac{\sqrt{65}}{4}\right\}$

57. $\{2 \pm \sqrt{11}\}$ **59.** $\{1, -7\}$ **61. a.** $t = \dfrac{\sqrt{d}}{4}$ **b.** 8 sec

63. $r = \sqrt{\dfrac{A}{\pi}}$ or $r = \dfrac{\sqrt{A\pi}}{\pi}$ **65.** $a = \sqrt{d^2 - b^2 - c^2}$

67. $r = \sqrt{\dfrac{3V}{\pi h}}$ or $r = \dfrac{\sqrt{3V\pi h}}{\pi h}$ **69.** The shelf extends 4.2 ft.

71. The sides are 7.1 in.   **73. a.** 4.5 thousand textbooks or 35.5
thousand textbooks **b.** Profit increases to a point as more books are
produced. Beyond that point, the market is "flooded," and profit
decreases. There are two points at which the profit is $20,000.
Producing 4.5 thousand books makes the same profit using fewer
resources as producing 35.5 thousand books.

Section 7.2 Practice Exercises, pp. 602–605

1. a. quadratic; $\dfrac{-b \pm \sqrt{b^2 - 4ac}}{2a}$ **b.** $ax^2 + bx + c = 0$

c. $8; -42; -27$ **d.** $7; 97$ **e.** $b^2 - 4ac$; discriminant **f.** imaginary
g. real **h.** less than

3. $4 - 2\sqrt{5}$ **5.** $2 - i\sqrt{3}$ **7.** linear   **9.** $\{-12, 1\}$

11. $\left\{\dfrac{1}{9} \pm \dfrac{2\sqrt{11}}{9}i\right\}$ **13.** $\left\{\dfrac{1}{6} \pm \dfrac{\sqrt{14}}{6}i\right\}$ **15.** $\{7, -5\}$

17. $\left\{\dfrac{-3 \pm \sqrt{41}}{2}\right\}$ **19.** $\left\{\dfrac{2}{5}\right\}$ **21.** $\{3 \pm i\sqrt{5}\}$

23. $\left\{\dfrac{1 - \sqrt{29}}{2}\right\}$ **25.** $\left\{\dfrac{-1 \pm \sqrt{13}}{4}\right\}$

27. $\left\{-\dfrac{2}{3} \pm \dfrac{2\sqrt{2}}{3}i\right\}$ **29.** $\left\{\dfrac{-5 \pm \sqrt{13}}{2}\right\}$

31. $\{-2, -4\}$ **33.** $\left\{\dfrac{-7 \pm \sqrt{109}}{6}\right\}$

35. a. $(z - 3)(z^2 + 3z + 9)$ **b.** $\left\{3, -\dfrac{3}{2} \pm \dfrac{3\sqrt{3}}{2}i\right\}$

37. a. $3x(x^2 - 2x + 2)$ **b.** $\{0, 1 \pm i\}$ **39.** The length of each
side is 3 ft.   **41.** The legs are approximately 1.6 in. and 3.6 in.
43. The legs are approximately 8.2 m and 6.1 m.
45. a. Approximately 4.5 fatalities per 100 million miles driven
b. Approximately 1 fatality per 100 million miles driven
c. Approximately 4.2 fatalities per 100 million miles driven
d. For drivers 26 yr old and 71 yr old

47. The times are $\dfrac{3+\sqrt{5}}{2}$ sec ≈ 2.62 sec or

$\dfrac{3-\sqrt{5}}{2}$ sec ≈ 0.38 sec.   **49. a.** $x^2 + 2x + 1 = 0$

b. 0   **c.** 1 rational solution   **51. a.** $19m^2 - 8m + 0 = 0$ **b.** 64
c. 2 rational solutions   **53. a.** $5p^2 + 0p - 21 = 0$ **b.** 420
c. 2 irrational solutions   **55. a.** $n^2 + 3n + 4 = 0$ **b.** -7
c. 2 imaginary solutions   **57.** Discriminant: 16; two x-intercepts
59. Discriminant: 0; one x-intercept
61. Discriminant: -39; no x-intercepts
63. x-intercepts: $\left(\dfrac{5+\sqrt{13}}{2}, 0\right), \left(\dfrac{5-\sqrt{13}}{2}, 0\right)$;
y-intercept: $(0, 3)$ **65.** x-intercepts: none; y-intercept: $(0, -1)$
67. x-intercepts: $\left(\dfrac{-5-\sqrt{41}}{4}, 0\right), \left(\dfrac{-5+\sqrt{41}}{4}, 0\right)$;
y-intercept: $(0, -2)$ **69.** $\{-1 \pm 3i\}$
71. $\left\{-\dfrac{1}{3}, 6\right\}$ **73.** $\left\{\dfrac{5}{2} \pm \dfrac{3\sqrt{2}}{2}i\right\}$ **75.** $\left\{-\dfrac{1}{2} \pm 2i\right\}$
77. $\left\{\dfrac{3 \pm \sqrt{7}}{2}\right\}$ **79.** $\{1 \pm 3i\sqrt{2}\}$ **81.** $\{2 \pm \sqrt{2}\}$
83. $\left\{-\dfrac{1}{8}, \dfrac{3}{4}\right\}$ **85.** $\left\{\pm\dfrac{\sqrt{21}}{2}\right\}$
87. a. $\{-3 \pm \sqrt{14}\}$ **b.** $\{-3 \pm \sqrt{14}\}$ **c.** Answers will vary

Section 7.2 Graphing Calculator Exercises, p. 605

89.   **90.**

91.   **92.**

Section 7.3 Practice Exercises, pp. 609–611

1. a. quadratic   **b.** $3x - 1$ **c.** $p^{1/3}$
3. $\left\{\dfrac{3}{2} \pm \dfrac{\sqrt{7}}{2}\right\}$ **5.** $\{-4\}$ **7.** $\{2 \pm \sqrt{26}\}$
9. a. $\{-4, -6\}$ **b.** $\{-4, -1, -2, -3\}$
11. $\{1 \pm i\sqrt{2}, 1 \pm \sqrt{2}\}$ **13.** $\{5, -1, 2\}$ **15.** $\{27, -8\}$
17. $\left\{-\dfrac{1}{32}, -243\right\}$ **19.** $\{4\}$ (The value 64 does not check.)
21. $\left\{\dfrac{1}{4}\right\}$ (The value 4 does not check.)
23. $\{-7\}$ **25.** $\{16\}$; yes   **27.** $\{\pm 3i, \pm\sqrt{5}\}$
29. $\left\{\pm\dfrac{\sqrt{3}}{2}, \pm i\sqrt{5}\right\}$ **31.** $\left\{\dfrac{-5 \pm \sqrt{13}}{2}\right\}$
33. $\left\{\dfrac{9 \pm \sqrt{73}}{4}\right\}$ **35.** $\{2 \pm \sqrt{3}\}$ **37.** $\{\pm 2, \pm 2i\}$
39. $\left\{-\dfrac{7}{4}, -\dfrac{3}{2}\right\}$ **41.** $\left\{\dfrac{1}{2}, -\dfrac{1}{2}, \sqrt{2}, -\sqrt{2}\right\}$
43. $\left\{2, 1, -1 \pm i\sqrt{3}, -\dfrac{1}{2} \pm \dfrac{\sqrt{3}}{2}i\right\}$ **45.** $\{5, 4\}$ **47.** $\{0, 2\}$
49. $\left\{1, \dfrac{17}{2}\right\}$ **51.** $\{64, -125\}$ **53.** $\{\pm\sqrt{2}, \pm 2i\}$
55. $\{\pm 4i, 1\}$ **57.** $\{4, \pm i\sqrt{5}\}$ **59.** $\left\{\dfrac{8}{3}, 2\right\}$

Section 7.3 Graphing Calculator Exercises, p. 611

61. a. $\{\pm i\sqrt{2}\}$ **b.** Two imaginary solutions; no real solutions
c. No x-intercepts
d.

62. a. $\{1, -1\}$ **b.** Two real solutions; no imaginary solutions
c. Two x-intercepts
d.

63. a. $\{0, 3, -2\}$ **b.** Three real solutions; no imaginary solutions
c. Three x-intercepts
d.

64. a. $\{\pm 1, \pm 3\}$ **b.** Four real solutions; no imaginary solutions
c. Four x-intercepts
d.

Chapter 7 Problem Recognition Exercises, pp. 611–612

1. $\{-5 \pm \sqrt{22}\}$ **2.** $\{8 \pm \sqrt{59}\}$ **3.** $\left\{-\dfrac{1}{6} \pm \dfrac{\sqrt{47}}{6}i\right\}$
4. $\left\{-\dfrac{3}{8} \pm \dfrac{\sqrt{71}}{8}i\right\}$ **5. a.** Quadratic   **b.** $\{2, -7\}$
6. a. Quadratic   **b.** $\{4, 5\}$ **7. a.** Quadratic form
b. $\{-3, -1, 1, 3\}$ **8. a.** Quadratic form   **b.** $\{2, -2, i, -i\}$
9. a. Quadratic form (or radical)   **b.** $\{16\}$ (The value 1 does not check.)   **10. a.** Quadratic   **b.** $\left\{\dfrac{9 \pm \sqrt{89}}{2}\right\}$
11. a. Linear   **b.** $\{4\}$ **12. a.** Linear   **b.** $\{-1\}$
13. a. Quadratic   **b.** $\{\pm i\sqrt{2}\}$ **14. a.** Quadratic
b. $\{\pm i\sqrt{6}\}$ **15. a.** Rational   **b.** $\{2, -1\}$
16. a. Rational   **b.** $\{-2, 4\}$
17. a. Quadratic   **b.** $\{10 \pm \sqrt{101}\}$
18. a. Quadratic   **b.** $\{-9 \pm \sqrt{77}\}$
19. a. Quadratic   **b.** $\left\{\dfrac{1 \pm \sqrt{17}}{2}\right\}$
20. a. Quadratic   **b.** $\left\{\dfrac{1 \pm \sqrt{13}}{2}\right\}$
21. a. Radical   **b.** $\{3\}$ (The value -1 does not check.)
22. a. Radical   **b.** $\{6\}$ (The value -1 does not check.)
23. a. Quadratic form (or radical)   **b.** $\{-125, 27\}$
24. a. Quadratic form (or radical)   **b.** $\{-64, -1\}$

Section 7.4 Practice Exercises, pp. 621–625

1. a. parabola   **b.** $>;<$ **c.** lowest; highest   **d.** (h, k); upward; downward   **e.** vertex; $x = h$

3. $\{3 \pm 2i\}$ **5.** $\left\{\dfrac{5 \pm \sqrt{10}}{5}\right\}$ **7.** $\{-27, -8\}$

9. The value of k shifts the graph of $f(x) = x^2$ vertically.

11.

13.

15.

17.

19.

21.

23.

25.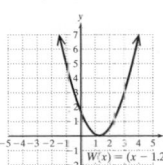

27. The value of a vertically stretches or shrinks the graph of $f(x) = x^2$.

29.

31.

33.

35.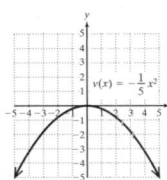

37. d   **39.** g   **41.** a   **43.** e

45.

47.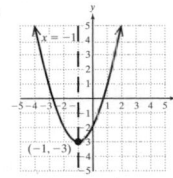

Domain: $(-\infty, \infty)$; range: $[2, \infty)$ Domain: $(-\infty, \infty)$; range: $[-3, \infty)$

49.

51.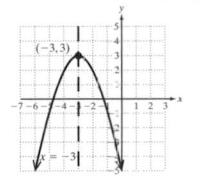

Domain: $(-\infty, \infty)$; range: $(-\infty, -2]$ Domain: $(-\infty, \infty)$; range: $(-\infty, 3]$

53.

55.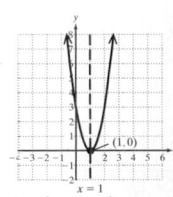

Domain: $(-\infty, \infty)$; range: $[1, \infty)$ Domain: $(-\infty, \infty)$; range: $[0, \infty)$

57.

59.

Domain: $(-\infty, \infty)$; range: $(-\infty, 3]$ Domain: $(-\infty, \infty)$; range: $[-1, \infty)$

61.

63.

Domain: $(-\infty, \infty)$; range: $(-\infty, 2]$ Domain: $(-\infty, \infty)$; range: $[1, \infty)$

65. a. $y = x^2 + 3$ is $y = x^2$ shifted up 3 units.
b. $y = (x + 3)^2$ is $y = x^2$ shifted left 3 units.
c. $y = 3x^2$ is $y = x^2$ with a vertical stretch.
67. Vertex $(6, -9)$; minimum point; minimum value: -9
69. Vertex $(2, 5)$; maximum point; maximum value: 5
71. Vertex $(-8, 0)$; minimum point; minimum value: 0
73. Vertex $\left(0, \dfrac{21}{4}\right)$; maximum point; maximum value: $\dfrac{21}{4}$
75. Vertex $\left(7, -\dfrac{3}{2}\right)$; minimum point; minimum value: $-\dfrac{3}{2}$
77. Vertex $(0, 0)$; minimum point; minimum value: 0
79. True   **81.** False   **83. a.** $(60, 30)$ **b.** 30 ft   **c.** 70 ft
85. a. The fireworks will explode at a height of 150 ft.
b. Yes, because the ordered pair $(3, 150)$ is the vertex.

Section 7.5 Practice Exercises, pp. 632–636

1. a. $\dfrac{-b}{2a}; \dfrac{-b}{2a}$ **b.** True   **c.** True   **d.** True   **e.** False

3. The graph of p is the graph of $y = x^2$ shrunk vertically by a factor of $\frac{1}{4}$.
5. The graph of r is the graph of $y = x^2$ shifted up 7 units.
7. The graph of t is the graph of $y = x^2$ shifted to the left 10 units.   **9.** $16; (x - 4)^2$ **11.** $\dfrac{49}{4}; \left(y + \dfrac{7}{2}\right)^2$

13. $\dfrac{1}{81}; \left(b + \dfrac{1}{9}\right)^2$ **15.** $\dfrac{1}{36}; \left(t - \dfrac{1}{6}\right)^2$

17. $g(x) = (x - 4)^2 - 11; (4, -11)$
19. $n(x) = 2(x + 3)^2 - 5; (-3, -5)$
21. $p(x) = -3(x - 1)^2 - 2; (1, -2)$
23. $k(x) = \left(x + \dfrac{7}{2}\right)^2 - \dfrac{89}{4}; \left(-\dfrac{7}{2}, -\dfrac{89}{4}\right)$
25. $F(x) = 5(x + 1)^2 - 4; (-1, -4)$
27. $P(x) = -2\left(x - \dfrac{1}{4}\right)^2 + \dfrac{1}{8}; \left(\dfrac{1}{4}, \dfrac{1}{8}\right)$ **29.** $(2, 3)$
31. $(-1, -2)$ **33.** $(-4, -15)$ **35.** $(-1, 2)$ **37.** $(1, 3)$
39. $\left(\dfrac{3}{2}, \dfrac{3}{4}\right)$ **41.** $(-4, -15)$ **43.** $(-1, 4)$

45. a. $(-1, -4)$ **b.** $(0, -3)$ **47. a.** $(\frac{1}{2}, \frac{7}{2})$ **b.** $(0, 4)$
c. $(1, 0), (-3, 0)$ **c.** No x-intercepts
d.

        d.

49. a. $(\frac{3}{2}, 0)$ **b.** $(0, -\frac{9}{4})$ **51. a.** $(-1, 4)$ **b.** $(0, 3)$
c. $(\frac{3}{2}, 0)$ **c.** $(1, 0), (-3, 0)$
d.

        d.

53. Mia must package 10 MP3 players.   **55. a.** 164.25 ft
b. 3.125 sec   **57. a.** 45 mph   **b.** 32 mpg   **59. a.** 48 hr
b. 2 g   **61.** $a = -9, b = 5, c = 4; y = -9x^2 + 5x + 4$
63. $a = 2, b = -1, c = -5; y = 2x^2 - x - 5$
65. $a = -3, b = 4, c = 0; y = -3x^2 + 4x$
67. a. The sum of the sides must equal the total amount of fencing.
b. $A = x(200 - 2x)$ **c.** 50 ft by 100 ft

Section 7.5 Graphing Calculator Exercises, p. 636

69.   **70.**

71.   **72.**

73.   **74.**

Section 7.6 Practice Exercises, pp. 645–647

1. a. quadratic   **b.** solutions; undefined   **c.** test point
d. true   **e.** rational   **f.** denominator   **g.** { }; $(-\infty, \infty)$
h. excludes; includes
3. $\left\{\dfrac{3}{4} \pm \dfrac{\sqrt{23}}{4}i\right\}$ **5.** $\{0, 4, -6\}$ **7.** $\{5\}$
9. a. $(-\infty, -2) \cup (3, \infty)$ **b.** $(-2, 3)$ **c.** $[-2, 3]$
d. $(-\infty, -2] \cup [3, \infty)$ **11. a.** $(-2, 0) \cup (3, \infty)$
b. $(-\infty, -2) \cup (0, 3)$ **c.** $(-\infty, -2] \cup [0, 3]$
d. $[-2, 0] \cup [3, \infty)$ **13. a.** $\{4, -\frac{1}{2}\}$ **b.** $(-\infty, -\frac{1}{2}) \cup (4, \infty)$
c. $(-\frac{1}{2}, 4)$ **15. a.** $\{-10, 3\}$ **b.** $(-10, 3)$ **c.** $(-\infty, -10) \cup (3, \infty)$
17. a. $\left\{-\dfrac{1}{2}, 3\right\}$ **b.** $\left[-\dfrac{1}{2}, 3\right]$ **c.** $\left(-\infty, -\dfrac{1}{2}\right] \cup [3, \infty)$

19. $(1, 7)$ **21.** $(-\infty, -2) \cup (5, \infty)$ **23.** $[-5, 0]$ **25.** $[4, 8]$
27. $\left(-\infty, \dfrac{1 - \sqrt{6}}{5}\right) \cup \left(\dfrac{1 + \sqrt{6}}{5}, \infty\right)$
29. $\left[\dfrac{-3 - \sqrt{33}}{2}, \dfrac{-3 + \sqrt{33}}{2}\right]$ **31.** $(-11, 11)$
33. $\left(-\infty, -\dfrac{1}{3}\right] \cup [3, \infty)$ **35.** $(-\infty, -3] \cup [0, 4]$
37. $(-2, -1) \cup (2, \infty)$ **39. a.** $\{7\}$ **b.** $(-\infty, 5) \cup (7, \infty)$
c. $(5, 7)$ **41. a.** $\{4\}$ **b.** $[4, 6)$ **c.** $(-\infty, 4] \cup (6, \infty)$
43. $(1, \infty)$ **45.** $(-\infty, -4) \cup (4, \infty)$
47. $\left(2, \dfrac{7}{2}\right)$ **49.** $(5, 7]$ **51.** $(-\infty, 0) \cup \left[\dfrac{1}{2}, \infty\right)$
53. $(0, \infty)$ **55.** $(-\infty, \infty)$ **57.** { }   **59.** $\{0\}$
61. { }   **63.** { }   **65.** $(-\infty, \infty)$
67. $(-\infty, -11) \cup (-11, \infty)$ **69.** $\left\{\dfrac{3}{2}\right\}$ **71.** Quadratic;
$[-4, 4]$ **73.** Quadratic; $(-\infty, \infty)$ **75.** Linear; $(-\frac{5}{2}, \infty)$
77. Rational; $(-\infty, -1)$ **79.** Polynomial (degree > 2);
$(0, 5) \cup (5, \infty)$ **81.** Polynomial (degree > 2); $(2, \infty)$
83. Polynomial (degree > 2); $(-\infty, 5]$
85. Quadratic; $(-\sqrt{2}, \sqrt{2})$
87. Quadratic; $\left(-\infty, \dfrac{-5 - \sqrt{33}}{2}\right] \cup \left[\dfrac{-5 + \sqrt{33}}{2}, \infty\right)$
89. Rational; $(-\infty, -2] \cup (5, \infty)$ **91.** Linear; $(-\infty, 5]$
93. Quadratic; $(-\infty, \infty)$ **95.** The fuses should be set for
between 1.3 sec and 5.4 sec after launch.

Section 7.6 Graphing Calculator Exercises, p. 648

96. $(-\infty, 0) \cup (2, \infty)$ **97.** $(0, 2)$

98. $(-1, 1)$ **99.** $(-\infty, -1) \cup (1, \infty)$

100. $\{-5\}$ **101.** $\{5\}$

102. { }   **103.** { }

Chapter 7 Problem Recognition Exercises, p. 648

1. a. Equation quadratic in form and polynomial equation **b.** $\{\pm 2\sqrt{2}, \pm 1\}$ **2. a.** Absolute value inequality **b.** $(-\frac{1}{2}, 1)$ **3. a.** Polynomial inequality **b.** $[-\frac{1}{2}, 5]$ **4. a.** Radical equation **b.** $\{1\}$
5. a. Absolute value equation **b.** $\{9, -1\}$
6. a. Rational equation **b.** $\{4 \pm 2\sqrt{6}\}$
7. a. Polynomial inequality **b.** $[-5, -2] \cup [2, \infty)$
8. a. Compound inequality **b.** $(-\infty, 1)$
9. a. Linear inequality **b.** $[-23, \infty)$
10. a. Absolute value equation **b.** $\{9, 1\}$
11. a. Rational inequality **b.** $(-\infty, 2) \cup [5, \infty)$
12. a. Absolute value inequality **b.** $(-\infty, -13) \cup (5, \infty)$
13. a. Radical equation **b.** $\{-4\}$ (The value -7 does not check.)
14. a. Quadratic equation **b.** $\left\{\frac{3}{4} \pm \frac{\sqrt{10}}{4}i\right\}$
15. a. Compound inequality **b.** $(-\infty, -6) \cup (4, \infty)$
16. a. Linear equation and rational equation **b.** $\left\{-\frac{11}{2}\right\}$
17. a. Polynomial inequality **b.** $\{5\}$
18. a. Rational inequality **b.** $\{\ \}$
19. a. Radical equation and equation quadratic in form **b.** $\{16, 81\}$ **20. a.** Polynomial equation and equation quadratic in form **b.** $\{\pm 2, \pm 3\}$

Chapter 7 Review Exercises, pp. 655–658

1. $\{\pm\sqrt{5}\}$ **2.** $\{\pm 2i\}$ **3.** $\{\pm 9\}$ **4.** $\left\{\pm\frac{\sqrt{57}}{3}i\right\}$
5. $\{2 \pm 6\sqrt{2}\}$ **6.** $\left\{\frac{5}{2} \pm \frac{3}{2}i\right\}$ **7.** $\left\{\frac{1 \pm \sqrt{3}}{3}\right\}$
8. $\{4 \pm \sqrt{5}\}$ **9.** $5\sqrt{3}$ in. ≈ 8.7 in. **10.** 9 in.
11. $5\sqrt{6}$ in. ≈ 12.2 in. **12.** $n = 64; (x + 8)^2$
13. $n = \frac{81}{4}; \left(x - \frac{9}{2}\right)^2$ **14.** $n = \frac{1}{16}; \left(y + \frac{1}{4}\right)^2$
15. $n = \frac{1}{25}; \left(z - \frac{1}{5}\right)^2$ **16.** $\{-2 \pm 3i\}$ **17.** $\{1 \pm \sqrt{6}\}$
18. $\{3 \pm 2\sqrt{3}\}$ **19.** $\{4 \pm 3i\}$ **20.** $\left\{\frac{1}{3}, -1\right\}$
21. $\left\{\frac{1}{2}, -4\right\}$ **22.** $r = \sqrt{\frac{V}{\pi h}}$ or $r = \frac{\sqrt{V\pi h}}{\pi h}$
23. $s = \sqrt{\frac{A}{6}}$ or $s = \frac{\sqrt{6A}}{6}$ **24.** See page 596.
25. Two rational solutions **26.** Two rational solutions
27. Two irrational solutions **28.** Two imaginary solutions
29. One rational solution **30.** Two imaginary solutions
31. $\{2 \pm \sqrt{3}\}$ **32.** $\left\{\frac{5}{2} \pm \frac{5\sqrt{3}}{2}i\right\}$ **33.** $\left\{2, -\frac{5}{6}\right\}$
34. $\left\{2, \frac{4}{3}\right\}$ **35.** $\left\{\frac{4}{5}, -\frac{1}{5}\right\}$ **36.** $\left\{\frac{1}{10}, -\frac{1}{2}\right\}$
37. $\{2 \pm 2i\sqrt{7}\}$ **38.** $\{4 \pm \sqrt{6}\}$ **39.** $\left\{\frac{2 \pm \sqrt{22}}{3}\right\}$
40. $\{8, -2\}$ **41.** $\{-7 \pm \sqrt{3}\}$ **42.** $\{-1 \pm \sqrt{11}\}$
43. a. 1822 ft **b.** 115 ft/s **44. a.** $\approx 53,939$ thousand **b.** 2021 **45.** The dimensions are approximately 3.1 ft by 7.2 ft.
46. The distance between Lincoln and Omaha is approximately 50 mi. **47.** $\{49\}$ (The value 9 does not check.)
48. $\{4, 16\}$ **49.** $\{\pm 3, \pm\sqrt{2}\}$ **50.** $\left\{\pm\frac{\sqrt{6}}{2}, \pm i\right\}$
51. $\{-243, 32\}$ **52.** $\{32, 1\}$ **53.** $\{3 \pm \sqrt{10}\}$
54. $\left\{\frac{1 \pm \sqrt{181}}{6}\right\}$ **55.** $\{\pm 3i, \pm i\sqrt{3}\}$ **56.** $\{\pm\sqrt{7}, \pm 2\}$

57.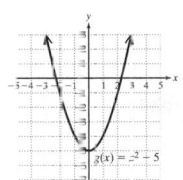
Domain: $(-\infty, \infty)$; range: $[-5, \infty)$
58.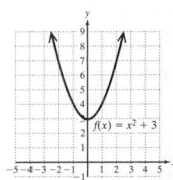
Domain: $(-\infty, \infty)$; range: $[3, \infty)$
59.
Domain: $(-\infty, \infty)$; range: $[0, \infty)$
60.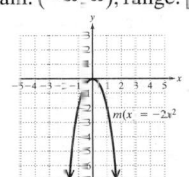
Domain: $(-\infty, \infty)$; range: $[0, \infty)$
61.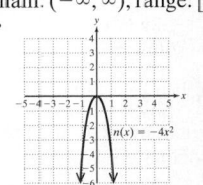
Domain: $(-\infty, \infty)$; range: $(-\infty, 0]$
62.
Domain: $(-\infty, \infty)$; range: $(-\infty, 0]$
63.
Domain: $(-\infty, \infty)$; range: $(-\infty, -5]$
64.
Domain: $(-\infty, \infty)$; range: $(-\infty, 3]$
65. $\left(4, \frac{5}{3}\right)$ is the minimum point. The minimum value is $\frac{5}{3}$.
66. $\left(1, -\frac{1}{7}\right)$ is the maximum point. The maximum value is $-\frac{1}{7}$ **67.** $x = -\frac{2}{11}$ **68.** $x = \frac{3}{16}$
69. $z(x) = (x - 3)^2 - 2; (3, -2)$
70. $b(x) = (x - 2)^2 - 48; (2, -48)$
71. $p(x) = -5(x + 1)^2 - 8; (-1, -8)$
72. $q(x) = -3(x + 4)^2 - 6; (-4, -6)$ **73.** $(1, -15)$
74. $(-1, 7)$ **75.** $\left(\frac{1}{2}, \frac{41}{4}\right)$ **76.** $\left(-\frac{1}{3}, -\frac{22}{3}\right)$
77. a. $(2, -3)$ **b.** $(0,0), (4,0)$
c.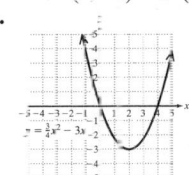
78. a. $(-2, 4)$ **b.** $(0,0)(-4,0)$
c.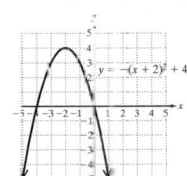
79. a. 3 sec **b.** 144 ft **80. a.** 150 meals **b.** $1200
81. $a = 1, b = 4, c = -1; y = x^2 + 4x - 1$
82. $a = 1, b = -1, c = 6; y = x^2 - x + 6$
83. a. $\{-2, 2\}$; The x-intercepts are $(-2, 0)$ and $(2, 0)$.
b. On the interval $(-2, 2)$ the graph is below the x-axis.
c. On the intervals $(-\infty, -2)$ and $(2, \infty)$ the graph is above the x-axis.
84. a. $x = 2$ **b.** $\{0\}$
c. $(-\infty, 0] \cup (2, \infty)$ **d.** $[0, 2)$ **85.** $(-2, 6)$

86. $(-\infty, \infty)$ **87.** $(-\infty, -2) \cup [0, \infty)$
88. $(-\infty, -1] \cup (1, \infty)$ **89.** $(-2, 0) \cup (5, \infty)$
90. $(-2, 0) \cup \left(\dfrac{5}{2}, \infty\right)$
91. $(-\infty, -2 - \sqrt{3}] \cup [-2 + \sqrt{3}, \infty)$
92. $(-\infty, -5) \cup (1, \infty)$ **93.** $(3, \infty)$
94. $(-3, \infty)$ **95.** $\{-5\}$ **96.** $(-\infty, -2) \cup (-2, \infty)$

Chapter 7 Test, pp. 658–660

1. $\{2, -8\}$ **2.** $\{2 \pm 2\sqrt{3}\}$ **3.** $\{-1 \pm i\}$
4. $n = \dfrac{121}{4}; \left(d + \dfrac{11}{2}\right)^2$ **5.** $\{-3 \pm 3\sqrt{3}\}$
6. $\left\{\dfrac{3}{4} \pm \dfrac{\sqrt{47}}{4}i\right\}$ **7. a.** $x^2 - 3x + 12 = 0$
b. $a = 1, b = -3, c = 12$ **c.** -39 **d.** Two imaginary solutions
8. a. $y^2 - 2y + 1 = 0$ **b.** $a = 1, b = -2, c = 1$ **c.** 0   **d.** One rational solution
9. $\left\{1, \dfrac{1}{3}\right\}$ **10.** $\left\{\dfrac{-7 \pm \sqrt{5}}{2}\right\}$ **11.** The height is approximately 4.6 ft and the base is 6.2 ft.   **12.** The radius is approximately 12.0 ft.   **13.** $\{9\}$ (The value 4 does not check.)   **14.** $\{8, -64\}$ **15.** $\left\{\dfrac{11}{3}, 6\right\}$
16. $\{\pm\sqrt{6}, \pm 3\}$ **17.** $\left\{\dfrac{5 \pm \sqrt{57}}{2}\right\}$ **18.** $\left\{5, -\dfrac{1}{2}\right\}$
19. $\{4 \pm \sqrt{15}\}$ **20.** $\{-7 \pm 2i\sqrt{6}\}$ **21.** $\{6 \pm \sqrt{23}\}$
22. The vertex is $(3, -17)$.
23.

24.

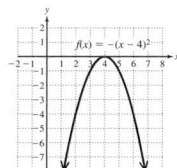

Domain: $(-\infty, \infty)$; range: $[-4, \infty)$ Domain: $(-\infty, \infty)$; range: $(-\infty, 0]$
25.

Domain: $(-\infty, \infty)$; range: $[-3, \infty)$
26. The rocket will hit the ground 256 ft away.
27. a. 1320 million   **b.** 1999   **28.** The graph of $y = x^2 - 2$ is the graph of $y = x^2$ shifted down 2 units.   **29.** The graph of $y = (x + 3)^2$ is the graph of $y = x^2$ shifted 3 units to the left.
30. The graph of $y = -4x^2$ is the graph of $y = 4x^2$ opening downward instead of upward.   **31. a.** $(4, 2)$
b. Downward   **c.** Maximum point   **d.** The maximum value is 2.
e. $x = 4$ **32. a.** $g(x) = 2(x - 5)^2 + 1; (5, 1)$
b. $(5, 1)$ **33. a.** $f(x) = (x + 2)^2 - 16$ **b.** Vertex: $(-2, -16)$
c. x-intercepts: $(-6, 0)$ and $(2, 0)$; y-intercept: $(0, -12)$
d. The minimum value is -16 **e.** $x = -2$
34. a. 200 ft   **b.** 20,000 ft^2 **35.** $\left[\dfrac{1}{2}, 6\right)$ **36.** $(-5, 5)$
37. $(-\infty, -3) \cup (-2, 2)$ **38.** $\left(-3, -\dfrac{3}{2}\right)$
39. $\{\ \}$ **40.** $\{-11\}$

Chapters 1–7 Cumulative Review Exercises, pp. 660–662

1. a. $\{2, 4, 6, 8, 10, 12, 16\}$ **b.** $\{2, 8\}$ **2.** $-3x^2 - 13x + 1$
3. -16 **4.** 1.8×10^{10} **5. a.** $(x + 2)(x + 3)(x - 3)$
b. Quotient: $x^2 + 5x + 6$; remainder: 0   **6.** $x - 2$
7. $\dfrac{2\sqrt{2x}}{x}$ **8.** $8000 in 12% account; $2000 in 3% account

9. $\{(8, 7)\}$ **10. a.** 720 ft   **b.** 720 ft   **c.** 12 sec
11. $\{3 \pm 4i\}$ **12.** $\left\{\dfrac{-5 \pm \sqrt{33}}{4}\right\}$ **13.** $n = 25; (x + 5)^2$
14. $2(x + 5)(x^2 - 5x + 25)$
15.

16. a. $\left(\dfrac{5}{2}, 0\right), (2, 0)$ **b.** $(0, 10)$ **17.** The domain element 3 has more than one corresponding range element.
18.

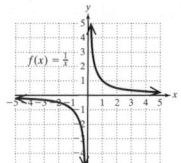

19. $y = 39$ **20. a.** Linear   **b.** $(0, 300,000)$. If there are no passengers, the airport runs 300,000 flights per year.
c. $m = 0.008$ or $m = \dfrac{8}{1000}$. There are eight additional flights per 1000 passengers.   **21. a.** $[-4, \infty)$ **b.** $(-\infty, \infty)$
22. a. $(-\infty, 2]$ **b.** $(-\infty, 4]$ **c.** 4   **d.** 3   **e.** $x = -1$
23. $\left\{\dfrac{1}{18}, \dfrac{1}{2}\right\}$ **24.** $f = \dfrac{pq}{p + q}$ **25.** $\{7\}$
26. $\dfrac{y + 1}{y - 3}$ **27. a.** $(3, 1)$ **b.** Upward   **c.** $(0, 19)$
d. No x-intercept
e.

28. Vertex: $(8, -62)$ **29. a.** $\left(-\infty, -\dfrac{5}{2}\right] \cup [2, \infty)$ **b.** On these intervals, the graph is on or above the x-axis.
30. $(-\infty, 3) \cup [5, \infty)$

Chapter 8

Chapter Opener Puzzle

5	1	B 4	C 3	6	2
3	F 6	2	1	4	5
G 1	5	3	6	D 2	4
A 2	4	6	5	1	3
6	2	5	E 4	3	1
4	3	H 1	2	5	6

Section 8.1 Practice Exercises, pp. 668–671

1. a. $f(x); g(x)$ **b.** $g(x)$ **c.** $f(g(x))$
3. $(f + g)(x) = 2x^2 + 5x + 4$
5. $(g - f)(x) = 2x^2 + 3x - 4$
7. $(f \cdot h)(x) = x^3 + 4x^2 + x + 4$
9. $(g \cdot f)(x) = 2x^3 + 12x^2 + 16x$
11. $\left(\dfrac{h}{f}\right)(x) = \dfrac{x^2 + 1}{x + 4}, x \neq -4$

13. $\left(\dfrac{f}{g}\right)(x) = \dfrac{x+4}{2x^2+4x}, x \neq 0, x \neq -2$

15. $(f \circ g)(x) = 2x^2 + 4x + 4$

17. $(g \circ f)(x) = 2x^2 + 20x + 48$ **19.** $(k \circ h)(x) = \dfrac{1}{x^2+1}$

21. $(k \circ g)(x) = \dfrac{1}{2x^2+4x}, x \neq 0, x \neq -2$ **23.** No

25. $(f \circ g)(x) = 25x^2 - 15x + 1; (g \circ f)(x) = 5x^2 - 15x + 5$

27. $(f \circ g)(x) = |x^3 - 1|; (g \circ f)(x) = |x|^3 - 1$

29. $(h \circ h)(x) = 25x - 24$ **31.** 0 **33.** -64 **35.** 2

37. 1 **39.** $\dfrac{1}{64}$ **41.** 0 **43.** Undefined

45. $\dfrac{4}{9}$ **47.** -2 **49.** 2 **51.** 0 **53.** 1

55. 0 **57.** Undefined **59.** -1 **61.** 2 **63.** 2

65. -1 **67.** -2 **69.** 3 **71.** -6 **73.** -1

75. 4 **77.** 0 **79.** 2 **81. a.** $P(x) = 3.78x - 1$

b. \$188 **83. a.** $F(t) = 0.2t + 6.4$; F represents the amount of child support (in billion dollars) not paid. **b.** $F(4) = 7.2$ means that in 2004, \$7.2 billion of child support was not paid.

85. a. $(D \circ r)(t) = 560t$; This function represents the total distance Joe travels as a function of time that he rides. **b.** 5600 ft

Section 8.2 Practice Exercises, pp. 677–680

1. a. $\{(2,1),(3,2),(4,3)\}$ **b.** one-to-one; y **c.** is not **d.** is
e. $y = x$ **f.** $x; x$ **g.** f^{-1} **h.** (b, a) **3.** Yes **5.** No

7. Yes **9.** $g^{-1} = \{(5,3),(1,8),(9,-3),(2,0)\}$

11. $r^{-1} = \{(3,a),(6,b),(9,c)\}$ **13.** The function is not one-to-one.

15. Yes **17.** No **19.** Yes

21. a. $(f \circ g)(x) = 6\left(\dfrac{x-1}{6}\right) + 1 = x$

b. $(g \circ f)(x) = \dfrac{(6x+1)-1}{6} = x$

23. a. $(f \circ g)(x) = \dfrac{\sqrt[3]{8x^3}}{2} = x$ **b.** $(g \circ f)(x) = 8\left(\dfrac{\sqrt[3]{x}}{2}\right)^3 = x$

25. a. $(f \circ g)(x) = (\sqrt{x-1})^2 + 1 = x$
b. $(g \circ f)(x) = \sqrt{(x^2+1)-1} = x$ **27.** $h^{-1}(x) = x - 4$

29. $m^{-1}(x) = 3(x+2)$ **31.** $p^{-1}(x) = -x + 10$

33. $n^{-1}(x) = \dfrac{5x-2}{3}$ **35.** $h^{-1}(x) = \dfrac{3x+1}{4}$

37. $f^{-1}(x) = \sqrt[3]{x-1}$ **39.** $g^{-1}(x) = \dfrac{x^3+1}{2}$

41. $g^{-1}(x) = \sqrt{x-9}$

43. a. 1.2192 m, 15.24 m **b.** $f^{-1}(x) = \dfrac{x}{0.3048}$ **c.** 4921.3 ft

45. False **47.** True **49.** False **51.** True **53.** $(b, 0)$
55. a. Domain $[1, \infty)$, range $[0, \infty)$ **b.** Domain $[0, \infty)$,
range $[1, \infty)$ **57. a.** $[-4, 0]$ **b.** $[0, 2]$ **c.** $[0, 2]$ **d.** $[-4, 0]$

59. a. $[0, 2]$ **b.** $[0, 4]$ **c.** $[0, 4]$ **d.** $[0, 2]$

61. $q^{-1}(x) = x^2 - 4, x \geq 0$ **63.** $z^{-1}(x) = x^2 - 4, x \leq 0$

65. $f^{-1}(x) = \dfrac{x+1}{1-x}$ **67.** $t^{-1}(x) = \dfrac{x+2}{x}$

69. $n^{-1}(x) = -\sqrt{x-9}$

Section 8.2 Graphing Calculator Exercises, p. 680

71. $f^{-1}(x) = x^3 - 5$

72. $k^{-1}(x) = \sqrt[3]{x+4}$

73. $g^{-1}(x) = \sqrt[3]{2x+4}$

74. $m^{-1}(x) = \dfrac{x+4}{3}$

Section 8.3 Practice Exercises, pp. 686–689

1. a. b^x **b.** is not; is **c.** increasing **d.** decreasing
e. $(-\infty, \infty); (0, \infty)$ **f.** $(0, 1)$ **g.** $y = 0$ **h.** is not

3. $-2x^2 + 2x - 3$ **5.** $\dfrac{3x-1}{2x^2+x+2}$ **7.** $6x^2 + 3x + 5$

9. 25 **11.** $\dfrac{1}{1000}$ **13.** 6 **15.** 8 **17.** 5.8731

19. 1385.4557 **21.** 0.0063 **23.** 0.8950
25. a. $x = 2$ **b.** $x = 3$ **c.** Between 2 and 3
27. a. $x = 4$ **b.** $x = 5$ **c.** Between 4 and 5

29. $f(0) = 1, f(1) = \dfrac{1}{5}, f(2) = \dfrac{1}{25}, f(-1) = 5, f(-2) = 25$

31. $h(0) = 1, h(1) = 3, h(-1) = \dfrac{1}{3}, h(\sqrt{2}) \approx 4.73,$

$h(\pi) \approx 31.54$ **33.** If $b > 1$, the graph is increasing. If $0 < b < 1$, the graph is decreasing.

35.

37.

39.

41.

43. a. 0.25 g **b.** ≈ 0.16 g **45. a.** 758,000 **b.** 379,000
c. 144,000 **47. a.** $P(t) = 153,000,000(1.0125)^t$
b. $P(41) \approx 255,000,000$ **49. a.** \$1640.67 **b.** \$2691.80
c. $A(0) = 1000$. The initial amount of the investment is \$1000.
$A(7) = 2000$ The amount of the investment doubles in 7 yr.

Section 8.3 Graphing Calculator Exercises, p. 689

51.   **52.**

53.   **54.**

55.   **56.**

57.   **58.**

Section 8.4 Practice Exercises, pp. 698–702

1. a. logarithmic; b **b.** logarithm; base; argument
c. $(0, \infty); (-\infty, \infty)$ **d.** exponential   **e.** common
f. increasing; decreasing   **g.** 2, 3, and 4   **h.** $y = x$ **3.** i

5. a. $s(-2) = \dfrac{25}{4}, s(-1) = \dfrac{5}{2}, s(0) = 1, s(1) = \dfrac{2}{5}, s(2) = \dfrac{4}{25}$

b.

7. $g(-2) = \dfrac{1}{9}, g(-1) = \dfrac{1}{3}, g(0) = 1, g(1) = 3, g(2) = 9$

9. $b^y = x$ **11.** $5^4 = 625$ **13.** $10^{-4} = 0.0001$ **15.** $6^2 = 36$

17. $b^x = 15$ **19.** $3^x = 5$ **21.** $\left(\dfrac{1}{4}\right)^{10} = x$ **23.** $\log_3 81 = x$

25. $\log_5 25 = 2$ **27.** $\log_7\left(\dfrac{1}{7}\right) = -1$ **29.** $\log_b y = x$

31. $\log_e y = x$ **33.** $\log_{1/3} 9 = -2$ **35.** 2   **37.** -1 **39.** $\dfrac{1}{2}$

41. 0   **43.** 5   **45.** 1   **47.** 3   **49.** $\dfrac{1}{2}$ **51.** 1   **53.** 3

55. 6   **57.** -2 **59.** 0.7782   **61.** 0.4971   **63.** -1.5051
65. -2.2676 **67.** 5.5315   **69.** -7.4202
71. a. Slightly less than 2   **b.** Slightly more than 1
c. $\log 93 \approx 1.9685, \log 12 \approx 1.0792$

73. a. $f\left(\dfrac{1}{64}\right) = -3, f\left(\dfrac{1}{16}\right) = -2, f\left(\dfrac{1}{4}\right) = -1, f(1) = 0,$
$f(4) = 1, f(16) = 2, f(64) = 3$

b.

75. $3^y = x$

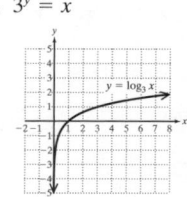

x	y
$\frac{1}{9}$	-2
$\frac{1}{3}$	-1
1	0
3	1
9	2

77. $\left(\dfrac{1}{2}\right)^y = x$

x	y
4	-2
2	-1
1	0
$\frac{1}{2}$	1
$\frac{1}{4}$	2

79. $(5, \infty)$ **81.** $(-\infty, 2)$ **83.** $\left(\dfrac{1}{3}, \infty\right)$ **85.** $(-1.2, \infty)$
87. $(-\infty, 2)$ **89.** $(-\infty, 0) \cup (0, \infty)$ **91.** ≈ 7.35
93. a.

t (months)	0	1	2	6	12	24
$S_1(t)$	91	82.0	76.7	65.6	57.6	49.1
$S_2(t)$	88	83.5	80.8	75.3	71.3	67.0

b. Group 1: 91; Group 2: 88   **c.** Method II

Section 8.4 Graphing Calculator Exercises, p. 702

95. Domain: $(-6, \infty)$;   **96.** Domain: $(-2, \infty)$;
asymptote: $x = -6$ asymptote: $x = -2$

97. Domain: $(2, \infty)$;   **98.** Domain: $(-8, \infty)$;
asymptote: $x = 2$ asymptote: $x = -8$

99. Domain: $(-\infty, 2)$;   **100.** Domain: $(-\infty, 3)$;
asymptote: $x = 2$ asymptote: $x = 3$

Chapter 8 Problem Recognition Exercises, p. 703

1. e   **2.** l   **3.** j   **4.** g   **5.** c   **6.** a
7. k   **8.** b   **9.** i   **10.** h   **11.** f   **12.** d

Section 8.5 Practice Exercises, pp. 709–712

1. a. $1; 0; x; x$ **b.** $\log_b x + \log_b y; \log_b x - \log_b y$ **c.** $p \log_b x$
d. False   **e.** False   **f.** False

3. 4   **5.** $\dfrac{1}{6}$ **7.** 0.9031   **9.** 5.0475   **11.** a   **13.** c

15. a, b, c   **17.** 1   **19.** 4   **21.** 11   **23.** 3   **25.** 0
27. 9   **29.** 0   **31.** 3   **33.** 1   **35.** $2x$ **37.** 4   **39.** 0
41. Expressions a and c are equivalent.
43. Expressions a and c are equivalent.
45. $\log_3 x - \log_3 5$ **47.** $\log 2 + \log x$ **49.** $4 \log_5 x$
51. $\log_4 a + \log_4 b - \log_4 c$

53. $\dfrac{1}{2}\log_b x + \log_b y - 3\log_b z - \log_b w$

55. $\log_2(x+1) - 2\log_2 y - \dfrac{1}{2}\log_2 z$

57. $\dfrac{1}{3}\log a + \dfrac{2}{3}\log b - \dfrac{1}{3}\log c$ **59.** $-5\log w$

61. $\dfrac{1}{2}\log_b a - 3 - \log_b c$ **63.** 3   **65.** 2   **67.** $\log_3\left(\dfrac{x^2 z}{y^3}\right)$

69. $\log_3\left(\dfrac{a^2 c}{\sqrt[4]{b}}\right)$ **71.** $\log_b x^2$ **73.** $\log_8(8a^5)$ or $\log_8 a^5 + 1$

75. $\log\left[\dfrac{(x+6)^2\sqrt[3]{y}}{z^5}\right]$ **77.** $\log_b\left(\dfrac{1}{x-1}\right)$ **79.** 1.792

81. 2.485   **83.** 4.396   **85.** 0.916   **87.** 13.812   **89.** 16.09

91. a. $B = 10\log I - 10\log I_0$ **b.** $10\log I + 160$

Section 8.5 Graphing Calculator Exercises, p. 712

93. a. Domain: $(-\infty, 0) \cup (0, \infty)$ **b.** Domain: $(0, \infty)$

c. They are equivalent for all x in the intersection of their domains, $(0, \infty)$.
94. a. Domain: $(-\infty, 1) \cup (1, \infty)$ **b.** Domain: $(1, \infty)$

c. They are equivalent for all x in the intersection of their domains, $(1, \infty)$.

Section 8.6 Practice Exercises, pp. 720–724

1. a. e **b.** e **c.** natural; $\ln x$ **d.** $0; 1; p, x$ **e.** $\ln x + \ln y$; $\ln x - \ln y$ **f.** $p \ln x$ **g.** $\log_a b$

3. $\log_3\left(\dfrac{a\sqrt{d}}{c^5}\right)$ **5.** $\dfrac{1}{4}\log_6 x + \dfrac{1}{2}\log_6 y - \dfrac{3}{4}\log_6 z$

7.

x	y
-4	0.05
-3	0.14
-2	0.37
-1	1
0	2.72
1	7.39

Domain: $(-\infty, \infty)$

9.

x	y
-2	2.14
-1	2.37
0	3
1	4.72
2	9.39
3	22.09

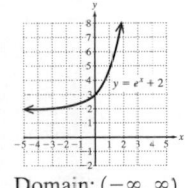

Domain: $(-\infty, \infty)$

11. a. \$12,209.97   **b.** \$13,488.50   **c.** \$14,898.46   **d.** \$16,050.09
An investment grows more rapidly at higher interest rates.
13. a. \$12,423.76   **b.** \$12,515.01   **c.** \$12,535.94   **d.** \$12,546.15
e. \$12,546.50   More money is earned at a greater number of compound periods per year.   **15. a.** \$6920.15   **b.** \$9577.70
c. \$13,255.84   **d.** \$18,346.48   **e.** \$35,143.44   More money is earned over a longer period of time.

17.

x	y
2.25	-1.39
2.50	-0.69
2.75	-0.29
3	0
4	0.69
5	1.10
6	1.39

Domain: $(2, \infty)$

19.

x	y
0.25	-2.39
0.5	-1.69
0.75	-1.29
1	-1.00
2	-0.31
3	0.10
4	0.39

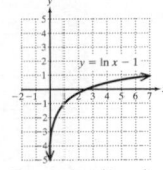

Domain: $(0, \infty)$

21. a.

b. Domain: $(-\infty, \infty)$; range: $(0, \infty)$
c. Domain: $(0, \infty)$; range: $(-\infty, \infty)$

23. 1   **25.** 0   **27.** -6 **29.** $2x + 3$ **31.** $\ln(p^6\sqrt[3]{q})$

33. $\ln\sqrt{\dfrac{x}{y^3}}$ **35.** $\ln\left(\dfrac{a^2}{b\sqrt[3]{c}}\right)$ **37.** $\ln\left(\dfrac{x^4}{y^3 z}\right)$

39. $2\ln a - 2\ln b$ **41.** $2\ln b + 1$ **43.** $4\ln a + \dfrac{1}{2}\ln b - \ln c$

45. $\dfrac{1}{5}\ln a - \dfrac{1}{5}\ln b - \dfrac{2}{5}\ln c$ **47. a.** 2.9570   **b.** 2.9570
c. They are the same.   **49.** 2.8074   **51.** 1.5283

53. -2.1269 **55.** 0 **57.** -3.3219 **59.** -3.8124
61. a. 15.4 yr   **b.** 6.9 yr   **c.** 13.8 yr   **63. a.** 19.8 yr
b. 13.9 yr   **c.** 27.8 yr

Section 8.6 Graphing Calculator Exercises, pp. 724–725

65. a–b

c. They appear to be the same.

66. a–b

c. They appear to be the same.

67.   **68.**

69.

Chapter 8 Problem Recognition Exercises, p. 725

	Exponential Form	Logarithmic Form
1.	$2^5 = 32$	$\log_2 32 = 5$
2.	$3^4 = 81$	$\log_3 81 = 4$
3.	$z^y = x$	$\log_z x = y$
4.	$b^c = a$	$\log_b a = c$
5.	$10^3 = 1000$	$\log 1000 = 3$
6.	$10^1 = 10$	$\log 10 = 1$
7.	$e^a = b$	$\ln b = a$
8.	$e^q = p$	$\ln p = q$
9.	$\left(\frac{1}{2}\right)^2 = \frac{1}{4}$	$\log_{1/2}\left(\frac{1}{4}\right) = 2$
10.	$\left(\frac{1}{3}\right)^{-2} = 9$	$\log_{1/3} 9 = -2$
11.	$10^{-2} = 0.01$	$\log 0.01 = -2$
12.	$10^x = 4$	$\log 4 = x$
13.	$e^0 = 1$	$\ln 1 = 0$
14.	$e^1 = e$	$\ln e = 1$
15.	$25^{1/2} = 5$	$\log_{25} 5 = \frac{1}{2}$
16.	$16^{1/4} = 2$	$\log_{16} 2 = \frac{1}{4}$
17.	$e^t = s$	$\ln s = t$
18.	$e^r = w$	$\ln w = r$
19.	$15^{-2} = \frac{1}{225}$	$\log_{15}\left(\frac{1}{225}\right) = -2$
20.	$3^{-1} = p$	$\log_3 p = -1$

Section 8.7 Practice Exercises, pp. 734–738

1. a. $x = y$ **b.** $x = y$ **3.** $\log_b [x(2x + 3)]$
5. $\log_b\left(\dfrac{x + 2}{3x - 5}\right)$ **7.** $\{9\}$ **9.** $\{10^{42}\}$ **11.** $\{e^{0.08}\}$
13. $\{10^{-9.2} - 40\}$ **15.** $\{5\}$ **17.** $\{10\}$ **19.** $\{25\}$ **21.** $\{59\}$
23. $\{1\}$ **25.** $\{0\}$ **27.** $\left\{-\dfrac{37}{9}\right\}$
29. $\{3\}$ (The value -3 does not check.)   **31.** $\{4\}$ **33.** $\{3\}$
35. $\{2\}$ **37.** $\{ \ \}$ (The value -3 does not check.)   **39.** $\{4\}$
41. $\{-6\}$ **43.** $\left\{\dfrac{1}{2}\right\}$ **45.** $\{2\}$ **47.** $\left\{\dfrac{11}{12}\right\}$ **49.** $\{1\}$
51. $\left\{\dfrac{4}{19}\right\}$ **53.** $\{-2, 1\}$ **55.** $\left\{\dfrac{\ln 21}{\ln 8}\right\}$ **57.** $\{\ln 8.1254\}$
59. $\{\log 0.0138\}$ **61.** $\left\{\dfrac{\ln 15}{0.07}\right\}$ **63.** $\left\{\dfrac{\ln 3}{1.2}\right\}$
65. $\left\{\dfrac{\ln 3}{\ln 5 - \ln 3}\right\}$ **67.** $\left\{\dfrac{2 \ln 2}{\ln 6 - \ln 2}\right\}$ **69.** $\left\{\dfrac{\ln 4}{0.04}\right\}$
71. $\left\{3 \ln\left(\dfrac{125}{6}\right)\right\}$ **73.** $\left\{\dfrac{\ln 20}{\ln 5} + 2\right\}$
75. a. ≈ 1285 million (or 1,285,000,000) people   **b.** ≈ 1466.5 million (or 1,466,500,000) people   **c.** The year 2049 ($t \approx 50.8$)
77. a. 500 bacteria   **b.** ≈ 660 bacteria   **c.** ≈ 25 min
79. It will take 9.9 yr for the investment to double.   **81.** 5 yr
83. a. 7.8 g   **b.** 18.5 days   **85.** The intensity of sound of heavy traffic is $10^{-3.07}$ W/m^2.   **87.** It will take 38.4 yr.   **89. a.** 1.42 kg
b. No   **91.** $\{10^5, 10^{-3}\}$ **93.** $\left\{\dfrac{1}{27}, \dfrac{1}{9}\right\}$

Section 8.7 Graphing Calculator Exercises, p. 738

95.   **96.**

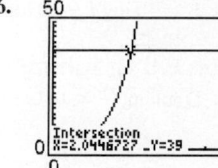

Chapter 8 Review Exercises, pp. 744–748

1. $(f - g)(x) = 2x^3 + 9x - 7$
2. $(f + g)(x) = -2x^3 - 7x - 7$
3. $(f \cdot n)(x) = \dfrac{x - 7}{x - 2}, x \neq 2$ **4.** $(f \cdot m)(x) = x^3 - 7x^2$
5. $\left(\dfrac{f}{g}\right)(x) = \dfrac{x - 7}{-2x^3 - 8x}, x \neq 0$
6. $\left(\dfrac{g}{f}\right)(x) = \dfrac{-2x^3 - 8x}{x - 7}, x \neq 7$
7. $(m \circ f)(x) = (x - 7)^2$ or $x^2 - 14x + 49$
8. $(n \circ f)(x) = \dfrac{1}{x - 9}, x \neq 9$ **9.** 100   **10.** $\dfrac{1}{8}$ **11.** -167
12. -10 **13. a.** $(2x + 1)^2$ or $4x^2 + 4x + 1$ **b.** $2x^2 + 1$
c. No, $f \circ g \neq g \circ f$ **14.** $\dfrac{1}{4}$ **15.** -3 **16.** 0
17. -1 **18.** 4   **19.** 1   **20.** No   **21.** Yes
22. $\{(5, 3), (9, 2), (-1, 0), (1, 4)\}$ **23.** $q^{-1}(x) = \dfrac{4}{3}(x + 2)$
24. $g^{-1}(x) = (x - 3)^5$ **25.** $f^{-1}(x) = \sqrt[3]{x} + 1$
26. $n^{-1}(x) = \dfrac{2x + 4}{x}$
27. $(f \circ g)(x) = 5\left(\dfrac{1}{5}x + \dfrac{2}{5}\right) - 2 = x + 2 - 2 = x$
$(g \circ f)(x) = \dfrac{1}{5}(5x - 2) + \dfrac{2}{5} = x - \dfrac{2}{5} + \dfrac{2}{5} = x$

28. The graphs are symmetric about the line $y = x$.

29. a. Domain: $[-1, \infty)$; range: $[0, \infty)$ **b.** Domain: $[0, \infty)$; range: $[-1, \infty)$ **30.** $p^{-1}(x) = (x-2)^2, x \geq 2$

31. 1024 **32.** $\frac{1}{36} \approx 0.028$ **33.** 2 **34.** 10

35. 8.825 **36.** 16.242 **37.** 1.627 **38.** 0.681

39.

40.

41.

42.

43. a. Horizontal **b.** $y = 0$ **44. a.** 15,000 mrem **b.** 3750 mrem **c.** Yes **45.** -3 **46.** 0 **47.** 1 **48.** 8 **49.** 4 **50.** 4 **51.** 5 **52.** -1

53.

54.

55. a. Vertical asymptote **b.** $x = 0$ **56. a.** 2.5 **b.** 9.5 **57.** 1 **58.** 6 **59.** 0 **60.** 7

61. a. $\log_b x + \log_b y$ **b.** $\log_b\left(\frac{x}{y}\right)$ **c.** $p \log_b x$

62. $\log_b\left(\frac{\sqrt[4]{x^3 y}}{z}\right)$ **63.** $\log_3\left(\frac{\sqrt{ab}}{c^2 d^4}\right)$ **64.** $\log 5$

65. 0 **66.** b **67.** a **68.** 148.4132 **69.** 14.0940 **70.** 32.2570 **71.** 57.2795 **72.** 1.7918 **73.** -2.1972 **74.** 1.3424 **75.** 1.3029 **76.** 3.3219 **77.** 1.9943 **78.** -0.8370 **79.** -3.6668 **80. a.** $33,361.92 **b.** $33,693.90 **c.** $33,770.48 **d.** $33,809.18

81. a. $S(0) = 95$; the student's score is 95 at the end of the course. **b.** $S(6) \approx 23.7$; the student's score is 23.7 after 6 months. **c.** $S(12) \approx 20.2$; the student's score is 20.2 after 1 yr. **82.** $(-\infty, \infty)$ **83.** $(-\infty, \infty)$ **84.** $(-\infty, \infty)$

85. $(0, \infty)$ **86.** $(-5, \infty)$ **87.** $(7, \infty)$ **88.** $\left(\frac{4}{3}, \infty\right)$

89. $(-\infty, 5)$ **90.** $\{125\}$ **91.** $\left\{\frac{1}{49}\right\}$ **92.** $\{216\}$

93. $\{3^{1/12}\}$ **94.** $\left\{\frac{1001}{2}\right\}$ **95.** $\{9\}$

96. $\{5\}$ (The value -2 does not check.) **97.** $\left\{\frac{190}{321}\right\}$

98. $\{-1\}$ **99.** $\left\{\frac{4}{7}\right\}$ **100.** $\left\{\frac{\ln 21}{\ln 4}\right\}$ **101.** $\left\{\frac{\ln 18}{\ln 5}\right\}$

102. $\{-\ln 0.1\}$ **103.** $\left\{-\frac{\ln 0.06}{2}\right\}$ **104.** $\left\{\frac{\log 1512}{2}\right\}$

105. $\left\{\frac{\log 821}{3}\right\}$ **106.** $\left\{\frac{3 \ln 2}{\ln 7 - \ln 2}\right\}$

107. $\left\{\frac{5 \ln 14}{\ln 14 - \ln 6}\right\}$ **108. a.** $1.09\,\mu g$ **b.** $0.15\,\mu g$ **c.** 16.08 days **109. a.** 150 bacteria **b.** ≈ 185 bacteria **c.** ≈ 99 min **110. a.** $V(0) = 15,000$; the initial value of the car is $15,000. **b.** $V(10) = 3347$; the value of the car after 10 years is $3347. **c.** 7.3 yr

Chapter 8 Test, pp. 748–750

1. $\left(\frac{f}{g}\right)(x) = \frac{x-4}{x^2+2}$ **2.** $(h \cdot g)(x) = \frac{x^2+2}{x}, x \neq 0$
3. $(g \circ f)(x) = x^2 - 8x + 18$
4. $(h \circ f)(x) = \frac{1}{x-4}, x \neq 4$
5. -48 **6.** $-\frac{3}{2}$ **7.** $\frac{1}{18}$ **8.** 18
9. $\left(\frac{g}{f}\right)(x) = \frac{x^2+2}{x-4}, x \neq 4$ **10.** A function is one-to-one if it passes the horizontal line test. **11.** b
12. $f^{-1}(x) = 4x - 12$ **13.** $g^{-1}(x) = \sqrt{x} + 1$
14.

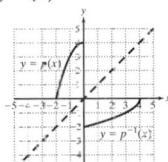

15. a. 4.5416 **b.** 32.2693 **c.** 687.2913
16.

17. a. $\log_{16} 8 = \frac{3}{4}$ **b.** $x^5 = 31$
18.

19. $\frac{\log_a n}{\log_a b}$ **20. a.** 1.3222 **b.** 1.8502 **c.** -2.5850
21. a. $1 + \log_3 x$ **b.** -5 **22. a.** $\log_b(\sqrt{x y^3})$
b. $\log\left(\frac{1}{a^3}\right)$ or $-\log a^3$ **23. a.** 1.6487 **b.** 0.0498
c. -1.0986 **d.** 1 **24. a.** $y = \ln x$ **b.** $y = e^x$
25. a. $p(4) \approx 59.8$; 59.8% of the material is retained after 4 months. **b.** $p(12) \approx 40.7$; 40.7% of the material is retained after 1 yr. **c.** $p(0) = 92$; 92% of the material is retained at the end of the course. **26. a.** 9762 thousand (or 9,762,000) people **b.** The year 2020 ($t \approx 20.4$)
27. a. $P(0) = 300$; there are 300 bacteria initially. **b.** 35,588 bacteria **c.** 1,120,537 bacteria **d.** 1,495,831 bacteria **28.** $\{25\}$ (The value -4 does not check.)
29. $\{32\}$ **30.** $\{e^{2.4} - 7\}$ **31.** $\{-7\}$ **32.** $\left\{\frac{\ln 50}{\ln 4}\right\}$
33. $\left\{\frac{\ln 250}{2.4}\right\}$ **34.** $\left\{\frac{\ln 58}{\ln 2} + 3\right\}$ **35.** $\left\{\frac{7 \ln 4}{\ln 5 - \ln 4}\right\}$
36. a. $P(2500) = 560.2$; at 2500 m the atmospheric pressure is 560.2 mm Hg. **b.** 760 mm Hg **c.** 1498.8 m
37. a. $2909.98 **b.** 9.24 yr to double

Chapters 1–8 Cumulative Review Exercises, pp. 750–751

1. $-\dfrac{5}{4}$ **2.** $-1 + \dfrac{p}{2} + \dfrac{3p^3}{4}$ **3.** Quotient:

$t^3 + 2t^2 - 9t - 18$; remainder: 0   **4.** $|x - 3|$

5. $\dfrac{2\sqrt[3]{25}}{5}$ **6.** $2\sqrt{7}$ in.   **7.** $\dfrac{4d^{1/10}}{c}$

8. $(\sqrt{15} - \sqrt{6} + \sqrt{30} - 2\sqrt{3})\,\text{m}^2$ **9.** $-\dfrac{7}{29} - \dfrac{26}{29}i$

10. $\left\{6, \dfrac{3}{2}\right\}$ **11.** 4.8 L   **12.** 24 min   **13.** $\{(7, -1)\}$

14. $\left\{-\dfrac{23}{11}\right\}$ **15.** $x = \dfrac{c + d}{a - b}$ **16.** $t = \dfrac{\sqrt{2sg}}{g}$

17. $\left\{\dfrac{4 \pm \sqrt{34}}{3}\right\}$ **18.** $(7, 0), (3, 0)$

19. a. $-30t$ **b.** $-10t^2$ **c.** $2t^2 + 5t$ **20.** $\{0\}$

21. a. $x = 2$ **b.** $y = 6$ **c.** $y = \dfrac{1}{2}x + 5$

22. $40°, 80°, 60°$ **23.** Infinitely many solutions;
$\{(x, y)\,|\,-2x + y = -4\}$; dependent equations

24. $\left[\dfrac{1}{2}, \infty\right)$ **25.** $f^{-1}(x) = \dfrac{1}{5}x + \dfrac{2}{15}$ **26.** $40\,\text{m}^3$

27. $\dfrac{-x^2 + 5x - 25}{(2x + 1)(x - 2)}$ **28.** $1 + x$ **29. a.** Yes;

$x \neq 4, x \neq -2$ **b.** $\{8\}$ **c.** $(-\infty, -2) \cup (4, 8]$

30. The numbers are $-\dfrac{3}{4}, 4$.   **31.** $\{-4\}$ (The value -9

does not check.)   **32.** $(-\infty, \infty)$
33. a. $P(6) = 2{,}000{,}000, P(12) = 1{,}000{,}000, P(18) = 500{,}000,$
$P(24) = 250{,}000, P(30) = 125{,}000$ **b.** 48 hr
34. a. 2   **b.** -3 **c.** 6   **d.** 3   **35. a.** 217.0723
b. 23.1407   **c.** 0.1768   **d.** 3.7293   **e.** -0.4005 **f.** 2.6047

36. $\left\{\dfrac{2}{3}\right\}$ **37.** $\{\ln(100)\}$ **38.** $\{3\}$ (The value -9

does not check.)   **39.** $\log\left(\dfrac{\sqrt{z}}{x^2 y^3}\right)$ **40.** $\dfrac{2}{3}\ln x - \dfrac{1}{3}\ln y$

Chapter 9

Chapter Opener Puzzle

$\dfrac{\text{C}}{5}\ \dfrac{\text{O}}{4}\ \dfrac{\text{N}}{6}\ \dfrac{\text{I}}{3}\ \dfrac{\text{C}}{5}\qquad \dfrac{\text{S}}{1}\ \dfrac{\text{E}}{7}\ \dfrac{\text{C}}{5}\ \dfrac{\text{T}}{2}\ \dfrac{\text{I}}{3}\ \dfrac{\text{O}}{4}\ \dfrac{\text{N}}{6}\ \dfrac{\text{S}}{1}$

Section 9.1 Practice Exercises, pp. 760–764

1. a. $\sqrt{(x_2 - x_1)^2 + (y_2 - y_1)^2}$ **b.** circle; center   **c.** radius

d. $(x - h)^2 + (y - k)^2 = r^2$ **e.** $\left(\dfrac{x_1 + x_2}{2}, \dfrac{y_1 + y_2}{2}\right)$

3. $3\sqrt{5}$ **5.** $\sqrt{34}$ **7.** $\dfrac{\sqrt{73}}{8}$ **9.** 4   **11.** 8

13. $4\sqrt{2}$ **15.** $\sqrt{42}$ **17.** Subtract 5 and -7. This becomes
$5 - (-7) = 12$.   **19.** $y = 13, y = 1$
21. $x = 0, x = 8$ **23.** Yes   **25.** No
27. Center $(4, -2); r = 3$ **29.** Center $(-1, -1); r = 1$

   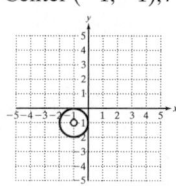

31. Center $(0, 2); r = 2$ **33.** Center $(3, 0); r = 2\sqrt{2}$

35. Center $(0, 0); r = \sqrt{6}$ **37.** Center $\left(-\dfrac{4}{5}, 0\right); r = \dfrac{8}{5}$

39. Center $(1, 3); r = 6$ **41.** Center: $(0, 3); r = 2$

43. Center $(0, -3); r = \dfrac{4}{3}$ **45.** Center: $(-1, -2); r = 3$

47. Center $(0, 0); r = 1$

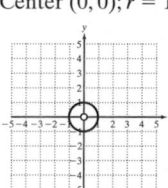

49. $x^2 + y^2 = 4$ **51.** $x^2 + (y - 2)^2 = 4$
53. $(x + 2)^2 + (y - 2)^2 = 9$ **55.** $x^2 + y^2 = 49$
57. $(x + 3)^2 + (y + 4)^2 = 36$ **59.** $(x - 5)^2 + (y - 3)^2 = 2.25$
61. $(1, 2)$ **63.** $(-1, 0)$ **65.** $(-1, 6)$

67. $(0, 3)$ **69.** $\left(-\dfrac{1}{2}, \dfrac{3}{2}\right)$ **71.** $(-0.4, -1)$

73. $\left(40, 7\tfrac{1}{2}\right)$; they should meet 40 mi east, $7\tfrac{1}{2}$ mi north of the
warehouse.   **75. a.** $(1, 3)$ **b.** $(x - 1)^2 + (y - 3)^2 = 5$
77. a. $(0, 3)$ **b.** $x^2 + (y - 3)^2 = 4$
79. $(x - 4)^2 + (y - 4)^2 = 16$

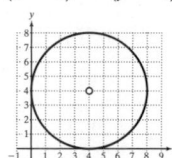

81. $(x - 1)^2 + (y - 1)^2 = 29$

Section 9.1 Graphing Calculator Exercises, p. 764

83.

84.

85.

86.

87.

88.

Section 9.2 Practice Exercises, pp. 771–774

1. a. conic; plane   **b.** parabola; directrix; focus   **c.** vertex; symmetry   **d.** $y = k$

3. 5

5. Center: $(0, -1)$; radius: 4

7. $\left(\dfrac{11}{2}, 1\right)$

9.

11.

13.

15.

17.

19.

21.

23.
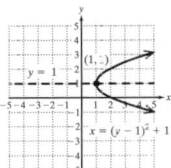

25. $(2, -1)$ **27.** $(5, -1)$ **29.** $\left(2, \dfrac{7}{4}\right)$ **31.** $\left(\dfrac{3}{2}, -\dfrac{1}{4}\right)$

33. $(10, -1)$ **35.** The maximum height of the water is 22 ft.

37. A parabola whose equation is in the form $y = a(x - h)^2 + k$ has a vertical axis of symmetry. A parabola whose equation is in the form $x = a(y - k)^2 + h$ has a horizontal axis of symmetry.

39. Vertical axis of symmetry; opens upward

41. Vertical axis of symmetry; opens downward

43. Horizontal axis of symmetry; opens right

45. Horizontal axis of symmetry; opens left

47. Vertical axis of symmetry; opens downward

49. Horizontal axis of symmetry; opens right

Section 9.3 Practice Exercises, pp. 779–783

1. a. ellipse; foci   **b.** ellipse   **c.** hyperbola; foci   **d.** transverse

e. $\dfrac{x^2}{a^2} - \dfrac{y^2}{b^2} = 1$ **f.** $\dfrac{y^2}{b^2} - \dfrac{x^2}{a^2} = 1$

3. Center: $(8, -6)$; radius: 10

5. Vertex: $(-3, -1)$; $x = -3$

7. $\left(x - \dfrac{1}{2}\right)^2 + \left(y - \dfrac{5}{2}\right)^2 = \dfrac{1}{4}$

9.

11.

13.

15.

17. Center: $(4, 5)$

19. Center: $(-1, 2)$

21. Center: $(2, -3)$

23. Center: $(0, 1)$
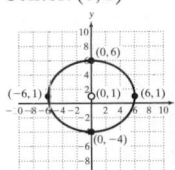

25. Vertical   **27.** Horizontal   **29.** Horizontal

31. Vertical

33.

35.

37.   **39.**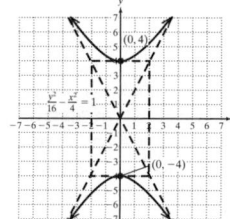

41. Hyperbola   **43.** Ellipse
45. Ellipse   **47.** Hyperbola
49. Ellipse   **51.** Hyperbola
53. The height 10 ft from the center is approximately 49 ft.
55.   **57.**

Section 9.3 Problem Recognition Exercises, pp. 783–784

1. Standard equation of a circle   **2.** Ellipse centered at the
origin   **3.** Distance between two points
4. Midpoint between two points   **5.** Parabola with vertical
axis of symmetry   **6.** Hyperbola with horizontal transverse axis
7. Hyperbola with vertical transverse axis
8. Parabola with horizontal axis of symmetry
9. Parabola   **10.** Hyperbola   **11.** Circle   **12.** Circle
13. Hyperbola   **14.** Ellipse   **15.** None of the above
16. Circle   **17.** Parabola   **18.** Hyperbola   **19.** Parabola
20. Circle   **21.** None of the above   **22.** Parabola
23. Ellipse   **24.** Circle   **25.** Parabola   **26.** Ellipse
27. Hyperbola   **28.** Circle   **29.** Ellipse   **30.** Parabola

Section 9.4 Practice Exercises, pp. 788–791

1. a. nonlinear   **b.** intersection

3. $7\sqrt{2}$ **5.** $(x + 5)^2 + (y - 3)^2 = 64$ **7.** $\left(-\dfrac{1}{2}, -\dfrac{7}{2}\right)$

9. Zero, one, or two   **11.** Zero, one, or two
13. Zero, one, two, three, or four   **15.** Zero, one, two, three, or
four
17. $\{(-3, 0), (2, 5)\}$ **19.** $\{(0, 1), (-1, 0)\}$

    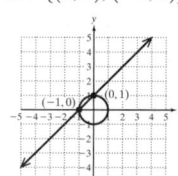

21. $\{(\sqrt{2}, 2), (-\sqrt{2}, 2)\}$

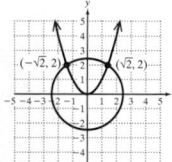

23. $\{(1, 1)\}$ **25.** $\{(0, 0)\}$ **27.** $\left\{\left(\dfrac{3}{2}, \dfrac{9}{4}\right)\right\}$

29. $\{(0, 0), (-2, -8)\}$ **31.** $\{(1, 4)\}$
33. $\{(1, 0), (-1, 0)\}$ **35.** $\{(2, 0), (-2, 0)\}$
37. $\{(3, 2), (-3, 2), (3, -2), (-3, -2)\}$
39. $\{(8, 6), (-8, -6)\}$ **41.** $\{(0, -2), (0, 2)\}$
43. $\{(2, 0), (0, 1)\}$ **45.** $\{(4, 5), (-4, -5)\}$

47. $\left\{\left(-\sqrt{2}, \dfrac{\sqrt{2}}{2}\right), \left(\sqrt{2}, -\dfrac{\sqrt{2}}{2}\right)\right\}$
49. The numbers are 3 and 4.
51. The numbers are 5 and $\sqrt{7}$, -5 and $\sqrt{7}$,
5 and $-\sqrt{7}$, or -5 and $-\sqrt{7}$.

Section 9.4 Graphing Calculator Exercises, p. 791

53.

54.

55.

56.

57. { }   **58.** { }

Section 9.5 Practice Exercises, pp. 795–799

1. k   **3.** e   **5.** i   **7.** j   **9.** a   **11.** d
13. True   **15.** False
17. a.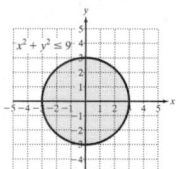

b. The set of points on and "outside" the circle $x^2 + y^2 = 9$
c. The set of points on the circle $x^2 + y^2 = 9$
19. a.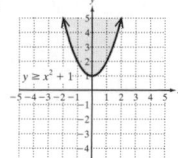

b. The parabola $y = x^2 + 1$ would be drawn as a dashed curve.
21. $(x - 3)^2 + (y + 4)^2 \leq 625$

23.

25.

27.

29.

31.

33.

35.

37.

39.

41.

43.

45.

47. { }

49.

51.

53.

55.
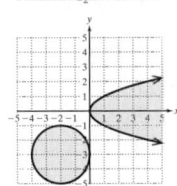

Chapter 9 Review Exercises, pp. 806–809

1. $2\sqrt{10}$ **2.** $\sqrt{89}$ **3.** $x = 5$ or $x = -1$
4. $x = -3$ **5.** Center $(12, 3)$; $r = 4$
6. Center $(-7, 5)$; $r = 9$ **7.** Center $(-3, -8)$; $r = 2\sqrt{5}$
8. Center $(1, -6)$; $r = 4\sqrt{2}$ **9. a.** $x^2 + y^2 = 64$
b. $(x - 8)^2 + (y - 8)^2 = 64$ **10.** $(x + 6)^2 + (y - 5)^2 = 10$
11. $(x + 2)^2 + (y + 8)^2 = 8$

12. $\left(x - \dfrac{1}{2}\right)^2 + (y - 2)^2 = 4$

13. $(x - 3)^2 + \left(y - \dfrac{1}{3}\right)^2 = 9$ **14.** $x^2 + y^2 = \dfrac{49}{4}$

15. $x^2 - (y - 2)^2 = 9$ **16.** $(-4, -2)$ **17.** $(-1, 8)$
18. Vertical axis of symmetry; parabola opens downward
19. Horizontal axis of symmetry; parabola opens right
20. Horizontal axis of symmetry; parabola opens left
21. Vertical axis of symmetry; parabola opens upward

22.

23.

24.

25.
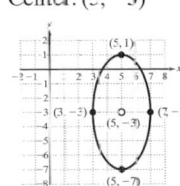

26. $y = (x - 3)^2 - 4$; vertex: $(3, -4)$;
axis of symmetry: $x = 3$
27. $x = (y + 2)^2 - 2$; vertex: $(-2, -2)$;
axis of symmetry: $y = -2$
28. $x = -4\left(y - \dfrac{1}{2}\right)^2 + 1$; vertex: $\left(1, \dfrac{1}{2}\right)$;

axis of symmetry: $y = \dfrac{1}{2}$

29. $y = -2\left(x + \dfrac{1}{2}\right)^2 + \dfrac{1}{2}$; vertex: $\left(-\dfrac{1}{2}, \dfrac{1}{2}\right)$;

axis of symmetry: $x = -\dfrac{1}{2}$

30.

31.
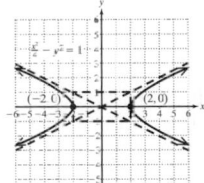

32. Center: $(5, -3)$ **33.** Center: $(0, 2)$

34. Horizontal   **35.** Vertical
36. Vertical   **37.** Horizontal
38.

39.

40. Hyperbola   **41.** Ellipse
42. Ellipse   **43.** Hyperbola

44. a. Line and parabola
b.

c. $\left\{\left(\dfrac{5}{2}, \dfrac{5}{4}\right), (-4, 11)\right\}$

45. a. Line and parabola
b.

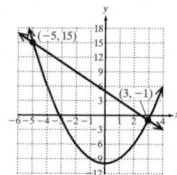

c. $\{(-5, 15), (3, -1)\}$

46. a. Circle and line
b.

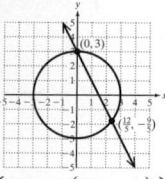

c. $\left\{(0, 3), \left(\dfrac{12}{5}, -\dfrac{9}{5}\right)\right\}$

47. a. Circle and line
b.

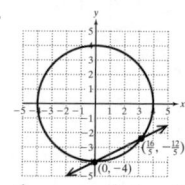

c. $\left\{(0, -4), \left(\dfrac{16}{5}, -\dfrac{12}{5}\right)\right\}$

48. $\left\{(0, -2), \left(\dfrac{16}{9}, \dfrac{14}{9}\right)\right\}$ **49.** $\left\{\left(-\dfrac{7}{5}, \dfrac{13}{5}\right), (-5, -1)\right\}$

50. $\{(8, 4), (2, -2)\}$ **51.** $\{(2, 4), (-2, 4), (\sqrt{2}, 2), (-\sqrt{2}, 2)\}$

52. $\{(3, 1), (3, -1), (-3, 1), (-3, -1)\}$

53. $\{(6, 5), (-6, 5), (6, -5), (-6, -5)\}$

54.

55.

56.

57.

58.

59.

60.

61.

Chapter 9 Test, pp. 809–810

1. $4\sqrt{2}$ **2.** Center: $\left(\dfrac{5}{6}, -\dfrac{1}{3}\right)$; $r = \dfrac{5}{7}$

3. Center: $(0, 2)$; $r = 3$ **4. a.** $\sqrt{5}$ **b.** $x^2 + (y - 4)^2 = 5$

5. The center is the midpoint $(3.8, 1.95)$.

6.

7.

8.

9.

10.

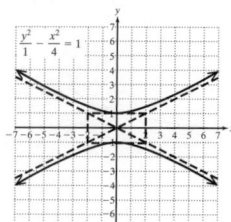

11. $\{(-3, 0), (0, 4)\}$; graph b   **12.** { }; graph a

13. The addition method can be used if the equations have corresponding *like* terms.   **14.** $\{(2, 0), (-2, 0)\}$

15.

16.

17.

18.

Chapters 1–9 Cumulative Review Exercises, pp. 811–812

1. $\{x \mid x \text{ is a real number}\}$

2. $\left(\dfrac{10}{3}, \infty\right)$

3. The integers are 10 and 15.   **4. a.** $(-5, 0), (0, 3)$ **b.** $m = \dfrac{3}{5}$

c.

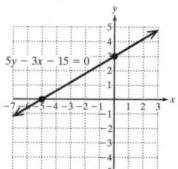

5. slope: $\dfrac{3}{4}$; y-intercept: $\left(0, -\dfrac{3}{2}\right)$

6. There are 12 dimes and 5 quarters.

7. $\{(2, -3, 1)\}$ **8.** $\left\{\left(\dfrac{5}{2}, \dfrac{3}{2}\right)\right\}$ **9.** $\{(6, 3)\}$

10. $f(0) = -12$; $f(-1) = -16$; $f(2) = -10$; $f(4) = -16$

11. $(-4, -2)$ **12.** $z = 32$ **13.** $(g \circ f)(x) = x + 7$; $x \geq -1$ **14. a.** -5 **b.** $(x + 1)(x^2 + 1)$; -5

c. They are the same.   **15.** $(x + y)(x - y - 6)$

16. $x^3 - 2x^2 - 7x - 4$ **17.** $\left\{6, \dfrac{3}{2}\right\}$ **18.** $\dfrac{a - 1}{a + 2}$

19. $\dfrac{-x^2 - x - 4}{(x + 3)(x - 2)}$ **20.** $\{0, -1\}$ **21. a.** { }

b. $\{-11\}$ **22.** $-30 + 24i$ **23.** $\dfrac{12}{41} + \dfrac{15}{41}i$

24. $2\sqrt{7}$ m   **25. a.** 17.6 ft; 39.6 ft; 70.4 ft   **b.** 8 sec

26. $\left\{-\dfrac{1}{5}, \dfrac{1}{10} \pm \dfrac{\sqrt{3}}{10}i\right\}$ **27.** $\{2 \pm \sqrt{11}\}$

28. $(-5, -36)$

29. a. $(-3, 0), (1, 0)$ **b.** $(0, 3)$ **c** $(-1, 4)$

30. $(-1, 19)$ **31.** $\left(-\infty, \frac{1}{2}\right] \cup \left[\frac{9}{2}, \infty\right)$ **32.** $\log_8 32 = \frac{5}{3}$

33. $\left\{\frac{2}{3}\right\}$ **34.** $h^{-1}(x) = \sqrt[3]{x + 1}$

35. $x^2 + (y - 5)^2 = 16$

36.

37. $\left(-\frac{1}{2}, 0\right)$ **38.** $\{(0, -4)\}$

39.

40.

Chapter 10

Chapter Opener Puzzle

1. $13, 16, 19$ **2.** $-26, -34, -42$ **3.** $48, 96, 192$
4. $\frac{1}{9}, \frac{1}{27}, \frac{1}{81}$ **5.** $25, 36, 49$ **6.** $-324, 972, -2916$
7. x^3y^4, x^2y^5, xy^6 **8.** a^2b^7, ab^8, b^9 **9.** $a^2 + 2ab + b^2$
10. $a^3 + 3a^2b + 3ab^2 + b^3$ **11.** $n + 1$

Section 10.1 Practice Exercises, pp. 819–820

1. a. $a^2 + 2ab + b^2$; $a^3 + 3a^2b + 3ab^2 + b^3$; binomial
b. $n(n - 1)(n - 2) \cdots (2)(1)$; factorial   **c.** $6; 2; 1; 1$
d. binomial   **e.** Pascal's   **3.** $a^3 + 3a^2b + 3ab^2 + b^3$
5. $1 + 4g + 6g^2 + 4g^3 + g^4$
7. $p^7 + 7p^6q^2 + 21p^5q^4 + 35p^4q^6 + 35p^3q^8 + 21p^2q^{10} + 7pq^{12} + q^{14}$
9. $s^5 - 5s^4t + 10s^3t^2 - 10s^2t^3 + 5st^4 - t^5$
11. $625 - 500u^3 + 150u^6 - 20u^9 + u^{12}$
13. $x^6 - 12x^4 + 48x^2 - 64$ **15.** 120 **17.** 1 **19.** False
21. True   **23.** $6! = 6 \cdot (5 \cdot 4 \cdot 3 \cdot 2 \cdot 1) = 6 \cdot 5!$ **25.** 1680
27. 6 **29.** 56 **31.** 1 **33.** $m^{11} + 11m^{10}n + 55m^9n^2$
35. $u^{24} - 12u^{22}v + 66u^{20}v^2$ **37.** 9 terms
39. $s^6 + 6s^5t + 15s^4t^2 + 20s^3t^3 + 15s^2t^4 + 6st^5 + t^6$
41. $b^3 - 9b^2 + 27b - 27$ **43.** $16x^4 + 32x^3y + 24x^2y^2 + 8xy^3 + y^4$ **45.** $c^{14} - 7c^{12}d + 21c^{10}d^2 - 35c^8d^3 + 35c^6d^4 - 21c^4d^5 + 7c^2d^6 - d^7$
47. $\frac{1}{32}a^5 - \frac{5}{16}a^4b + \frac{5}{4}a^3b^2 - \frac{5}{2}a^2b^3 + \frac{5}{2}ab^4 - b^5$
49. $x^4 + 16x^3y + 96x^2y^2 + 256xy^3 + 256y^4$
51. $-462m^6n^5$ **53.** $495u^{16}v^4$ **55.** g^9

Section 10.2 Practice Exercises, pp. 826–828

1. a. infinite; finite   **b.** terms; nth   **c.** alternating   **d.** series
e. summation   **f.** index; $(3)^2 + (4)^2 + (5)^2$
3. 28 **5.** $16x^4 + 32x^3z + 24x^2z^2 + 8xz^3 + z^4$
7. $4, 7, 10, 13, 16$ **9.** $\sqrt{3}, 2, \sqrt{5}, \sqrt{6}$ **11.** $-\frac{2}{3}, \frac{3}{4}, -\frac{4}{5}, \frac{5}{6}$
13. $0, -3, 8$ **15.** $0, 2, 6, 12, 20, 30$ **17.** $-3, 9, -27, 81$
19. When n is odd, the term is negative. When n is even, the term is positive.   **21.** $a_n = 2n$ **23.** $a_n = 2n - 1$

25. $a_n = \frac{1}{n^2}$ **27.** $a_n = (-1)^{n+1}$ **29.** $a_n = (-1)^n 2^n$

31. $a_n = \frac{3}{5^n}$ **33.** $\$60, \$58.80, \$57.62, \$56.47$

35. $25{,}000; 50{,}000; 100{,}000; 200{,}000; 400{,}000; 800{,}000; 1{,}600{,}000$
37. A sequence is an ordered list of terms. A series is the sum of the terms of a sequence.

39. 90 **41.** $\frac{31}{16}$ **43.** 30 **45.** 10 **47.** $\frac{73}{12}$ **49.** 38

51. -1 **53.** 55 **55.** $\sum_{n=1}^{6} n$ **57.** $\sum_{i=1}^{5} i$ **59.** $\sum_{j=1}^{5} 4j$

61. $\sum_{k=1}^{4} (-1)^{k+1} \frac{1}{3^k}$ **63.** $\sum_{i=1}^{6} \frac{i + 4}{11i}$ **65.** $\sum_{n=1}^{5} x^n$

67. $-3, 2, 7, 12, 17$ **69.** $5, 21, 85, 341, 1365$
71. $1, 1, 2, 3, 5, 8, 13, 21, 34, 55$

Section 10.3 Practice Exercises, pp. 833–834

1. a. arithmetic   **b.** difference   **c.** $a_1 + (n - 1)d$; d
d. series   **e.** $\frac{n}{2}(a_1 + a_n)$

3. $-1, 1, -1, 1$ **5.** -55 **7.** $3, 11, 19, 27, 35$
9. $80, 60, 40, 20, 0$ **11.** $3, \frac{15}{4}, \frac{9}{2}, \frac{21}{4}, 6$ **13.** 2 **15.** -3

17. -2 **19.** $3, 8, 13, 18, 23$ **21.** $2, \frac{5}{2}, 3, \frac{7}{2}, 4$

23. $2, -2, -6, -10, -14$ **25.** $a_n = -5 + 5n$

27. $a_n = -2n$ **29.** $a_n = \frac{3}{2} + \frac{1}{2}n$ **31.** $a_n = 25 - 4n$

33. $a_n = -14 + 6n$ **35.** $a_6 = 17$ **37.** $a_9 = 47$
39. $a_7 = -30$ **41.** $a_{11} = -48$ **43.** 19 **45.** 22
47. 23 **49.** 11 **51.** $a_1 = -2, a_2 = -5$ **53.** 670
55. 290 **57.** -15 **59.** 95 **61.** 924 **63.** 300
65. -210 **67.** 5050 **69.** 980 seats; $\$14{,}700$

Section 10.4 Practice Exercises, pp. 840–843

1. a. geometric   **b.** ratio   **c.** $a_1 r^{n-1}$ **d.** $\frac{a_1(1 - r^n)}{1 - r}$

e. $\frac{a_1}{1 - r}$; 1 **3.** $a_n = 4n - 8$ **5.** 25 **7.** $1, 10, 100, 1000$

9. $64, 32, 16, 8$ **11.** $8, -2, \frac{1}{2}, -\frac{1}{8}$ **13.** 2 **15.** $-\frac{1}{4}$

17. -2 **19.** $-3, 6, -12, 24, -48$ **21.** $6, 3, \frac{3}{2}, \frac{3}{4}, \frac{3}{8}$

23. $-1, -6, -36, -216, -1296$ **25.** $a_n = 3(4)^{n-1}$

27. $a_n = -5(-3)^{n-1}$ **29.** $a_n = \frac{1}{2}(4)^{n-1}$ **31.** $\frac{1}{64}$

33. $-\frac{243}{8}$ **35.** -48 **37.** -9 **39.** $\frac{1}{8}$ **41.** 4

43. A geometric sequence is an ordered list of numbers in which the ratio between each term and its predecessor is constant. A geometric series is the sum of the terms of such a sequence.
45. a. $3 + 12 + 48 + 192$ **b.** 255

47. $\frac{1562}{125}$ **49.** $-\frac{11}{8}$ **51.** $\frac{3124}{27}$ **53.** $\frac{665}{243}$

55. -172 **57. a.** $\$1050.00, \$1102.50, \$1157.63, \$1215.51$
b. $a_{10} = \$1628.89; a_{20} = \$2653.30; a_{40} = \$7039.99$

59. $r = \frac{1}{6}$; sum is $\frac{6}{5}$ **61.** $r = -\frac{1}{4}$; sum is $\frac{4}{5}$

63. $r = -\frac{3}{2}$; sum does not exist   **65.** The sum is $\$800$ million.

67. The total vertical distance traveled is 28 ft.

69. a. $\frac{7}{10}$ **b.** $\frac{1}{10}$ **c.** $\frac{7}{9}$ **71. a.** $\$1{,}429{,}348$

b. $\$1{,}505{,}828$ **c.** $\$76{,}480$

Section 10.4 Problem Recognition Exercises, p. 843

1. Geometric, $r = -\dfrac{1}{2}$ **2.** Arithmetic, $d = \dfrac{1}{2}$

3. Geometric, $r = 2$ **4.** Neither   **5.** Arithmetic, $d = \dfrac{2}{3}$

6. Geometric, $r = -\dfrac{3}{2}$ **7.** Neither   **8.** Geometric, $r = -1$

9. Arithmetic, $d = -2$ **10.** Neither   **11.** Neither

12. Arithmetic, $d = 4$ **13.** Arithmetic, $d = \dfrac{1}{4}\pi$

14. Geometric, $r = -1$ **15.** Neither

16. Arithmetic, $d = \dfrac{1}{3}e$ **17.** Neither   **18.** Geometric, $r = 3$

Chapter 10 Review Exercises, pp. 848–849

1. 40,320   **2.** 120   **3.** 66   **4.** 84

5. $x^{10} + 20x^8 + 160x^6 + 640x^4 + 1280x^2 + 1024$

6. $c^4 - 12c^3d + 54c^2d^2 - 108cd^3 + 81d^4$

7. $a^{11} - 22a^{10}b + 220a^9b^2$ **8.** $1512x^{10}y^3$

9. $-15,000x^3y^{21}$ **10.** $160a^3b^3$ **11.** $1, -2, -5, -8, -11$

12. $-2, -16, -54$ **13.** $\dfrac{1}{3}, -\dfrac{1}{2}, \dfrac{3}{5}, -\dfrac{2}{3}$ **14.** $-\dfrac{2}{3}, \dfrac{4}{9}, -\dfrac{8}{27}, \dfrac{16}{81}$

15. $a_n = \dfrac{n}{n+1}$ **16.** $a_n = (-1)^n \cdot \dfrac{3}{2^n}$ **17.** k **18.** 5

19. 5   **20.** -22 **21.** $\displaystyle\sum_{i=1}^{7} \dfrac{3+i}{i}$ **22.** $a_n = n^2 + \left(\dfrac{n}{2}\right)h$

23. $-12, -10.5, -9, -7.5, -6$ **24.** $\dfrac{2}{3}, 1, \dfrac{4}{3}, \dfrac{5}{3}, 2$

25. $a_n = 10n - 9$ **26.** $a_n = -14n + 20$ **27.** $a_{17} = \dfrac{9}{2}$

28. $a_{25} = -155$ **29.** 24 terms   **30.** 19 terms

31. $d = 10$ **32.** $a_n = -5n + 19$ **33.** -620

34. 92.5   **35.** 294   **36.** 5989   **37.** $r = 3$

38. $r = 1.2$ **39.** $-1, \dfrac{1}{4}, -\dfrac{1}{16}, \dfrac{1}{64}$ **40.** 10, 20, 40, 80

41. $a_n = -4(2)^{n-1}$ **42.** $a_n = 6\left(-\dfrac{1}{3}\right)^{n-1}$ **43.** $a_6 = \dfrac{128}{243}$

44. $a_4 = 270$ **45.** $a_1 = 4$ **46.** $a_1 = 1$ **47.** 1275

48. $-\dfrac{182}{3}$ **49.** 18   **50.** $\dfrac{25}{2}$

51. a. \$10,700, \$11,449, \$12,250.43   **b.** The account is worth \$12,250.43 after 3 yr.   **c.** \$19,671.51

Chapter 10 Test, pp. 849–850

1. 1   **2.** 210   **3.** $a^4 + 4a^3b + 6a^2b^2 + 4ab^3 + b^4$

4. $81y^4 - 216y^3x^2 + 216y^2x^4 - 96yx^6 + 16x^8$

5. $-56a^3c^{15}$ **6.** $-1, -\dfrac{3}{4}, -\dfrac{3}{5}, -\dfrac{1}{2}$ **7.** 65

8. a. $8, 9.5, 11, 12.5, 14$ **b.** $a_n = 1.5n + 6.5$ **9.** $\displaystyle\sum_{n=1}^{4} x^{3n}$

10. $d = \dfrac{1}{4}$ **11.** $r = \dfrac{1}{3}$ **12. a.** $-15, -18, -21, -24$

b. $a_n = -3n - 12$ **c.** $a_{40} = -132$ **13. a.** $4, -8, 16, -32$

b. $a_n = 4(-2)^{n-1}$ **c.** -2048 **14.** $a_n = \dfrac{3}{5} \cdot \left(\dfrac{1}{2}\right)^{n-1}$

15. $a_n = -2n + 22$ **16.** 31 terms   **17.** 7 terms

18. 10,900   **19.** $\dfrac{1023}{64}$ **20.** 8   **21.** $a_1 = \dfrac{1}{27}$

22. $a_{18} = -202$ **23. a.** \$1095   **b.** \$1160.70
c. \$5933.13   **d.** \$86,568.71

Chapters 1–10 Cumulative Review Exercises, pp. 851–852

1. $x^2 - x - 13$ **2.** $\dfrac{20a^6}{b^{10}}$ **3.** $\dfrac{1}{16}$ **4.** $\dfrac{\sqrt{10}}{2}$

5. $\dfrac{2c+5}{2(c+4)}$ **6.** $-\dfrac{xy}{x+y}$ **7.** 4   **8.** $3xy^4\sqrt[3]{2x^2z^2}$

9. $-\dfrac{a+1}{a}$ **10.** $\dfrac{2x+5}{(x-5)(x-2)(x+10)}$ **11.** $7x^2\sqrt{2x}$

12. $x - 9y$ **13.** $2x^3 + 3x^2 + 6x + 17 + \dfrac{27}{x-2}$

14. $(2a-7)(3a+2)$ **15.** $3c(5c-2)(5c+2)$

16. $(3x + 4y^2)(9x^2 - 12xy^2 + 16y^4)$

17. $(w+9)(w-2)(w+2)$ **18.** $\{4\}$ **19.** $\{0\}$

20. $\left\{\dfrac{5}{4}\right\}$ **21.** $\left\{\dfrac{2}{5} \pm \dfrac{\sqrt{3}}{5}i\right\}$ **22.** $\{4, -4, 1, -1\}$

23. $\left\{\dfrac{7}{2}, -\dfrac{3}{4}\right\}$ **24.** $\{1\}$ **25.** $\{2\}$ (The value -5 does not check.)   **26.** $\ln\left(\dfrac{x\sqrt{z}}{y^5}\right)$ **27.** 3.4929

28. $[-2, 5)$ **29.** $[-5, 0] \cup [1, \infty)$ **30.** $\{5\}$

31. $(-\infty, -4) \cup [-1, \infty)$ **32.** $(-\infty, 1) \cup (3, \infty)$

33.   **34.**

35.   **36.**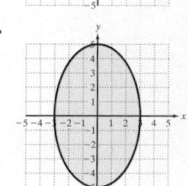

37. a. $(-1, 0), (-5, 0)$ **b.** $(0, 5)$ **c.** $(-3, -4)$

38. slope -4; y-intercept $(0, 6)$ **39.** $y = -5x + 18$

40. a. 2   **b.** undefined   **c.** $x = 2$ **d.** $(-\infty, 4)$ **e.** $(-\infty, 3]$

41. $(-\infty, 5]$ **42.** $\{(3, -1)\}$ **43.** $\{(1, 2, 3)\}$

44. a. $2\sqrt{13}$ **b.** $(6, -2)$ **45.** $b_1 = \dfrac{2A - hb_2}{h}$

46. $t(9) = 4.2$; This means that it will take the rat 4.2 min if the rat has practiced 9 times.   **47.** The car is moving 75 mph.   **48.** The angles are 57° and 123°.   **49.** She invested \$8000 at 6.5% and \$5000 at 5%.   **50.** The plane travels 500 mph in still air. The wind speed is 50 mph.

Additional Topics Appendix

Section A.1 Practice Exercises, pp. A9–A12

1. a. determinant; $ad - bc$ **b.** minor   **c.** $a_2; \begin{vmatrix} b_1 & c_1 \\ b_2 & c_2 \end{vmatrix}$

3. 16   **5.** 1   **7.** 4   **9.** 6   **11.** 26   **13.** 6

15. 46   **17. a.** 30   **b.** 30   **19.** Choosing the row or column with the most zero elements simplifies the arithmetic when evaluating a determinant.

21. 12   **23.** -15 **25.** 0   **27.** $8a - 2b$

29. $4x - 3y + 6z$ **31.** 0

33. $\mathbf{D} = 16; \mathbf{D}_x = -63; \mathbf{D}_y = 66$ **35.** $\{(2, -1)\}$ **37.** $\{(4, -1)\}$

39. $\left\{\left(\dfrac{23}{13}, \dfrac{9}{13}\right)\right\}$ **41.** $x = 1$ **43.** $z = \dfrac{1}{2}$ **45.** $y = \dfrac{16}{41}$

47. Cramer's rule does not apply when the determinant $\mathbf{D} = 0$.

49. { }; inconsistent system   **51.** $\{(0, 0)\}$

53. Infinitely many solutions; $\{(x, y)\,|\,x + 5y = 3\}$; dependent equations   **55.** $\{(3, 2, 1)\}$

57. { }; inconsistent system   **59.** $x = -19$

61. $w = 1$ **63.** 36   **65. a.** 2   **b.** -2 **c.** $x = -1$

67. The measures are 37.5° and 52.5°.

69. 40 iPods, 15 iPads and 20 iPhones were sold.

71. There were 560 women and 440 men in the survey.

Application Index

BIOLOGY/HEALTH/LIFE SCIENCES

Age versus height of boys, 84
Amount of water needed by hiker, 477
Antacid tablets required per day for women, 315
Average weights of boys and girls based on age, 156
Bacteria population in culture over time, 339, 689, 736, 748–749, 751, 827
Bacterium maximum population, 635
Bison population approximation, 471
Cages available for cats and dogs, 276
Calcium content in selected foods, 291
Calories in box of candy, from ratio, 470
Calories in fast foods, 267
Cats to dogs ratio at animal rescue facility, 464
Cholesterol levels for adult humans, 124
Deaths from cholera, over time, 724
Fat content of ice cream, 267, 488
Fat in box of candy, from ratio, 470
Fiber content in selected foods, 291
Glasses of milk required per day for women, 315
Grams of medicine prescribed according to weight of patient, 481
Half-life of iodine, 747–748
HDL cholesterol level, 464
Hemoglobin levels in human blood, 96
Length of woman's femur and her height, 183, 187
Life expectancy for females by birth year, 232
Longevity of selected animals, 184
Manatee population approximation, 471
Maximum heart rate, 124, 190
pH level of antacid tablet, 746
pH levels in blood, 701
Range of adenosine deaminase (ADA) in humans, 126
Sodium content of selected foods, 287–288
Tetanus toxin yield in a culture, 635
Time required for bacteria in culture to decrease, 751
Tomato seedling growing heights each day, 849
White blood cell numbers in humans, 96

BUSINESS/ECONOMICS

Advertising effect on sales, 702
Airline cost for selected number of passengers, 555
Apartment rental fees, 253
Car rental charges, 176–177, 269, 310
Commercial time length on TV, A-12
Commission/sales calculations, 66, 84, 143, 177, 231–232
Cost calculations, 234, 261–262, 266, 634
Currency exchange rate, 491
Delivery times, 282
Depreciation, 135–136, 143, 748
Dollar amounts spent on video games, 178
DVD rental fees, 317
Fish caught commercially, from weight ratio, 470
Income after taxes, 122
Income tax, 122
Manufacturing times, 282
Markup/markdown, 66, 68
Measurement error for bag of potato chips, 112
Measurement error for filling jug of juice, 112
Minimum hourly wage, 173–174
Monthly salary needed to qualify for loan, 81
Motel charges, 253
Number of airline passengers, from total cost, 555
Number of items produced for maximum profit, 657
Number of items to be packaged to minimize cost, 634
Orders processed by each person in business, 315
Original wholesale cost, 60–61
Per capita income in U.S., 222

Phone/cell-phone plan charges, 174, 269, 311
Price/cost of items and number of items sold, 180–181, 227, 481
Production costs, 481
Profit calculations, 85, 338, 406, 555, 591, 657, 670
Property taxes, 177
Rental charges, 310
Revenue calculations, 834
Salaries, 68
Salaries over time after annual raises, 843
Sales made by each person in business, 288
Sales tax calculations, 66
Taxi charges, 143
Time for two persons to work together, 468–469, 473, 491, 493
Tips calculations, 229
Tourist money spent and respent in community, 842
Weekly earnings for women over 16, 216
Weight of car and its gas mileage, 187

CHEMISTRY

Acid solution, 122, 125, 262, 267, 299, 315, 420
Alcohol mixture, 63, 67, 750
Antifreeze mixture, 62–63, 67
Avogadro's number of atoms, 329
Bleach mixture, 67, 267
Carbon 14 decay to determine age, 719
Disinfectant mixture, 268
Face cream mixture, 267
Fertilizer mixture, 67, 267
Gold mixture, 268
Hydrogen and oxygen molecules in a substance, 328
pH calculations, 696–697, 701, 746
Saline mixture, 67, 262, 311
Water to alcohol ratio in mixture, 471

CONSTRUCTION/DESIGN

Area of metal washer, 391
Area of window, 26
Board length measurement error, 112
Bolt length and width measurement errors, 112
Cost to fence a pasture, 522
Dimensions of circular fountain, 70–71
Dimensions of corral, 68–69, 635
Dimensions of dog run, 281
Dimensions of one pen within an area of pens, 636
Dimensions of rabbit pen, 74
Dimensions of room, 69, 405
Distance between base of ladder and wall, 405
Elevator weight capacity, 126
Fencing needed for dog run, 74
Height of arched bridge, 625, 783
Height of arched tunnel, 783
Height of bridge buckling in heat, 575
Heights of bridge suspension cable and towers, 625
Length and width of swimming pool, 493
Length of roof, from dimensions of room, 578
Length of screw, margin of error, 125
Length of stabilizing wires of tower, 576
Maximum load of wooden beam, 482
Perimeter of rectangular room, 338
Pumping water from construction site, 513
Slope of ladder leaning against wall, 146, 153
Slope of roof, 147, 153, 157
Slope of stairway, 226
Slope of telephone pole leaning after storm, 420
Slope of treadmill, 153
Thickness of piece of material cut by machine, 96, 110
Time for two persons to work together, 468–469, 473
Volume of Pyramid of Khufu, 551–552

CONSUMER APPLICATIONS

Amount of gasoline used versus miles driven, 30
Amount spent on popcorn and drinks, 234
Amounts spent on rent and car payments, from ratio, 471
Apartment rental fees, 253
Appreciation of home value, 842
Car rental charges, 176–177, 269, 310
City and highway mileages of vehicles, 260
Coin collection, 269, 811
Cost of each item in group, 261–262, 266, 291
Cost per night at each hotel, 292
Cost to carpet square room, 671
Currency exchange rate, 491
Dollar amounts spent on video games, 178
DVD rental fees, 317
Food mixture calculations, 67–68, 267, 291
Fuel remaining in truck after running certain number of hours, 136
Gallons of gas used in city driving, 471
Home prices, 68
Hours needed to cook a turkey, 556
Income after taxes, 122
Income tax, 122
Laundry detergent cost, from weight ratio, 470
Maximum gas mileage of cars, 635
Monthly loan payments, 827
Motel charges, 253
Number of e-readers used, A-12
Number of iPods, iPads, and iPhones sold, A-12
Number of people served by weight of a turkey, 481
Phone/cell-phone plan charges, 174, 269, 311
Picture frame dimensions, 73
Power company charges and kilowatt-hours used, 56
Property taxes, 177
Rental charges, 310
Taxi charges, 143
Tourist money spent and respent in community, 842
Weight of turkey cooked for 4 hours, 556
Yearly mileage for passenger cars in U.S., 178

DISTANCE/SPEED/TIME

Amount of gasoline used versus miles driven, 30
Area of picture on wall versus distance from projector, 481
Depth of ocean wave when traveling certain speed, 576
Distance apart for two persons leaving from same location, 595
Distance apart for two vehicles/persons, over time, 502, 506
Distance between base of ladder and wall, 405
Distance between cities, 656
Distance between storm and observer, 178
Distance from home plate to second base, 522
Distance seen to horizon from a height, 492
Distances of direct and indirect driving routes, 405, 522, 594–595
Distance traveled, from revolutions of wheels, 671
Distance traveled by accelerating car over time, 812
Distance traveled vertically by rebounding ball, 839, 842
Driving speed, from total walking/driving speed, 491
Driving speed during clear weather, 64, 472
Driving speed during stormy weather, 472
Halfway point for two vehicles traveling toward each other, 764
Height and horizontal distance of object shot/thrown from ground, 339, 418, 634, 659
Height for fireworks to explode, 625, 630–631, 634

Height of dropped object, over time, 203, 403, 553
Height of kite, 522
Height of object shot upward, over time, 402–403, 406, 417, 595–596, 631, 657, 660
Height of thrown ball, over time, 302–303, 649
Height of tree, from shadow cast, 465–466
Height of water from fountain jet, 774
Heights of rebounding dropped ball, 824, 839
Hours needed to cook a turkey, 556
Initial height of dropped object, from velocity at impact, 555
Landing distance of airplane, 655–656
Landing speed of airplane, 655–656
Location of campground for hikers, 764
Map distances, from inches, 471, 493, 494
Measurement conversions, 679
Skid marks and speed of vehicle, 552–553
Slope of airplane gaining altitude over distance, 153
Speed conversions, 679
Speed of airplane in still air, 264–265, 267–268, 311, 852
Speed of bike in still air, 290
Speed of bike with wind and against wind, 493
Speed of boat in still water, 267–269
Speed of boat to island and back, from total time, 472
Speed of canoe of specific length, 477
Speed of car in race, 75
Speed of car varied by time and distance, 493, 852
Speed of current, 267–269
Speed of cyclist against wind, from total time, 472
Speed of cyclist each day, 491
Speed of hiking uphill and downhill, 64, 67
Speed of maximum gas mileage in cars, 635
Speed of moving sidewalk, 268
Speed of two vehicles traveling at different times, 473
Speed of vehicle, from braking distance, 604
Speed of wind, 264–265, 267–268, 290, 311, 852
Speeds of airplane trips, 67
Speeds of boats traveling in same direction, 68
Speeds of cars traveling toward each other, 67, 122
Speeds of two skiers, 467
Speeds of two vehicles, from total time traveled, 467–468
Speeds of two vehicles in same amount of time, 472
Speeds running/walking and biking, 466–467, 472–473
Stopping distance of car, 482
Time at which thrown/launched object reaches a height, 604, 631, 657
Time for accelerating car to travel selected distance, 812
Time for dropped object to hit ground, 545, 553, 555, 588, 590, 591
Time for fireworks to reach maximum height, 625, 630–631, 634, 647
Time for object launched upward to hit ground, 596
Time for one person to complete a job, 473
Time for rat to run through maze, 852
Time for two persons to work together, 468–469, 473, 491, 493
Time for two pumps to fill container, 473, 488, 491
Time for vehicle to overtake another, 750
Time of car in race, 75
Time required to produce bacterium population, 635
Time required to produce tetanus toxin yield, 635
Vertical distance traveled by bungee jumper, 842
Walking distance and time, 125
Walking/running speeds of two persons, 472–473
Walking speed on and off moving walkway, 472
Walking speed on nonmoving ground, 268

EDUCATION/SCHOOL

Adults on staff of preschool, from ratio, 470
Associate degrees conferred in U.S., 180
College enrollment by year, 226
College textbook costs, 175
Credit-hours taken at college, 56
Dormitory charges, 339
Grade point average versus hours studying, 181

Memory retention of students, 697–698, 747, 749
Men enrolled in a course, from ratio, 471
Number of children enrolled, from ratio, 470
Numbers of men and women single college students, 471
Overall course grades, 216
Passing to failing students ratio, 465
Phone rates for students, 311
Salary with college degree, 68
School supplies cost, 266
Study time spent on each course, 275–276
Test scores required for A, B, or C in course, 81, 84, 97
Test scores required for final numerical average, 123, 420
Test scores versus hours studying, 182, 204
Vocabulary growth and teaching techniques, 701–702
Women graduating from law school in U.S., 174–175

ENVIRONMENT/EARTH SCIENCE/GEOGRAPHY

Air temperature and altitude, 143
Annual rainfall amounts, 234
Atmospheric pressure and altitude, 750
Average high and low temperatures, 29
Average January temperatures, 678
Average precipitation per month, 190
Cost of earthquakes yearly, 329I-2
Distance between storm and observer, 178
Distance seen to horizon from a height, 492
Earthquake intensity, 712
Nautical mile range of ship's radar, 763
pH levels in streams and rivers, 701
Points on map where earthquake is felt, 796
Range of cell tower signals, 763, 796
Slope of a hill, 153
Snow depths over time, 172
Temperature conversions, 29, 97
Tons of air pollution versus population density, 481
Water level in pond versus days of drought, 81–82, 179
Windchill and wind speed, 179

GARDENING/LANDSCAPING

Acres available for selected fruit trees, 312–313
Area of garden and walk surrounding it, 348, 415
Area of square garden, 415
Area of walkway around garden, 415
Base and height of triangular garden, 405
Cost of edging a garden, 527
Dimensions of garden, from amount of fencing, 74, 282
Dimensions of garden, from dimensions of border, 117
Edging around tree, 74
Fertilizer mixture, 67, 267
Garden side lengths, 73
Height of tree, from shadow cast, 465–466
Perimeter of rectangular garden, 338
Radius of circular garden, 658

GEOMETRY

Angle measurement, 122–123
Angle measures of complementary angles, 69–70, 74, 268, 311, 315
Angle measures of supplementary angles, 70, 74, 265, 268, 852
Angle measures of triangle, 74, 265, 268, 269, 286–287, 291, 313, 750
Area of frame around photograph, 348
Area of geometric figure, from lengths of sides, 482, 552, 660, 671
Area of photograph, 346
Area of trapezoid, 26, 71
Base and height of triangle, 405, 594–595, 658
Base and height of triangle, from area, 405
Consecutive integers, 66, 122, 125

Depth of parabolic satellite dish, 774
Dimensions of rectangle, 69, 74, 269, 334, 399, 405, 656
Dimensions of rectangle/box, from volume, 417, 590, 603
Height of tree, from shadow cast, 465–466
Height of triangle, 655
Original base and height of triangle before increasing, 405
Perimeter of geometric shape, 334, 338, 527
Radius and diameter of circle, 70–71
Radius of circle, from area, 406, 588, 658
Radius of sphere, from volume, 514
Reciprocal numbers, 472, 493
Side lengths of cube, from volume, 603
Side lengths of geometric shape, from area, 514, 591, 655, 656
Side lengths of geometric shape, from perimeter, 73–74, 122, 125, 287, 291, 313, 317, 407, 493, 527
Side lengths of triangle, 400, 405, 576, 590, 603–604, 656
Side lengths of triangle, from similar triangle, 466, 471
Surface area of cube, 482
Surface area of right circular cylinder, 407
Surface area of sphere, 71
Values of π, 123
Volume of Pyramid of Khufu, 551–552
Volume of rectangular box, 345–346, 348–349
Volume of right circular cylinder, 71

INVESTMENT

Amount borrowed, 62, 67, 116, 267, 306, 315
Amount deposited, 67–68, 269, 292, 329
Amount invested, 60–62, 67, 122, 125, 263–264, 267, 269, 292, 311, 368, 660, 852
Annual rate of return, 512, 514, 849
Appreciation of home value, 842
Compound interest calculations, 713–714, 721–722, 737, 747, 750, 827, 842, 850
Interest rate calculations, 514, 591, 721–722, 850
Investment growth, over time, 689, 849
Monthly salary earned to qualify for loan, 81
Savings account calculations, 173, 483–484, 850
Simple interest calculations, 60, 66–67, 75, 84, 267, 479, 481, 494
Time required for investment to grow, 84, 724, 736–737, 750
Yield on a bond, 478

POLITICS/PUBLIC POLICY

Money per person lent by U.S. government, 325
Percentage of votes for a candidate, 96, 113
Representation in U.S. House by state, 184

SCIENCE/TECHNOLOGY

Background radiation levels, 746
Current in a wire, 481
Decibel levels of sound, 477–478
Depth of ocean wave when traveling certain speed, 576
Distance between Earth and Alpha Centauri, 328
Distance between Earth and Andromeda Galaxy, 325–326
Distance between Earth and Barnard's Star, 326
Distance between storm and observer, 178
Earthquake intensity, 712
Exponential decay of radioactive element, 684–685, 688
Force required to stretch a spring, 178, 482, 492
Frequency of vibrating string, 481
Half-life of radioactive elements, 733–734, 737–738, 746–748, 827
Intensity of light, 481
Intensity of sound, 712, 737
Kinetic energy, 478–479
Mass of an object, 75, 126

Period of pendulum, 482, 493, 545, 555, 824
Power in electric circuit, 482
Radioactive decay, 733–734, 737–738, 827
Resistance of a wire, 481
Strength of wooden beam, 482
Thickness of a penny, 326
Velocity of ocean wave, 576
Volume of a gas, 751
Weight of medicine ball, from radius, 482

SPORTS/EXERCISE/ENTERTAINMENT

Amount of water needed by hiker, 477
Basketball basket types, 269
Dimensions of basketball court, 399
Dimensions of volleyball court, 73
Distance from home plate to second base, 522
Dollar amounts spent on video games, 178
Dominoes in triangular array, 834
Football passes completed, 491
Football points scored, 419

Football yards rushed and margin of victory, 227–228
Gender in workout survey, A-12
Golf fees, 266
Goofie Golf par holes, 291
Lengths of rope needed by magician, 122
Speed of bike in still air, 290
Theater seating arrangement, 834
Ticket prices, 266, 269, 311
Women's track and field records, 191

STATISTICS/DEMOGRAPHICS

Alcohol-related deaths in U.S., 122
Amount of child support due in U.S., 339
Associate degrees conferred in U.S., 180
Bottled water consumption in U.S., 230, 414
Cellular phone subscriptions in U.S., 155
Child support amounts due and paid in U.S., 670
Deaths from cholera, over time, 734
Fatality rate for drivers of selected ages, 604
Male and female police officer proportions, 464–465
Males over 20 employed full-time in U.S., 151–152

Margin of error in survey results, 566–567
Men in college in U.S., 122
Number of flights from large airport, 661
Number of women paid child support, 339
Numbers of men and women single college
     students, 471
Per capita income in U.S., 222
Percent of male smokers in U.S., 335
Population density, 152, 329
Population estimation, 733, 736, 738–739, 749
Population growth, 155, 190, 335, 419, 656, 659, 686,
     689, 732–733
Population of Alaska over time, 82
Prisoners in federal and state facilities by year, 180
Representatives in House, 678
Rural and urban populations in southern U.S., 670
Soft drink consumption in U.S., 232
Television viewers watching *American Idol*, 125
Weekly earnings for women over 16, 216
Women graduating from law school in U.S.,
     174–175
Yearly mileage for passenger cars in U.S., 178

Subject Index

Absolute value
     on calculators, 17
     equality of, 101
     evaluation of, 17
     explanation of, 16, 17, 39, 103, 105
     to express measurement error, 110
     translating to/from words, 109–110
Absolute value bars, 16, 17
Absolute value equations
     containing two absolute values, 101–102
     explanation of, 97, 120
     method to solve, 97–102, 104
Absolute value functions, 205, 207
Absolute value inequalities
     explanation of, 103
     methods to solve, 103–109, 120–121
     test point method to solve, 107–109
AC-method, to factor trinomials, 368–370, 411
Addition
     associative property of, 32
     commutative property of, 32, 35
     of complex numbers, 561
     distributive property of multiplication over,
          32–34, 40–41, 341, 342
     of functions, 664–665
     identity property of, 32
     inverse property of, 32
     of *like* terms, 33–35, 41, 523
     of polynomials, 330–331, 409
     of radicals, 522–525, 570
     of rational expressions, 437, 441–444, 486
     of real numbers, 17–18, 39
Addition method
     explanation of, 253, 255
     to solve nonlinear systems of equations, 787–788
     to solve systems of linear equations,
          253–257, 305
Addition property of equality, 45, 77
Addition property of inequality, 77

Algebraic expressions
     explanation of, 31
     method to simplify, 34–35, 40–41, 320
     writing English statements as, 57
Alternating sequences, 822–823
Angles, 69–70, 265
Applications. *See also* Applications Index
     of binomial theorem, 817–818
     of compound inequalities, 92–93
     of consecutive integers, 58–59, 358
     of distance, rate, and time, 63–64
     of exponential equations, 733–734
     of exponential functions, 684–686, 732–733
     of geometric series, 839
     geometry, 68–70, 117, 286–287, 551–552
     of linear equations in one variable, 57–64,
          68–70, 116
     of linear equations in three variables, 286–288
     of linear equations in two variables, 171–173, 222
     of linear inequalities, 81–82
     of literal equations, 70–71, 117
     of mixtures, 62–63
     of natural logarithmic functions, 718–719
     of percents and rates, 59–61, 335
     of polynomial functions, 334–335
     of principal and interest, 61–62, 263–264,
          713–714
     of product of polynomials, 345–346
     of proportions, 463–465
     of Pythagorean theorem, 399, 502
     of quadratic equations, 398–400
     of quadratic formula, 594–596
     of quadratic functions, 402–403, 630–631
     of radical equations, 551–552
     of radical functions, 552–553
     of rational equations, 466–469
     of rational exponents, 512
     of relations, 187
     of sequences, 824

of similar triangles, 465–466
of slope, 146–147, 151–152
of systems of linear equations in two variables,
     261–265, 306
of variation, 476–479
of work, 468–469
Approximately equal to, 9
Approximation, of exponential expressions,
     681–682
Area, of circle, 588
Argument, of logarithmic expressions, 690
Arithmetic sequences
     explanation of, 829
     finding number of terms in, 830–831
     finding specified term of, 830
     *n*th term of, 829–830
Arithmetic series
     explanation of, 831–832, 847
     sum of, 832
Associative property of addition, 32
Associative property of multiplication, 32
Asymptotes, 206, 777
Augmented matrix, 294–298, 309
Axis of symmetry, of parabolas, 612, 765–768, 802

Base
     of exponential expression, 22, 40
     of logarithmic expressions, 690
Binomial expansions
     binomial theorem for, 816–818, 845
     explanation of, 814–815, 845
     factorial notation to find coefficients of, 815–816
     finding specific term in, 818–819
     Pascal's triangle and, 815, 845
Binomial factors, 362
Binomials. *See also* Polynomials
     difference of squares to factor, 382–385, 387, 412
     explanation of, 330, 409
     factoring methods for, 382–388

Binomials. *See also* Polynomials—*(continued)*
  of form $x^6 - x^6$, 388
  product of, 342–343, 368, 371
  square of, 343, 814
  sum and difference of cubes to factor, 385–387, 413
Binomial theorem
  applications of, 817–818
  explanation of, 816–817, 845
Boundary points, 638, 640–643
Braces, 6, 24, 44
Brackets, 24

Calculator features
  absolute value function, 17
  Dot Mode, 423
  Graph, 138, 196
  Intersect, 240–241
  Minimum and *Maximum,* 620
  Seq, 822
  square window option, 165
  Table, 138, 196
  Value, 161
  Zero, 402
Calculators
  change-of-base formula on, 718
  circles on, 757
  exponential equations on, 730
  exponential expressions on, 23, 681–682, 713
  factorials on, 815, 816
  logarithms on, 693, 696, 715
  matrices on, 299
  minimum and maximum values on, 620
  nonlinear systems of equations on, 786
  *n*th roots on, 499
  order of operations on, 25
  parabolas on, 614, 769
  quadratic inequalities on, 639
  radical expressions on, 509, 518
  radical functions on, 503
  scientific notation on, 325, 326
  sequences on, 822, 825
  square roots on, 497
  x- and *y*-intercepts of functions on, 210
Center, of circle, 755, 757
Change-of-base formula, 717–718
Circles
  area of, 588
  on calculators, 756
  center of, 755, 757
  explanation of, 755
  graphs of, 755–757
  radius of, 755, 757
  standard equation of, 755, 757, 758, 801
  writing equation of, 757, 758
Clearing decimals, 49, 51
Clearing fractions, 49–51, 455, 457
Clearing parentheses, 33–35
Coefficient matrix, 293
Coefficients
  explanation of, 31, 40
  leading, 330, 375–376, 409
  of polynomial, 329, 330, 409
Column matrices, 293
Commission, 59
Common difference, of arithmetic sequence, 829–830, 846
Common logarithmic functions
  applications of, 696–697
  evaluation of, 692–693
  explanation of, 692
Common ratio, of geometric sequence, 835, 847
Commutative property of addition, 32, 35
Commutative property of multiplication, 32
Completing the square
  explanation of, 584–585, 650
  to graph parabolas, 766–767, 769–770
  to solve quadratic equations, 585–587, 599

Complex conjugates, 560, 573
Complex fractions
  explanation of, 447, 487
  methods to simplify, 447–451
Complex numbers
  addition of, 561
  division of, 562–563
  explanation of, 559–560, 573
  method to simplify, 563
  multiplication of, 562
  real and imaginary parts of, 560–561
  subtraction of, 561
Composition
  of functions, 665–666, 739
  of functions with inverse, 674
Compound inequalities
  applications of, 92–93
  explanation of, 118–119
  of form $a < x < b$, 89–90, 119
  joined by word *and,* 87–89
  joined by word *or,* 91–92
  linear in two variables, 273–274
  union and intersection of sets and, 85–87, 118–119
Compound interest, 713–714
Concentration rate, in mixture problem, 62
Conditional equations, 52–53
Conic sections, 765. *See also* Circles; Ellipse; Hyperbolas; Parabolas
Conjugate radical expressions, 531–532
Conjugates
  complex, 560, 573
  explanation of, 342, 410
Consecutive integers, 58–59, 398
Consistent systems of linear equations, 237
Constant functions, 204, 205, 224
Constant terms, 31, 40, 171
Continuously compounded interest, 714
Contradictions, 52–53
Cost problems, 261
Cube roots, 498
Cubes
  perfect, 388, 499, 500
  sum and difference of, 385–387
Cubic functions, 205, 207

Decimals, clearing, 49, 51
Degree
  of polynomial, 330, 409
  of the term, 329, 409
Denominators
  least common, 438–439
  rational expressions with like, 437
  rational expressions with unlike, 441–444
  rationalizing, 538–543, 571
Dependent systems of linear equations
  explanation of, 237
  Gauss-Jordan method to solve, 298
  in three variables, 288–289
  in two variables, 237, 250, 257–258
Difference of cubes, 385–387
Difference of squares
  explanation of, 342, 343, 410
  to factor binomials, 382–383, 387, 412
  in grouping, 383–385
Directrix, of parabola, 765
Direct variation
  applications of, 476, 477
  explanation of, 474, 489
Discriminant
  to determine number and type of solutions to quadratic equation, 596–598
  explanation of, 596, 651
  to find *x*- and *y*-intercepts of quadratic function, 598–599
Distance, between two points, 754, 755
Distance, rate, and time problems, 63–64, 264–265, 466–468
Distance formula, 63, 754–755, 801

Distributive property of multiplication over addition
  application of, 33, 34, 341, 342
  explanation of, 32, 40–41
Division
  of complex numbers, 562–563
  of fractions, 433
  of functions, 664–665
  long, 350–353
  of polynomials, 350–356, 410
  of rational expressions, 433–434, 485
  of real numbers, 21–22, 40
  synthetic, 354–356, 411
Division algorithm, 352
Division property of equality, 45
Division property of inequality, 78
Division property of radicals, 537–538, 571
Domain
  of functions, 197–198, 207
  of inverse functions, 676
  of logarithmic functions, 693, 695–696
  of radical functions, 502
  of rational functions, 422–424
  of relations, 183–186, 223
Dot Mode (calculator feature), 423

e (irrational number), 712–713, 743
Elementary row operations, 295
Elements, 6
Elimination method. *See* Addition method
Ellipse
  explanation of, 774–775
  foci of, 774, 803
  graphs of, 775–776, 803
  standard form of equation of, 775, 803
Empty sets, 86
Equality
  of absolute values, 101
  addition property of, 45, 77
  division property of, 45
  multiplication property of, 45
  subtraction property of, 45, 77
Equations. *See also* Linear equations in one variable; Linear equations in two variables; Nonlinear systems of equations; Quadratic equations; *specific types of equations*
  absolute value, 97–102, 120
  of circles, 755, 757, 758, 801
  conditional, 52–53
  contradictions as, 52–53
  of ellipse, 775, 803
  explanation of, 44
  exponential, 729–734, 744
  of hyperbola, 777, 792, 803
  identities as, 52–53
  of inverse of functions, 674–676
  of lines, 157–165, 221
  literal, 70–73, 588
  logarithmic, 726–729, 743
  of parabola, 768
  in quadratic form, 606–609, 651
  radical, 546–552, 572
  rational, 454–460, 466–469, 488
  solutions to, 44
  systems of dependent, 237, 250, 257–258
  systems of independent, 237
Equivalent rational expressions, 440
Exponential equations
  applications of, 733–734
  on calculators, 730
  explanation of, 729, 744
  method to solve, 729–732
Exponential expressions
  on calculators, 23, 681–682
  evaluation of, 23
  explanation of, 22
Exponential form, logarithmic form converted to, 690–691
Exponential functions
  applications of, 684–686, 732–733

decay, 683–685
  explanation of, 680–681, 741
  graphs of, 682–684, 695
  growth, 683, 686
Exponents
  explanation of, 22, 40, 320
  integer, 320–323, 409
  properties of, 320–321
  rational, 508–512, 569
  simplifying expressions with, 321–323
Expressions. See Algebraic expressions; Exponential
    expressions; Radical expressions; Rational
    expressions
Extraneous solutions, 546

Factorial notation, 815–816
Factors/factoring
  binomial, 362, 382–388, 412–413
  difference of squares, 382–385, 387
  explanation of, 360
  greatest common, 361–365, 370
  by grouping, 363–365, 383–385, 411
  negative, 361–362
  to solve quadratic equations, 395–398
  sum and difference of cubes, 385–387
  trinomials, 368–379
  using substitution, 378–379
Feasible regions, 275–276
Finite sequences, 821, 846
Finite series, 825
First-degree polynomial equations. See Linear
    equations in one variable
Focus
  of ellipse, 774, 803
  of hyperbola, 777, 803
  of parabola, 765
Form fitting, 72
Formulas. See also specific formulas
  evaluation of, 26
  explanation of, 26
  involving rational expressions, 459–460
  as literal equations, 70
Fractions
  clearing, 49–51, 455, 457
  complex, 447–451, 487
  division of, 433
  multiplication of, 432
Function notation
  explanation of, 194–196, 205, 224
  finding intercepts using, 209–211
Functions. See also specific types of functions
  absolute value, 205, 207
  algebra of, 664–665, 739
  composition of, 665–666, 739
  constant, 204, 205, 224
  cubic, 205, 207
  domain of, 197–198
  evaluation of, 194–196
  explanation of, 192, 223
  exponential, 680–686, 741
  graphs of, 197, 205–208, 210, 211, 224
  identity, 205, 207
  inverse, 671–676, 740
  linear, 205, 224
  logarithmic, 690–697, 741
  method to combine, 666–667
  natural logarithmic, 714–717
  one-to-one, 671–672
  polynomial, 333–335
  quadratic, 205, 207–209, 224, 400–403, 595–596,
      612–620, 626–627, 630–631, 636, 652
  radical, 502–504
  range of, 197
  rational, 422–424, 484
  reciprocal, 205, 207
  relations vs., 192, 197
  square root, 205, 207
  vertical line test for, 193–194, 223
Function values, from graphs, 197, 667–668

Fundamental principle of rational expressions,
    425, 427
Future value, 483–484

Gauss-Jordan method
  explanation of, 295
  to solve system of linear equations, 296–298, 309
GCF. See Greatest common factor (GCF)
Geometric sequences
  explanation of, 835, 847
  finding specified term of, 836
  nth term of, 835–836
Geometric series
  applications of, 839
  explanation of, 836–837, 847
  infinite, 838–839
  sum of, 837–839
Geometry applications, 68–70, 117, 265, 286–287,
    551–552
Graph (calculator feature), 138
Graphing calculators. See Calculator features;
    Calculators
Graphs. See also Rectangular coordinate system
  of basic functions, 205–208, 224
  of circles, 755–757
  of ellipse, 775–776, 803
  of exponential functions, 682–684, 695
  of feasible region, 275–276
  of function values, 197, 667–668
  of horizontal line, 137
  of hyperbolas, 777–779, 792–793
  of linear equations in two variables, 130–133
  of linear inequalities in two variables, 270–272, 307
  of logarithmic functions, 693–696, 715, 718
  of nonlinear functions, 207
  of nonlinear inequalities in two variables,
      792–793
  of nonlinear systems of equations, 785, 786
  of nonlinear systems of inequalities, 793–794
  of parabolas, 401, 612–620, 652, 766–770
  of quadratic functions, 208–209, 401, 612–620,
      636, 652
  of radical functions, 503–504
  to solve systems of linear equations, 237–240
  of vertical line, 136–137
  x- and y-intercepts on, 209–211
Greater than, 9
Greater than or equal to, 9
Greatest common factor (GCF)
  explanation of, 361, 411
  factoring out, 361–365, 370
Grouping
  difference of squares used in, 383–385
  factoring by, 363–365, 411

Horizontal asymptotes, 206
Horizontal lines
  explanation of, 136, 137, 165, 219, 221
  slope of, 148
Horizontal line test, 672, 740
Horizontal transverse axis, of hyperbola, 777, 803
Hyperbolas
  explanation of, 777, 803
  focus of, 777, 803
  graphs of, 777–779, 792–793
  standard form of equation of, 777, 792, 803
  vertex of, 777

Identities, 52–53
Identity functions, 205, 207
Identity property of addition, 32
Identity property of multiplication, 32
Imaginary numbers (i)
  explanation of, 556
  powers of, 558–559
  simplifying expressions in terms of, 557–558
Inconsistent systems of linear equations
  addition method to solve, 257
  explanation of, 237
  Gauss-Jordan method to solve, 298

method to solve, 249, 289
  in three variables, 289
Independent systems of linear equations, 237
Index of radical, 498, 532–533
Index of summation, 825, 832
Inequalities. See also Nonlinear systems of
    inequalities; specific types of inequalities
  absolute value, 103–110, 120–121
  addition property of, 77
  applications of, 81–82
  compound, 85–93, 118–119
  division property of, 78
  explanation of, 9–10
  linear in one variable, 76–82
  linear in two variables, 270–274
  multiplication property of, 78
  nonlinear, 791–793
  notation for, 9, 78, 80, 270, 271, 637
  polynomial, 639–641, 654
  quadratic, 636–639
  rational, 641–643, 654
  with "special case" solution sets, 644
  subtraction property of, 77
  translating to/from words in, 12
Infinite sequences, 821, 846
Infinite series, 825, 838–839
Integer exponents, 320–323, 409
Integers
  consecutive, 58–59, 398
  explanation of, 6, 39
Interest
  applications of, 61–62, 253–264, 713–714
  compound, 713–714
  compounded continuously, 714
  simple, 59
Intersect (calculator feature), 240–241
Intersection, of sets, 85–86
Interval notation
  domain in, 198
  explanation of, 10, 12, 39
  expressing sets by using, 10–11, 39, 53
  inequalities and, 77
  union and intersection in, 86, 87
Inverse functions
  explanation of, 671, 673, 674, 740
  finding equation of, 674–676
  one-to-one, 671–672
  property of, 673
Inverse of f, 671
Inverse property of addition, 32
Inverse property of multiplication, 32
Inverse variation
  applications of, 477–478
  explanation of, 474, 489
Irrational number
  e, 712–713, 743
  explanation of, 7–8, 39
Joint variation
  applications of, 478–479
  explanation of, 475, 489

LCD. See Least common denominator (LCD)
Leading coefficient, 330, 375–376, 409
Leading term, of polynomial, 330, 375, 409
Least common denominator (LCD)
  explanation of, 49, 50
  of rational expressions, 438–439, 441–444
Less than, 9
Less than or equal to, 9
Like radicals, 522, 523
Like terms
  addition and subtraction of, 33–35, 41, 523
  explanation of, 31, 40, 330
Linear equations in one variable
  applications of, 57–64, 68–70, 116
  clearing fractions and decimals to solve, 49–51
  conditional, 52–53
  with consecutive integers, 58–59
  contradictions as, 52–53

Linear equations in one variable—(*continued*)
  explanation of, 44, 115
  identities as, 52–53
  methods to solve, 45–49
Linear equations in three variables
  applications of, 286–288
  explanation of, 308
Linear equations in two variables. *See also* Systems
    of linear equations
  applications of, 171–173, 222
  explanation of, 129, 219
  graphs of, 130–133
  horizontal and vertical lines and, 136–137
  solutions to, 130
  x- and *y*-intercepts of, 133–136, 219
Linear functions, 205, 224
Linear inequalities in one variable
  applications of, 81–82
  explanation of, 76
  methods to solve, 76–80
  properties of, 77, 78
Linear inequalities in two variables
  compound, 273–274
  explanation of, 270, 307
  graphs of, 270–272, 307
Linear models
  explanation of, 171, 222
  interpretation of, 173–174
  method to write, 171–173
  from observed data points, 174–175
Lines
  equations of, 157–165, 221
  horizontal, 136, 137, 165, 219, 221
  slope of, 145–152
  vertical, 136–137, 165, 219, 221
Literal equations
  application of, 70–71, 117
  explanation of, 70, 588
  method to solve, 71–72
Logarithmic equations
  explanation of, 726, 743
  method to solve, 726–729
Logarithmic expressions
  equivalence of, 728
  evaluation of, 691–692
  in expanded form, 706–708, 717
  as single logarithm, 708–709
Logarithmic form, converted to exponential
    form, 690–691
Logarithmic functions
  applications of, 696–697
  common, 692–693, 696–697
  conversions to exponential form, 690–691
  domain of, 693, 695–696
  explanation of, 690, 741
  graphs of, 693–696, 718
  natural, 714–717
Logarithms
  on calculators, 693, 696, 715
  change-of-base formula and, 717–718
  explanation of, 690
  power property of, 705, 706, 709, 715
  product property of, 704–706, 715
  quotient property of, 705, 706, 715
  review of properties of, 742
  sum or difference of logarithms as single,
    708–709
Long division
  of polynomials, 350–353
  synthetic division vs., 356

Mathematical models, 171, 172. *See also*
    Linear models
Matrices
  augmented, 294–298, 309
  on calculators, 299
  coefficient, 293
  column, 293
  explanation of, 293

order of, 293–294
  row, 293
  to solve systems of linear equations, 295–299, 309
  square, 293
Maximum (calculator feature), 620
Maximum value, 619
Mean, 844
Midpoint formula, 758–759, 801
Minimum (calculator feature), 620
Minimum value, 619
Mixture problems, 62–63, 262
Models. *See* Linear models
Monomials. *See also* Polynomials
  division by, 350
  explanation of, 330, 409
  multiplication of, 340
Motion problems, 264–265
Multiplication
  associative property of, 32
  commutative property of, 32
  of complex numbers, 562
  distributive property of, over addition, 32–34
  of fractions, 432
  of functions, 664–665
  identity property of, 32
  inverse property of, 32
  of polynomials, 340–344, 410
  of radical expressions, 528–533, 570
  of rational expressions, 432–433, 485
  of real numbers, 20, 21, 40
Multiplication property of equality, 45
Multiplication property of inequality, 78
Multiplication property of radicals, 516–518, 528–530

Natural logarithmic functions
  applications of, 718–719
  in expanded form, 717
  explanation of, 714
  graphs of, 715
  properties of, 715
  simplifying expressions with, 716
  writing sum or difference of, 716
Natural numbers, 6, 39
Negative factors, 361–362
Negative infinity symbol, 10
Negative slope, 148
Nonlinear inequalities
  graphs of, 792–793, 805
  systems of, 793–794
  test point method for, 791, 792, 805
  in two variables, 791–793
Nonlinear systems of equations
  addition method to solve, 787–788
  on calculators, 786
  explanation of, 784, 804
  graphs of, 785, 786
  substitution method to solve, 784–787, 804
Nonlinear systems of inequalities
  explanation of, 793
  graphs of, 793–794, 805
Notation/symbols
  approximately equal to, 9
  empty set, 86
  factorial, 815–816
  function, 194–196, 205, 209–211, 224
  greater than, 9
  greater than or equal to, 9
  grouping, 24
  inequality, 9, 78, 80, 270, 271, 637
  interval, 10–12, 77, 86, 87
  less than, 9
  less than or equal to, 9
  negative infinity, 10
  radical, 23, 496, 510
  scientific, 324–326, 409
  set-builder, 6, 77, 240
  summation, 825–826
*n*th roots
  on calculators, 499

explanation of, 498–499, 568
  identification of, 499
  method to simplify, 499, 501
  of real numbers, 499
  of variable expressions, 499–501
*n*th term of sequence
  on calculators, 822
  explanation of, 821
  method to find, 823–824
Number line. *See* Real number line
Numbers. *See also* Complex numbers; Integers;
    Real numbers
  imaginary, 556–559
  irrational, 7–8, 39
  natural, 6, 39
  rational, 6–7, 39
  whole, 6, 39

Observed data points, linear models from, 174–175
One-to-one functions
  explanation of, 671–672, 740
  inverse of, 674–676
Opposite, of real numbers, 16, 39
Ordered pairs
  explanation of, 130
  graphs of, 129, 131
  relations as set of, 183, 184
  as solution to system of linear equations, 23, 62
Ordered triples, 283, 308
Order of matrix, 293, 309
Order of operations
  application of, 25
  on calculators, 25
  explanation of, 24, 40
  to simplify radicals, 518–519
Origin, on rectangular coordinate system, 128

Parabolas. *See also* Quadratic equations
  axis of symmetry of, 612, 765–768, 802
  on calculators, 614, 769
  directrix of, 765
  explanation of, 208, 224, 400, 612, 765, 802
  focus of, 765
  graphs of, 401, 612–620, 652, 766–770
  standard form of equation of, 766, 768, 802
  vertex of, 612, 628–629, 653, 765, 766, 770–771, 802
Parallel lines
  explanation of, 149
  in slope-intercept form, 159
  slope of, 149, 150
Parentheses
  clearing, 33–35
  explanation of, 24, 77
Pascal, Blaise, 815
Pascal's triangle, 408, 815, 818, 845
Percents, applications of, 59–61, 335
Perfect cubes, 388, 499, 500
Perfect fifth powers, 499
Perfect fourth powers, 499
Perfect squares, 388, 497, 500
Perfect square trinomials
  explanation of, 343, 400, 410, 412
  factoring, 376–377
Perimeter
  application involving, 68–69, 334
  formula for, 70
Perpendicular lines
  explanation of, 149
  in slope-intercept form, 159
  slope of, 149, 150
π (pi), 7
Plotting points, 129
Point-slope formula
  explanation of, 161–162, 165, 221
  to find equation of line, 162–164
Polynomial functions
  in applications, 334–335
  evaluation of, 333–334
  explanation of, 333

Polynomial inequalities, 639–641, 654
Polynomials. *See also* Binomials; Factors/factoring;
    Monomials; Trinomials
  addition of, 330–331, 409
  applications of product of, 345–346
  degree of, 330
  division of, 350–356, 410
  explanation of, 329–330, 409
  finding opposite of, 331
  leading coefficient of, 330, 375–376
  leading term of, 330, 375
  multiplication of, 340–344, 410
  prime, 369, 372, 383
  special case products and, 342–344
  subtraction of, 331–333, 409
  translations involving, 344–345
Population models, 738–739
Positive slope, 148
Power, 22
Power property of logarithms, 705, 706, 709, 715
Prime polynomials, 369, 372, 383
Principal and interest problems, 61–62, 263–264,
    713–714
Principal square roots, 23, 40, 496, 568
Problem solving, 57–58, 116. *See also* Translating
    to/from English form; Word problems
Product property of logarithms, 704–706, 715
Proportions. *See also* Variation
  in applications, 464–465, 488
  explanation of, 463
  methods to solve, 463–465
  with similar triangles, 465–466
Pythagorean theorem
  applications of, 399, 502
  explanation of, 399, 501, 568

Quadrants, on rectangular coordinate system, 128
Quadratic equations. *See also* Parabolas
  applications of, 398–400
  completing the square to solve, 585–587
  discriminant and, 596–599
  explanation of, 394
  factoring to solve, 395–398
  high-degree, 397, 398
  methods to solve, 599–602
  in one variable, 394
  quadratic formula to solve, 593–594
  square root property to solve, 582–583
  zero product rule to solve, 394–398, 413
Quadratic formula
  applications using, 594–596
  derivation of, 592
  explanation of, 592, 651
  to solve quadratic equations, 593–594, 599
Quadratic functions
  analysis of, 595–596, 627
  applications of, 402–403, 630–631
  explanation of, 205, 207–209, 224, 400
  in form $f(x) = a(x - h)^2 + k$, 626–627
  graphs of, 208, 401, 612–620, 636, 652
  x- and y-intercepts of, 401
Quadratic inequalities
  explanation of, 636
  graphs to solve, 637–638
  test point method to solve, 638–639
Quadratic in form, equations, 606–609, 651
Quadratic models, 302, 631–632, 649
Quotient property of logarithms, 705, 706, 715
Quotients, negative and positive, 21, 22

Radical equations
  applications of, 551–553
  explanation of, 546
  method to solve, 546–551, 572
Radical expressions
  addition of, 522–525, 570
  on calculators, 509, 518
  conjugate, 531–532
  division property of, 537–538, 571
  like, 522, 523

  method to simplify, 497–498, 500, 515–519, 530,
    537–538, 569
  multiplication of, 528–533, 570
  multiplication property of, 515–518, 528–530
  order of operations to simplify, 518–519
  rationalizing the denominator of, 538–543, 571
  simplified form of, 516, 536–537
  special case products and, 531–532
  subtraction of, 522–525, 570
Radical functions
  applications, 552–553
  on calculators, 503
  domain of, 502
  explanation of, 502–503
  graphs of, 503–504
Radical notation, 23, 496, 510
Radicand, 498
Radius, of circle, 755, 757
Range
  of functions, 197, 207
  of relations, 183–186, 223
Rate of change, 171, 220
Rate problems, 63–64
Rational equations
  applications of, 466–469
  explanation of, 454
  formulas involving, 459–460
  methods to solve, 455–459, 488
  steps to solve, 457
Rational exponents
  applications of, 512
  converting between radical notation and, 510–511
  evaluating expressions with, 508–509
  explanation of, 508–509, 569
  properties of, 511
Rational expressions
  addition of, 437, 441–444, 486
  division of, 433–434, 485
  equivalent, 440
  explanation of, 422
  fundamental principle of, 425, 427
  least common denominator of, 438–439, 441–444
  with like denominators, 437
  method to simplify, 424–427
  multiplication of, 432–433, 485
  subtraction of, 437, 441–444, 486
  with unlike denominators, 441–444
Rational functions
  domain of, 422–424
  evaluation of, 422–423
  explanation of, 422, 484
Rational inequalities
  explanation of, 641
  test point method to solve, 641–643, 654
Rationalizing the denominator
  explanation of, 538–539
  one term, 540–541, 571
  two terms, 541–543, 571
Rational numbers
  explanation of, 39
  identification of, 7
  set of, 6–7
Ratios
  of -1, 427–428
  common, 835
  explanation of, 463
  proportions and, 463
Real number line
  absolute value on, 16, 17
  explanation of, 6, 10
  inequalities on, 9, 76, 78
Real numbers
  absolute value of, 16–17, 105
  addition of, 17–18, 39
  division of, 21–22, 40
  explanation of, 6, 7, 39
  multiplication of, 20, 21, 40
  opposite of, 15
  ordering, 9–10

  properties of, 32–34
  set of, 5–8
  subtraction of, 19, 40
Reciprocal functions, 205, 207
Reciprocals, 20, 39
Rectangles, perimeter of, 70, 334
Rectangular coordinate system, 128–129.
    See also Graphs
Reduced row echelon form, 295–298, 309
Relations
  applications of, 187
  domain and range of, 183–186, 223
  explanation of, 183, 223
  functions vs., 192, 197
Right circular cylinders, 71
Right triangles, 399, 501, 502
Roots
  nth, 498–499
  square, 23–24, 496–498
  of variable expressions, 499–501

Sales tax, 59
Sample mean, 844
Sample standard deviation, 844
Scientific notation
  application of, 325–326
  on calculators, 325, 326
  explanation of, 324, 409
  writing numbers in, 324–325
Second-degree polynomial equations.
    See Quadratic equations
Seq (calculator feature), 822
Sequences
  alternating, 822–823
  in applications, 824
  arithmetic, 829–831, 846
  on calculators, 822, 825
  explanation of, 820–821, 846
  finite, 821, 8446
  geometric, 835–836, 847
  infinite, 821, 846
  nth term of, 823
  terms of, 821–824
Series
  arithmetic, 831–832, 847
  explanation of, 824–826, 846
  finite, 825
  geometric, 837–839, 847
  infinite, 825
Set-builder notation, 6, 77, 240
Sets
  classifying numbers by, 8
  explanation of, 6, 39
  in interval notation, 10–11, 39, 53
  of real numbers, 5–8
  solution, 44
  union of, 85–87, 274
Similar triangles, 465–466
Simple interest, 59
Simplification
  of algebraic expressions, 34–35, 40–41, 320
  of complex fractions, 447–451
  of complex numbers, 563
  of expressions in terms of i, 557–558
  with natural logarithms, 716
  of nth roots, 499, 501
  of radical expressions, 497–498, 500, 515–519,
    530, 536–538, 569
  with rational exponents, 511
  of rational expressions, 424–427
  of ratios of -1, 427–428
Slope
  applications of, 146–147, 151–152
  explanation of, 145–146, 165, 220
  formula for, 147–149
  interpretation of, 151–152
  methods to find, 146, 158
  negative, 148
  of parallel lines, 149–151

Slope—(continued)
    of perpendicular lines, 149–151
    positive, 148
    as rate of change, 171
    undefined, 148
    zero, 148
Slope-intercept form
    equations in, 171, 238
    explanation of, 157–158, 165, 221
    to find equation of line, 160–161
    uses of, 159–160
Solutions
    to equations, 44–49
    extraneous, 546
    to systems of linear equations, 236–237
Solution sets
    explanation of, 44
    inequalities with "special case," 644
Special case products, 342–344, 531–532
Spheres, surface area of, 71
Square matrices, 293
Square root functions, 205, 207
Square root property
    explanation of, 582, 650
    to solve quadratic equations, 582–583, 585–587
Square roots
    evaluation of, 24
    explanation of, 23, 496
    identification of, 496
    method to simplify, 497–498
    of negative number, 24
    positive and negative, 496
    principal, 23, 40, 496, 568
Squares. See also Completing the square;
        Difference of squares; Sum of squares
    of binomials, 343, 814
    perfect, 388, 497, 500
Square window option (calculator feature), 165
Standard deviation, 844
Standard form
    of ellipse equation, 775, 803
    of equation of circle, 755
    of equation of line, 157, 165, 221
    of hyperbola equation, 777, 792
    of parabola equation, 766, 768, 802
Study skills, 2–3
Subscripts, 26
Subsets, 6
Substitution method
    explanation of, 246
    to factor polynomials, 378–379
    to solve inconsistent systems, 249
    to solve nonlinear systems of equations,
        784–787, 804
    to solve systems of dependent equations, 250
    to solve systems of linear equations, 246–248, 304
Subtraction
    of complex numbers, 561
    of functions, 664–665
    of like terms, 33–35, 41, 523
    of polynomials, 331–333, 409
    of radicals, 522–525, 570
    of rational expressions, 437, 441–444, 486
    of real numbers, 19, 40
Subtraction property of equality, 45, 77
Subtraction property of inequality, 77

Sum formula, 832
Summation notation, 825–826
Sum of cubes, 385–386
Sum of squares, 383
Superscripts, 26
Surface area, of sphere, 71
Symbols. See Notation/symbols
Synthetic division
    to divide polynomials, 354–356
    explanation of, 354, 411
    long division vs., 356
Systems of linear equations in three variables
    applications of, 286–288
    with dependent equations, 288–289
    explanation of, 283, 289, 308
    methods to solve, 284–286
    solutions to, 283–284
Systems of linear equations in two variables
    addition method to solve, 253–257, 305
    applications of, 261–265, 306
    consistent, 237
    dependent, 237, 250, 257–258
    explanation of, 236
    Gauss-Jordan method to solve, 295–298
    graphs to solve, 237–240, 303
    inconsistent, 237, 249, 289
    independent, 237
    matrices to solve, 295–298, 309
    solutions to, 236, 237, 303
    substitution method to solve, 246–248, 304
Systems of nonlinear equations. See Nonlinear
        systems of equations
Systems of nonlinear inequalities. See Nonlinear
        systems of inequalities

Table (calculator feature), 138, 196
Terms
    coefficient of, 31, 40
    constant, 30, 31, 171
    degree of, 329
    explanation of, 31, 40
    leading, 330, 375, 409
    like, 31, 33–35, 40, 330
    of sequences, 821–824
    unlike, 31
    variable, 31, 40, 171
Test point method
    for linear inequalities in one variable, 80
    for nonlinear inequalities, 791, 792
    for polynomial inequalities, 639–641, 654
    for quadratic inequalities, 638–639
    for rational inequalities, 641–643, 654
Time problems, 63–64
Translating to/from English form. See also
        Problem solving; Word problems
    for absolute value, 109–110
    for inequalities, 12
    for polynomials, 344–345
    for variation, 474–475
Transverse axis, of hyperbola, 777, 803
Trial-and-error method, to factor trinomials,
        371–376, 412
Triangles
    Pascal's, 408, 815, 818, 845
    right, 399
    similar, 465–466

Trinomials. See also Polynomials
    AC-method to factor, 368–370, 411
    explanation of, 330, 409
    guidelines to factor, 373, 379
    perfect square, 343, 376–377, 400, 410, 412
    substitution to factor, 378–379
    trial-and-error method to factor, 371–376, 412

Undefined slope, 148
Union, of sets, 85–87, 274
Unlike terms, 31

Value (calculator feature), 161, 196
Variable expressions, 499–501
Variables, 26
Variable terms, 31, 40, 171
Variation. See also Proportions
    applications of, 476–479
    direct, 474, 476, 477, 489
    inverse, 474, 477–478, 489
    joint, 475, 478–479, 489
    translations involving, 474–475
Vertex
    of hyperbola, 777
    of parabola, 612, 628–629, 653, 765, 766,
        770–771, 802
    of quadratic function, 629–630
Vertex formula, 628–629, 770, 771
Vertical asymptotes, 206
Vertical lines
    explanation of, 136–137, 165, 219, 221
    slope of, 148
Vertical line test, 193–194, 223
Vertical transverse axis, of hyperbola, 777, 803
Volume, of right circular cylinder, 71

Whole numbers, 6, 39
Word problems, 57–58, 116. See also Translating
        to/from English form
Work problems, 468–469

x-axis, 128
x-coordinates
    explanation of, 129
    in one-to-one functions, 671
x-intercepts
    explanation of, 133, 219
    function notation to find, 209–211
    of linear equation, 133–136
    of quadratic functions, 401, 598–599

y-axis, 128
y-coordinates
    explanation of, 129
    in one-to-one functions, 671
y-intercepts
    explanation of, 133, 219
    function notation to find, 209–211
    of linear equation, 133–136
    of quadratic functions, 401, 598–599
    slope-intercept form to find, 158, 159

Zero (calculator feature), 402
Zero product rule
    explanation of, 394–398, 413, 582
    to solve quadratic equations, 599
Zero slope, 148

Perimeter and Circumference

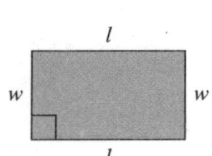

Rectangle

$P = 2l + 2w$

Square

$P = 4s$

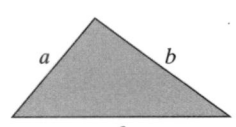

Triangle

$P = a + b + c$

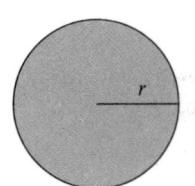

Circle

Circumference: $C = 2\pi r$

Area

Rectangle

$A = lw$

Square

$A = s^2$

Parallelogram

$A = bh$

Triangle

$A = \frac{1}{2}bh$

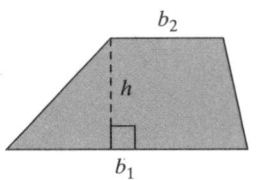

Trapezoid

$A = \frac{1}{2}(b_1 + b_2)h$

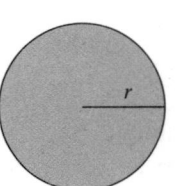

Circle

$A = \pi r^2$

Volume

Rectangular Solid

$V = lwh$

Cube

$V = s^3$

Right Circular Cylinder

$V = \pi r^2 h$

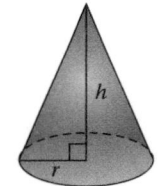

Right Circular Cone

$V = \frac{1}{3}\pi r^2 h$

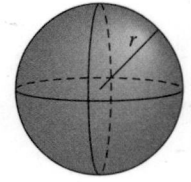

Sphere

$V = \frac{4}{3}\pi r^3$

Angles

- Two angles are **complementary** if the sum of their measures is 90°.

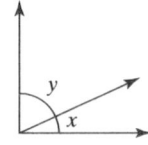

- Two angles are **supplementary** if the sum of their measures is 180°.

- $\angle a$ and $\angle c$ are vertical angles, and $\angle b$ and $\angle d$ are vertical angles. The measures of vertical angles are equal.

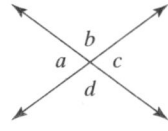

- The sum of the measures of the angles of a triangle is 180°.

$x° + y° + z° = 180°$